Machining Fundamentals

Ninth Edition

by

John R. Walker
Bob Dixon

Publisher
The Goodheart-Willcox Company, Inc.
Tinley Park, IL
www.g-w.com

Manufactured in the United States of America.

Library of Congress Catalog Card Number 2012050405

ISBN 978-1-61960-209-0

2 3 4 5 6 7 8 9 – 14 – 17 16 15 14

The Goodheart-Willcox Company, Inc. Brand Disclaimer: Brand names, company names, and illustrations for products and services included in this text are provided for educational purposes only and do not represent or imply endorsement or recommendation by the author or the publisher.

The Goodheart-Willcox Company, Inc. Safety Notice: The reader is expressly advised to carefully read, understand, and apply all safety precautions and warnings described in this book or that might also be indicated in undertaking the activities and exercises described herein to minimize risk of personal injury or injury to others. Common sense and good judgment should also be exercised and applied to help avoid all potential hazards. The reader should always refer to the appropriate manufacturer's technical information, directions, and recommendations; then proceed with care to follow specific equipment operating instructions. The reader should understand these notices and cautions are not exhaustive.

The publisher makes no warranty or representation whatsoever, either expressed or implied, including but not limited to equipment, procedures, and applications described or referred to herein, their quality, performance, merchantability, or fitness for a particular purpose. The publisher assumes no responsibility for any changes, errors, or omissions in this book. The publisher specifically disclaims any liability whatsoever, including any direct, indirect, incidental, consequential, special, or exemplary damages resulting, in whole or in part, from the reader's use or reliance upon the information, instructions, procedures, warnings, cautions, applications, or other matter contained in this book. The publisher assumes no responsibility for the activities of the reader.

The Goodheart-Willcox Company, Inc. Internet Disclaimer: The Internet resources and listings in this Goodheart-Willcox Publisher product are provided solely as a convenience to you. These resources and listings were reviewed at the time of publication to provide you with accurate, safe, and appropriate information. Goodheart-Willcox Publisher has no control over the referenced websites and, due to the dynamic nature of the Internet, is not responsible or liable for the content, products, or performance of links to other websites or resources. Goodheart-Willcox Publisher makes no representation, either expressed or implied, regarding the content of these websites, and such references do not constitute an endorsement or recommendation of the information or content presented. It is your responsibility to take all protective measures to guard against inappropriate content, viruses, or other destructive elements.

Cover image credit: sspopov / Shutterstock.com

Library of Congress Cataloging-in-Publication Data

Walker, John R.,
Machining fundamentals / by John R. Walker, Bob A. Dixon.
—Ninth edition.
pages cm
Includes index.
ISBN 978-1-61960-209-0
1. Machine-shop practice. 2. Machining. I. Title. II. Author
TJ1160 .W25 2014
671.3'5--dc23

2012050405

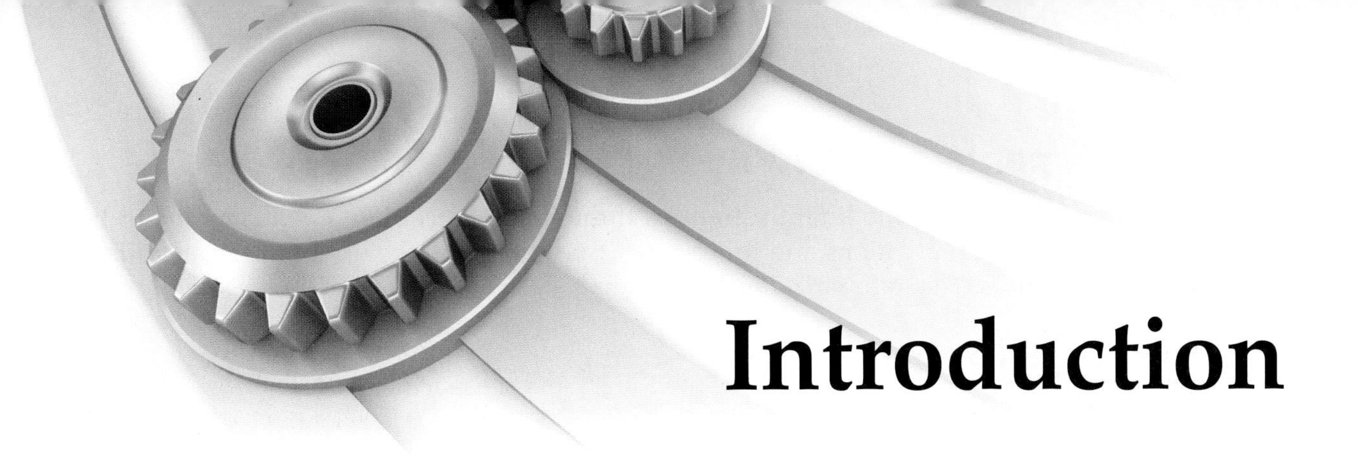

Introduction

Machinists are highly skilled men and women. They use drawings, hand tools, precision measuring tools, drilling machines, grinders, lathes, milling machines, and other specialized machine tools to shape and finish metal and nonmetal parts. Machinists must have a sound understanding of basic and advanced machining technology, which includes:

- Proficiency in safely operating machine tools of various types (manual, automatic, and computer controlled).
- Knowledge of the working properties of metals and nonmetals.
- The academic skills (such as math, science, English, print reading, and metallurgy) needed to make precision layouts and machine setups.

Machining Fundamentals provides an introduction to these important areas of manufacturing technology. The text explains the "how, why, and when" of numerous machining operations, setups, and procedures. Through it, you will learn how machine tools operate and when to use one particular machine instead of another. The advantages and disadvantages of various machining techniques are discussed, along with their suitability for particular applications.

Machining Fundamentals details the many common methods of machining and shaping parts to meet given specifications. It also covers more advanced processes, such as laser machining and welding, water-jet cutting, high-energy-rate forming (HERF), cryogenics, chipless machining, electrical discharge machining (EDM), electrochemical machining (ECM), robotics, and rapid prototyping. The importance of computer numerical control (CNC) in the operation of most machine tools and its role in automated manufacturing is explored thoroughly.

Machining Fundamentals has many features that make it easy to read and understand. The heads in each chapter are numbered to make it easier to locate specific information within a chapter. Learning objectives are presented at the beginning of each chapter, along with a list of selected technical terms important to understanding the material in that chapter. Throughout the text, technical terms are highlighted in bold italic type as they are introduced and defined. These terms are also listed and defined in a *Glossary of Technical Terms* at the end of this text. Review questions that reinforce key learning objectives are presented at the end of each chapter.

Machining Fundamentals is a valuable guide to anyone interested in machining, since the procedures and techniques presented have been drawn from all areas of machining technology.

Machining Fundamentals Color Key

A consistent color code is used in the line illustrations throughout ***Machining Fundamentals*** to help you better visualize the machining operations and procedures. Specific colors are used to indicate different materials and equipment features. The following key shows what each color represents.

- Metals (surfaces)
- Metals (in section)
- Machines/machine parts
- Tools
- Cutting edges
- Work-holding and tool-holding devices
- Rulers and measuring devices
- Direction or force arrows, dimensional information
- Fasteners
- Abrasives
- Fluids
- Miscellaneous

About the Authors

John R. Walker is the author of thirteen textbooks and has written numerous magazine articles. Mr. Walker completed his undergraduate studies at Millersville University and has a master's degree in Industrial Education from the University of Maryland. He taught industrial arts and vocational education for more than thirty-two years, including five years as Supervisor of Industrial Education. He also worked as a machinist for the US Air Force and as a draftsman at the US Army Aberdeen Proving Grounds.

Bob Dixon is an Associate Professor and Head of the Engineering Technology Department at Walters State Community College in Morristown, Tennessee. Bob holds bachelor's and master's degrees in Engineering Technology from East Tennessee State University and a master's degree in Industrial Engineering from the University of Tennessee. Prior to entering the education field, Bob spent over 20 years in industry working in a variety of machining, manufacturing, and engineering positions. Bob is an ATMAE Certified Senior Technology Manager and recipient of the 2005 ATMAE Outstanding Faculty of Industrial Technology Award for Region 3.

Reviewers

The authors and publisher would like to thank the following individuals for their valuable input in the development of ***Machining Fundamentals***.

Terry Anselmo
Precision Machining Instructor
Sun Area Technical Institute
Northumberland, Pennsylvania

Marte Arreola
Machinist Instructor
British Columbia Institute of Technology
Burnaby, British Columbia

George Beauchemin
CAD-CAM-CNC Instructor
Bay Path Regional Vocational High School
Worcester, Maryland

Paul Bois
Precision Machining Teacher
Milford High School and Applied Technology Center
Somersworth, New Hampshire

Mark Bosworth
Precision Machining Technology Coordinator
Southwestern Illinois College
Belleville, Illinois

Craig Brazil
Professor, IMM Program
Sheridan College
Oakville, Ontario

Dr. Chesley Chambers
Metals Instructor
Floyd County College & Career Academy
Cedartown, Georgia

Donald Dean
Manufacturing Technologies Coordinator
C-TEC
Newark, Ohio

Bruce Dirks
Technology Education Instructor
Owatonna High School
Owatonna, Minnesota

David Eledge
INCT Instructor
Metropolitan Community College
Council Bluffs, Iowa

Gottfried Georgi
WI Technology Ed. Instructor, retired
Wm. Horlick High School
Mount Pleasant, Wisconsin

John Glenn
Machine Tool Technology Instructor
Clinton Technical School
Clinton, Missouri

Howard Gray
Coordinator, Mechanical Technology Programs
Sault College
Sault Ste. Marie, Ontario

Richard Grossen
CNC Instructor/Department Chair
Blackhawk Technical College
Cottage Grove, Wisconsin

Henry Hatem
Instructor of Precision Machining Technologies
Renton Technical College
Carnation, Washington

Michael Hilber
Engineering/Manufacturing Teacher
Secondary Technical Education Program/Anoka Hennepin
East Bethel, Minnesota

Jeremy Hodkiewicz
Technology Education Teacher
Shawano High School
Clintonville, Wisconsin

Roger Holland
Instructor, Computerized CNC Precision Machining Technology
Minnesota State College—Southeast Technical
La Crescent, Minnesota

Bruce Jacobs
Industrial Arts Technology/STEM/CTE, Visual Arts & Yearbook Teacher
Mt. Bauer Middle School
Puyallup, Washington

Steve Jenkins
Wor-Wic Community College & Parkside High School
High Performance Mfg. Instructor
Salisbury, Maryland

Bart Jenkins
Director/Instructor Machine Tool Technology
Georgia Northwestern Technical College
Rome, Georgia

Mike Koppy
Machine Technology Instructor
Lake Superior College
Hermantown, Minnesota

Danny Lester
Associate Professor
Somerset Community College
Somerset, Kentucky

Tom Livingstone
Associate Professor
The Pennsylvania College of Technology
Williamsport, Pennsylvania

Dale Miller
Industrial Machine Technology Instructor
Lebanon County Career and Technology Center
Lebanon, Pennsylvania

Brian Nelson
Senior Instructor
Dunwoody College of Technology
West Saint Paul, Minnesota

Shan Packer
Precision Machining Instructor
Central Mountain High School
Howard, Pennsylvania

Samuel Ray
Machine Tool Technology Instructor, retired
The Williamson Free School of Mechanical Trades
Saint Michaels, Maryland

Edward Schmidt
Career Technical Educator (C.T.E.)
Trenton Central High School
Philadelphia, Pennsylvania

Preet Singh
Dual Enrollment Teacher
Donna High School—CTE Department
Pharr, Texas

Bruce Smith
Instructor
Tulsa Technology Center
Claremore, Oklahoma

Marty Snagg
Engineering Instructor
Heritage High School
Conyers, Georgia

Paul Sorenson
Computerized CNC Precision Machining Technology Instructor
Minnesota State College—Southeast Technical
Winona, Minnesota

James Standing
Professor, Toolmaking/Machine Shop
Algonquin College
Ottawa, Ontario

Steve Stayton
Director, Career Technical Education
Colbert County Schools
Tuscumbia, Alabama

Daniel Wagner
Managing Director of Workforce Training
HACC–Central Pennsylvania's Community College
Hegins, Pennsylvania

Roger Wensink
Machine Tool and Apprenticeship Instructor
Lakeshore Technical College
Sheboygan Falls, Wisconsin

Olaf Wick
Instructor, Machine Tooling Technics
Wisconsin Indianhead Technical College
New Richmond, Wisconsin

Dr. John Wren
Professor
Fanshawe College, St. Thomas Campus
St. Thomas, Ontario

Dr. John Wyatt
Associate Professor
Mississippi State University
Starkville, Mississippi

Nick Yates
Project Lead the Way Engineering Teacher
Patterson High School
Baltimore, Maryland

Jim Zeleny
Confederation College
Program Coordinator—Aerospace Manufacturing Technology, retired
Thunder Bay, Ontario

Andy Zwanch
Precision Machining Technology Department Chair
Johnson College
Scranton, Pennsylvania

Brief Table of Contents

Contents

CHAPTER 15

CHAPTER 16

CHAPTER 17

CHAPTER 18

CHAPTER 19

CHAPTER 20

CHAPTER 21

CHAPTER 30

CHAPTER 31

CHAPTER 32

CHAPTER 33

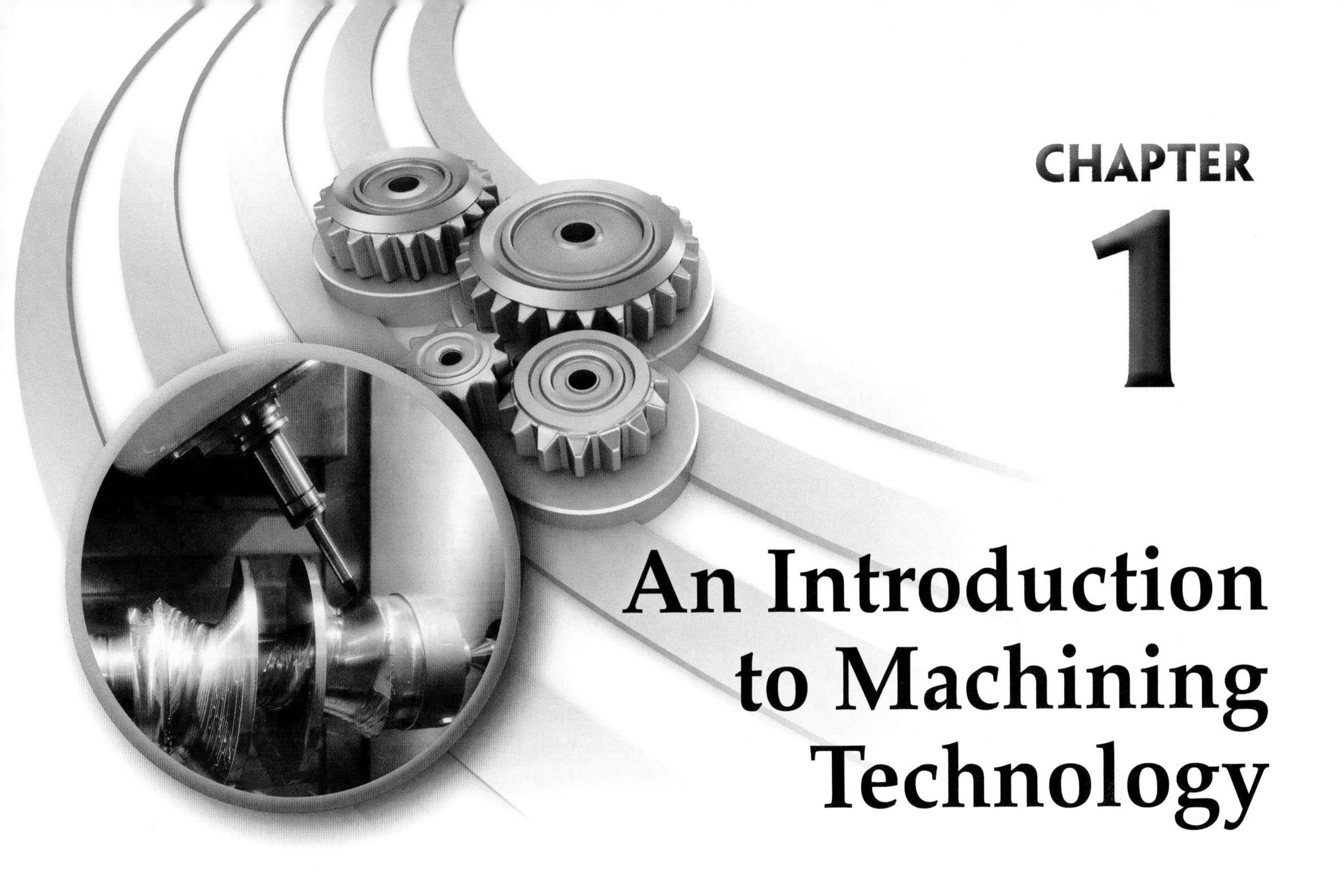

CHAPTER

1

An Introduction to Machining Technology

Learning Objectives

After studying this chapter, you will be able to:

- Discuss how modern machine technology affects the workforce.
- Give a brief explanation of the evolution of machine tools.
- Provide an overview of machining processes.
- Explain how CNC machining equipment operates.
- Describe the role of the machinist.

Technical Terms

broaching machine
computer numerical control (CNC)
drill press
grinding machine
lathe
machine tools
machinist
milling machine
numerical control (NC)
sawing machine
skill standards
turning

A study of technology will show that industry has progressed from the time when everything was made by hand to the present fully automated manufacturing of products. Machine tools have played an essential role in all technological advances.

Without machine tools, **Figure 1-1**, there would be no airplanes, automobiles, television sets, or computers. Many of the other industrial, medical, recreational, and domestic products we take for granted would not have been developed. For example, if machine tools were not available to manufacture tractors and farming implements, farmers might still be plowing with oxen and hand-forged plowshares.

It is difficult to name a product that does not require, either directly or indirectly, the use of a machine tool somewhere in its manufacture. Today, no country can hope to compete successfully in a global economy without making use of the most advanced machine tools available.

Figure 1-1. Machine tools have made it possible to manufacture parts with the precision and speed necessary for low-cost mass production. Without machine tools, most products on the market today would not be available or affordable.

And no industry or country can hope to take advantage of the most advanced machine tools without the aid of a ***machinist***—a person highly skilled in the use of machine tools and capable of creating the complex machine setups required of modern manufacturing.

These high-paying skilled jobs in manufacturing, such as tool and die making and precision machining, *require aptitudes comparable to those of college graduates*. Jobs that require few or no skills have almost disappeared.

1.1 The Evolution of Machine Tools

Machine tools are the class of machines which, taken as a group, can reproduce themselves (manufacture other machine tools). There are many variations of each type of machine tool, and they are available in many sizes. Tools range from those small enough to fit on a bench top to machines weighing several hundred tons.

The evolution of machine tools is somewhat akin to the old question, "Which came first, the chicken or the egg?" You could also ask, "How could there be machine tools when there were no machine tools to make them?"

1.1.1 Early Machine Tools

The first machine tools, the bow lathe and bow drill, were handmade and human-powered. They have been dated back to about 1200 BC. Until the end of the seventeenth century, the lathe could only be used to turn softer materials, such as wood, ivory, or at most, soft metals like lead or copper. Eventually, the bow lathe with its reciprocating (back-and-forth) motion gave way to treadle power, which made possible work rotation that was continuous in one direction. Later, machines were powered by a "great wheel" turned by flowing water or by a person or animal walking on a treadmill. Power was transmitted from the wheel to one or more machines by a belt and pulley system.

When inventor James Watt first experimented with his steam engine, the need for perfectly bored cylinders soon became apparent. This brought about the development of the first true machine tool. It was a form of the lathe and was called a "boring mill," **Figure 1-2**. The water-powered tool was developed in 1774 by Englishman John Wilkinson.

This machine was capable of turning a cylinder 36″ in diameter to an accuracy of a "thin-worn shilling"

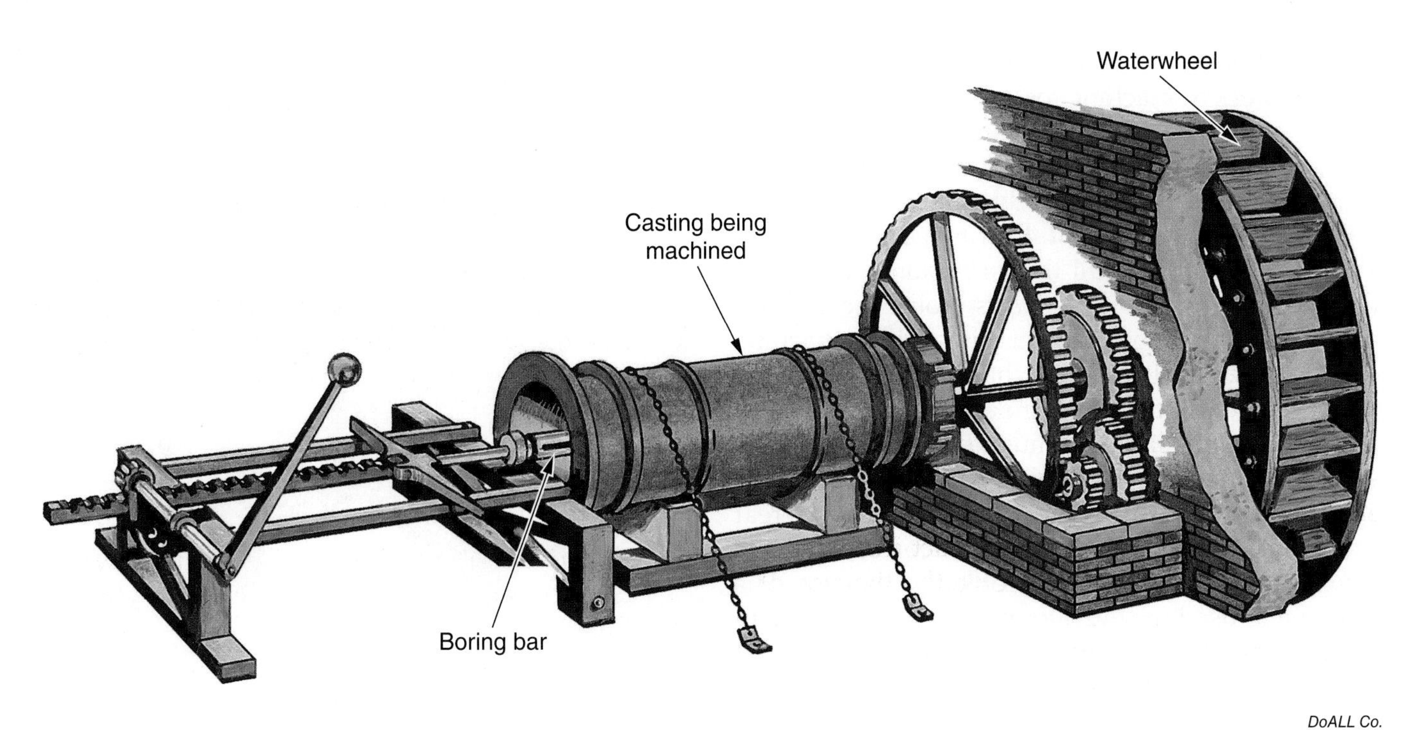

DoALL Co.

Figure 1-2. The first true machine tool is thought to be the boring mill invented by John Wilkinson in 1774. It enabled James Watt to complete the first successful steam engine. The boring bar was rigidly supported at both ends, and was rotated by waterpower. It could bore a 36″ diameter cylinder to an accuracy of less than 1/16″.

(an English coin about the size of a modern US quarter). However, operation of the boring mill, like all metal cutting lathes at the time, was hampered by the lack of tool control. The "mechanic" (the first machinist) had to unbolt and reposition the cutting tool after each cut.

About 1800, the first lathe capable of cutting accurate screw threads was designed and constructed by Henry Maudslay, an English master mechanic and machine toolmaker. As shown in **Figure 1-3**, a handmade screw thread was geared to the spindle and moved a cutting tool along the work. Maudslay also devised a slide rest and fitted it to his lathe. It allowed the cutting tool to be accurately repositioned after each cut. Maudslay's lathe is considered the "granddaddy" of all modern chip-making machine tools.

In retrospect, the Industrial Revolution could not have taken place if there had not been a cheap, convenient source of power: the steam engine. Until the advent of the steam engine, industry had to locate near sources of water power. This was often some distance from raw materials and workers. With cheap power, industry could locate where workers were plentiful and where the products they produced were needed. The steam engine, in turn, would not have been possible without machine tools.

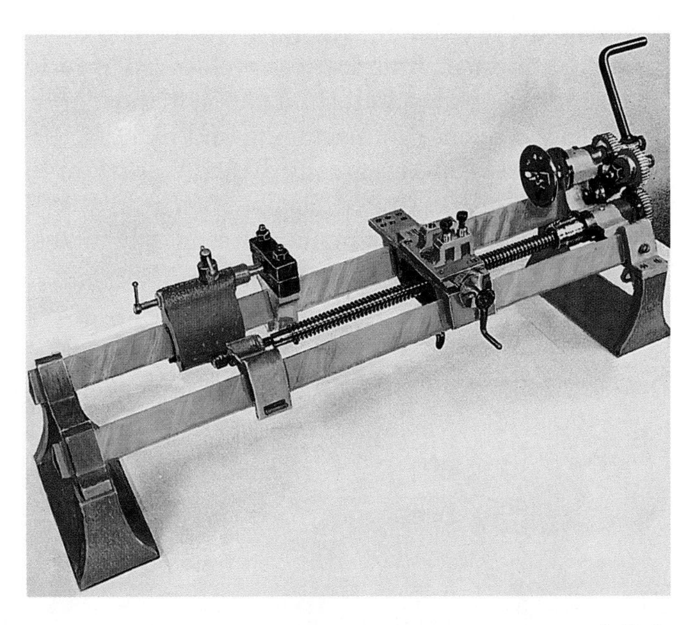

DoALL Co.

Figure 1-3. Henry Maudslay's screw-cutting lathe. This machine tool, constructed on a heavy frame, combined a master lead screw and a movable slide rest. The lead screw had to be changed when a different thread pitch was required.

Until the boring mill and lathe were developed to the point where metal could be machined with some degree of accuracy, there could be no steam engine.

The milling machine was the next important development in machine tools. It also evolved from the lathe. In 1820, Eli Whitney, an American inventor and manufacturer, devised a system to mass-produce muskets (guns). Whitney began using a milling machine, **Figure 1-4**, to make interchangeable musket parts. Until then, muskets were made individually by hand, so parts from one musket would not fit in another. Whitney's milling machine even had power feed, but it had one defect. There was no provision to raise the worktable. The part had to be raised by shimming after each cut. Since each machine was used to produce the same part again and again, this shortcoming was not a great problem. This problem was quickly corrected.

Whitney had another problem, however. His ideas were used in several armories producing gun parts. There was no standard of measurement at that time, so parts made in one armory were not interchangeable with parts made in another armory. It was not until the mid-1860s that the United States adopted a standard measuring system.

By 1875, basic machine tools such as the lathe, the milling machine, and the drill press, **Figure 1-5**, were capable of attaining accuracies of one one-thousandth of an inch. America was well on its way to becoming the greatest industrial nation in the world.

This proficiency in machining and manufacturing would help America greatly during World War II. A large part of the United States' success in WWII was due to its distinct manufacturing advantage.

DoALL Co.

Figure 1-4. One of the first practical milling machines manufactured in America. Eli Whitney used it and similar machines to mass-produce musket parts that were interchangeable.

Factories were rapidly turned over from producing consumer goods to military hardware. Of special importance to the war effort was the opening up of heavy industry professions to women. This supplied the labor needed to produce the large quantities of guns, ammunition, tanks, planes, and ships necessary to win the war, **Figure 1-6**.

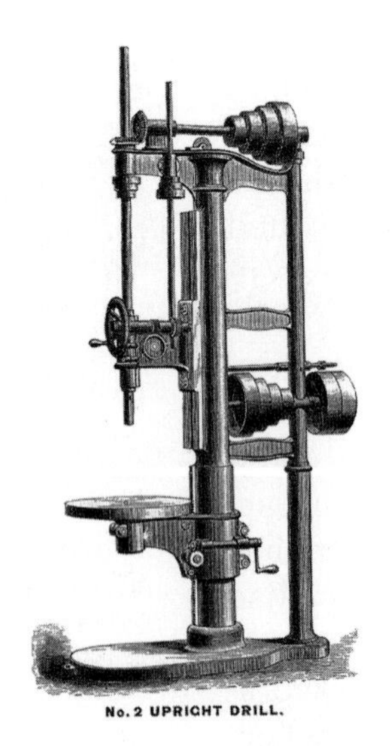

Goodheart-Willcox Publisher

Figure 1-5. Illustrations of Pratt & Whitney machine tools from an 1876 advertisement. Built from heavy iron castings, the machines were driven by overhead pulleys and belting. A central steam engine or large electric motor powered the overhead pulleys in factories until the 1920s.

Library of Congress, Prints & Photograph Division, FSA-OWI Collection, LC-DIG-fsac-1a34951

Figure 1-6. Lathe operator machining parts for transport planes at the Consolidated Aircraft Corporation plant, Fort Worth, Texas.

1.1.2 Power Sources

As machine tools were improved, so was the way they were powered. At first, the changes were very slow, taking hundreds of years. The greatest changes have come only in the last 150 years or so. The following are the various power sources used by machine tools throughout history in the order they evolved:

- **Hand power.** The bow lathe and bow drill are examples. The direction of rotation changed at each stroke of the bow.
- **Foot power.** A treadle or a treadmill made possible continuous rotation of the work in one direction.
- **Animal power.** Treadmills were used to power early devices for boring cannon barrels. Human foot power was not sufficiently strong for this work.
- **Water power.** Not always dependable as a power source, because of lack of water during dry seasons.
- **Steam power.** The first real source of dependable power. A centrally located steam engine turned shafts and overhead pulleys that were belted to the individual machines.
- **Central electrical power.** Large electric motors simply replaced the steam engines. Power transmission to the machines did not change.
- **Individual electrical power.** Motors were built into the individual machine tools. Overhead belting was eliminated.

1.2 Basic Machine Tool Operation

Almost all machine tools have evolved from the ***lathe***, **Figure 1-7**. This machine tool performs one of the most important machining operations, ***turning***. It operates on the principle of work being rotated against the edge of a cutting tool, as shown in **Figure 1-8**.

Photo courtesy of Grizzly Industrial, Inc. www.grizzly.com

Figure 1-7. A modern lathe featuring chuck safety guard, foot brake, coolant system, inch/metric dials, and a universal gearbox capable of cutting inch, metric, and diametral threads. Except those tools that perform nontraditional machining operations, all machine tools have evolved from the lathe.

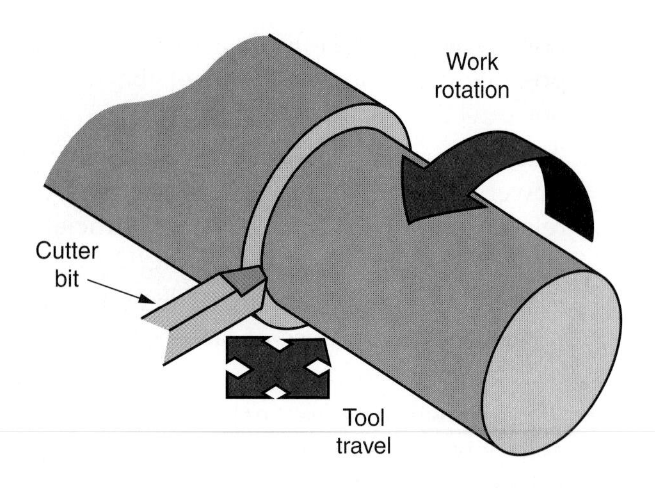

Goodheart-Willcox Publisher

Figure 1-8. The lathe operates on the principle of the work being rotated against the edge of a cutting tool.

Many other operations—drilling, boring, threadcutting, milling, and grinding—can also be performed on a lathe. The most advanced version of the lathe is the CNC turning center.

1.2.1 Sawing Machines

A ***sawing machine***, **Figure 1-9**, or saw, makes use of a multitoothed saw blade to cut away material. Sawing machines come in a variety of forms. All sawing machines perform one of two basic operations:

- **Cutoff sawing.** Sawing and cutoff machines cut stock material into more manageable lengths in preparation for other machining operations.
- **Band machining.** A vertical band saw uses a continuous saw blade. Chip removal is rapid and accuracy can be held to close tolerances, eliminating or minimizing many secondary machining operations.

Photo courtesy of Grizzly Industrial, Inc. www.grizzly.com

Figure 1-9. Sawing machines, like this horizontal band saw, make use of a continuous saw blade, with each tooth functioning as a precision cutting tool.

1.2.2 Drill Press

A ***drill press***, **Figure 1-10**, rotates a cutting tool (drill) against the material with sufficient pressure to cause the tool to penetrate the material. It is primarily used for cutting round holes. See **Figure 1-11**. Drill presses are available in many versions. Some are designed to machine holes as small as 0.0016″ (0.04 mm) in diameter.

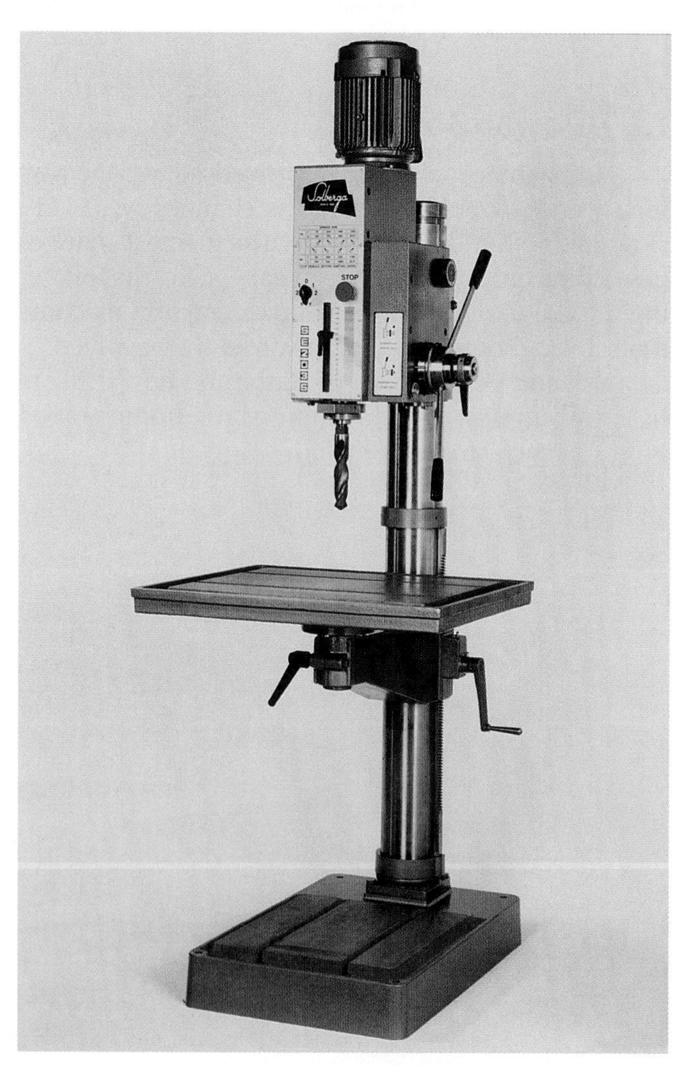

Willis Machinery and Tools Corp.

Figure 1-10. A typical 20″ variable-speed gear head drill press with power feed. It can drill holes up to 1½″ in diameter in cast iron.

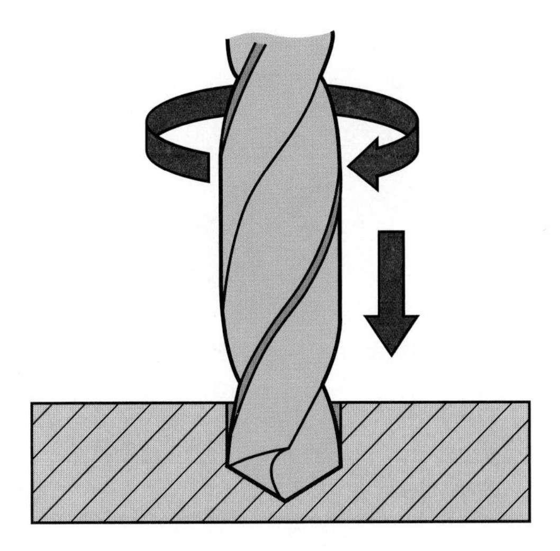

Goodheart-Willcox Publisher

Figure 1-11. A drill press operates by rotating a cutting tool (drill) against the material with sufficient pressure to cause the tool to penetrate the material.

1.2.3 Grinding Machines

A ***grinding machine***, **Figure 1-12**, or grinder, removes metal by rotating a grinding wheel or abrasive belt against the work. The process falls into two basic categories:

- **Offhand grinding.** Work that does not require great accuracy is handheld and manipulated until ground to the desired shape.
- **Precision grinding.** Only a small amount of material is removed with each pass of the grinding wheel, so that a smooth, accurate surface is generated. Precision grinding is a finishing operation.

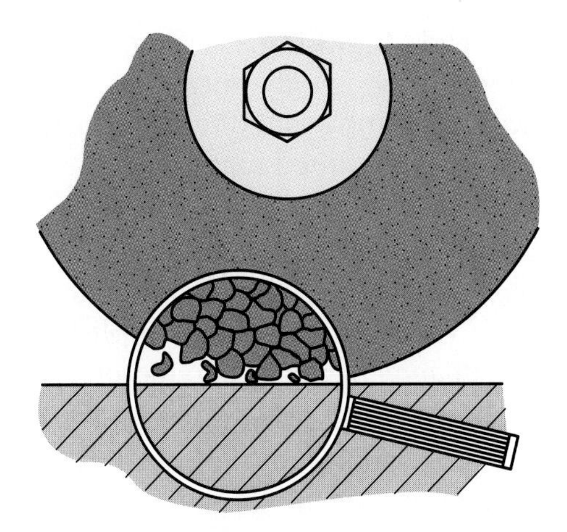

Goodheart-Willcox Publisher

Figure 1-12. Grinding is a cutting operation, like turning, drilling, milling, or sawing. However, instead of the one, two, or multiple-edge cutting tools used in other applications, grinding uses an abrasive tool composed of thousands of cutting edges.

1.2.4 Milling Machine

A ***milling machine*** rotates a multitoothed cutter into the work, **Figure 1-13**. A wide variety of cutting operations can be performed on milling machines, including machining flat, or contoured surfaces, slots, grooves, recesses, threads, gears, and spirals. Milling machines are available in more variations than any other family of machine tools, **Figure 1-14**, and are well suited to computer-controlled operation. The most advanced version of a milling machine is the CNC milling center.

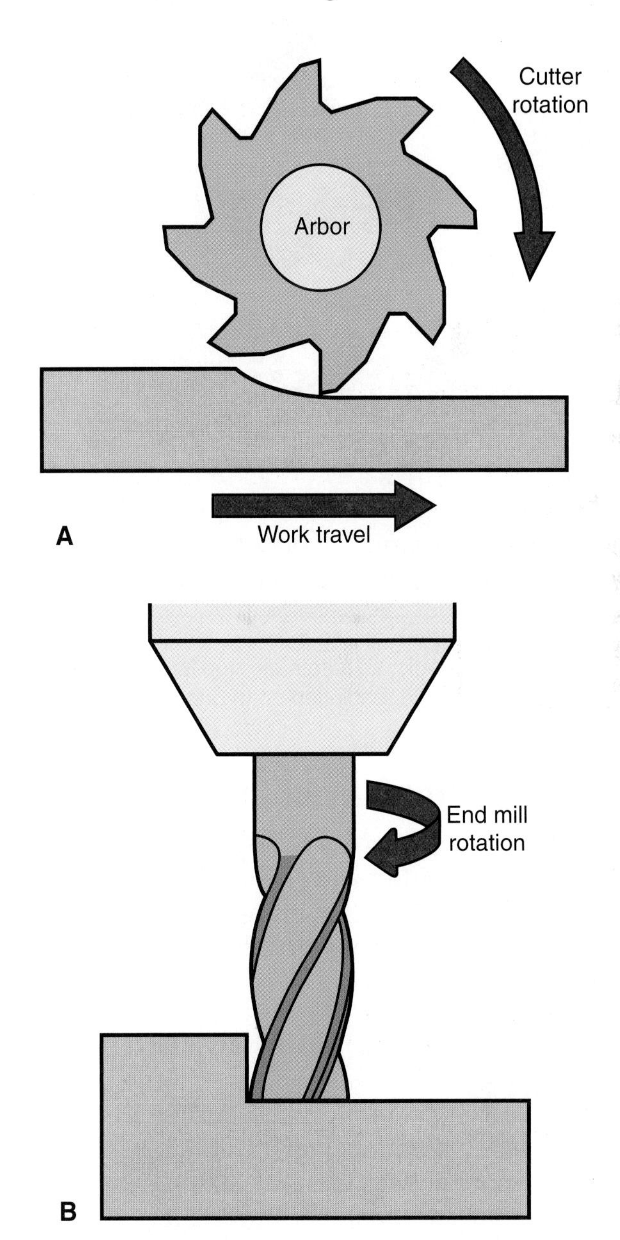

Goodheart-Willcox Publisher

Figure 1-13. Milling removes material by rotating a multitoothed cutter into the work. A—With peripheral milling, the surface being machined is parallel to periphery of the cutter. B—End mills have cutting edges on the circumference and the end.

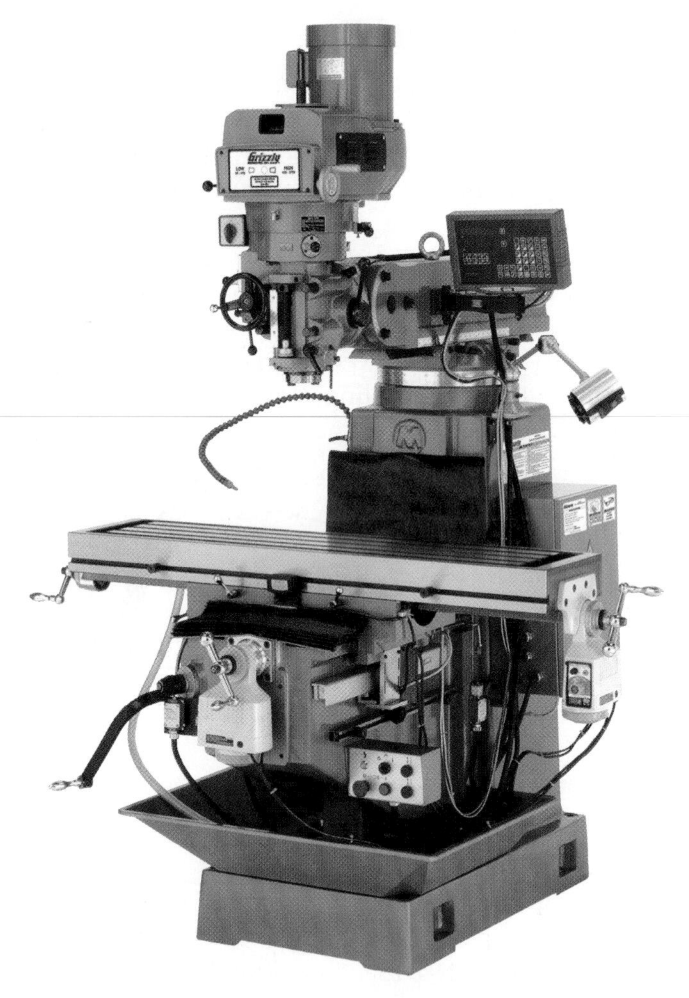

Photo courtesy of Grizzly Industrial, Inc. www.grizzly.com

Figure 1-14. A modern milling machine featuring power feed, variable speed controls, an automatic stop function, coordinate display, and selectable resolution up to one micrometer.

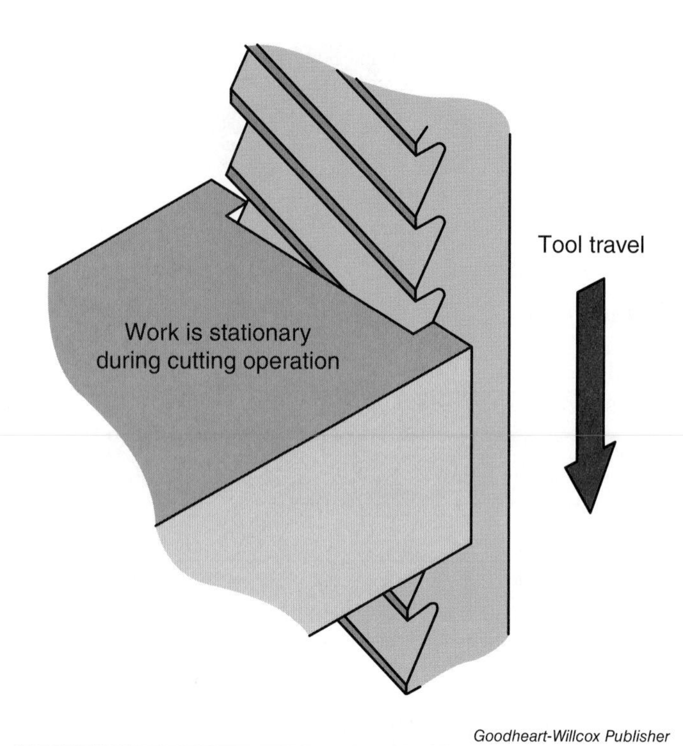

Goodheart-Willcox Publisher

Figure 1-15. A broach is a multitoothed cutting tool that moves against the work. Each tooth removes only a small portion of the material being machined. The cutting operation may be on a vertical or horizontal plane.

1.2.5 Broaching Machines

A ***broaching machine*** is designed to push or pull a multitoothed cutter across the work, **Figure 1-15**. Each tooth of the broach (cutting tool) removes only a small amount of the material being machined.

1.3 Nontraditional Machining Processes

There are a number of machining operations that have not evolved from the lathe. They are classified as nontraditional machining processes. These processes include the following:

- **Electrical discharge machining (EDM).** An advanced machining process that uses a fine, accurately controlled electrical spark to erode metal.
- **Electrochemical machining (ECM).** A method of material removal that shapes a workpiece by removing electrons from its surface atoms. In effect, ECM is exactly the opposite of electroplating.
- **Chemical milling.** A process in which chemicals are used to etch away selected portions of metal.
- **Chemical blanking.** A material removal method in which chemicals are used to produce small, intricate, ultrathin parts by etching away unwanted material.
- **Hydrodynamic machining (HDM).** A computer-controlled technique that uses a 55,000 psi water jet to cut complex shapes with minimum waste. The work can be accomplished with or without abrasives added to the jet.
- **Ultrasonic machining.** A method that uses ultrasonic sound waves and an abrasive slurry to remove metal.
- **Electron beam machining (EBM).** A thermoelectric process that focuses a high-speed beam of electrons on the workpiece. The heat that is generated vaporizes the metal.

- **Laser machining.** The laser produces an intense beam of light that can be focused onto an area only a few microns in diameter. It is useful for cutting and drilling.

1.4 Automating the Machining Process

In the late 1940s, the United States Air Force was searching for ways to increase production on complex parts for the new jet aircraft and missiles then going into production. The Parsons Corporation, a manufacturer of aircraft parts, had developed a two-axis technique for generating data to check helicopter blade airfoil patterns. This system used punched-card tabulating equipment. To determine the accuracy of the data, a pattern was mounted on a Bridgeport milling machine. With a dial indicator in place, the X and Y points were called out to a machinist operating the machine's X-axis handwheel and another machinist who controlled the Y-axis handwheel. With enough reference points established, the generated data proved accurate to ±0.0015″ (0.038 mm).

1.4.1 The Development of Numerical Control

Parsons realized that the technique might also be developed into a two-axis, or even three-axis, machining system. With an Air Force contract to manufacture a contoured integrally stiffened aircraft wing section, the Parsons Corporation subcontracted with the Servomechanism Laboratory at the Massachusetts Institute of Technology to design a three-axis machining system. MIT eventually took over the entire development project.

By 1952, MIT had designed a control system and mounted it on a vertical spindle machine tool. The system operated on instructions coded in the binary number system on punched (perforated) tape. Programming required the use of an early computer on which MIT was also experimenting.

Later in that year, MIT demonstrated the first machine tool capable of executing simultaneous cutting tool movement on three axes. Since mathematical information was the basis of the concept, MIT coined the term ***numerical control (NC)***. The first NC machines became available to industry in 1955.

1.4.2 Computer Numerical Control

In the mid-1970s, with the introduction of the microchip, the use of onboard computers on individual machine tools became possible. This led to the introduction of ***computer numerical control (CNC)***, **Figure 1-16**.

CNC machine tools are much easier to use than manually controlled machines. They have menu-selectable displays, advanced graphics (the multifunction screen displays the full operational data as a part is being machined), and a word address format for programming. The program is made up of sentence-like commands. Programs can be entered at the machine or downloaded from an external computer. Programs on punched tapes are no longer used. A modern CNC machining center is shown in **Figure 1-17**.

A CNC machine tool offers several benefits, including:

- **Accuracy.** It is capable of producing consistent and accurate workpieces.
- **Repeatability.** It is able to produce any number of identical workpieces once a program is verified.
- **Flexibility.** Changeover to running another type of part requires only a short period of nonproductive machine downtime.

AMT—The Association for Manufacturing Technology

Figure 1-16. CNC machine tools are equipped with onboard computers that permit computer-aided or manual programming. All controls needed for complete machine operation are in one location.

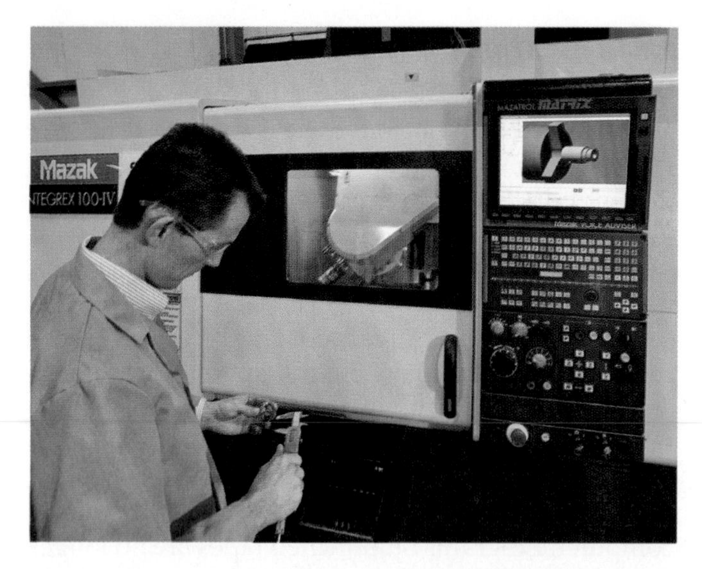

Mack Molding Co., Arlington, Vermont

Figure 1-17. A modern CNC turning and machining center with multiaxis capabilities. Its advanced multitasking technology allows turning and milling operations to be performed with a single setup.

The use of robotic systems for loading and unloading permits some machine tools to operate unattended during the entire machining cycle. Robots, **Figure 1-18**, also have many abilities useful in industrial applications:

- Operating in hazardous and harsh environments.
- Performing operations that would be tedious for a human operator.

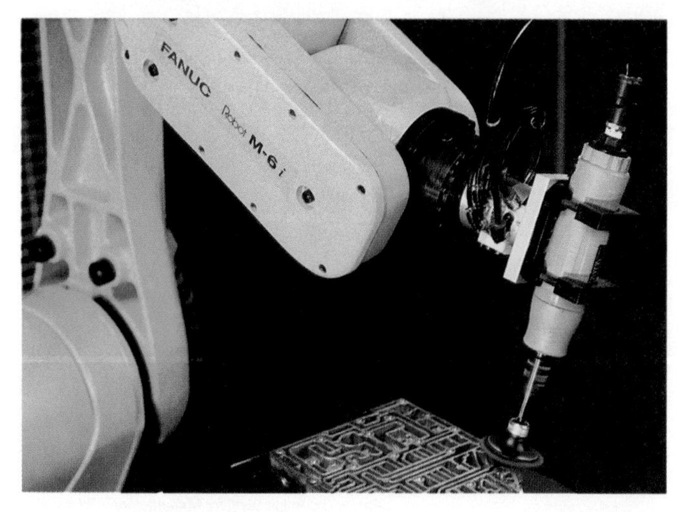

Fanuc Robotics

Figure 1-18. A robot is a programmable, multifunctional manipulator designed to move material, tools, or specialized devices through programmed motions for the performance of a variety of tasks. This robot is deburring a complex part following machining operations.

- Handling heavy materials.
- Positioning parts with great repetitive precision.

The automotive industry makes extensive use of robots in the manufacture and assembly of motor vehicles, **Figure 1-19**.

1.5 The Evolving Role of the Machinist

In recent years, the number of highly skilled machinists has been in decline. CNC machine tools have compensated for this trend to some degree. Since these machines operate under programmed control, the men and women who use them do not require the same level of skill or training as a skilled machinist.

However, because of these same CNC machine tools, the demand for machinists has not diminished. Machinists understand machining technology and what machine tools are capable of accomplishing. For these reasons, they make the best programmers and setup personnel.

There is still another reason for the high demand for machinists: although CNC equipment is found in almost all machine shops, surveys consistently show that there is still considerable work being produced on conventional manually operated machine tools.

Whether planning a CNC program or preparing to produce work on a conventional machine tool, a machinist must make many decisions on how to manufacture a part in the most economical way. A machinist must be able to perform the following activities:

- Make a thorough study of the print.
- Determine the machining that must be done.
- Ascertain tolerance requirements.
- Plan the machining sequence.
- Determine how the setup will be made.
- Select the machine tool, cutter(s), and other tools and equipment that will be needed.
- Calculate cutting speeds and feeds.
- Select a proper cutting fluid for the material being machined.

All of this is possible because of the skill, knowledge, and experience of the machinist. Essentially, a machinist is able to visualize the machining program.

Rainer Plendl/Shutterstock.com

Figure 1-19. The automotive industry makes extensive use of robots for positioning parts, welding, painting, and performing quality control tasks. Many production operations include computer-controlled robotic assembly lines like this one.

1.6 Acquiring Machining Skills and Knowledge

The skills and knowledge needed by the machinist are not acquired in a short time. It normally requires taking part in a multiyear salaried apprentice program. In addition to machine tool training under an experienced machinist, the program also involves related subjects such as English, algebra, geometry, trigonometry, print reading, safety, production techniques, and CNC principles and programming.

The National Institute for Metalworking Skills (NIMS), with the aid of the metalworking industry, developed a set of ***skills standards***, industry requirements for skilled workers. NIMS uses these standards to certify individuals through performance testing and accredited training programs that meet their standards. The standards provide skilled workers with certification that will afford them industry recognition.

Chapter Review

Summary

- Machine tools and machinists are vital to our modern industrialized world.
- Modern machine technologies require highly educated machinists.
- Machine tools are the foundation of industry and precision metalworking.
- Evolving power sources have changed the way machine tools operate and what materials they can machine.
- Basic machine tool processes include turning, sawing, drilling, grinding, milling, and broaching.
- There are a growing number of nontraditional machining processes.
- Automation and computers have dramatically affected machine tool operations.
- Through skill, knowledge, and experience, a machinist determines the best possible way to manufacture a part, whether using manual or CNC machining techniques.

Review Questions

Answer the following questions using the information provided in this chapter.

1. Jobs such as tool and die making and precision machining require aptitudes comparable to those of _____.
   A. high school graduates
   B. college graduates
   C. high school equivalency graduates
   D. All of the above.
   E. None of the above.
2. One of the first machine tools, the bow lathe _____.
   A. could only turn softer materials
   B. has been dated back to about 1200 BC
   C. eventually gave way to treadle power
   D. All of the above.
   E. None of the above.
3. The Industrial Revolution could not have taken place without the cheap, convenient power of the _____.
4. Eli Whitney's mass-production system for muskets had a major problem because _____.
   A. there were no skilled workers
   B. there was no good source of power
   C. there was no standard of measurement
   D. All of the above.
   E. None of the above.
5. What occurred in the mid-1860s that was very important to the development of machining technology in the United States?
6. List seven power sources in the order they have evolved.
7. Almost all machine tools have evolved from the _____.
8. List three types of basic machine tools and briefly describe their operation.
9. List four types of nontraditional machining processes and briefly describe their operation.
10. The introduction of the microchip in the mid-1970s led to the introduction of _____ machine tools.
11. CNC machine tools operate according to a(n) _____ made up of sentence-like commands.
12. List four industrial applications of robots.
13. When planning the manufacture of a part, a machinist must _____.
    A. ascertain tolerance requirements
    B. select the machine tool, cutter(s), and other tools and equipment that will be needed
    C. select a proper cutting fluid for the material being machined
    D. All of the above.
    E. None of the above.

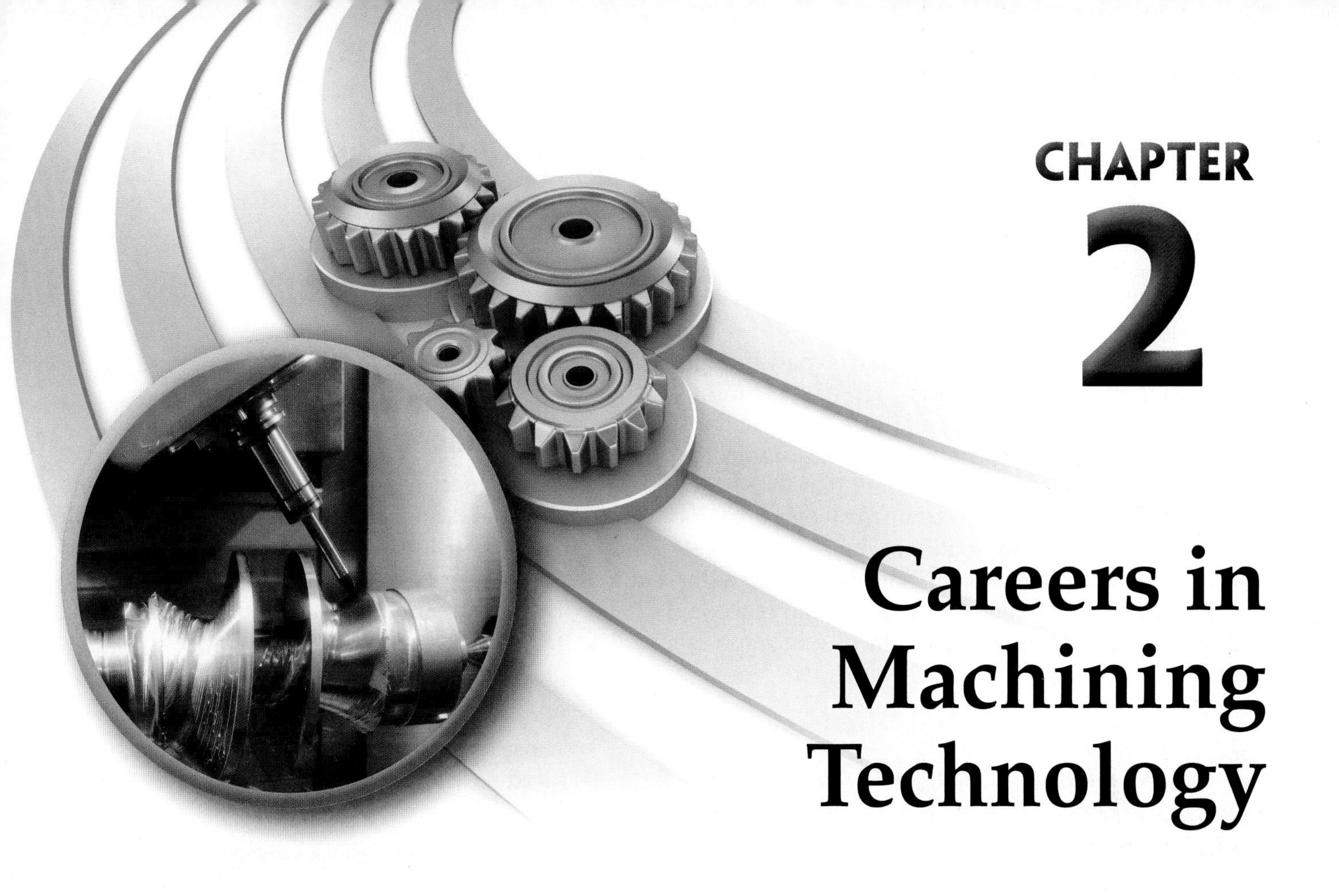

CHAPTER 2

Careers in Machining Technology

Learning Objectives

After studying this chapter, you will be able to:

- List the requirements for the various machining technology occupations.
- Explain where to obtain information on occupations in machining technology.
- State what industry expects of an employee.
- Describe what an employee should expect from industry.
- Summarize the information given on a résumé.

Technical Terms

apprentice
associate's degree
bachelor's degree
career
engineer
job shop
part programmer
résumé
semiskilled worker
skilled worker
technician
toolmaker

A ***career*** is a job that requires specialized training and commitment to the profession. Many people want a career that is both challenging and interesting. Their philosophy is that if you enjoy your job, you will never "work" a day in your life. Others are satisfied with whatever job comes along. With which viewpoint do you agree?

If you are looking for a career that is challenging, interesting, and rewarding, the field of machining offers many opportunities. See **Figure 2-1**. Whether you choose one of the machine shop areas or select a career in a related field, you will find that the study of *Machining Fundamentals* is basic to all of them.

No matter which occupational choice you make, you should realize that you will have to keep up with technical progress. To be successful and advance in your career, a continuing program of education is usually necessary.

Dmitry Kalinovsky/Shutterstock.com

Figure 2-1. The field of metal machining offers many opportunities for semiskilled and skilled workers, technicians, and professional personnel.

2.1 Machining Job Categories

Jobs in material machining fall into four general categories:

- Semiskilled worker.
- Skilled worker.
- Technicians.
- Professionals.

2.1.1 Semiskilled Workers

A ***semiskilled worker*** performs basic, routine operations that do not require a high degree of skill or training. Semiskilled workers may be classified into the following general groups:

- Those who are helpers for skilled workers.
- Those that operate machines and equipment used in making things. The machines are set up by skilled workers.
- Those who assemble the various manufactured parts into final products, **Figure 2-2**.

Dmitry Kalinovsky/Shutterstock.com

Figure 2-2. Many semiskilled workers are employed in assembly industries, where they assemble manufactured parts into complete units. Training periods to learn these job skills are relatively short.

There is little chance for advancement out of semiskilled jobs without additional schooling and training. Most semiskilled work is found in production shops where there are great numbers of repetitive operations. In general, semiskilled workers are told what to do and how the work is to be done. They are often the first to lose their jobs when there is a downturn in the economy.

2.1.2 Skilled Workers

A ***skilled worker*** has been trained to do more complex tasks. Skilled workers are found in all areas of material machining. Many skilled workers obtain their training as an ***apprentice*** in an apprenticeship program (on-the-job training while working with skilled machinist), **Figure 2-3**.

Four or more years of instruction under an experienced machinist is generally required. In addition to working in the shop, an apprentice usually studies related subjects, such as math, science, English, print reading, metallurgy, safety, and production techniques. Upon completion of an apprenticeship program, the worker is capable of performing the precise work essential to the trade.

In recent years, the demand for skilled workers has grown tremendously. Workers entering the field can receive their training through the armed forces, **Figure 2-4**, or in career and technical education

U.S. Army

Figure 2-4. The Army, and other branches of the armed forces, offers excellent opportunities for learning a trade. These service women are learning to machine parts on a manual lathe. The coursework they complete can be used toward obtaining National Institute of Metalworking Skills (NIMS) credentials.

(Use of military imagery does not imply or constitute endorsement of Goodheart-Willcox Publisher, its products, or services by the U.S. Department of Defense.)

Alan Poulson Photography/Shutterstock.com

Figure 2-3. The apprentice studies under an experienced machinist for a period of four or more years. The training program also includes the study of related subjects like math, English, and science.

programs offered in high schools and community colleges. Many community college programs are offered in conjunction with local industry.

There are several areas in which a machinist may concentrate. Some of these areas of specialization are discussed in the following sections.

All-Around Machinist

An all-around machinist is skilled in the setup and operation of most types of machine tools. He or she must be familiar with both manual and computer-controlled machine tools and how they are programmed, **Figure 2-5**. An all-around machinist is expected to plan and carry out all of the operations needed to machine a job.

Many all-around machinists work in ***job shops***. Some job shops specialize in creating custom or experimental machining projects for various clients. Other job shops specialize in manufacturing products with very small production runs.

Tool and Die Maker

A ***toolmaker*** is a highly skilled machinist who specializes in producing the tools and tooling needed for machining operations. These include the following:

- **Dies.** Special tools for shaping, forming, stamping, or cutting metal or other materials.
- **Jigs.** Devices that position work and guide cutting tools.
- **Fixtures.** Devices to hold work while it is machined.

John-james Gerber/Shutterstock.com

Figure 2-5. The all-around machinist can set up and operate most types of machine tools, whether manual or CNC.

These tools are necessary for modern mass production techniques. Toolmakers must have a broader background in machining operations and mathematics than most other skilled workers in the trade. See **Figure 2-6**.

A diemaker specializes in making the punches and dies needed to stamp out such parts as auto body panels, electrical components, and similar products. He or she will also produce the dies for making extrusions (metal shaped by being pushed through an opening in a metal disc of proper configuration) and die castings (parts made by forcing molten metal into a mold). Like the toolmaker, a diemaker is a highly skilled machinist.

Specialist

The layout specialist is a machinist who interprets the drawings and uses precision measuring tools to mark off where metal must be removed by machining from castings, forgings, and metal stock. This person must be very familiar with the operation and capabilities of machine tools. He or she is well-trained in mathematics and print reading.

A setup specialist is a person who locates and positions ("sets up") tooling and work-holding devices on a machine tool for use by a machine tool operator. This worker may also show the machine tool operator how to do the job, and often checks the accuracy of the machined part. See **Figure 2-7**.

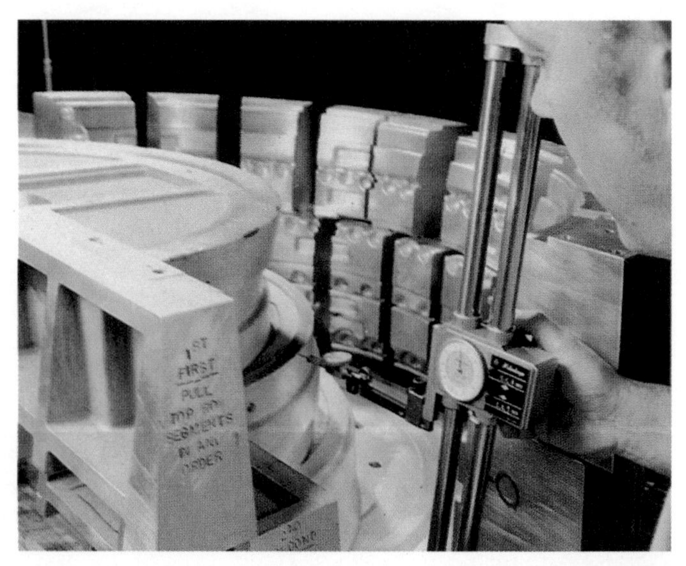

Precision Castparts Corp.

Figure 2-6. A tool-and-die maker checking the dies used for molding a plastic pattern used to cast a jet engine component. Master tooling ensures that other sections of the engine, made elsewhere in the United States, Israel, and Europe, will fit together perfectly.

Hydromat, Inc.

Figure 2-7. A setup specialist is a master machinist who prepares machine tools for operation by less highly trained personnel. After thoroughly checking the machined part to be sure it will meet specifications, the setup specialist will turn the machine tool over to a machine operator.

Tri-Tool, Inc.

Figure 2-8. The programmer prepares the information such as tools, tool paths, and machining sequence for a CNC program that will direct the entire machining process of a specific part. The programmer must thoroughly understand machining technology.

Part Programmer

A ***part programmer*** inputs data into a computer-controlled (CNC) machine tool for machining a product. CNC machine tools have revolutionized the fields of machining and manufacturing. However, computers have no inherent intelligence and cannot think or exercise judgment. They must be programmed by a highly skilled part programmer who studies the drawings and determines the sequences, tools, and motions the machine tool must carry out to machine the part, **Figure 2-8**.

To perform this task, a part programmer must have a background that includes the following:

- Formal training in computer hardware as it relates to machine tool operation.
- Formal training using computer-aided design (CAD) and computer-aided manufacturing (CAM) software.
- Experience in reading and interpreting drawings.
- A thorough grounding in machining technology and procedures.
- A working knowledge of cutting speeds and feeds for various tools and materials.
- Extensive training in mathematics.

Many community colleges and career and technical education centers offer programs in CNC programming, computer-aided design, and computer-aided manufacturing.

Supervisor or Manager

A supervisor or manager is usually a skilled machinist who has been promoted to a position of greater responsibility. This person will direct other workers in the shop and is responsible for meeting production deadlines and keeping work quality high, **Figure 2-9**. In many shops, the manager may also be responsible for training and other tasks.

2.1.3 Technicians

The ***technician*** is a member of the production team who operates in the realm between the shop and engineering departments. The position is an outgrowth of today's highly technological and scientific world. The job usually requires at least two years of college, with a program of study centered on math, science, English, computer science, quality control, manufacturing, and production processes. Many state and community colleges offer two year ***associate's degree*** programs devoted to preparing students for such technical positions.

Steve Good/Shutterstock.com

Figure 2-9. The supervisor or manager of the production department works very closely with machinists, engineers, metallurgists, and other staff. Supervisors may also be responsible for ordering stock, meeting production deadlines, and ensuring high quality work.

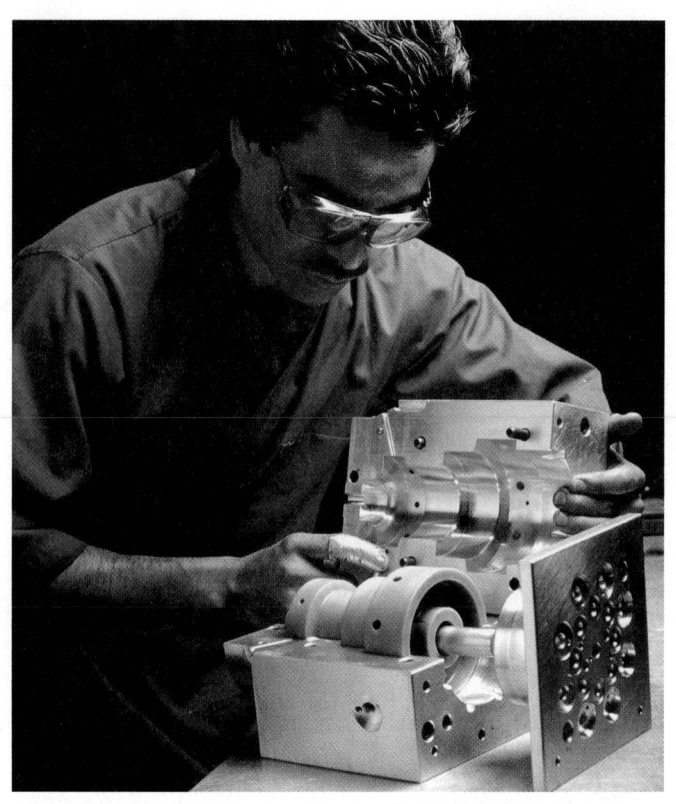

Chuck Rausin/Shutterstock.com

Figure 2-10. This quality control technician is inspecting a die used to make plastic or wax patterns to be used in a metal casting operation. The technician must ensure that the die meets engineering specifications before being shipped to customers.

The technician assists the engineer by testing various experimental devices and machines, compiling statistics, making cost estimates, and preparing technical reports. Many inspection and quality control programs are managed by technicians, **Figure 2-10**. Technicians also repair and maintain computer controlled machine tools and robots.

2.1.4 Professionals

Several professions offer many excellent opportunities in the fields of machining, metalworking, and manufacturing. Teaching is a challenging profession that offers a freedom not found in most other professions. Teaching can be a very personally satisfying profession, though it is a field that students often overlook, **Figure 2-11**. Teachers of industrial arts, industrial technology, vocational education, and career and technical education are in a fortunate position. It is not an overcrowded profession and it appears there will be a demand for teachers for many years to come.

Goodheart-Willcox Publisher

Figure 2-11. The teaching profession is a challenging one. Many skilled educators will be needed in machining technology if the United States is to maintain its position as a world leader in that industry.

To teach machining, four years of college training are usually needed, **Figure 2-12**. While industrial experience is ordinarily not required, it will prove very helpful.

Engineering is a fast growing and challenging profession. An ***engineer*** uses mathematics, science, and knowledge of manufacturing principles to develop new products and processes for industry, **Figure 2-13**.

A four year ***bachelor's degree*** program in advanced theoretical education and training is usually the minimum requirement for entering the engineering profession. Some men and women have been able to enter the profession without a degree after a number of years experience as machinists, drafters, or engineering technicians. However, they are usually required to take additional college-level training.

Industrial engineers are primarily concerned with the safest and most efficient use of machines, materials, and personnel, **Figure 2-14**. In some instances, he or she may be responsible for the design of special machinery and equipment to be utilized in manufacturing operations.

Mechanical engineers are normally responsible for the design and development of new machines, devices, and ideas. This engineering specialty is also involved with the redesign and improvements of

William Schotta, Millersville University

Figure 2-12. This college student may someday teach machine technology. During four or more years of training, she will learn all phases of machine tool operation and programming.

Corepics VOF/Shutterstock.com

Figure 2-13. An engineer inspects the rotors of a huge industrial wind turbine that will be used to test the aerodynamic qualities of everything from aircraft to automobiles.

Minerva Studio/Shutterstock.com

Figure 2-14. Industrial engineers have many duties. Some industrial engineers are responsible for maintaining a safe work environment, managing work flow, and making the most efficient use of the machines and materials.

Chuck Rausin/Shutterstock.com

Figure 2-15. Mechanical engineers are responsible for the design and development of new machines, devices, and ideas.

existing equipment, **Figure 2-15**. Some mechanical engineers specialize in different types of mechanical systems, such as heating and cooling systems, automotive vehicles, and robotics.

Tool and manufacturing engineers often work with the other engineers. A principal concern of the mechanical engineer is the design and development of the original or prototype model. When this model has been thoroughly tested and has met design requirements, the product is turned over to the tool and manufacturing engineer to devise methods and means required to manufacture and assemble the item, **Figure 2-16**.

Metallurgical engineers are involved in the development and testing of metals that are used in products and manufacturing processes.

2.2 Preparing to Find a Job in Machining Technology

Machining technology is a technical area with constantly developing new ideas, materials, processes, and manufacturing techniques.

OSG Tap & Die, Inc.

Figure 2-16. Tool and manufacturing engineers devise new methods to manufacture complex products. This engineer is setting up a CNC machine for a new cycle.

This means that occupational opportunities are created that were not previously available. One study reported that the average graduate will be employed in at least five different jobs in his or her lifetime, and three of them do not even exist yet!

2.2.1 Obtaining Information on Machining Occupations

There are many sources of occupational information. The most accessible for today's students are the school's career center, technical education instructors, and the Internet.

State employment services are also excellent sources for getting information on local and state employment opportunities, as are the various trade unions concerned with the metalworking trades.

The field and regional offices of the Office of Apprenticeship of the United States Department of Labor may also be contacted for information on apprenticeship programs in your area. The *Occupational Outlook Handbook*, available on the US Department of Labor's website and in published formats, describes many specific job categories and estimates the future demand for workers in each occupation.

Information on technical occupations is also available from community colleges. Many of them offer associate's degrees in technical areas.

2.2.2 Traits Employers Look for in an Employee

Industry is always on the lookout for bright people who are not afraid to work and assume responsibility. Employers also look for the following traits in an employee, often referring to scholastic records, references, and previous employers to obtain the necessary information:

- **Skills and knowledge.** Has the technical skills and knowledge necessary for an entry position. Does work neatly and accurately. Pays attention to details.
- **Integrity and honesty.** This trait is on the same level of importance as technical skills and knowledge.
- **Comprehension.** Is able to understand oral and written instructions and to read and interpret prints.
- **Dependability.** Has a good attendance and punctuality record in class and at former jobs.
- **Teamwork.** Shows the ability to work well with peers and supervisors.
- **Communication.** Is able to communicate ideas and suggestions orally and in writing.
- **Self-confidence.** Takes pride in work and will not knowingly turn out inferior or substandard material.
- **Accountability.** Is able to assume responsibility and be accountable for his or her actions.
- **Initiative.** Volunteers ideas. Demonstrates leadership.
- **Grooming and dress.** Presents a positive personal appearance.

2.2.3 Factors for Rejection for Employment

There are many factors and traits that can cause a person to be rejected for employment. They include the following:

- Poor personal appearance.
- Poor scholastic performance.
- Poor attendance record.
- Lack of maturity.
- Lack of interest or enthusiasm for the job being sought.
- Knowing little or nothing about the company where employment is sought.
- Unrealistic salary demands.
- Lack of ability to communicate.

2.3 How to Get a Job

Securing a job is a very important task. To be successful, you will have to spend as much time looking for this position as you would working at a regular job. There are several other things that you can do to make this task easier.

You will have to decide what type of work you would like to do. Most schools and state employment services administer tests that will help you determine the areas of employment where you will have a good chance of succeeding.

Answering the following questions will give you additional help:

- What can I do with some degree of success?
- What have I done that others have commended me for doing well?
- What are the things I really like to do?
- What are the things I do not like to do?

- What jobs have I held? Why did I leave them?
- What skills have I acquired while in school?

You will probably have two or more areas of interest. After listing them, start gathering information on these areas of interest. Use as many different sources as time permits. This may include searching the Internet, reading publications about the profession, talking with persons doing this type of work, and visiting industry. Plan your educational program to prepare for entry into a specific job or for advanced schooling.

The next problem is figuring out how you go about getting that job. Jobs are always available. Workers get promoted or they retire. Some quit, die, or get fired. Technological progress also creates new jobs. However, you must "track down" these jobs. There is no easy way to get a challenging job.

Concentrate on getting the job. If possible, always make your initial request for a job in person. Always be specific on the type of job you are seeking. Make sure you are qualified for that job. Never ask for "just any job" or inquire, "What openings do you have?"

Dress appropriately. Job hunting is not the time to wear old clothes or torn and beat-up shoes. Be clean and well-groomed.

When filling out a job application, avoid leaving any spaces blank. The employer may think there is something you do not want to answer. If the question is not applicable to you, write in "Not applicable," "Does not apply," or "NA." For example, there might be a question on the application that asks, "What was your highest rank in the armed forces?" If you were not in the armed forces, you would write in, "Does not apply," or "NA," rather than leaving the answer space blank.

Last, but not least, know where to look for a job. There are many websites dedicated to listing job opportunities. Schools often host job fairs where potential job seekers can meet with employers and companies. Even check the classified advertising section of local newspapers. Talk with friends and relatives who are employed. They may be aware of job openings at their places of employment before the jobs are advertised.

A new office or factory building may indicate potential job openings. It would also be to your advantage to prepare a list of desirable employers in your community and visit their personnel offices. Plan these visits on a routine basis when jobs are not readily available. The office staff will then know you are interested in working for their firm and may give you preference.

One thing you must remember: The job will *not* come to you, you must search for it!

2.3.1 Preparing a Résumé

To speed the tedious task of filling out job applications, prepare a ***résumé*** in advance. A résumé is a summary of your educational and employment background. It will ensure uniform information with little chance for confusing responses. The résumé is submitted when applying for a job. Your résumé should include the following items:

- Your full name.
- Your telephone number, e-mail address, and full address. Do not forget area and zip codes.
- Education and any special training. Include dates of all educational attendance.
- The types of equipment that you can safely operate.
- Names of previous employers. List the places you have worked, starting with the most recent. Include the items that follow (for each previous employer):
  - Company name and address.
  - Dates employed.
  - Immediate supervisor's name.
  - Salary or pay rate.
  - Reason for leaving.
- Names and addresses of references. Do not include relatives unless you have worked for them. Make sure you secure permission before using a person for a reference. (Today, many job seekers note "References available on request," instead of listing the names and addresses on the résumé.)

2.3.2 What an Employee Should Expect from Industry

From the preceding sections, you have some idea of what industry expects from an employee. However, are you aware of what an employee should expect from industry? Over and above salary and fringe benefits, what should you expect from an employer?

The following are a few questions you can ask when selecting a place of employment:

- Is a relatively safe and clean work area provided? Obviously, some areas can never be made as safe as others. For example, tapping a blast furnace is inherently more dangerous than working on a small lathe or drill press.
- Are work areas adequately lighted, heated, and ventilated? Are noxious fumes and dust particles filtered from the air?
- Is proper safety clothing and equipment available for all dangerous work? Safety items such as goggles, hearing protectors, and steel-tipped safety shoes may be provided free or at minimum cost.
- Are all necessary precautions observed when hazardous materials are involved?
- Is there a preventative safety program, and are safety regulations and precautions rigorously enforced?

2.3.3 Factors That Can Lead to Job Termination

The following factors can lead to failure to get a promotion, or possibly being terminated (fired) from a job. They include the following:

- Alcohol or illegal drug abuse on the job.
- Inability or refusal to perform the work required.
- Being habitually tardy or missing work repeatedly without adequate reasons.
- Inability to work with supervisors or peers.
- Fighting with or making threats to fellow workers or supervisors.
- Inability to work as a team member.

2.4 Keeping Your Skills Current

The completion of your formal schooling does not mean the end of your training and study. To keep a job and advance in it, you will have to keep up-to-date with the knowledge and new skills that advanced technologies demand. Keen competition from foreign-made products and the ever changing nature of technology make this a very real necessity.

Chapter Review

Summary

- Machining careers are divided into four main categories—semiskilled, skilled, technical, and professional—based on training and level of education.
- Information on machining-related careers can be found through a school's career center, from instructors, the Internet, community colleges or career and technical education centers, state employment services, the armed forces, or the Department of Labor.
- Employers are looking for very specific traits in their employees including work-related skills, knowledge, and ability, but also honesty, dependability, teamwork and communication, initiative, and proper grooming and dress.
- What is on your résumé may get you hired, but in order to keep and advance in your chosen career it is very important to keep up-to-date with new skills and advances in technology.

Review Questions

Answer the following questions using the information provided in this chapter.

1. List the four general categories of metalworking occupations.
2. _____ workers are those who perform operations that do *not* require a high degree of skill or training.
3. The _____ worker usually starts his or her career as an apprentice.
4. Other than apprenticeship programs, where can specialized career training now be obtained?
5. Describe what an all-around machinist is expected to do.
6. What does a layout specialist do?
7. List five areas of knowledge a part programmer should have to perform his or her job properly.
8. What are some of the duties of a technician?
9. List the areas of study usually included in a technician's educational program.
10. List three sources of information on metalworking occupations.
11. List five traits an employer wants in a prospective employee.
12. List five factors that can lead to rejection for employment.
13. What is a job résumé?
14. Why should a résumé be prepared in advance?
15. List three questions an employee should ask when selecting a place of employment?
16. What are three factors that can lead to job termination?
17. What is necessary if you want to keep and advance in your chosen career?

CHAPTER 3

Shop Safety

Learning Objectives

After studying this chapter, you will be able to:

- Explain why it is important to develop safe work habits.
- Dress in the proper safety equipment and clothing for a machine shop.
- Recognize and correct unsafe work practices.
- Apply safe work practices when employed in a machine shop.
- Select the appropriate fire extinguisher for a particular type of fire.

Technical Terms

approved respirator
combustible material
machine shield
OSHA
protective clothing
safety equipment
spontaneous combustion
ventilation

Shop safety is not something to be studied at the start of a training program and then forgotten; most accidents are caused by carelessness or by not observing safety rules. Remember this when your instructor insists on safe work practices. If you are diligent and follow instructions with care, machining operations can be safe and enjoyable. Safe work practices should become a force of habit.

Since it is not possible to include every safety precaution, the safety practices in this chapter are general. Safety precautions for specific tools and machines are described in the text where they apply, along with the description and operation of the equipment. Refer to **Figure 3-1**.

Study all safety rules carefully and constantly apply them. When in doubt about any task, get help. *Do not* take chances.

SAFETY NOTE

Throughout the text, 'Safety Notes' are set aside to emphasize important tips and warn against potential hazards.

3.1 Safety in the Shop

Keep the shop clean. Metal scraps should be placed in the scrap bin. Never allow them to remain on the bench or floor. These metal scraps can be very sharp and cause injury if left on the bench or floor.

Christopher Parypa/Shutterstock.com

Figure 3-1. None of the pilots in this precision flying team would think of taking off to give a flight demonstration until all the plane's systems were in safe operating condition. In the same way, you should never operate a machine tool until you have determined that it is in safe operating condition.

Exercise extreme care when you are machining unfamiliar materials. For example, magnesium chips burn with great intensity under certain conditions. Applying water to the burning magnesium chips only intensifies the fire. Machining equipment can be damaged beyond repair and very serious burns can result.

Inhaling fumes or dust from some of the more advanced and exotic materials can cause serious respiratory ailments. *Do not* machine a material until you know what it is and how it should be handled.

An approved respirator and special ***protective clothing*** must be worn when machining some materials. Machines must be fitted with effective vacuum systems as needed.

The shop is a place to work, not play. It is not a place for horseplay. A "joker" in a machine shop is a walking hazard to everyone. Daydreaming also increases your chances of injury.

If you have been ill and are using medication, check with your doctor or school clinic to determine whether it is safe for you to operate machinery. For example, many cold remedies recommend that you do not operate machinery while taking the medication because of possible drowsiness.

SAFETY NOTE

Avoid using compressed air to remove chips and cutting oil from machines. Flying chips can cause serious eye injuries. Also, oil that has been vaporized by the stream of air can ignite, resulting in painful burns and property damage.

Oily rags must be placed in an approved safety container. See **Figure 3-2**. Rags or waste used to clean machines will also have metal slivers embedded in them, posing an additional hazard. Placing them in a safety container will help make sure they will not be used again. Dispose of the rags daily. This will minimize the possibility of ***spontaneous combustion*** (ignition by rapid oxidation or burning of oil without an external source of heat).

Keep hand tools in good condition. Store tools in such a way that people cannot be injured while they are removing the tools from the tool panel or storage rack.

Use care when handling long sections of metal stock. Accidentally contacting a light fixture with the stock, for example, could cause severe electrical burns or even death.

Justrite Manufacturing Company

Figure 3-2. Oily rags used for cleaning machines or soaking up spills should be placed in an approved safety container like this one to minimize fire hazards. The container should be emptied daily and contents disposed of properly.

When moving heavy machine accessories or large pieces of metal stock, always ask for help. The back injuries that can result from improper lifting are usually long-term injuries.

Dress properly for working around machinery. Severe injuries or even death can result if clothing, hair, or jewelry gets caught in moving parts. Avoid wearing loose-fitting clothing that could catch in machinery. A snug-fitting shop coat or apron can be worn to protect your street clothes, **Figure 3-3**. Keep sleeves rolled up. Rings and other jewelry should be removed before working around machinery. If you have long hair, wear a cap or use other means of containing it. When operating a machine with a rotating spindle, do not wear gloves. Crushing of the hands can result from gloves getting caught in the spindle.

For jobs where dust and fumes are a hazard, ensure that there is adequate ***ventilation***. Return solvents and oils to proper storage place after use. Wipe up spilled oil or solvent right away. If the spill area is extensive, use an approved-type oil absorbent. See **Figure 3-4**.

Millersville University

Figure 3-3. This trainee is properly dressed for the job she is doing. She is wearing approved eye protection and a snug-fitting apron. The machine was carefully checked before she began to operate it.

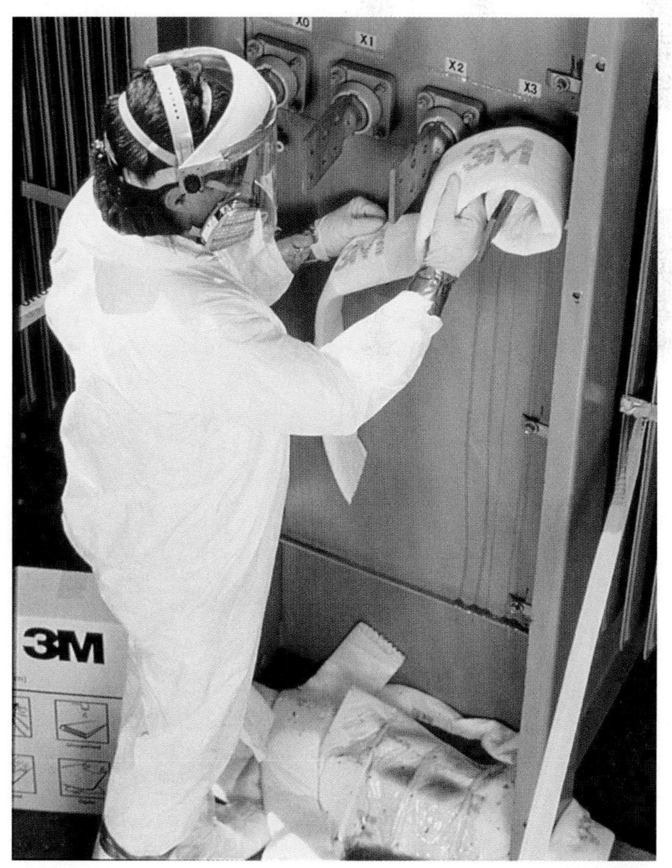

3M Company

Figure 3-4. This worker in a manufacturing plant is using a special oil-absorbent material to wrap a leaking machine component. Note that the material, which is packaged in roll form, has also been used to soak up oil on the machine base and floor.

Wear appropriate ***safety equipment***, **Figure 3-5**. In noisy areas, use earplugs or another type of hearing protection. Disposable plastic gloves will protect your hands when handling oils, cutting fluids, or solvents. Wear a dust mask when machining produces airborne particles, such as those from sand castings, plastics, and some grinding operations. An ***approved respirator*** must be worn in areas where machining operations produce a mist of oil or coolants. See **Figure 3-6**. Suitable personal protective equipment must also be worn when handling sharp, hot, or contaminated materials.

Take no chances. Always protect your eyes. Eyesight that has been damaged or destroyed cannot be replaced. Wear safety glasses, goggles, or face shields approved by ***OSHA*** (the United States Occupational Safety and Health Administration).

Wear eye protection whenever you are in the shop. Protective eyewear should be used when conditions call for it. It is good practice to have your own personal safety glasses. The cost is reasonable. Your instructor can help you determine the style best suited for your needs, **Figure 3-7**. If you wear glasses, special safety lenses are available that can be ground to your prescription. Your eye doctor or optician can help get them.

Know your job. It is foolish and disastrous to operate machines without first receiving proper instruction. If you are not sure what must be done, or how a task should be performed, get help.

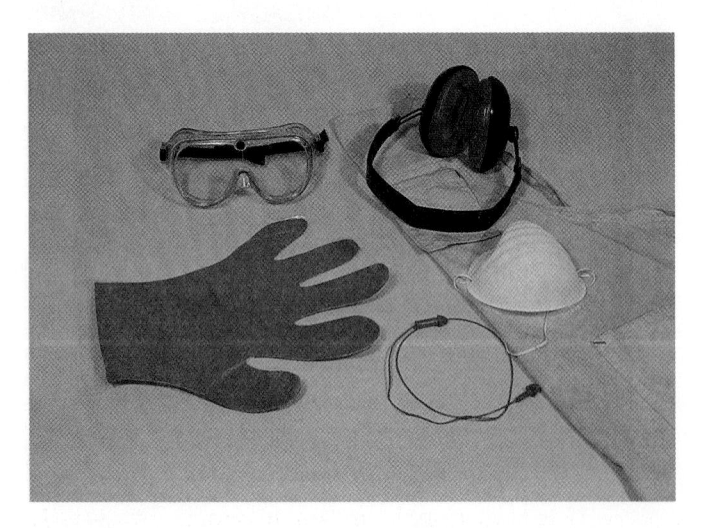

Goodheart-Willcox Publisher

Figure 3-5. Wear appropriate safety equipment. Shown are approved eye protection, an apron to protect clothing, plastic gloves for handling oils and solvents, a hearing protector, earplugs, and a dust mask.

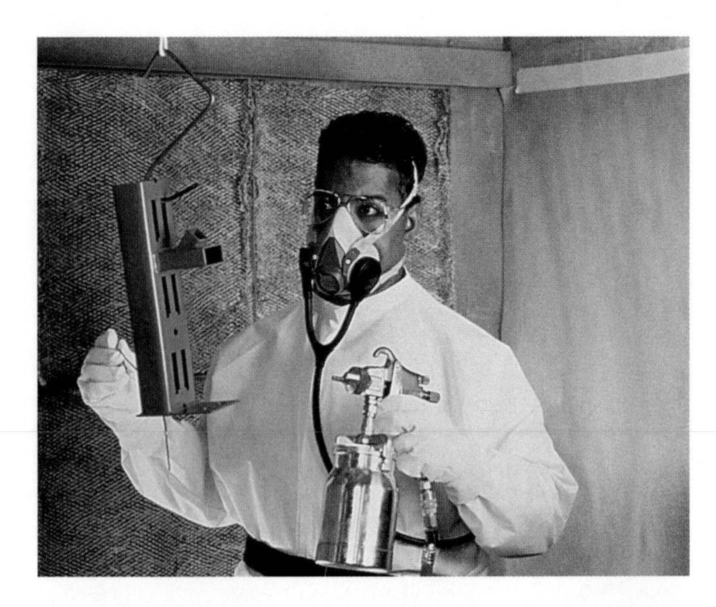

3M Company

Figure 3-6. Whenever fine airborne mists of oil, coolant, or other materials are present, an approved respirator is required. This spray painter is wearing a respirator supplied with clean air through a tube from a central source. It is also important to use proper eye protection, such as the safety glasses worn by this worker.

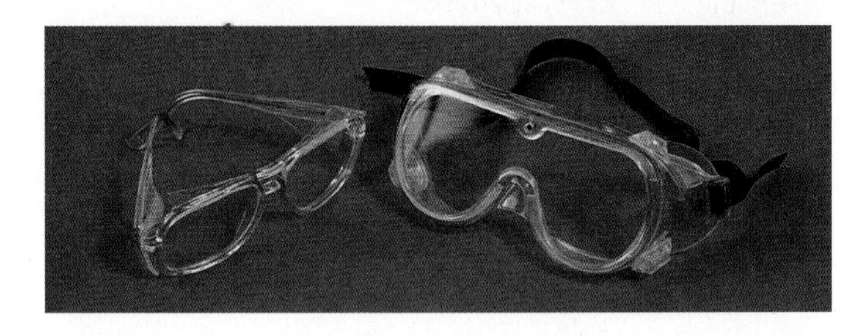

Goodheart-Willcox Publisher

Figure 3-7. Safety glasses are available in a number of styles. The model at left is similar to regular eyeglasses, but has "wings" at each side and on top to guard against flying particles. The goggle-style model at right fits tightly against the face and can be worn over regular eyeglasses.

3.1.1 Safety Aids

Not all barriers provide physical protection from machine hazards. Awareness barriers serve to remind the operator of an area that is dangerous. In its simplest form, a barrier may be nothing more than red or yellow lines painted on the floor. More complex barriers stop the machine when a light beam or electronic beam is broken by someone entering the danger area.

Machine shields, **Figure 3-8**, provide protection from flying chips and splashing cutting fluids or coolants. Many CNC machine tools are fitted with large sliding shields that cover the entire machining area, **Figure 3-9**.

Goodheart-Willcox Publisher

Figure 3-8. Specially designed safety shields are available for almost all machine tools. They should always be in place before the machine is operated.

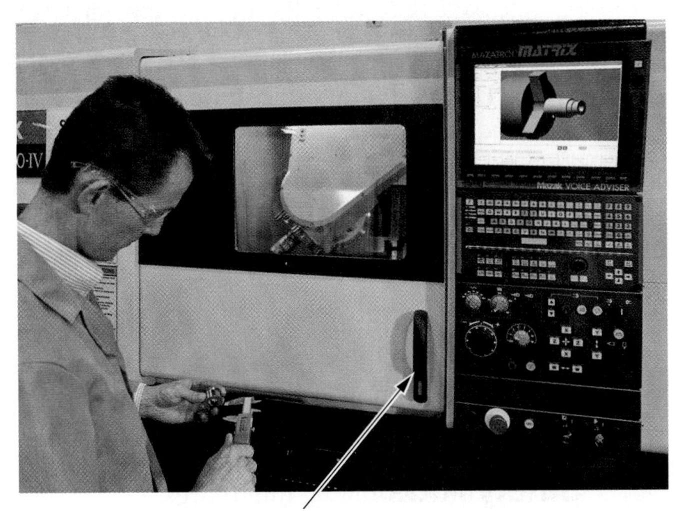

Mack Molding Co., Arlington, Vermont

Figure 3-9. Many CNC machine tools use sliding shields over the machining area to protect the machinist from flying chips and vaporized coolant.

Many different types of warning signs, **Figure 3-10**, are used to notify workers of potential hazards. *No Smoking* signs must be posted in areas where flammable or combustible materials are used and stored.

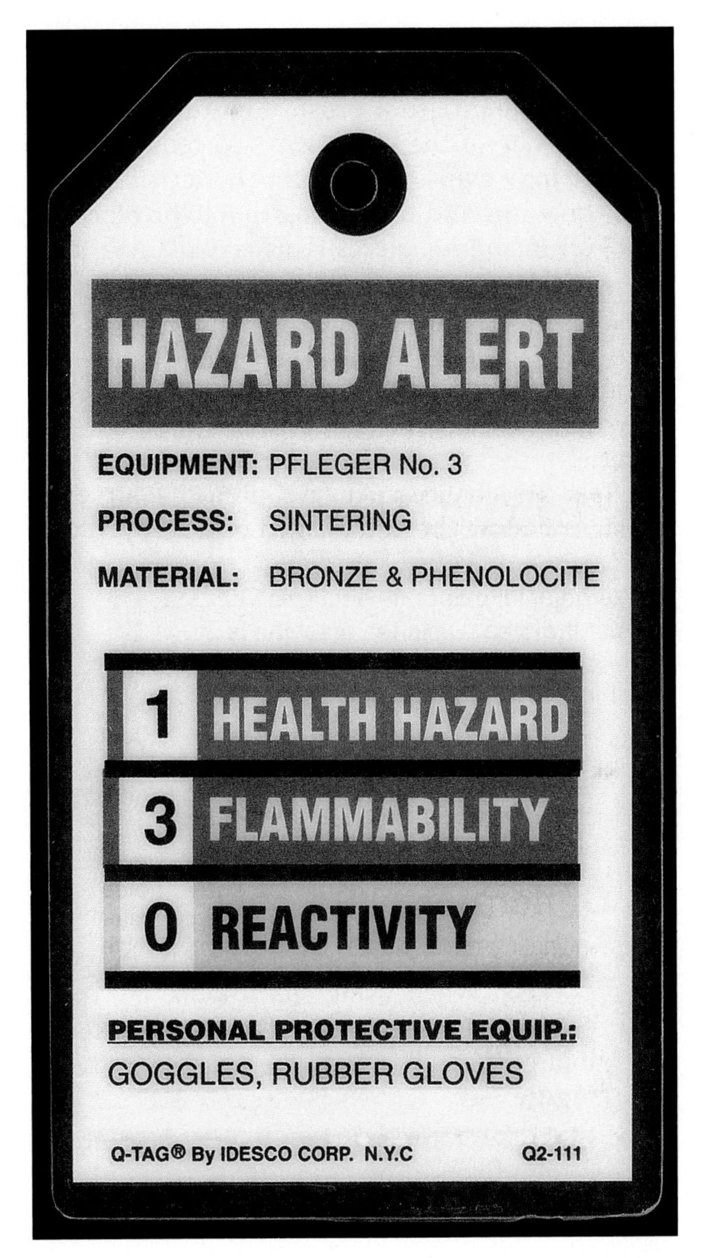

Idesco Corp.

Figure 3-10. Signs can remind workers to be alert for potential hazards.

3.2 General Machine Safety

- Never operate a machine until all guards are in place.
- Keep the floor around your machine clear of oil, chips, and metal scrap.
- It is considered an unsafe practice to talk to anyone while you are operating a machine. You might become distracted and injure yourself, or someone else.

- *Never* attempt to remove chips or cuttings with your hands or while the machine is operating. Use a brush, **Figure 3-11**. Pliers are one of the safest ways to remove long, stringy chips from a lathe. Better still, learn how to grind the cutting tool to break chips off in shorter pieces. This is explained later in the text.
- Secure prompt medical attention for any cut, bruise, scratch, burn, or other injury. No matter how minor the injury may appear, report it to your instructor.

SAFETY NOTE

Always stop your machine to make adjustments or measurements. Resist the urge, while the machine is running, to touch a surface that has been machined. Severe lacerations can result.

3.3 General Tool Safety

- Never carry sharp-pointed tools in your pockets. When using sharp tools, lay them on the bench in such a way that you will not injure yourself when you reach for them, **Figure 3-12**.
- Make sure tools are properly sharpened, in good condition, and fitted with suitable handles.

Photo courtesy of Weiler Corporation

Figure 3-11. Use a brush, not your hands, to remove accumulated chips.

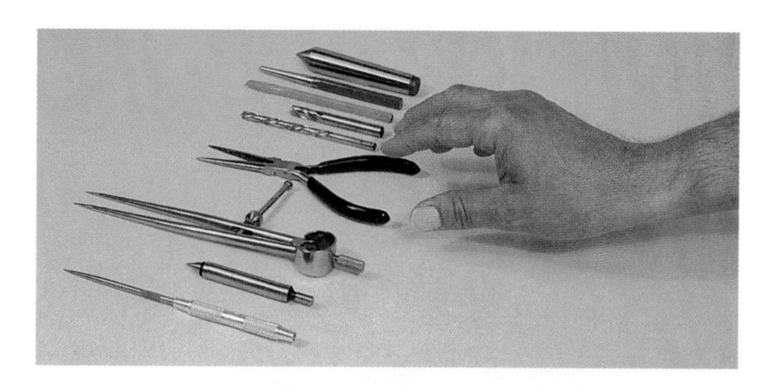

Goodheart-Willcox Publisher

Figure 3-12. Arrange sharp-pointed tools on the bench in a way that they will not injure you when you reach to pick them up.

3.4 Fire Safety

Combustible materials are classified into four categories, **Figure 3-13**. Extinguishers should have color-coded symbols to identify their appropriateness for a particular type of fire.

- **Class A fires.** Those involving ordinary combustible materials—such as paper, wood, and textiles. They require the cooling and quenching effect of water, or solutions containing a large percentage of water. Do not use Class A extinguishers on Class C and D fires.
- **Class B fires.** Flammable liquid and grease fires require the blanketing or smothering effects of dry chemicals or carbon dioxide.
- **Class C fires.** Electrical equipment fires require nonconducting extinguishing agents that will smother the flames. Do not use Class A extinguishers on electrical fires.
- **Class D fires.** Extinguishers containing specially prepared heat-absorbing dry powder are used on flammable metals, such as magnesium and lithium. Do not use Class A extinguishers on flammable metal fires.

Know what to do in case of a fire. Be familiar with the location of your building's fire exits and how they are opened. Be aware of alternate escape routes.

In some situations, students are trained in the use of fire extinguishers by the local fire department. If you are one of those students, make sure that you know where the fire extinguishers are located. If you have not received such training, get out of the fire area immediately.

Think before acting. It costs nothing, and you may be saved from painful injury that could result in a permanent disability.

Fires
Class A Fires Ordinary Combustibles (Materials such as wood, paper, textiles.) *Requires... cooling-quenching* Old: A   New 
Class B Fires Flammable Liquids (Liquids such as grease, gasoline, oils, and paints.) *Requires...blanketing or smothering.* Old: B   New 
Class C Fires Electrical Equipment (Motors, switches, etc.) *Requires... a nonconducting agent.* Old: C   New 
Class D Fires Combustible Metals (Flammable metals such as magnesium and lithium.) *Requires...blanketing or smothering.* D

Type	Use	Operation
Soda-acid Bicarbonate of soda solution and sulfuric acid 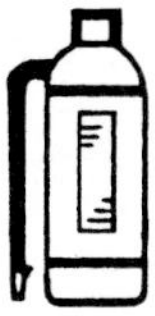	Okay for use on: A Not for use on: B C D	Direct stream at base of flame.
Pressurized Water Water under pressure 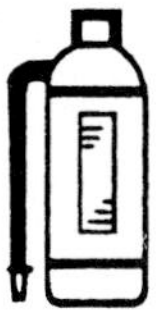	Okay for use on: A Not for use on: B C D	Direct stream at base of flame.
Carbon Dioxide (CO_2) Carbon dioxide (CO_2) gas under pressure 	Okay for use on: B C Not for use on: A D	Direct discharge as close to fire as possible, first at edge of flames and gradually forward and upward.
Foam Solution of aluminum sulfate and bicarbonate of soda 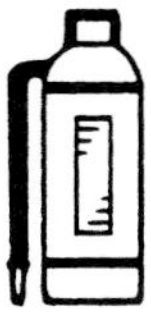	Okay for use on: A B Not for use on: C D	Direct stream into the burning material or liquid. Allow foam to fall lightly on fire.
Dry Chemical 	Multi-purpose type — Okay for: A B C; Not okay for: D Ordinary BC type — Okay for: B C; Not okay for: A D	Direct stream at base of flames. Use rapid left-to-right motion toward flames.
Dry Chemical Granular type material	Okay for use on: D Not for use on: A B C	Smother flames by scooping granular material from bucket onto burning metal.

Goodheart-Willcox Publisher

Figure 3-13. Symbols coded by color and shape identify the four classifications of fire and the extinguishers that can be used on them.

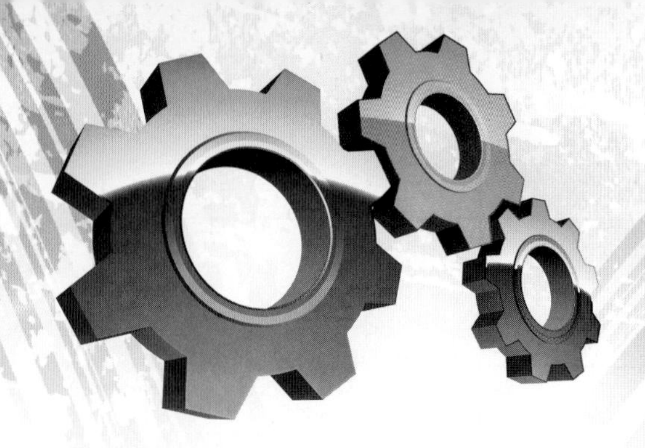

Chapter Review

Summary

- Shop safety should be maintained throughout your machining career, not just in the training program, to help prevent injuries.
- Care should be taken to wear proper clothing that will not be caught in the machine.
- Safety equipment should be used when working with machines to protect yourself from injury.
- Awareness barriers and machine shields are put in place to protect machine users.
- Get proper instruction before operating any machine.
- Warning signs are placed around a shop to identify potential hazards.
- Combustible materials are classified into four different categories (A, B, C, and D) along with the type of extinguisher used to put them out.

Review Questions

Answer the following questions using the information provided in this chapter.

1. Why is shop safety so important?
2. Most shop accidents are caused by _____.
3. What should you do before operating a machine tool if you are taking medication of any sort?
4. Why should compressed air *not* be used to clean chips from machine tools?
5. Oily rags should be placed in a safety container to prevent _____.
6. Why is it necessary to take special precautions when handling long sections of metal stock?
7. Get help when moving _____.
8. When working in an area contaminated with dust or solvent fumes, be sure to wear a dust mask and that there is _____.
9. Safety glasses should be worn _____.
   A. most of the time
   B. only when working on machines
   C. the entire time you are in the shop
   D. None of the above.
10. Under what conditions should you never attempt to operate a machine?
11. Always stop machine tools before making adjustments and _____.
12. Use a(n) _____ to remove chips and shavings, *not* your hand.
13. Secure prompt _____ for any cut, bruise, scratch, or burn.

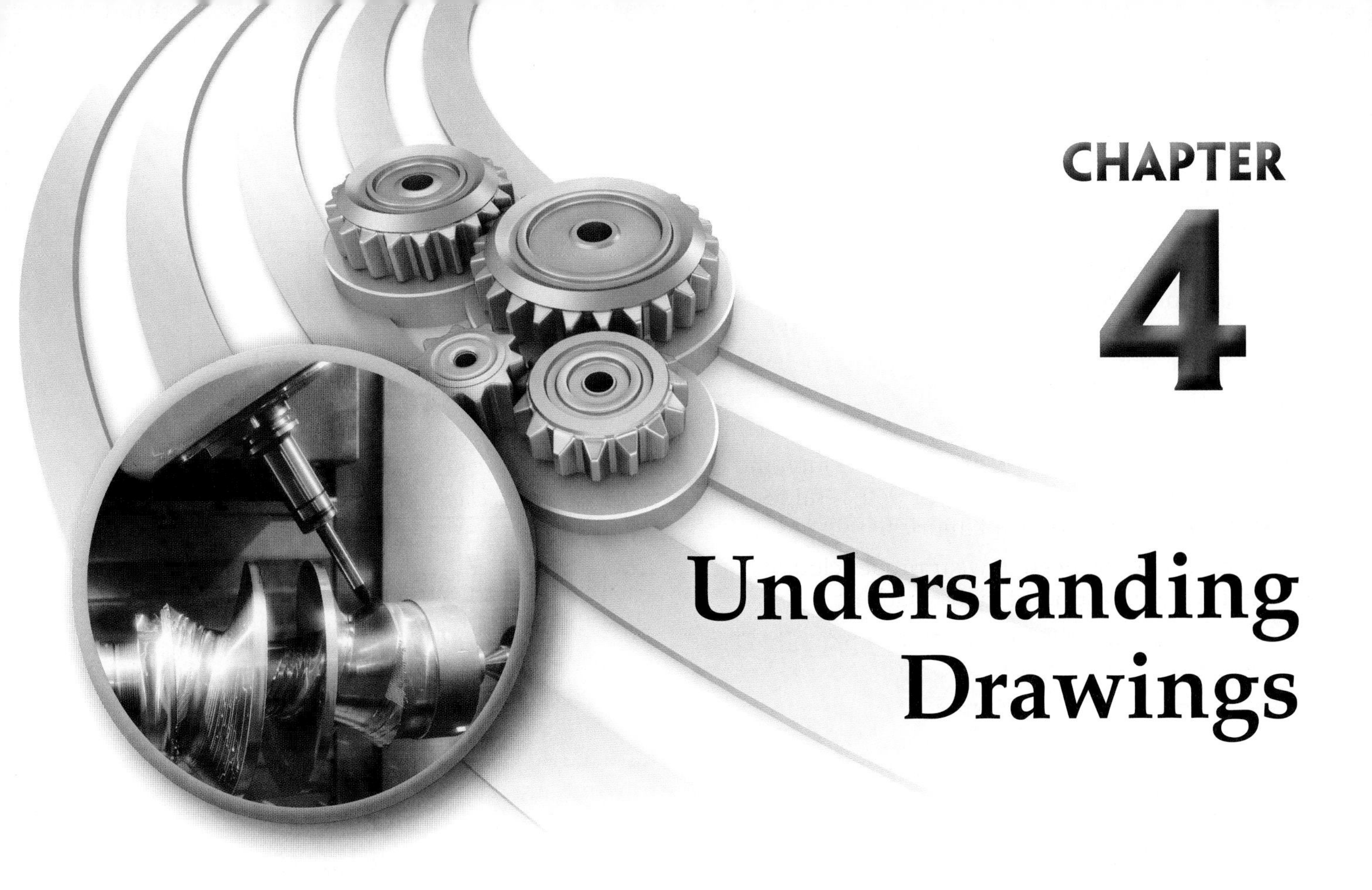

CHAPTER 4
Understanding Drawings

Learning Objectives

After studying this chapter, you will be able to:

- Read drawings that are dimensioned in fractional inches, decimal inches, and in metric units.
- Explain the information found on a typical drawing.
- Describe how detail, subassembly, and assembly drawings differ.
- Indicate why drawings are numbered.
- Explain the basics of geometric dimensioning and tolerancing.

Technical Terms

actual size
American National Standards Institute (ANSI)
angularity
bill of materials
circular runout
circularity
concentricity
cylindricity
datum
dimensions
dual dimensioning
feature
feature control frame
flatness
form geometric tolerances
geometric dimensioning and tolerancing
geometric tolerance
least material condition (LMC)
location geometric tolerances
maximum material condition (MMC)
orientation geometric tolerances
parallelism
perpendicularity
position tolerance
profile geometric tolerance
profile of a line tolerance
profile of a surface tolerance
profilometer
runout geometric tolerance
scale drawing
SI Metric system
straightness
symmetry
tolerance
total runout
US Conventional system
working drawing

Many products manufactured today are an assembly of parts supplied by a number of different industries. These industries may be in distant geographic locations, **Figure 4-1**.

It would not be possible for industry to manufacture a complex product without using drawings. Drawings show the machinist what to make and identify the standards that must be followed so the various parts will fit together properly. The resulting parts will also be interchangeable with similar components on equipment already in service.

Drawings range from a simple freehand sketch, **Figure 4-2**, to detailed drawings for complex products, **Figure 4-3**.

Anatoliy Lukich/Shutterstock.com

Figure 4-1. Thousands of drawings were required in the design and construction of this vertical takeoff and landing aircraft. Standards and specifications had to be exact because components were manufactured in several geographic locations.

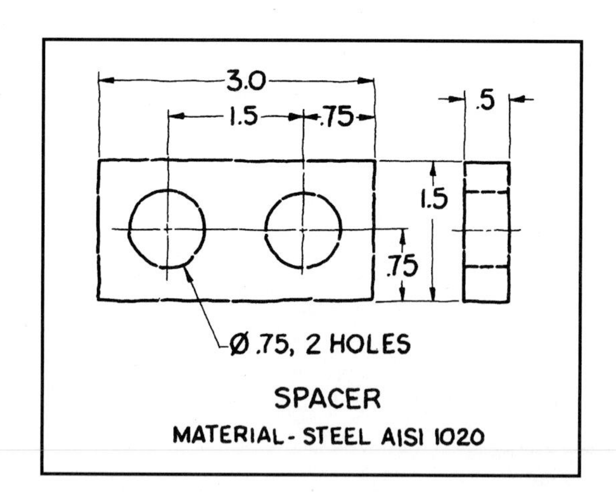

Goodheart-Willcox Publisher

Figure 4-2. Some drawings are as simple as this freehand sketch.

Symbols, lines, and figures are used to give drawings meaning, **Figure 4-4**, and are known as the "language of industry." They have been standardized so they have the same meaning wherever drawings are made and used.

These symbols, lines, and figures have been devised by the ***American National Standards Institute***, better known as ***ANSI***, an organization that promotes and assists in the national adoption of standardized definitions, terminology, symbols, materials, performance characteristics, procedures, and testing methods for use in manufacturing and business.

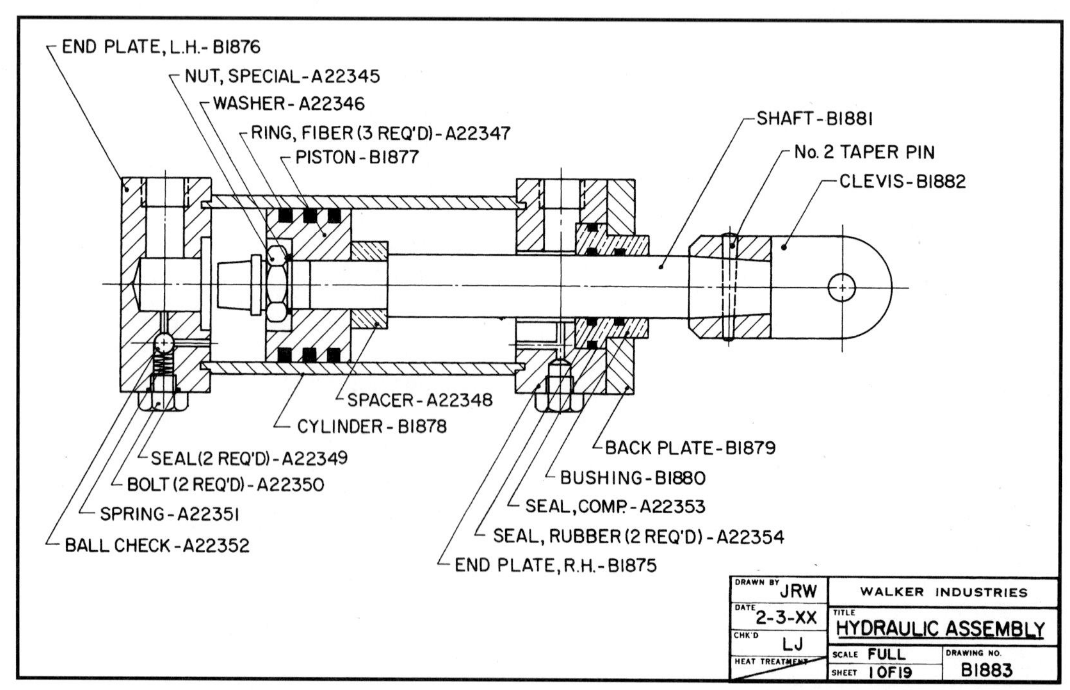

Goodheart-Willcox Publisher

Figure 4-3. Each manufactured product may require dozens of drawings, one for each part. Even the smallest screw, washer, or pin may require a drawing.

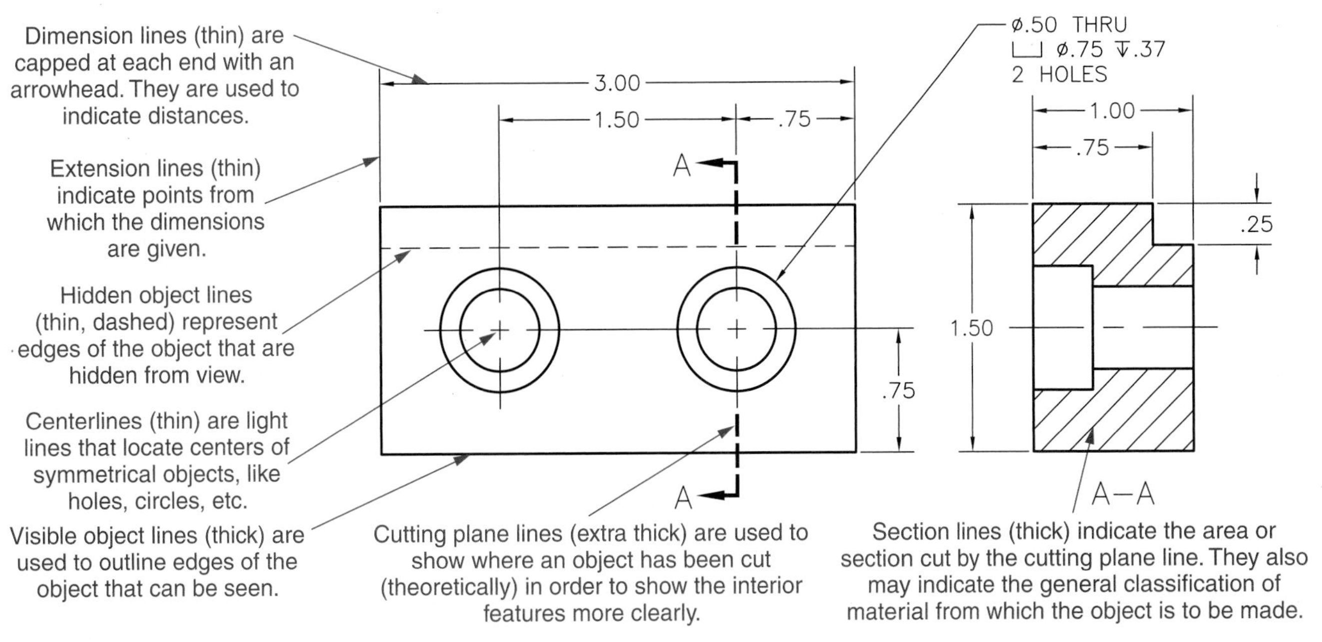

Figure 4-4. Many types of lines, symbols, and figures are used to give a drawing exact meaning.

Periodically, ANSI changes standard drawing symbols. Machinists must be familiar with past and present practices because only recently made drawings will follow the new standards. It is too expensive to revise the millions of drawings made before the new standards were devised. **Figure 4-5** shows past and present metalworking symbols.

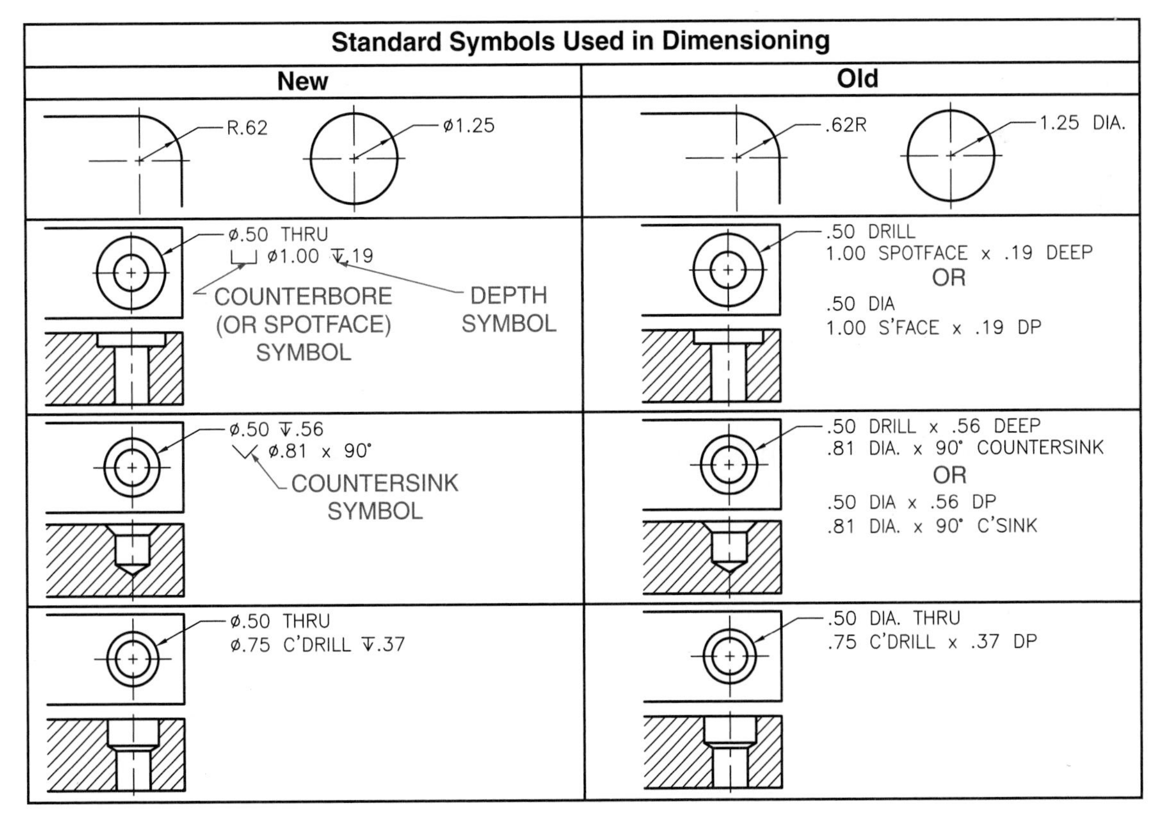

Figure 4-5. Standard ANSI symbols are changed periodically. You must be familiar with both the old and new symbols because either may be used on the drawings. Compare these examples.

Lines are used to draw views that fully describe the object to be manufactured. In addition, the drawing usually includes other information needed to make the product. Details often show threads, for example. **Figure 4-6** shows several methods of showing threads on a drawing.

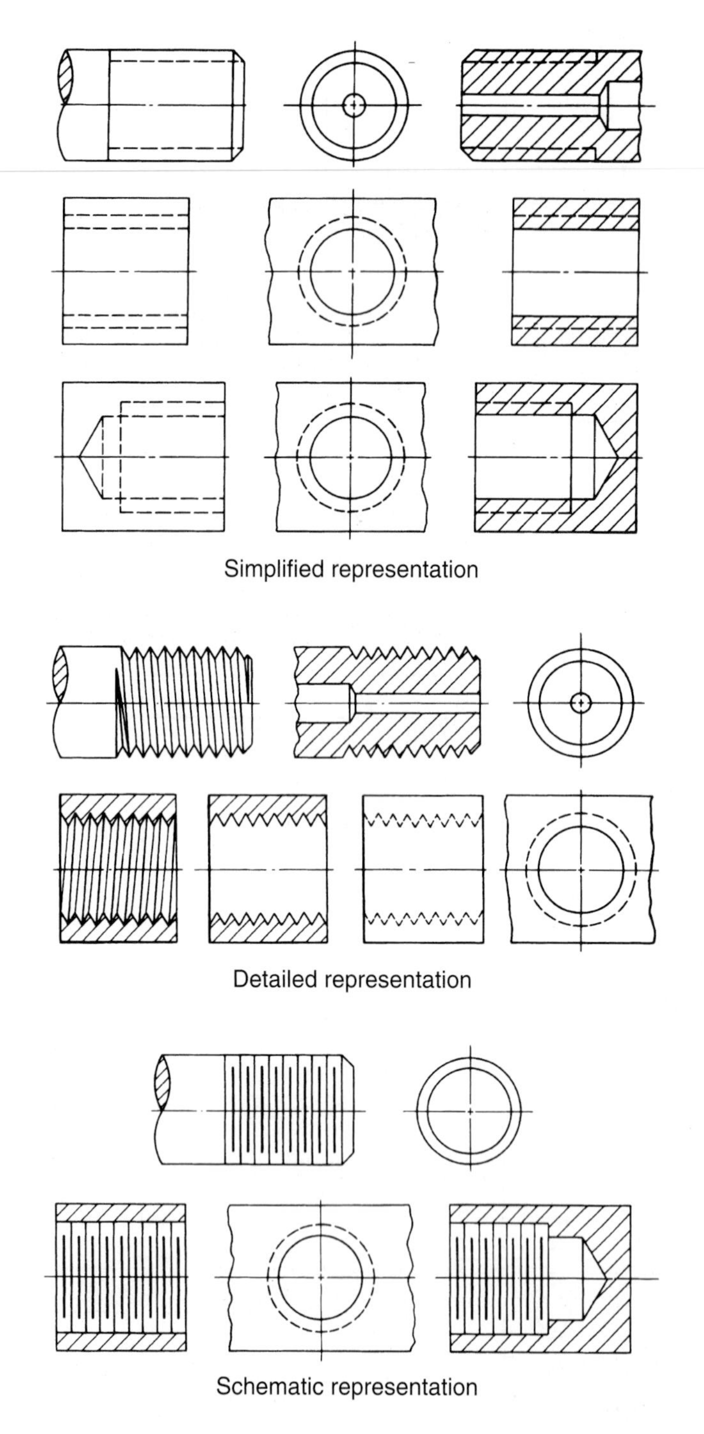

Goodheart-Willcox Publisher

Figure 4-6. Methods used to depict threads on drawings. Only one type will be found on a drawing. The simplified version is the most common style.

4.1 Dimensions

A proper drawing includes all ***dimensions*** (sizes or measurements) in proper relation to one another. The dimensions are needed to produce the part or object.

Until recently, drawings were only dimensioned in decimal or fractional parts of an inch, **Figure 4-7**. However, some industries in the United States are in the process of converting to the metric system of measurement. During this transition period, machinists will need to understand drawings dimensioned in more than one system.

4.1.1 Fractional Dimensioning

Drawings using fractional dimensioning usually show objects that do not require a high degree of precision in their manufacture. Greater precision is indicated when dimensions are given in decimal parts of an inch. Fractional dimensioning is rarely used today, but older drawings may still use this format.

4.1.2 Dual Dimensioning

Dual dimensioning is a system that uses the ***US Conventional system*** of fraction or decimal

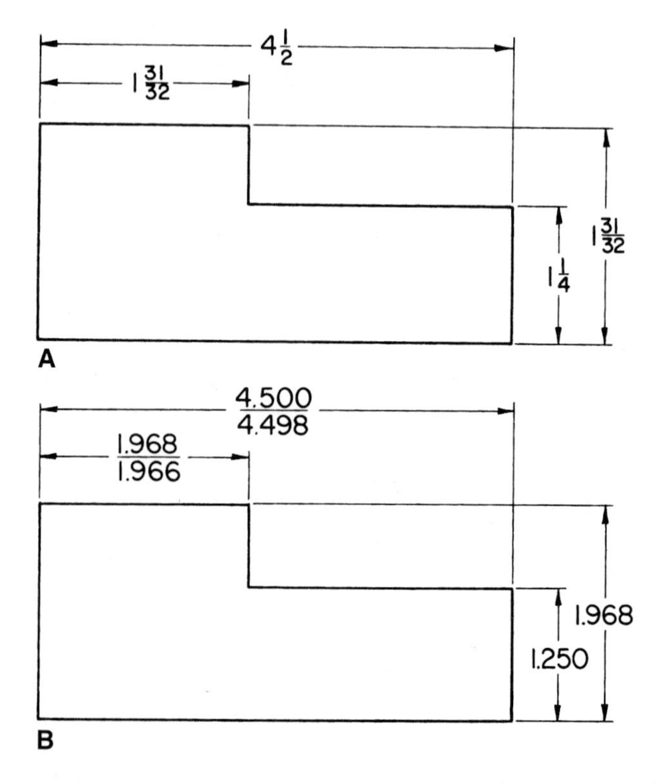

Goodheart-Willcox Publisher

Figure 4-7. Dimensional styles. A—Fractional dimensions do not require tolerances closer than ±1/64″. B—Decimal dimensions normally have tolerances of ±0.010″ or 0.015″.

dimensions and metric dimensions on the same drawing, **Figure 4-8**. If the drawing is intended primarily for use in the United States, the decimal inch will appear above the metric dimension, as in **Figure 4-9A**. The reverse is true if the drawing is to be used in a metric-oriented country, as in **Figure 4-9B**. Some companies place the metric dimension within brackets, as in **Figure 4-9C**.

4.1.3 Metric Dimensioning

With metric dimensioning, all of the dimensions on the drawing are in the ***SI Metric system*** (*SI* stands for the French words *Systeme Internationall*), usually in millimeters.

4.2 Information Included on Drawings

Drawings contain additional information to inform the machinist of the material to be used, required surface finish, tolerances, quantity of units,

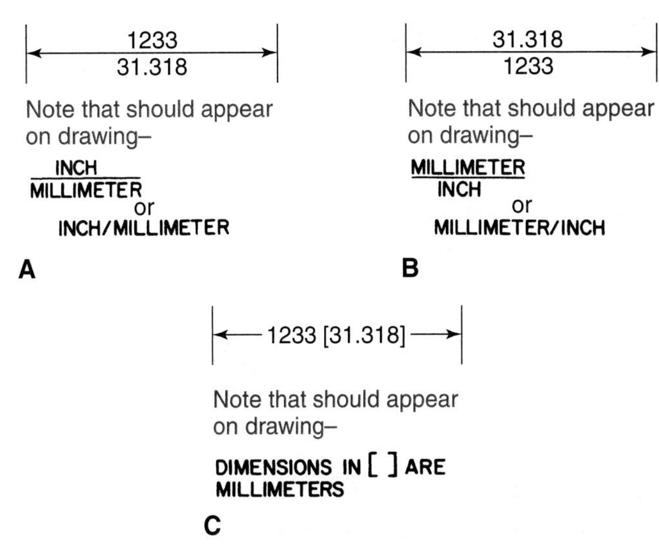

Goodheart-Willcox Publisher

Figure 4-9. Indicating inch and millimeter dimensions on a dual-dimensioned drawing. A—When the drawing is to be used in the United States, the inch value appears on top. B—When the drawing is to be used primarily in a country that uses the metric system, the millimeter value appears on top. C—Sometimes, brackets are used to indicate the metric equivalent on a drawing to be used in the United States.

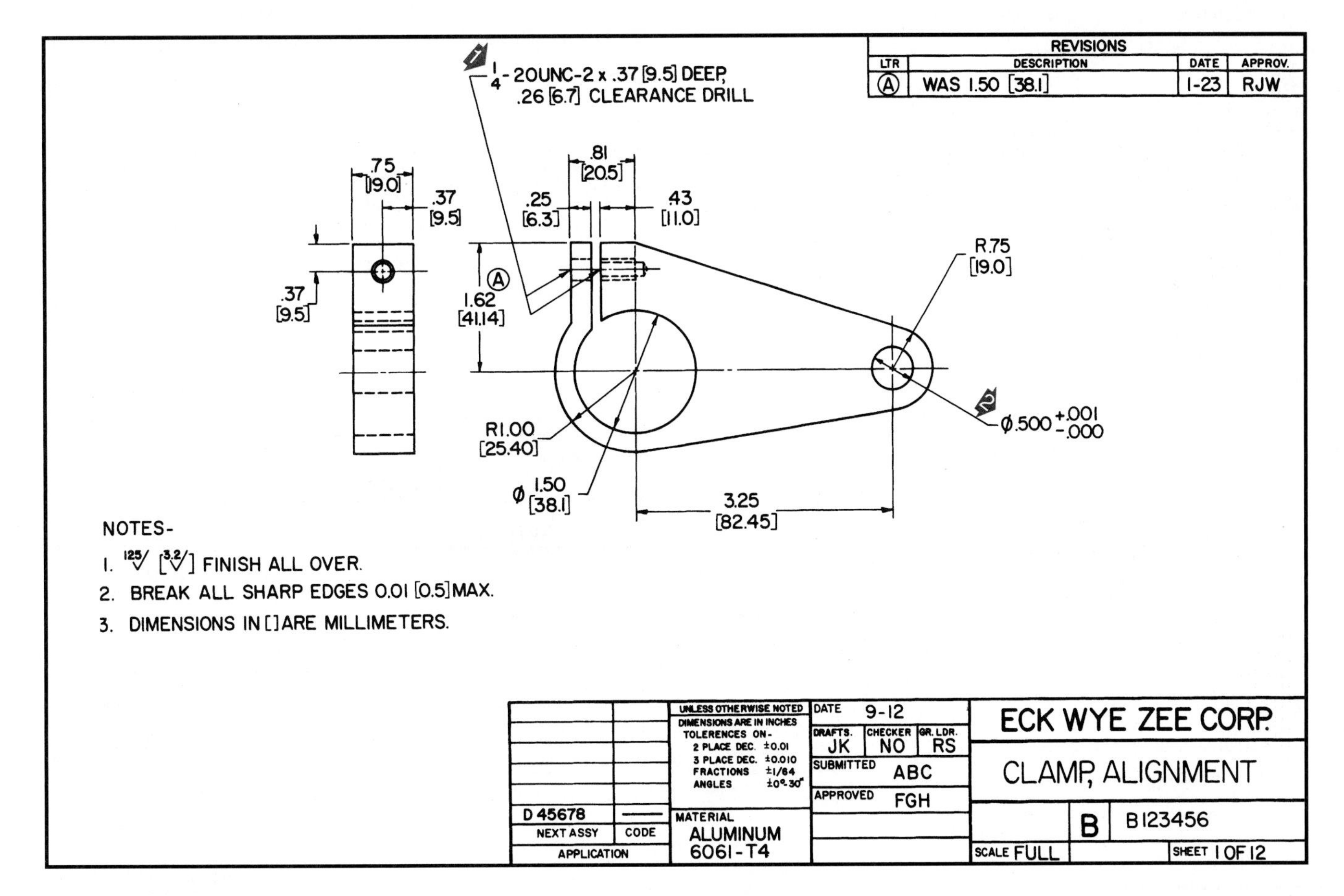

Goodheart-Willcox Publisher

Figure 4-8. Dual-dimensioned drawing. 1—A metric thread size has not been given. There is no metric thread that is equal to this size fractional thread. 2—There is no metric reamer equal to this size.

scale, assembly and subassembly instructions, past revisions, and the name of the object. It is important for you to be familiar with this information.

4.2.1 Materials

The general classification of materials to be used in the manufacture of an object may be indicated by the type of section line on the drawing or plan, **Figure 4-10**. Exact material specification is included in a section of the title block, **Figure 4-11A**. Sometimes, the material specification may be found in the notes shown elsewhere on the drawing.

4.2.2 Surface Finishes

The quality of the surface finish (degree of surface smoothness) is important in the manufacture of many products. The smoothness of the bore of an engine cylinder is an example. Usually, the superior (smoother) the finish of a machined surface, the more expensive it is to manufacture.

In the past, symbols were used to indicate machined surfaces, **Figure 4-12**. These symbols may still be found on some older drawings. With so many machining techniques now in use, symbols such as these do not indicate, in sufficient detail, the quality of the surface finish required on a part.

The method presently used provides more complete surface information. Shown in **Figure 4-13**, a check mark and number are used to indicate surface roughness in microinches or micrometers. A microinch is one-millionth of an inch (0.000001″). A micrometer (micron) is one-millionth of a meter (0.000001 m) and is abbreviated μm.

A machinist compares surface finishes to required specifications by using a surface roughness comparison standard as a guide, **Figure 4-14**. If the surface finish is critical, as it is in some jet engine components, the finish is measured electronically with a device called a ***profilometer***, **Figure 4-15**.

4.2.3 Tolerances

The control of dimensions to achieve interchangeable manufacturing is known as tolerancing. This controls the size of the features of a part. A standard system of geometric dimensioning and tolerancing has been established.

A ***tolerance*** is an allowance, either oversized or undersized, that is permitted when machining or making a part. Refer to **Figure 4-11B**. Acceptable tolerances are shown on drawings in several different ways.

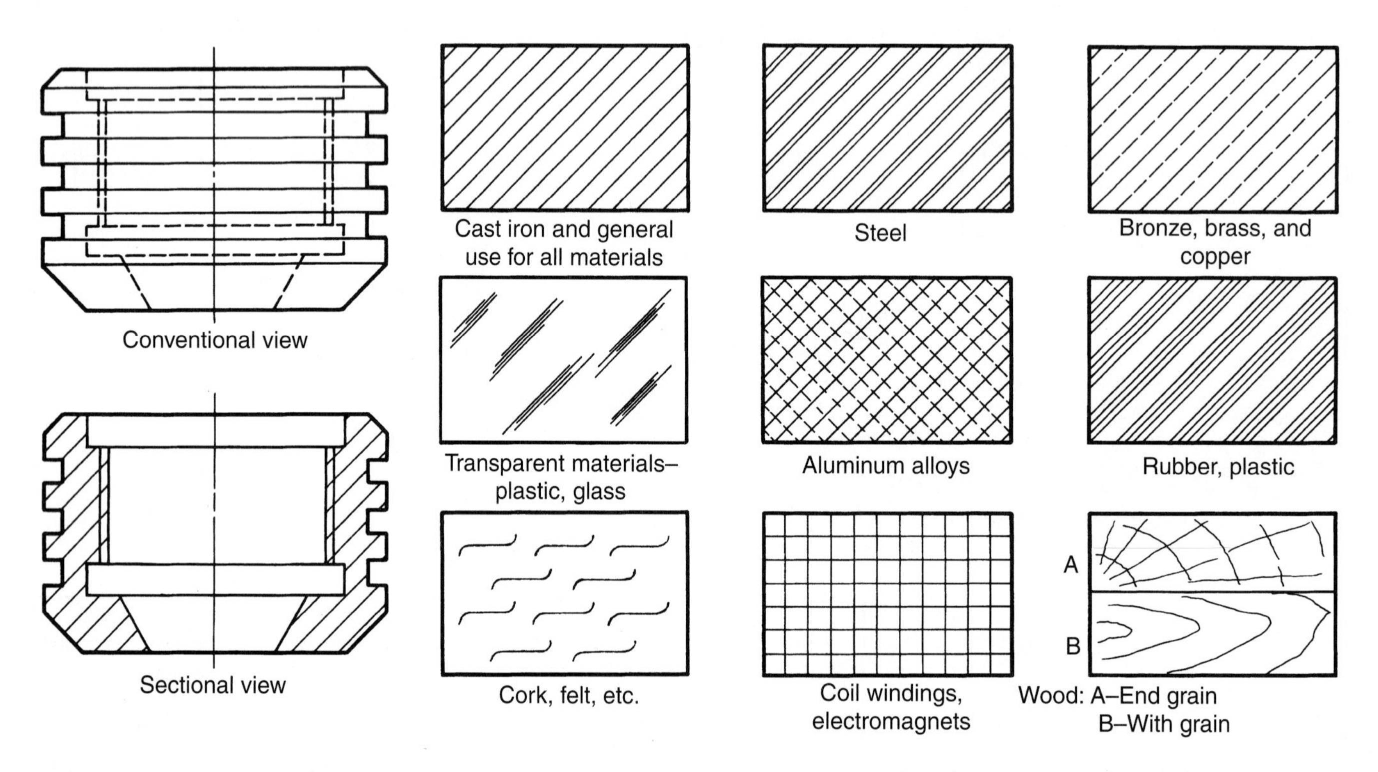

Figure 4-10. Sectional views make a drawing easier to understand because internal details are shown more clearly. Various materials are identified with unique section lines. However, many sectional views use general section lining regardless of the material being used.

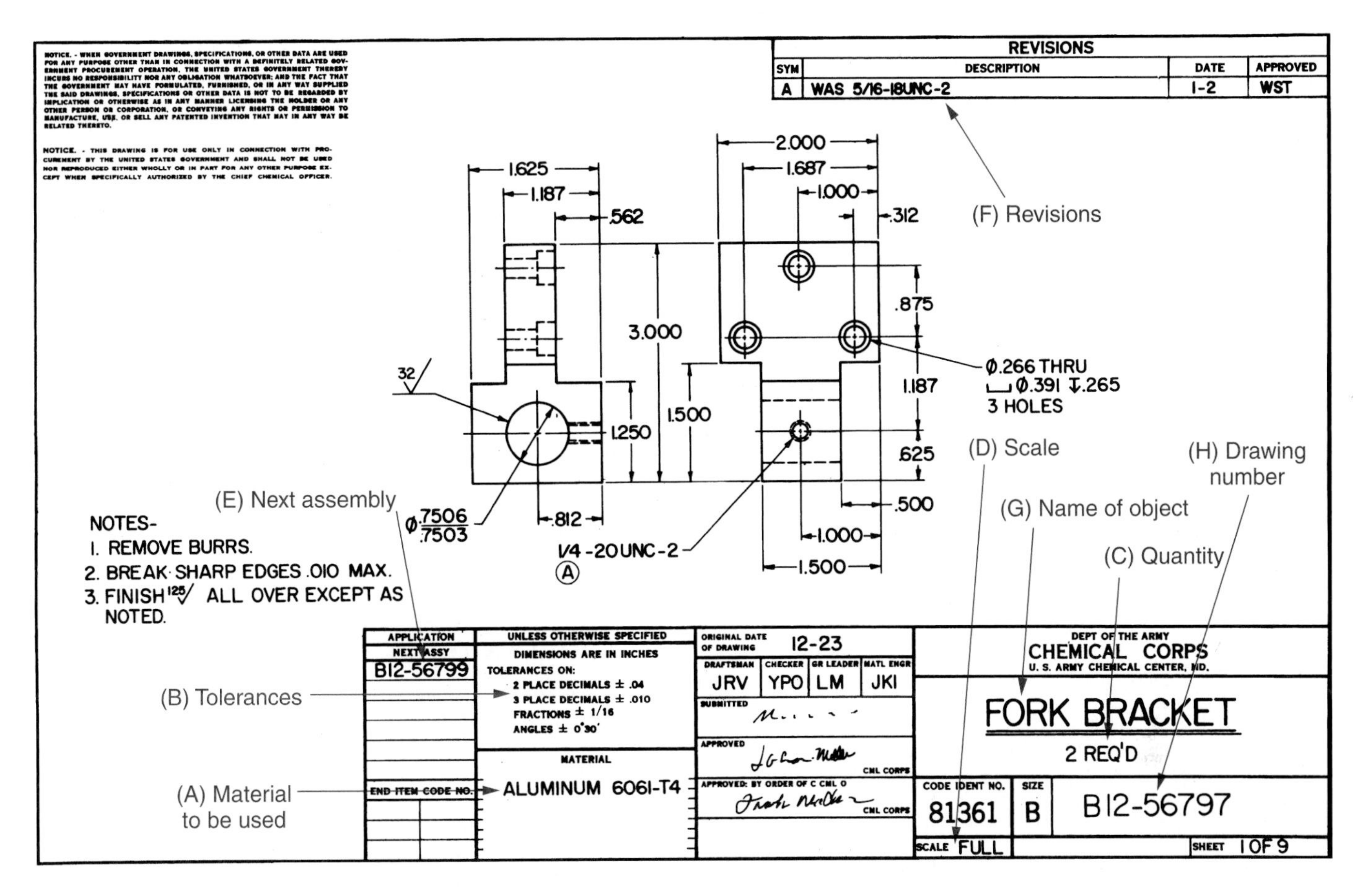

Figure 4-11. A great deal of information is contained in the drawing's title block. The components highlighted here are standard on most drawings.

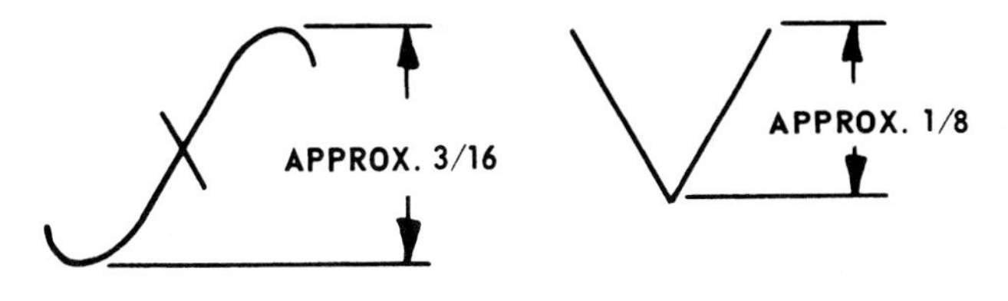

Figure 4-12. These are older-style finish marks. They do not indicate the degree of smoothness required; they simply specify that the surface is machined. Finish marks of these types are still found on older drawings.

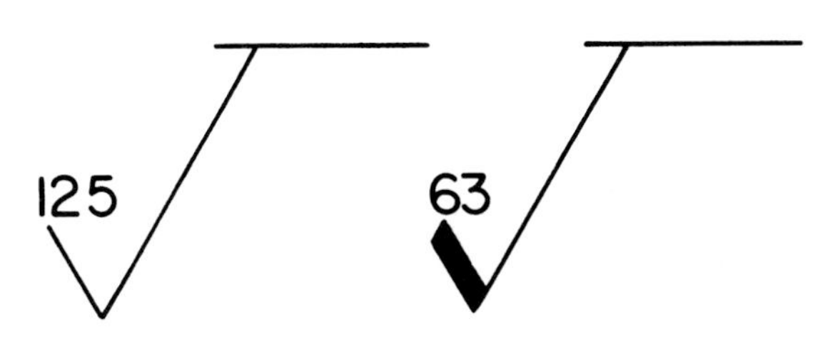

Figure 4-13. Current surface finish marks. The number indicates the degree of smoothness in microinches—the larger the number, the rougher the finish.

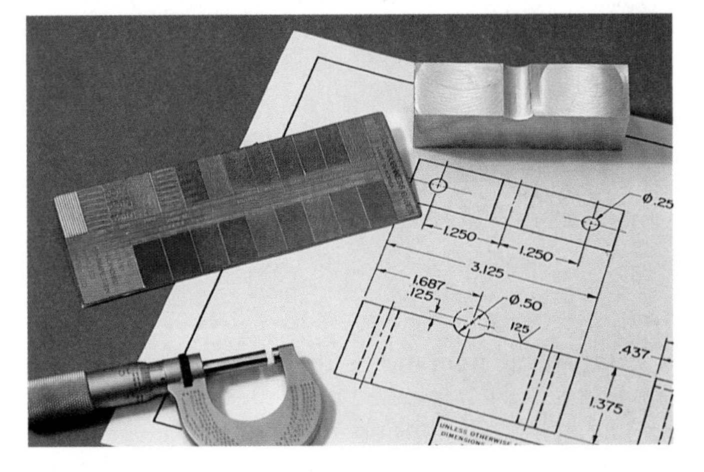

Figure 4-14. A surface roughness comparison standard is used to check whether a milled surface meets the required specifications.

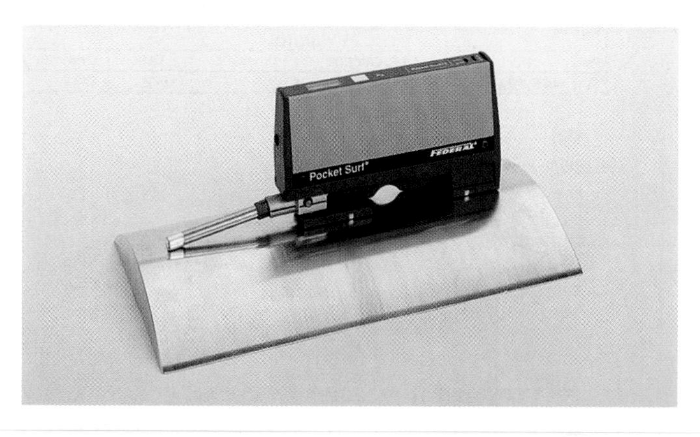

Federal Products Co.

Figure 4-15. Surface roughness is best determined with a profilometer or electronic surface roughness gage. The probe on the unit is moved across the work and measures surface roughness electronically. A digital display presents the measured roughness value in microinches or micrometers.

When the dimension is given in fractional inch units, the permissible tolerances can be assumed to be ±1/64″, unless otherwise indicated.

The symbol "±" means that the machined surface can be *plus* (larger) or *minus* (smaller) by the dimension that follows and still be acceptable. In the example above, the dimension may be up to 1/64″ larger or smaller than the dimension given on the drawing. This "plus and minus" tolerance is called a bilateral tolerance.

If it is permissible to machine the part larger, but not smaller, the dimension on the drawing might read as follows:

$$2\ 1/2^{+1/64}$$

If only a minus tolerance is permitted, the dimension might read as follows:

$$2\ 1/2^{-1/64}$$

When the tolerance is only plus or only minus (one direction), it is called a unilateral dimension.

Drawings dimensioned in decimal inches usually indicate that the work must be machined more precisely than dimensioning in fractional inches. The part can be used as long as the machined dimensions measure within these limits. Unless otherwise indicated, the tolerances can be assumed to be ±0.001″.

A *plus* tolerance may be shown as follows:

$$2.500^{+.001} \text{ or } \frac{2.501}{2.500}$$

A *minus* tolerance may be shown as follows:

$$2.500^{-.001} \text{ or } \frac{2.500}{2.499}$$

Metric tolerances are presented in the same way as decimal tolerances, **Figure 4-16**.

4.2.4 Quantity of Units

Also shown on the drawing is the number of parts (quantity) needed in each assembly. Refer to **Figure 4-11C**. A work order, included with the job information received by the shop, gives the total number of units to be manufactured. This facilitates ordering the necessary materials, and will help in determining the most economical way to manufacture the pieces.

4.2.5 Drawing Scale

Drawings made other than actual size (1:1) are called ***scale drawings***. The scale is usually shown in a section of the title block, **Figure 4-11D**. A drawing made one-half size would have a scale of 1:2 (one-to-two). A scale of 2:1 (two-to-one) would mean that the drawing is twice the size of the actual part.

4.2.6 Assembly or Subassembly

Assembly or subassembly information is necessary to correctly fit the various parts together. The term *application* is sometimes used in place of the term *next assembly*, **Figure 4-11E**.

4.2.7 Revisions

Revisions indicate what changes were made to the original drawing and when they were made. Refer to **Figure 4-11F**.

4.2.8 Name of the Object

A portion of the title block provides this information. It tells the machinist the correct name of the piece, **Figure 4-11G**.

4.3 Prints

On many jobs, several sets of plans are required. In the past a variety of techniques existed to reproduce drawings. Today, most drawings are created using computer-aided design (CAD) software. Both 2D and 3D models can be created using CAD software. Prints can be generated on a plotter (automatic drafting machine),

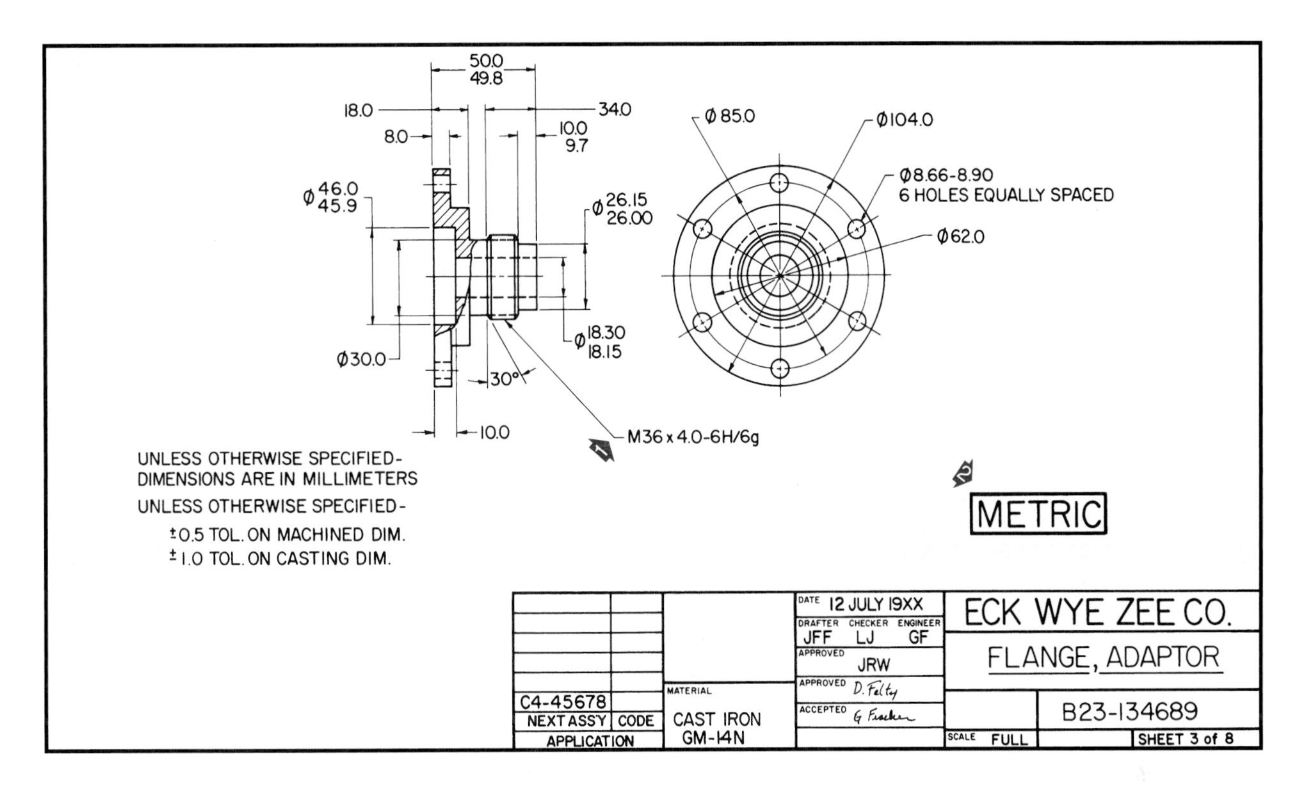

Goodheart-Willcox Publisher

Figure 4-16. A metric detail drawing. 1—Note that metric thread specifications are different from the more familiar UNC (coarse) and UNF (fine) series threads. The letter "M" denotes standard metric screw threads. The 36 indicates the nominal thread diameter in millimeters. The 4.0 denotes thread pitch in millimeters. The 6H and 6g are tolerance class designations. 2—To avoid possible misunderstanding, metric is shown on the drawing in large letters.

using either the 2D or 3D models, or viewed electronically, see **Figure 4-17**. This same information can also be used to control machine tools, using computer-aided manufacturing (CAM) software. When these methods are used, the overall manufacturing technique is called computer-integrated manufacturing (CIM). Computer numerical control (CNC) machining uses CAD models to precisely carry out the sequence of operations needed to produce a part. More information on computers in manufacturing is provided in later chapters of this book.

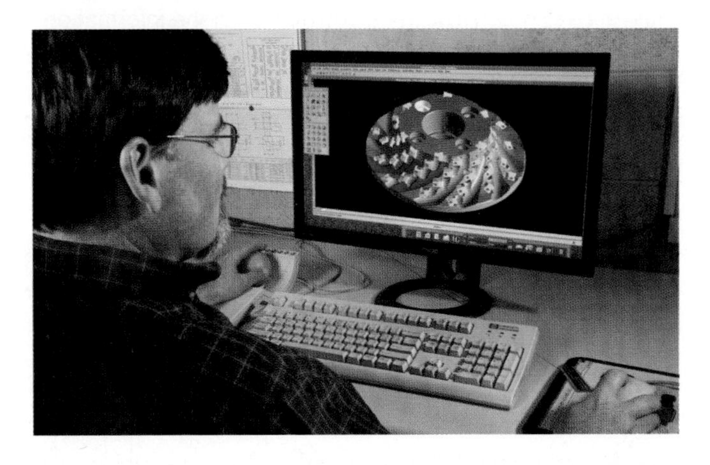

Lovejoy Tool Company, Inc

Figure 4-17. A computer generated 3D model of an indexable insert cutting tool.

4.4 Types of Drawings Used in the Shop

Working drawings, also called prints, establish the standards for the product and show the machinist what to make. There are two major kinds of working drawings:

- **Detail drawings.** These consist of a drawing (usually multiview) of the part with dimensions and other information for making the part, **Figure 4-18**.
- **Assembly drawings.** These drawings show where and how the parts, described on detail drawings, fit into the completed assembly. See **Figure 4-19**.

On large or complex products, subassembly drawings are used to show the assembly of a small portion of the completed object, **Figure 4-20**.

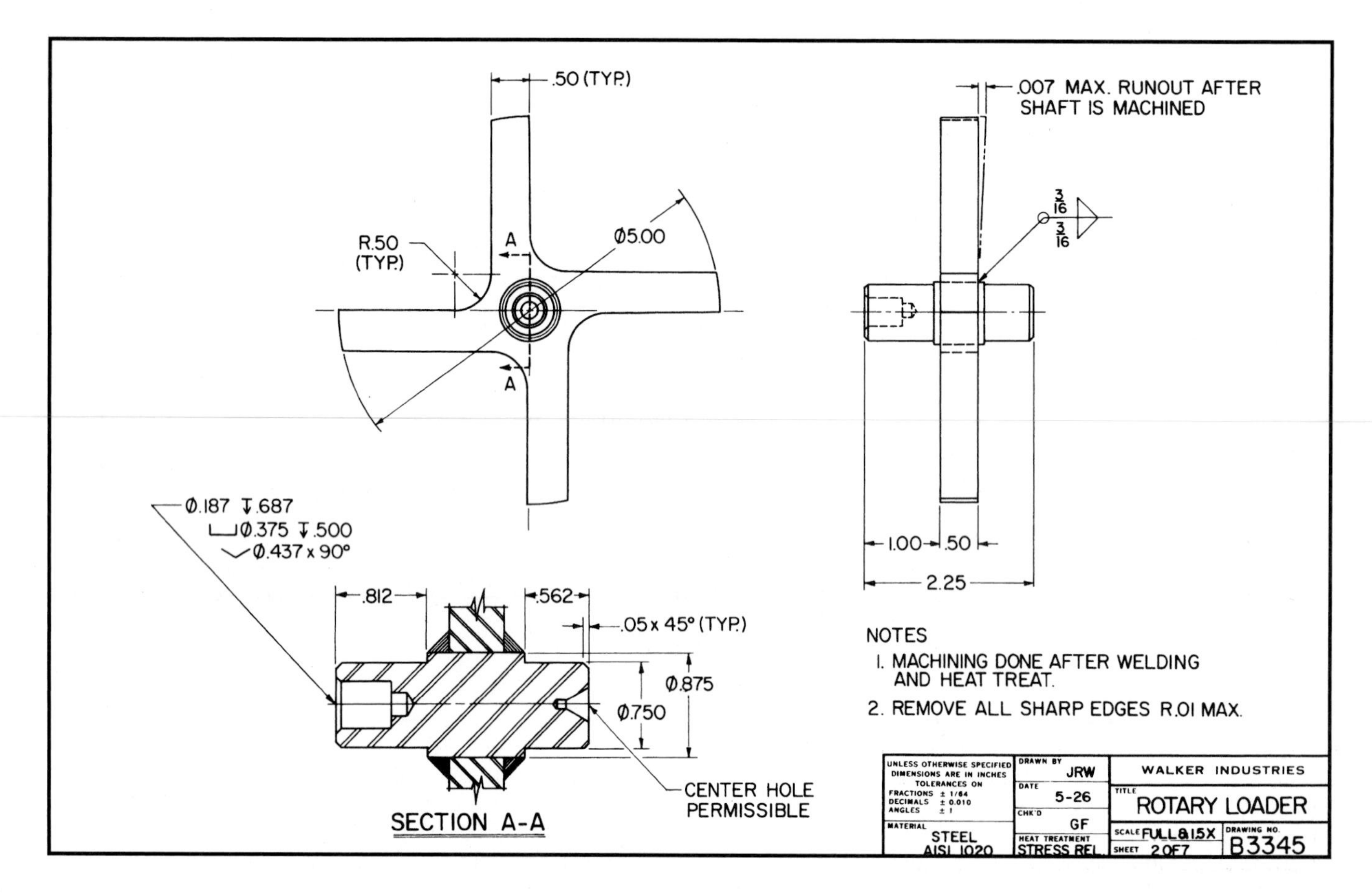

Goodheart-Willcox Publisher

Figure 4-18. A detail drawing contains all of the information needed to produce the part.

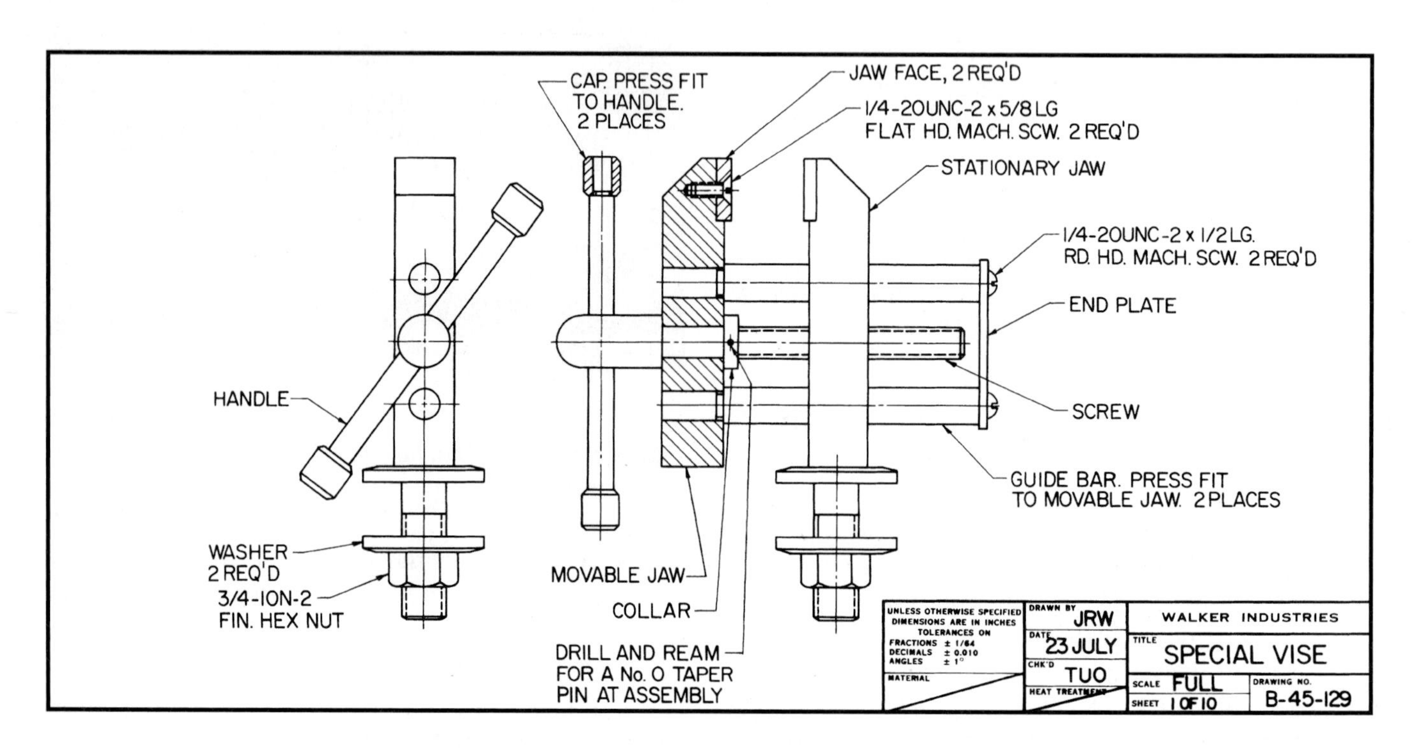

Goodheart-Willcox Publisher

Figure 4-19. An assembly drawing shows how various parts fit together.

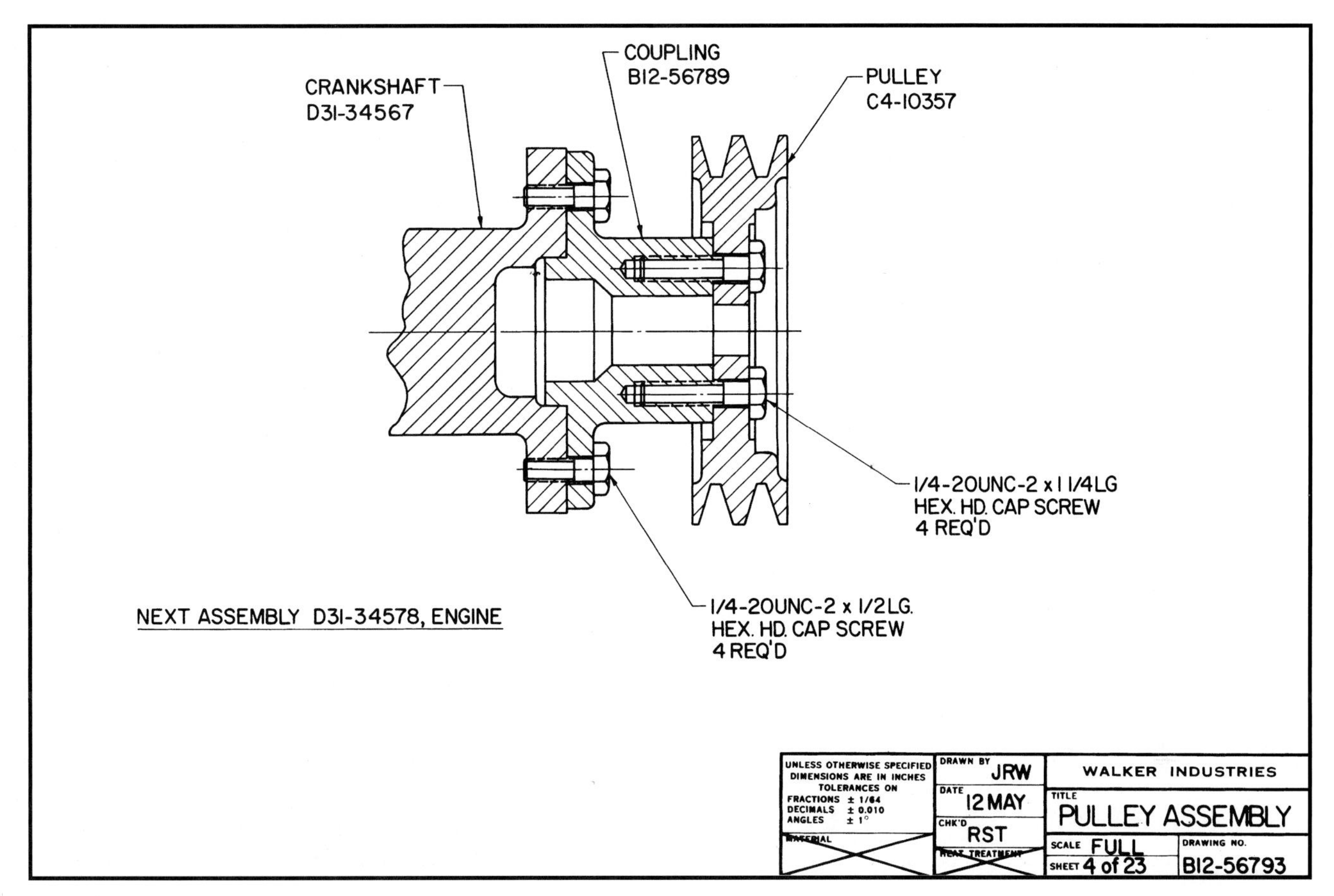

Goodheart-Willcox Publisher

Figure 4-20. A subassembly drawing contains the assembly of only a portion of the entire product.

Some assembly and subassembly drawings are shown as exploded pictorial drawings (a drawing with parts separated, but in proper relationship). One is shown in **Figure 4-21**.

In most instances, a detail drawing provides information on just one item. However, if the mechanism is small in size or if it is composed of only a few parts, the detail and assembly drawings may appear on the same sheet, **Figure 4-22**.

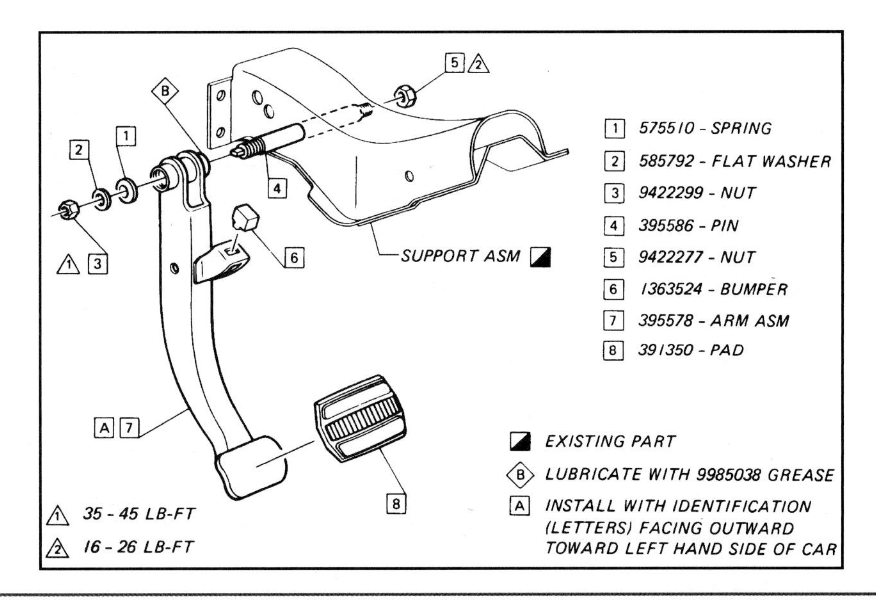

General Motors Corp.

Figure 4-21. Exploded pictorial drawings are often used with semiskilled workers who have received a minimum of training in print reading.

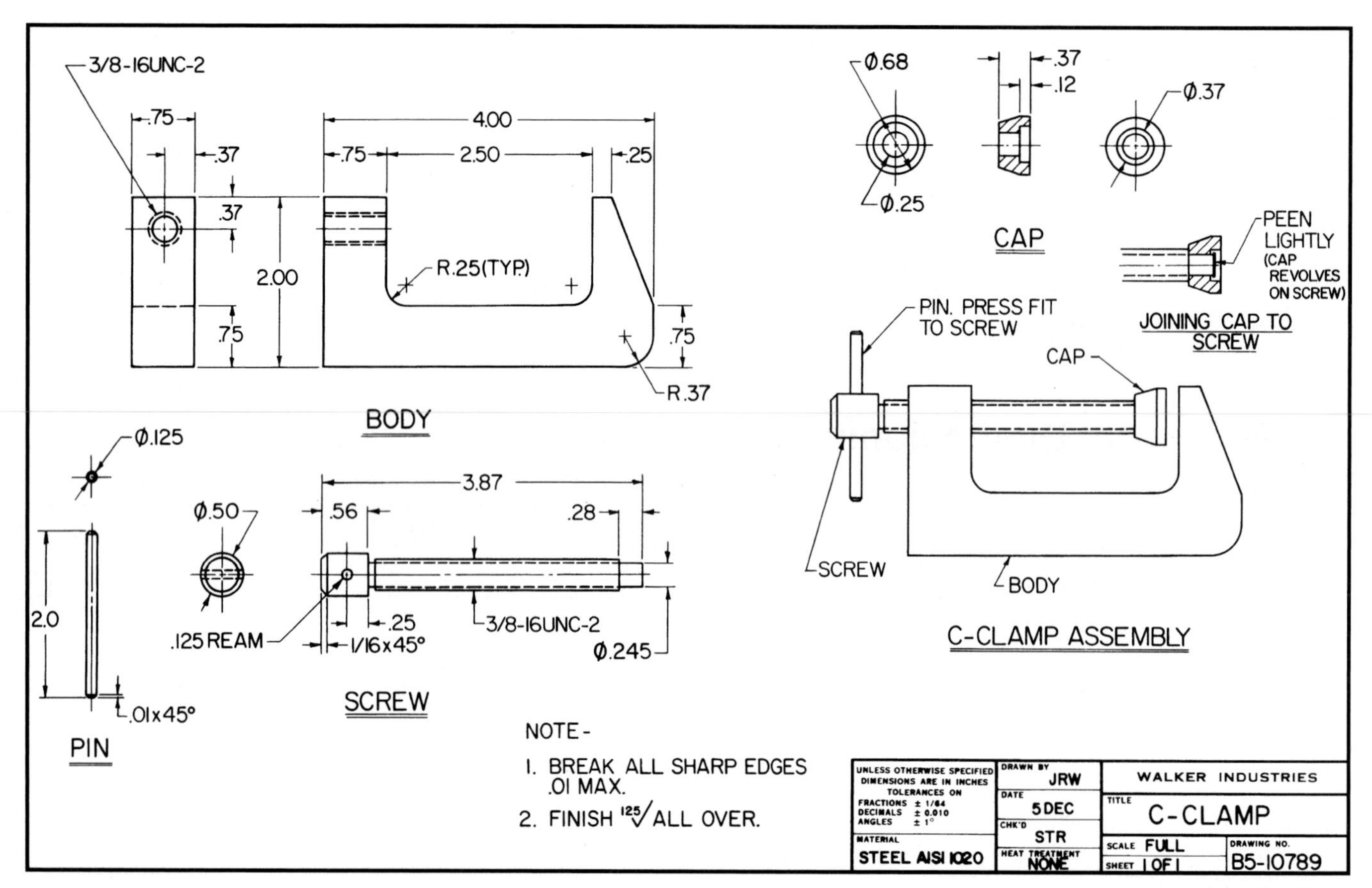

Goodheart-Willcox Publisher

Figure 4-22. A detail and assembly drawing on the same sheet.

4.5 Parts List

Parts are identified by circled numbers and are listed in a note. Some drawings will also include a parts list or ***bill of materials*** listing all of the parts used in the assembly. See **Figure 4-23**.

4.6 Drawing Sizes

Most firms centralize the preparation and storage of drawings in the engineering department. Generally, engineers and drafters prepare drawings on standard-size sheets. This simplifies the stocking, handling, and storage of the completed drawings.

Standard sizes for drawing sheets include the following:

US Conventional Sheet Sizes

A size = 8 1/2″ × 11″
B size = 11″ × 17″
C size = 17″ × 22″
D size = 22″ × 34″
E size = 34″ × 44″

A

PARTS LIST		
No.	*Name*	*Quan.*
1	CRANKCASE	1
2	CRANKSHAFT	1
3	CRANKCASE COVER	1
4	CYLINDER	2
5	PISTON	2

B

A1776	NUT	BRASS	6
A1985	BOLT	BRASS	6
B1765	PLATE	ALUMINUM	2
B1767	CYLINDER	CAST IRON	2
Pt. No.	*Name*	*Material*	*Quan.*
BILL OF MATERIALS			

Goodheart-Willcox Publisher

Figure 4-23. A list of parts and materials is normally included with the drawings for a project. A—A typical, but partial, parts list. B—An example of a partial bill of materials.

SI Metric Sheet Sizes		
A4 size	=	210 × 297 mm
A3 size	=	297 × 420 mm
A2 size	=	420 × 594 mm
A1 size	=	594 × 841 mm
A0 size	=	841 × 1189 mm

Also, for convenience in filing and locating drawings in storage, each drawing has an identifying number, **Figure 4-11H**.

4.7 Geometric Dimensioning and Tolerancing

Conventional tolerancing is appropriate for many products. However, for accurately machined parts, the amount of variation (tolerances) in form (shape and size) and position (location) may need to be more strictly defined. This definition provides the precision needed to allow for the most economical manufacture of parts. See **Figure 4-24**.

Geometric dimensioning and tolerancing is a system that provides additional precision compared to conventional dimensioning. It ensures that parts can be easily interchanged.

Only a brief introduction to geometric dimensioning and tolerancing is included in this text. Detailed information can be found in the publication ASME Y14.5M.

4.7.1 Definitions

Geometric characteristic symbols are used to provide clarity and precision in communicating design specifications. See **Figure 4-25**. These symbols are

Ford Motor Co.

Figure 4-24. A high degree of precision is needed to produce the parts used in this transmission. The tolerances allowed for the shape and location of features on the parts must not be exceeded.

Symbol for:	ASME Y14.5
Straightness	—
Flatness	⏥
Circularity	○
Cylindricity	⌭
Profile of a line	⌒
Profile of a surface	⌓
All-around profile	◀—⊖—
Angularity	∠
Perpendicularity	⊥
Parallelism	//
Position	⌖
Concentricity/coaxiality	◎
Symmetry	⌯
Circular runout	* ↗
Total runout	* ⌰
At maximum material condition	Ⓜ
At least material condition	Ⓛ
Regardless of feature size	NONE
Projected tolerance zone	Ⓟ
Diameter	Ø
Basic dimension	[30]
Reference dimension	(30)
Datum feature	[A]—◀
Datum target	Ø6 / A1
Target point	X
Dimension origin	⌱
Feature control frame	[⌖ | Ø0.5Ⓜ | A | B | C]
Conical taper	⌲
Slope	⌳
Counterbore/spotface	⌴
Countersink	⌵
Depth/deep	↧
Square (shape)	□
Dimension not to scale	15 (underlined)
Number of times/places	8X
Arc length	⌒105
Radius	R
Spherical radius	SR
Spherical diameter	SØ

* May be filled

American National Standards Institute

Figure 4-25. Symbols used to specify positional and form tolerances in geometric dimensioning.

standardized by the American Society of Mechanical Engineers (ASME). ***Geometric tolerance*** is a general term that refers to tolerances which control form, profile, orientation, location, and runout.

A basic dimension is a numerical value denoting the exact size, profile, orientation, or location of a feature. The true position of a feature is its theoretically exact location as established by basic dimensions. A reference dimension is a dimension provided for information only. It is not used for production or inspection purposes. See **Figure 4-26**.

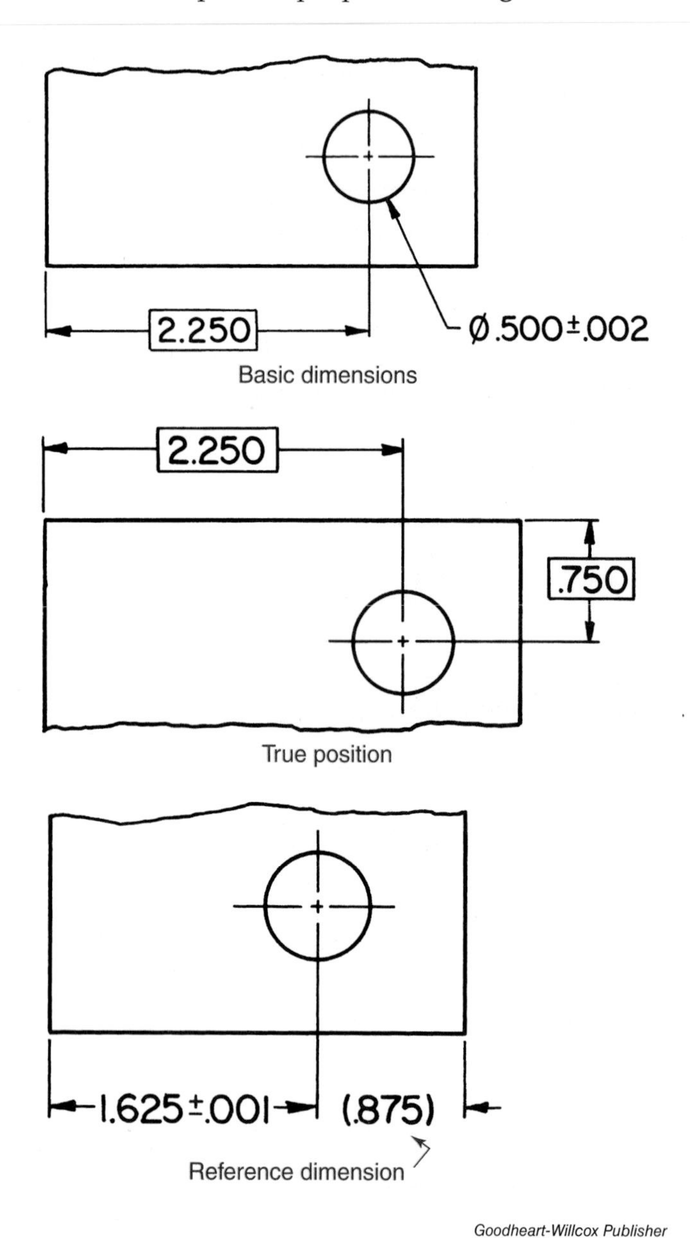

Goodheart-Willcox Publisher

Figure 4-26. Basic dimensions are usually indicated by being enclosed in a rectangular frame. They are not toleranced. True position is the theoretical exact location of feature. It is established by basic dimensions. Reference dimensions are not used for production or inspection purposes. On a drawing, they are shown enclosed in parentheses.

Datum is an exact point, axis, or plane. It is the origin from which the location or geometric characteristic of features of a part is established. It is identified by a solid triangle with an identifying letter. See **Figure 4-27**. ***Feature*** is a general term applied to a physical portion of a part, such as a surface, pin, hole, or slot. A datum feature is the actual feature of a part used to establish a datum. See **Figure 4-28.**

Maximum material condition (MMC) is the condition in which the size of a feature contains the maximum amount of material within the stated limits of size. Examples include a minimum hole diameter and maximum shaft diameter, both of which result in the greatest possible amount of material being used. See **Figure 4-29**. MMC is indicated by an *M* within a circle.

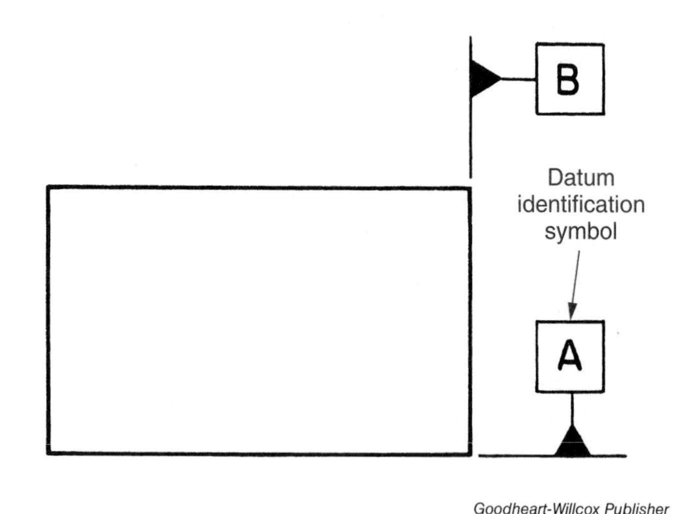

Goodheart-Willcox Publisher

Figure 4-27. Datums are exact points, axes, or planes from which features of a part are located.

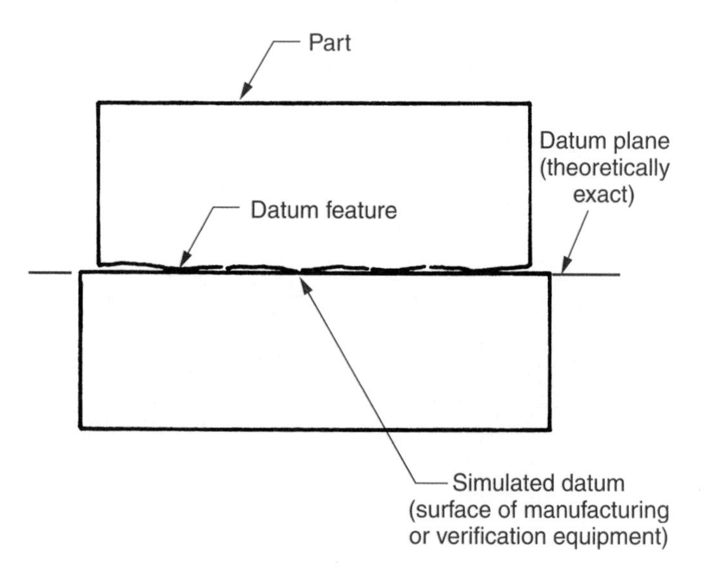

Goodheart-Willcox Publisher

Figure 4-28. A datum feature is a physical feature on a part used to establish a datum.

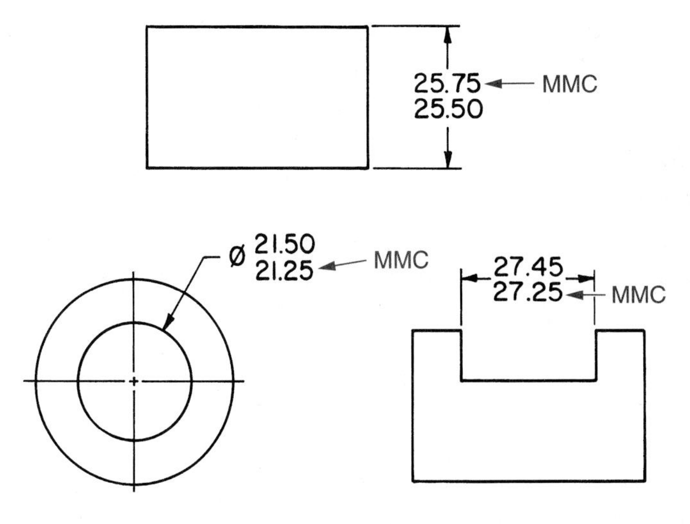

Goodheart-Willcox Publisher

Figure 4-29. Maximum material condition (MMC) indicates that the size of a feature contains the maximum amount of material within the stated tolerance limits.

Least material condition (LMC) is the condition in which the size of a feature contains the least amount of material within the stated tolerance limits. Examples include a maximum hole diameter and a minimum shaft diameter. See **Figure 4-30**. LMC is indicated by an *L* within a circle.

Regardless of feature size (RFS) specifies that the size of a feature tolerance must not be exceeded. RFS is assumed for all geometric tolerances unless otherwise specified.

The maximum and minimum sizes of a feature are called the limits of size. See **Figure 4-31**. The measured size of a part after it is manufactured is the ***actual size***.

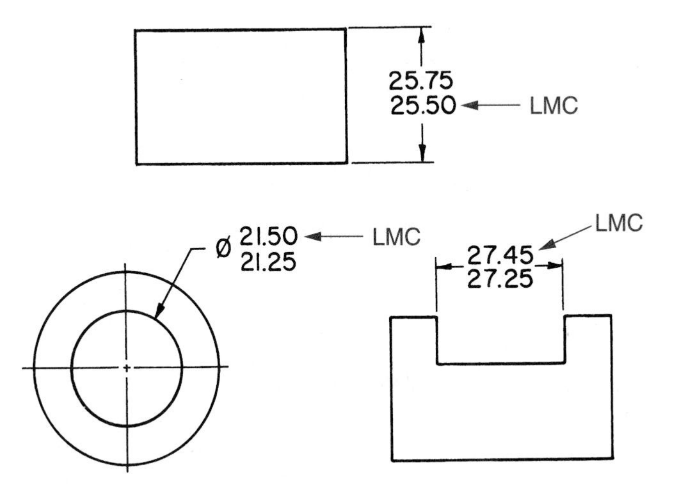

Goodheart-Willcox Publisher

Figure 4-30. Least material condition (LMC) indicates that the size of a feature contains the least amount of material within the stated limits of size.

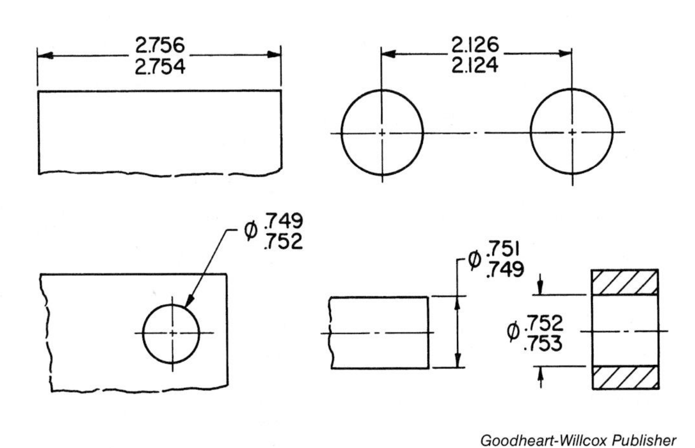

Goodheart-Willcox Publisher

Figure 4-31. Limits of size are the maximum and minimum sizes of a feature.

4.7.2 Application of Geometric Dimensioning and Tolerancing

A datum identification symbol, **Figure 4-32**, consists of a square frame that contains the datum reference letter. All letters but *I*, *O*, and *Q* may be used. A rectangular frame with the datum reference letter preceded and followed by a dash may be found on older drawings.

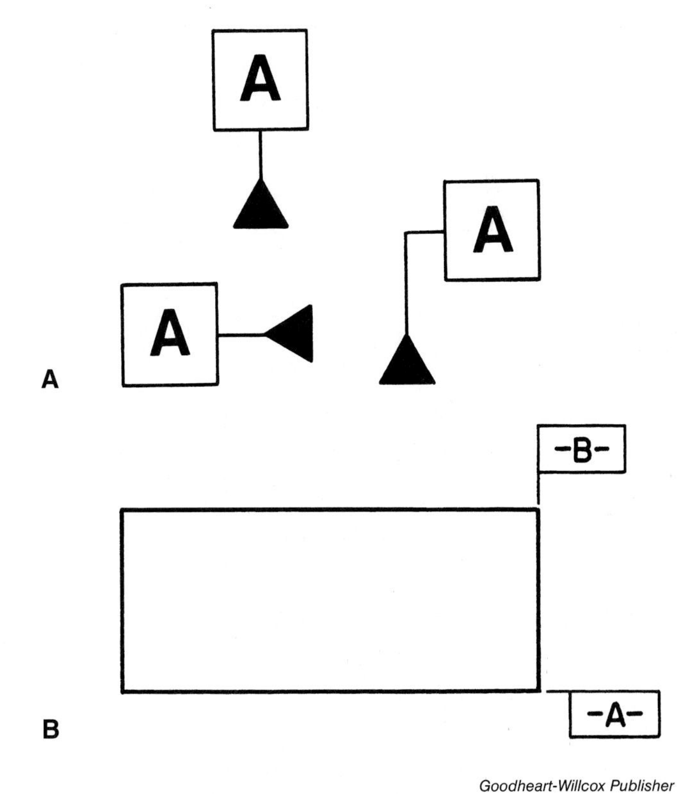

Goodheart-Willcox Publisher

Figure 4-32. Datum points and surfaces are identified by a datum identification symbol. A—Datum identification symbols used on new drawings. B—This type of datum symbol is not used currently, but is still found on old drawings.

A ***feature control frame*** is used to define the geometric tolerancing characteristics of a feature. It contains the geometric symbol, allowable tolerance, and the datum reference letter(s). It is connected to an extension line of the feature, a leader running to the feature, or below a leader-directed note of the feature, **Figure 4-33**.

Datum references indicated on the right end of the feature control frame are read from left to right. The letters signify datum preference. They establish up to three mutually perpendicular planes, **Figure 4-34**.

4.7.3 Form Geometric Tolerances

Form geometric tolerances control flatness, straightness, circularity (roundness), and cylindricity. They are indicated by the symbols shown in **Figure 4-35**. Form tolerances control only the variation permitted on a single feature and are used when form variation is less than that permitted by size tolerance.

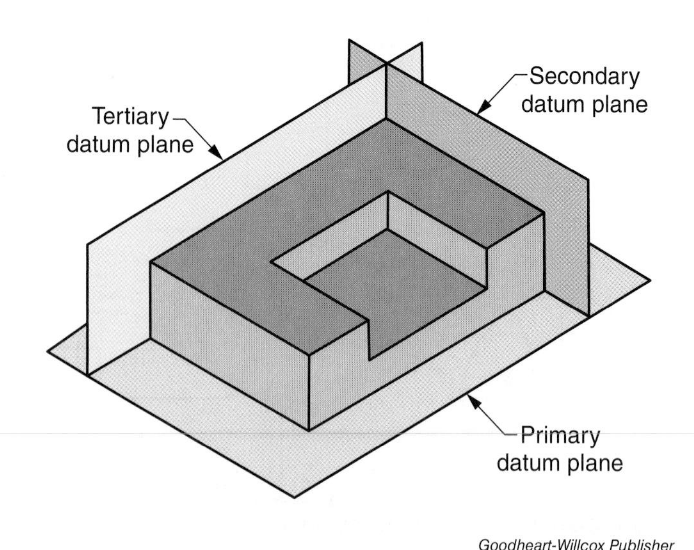

Goodheart-Willcox Publisher

Figure 4-34. Datum references are perpendicular planes. The first datum referenced is the primary datum, followed by the secondary and tertiary datums.

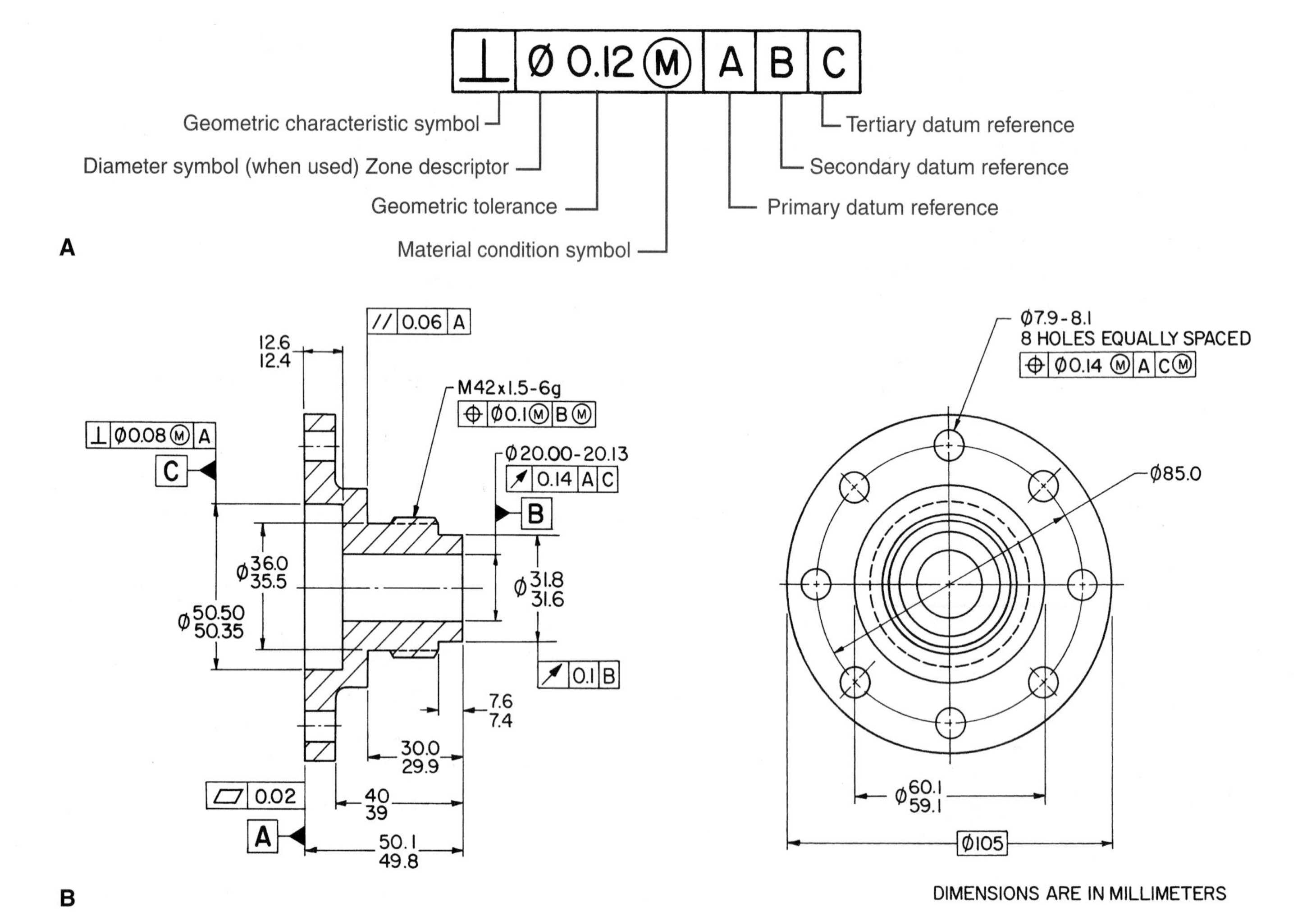

Goodheart-Willcox Publisher

Figure 4-33. A feature control frame is used when a location or form tolerance is related to a datum. A—Components of a feature control frame. B—Feature control frames are used to specify tolerances on this drawing.

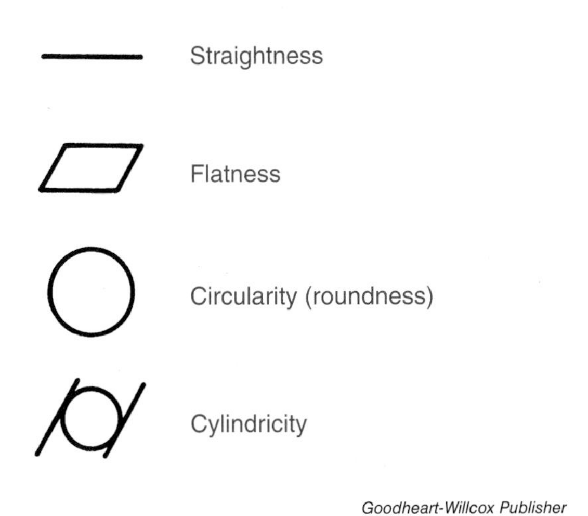

Figure 4-35. Form geometric symbols.

Flatness is a measure of the variation of a surface perpendicular to its plane. The flatness tolerance specifies the two parallel planes within which all points of a surface must lie, **Figure 4-36**.

Straightness describes how closely the surface of an object is to a line. A straightness tolerance establishes a tolerance zone of uniform width along a line. All elements of the surface must lie within this zone, **Figure 4-37**.

Circularity (roundness) is characterized by any given cross section taken perpendicular to the axis of a cylinder or a cone, or through the common center of a sphere. A circularity tolerance specifies a tolerance zone bounded by two concentric circles, indicated on a plane perpendicular to the axis of a cylinder or a cone, within which each circular element must lie. It is a single cross-sectional tolerance. See **Figure 4-38**.

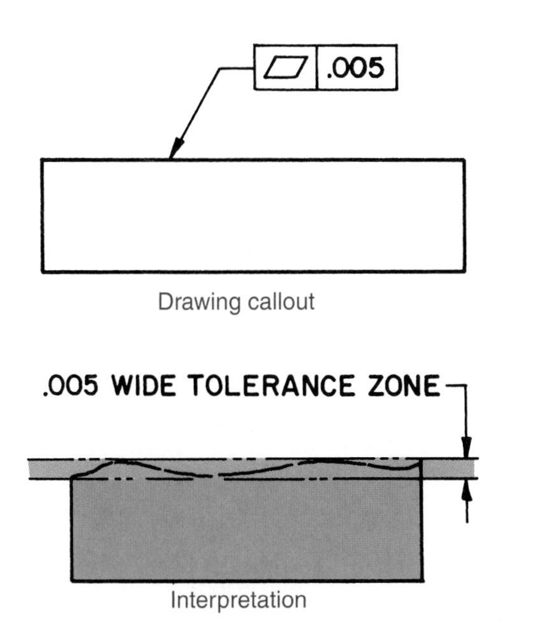

Figure 4-36. The flatness geometric form tolerance specifies the two parallel planes within which a surface must lie.

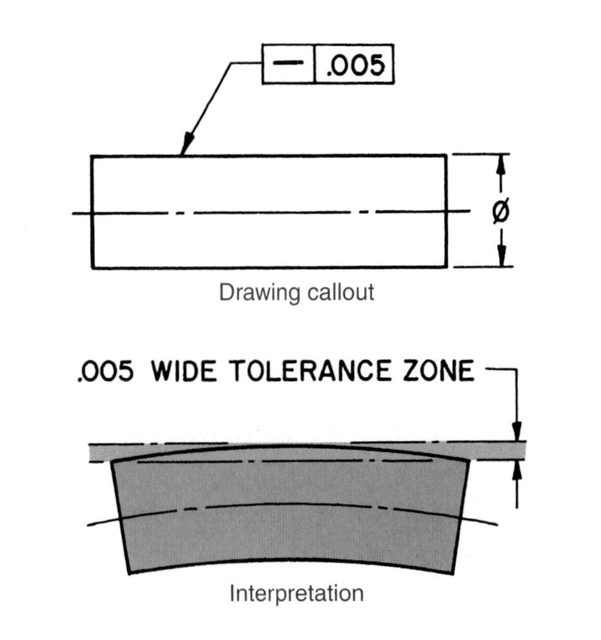

Figure 4-37. A straightness geometric form tolerance establishes a tolerance zone of uniform width along a straight line. All elements of the surface must lie within this zone.

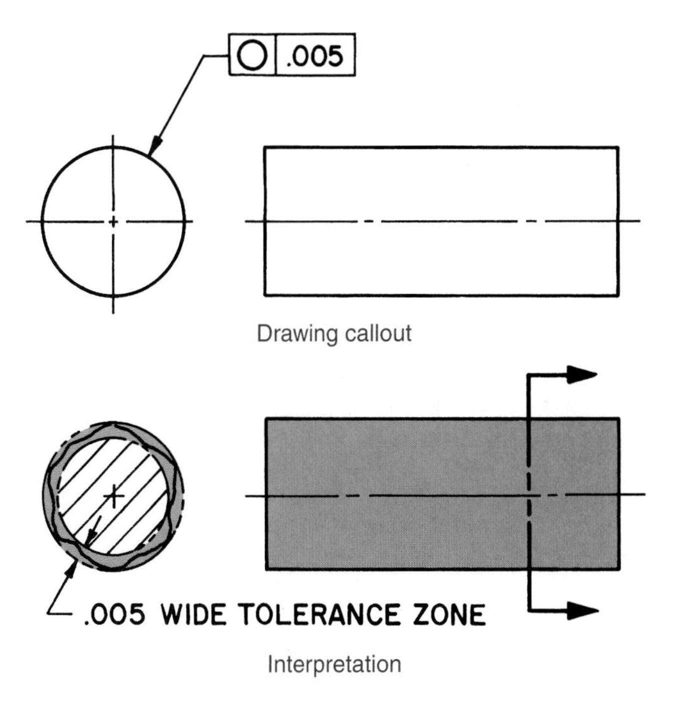

Figure 4-38. A circularity geometric tolerance specifies a tolerance zone bounded by two concentric circles on a plane perpendicular to the axis of a cylinder or cone, within which each circular element must lie.

Cylindricity represents a surface in which all points are an equal distance from a common center. The cylindricity tolerance establishes a tolerance zone that controls the diameter of a cylinder throughout its entire length. It consists of two concentric cylinders within which the actual surface must lie. This tolerance covers both the circular and longitudinal elements. See **Figure 4-39**.

4.7.4 Profile Geometric Tolerances

A ***profile geometric tolerance*** controls the outline or contour of an object and can be represented by an external view or by a cross section through the object. It is a boundary along the true profile in which elements of the surface must be contained. The symbols used to indicate profile tolerances are shown in **Figure 4-40**.

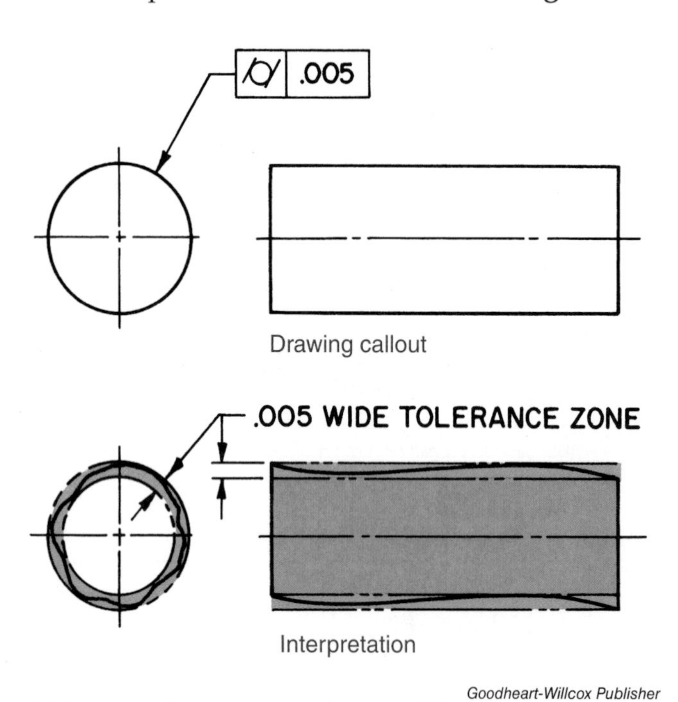

Figure 4-39. The cylindricity geometric tolerance establishes a tolerance zone that controls the diameter of a cylinder throughout its entire length.

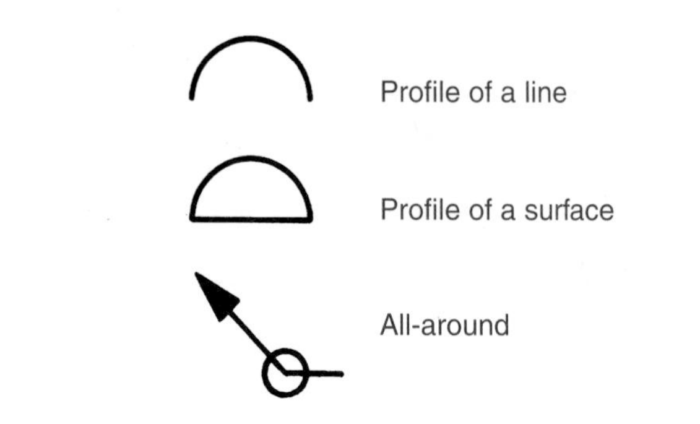

Figure 4-40. Profile geometric tolerance symbols. When a tolerance is specified for all sides of an object, the "all-around" symbol is used.

A ***profile of a line tolerance*** is a two-dimensional (cross-sectional) tolerance zone extending along the length of the element. It is located using basic dimensions, **Figure 4-41**.

The ***profile of a surface tolerance*** is three-dimensional and extends along the length and width of the surface. For proper orientation of the profile, a datum reference is usually required, **Figure 4-42**.

4.7.5 Orientation Geometric Tolerances

Orientation geometric tolerances control the degree of parallelism, perpendicularity, or angularity of a feature with respect to one or more datums. There are three orientation tolerances, **Figure 4-43**.

Angularity is concerned with the position of a surface or axis at a specified angle to a datum plane or axis. The specified angle must be other than 90°. An angularity tolerance establishes a tolerance zone defined by two parallel lines, planes, or a cylindrical zone at a specified basic angle other than 90°. The line elements, surface, or axis of the considered feature must lie within this zone, **Figure 4-44**.

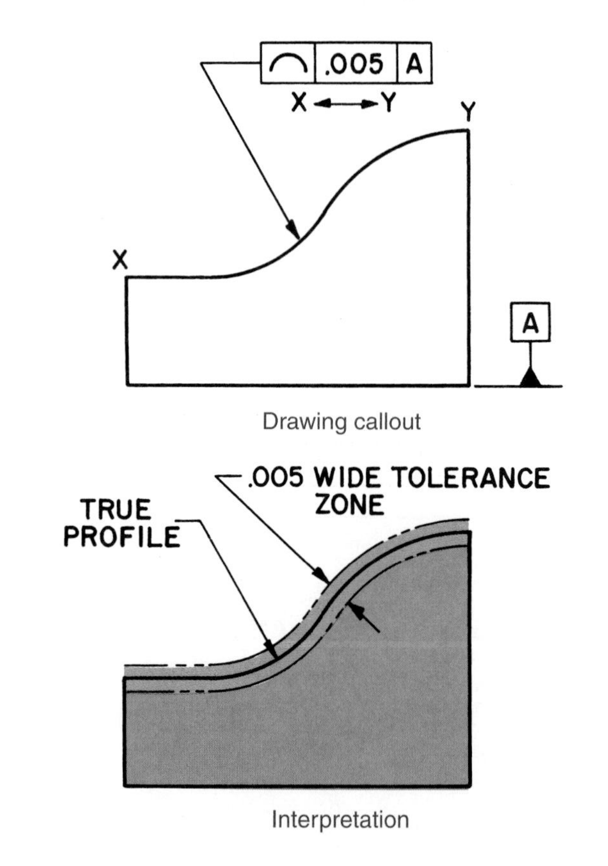

Figure 4-41. A profile line geometric tolerance is a two-dimensional tolerance zone extending along the length of the considered element.

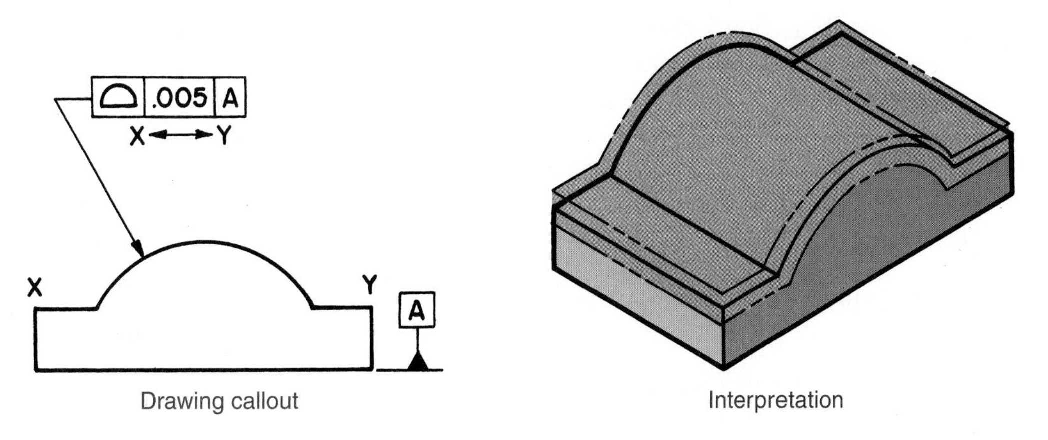

Goodheart-Willcox Publisher

Figure 4-42. The profile surface geometric tolerance is three-dimensional and extends along the length and width of the surface.

Goodheart-Willcox Publisher

Figure 4-43. Orientation geometric tolerance symbols.

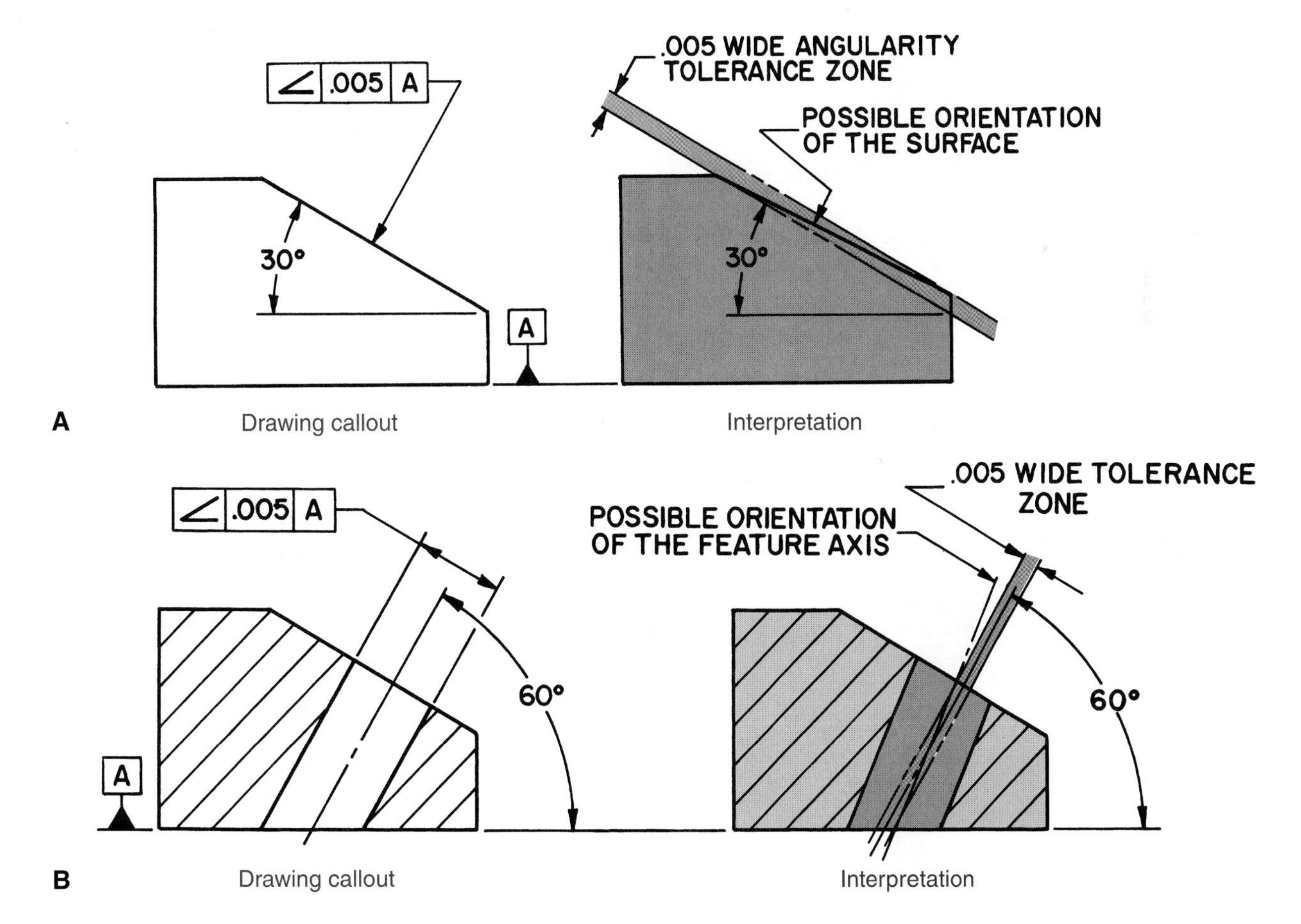

Goodheart-Willcox Publisher

Figure 4-44. An angularity geometric tolerance establishes a tolerance zone defined by two parallel lines, planes, or a cylindrical zone at a specified basic angle other than 90°. A—Angularity of a surface. B—Angularity of an axis.

Perpendicularity describes how closely two lines or surfaces are to being perpendicular (at 90° to each other). A perpendicularity tolerance specifies a tolerance zone at right angles to a given datum or axis. It is described by two parallel lines, planes, or a cylindrical tolerance zone. The line, surface, or axis of the considered feature must lie within this zone, **Figure 4-45**.

Parallelism describes how close all elements of a line or surface are to being parallel (equidistant) to a given datum plane or axis. A parallelism tolerance is a tolerance zone defined by two lines parallel to a datum within which the elements of a surface or axis must lie, **Figure 4-46**.

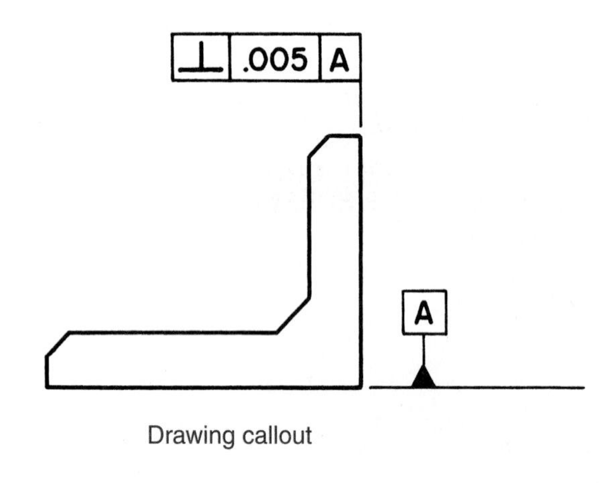

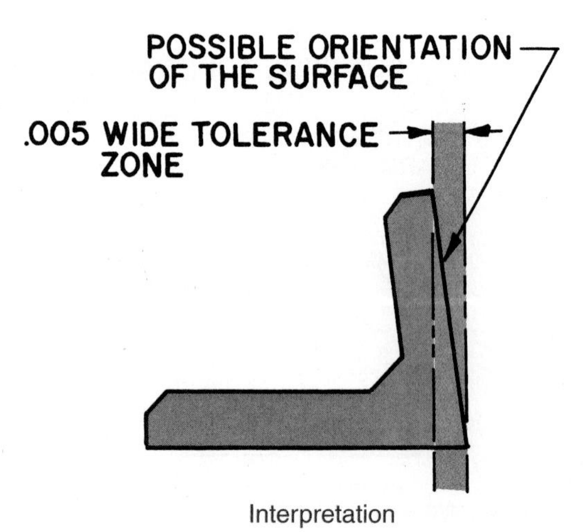

Goodheart-Willcox Publisher

Figure 4-45. The line, surface, or axis of a considered feature must lie within the perpendicularity geometric tolerance zone.

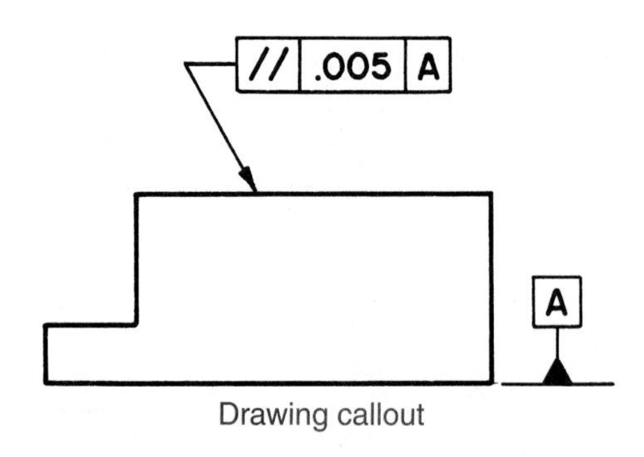

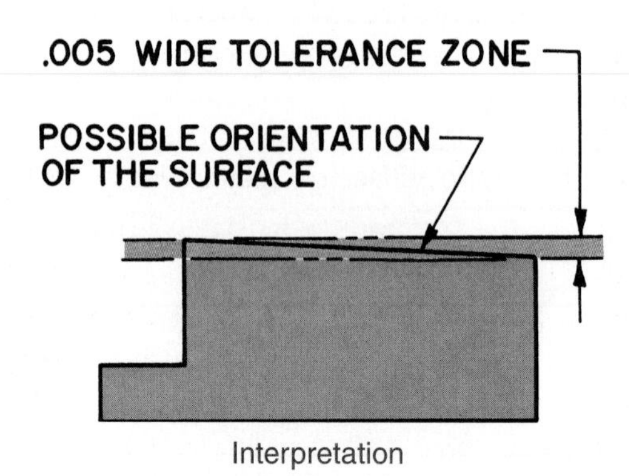

Goodheart-Willcox Publisher

Figure 4-46. A parallelism geometric tolerance is a tolerance zone defined by two lines parallel to a datum within which the elements of a surface or axis must lie.

4.7.6 Location Geometric Tolerances

Location geometric tolerances are used to establish the location of features and datums. They define the zone within which the center, axis, or center plane of a feature may vary from a true (theoretically exact) position. Location tolerances are also known as positional tolerances and include position, concentricity, and symmetry. See **Figure 4-47**.

Basic dimensions establish the true position of a feature from specified datums and related features. A ***position tolerance*** establishes how far a feature may vary from its true position, **Figure 4-48**.

Concentricity defines the relationship between the axes of two or more of an object's cylindrical features. A concentricity tolerance is expressed as a cylindrical tolerance zone. The axis or center point of this zone coincides with a datum axis, **Figure 4-49**.

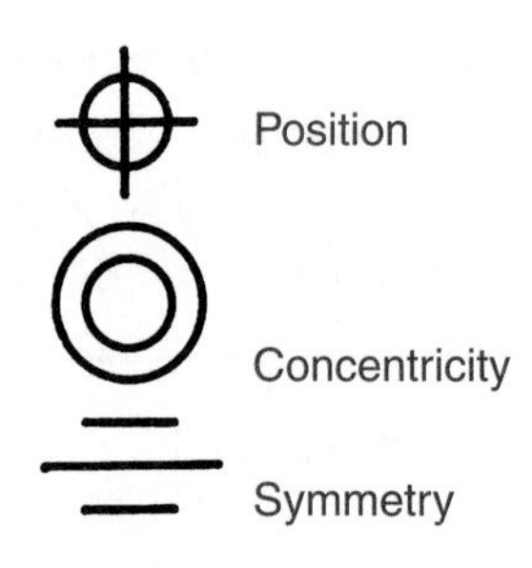

Goodheart-Willcox Publisher

Figure 4-47. Location or positional tolerance symbols.

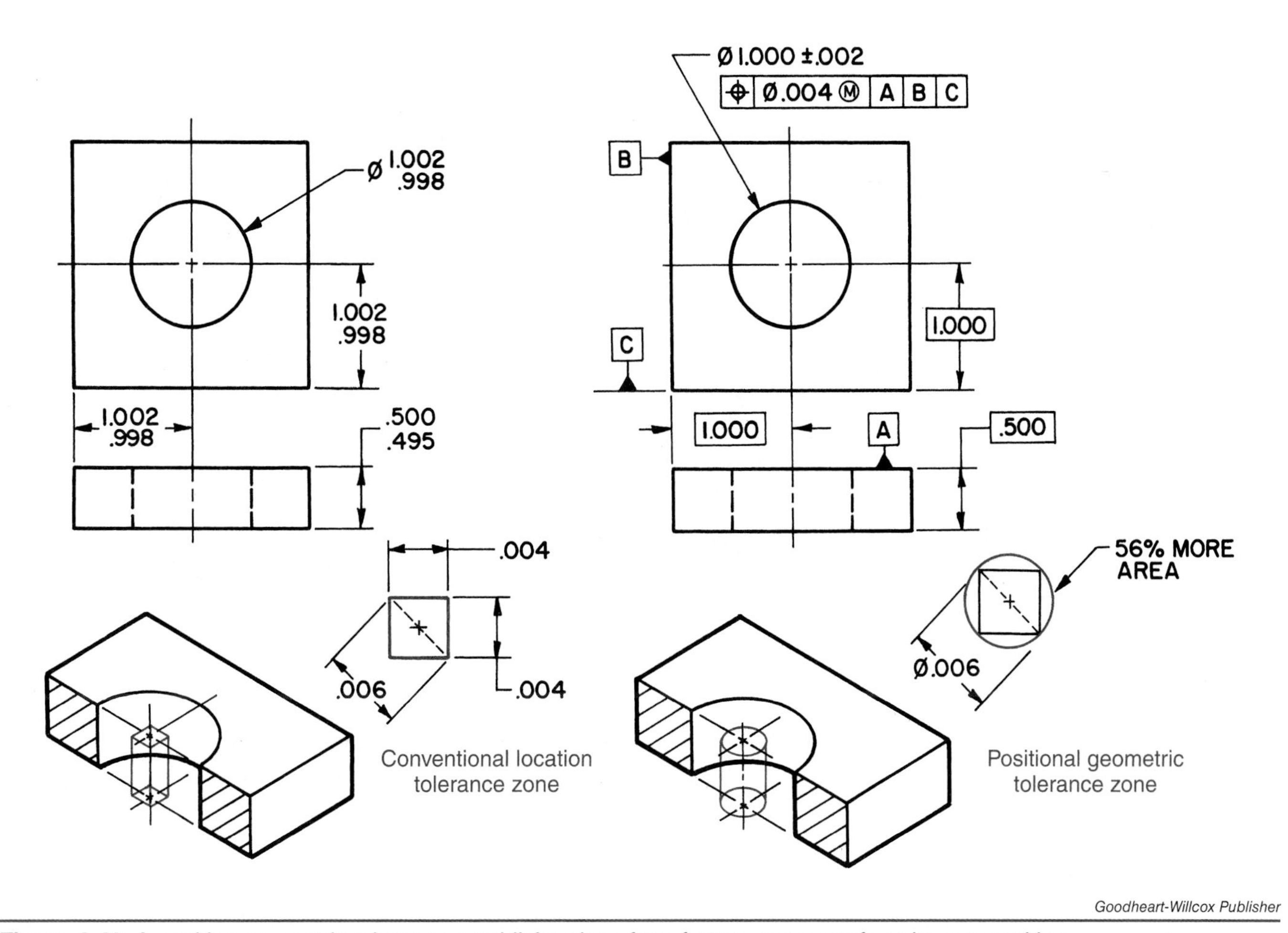

Goodheart-Willcox Publisher

Figure 4-48. A position geometric tolerance establishes how far a feature may vary from its true position.

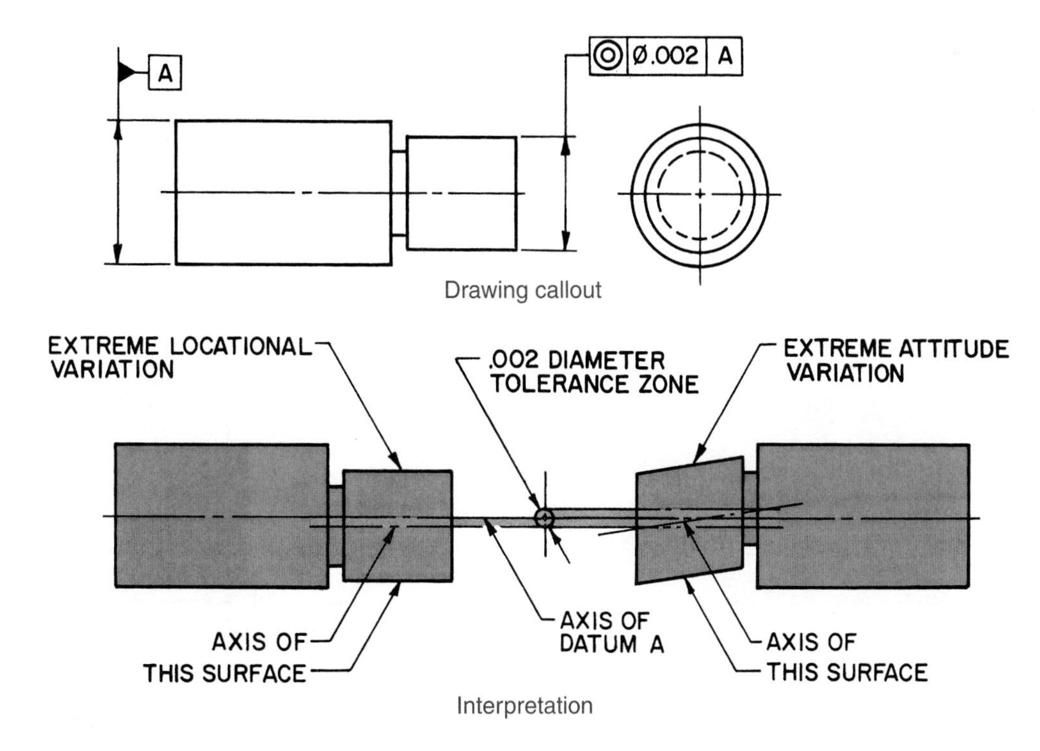

Goodheart-Willcox Publisher

Figure 4-49. A concentricity geometric tolerance is expressed as a cylindrical tolerance zone. The axis or center point of this zone coincides with a datum axis.

Since this tolerance is sometimes difficult and time-consuming to verify, runout or position geometric tolerances are often used instead.

Symmetry indicates equal or balanced proportions on either side of a central plane or datum, **Figure 4-50**. A symmetry tolerance is a zone within which the symmetrical surfaces align with the datum of a center plane or axis, **Figure 4-51**.

4.7.7 Runout Geometric Tolerances

A ***runout geometric tolerance*** is a combination of geometric tolerances used to control the relationship between one or more features of a part and an associated datum axis. There are two types of runout geometric tolerances—total runout and circular runout. These tolerances are indicated by the symbols shown in **Figure 4-52**. Runout tolerances are used to control runout of surfaces around or perpendicular to a datum axis.

Total runout controls circularity, straightness, angularity, and cylindricity of a part when applied

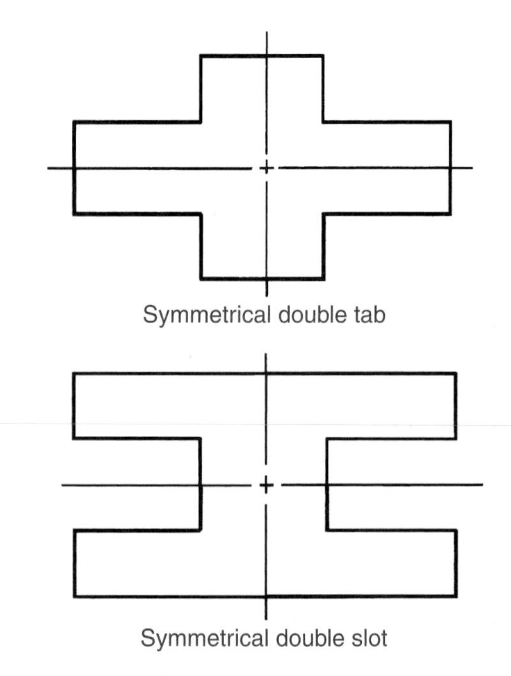

Goodheart-Willcox Publisher

Figure 4-50. Symmetry indicates equal or balanced proportions on either side of a central plane.

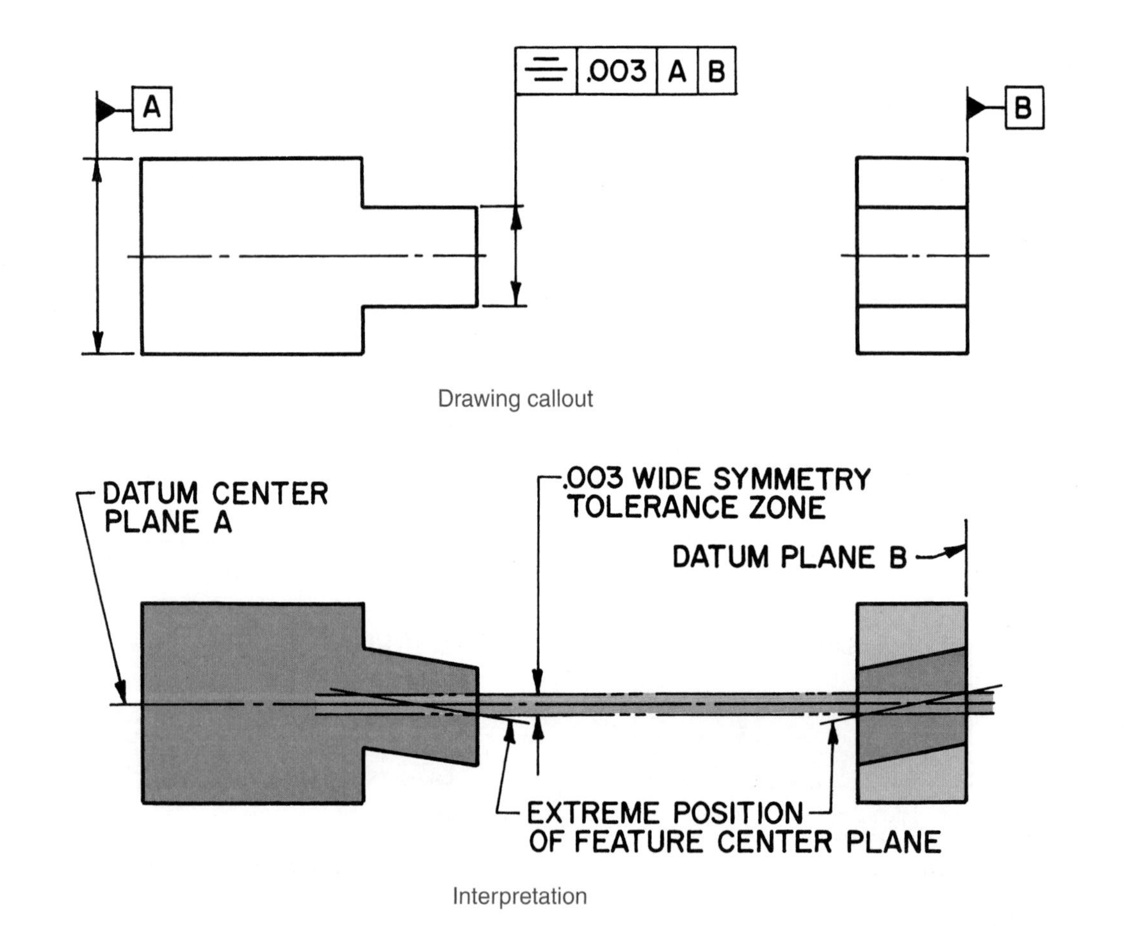

Goodheart-Willcox Publisher

Figure 4-51. A symmetry geometric tolerance is a zone within which the symmetrical surfaces align with the datum of a center plane or axis.

Goodheart-Willcox Publisher

Figure 4-52. Runout geometric tolerance symbols. Arrows may be filled or unfilled.

to surfaces rotated around a datum axis, **Figure 4-53.** The entire surface must lie within the tolerance zone.

Circular runout is applied to features independently and controls circularity of a single circular cross section, **Figure 4-54.** The tolerance is measured by the full indicator movement (FIM) of a dial indicator when it is placed at several positions as the part is rotated.

4.7.8 Geometric Dimensioning and Tolerancing Further Education

Geometric dimensioning and tolerancing is far more involved than described on the preceding pages. As you progress in machining technology, you should consider purchasing a text on the subject, studying a copy of ASME Y14.5M, or enrolling in a class on geometric dimensioning and tolerancing.

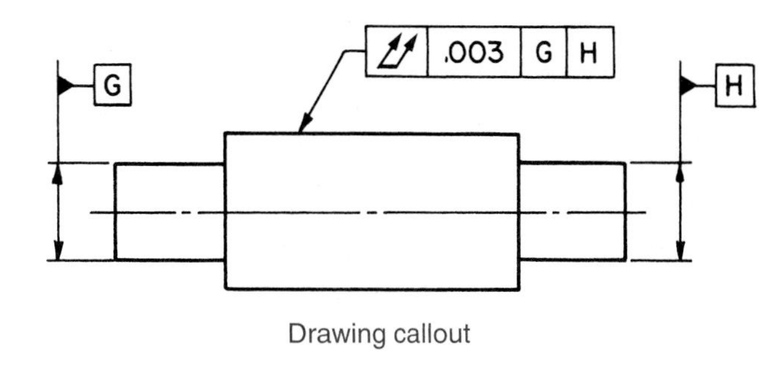

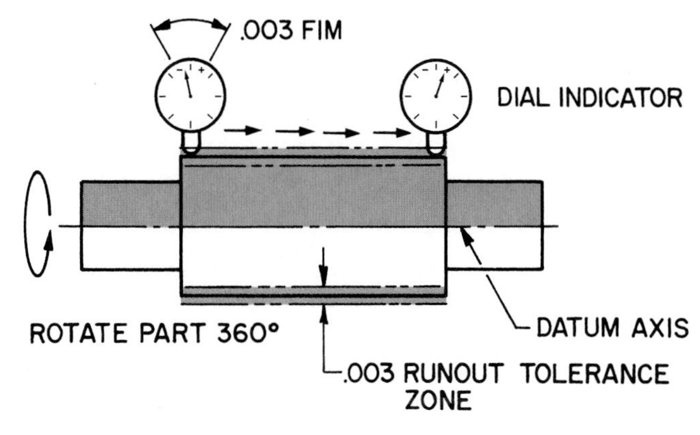

Goodheart-Willcox Publisher

Figure 4-53. Total runout controls circularity, straightness, angularity, and cylindricity of a part when applied to surfaces rotated around a datum. The entire surface must lie within the tolerance zone.

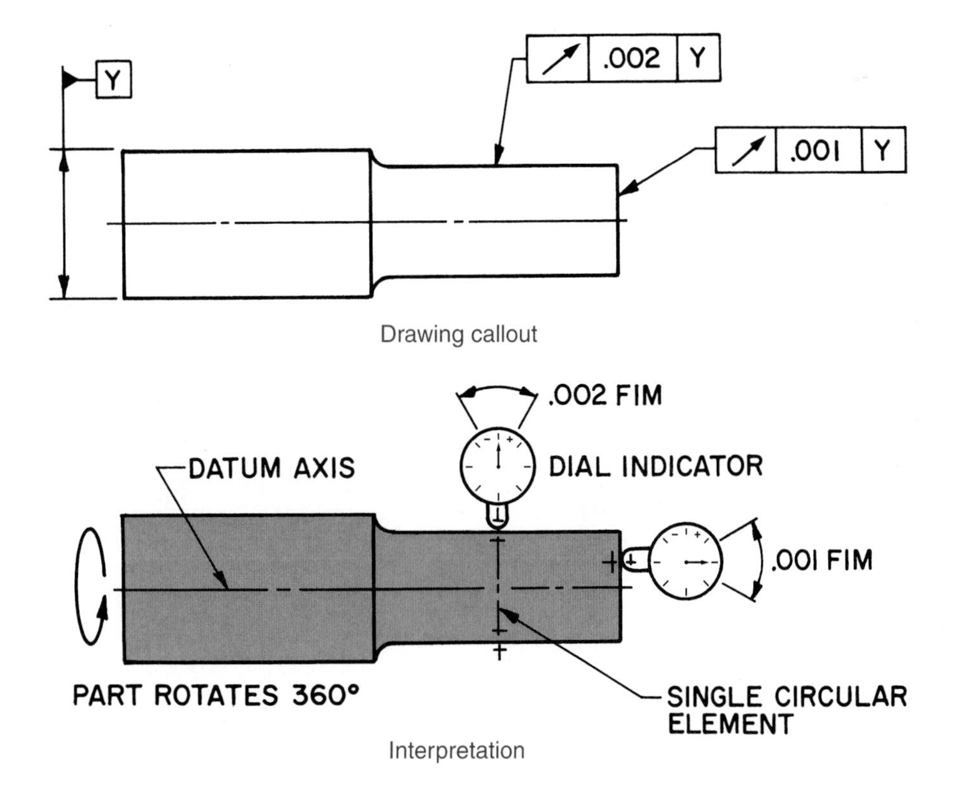

Goodheart-Willcox Publisher

Figure 4-54. Circular runout controls circularity of a single circular cross section.

Chapter Review

Summary

- Drawings are the foundation of parts manufacturing.
- The American National Standards Institute's (ANSI) standardized system of symbols, lines, and figures are the language of industry.
- Drawings include a wealth of information and come in a variety of forms.
- The geometric dimensioning and tolerancing system provides additional precision.

Review Questions

Answer the following questions using the information provided in this chapter.

1. Drawings are used to _____.
   A. show, in multiview, what an object looks like before it is made
   B. standardize parts
   C. show what to make and the sizes to make it
   D. All of the above.
   E. None of the above.
2. The symbols, lines, and figures that make up a drawing are frequently called the _____.
3. A microinch is _____ of an inch.
4. A micrometer, one millionth of a meter, is also known as a(n) _____.
5. How can surface roughness of a machined part be checked against specifications on the drawing? How can it be measured electronically?
6. Tolerances are _____.
   A. the different materials that can be used
   B. acceptable allowances, either oversized or undersized
   C. dimensions
   D. All of the above.
   E. None of the above.
7. When tolerances are plus and minus, it is called a(n) _____ tolerance.
8. When tolerances are only plus or only minus, it is called a(n) _____ tolerance.
9. Drawings made other than actual size are called _____.
10. A subassembly drawing differs from an assembly drawing by _____.
    A. showing only a small portion of the complete object
    B. making it possible to use smaller drawings
    C. showing the object without all needed dimensions
    D. All of the above.
    E. None of the above.
11. The machinist is given all of the information needed to make a part on a(n) _____ drawing.
12. What does an assembly drawing show?
13. Why are standard size drawing sheets used?
14. A dimension providing the theoretical exact location of a feature is a(n) _____ dimension.
15. Dimensions placed between parentheses are _____ dimensions.
16. When is a feature control frame used?
17. List the four form geometric tolerances and describe the symbol used to represent each tolerance.
18. Define the term maximum material condition (MMC). Use a sketch if necessary.
19. Define the term least material condition (LMC). Use a sketch if necessary.

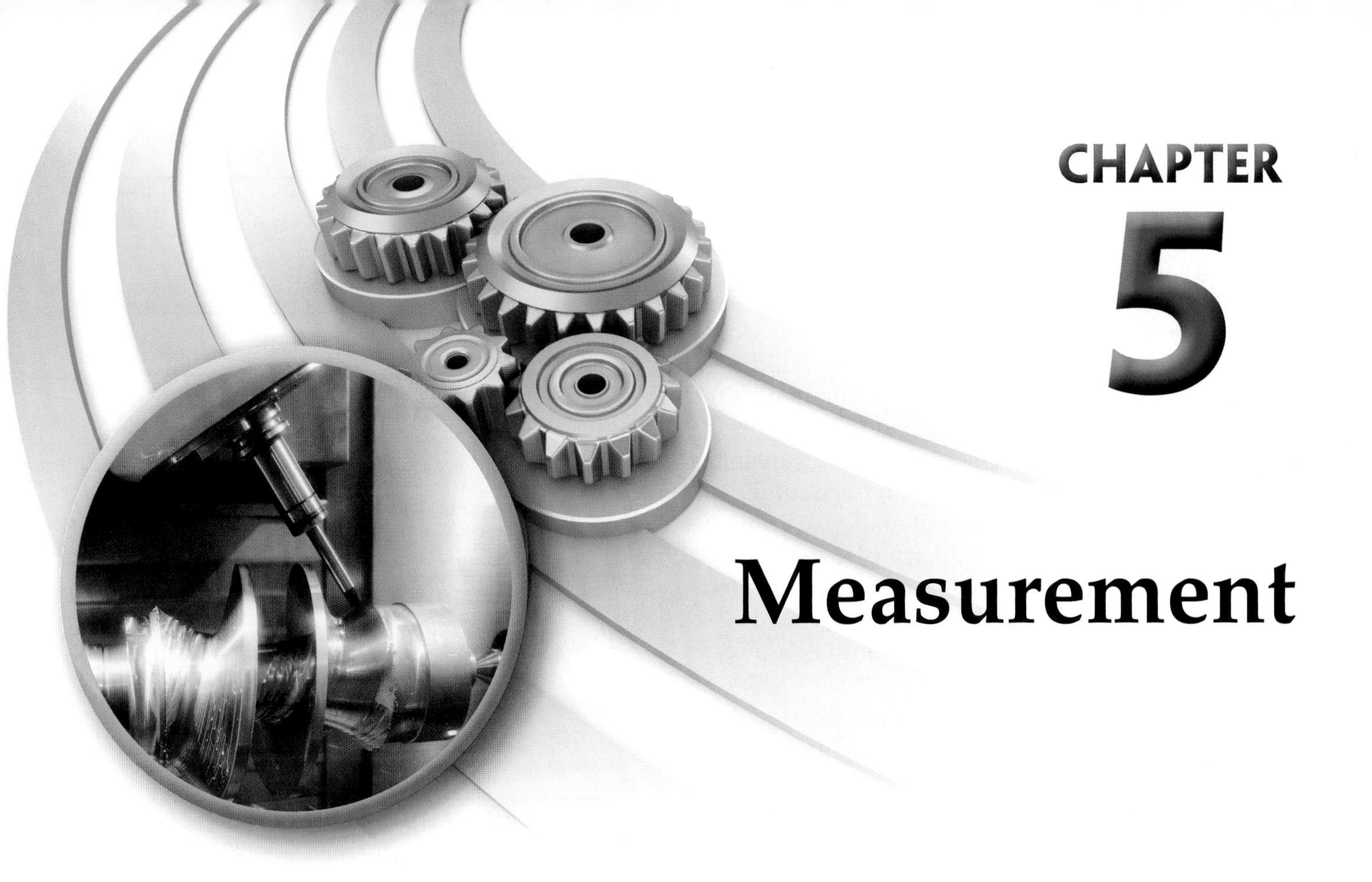

CHAPTER 5

Measurement

Learning Objectives

After studying this chapter, you will be able to:

- Measure to 1/64″ (0.5 mm) with a steel rule.
- Measure to 0.0001″ (0.002 mm) using a vernier micrometer caliper.
- Measure to 0.001″ (0.02 mm) using vernier measuring tools.
- Measure angles to 0°5′ using a universal vernier bevel.
- Identify and use various types of gages found in a machine shop.
- Use a dial indicator.
- Use the various helper measuring tools found in a machine shop.

Technical Terms

dial indicator
gage blocks
gaging
graduations
International System of Units (SI)
metrology
micrometer
micrometer caliper
micron
steel rule
universal bevel protractor
vernier caliper

Without some form of accurate measurement, modern industry could not exist. The science that deals with systems of measurement is called ***metrology***. Today, industry can make measurements accurate to one microinch (one-millionth of an inch).

If a microinch were as thick as a dime, one inch would be as high as *four* Empire State Buildings (about 5000′ total). An engineer once estimated, with tongue-in-cheek, that a steel railroad rail supported at both ends would sag one-millionth of an inch when a "fat horsefly" landed on it in the middle.

In addition to using US Conventional units of measure (inch and foot), industry also uses metric units of measure (millimeter, meter, etc.), called the ***International System of Units*** (abbreviated SI). A ***micrometer*** or ***micron*** is one-millionth of a meter (0.000001 m).

All of the familiar measuring tools are available with scales graduated in metric units, **Figure 5-1**. A metric (millimeter) rule is compared with conventional fractional and decimal rules in **Figure 5-2**. Metric-based measuring tools should offer no problems for the user. As a matter of fact, they are often easier to read than inch-based measuring tools.

Although you will measure in very tiny units when you go to work in industry, you must first learn to read a rule to 1/64″ and 0.5 mm. Then, you can progress through 1/1000″ (0.001″) and 1/100 mm (0.01 mm) by learning to use micrometer and vernier-type measuring tools. Finally, you can progress to 1/10,000″ (0.0001″) and 1/500 mm (0.002 mm) by using the vernier scale on some micrometers.

5.1 The Rule

The ***steel rule***, often incorrectly called a scale, is the simplest of the measuring tools found in the shop. **Figure 5-2** shows the three basic types of rule graduations. A few of the many rule styles are shown in **Figure 5-3**.

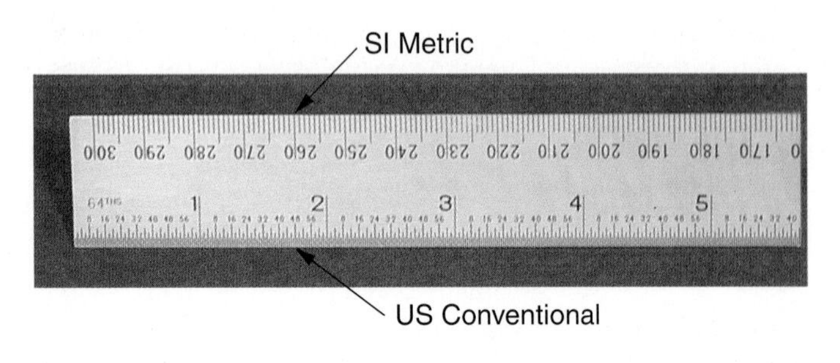

Goodheart-Willcox Publisher

Figure 5-1. This rule can be used to make measurements in both US Conventional and SI Metric units.

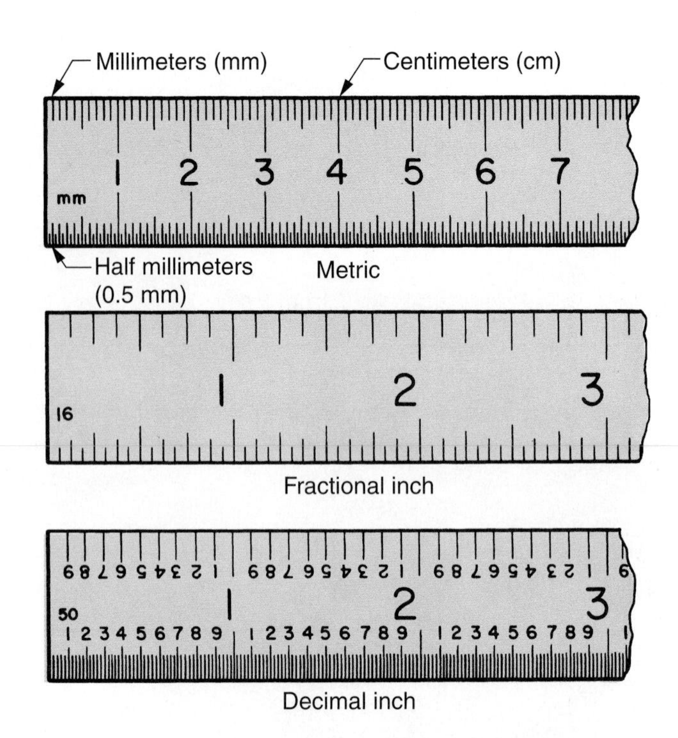

Goodheart-Willcox Publisher

Figure 5-2. Compare the metric (millimeter-graduated) rule with the rules graduated in fractional and decimal inch units.

5.1.1 Reading the Rule (US Conventional)

A careful study of the enlarged rule section will show the different fractional divisions of the inch from 1/8 to 1/64, **Figure 5-4**. The lines representing the divisions are called ***graduations***. On many rules, every fourth graduation is numbered on the 1/32 edge, and every eighth graduation on the 1/64 edge.

To become familiar with the rule, begin by measuring objects on the 1/8 and 1/16 scales. Once you become comfortable with these scales, begin using the 1/32 and 1/64 scales. Practice until you can quickly and accurately read measurements. Some rules are graduated in 10ths, 20ths, 50ths, and 100ths. Additional practice will be necessary to read these rules.

Fractional measurements are always reduced to the lowest terms. A measurement of 14/16″ is reduced to 7/8″, 2/8″ becomes 1/4″, and so on.

5.1.2 Reading the Rule (Metric)

Most metric rules are divided into millimeter or half-millimeter graduations. They are numbered every 10 mm. See **Figure 5-5**. The measurement is determined by counting the number of millimeters.

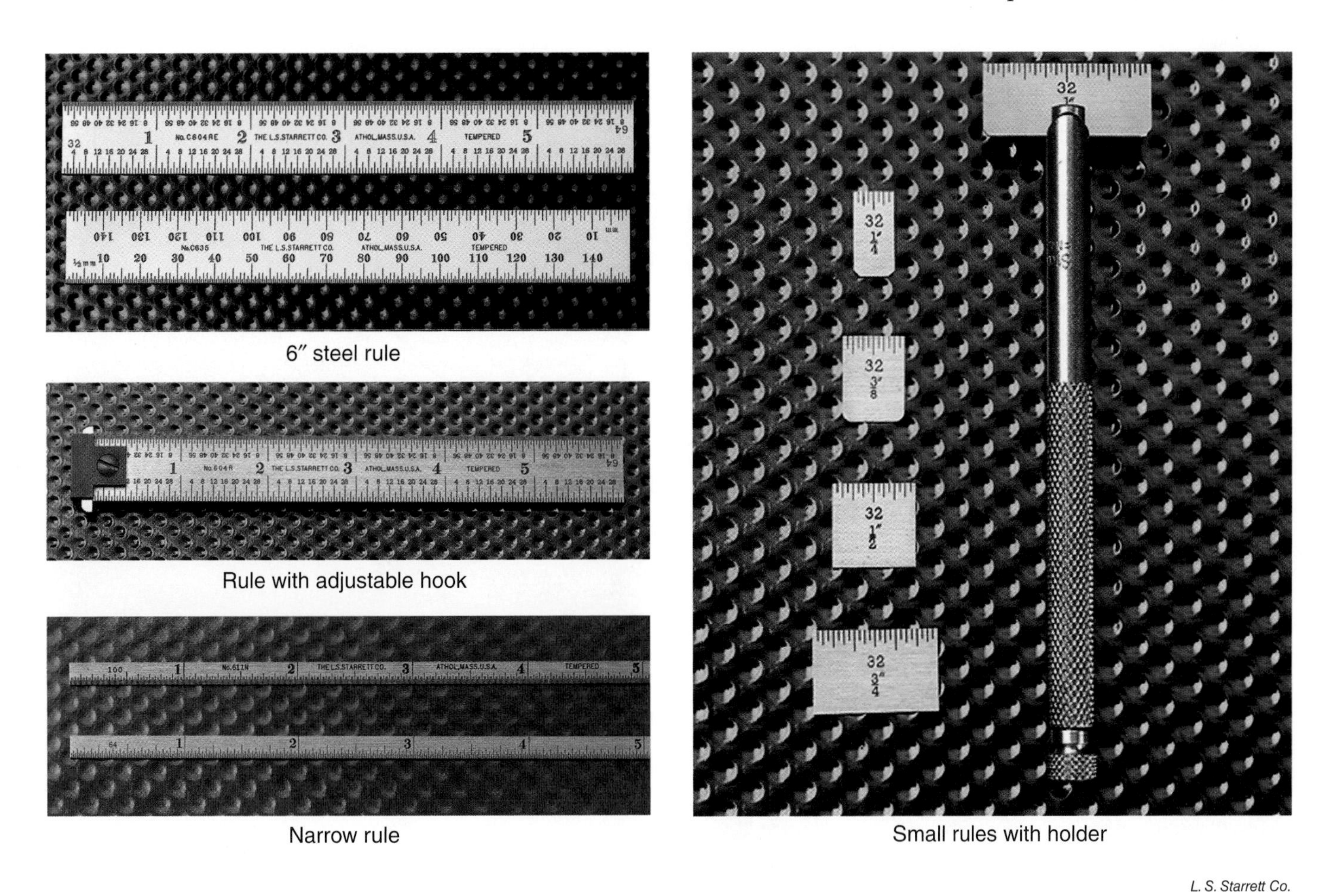

Figure 5-3. Many different types of rules are used to make measuring quicker and more accurate.

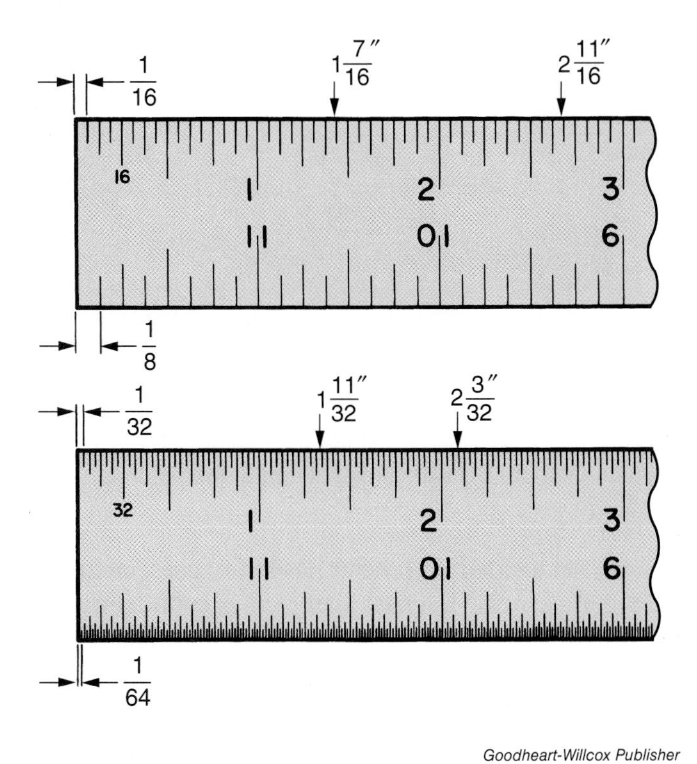

Figure 5-4. These are the fractional graduations found on a rule. Measurements are taken by counting the number of graduations.

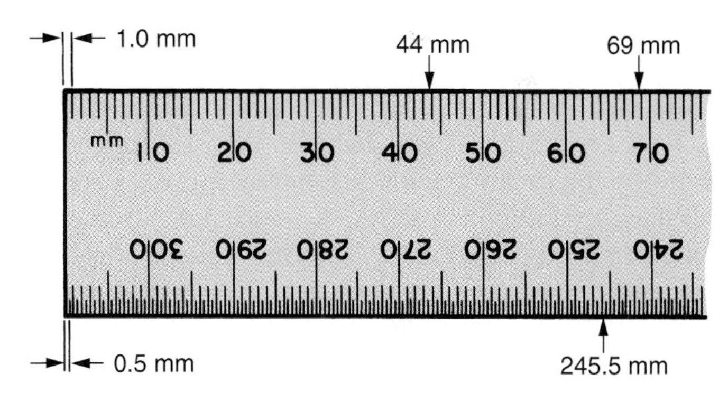

Figure 5-5. Most metric rules are graduated in millimeters and half-millimeters. They are available in a variety of sizes.

5.1.3 Care of the Rule

The steel rule is precision-made and, like all tools, its accuracy depends on the care it receives. Here are a few suggestions:

- Use the rule for measurements only. Do not adjust screws or open paint cans with it. Be careful not to bend your rule.

- Avoid laying other tools on the rule.
- Wipe steel rules with an oily cloth before storing. This will prevent rust. If the rule is to be stored for a prolonged period, coat it with wax or rust preventative.
- Clean the rule with steel wool to keep the graduations legible.
- Make measurements and tool settings from the 1″ line (10 mm line on a metric rule) or other major graduations, rather than from the end of the rule.
- Store rules separately. Do not throw them in a drawer with other tools.
- Use the rule with care to protect the ends from nicks and wear.
- Use the correct rule for the job being done.

SAFETY NOTE

Keep the rule clear of moving machinery. Never use it to clean metal chips as they form on the cutting tool. This is extremely dangerous and will ruin the rule.

5.2 The Micrometer Caliper

A Frenchman, Jean Palmer, devised and patented a measuring tool that made use of a screw thread, making it possible to read measurements quickly and accurately without calculations. It incorporated a series of engraved lines on the sleeve and around the thimble. The device, called *Systeme Palmer*, is the basis for the modern micrometer caliper, **Figure 5-6**.

The ***micrometer caliper***, also known as a "mike," is a precision tool capable of measuring to 0.001″ or 0.01 mm. When fitted with a vernier scale, it will read to 0.0001″ or 0.002 mm.

5.2.1 Types of Micrometers

Micrometers are produced in a wide variety of models. Digital display is included in many micrometers, making measuring easier. Some of the most popular models are the following:

- An outside micrometer measures external diameters and thickness, **Figure 5-7**.

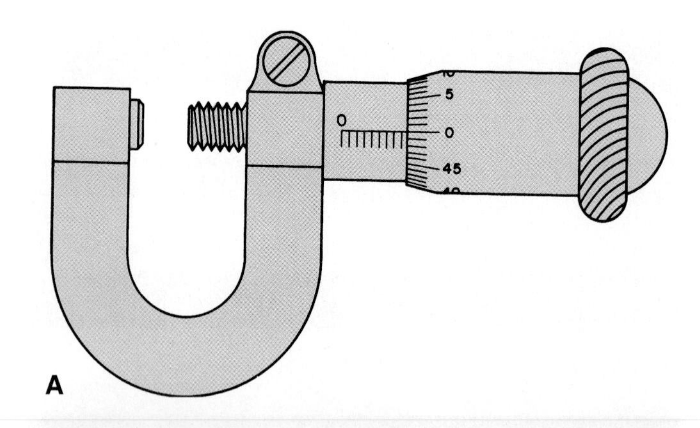

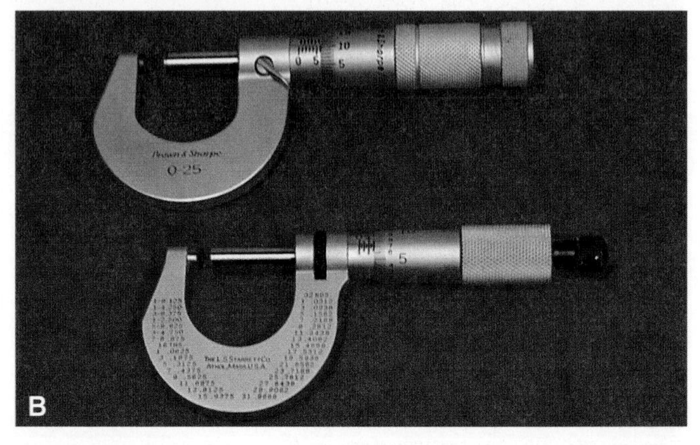

Goodheart-Willcox Publisher

Figure 5-6. The micrometer caliper, past and present. A—A drawing of the Systeme Palmer measuring device. B—These modern micrometer calipers operate on the same principle as the original 1848 invention.

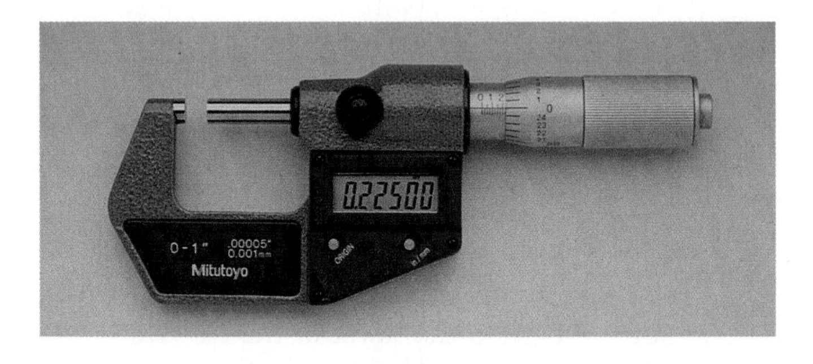

Mitutoyo/MTI Corp.

Figure 5-7. This digital outside micrometer can be used to measure in both US Conventional and SI Metric units.

- An inside micrometer has many uses, including measuring internal diameters of cylinders, rings, and slots. The range of a conventional inside micrometer can be extended by fitting longer rods to the micrometer head. The range of a jaw-type inside micrometer is limited to 1″ or 25 mm. The jaw-type inside micrometer has a scale graduated from right to left. See **Figure 5-8**.

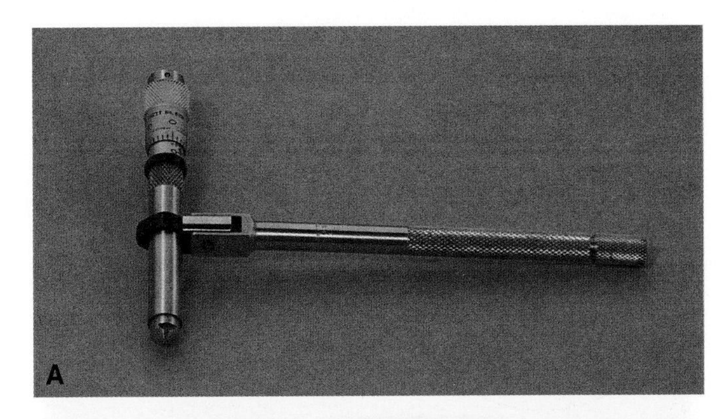

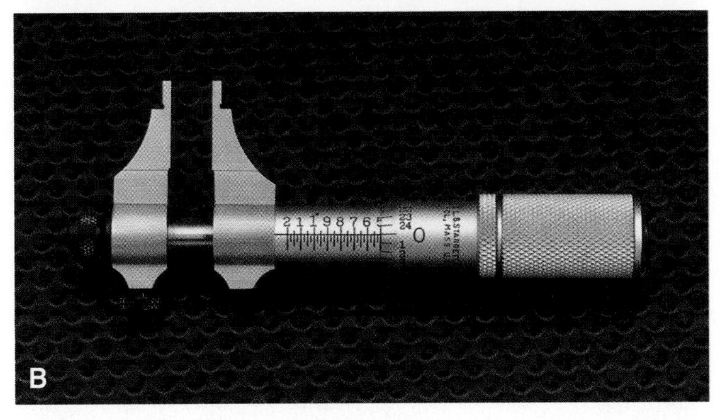

L. S. Starrett Co.

Figure 5-8. Inside micrometers. A—A conventional inside micrometer. B—The caliper jaws on this inside micrometer allow quick and accurate measurements. The divisions on the sleeve are numbered in the reverse order of a conventional outside micrometer.

- A micrometer depth gage measures the depths of holes, slots, and projections. See **Figure 5-9**. The measuring range can be increased by changing to longer spindles. Measurements are read from right to left.
- A screw thread micrometer has a pointed spindle and a double-V anvil shaped to contact the screw thread, **Figure 5-10**. It measures the pitch diameter of the thread, which equals the outside (major) diameter of the thread minus the depth of one thread. Since each thread micrometer is designed to measure only a limited number of threads per inch, a set of thread micrometers is necessary to measure a full range of thread pitches.
- A chamfer micrometer will accurately measure countersunk holes and other chamfer-type measurements. With fastener tolerances so critical on some aerospace and other applications, it is important that countersunk holes and tapers on fasteners meet specifications. A chamfer micrometer makes it possible to check these critical areas.

Goodheart-Willcox Publisher

Figure 5-9. A standard micrometer depth gage.

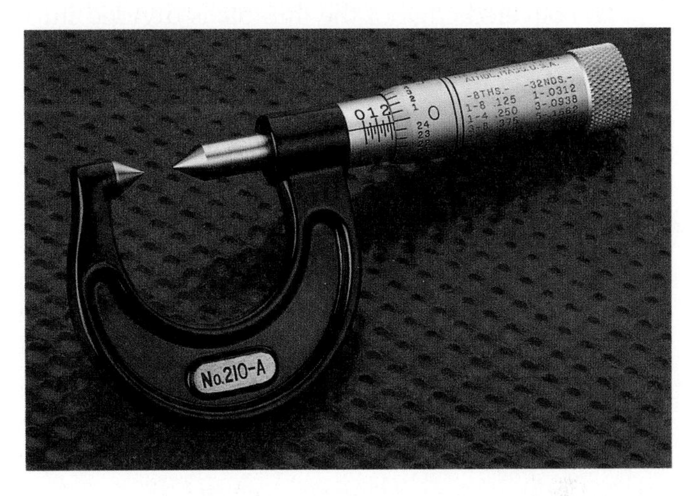

L. S. Starrett Co.

Figure 5-10. This screw thread micrometer can measure threads as wide as 7/8″.

Special micrometers are available for other applications. These micrometers are devised to handle nonstandard measurement tasks.

5.2.2 Reading an Inch-Based Micrometer

A micrometer uses a very precisely made screw thread that rotates in a fixed nut. The screw thread is ground on the spindle and is attached to the thimble. The spindle advances or recedes from the anvil as the thimble is rotated. See **Figure 5-11**. The threaded section has 40 threads per inch; therefore, each revolution of the thimble moves the spindle 1/40″ (0.025″).

The line engraved lengthwise on the sleeve is divided into 40 equal parts per inch (corresponding to the number of threads per inch on the spindle).

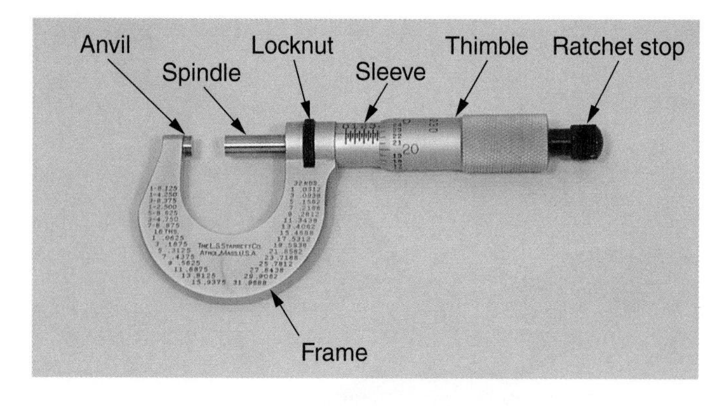

Goodheart-Willcox Publisher

Figure 5-11. Basic parts of a micrometer caliper.

Each vertical line equals 1/40″, or 0.025″. Every fourth division is numbered, representing 0.100″, 0.200″, etc.

The beveled edge of the thimble is divided into 25 equal parts around its circumference. Each division equals 1/1000″ (0.001″). On some micrometers, every division is numbered, while every fifth division is numbered on others.

The micrometer is read by recording the highest number on the sleeve (1 = 0.100, 2 = 0.200, etc.). To this number, add the number of vertical lines visible between the number and thimble edge (1 = 0.025, 2 = 0.050, etc.). To this total, add the number of thousandths indicated by the line that coincides with the horizontal sleeve line.

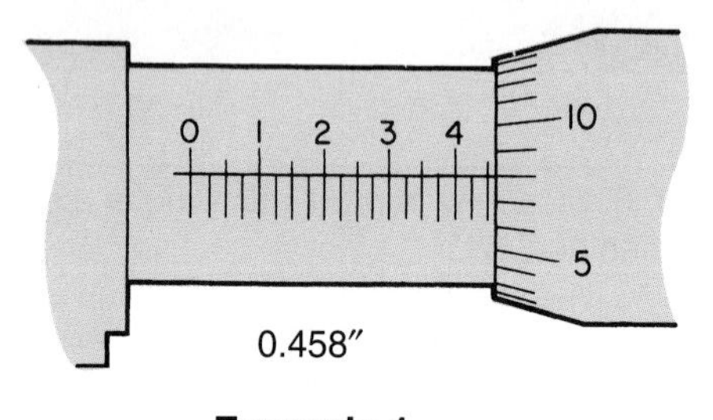

Example 1

Add the readings from the sleeve and the thimble:

4 large graduations:	4 × 0.100 = 0.400
2 small graduations:	2 × 0.025 = 0.050
8 thimble graduations:	8 × 0.001 = 0.008
Total mike reading	= 0.458″

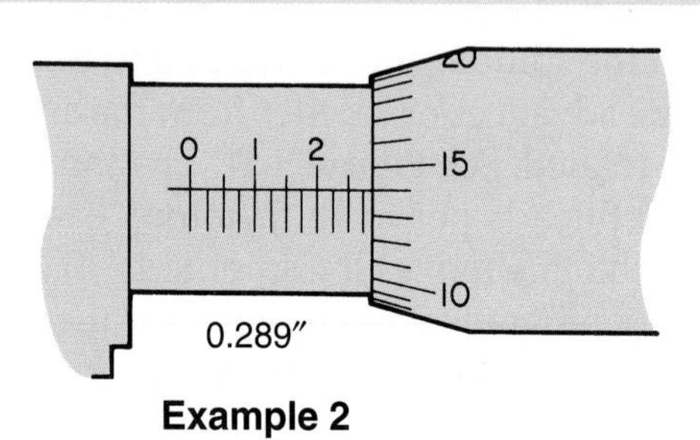

Example 2

Add the readings from the sleeve and the thimble:

2 large graduations:	2 × 0.100 = 0.200
3 small graduations:	3 × 0.025 = 0.075
14 thimble graduations:	14 × 0.001 = 0.014
Total mike reading	= 0.289″

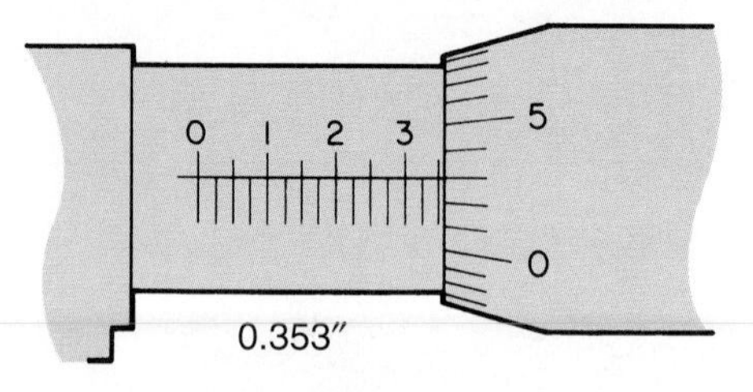

Example 3

Add the readings from the sleeve and the thimble:

3 large graduations:	3 × 0.100 = 0.300
2 small graduations:	2 × 0.025 = 0.050
3 thimble graduations:	3 × 0.001 = 0.003
Total mike reading	= 0.353″

5.2.3 Reading a Vernier Micrometer

On occasion, it is necessary to measure more precisely than 0.001″. A vernier micrometer caliper is used in these situations. This micrometer has a third scale around the sleeve that will furnish the 1/10,000″ (0.0001″) reading. See **Figure 5-12.**

The vernier scale has 11 parallel lines that occupy the same space as 10 lines on the thimble. The lines around the sleeve are numbered 1 to 10. The difference between the spaces on the sleeve and those on the thimble is one-tenth of a thousandth of an inch.

To read the vernier scale, first obtain the thousandths reading, then observe which of the lines on the vernier scale coincides (lines up) with a line on the thimble. Only one of them can line up. If the line is 1, add 0.0001 to the reading; if line 2, add 0.0002 to the reading, etc. See **Figure 5-13.**

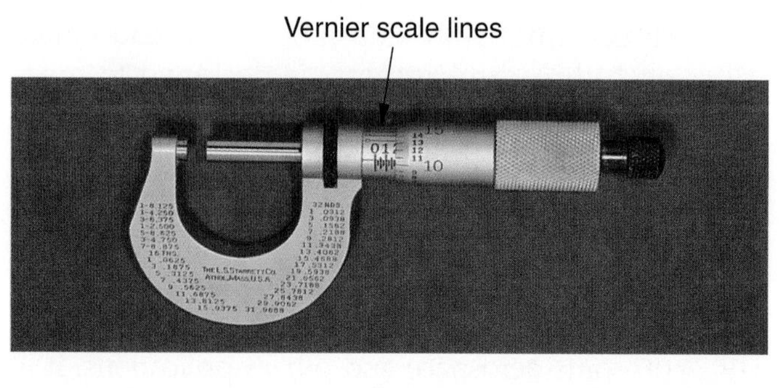

Goodheart-Willcox Publisher

Figure 5-12. A vernier micrometer caliper includes a vernier scale on the sleeve.

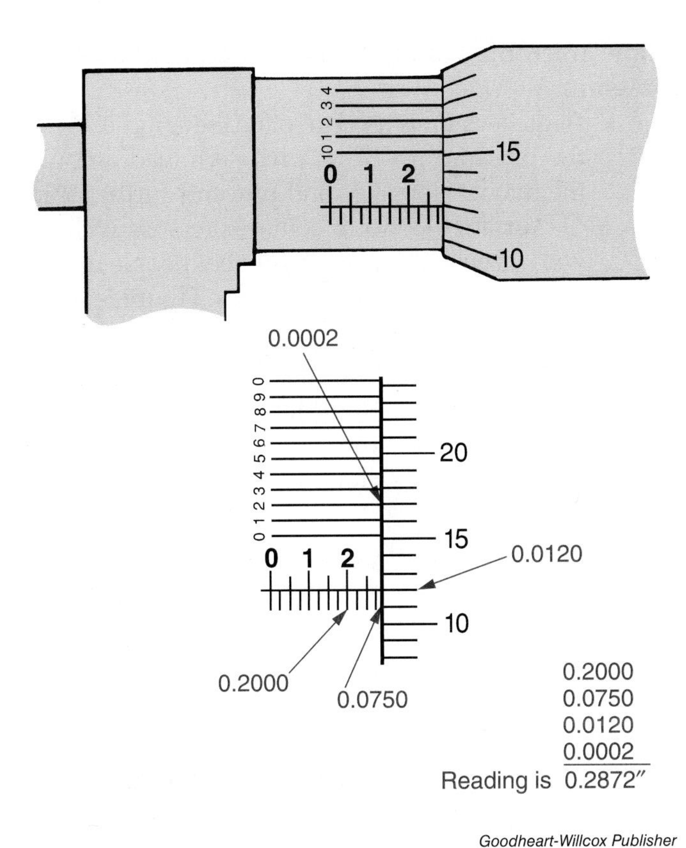

Goodheart-Willcox Publisher

Figure 5-13. How to read a vernier micrometer caliper. Add the total reading in thousandths, then observe which of the lines on the vernier scale coincides with a line on the thimble. In this case, it is the second line, so 0.0002 is added to the reading.

5.2.4 Reading a Metric Micrometer

The metric micrometer is read as shown in **Figure 5-14**. If you are able to read the conventional inch-based micrometer, reading the metric tool will offer no difficulties.

5.2.5 Reading a Metric Vernier Micrometer

Metric vernier micrometers are read in the same way as standard metric micrometers. However, using the vernier scale on the sleeve, an additional reading of two-thousandths of a millimeter can be obtained, **Figure 5-15**.

5.2.6 Using the Micrometer

The proper way to hold a micrometer when making a measurement is shown in **Figure 5-16**. The work is placed into position, and the thimble rotated until the part is clamped lightly between the anvil and spindle. Guard against excessive pressure, which will cause an erroneous reading.

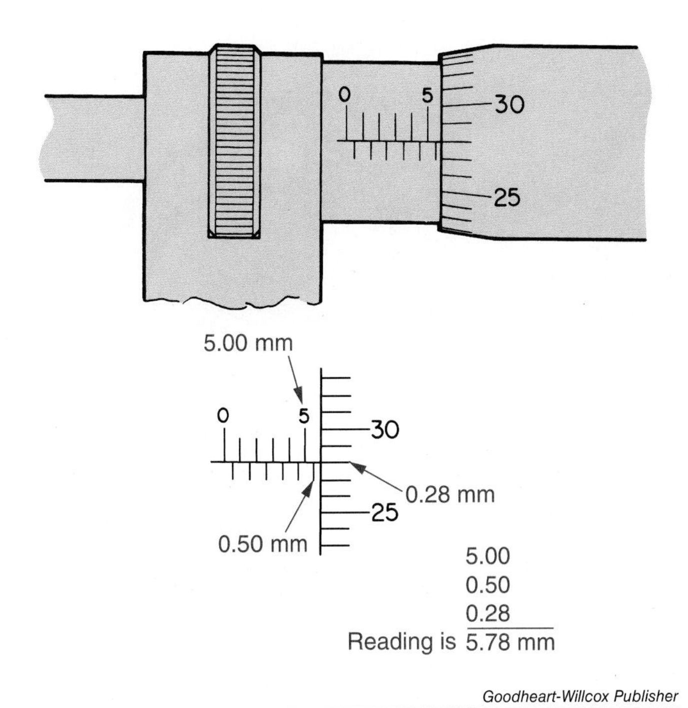

Goodheart-Willcox Publisher

Figure 5-14. To read a metric micrometer, add the total reading in millimeters visible on the sleeve to the reading of hundredths of a millimeter, indicated by the graduation on the thimble. Note that the thimble reading coincides with the longitudinal line on the micrometer sleeve.

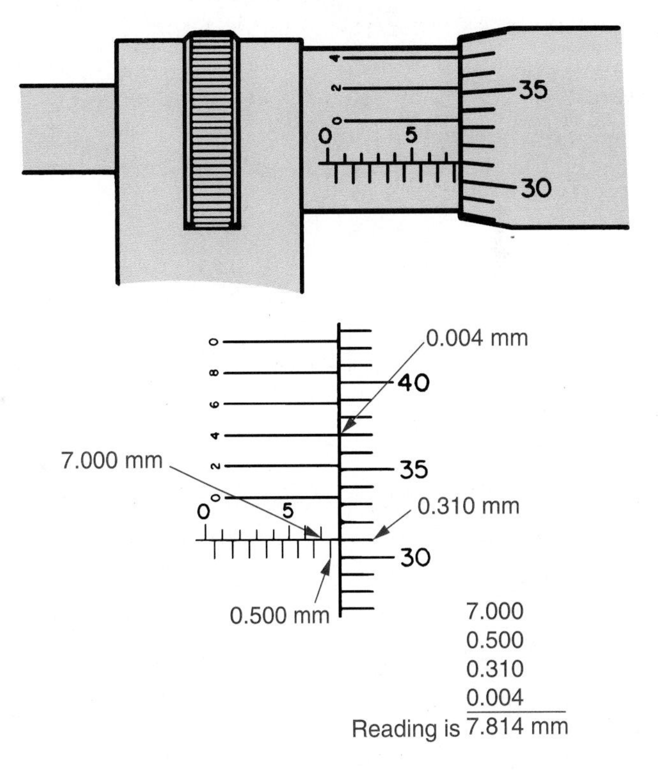

Goodheart-Willcox Publisher

Figure 5-15. Reading a metric vernier micrometer caliper. To the regular reading in hundredths of a millimeter (0.01), add the reading from the vernier scale that coincides with a line on the thimble. Each line on the vernier scale is equal to two thousandths of a millimeter (0.002 mm).

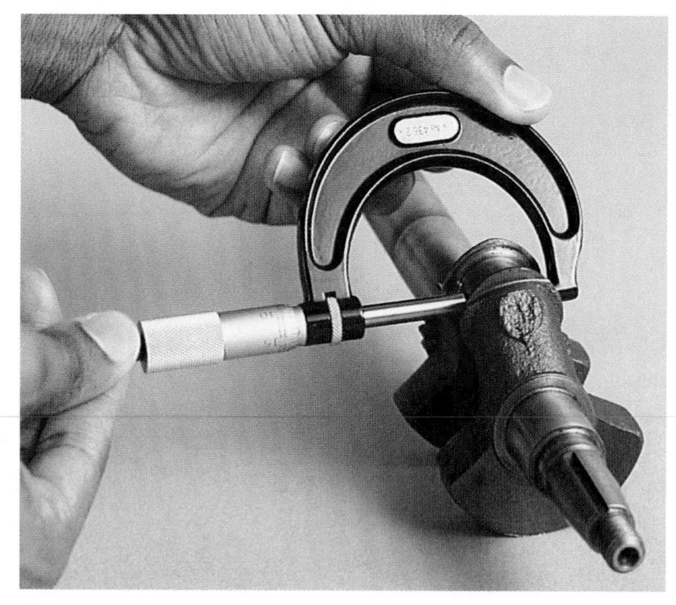

A

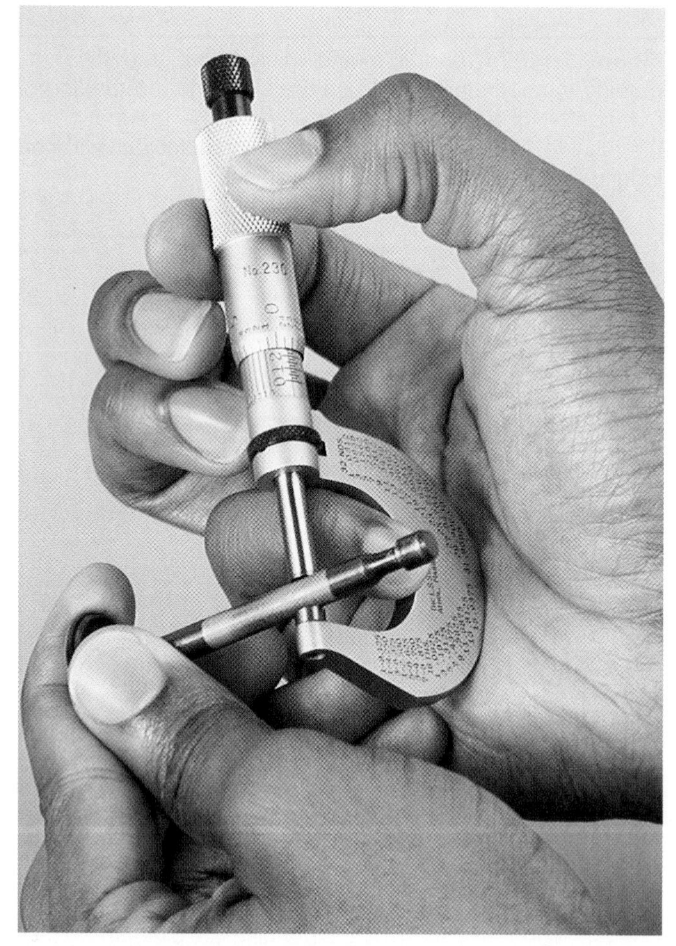

B

Goodheart-Willcox Publisher

Figure 5-16. Proper technique of handling a micrometer. A—Use very light pressure when turning the thimble. B—When the piece being measured must also be held, position the micrometer as shown, with a finger in the micrometer frame.

Some micrometers have features to help regulate pressure:

- A ratchet stop is used to rotate the spindle. When the pressure reaches a predetermined amount, the ratchet stop slips and prevents further spindle turning. Uniform contact pressure with the work is ensured, even if different people use the same micrometer. Refer again to **Figure 5-11**.
- A friction thimble may be built into the upper section of the thimble. This produces the same results as the ratchet stop but permits one-handed use of the micrometer.
- A locknut is used when several identical parts are to be gaged. Refer again to **Figure 5-11**. The nut locks the spindle into place. Gaging parts with a micrometer locked at the proper setting is an easy way to determine whether the pieces are sized correctly.

5.2.7 Reading an Inside Micrometer

To get a correct reading with an inside micrometer, it is important that the tool be held square across the diameter of the work. It must be positioned so that it will measure across the diameter on the exact center, **Figure 5-17**.

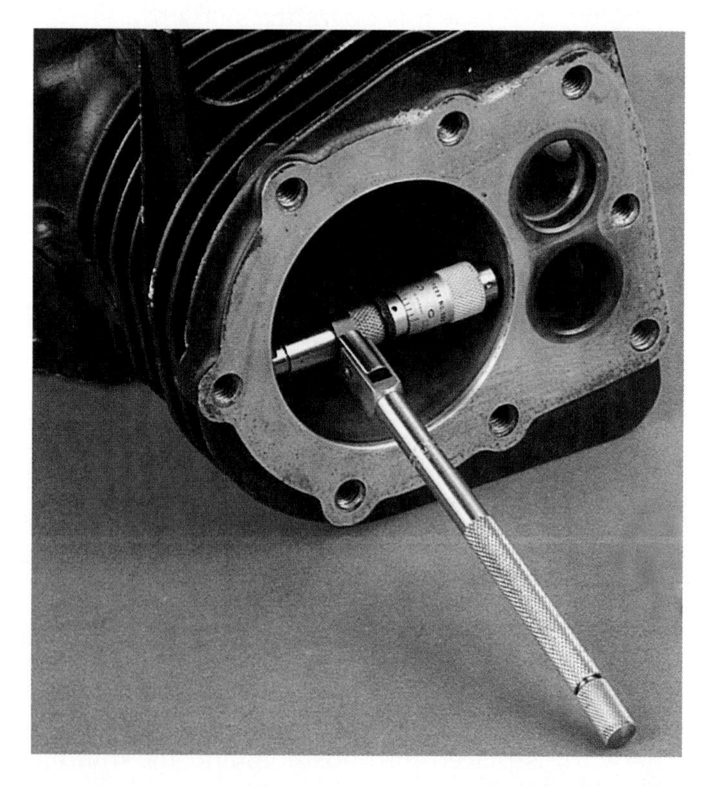

Goodheart-Willcox Publisher

Figure 5-17. Using an inside micrometer. Extension rods can be added to increase the tool's measuring range.

Measurement is made by holding one end of the tool in place and then "feeling" for the maximum possible setting by moving the other end from left to right, and then in and out of the opening. The measurement is made when no left or right movement is felt, and a slight drag is noticeable on the in-and-out swing. It may be necessary to take several readings and average them.

5.2.8 Reading a Micrometer Depth Gage

Be sure to read a micrometer depth gage correctly. The graduations on this measuring tool are in reverse order of the graduations on an outside micrometer. See **Figure 5-18**. The graduations under the thimble must be read, rather than those that are exposed.

5.2.9 Care of a Micrometer

Micrometers are precision instruments and must be handled with care. The following techniques are recommended:

- Place the micrometer on the work carefully so the faces of the anvil and spindle will not be damaged. The same applies when removing the tool after a measurement has been made.
- Keep the micrometer clean. Wipe it with a slightly oiled cloth to prevent rust and tarnish. A drop of light oil on the screw thread will keep the tool operating smoothly.
- Avoid "springing" a micrometer by applying too much pressure when you are making a measurement.
- Clean the anvil and spindle faces before use. This can be done with a soft cloth or by lightly closing the jaws on a clean piece of paper and drawing the paper out.
- Check for accuracy by closing the spindle gently on the anvil and note whether the zero line on the thimble coincides with the zero on the sleeve. If they are not aligned, follow the manufacturer's recommended adjustments.
- Avoid placing a micrometer where it may fall on the floor or have other tools placed on it.
- If the micrometer must be opened or closed a considerable distance, do not "twirl" the frame; gently roll the thimble with your palm. See **Figure 5-19**.
- Never attempt to make a micrometer reading until a machine has come to a complete stop.
- Clean and oil the tool if it is to be stored for some time. If possible, place the micrometer in a small box for protection.

5.3 Vernier Measuring Tools

The vernier principle of measuring was named for its inventor, Pierre Vernier, a French mathematician. The ***vernier caliper*** can make accurate measurements to 1/1000″ (0.001″) and 1/50 mm (0.02 mm). See **Figure 5-20**.

The design of the tool permits measurements to be made over a large range of sizes. It is manufactured as a standard item in 6″, 12″, 24″, 36″, and 48″ lengths. SI Metric vernier calipers are available in 150 mm, 300 mm, and 600 mm lengths. The 6″, 12″, 150 mm, and 300 mm sizes are most commonly used. Unlike the micrometer caliper, the vernier caliper can be used for both inside and outside measurements, **Figure 5-21**.

The following measuring instruments may include a vernier scale:

- Height and depth gages are used for layout work and to inspect the locations of features. See **Figure 5-22**.

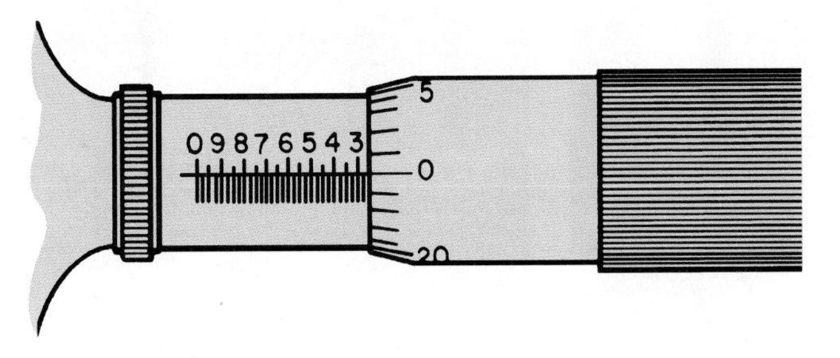

Goodheart-Willcox Publisher

Figure 5-18. A micrometer depth gage. When making measurements with a depth gage, remember that the graduations are in reverse order. This gage indicates a depth of 0.250″.

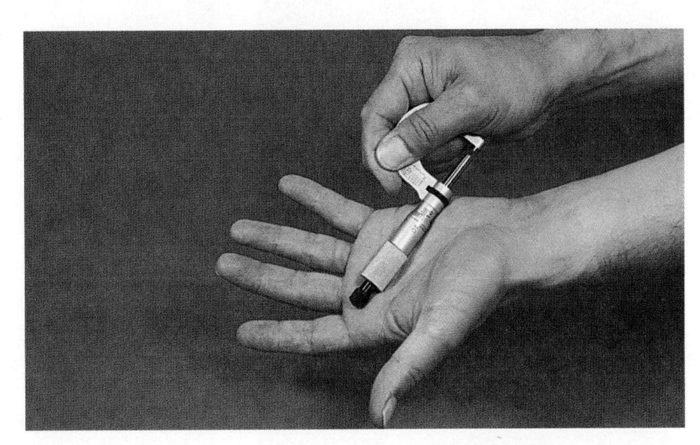

Goodheart-Willcox Publisher

Figure 5-19. Micrometers must be treated carefully. Roll the micrometer thimble on the palm of your hand if the instrument must be opened or closed a considerable distance.

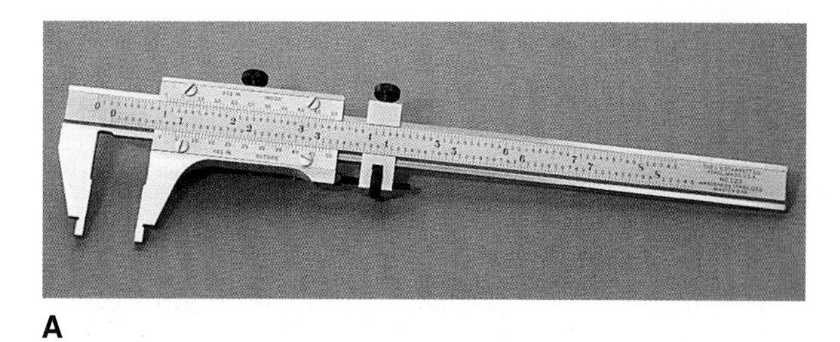

A

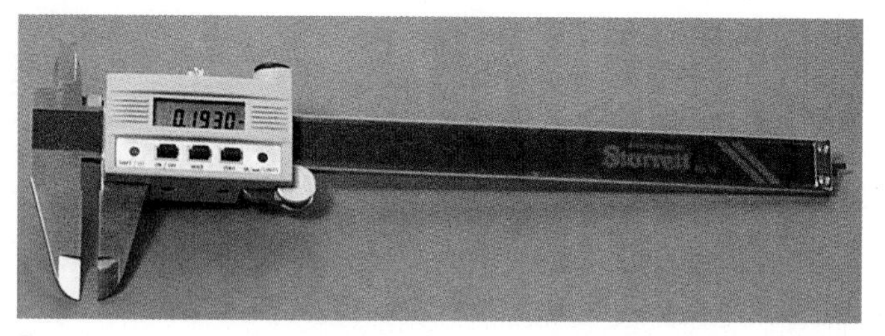

B

Goodheart-Willcox Publisher

Figure 5-20. Vernier calipers can be used to make very accurate measurements. A—Standard vernier caliper. B—Modern digital calipers are easier to read than mechanical instruments.

Mahr Federal Inc.

Figure 5-21. Vernier calipers can be used to make both internal and external measurements.

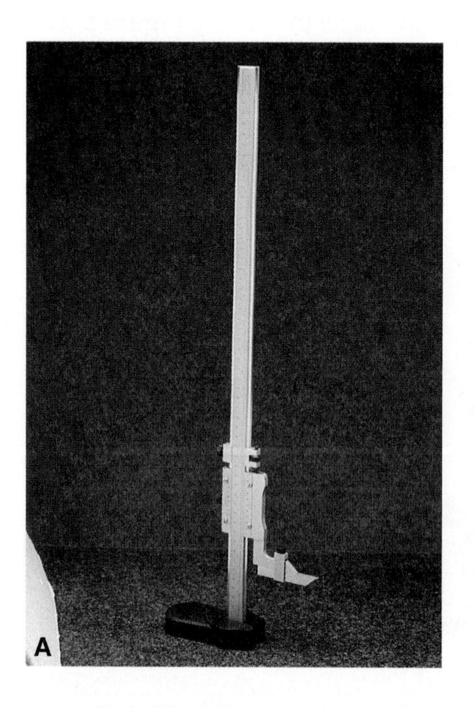

A

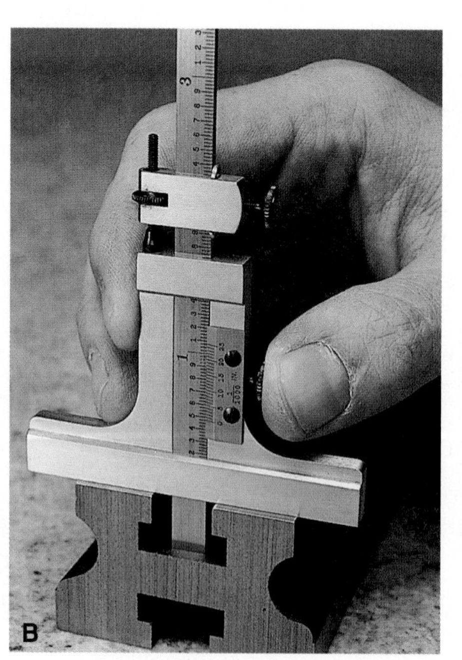

B

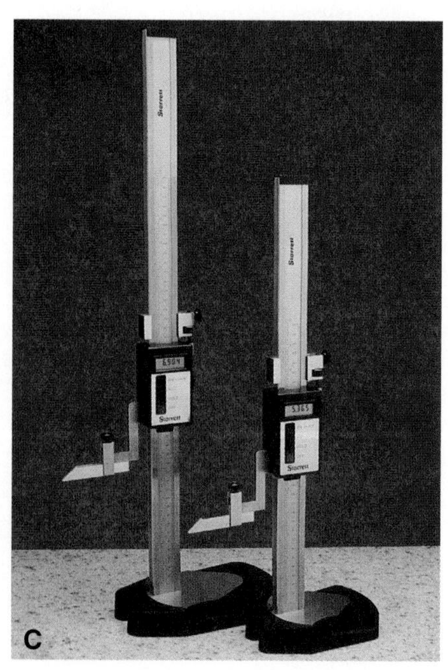

C

L. S. Starrett Co.

Figure 5-22. Many instruments are equipped with a vernier scale. A—Height gage. B—Depth gage. C—The digital readout on this type of height gage serves the same function as a standard vernier scale.

- Gear tooth calipers are used to measure gear teeth and threading tools, **Figure 5-23**.
- Universal vernier bevel protractors are used for the layout and inspection of angles, **Figure 5-24**.

Vernier measuring tools, with the exception of the vernier bevel protractor, consist of a graduated beam with fixed jaw or base and a vernier slide assembly. The vernier slide assembly is composed of a movable jaw or scribe, vernier plate, and clamping screws. The slide moves as a unit along the beam.

Unlike other vernier measuring tools, the beam of the vernier caliper is graduated on both sides. One side is for making outside measurements, the other for inside measurements. Many of the newer vernier measuring tools are graduated to make both inch and millimeter measurements.

L. S. Starrett Co.

Figure 5-23. Gear tooth vernier calipers are used to measure gear teeth, form tools, and threaded tools.

Mahr Federal Inc.

Figure 5-24. A digital universal vernier bevel protractor is used to accurately measure angles.

5.3.1 Reading an Inch-Based Vernier Scale

These measuring tools are available with either 25-division or 50-division vernier plates. Both plates can be read to 0.001″.

On measuring tools using the 25-division vernier plate, every inch section on the beam is graduated into 40 equal parts. Each graduation is 1/40″ (0.025″). Every fourth division, representing 0.100″, is numbered.

There are 25 divisions on the vernier plate. Every fifth line is numbered: 5, 10, 15, 20, and 25. The 25 divisions occupy the same space as 24 divisions on the beam. This slight difference, equal to 0.001 (1/1000″) per division, is the basis of the vernier principle of measuring.

To read a 25-division vernier plate measuring tool, note how many inches (1, 2, 3, etc.), tenths (0.100, 0.200, etc.), and fortieths (0.025, 0.050, or 0.075) there are between the "0" on the vernier scale and the "0" line on the beam, then add them. Then count the number of graduations (each graduation equals 0.001″) that lie between the "0" line on the vernier plate and the line that coincides (corresponds exactly) with a line on the beam. Only one line will coincide. Add this to the above total for the reading.

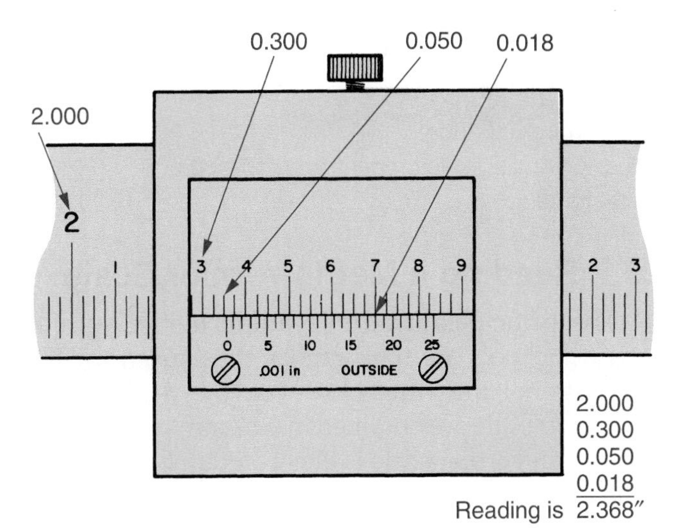

The "0" line on the vernier plate is:

Past the 2:	2 × 1	= 2.000
Past the 3:	3 × 0.100	= 0.300
Plus 2 graduations:	2 × 0.025	= 0.050
Plus 18 vernier scale graduations:	18 × 0.001	= 0.018
Total reading		= 2.368″

On the 50-division vernier plate, every second graduation between the inch lines is numbered, and equals 0.100″. The unnumbered graduations equal 0.050″.

The vernier plate is graduated into 50 parts, each representing 0.001″. Every fifth line is numbered: 5, 10, 15 . . . 40, 45, and 50.

To read a 50-division vernier measuring tool, first count how many inches, tenths (0.100), and twentieths (0.050) there are between the "0" line on the beam, and the "0" line on the vernier plate. Then add them. Then count the number of 0.001 graduations on the vernier plate from its "0" line to the line that coincides with a line on the beam. Add this to the above total.

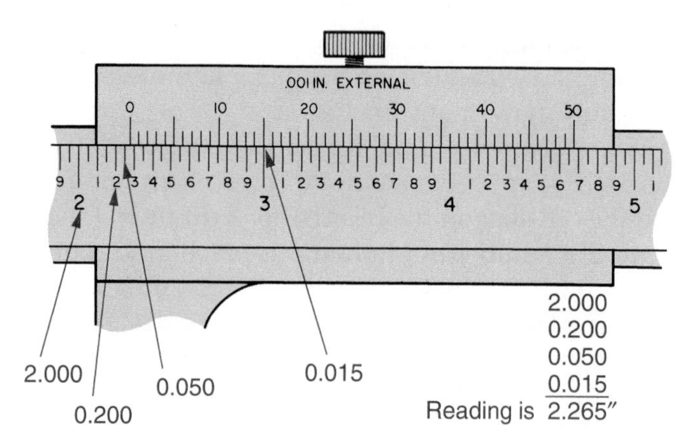

The "0" line on the vernier plate is:

Past the 2:	2 × 1.000	= 2.000
Past the 2:	2 × 0.100	= 0.200
Plus one graduation:	1 × 0.050	= 0.050
Plus 15 vernier scale graduations:	15 × 0.001	= 0.015
Total reading		= 2.265″

5.3.2 Reading a Metric Vernier Scale

The principles used in reading metric vernier measuring tools are the same as those used for US Conventional measure. However, the readings on the vernier scale are obtained in 0.02 mm precision. A 25-division vernier scale is illustrated in **Figure 5-25**, while a 50-division scale is described in **Figure 5-26**.

5.3.3 Using the Vernier Caliper

As with any precision tool, a vernier caliper must not be forced on the work. Slide the vernier assembly until the jaws nearly contact the section being measured. Lock the clamping screw. Make the tool adjustment with the fine adjusting nut. The jaws must contact the work firmly, but not tightly.

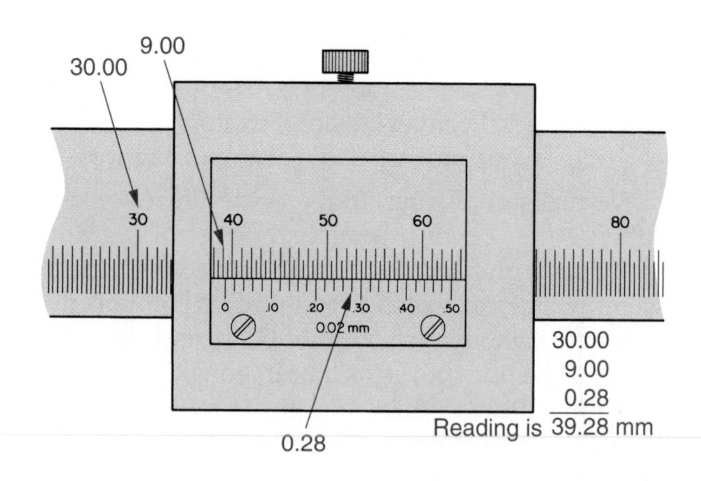

Goodheart-Willcox Publisher

Figure 5-25. How to read a 25-division metric vernier scale. Readings on the scale are obtained in units of two hundredths of a millimeter (0.02 mm).

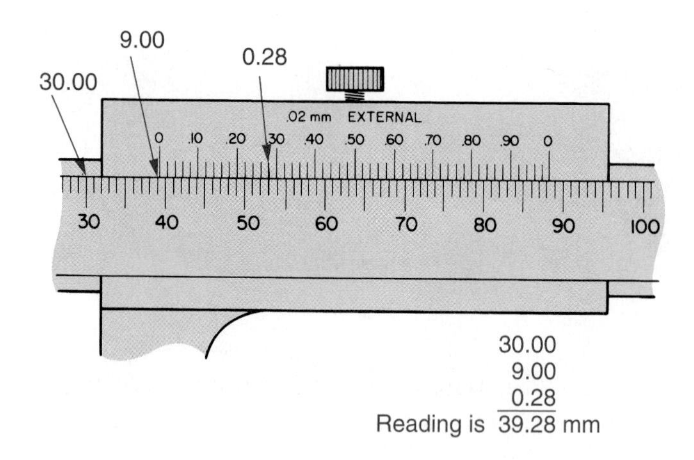

Goodheart-Willcox Publisher

Figure 5-26. How to read a 50-division metric vernier scale. Each division equals two hundredths of a millimeter (0.02 mm).

Lock the slide on the beam. Carefully remove the tool from the work and make your reading. For precise layout work, divider and trammel point settings are located on the outside measuring scale and on the slide assembly.

5.3.4 Using a Dial Caliper

A dial caliper is a direct-reading instrument that resembles a vernier caliper. It can be used to make outside, inside, and depth measurements (with the addition of a depth attachment). A lock permits the tool to be employed for repetitive measurements. See **Figure 5-27**.

The beam is graduated into 0.10″ increments. The caliper dial is graduated into 100 divisions. The reading is made by combining the division on the beam and the dial reading.

L.S. Starrett Co.

Figure 5-27. Dial calipers provide direct readings of measurements.

The dial hand makes one full revolution for each 0.10″ movement. Each dial graduation, therefore, represents 1/100 of 0.10″, or 0.001″. On the metric version, each dial graduation represents 0.02 mm.

5.3.5 Universal Vernier Bevel Protractor

A quick review of the circles, angles, and units of measurement associated with them will help in understanding how to read a universal vernier bevel protractor:

- **Degree (°).** Regardless of its size, a circle contains 360°. Angles are also measured by degrees.
- **Minute (′).** A minute represents a fractional part of a degree. If a degree is divided into 60 equal parts, each part is one minute. A foot mark (′) is used to signify minutes (for example, 30°15′).
- **Second (″).** Minutes are divided into smaller units known as seconds. There are 60 seconds in one minute. An angular measurement written in degrees, minutes, and seconds appears as 36°18′22″. This would read "36 degrees, 18 minutes, and 22 seconds."

A ***universal bevel protractor*** has several parts: a dial, a base or stock, and a sliding blade. The dial is graduated into degrees, and the blade can be extended in either direction and set at any angle to the stock. The blade can be locked against the dial by tightening the blade clamp nut. The blade and dial can be rotated as a unit to any desired position, and locked by tightening the dial clamp nut.

The protractor dial is graduated into 360° and reads from 0° to 90° and then back down to 0°. Every ten degree division is numbered, and every five degrees is indicated by a fine line longer than those on either side. The vernier scale is divided into twelve equal parts on each side of the "0." Every third graduation is numbered (0, 15, 30, 45, 60), representing minutes. Each division equals five minutes. Since each degree is divided into 60 minutes, one division is equal to 5/60 of a degree.

To read the protractor, note the number of degrees that can be read up to the "0" on the vernier plate. To this, add the number of minutes indicated by the line beyond the "0" on the vernier plate that aligns exactly with a line on the dial.

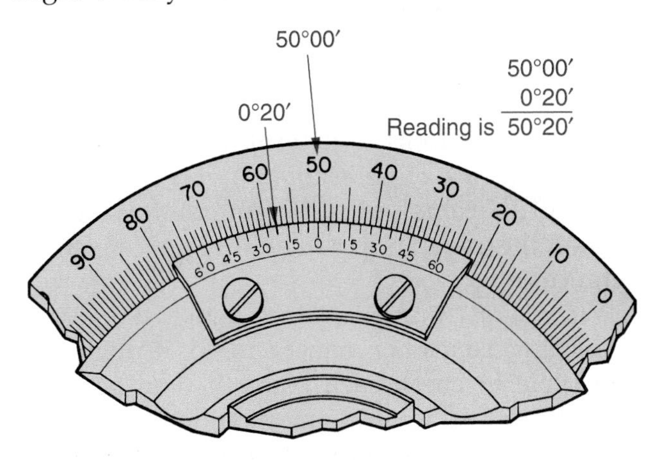

In this example the "0" is past the 50° mark, and the vernier scale aligns at the 20′ mark. Therefore, the measurement is 50°20′.

5.3.6 Care of Vernier Tools

Reasonable care in handling these expensive tools will maintain their accuracy:

- Wipe the instrument with a soft, lint-free cloth before using. This will prevent dirt and grit from being ground in, which could eventually affect the accuracy of the tool.
- Wipe the tool with a lightly oiled, soft cloth after use and before storage.
- Store the tool in its case.

- Never force the tool when you are making measurements.
- Use a magnifying glass or a jeweler's loupe to make vernier readings. Hold the tool so the light is reflected on the scale.
- Handle the tool as little as possible. Sweat and body acids cause rusting and staining.
- Periodically check for accuracy. Use a measuring standard, gage block, or ground parallel. Return the tool to the manufacturer for adjustments and repairs.
- Lay vernier height gages on their side when not in use. Then there will be no danger that they will be knocked over and damaged.

5.4 Gages

It is impractical to check every dimension on every manufactured part with conventional measuring tools. Specialized tools, such as plug gages, ring gages, and optical gages are used instead. These gaging devices can quickly determine whether the dimensions of a manufactured part are within specified limits or tolerances.

Measuring requires the skillful use of precision measuring tools to determine the exact geometric size of the piece. ***Gaging*** involves checking parts with various gages. Gaging simply shows whether the piece is made within the specified tolerances.

When great numbers of an item with several critical dimensions are manufactured, it might not be possible to check each piece. It then becomes necessary to decide how many randomly selected pieces must be checked to ensure satisfactory quality and adherence to specifications. This technique is called statistical quality control.

Always handle gages carefully. If dropped or mishandled, the accuracy of the device could be affected. Gages provide a method of checking your work and are very important tools.

5.4.1 Plug Gage

Plug gages are used to check whether hole diameters are within specified tolerances. The double-end cylindrical plug gage has two gaging members known as go and no-go plugs, **Figure 5-28**. The go plug should enter the hole with little or no interference. The no-go plug should not fit.

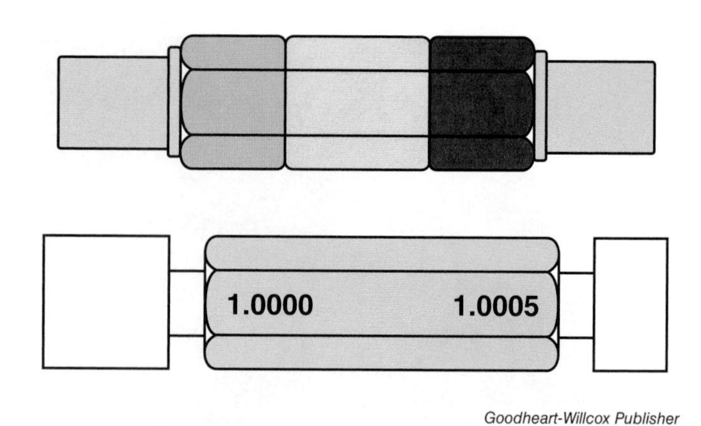

Goodheart-Willcox Publisher

Figure 5-28. A double-end cylindrical plug gage.

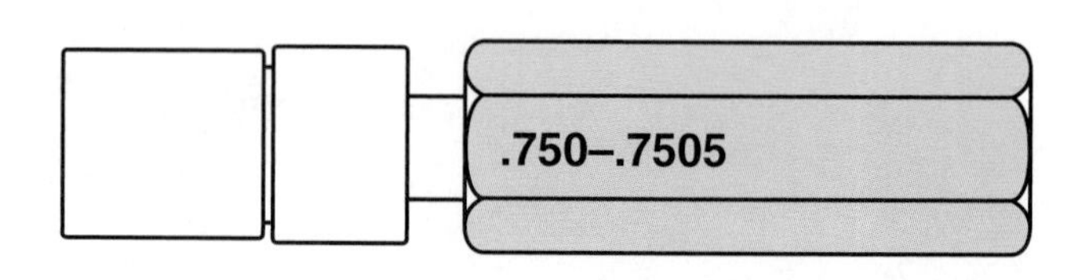

Goodheart-Willcox Publisher

Figure 5-29. A step plug gage can check for oversize and undersize in a single test.

The go plug is longer than the no-go plug. A progressive plug gage, or step plug gage, has the go and no-go plugs on the same end. This gage is able to check the dimensions in one motion. See **Figure 5-29.**

5.4.2 Ring Gage

External diameters are checked with ring gages. The go and no-go ring gages are separate units, and can be distinguished from each other by a groove cut on the knurled outer surface of the no-go gage.

On ring gages, the gage tolerance is the reverse of plug gages. The opening of the go gage is larger than the opening for the no-go gage.

5.4.3 Snap Gage

A snap gage serves the same purpose as a ring gage. Snap gages are designed to check internal diameters, external diameters, or both. There are three general types:

- An adjustable snap gage can be adjusted through a range of sizes. See **Figure 5-30.**
- A nonadjustable snap gage is made for one specific size.
- A dial indicator snap gage measures the amount of variation in the part measurement. The dial face has a double row of graduations reading in opposite directions from zero.

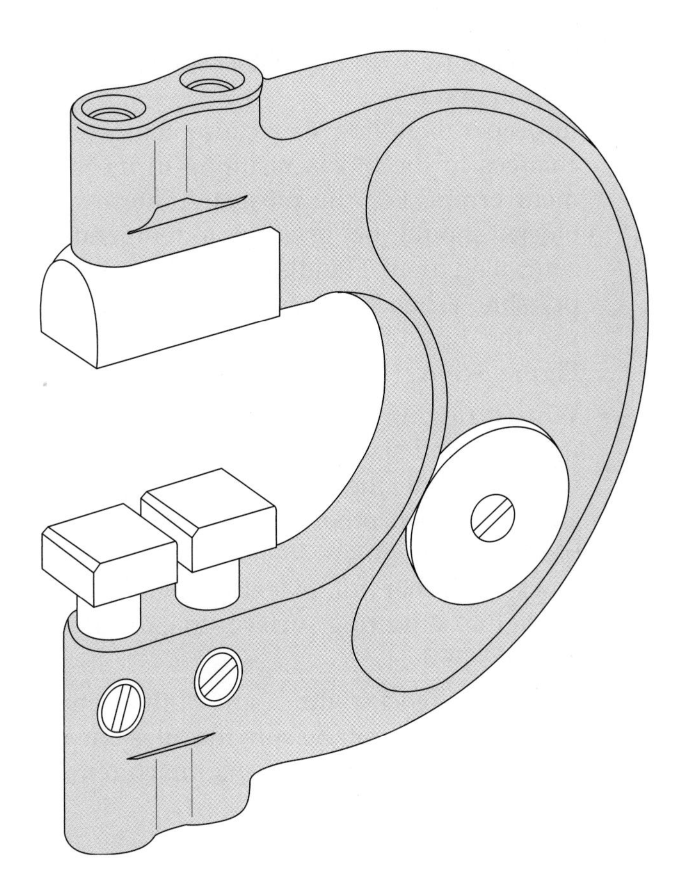

Goodheart-Willcox Publisher

Figure 5-30. An adjustable snap gage.

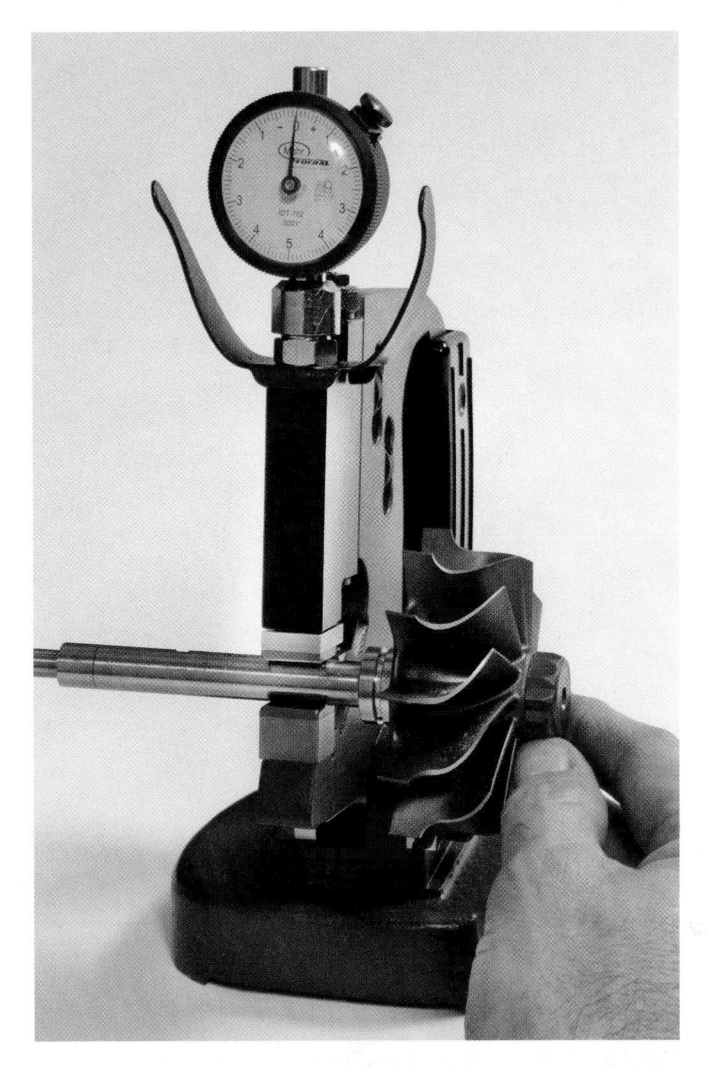

Mahr Federal Inc.

Figure 5-31. A dial indicator snap gage.

Minus graduations are red and plus graduations are black. Both adjustable and nonadjustable indicating snap gages are available. See **Figure 5-31**.

On snap gages, the anvils should be narrower than the work being measured. This will avoid uneven wear on the measuring surfaces.

5.4.4 Thread Gages

Several types of gages are used to check screw thread fits and tolerances. These gages are similar to the gages already discussed:

- A thread plug gage is similar to a normal plug gage except that the plugs are threaded. Threaded plug gages are used to check internal threads. The go plug should enter the threaded hole, while the no-go plug should not.
- A thread ring gage is similar to a ring gage with internal threads. Thread ring gages are used to check external threads. The no-go ring gage features a groove cut into the knurled surface to distinguish it from the go ring gage.
- A thread snap gage features thread shaped anvils or rollers. Thread snap gages function similar to normal snap gages. Dial thread snap gages are also available.

5.4.5 Gage Blocks

Gage blocks, commonly known as Jo-blocks or Johansson blocks, are precise steel measuring standards. Gage blocks can be purchased in various sets ranging from a few commonly used block sizes to more complete sets. See **Figure 5-32**.

Gage blocks are used to verify the accuracy of master gages. They are also used as working gages and for setting up machining work requiring great accuracy. The Federal Accuracy Grades for gage blocks are shown in **Figure 5-33**.

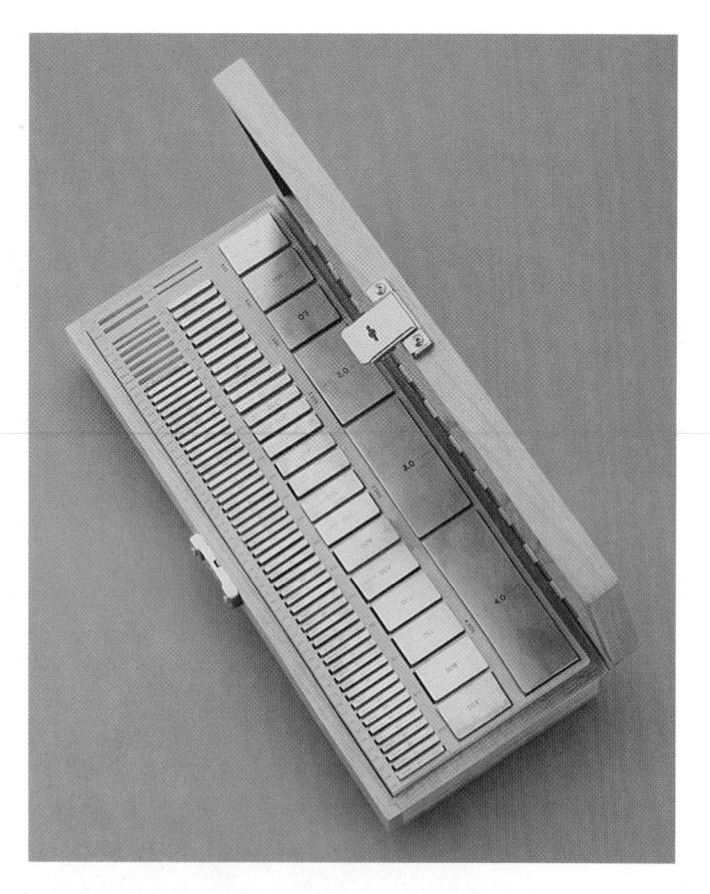

Federal Products Co.

Figure 5-32. A typical set of gage blocks.

Federal Accuracy Grades			
		Tolerance	
Accuracy grade	Former designation	US Conventional system (inch)	Metric system (millimeter)
0.5	AAA	±.000001″	±.00003 mm
1	AA	±.000002″	±.00005 mm
2	A+	+.000004″ –.000002″	+.0001 mm –.00005 mm
3	A&B	+.000006″ –.000002″	+.00015 mm –.00005 mm

Reference temperature: 68°F (20° C)
One inch = 25.4 millimeters exactly

Goodheart-Willcox Publisher

Figure 5-33. Federal Accuracy Grades for gage blocks.

When working with gage blocks, keep the following tips in mind:

- Improper handling can cause temperature changes in the block, resulting in measurement errors. For the most accurate results, blocks should be used in a temperature-controlled room. Handle the blocks as little as possible. When you must handle the blocks, use the tips of your fingers, as shown in **Figure 5-34A**.
- When wringing gage blocks together to build up to desired size, wipe the blocks and then carefully slide them together. They should adhere to each other strongly. Separate the blocks when you are finished. Leaving gage blocks together for extended periods may cause the contacting surfaces to corrode. See **Figure 5-34B**.
- Wipe gage blocks with a soft cloth or chamois treated with oil. Be sure the oil is one recommended by the gage manufacturer. See **Figure 5-34C**.

5.5 Dial Indicators

Industry is constantly searching for ways to reduce costs without sacrificing quality. Inspection has always been a costly part of manufacturing. To speed up this phase of production without sacrificing accuracy, dial indicators and electronic gages are receiving increased attention.

Dial indicators are designed with shockproof movements and have jeweled bearings (similar to fine watches). There are two types of indicators: balanced and continuous. Balanced indicators can take measurements on either side of a zero line. Continuous indicators read from "0" in a clockwise direction. See **Figure 5-35**.

Dial faces are available in a wide range of graduations. They usually read in the following increments:

- 1/1000″ (0.001″).
- 1/100 mm (0.01 mm).
- 1/10,000″ (0.0001″).
- 2/1000 mm (0.002 mm).

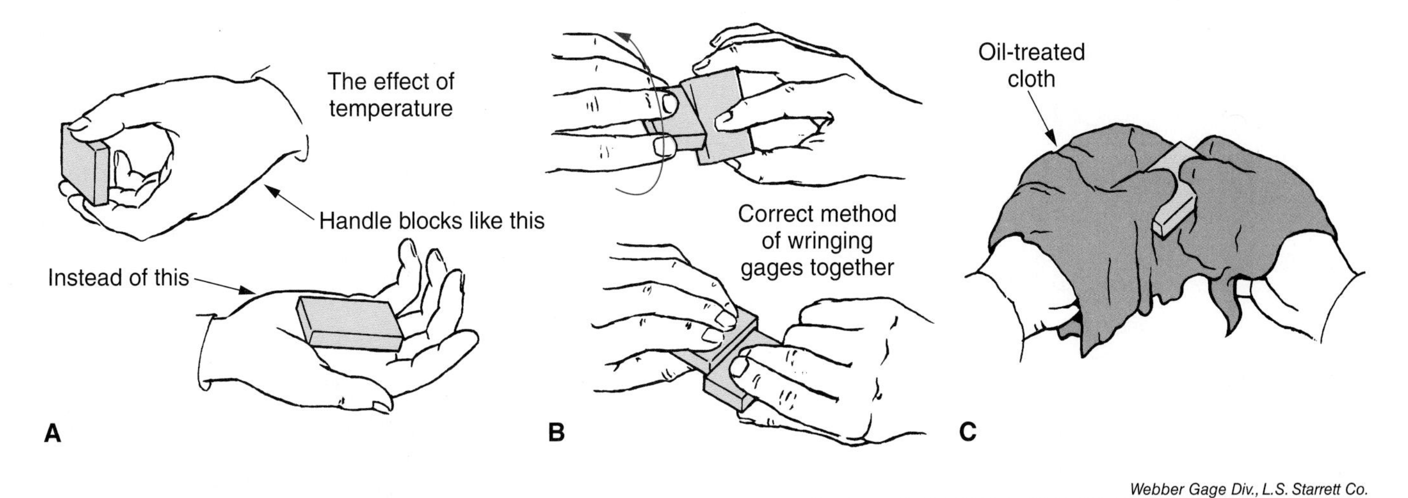

Webber Gage Div., L.S. Starrett Co.

Figure 5-34. Proper care of gage blocks. A—Handling gage blocks. B—Wipe blocks and slide them together. Do not leave blocks together for extended periods. C—Wipe blocks with a soft cloth before storing.

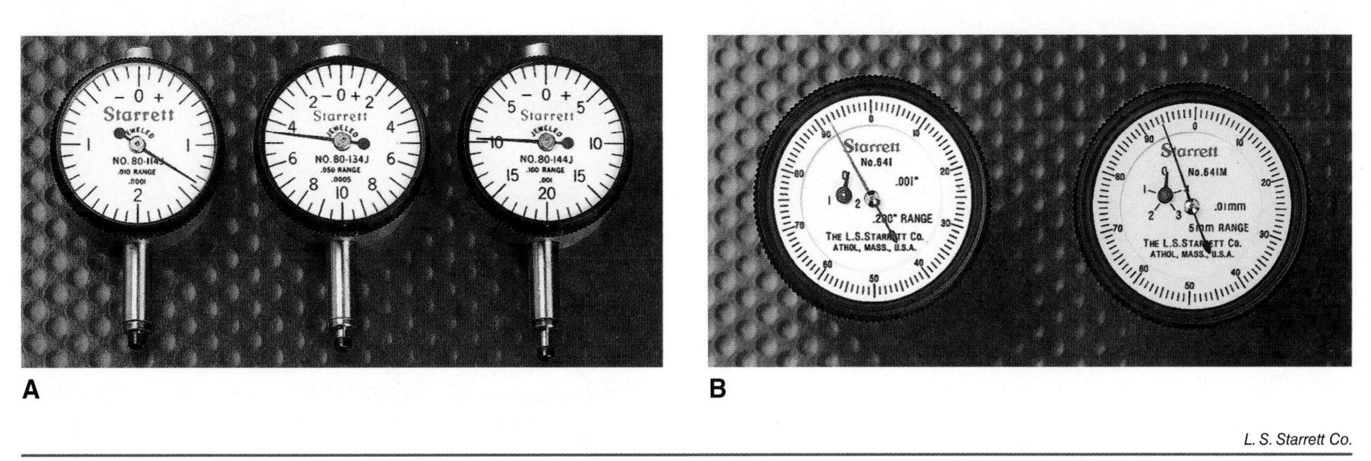

L. S. Starrett Co.

Figure 5-35. The two basic varieties of dial indicators. A— Balanced indicators. B—Continuous indicators.

Much use is made of dial indicators for centering and aligning work on machine tools, checking for eccentricity, and inspecting work. Dial indicators must be mounted to rigid holding devices, **Figure 5-36**.

A digital electronic indicator, **Figure 5-37**, features direct digital readouts and a traditional graduated dial for fast, accurate reading. These indicators are available as both self-contained and remote readout units.

5.5.1 How to Use a Dial Indicator

The hand on the dial is actuated by a sliding plunger. Place the plunger lightly against the work until the hand moves. The dial face is turned until the "0" line coincides with the hand. As the work touching the plunger is slowly moved, the indicator hand will measure movement.

The dial indicator can show the difference between the high and low points, or the total runout of the piece in a lathe. When machining, adjustments are made until there is little or no indicator movement.

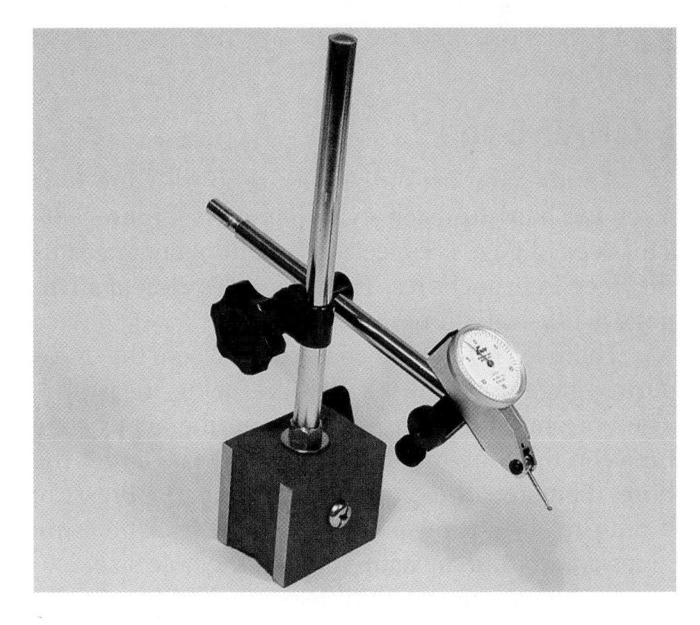

Goodheart-Willcox Publisher

Figure 5-36. Mounting this dial indicator on a magnetic base permits it to be attached to any ferrous metal surface. A push button releases the magnet.

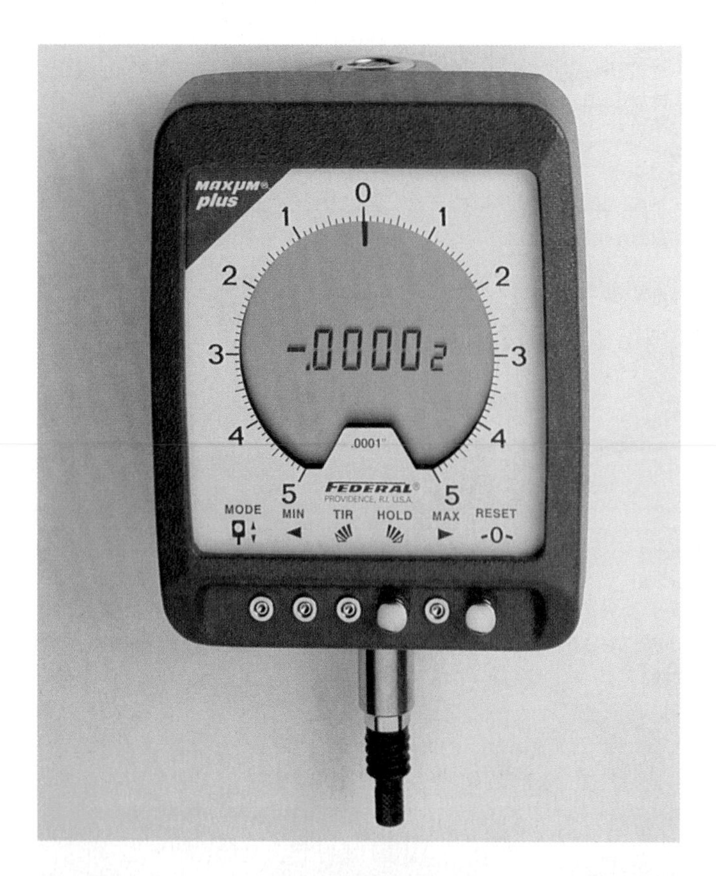

Federal Products Company

Figure 5-37. This digital electronic indicator has numeric readouts and a conventional graduated dial.

5.6 Other Gaging Tools

Industry makes wide use of other types of gaging tools. Most of these tools are used for special purposes.

5.6.1 Air Gage

An air gage uses air pressure to measure hole sizes and hard-to-reach shaft diameters, **Figure 5-38**. This type of gage is especially helpful when measuring deep internal bores. The basic operation of an air gage is illustrated in **Figure 5-39**.

There is no actual contact between the measuring gage and wall of the bore being measured. The bore measurement depends on the air leakage between the plug and the hole wall. (The larger the bore diameter, the greater the leakage.) Pressure builds up and the measurement of the back pressure gives an accurate measurement of the hole size.

Change in pressure (air leakage) is measured by a dial indicator, a cork floating on the air stream, or by a manometer (U-shaped tube in which the height of fluid in the tube indicates pressure).

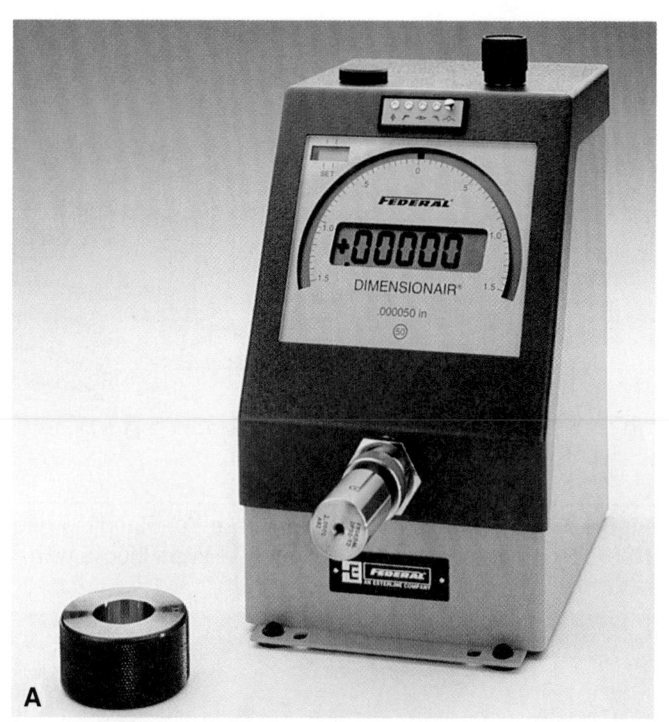

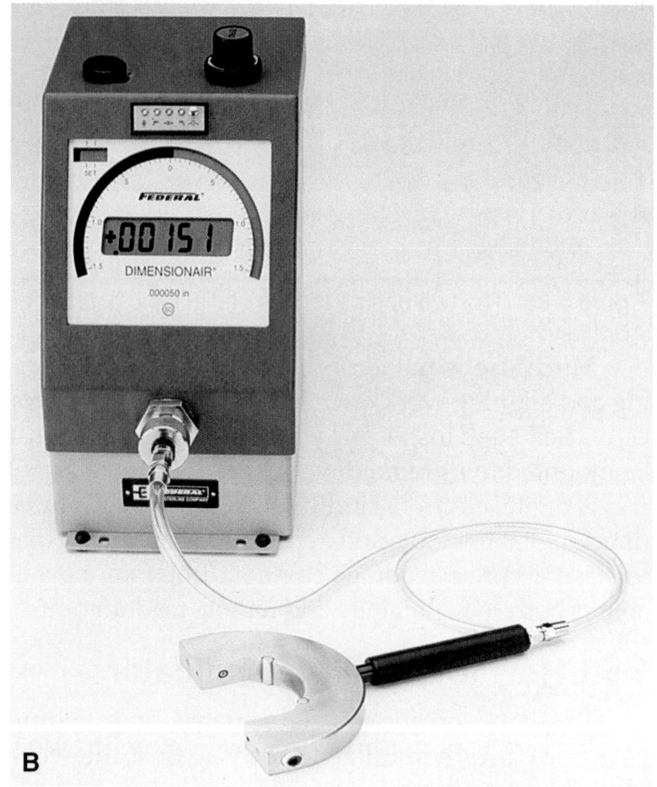

Federal Products Company

Figure 5-38. Digital air gages are available with either US Conventional or SI Metric readouts. They can check either inside or outside diameters. A—An air gage set up to inspect an internal dimension. The master ring shown with the gage is used to set zero on the readout. B—This gage has an air fork, which is used to check hard-to-reach diameters, such as crankshaft journals.

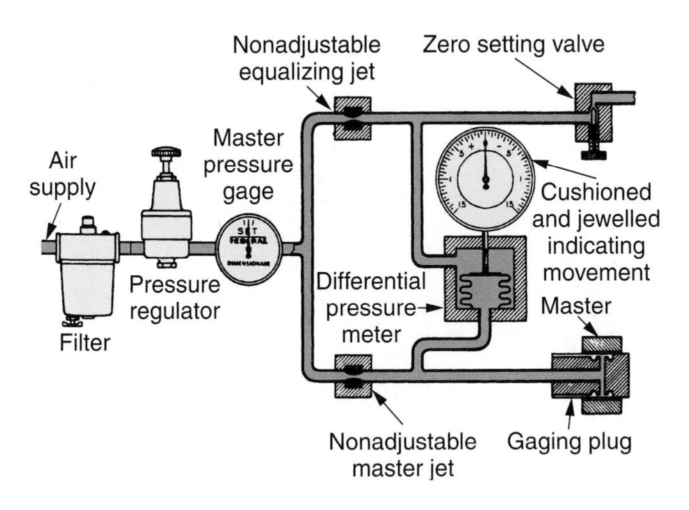

Goodheart-Willcox Publisher

Figure 5-39. This diagram illustrates the operation of an air gage.

5.6.2 Electronic Gage

An electronic gage, **Figure 5-40**, is another type of gaging tool used to make extremely precise measurements. Electronic gages are comparison gages: they compare the size of the work to a reference size. Some are calibrated by means of master gage blocks and others use replaceable gaging probes. These instruments measure in both US Conventional and SI Metric units.

5.6.3 Laser Gaging

A laser is a device that produces a very narrow beam of extremely intense light. Lasers are used in communication, medical, and industrial applications. Laser is an acronym for light amplification by stimulated emission of radiation.

The laser is another area of technology that has moved from the laboratory into the shop. When employed for inspection purposes, it can check the accuracy of critical areas in machined parts quickly and accurately.

5.6.4 Optical Comparator

The optical comparator uses magnification as a means for inspecting parts, **Figure 5-41**. An enlarged image of the part is projected on a screen for inspection. The part image is superimposed on an enlarged, accurate drawing of the correct shape and size. The comparison is made visually. Variations as small as 0.0005″ (0.012 mm) can be noted by a skilled operator.

Sunnen Products Company

Figure 5-40. This electronic bore gaging system can deliver electronic resolution as fine as 0.00001″ (0.0002 mm). Using replaceable gaging probes, the self-contained unit measures diameters ranging from 0.370″ to 2.900″. It also measures in millimeters. It can be linked to a computer for statistical process control (SPC) data collection.

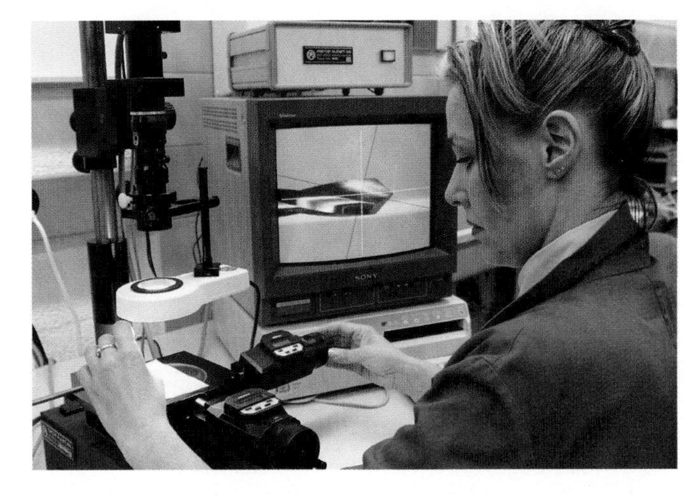

BIG Kaiser Precision Tooling Inc.

Figure 5-41. A technician uses an optical comparator to quickly and efficiently inspect a microdrill.

5.6.5 Optical Flats

Optical flats are precise measuring instruments that use light waves as a measuring standard, **Figure 5-42**. The flats are made of quartz and have one face ground and polished to optical flatness. When this face is placed on a machined surface and a special light passed through it, light bands appear on the surface, **Figure 5-43**. The shape of these bands indicate to the inspector the accuracy of the part. See **Figure 5-44**.

L. S. Starrett Co.

Figure 5-42. Optical flats are used for precision flatness, parallelism, size, and surface variations.

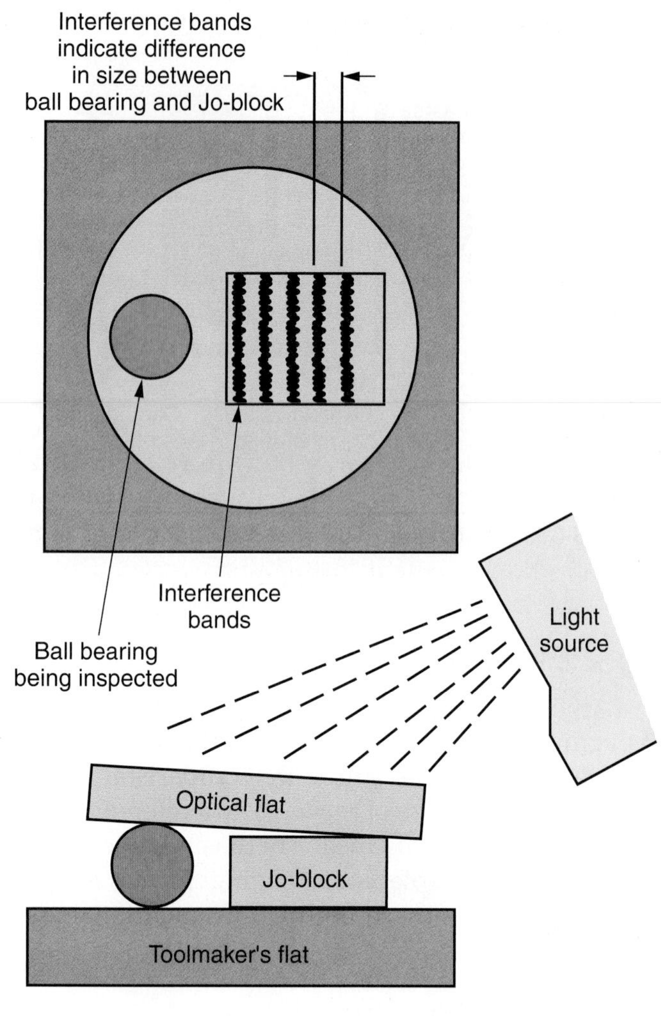

Goodheart-Willcox Publisher

Figure 5-43. Optical flat set-up. Optical flat is placed on top of the work and light is positioned above the flat.

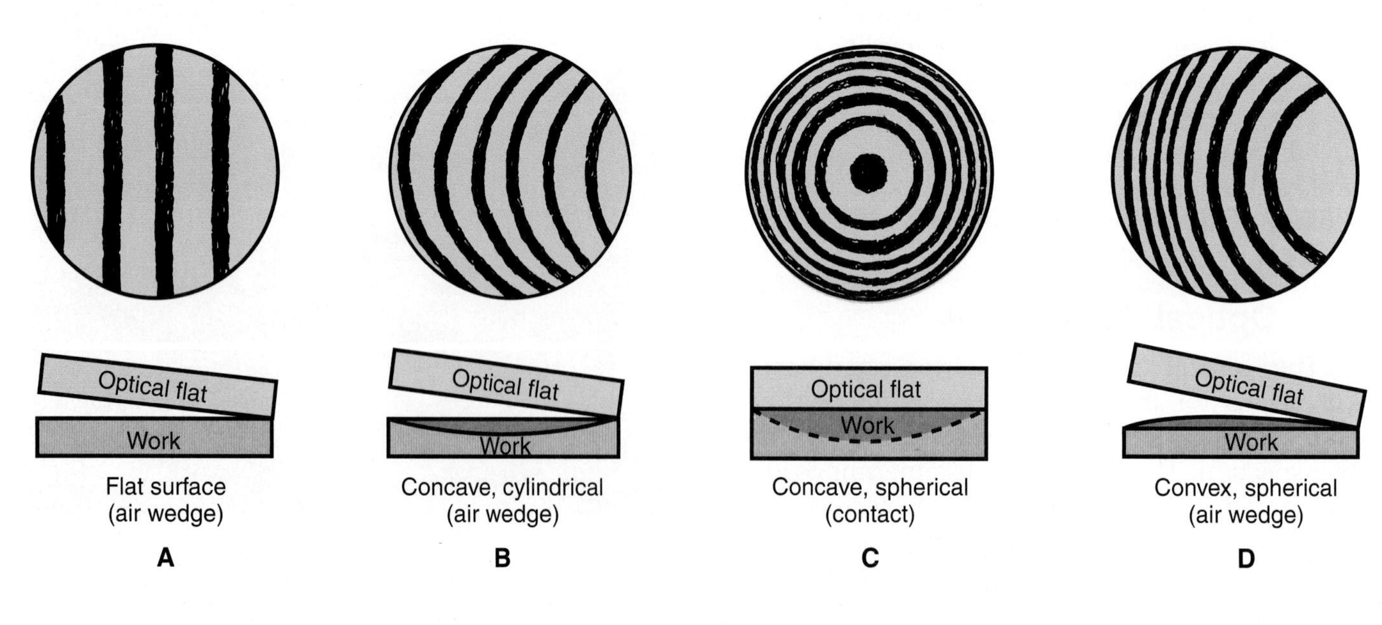

Goodheart-Willcox Publisher

Figure 5-44. Interference band patterns indicate surface flatness and variations.

5.6.6 Thickness (Feeler) Gage

Thickness gages are pieces or leaves of metal manufactured to precise thickness, **Figure 5-45**. Thickness gages are made of tempered steel and are usually 1/2″ (12.7 mm) wide.

Thickness gages are ideal for measuring narrow slots, setting small gaps and clearances, determining fit between mating surfaces, and checking flatness of parts in straightening operations. See **Figure 5-46**.

5.6.7 Screw Pitch Gage

Screw pitch gages are used to determine the pitch or number of threads per inch on a screw, **Figure 5-47**. Each blade is stamped with the pitch or number of threads per inch. Screw pitch gages are available in US Conventional and SI Metric thread sizes.

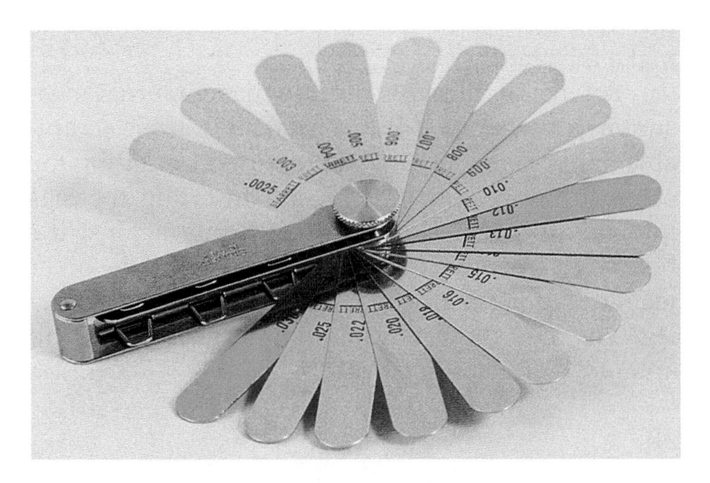

Goodheart-Willcox Publisher

Figure 5-45. Thickness or feeler gages.

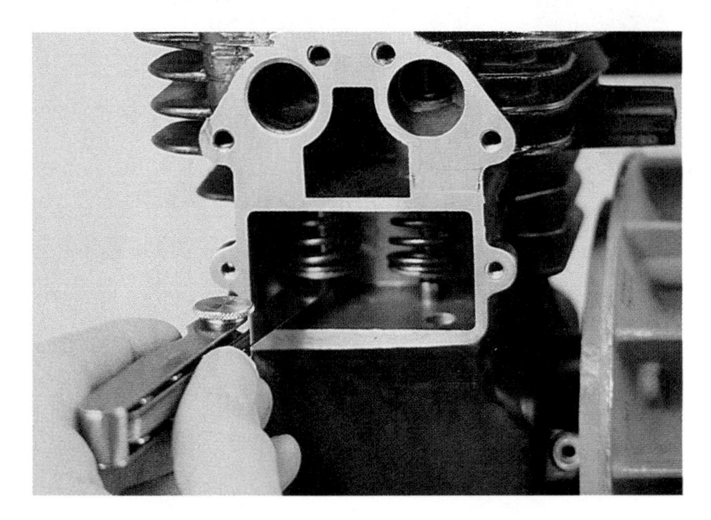

Goodheart-Willcox Publisher

Figure 5-46. A thickness gage is used to check part clearance.

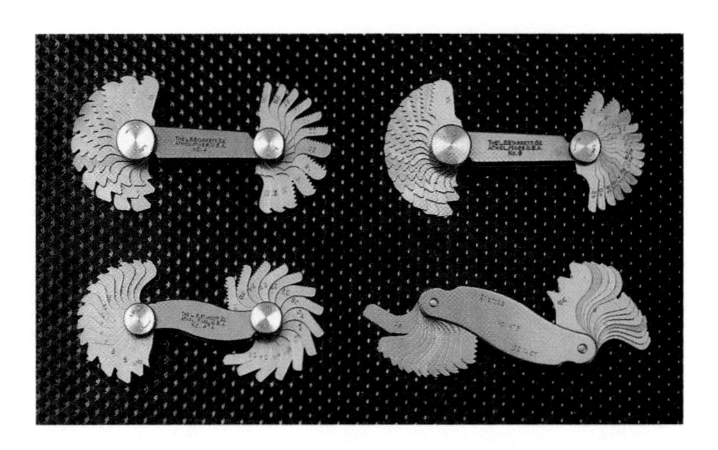

L. S. Starrett Co.

Figure 5-47. Screw pitch gages are made for both inch-based and metric threads.

5.6.8 Fillet and Radius Gage

The thin steel blades of a fillet and radius gage, **Figure 5-48**, are used to check concave and convex radii on corners or against shoulders. The gage is used for layout work and inspection, and as a template when grinding form cutting tools. See **Figure 5-49**. The gages increase in radius in 1/64″ (0.5 mm) increments.

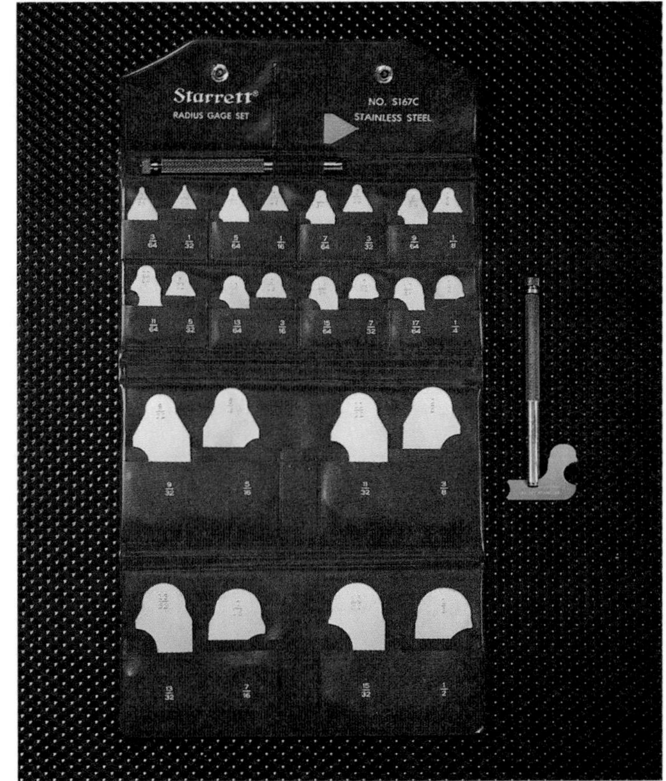

L. S. Starrett Co.

Figure 5-48. A set of radius and fillet gages.

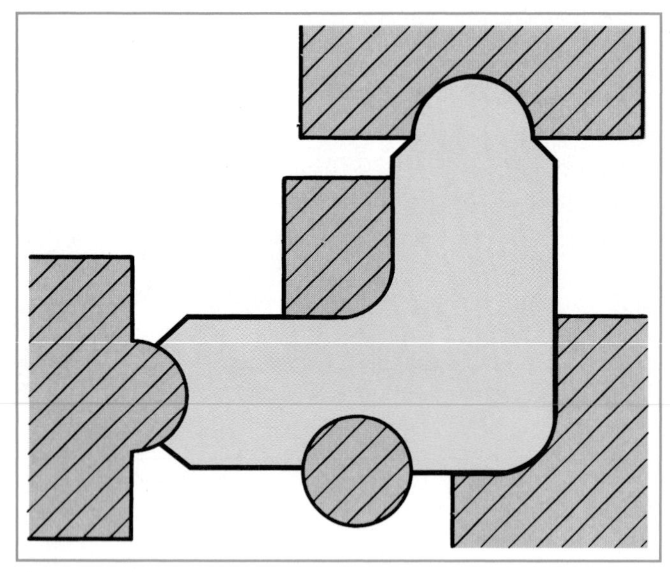
A

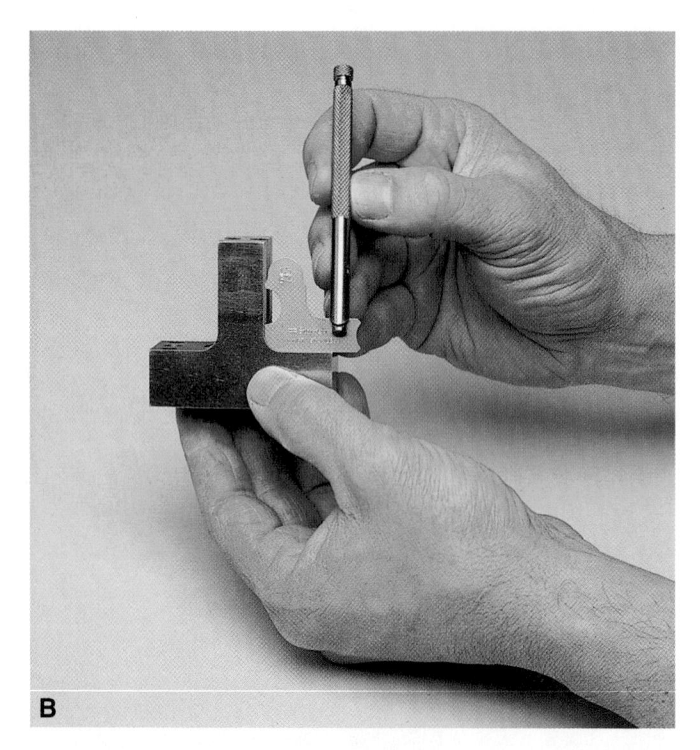
B

L. S. Starrett Co.

Figure 5-49. Using a radius gage. A—Various ways a radius gage can be used. B—Using a radius gage holder.

5.6.9 Drill Rod

Drill rods are steel rods manufactured to close tolerances to twist drill diameters. They are used to inspect hole alignment, location, and diameter. Drill rods are available in both US Conventional and SI Metric sizes.

5.7 Helper Measuring Tools

Some measuring tools are not direct reading and require the help of a rule, micrometer, or vernier caliper to determine the size of the measurement taken. These are called helper measuring tools.

5.7.1 Calipers

External or internal measurements of 1/64″ (0.4 mm) can be made with calipers, **Figure 5-50**. A caliper does not have a dial or scale that shows a measurement; the distance between points must be measured with a steel rule.

Round stock is measured by setting the caliper square with the work and moving the caliper legs down on the stock. Adjust the tool until the caliper point bears lightly on the center line of the stock. Caliper weight should cause the caliper to slip over the diameter. Hold the caliper next to the rule to make the reading.

An inside caliper is used to make internal measurements where 1/64″ (0.4 mm) accuracy is acceptable. Hole diameter can be measured by setting the caliper to approximate size and inserting the legs into the opening. Hold one leg firmly against the hole wall, and adjust the thumbscrew until the other leg lightly touches the wall exactly opposite the first leg. The legs should drag slightly when moved in and out, or from side to side.

L. S. Starrett Co.

Figure 5-50. Inside and outside calipers.

Considerable skill is required to make accurate measurements with a caliper. See **Figure 5-51**. Much depends on the machinist's sense of touch. With practice, measurements with accuracy of 0.003″ (0.07 mm) can be made. However, a micrometer or vernier caliper is preferred and must be used when greater accuracy is required.

5.7.2 Telescoping Gage

A telescoping gage is intended for use with a micrometer to determine internal dimensions, **Figure 5-52**. Sets of telescoping gages with varying ranges are available, **Figure 5-53**.

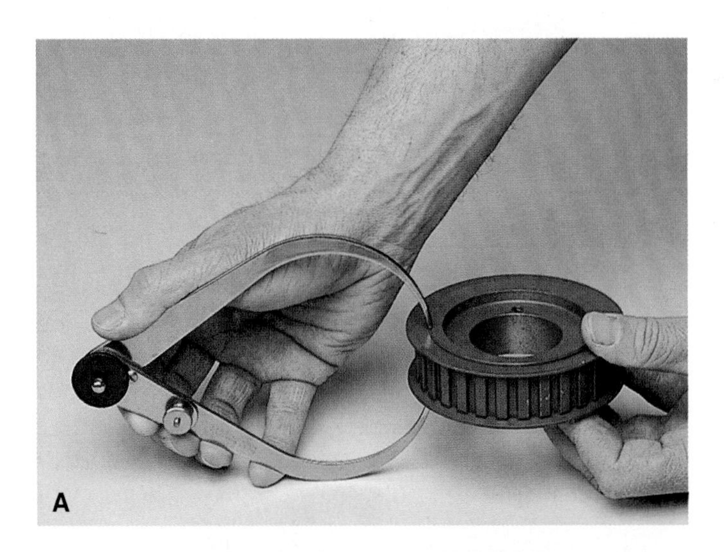

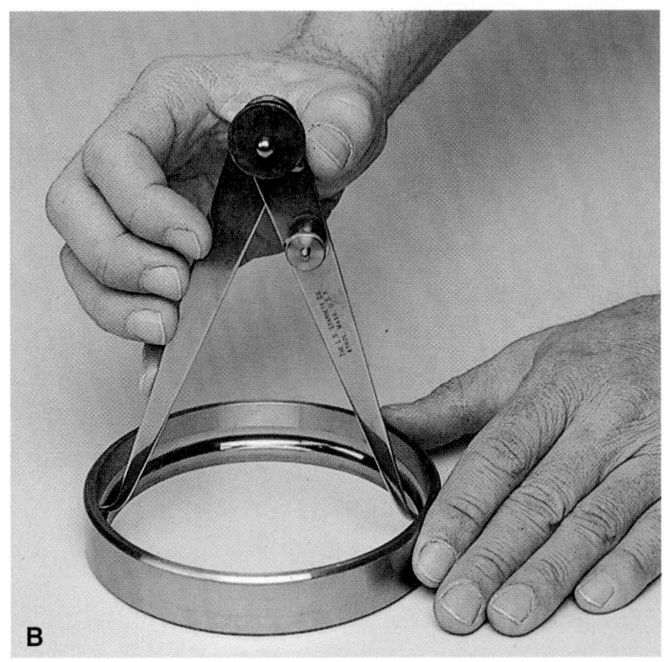

L. S. Starrett Co.

Figure 5-51. Using outside and inside calipers.

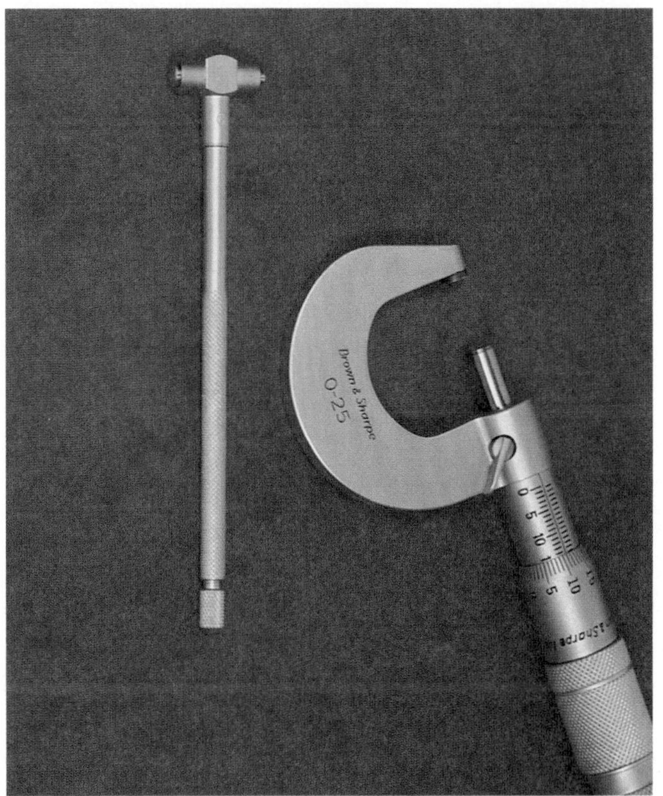

Goodheart-Willcox Publisher

Figure 5-52. A telescoping gage is used with a micrometer.

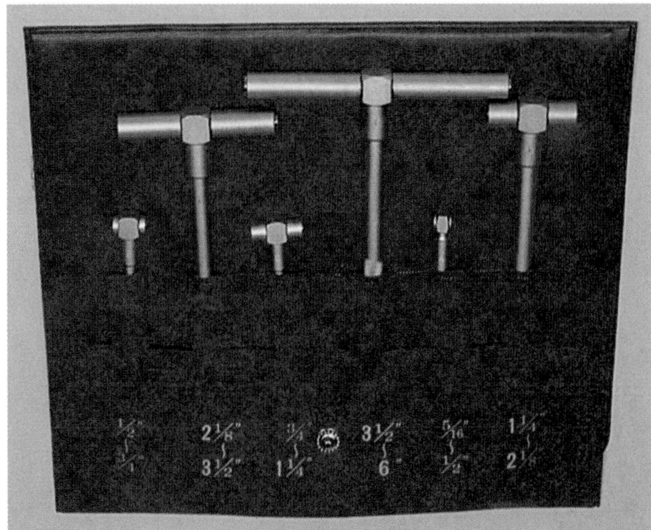

Goodheart-Willcox Publisher

Figure 5-53. A typical set of telescoping gages.

To use a telescoping gage, compress the contact legs. The legs collapse within one another under spring tension. Insert the gage into the hole and allow the legs to expand, **Figure 5-54**.

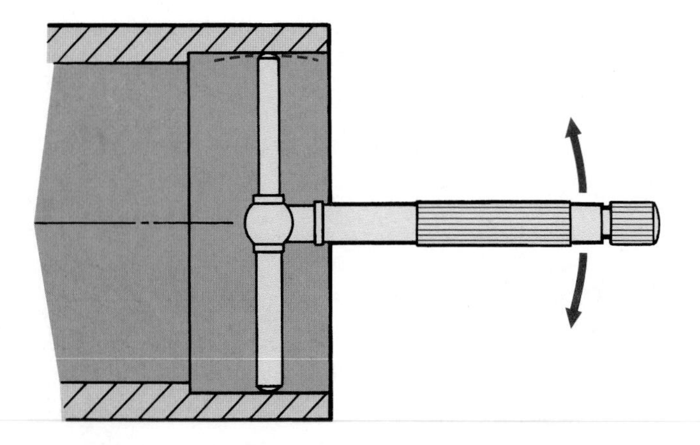

Goodheart-Willcox Publisher

Figure 5-54. Positioning a telescoping gage to measure an inside diameter.

After the proper fitting is obtained, lock the contacts into position. Remove the gage from the hole and make your reading with a micrometer, **Figure 5-55**.

5.7.3 Small Hole Gage

A small hole gage is used to measure openings that are too small for a telescoping gage, **Figure 5-56**. The contacts are designed to allow accurate measurement of shallow grooves and small diameter holes.

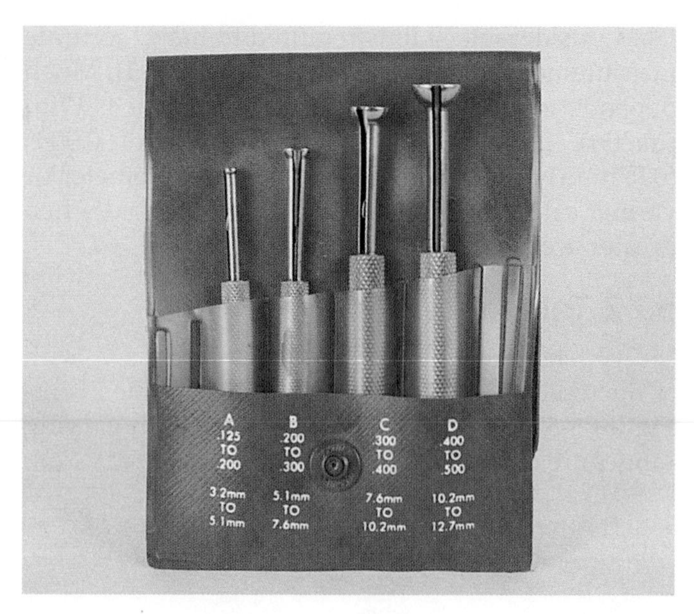

Goodheart-Willcox Publisher

Figure 5-56. Small hole gages are used to measure the diameter of holes that are too small for telescoping gages.

They are adjusted to size by the knurled knob at the end of the handle. Measurement is made over the contacts with a micrometer, **Figure 5-57**.

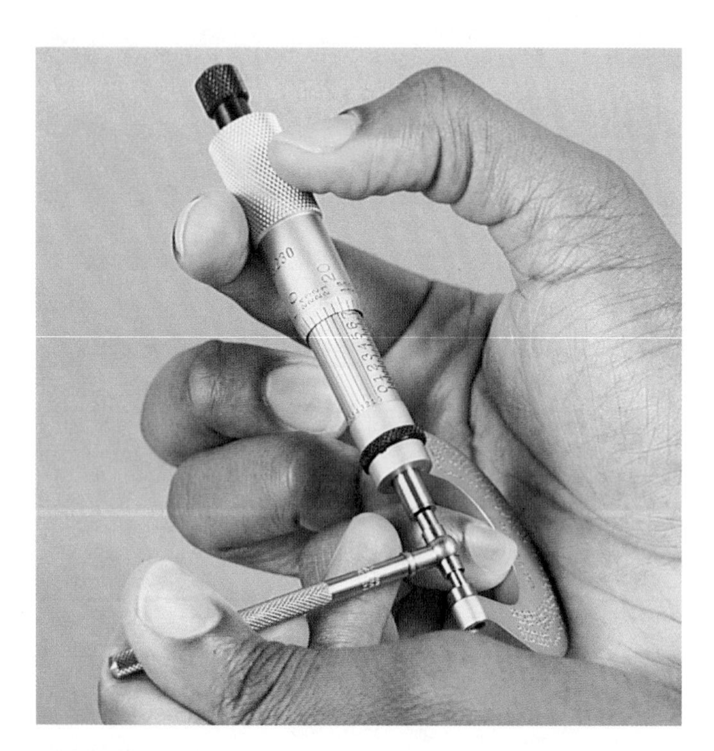

Goodheart-Willcox Publisher

Figure 5-55. After removing the locked telescoping gage, measure it with a micrometer.

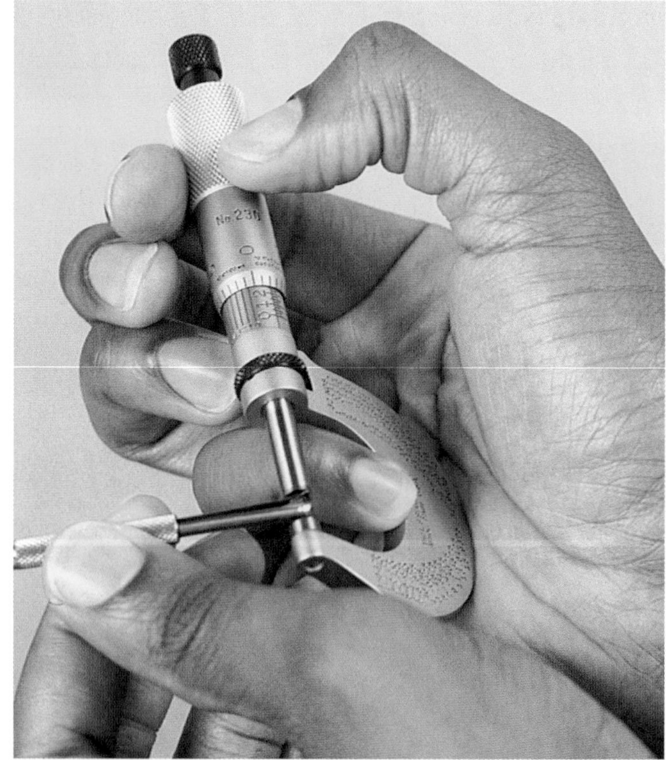

Goodheart-Willcox Publisher

Figure 5-57. The correct way to measure a small hole gage with a micrometer.

Chapter Review

Summary

- Today's machinists need precise measurements, accurate to a microinch or micrometer.
- The steel rule is the most basic measuring tool and is the first to be mastered.
- Both the micrometer caliper and the vernier caliper are used to measure distances very accurately.
- Different gage tools can be used to measure less conventional measurements.
- All measuring tools must be taken care of to maintain the accuracy of their measurements.

Review Questions

Answer the following questions using the information provided in this chapter.

1. One-millionth part of a standard inch is known as a(n) _____.
2. One-millionth part of a meter is known as a _____.
3. The micrometer caliper is nicknamed _____.
4. A micrometer is capable of measuring accurately to 0.0001″ or 0.002 mm when fitted with a(n) _____.
5. The vernier caliper has several advantages over the micrometer. List two of them.
6. List four suggestions for keeping a micrometer or vernier caliper in good condition.
7. The vernier-type tool for measuring angles is called a(n) _____.
8. How does a double-end cylindrical plug gage differ from a step plug gage?
9. A ring gage is used to check whether _____ are within the specified _____ range.
10. Gage blocks are often referred to as _____ blocks.
11. The dial indicator is available in two basic types. List them.
12. What are some uses for the dial indicator?
13. An air gage employs air pressure to measure deep internal openings and hard-to-reach shaft diameters. It operates on the principle of _____.
    A. air pressure leakage between the plug and hole walls
    B. the amount of air pressure needed to insert the tool properly in the hole
    C. amount of air pressure needed to eject the gage from the hole
    D. All of the above.
    E. None of the above.
14. The _____ is used for production inspection. An enlarged image of the part is projected on a screen where it is superimposed on an accurate drawing.
15. Name the measuring device that employs light waves as a measuring standard.
16. The pitch of a thread can be determined with a(n) _____.
17. What use are fillet and radius gages?
18. What are helper measuring tools?
19. How is a telescoping gage used?
20. Make readings from the rules.

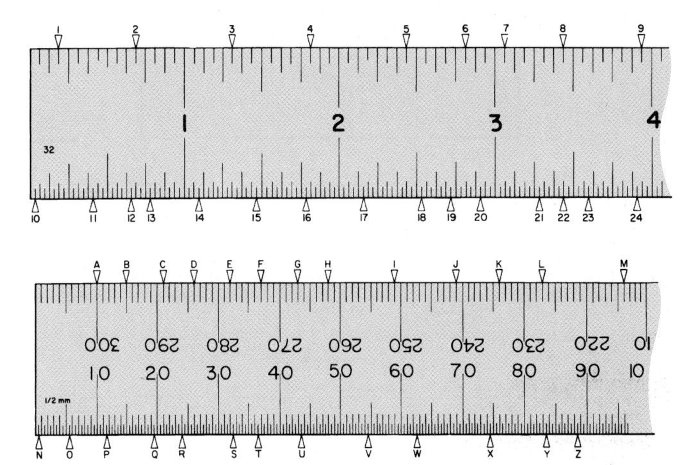

21. Make readings from the vernier scales shown below.

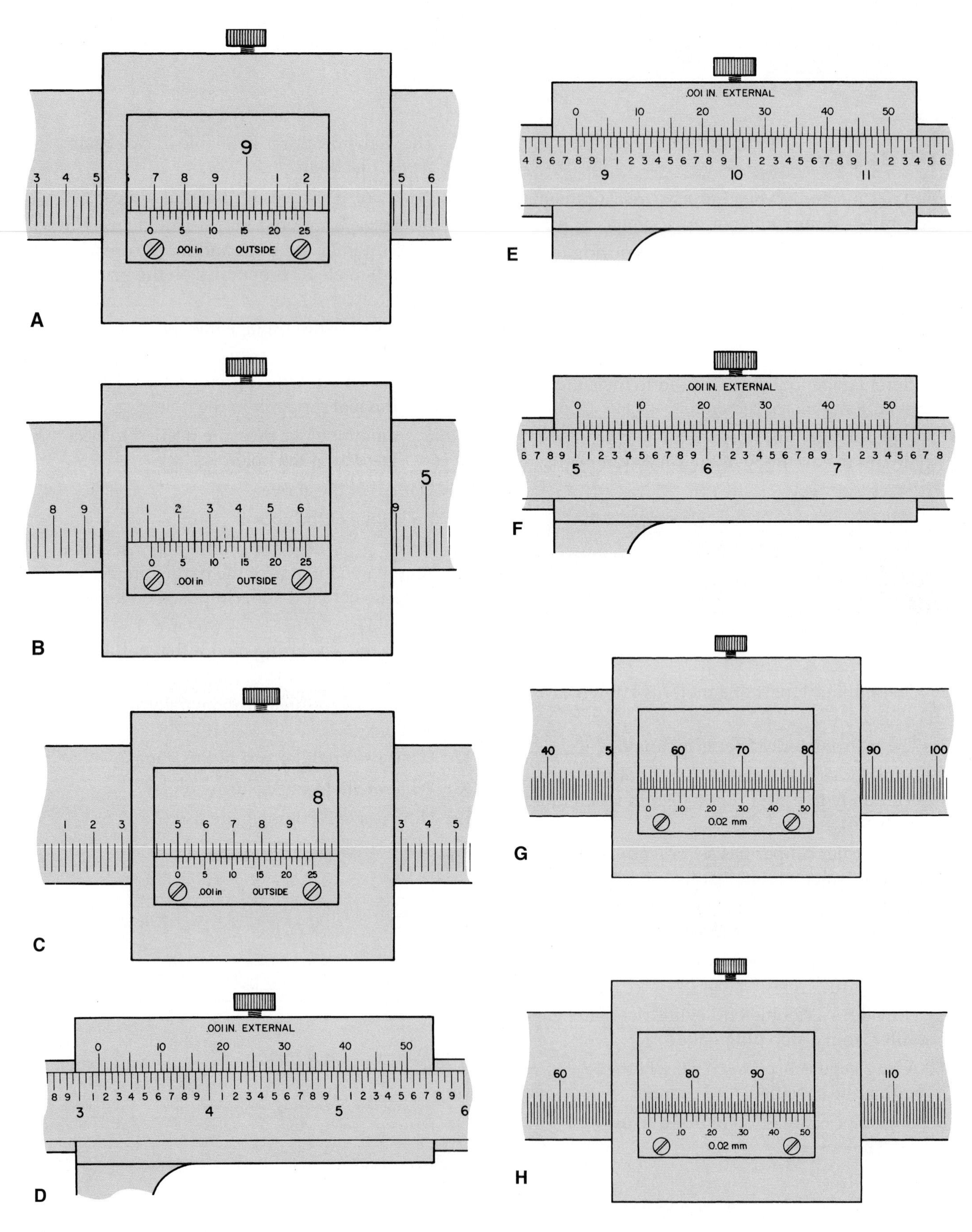

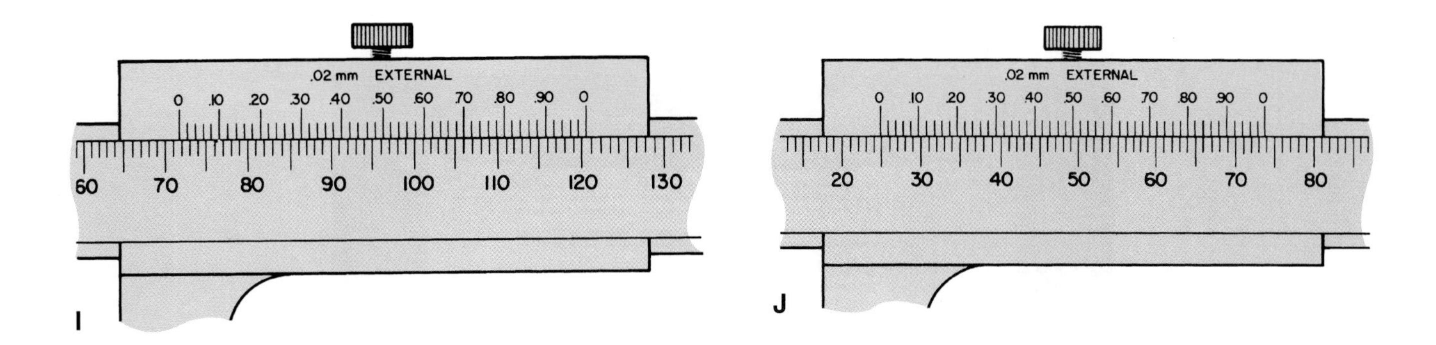

22. Make readings from the micrometer illustrations.

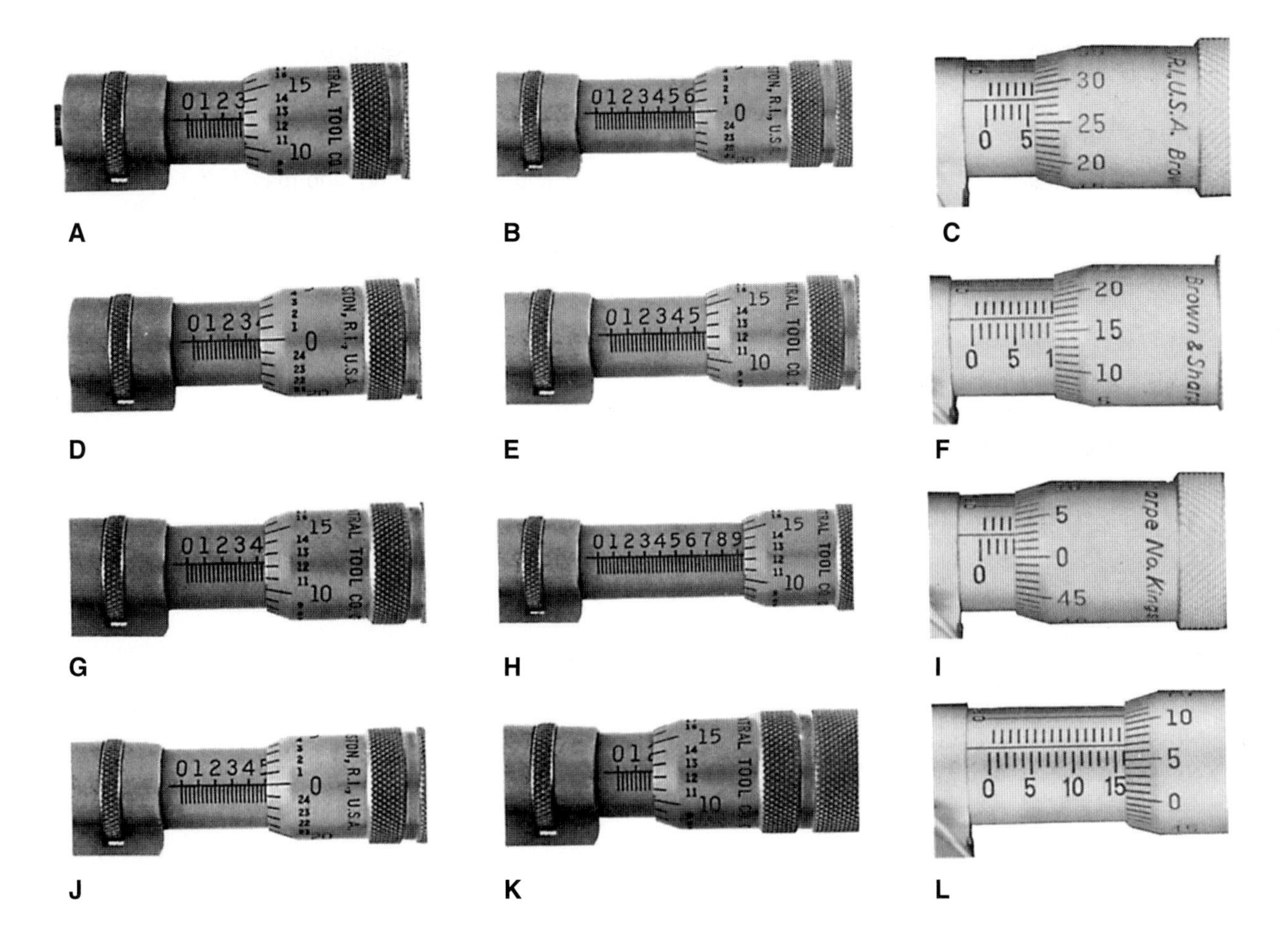

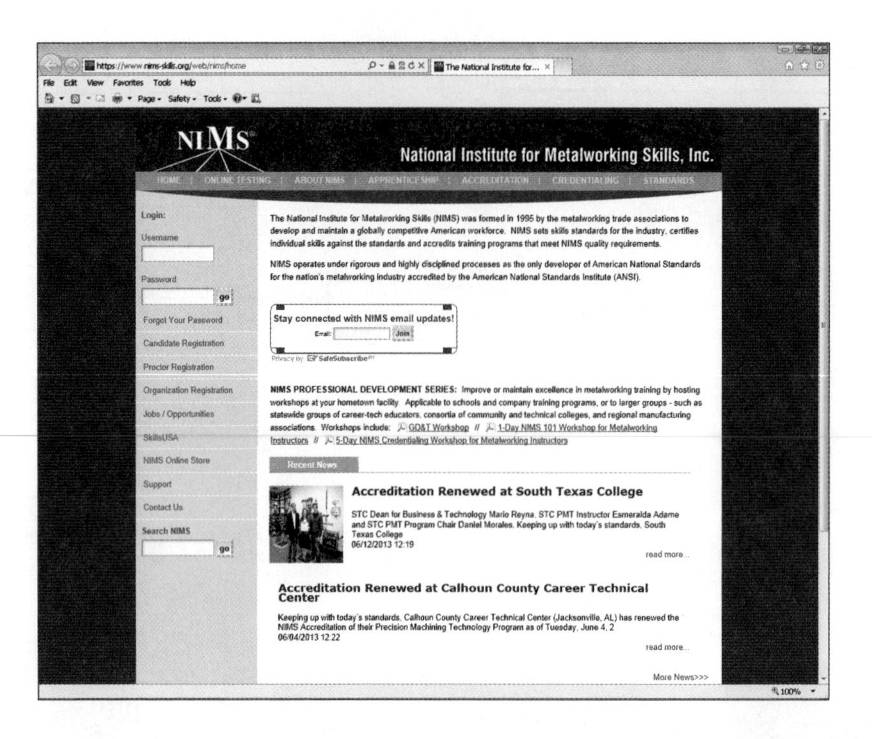

Goodheart-Willcox Publisher

The National Institute for Metalworking Skills (NIMS) works with the machining, metalworking, and manufacturing industries to promote careers in these fields and set skills standards to certify individuals through performance testing and accredited training programs.

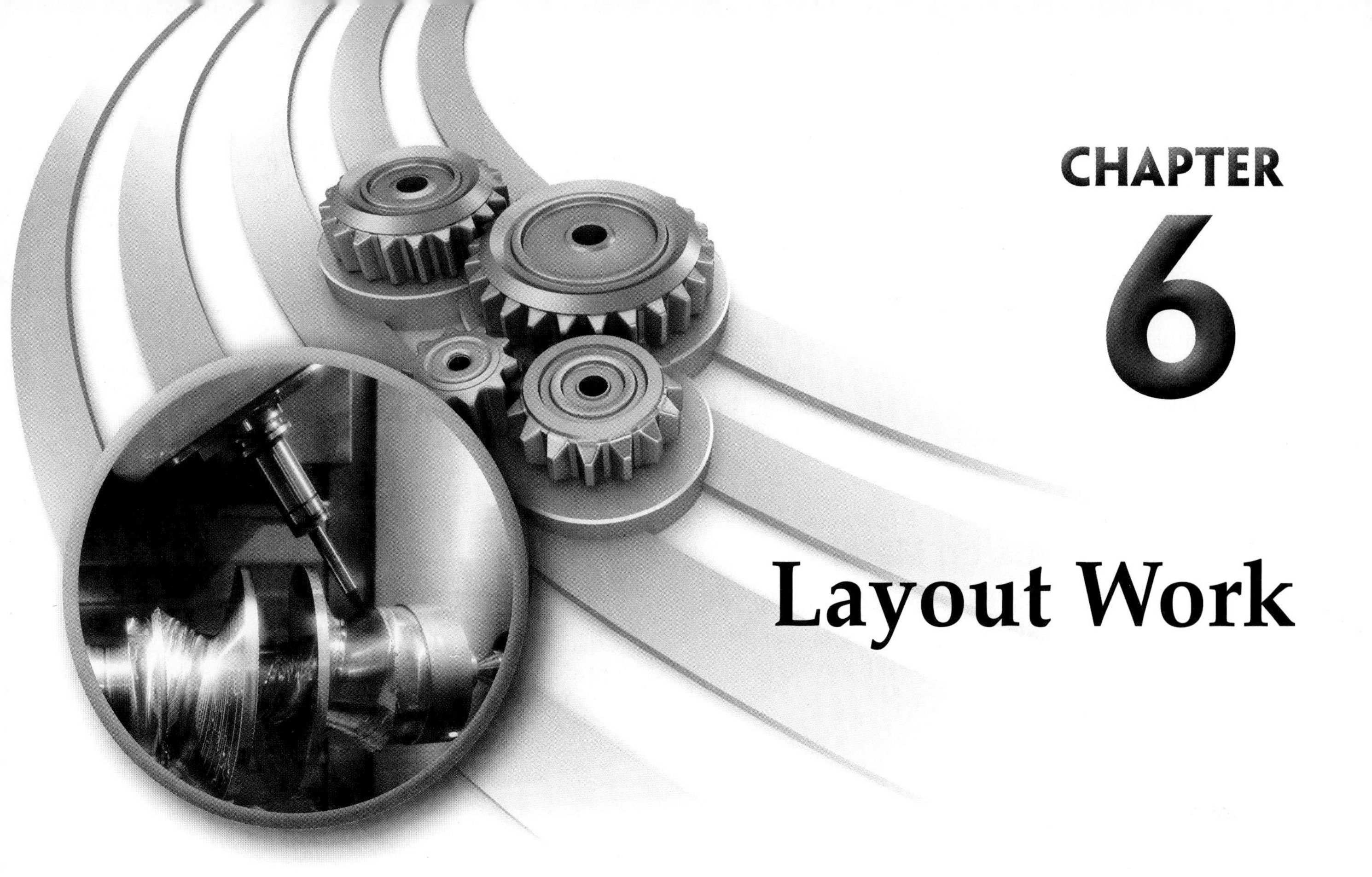

CHAPTER 6

Layout Work

Learning Objectives

After studying this chapter, you will be able to:

- Explain why layouts are needed.
- Identify common layout tools.
- Use layout tools safely.
- Make basic layouts.
- List safety rules for layout work.

Technical Terms

divider
lay out
layout dye
plain protractor
reference line
scriber
square
straightedge
surface gage
surface plate
V-block
vernier protractor

Lay out is the term used to describe the locating and marking of lines, circles, arcs, and points for drilling holes or making cuts. These lines and reference points on the metal show the machinist where to machine.

The tools for this work are known as layout tools. Many common hand tools fall into this category. The accuracy of the job will depend on the proper and careful application of these tools. See **Figure 6-1**. A good layout job is determined by its neatness, accuracy, and legibility.

6.1 Making Lines on Metal

The shiny finish of most metals makes it difficult to distinguish layout lines. For this reason, a coating must be placed on the metal before layout.

6.1.1 Layout Dye

There are many coatings used to make layout lines stand out better. Of these coatings, ***layout dye*** is probably the easiest to use. When applied to the metal, this, usually blue-colored, fluid offers an excellent contrast between the metal and the layout lines. All dirt, grease, and oil must be removed before applying the dye. If these substances are present on the surface, the dye will not adhere properly.

Chalk will also work on hot rolled steel as a layout background. A pencil should not be used because it marks too wide and rubs off.

6.1.2 Scriber

An accurate layout requires fine lines that must be scribed (scratched) into the metal. A ***scriber*** will produce these lines, **Figure 6-2**. The point is made of hardened steel, and is kept needle-sharp by frequent honing on a fine oilstone. Many styles of scribers are available.

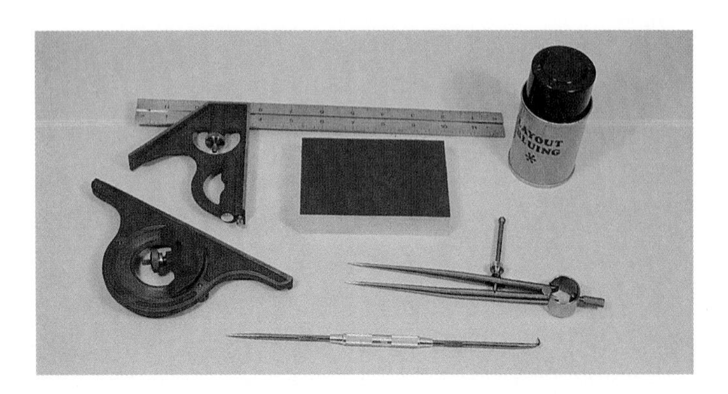

Goodheart-Willcox Publisher

Figure 6-1. A few of the tools needed to make a simple layout.

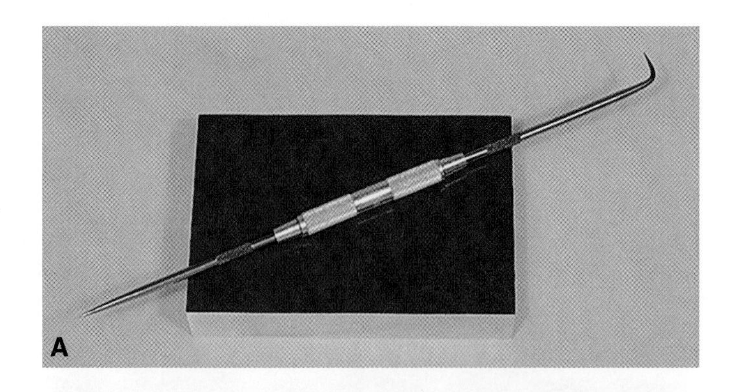

Goodheart-Willcox Publisher

Figure 6-2. Scribers are used to mark parts during layout. A—The long bent point on this scriber can reach through holes. B—This pocket scriber has a removable point. The point can be reversed when the scriber is not being used, protecting the tip and making the work area safer.

6.1.3 Divider

The scriber is used to draw straight lines. A ***divider*** is used to draw circles and arcs, **Figure 6-3**. It is essential that both legs of the tool be equal in length and kept pointed. Measured distances can be laid out with a divider, **Figure 6-4**. To set the tool to the correct distance, set one point on the inch or centimeter mark of a steel rule, and open the divider until the other leg is set to the proper measurement, **Figure 6-5**.

Circles and arcs that are too large to be made with a divider are drawn with a trammel, **Figure 6-6**. This consists of a long thin rod, called a beam, on which two sliding heads with scriber points are mounted. One head is equipped with an adjusting screw. Extension rods can be added to the beam to increase the capacity of the tool.

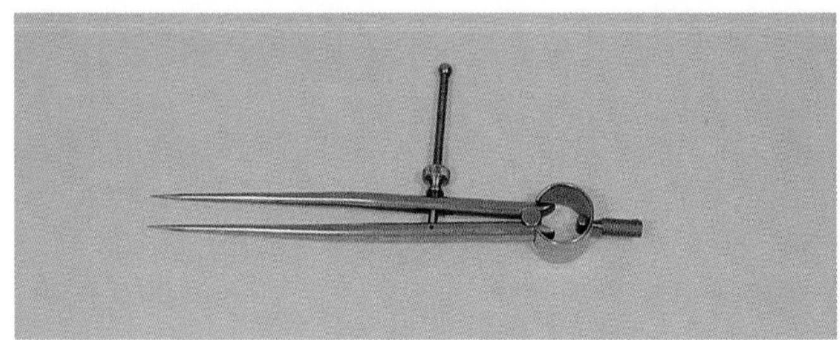

Goodheart-Willcox Publisher

Figure 6-3. A divider is used to mark lines, arcs, and circles.

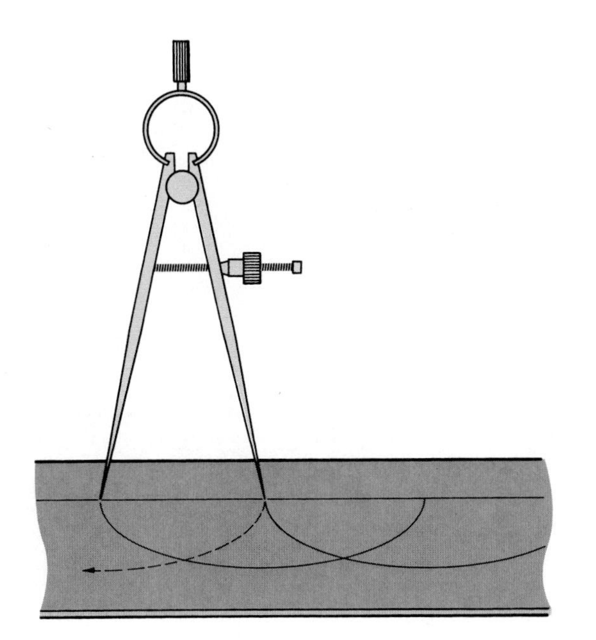

Goodheart-Willcox Publisher

Figure 6-4. Equal spaces can be laid out by "walking" the divider.

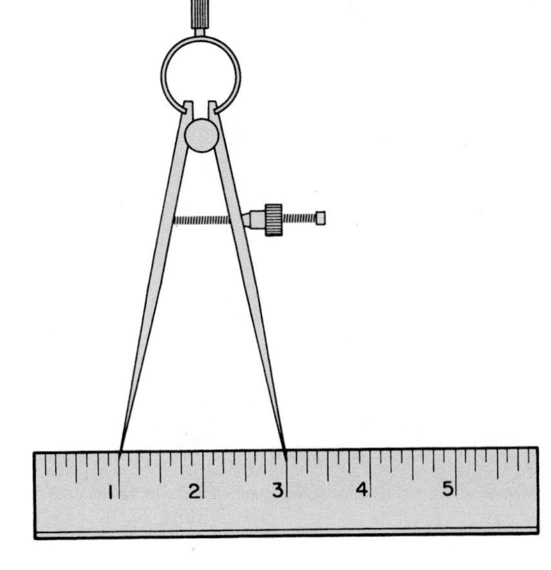

Goodheart-Willcox Publisher

Figure 6-5. To set a divider to a desired dimension, place it on a rule as shown.

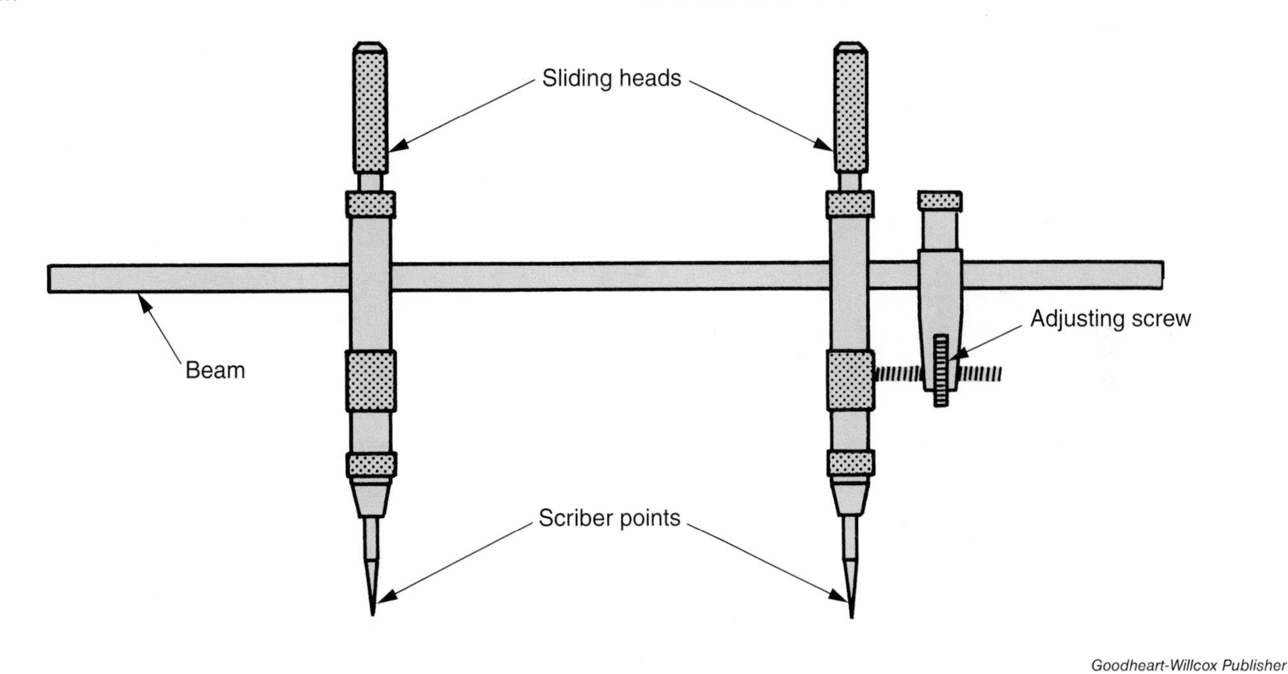

Goodheart-Willcox Publisher

Figure 6-6. Large circles and arcs are drawn with a trammel.

The hermaphrodite caliper is a layout tool with one leg that is shaped like a caliper and the other pointed like a divider, **Figure 6-7**. Lines parallel to the edge of the material, either straight or curved, can be drawn with the tool, **Figure 6-8**. It can also be used to locate the center of irregularly shaped stock.

SAFETY NOTE

- Never carry an open scriber, divider, trammel, or hermaphrodite caliper in your pocket. The sharp points of these tools can puncture the skin easily.
- Always cover sharp points with a cork when the tool is not being used.
- Wear goggles when grinding scriber points.

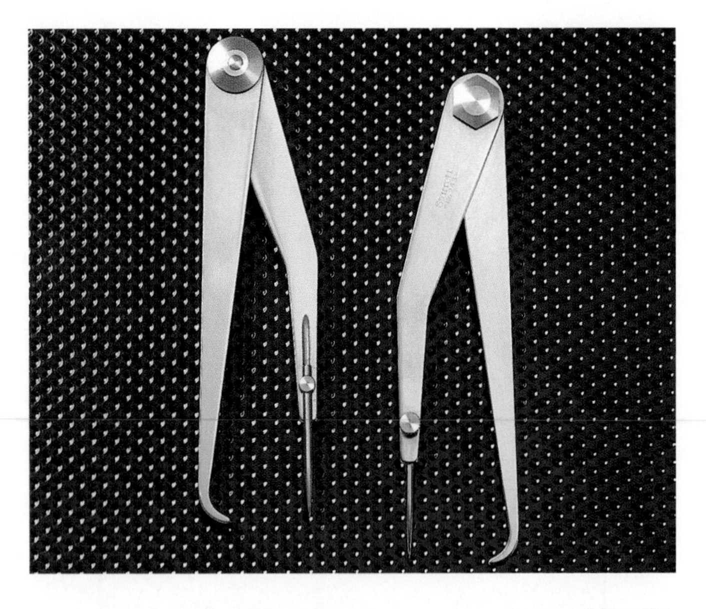

L. S. Starrett Co.

Figure 6-7. A hermaphrodite caliper has a blunt end for the sliding surface and a point for scribing.

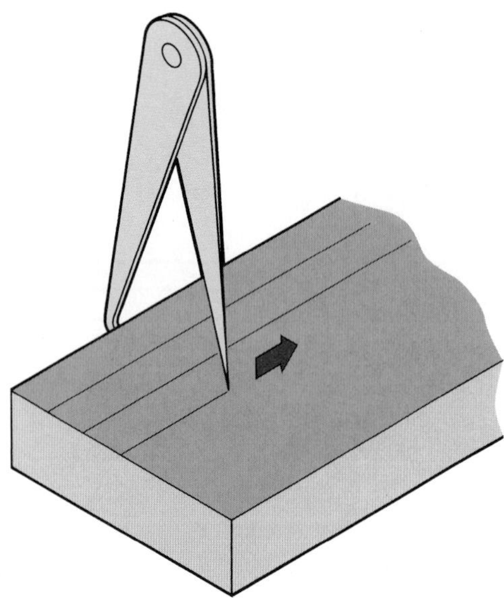

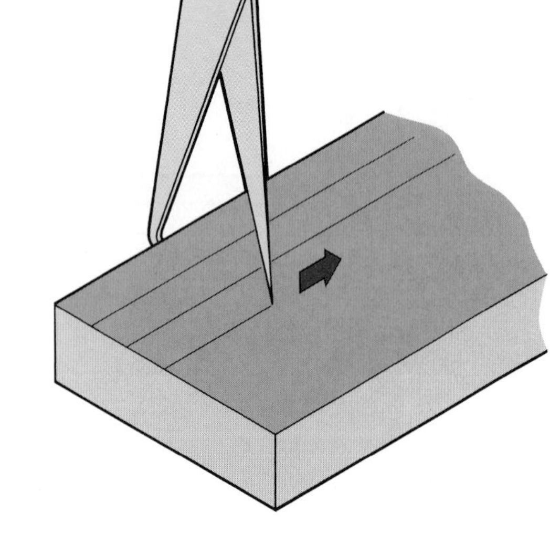

Goodheart-Willcox Publisher

Figure 6-8. Scribing lines parallel to an edge with a hermaphrodite caliper.

6.1.4 Surface Gage

A ***surface gage*** has many uses, but is most frequently used for layout work, **Figure 6-9**. It consists of a base, spindle, and scriber. An adjusting screw is fitted for making fine adjustments. The scriber is mounted in a way that allows it to be pivoted into any position. A surface gage can be used to scribe lines at a given height and parallel to the surface, **Figure 6-10**. A V-slot in the base permits the tool to also be used on a curved surface.

L. S. Starrett Co.

Figure 6-9. This small surface gage is designed for light work.

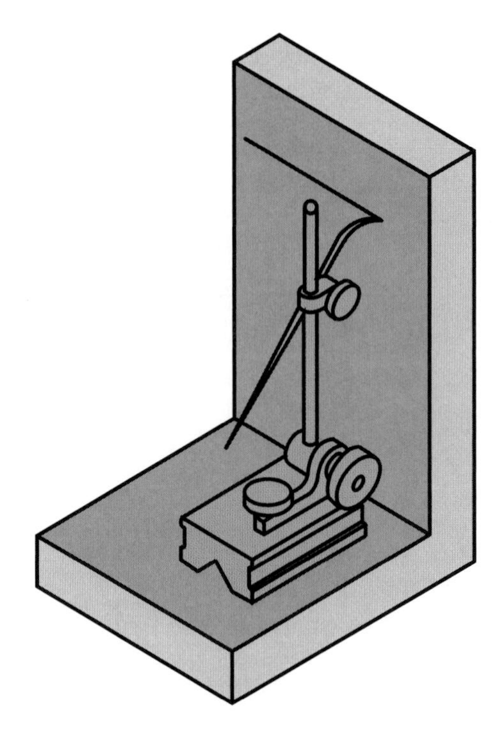

Goodheart-Willcox Publisher

Figure 6-10. Carefully slide the surface gage to scribe lines parallel to the base. Handle the gauge carefully because sharp points can cause injury.

To check whether a part is parallel to a given surface, fit the surface gage with a dial indicator. Set the indicator to the required dimension with the aid of gage blocks. The tool is then moved back and forth along the work. A height gage can be used in the same manner as a surface gage, **Figure 6-11**.

Tibor Machine Products

Figure 6-11. A digital vernier height gage with a dial indicator being used to check whether the V-blocks are parallel to the surface plate.

6.1.5 Surface Plate

Every linear measurement depends on an accurate reference surface. ***Surface plates*** provide a reference surface (plane) for layout and inspection.

Surface plates can be purchased in sizes up to 72″ by 144″ (1800 mm by 3600 mm) and in various grades. Surface plate grade differences are given in degrees of flatness:

- Grade AA for laboratories.
- Grade A for inspection.
- Grade B for tool room and layout applications.

Most surface plates made today are granite, **Figure 6-12**, but some are cast iron. Granite is more stable. Cast iron surface plates are more affected by temperature changes.

L. S. Starrett Co.

Figure 6-12. Surface plates are available in various grades and materials. Granite surface plates are the most common.

Surface plates are used primarily for layout and inspection work. They should never be used for any job that could mar or nick the surface.

When square reference surfaces are needed, a right angle plate is used, **Figure 6-13**. The plates can be placed in any position with the work clamped to the face for layout, measurement, or inspection.

An accurate surface parallel to the surface plate can be obtained using box parallels, **Figure 6-14**. All surfaces are precision-ground to close tolerances.

6.1.6 V-Blocks

V-blocks support round work for layout and inspection, **Figure 6-15**. They are furnished in matched pairs with surfaces that are ground square to close tolerances. Ribs are cast into the body for weight reduction. The ribs also can be used as clamping surfaces.

Goodheart-Willcox Publisher

Figure 6-13. Right angle plates are often used to check perpendicular surfaces.

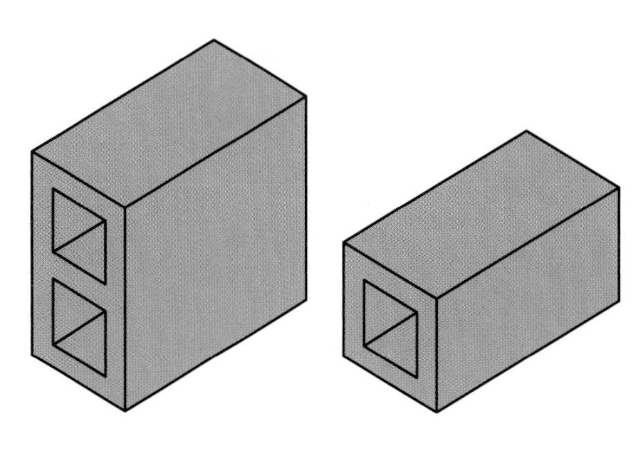

Goodheart-Willcox Publisher

Figure 6-14. Box parallels are available in a number of sizes.

Goodheart-Willcox Publisher

Figure 6-15. V-blocks can be used to hold round stock for layout and measurement work.

SAFETY NOTE

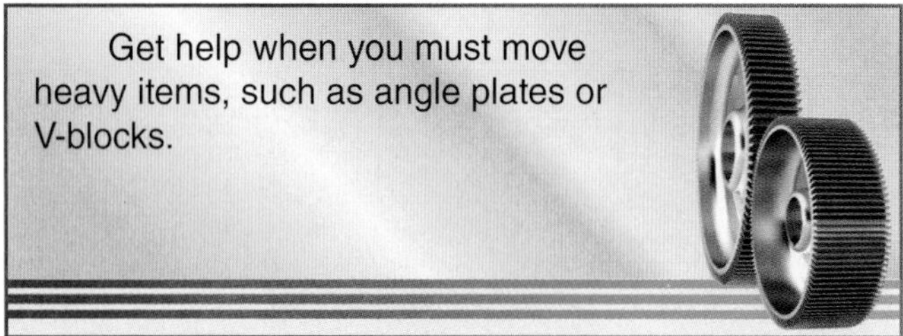

Get help when you must move heavy items, such as angle plates or V-blocks.

6.1.7 Straightedge

Long flat surfaces are checked for accuracy with a ***straightedge***, **Figure 6-16**. This tool is also used for laying out long straight lines. Straightedges can be made from steel or granite, with steel being more common.

6.2 Squares

The ***square*** is used to check 90° (square) angles. The tool is also used for laying out lines that must be at right angles to a given edge or parallel to another edge. Some simple machine setups can be made quickly and easily with a square.

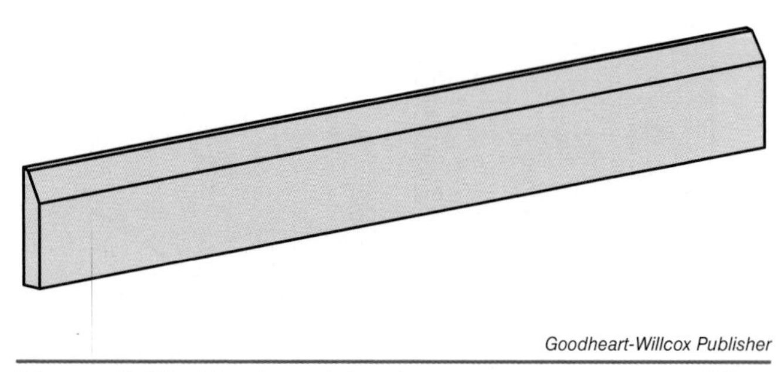

Goodheart-Willcox Publisher

Figure 6-16. Steel straightedges are very common. They are also available in granite.

A hardened steel square, also commonly called a machinist's square, is recommended when extreme accuracy is required. See **Figure 6-17**. The square has true right angles, both inside and outside. It is accurately ground and lapped for straightness and parallelism. The tool comes in sizes up to 36" (910 mm).

The double square is more practical than the steel square for many jobs because the sliding blade is adjustable and interchangeable with other blades, **Figure 6-18**. The tool should not be used where great precision is required. The bevel blade has one angle for checking octagons (45° angles), and another for checking hexagons (60° angles).

A drill grinding blade is also available for the double square. One end is beveled to 59° for drill grinding and the other end is beveled at 41° for checking the cutting angles of machine screw countersinks. Both ends are graduated for measuring the length of the cutting lips, to ensure that the cutting tools are sharpened on center.

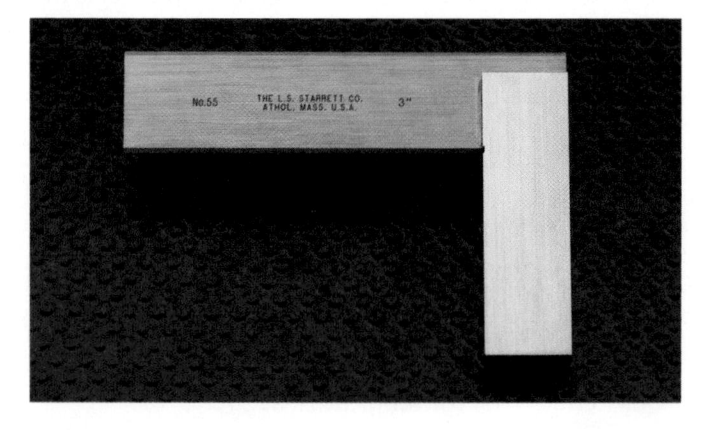

L. S. Starrett Co.

Figure 6-17. The hardened steel square is handy during layout work.

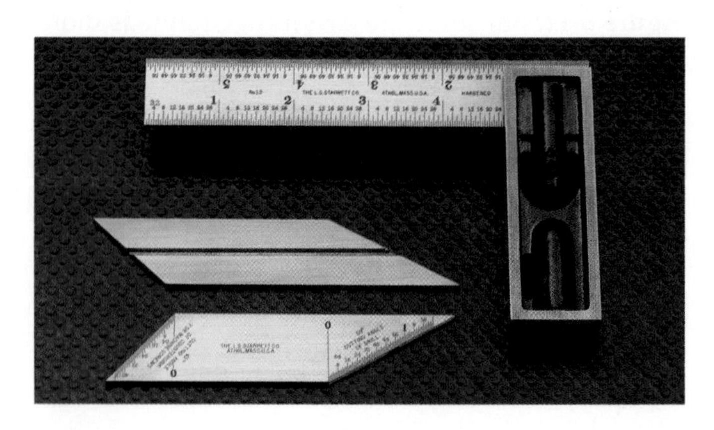

L. S. Starrett Co.

Figure 6-18. This double square has a graduated blade, beveled blade, and a drill grinding blade.

The combination set consists of a hardened blade, square head, center head, and bevel protractor. The blade fits all three heads, **Figure 6-19**. Combination sets are adaptable to a large variety of operations, making them especially valuable in the shop.

The square head, which has one 45° edge, makes it possible for the tool to serve as a miter square. By projecting the blade the desired distance below the edge, it also serves as a depth gage, **Figure 6-20**. The spirit level, fitted in one edge, allows it to be used as a simple level.

With the rule properly inserted, the center head can be used to quickly find the center of round stock. This is illustrated in **Figure 6-21**.

The protractor head can be rotated through 180° and is graduated accordingly. The head can be locked with a locking nut, making it possible to accurately determine and scribe angles, **Figure 6-22**. The head also has a level built in, making it possible to use it as a level for positioning angles for inspection, layout, or machining.

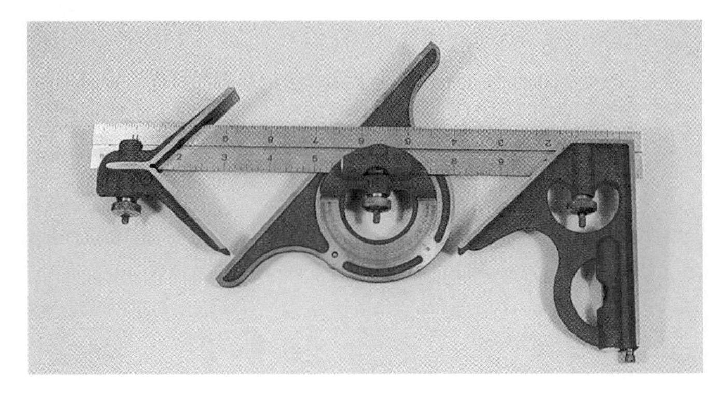

Goodheart-Willcox Publisher

Figure 6-19. A combination set will perform various layout tasks.

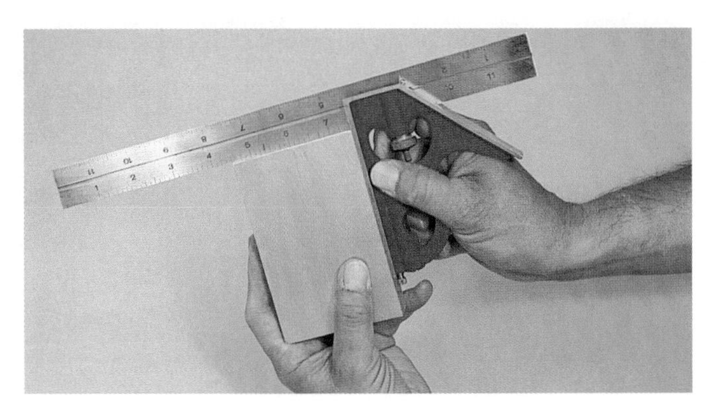

Goodheart-Willcox Publisher

Figure 6-20. The combination set can be used to check squareness and to measure like a depth gage.

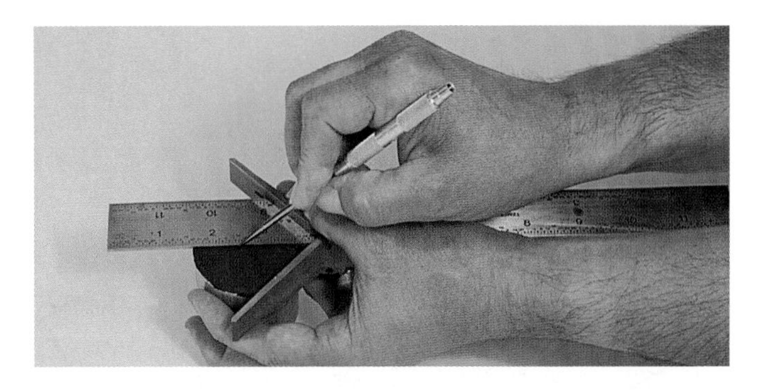

Goodheart-Willcox Publisher

Figure 6-21. Using a center head and rule to locate the center of a piece of round stock.

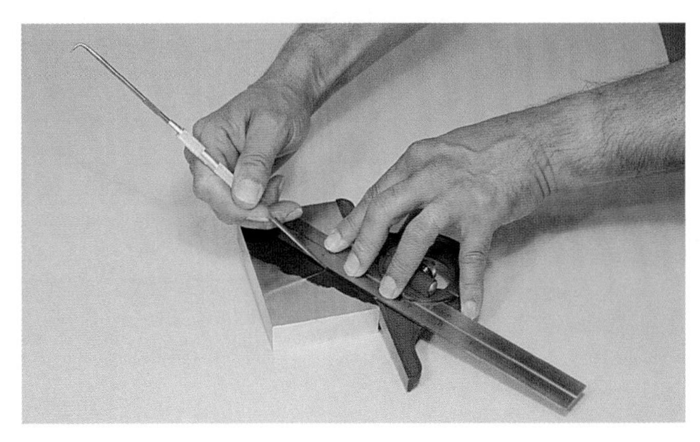

Goodheart-Willcox Publisher

Figure 6-22. Angular settings on layouts can be made with the protractor head and rule of the combination set.

Handle a square with care. The blade is mounted solidly, but if the tool is dropped, the blade can be "sprung," ruining the square.

6.3 Measuring Angles

In addition to the protractor head of the combination set, other angle measuring tools are used in layout work. The accuracy required by a job will determine which tool must be used.

When angles do not need to be laid out or checked to extreme accuracy, a ***plain protractor*** can be used, **Figure 6-23**. The head is graduated from 0° to 180° in both directions for easy reading.

A protractor depth gage is suitable for checking angles and measuring slot depths, **Figure 6-24**.

A universal bevel is useful for checking, laying out, and transferring angles, **Figure 6-25**. Both blade and stock are slotted, making it possible to adjust the blade into any desired position. A thumbscrew locks it tightly in place.

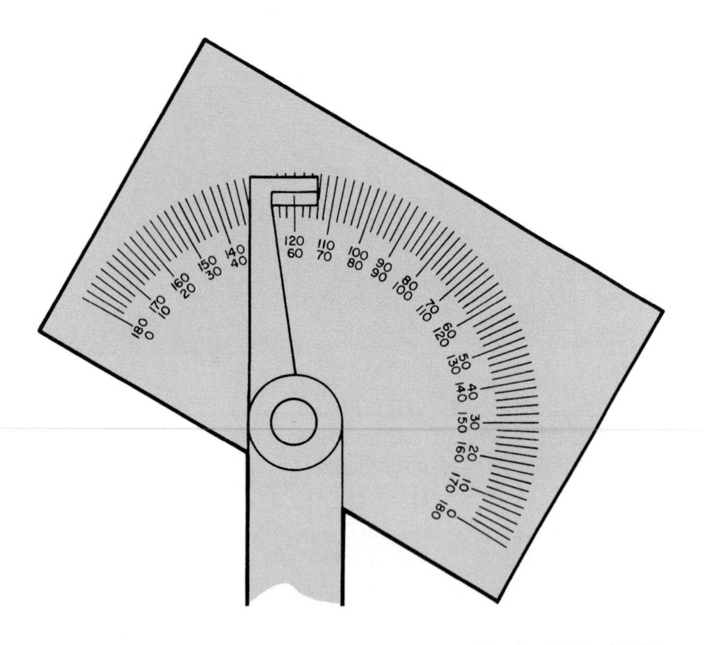

Goodheart-Willcox Publisher

Figure 6-23. A plain steel protractor will show angles with moderate precision.

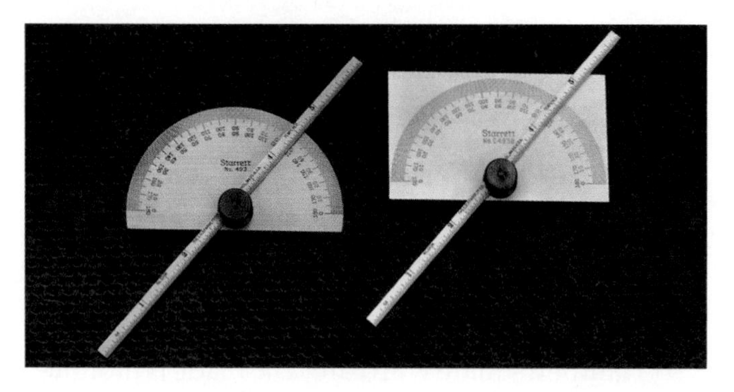

L. S. Starrett Co.

Figure 6-24. A protractor depth gage.

L. S. Starrett Co.

Figure 6-25. A universal bevel can be locked at various angles.

When a job requires extreme accuracy, the machinist uses a ***vernier protractor***, **Figure 6-26**. With this tool, angles of 1/12 of a degree (5 minutes) can be accurately measured.

6.4 Simple Layout Steps

Each layout job requires planning before the operation can be started. **Figure 6-27** shows a typical job. Use the following planning procedure:

1. Carefully study the drawings.
2. Cut stock to size.

SAFETY NOTE

Remove all burrs and sharp edges from stock before starting layout work.

3. Clean all dirt, grease, and oil from the work surface. Apply layout dye.
4. Locate and scribe a ***reference line*** (base line). You will make all measurements from this line. If the material has one true edge, it can be used in place of the base line.
5. Locate and center points of all circles and arcs.

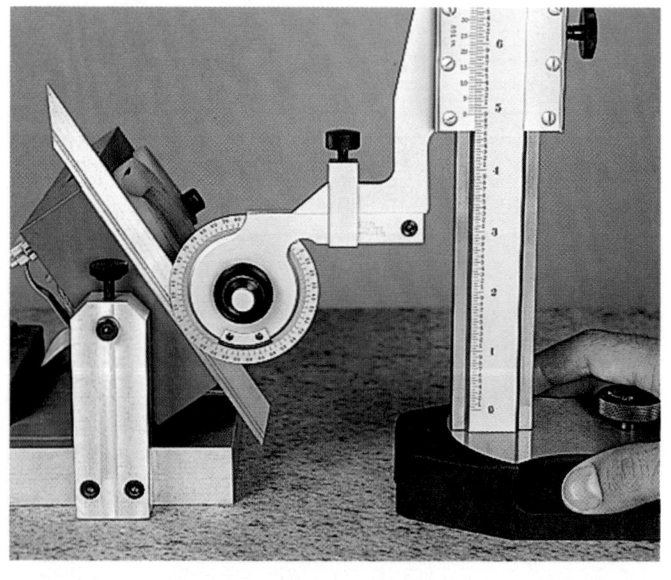

L. S. Starrett Co.

Figure 6-26. Precise angular measurements are made with a vernier protractor. In this view, the protractor is mounted on a height gage.

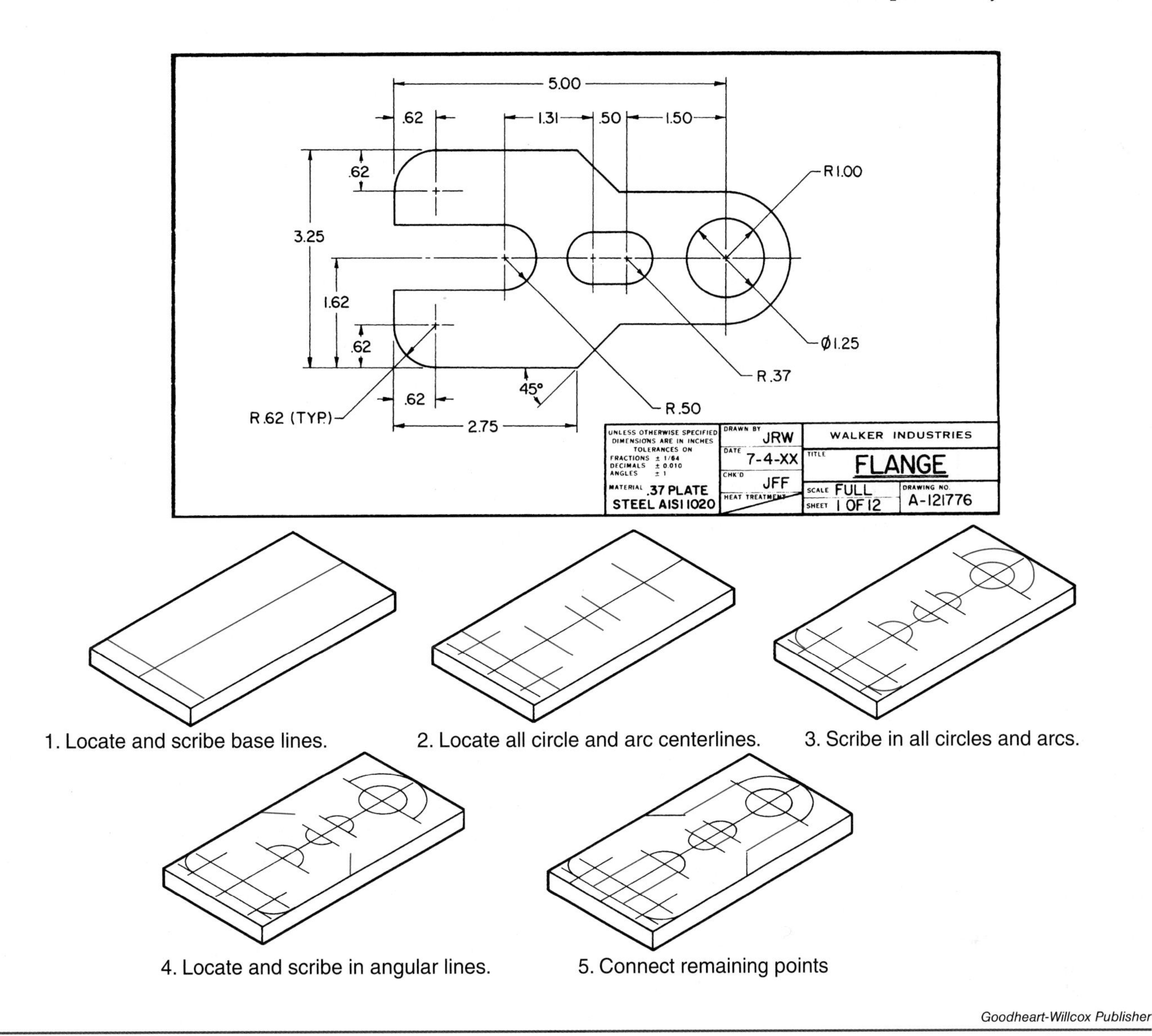

Figure 6-27. Compare the part drawing with steps involved in laying out the job.

6. Use a prick punch to mark the point where centerlines intersect. The sharp point (30° to 60°) of this punch makes it easy to locate the position. After the prick punch mark has been checked, it is enlarged slightly with a center punch, **Figure 6-28.**
7. Scribe in all circles and arcs with a divider or trammel.
8. If angular lines are necessary, scribe them using the proper layout tools. You can also locate the correct points by measuring and connecting them using a rule or straightedge and a scribe.
9. Scribe in all other internal openings.
10. Lines should be clean and sharp. Any double or sloppy line work should be removed by cleaning it off with a solvent. Then apply another coat of dye before scribing the line again.

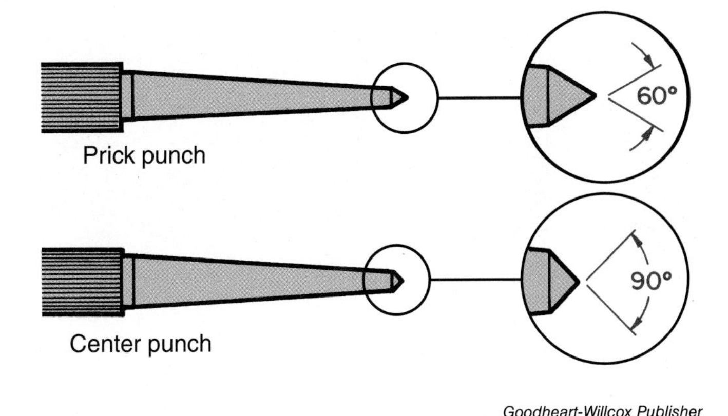

Figure 6-28. A prick punch has a more sharply angled point than a center punch. The prick punch is used to mark a location. After the prick punch mark is checked, it is enlarged with a center punch.

Chapter Review

Summary

- Layout work is an important first step to the machining process.
- Machinists use a variety of layout tools including layout dye, scribers, dividers, surface gages, surface plates, V-blocks, and straightedges.
- A variety of squares and protractors are used to measure angles.
- Basic layouts can be created following ten simple steps.

Review Questions

Answer the following questions using the information provided in this chapter.

1. Why are layout lines used?
2. What is used to make layout lines easier to see?
3. What is wrong with using a pencil to make layout lines on metal?
4. Straight layout lines are drawn with a(n) _____.
5. Circles and arcs are drawn on work with a(n) _____.
6. Large circles and arcs are drawn with a(n) _____.
7. A(n) _____ is the flat granite or cast iron surface used for layout and inspection work.
8. Round stock is usually supported on _____ for layout and inspection.
9. Long flat surfaces can be checked for trueness with a(n) _____.
10. List four layout operations that can be performed with a combination set.
11. The center of round stock can be found quickly with the _____ and rule of a combination set.
12. Angular lines that must be very accurate should be laid out with a(n) _____.
13. The _____ punch has a sharper point than the center punch.
14. List three safety precautions that you should observe when doing layout work.

CHAPTER 7

Hand Tools

Learning Objectives

After studying this chapter, you will be able to:

- Identify the most commonly used machine shop hand tools.
- Select the proper hand tool for the job.
- Maintain hand tools properly.
- Explain how to use hand tools safely.

Technical Terms

abrasive
American National Thread System
blind hole
class of fit
foot-pound
newton-meter
reamer
safe edge
torque
Unified System

Selecting and using hand tools correctly will help you do a job safely, with a minimum expenditure of time. When a hand tool is used incorrectly, it can be damaged or someone may be injured. It is to your advantage to learn to work properly with hand tools.

7.1 Clamping Devices

Clamping devices are used to hold and position material while it is being worked on. Several types of clamping devices are used in machining.

7.1.1 Vises

The bench vise is used for many holding tasks. It should be mounted on the bench edge far enough out to permit clamping long work in a vertical position. A vise may be a solid base type or may have a swivel base, which allows the vise to be rotated. See **Figure 7-1**.

Small precision parts may be held in a small bench vise or toolmaker's vise, **Figure 7-2**. This type of vise can be rotated and tilted to any desired position. Vise size is determined by the width of the jaws, **Figure 7-3**.

A vise's clamping action is obtained from a heavy screw turned by a handle. The handle is long enough to apply ample pressure for any work that will fit the vise. Under no circumstances should the vise handle be hammered tight, nor should additional pressure be applied using a length of pipe on the handle for leverage.

Vise jaws are hardened. When clamping work that could be damaged or marred by the jaw serrations, cover the jaws with soft copper, brass, or aluminum caps, **Figure 7-4**.

ekipaj/Shutterstock.com

Figure 7-1. A swivel base bench vise. The base is made in two parts so that the body can be rotated to any desired position.

Birgit Reitz-Hofmann/Shutterstock.com

Figure 7-2. A small vise used by a toolmaker. It can be rotated and pivoted to desired working position.

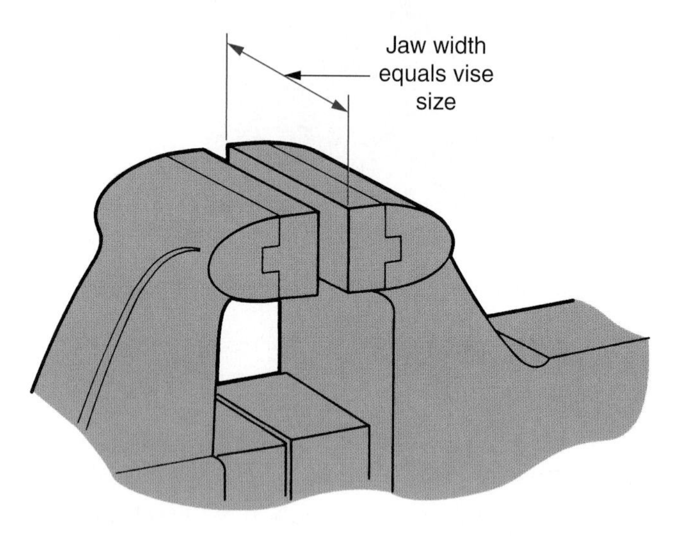

Goodheart-Willcox Publisher

Figure 7-3. Vise size is determined by width of vise jaws.

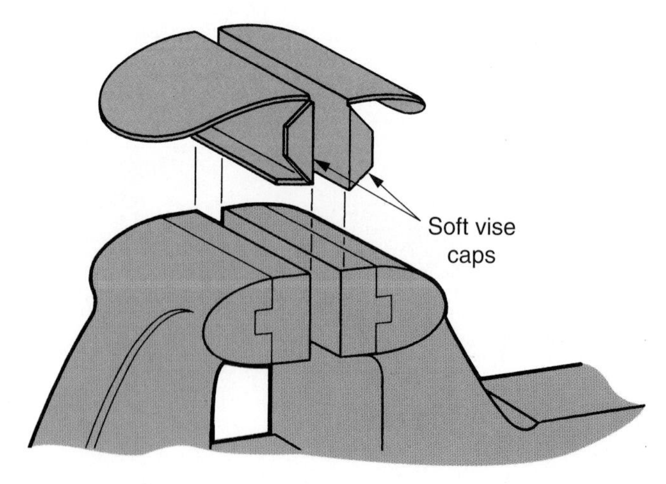

Goodheart-Willcox Publisher

Figure 7-4. Caps made of copper, lead, or aluminum are slipped over hardened vise jaws to protect work from becoming marred or damaged by jaw serrations.

SAFETY NOTE

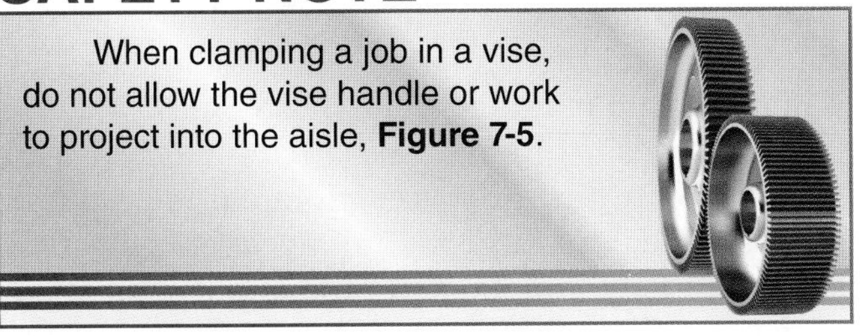

7.1.2 Clamps

The C-clamp and the parallel clamp hold parts together while they are worked on. The C-clamp, **Figure 7-6**, is made in many sizes. Jaw opening determines clamp size.

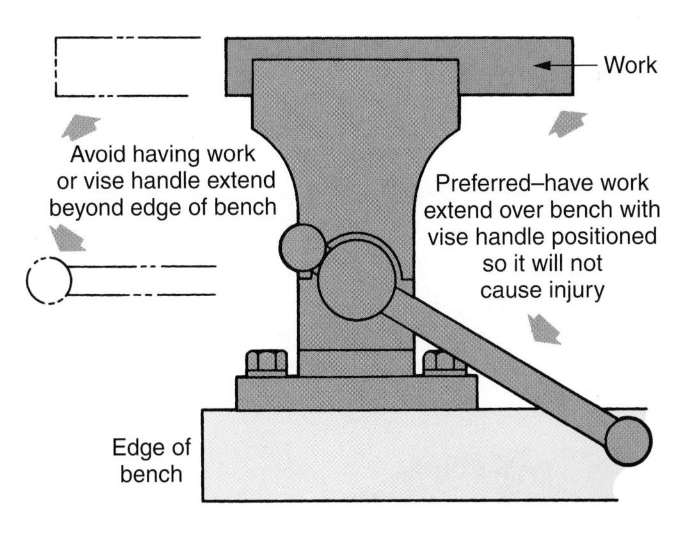

Goodheart-Willcox Publisher

Figure 7-5. To prevent injury, avoid letting the vise handle or work project into the aisle.

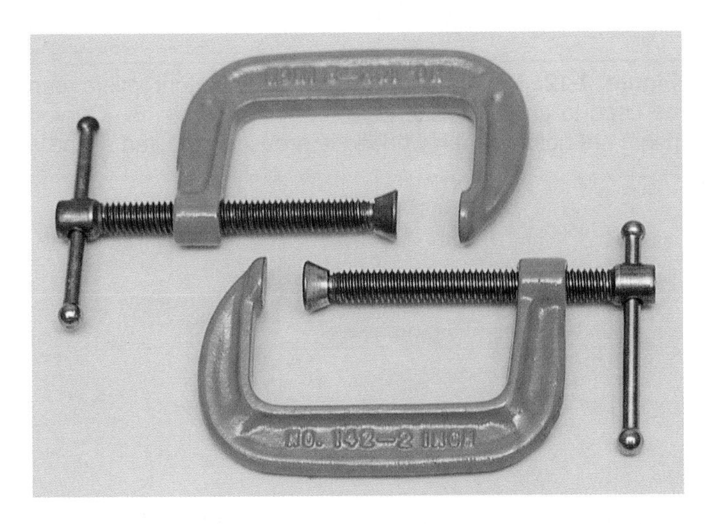

Goodheart-Willcox Publisher

Figure 7-6. C-clamps are available in a range of sizes.

A parallel clamp is ideal for holding small work. For maximum clamping action, the jaw faces must be parallel. See **Figure 7-7**. Placing strips of paper the width of the clamp jaw between the work and the jaws will improve clamping action.

7.2 Pliers

Slip joint or combination pliers are widely used for holding tasks. See **Figure 7-8**. The slip joint permits the pliers to be opened wider at the hinge pin to grip larger size work. They are made in 5″, 6″, 8″, and 10″ sizes. The plier size is measured by the overall length of the tool.

Some slip joint pliers are made with cutting edges for clipping wire and small metal sections to needed lengths. The better grade pliers are of forged construction.

Diagonal pliers are another widely used tool for light cutting tasks, **Figure 7-9**. The cutting edges are

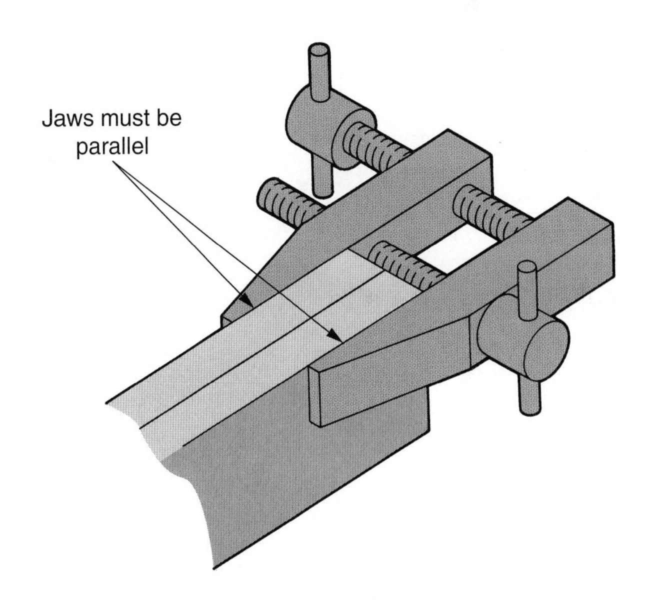

Goodheart-Willcox Publisher

Figure 7-7. For maximum clamping action with a parallel clamp, adjust the jaws until they are parallel.

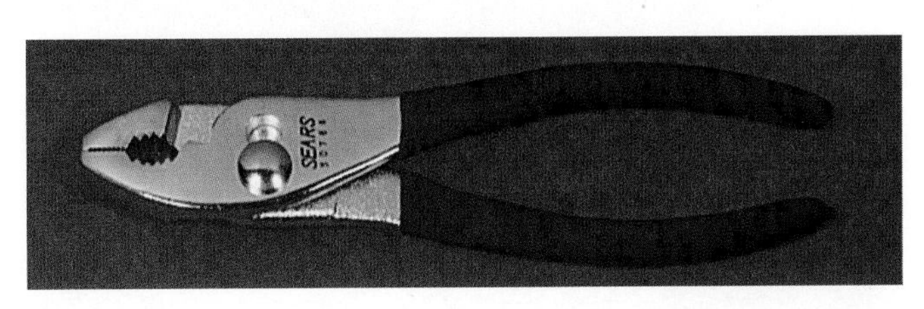

Goodheart-Willcox Publisher

Figure 7-8. Combination or slip-joint pliers are used for holding tasks.

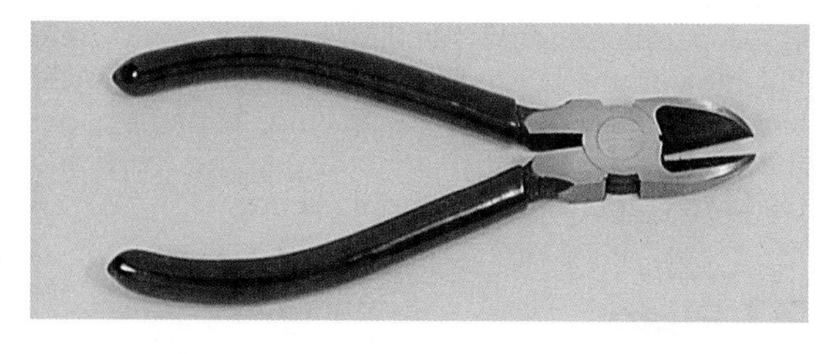

Goodheart-Willcox Publisher

Figure 7-9. Diagonal pliers will cut flush with a surface.

at an angle to permit the pliers to cut flush (even) with the work surface. Diagonal pliers are made in 4″, 5″, 6″, and 7″ lengths.

Side-cutting pliers are capable of cutting heavier wire and pins, **Figure 7-10**. Some of these pliers have a wire stripping groove and insulated handles. They are made in 6″, 7″, and 8″ lengths.

Round-nose pliers, **Figure 7-11**, are helpful when forming wire and light metal. Their jaws are smooth and will not mar the metal being grasped. Round-nose pliers are available in 4″, 4 1/2″, 5″, and 6″ sizes.

Goodheart-Willcox Publisher

Figure 7-10. Side-cutting pliers have square jaws for holding and cutting tasks.

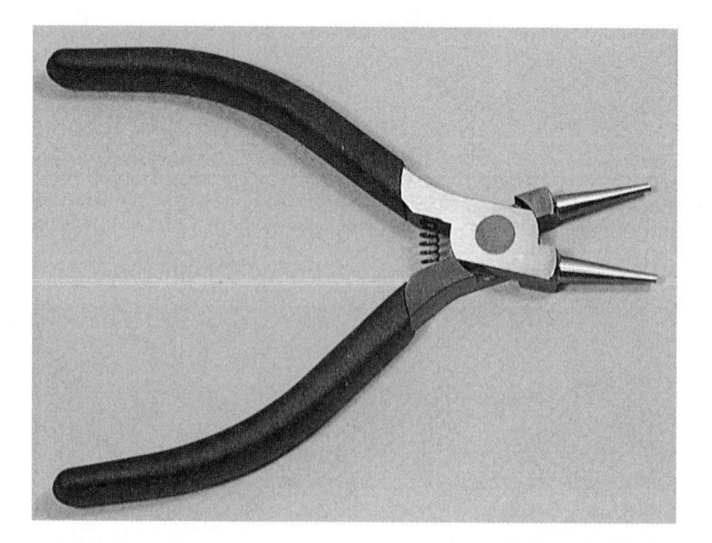

Goodheart-Willcox Publisher

Figure 7-11. Round-nose pliers have smooth jaws.

Needle-nose pliers are available in both straight and curved-nose types. They are handy for holding small work and when work space is limited. They will reach into cramped places. See **Figure 7-12**.

Tongue and groove pliers have aligned teeth for flexibility in gripping different size work, **Figure 7-13**. The size of the jaw opening can be adjusted easily. Tongue and groove pliers are made in many different sizes. The 6″ size usually has five adjustments, while the larger 16″ size has eleven adjustments.

Locking pliers have jaw openings that can be adjusted through a range of sizes by using a threaded mechanism on one handle. See **Figure 7-14**. After adjustment, a squeeze of the hand can lock the jaws onto the work with more than a ton of pressure. Jaw pressure can be relieved by using the quick release on the handle. These pliers are made

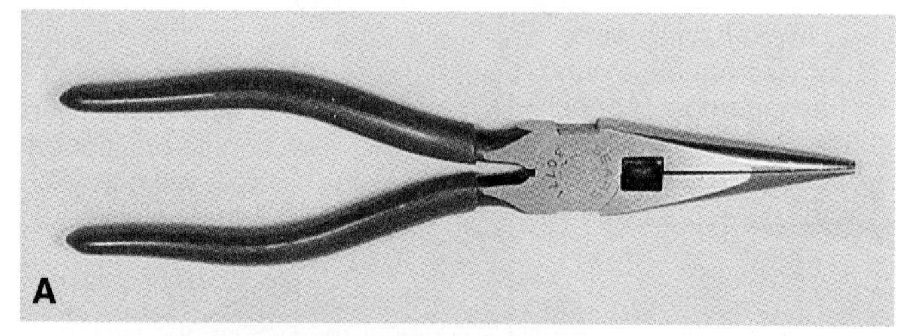

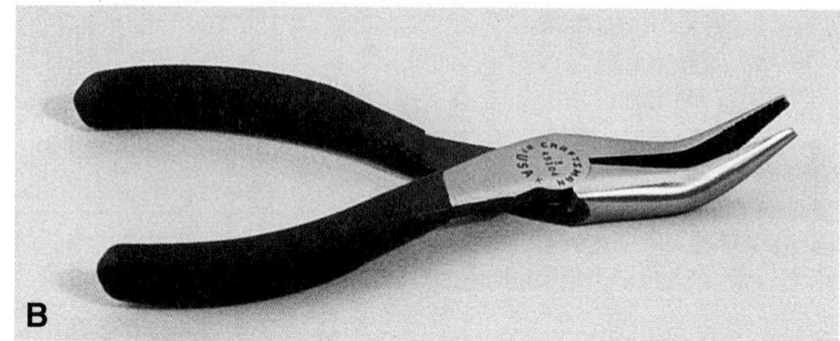

Goodheart-Willcox Publisher

Figure 7-12. Needle-nosed pliers. A—Straight pliers can be used to grasp smaller hard-to-reach objects. B—Curved pliers are helpful when working in areas with limited space.

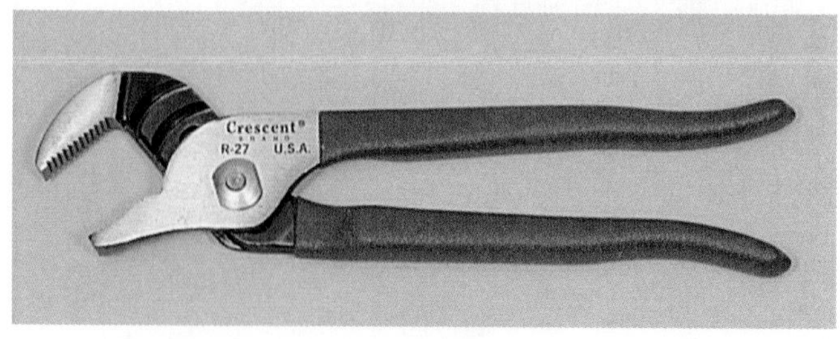

Goodheart-Willcox Publisher

Figure 7-13. Tongue and groove pliers allow the jaws to expand to hold large objects.

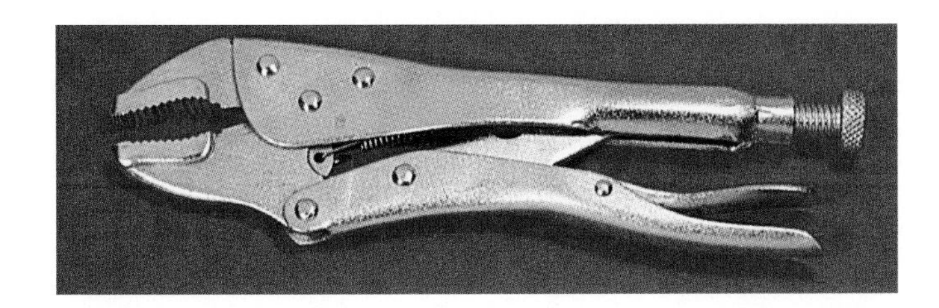

Goodheart-Willcox Publisher

Figure 7-14. Adjustable clamping pliers can be locked on work of different sizes.

in many sizes with straight, curved, or long-nose jaws. They are known by several names, including adjustable clamping pliers, Vise Grip® pliers, and Tag-L-Lock® pliers.

Another type of adjustable pliers, called Robo-Grip® pliers, permits one-handed jaw-size adjustment by merely squeezing the handles. See **Figure 7-15**. This type of adjustable pliers does not have a locking feature.

7.2.1 Care of Pliers

Like many tools, pliers will give long, useful service if a few simple precautions are taken:

- Never use pliers as a substitute for a wrench.
- Do not try to cut metal sizes that are too large, or work that has been heat-treated. Pliers with cutters will deform or break if used in this way. Breakage will also occur if additional leverage is applied to the handles.
- Occasionally, clean and oil pliers to keep them in good working condition.
- Store pliers in a clean, dry place. Avoid throwing them in a drawer or toolbox with other tools.
- Use pliers that are large enough for the job.

Goodheart-Willcox Publisher

Figure 7-15. The newest type of adjustable pliers offers one-handed operation.

7.3 Wrenches

Wrenches comprise a family of tools designed for use in assembling and disassembling many types of threaded fasteners. They are available in a vast number of types and sizes. Only the most commonly used wrenches will be covered.

7.3.1 Torque-Limiting Wrenches

Torque is the amount of turning or twisting force applied to a threaded fastener or part. It is measured in force units of ***foot-pounds*** (ft·lb) or the metric equivalent, ***newton-meters*** (N·m). Torque is the product of the force applied times the length of the lever arm. See **Figure 7-16**.

A torque-limiting wrench allows you to measure the tightening of a threaded fastener in foot-pounds or newton-meters. This provides maximum holding power, without danger of the fastener or part failing or causing the work to warp or spring out of shape. See **Figure 7-17**.

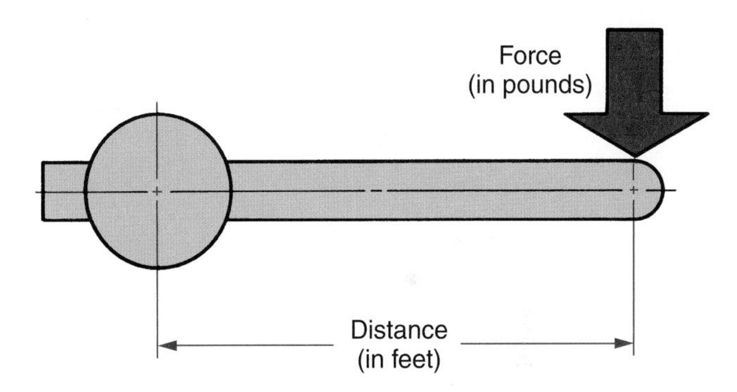

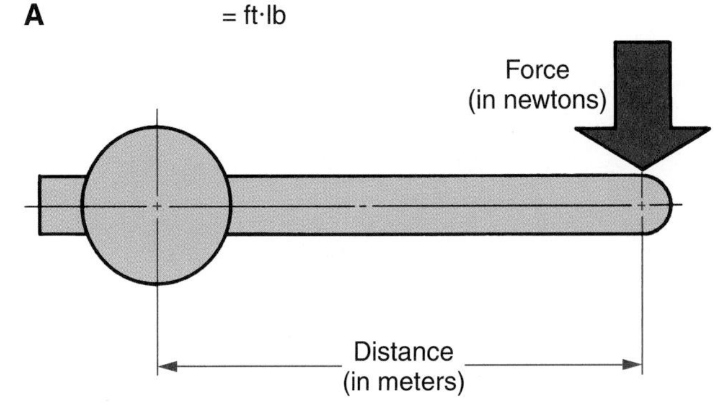

Goodheart-Willcox Publisher

Figure 7-16. Torque measurement. A—In the US Conventional system, torque is measured in foot-pounds. B—Torque values in SI Metric are given in newton-meters.

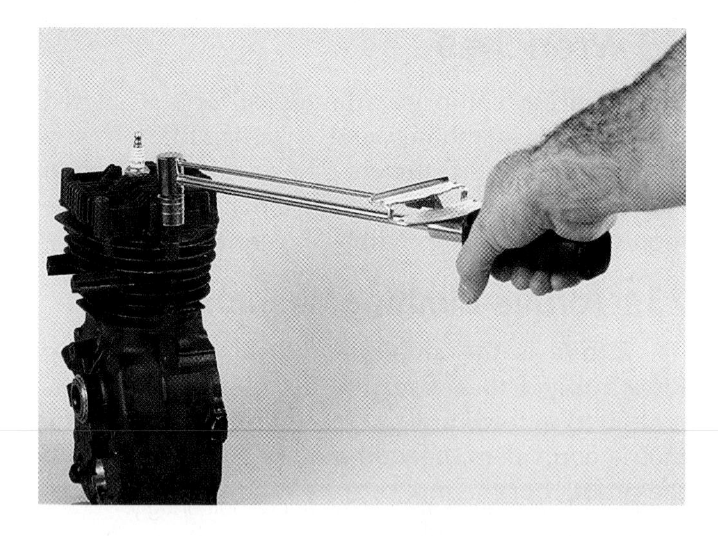

Goodheart-Willcox Publisher

Figure 7-17. Torque-limiting wrenches are used when fasteners must be tightened to within certain limits to prevent undue stresses and strains from developing in the part.

There are many types of torque-limiting wrenches, **Figure 7-18**. It is possible to obtain torque wrenches that are direct reading, or that feature a sensory signal (clicking sound or momentary release) when a preset torque is reached.

The right and wrong methods of gripping the wrench handle are shown in **Figure 7-19**. You should never lengthen the handle for additional leverage. These tools are designed to take a specific maximum force load. Any force over this amount will destroy the accuracy of the wrench.

Goodheart-Willcox Publisher

Figure 7-19. The right and wrong ways to apply pressure to a torque-limiting wrench handle.

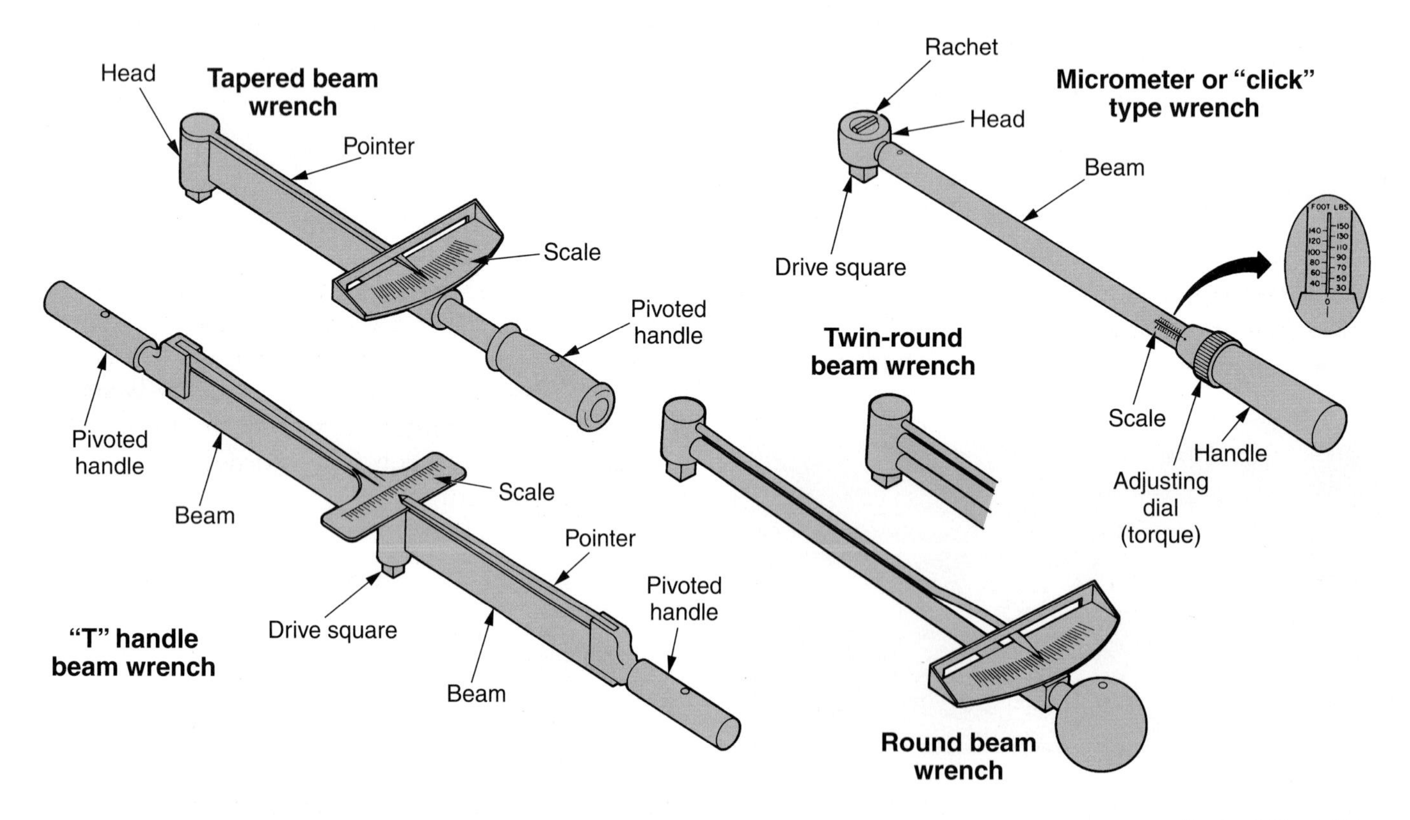

Goodheart-Willcox Publisher

Figure 7-18. Several types of torque-limiting wrenches.

Torque-limiting wrenches will provide accurate measurements whether they are pushed or pulled. However, to prevent hand injury, the preferred method is to pull on the wrench handle.

7.3.2 Adjustable Wrenches

The term "adjustable wrench" could be used to describe several different types of wrenches, such as the "monkey wrench" and pipe wrench. However, the wrench that is somewhat like an open-end wrench, but with an adjustable jaw, is commonly referred to as an adjustable wrench, **Figure 7-20**.

As the name implies, the wrench can be adjusted to fit a range of bolt-head and nut sizes. Although it is convenient at times, the adjustable wrench is not intended to take the place of open-end, box, and socket wrenches.

When using the adjustable wrench, keep the following tips in mind:

- The wrench should be placed on the bolt head or nut so that the movable jaw faces the direction the fastener is to be rotated, **Figure 7-21**.
- Adjust the thumbscrew so the jaws fit the bolt head or nut snugly, **Figure 7-22**.
- Do not place an extension on the wrench handle for additional leverage.

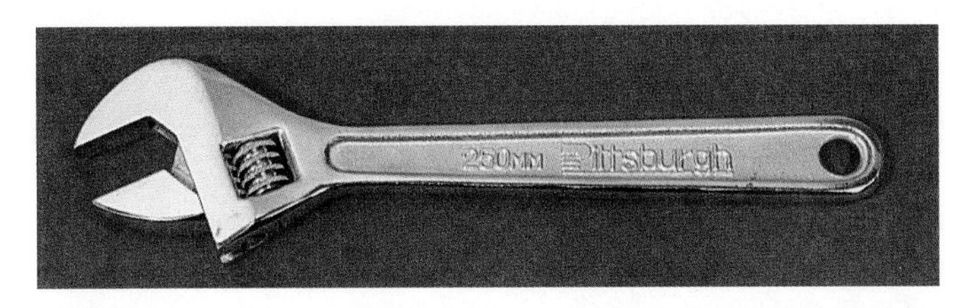

Goodheart-Willcox Publisher

Figure 7-20. An adjustable wrench is handy when a full wrench set is not available.

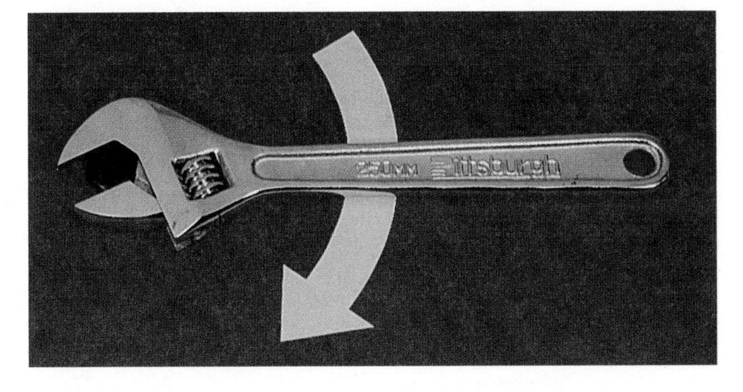

Goodheart-Willcox Publisher

Figure 7-21. The movable jaw of the wrench should always face the direction of rotation.

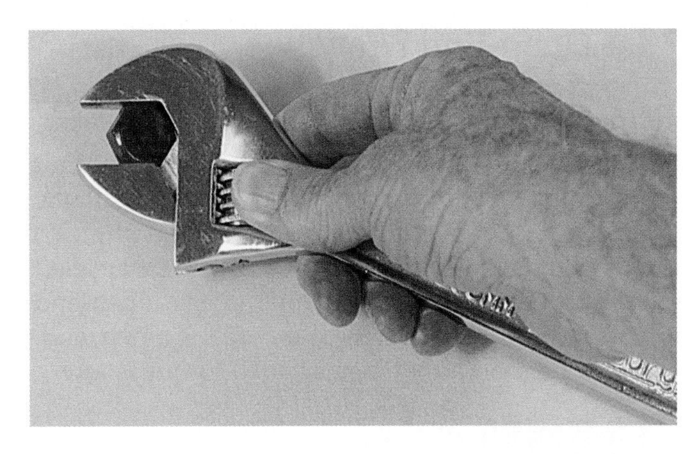

Goodheart-Willcox Publisher

Figure 7-22. A wrench must fit the nut or bolt snugly.

- Never hammer on the handle to loosen a stubborn fastener.
- Use the smallest wrench that will fit the fastener on which you are working. This will minimize the possibility of twisting off the fastener.

SAFETY NOTE

It is dangerous to push on, rather than pull, any wrench. If the fastener fails or loosens unexpectedly, you will almost always strike and injure your knuckles on the work. This operation is commonly known as "knuckle dusting."

7.3.3 Pipe Wrenches

The pipe wrench is designed to grip round stock, **Figure 7-23**. However, the jaws will leave marks on the work. Do not use a pipe wrench on bolt heads or nuts unless they cannot be turned with another type of wrench. For instance, you might need a pipe wrench to remove a bolt if the corners of its head have been rounded.

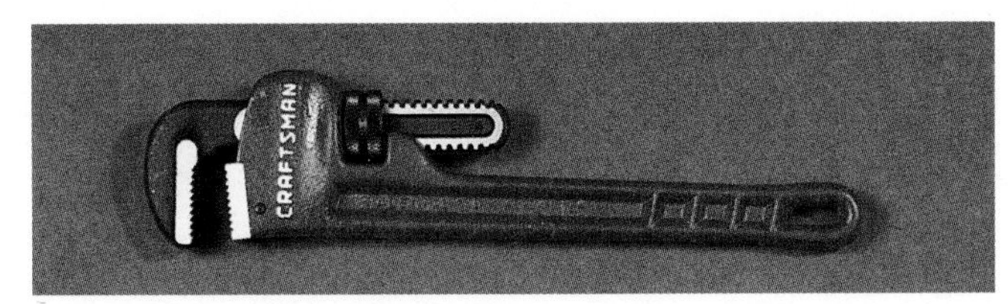

Goodheart-Willcox Publisher

Figure 7-23. The pipe wrench has jaws that will grasp round objects.

7.3.4 Open-End Wrenches

Open-end wrenches are usually double-ended, with two different size openings, **Figure 7-24**. They are made about 0.005″ (0.13 mm) oversize to permit them to easily slip on bolt heads and nuts of the specified wrench size. Openings are at an angle to the wrench body, so that the wrench can be applied in close quarters. Standard and metric open-end wrenches are available. Because of the open end, they can be used only when applied torque is low.

7.3.5 Box Wrenches

The body or jaw of the box wrench completely surrounds the bolt head or nut, so it can be used when higher torque must be applied than is possible with an open-end wrench. See **Figure 7-25**. A properly fitted box wrench will not normally slip. It is preferred for many jobs. Box wrenches are available in the same sizes as open-end wrenches and with straight and offset handles.

7.3.6 Combination Open-End and Box Wrenches

A combination open-end and box wrench has an open-end wrench at one end of the handle and a box wrench at the other end. These wrenches are made in standard and metric sizes, **Figure 7-26**.

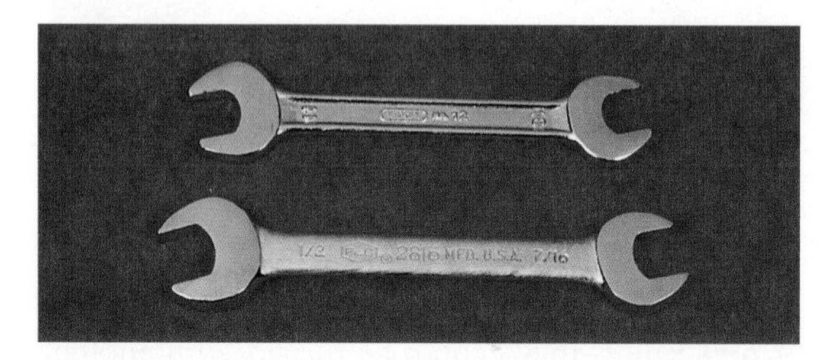

Goodheart-Willcox Publisher

Figure 7-24. An open-end wrench is acceptable when the torque applied is low.

Goodheart-Willcox Publisher

Figure 7-25. A box wrench can handle more torque than an open-end wrench.

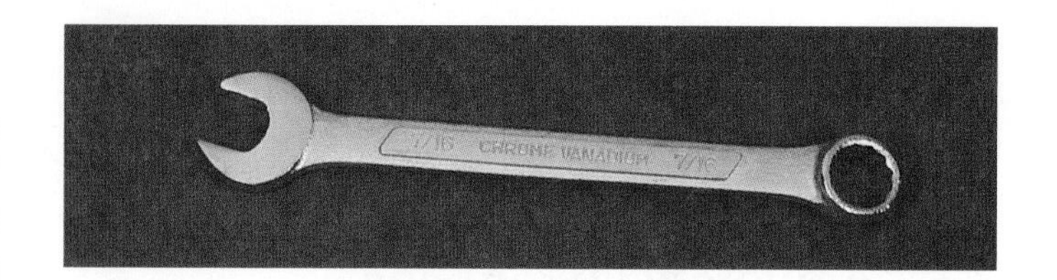

Goodheart-Willcox Publisher

Figure 7-26. Combination wrenches are handy because of two end configurations.

7.3.7 Socket Wrenches

Socket wrenches are like box wrenches and are made with a tool head-socket (opening) that fits many types of handles (either solid bar or ratchet type). A typical socket wrench set contains various handles and a wide range of socket sizes, **Figure 7-27**. Many sets include both standard and metric sockets. Various types of socket openings are shown in **Figure 7-28**.

7.3.8 Spanner Wrenches

Spanner wrenches are special wrenches with drive lugs, and are designed to turn flush- and recessed-type threaded fittings. The fittings have slots or holes to receive the wrench end. They are usually furnished with machine tools and attachments. See **Figure 7-29**.

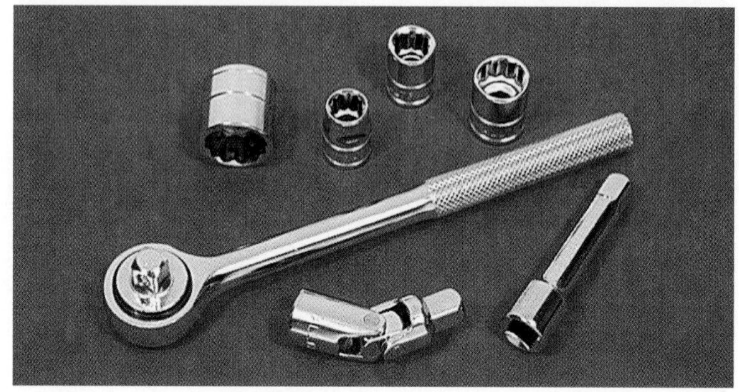

Goodheart-Willcox Publisher

Figure 7-27. A typical socket wrench and sockets. The wrench has a right- and left-hand ratchet mechanism.

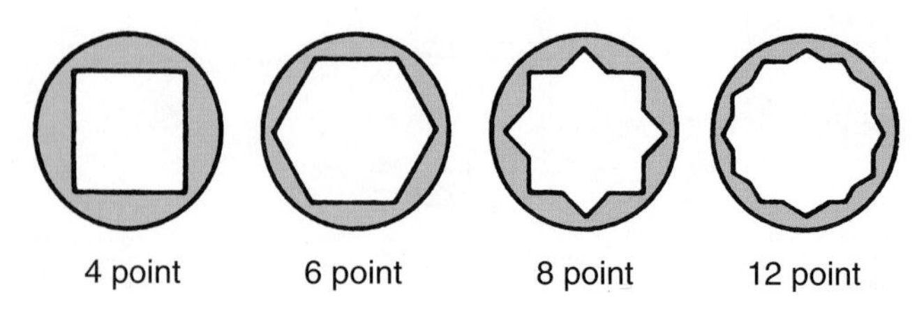

Goodheart-Willcox Publisher

Figure 7-28. Types of socket openings available. The 12-point socket can be used with both square and hex head fasteners.

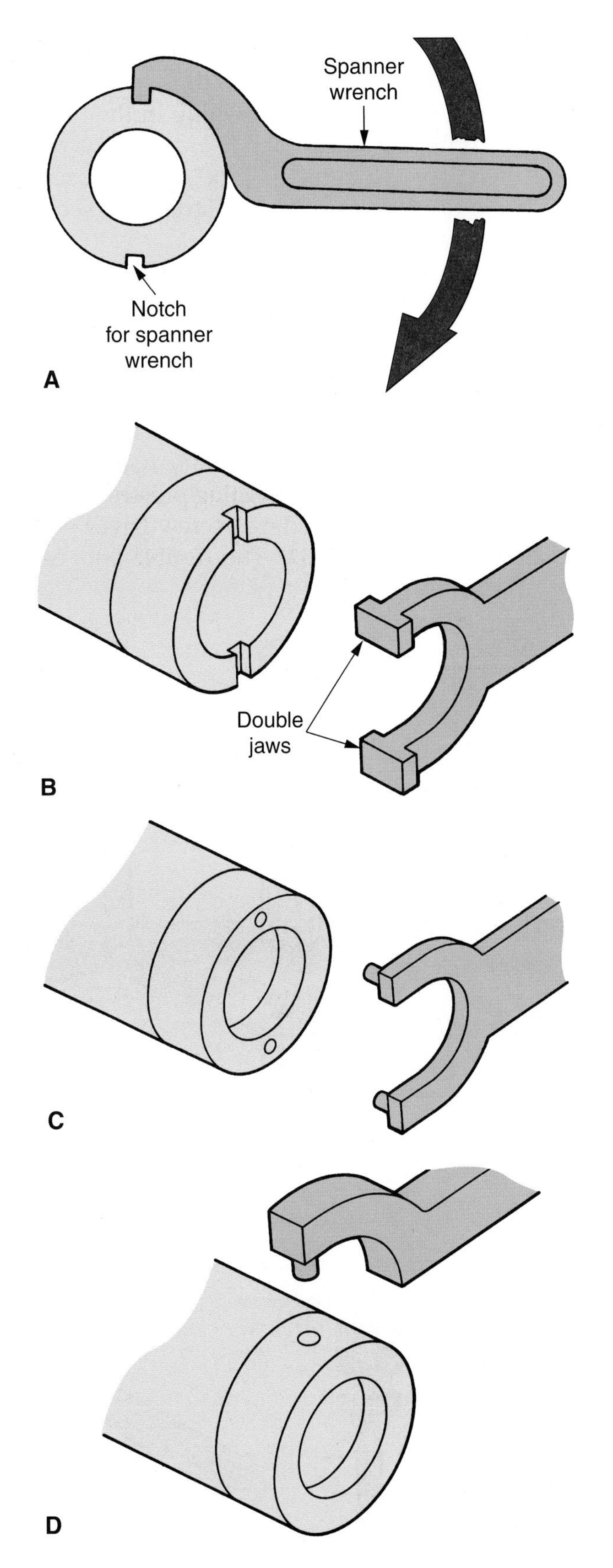

Figure 7-29. Spanner wrenches. A—Hook-type spanner wrench. Some can be adjusted to fit different size fasteners. B—End spanner wrench. C and D—Two types of pin spanner wrenches.

A hook spanner is equipped with a single lug that is placed in a slot or notch cut in the fitting. An end spanner has lugs on both faces of the wrench for better access to the fitting. The lugs fit notches or slots machined into the face of the fitting. On pin spanner wrenches, the lugs are replaced with pins that fit into holes on the fitting, rather than into notches.

7.3.9 Allen Wrenches

The wrench that is used with socket-headed fasteners is commonly known as an Allen wrench, **Figure 7-30**. It is manufactured in many sizes to fit fasteners of various standard and metric dimensions.

7.3.10 Wrench Safety

When using a wrench, take care to keep the following safety precautions in mind:

- Always pull on a wrench; never push. You have more control over the tool and there is less chance of injury.
- Select a wrench that fits properly. A loose-fitting wrench, or one with worn jaws, may slip and cause injury. It can also round off and ruin the bolt or nut on which it is being used.
- Never hammer on a wrench to loosen a stubborn fastener.
- Lengthening a wrench handle for additional leverage is a dangerous practice. Use a larger wrench.
- Before using a wrench, clean any grease or oil off the handle and the floor in the work area. This will reduce the possibility of your hands or feet slipping.
- Never try to use a wrench on moving machinery.

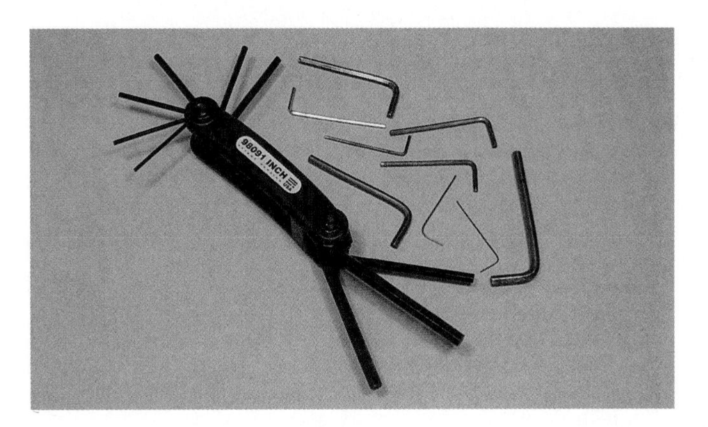

Figure 7-30. Allen wrenches are used with socket-headed fasteners. They are made in both inch and metric sizes.

7.4 Screwdrivers

Screwdrivers are manufactured with many different tip shapes, **Figure 7-31**. Each shape has been designed for a particular type of fastener. The standard and Phillips type screwdrivers are familiar to all shop workers. The other shapes may not be as well-known.

The Phillips screwdriver has a +-shaped tip for use with Phillips recessed head screws. Four sizes (#1, #2, #3, and #4) handle the full range of this type fastener. They are manufactured in the same general styles as the standard screwdriver.

The Pozidriv® screwdriver tip is similar in appearance to the Phillips tip but has a slightly different shape. This type has been designed for Posidriv® screws used extensively in the aircraft, automotive, electronic, and appliance industries.

The tip of this screwdriver has a black oxide finish to distinguish it from the Phillips tool. Using a Phillips tip will damage the opening in the head of the Pozidriv® screw.

Clutch head, Robertson, Torx®, and hex screwdrivers are used for special industrial and security applications.

A standard screwdriver has a flattened wedge-shaped tip that fits into the slot in a screw head. This tool is made in 3″ to 12″ lengths. The shank diameter and the width and thickness of the tip are proportional with the length. Screwdriver length is measured from blade tip to the bottom of the handle. The blade is heat-treated to provide the necessary hardness and toughness to withstand the twisting pressures.

A few of the standard screwdriver types are shown in **Figure 7-32**. The double-end offset

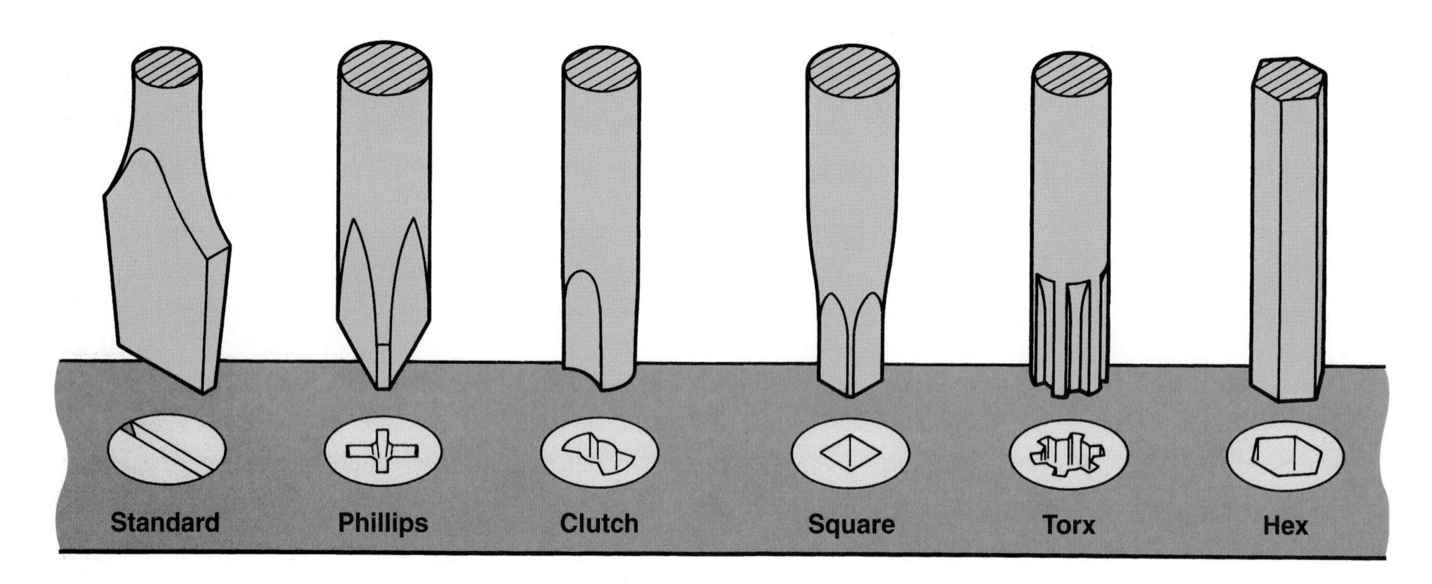

Goodheart-Willcox Publisher

Figure 7-31. Types of screwdriver tips.

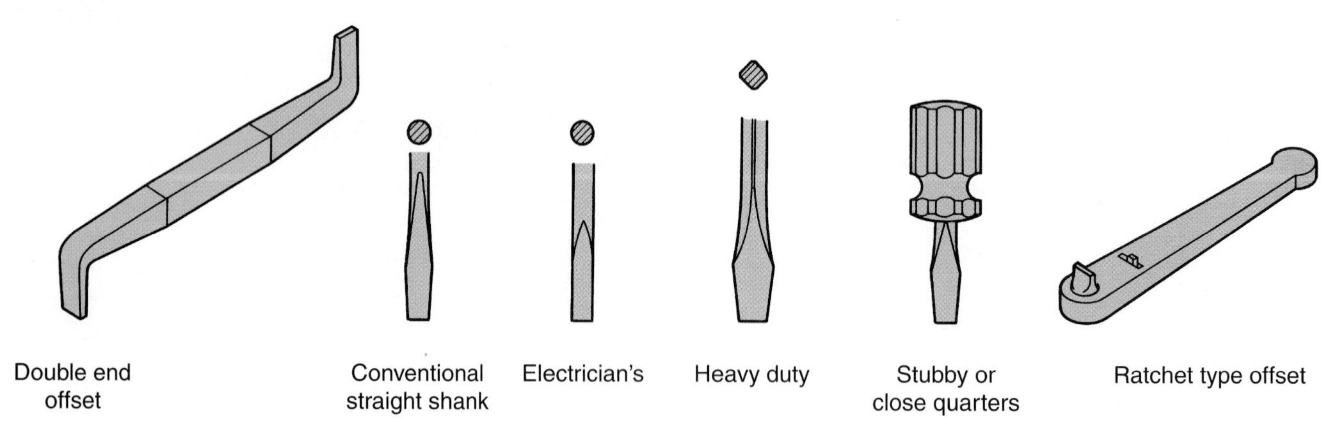

Goodheart-Willcox Publisher

Figure 7-32. Styles and types of standard screwdrivers.

screwdriver can be used where there is not enough space for a conventional straight shank tool. The conventional straight shank screwdriver is widely used for a variety of work. The electrician's screwdriver has a long thin blade and an insulated handle. The long thin blade will reach into tight areas. A heavy-duty screwdriver has a thick, square shank that permits a wrench to be applied for driving or removing large or stubborn screws. The stubby, or close quarters screwdriver, is designed for use where work space is limited. The ratchet screwdriver moves the screw on the power stroke, but not on the return stroke. It can be set for right-hand or left-hand operation.

7.4.1 Using a Screwdriver

Always select the correct size screwdriver for the screw being driven, **Figure 7-33**. A poor fit will damage the screw slot and often will damage the tool's tip. Damaged screw heads are dangerous, and are often difficult to drive or remove. They should be replaced.

When driving or removing a screw, hold the screwdriver square with the fastener. Guide the tip with your free hand.

A worn screwdriver tip, such as the one shown at right in **Figure 7-34**, must be reground. A fine grinding wheel and light pressure is required. Avoid overheating the tip during the grinding operation. It will destroy the tool. Check the tip during the grinding operation by fitting it to a screw slot. A properly ground tip will fit snugly and hold the head firmly in the slot.

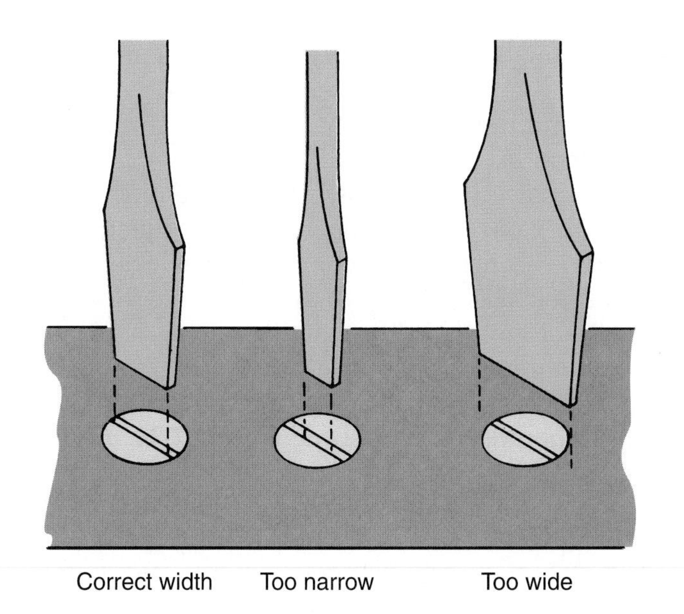

Figure 7-33. Use the correct screwdriver tip for the job.

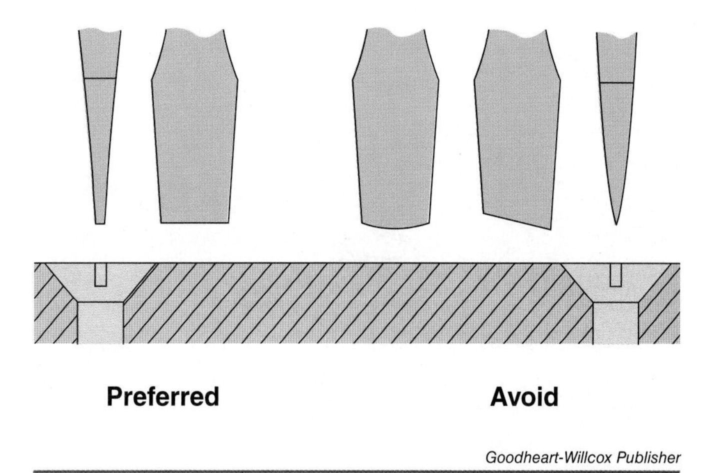

Figure 7-34. Tips on the right are to be avoided. They are worn or improperly sharpened. The tip at left is ground correctly. The sides are concave to help hold the tip in the slot when pressure is applied.

7.4.2 Screwdriver Safety

When using a screwdriver, take care to keep the following safety precautions in mind:

- A screwdriver is not a substitute for a chisel, nor is it made to be hammered on or used as a pry bar.
- Wear safety goggles when regrinding screwdriver tips.
- Screws with burred heads are dangerous. They should be replaced or the burrs removed with a file or abrasive cloth.
- Always turn electric power off before working on electrical equipment. The screwdriver should have an insulated handle specifically designed for electrical work.
- Avoid carrying a screwdriver in your pocket. It is a dangerous practice that can cause injury to you or to someone else.

7.5 Striking Tools

The machinist's ball-peen hammer, **Figure 7-35**, is the most commonly used shop hammer. It has a hardened striking face and is used for all general purpose work that requires a hammer.

Ball-peen hammer sizes are classified according to the weight of the head, without the handle. They are available in weights of 2, 4, 8, and 12 ounces, and 1, 1 1/2, 2, and 3 pounds.

A soft-face hammer or mallet allows heavy blows to be struck without damaging the part or surface.

Goodheart-Willcox Publisher

Figure 7-35. The ball-peen hammer is most common type in a machine shop.

A steel face hammer would damage or mar the work surface. Soft-face hammers are especially useful for setting work tightly on parallels (steel bars) when mounting material in a vise. Soft-face hammers are made of many different materials: copper, brass, lead, rawhide, and plastic. See **Figure 7-36**.

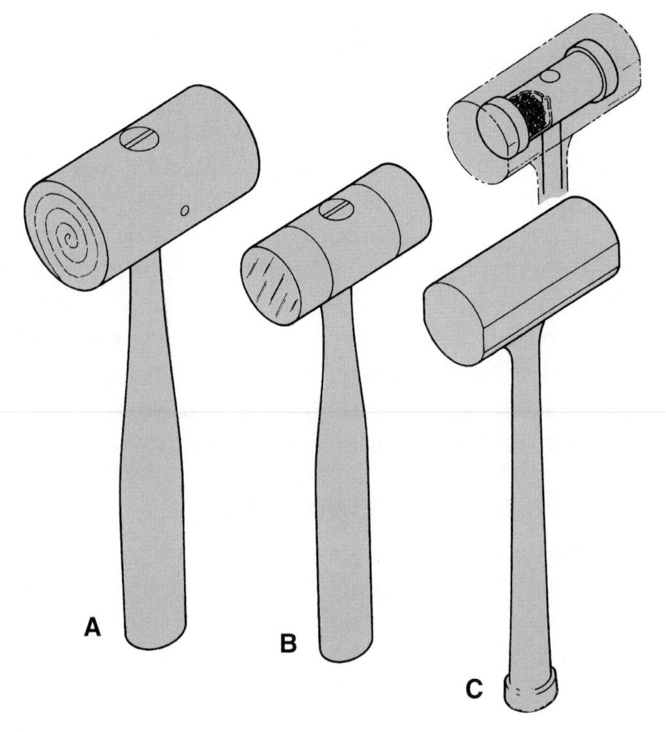

Goodheart-Willcox Publisher

Figure 7-36. Soft-face hammers and mallets. A—Rawhide mallet. B—Plastic-face hammer. C—Dead blow hammer. The head of the dead blow hammer contains tiny steel shot encased in plastic. This provides a hammer that has striking power but will not rebound (bounce back) as will other mallets and soft-face hammers.

7.5.1 Striking Tool Safety

When using a hammer, take care to keep the following safety precautions in mind:

- Never strike two hammers together. The faces are very hard; the blow might cause a chip to break off and fly out at high speed.
- Do not use a hammer unless the head is on tightly and the handle is in good condition.
- Do not hold the handle too close to the head when striking a blow, or you may injure your knuckles.
- Strike each blow squarely, or the hammer may glance off of the work and injure you or someone working nearby.
- Place a hammer on the bench carefully. A falling hammer can cause a painful foot injury, or damage precision tools on the bench.

7.6 Chisels

Not all cutting in metalworking is done by machine. Chisels are one of several basic hand tools, along with hacksaws and files, that are considered cutting implements. These tools, when in good condition, sharpened, and properly handled, are safe to use.

The chisel is used mostly to cut cold metal, hence the term "cold" chisel. The four chisels illustrated in **Figure 7-37** are the most common types. The general term cold chisel is used when referring to these chisels. Other chisels in this category are variations or combinations of these chisels.

The work to be cut will determine how the chisel should be sharpened, **Figure 7-38**. A chisel with a slightly curved cutting edge will work better when cutting on a flat plate. The curved edge will help prevent the chisel from cutting unwanted grooves in the surrounding metal, as when shearing rivet heads. If the chisel is to be used to shear metal held in a vise, the cutting edge should be straight.

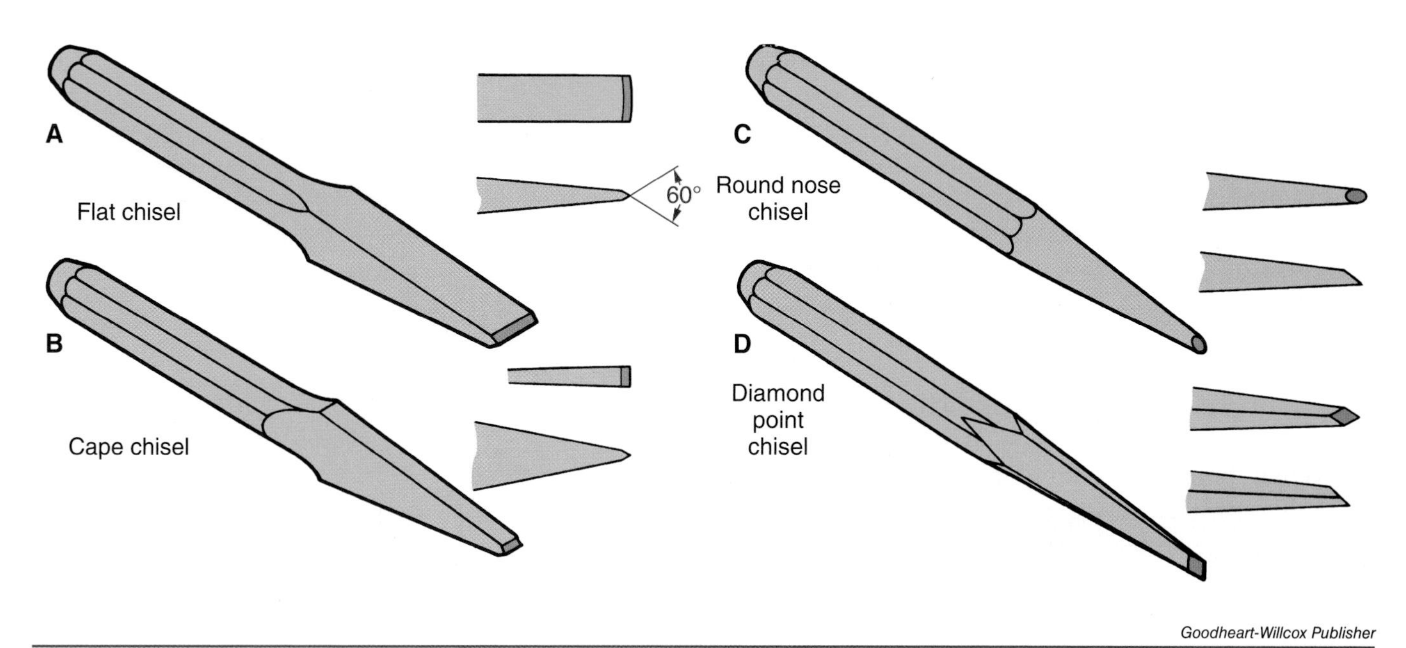

Goodheart-Willcox Publisher

Figure 7-37. Cold chisels. A—Flat chisel is used for general cutting and chipping work. B—Cape chisel has a narrower cutting edge than the flat chisel and is used to cut grooves. C—Round nose chisel can cut radii and round grooves. D—Diamond point chisel is principally used for squaring corners.

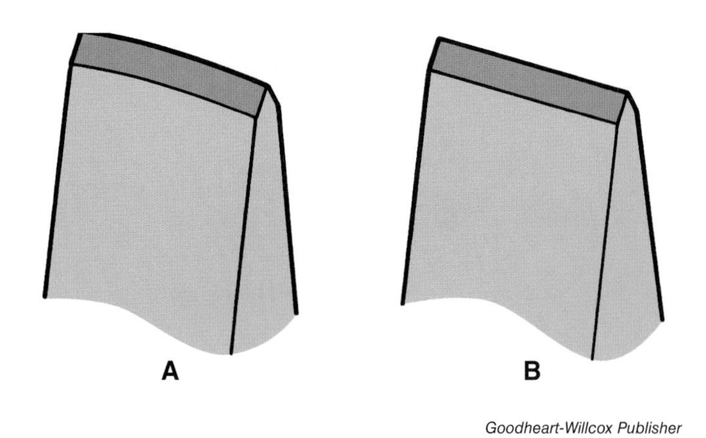

Goodheart-Willcox Publisher

Figure 7-38. The work to be done determines how a chisel should be sharpened. A—Slightly rounded edge for cutting on flat plate. B—Straight edge for shearing.

The chisel is frequently used to chip surplus metal from castings. Chipping is started by holding the chisel at an angle, as shown in **Figure 7-39**. The angle must be great enough to cause the cutting edge to enter the metal.

After the chisel cut has been started and the proper depth reached, the chisel angle can be decreased enough to keep the cutting action at the proper depth. Cut depth can be reduced by decreasing the chisel angle. However, if the cutting angle is decreased too much, the chisel will ride on the heel of the cutting edge and lift out of the cut.

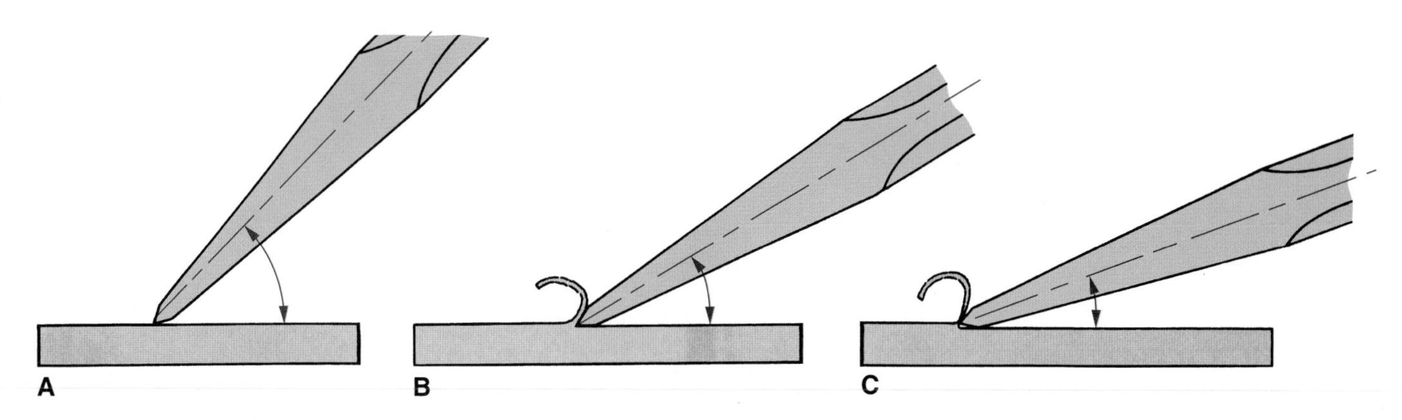

Goodheart-Willcox Publisher

Figure 7-39. Proper chisel angles for various cutting operations. A—Starting the cut. B—Maintaining cut at desired depth. C—Reducing the cutting angle too much will cause chisel to lift out of cut.

When shearing metal in a vise, position it so the layout line is just below the vise jaws. This will leave sufficient metal to finish by filing or grinding. When cutting, it is usually best to hold the metal in a vise without using jaw caps. This provides a better shearing action between the vise jaws and chisel. Advance the chisel after each blow so the cutting is done by the center of the cutting edge.

A chisel is an ideal tool for removing rivets. The head can be sheared off and the rivet punched out. A variation of the conventional cold chisel for removing rivet heads is called a "rivet buster," **Figure 7-40**.

When there is not enough room to swing a hammer with sufficient force to cut a rivet, an alternate procedure can be used, **Figure 7-41A**. Drill a hole, about the size of the rivet body, almost through the head. The head can then be removed easily with the chisel.

If the head is so large that it cannot be removed in one piece, make a saw cut almost through the head, then use the chisel to cut away half the head at a time. **Figure 7-41B** shows how this is done. Rivets can also be removed by drilling and using the narrow cape chisel, **Figure 7-41C**.

7.6.1 Chisel Safety

When using a chisel, take care to keep the following safety precautions in mind:

- Flying chips are dangerous. When cutting metal with a chisel, wear safety goggles and erect a shield around the work. These steps will protect you and people working nearby.
- Hold a chisel so that if you miss it with the hammer, you will not strike and injure your hand. If one is available, use a chisel holder.

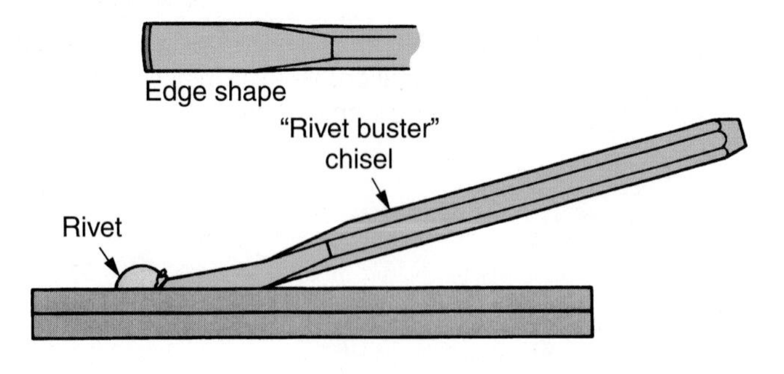

Figure 7-40. This variation of the flat chisel is often referred to as a "rivet buster." Upper drawing shows how it is sharpened.

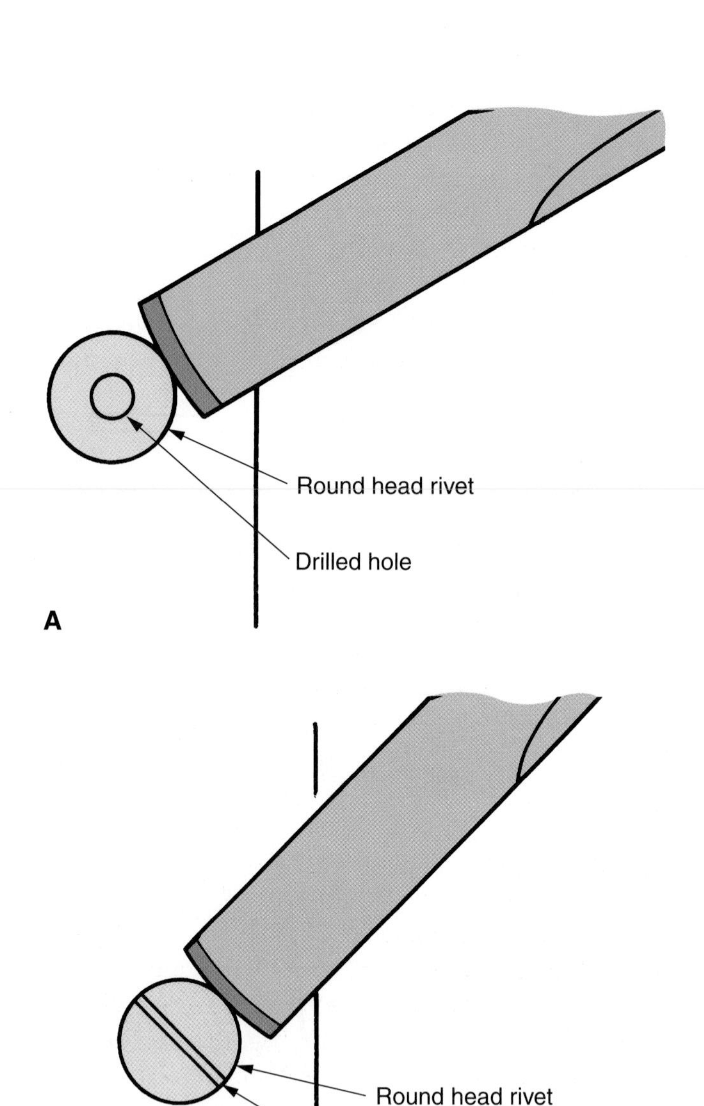

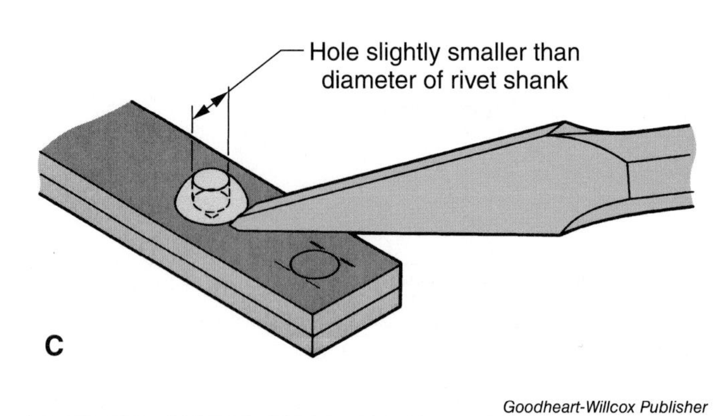

Figure 7-41. Alternate methods for removing rivet heads. A—When there is not enough room to swing hammer with sufficient force. B—When rivet head is too large to be removed as one piece. C—A cape chisel may also be used to remove rivets.

- A mushroomed chisel head is extremely dangerous, since jagged metal can be knocked or chipped off and cause serious injury. This mushroom effect can occur after repeated use with a hammer. Remove this hazardous condition by grinding away the excess metal. See **Figure 7-42**.
- Edges on metal cut with a chisel are sharp and can cause bad cuts. Remove them by grinding or filing.

7.7 Hacksaw

The typical hacksaw is composed of a frame with a handle and a replaceable blade, **Figure 7-43**. Almost all hacksaws made today are adjustable to accommodate several different blade lengths. They are also made so the blade can be installed in either a vertical or horizontal position, **Figure 7-44**.

When placing a blade in the saw frame, make sure the frame is adjusted for the blade length being inserted. There should be sufficient adjustment remaining to permit tightening the blade until it "pings" when snapped with your finger. Frequently, a new blade must be retightened after a few strokes because it will stretch slightly from the heat produced while cutting.

The hacksaw blade must be positioned with the teeth pointing away from the handle, **Figure 7-45**. This will make it cut on the forward (push) stroke.

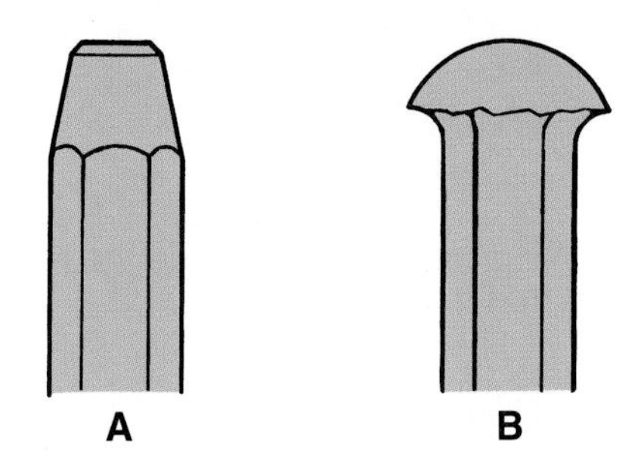

Goodheart-Willcox Publisher

Figure 7-42. Chisel safety. A—Chisel head ground to a safe condition. B—A dangerous mushroomed head.

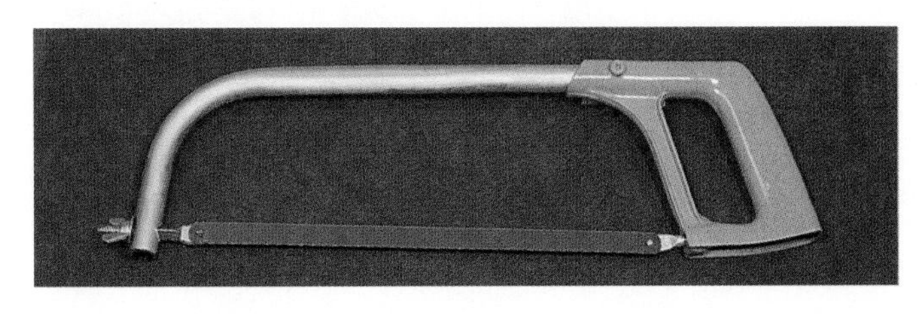

Goodheart-Willcox Publisher

Figure 7-43. A typical hacksaw.

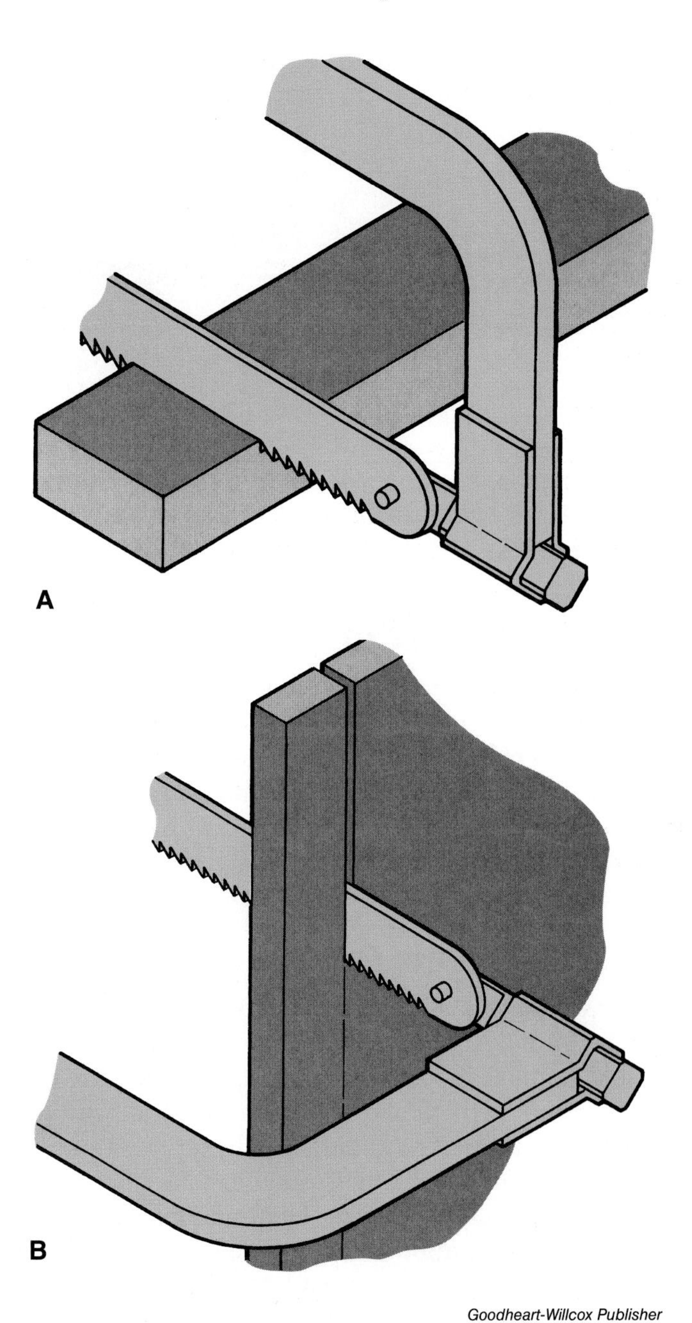

Goodheart-Willcox Publisher

Figure 7-44. Blade positions. A—The blade is set to cut in a conventional vertical position. B—The blade has been pivoted 90° to cut a long narrow strip of stock.

7.7.1 Holding Work for Sawing

The work must be held securely, with the point to be cut as close to the vise as practical. This helps to eliminate "chatter" and vibration that will dull the saw teeth.

Figure 7-46 shows some methods preferred for holding work that is irregular in shape. The work is clamped so the cut is started on a flat side rather than on a corner or edge. This lessens the possibility of ruining the teeth or breaking the blade.

Goodheart-Willcox Publisher

Figure 7-45. A hacksaw blade must be inserted with the teeth pointing away from the handle. This positions it to cut on the forward stroke of the hacksaw.

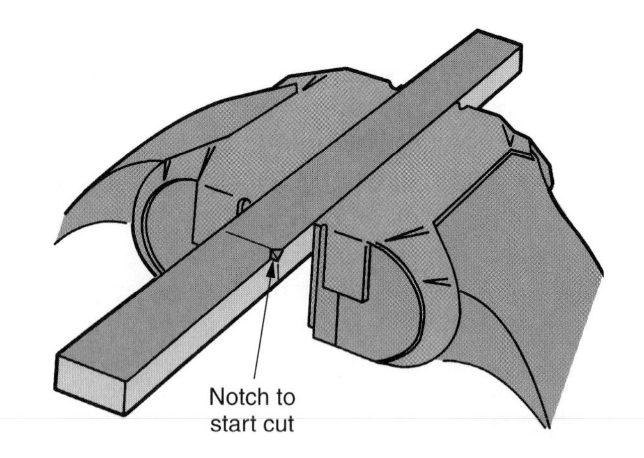

Goodheart-Willcox Publisher

Figure 7-47. Using a file to notch or nick the edge of a piece to be cut permits easier starting of the hacksaw cut.

7.7.2 Starting a Cut

When starting a cut to a marked line, it is best to notch the work with a file, **Figure 7-47**. You can also use the thumb of your left hand to guide the blade until it starts the cut, **Figure 7-48**. Work carefully to avoid injury. Some hacksaw blades are manufactured with very fine teeth at the front to make starting a cut easier. Use enough pressure so the blade will begin to cut immediately.

7.7.3 Making the Cut

Grasp the hacksaw firmly by the handle and front of the frame. Apply enough pressure on the forward stroke to make the teeth cut. Insufficient pressure will permit the teeth to slide over the material, dulling the teeth. Lift the saw slightly on the return stroke, to avoid dragging the teeth across the material.

Cut the full length of the blade and make about 40 to 50 strokes per minute. More strokes per minute may generate enough heat to draw the blade temper and dull the teeth. Keep the blade moving in a straight line.

Goodheart-Willcox Publisher

Figure 7-48. Using your thumb to guide the hacksaw blade as the cut is started.

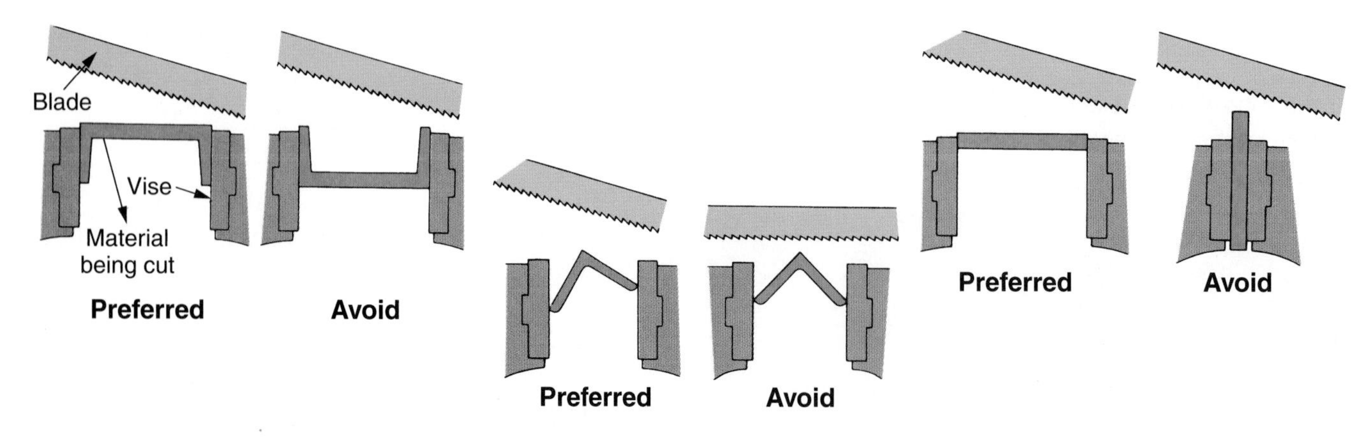

Goodheart-Willcox Publisher

Figure 7-46. Preferred methods of holding irregular stock for sawing.

Avoid any twisting or binding, which can bend or break the blade.

7.7.4 Dull or Broken Blade

If you start a cut with an old blade and the blade breaks or dulls, do not continue in the same cut with a new blade. As a blade becomes dull, the kerf (slot made by blade) becomes narrower. If you try to continue the cut in the same slot, the new blade will usually bind and be ruined in the first few strokes. If possible, rotate the work and start a new cut on the other side.

7.7.5 Finishing a Cut

When the blade has cut almost through the material, saw carefully. Support the stock being cut off with your free hand to prevent it from dropping when the cut is completed.

7.7.6 Saw Blades

All hacksaw blades are heat-treated to provide the hardness and toughness needed to cut metal. The shape and kind of material to be cut has an important bearing on blade choice, in terms of the number of teeth per inch, **Figure 7-49**.

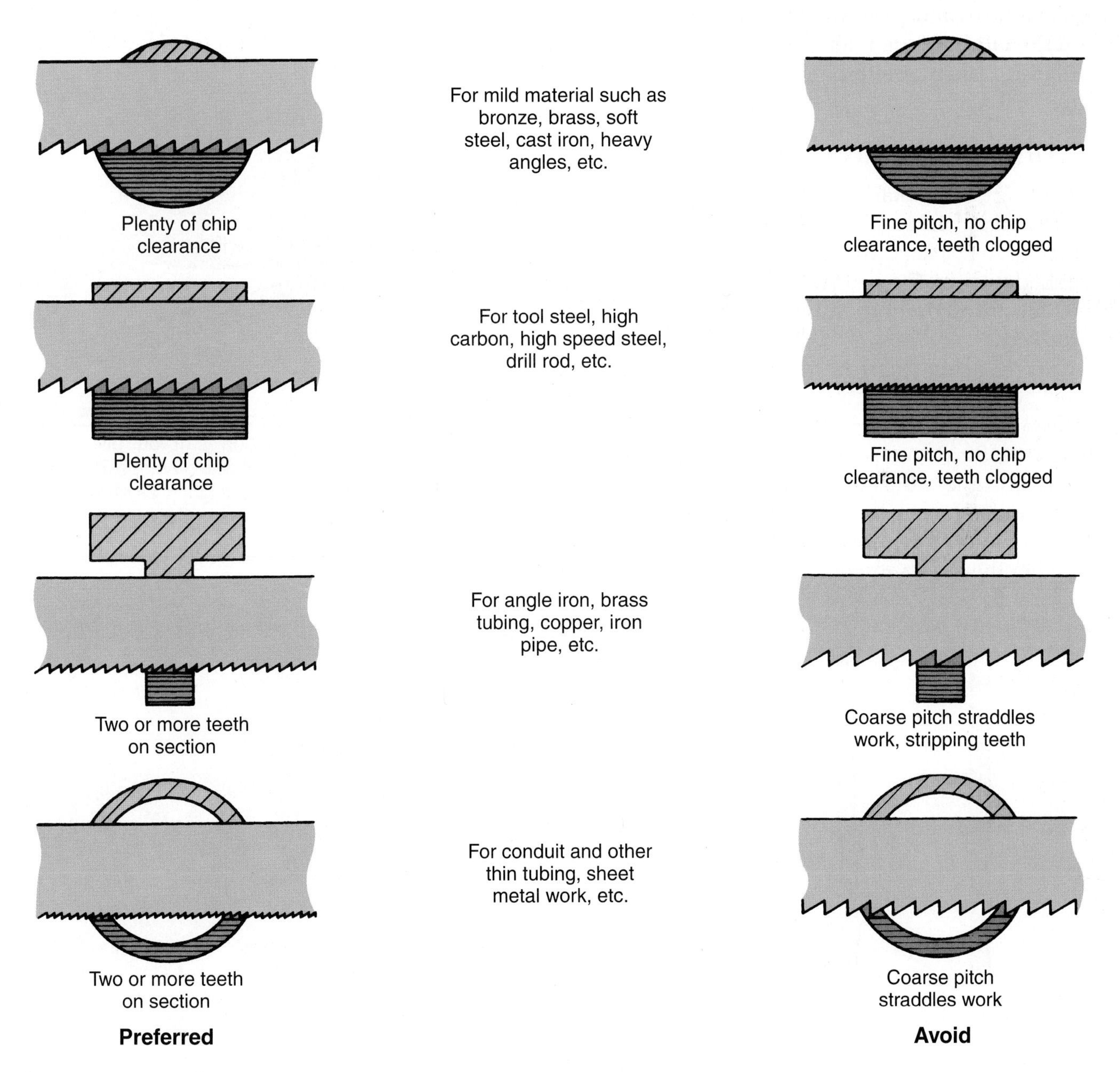

Figure 7-49. The proper hacksaw blade should be used for each job to ensure long blade life and rapid cutting action.

The flexible back blade has only the teeth hardened. The all-hard blade is hardened throughout. The hardness is reduced near the end holes, however, to reduce the possibility of breakage at these points. Flexible back blades are best for sawing soft materials or materials with thin cross sections. An all-hard blade is best for cutting hard metals. It does not buckle when heavy pressure is applied.

At least three teeth should be cutting at all times; otherwise, the teeth will straddle the section being cut and snap off when cutting pressure is applied. The set of the blade provides the necessary clearance, and prevents the blade from binding in the cut. A blade may have one of three sets: undulating, raker, or alternate. These are shown in **Figure 7-50**.

7.7.7 Unusual Cutting Situations

Cutting soft metal tubing can be a problem. The blade may bind and tear the tubing or the tubing may flatten. This can be eliminated by inserting a wood dowel of the proper size into the tubing. Then cut through both tubing and dowel. See **Figure 7-51**.

Cutting a long narrow strip from thin metal can be done by setting the blade at right angles to the frame. Make the cut in the usual way, as in **Figure 7-52**. Strips of any width, up to the capacity of the saw frame, can be made in this manner. Thin metal can also be cut more easily and precisely by putting it between two pieces of wood, and cutting through both wood and metal, **Figure 7-53**.

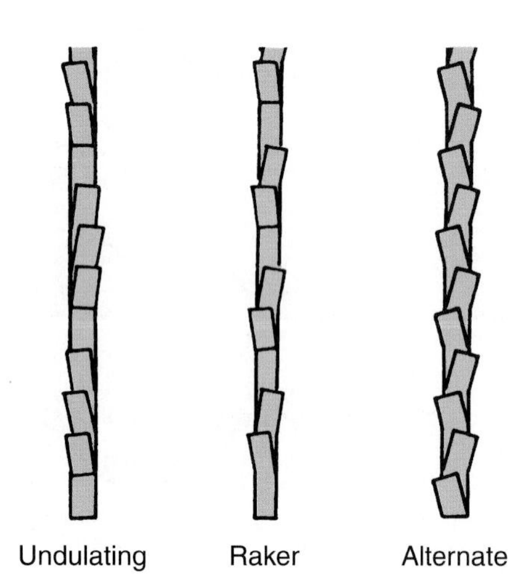

Goodheart-Willcox Publisher

Figure 7-50. Types of sets in hacksaw teeth.

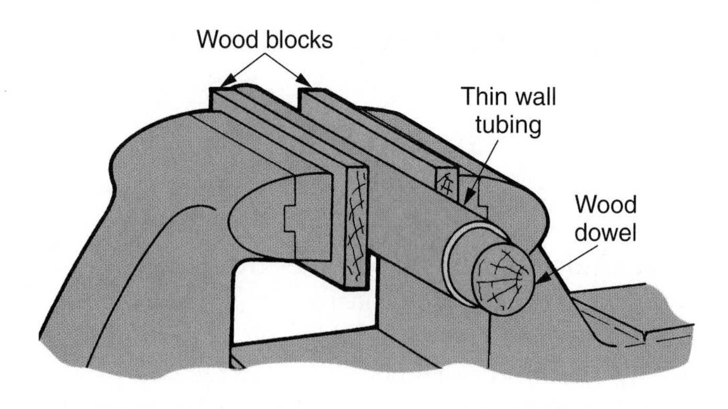

Goodheart-Willcox Publisher

Figure 7-51. A snug-fitting dowel slid into thin-wall tubing will make cutting the tubing easier. If tubing is to be held in a vise for cutting, place soft wood blocks between the vise jaws and the work to prevent marring the exterior surface of the tubing.

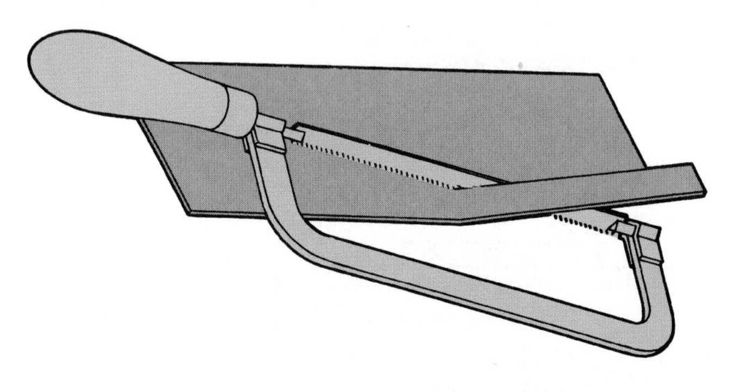

Goodheart-Willcox Publisher

Figure 7-52. The hacksaw blade can be pivoted to a horizontal position for cutting long narrow strips. Best results can be obtained if strip is bent up slightly, as shown, during the sawing operation.

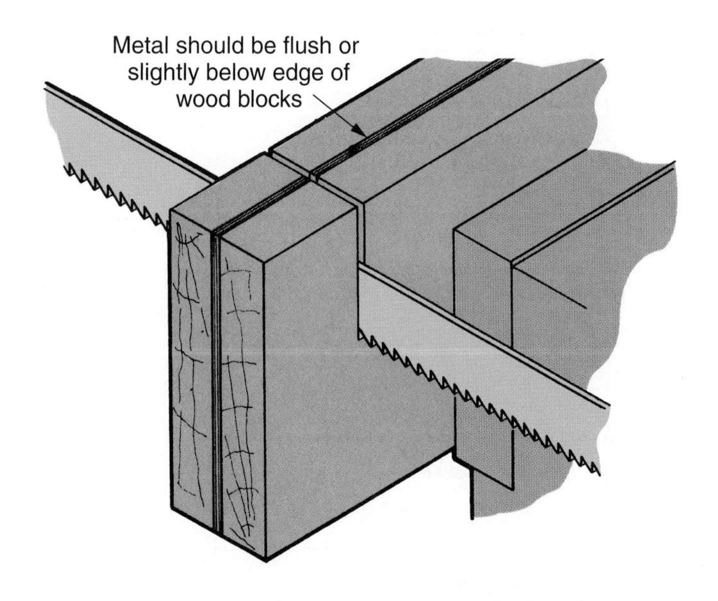

Goodheart-Willcox Publisher

Figure 7-53. Sandwiching thin metal between two pieces of wood will make sawing easier and more precise.

7.7.8 Hacksaw Safety

When using a hacksaw, take care to keep the following safety precautions in mind:

- Never test the sharpness of a blade by running your fingers across its teeth.
- Store saws in a way that will prevent accidentally grasping the teeth when you pick up a saw.
- Burrs formed on the cut edge of metal are sharp and can cause a serious cut.
- Do not brush away chips with your hand; use a brush.
- Always wear safety goggles while using a hacksaw. All-hard blades can shatter and produce flying chips.
- Be sure the hacksaw blade is properly tensioned. If it should break while you are on the cutting stroke, your hand may strike the work, causing a painful injury.

7.8 Files

A file is used for hand smoothing and shaping operations. The modern file is made from high-grade carbon steel and is heat-treated to provide the necessary hardness and toughness.

In manufacturing a file, the first production step is to cut the blank to approximate shape and size. The tang (the part of the file inserted into the handle) and point are formed next. Then, the blank is annealed and straightened. The point and tang are trimmed after the sides and faces have been ground and the teeth cut. After another straightening, the file is heat-treated, cleaned, and oiled. Tests are made continually to ensure a quality tool.

7.8.1 File Classifications

Files are classified by their shape. The shape is the general outline and cross section. The outline is either tapered or blunt, **Figure 7-54.**

Files are also classified according to the cut of the teeth: single-cut, double-cut, curved tooth, and rasp, **Figure 7-55**, and to the coarseness of the teeth: rough, coarse, bastard, second-cut, smooth, and dead smooth.

7.8.2 File Care

SAFETY NOTE

A file should never be used without a handle. It is too easy to drive the unprotected tang into your hand.

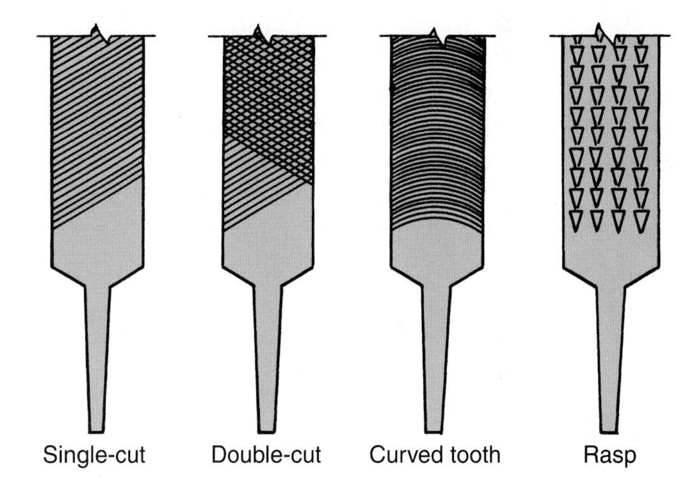

Goodheart-Willcox Publisher

Figure 7-55. File cut classifications.

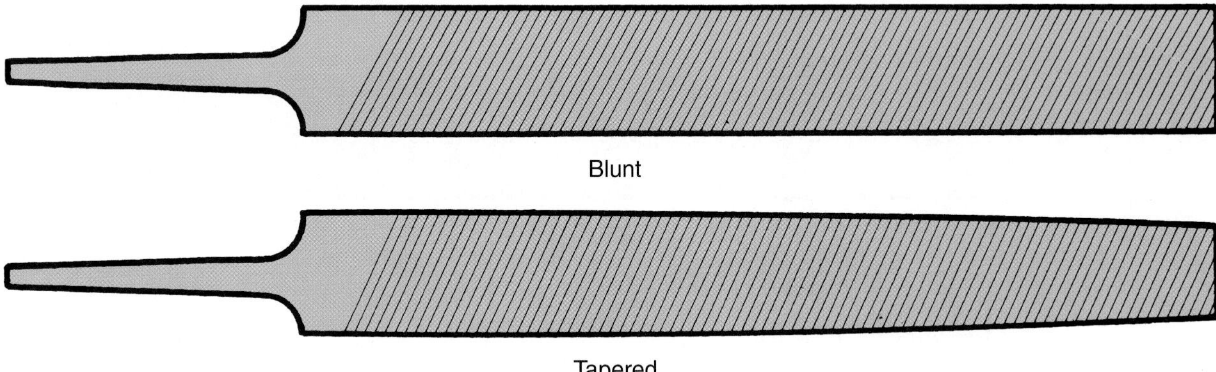

Goodheart-Willcox Publisher

Figure 7-54. File shape classifications.

A handle can be fit to the file by drilling a hole in the handle. The hole should equal in diameter the width of the tang at its midpoint, **Figure 7-56**. Mate the file and handle by inserting the tang in the hole, then sharply striking the handle on a solid surface.

Store files so that they are always separated, **Figure 7-57**. Never throw files together in a drawer or store them in a damp place. Clean files frequently with a file card or brush, **Figure 7-58**. Some soft metals cause pinning, a condition in which the teeth become loaded or clogged with material the file has removed. Pinning will cause gouging and scratching of the work surface. The particles can be removed from the file with a pick or scorer. A file card combines the card, brush, and pick for file cleaning.

7.8.3 Selecting a File

The number of different kinds, shapes, and cuts of files that are manufactured is almost unlimited. For this reason, only the general classification of files will be covered.

Files have three distinct characteristics: length, kind or type, and cut. The file length is always measured from the heel to the point, **Figure 7-59**. The tang is not included in the measurement.

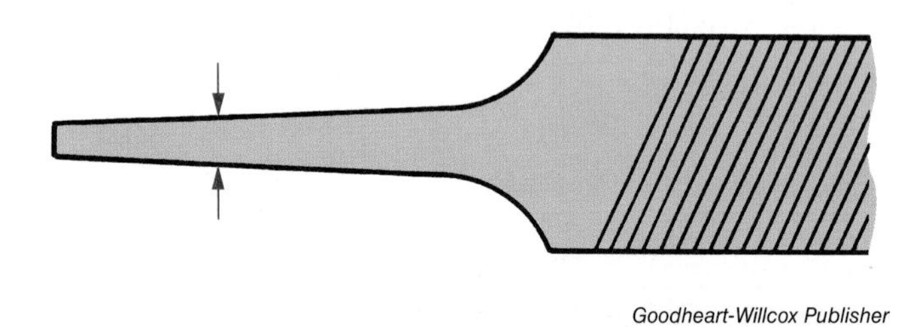

Goodheart-Willcox Publisher

Figure 7-56. File handle hole should be equal in diameter to width of file tang at the point indicated.

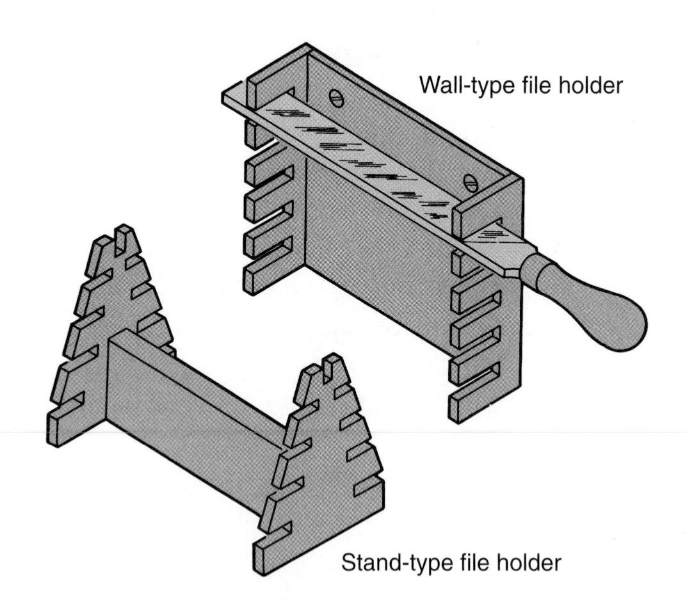

Goodheart-Willcox Publisher

Figure 7-57. Storing files properly, in holders like these, will greatly extend their useful life.

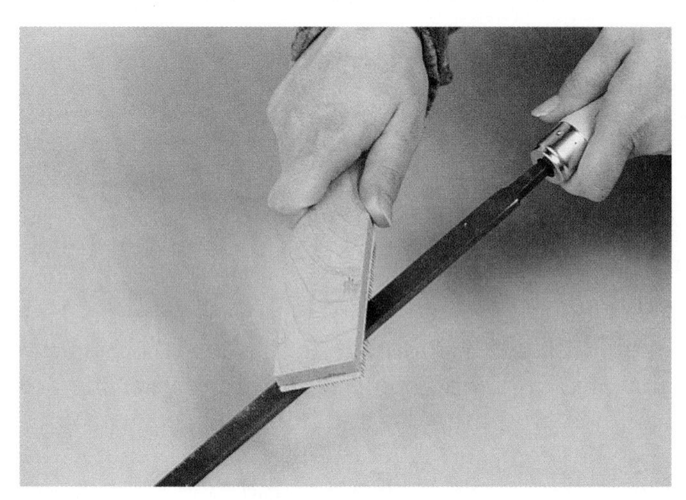

Goodheart-Willcox Publisher

Figure 7-58. Using file card to clean a file.

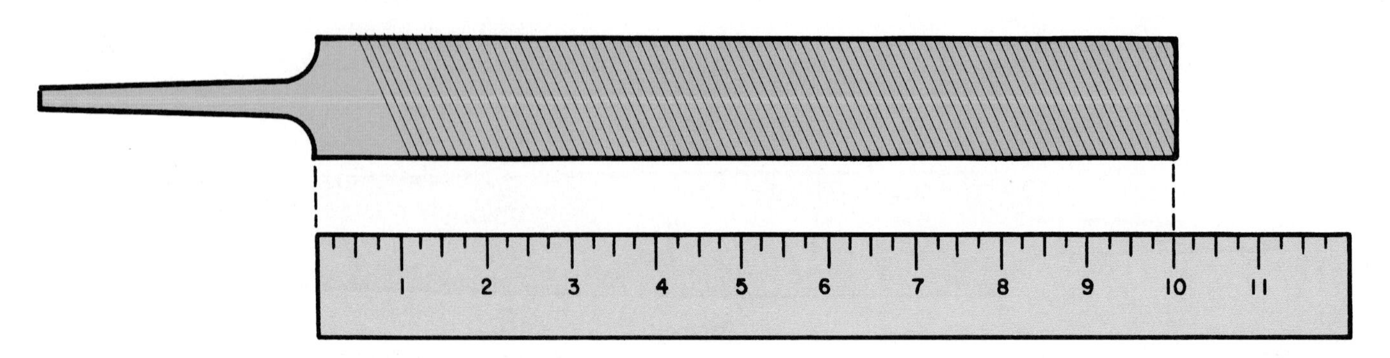

Goodheart-Willcox Publisher

Figure 7-59. How a file is measured.

The file type refers to its shape, such as flat, mill, half-round, or square. The file cut indicates the relative coarseness of the teeth:

- Single-cut files are usually used to produce a smooth surface finish. They require only light pressure to cut.
- Double-cut files remove metal much faster than single-cut files. They require heavier pressure and they produce a rougher surface finish.
- Rasps are best for working wood or other soft materials where a large amount of stock must be removed in a hurry.
- Curved-tooth files are used to file flat surfaces of aluminum and sheet steel.

Some files have ***safe edges***. This denotes that the file has one or both edges without teeth, **Figure 7-60**. This permits filing corners without danger to the portion of the work that is not to be filed.

In selecting the file, many factors must be considered if maximum cutting efficiency is to be attained:

- The nature of the work (flat, concave, convex, notched, etc.).
- Kind of material.
- Amount of material to be removed.
- Surface finish and accuracy demanded.

Of the many file shapes available, the most commonly used are flat, pillar, square, 3-square, knife, half-round, crossing, and round, **Figure 7-61**. Each shape is available in many sizes and degrees of coarseness: rough, coarse, bastard, second-cut, smooth, and dead smooth. Because file coarseness is related to length, **Figure 7-62**, a small (4″) rough cut file may be as fine as a large (16″) second-cut file.

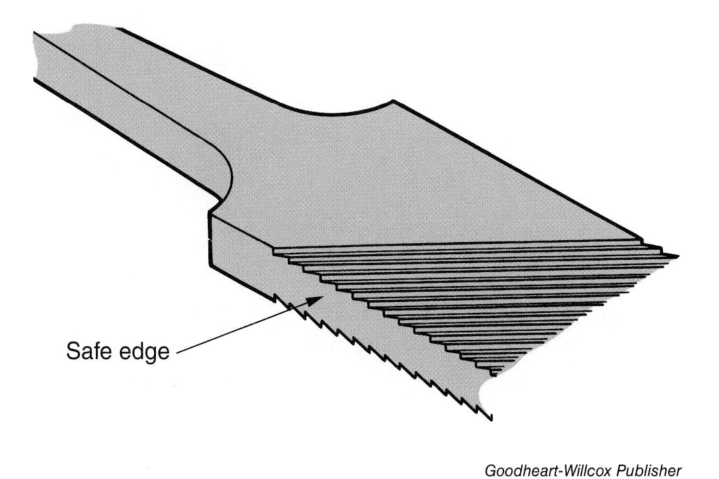

Goodheart-Willcox Publisher

Figure 7-60. The safe edge of a file does not have teeth.

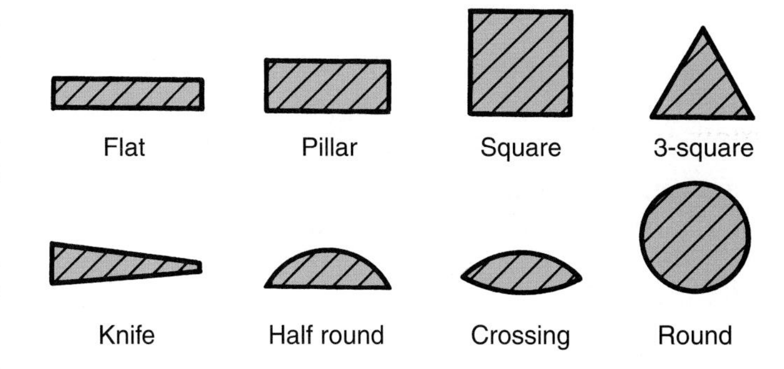

Goodheart-Willcox Publisher

Figure 7-61. Cross-sectional views of the most widely used file shapes.

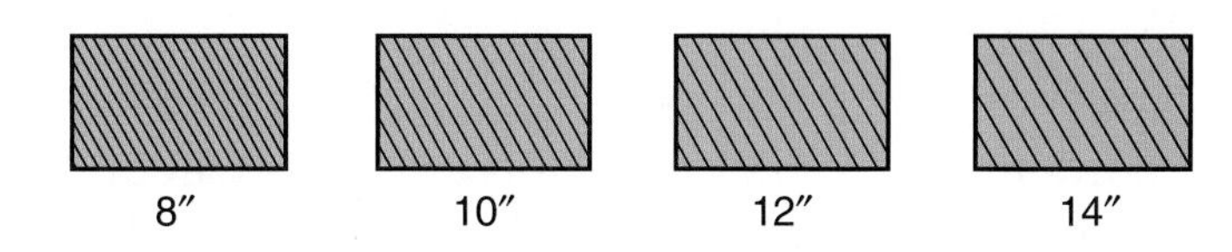

Goodheart-Willcox Publisher

Figure 7-62. File coarseness is directly related to file length. The longer the file, the coarser the teeth.

7.8.4 Types of Files

The wide variety of files can be divided into five general groups:

- The machinists' file is used whenever metal must be removed rapidly, and the finish is of secondary importance. It is made in a large range of shapes and sizes, and is double-cut.
- The mill file is a single-cut and tapers for the last third of its length away from the tang. It is suitable for general filing when a smooth finish is required. A mill file works well for draw filing, lathe work, and working on brass and bronze.
- Swiss pattern and jewelers' files are manufactured in more than a hundred different shapes. They are used primarily by tool-and-die makers, jewelers, and others who do precision filing.
- The rasp has teeth that are individually formed and disconnected from each other. It is used for relatively soft materials (plastic for example) when large quantities of the material must be removed.
- The special purpose files group includes those specifically designed to cut one type of metal or for one kind of operation. An example is the long-angle lathe file, which does an efficient filing job on the lathe.

7.8.5 Using a File

Efficient filing requires that the work be held solidly. Where practical, hold the work at about elbow height for general filing, **Figure 7-63**. If large quantities of metal must be removed by heavy filing, mount the work slightly lower.

Straight or cross filing consists of pushing the file lengthwise across the work, either straight ahead or at a slight angle. Grasp the file as shown in **Figure 7-64A**. Heavy-duty filing requires heavy pressure and can best be done if the file is held as in **Figure 7-64B**.

A file can be ruined by using either too much pressure or too little pressure on the cutting stroke. Apply just enough pressure to permit the file to cut on the entire forward stroke. Too little pressure allows the file to slide over the work. This will dull the file. Too much pressure "overloads" the file, causing the teeth to clog and chip.

Lift the file from the work on the reverse stroke, except when filing soft metal. The pressure on the return stroke when filing soft metal should be no more than the weight of the file itself.

Draw filing, when properly done, will produce a finer finish than straight filing. Hold the file as shown in **Figure 7-65**. Do not use a short-angle file for draw filing. The short-angle file can cause scoring or scratching, instead of the desired shaving and shearing action, as the file is pushed and pulled across the work. Use a double-cut file to "rough down" the surface, then a single-cut file to produce the final finish.

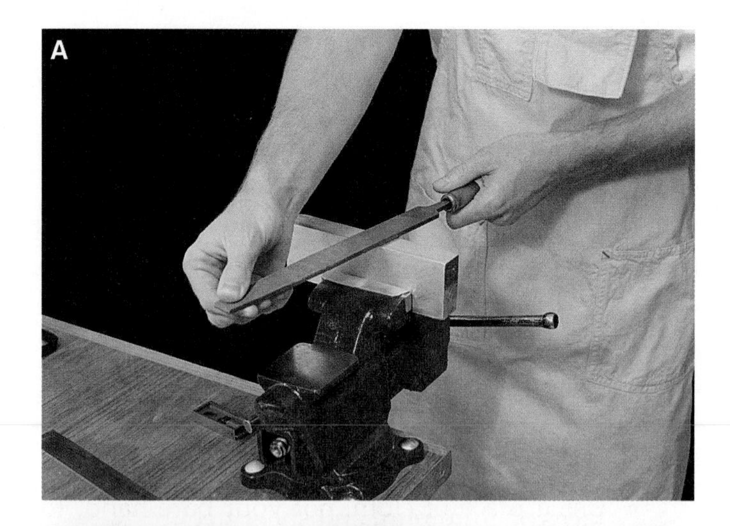

Goodheart-Willcox Publisher

Figure 7-64. Holding a file. A—The proper way to hold a file for straight or cross filing. B—Holding method used to apply the additional pressure required when a considerable quantity of metal must be removed.

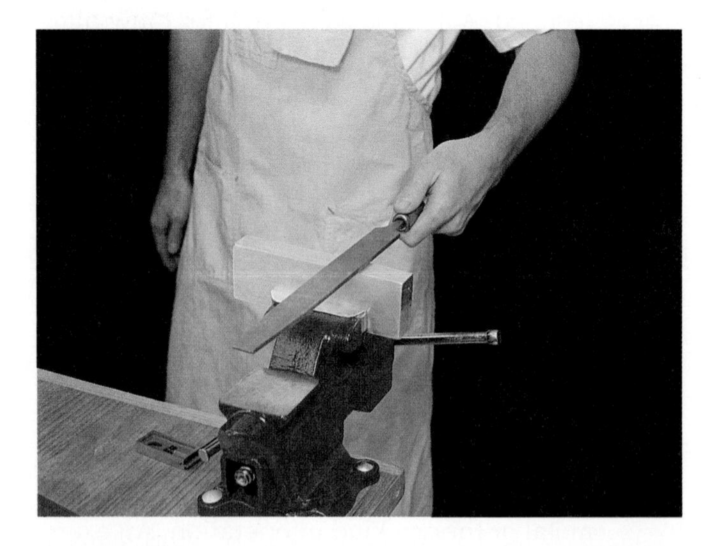

Goodheart-Willcox Publisher

Figure 7-63. Mount work at elbow height for general filing.

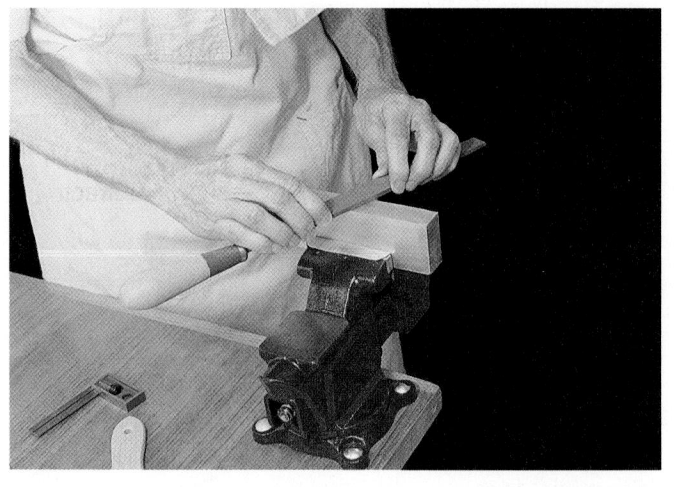

Goodheart-Willcox Publisher

Figure 7-65. Draw filing, when done properly, improves the surface finish.

7.8.6 File Safety

When using a file, take care to keep the following safety precautions in mind:

- Never use a file without a handle. Painful injuries may result.
- Clean files with a file card, not your hand. The chips can penetrate your skin and result in an infection.
- Do not clean a file by slapping it on the bench, this may cause the file to shatter.
- Files are very brittle. Never use one for prying tasks.
- Use a piece of cloth, not your bare hand, to clean the surface being filed. Sharp burrs are formed in filing and can cause serious cuts.
- Never hammer on or with a file. It can shatter, causing chips to fly in all directions.

7.9 Reamers

A drill does not produce a smooth or accurate enough hole for a precision fit. A ***reamer*** is a tool that will produce smoothness and accuracy. Ordinarily, hand reaming is used only for final sizing of a hole.

7.9.1 Hand Reamer

A hand reamer has a square shank end so that it can be held in a tap wrench. See **Figure 7-66A**. Reamers may be made of high-speed steel or carbon steel, and are available in sizes from 1/8″ to 1 1/2″ (3.175 mm to 38.1 mm). The cutting end is ground with a slight taper to provide easy starting in the hole.

Straight-fluted reamers are suitable for most work. However, when reaming a hole with a keyway or other interruption, it is better to have a spiral-fluted reamer. When preparing a piece to be reamed by hand, 0.005″ to 0.010″ (0.15 mm to 0.25 mm) of stock should be left in the hole for removal by the reaming tool.

An expansion hand reamer is used when a hole must be cut a few thousandths inch over nominal size for fitting purposes, **Figure 7-66B**. Slots are cut into the center of the tool. The center opening is machined on a slight taper. The reamer is expanded by tightening a taper screw into this opening. The amount of expansion is limited; the reamer may be broken if expanded too much. Because of the danger of producing oversize holes, do not use an expansion reamer instead of a solid reamer unless absolutely necessary.

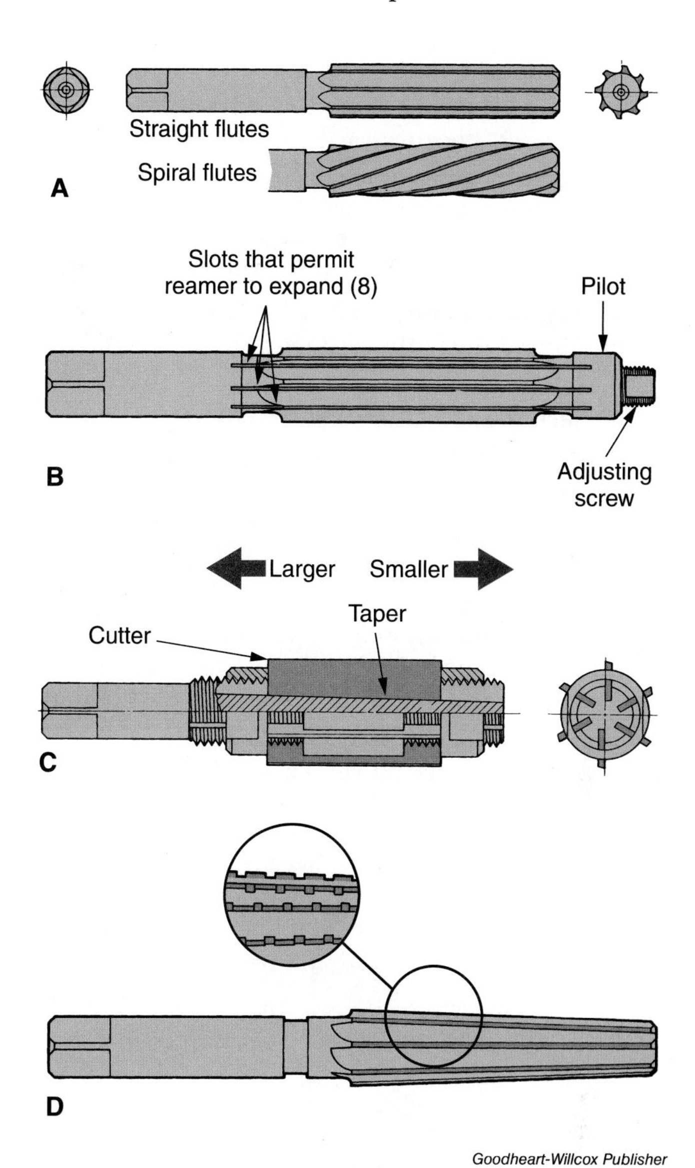

Figure 7-66. Hand reamers. A—Straight flute and spiral flute solid reamers are used for different applications. B—The expansion hand reamer and how its size is adjusted. C—The adjustable hand reamer can be set for odd sizes. D—A taper hand reamer. Enlarged section shows how the cutting edges are notched on a roughing taper reamer.

The adjustable hand reamer is threaded its entire length and is fitted with tapered slots to receive the adjustable blades, **Figure 7-66C**. The blades are tapered along one edge to correspond with the taper slots in the reamer body, so that the cutting edges of the blades remain parallel.

Reamer diameter is set by loosening one adjusting nut and tightening the other. The blades can be moved in either direction. This type of reamer is manufactured in sizes ranging from 3/8″ to 3 1/2″ (9.5 mm to 85 mm). Each reamer has sufficient adjustment to increase its diameter to the next larger reamer size.

The taper reamer will finish a tapered hole accurately and with a smooth finish for taper pins, **Figure 7-66D**. Because of their long cutting edges, taper reamers are somewhat difficult to operate.

To provide for easier removal of surplus metal, a roughing reamer is first rotated into the hole. This reamer is slightly smaller (0.010″ or 0.25 mm) than the finish reamer. Left-hand spiral grooves are ground along the cutting edges to break up chips.

7.9.2 Using a Hand Reamer

A two-handle tap wrench is commonly used with a reamer, because it permits an even application of pressure. It is virtually impossible to secure a satisfactory hole using an adjustable wrench to turn the reamer.

To start reaming, rotate the tool slowly to allow it to align with the hole. Check several points around the reamer's circumference, to make sure that the reamer has started square, **Figure 7-67**.

Rotation should be steady and rapid. Keep the reamer cutting, or it will start to "chatter," producing a series of tool marks in the surface of the hole. This could also cause the hole to be out-of-round.

Turning pressure is applied evenly with both hands, always in a clockwise direction, **Figure 7-68**. Never turn a reamer in a counterclockwise direction, since this will dull the cutting edges.

Feed the reamer deeply enough into the hole to take care of the starting taper. The choice of cutting fluid to be applied will depend on the metal being reamed.

7.9.3 Reaming Safety

When using a hand reamer, take care to keep the following safety precautions in mind:

- To prevent injury, remove all burrs from holes.
- Never use your hands to remove chips and cutting fluid from the reamer. Use a piece of cotton waste.
- Store reamers carefully so they do not touch one another. Never store reamers loose or throw them into a drawer with other tools.
- Clamp work solidly before starting to ream.
- Do not use compressed air to remove chips and cutting fluid or to clean a reamed hole.

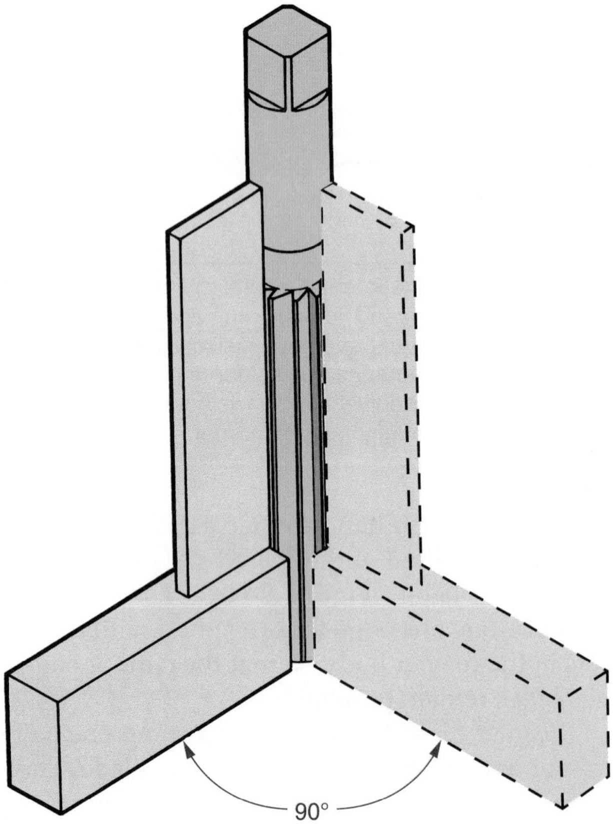

90°

Goodheart-Willcox Publisher

Figure 7-67. Always make sure that the reamer is square with the work.

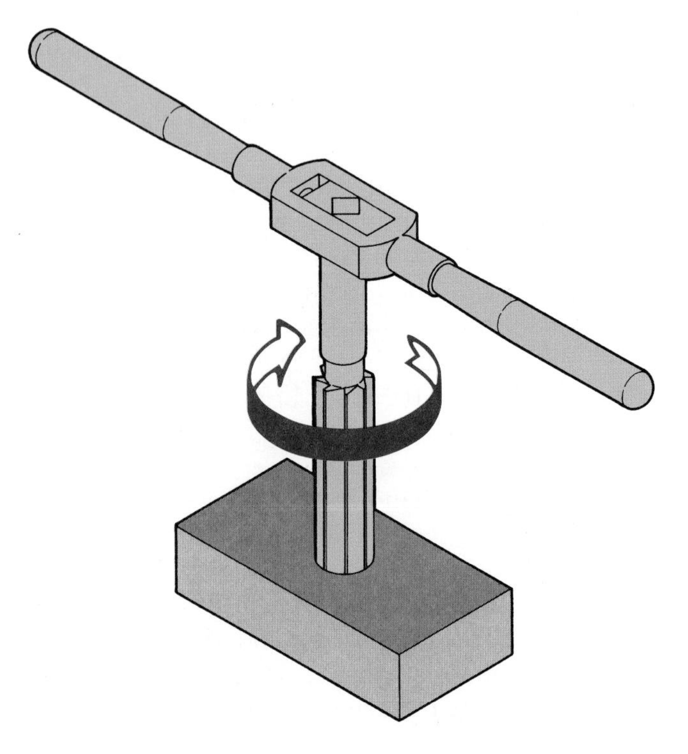

Goodheart-Willcox Publisher

Figure 7-68. Always turn a hand reamer in a clockwise direction.

7.10 Hand Threading

Threaded sections have many applications in our everyday life. A thread is a spiral or helical ridge found on nuts and bolts. When required on a job, threads are indicated on the plans and drawings in a special way, **Figure 7-69**. They are specified by diameter and number of threads per inch. Metric threads are specified by diameter and thread pitch is given in millimeters.

The ***American National Thread System*** was adopted in 1911. It is the common thread form used in the United States and is characterized by the 60° angle formed by the sides of the thread.

The National Coarse (NC) Thread is for general purpose work; the National Fine (NF) Thread is for precision assemblies. These are the most widely used thread groups in the American National series. The NF group has more threads per inch for a given diameter than the NC group.

A considerable amount of confusion resulted during World War II from the many different forms and kinds of threads used by the Allied nations. As a result, the powers that make up NATO (the North Atlantic Treaty Organization) adopted a standard thread form. It is referred to as the ***Unified System***, **Figure 7-70**. It is very similar to the American National Thread System. It differs only in the thread shape. The thread root is rounded and the crest may be flat or rounded. The threads are identified by UNF (Unified National Fine) and UNC (Unified National Coarse), **Figure 7-71**. Fasteners using this thread series are interchangeable with fasteners using the American National thread.

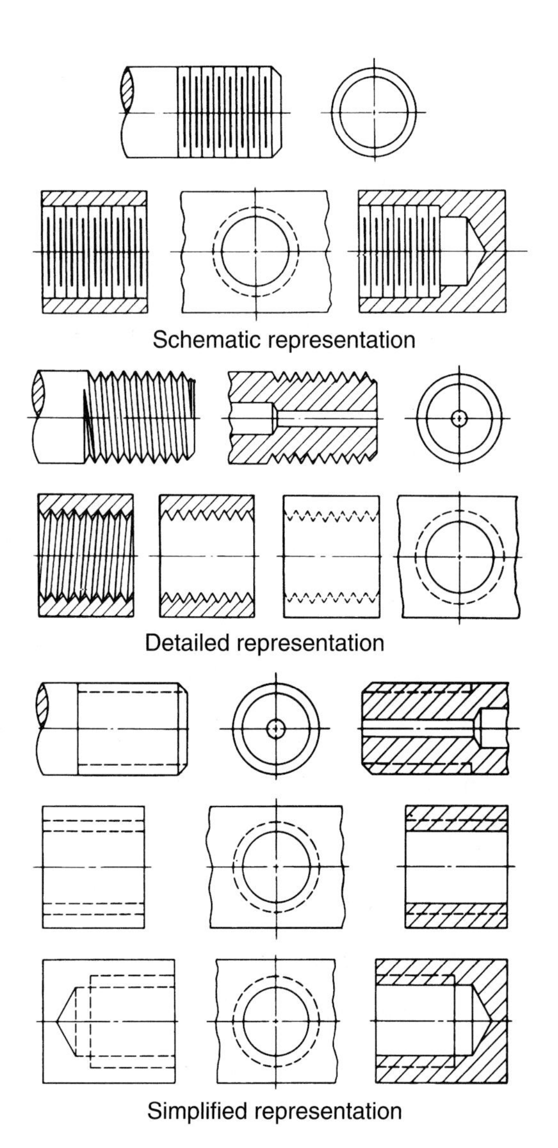

Figure 7-69. Methods used to visually depict threads on drawings. Only one type will be used on a given drawing.

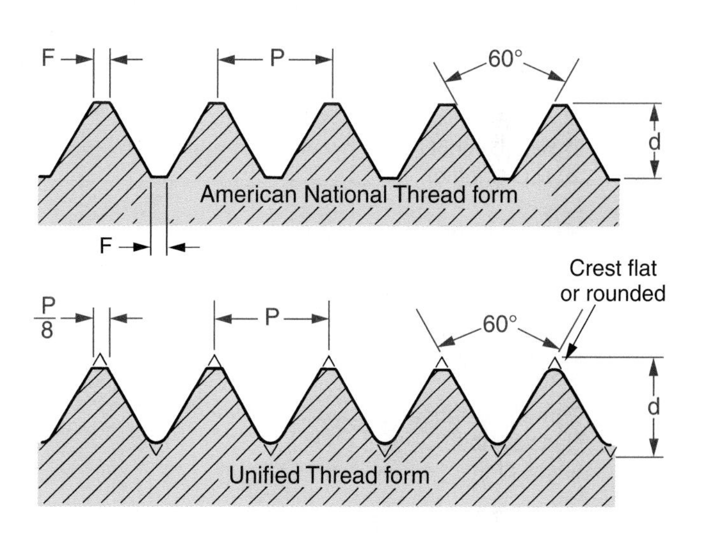

Figure 7-70. Drawings illustrate similarities and differences of the American National Thread form and the Unified Thread form.

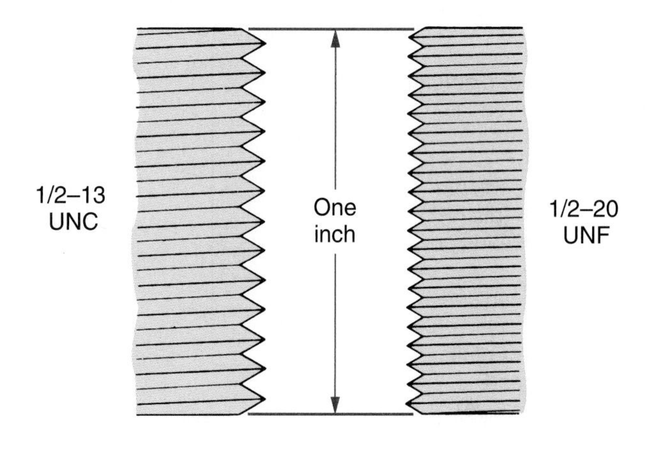

Figure 7-71. Comparison of Unified Coarse (UNC) and Unified Fine (UNF) threads. Both have the same geometric shape.

In addition to those just described, there are several other thread groups. Included are the Unified National Extra Fine (UNEF), Unified National 8 Series, and Unified National 12 Series. The 8 Series has 8 threads per inch and is used on diameters ranging from 1″ to 6″ in 1/8″ and 1/4″ increments. The 12 Series has 12 threads per inch and is used on diameters that range from 1″ to 6″.

Metric unit threads have the same shape as the Unified Thread, but are specified in a different manner, **Figure 7-72**. Metric threads and Unified National Series threads are not interchangeable. See **Figure 7-73**.

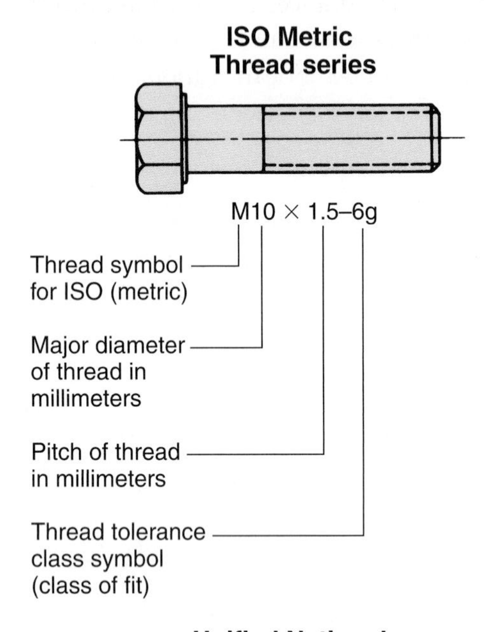

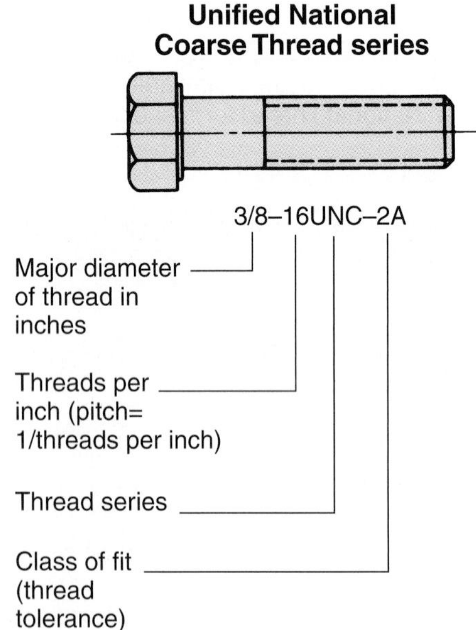

Goodheart-Willcox Publisher

Figure 7-72. How thread size is noted and what each term means.

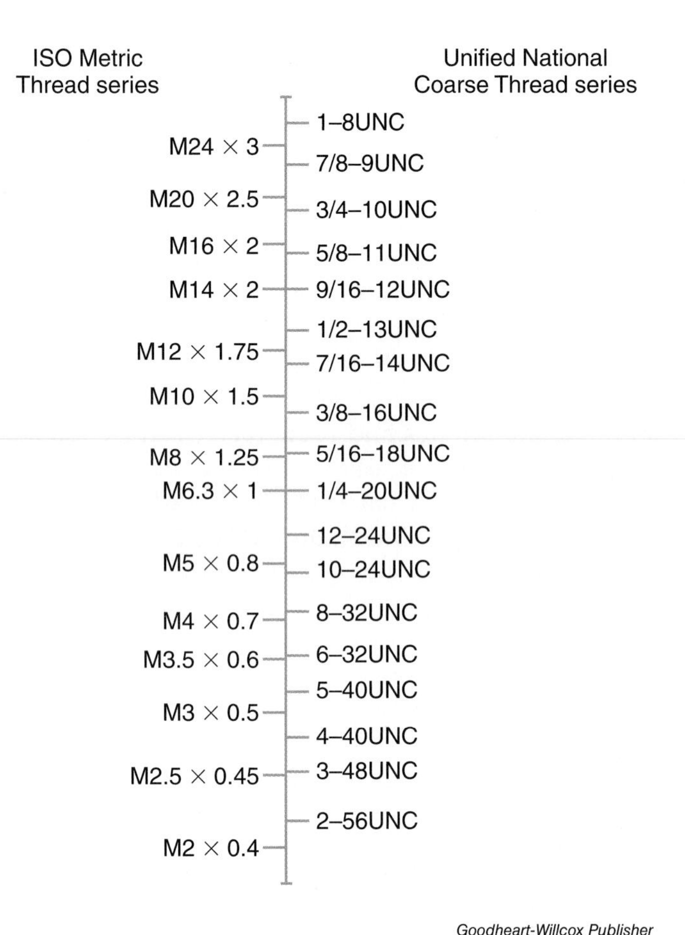

Goodheart-Willcox Publisher

Figure 7-73. While ISO Metric threads may appear to be similar in diameter to the Unified National Thread series, the two are not interchangeable.

7.10.1 Thread Size

Threads of the Unified National system smaller than 1/4″ diameter are not measured as fractional sizes. They are given by number sizes that range from #0 (approximately 1/16″ or 0.060″ diameter) to #12 (just under 1/4″ or 0.216″ diameter). Both UNC and UNF series are available.

Care must be taken so the number denoting the thread diameter and the number of threads are not mistaken for a fraction. For example: a #8-32 UNC thread would be a thread that is a #8 (0.164″) diameter and 32 threads per inch, not an 8/32 (1/4″) diameter fastener with a UNC series thread.

7.10.2 Cutting Threads

Because thread dimensions have been standardized, the use of taps to cut internal threads, and dies to cut external threads have become universal practice whenever threads are to be cut by hand. See **Figure 7-74**.

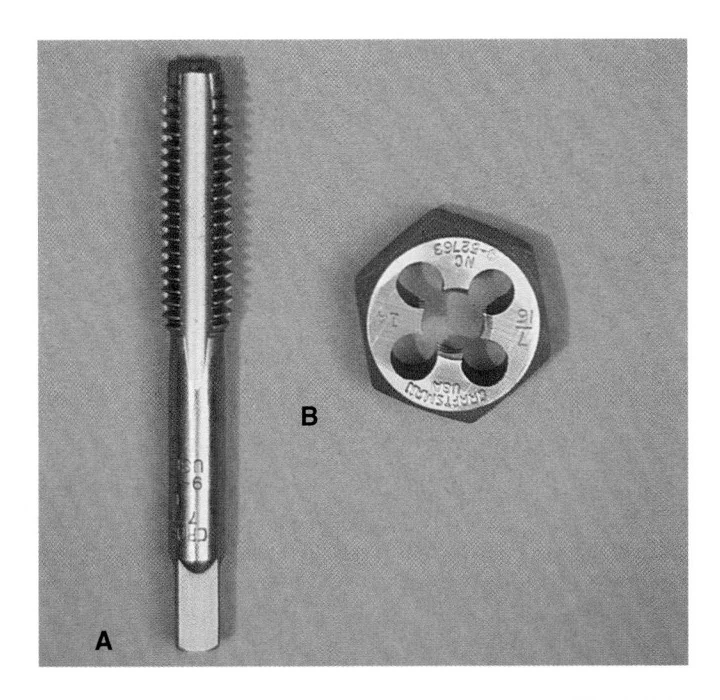

Goodheart-Willcox Publisher

Figure 7-74. Thread cutting. A—Tap is for cutting internal threads. B—Die is for cutting external threads.

7.10.3 Internal Threads

Internal threads are made with a tap, **Figure 7-75**. Taps are made of carbon steel or high-speed steel (HSS) and are carefully heat-treated for long life. Taps are quite brittle and are easily broken if not handled properly.

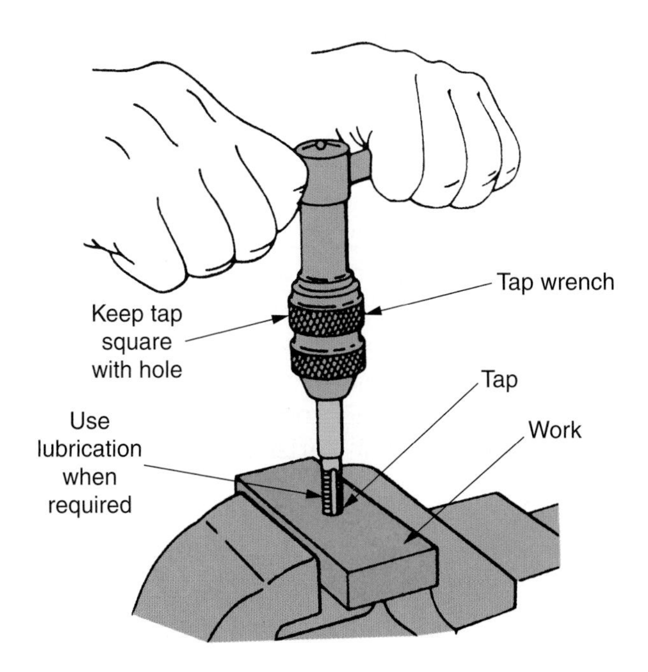

Goodheart-Willcox Publisher

Figure 7-75. A machinist is cutting internal threads with a tap.

To meet demands for varying degrees of thread accuracy, it became necessary for industry to adopt standard working tolerances for threads. Working tolerances for threads have been divided into ***classes of fits***, which are indicated by the last number on the thread description (1/2-13 UNC-2).

Fits for inch-based threads are as follows:

- Class 1—Loose fit.
- Class 2—Free fit.
- Class 3—Medium fit.
- Class 4—Close fit.

Under revised ISO standards, there are two classes of thread tolerances for external threads: 6g for general-purpose threads and 5g6g for close tolerance threads. There is only one tolerance class for internal threads, 6H. A lowercase letter indicates the tolerance on a bolt and a capital letter is used for the nut.

7.10.4 Taps

Standard hand taps are made in sets of three. They are known as taper, plug, and bottoming taps, **Figure 7-76**:

- Threads are started with a taper tap. It is tapered back from the end 6 to 10 threads before full thread diameter is reached.

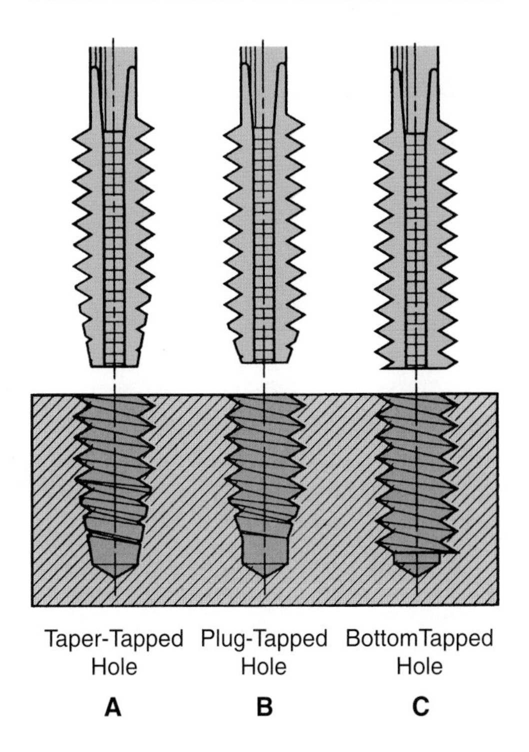

Goodheart-Willcox Publisher

Figure 7-76. Standard hand taps. A—Taper tap for starting the thread. B—Plug tap for continuing thread after taper tap has cut as far into hole as it can. C—Bottoming tap for continuing threads to bottom of a blind hole.

- The plug tap is used after the taper tap has cut threads as far into the hole as possible. It tapers back 3 or 4 threads before full thread diameter is reached.
- Threads are cut to the bottom of a ***blind hole*** (one that does not go through the part) with a bottoming tap. This tap tapers back 1 or 20 threads before full thread diameter is reached. It is necessary to use the full set of taps only when a blind hole is to be tapped, **Figure 7-77**.

Another tap used in the shop is a pipe tap. A pipe tap cuts a tapered thread, so there is a "wedging" action set up to make a leak-tight joint. The fraction that indicates pipe tap size may be confusing at first because it indicates pipe size and not the thread diameter. See **Figure 7-78**.

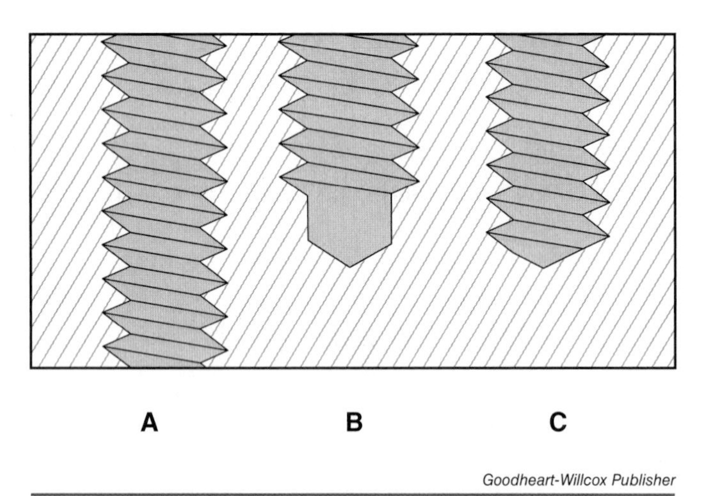

Goodheart-Willcox Publisher

Figure 7-77. Cutaway of metal block illustrates three types of threaded holes. A—Open or through hole. B—Blind hole that has been drilled deeper than desired threads. C—Blind hole with threads tapped to bottom.

Goodheart-Willcox Publisher

Figure 7-78. There is considerable difference between a 1/8″ standard thread tap and a 1/8″ pipe thread tap.

A pipe thread is indicated by the abbreviation NPT (National Pipe Thread). To obtain the wedging action needed for leakproof joints, the threads taper 3/4″ per foot of length.

7.10.5 Tap Drill

The drill used to make the hole prior to tapping is called a tap drill. Theoretically, it should be equal in diameter to the minor diameter of the screw that will be fitted into the tapped hole. See **Figure 7-79**. This situation would cause the tap to cut a full thread, however. The pressure required to rotate the tap would be so great that tap breakage could occur. Full-depth threads are not necessary because three-quarter-depth threads are strong enough that the fastener usually breaks before the threads strip. Drill sizes can be secured from a tap drill chart, **Figure 7-80** and **Figure 7-81**.

7.10.6 Tap Wrenches

Two types of tap wrenches are available, **Figure 7-82**. The type used will depend on the size of the tap. A T-handle tap wrench should be used

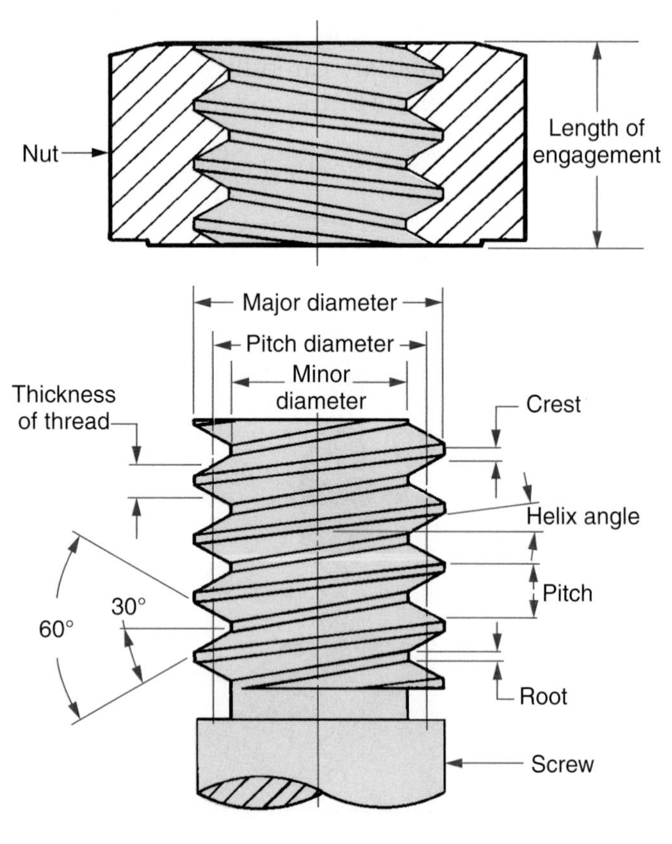

Goodheart-Willcox Publisher

Figure 7-79. Nomenclature of a fastener thread.

National Coarse and National Fine threads and tap drills								
Size	Threads per inch	Major dia.	Minor dia.	Pitch dia.	Tap drill 75% thread	Decimal equivalent	Clearance drill	Decimal quivalent
2	56	.0860	.0628	.0744	50	.0700	42	.0935
	64	.0860	.0657	.0759	50	.0700	42	.0935
3	48	.099	.0719	.0855	47	.0785	36	.1065
	56	.099	.0758	.0874	45	.0820	36	.1065
4	40	.112	.0795	.0958	43	.0890	31	.1200
	48	.112	.0849	.0985	42	.0935	31	.1200
6	32	.138	.0974	.1177	36	.1065	26	.1470
	40	.138	.1055	.1218	33	.1130	26	.1470
8	32	.164	.1234	.1437	29	.1360	17	.1730
	36	.164	.1279	.1460	29	.1360	17	.1730
10	24	.190	.1359	.1629	25	.1495	8	.1990
	32	.190	.1494	.1697	21	.1590	8	.1990
12	24	.216	.1619	.1889	16	.1770	1	.2280
	28	.216	.1696	.1928	14	.1820	2	.2210
1/4	20	.250	.1850	.2175	7	.2010	G	.2610
	28	.250	.2036	.2268	3	.2130	G	.2610
5/16	18	.3125	.2403	.2764	F	.2570	21/64	.3281
	24	.3125	.2584	.2854	I	.2720	21/64	.3281
3/8	16	.3750	.2938	.3344	5/16	.3125	25/64	.3906
	24	.3750	.3209	.3479	Q	.3320	25/64	.3906
7/16	14	.4375	.3447	.3911	U	.3680	15/32	.4687
	20	.4375	.3725	.4050	25/64	.3906	29/64	.4531
1/2	13	.5000	.4001	.4500	27/64	.4219	17/32	.5312
	20	.5000	.4350	.4675	29/64	.4531	33/64	.5156
9/16	12	.5625	.4542	.5084	31/64	.4844	19/32	.5937
	18	.5625	.4903	.5264	33/64	.5156	37/64	.5781
5/8	11	.6250	.5069	.5660	17/32	.5312	21/32	.6562
	18	.6250	.5528	.5889	37/64	.5781	41/64	.6406
3/4	10	.7500	.6201	.6850	21/32	.6562	25/32	.7812
	16	.7500	.6688	.7094	11/16	.6875	49/64	.7656
7/8	9	.8750	.7307	.8028	49/64	.7656	29/32	.9062
	14	.8750	.7822	.8286	13/16	.8125	57/64	.8906
1	8	1.0000	.8376	.9188	7/8	.8750	1- 1/32	1.0312
	14	1.0000	.9072	.9536	15/16	.9375	1- 1/64	1.0156
1-1/8	7	1.1250	.9394	1.0322	63/64	.9844	1- 5/32	1.1562
	12	1.1250	1.0167	1.0709	1- 3/64	1.0469	1- 5/32	1.1562
1-1/4	7	1.2500	1.0644	1.1572	1- 7/64	1.1094	1- 9/32	1.2812
	12	1.2500	1.1417	1.1959	1-11/64	1.1719	1- 9/32	1.2812
1-1/2	6	1.5000	1.2835	1.3917	1-11/32	1.3437	1-17/32	1.5312
	12	1.5000	1.3917	1.4459	1-27/64	1.4219	1-17/32	1.5312

Goodheart-Willcox Publisher

Figure 7-80. Thread and tap drill chart for Unified National threads.

Nominal size	Internal thread minor diameter		Tap drill diameter
	Max.	Min.	
M1.6×0.35	1.321	1.221	1.25
M2×0.4	1.679	1.567	1.6
M2.5×0.45	2.138	2.013	2.05
M3×0.5	2.599	2.459	2.5
M3.5×0.6	3.010	2.850	2.9
M4×0.7	3.422	3.242	3.3
M5×0.8	4.334	4.134	4.2
M6.3×1	5.553	5.217	5.3
M8×1.25	6.912	6.647	6.8
M10×1.5	8.676	8.376	8.5
M12×1.75	10.441	10.106	10.2
M14×2	12.210	11.835	12.0

Nominal size	Internal thread minor diameter		Tap drill diameter
	Max.	Min.	
M16×2	14.210	13.885	14.0
M20×2.5	17.744	17.294	17.5
M24×3	21.252	20.752	21.0
M30×3.5	26.771	26.211	26.5
M36×4	32.370	31.670	32.0
M42×4.5	37.799	37.129	37.5
M48×5	43.297	42.587	43.0
M56×5.5	50.796	50.046	50.5
M64×6	58.305	57.505	58.0
M72×6	66.305	65.505	66.0
M80×6	74.305	73.505	74.0
M90×6	84.305	83.505	84.0
M100×6	94.305	93.505	94.0

Goodheart-Willcox Publisher

Figure 7-81. Thread and tap drill chart for metric threads.

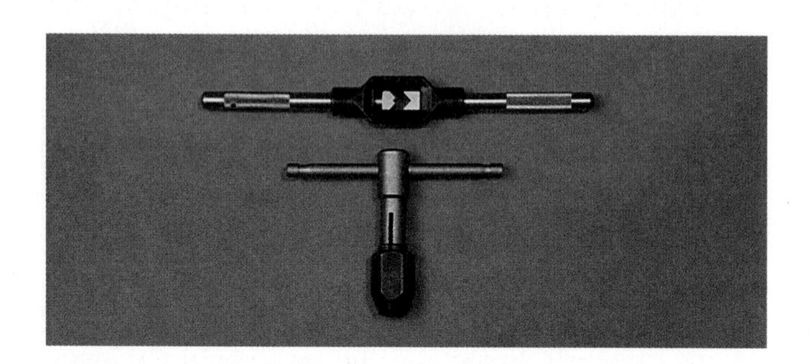

Goodheart-Willcox Publisher

Figure 7-82. Tap wrenches. Top—Hand tap wrench. Bottom—T-handle tap wrench.

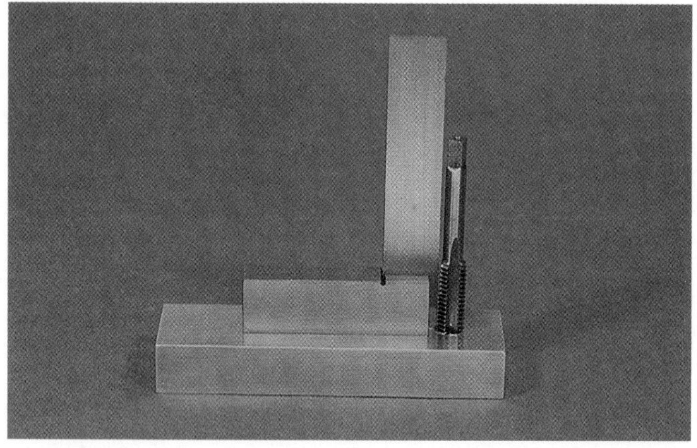

Goodheart-Willcox Publisher

Figure 7-83. A tap must be started squarely with the hole. A quick way to check this is to use a machinist's square.

with all small taps. It allows a more sensitive "feel" when tapping. The hand tap wrench is best suited for large taps where more leverage is required.

When tapping by hand, the chief requirement is to make sure that the tap is started straight, and remains square during the entire tapping operation, **Figure 7-83**. The tap must be backed off (reversed in rotation) for one-half of a turn every one or two cutting turns. This will break the chips free and allow them to drop through the tap flutes. Backing off prevents chips from jamming the tap and damaging the threads. Use a cutting fluid designed for the particular metal you are tapping.

When tapping a blind hole, some machinists place a dab of grease, or a piece of grease pencil or wax crayon in the hole. As the tap cuts the threads, the grease is forced up and out of the hole, carrying the chips along.

7.10.7 Power Tapper

Tapping can also be done using an articulating arm power tapper which can be fitted with an auto reverse (the tap is run out of the hole automatically when the threads are at the proper depth), automatic tap lubrication, and digital depth readout. Holes can be tapped quickly with little chance of tap breakage.

7.10.8 Care in Tapping

Considerable care must be exercised when tapping:

- Use the correct size tap drill. Secure this information from a tap drill chart.
- Use a sharp tap and apply sufficient quantities of cutting fluid. With some cutting fluids, the area is flooded with fluid; with others, a few drops are sufficient. Read the container label.
- Start the taper tap square with the work.
- Do not force the tap to cut. Remove the chips using a piece of cloth or cotton waste, not your fingers.
- Avoid running a tap to the bottom of a blind hole and continuing to apply pressure. Do not allow the hole to fill with chips and jam the tap. Either condition can cause the tap to break (especially if the tap is small).
- Remove burrs on the tapped hole with a smooth file.

7.10.9 Dealing with Broken Taps

Taps sometimes break off in a hole. Several tools and techniques have been developed for removing them without damaging the threads already cut. These methods do not always work and the part may have to be discarded.

Frequently, a tap will shatter in the hole. It may then be possible to remove the fragments with a pointed tool such as a scribe.

Broken carbon steel taps can sometimes be removed from steel if the work can be heated to annealing temperature. The tap can then be drilled out. This cannot be done with high-speed steel taps. If the HSS tap is large enough, it can be ground out with a hand grinder.

A tap extractor can sometimes be used to remove a broken tap. See **Figure 7-84**. Penetrating oil should be applied and allowed to "soak in" for a short time before the fingers of the tap extractor are fitted into the flutes of the broken tap. The collar on the extractor is slipped down flush with the work surface.

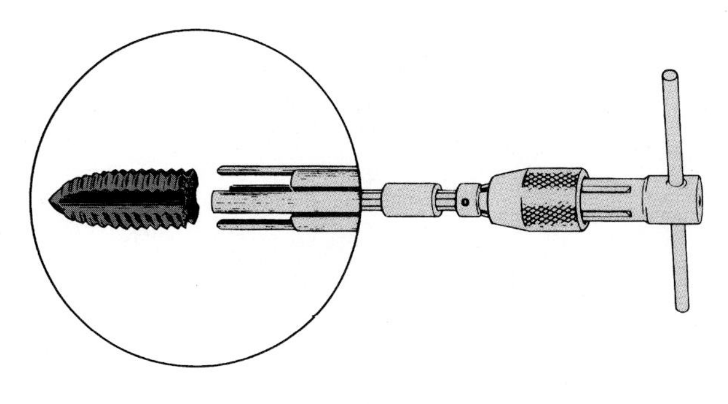

Goodheart-Willcox Publisher

Figure 7-84. Tap extractor will help remove a broken tap. Close-up shows fingers of extractor and how they fit into flutes of broken tap. It does not always work.

A tap wrench is fitted on the extractor. The tap extractor is then carefully twisted back and forth to loosen the tap segments. After the broken parts have been loosened, it is a simple matter to remove them.

In some shops, a tap disintegrator is used to remove broken taps. This device makes use of an electric arc to cause the tap to disintegrate. If used properly, it will break up the tap without affecting the metal surrounding the broken tool.

7.10.10 External Threads

External threads are cut with a die, **Figure 7-85**. Solid dies are not adjustable and for that reason are not often used. The adjustable die, and the two-part adjustable die, are preferred. The two-part die has a wide range of adjustment and is fitted with guides to keep it true and square on the work. Dies are available for cutting most standard threads.

7.10.11 Die Stocks

A die stock holds the die and provides leverage for turning the die on the work. See **Figure 7-86**. When cutting external threads, it is necessary to remember the following:

- Material diameter is the same size as the desired thread diameter. That is, 1/2-13 UNC threads are cut on a 1/2″ diameter shaft.
- Mount work solidly in a vise.
- Set the die to the proper size. Make trial cuts on a piece of scrap until the proper adjustment is found.
- Grind a small chamfer on the shaft end, as shown in **Figure 7-87**. This permits a die to start easily.
- Start the cut with the tapered end of the die.

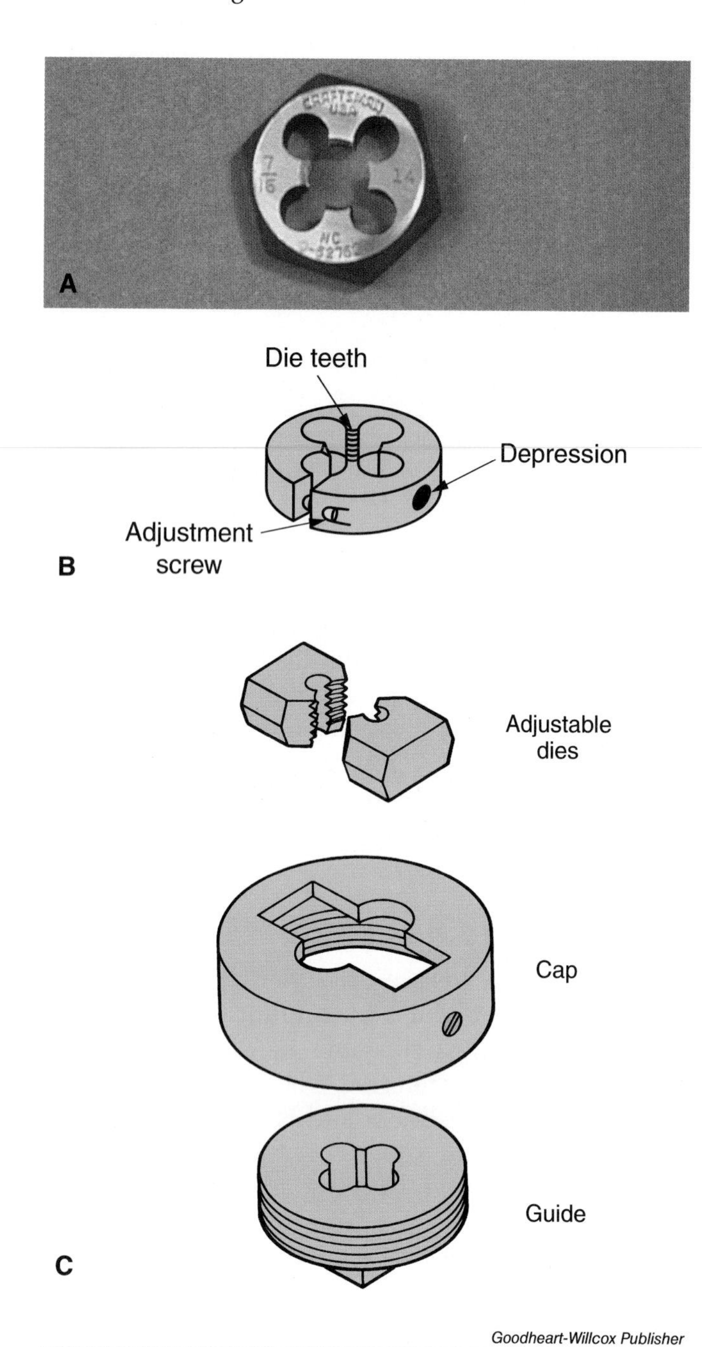

Figure 7-85. Types of dies. A—Solid dies used to cut external threads by hand. They cannot be adjusted. B—A small screw on one side of the split permits small changes in size of an adjustable die. C—Construction of a multi-part adjustable die.

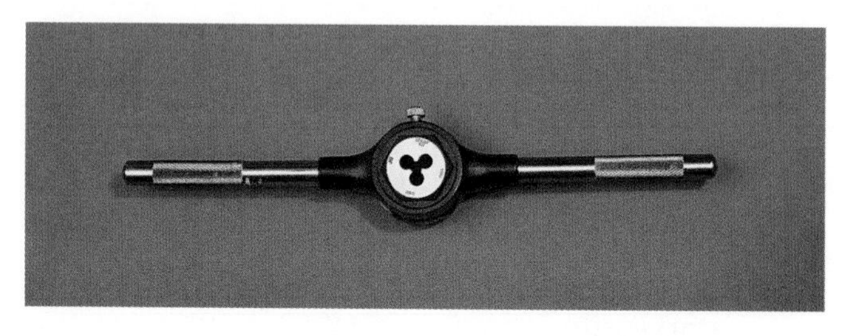

Figure 7-86. The die is held in a die stock.

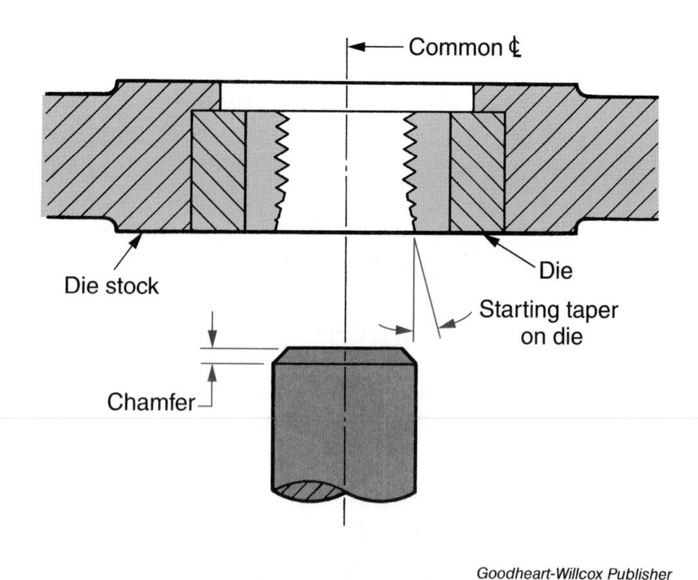

Figure 7-87. A die will start more easily if a small chamfer is cut or ground on the end of the shaft to be threaded. Section through die and die stock shows proper way to start threads.

- Back off the die every one or two turns to break the chips.
- Use cutting oil. Place a paper towel down over the work to absorb excess cutting oil. The towel will also prevent the oil from getting on the floor.
- Remove any burrs from the finished thread with a fine cut file.

7.10.12 Threading to a Shoulder

When a thread must be cut by hand to a shoulder, start and run the threads as far as possible in the usual manner, **Figure 7-88**. Remove the die. Turn it over with the taper up. Run the threads down again to the shoulder. Never try this operation without first starting the threads in the usual manner.

7.10.13 Problems in Cutting External Threads

The most common problem encountered when cutting external threads with a die is ragged threads. Ragged threads can be caused by any of the following:

- Applying little or no cutting oil.
- Dull die cutters.
- Stock too large for the threads being cut.
- Die not started square.
- One set of cutters upside down when using a two-part die.

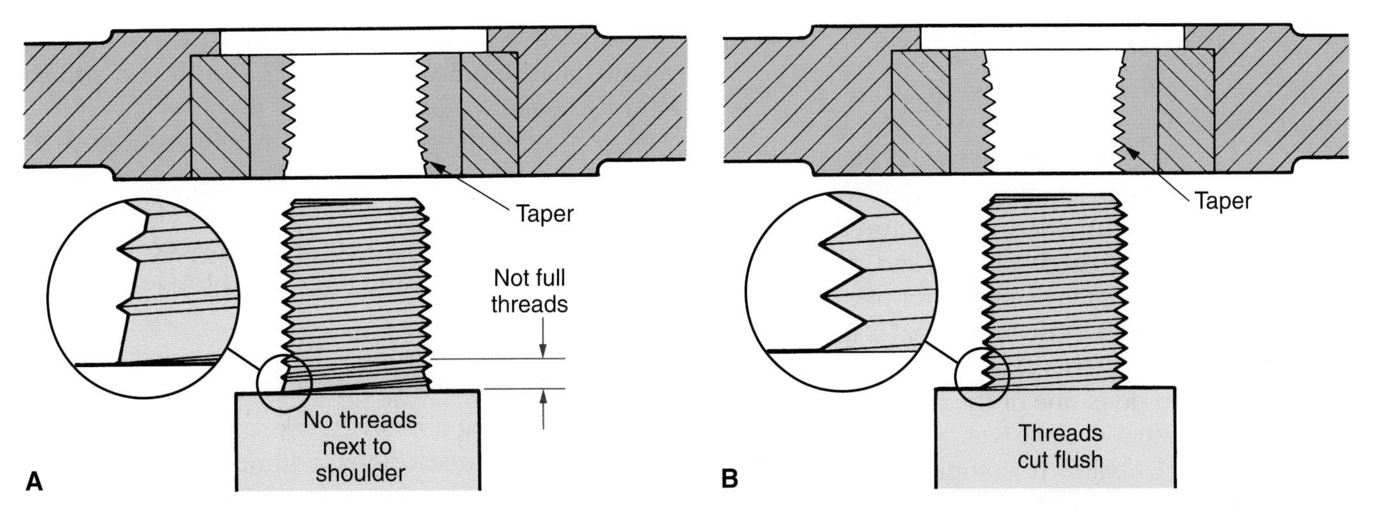

Goodheart-Willcox Publisher

Figure 7-88. How to cut threads to a shoulder. After die has been run down as far as possible, the die is reversed. When rotated down the shaft, it will cut threads almost flush with shoulder. A—Running die down normally. B—Reversing die to cut flush.

7.10.14 Hand Threading Safety

When threading, take care to keep the following safety precautions in mind:

- If a tap or threaded piece must be cleaned of chips with compressed air, protect your eyes from flying chips by wearing goggles. Take care not to endanger persons working in the area near you.
- Chips produced by hand threading are sharp. Use a brush or piece of cloth, not your hand, to remove them.
- Newly-cut external threads are very sharp. Again, use a brush or cloth to clean them.
- Wash your hands after using cutting fluids or oils. Some cause skin rash. This can develop into a serious skin disorder if the oils are left on the hands for an extended period.
- Have cuts treated by a qualified person. Infections can occur when cuts and other injuries are not properly treated.
- Be sure the die is clamped firmly in the die stock. If not, it can fall from the holder and cause injury.
- Broken taps have very sharp edges and are very dangerous. Handle them as you would broken glass.

7.11 Hand Polishing

An ***abrasive*** is commonly thought of as any hard substance that will wear away another material. The substance, grain size, backing material, and the manner in which the substance is bonded to the backing material determine the performance and efficiency of an abrasive. The table in **Figure 7-89** shows a comparison of abrasive grain size and indicates how the various abrasives are graded.

Technical grades		Simplified grades	Other grades
Mesh	Aluminum oxide Silicon carbide	Emery	Emery polishing
600			4/0
			3/0
500			2/0
400	10/0		0
360			
320	9/0		1/2
280	8/0		
240	7/0		1 G
220	6/0		2
180	5/0		3
150	4/0	Fine	
120	3/0		
100	2/0	Medium	
80	0	coarse	
60	1/2		
50	1	Extra coarse	
40	1 1/2		
36	2		
30	2 1/2		
24	3		
20	3 1/2		
16	4		
12	4 1/2		

Coated Abrasive Manufacturers Institute

Figure 7-89. A comparative grading chart for abrasives.

7.11.1 Abrasive Materials

Emery is a natural abrasive. It is black in color and cuts slowly, with a tendency to polish.

Aluminum oxide has replaced emery as an abrasive when large quantities of metal must be removed. It is a synthetic (manufactured) material that works best on high-carbon and alloy steels. Aluminum oxide that is designed for use on metal is manufactured with a grain shape that is not as sharp as that made for woodworking.

Silicon carbide is one of the hardest and sharpest of the synthetic abrasives. Silicon carbide is greenish-black in color. It is superior to aluminum oxide in its ability to cut fast under light pressure. It is ideal for "sanding" metals like cast iron, bronze, and aluminum.

Crocus may be synthetic or natural iron oxide. It is bright red in color, very soft, and is used for cleaning and polishing when a minimum of stock is to be removed.

Diamonds are the hardest natural substance known. However, they can also be manufactured. Synthetic diamonds have no value as gems; they are used almost exclusively by industry for polishing and grinding. Diamond dust polishing compound is made by crushing synthetic diamonds. It is the only abrasive hard enough to polish the newer heat-treated, exotic alloy steels used by industry.

7.11.2 Coated Abrasives

A coated abrasive is cloth or paper with abrasive grains bonded to one surface. Because of its flexibility, cloth is used as a backing material for abrasives found in metalworking or machining. It is available in 9″ × 11″ (210 mm × 280 mm) sheets. It is also available in rolls starting at 1/2″ (12.5 mm) in width, and is called abrasive cloth.

7.11.3 Using Abrasive Cloth

Abrasive cloth is quite expensive. Use only what you need. Tear the correct amount from the sheet or roll. Do not discard abrasive cloth unless it is completely worthless. Used cloth is excellent for polishing.

Avoid using abrasives on machined surfaces. If a job has been filed properly, only a fine-grain cloth will be needed to polish the surface. However, if scratches are deep, start the polishing operation by using a coarse-grain cloth first. Change to a medium-grain cloth next, and finally a fine-grain abrasive. A few drops of oil will speed the operation. For a high polish, leave the oil on the surface after the scratches have been removed. Reverse the cloth and rub the smooth backing over the work.

Abrasive cloth must be properly supported to work efficiently. To get such support, wrap the cloth around a block of wood or file, **Figure 7-90**. Apply pressure and rub the abrasive back and forth in a straight line, parallel to the long side of the work.

7.11.4 Abrasives Safety

When using abrasives, take care to keep the following safety precautions in mind:

- Avoid rubbing your fingers or hand over polished surfaces or surfaces to be polished. Burrs on the edges of the metal can cause painful cuts.
- Wash your hands thoroughly after polishing operations.
- Treat all cuts immediately, no matter how small.
- Place all oily rags in a closed container. Never put them in your apron/shop coat or in a locker.
- Wipe up any oil dropped on the floor during polishing operations.
- If a lathe is used for polishing operations, make sure the machine is protected from the abrasive grains that fall from the polishing cloth. Stop the machine when inspecting your work.

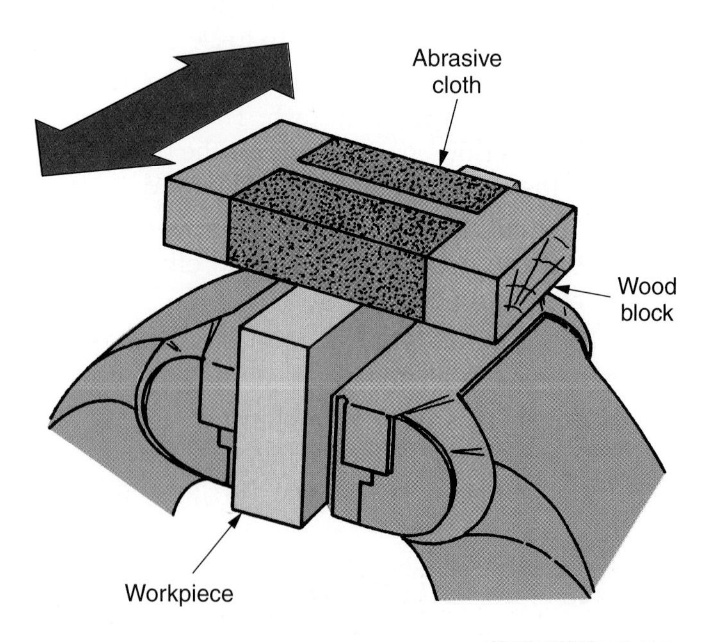

Goodheart-Willcox Publisher

Figure 7-90. Abrasive cloth should be supported with a block of wood or a file. Do not support it with your fingers alone.

Chapter Review

Summary

- There are several different vises and clamps used to mount a piece being worked on.
- Wrenches and screwdrivers come in many different shapes and sizes and serve many different purposes.
- Great care must be taken when working with chisels and hacksaws to avoid injury.
- It is important to know how to select the right tool for the project.
- Threading is a precise process that has different classes of fits.
- There are many different standards of thread measurement to be learned.
- Various materials can be used as abrasives for polishing and finishing a surface.

Review Questions

Answer the following questions using the information provided in this chapter.

1. List two variations of the bench vise.
2. How is vise size determined?
3. Work held in a vise can be protected from damage by the jaw serrations if _____ are placed over the jaws.
4. To prevent injuries, what should be avoided when mounting work in a vise?
5. A(n) _____ is ideal for holding small work.
6. How do slip-joint pliers have an advantage over many other types of pliers?
7. Why are the cutting edges on diagonal pliers set at an angle?
8. What are adjustable clamping pliers?
9. List three ways of extending the working life of pliers.
10. What use are torque-limiting wrenches?
11. Do torque-limiting wrenches give a more accurate reading when they are pushed or when they are pulled?
12. Several different wrenches can be classified as adjustable wrenches. Name two.
13. List three points that should be observed when using an adjustable wrench.
14. Round work can be gripped with a(n) _____ wrench.
15. Describe socket wrenches.
16. The wrench that is used with socket-headed fasteners is commonly known as a(n) _____.
17. When using a wrench, is it better to push or pull on the handle?
18. List five safety precautions that should be observed when using a wrench.
19. What is the difference between a standard screwdriver tip and a Phillips screwdriver tip?

Match each phrase with the correct screwdriver name listed below.

20. _____ Pozidriv®
21. _____ Standard
22. _____ Electrician
23. _____ Heavy-duty
24. _____ Stubby
25. _____ Ratchet

A. Has a flattened wedge-shaped tip.
B. Moves the fastener on the power stroke, but not on the return stroke.
C. Has a square shank to permit additional force to be applied with a wrench.
D. Useful when handling small screws.
E. Tip is similar to that of a Phillips head screwdriver.
F. Has an insulated handle.
G. Is short and is used when space is limited.

26. List three safety precautions that should be observed when using a screwdriver.
27. How is the size of a ball-peen hammer determined?
28. Why are soft-face hammers and mallets used in place of a ball-peen hammer?
29. List three safety precautions that should be observed when using striking tools.
30. List the four general types of cold chisels.
31. The chisel is an ideal tool for removing _____.
32. A dangerous condition that can cause injury if not fixed is a(n) _____ chisel.
33. Why are almost all hacksaws adjustable?
34. Why should the work be mounted solidly and close to the vise before cutting with a hacksaw?
35. A hacksaw cuts best at about _____ strokes per minute.
36. If a blade breaks or dulls before completing a cut, why should you not continue in the same cut with a new blade?
37. It is important to choose a hacksaw blade that will have at least _____ teeth cutting at all times.
38. What is the best way to hold soft metal tubing for hacksawing?
39. What is the best way to hold thin metal for hacksawing?
40. List the four cuts of files.
41. Files are cleaned with a(n) _____, never with your hand.
42. What are the most commonly used file shapes?
43. List three safety precautions that should be observed when files are used.
44. What is hand reaming used for?
45. How much stock should be left in a hole for hand reaming?
46. How does the UNC thread series differ from the UNF thread series?
47. A(n) _____ is used to cut internal threads.
48. External threads are cut with a(n) _____.
49. A hole that does not go entirely through the part is called a(n) _____ hole.
50. List the correct sequence of taps used to form threads to the full depth of a blind hole.
51. The drill used to make the hole prior to tapping is called a(n) _____.
52. Taps are turned in with a(n) _____.
53. What is an abrasive?
54. Avoid polishing _____ surfaces with an abrasive.

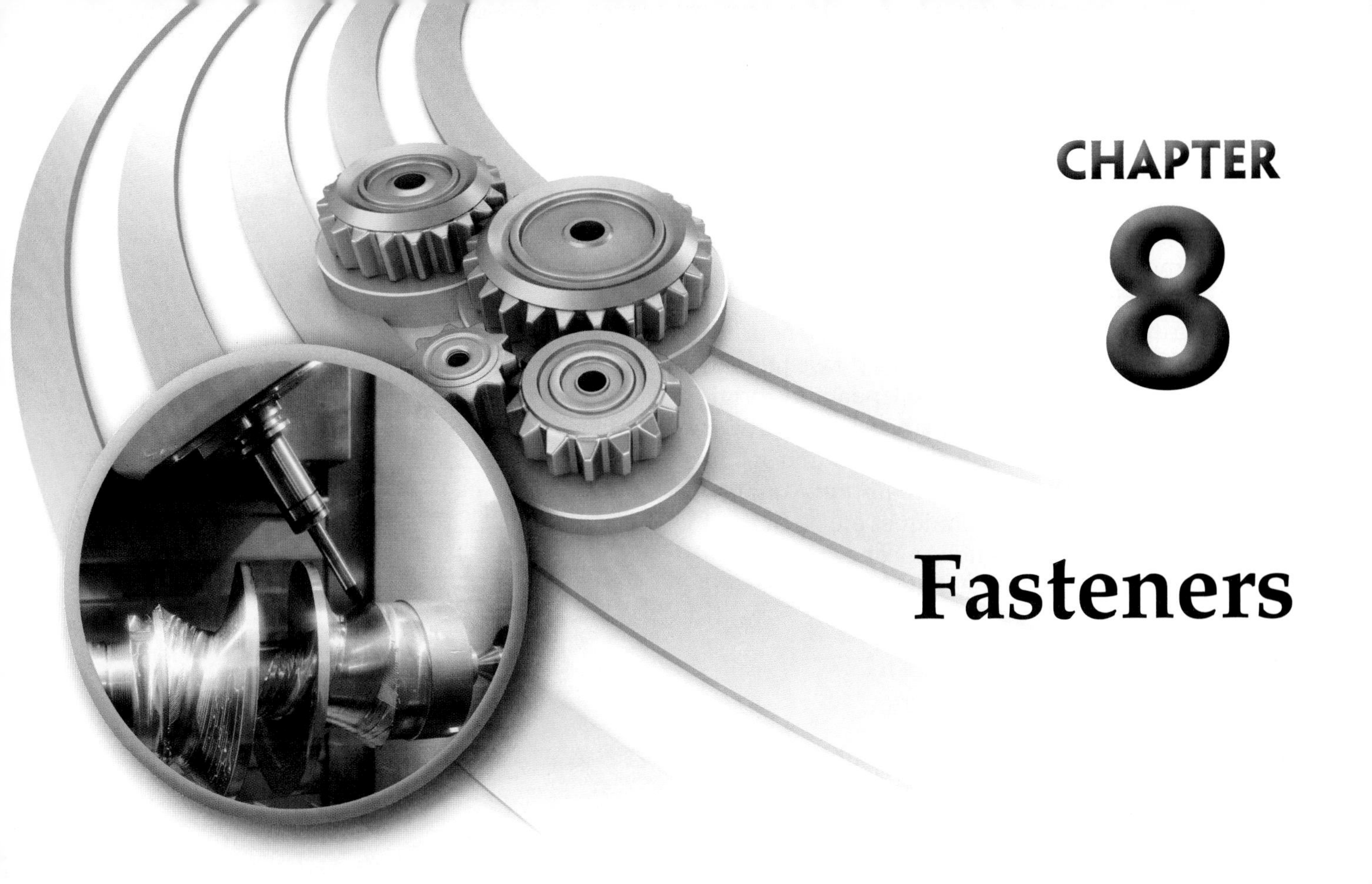

CHAPTER 8

Fasteners

Learning Objectives

After studying this chapter, you will be able to:

- Identify several types of fasteners.
- Explain why inch-based fasteners are not interchangeable with metric-based fasteners.
- Describe how some fasteners are used.
- Select the proper fastening technique for a specific job.
- Describe adhesive fastening techniques.

Technical Terms

adhesive
cyanoacrylate quick setting adhesive
fastener
key
threaded fastener

A ***fastener*** is any device used to hold two objects or parts together. This definition would include bolts, nuts, screws, pins, keys, rivets, and even chemical bonding agents or adhesives. The most common types of fasteners will be explained and illustrated in this chapter.

It is critical to choose the proper fasteners for each job, **Figure 8-1**. A poorly selected fastener can greatly reduce the safety and dependability originally designed into a product. Choosing improper fasteners could increase assembly costs and result in an inferior or faulty product. To improve quality, several different fastening techniques are often used in the same or related assemblies. For example, one auto manufacturer uses more than 11,000 kinds and sizes of fasteners.

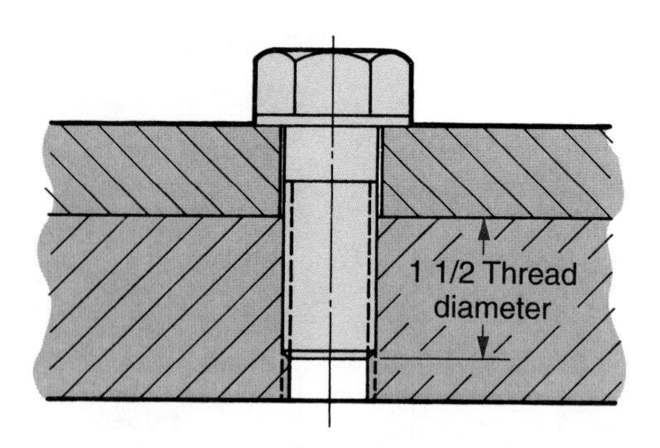

Goodheart-Willcox Publisher

Figure 8-2. For maximum strength, a threaded fastener must screw into the mating part a distance equal to 1 1/2 times the diameter of the thread.

8.1 Threaded Fasteners

Threaded fasteners make use of the wedging action of the screw thread to clamp parts together. To achieve maximum strength, a threaded fastener should screw into its mating part at least a distance equal to one and one-half times the thread diameter. See **Figure 8-2**.

Threaded fasteners vary in cost from thousands of dollars for special bolts that attach the wings to the fuselage of large aircraft, to a fraction of a penny for small machine screws. See **Figure 8-3**.

Shutterstock.com

Figure 8-3. There is a wide range of threaded fasteners available, from tiny machine screws used in precision instruments to large bolts used in building construction. This 1″ anchor bolt is being used to mount a steel column on a concrete foundation pier.

Steve Mann/Shutterstock.com

Figure 8-1. Complex assemblies, such as this helicopter tail rotor, require the use of many types and sizes of fasteners. Reliability of the product, and the safety of persons using it, can be greatly affected if improper fasteners are selected in the design or assembly phases.

Most threaded fasteners are available in metric sizes. Many American manufacturers now use metric-sized fasteners in their products, which has led to some problems. Metric threads and the common unified (inch-based) threads have the same basic profile (shape), but are *not* interchangeable. See **Figure 8-4**.

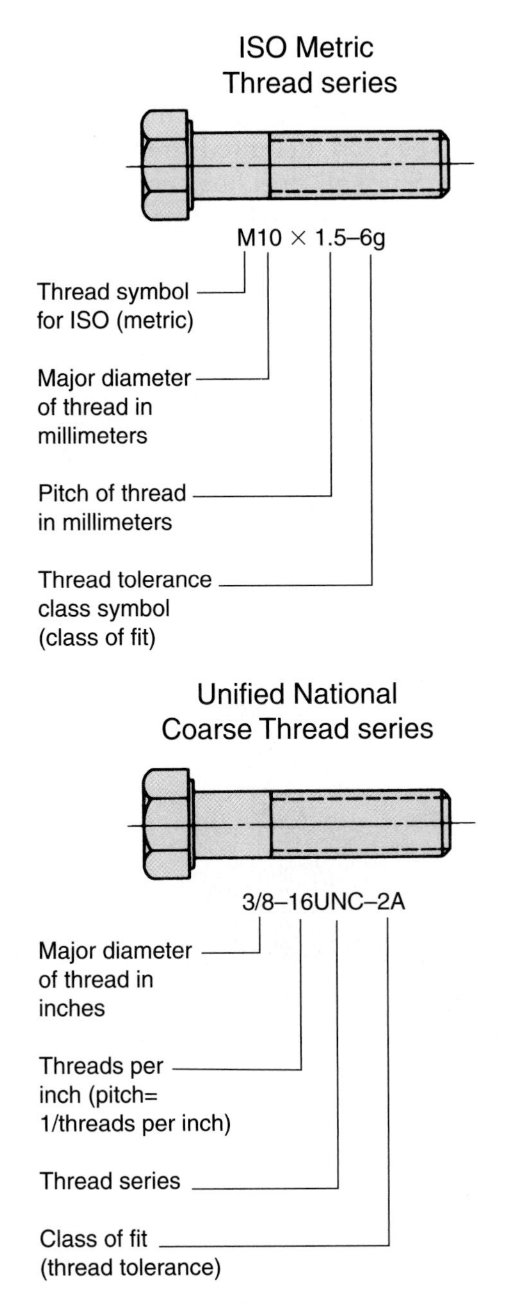

Goodheart-Willcox Publisher

Figure 8-4. Metric threads have the same basic profile (shape) as the Unified Thread series. However, the Unified and Metric threads are not interchangeable.

8.1.1 Machine Screws

Machine screws are widely used in general assembly work. They have slotted or recessed heads, and are made in a number of head styles, **Figure 8-5**.

Machine screws are available in body diameters ranging from #0000 (0.021″) to 3/4″ and in lengths from 1/8″ to 3″. Metric sizes are also manufactured. Nuts, in either square or hexagonal shapes, are purchased separately.

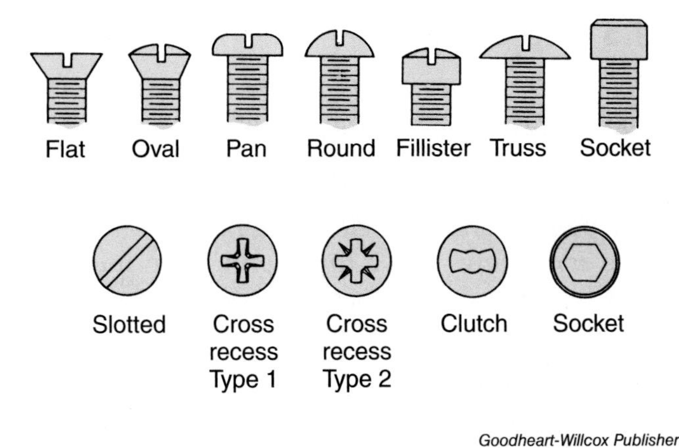

Goodheart-Willcox Publisher

Figure 8-5. A sampling of the many types of machine screws available.

8.1.2 Machine Bolts

Machine bolts are used to assemble parts that do not require close tolerances. They are manufactured with square and hexagonal heads, in diameters ranging from 1/2″ to 3″. The nuts are similar in shape to the bolt head. They are usually furnished with the machine bolts. Tightening the nut produces a clamping action to hold parts together, **Figure 8-6**.

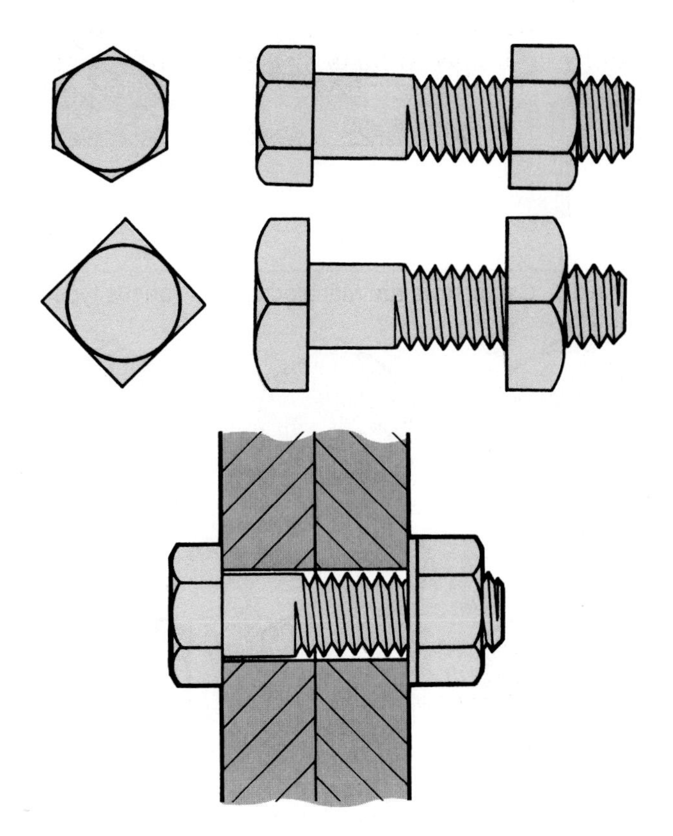

Goodheart-Willcox Publisher

Figure 8-6. Machine bolts use a nut to produce clamping force.

8.1.3 Cap Screws

Cap screws are found in assemblies requiring a higher quality and a more finished appearance, **Figure 8-7**. Instead of tightening a nut to develop clamping action, as with the machine bolt, the cap screw passes through a clearance hole in one of the pieces and screws into a threaded hole in the other part. Clamping action is accomplished by tightening the bolt into the threaded part.

Cap screws are held to much closer tolerances in their manufacture than machine screws. They are provided with a machined or semifinished bearing surface under the head. Some types of cap screws are heat-treated.

The application determines the strength of the cap screw to be used. Required strength is indicated on the print. Since all steel hex head cap screws are similar in appearance, a series of markings on the bolt head indicate strength capabilities. See **Figure 8-8**. The stronger the cap screw, the more expensive it is.

Cap screws are stocked in coarse and fine thread series and in diameters from 1/4″ to 2″. Lengths from 3/8″ to 10″ are available. Metric sizes can also be supplied.

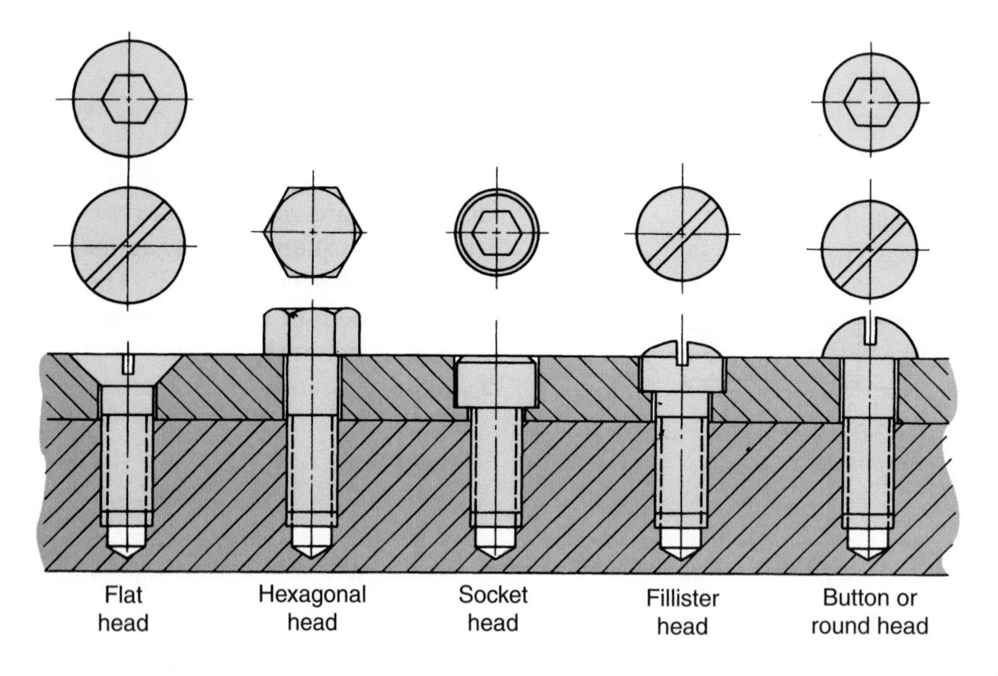

Figure 8-7. Cap screws are manufactured in various types.

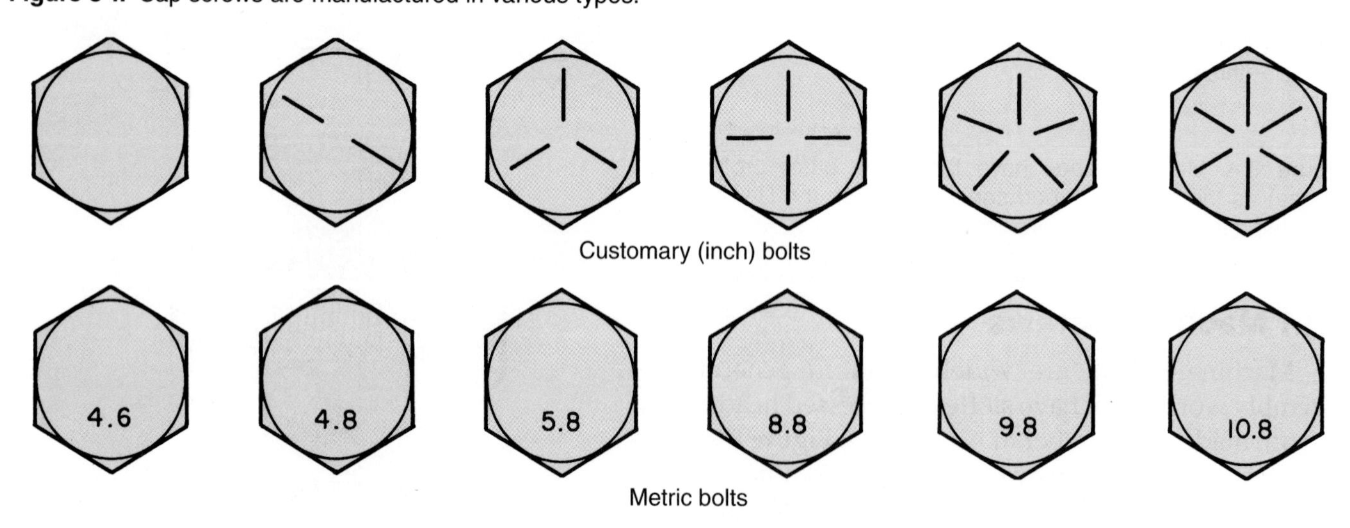

Figure 8-8. Identification marks (inch size) and class numbers (metric size) are used to indicate the relative strength of hex head cap screws. As identification marks increase in number, or class numbers become larger, increasing strength is indicated.

8.1.4 Setscrews

Setscrews are semipermanent fasteners that are used for such applications as preventing pulleys from slipping on shafts, holding collars in place on assemblies, and positioning shafts on assemblies. See **Figure 8-9**. Setscrews are usually made of heat-treated steel. They are classified in two ways, by head style, and by point style, **Figure 8-10**.

The thumbscrew is a variation of the setscrew that can be turned by hand. It is typically used in place of a setscrew for assemblies that require rapid or frequent disassembly, **Figure 8-11**. Thumbscrews are available with points similar to those on setscrews.

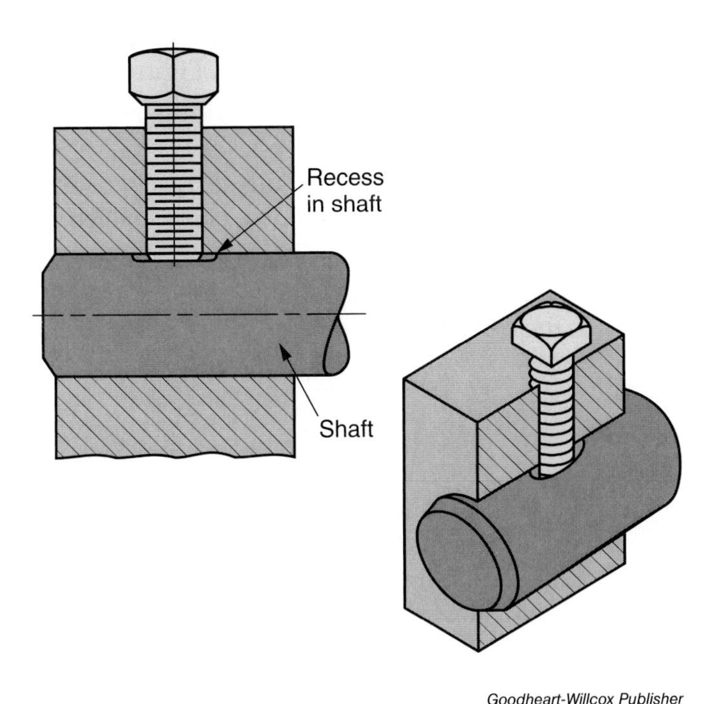

Goodheart-Willcox Publisher

Figure 8-9. Setscrews lock pulleys and gears to shafts to prevent rotation of the shaft in the hole.

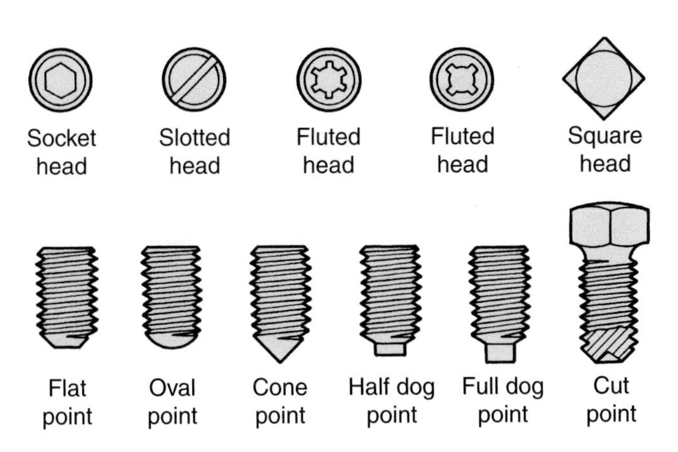

Goodheart-Willcox Publisher

Figure 8-10. Setscrew head and point designs.

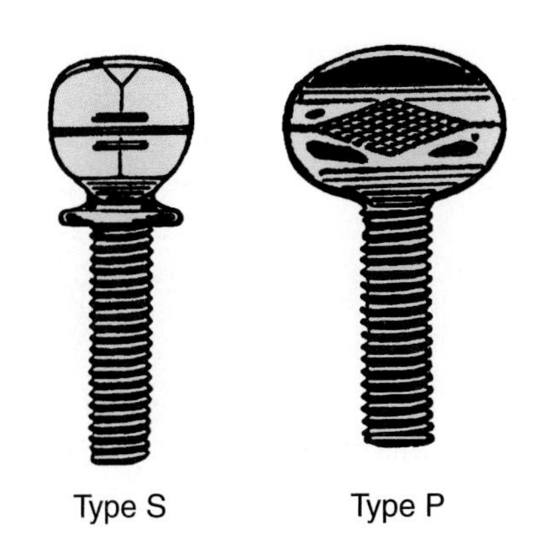

Parker-Kalon

Figure 8-11. Thumb screws can be removed or installed by hand.

8.1.5 Stud Bolts

Stud bolts are headless bolts that are threaded for their entire length, or (more commonly) on both ends, **Figure 8-12**. One end is designed for semipermanent installation in a tapped hole, while the other end is threaded for standard nut assembly to clamp the pieces together.

8.1.6 Removing Broken or Sheared Bolts

Bolts that have broken or sheared off can be hard to remove without proper tools. The drill-out power extractor, **Figure 8-13**, is available in several sizes. It is used with a 3/8″ capacity variable speed/reversing power drill. The built-in drill cuts the proper size hole for the extractor unit to fit. After the

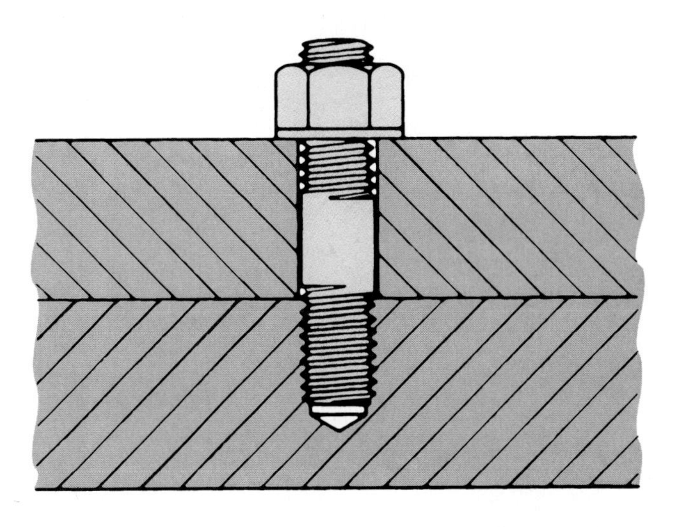

Goodheart-Willcox Publisher

Figure 8-12. One end of a stud bolt usually threads into a part, while the other end accepts a nut.

Alden Corp.

Figure 8-13. The Drill-Out Power Extractor™ combines a drill with an adjustable extractor collar. The drill makes the required size hole, and then the power drill is reversed. This causes the extractor to grip the broken bolt and torque it out.

hole is drilled and the extractor is placed into position, drill speed is reduced and reversed. This will remove the broken bolt.

A spiral fluted bolt extractor, **Figure 8-14**, is also available in several sizes. A chart furnished with the extractor indicates drill size to be used. After the hole has been drilled, the extractor is inserted and turned counterclockwise with an appropriate size tap wrench.

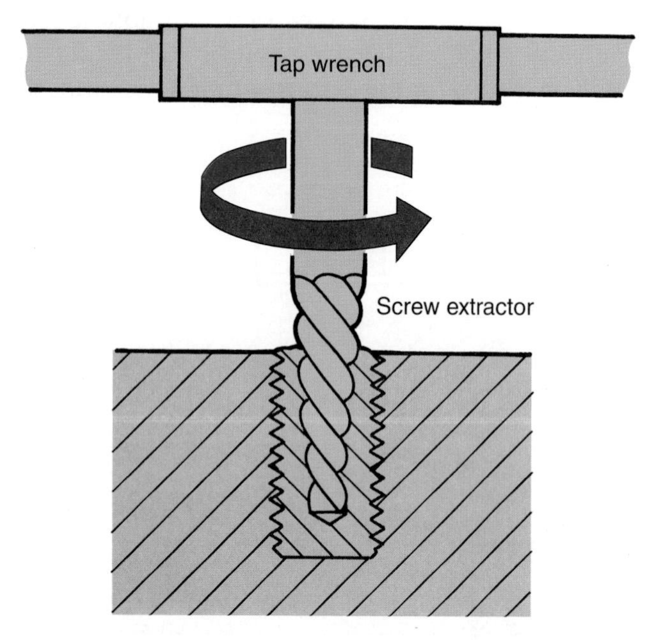

Goodheart-Willcox Publisher

Figure 8-14. Spiral-flute broken bolt extractor. A hole of proper size is drilled, a tap wrench applied, and the broken bolt is turned out.

Extractors of the type shown in **Figure 8-15** are designed to remove sheared machine screws. The blade is made of heat-treated tool steel. A hole is drilled in the screw, and then the tapering blade is lightly driven into the hole. The extractor is turned counterclockwise to remove the screw.

Treatment with penetrating oil will often make it easier to remove stubborn sheared bolts and machine screws.

8.1.7 Nuts

For most threaded fasteners, nuts with hexagonal or square shapes are used with bolts having the same-shape head. Nuts are usually manufactured of the same materials as their mating bolts. Nuts are available in various degrees of finish, **Figure 8-16**:

- A regular nut is unfinished (not machined) except on the threads.
- A regular semifinished nut is machined on the bearing face to provide a truer surface for the washer.
- A heavy semifinished nut is identical in finish to the regular semifinished nut. However, the body is thicker for additional strength.

Nuts vary in shape and size depending on their intended function. Plain hexagonal nuts are the most common. Square nuts are no longer common, but can be found in older assemblies. The jam nut is thinner than the standard nut. It is frequently used to lock a full-size nut in place, **Figure 8-17**.

Castle and slotted nuts have slots across the flats so they can be locked in place with a cotter pin or safety wire. A hole is drilled in the bolt or stud, and a cotter pin or wire is inserted through the slot and hole to prevent the nut from turning loose, **Figure 8-18**.

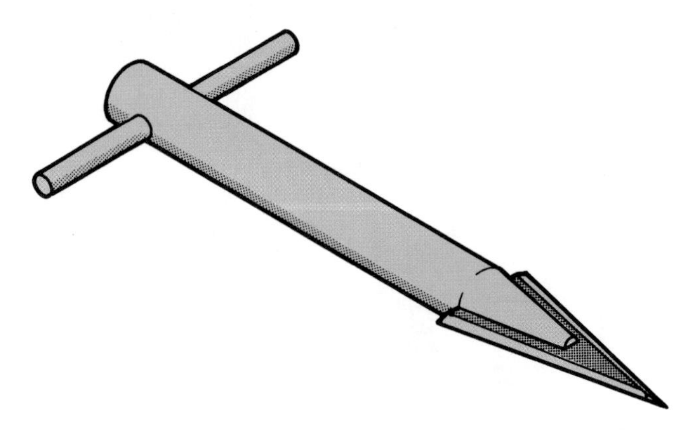

Goodheart-Willcox Publisher

Figure 8-15. An extractor designed to remove machine screws. A hole is drilled in the broken-off screw. The extractor blade is tapped into place and carefully turned to remove broken screw.

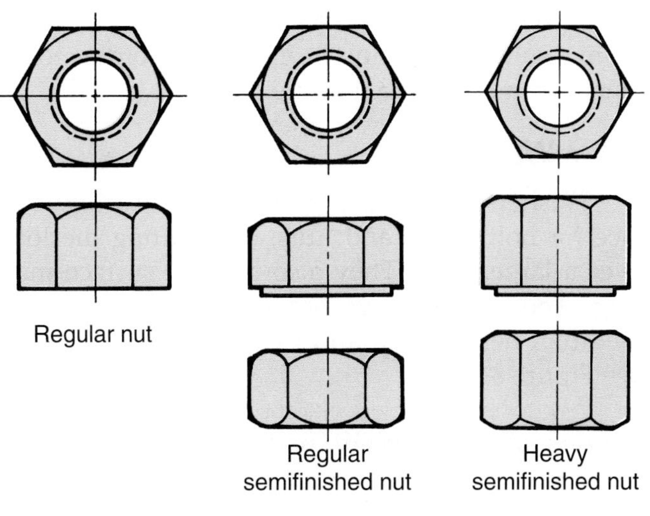

Figure 8-16. Degrees of finished nuts. Only the threads of regular nuts are machined. Regular semifinished nuts have a machined bearing surface. Heavy semifinished nuts are thicker than regular nuts.

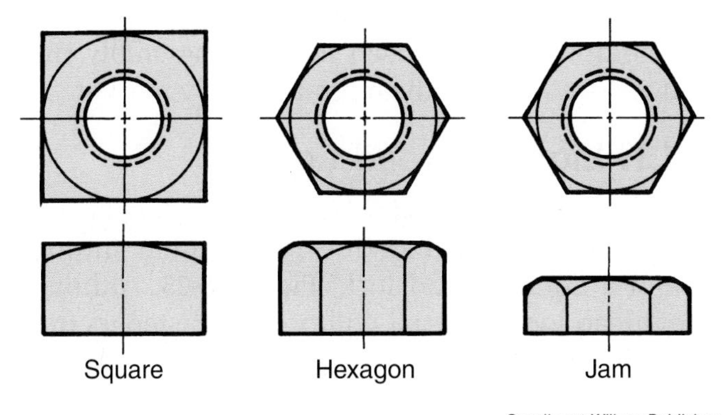

Figure 8-17. Square nuts are no longer common. A hexagon nut has six vertical sides and can be loosened or tightened with standard sockets and wrenches. A jam nut can lock a full-size nut in place to prevent loosening of the bolt.

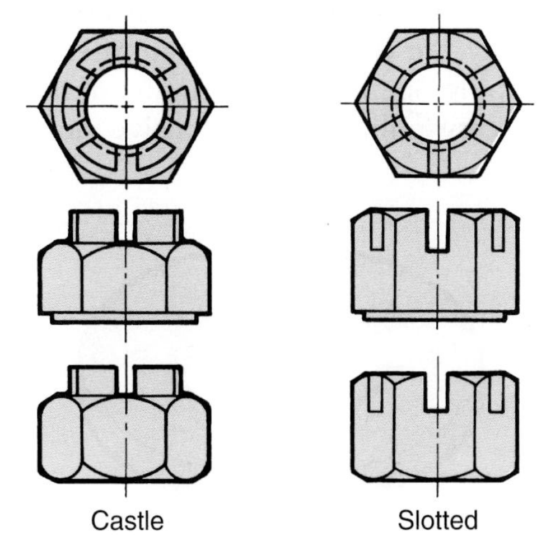

Figure 8-18. Castle and slotted nuts can be locked in place with a cotter pin or wire, but a hole must be drilled through the bolt first.

These types of locking nuts are being replaced on many applications by self-locking nuts, **Figure 8-19.** Self-locking nuts are slightly deformed to produce a friction fit, or have a nylon insert, so they cannot vibrate loose. No hole through the bolt is required when self-locking nuts are used in an assembly.

In critical assemblies, use a new self-locking nut to replace one that has been removed for any reason. The used nut may not have adequate locking action remaining and may loosen in service.

Acorn nuts are used when appearance is of primary importance, or where projecting threads must be protected. They are available in high or low crown styles. See **Figure 8-20.**

The wing nut is found where frequent adjustment or frequent removal is necessary. It can be loosened and tightened rapidly without the need of a wrench. Refer to **Figure 8-21.**

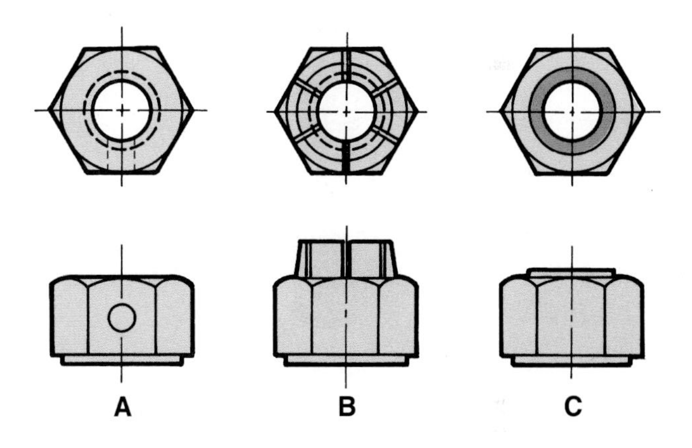

Figure 8-19. Self-locking nuts. A—The threads of center-lock nuts are deformed in the middle of the nut. B—Split-beam nuts have slots cut into the top of the nut that are bent slightly inward. C—Nylon insert lock nut.

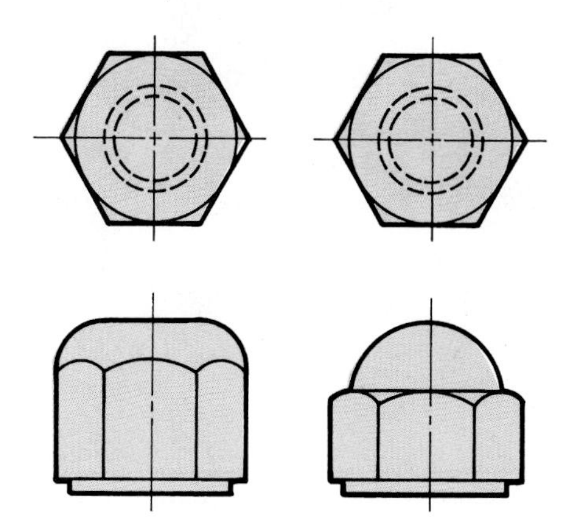

Figure 8-20. Acorn or cap nuts look good and will protect threads.

Parker-Kalon

Figure 8-21. Wing nut looks as if it has "wings." Like the thumb-screw, it is turned by hand.

8.1.8 Inserts

An insert is a special form of nut or internal thread. Inserts are designed to provide higher strength threads in soft metals and plastics. The types shown in **Figure 8-22** are frequently used to replace damaged or stripped threads. The threaded hole is drilled and tapped. The insert is then screwed into the hole.

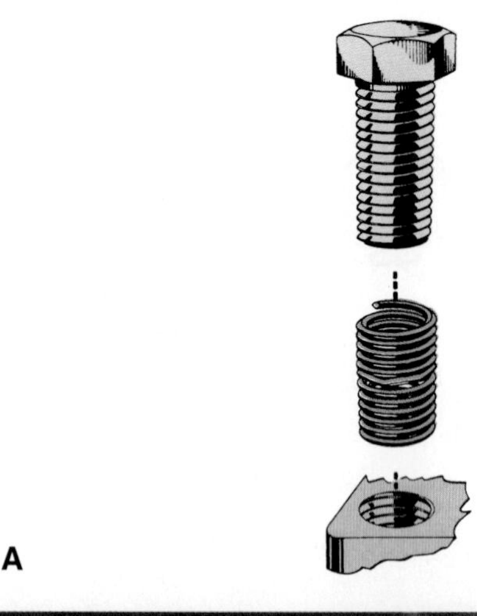

Heli-Coil Corp.; Jergens, Inc.

Figure 8-22. Thread repair inserts. A—An insert is frequently used to replace damaged or stripped threads in a part. B—These key-locking inserts can be easily installed or removed without special tools. They are available in both carbon steel and stainless steel.

Its internal thread is standard size and form. For optimum results, inserts must be installed according to the manufacturer's instructions.

8.1.9 Washers

Washers provide an increased bearing surface for bolt heads and nuts, distributing the load over a larger area. They also prevent surface marring. The standard washer is produced in light, medium, heavy-duty, and extra heavy-duty series. See **Figure 8-23**.

8.1.10 Lock Washers

A lock washer will prevent a bolt or nut from loosening under vibration. The split-ring lock washer is rapidly being replaced by the tooth-type lock washer, which has greater holding power on most applications. See **Figure 8-24**.

Preassembled lock washer and screw nuts and lock washer and nut units have a washer mounted on the nut. They are used to lower assembly time and reduce waste in the mass-assembly market.

8.1.11 Liquid Thread Lock

Nuts, bolts, and machine screws can be prevented from loosening due to vibration through use of a liquid thread lock, **Figure 8-25**. Although the thread lock material will prevent fasteners from vibrating loose, it allows easy removal of the fastener should disassembly be necessary. When using a liquid thread lock, follow the manufacturer's recommendations for maximum effectiveness.

8.1.12 Thread-Forming Screws

Thread-forming screws produce a thread in the part as they are driven. This feature eliminates a costly tapping operation. A variation of the thread-forming screw eliminates expensive hole-making

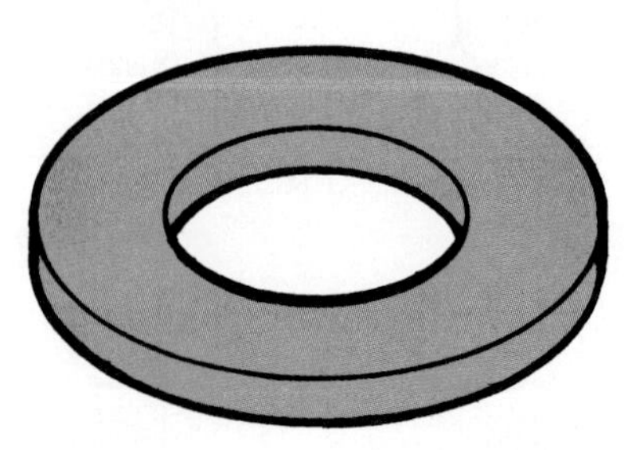

Goodheart-Willcox Publisher

Figure 8-23. The standard flat washer provides a bearing surface for a fastener.

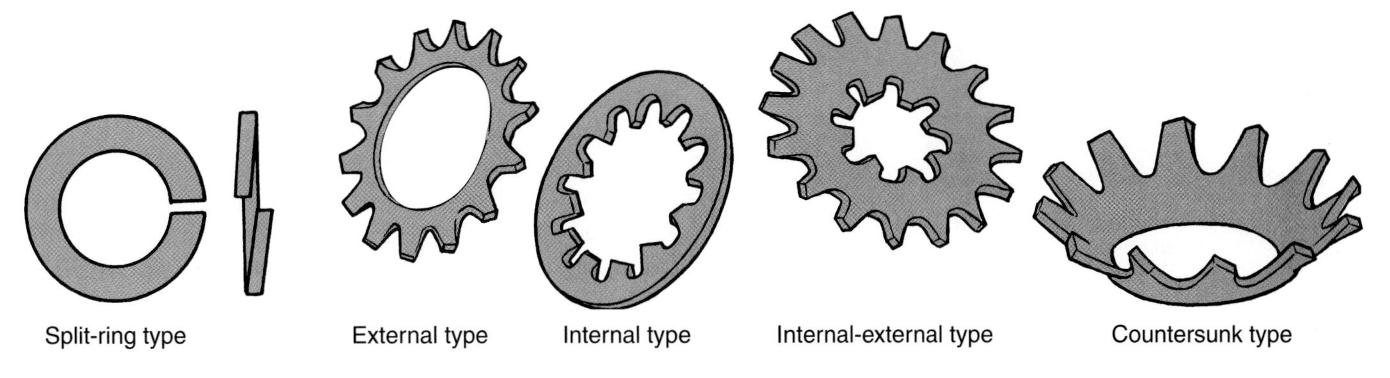

Goodheart-Willcox Publisher

Figure 8-24. Lock washer variations. External typelock washers should be used whenever possible, because it provides greatest resistance to turning. The internal type lock washer is used with small head screws and to hide teeth, either for appearance or to prevent snagging. Internal-external type lock washers are used when mounting holes are oversize. The countersunk type is used with flat or oval head screws.

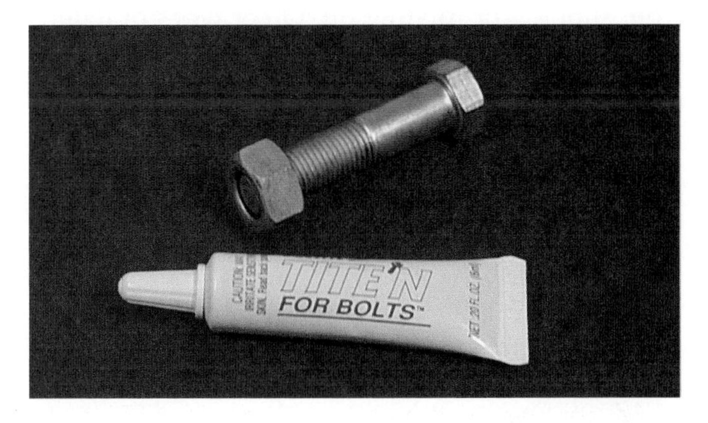

Goodheart-Willcox Publisher

Figure 8-25. There are many liquid thread locks available. They prevent a bolt, nut, or screw from vibrating loose, but allow easy removal should disassembly be required.

(drilling or punching) and aligning operations because the screw drills its own hole as it is driven into place. See **Figure 8-26**.

8.1.13 Thread Cutting Screws

Thread cutting screws differ from thread forming screws because they actually cut threads into the material when driven. Refer to **Figure 8-27**. Thread cutting screws are hardened. They are used to join heavy-gage sheet metal, and to thread into nonferrous metal assemblies.

8.1.14 Drive Screws

Drive screws are simply hammered into a drilled or punched hole of the proper size. A permanent assembly (one that does not have to be taken apart) results, **Figure 8-28**.

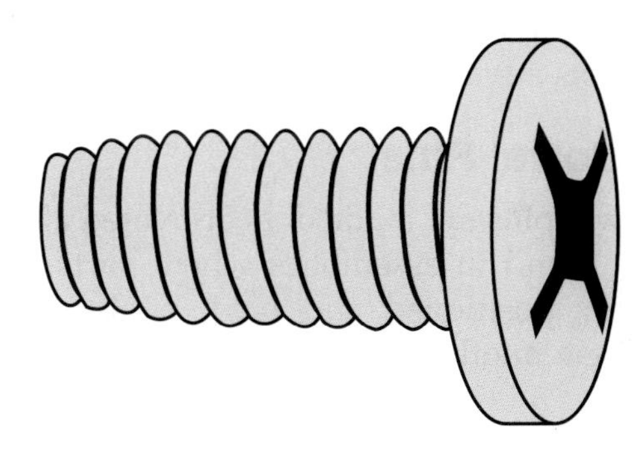

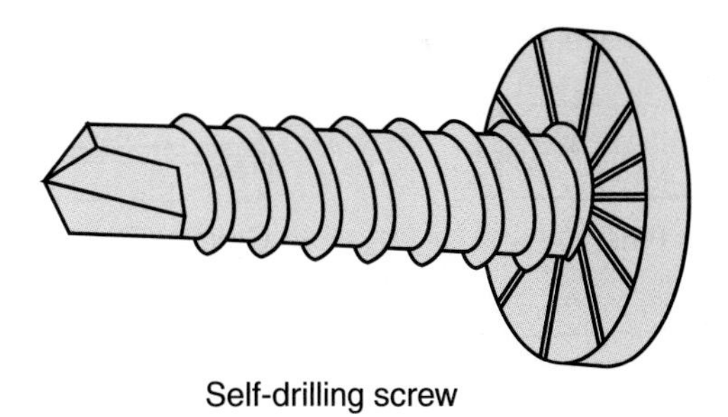

Goodheart-Willcox Publisher

Figure 8-26. Thread-forming screws.

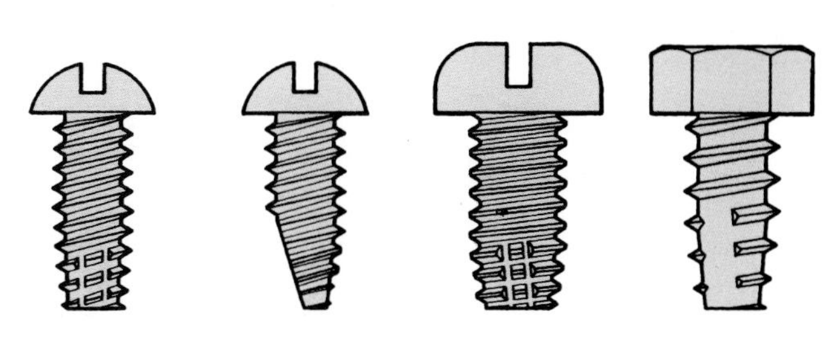

Goodheart-Willcox Publisher

Figure 8-27. Variations among thread cutting screws.

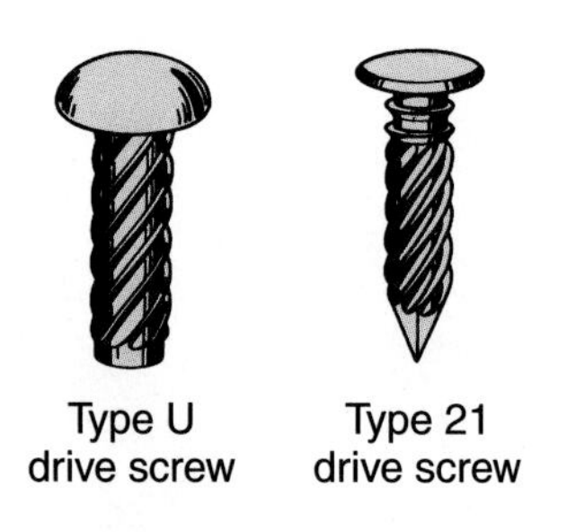

Parker-Kalon

Figure 8-28. Drive screws are hammered or forced into place in a presized hole.

8.2 Nonthreaded Fastening Devices

Nonthreaded fasteners comprise a large group of mechanical holding devices. These include dowel pins, cotter pins, retaining rings, rivets, and keys. Each has its advantages.

8.2.1 Dowel Pins

Dowel pins are made of heat-treated alloy steel and are found in assemblies where parts must be accurately positioned and held in absolute relation to one another. See **Figure 8-29**. They ensure perfect alignment and facilitate quicker disassembly and reassembly of parts in exact relationship to each other. They are fitted into reamed holes and are available in diameters from 1/16″ to 1″. They are also available in metric sizes.

Dowel pins are normally 0.0002″ (0.005 mm) oversize (identified by a plain steel finish) but are available in 0.001″ (0.025 mm) oversize (identified by a black finish) for repairs.

Taper pins are made with a uniform taper of 1/4″ per foot in lengths up to 6″, with diameters as small as 5/32″ at the large end, **Figure 8-30**.

8.2.2 Cotter Pins

A cotter pin is fitted into a hole drilled crosswise through a shaft, **Figure 8-31**. The pin prevents parts from slipping or rotating off. Other types of retaining devices are replacing the cotter pin.

8.2.3 Retaining Rings

The retaining ring, **Figure 8-32**, has been developed for both internal and external applications.

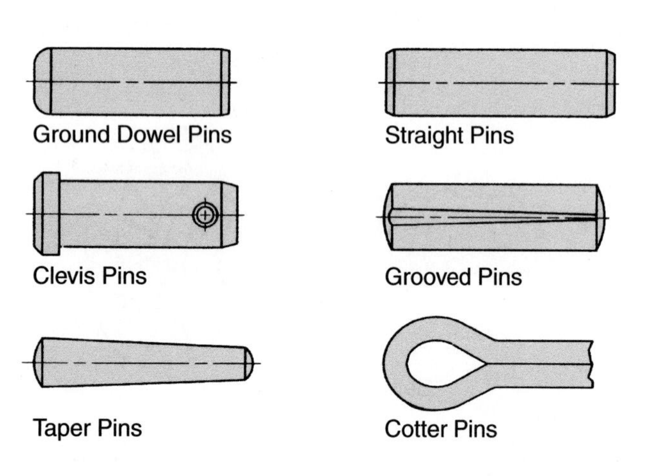

Goodheart-Willcox Publisher

Figure 8-29. Types of dowel pins. They are made in a wide range of sizes and types.

Goodheart-Willcox Publisher

Figure 8-30. A taper pin is often used to lock a handle to its shaft.

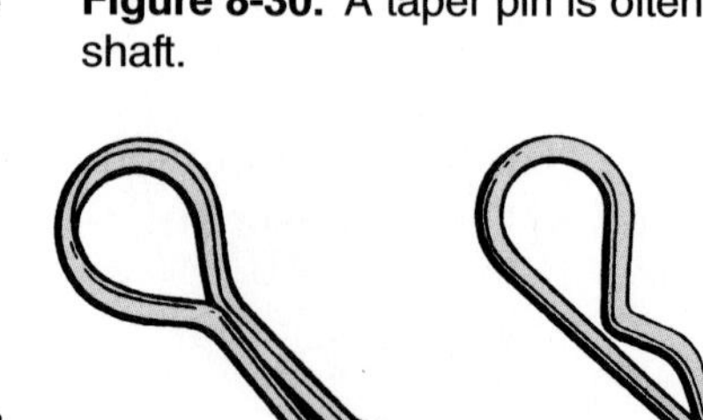

Goodheart-Willcox Publisher

Figure 8-31. Types of cotter pins.

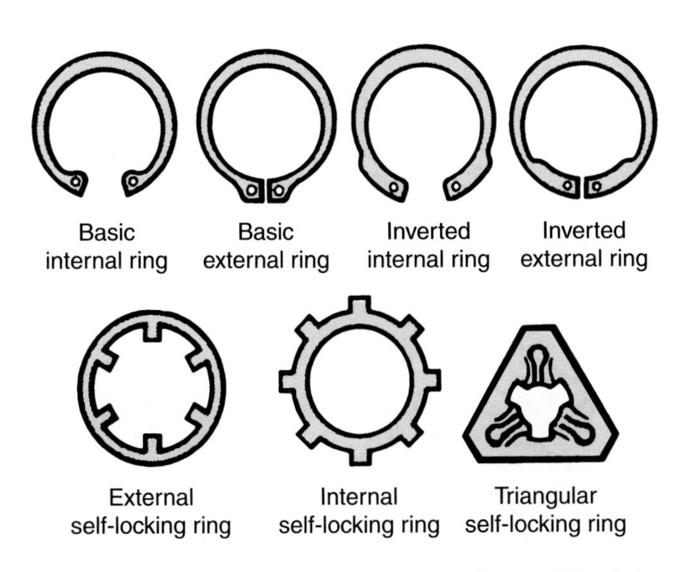

Goodheart-Willcox Publisher

Figure 8-32. Retaining rings.

Retaining rings reduce both cost and weight of the product on which they are used. While most retaining rings must be seated in grooves, **Figure 8-33**, a self-locking type does not require this special recess. Special pliers are needed for rapid installation and removal of the retaining rings, **Figure 8-34**.

8.2.4 Rivets

Permanent assemblies can be made with rivets, **Figure 8-35**. Solid rivets can be set, or deformed to become larger on one end, by hand or machine methods.

Blind rivets are mechanical fasteners that have been developed for applications where the joint is accessible from only one side. They require special tools for installation, **Figure 8-36**. Blind rivet types are shown in **Figure 8-37**.

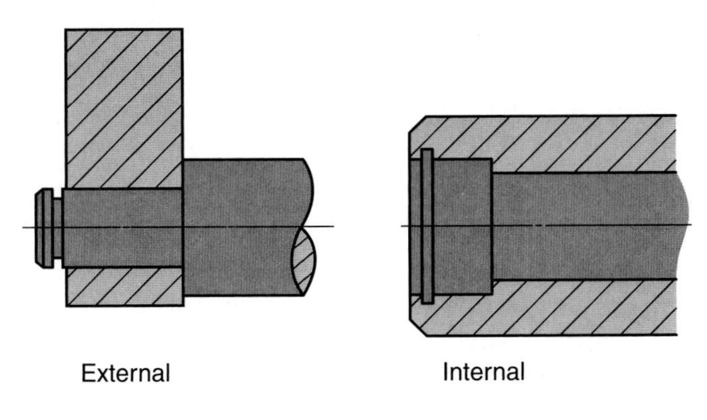

Goodheart-Willcox Publisher

Figure 8-33. Grooves are machined into parts to receive retaining rings. They eliminate many other expensive machining operations.

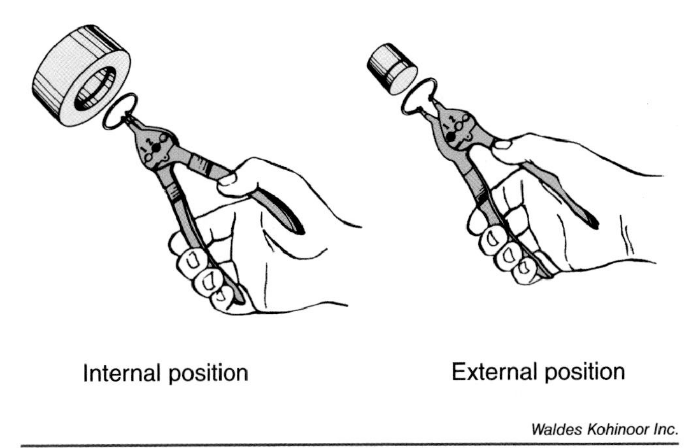

Waldes Kohinoor Inc.

Figure 8-34. Special pliers are used to install some types of retaining rings.

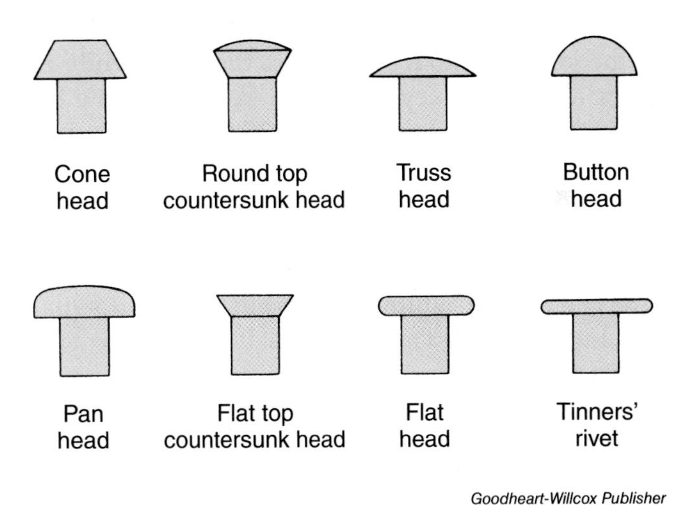

Goodheart-Willcox Publisher

Figure 8-35. Rivet head styles.

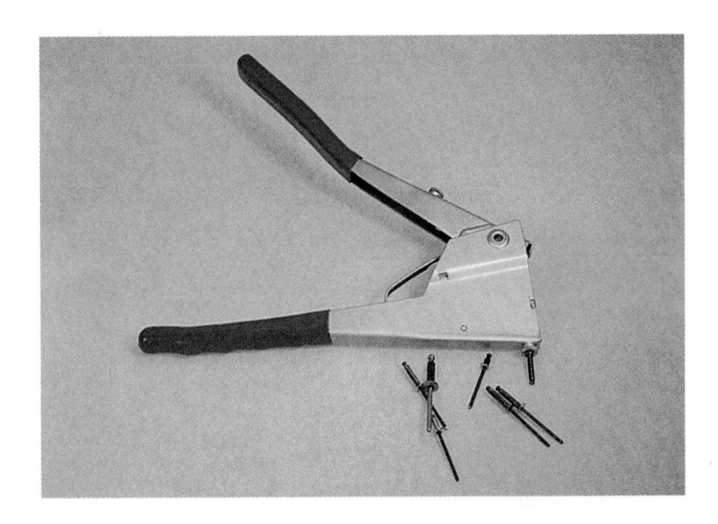

Goodheart-Willcox Publisher

Figure 8-36. A pliers-type rivet gun is used to insert one type of blind rivet.

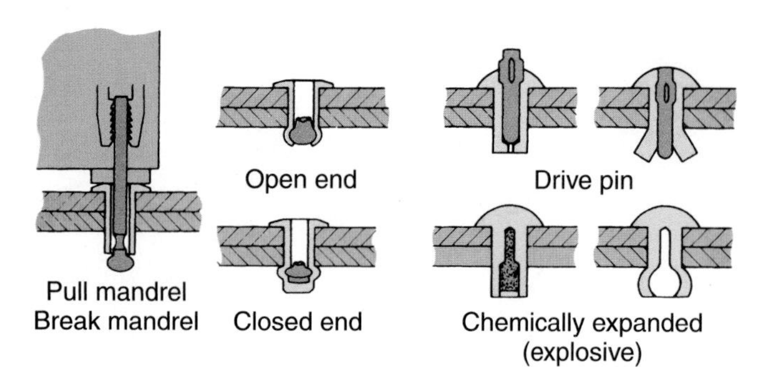

Figure 8-37. Types of blind rivets.

8.2.5 Keys

A ***key*** is a small piece of metal that prevents a gear or pulley from rotating on its shaft. One-half of the key fits into a keyseat on the shaft while the other half of the key fits into a keyway in the hub of the gear or pulley, **Figure 8-38**. Commonly used keys are shown in **Figure 8-39**.

A square key is usually one-fourth the shaft diameter. It may be slightly tapered on the top to make it easier to install.

The Pratt & Whitney key is similar to the square key, but is rounded at both ends. It fits into a keyseat of the same shape.

The gib head key is interchangeable with the square key. The head design permits easier removal from the assembly.

A Woodruff key is semicircular and fits into a keyseat of the same shape. The top of the key fits into the keyway of the mating part.

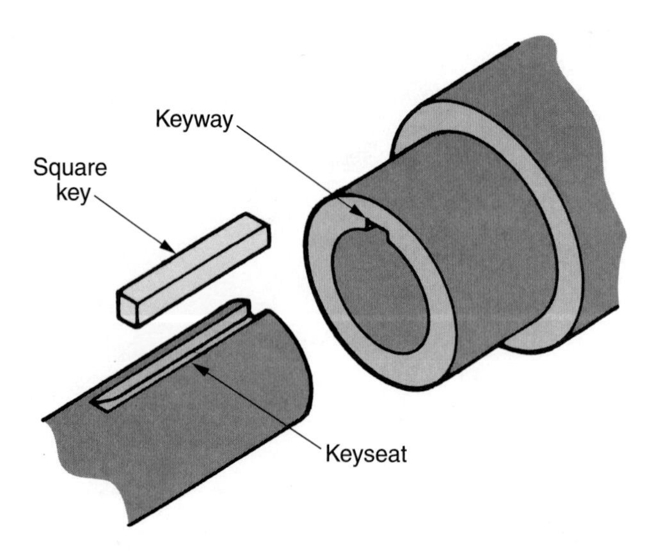

Figure 8-38. A square key is used to prevent a pulley or gear from rotating on its shaft.

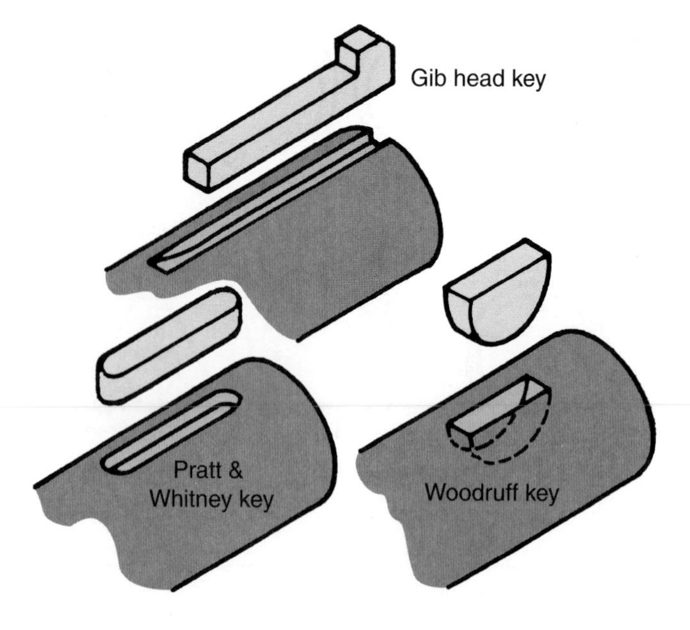

Figure 8-39. Three types of keys.

8.3 Adhesives

Adhesives provide another way to join metals and to keep threaded fasteners from vibrating loose. In some applications, the resulting joints are stronger than the metal itself. Adhesive bonded joints do not require costly and time-consuming operations such as drilling, countersinking, riveting, or welding.

The major drawback to the use of adhesives is heat. While some adhesives retain their strength at temperatures as high as 700°F (371°C), most should not be used for assemblies that will be exposed to temperatures above 150–200°F (66–93°C).

Adhesives for locking threaded fasteners in place are made in a number of chemical formulations. The desired permanence of the threaded joint will determine the type of adhesive to be used.

Adhesive-bonded assemblies offer many advantages over other fastening techniques:

- The load is distributed evenly over the entire joined area, **Figure 8-40**.
- There is continuous contact between the mating surfaces, **Figure 8-41**.
- Full strength of the mating parts is maintained, since holes do not have to be made to insert fasteners. The extreme heat required for joining methods like welding is not necessary with adhesive bonding. This means there is no danger of the work becoming distorted or having its heat treatment affected.

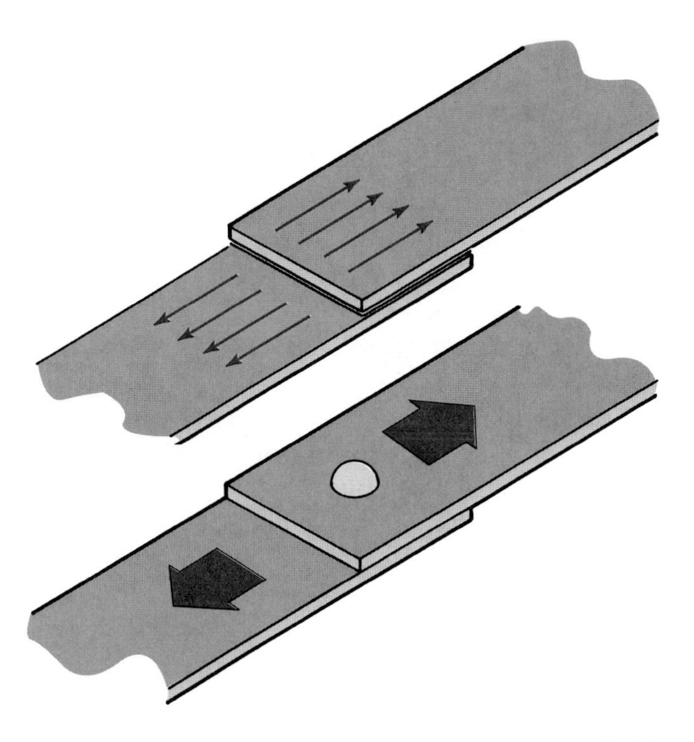

Goodheart-Willcox Publisher

Figure 8-40. When an adhesive is used to join metal, the load is distributed evenly over the entire joint. A rivet or conventional threaded fastener localizes the load in a small area.

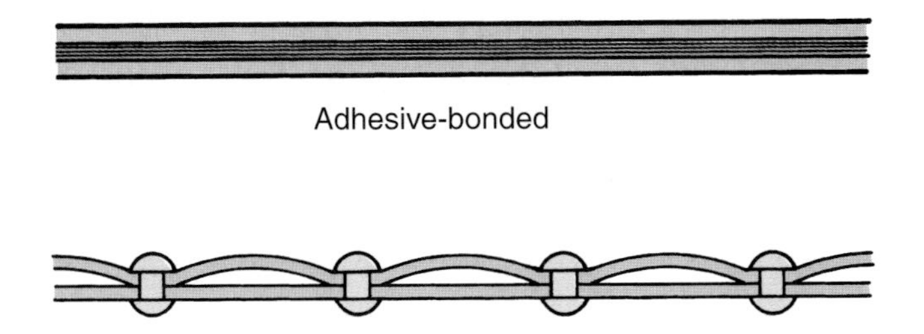

Goodheart-Willcox Publisher

Figure 8-41. On parts joined with an adhesive, the mating surfaces are in continuous contact.

- Smooth surfaces result from adhesive bonding—there are no external projections (as with rivets or bolts), and the surface is not marred by the heat and pressure necessary to join pieces with spot welds.

Many commercial adhesives are sold in small quantities. They are suitable for use in training areas and the home, **Figure 8-42**.

Adhesives are available in liquid, paste, or solid form. Many can be applied directly from the container. Others must be mixed with a catalyst or hardener. A few pressure-sensitive adhesives are manufactured in sheet form.

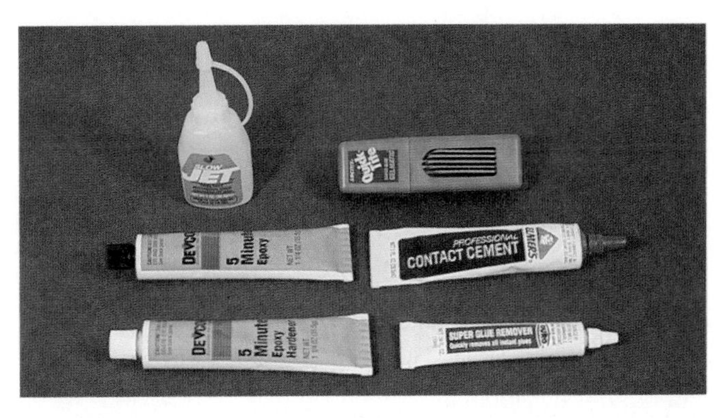

Goodheart-Willcox Publisher

Figure 8-42. Adhesives for joining metal to metal and metal to other materials are available in good hardware stores. They are similar to those found in industry.

One type of adhesive has found growing use in machining technology to make temporary bonds. ***Cyanoacrylate quick setting adhesives*** (known by such trade names as Eastman 910™, Super Glue™, and Crazy Glue™) are used to hold matching metal sections together while they are being machined. Round stock too small for existing collets can be glued into larger stock for turning, milling, or grinding. Some fragile parts have been glued to backup blocks for machining, and parts like the gun sight casting shown in **Figure 8-43** have been glued to a fixture for such operations as machining the sighting and bottom grooves.

After machining, the parts can be removed from the holding device by an application of heat (175°F or 79°C maximum). Very small parts can be removed by applying a cyanoacrylate debonder.

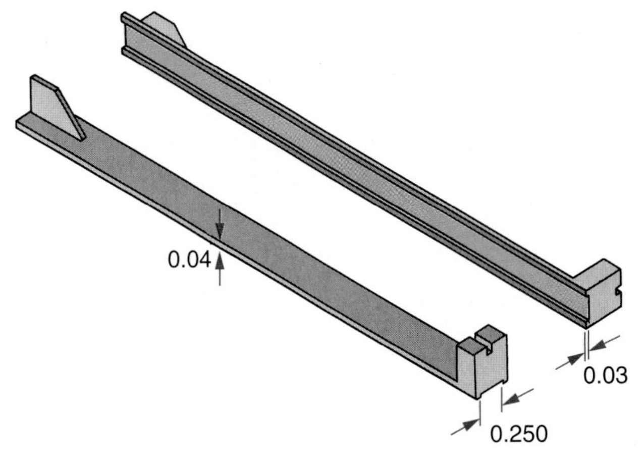

Goodheart-Willcox Publisher

Figure 8-43. This gun sight was held on a fixture with cyanoacrylate adhesive to allow machining. The thin base section of the part made it difficult and expensive to mount to a fixture by other means.

For successful use of cyanoacrylate adhesives, the part and mounting surface must be prepared according to the adhesive manufacturer's directions.

SAFETY NOTE

When using cyanoacrylate adhesives, always wear approved eye protection and keep fingers away from your eyes and mouth. Since this adhesive can instantly bond fingers to each other or to other surfaces, always have a debonder available for immediate use. Unless a suitable solvent is available, surgery might be needed to separate the joined fingers. Should you get adhesive in your eyes, see a physician immediately.

8.4 Using Adhesives

Most adhesives require following a five-step process to produce solidly bonded joints:

1. Surface preparation is critical! All adhesives require clean surfaces to produce full-strength bonds. Preparation may range from simply wiping surfaces with a solvent to performing multistage cleaning and chemical treatment.

SAFETY NOTE

The chemicals in adhesives for joining metals and other materials can cause severe skin irritation. To be safe, wear disposable plastic gloves when preparing or applying all types of adhesives.

2. Adhesive preparation must be done properly. Mixing, delivery to the work area, and setting up equipment must all be done according to the manufacturer's directions.

SAFETY NOTE

Carefully follow all instructions on the adhesive container when mixing and using adhesives. Only mix the amount you will need. Promptly remove any adhesive from your skin by washing in water.

3. Adhesive application may be done by brushing, rolling, spraying, dipping, or methods designed for a specific assembly.

SAFETY NOTE

Do not apply adhesives near areas where there are open flames. Solvents used in some adhesives can be highly flammable or toxic. Apply them only in well-ventilated areas, and wear a suitable respirator.

4. Assembly involves positioning of materials to be joined. This often requires the use of jigs or fixtures for alignment.
5. Bond development is the process of evaporation of solvents and curing of the adhesive. It may involve application of pressure or heat.

Chapter Review

Summary

- Threaded fasteners include machine screws, machine bolts, cap screws, setscrews, and stud bolts.
- Nuts, inserts, and washers are commonly used with threaded fasteners.
- English and metric fasteners are *not* interchangeable.
- Nonthreaded fasteners include dowel pins, cotter pins, retaining rings, rivets, and keys.
- Adhesives are used to join metal and have many advantages, but also several unique disadvantages.

Review Questions

Answer the following questions using the information provided in this chapter.

1. For maximum strength, a threaded fastener should screw into its mating part a distance equal to _____ times the diameter of the thread.
2. List four types of threaded fasteners. Briefly describe how each is used.
3. _____ screws are used for general assembly work.
4. How is the strength of hex-head cap screws indicated?
5. To prevent a pulley from slipping on a shaft, a(n) _____ is often used.
6. The _____ bolt is threaded at both ends.
7. When removing stubborn sheared bolts, what can be done to make their removal easier?
8. When is a jam nut used?
9. The shape of the _____ nut permits it to be loosened and tightened without a wrench.
10. Why are lock washers used?
11. Drive screws and rivets can be used to create a(n) _____.
12. While most _____ must be seated in grooves, a self-locking type does not require the special recess.

Match each word in the left column with the most correct sentence in the right column. Place the appropriate letter in the blank.

13. _____ Rivet.
14. _____ Jam nut.
15. _____ Drive screw.
16. _____ Thread-cutting screw.
17. _____ Acorn nut.
18. _____ Dowel pin.
19. _____ Blind rivet.
20. _____ Keyway.
21. _____ Keyseat.
22. _____ Key.

A. Developed for use in confined area, where a joint is only accessible from one side.
B. Used where parts must be aligned accurately and held in absolute relation with one another.
C. Prevents a pulley or gear from slipping on a shaft.
D. Protects projecting threads.
E. Is hammered into a drilled or punched hole.
F. Used to make permanent assemblies.
G. Slot cut in gear or pulley to receive a key.
H. Locks a regular nut in place.
I. Eliminate costly tapping operations.
J. Slot cut in shaft to receive a key.

23. What advantages do adhesives offer over other fastening techniques?
    A. The load is distributed evenly over the entire area.
    B. There is continuous contact between the mating surfaces.
    C. Full strength of the mating parts is maintained.
    D. No external projections result in smooth surfaces.
    E. All of the above.
24. Briefly describe, in order, the steps that must be used to join metals with adhesives.
25. List at least three safety precautions that must be observed when using adhesives.

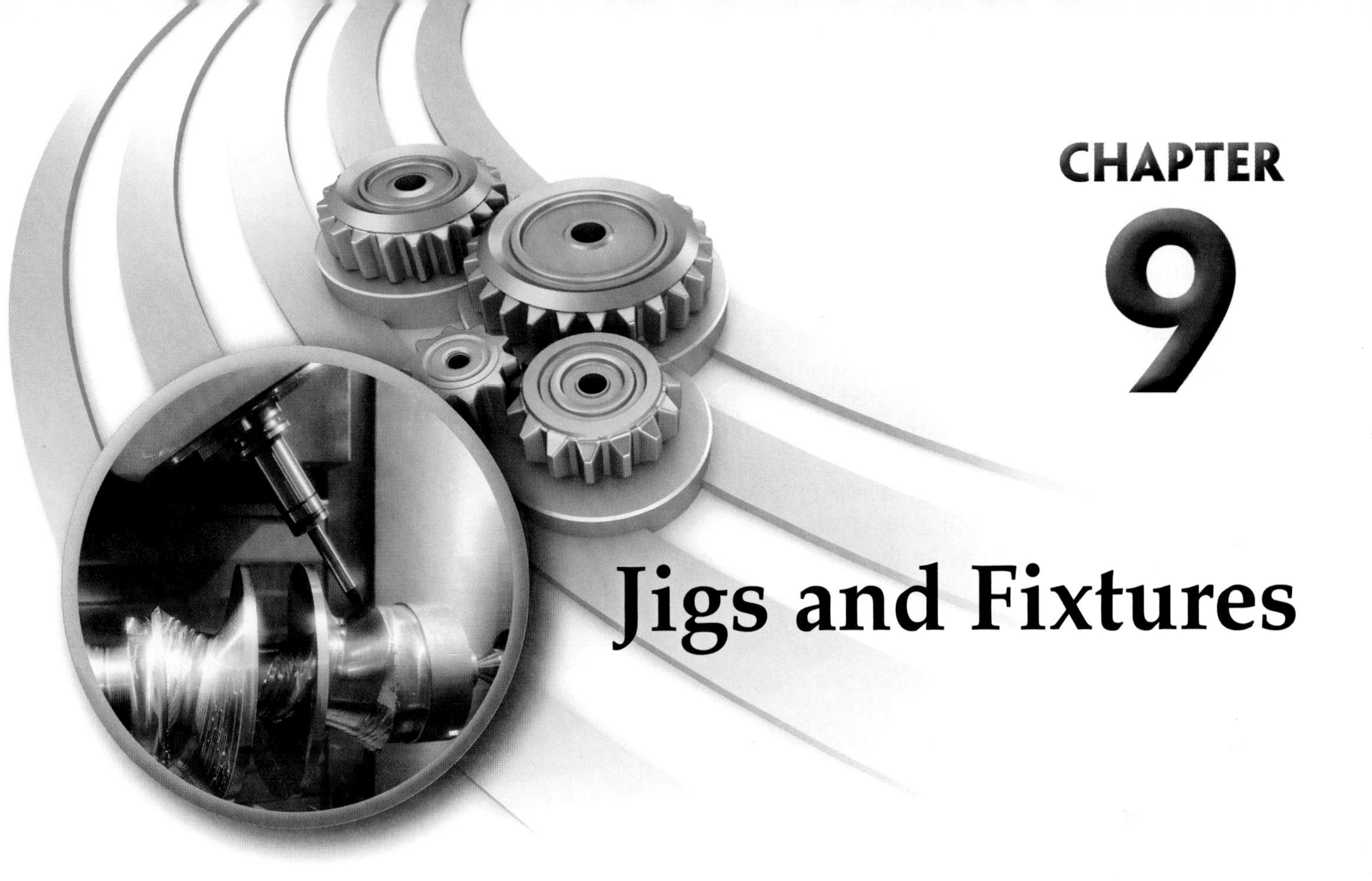

CHAPTER 9

Jigs and Fixtures

Learning Objectives

After studying this chapter, you will be able to:

- Describe a jig.
- Describe a fixture.
- Elaborate on the classifications of jigs and fixtures.
- Explain why jigs and fixtures are used.
- Describe a tombstone.

Technical Terms

bushing
fixture
jig
tombstone

Jigs and fixtures are devices that are used extensively in production machine shops to hold work while machining operations are performed. They position the work so that all of the parts produced are uniform and within specifications. When large numbers of identical and interchangeable parts must be produced, the use of jigs and fixtures helps to reduce manufacturing costs. The use of jigs and fixtures is often justified when limited production is required because they allow relatively unskilled workers to operate the machines.

Jigs and fixtures are also used in assembly work for operations such as welding or riveting. They position and hold work to make the fabrication of standardized parts feasible.

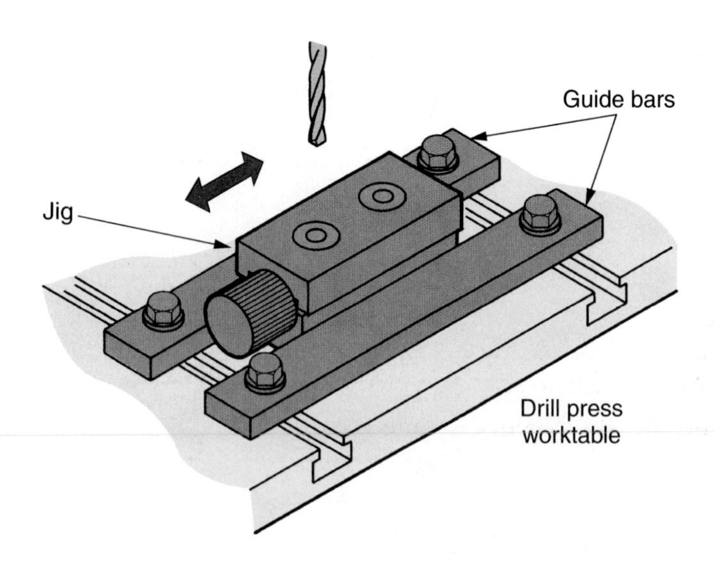

Goodheart-Willcox Publisher

Figure 9-2. A drill jig is nested between guide bars to prevent dangerous and undesirable "merry-go-round" rotation.

9.1 Jigs

A ***jig*** is a device that holds a workpiece in place and guides the cutting tool during a machining operation such as drilling, reaming, or tapping. Hardened steel ***bushings*** are used to guide the drill or cutting tool, **Figure 9-1**.

The jig is seldom mounted solidly to the drill press table. For safety, however, it is usually nested between guide bars that are mounted solidly to the table, **Figure 9-2**.

Drill jigs fall into two general types: open jigs and box (or closed) jigs. The drill template or plate jig is the simplest form of the open jig. It consists of a plate with holes to guide the drill. The jig fits over the work, **Figure 9-3**.

In a more elaborate form of an open drill jig, **Figure 9-4**, clamps are used to hold the work in place. Drill jigs may be fitted with a base plate to provide clearance for the drill as it breaks through the work.

The box jig encloses the work, **Figure 9-5**. This type is more costly to make than an open jig, but is often used when holes must be drilled in several directions. **Figure 9-6** illustrates a box jig in its simplest form. The work is fitted into the jig through a hinged or swinging cover. The clamps that hold the work in place are permanently mounted to the jig.

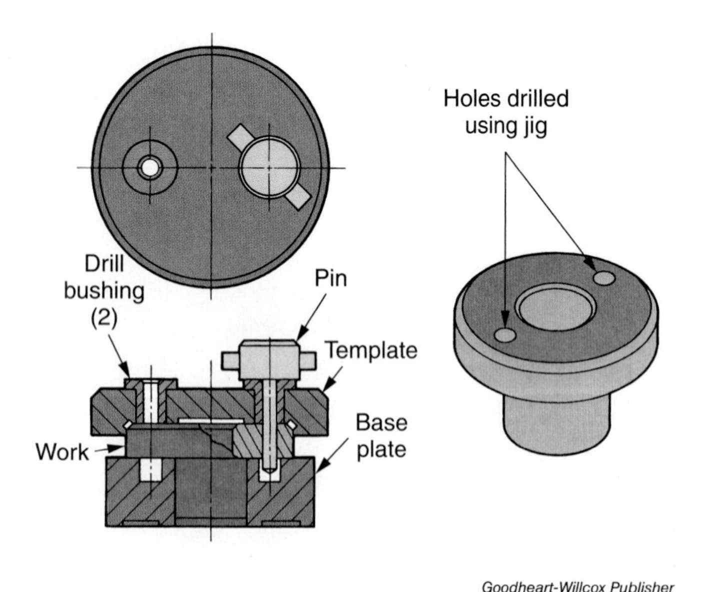

Goodheart-Willcox Publisher

Figure 9-1. With this circular type drill jig, a pin is placed into the first hole after it is drilled. This holds the workpiece in position when drilling the second hole. The base plate provides clearance for the drill as it breaks through the workpiece.

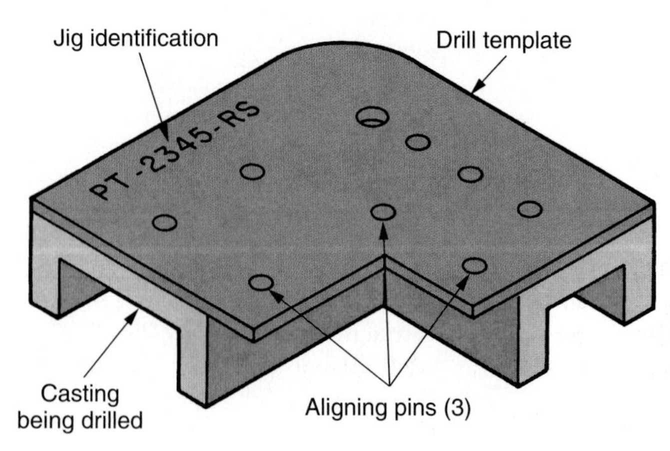

Goodheart-Willcox Publisher

Figure 9-3. A simple drill template. Identification numbers on jigs and fixtures allow these devices to be located easily when stored away between uses.

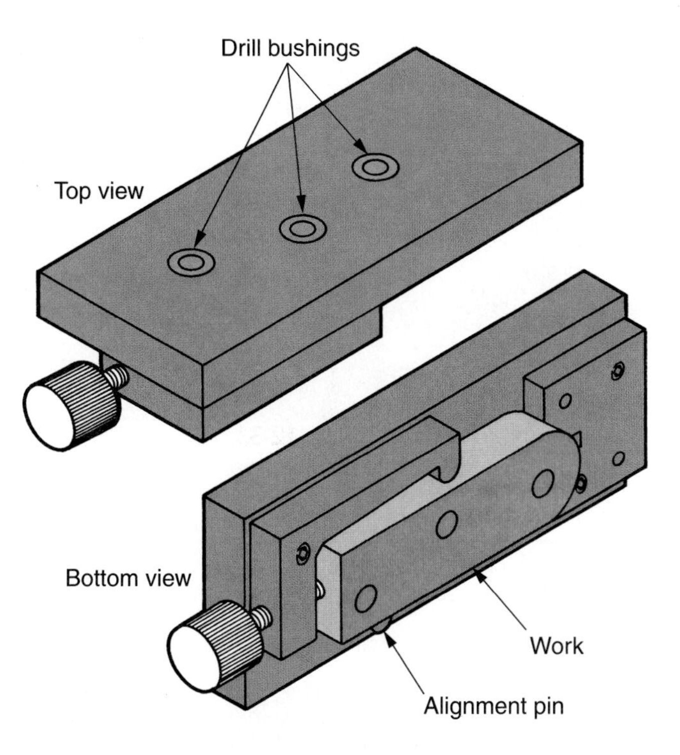

Goodheart-Willcox Publisher

Figure 9-4. This open jig has a clamp to hold the work in position for drilling. A V-notch at one end and an alignment pin at the other end position the work properly in the jig.

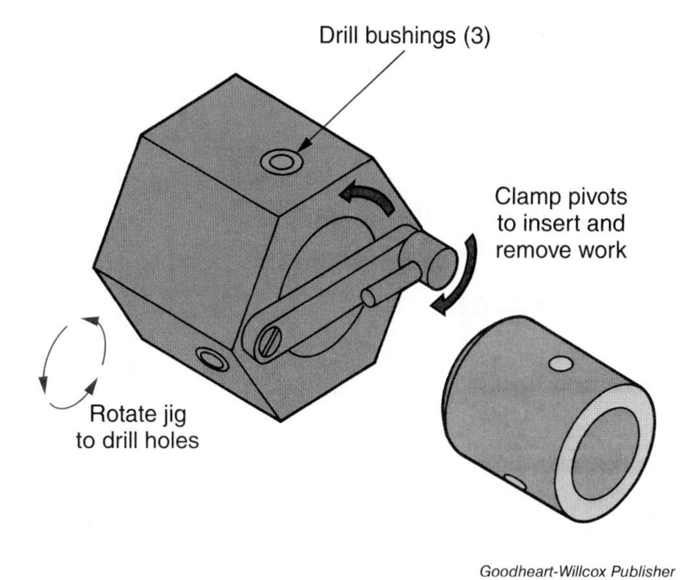

Goodheart-Willcox Publisher

Figure 9-6. A light box jig used to drill three equally spaced holes in a base end cap. Since only a limited production run was required, it was not necessary to construct a more elaborate jig.

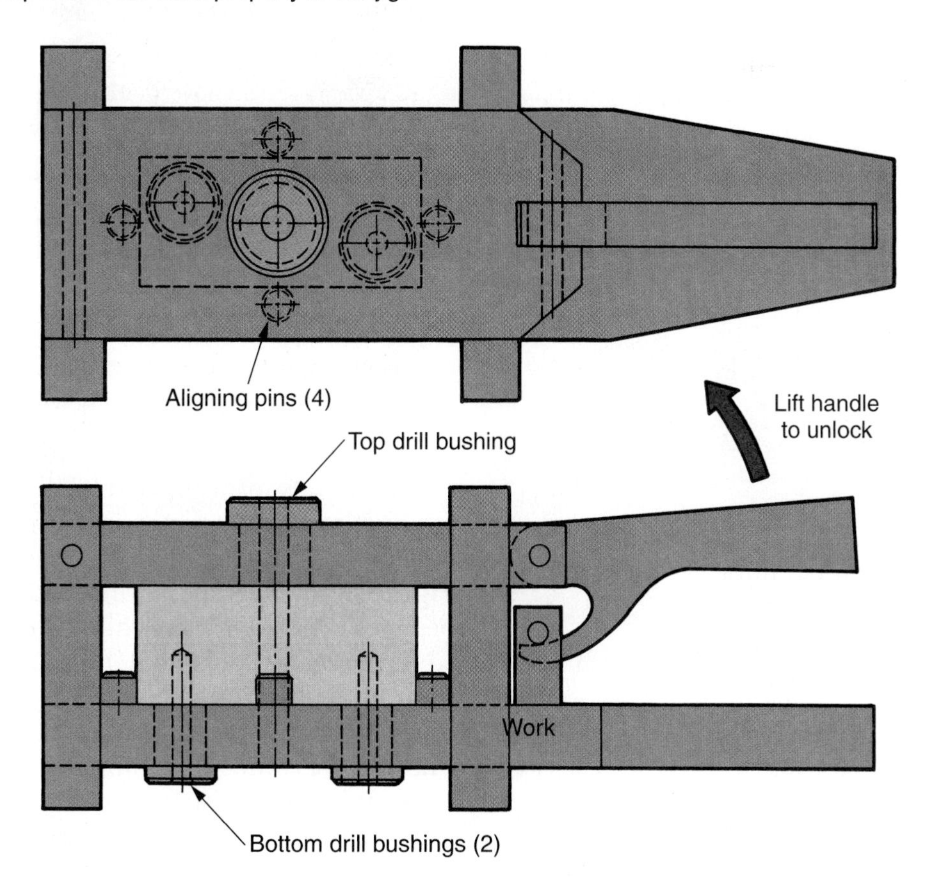

Goodheart-Willcox Publisher

Figure 9-5. A box drill jig. Lowering the handle locks work in the jig.

When several different operations must be performed on a job, a combination of open and box jigs is often used. Slip bushings are used to guide the drills. The slip bushings are then removed for subsequent operations such as reaming, tapping, countersinking, counterboring, or spot facing.

9.2 Fixtures

A ***fixture*** is used to position and hold a workpiece while machining operations are performed on it, **Figure 9-7**. Unlike a jig, a fixture does not guide the cutting tools.

Fixtures fall into many classifications. The class is determined by the type of machine tool on which the fixture is used, such as a machining center, milling machine (vertical or horizontal), lathe, band saw, or grinder. Fixture designs range from simple vise jaw modifications, **Figure 9-8**, to very large and complex devices used by the aerospace industry, **Figure 9-9**.

9.3 Jig and Fixture Construction

Jigs and fixtures are devices that are designed for specific jobs. Their complexity is determined by the number of pieces to be produced, the degree of accuracy required, and the kind of machining operations that must be performed.

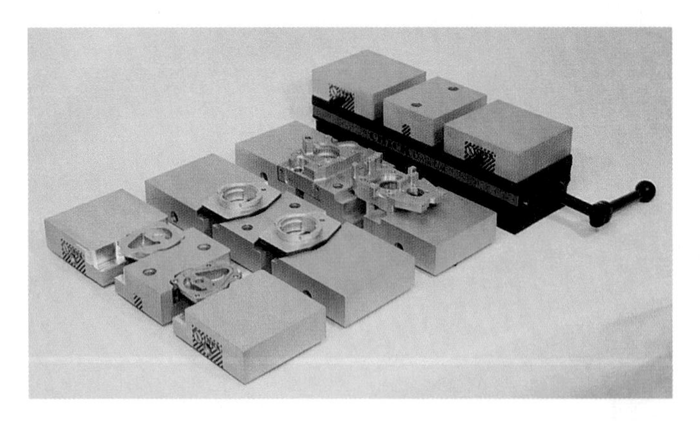

Chick Machine Tool, Inc.

Figure 9-7. Machining centers often require special fixture holding devices. The pockets to hold the work are machined directly into the body of the jaws. Three examples are shown in the foreground. The jaws, which snap on and off the work-holding system, are shown in the background. New setups can be made in a very short period of time.

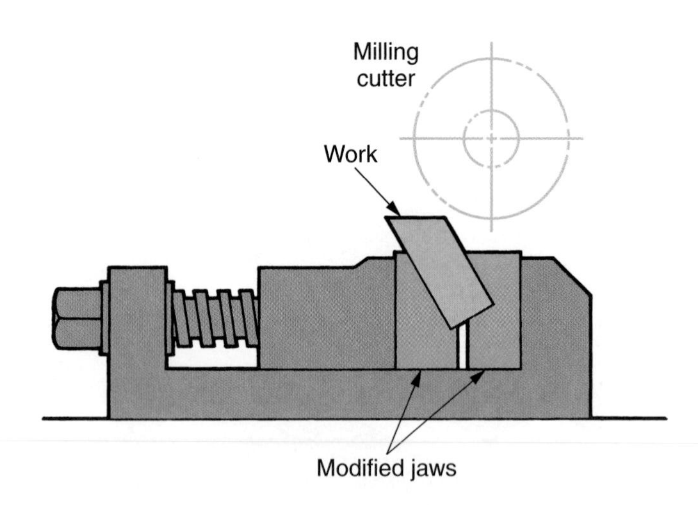

Goodheart-Willcox Publisher

Figure 9-8. This simple fixture consists of vise jaws that have been modified to position a workpiece so that an angular cut can be made on it.

Peter R Foster IDMA/Shutterstock.com

Figure 9-9. Many fixtures are used in the assembly of an aircraft fuselage. In fixture construction, extreme accuracy is critical, requiring use of lasers to ensure precise alignment.

The body of a jig or fixture may be built-up, welded, or cast. Commercial components are available in a wide range of sizes, types, and shapes. See **Figure 9-10**. Special fixture-holding devices have been developed for machining centers and other CNC machine tools that permit multiple setups, **Figure 9-11**. A ***tombstone*** is a common fixture-holding device used with CNC machine tools, **Figure 9-12**. Tombstones are made from heavy castings and are precisely machined. Tombstones are mounted on a machine's worktable. Fixtures are mounted directly onto the tombstones. The tombstone may be blank, have drilled and tapped mounting holes, or use T-slots for mounting fixtures.

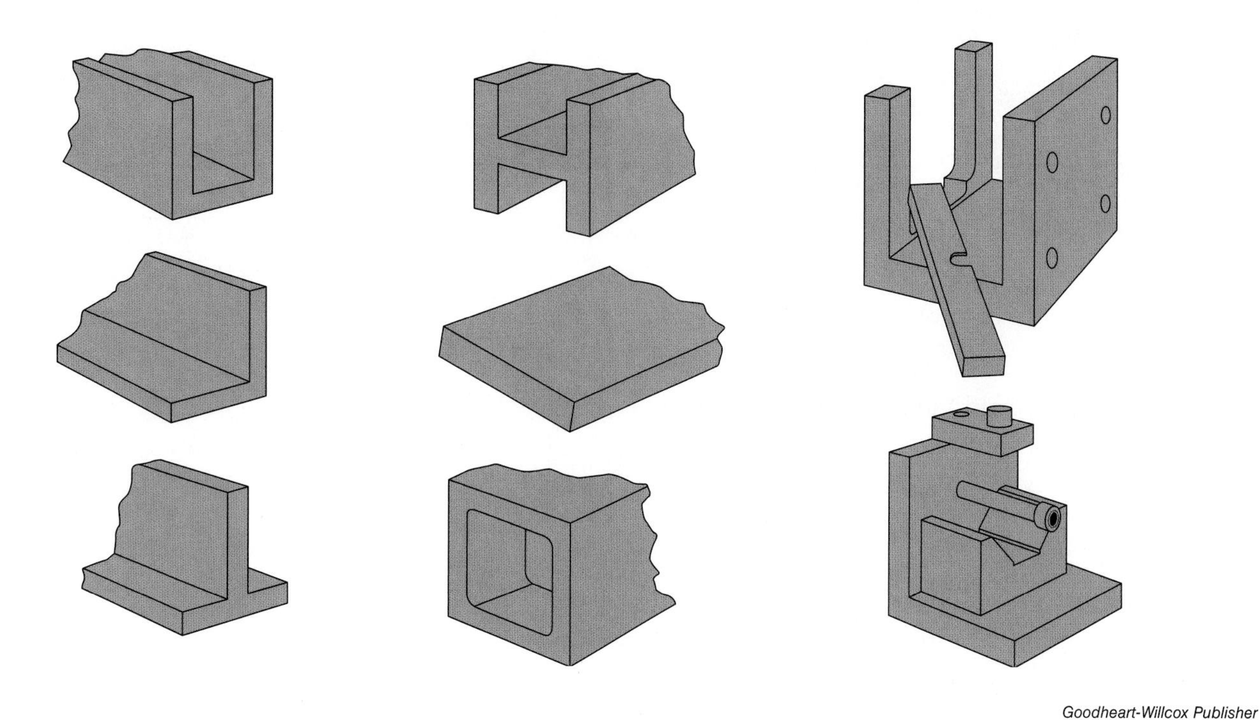

Goodheart-Willcox Publisher

Figure 9-10. Standard cast iron shapes are machined parallel and square to save time and money in both designing and building jigs and fixtures. Sections of different shapes can be bolted together to form complex holding devices. Two completed units are shown at the right of the illustration.

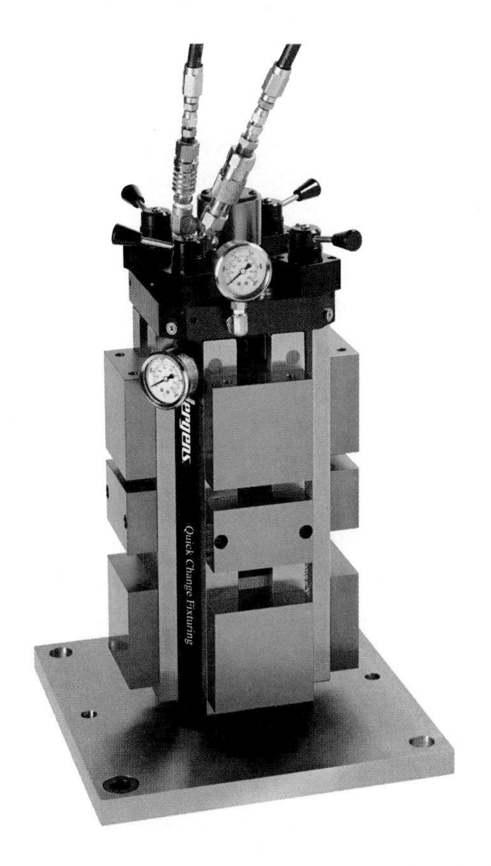

Chick Machine Tool, Inc.

Figure 9-11. Special fixture-holding devices for machining centers and other CNC machine tools. Work-holding pockets are cut directly into the jaw blocks. Pivoting the vertical setup will bring the next set of workpieces into position.

Tibor Machine Products

Figure 9-12. A typical tombstone with drilled and tapped mounting holes. The work-holding fixtures have been mounted directly to the tombstone. A well designed fixture holds the workpiece securely and accurately, while allowing quick part changes.

Chapter Review

Summary

- Jigs and fixtures hold the workpiece during machining operations.
- Open and box type jigs guide drills or cutting tools.
- Fixtures are used with many machine tools.
- Special fixture holding devices for CNC machine tools allow multiple setups.

Review Questions

Answer the following questions using the information provided in this chapter.

1. Jigs and fixtures are devices used by _____ machine shops.
2. Jigs and fixtures _____.
   A. hold the work
   B. position the work
   C. are used in assembly work
   D. All of the above.
   E. None of the above.
3. *True or False?* Jigs and fixtures help to reduce manufacturing costs when large numbers of identical and interchangeable parts need to be produced.
4. *True or False?* Jigs and fixtures allow relatively unskilled workers to operate machines.
5. What is a jig?
6. Hardened steel _____ guide the drill or cutting tool during machine operation.
7. Jigs fall into two general types. List and briefly describe each type.
8. When using a combination of open and box jigs, _____ bushings are used to guide the drills and then removed for subsequent operations.
9. What is a fixture?
10. Special _____ that permit multiple setups have been developed for machining centers and other CNC machine tools.
11. The common fixture-holding devices made of heavy, precisely machined castings are known as _____.

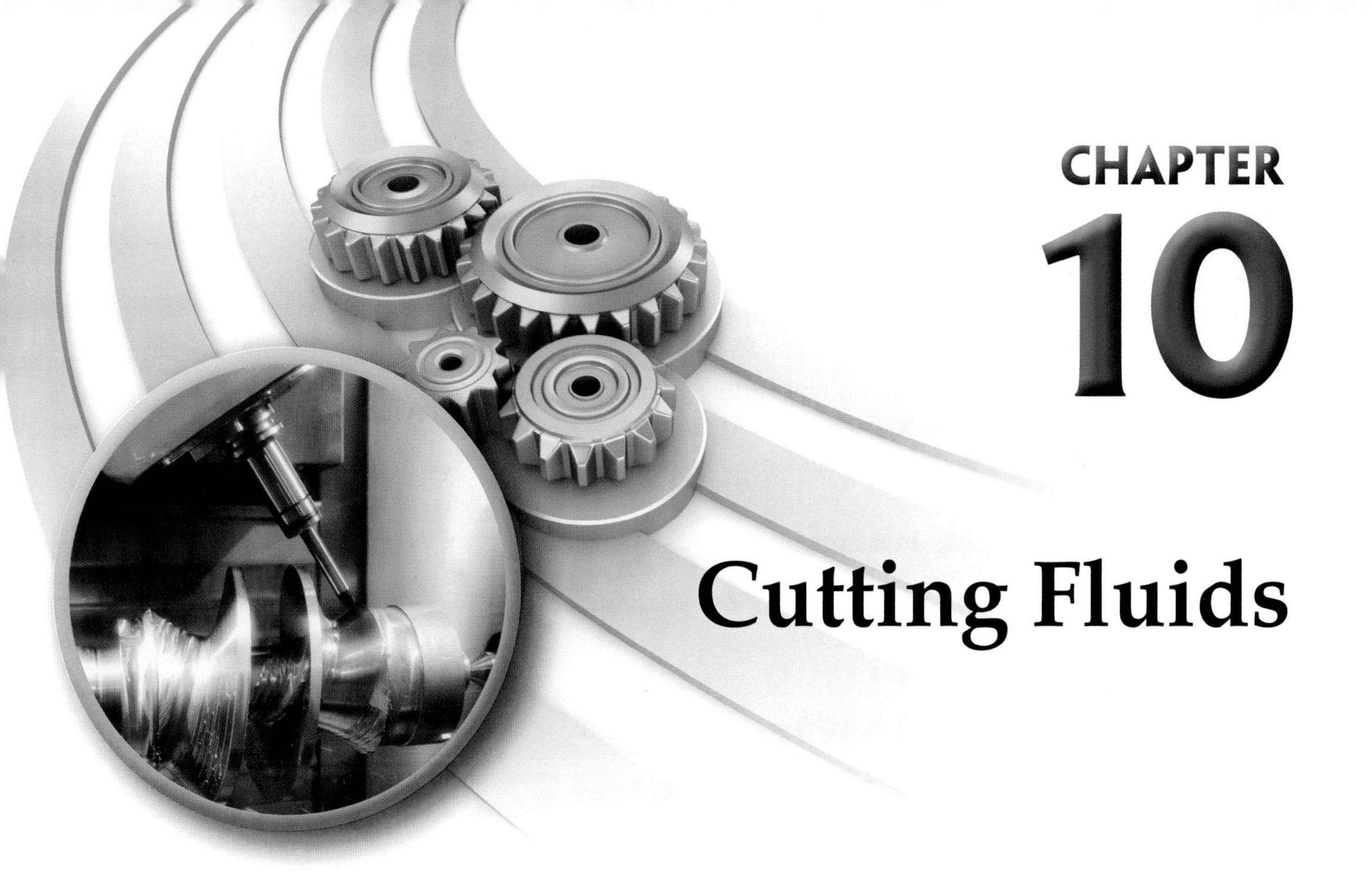

CHAPTER 10

Cutting Fluids

Learning Objectives

After studying this chapter, you will be able to:

- Understand why cutting fluids are necessary.
- List the types of cutting fluids.
- Describe each type of cutting fluid.
- Discuss how cutting fluids should be applied.

Technical Terms

bacteriostat
chemical cutting fluid
cutting fluid
emulsifiable oil
gaseous fluid
mineral oil
semichemical cutting fluid

Cutting fluids are required to do many things simultaneously. These important functions include the following:

- Cooling the work and cutting tool, **Figure 10-1**.
- Improving surface finish quality.
- Lubricating to reduce friction and cutting forces, thereby extending tool life.
- Minimizing material buildup on tool cutting edges.
- Protecting machined surface against corrosion.
- Flushing away chips, **Figure 10-2**.

In addition, cutting fluids must comply with all federal, state, and local regulations for human safety, air and water pollution, waste disposal, and shipping restrictions.

10.1 Types of Cutting Fluids

Cutting fluids fall into the following four basic types:

- Mineral oils.
- Emulsifiable (water-based) oils.
- Chemical and semichemical fluids.
- Gaseous fluids.

BIG Kaiser Precision Tooling Inc.

Figure 10-1. For maximum results, coolant fluid should flood the area being machined and cutting tool to provide the most efficient removal of the heat generated.

MAG IAS, LLC

Figure 10-2. An additional function of cutting fluids is to flush away chips from the area where cutting is taking place.

10.1.1 Mineral Cutting Oils

Cutting fluids made from mineral oil may be used straight or combined with additives. Straight ***mineral oils*** are best suited for light-duty (low speed, light feed) operations where high levels of cooling and lubrication are not required.

They are noncorrosive (will not wear away material gradually through chemical action) and are usually used with high machinability metals, such as aluminum, magnesium, brass, and free-machining steel.

Mineral oils are often combined with animal and vegetable oils and contain sulfur, chlorine, or phosphorus. Their use is limited by high cost, operator health problems, and danger from smoke and fire. Mineral oils also stain some metals. They have a tendency to become rancid, so the tank containing them must be cleaned periodically and the fluid replaced. ***Bacteriostats*** may be added to regulate and control the growth of bacteria.

SAFETY NOTE

When working in situations where cutting fluid mists or vapors are present, always wear an approved respirator. A simple dust mask is not sufficient protection.

10.1.2 Emulsifiable Oils

Emulsifiable oils, also known as soluble oils, are composed of oil droplets that are suspended in water by blending the oil with emulsifying agents and other materials. Emulsifiable oils range in appearance from milky to translucent. They are available in many variations for metal removal applications that generate considerable heat.

Emulsifiable oils offer a number of advantages over straight cutting oils. They provide increased cooling capacity in some applications. They are cleaner to work with than other cutting fluids, and provide cooler and cleaner parts for the machinist to handle. These oils reduce the misting and fogging that are health hazards for machine operators. Because they are diluted with water, they offer increased economy and present no fire hazard.

Emulsifiable oils can be used in most light- and moderate-duty machining operations. For economy and best machining results, these oils must be mixed according to the manufacturer's recommendations. The manufacturer's recommendations take into account the material being machined and the machining operation performed. Fluid maintenance must be performed on a routine basis to control rancidity. Bacteriostats can be added to control bacterial growth.

SAFETY NOTE

Water-based cutting fluids must never be used when machining magnesium. Magnesium dust and chips are flammable, and burning magnesium reacts violently with water.

10.1.3 Chemical and Semichemical Cutting Fluids

Chemical cutting fluids generally contain no oil and are not actually liquids. Examples include graphite, mica, and white lead. They have various rates of dilution depending upon use. A wetting agent is often added to provide moderate lubricating qualities.

Semichemical cutting fluids may have a small amount of mineral oil added to improve the fluid's lubricating qualities. Semichemical cutting fluids incorporate the best qualities of both chemical and emulsifiable oil cutting fluids.

Chemical and semichemical cutting fluids offer the following advantages:

- Fluids dissipate heat rapidly.
- They are clean to use.
- After machining, residue is easy to remove.
- The fluids are easy to mix and do not become rancid.

Their disadvantages include the following:

- Some formulas have minimal lubricating qualities.
- Fluids may cause skin irritation in some workers. Material safety information should always be kept in the shop when using chemical or semichemical cutting fluids.
- When they become contaminated with other oils, disposal can be a problem.

10.1.4 Gaseous Fluids

Compressed air is the most commonly used ***gaseous fluid*** coolant. It cools by forced convection, the transfer of heat through the mechanical movement of a gas or liquid.

In addition to cooling the workpiece and tool, compressed air also blows chips away at high velocity. Workers in surrounding areas must be shielded from the flying chips.

SAFETY NOTE

Compressed air should never be directed at yourself or another person. Compressed air can enter the body through a cut or opening and cause an air bubble to form in the blood stream leading to a potentially fatal blockage of a blood vessel.

10.2 Application of Cutting Fluids

Machining and grinding applications require a continuous flooding of fluid around the cutting tool and work to provide efficient removal of the heat generated, **Figure 10-3**. Coolant nozzles must be positioned carefully so that, in addition to cooling the work area, the cutting fluid will also carry the chips away. In some machining operations, a conveyor system, **Figure 10-4**, is used to remove chips and cutting fluid from the cutting area. The cutting fluid is filtered to remove contaminants (unwanted debris) and returned to the machine's coolant tank for reuse.

Goodheart-Willcox Publisher

Figure 10-4. This powered conveyor system moves chips away from a machine's cutting area. A rotating drum filter is used to separate chips and metal particles from the cutting fluid. The filtered fluid is returned to the machine for reuse.

Lockwood Products, Inc.

Figure 10-3. Coolant fluid must surround the cutting area for maximum effect. Shown is a modular hose system for applying air and liquids. The units snap together and can be shaped to fit any job.

CNC Software, Inc. © Copyright 1983-2013. All rights reserved.

Figure 10-5. Carbide cutting tools, like this endmill, are widely used on machine tools, but require coolants that can remove heat at higher rates.

10.3 Evaluation of Cutting Fluids

It is not possible in this text's limited space to cover every cutting fluid, nor does space permit recommending specific cutting fluids for every machining operation. This information can be obtained from data published by cutting fluid manufacturers. Recommendations for cutting fluid use are included in the chapters of this text dealing with each type of machine tool. In general, however, cutting fluids (excepting gaseous fluids) are compatible with HSS (high-speed steel) and carbide tooling. See **Figure 10-5**. Since carbide tooling operates at higher cutting speeds and generates higher cutting temperatures, cutting fluids that have high cooling rates should be used in such applications. Machining with ceramic tooling is usually accomplished without the use of cutting fluids, **Figure 10-6**.

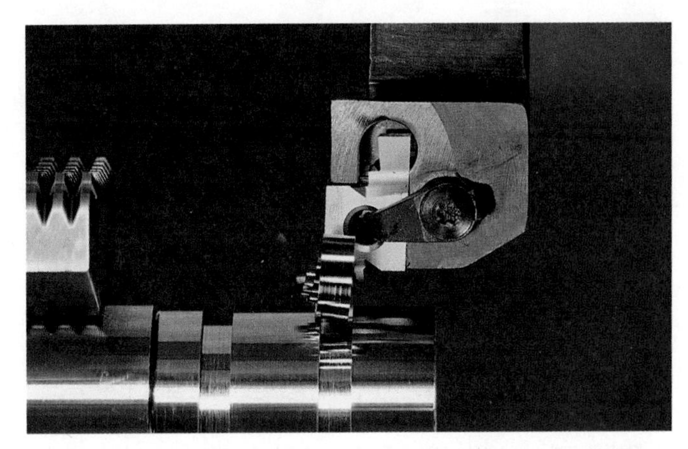

Carboloy

Figure 10-6. Standard turning operation on a vertical turning center. Cutting fluids are not necessary with ceramic tooling.

Chapter Review

Summary

- Cutting fluids cool the work and cutting tool, improve surface quality, lubricate, minimize material build up, protect against corrosion, and flush away chips.
- The four types of cutting fluids are mineral oils, emulsifiable oils, chemical and semichemical cutting fluids, and gaseous fluids.
- Selection and application of cutting fluids depends upon the material being machined, specific machine operation, the type of cutting tool, and cutting speed and feed.

Review Questions

Answer the following questions using the information provided in this chapter.

1. Cutting fluids must do many things simultaneously. List five functions of cutting fluids.
2. List the four basic types of cutting fluids.
3. What type cutting fluid is recommended for machining aluminum, magnesium, brass, and free-machining steels?
4. Why do mineral oil cutting fluids have limited use?
5. A(n) _____ is a substance that controls the growth of bacteria.
6. _____ cutting fluids are also known as soluble oils.
7. What advantages do the emulsifiable oil cutting fluids have over mineral oil cutting fluids?
8. _____ cutting fluids contain no oils.
9. When small amounts of mineral oil are added to chemical cutting fluids, it is known as a(n) _____ cutting fluid.
10. What are the advantages of chemical and semichemical cutting fluids?
11. What is dangerous about using compressed air to cool the area being machined?
12. What does continuous flooding of the work area and cutting tool with cutting fluid accomplish?

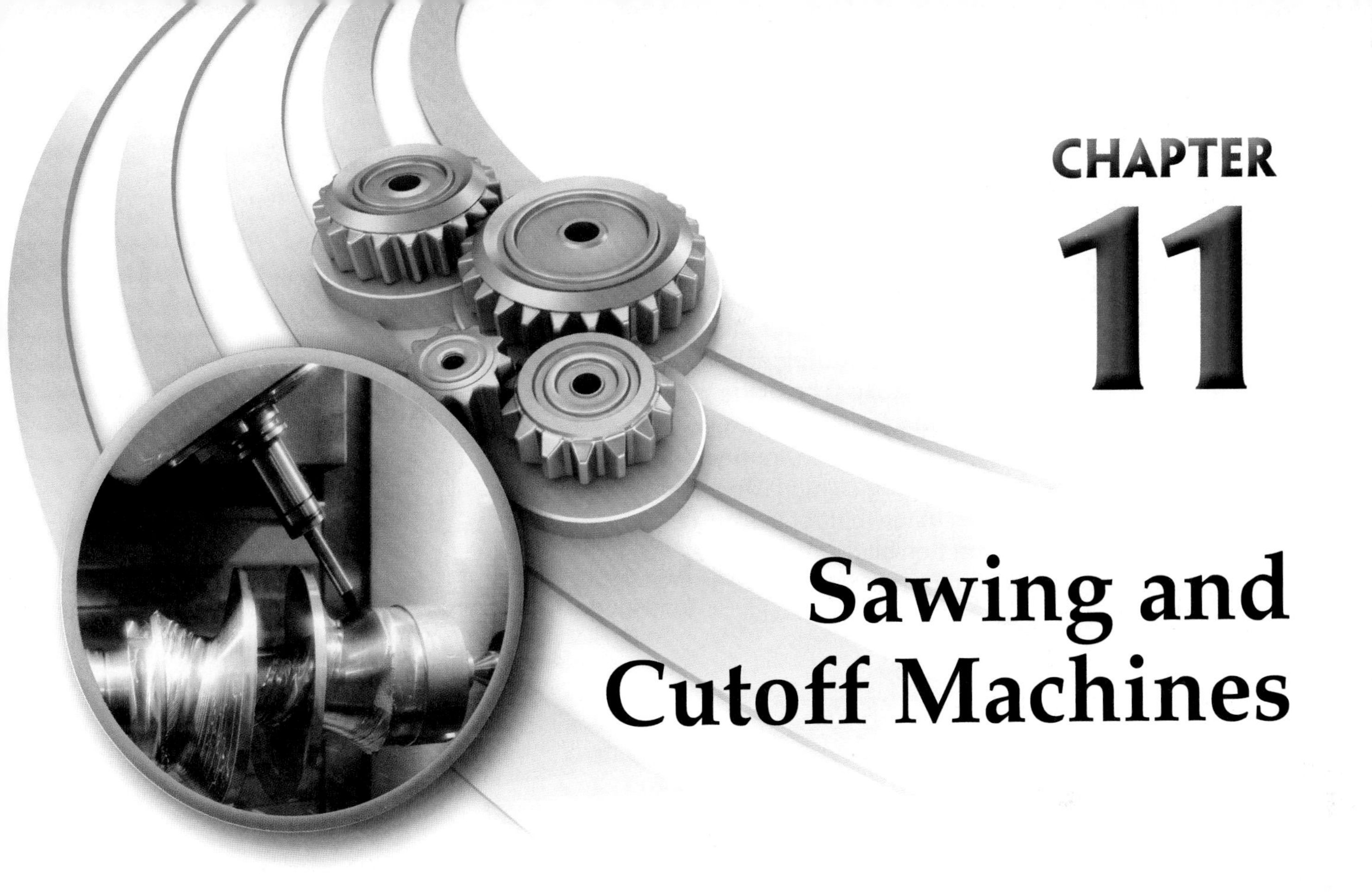

CHAPTER 11

Sawing and Cutoff Machines

Learning Objectives

After studying this chapter, you will be able to:

- Identify the various types of sawing and cutoff machines.
- Select the correct machine for the job.
- Mount a blade and prepare the machine.
- Position the work for the most efficient cutting.
- Safely operate sawing and cutoff machines.

Technical Terms

abrasive cutoff saw
all-hard blade
cold circular saw
flexible-back blade
friction saw
horizontal band saw
three-tooth rule

11.1 Metal-Cutting Power Saws

The first step in most machining jobs is to cut the stock to required length. This can be done using power saws, **Figure 11-1**.

There are three principal types of metal-cutting saws, **Figure 11-2**. Reciprocating-type power saws use a back-and-forth (reciprocating) cutting action. The cutting is done on the backstroke. The blade is similar to that found on a hand hacksaw, only larger and heavier. Band-type power saws have a continuous blade that moves in one direction. Circular-type power saws have a round, flat blade that rotates into the work. A toothed blade, friction blade, or abrasive blade may be used, depending on the material and the operation.

alterfalter/Shutterstock.com

Figure 11-1. The first step in most machining jobs is to cut the stock to the desired length. Measure the cutoff length carefully and observe all safety precautions.

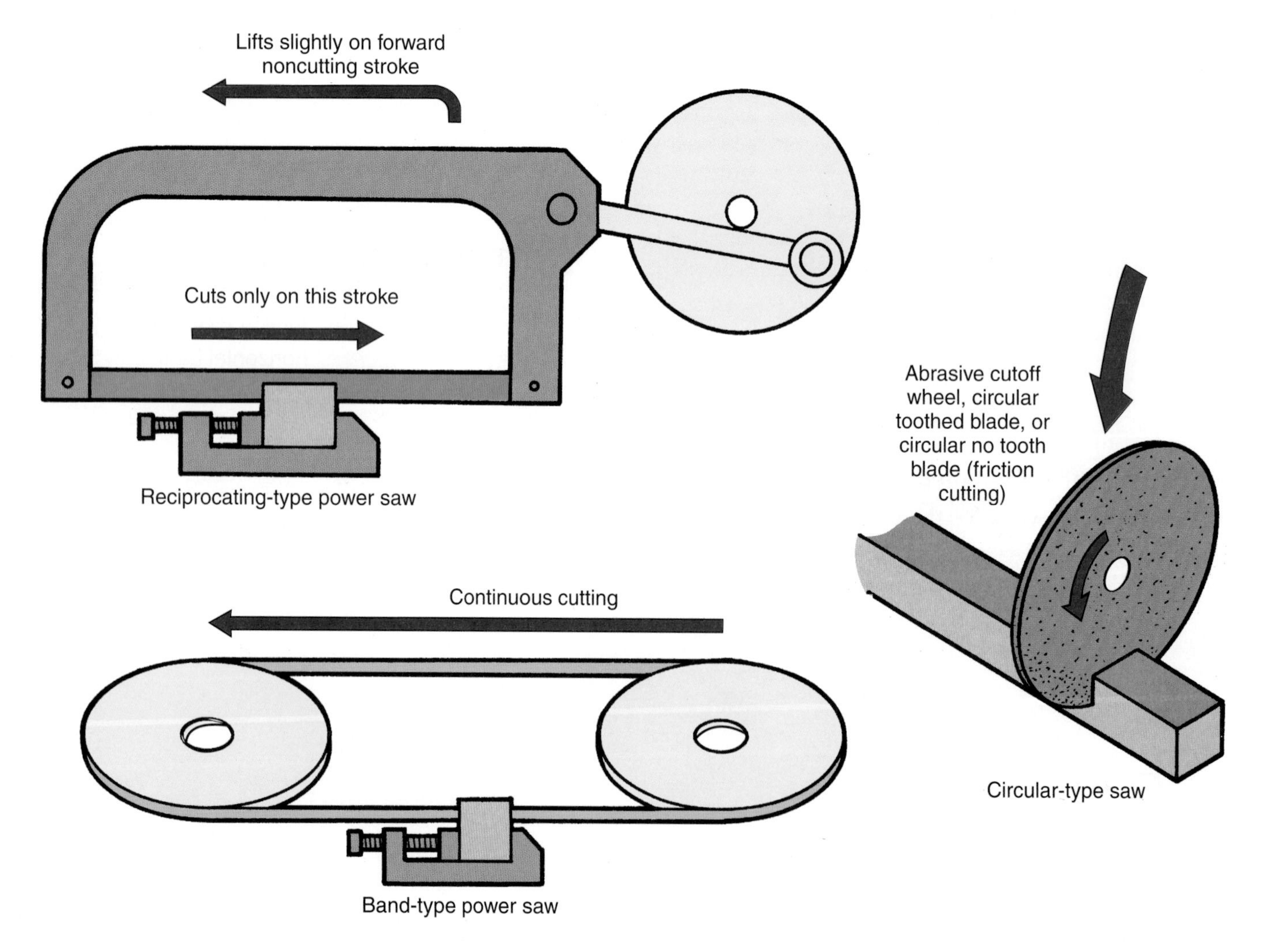

Goodheart-Willcox Publisher

Figure 11-2. The three principal types of cutoff saws.

11.2 Power Hacksaw

A power hacksaw, **Figure 11-3**, uses a reciprocating motion to move the blade across the work. The blade cuts on the backstroke. There are several types of feeds available.

Positive feed produces an exact depth of cut on each stroke. The pressure on the blade varies with the number of teeth in contact with the work.

Definite pressure feed yields a pressure on the blade that is uniform regardless of the number of teeth in contact with the work. The depth of the cut varies with the number of teeth contacting the work. This condition prevails with gravity feed.

Feed can be adjusted to meet varying conditions. For best performance, the blade and feed must be selected to permit high-speed cutting and heavy feed pressure with minimum blade bending and breakage.

Standard power hacksaws are available in sizes from 6″ × 6″ (150 mm × 150 mm) to 24″ × 24″ (900 mm × 900 mm). The saws can be fitted with many accessories. A swivel vise permits angular cuts to be made quickly, **Figure 11-4**. Quick-acting vises allow faster manual clamping of the workpiece. Power stock feed, power clamping of work, and automatic cycling can automate the cutting operation. Automatic cycling moves the work out the required distance, clamps it, and makes the cut automatically. The cycle is repeated on completion of the cut.

High-speed cutting requires use of a coolant. Coolant reduces friction, increases blade life, and prevents chip-clogged teeth. Cast iron and some brass alloys, unlike most materials, do not require coolant.

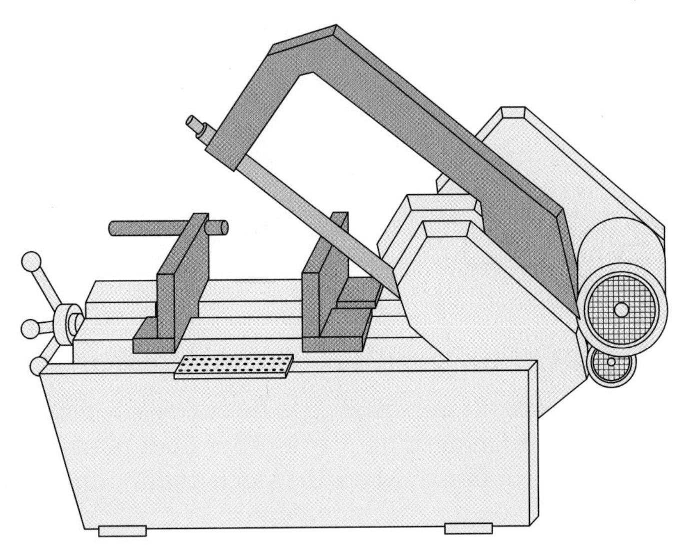

Goodheart-Willcox Publisher

Figure 11-3. An industrial reciprocating power hacksaw.

Worakit Sirijinda/Shutterstock.com

Figure 11-4. A swivel vise permits angular cuts.

11.2.1 Selecting a Power Hacksaw Blade

Proper blade selection is important. Use the ***three-tooth rule***—at least three teeth must be in contact with the work. Large sections and soft materials require a coarse-tooth blade. Small or thin work and hard materials require a fine-tooth blade.

For best cutting action, apply heavy feed pressure on soft materials and large work. Use light feed pressure on hard materials and work with small cross sections, **Figure 11-5**.

Blades are made in two principal types: flexible-back and all-hard. The choice depends on use.

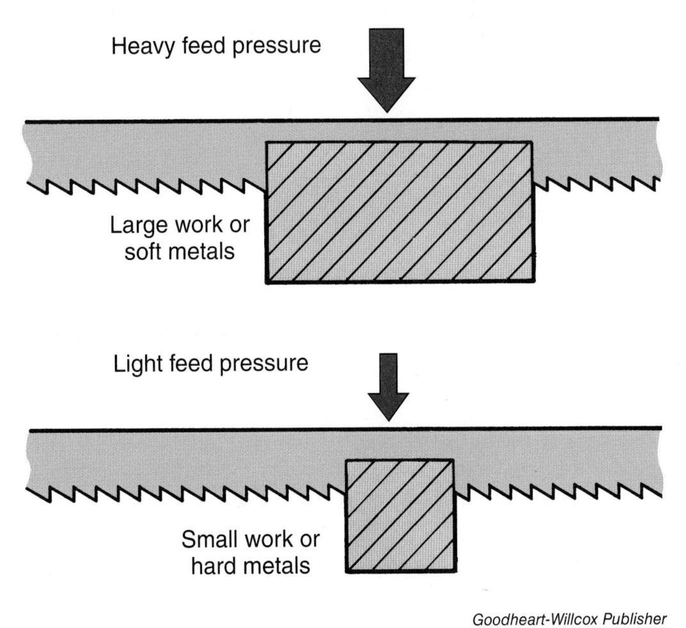

Goodheart-Willcox Publisher

Figure 11-5. Apply heavy feed pressure on soft metals and large work. Use light pressure on hard metals and work with small cross sections.

A ***flexible-back blade*** should be used when safety requirements demand a shatterproof blade. These blades should also be used for cutting odd-shaped work if there is a possibility of the work coming loose in the vise.

For a majority of cutting jobs, the ***all-hard blade*** is best for straight, accurate cutting under a variety of conditions.

SAFETY NOTE

When starting a cut with an all-hard blade, be sure the blade does not drop on the work when cutting starts. If it falls, the blade could shatter and fly apart, causing injuries.

Blades are also made from tungsten and molybdenum steels, and with tungsten carbide teeth on steel alloy backs. The following "rule-of-thumb" can be used for selecting the correct blade:

- Use a 4-tooth blade for cutting large sections or readily machined metals.
- Use a 6-tooth blade for cutting harder alloys and miscellaneous cutting.
- Use 10- and 14-tooth blades primarily on light-duty machines where work is limited to small sections requiring moderate or light feed pressure.

11.2.2 Mounting a Power Hacksaw Blade

The blade must be mounted to cut on the power (back) stroke. The blade must also lie perfectly flat against the mounting plates, **Figure 11-6**. If long life and accurate cuts are to be achieved, the blade must be properly tensioned.

Many techniques have been developed for properly mounting and tensioning blades. Use a torque wrench and consult the manufacturer's literature. If the information (proper torque for a given blade on a given machine) is not available, the following methods can be used:

- Tighten the blade until a low musical ring is heard when the blade is tapped lightly. A high-pitched tone indicates that the blade is too tight. A dull thud means the blade is too loose.
- The shape of the blade pin hole can serve as an indicator of whether the blade is tensioned properly. When proper tension is achieved, the pin holes will become slightly elongated, **Figure 11-7**.

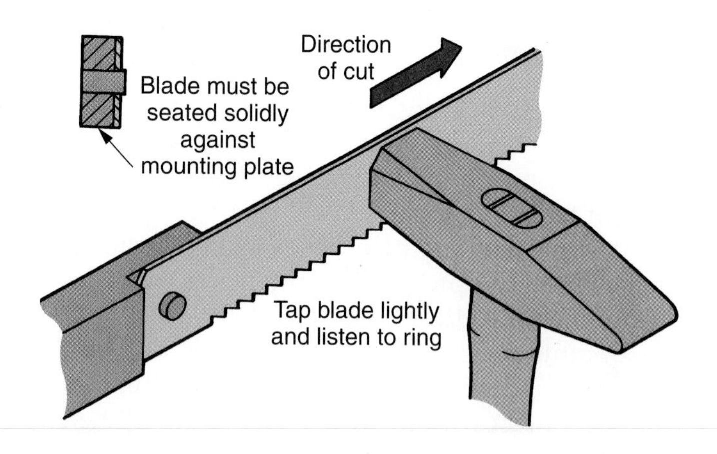

Figure 11-6. The blade must be adjusted to cut on the back stroke. Make sure it is perfectly flat against the mounting plates before tensioning. Tighten the blade until a low musical ring is heard when the blade is tapped with a small hammer. Since blades have a tendency to stretch slightly after making a few cuts, tension should be checked and, if necessary, adjusted.

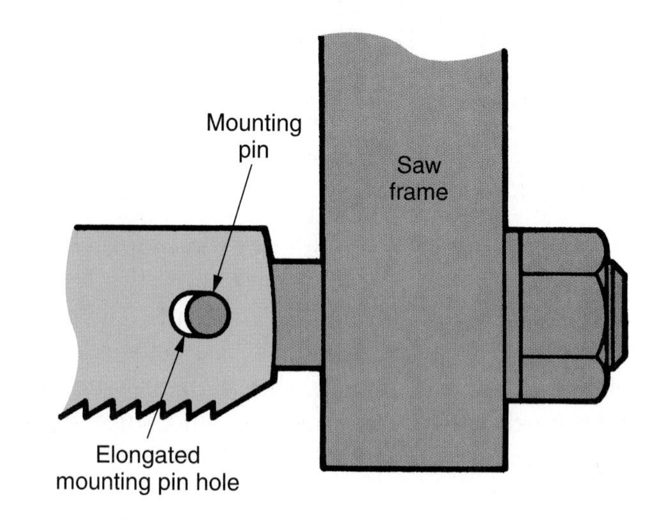

Figure 11-7. Pin holes on a properly tensioned blade will be slightly elongated, rather than round.

The blade will become more firmly seated after the first few cuts and will stretch slightly. The blade will require retensioning (retightening) before further cutting can be done.

11.2.3 Cutting with a Power Hacksaw

Measure off the distance to be cut. Allow ample material for facing if the work order does not specify the length of cut. Mark the stock and mount the work firmly on the machine, **Figure 11-8**.

If several sections are to be cut, use a stop gage, **Figure 11-9**. Apply an ample supply of coolant if the machine has a built-in coolant system.

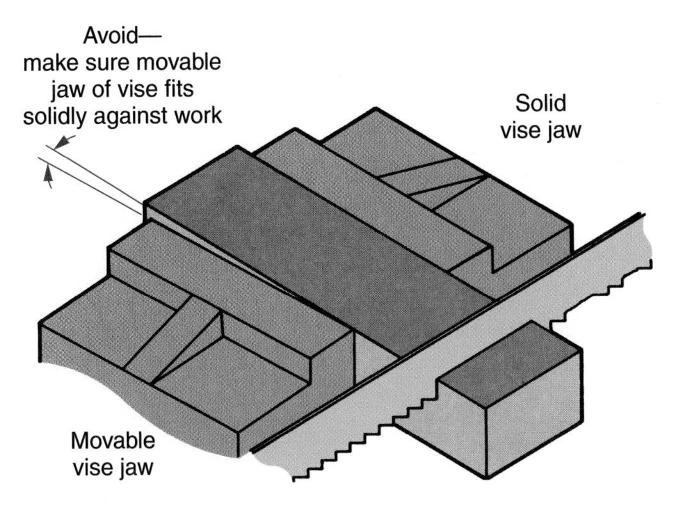

Goodheart-Willcox Publisher

Figure 11-8. If the work is not clamped solidly, it will twist and the blade will bind and be ruined in the first few seconds of use.

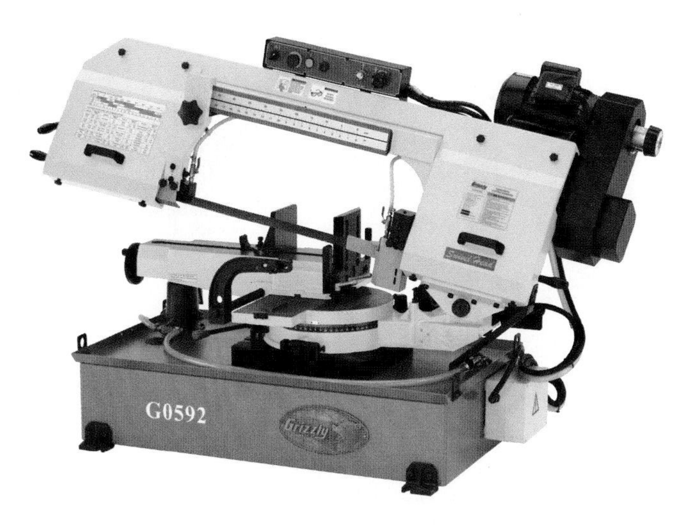

Photo courtesy of Grizzly Industrial, Inc. www.grizzly.com

Figure 11-10. Horizontal band saw with built-in coolant system, adjustable hydraulic down feed, automatic shutoff, quick positioning vise, and –45° to +60° swivel mount.

11.3 Power Band Saw

The ***horizontal band saw***, **Figure 11-10**, is frequently referred to as the cutoff machine. It offers the following advantages over the power hacksaw:

- **Greater precision.** The blade on a band saw can be guided more accurately than the blade on the reciprocating power saw. It is common practice to cut directly "on the line" when band sawing, because finer blades can be used.
- **Faster speed.** The long, continuous blade moves in only one direction, so cutting is also continuous. The blade can run at much higher speeds because it rapidly dissipates the cutting heat.
- **Less waste.** The small cross section of the band saw blade makes smaller and fewer chips than the thicker blade of the reciprocating power hacksaw, **Figure 11-11**.

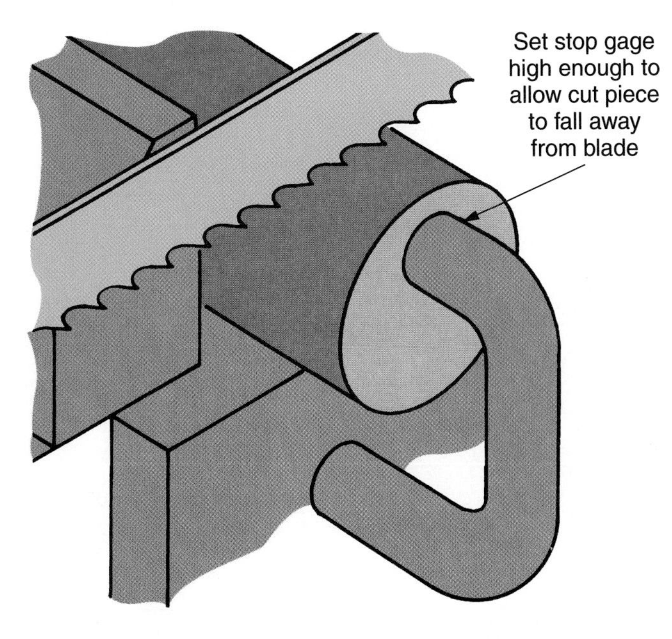

Goodheart-Willcox Publisher

Figure 11-9. A stop gage is used when several pieces of the same length must be cut. Set it high to permit the work to fall free when completely cut.

11.3.1 Selecting a Band Saw Blade

Band saw blades are made with raker teeth or wavy teeth, **Figure 11-12**. Most manufacturers also make variations of these sets. The raker set is preferred for general use.

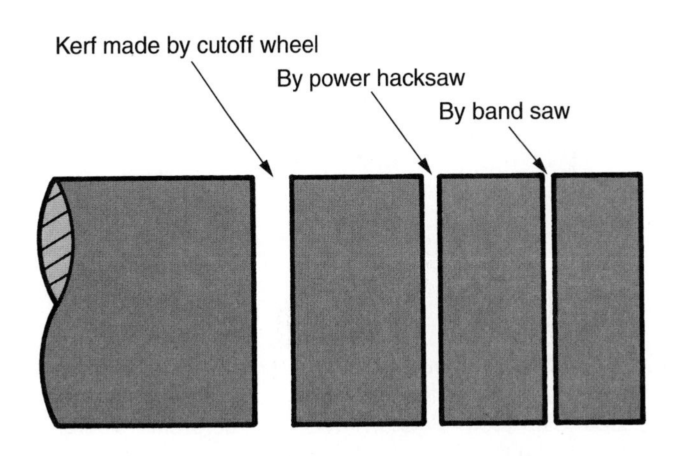

Goodheart-Willcox Publisher

Figure 11-11. Differences in the amount of metal converted to chips (waste) by each cutoff machine.

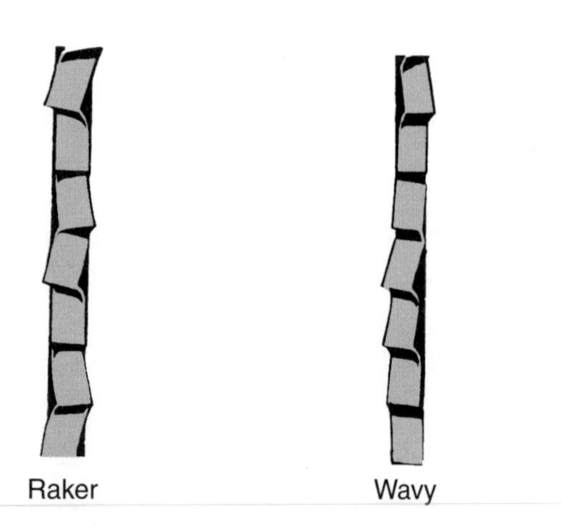

Goodheart-Willcox Publisher

Figure 11-12. Saw blades commonly have raker or wavy teeth. Raker teeth are preferred for general use, cutting large solid sections, and cutting thick plate.

Tooth pattern determines the efficiency of a blade in various materials. The standard tooth blade pattern is best suited for cutting most ferrous metals. A skip tooth blade pattern is preferred for cutting aluminum, magnesium, copper, and soft brasses. The hook tooth blade pattern is also recommended for most nonferrous metallic materials. See **Figure 11-13**.

For best results, consult the blade manufacturer's chart or manual for the proper blade characteristics (set, pattern, and number of teeth per inch) for the particular material being cut.

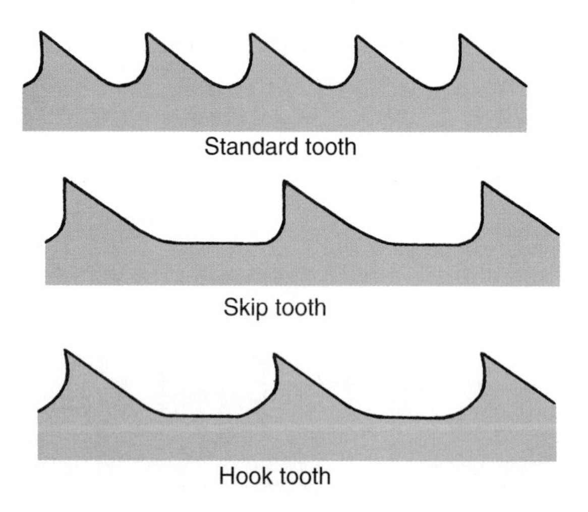

Goodheart-Willcox Publisher

Figure 11-13. Standard tooth blades, with rounded gullets, are usually best for most ferrous metals, hard bronzes, and hard brasses. Skip tooth blades provide for more chip clearance without weakening the blade body. They are recommended for cutting aluminum, magnesium, copper, and soft brasses. Hook tooth blades offer two advantages over skip tooth blades—easier feeding and less "gumming up."

11.3.2 Installing a Band Saw Blade

If the saw is to work at top efficiency, the blade must be installed carefully.

SAFETY NOTE

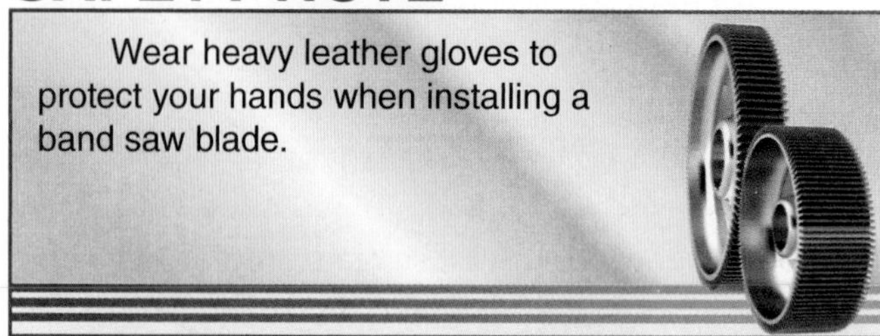

Wear heavy leather gloves to protect your hands when installing a band saw blade.

Blade guides should be adjusted to provide adequate support, **Figure 11-14**. Proper blade support is required to cut true and square with the holding device.

Follow the manufacturer's instructions for adjusting blade tension. Improper blade tension ruins blades and can cause premature failure of bearings in the drive and idler wheels.

Cutting problems encountered with the band saw are similar to those of the reciprocating hack saw. Most problems are caused by poor machine condition. Problems can be kept to a minimum if a maintenance program is followed on a regular basis. This typically includes checking wheel alignment, guide alignment, feed pressure, and hydraulic systems.

AMT—The Association for Manufacturing Technology

Figure 11-14. Adjust blade guides to provide adequate blade support. Otherwise, the blade will not cut true.

11.4 Using Power Hacksaws and Band Saws

Most sawing problems can be prevented by careful planning and observing a few rules. These rules apply to both power hacksaws and band saws.

11.4.1 Blades Breaking

Blades are normally broken when they are dropped on the work. A loose blade or excessive feed can cause the blade to fracture. Loose work can also cause blade damage, as will making a cut on a corner or sharp edge where the three-tooth rule is not observed. Broken blades can normally be avoided with proper machine setup.

11.4.2 Crooked Cutting

Crooked cutting is usually the result of a worn blade. Remember to reverse the work after replacing a blade, and start a new cut on the opposite side. See **Figure 11-15**. A loose blade or a blade rubbing on a clamping fixture will cause the same problem. It can also be caused by excessive blade pressure on the work or by worn saw guides.

11.4.3 Blade Pin Holes Breaking Out

This reciprocating blade problem can be caused by dirty mounting plates or too much tension on the blade. Worn mounting plates can cause a blade to twist and strain in such a way that the pin hole will break out.

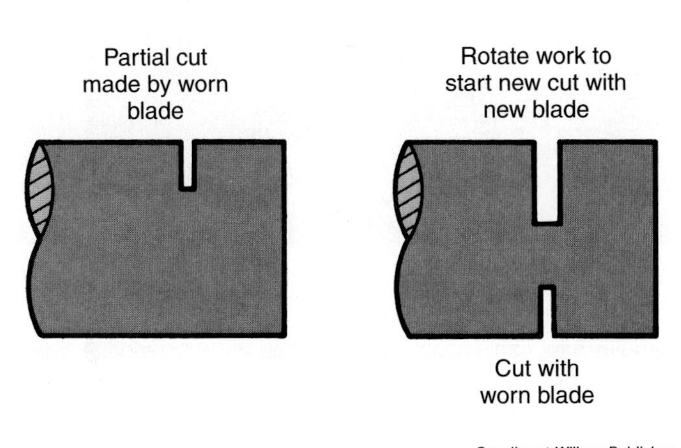

Figure 11-15. Never attempt to start a new blade in a cut made by a worn blade. Reverse the work and start another cut on the opposite side. Cut through to the old cut.

11.4.4 Premature Blade Tooth Wear

When this problem occurs, the teeth become rounded and dull quickly. Insufficient feed pressure (indicated by light, powdery chips) is one of the major causes of premature blade tooth wear. Excessive pressure (indicated by burned chips) causes the same problem.

Insufficient pressure can be corrected by increasing cutting pressure until a full curled chip is produced. If too much pressure is the culprit, reduce feed pressure until a full curled chip is formed.

Lack of coolant or a poorly adjusted machine can also cause rapid wear. Correct by following the manufacturer's recommendations.

11.4.5 Teeth Strip Off

This failure results when the teeth snap off the blade. Starting a cut on a sharp corner is a major cause of this problem. A machine setup with a flat starting surface will greatly reduce tooth stripping. Be sure the work is clamped securely. Loose work can also cause the teeth to strip, **Figure 11-16**.

Check the manufacturer's chart to determine the proper blade for the job to be done. A blade with

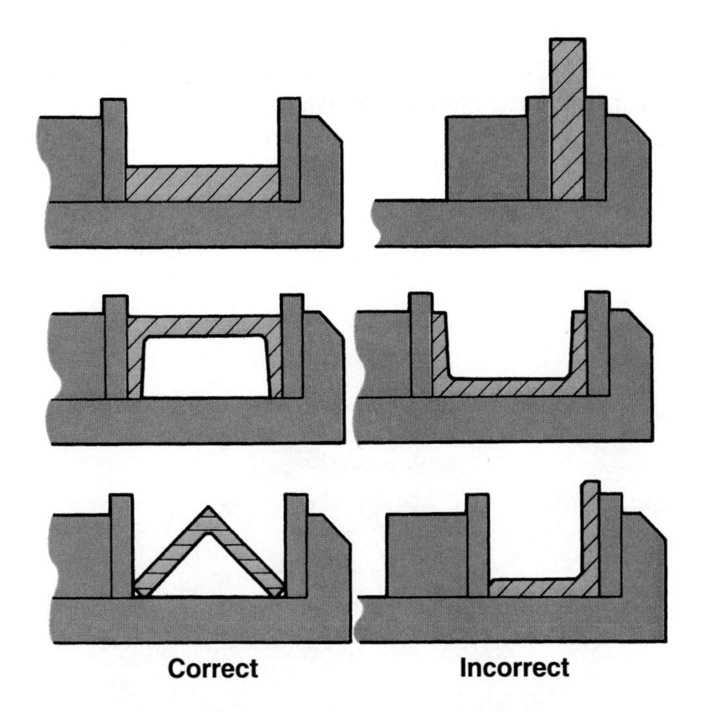

Figure 11-16. Recommended ways to hold sharp-cornered work for cutting. A carefully planned setup will ensure that at least three teeth will be cutting, greatly extending blade life.

teeth too fine will clog (load) and jam, causing the teeth to shear off. A blade that is too coarse (less than three teeth cutting) will cause the same problem. Make sure the blade is properly mounted and cutting on the power stroke.

11.5 Metal-Cutting Circular Saws

Metal-cutting circular saws are found in many areas of metalworking. Primarily production machines, these saws are divided into the following three classifications:

- Abrasive cutoff saw.
- Cold circular saw.
- Friction saw.

An ***abrasive cutoff saw***, **Figure 11-17**, cuts material using a rapidly revolving, thin abrasive wheel. Most materials—glass, ceramics, and metals—can be cut to close tolerances. Hardened steel does not require annealing to be cut. Special heat-resistant abrasive wheels are available for high-speed cutoff of hot stock.

Abrasive cutting falls into two classifications, dry and wet. Wet abrasive cutting, while not quite as rapid as dry cutting in some applications, produces a finer surface finish and permits cutting to close tolerances. The cuts are burn-free and have few or no burrs. Dry abrasive cutting does not use a coolant and is used for rapid, less-critical cutting.

A ***cold circular saw***, **Figure 11-18**, makes use of a circular, toothed blade capable of producing very accurate cuts. Large cold circular saws can sever round metal stock up to 27″ (675 mm) in diameter.

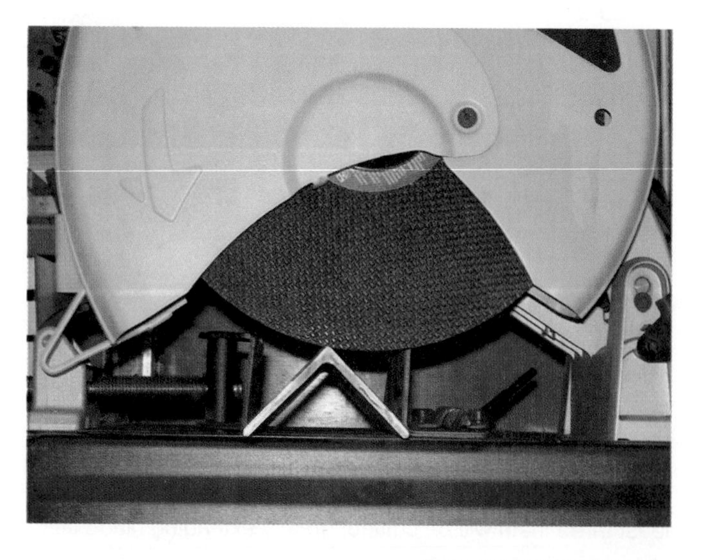

Photo courtesy of Weiler Corporation

Figure 11-17. An industrial abrasive cutoff saw.

W.J. Savage Co.

Figure 11-18. Cold circular saws. A—This automated cold circular saw can accept a piece up to 2500 pounds. A laser guide light marks the position of the cut. B—This machine can be fitted with carbide-tipped blades or abrasive disks.

A ***friction saw*** operates at very high speeds (20,000 surface feet per minute or 6000 meters per minute) and actually melts its way through the metal. The friction saw blade may or may not have teeth.

The teeth are used primarily to carry oxygen to the cutting area. These machines find many applications in steel mills to cut red-hot billets (sections of semifinished steel).

11.6 Power Saw Safety

- Never attempt to operate a sawing machine while your senses are impaired by medication or other substances.
- Get help when you are lifting and cutting heavy material.
- Clean oil, grease, and coolant from the floor around the work area.
- Burrs on cut pieces are sharp. Use special care when handling pieces with burrs.
- Follow the manufacturer's instructions for tensioning a blade. Too much tension can shatter the blade.
- Handle band saw blades with extreme care. They are long and springy and can uncoil suddenly.
- Be sure the work is mounted solidly before starting a cut.
- Be sure all guards are in place before using the saw.
- Always wear a dust mask and full face shield when cutting stock with a dry-type abrasive cutoff saw.
- Avoid standing directly in line with the blade when operating a circular cutoff saw.
- Use a brush to clean chips from the machine. Do not use your hands. Wait for the machine to come to a complete stop before cleaning.
- Keep your hands out of the way of moving parts.
- Stop the machine before making adjustments.
- Have all cuts, bruises, and scratches, even minor ones, treated immediately.

Chapter Review

Summary

- Reciprocating power hacksaws, power band saws, and circular metal-cutting saws are the principal types of saws used in machining operations.
- Power band saws offer many advantages over other metal-cutting saws.
- Proper blade tension is important for both power hacksaws and band saws.
- Follow the three-tooth rule for the best cutting action.
- Select the correct feed pressure, tooth set, and blade pattern to ensure efficient and safe cutting.
- Abrasive cutoff saws, cold circular saws, and frictions saws are the three classifications of metal-cutting circular saws.
- Always follow proper safety procedures when operating a power saw.

Review Questions

Answer the following questions using the information provided in this chapter.

1. List the three basic types of metal-cutting saws.
2. The _____ type saw has a back-and-forth cutting action.
3. The power hacksaw only cuts on the _____ stroke.
4. When using a power sawing machine, with what materials should you *not* use coolant?
5. What is the "three-tooth rule" for sawing?
6. List the two principal types of hacksaw blades.

For questions 7–9, give the "rule-of-thumb" for selecting the correct blade.

7. _____ teeth per inch for cutting large sections or readily machined materials.
8. _____ teeth per inch for cutting harder alloys and miscellaneous cutting.
9. _____ teeth per inch for cutting on the majority of light-duty machines, where work is limited to small sections and moderate to light feed pressures.
10. List three methods used to put proper tension on a power hacksaw blade.
11. When is a stop gage used?
12. What three advantages does the band sawing machine offer over other types of power saws?
13. Band saw blades are made with two types of teeth. List them.

For questions 14–16, give the tooth pattern best suited for cutting the given materials.

14. The _____ tooth patten is best suited for cutting most ferrous metals.
15. The _____ tooth pattern is preferred for cutting aluminum, magnesium, copper, and soft brass.
16. The _____ tooth pattern is also recommended for most nonferrous metallic materials.
17. List the three types of circular metal-cutting saws.
18. List five safety precautions to be observed when operating a power saw.

CHAPTER 12

Drills and Drilling Machines

Learning Objectives

After studying this chapter, you will be able to:

- Select and safely use the correct drills and drilling machine for a given job.
- Explain the safety rules that pertain to drilling operations.
- Identify and describe common drills and drill-holding devices.
- Make safe setups on a drill press.
- Sharpen a twist drill.
- Describe basic drilling operations.

Technical Terms

blind hole
center finder
counterboring
countersinking
cutting speed
drill gage
drilling machine
drill point gage
feed
flutes
pilot hole
reaming
spotfacing
tapping
twist drill

12.1 Drilling Machines

A ***drilling machine*** is a power-driven machine that holds both the material and cutting tool, bringing them together to create a round hole in the material. Many different types of drilling machines are used in industry. The type of machine used depends on the operation being performed, the size of the workpiece, and the variety of operations required of the machine.

12.1.1 Types of Drilling Machines

The most common drilling machine is the drill press, **Figure 12-1**. A drill press operates by rotating a cutting tool, or drill, against the material with sufficient pressure to cause the drill to penetrate the material. See **Figure 12-2**.

The size of a drill press is determined by the largest diameter circular piece that can be drilled on center, **Figure 12-3**. A 17″ drill press can drill to the center of a 17″ diameter piece. The centerline of the drill is 8 1/2″ from the column.

Photo courtesy of Grizzly Industrial, Inc. www.grizzly.com

Figure 12-1. A twelve-speed 20″ drill press.

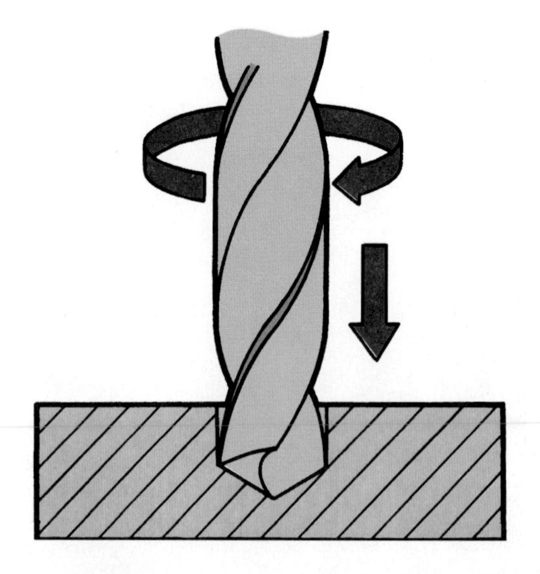

Goodheart-Willcox Publisher

Figure 12-2. Drilling is the operation most often performed on a drill press. Both rotating force and a downward pushing force are needed for drilling.

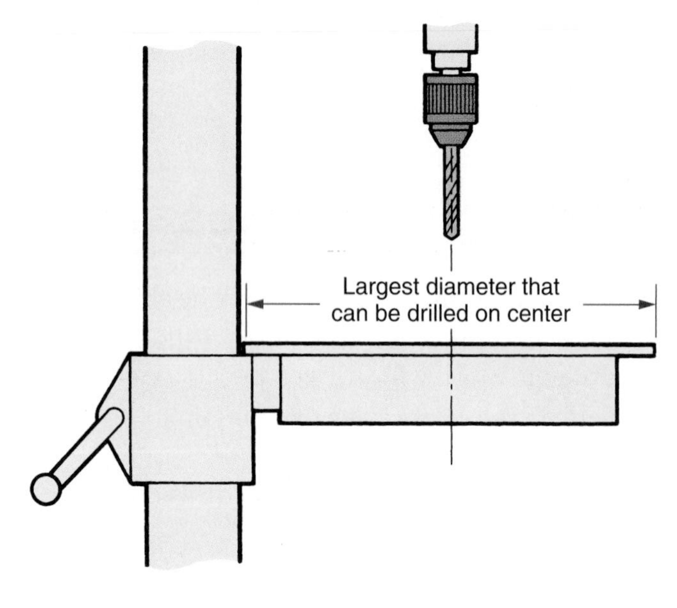

Goodheart-Willcox Publisher

Figure 12-3. How drill press size is determined.

Bench drill presses can be used to drill holes in small workpieces, **Figure 12-4**. These presses do not have as many capabilities as the floor models.

Electric hand drills are used to drill small holes in relatively thin material. They are reasonably priced and convenient to use. See **Figure 12-5**.

A radial drill press is designed to handle very large drilling work. The drill head is mounted in a way that allows it to be moved back and forth on an arm that extends from the massive machine column.

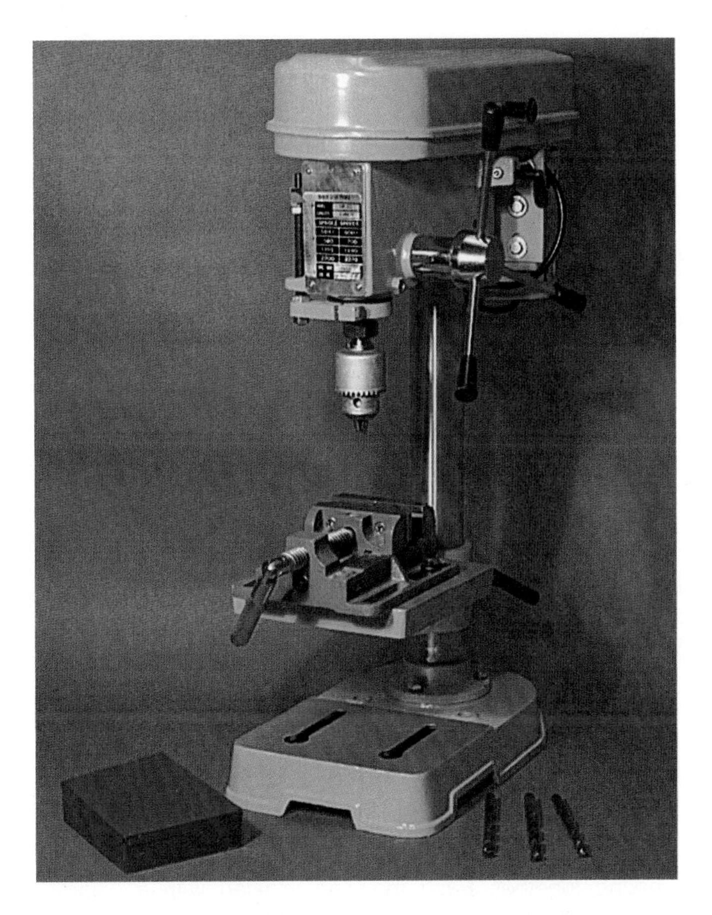

Goodheart-Willcox Publisher

Figure 12-4. A small bench drill press can be very useful.

Makita USA, Inc.

Figure 12-5. Portable electric drills are manufactured in a wide range of sizes. This model is battery powered.

The arm can be moved up and down and pivoted on the column, **Figure 12-6**.

Often, a large pit is located along one side of the machine to permit the positioning of large, odd-shaped work. The pit is covered when not in use.

Sharp Industries, Inc.

Figure 12-6. This radial drill press can drill large diameter holes in large workpieces.

Smaller bench radial drill presses are used to drill smaller holes. See **Figure 12-7**. These units are not as expensive as a full-size radial drill press.

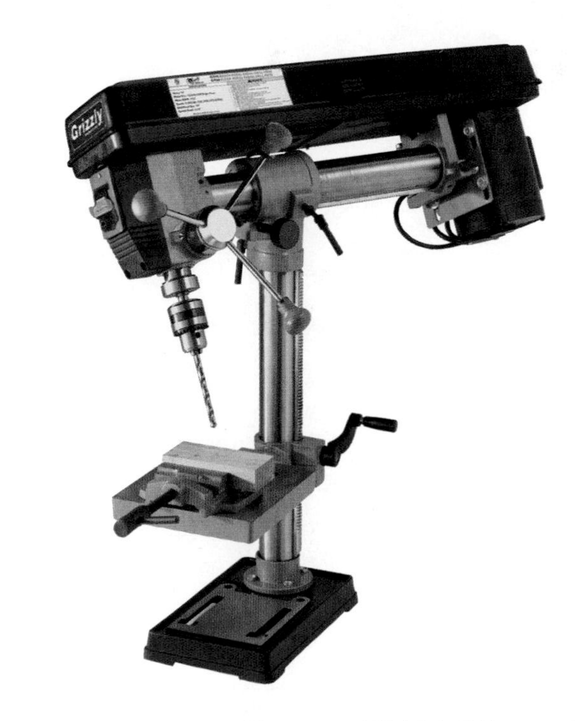

Photo courtesy of Grizzly Industrial, Inc. www.grizzly.com

Figure 12-7. Bench radial drill press.

Portable magnetic drills can be used in the shop and on the job site. These machines can be positioned in an upright, horizontal, or vertical position when drilling. See **Figure 12-8**.

Gang drilling machines consist of several drill assemblies, **Figure 12-9**. The workpiece is moved from one assembly to another. A different operation is performed at each stage.

Multiple-spindle drilling machines have several drilling heads. Several operations can be performed without changing drills.

Machining centers operate very efficiently and accurately. The center operates under computer numerical control (CNC). The center can be programmed to drill holes as part of a machining sequence. The drill position and speed are programmed into the center. No drill jigs are required. See **Figure 12-10**.

Robotic drilling machines are basically programmable, mobile drilling machines. The machine is programmed to move along one workpiece or between several workpieces, drilling at specified locations.

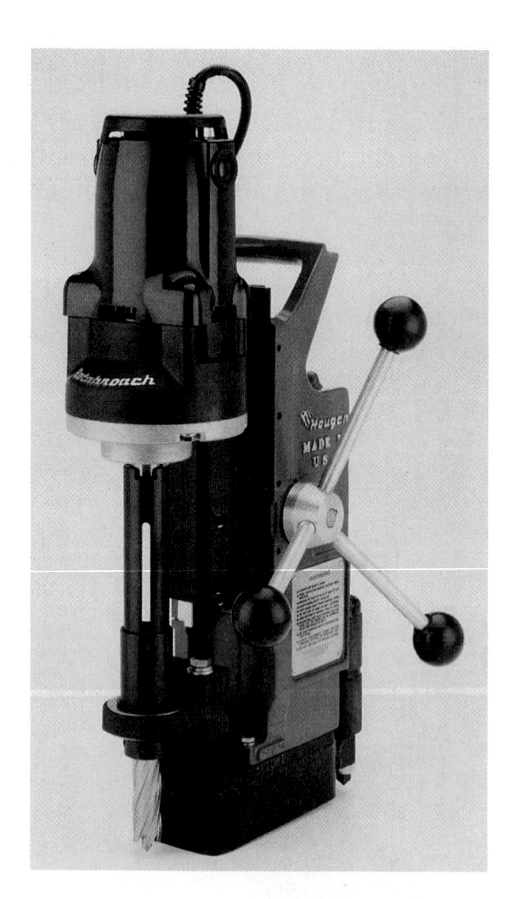

Hougen Manufacturing, Inc.

Figure 12-8. Portable magnetic drill. The magnetic base locks the machine to ferrous metals and enables it to be used in situations where a conventional drill press cannot be used.

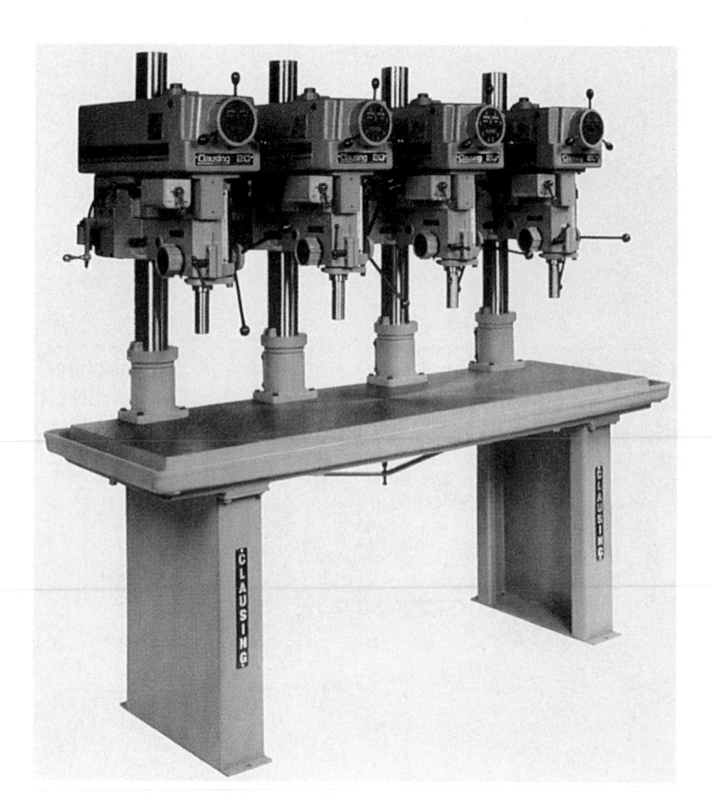

Clausing Industrial, Inc.

Figure 12-9. This gang drilling machine has four drills working together. Each machine is fitted with a different cutting tool. The work is held in a drill jig that moves from position to position as each operation is performed.

Haas Automation, Inc.

Figure 12-10. This CNC machine has a tool magazine capable of holding twenty tools.

12.1.2 Uses of Drilling Machines

Drilling machines are primarily used for cutting round holes. They can also be used for many different machining operations, **Figure 12-11**, including the following:

- ***Countersinking.*** Enlarging a hole at the workpiece surface along an angle to allow a screw head to be flush with the surface.

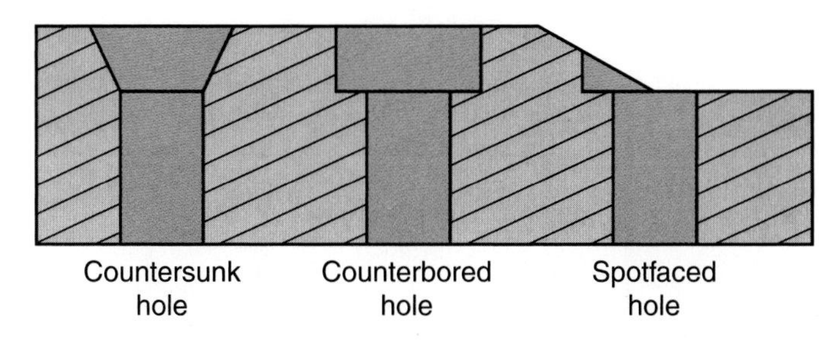

Goodheart-Willcox Publisher

Figure 12-11. Sectional view of three drilled and machined holes. Countersinking cuts a chamfer in a hole. Counterboring is done to prepare a hole to receive a fillister- or socket-head screw. Spotfacing machines a surface to permit a nut or bolt head to bear uniformly.

- ***Counterboring.*** Similar to countersinking, this operation cuts a cylindrical enlargement at the surface of a hole to allow bolt heads to be flush with the surface of the workpiece.
- ***Spotfacing.*** An operation performed to put a smooth finish on a raised area surrounding a hole.
- ***Tapping.*** This operation cuts screw threads into an existing hole.
- ***Reaming.*** An operation performed on an existing hole. The hole is enlarged and finished to very accurate dimensions.

12.2 Drill Press Safety

- Wear goggles when working on a drill press.
- Remove any jewelry and tuck in loose clothing so they do not become entangled in the rotating drill.
- Check the operation of the machine.
- Be sure all guards are in place.
- Clamp the work solidly. Do not hold work with your hand.
- When removing a drill, place a piece of wood below it. Small drills can be damaged if dropped and larger drills can cause injuries.
- Never attempt to operate a drilling machine while your senses are impaired by medication or other substances.
- Use sharp tools.
- Always remove the key from the chuck before turning on the power.
- Let the drill spindle come to a stop after completing the operation. Do not stop it with your hand.
- Clean chips from the work with a brush, not your hands.
- Keep the work area clear of chips. Place them in an appropriate container. Do not brush them onto the floor.
- Wipe up all cutting fluid that spills on the floor right away.
- Place all oily and dirty waste in a closed container when the job is finished.

12.3 Drills

Common drills are known as ***twist drills*** because most are made by forging or milling rough flutes and then twisting them to a spiral shape. After twisting, the drills are milled and ground to approximate size, **Figure 12-12**. Then, they are heat-treated and ground to exact size.

Most drills are made of high-speed steel (HSS) or carbon steel. High-speed steel drills can be operated at much higher cutting speeds than carbon steel drills without danger of burning and drill damage.

Most drills are available with straight or taper shanks and with tungsten carbide tips. Coating drills with titanium nitride greatly increases tool life.

12.3.1 Parts of a Drill

The twist drill is an efficient cutting tool. It is composed of three principal parts: point, shank, and body, **Figure 12-13**.

Point

The point is the cone-shaped end that does the cutting. The point consists of the following components:

- The dead center refers to the sharp edge at the extreme tip of the drill. This should always be in the exact center of the drill axis.
- The lips are the cutting edges of the drill.
- The heel is the portion of the point back from the lips.
- Lip clearance is the amount by which the surface of the point is relieved back from the lips.

Klein Tools, Inc.

Figure 12-12. A common twist drill.

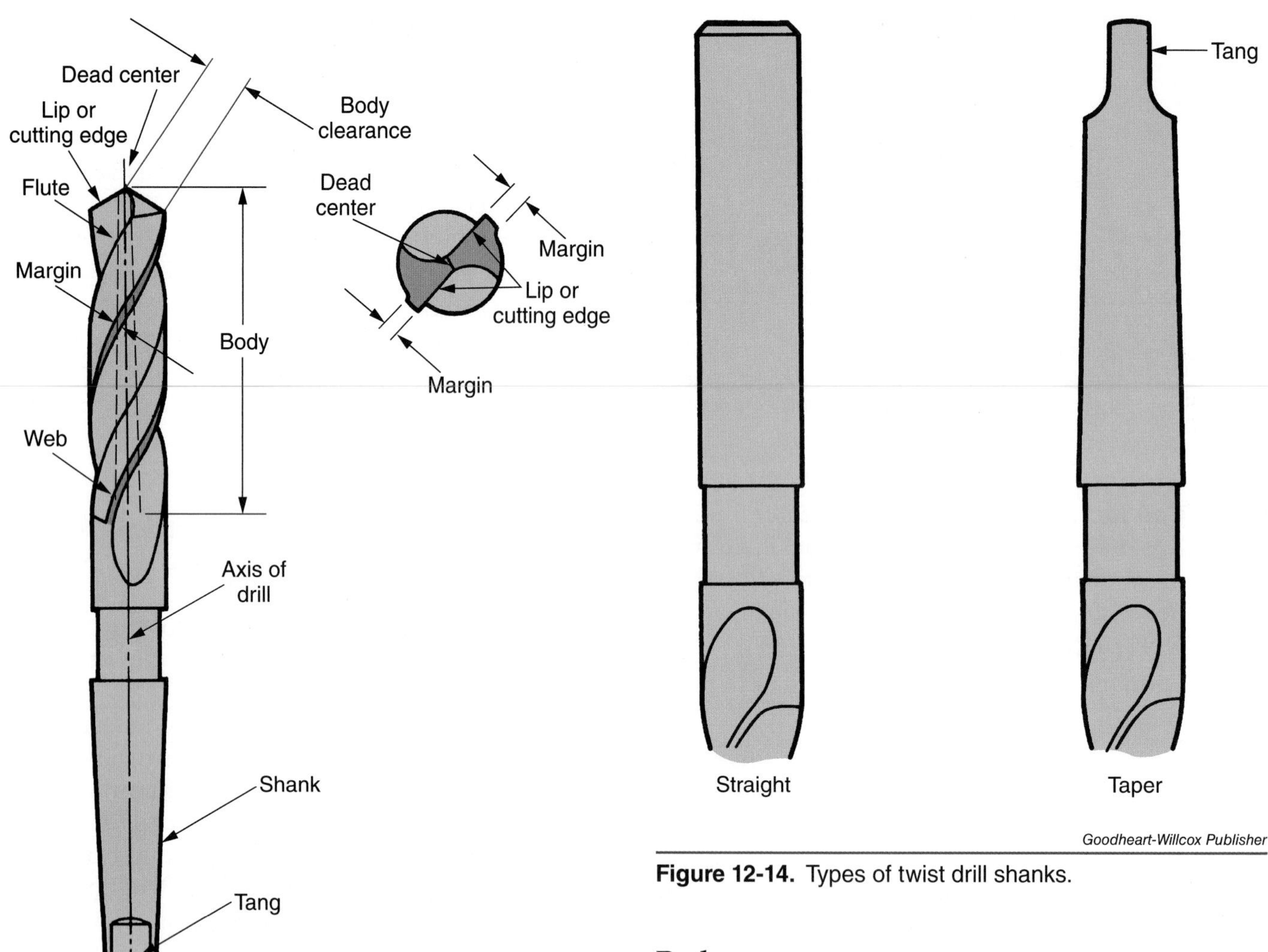

Goodheart-Willcox Publisher

Figure 12-13. Parts of a twist drill.

Goodheart-Willcox Publisher

Figure 12-14. Types of twist drill shanks.

Shank

The shank is the portion of the drill that mounts into the chuck or spindle. Twist drills are made with shanks that are either straight or tapered, **Figure 12-14**. Straight shank drills are used with a chuck. Taper shank drills have self-holding tapers (No. 1 to No. 5 Morse taper) that fit directly into the drill press spindle.

A tang is located on the end of the taper shank. It fits into a slot in the spindle, sleeve, or socket, and assists in driving the tool. The tang also provides a means of separating the taper from the holding device.

Body

The body is the portion of the drill between the point and the shank. It consists of the following components:

- The ***flutes*** are two or more spiral grooves that run along the length of the drill body. The flutes serve four purposes:
  - Help form the cutting edges of the drill point.
  - Curl the chip tightly for easier removal.
  - Form channels through which the chips can escape as the hole is drilled.
  - Allow coolant and lubricant to flow down to the cutting edges.
- The margin is the narrow strip extending back along the entire length of the drill body.
- Body clearance refers to the part of the drill body that has been reduced in order to lower friction between the drill and the wall of the hole.
- The web is the metal column that separates the flutes. It gradually increases in thickness toward the shank for added strength.

12.3.4 Drill Size

Drill sizes are expressed by the following series:

- **Numbers.** #80 to #1 (0.0135″ to 0.2280″ diameters).
- **Letters.** A to Z (0.234″ to 0.413″ diameters).
- **Inches and fractions.** 1/64″ to 3 1/2″ diameters.
- **Metric.** 0.15 mm to 76.0 mm diameters.

The size ranges listed are the most common drill sizes within each series. Larger and smaller size drills exist, but are uncommon or specialized. The drill size chart will give an idea of this vast array of drill sizes, **Figure 12-15**.

12.3.5 Drill Measurements

Most drills, with the exception of small drills in the number series, have the diameter stamped on the shank. These figures frequently become obscured, making it necessary to determine the diameter by measuring.

When a micrometer is used for measuring, the measurement is made across the drill margins. However, if the drill is worn, the measurement is made on the shank at the end of the flutes. See **Figure 12-16** for both techniques.

Inch	mm	Wire gage	Decimals of an inch
		80	.0135
		79	.0145
1/64			.0156
	.4		.0157
		78	.0160
		77	.0180
	.5		.0197
		76	.0200
		75	.0210
	.55		.0217
		74	.0225
	.6		.0236
		73	.0240
		72	.0250
	.65		.0256
		71	.0260
	.7		.0276
		70	.0280
		69	.0293
	.75		.0295
		68	.0310
1/32			.0313
	.8		.0315
		67	.0320
		66	.0330
	.85		.0335
		65	.0350
	.9		.0354
		64	.0360
		63	.0370
	.95		.0374
		62	.0380
		61	.0390
	1		.0394
		60	.0400
		59	.0410
	1.05		.0413
		58	.0420
		57	.0430
	1.1		.0433
	1.15		.0453
		56	.0465

Inch	mm	Wire gage	Decimals of an inch
3/64			.0469
	1.2		.0472
	1.25		.0492
	1.3		.0512
		55	.0520
	1.35		.0531
		54	.0550
	1.4		.0551
	1.45		.0571
	1.5		.0591
		53	.0595
	1.55		.0610
1/16			.0625
	1.6		.0630
		52	.0635
	1.65		.0650
	1.7		.0669
		51	.0670
	1.75		.0689
		50	.0700
	1.8		.0709
	1.85		.0728
		49	.0730
	1.9		.0748
		48	.0760
	1.95		.0768
5/64			.0781
		47	.0785
	2		.0787
	2.05		.0807
		46	.0810
		45	.0820
	2.1		.0827
	2.15		.0846
		45	.0860
	2.2		.0866
	2.25		.0886
		43	.0890
	2.3		.0906
	2.35		.0925
		42	.0935

Inch	mm	Wire gage	Decimals of an inch
3/32			.0938
	2.4		.0945
		41	.0960
	2.45		.0966
		40	.0980
	2.5		.0984
		39	.0995
		38	.1015
	2.6		.1024
		37	.1040
	2.7		.1063
		36	.1065
	2.75		.1083
7/64			.1094
		35	.1100
	2.8		.1102
		34	.1110
		33	.1130
	2.9		.1142
		32	.1160
	3		.1181
		31	.1200
	3.1		.1220
1/8			.1250
	3.2		.1260
	3.25		.1280
		30	.1285
	3.3		.1299
	3.4		.1339
		29	.1360
	3.5		.1378
		28	.1405
9/64			.1406
	3.6		.1417
		27	.1440
	3.7		.1457
		26	.1470
	3.75		.1476
		25	.1495
	3.8		.1496
		24	.1520
	3.9		.1535
		23	.1540

Inch	mm	Wire gage	Decimals of an inch
5/32			.1563
		22	.1570
	4		.1575
		21	.1590
		20	.1610
	4.1		.1614
	4.2		.1654
		19	.1660
	4.25		.1673
	4.3		.1693
		18	.1695
11/64			.1719
		17	.1730
	4.4		.1732
		16	.1770
	4.5		.1772
		15	.1800
	4.6		.1811
		14	.1820
		13	.1850
	4.7		.1850
	4.75		.1870
3/16			.1875
	4.8		.1890
		12	.1890
		11	.1910
	4.9		.1929
		10	.1935
		9	.1960
	5		.1969
		8	.1990
	5.1		.2008
		7	.2010
13/64			.2031
		6	.2040
	5.2		.2047
		5	.2055
	5.25		.2067
	5.3		.2087
		4	.2090
	5.4		.2126
		3	.2130
	5.5		.2165
7/32			.2188
	5.6		.2205
		2	.2210
	5.7		.2244
	5.75		.2264
		1	.2280
	5.8		.2883

Goodheart-Willcox Publisher

Figure 12-15. Decimal equivalents of drill sizes.

Inch	mm	Letter sizes	Decimals of an inch
	5.9		.2323
		A	.2340
15/64			.2344
	6		.2362
		B	.2380
	6.1		.2402
		C	.2420
	6.2		.2441
		D	.2460
	6.25		.2461
	6.3		.2480
1/4		E	.2500
	6.4		.2520
	6.5		.2559
		F	.2570
	6.6		.2598
		G	.2610
	6.7		.2638
17/64			.2656
	6.75		.2657
		H	.2660
	6.8		.2677
	6.9		.2717
		I	.2720
	7		.2756
		J	.2770
	7.1		.2795
		K	.2810
9/32			.2812
	7.2		.2835
	7.25		.2854
	7.3		.2874
		L	.2900
	7.4		.2913
		M	.2950
	7.5		.2953
19/64			.2969
	7.6		.2992
		N	.3020
	7.7		.3031
	7.75		.3051
	7.8		.3071
	7.9		.3110
5/16			.3125
	8		.3150
		O	.3160
	8.1		.3189
	8.2		.3228
		P	.3230
	8.25		.3248
	8.3		.3268

Inch	mm	Letter sizes	Decimals of an inch
21/64			.3281
	8.4		.3307
		Q	.3320
	8.5		.3346
	8.6		.3386
		R	.3390
	8.7		.3425
11/32			.3438
	8.75		.3345
	8.8		.3465
		S	.3480
	8.9		.3504
	9		.3543
		T	.3580
	9.1		.3583
23/64			.3594
	9.2		.3622
	9.25		.3642
	9.3		.3661
		U	.3680
	9.4		.3701
	9.5		.3740
3/8			.3750
		V	.3770
	9.6		.3780
	9.7		.3819
	9.75		.3839
	9.8		.3858
		W	.3860
	9.9		.3898
25/64			.3906
	10		.3937
		X	.3970
		Y	.4040
13/32			.4063
		Z	.4130
	10.5		.4134
27/64			.4219
	11		.4331
7/16			.4375
	11.5		.4528
29/64			.4531
15/32			.4688
	12		.4724
31/64			.4844
	12.5		.4921
1/2			.5000
	13		.5118
33/64			.5156
17/32			.5313
	13.5		.5315

Inch	mm	Decimals of an inch
35/64		.5469
	14	.5512
9/16		.5625
	14.5	.5709
37/64		.5781
	15	.5906
19/32		.5938
39/64		.6094
	15.5	.6102
5/8		.6250
	16	.6299
41/64		.6406
	16.5	.6496
21/32		.6563
	17	.6693
43/64		.6719
11/16		.6875
	17.5	.6890
45/64		.7031
	18	.7087
23/32		.7188
	18.5	.7283
47/64		.7344
	19	.7480
3/4		.7500
49/64		.7656
	19.5	.7677
25/32		.7812
	20	.7874
51/64		.7969
	20.5	.8071
13/16		.8125
	21	.8268
53/64		.8281
27/32		.8438
	21.5	.8465
55/64		.8594
	22	.8661
7/8		.8750
	22.5	.8858
57/64		.8906
	23	.9055
29/32		.9063
59/64		.9219
	23.5	.9252
15/16		.9375
	24	.9449
61/64		.9531
	24.5	.9646
31/32		.9688
	25	.9843
63/64		.9844

Inch	mm	Decimals of an inch
1		1.0000
	25.5	1.0039
1 1/64		1.0156
	26	1.0236
1 1/32		1.0313
	26.5	1.0433
1 3/64		1.0469
1 1/16		1.0625
	27	1.0630
1 5/64		1.0781
	27.5	1.0827
1 3/32		1.0938
	28	1.1024
1 7/64		1.1094
	28.5	1.1220
1 1/8		1.1250
1 9/64		1.1406
	29	1.1417
1 5/32		1.1562
	29.5	1.1614
1 11/64		1.1719
	30	1.1811
1 3/16		1.1875
	30.5	1.2008
1 13/64		1.2031
1 7/32		1.2188
	31	1.2205
1 15/64		1.2344
	31.5	1.2402
1 1/4		1.2500
	32	1.2598
1 17/64		1.2656
	32.5	1.2795
1 9/32		1.2813
1 19/64		1.2969
	33	1.2992
1 5/16		1.3125
	33.5	1.3189
1 21/64		1.3281
	34	1.3386
1 11/32		1.3438
	34.5	1.3583
1 23/64		1.3594
1 3/8		1.3750
	35	1.3780
1 25/64		1.3906
	35.5	1.3976
1 13/32		1.4063
	36	1.4173
1 27/64		1.4219
	36.5	1.4370

Goodheart-Willcox Publisher

Figure 12-15. (continued).

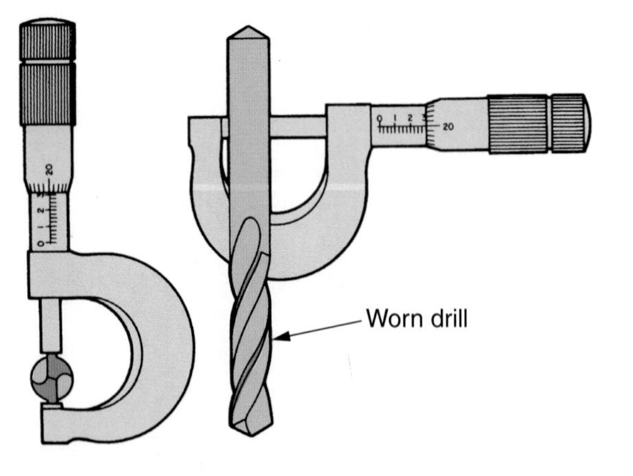

Goodheart-Willcox Publisher

Figure 12-16. Measuring drill size with a micrometer.

Diameter can also be checked with a ***drill gage,*** **Figure 12-17**. Drill gages are made for various drill series. However, 1/2″ drills are the largest that can be checked. New drills are checked at the points. Worked drills are checked at the end of the flutes.

Always check the drill diameter before using it. Using the wrong size drill can be a very expensive and time-consuming mistake.

12.3.6 Types of Drills

Industry uses special drills to improve the accuracy of the drilled hole, to speed production, and to improve drilling efficiency. Other special drills are designed for very specific operations or drilling in specific materials.

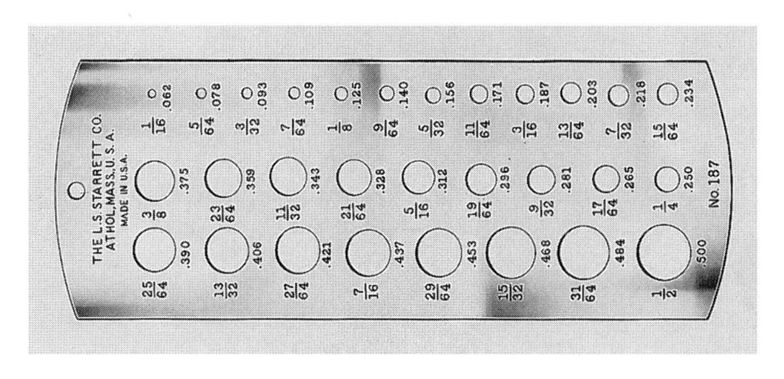

Goodheart-Willcox Publisher

Figure 12-17. This drill gage is used to measure fractional size drills. Similar gages are available for measuring letter, number, and millimeter size drills.

A reduced-shank drill has a shank with a smaller diameter than the body. Reduced-shank drills allow larger holes to be drilled using machine tools with smaller chuck or spindle sizes.

Microdrills, **Figure 12-18,** have diameters smaller than 0.020″ (0.508 mm). They require special drilling equipment.

A coolant-hole drill has coolant holes through the body, which permit fluid or air to remove heat from the point, **Figure 12-19**. The pressure of the fluid or air also ejects the chips from the hole while drilling.

BIG Kaiser Precision Tooling Inc.

Figure 12-18. Microdrills can be made of either HSS or tungsten carbide. Microdrills can drill holes as small as 0.0012″ (0.03 mm) through materials ranging from steel to cast iron to nonferrous alloys.

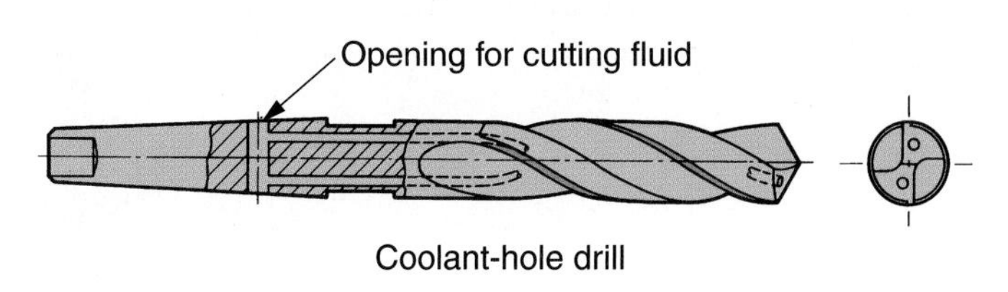

Goodheart-Willcox Publisher

Figure 12-19. A taper shank twist drill with holes to direct coolant to the cutting edges.

Special step drills eliminate drilling operations in production work. **Figure 12-20** illustrates this type of drill. A combination drill and reamer speeds up production by performing two operations at once, **Figure 12-21**.

Core drills are three- or four-flute drills used to enlarge core holes in a casting. See **Figure 12-22**. Drills with three or four flutes provide a better finish than drills with only two flutes.

Half-round straight-flute drills, **Figure 12-23**, are designed for producing holes in brass, copper alloys, and other soft nonferrous materials. Heavy duty carbide-tipped versions are available for drilling hardened steels. These drills are manufactured in fractional and number sizes.

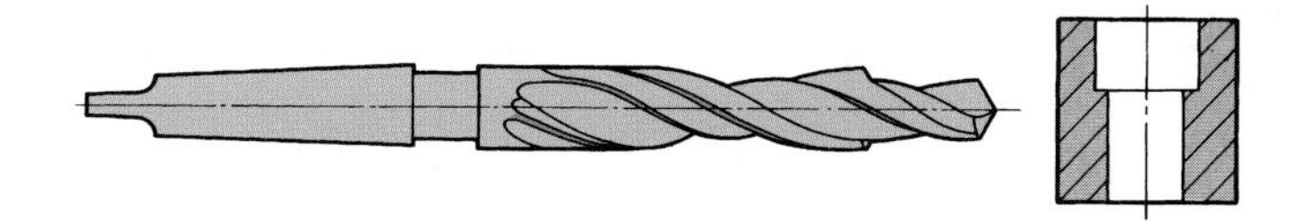

Goodheart-Willcox Publisher

Figure 12-20. A step drill designed to drill a hole and counterbore the hole all in the same operation.

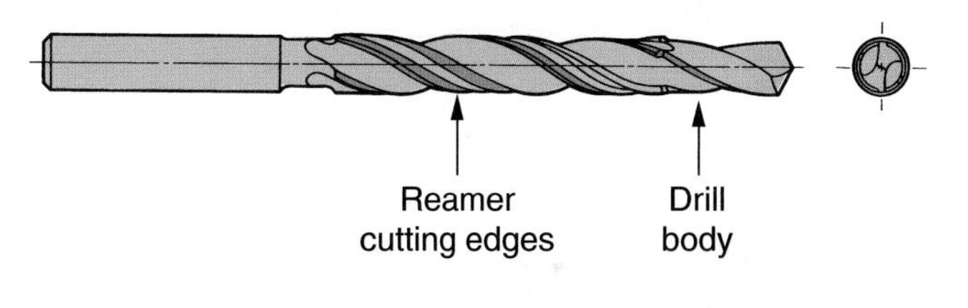

Goodheart-Willcox Publisher

Figure 12-21. Combination drill and reamer.

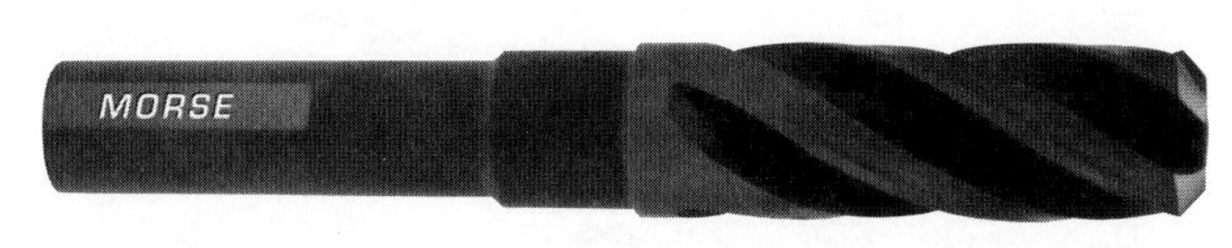

Morse

Figure 12-22. This four-flute core drill is used to enlarge cored and drilled holes.

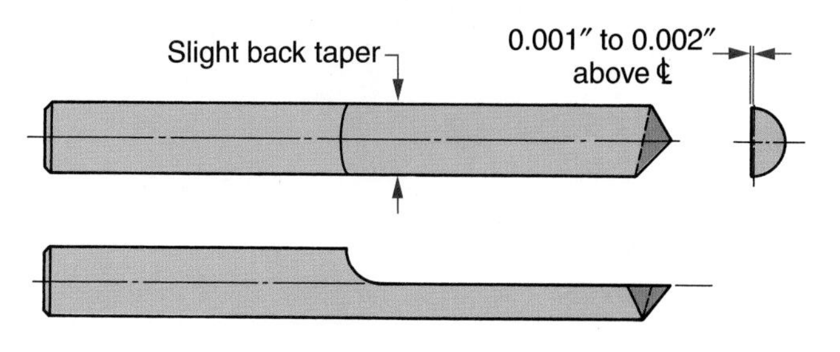

Goodheart-Willcox Publisher

Figure 12-23. A half-round drill.

The straight-flute gun drill is designed for ferrous and nonferrous metals, **Figure 12-24**. It is usually fitted with a carbide cutting tip.

Indexable-insert drills, **Figure 12-25**, are capable of drilling at much higher speeds than high-speed steel drills. The low-cost carbide inserts with multiple cutting edges eliminate costly sharpening. The entire drill does not need to be replaced when the cutting edges are worn—only the insert is changed.

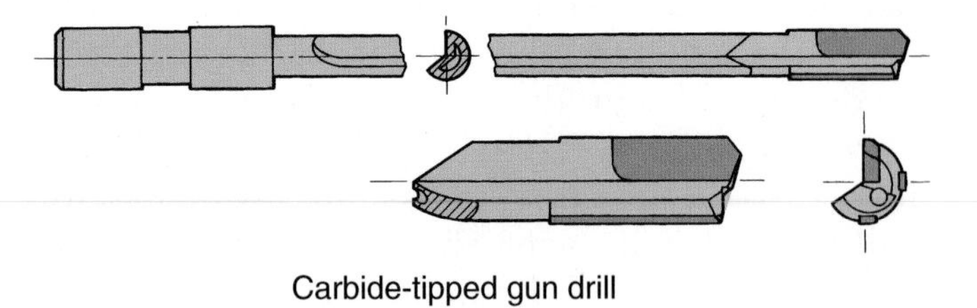

Goodheart-Willcox Publisher

Figure 12-24. A straight-flute gun drill. The tip is shown in larger scale. The light-colored portions are tungsten carbide. The larger section does the cutting, while the smaller sections act as wear surfaces.

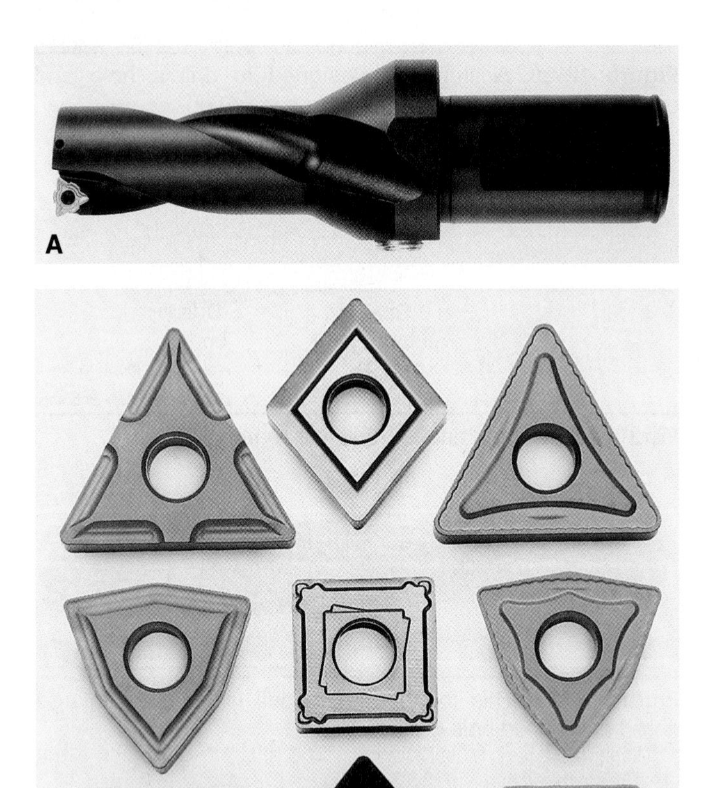

Iscar Metals, Inc.; Hertel Cutting Technologies, Inc.

Figure 12-25. Indexable-insert drill. A—When the carbide insert becomes worn, it can be replaced with a new insert. B—Different insert shapes are used with different materials and operations.

Indexable-insert drills do have limitations. Hole depth is limited to approximately four times the hole diameter. The smallest size available is 5/8″ diameter.

Spade drills have replaceable cutting tips that are normally made of tungsten carbide, **Figure 12-26**. These drills are available in sizes from 1″ to 5″ (25 mm to 125 mm). They are less expensive than twist drills of the same size.

12.4 Drill-Holding Devices

A drill is held in the drill press by one of two of the following methods:

- **Chuck.** A movable jaw mechanism for drills with straight shanks. See **Figure 12-27**.
- **Tapered spindle.** A tapered opening for drills with taper shanks, **Figure 12-28**.

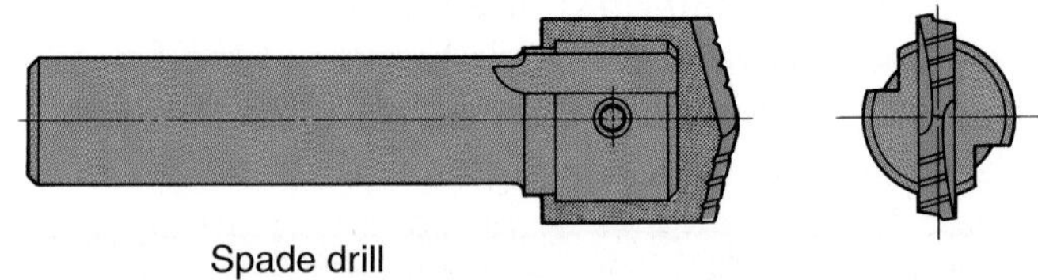

Goodheart-Willcox Publisher

Figure 12-26. This spade drill has a replaceable carbide insert cutting tip. Replaceable inserts are an economical way of drilling large holes.

Yukiwa Seiko USA, Inc.

Figure 12-27. A drill chuck can be tightened with a key to hold the drill. Keyless chucks are also available.

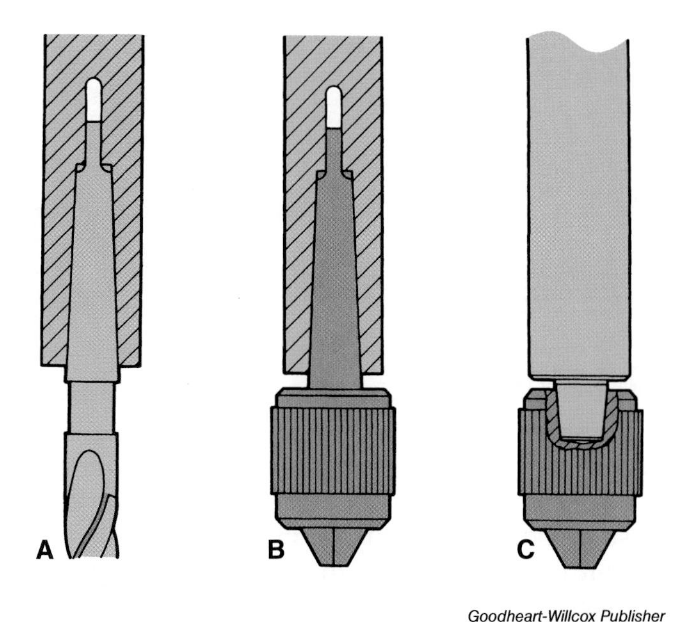

Figure 12-28. Tapered spindles. A—Taper shank drill. B—Drill chuck with a taper shank. C—This is a solid spindle with a short external taper that fits into a drill chuck. The chuck is permanently attached to the spindle.

A drill chuck with a taper shank makes it possible to use straight-shank drills when the drill press is fitted with a tapered spindle. When using a chuck, first insert the drill and tighten the chuck jaws by hand. If the chuck is centered and running true, tighten the chuck with a chuck key.

SAFETY NOTE

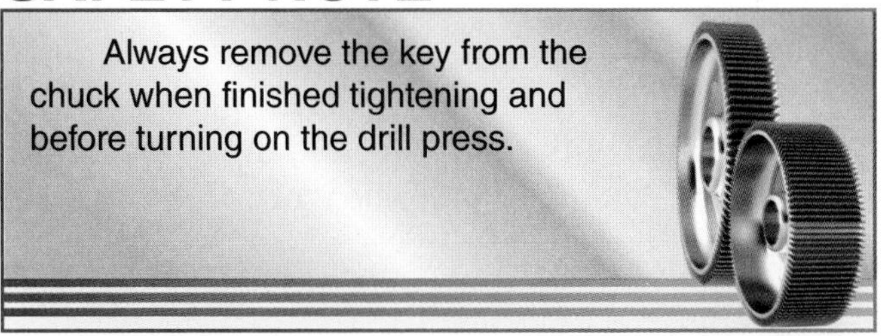

Taper shank drills must be wiped clean before inserting the shank into the spindle. Nicks in the shank must be removed with an oilstone otherwise the shank will not seat properly.

SAFETY NOTE

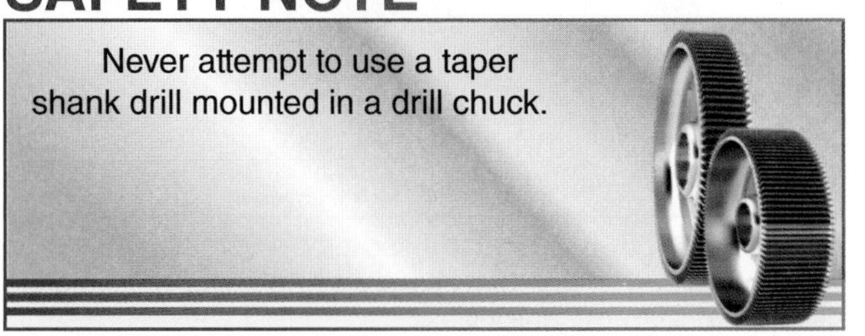

Most drill press spindles are made with a No. 2 or No. 3 Morse taper (often indicated as "MT"). A drill with a shank smaller than the spindle taper must be enlarged to fit by using a sleeve, **Figure 12-29**. Drills with shanks larger than the spindle opening can be fit by using a socket, **Figure 12-30**. The taper opening in the socket is larger than the taper on its shank. A socket should only be used when a larger drill press is not available.

SAFETY NOTE

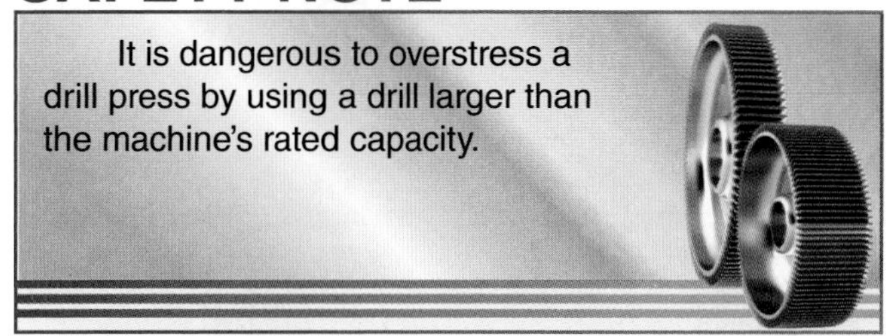

Sleeves, sockets, and taper shank drills are separated with a drift, **Figure 12-31**. To use a drift, insert it in the slot with the round edge up, **Figure 12-32**.

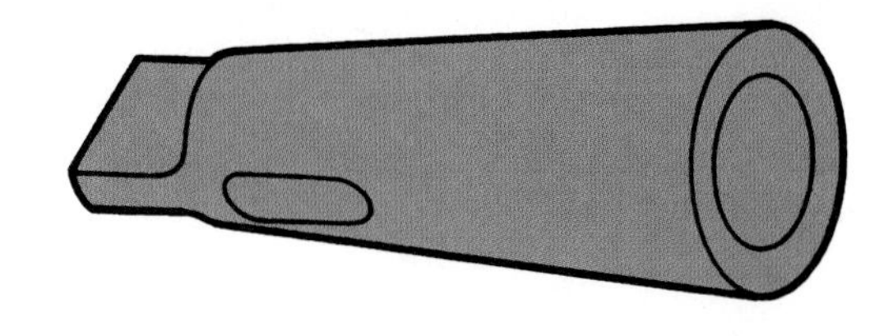

Figure 12-29. A drill sleeve is needed when the drill is too small for the spindle.

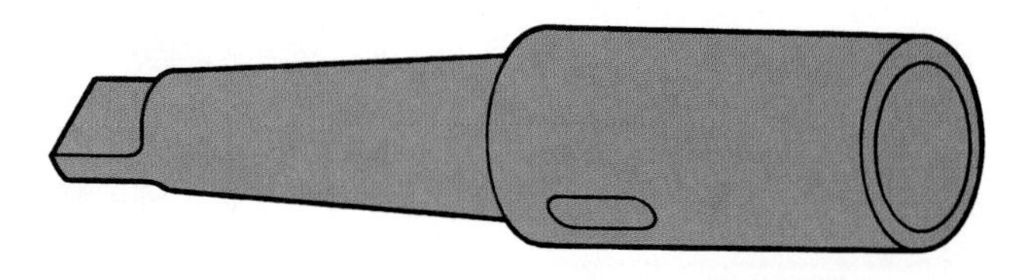

Figure 12-30. A drill socket is used when the spindle is too small for the drill.

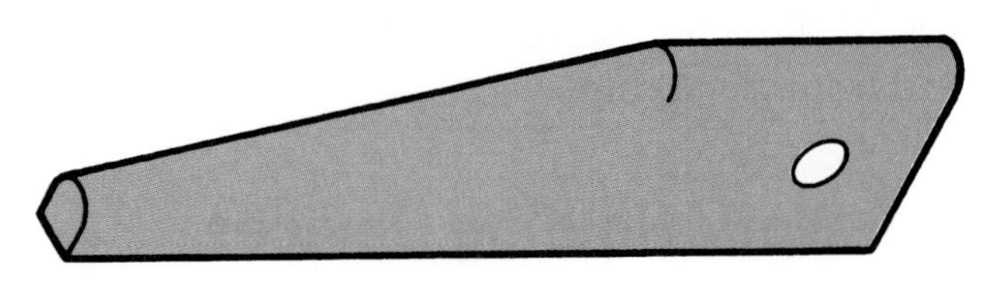

Figure 12-31. A drift is used to remove taper shank tools from the drill spindle. Never use a file tang.

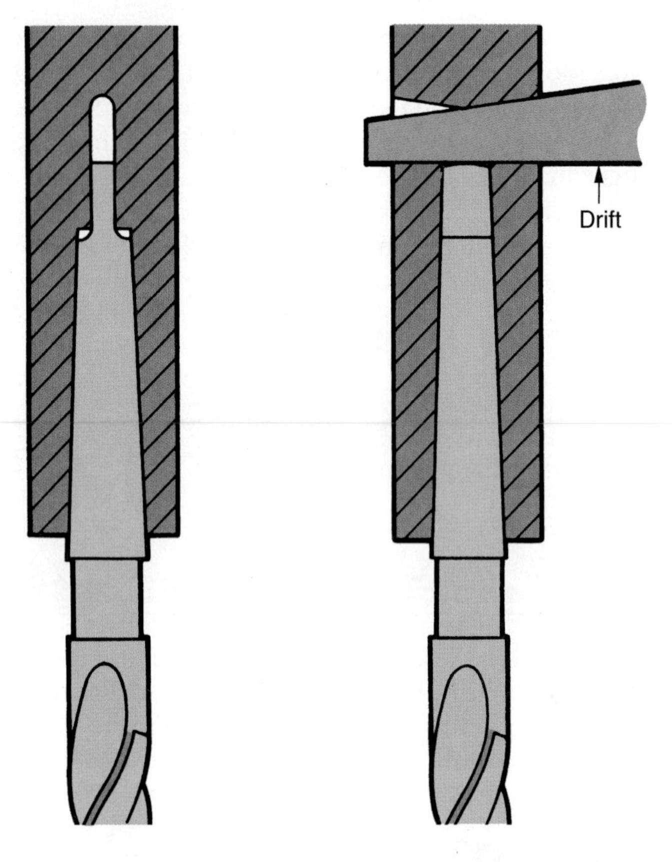

Goodheart-Willcox Publisher

Figure 12-32. Using a drift. A—The drill is locked in the spindle. B—Using a drift to remove the drill from the spindle.

A sharp rap with a lead hammer will cause the parts to separate.

SAFETY NOTE

Never use a file tang in place of a drift. It will damage the drill shank and machine spindle. Then other drills will not fit properly. The file could also shatter.

When removing a drill from the spindle, hold the drill to prevent it from falling to the floor. Dropping the drill may damage the drill point. Wrap a piece of clean cloth around the drill to protect your hand from metal chips.

12.5 Work-Holding Devices

Work must be mounted solidly on the drilling machine. If work is mounted improperly, it may spring or move, causing drill damage or breakage.

SAFETY NOTE

Serious injury can result from work that becomes loose and spins on a drill press. This dangerous situation is nicknamed a "merry-go-round."

12.5.1 Vises

Vises are widely used to hold work, **Figure 12-33.** For best results, the vise must be bolted to the drill table.

Parallels are often used to level the work and raise it above the vise base, **Figure 12-34.** This will permit the drill to come through the work and not damage the vise. Parallels can be made from stock steel bars or from special heat-treated steel. Heat-treated parallels are ground to size.

Seat the work on the parallels by tightening the vise and tapping the work with a mallet until the parallels do not move. Loose parallels indicate that the work is not seated properly.

An angular vise permits angular drilling without tilting the drill press table. See **Figure 12-35.** A cross-slide permits rapid alignment of the work. Some cross-slides are fitted with a vise, **Figure 12-36.** Others have a series of tapped holes for mounting a vise or another work-holding device.

12.5.2 V-Blocks

V-blocks support round work for drilling, **Figure 12-37.** These blocks are made in many sizes. Some are fitted with clamps to hold the work. Larger sizes must be clamped with the work, **Figure 12-38.**

Photo courtesy of Grizzly Industrial, Inc. www.grizzly.com

Figure 12-33. A typical vise used on drill press. The swivel base permits the vise body to pivot 180°. The quick-acting vise jaw locks parts quickly.

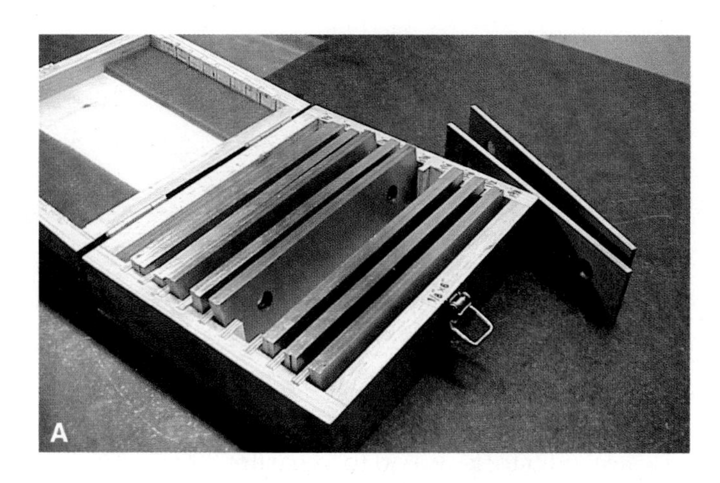

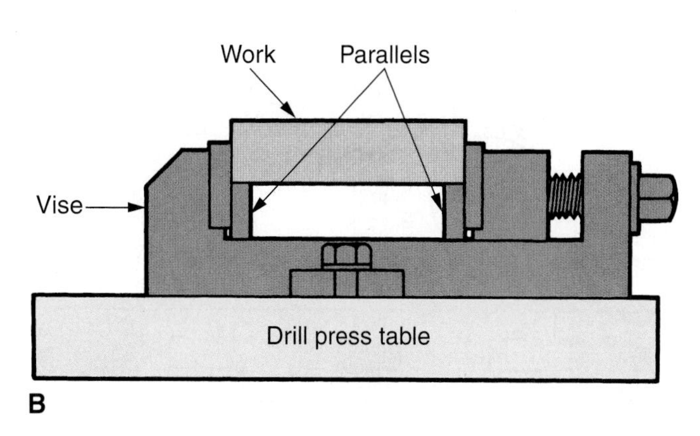

Goodheart-Willcox Publisher

Figure 12-34. Parallels. A— Steel parallels are available in a large variety of sizes. B—Parallels are often used to raise work above the vise base. This will prevent the drill from cutting into the vise as it goes through the work.

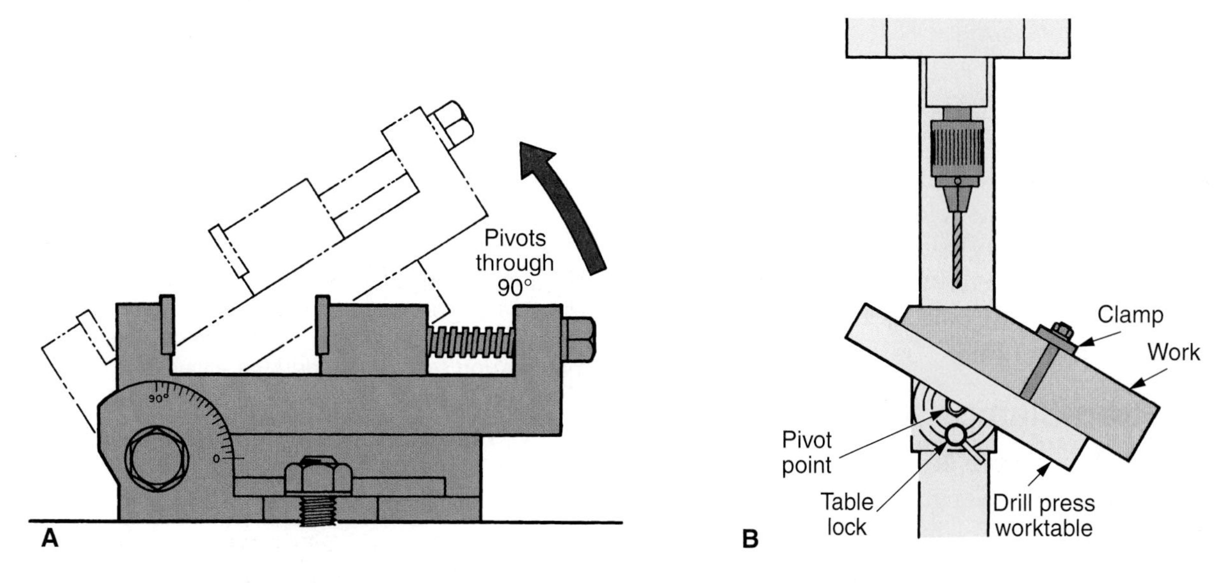

Goodheart-Willcox Publisher

Figure 12-35. Angular vise. A—An angular vise can be adjusted through 90° to permit drilling on an angle without tilting the entire vise or drill table. B—Angular drilling can also be done by tilting the drill table. Be sure the table is locked tightly before starting to drill.

Photo courtesy of Grizzly Industrial, Inc. www.grizzly.com

Figure 12-36. A cross-slide vise permits rapid alignment of work for drilling.

Photo courtesy of Grizzly Industrial, Inc. www.grizzly.com

Figure 12-37. V-blocks fitted with clamps to hold round workpieces.

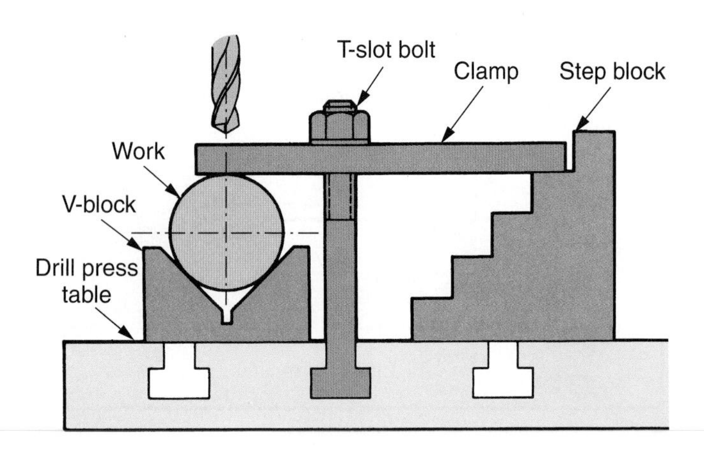

Goodheart-Willcox Publisher

Figure 12-38. One method of clamping large diameter stock for drilling. Always check to be sure that the drill will clear the V-block when it comes through the work.

12.5.3 T-Bolts

T-bolts fit into the drill press table slots and fasten the work or clamping devices to the machine, **Figure 12-39**. A washer should always be used between the nut and the holding device. For convenience, it is desirable to have an assortment of different length T-bolts. To reduce the chance of a setup working loose, place the bolts as close to the work as possible. See **Figure 12-40**.

12.5.4 Strap Clamps

Strap clamps, **Figure 12-41**, make the clamping operation easier. The elongated slot permits some adjustment without removing the washer and nut.

Photo courtesy of Grizzly Industrial, Inc. www.grizzly.com

Figure 12-39. A few of the many types and sizes of T-bolts available.

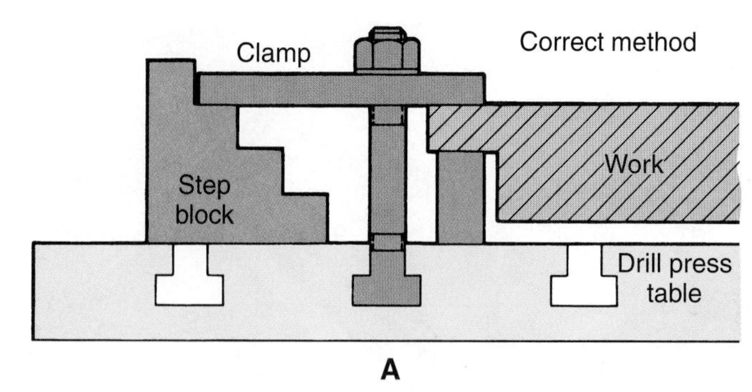

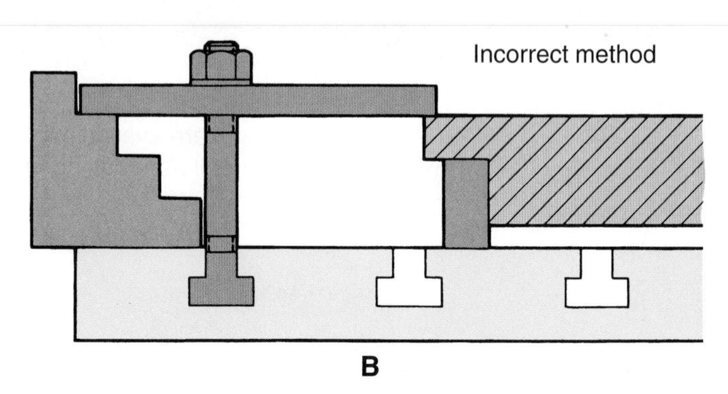

Goodheart-Willcox Publisher

Figure 12-40. Examples of clamping techniques. A—Correct clamping technique. The clamp is parallel to the work. Clamp slippage can be reduced by placing a piece of paper between the work and the clamp. B—Incorrect clamping technique. T-bolt is too far from work. This allows the clamp to spring under pressure.

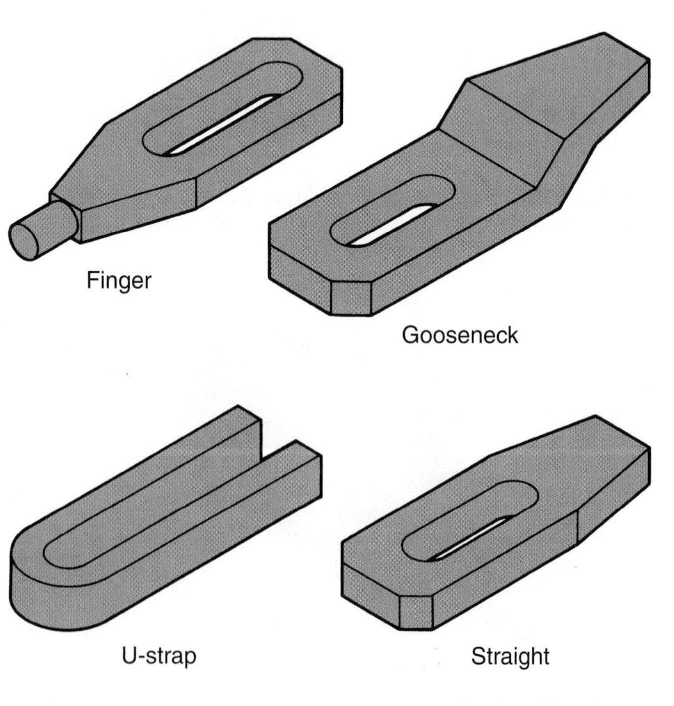

Goodheart-Willcox Publisher

Figure 12-41. Types of strap clamps.

Use a strip of copper or aluminum to protect a machined surface that must be clamped.

A U-strap clamp is used when the clamp must bridge the work. It can straddle the drill and not interfere with the drilling operation. The small, round section that projects from a finger clamp permits the use of small holes or openings in the work for clamping.

12.5.5 Step Blocks

A step block supports the strap clamp opposite the work, **Figure 12-42**. The steps allow the adjustments necessary to keep the strap parallel with the work.

12.5.6 Angle Plate

An angle plate is often used when work must be clamped to a support. The angle plate is then bolted to the machine table, **Figure 12-43**.

12.5.7 Drill Jig

A drill jig permits holes to be drilled in a number of identical pieces, **Figure 12-44**. This clamping device supports and locks the work in the proper position. With the use of bushings, the jig guides the drill to the correct location. This makes it unnecessary to lay out each individual piece for drilling.

12.6 Cutting Speeds and Feeds

The ***cutting speed*** is the speed at which the drill rotates. The ***feed*** is the distance the drill is moved into the work with each revolution. Both are important considerations because they determine the time required to produce the hole.

Photo courtesy of Grizzly Industrial, Inc. www.grizzly.com

Figure 12-42. Step blocks are used to support strap clamps.

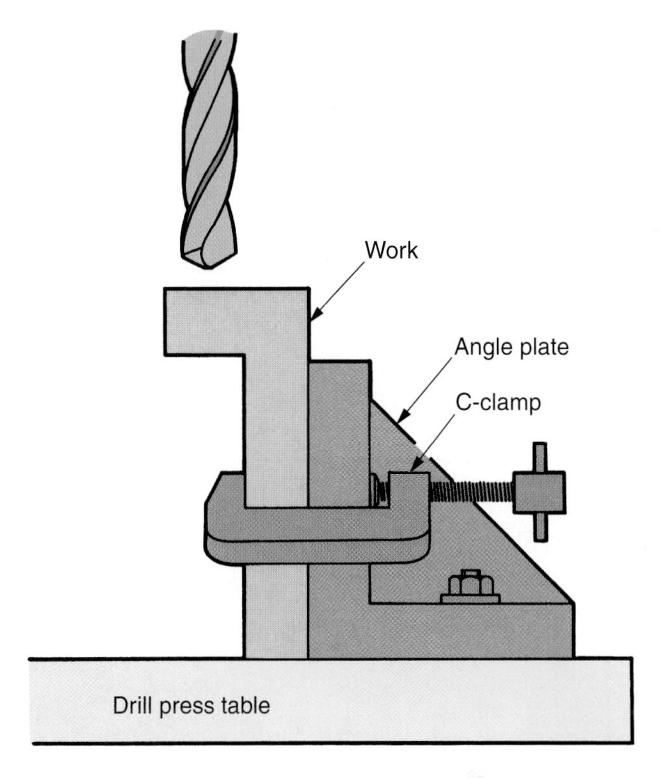

Goodheart-Willcox Publisher

Figure 12-43. Work must sometimes be mounted against an angle plate for adequate support for drilling.

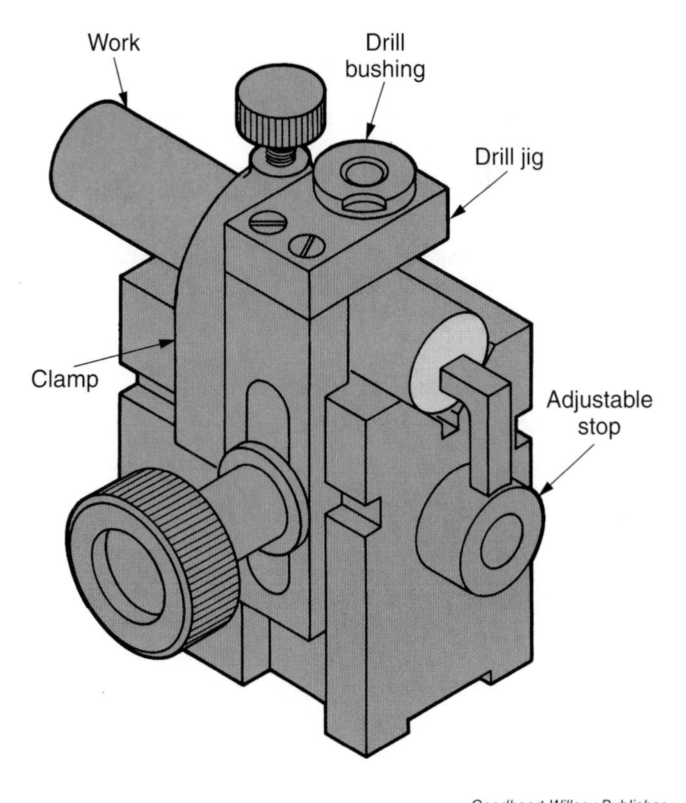

Goodheart-Willcox Publisher

Figure 12-44. A typical drill jig for holding round stock for drilling through center.

Drill cutting speed, also known as peripheral speed, does not refer to the revolutions per minute (rpm) of the drill, but rather to the distance that the drill cutting edge circumference travels per minute.

12.6.1 Feed

Contrary to popular belief, the spiral shape of a drill flute does not cause the drill to pull itself into the work. Constant pressure must be applied and maintained to advance the drill point at a given rate. This advance is called feed and is measured in either decimal fractions of an inch or millimeters.

Because so many variables affect results, there can be no hard and fast rule to determine the exact cutting speed and feed for a given material. For this reason, the drill speed and feed table indicates only recommended speeds and feeds, **Figure 12-45.** They are a starting point and can be increased or decreased for optimum cutting.

Material	Cutting Fluid	Speed feet per minute	Feeds per revolution Over .040 diameter**				
			Under 1/8	1/8 to 1/4	1/4 to 1/2	1/2 to 1	Over 1
Aluminum & aluminum alloys	Sol. oil, ker. & lard oil, lt. oil	200-300	.0015	.003	.006	.010	.012
Aluminum & bronze	Sol. oil, ker. & lard oil, lt. oil	50-100	.0015	.003	.006	.010	.012
Brass, free machining	Dry, sol. oil, ker. & lard oil, lt. min. oil	150-300	.0025	.005	.010	.020	.025
Bronze, common	Dry, sol. oil, lard oil, min. oil	200-250	.0025	.005	.010	.020	.025
Bronze, soft and medium hard	Min. oil with 5%-15% lard oil	70-300	.0025	.005	.010	.020	.025
Bronze, phosphor, 1/2 hard	Dry, sol. oil, lard oil, min. oil	110-180	.0015	.003	.006	.010	.012
Bronze, phosphor, soft	Dry, sol. oil, lard oil, min. oil	200-250	.0025	.005	.010	.020	.025
Cast iron, soft	Dry or airjet	100-150	.0025	.005	.010	.020	.025
Cast iron, medium	Dry or airjet	70-120	.0015	.003	.006	.010	.012
Cast iron, hard	Dry or airjet	30-100	.001	.002	.003	.005	.006
*Cast iron, chilled	Dry or airjet	10-25	.001	.002	.003	.005	.006
*Cast steel	Soluble oil, sulphurized oil. min. oil	30-60	.001	.002	.003	.005	.006
Copper	Dry, soluble oil, lard oil, min. oil	70-300	.001	.002	.003	.005	.006
Magnesium & magnesium alloys	Mineral seal oil	200-400	.0025	.005	.010	.020	.025
Manganese copper, 30% mn.	Soluble oil, sulphurized oil	10-25	.001	.002	.003	.005	.006
Malleable iron	Dry, soluble oil, soda water, min. oil	60-100	.0025	.005	.010	.020	.025
Monel metal	Sol. oil, sulphurized oil, lard oil	30-50	.0015	.003	.006	.010	.012
Nickel, pure	Sulphurized oil	60-100	.001	.002	.003	.005	.006
Nickel, steel 3-1/2%	Sulphurized oil	40-80	.001	.002	.003	.005	.006
Plastics, thermosetting	Dry or airjet	100-300	.0015	.003	.006	.010	.012
Plastics, thermoplastic	Soluble oil, soapy water	100-300	.0015	.003	.006	.010	.012
Rubber, hard	Dry or airjet	100-300	.001	.002	.003	.005	.006
Spring steel	Soluble oil, sulphurized oil	10-25	.001	.002	.003	.005	.006
Stainless steel, free mach'g.	Soluble oil, sulphurized oil	60-100	.0025	.005	.010	.020	.025
Stainless steel, tough mach'g.	Soluble oil, sulphurized oil	20-27	.0025	.005	.010	.020	.025
Steel, free machin'g SAE 1100	Soluble oil, sulphurized oil	70-120	.0015	.003	.006	.010	.012
Steel, SAE-AISI, 1000-1025	Soluble oil, sulphurized oil	60-100	.0015	.003	.006	.010	.012
Steel, .30-.60% carb., SAE 1000-9000							
Steel, annealed 150-225 Brinn.	Soluble oil, sulphurized oil	50-70	.0015	.003	.006	.010	.012
Steel, heat treated 225-283 Brinn.	Sulphurized oil	30-60	.0025	.005	.010	.020	.025
Steel, tool hi. car. & hi. speed	Sulphurized oil	25-50	.0025	.005	.010	.020	.025
Titanium	Highly activated sulphurized oil	15-20	.0025	.005	.010	.020	.025
Zinc, alloy	Soluble oil, kerosene & lard oil	200-250	.0015	.003	.006	.010	.012

*Use specially constructed heavy duty drills.
**For drill under .040, feeds should be adjusted to produce chips and not powder with ability to dispose of same without packing.

A

Chicago-Latrobe

Figure 12-45. A—Drill speed and feed table. B—Drill speeds in rpm. These tables are starting points for drilling different materials. Feeds and speeds should be increased or decreased depending on the specific metal being drilled and the condition of the drill press.

Drill Speeds in R.P.M.								
Diameter of drill	Soft metals 300 fpm	Plastics and hard rubber 200 fpm	Annealed cast from 140 fpm.	Mild steel 100 fpm	Malleable iron 90 fpm	Hard cast iron 80 fpm	Tool or hard steel 60 fpm	Alloy steel cast steel 40 fpm
1/16 (No. 53 to 80)	18320	12217	8554	6111	5500	4889	3667	2445
3/32 (No. 42 to 52)	12212	8142	5702	4071	3666	3258	2442	1649
1/8 (No. 31 to 41)	9160	6112	4278	3056	2750	2445	1833	1222
5/32 (No. 23 to 30)	7328	4888	3420	2444	2198	1954	1465	977
3/16 (No. 13 to 22)	6106	4075	2852	2037	1833	1630	1222	815
7/32 (No. 1 to 12)	5234	3490	2444	1745	1575	1396	1047	698
1/4 (A to E)	4575	3055	2139	1527	1375	1222	917	611
9/32 (G to K)	4071	2712	1900	1356	1222	1084	814	542
5/16 (L, M, N)	3660	2445	1711	1222	1100	978	733	489
11/32 (O to R)	3330	2220	1554	1110	1000	888	666	444
3/8 (S, T, U)	3050	2037	1426	1018	917	815	611	407
13/32 (V to Z)	2818	1878	1316	939	846	752	563	376
7/16	2614	1746	1222	873	786	698	524	349
15/32	2442	1628	1140	814	732	652	488	326
1/2	2287	1528	1070	764	688	611	458	306
9/16	2035	1357	950	678	611	543	407	271
5/8	1830	1222	856	611	550	489	367	244
11/16	1665	1110	777	555	500	444	333	222
3/4	1525	1018	713	509	458	407	306	204

B Figures are for high-speed drills. The speed of carbon drills should be reduced one-half. Use drill speed nearest to figure given.

Goodheart-Willcox Publisher

Figure 12-45. (continued).

Feed cannot be controlled accurately on a hand-fed drill press. A machinist must be aware of the cutting characteristics (such as uniform chips) that indicate whether the drill is being fed at the correct rate.

A feed that is too light will cause the drill to scrape, "chatter," and dull rapidly. Chipped cutting edges, drill breakage, and drill heating (despite the application of coolant) usually indicate that the rate is too great.

12.6.2 Speed Conversion

A problem arises in setting a drill press to the correct speed because its speed is given in revolutions per minute (rpm), while recommended drill cutting speed (CS) is given in feet or meters per minute (fpm or mpm).

The following simple formula determines the rpm to operate any diameter drill (D) at any specified speed:

$$\text{rpm} = \frac{4 \times \text{CS}}{\text{D}}$$

Drill speed problem. At what speed (rpm) must a 1/2″ diameter high-speed steel drill rotate when drilling aluminum?

To solve this problem, follow the steps below:

1. Refer to the speed and feed table, **Figure 12-45**. It gives the recommended cutting speed for aluminum (250 fpm).
2. Convert drill diameter (1/2″) to decimal fraction (0.5).
3. Substitute values into the formula

$$\text{rpm} = \frac{4 \times \text{CS}}{\text{D}}$$

$$= \frac{4 \times 250}{0.5}$$

$$= 2000 \text{ rmp}$$

Metric problems are solved in a similar manner using the following formula:

$$rpm = \frac{CS \times 1000}{D \times \pi}$$

Where:

CS = Cutting speed (mpm)
D = Drill diameter (mm)
π = 3 (rounded)

12.6.3 Drill Press Speed Control Mechanisms

With some drill presses, it is possible to set a dial to the desired rpm. However, on most conventional drilling machines, it is not possible to set the machine at the exact speed desired. The machinist must settle for the available speed nearest the desired speed.

The number of speed settings is limited by the number of pulleys in the drive mechanism, **Figure 12-46**. A decal or an engraved metal chart showing spindle speeds at various settings is attached to many machines. Information on spindle speeds can be found in the operator's manual, or it can be calculated if motor speed and pulley diameters are known.

Photo courtesy of Grizzly Industrial, Inc. www.grizzly.com

Figure 12-46. With step-pulley speed control, the belt is transferred to different pulley ratios to change drill speed.

12.7 Cutting Fluids

Drilling at the recommended cutting speeds and feeds generates considerable heat at the cutting point. This heat must be dissipated (carried away) as fast as it is generated, or it will destroy the drill's temper and cause it to dull rapidly.

Cutting fluids are applied to absorb the heat. They cool the cutting tool, serve as a lubricant to reduce friction at the cutting edges, and minimize the tendency for the chips to weld to the lips. Cutting fluids also improve hole finish and aid in the rapid removal of chips from the hole, **Figure 12-47**.

There are many kinds of cutting fluids. Many cutting fluids must be applied liberally. However, some newer fluids should be applied sparingly. Carefully read the instructions provided by the manufacturer.

Avoid using cutting fluids when drilling cast iron or other brittle materials. The fluids tend to cause the chips to pack and glaze the opening. Compressed air, used with care, will work when drilling these materials.

BIG Kaiser Precision Tooling Inc.

Figure 12-47. Cutting fluids minimize overheating and distortion of workpieces. Reduced heat also greatly extends the life of the tool. A coolant-hole drill allows the cutting fluid to flow through the cutting tool to provide faster metal removal by flushing chips out of the hole.

SAFETY NOTE

Always wash your hands thoroughly with soap and warm water after using cutting compounds and fluids. Before using the cutting compound, check the container to determine what should be done if you get any in your eye.

12.8 Sharpening Drills

A drill becomes dull with use and must be resharpened. Continued use of a dull drill may result in drill breakage or burning. Improper sharpening will cause the same problems.

Remove the entire point if it is badly worn or if the margins are burned, chipped, or worn off near the point. If the drill becomes overheated during grinding, do not plunge it into water. Allow it to cool in still air. The shock of sudden cooling may cause it to crack.

Three factors must be considered when repointing a drill: lip clearance, length and angle of the lips, and proper location of dead center.

Lip clearance. The two cutting edges, or lips, are comparable to chisels, **Figure 12-48**. To cut effectively, the heel (part of the point back of the cutting edge) must be relieved. Without this lip clearance, it is impossible for the lips to cut. If there is too much clearance, the cutting edges will be weakened. Too little clearance results in the drill point merely rubbing without penetrating the material.

Gradually increase lip clearance toward the center until the line across dead center stands at an angle of 120° to 135° with the cutting edge. See **Figure 12-49**.

Length and angle of lips. The material being drilled determines the proper point angle, **Figure 12-50**. The angles, in relation to the axis, must be the same (a 59° angle is satisfactory for most metals). If the angles are unequal, only one lip will cut and the hole will be oversized.

Proper location of dead center. The drill dead center must be accurate. Lips of different lengths will result in oversized holes, causing "wobble." This places tremendous pressures on the drill press spindle and bearings. See **Figure 12-51**.

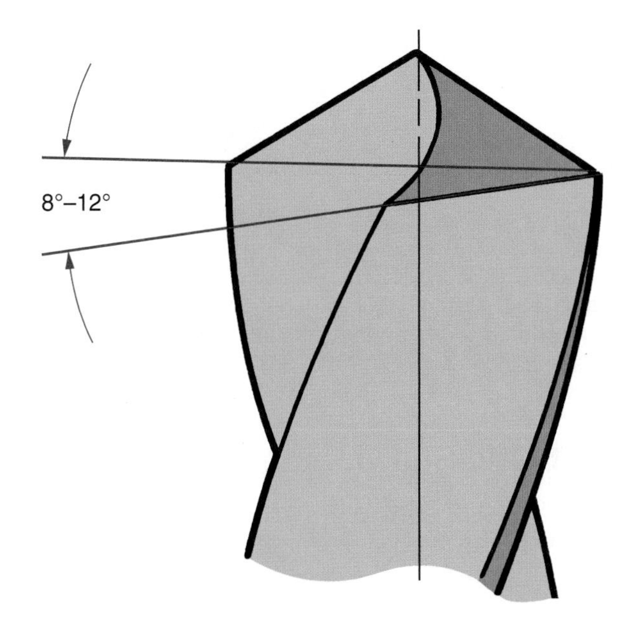

Goodheart-Willcox Publisher

Figure 12-48. Lip clearance of 8° to 12° is satisfactory for most drilling.

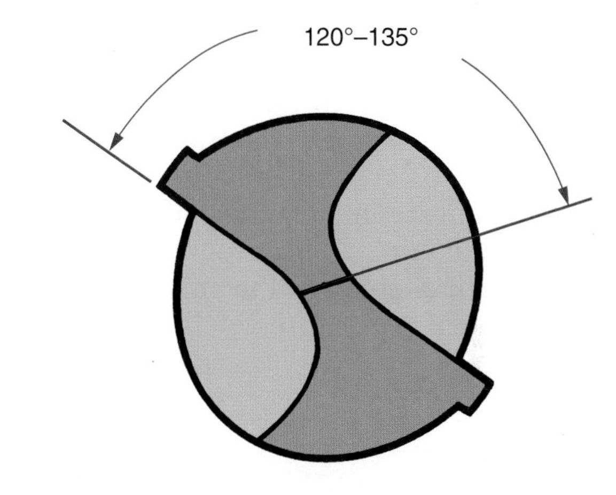

Goodheart-Willcox Publisher

Figure 12-49. Proper angle of drill dead center.

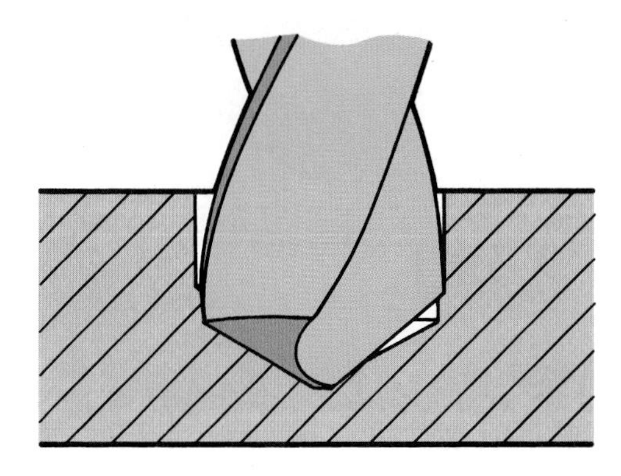

Goodheart-Willcox Publisher

Figure 12-50. Unequal drill point angles will produce a drilled hole that is oversize.

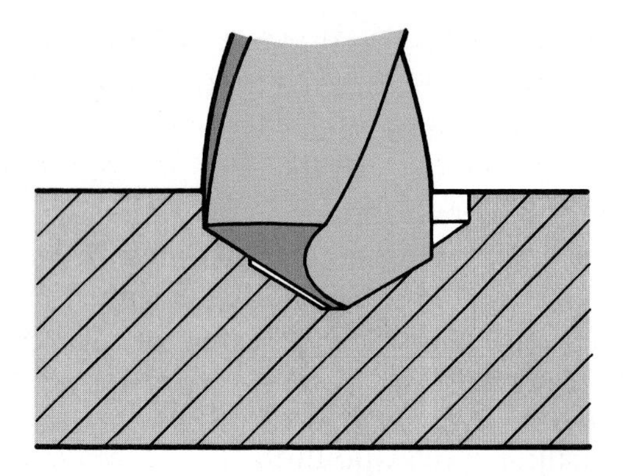

Goodheart-Willcox Publisher

Figure 12-51. Hole produced when drill is sharpened off center.

A combination of lip and dead center faults can result in a broken drill. If the drill is very large, permanent damage to the drilling machine can result. The hole produced will be oversized and often out-of-round. Refer to **Figure 12-52**.

The web of a drill increases in thickness toward the shank, **Figure 12-53**. When a drill has been shortened by repeated grindings, the web must be thinned to minimize the pressure required to make the drill penetrate the material. The thinning must be done equally on both sides of the web and care must be taken to ensure that the web is centered.

A ***drill point gage*** is used to check a drill point while sharpening. Its use is shown in **Figure 12-54**.

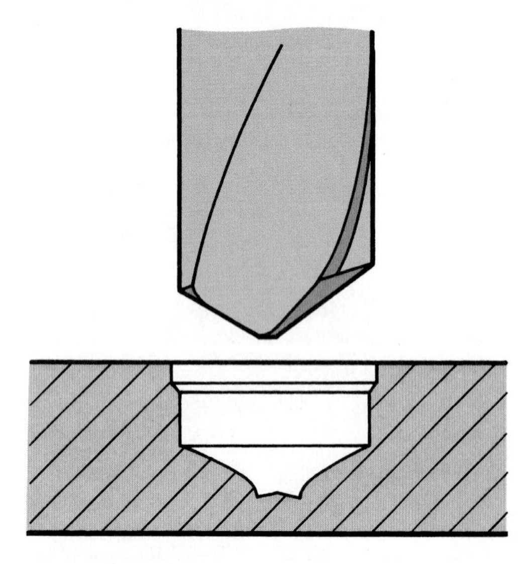

Goodheart-Willcox Publisher

Figure 12-52. This is the type of hole produced when the drill point has unequal point angles and is sharpened off center.

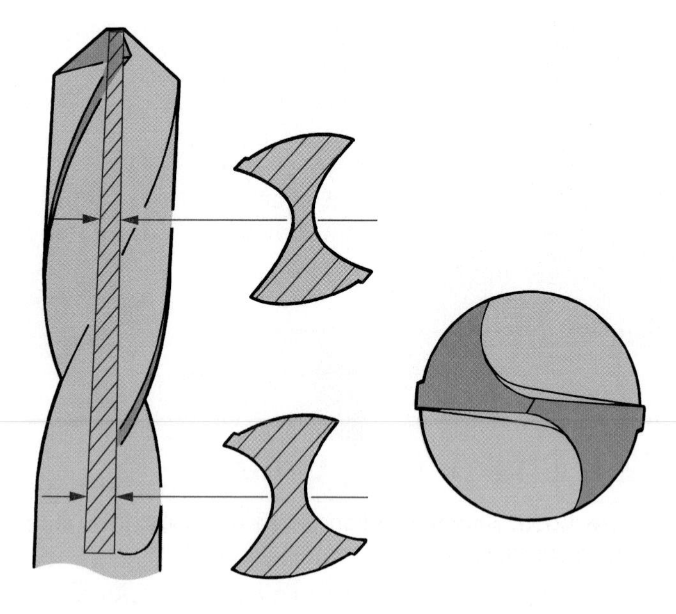

Goodheart-Willcox Publisher

Figure 12-53. The drill web tapers down toward the tip. The point is sometimes relieved to improve cutting action.

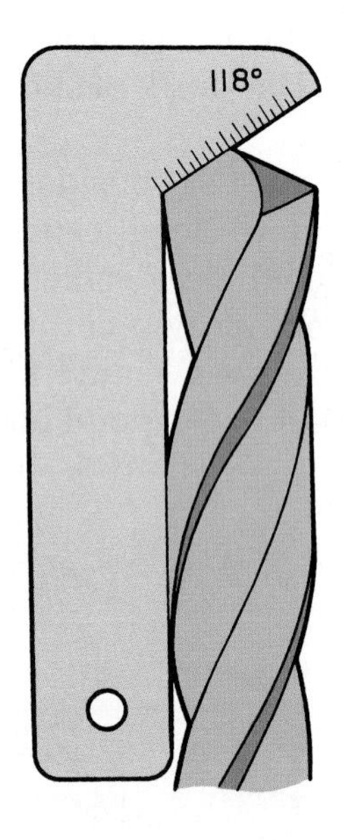

Goodheart-Willcox Publisher

Figure 12-54. Using a drill point gage will help ensure proper drill sharpening. For general drilling, an included angle of 118° is used.

12.8.1 Drill Sharpening Procedures

Use a coarse grinding wheel for roughing out the drill point if a large quantity of metal must be removed. Complete the operation on a fine wheel.

Many hand sharpening techniques have been developed. The following technique is suggested:

1. Grasp the drill shank with your right hand and the rest of the drill with your left hand. See **Figure 12-55**.

OSG Tap & Die, Inc.

Figure 12-55. One recommended way to hold a drill when it is being sharpened.

2. Place your left-hand fingers that are supporting the drill on the grinder tool rest. The tool rest should be slightly below center (about 1″ down on a 7″ diameter wheel, for example).
3. Stand so the centerline of the drill will be at a 59° angle to the centerline of the wheel, **Figure 12-56**. Lightly touch the drill lip to the wheel in a horizontal position.
4. Use your left hand as a pivot point and slowly lower the shank with your right hand. Increase pressure as the heel is reached to ensure proper clearance.
5. Repeat the operation on each lip until the drill is sharpened. Do not quench high-speed steel drills in water to cool them. Allow them to cool in air.
6. Check the drill tip frequently with a drill point gage to ensure a correctly sharpened drill.

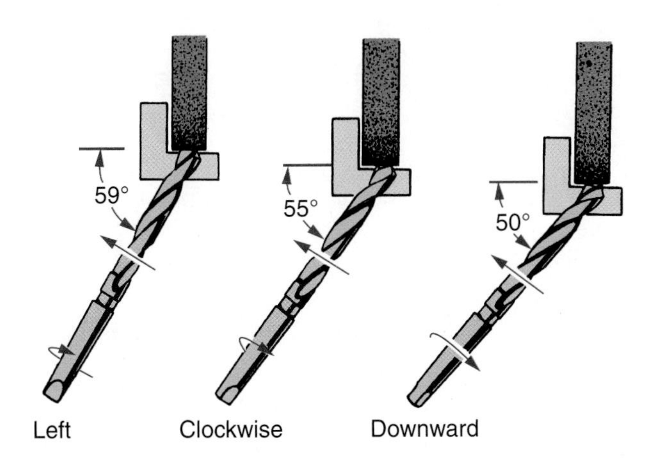

Goodheart-Willcox Publisher

Figure 12-56. One drill sharpening technique. Hold the point lightly against the rotating wheel and use three motions of the shank: to the left, clockwise rotation, and downward.

Sharpening a drill is not as difficult as it may first appear. However, before attempting to sharpen a drill, secure a properly sharpened drill and run through the motions explained above. When you have acquired sufficient skill, sharpen a dull drill.

To test, drill a hole in soft metal and observe the chip formation. When properly sharpened, chips will come out of the flutes in curled spirals of equal size and length. Tightness of the chip spiral is governed by the rake angle, **Figure 12-57**.

A standard drill point has a tendency to stick when used to drill brass. When brass is drilled, sharpen the drill as shown in **Figure 12-58**.

12.8.2 Drill Grinding Attachments

A drill sharpening device ensures that cutting edges of the drill will be uniform. An attachment for conventional grinders is shown in **Figure 12-59**. In the machine shop where a high degree of hole accuracy is required and a large amount of sharpening must be done, these devices are a must.

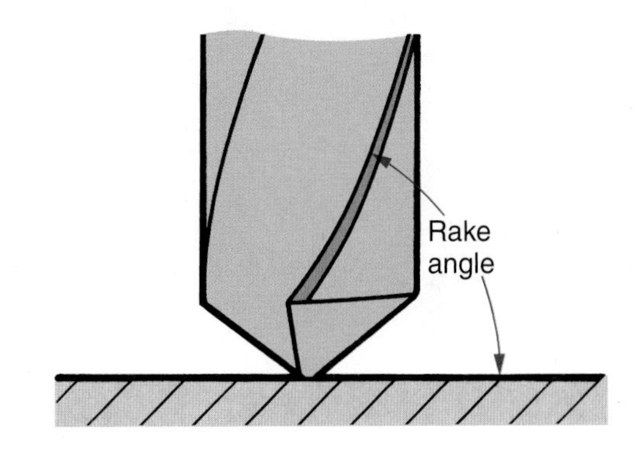

Goodheart-Willcox Publisher

Figure 12-57. Rake angle of the drill.

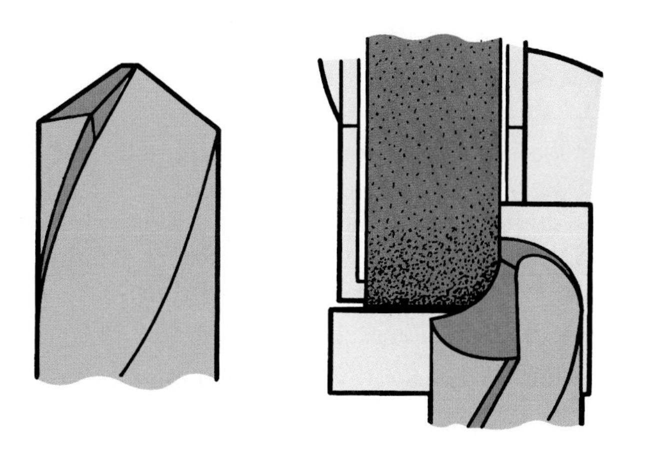
Goodheart-Willcox Publisher

Figure 12-58. Modified rake angle for drilling brass.

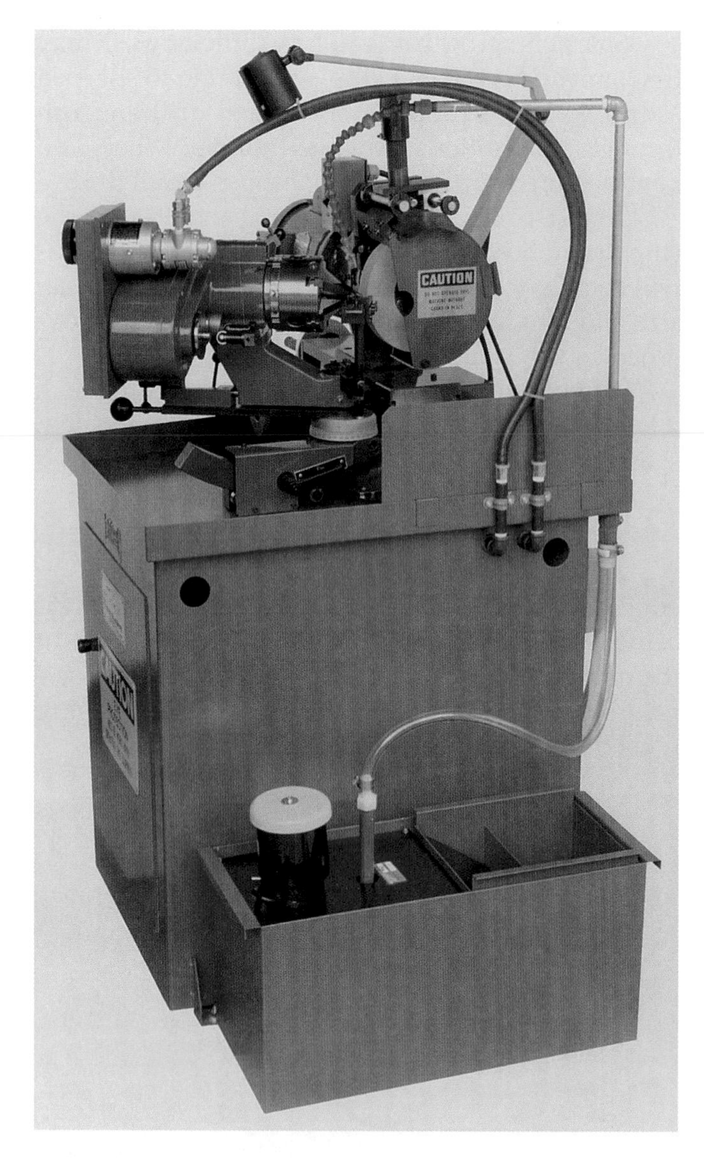

Rush Machinery, Inc.

Figure 12-59. A drill and tool grinder with built-in point splitting and web thinning capabilities. The user can choose from a wide variety of drill points. Machines of this type are available in semiautomatic and automatic versions.

12.9 Drilling

Obeying a few simple rules will help you drill accurately. Use the following procedure:

1. Carefully study the drawing to determine hole locations. Lay out the positions and mark the intersecting lines with a prick punch.
2. Secure a drill and check its size.
3. Mount work solidly on the drilling machine.

SAFETY NOTE

Never hold the work by hand. The workpiece could whip out of your hand and cause serious injuries.

4. Insert a ***center finder***, also known as a wiggler, into the drill chuck and position the point to be drilled directly under the chuck or spindle. Turn on the power and center the center finder point with your fingers. Position the work until the revolving center finder point does not wiggle when it is lightly dropped into the punched hole location and removed. If there is any point movement, additional alignment is necessary because the work is not positioned properly, **Figure 12-60**.
5. Remove the center finder and insert a center drill. Hand tighten the chuck, **Figure 12-61**. Check to be sure the drill runs true. If it does, tighten the chuck with a chuck key. Remember to remove the key before starting the machine.
6. After center drilling, replace the tool with the required drill. Hand tighten it in the chuck. Turn on the machine. If it does not run true, the drill may be bent or may have been placed in the chuck off center. Also check that it will drill to the required depth.

Goodheart-Willcox Publisher

Figure 12-60. It is difficult to align a workpiece with centerlines by eye. To assist in this job, a center finder, or wiggler, is used.

Goodheart-Willcox Publisher

Figure 12-61. After alignment with the center finder, center drilling will ensure that the drill will make the hole in the proper location.

7. Calculate the correct cutting speed and feed if you plan to use a power feed. Adjust the machine to operate as closely as possible to this speed.
8. Turn on the power and apply cutting fluid. Start the cut. Even pressure on the feed handle will keep the drill cutting freely.
9. Watch for the following signs that indicate a poorly cutting drill:
   - A dull drill will squeak and overheat. Chips will be rough and blue, and cause the machine to slow down. Small drills will break.
   - Infrequently, a chip will get under the dead center and act as a bearing, preventing the drill from cutting. Remove it by raising and lowering the drill several times.
   - Chips packed in the flutes will cause the drill to bind and slow the machine or cause the drill to break. Remove the drill from the hole and clean it with a brush that has been dipped in cutting fluid. Do not use cutting fluid when drilling cast iron.
10. Clear chips and apply cutting fluid as needed.
11. The most critical time of the drilling operation occurs when the drill starts to break through the work. Ease up on feed pressure at this point to prevent the drill from "digging in."
12. Remove the drill from the hole and turn off the power. Clear the chips with a brush. Unclamp the work and use a file to remove all burrs.

SAFETY NOTE

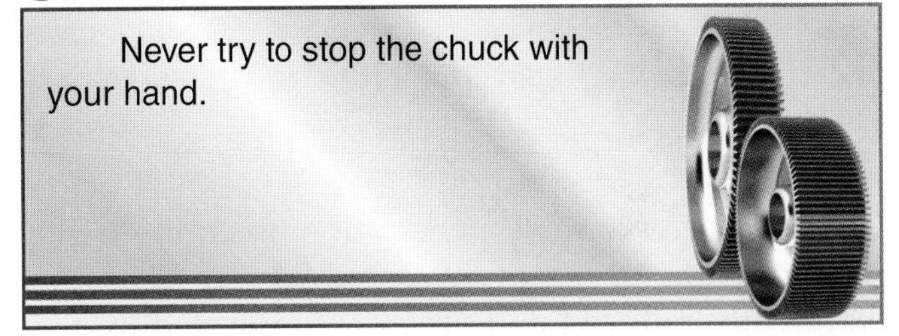

Never try to stop the chuck with your hand.

13. Clean chips and cutting fluid from the machine. Wipe it down with a soft cloth. Return equipment to storage after cleaning.

Observe extreme care in positioning the piece for drilling. A poorly planned setup may permit the drill to cut into the vise or drill table when it breaks through the work.

If a hole must be located precisely, certain additional precautions can be taken to ensure that the hole will be drilled where it is supposed to be drilled. After the center point has been determined, a series of proof

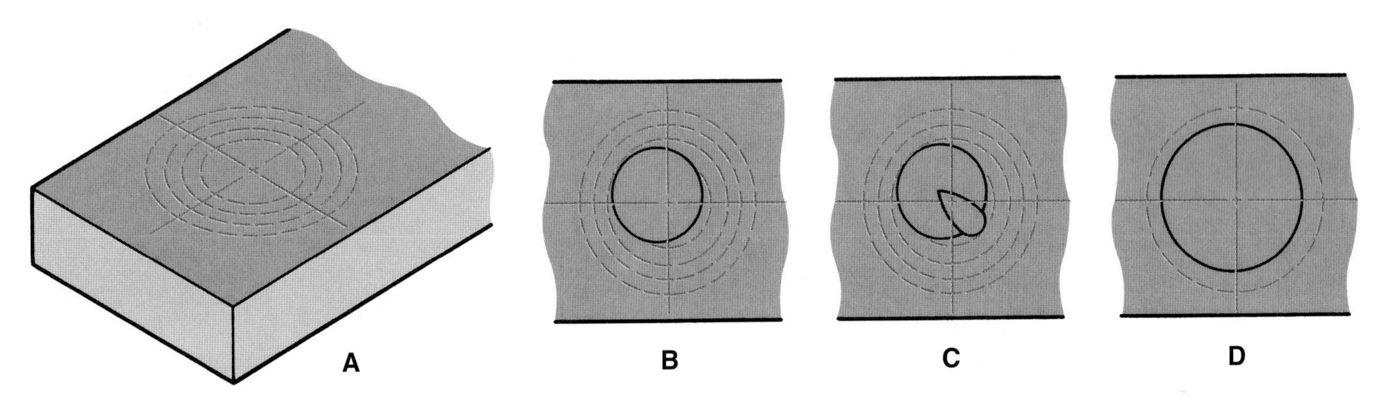

Goodheart-Willcox Publisher

Figure 12-62. How to bring a drill back on center. A—Proof circles. B—Drill has been started off center (exaggerated). C—Groove cut to bring drill back on center. D—Drill back on center. This operation will only work if the drill has not begun to cut to its full diameter.

circles are scribed, **Figure 12-62A**. They will serve as reference points to help check whether the drill remains on center as it starts to penetrate the material.

Even when work is properly centered, the drill may "drift" when starting a hole. Various factors can cause this, such as hard spots in the metal or an improperly sharpened drill. The drill cannot be brought back on center by moving the work, because it will still try to follow the original hole. This condition, **Figure 12-62B**, must be corrected before the full diameter of the drill is reached.

The drill is brought back on center by using a round-nose cape chisel to cut a groove on the side of the hole where the drill must be drawn, **Figure 12-62C**. This groove will "pull" the drill point to the center. Repeat the operation until the hole is centered in the proof circles, **Figure 12-62D**.

12.9.1 Drilling Larger Holes

On drills larger than 1/2″ (12.5 mm), a conventional dead center requires more than 50% of the force required for drill penetration. Even then, they have a tendency to run off center. Feed pressure can be greatly reduced and accuracy improved by first drilling a ***pilot hole*** (lead hole). See **Figure 12-63**.

The small pilot hole permits pressure to be exerted directly on the cutting edges of the large drill, causing it to drill faster. The pilot hole should have a diameter as large as, or slightly larger, than the width of the dead center.

There are additional methods to improve drilling efficiency without having to first drill a pilot hole. A pilot hole can often be eliminated by changing the drill point geometry as shown in **Figure 12-64**.

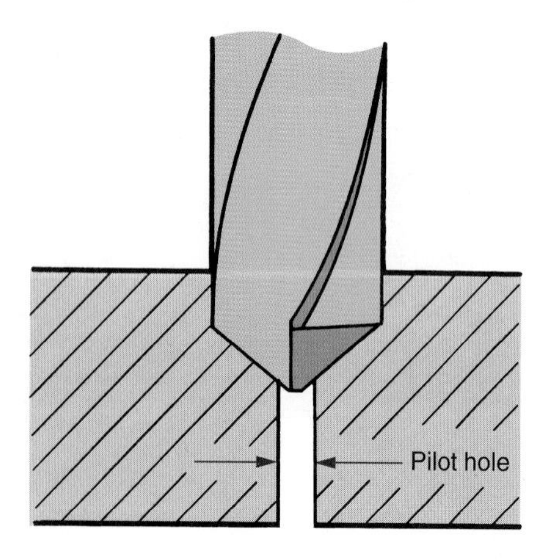

Goodheart-Willcox Publisher

Figure 12-63. A pilot hole makes drilling a large hole much easier.

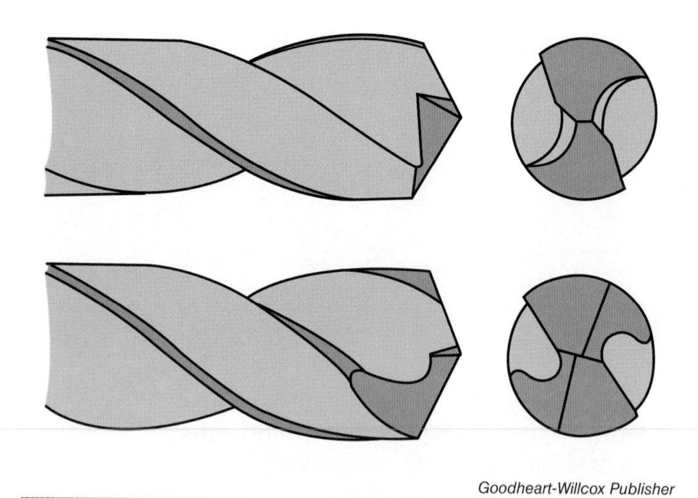

Goodheart-Willcox Publisher

Figure 12-64. Pilot holes can often be eliminated by changing drill point geometry. Two designs are shown.

However, they require specialized grinding skills or dedicated drill sharpening equipment.

The Winslow-helical drill point is another design that eliminates the pilot hole and improves drilling efficiency. See **Figure 12-65**. It has a raised S-shaped dead center that allows it to immediately cut metal into chips instead of "pushing" or extruding metal as a conventional straight dead center does.

Because of the limited room for chips in drill flutes, it is always desirable to have them broken into small pieces. Coiling chips must be avoided, especially in deep hole drilling. Coiling tends to clog the flutes and prevent cutting fluid from reaching the cutting edges.

Ductile materials, like many aluminum alloys, require chip breakers to produce proper chip breakup. **Figure 12-66** illustrates several chip breaker designs that have proven successful.

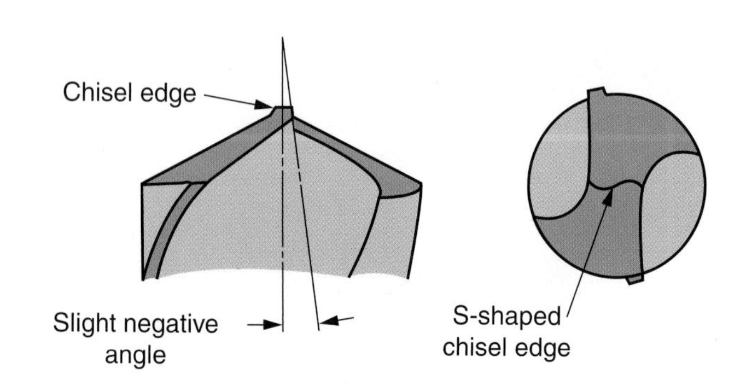

Goodheart-Willcox Publisher

Figure 12-65. The Winslow-helical drill point. It requires a specially designed drill point grinder.

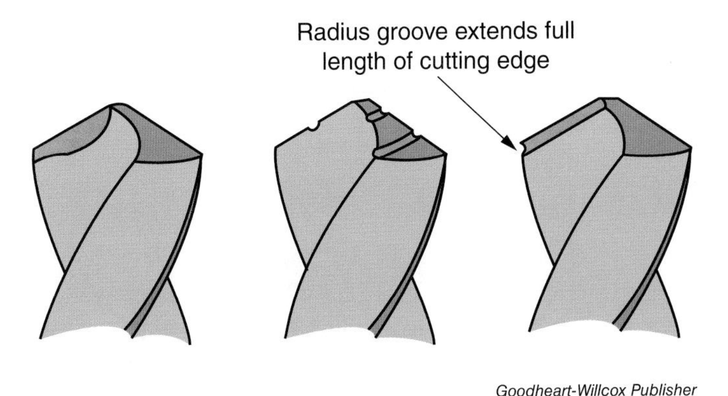

Figure 12-66. Chip breaker designs that have proven successful in preventing metal coiling.

12.9.2 Drilling Round Stock

Holes are more difficult to drill in the curved surface of round stock. Many difficulties can be eliminated by holding the round material in a V-block, **Figure 12-67**. A V-block can be held in a vise or clamped directly to the table.

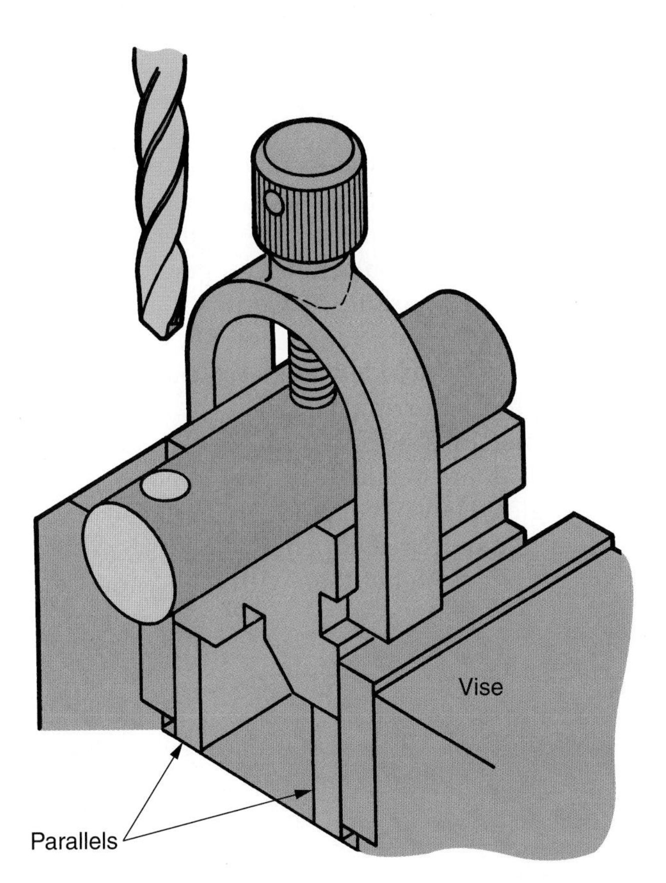

Figure 12-67. The V-block eliminates many difficulties when drilling round stock. Be sure the drill will clear the V-block when it comes through material.

Use the following procedure to center round stock in a V-block:

1. Locate the hole position on the stock. Prick punch the intersection of the layout lines. Place the stock in a V-block. If the hole is to go through the piece, make certain that the drill will clear the V-block. Also be sure there is ample clearance between the clamp and drill chuck.
2. To align the hole for drilling through exact center, place the work and V-block on the drill press table or on a surface plate. Rotate the punch mark until it is upright. Place a steel square on the flat surface with the blade against the round stock as shown in **Figure 12-68**. Measure from the square blade to the punch mark, and rotate the stock until the measurement is the same when taken from both sides of the stock.
3. From this point, the drilling sequence is identical to that previously described.

If a large number of identical parts must be drilled, it may be desirable to make a drill jig, **Figure 12-69**. The drill jig automatically positions and centers each piece for drilling.

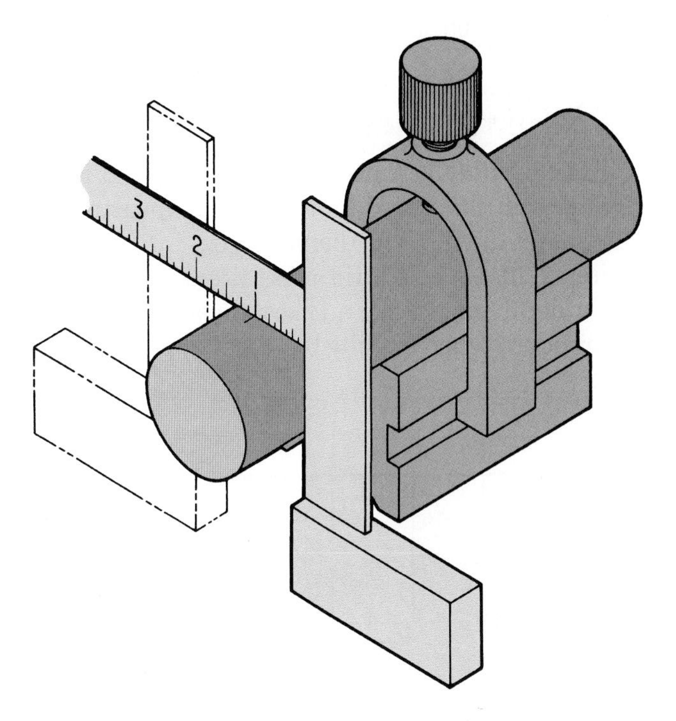

Figure 12-68. Using a square to center round stock in a V-block.

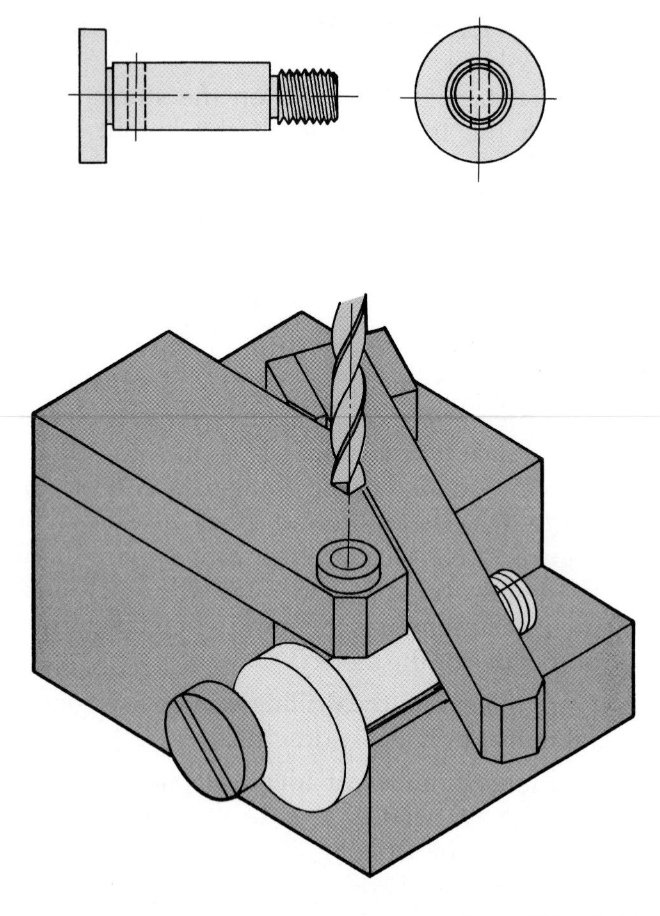

Goodheart-Willcox Publisher

Figure 12-69. This typical drill jig has an arm that lifts to allow easy insertion and removal of the part being drilled.

12.9.3 Blind Holes

A ***blind hole*** is a hole that is not drilled all the way through the work. Hole depth is measured by the distance the full hole diameter goes into the work, **Figure 12-70**. Using a drill press fitted with a depth stop or depth gage is the quickest means of achieving proper depth when drilling blind holes, **Figure 12-71**.

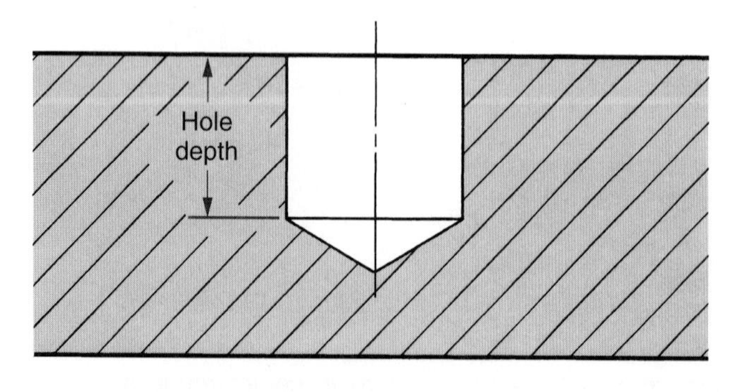

Goodheart-Willcox Publisher

Figure 12-70. Measuring the depth of a blind hole.

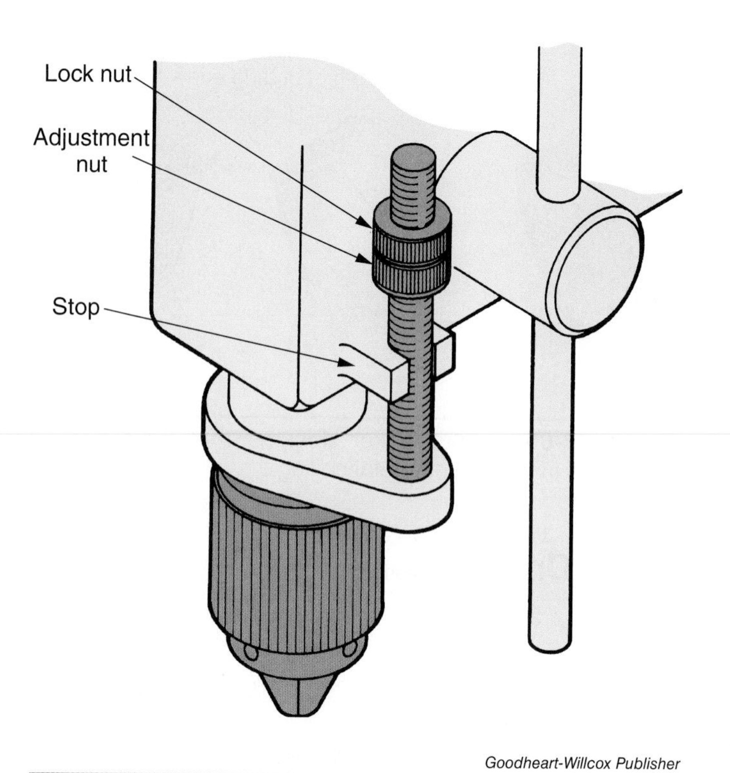

Goodheart-Willcox Publisher

Figure 12-71. The depth gage attachment provides easy adjustment of how far the drill moves into the work.

12.10 Countersinking

Countersinking is the operation that cuts a chamfer in a hole to permit a flat-headed fastener to be inserted with the head flush to the surface, **Figure 12-72**.

The tool used to machine countersinks is called a countersink, **Figure 12-73**. Countersinks are available with cutting edge angles of 60°, 82°, 90°, 100°, 110°, and 120° included angles. Countersinks are also used for deburring holes.

Countersinks with indexable carbide inserts, **Figure 12-74**, are available in a number of sizes and point angles. They have two cutting edges per insert and do not require resharpening. Cutting speeds are five to ten times higher than with HSS countersinks.

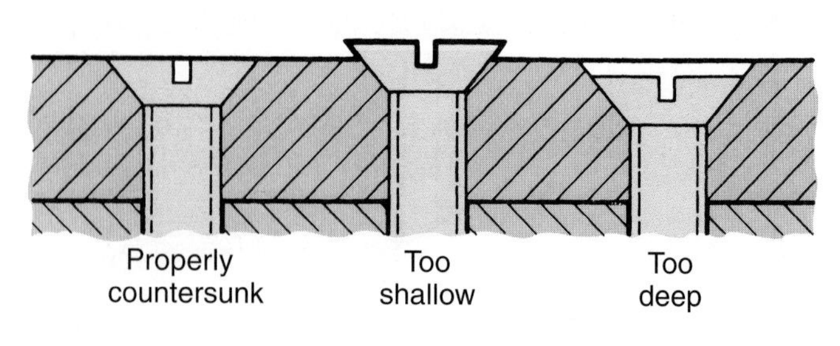

Goodheart-Willcox Publisher

Figure 12-72. Correctly and incorrectly countersunk holes. The countersink angle must match the fastener head angle.

Photo courtesy of Grizzly Industrial, Inc. www.grizzly.com

Figure 12-73. These six-fluted countersinks come in various sizes.

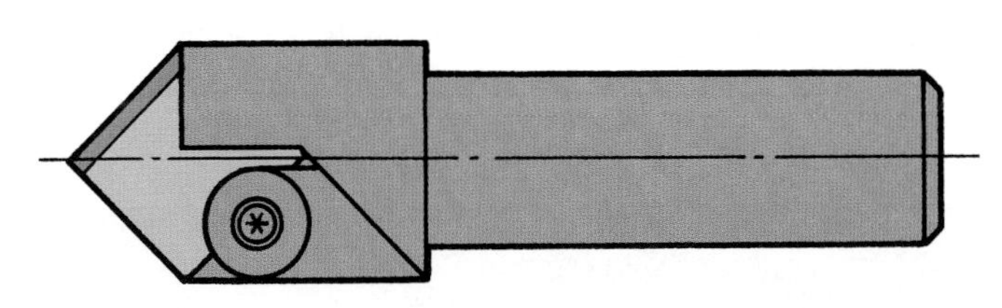

Goodheart-Willcox Publisher

Figure 12-74. Countersinks with indexing carbide inserts have a life five to ten times longer than similar HSS countersinks.

A single cutting edge countersink, **Figure 12-75**, is free cutting and produces minimum chatter. Chips produced by the cutting edge pass through the hole and are ejected.

Refer to the following procedures for using a countersink:

1. The cutting speed should be about one-half that recommended for a similar size drill. This will minimize the probability of chatter.
2. Feed the tool into the work until the chamfer is large enough for the fastener head to be flush.
3. Use the depth stop on the drill press if a number of similar holes must be countersunk.

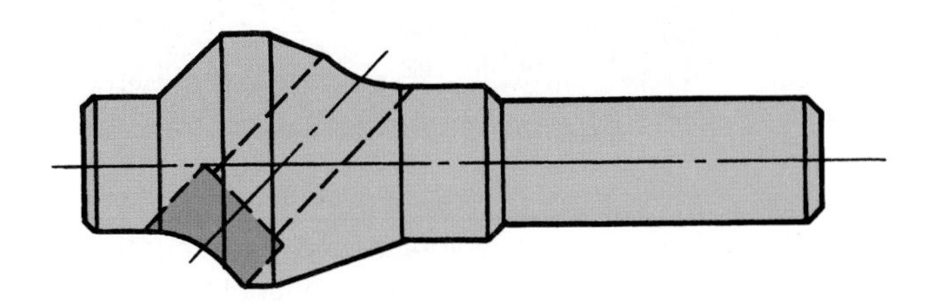

Goodheart-Willcox Publisher

Figure 12-75. Countersink with a single cutting edge and pilot.

12.11 Counterboring

The heads of fillister-head and socket-head screws are usually set below the work surface. A counterbore is used to enlarge the drilled hole to the proper depth and machine a square shoulder on the bottom to secure maximum clamping action from the fastener, **Figure 12-76**.

The counterbore tool has a guide, called a pilot, which keeps it positioned correctly in the hole. Solid counterbores are available. However, counterbores with interchangeable pilots and cutters are commonly used, **Figure 12-77**.

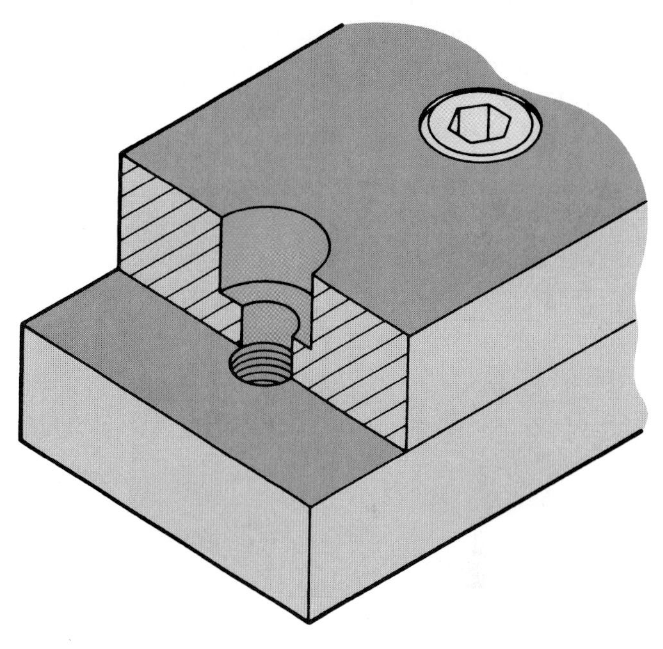

Goodheart-Willcox Publisher

Figure 12-76. A sectional view of a hole that has been drilled and counterbored to receive a socket-head screw.

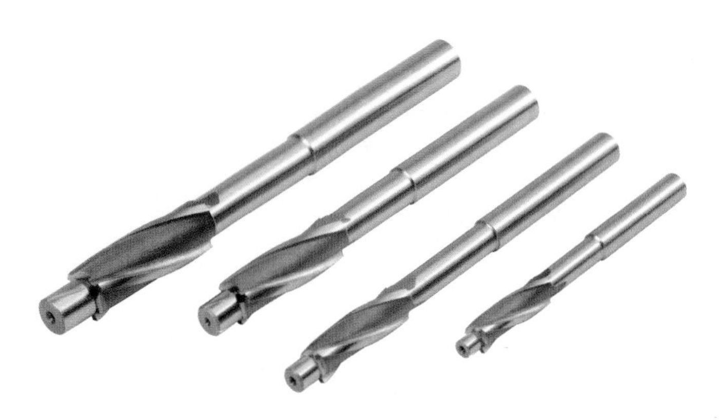

Photo courtesy of Grizzly Industrial, Inc. www.grizzly.com

Figure 12-77. Different sizes of straight shank counterbores.

They can be changed easily from one size cutter or pilot to another size. A drop of oil on the pilot will prevent it from binding in the drilled hole.

Counterbores with indexable carbide inserts, **Figure 12-78**, are also available. When the cutting edges become dull, new edges can be indexed into place without affecting opening diameter. Costly sharpening is eliminated.

12.12 Spotfacing

Spotfacing is the operation during which a circular spot is machined on a rough surface (such as a casting or forging) to provide a bearing surface for a bolt, washer, or nut. A counterbore may be used for spotfacing, although a special tool manufactured for inverted spotfacing is available, **Figure 12-79**.

Special backspotface and backcounterbore tools are required to perform operations in areas where conventional tools cannot be used. See **Figure 12-80**.

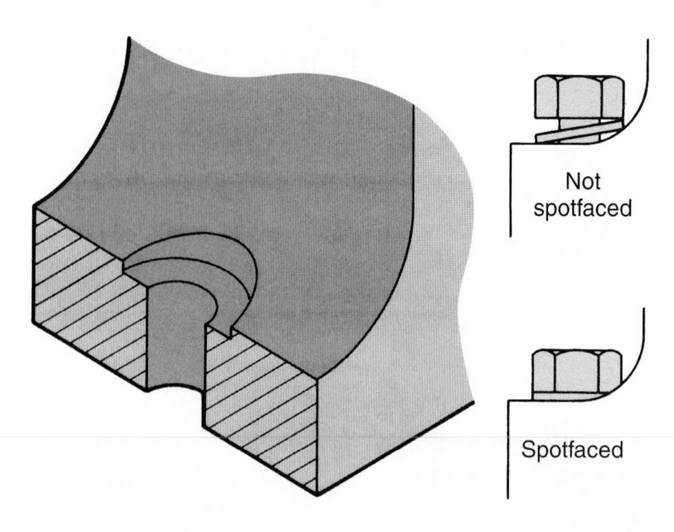

Goodheart-Willcox Publisher

Figure 12-79. Sectional view of a casting with a mounting hole that has been spotfaced. Profile drawings show the casting before and after spotfacing. The bolt head cannot be drawn down tightly until the mounting hole is spotfaced.

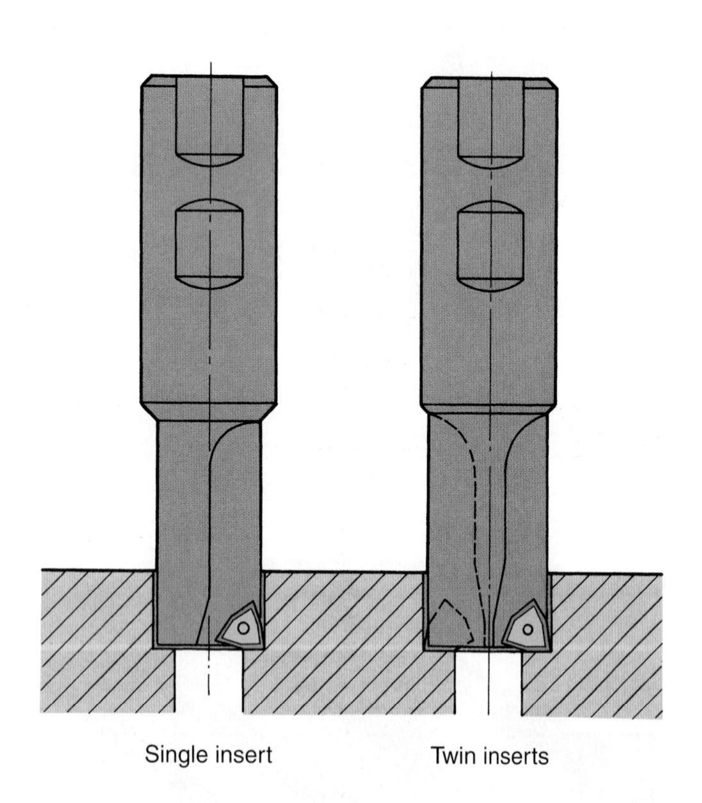

Goodheart-Willcox Publisher

Figure 12-78. Counterbores with carbide indexable inserts. The inserts are rotated when a cutting edge becomes dull.

Parlec, Inc.

Figure 12-80. Special backspotface and backcounterbore tools are used in situations where conventional tools cannot be inserted. Blade setting can be manual or automatic. Standard tools are available for bores and spotfaces of 0.250″ and larger.

The cutting point is lifted up into the workpiece, rather than being pushed down into it. See **Figure 12-81**.

Large diameter openings can be counterbored, spotfaced, or drilled with the APT Multi-Tool™, **Figure 12-82**. A pilot hole is drilled with a conventional twist drill. The required size pilot and blade is inserted and the opening is made. For very large openings, it may be necessary to use a smaller-size blade before machining the specified size. Multi-Tool blades are available in sizes from 1 1/8″ diameter to 4″ diameter.

Counterboring

Spotfacing

Goodheart-Willcox Publisher

Figure 12-81. These section drawings show how backcounterboring and backspotfacing tools work.

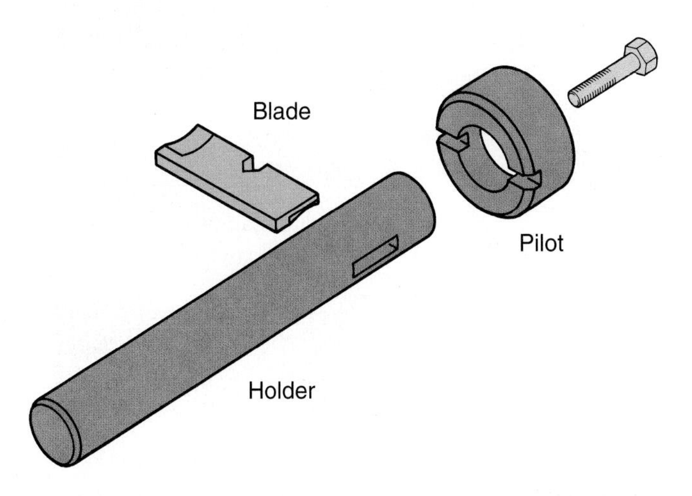

Goodheart-Willcox Publisher

Figure 12-82. Interchangeable drilling, spotfacing, and counterboring tool. Blades for producing holes up to 4″ in diameter are available.

12.13 Tapping

Tapping may be done by hand on a drill press using the following steps:

1. Drill the correct size hole for the tap, **Figure 12-83**.
2. With the work clamped in the machine, insert a small 60° center in the chuck. The center holds the tap vertically.
3. Place the center point in the tap's center hole.
4. Feed the tap into the work by holding down on the feed handle and turning the tap with a tap wrench.

SAFETY NOTE

Never insert a tap into the drill chuck and attempt to use the drill press power to run the tap into the work. The tap will shatter when power is applied. Turn the tap by hand.

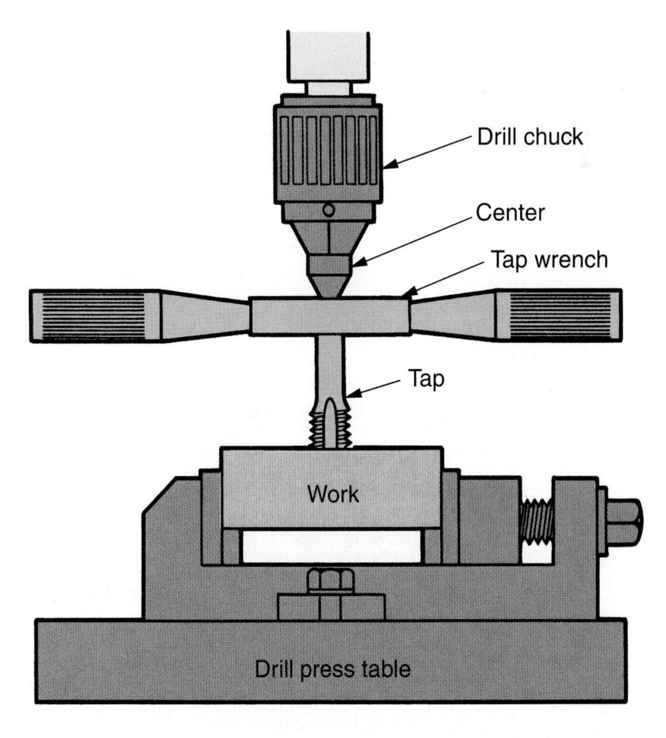

Goodheart-Willcox Publisher

Figure 12-83. Setup for hand tapping on a drill press. Work must be mounted solidly.

Tapping can only be done with power through the use of a tapping attachment, **Figure 12-84**. This device fits the standard drill press. It has reducing gears that slows the tap to about one-third of the drill press speed. A table provided with the attachment gives recommended spindle speeds for tapping.

A clutch arrangement drives the tap until it reaches the predetermined depth, at which time the tap stops rotating. Raising the feed handle causes the tap to reverse direction and back out of the hole.

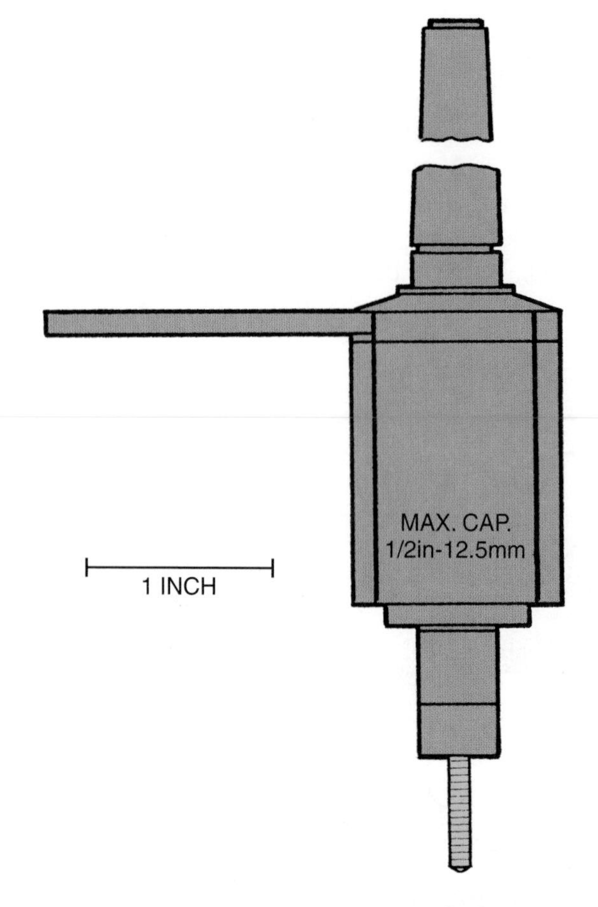

Goodheart-Willcox Publisher

Figure 12-84. Tapping attachment on a drill press.

12.14 Reaming

Reaming produces holes that are extremely accurate in diameter and have an exceptionally fine surface finish. Machine reamers are made in a variety of sizes and styles. They are usually manufactured from high-speed steel. Some are fitted with carbide cutting edges.

12.14.1 Types of Machine Reamers

Descriptions of a few common types of machine reamers follow. Refer to **Figure 12-85**.

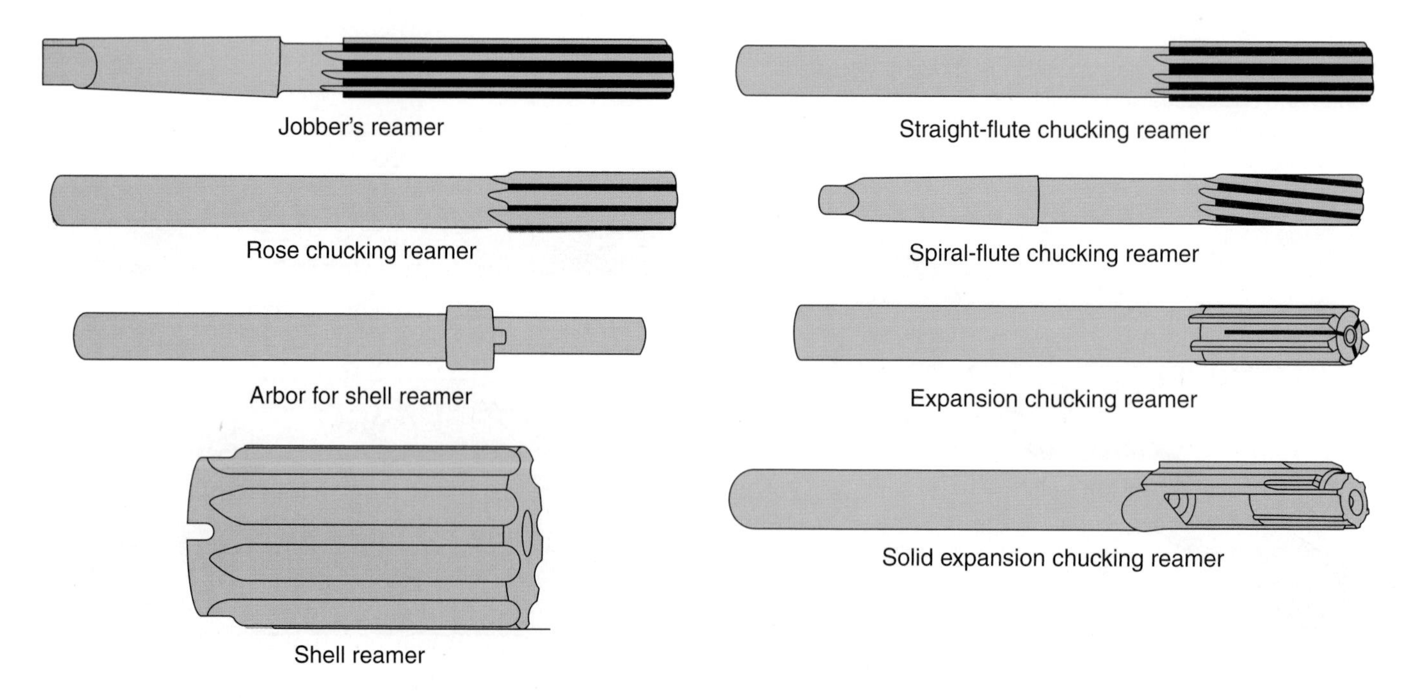

Goodheart-Willcox Publisher

Figure 12-85. Each type of reamer is suited for a different application. The solid expansion reamer has tungsten carbide cutting edges for extended cutting life.

- A jobber's reamer, also called a machine reamer, is identical to a hand reamer except that a taper shank is available and the tool is designed for machine operation.
- A chucking reamer is manufactured with both straight and taper shanks. It is similar to a jobber's reamer but its flutes are shorter and deeper. It is available with straight or spiral flutes.
- A rose chucking reamer is designed to cut on its end. The flutes provide chip clearance and are ground to act only as guides. This type of reamer is best used when considerable metal must be removed and the finish is not critical.
- A shell reamer is mounted on a special arbor that can be used with several reamer sizes. The arbor can have straight or spiral flutes and is also made in the rose style. The arbor shank may be straight or tapered. A hole in the reamer is tapered to fit the arbor, which is fitted with drive lugs.
- An expansion chucking reamer is available with straight flutes and either a straight or taper shank. Slots are cut into the body to permit the reamer to expand when an adjusting screw in the end is tightened.

A regular expansion reamer has several drawbacks. The slots, which are necessary for the reamer to expand, reduce tool rigidity. This diminishes accuracy and surface finish. Also, cutting-edge clearance is reduced as the reamer expands, creating a "drag." This often causes the tool to chatter with a resulting decrease in finish quality.

A solid expansion reamer provides rigidity and accuracy not possible with conventional expansion reamers. To expand this type, a tapered plug is forced into the reamer end. The tool body expands well beyond the tip, and ensures uniform parallel expansion across the full length of the carbide cutting lips. Clearance is automatically provided. The plug can be removed for shimming to a larger size. Once expanded, the reamer diameter cannot be reduced without grinding.

12.14.2 Using Machine Reamers

Reamers are expensive precision tools. The quality of the finish and accuracy of the reamed hole will depend on how the tool is used. Obey the following reaming rules:

- Carefully check the reamer diameter before use. If the hole diameter is critical, drill and ream a hole in a piece of similar material to check tool accuracy.
- Use a sharp reamer.
- Mount the reamer solidly.
- Cutting speed for a high-speed steel reamer should be about two-thirds that of a similar size drill.
- Feed should be as high as possible while still providing a good finish and accurate hole size.
- Allow enough material in the drilled hole to permit the reamer to cut rather than burnish (smooth and polish). The following allowances are recommended:
  - Up to 1/4″ (6.3 mm) diameter, allow 0.010″ (0.25 mm).
  - 1/4″ to 1/2″ (6.3 mm to 12.5 mm) diameter allow 0.015″ (0.4 mm).
  - 1/2″ to 1.0″ (12.5 mm to 25.0 mm) diameter allow 0.020″ (0.5 mm).
  - 1.0″ to 1.5″ (25.0 mm to 38.0 mm) diameter allow 0.025″ (0.6 mm).
- Use an ample supply of cutting fluid.
- Remove the reamer from the hole before stopping the machine.
- When not being used, reamers should be stored in separate containers or storage compartments. This will minimize chipping and dulling of the cutting edges.

12.15 Microdrilling

Specially designed machines, toolholding devices, and drills are required for drilling holes smaller than 0.020″ (0.508 mm) to within close tolerances, **Figure 12-86.** Microdrilling operations require very accurate spindles and collets to reduce drill flexing and breakage. Many microdrilling machines are controlled by computer numerical control (CNC) systems.

Royal Products, Division of Curran Manufacturing Corporation

Figure 12-86. This sensitive drill feed eliminates the need for a specialized microdrilling machine. It makes drilling small holes on a normal drill press easier, reduces tool breakage, and decreases the tendency for microdrills to "walk" off center.

Microdrilling uses a "pecking" technique to cut these small diameter holes. In this technique, the drill is repeatedly inserted and removed from the hole. The drills have flutes to pull chips out of the hole, but because they are so small, pecking is necessary for chip removal. The depth of each peck is determined by the drill diameter and the material being drilled.

Small-size holes with other geometric shapes (such as square, rectangular, or hexagonal) are made by electrical discharge machining (EDM). EDM operations are discussed in more detail in a later chapter.

Chapter Review

Summary

- Selection of the proper drilling machine, drill, and drill-holding device for specific drilling operations is important.
- Twist drills are the most common form of drill, but there are also a large variety of special purpose drills available.
- Drills are held in the drill press by a chuck or tapered spindle.
- Proper work-holding devices, cutting speeds, feed rates, and cutting fluids are important for safe, efficient and accurate drilling operations.
- A properly sharpened drill is critical to efficient and safe drilling operations.
- A center finder can be used to help position the chuck or spindle directly below the point to be drilled.
- Drilling machines are used for a variety of operations other than drilling including countersinking, counterboring, spotfacing, tapping, and reaming.
- Microdrills permit the drilling of very small holes, but require special tooling and the "pecking" technique.

Review Questions

Answer the following questions using the information provided in this chapter.

1. A drill works by _____.
   A. being forced into material
   B. rotating against material and being pulled through by the spiral flutes
   C. rotating against material with sufficient pressure to cause penetration
   D. All of the above.
   E. None of the above.
2. How is drill press size determined?
3. Drills are made from _____.
   A. high-speed steel (HSS)
   B. carbon steel
   C. titanium
   D. Both A and B.
   E. None of the above.
4. List the two types of drill shanks.
5. _____ shank drills are used with a chuck.
6. _____ shank drills fit directly into the drill press spindle.
7. The spiral grooves that run the length of the drill body are called _____.
8. The spiral grooves in a drill body are used to _____.
   A. help form the cutting edge of the drill point
   B. curl chips for easier removal
   C. form channels through which the chips can escape from the hole
   D. All of the above.
   E. None of the above.
9. Drill sizes are expressed by what four series?
10. What are two techniques used to determine a drill's size?

11. Name the device used to enlarge a taper shank drill so it will fit the spindle opening.
12. The device used to permit a drill with a taper shank too large to fit the spindle opening is called a(n) _____.
13. What is the name of the tool used to separate a taper shank drill from a drill-holding device?
14. Cutting fluids are used to _____.
    A. cool the drill
    B. improve the finish of a drilled hole
    C. aid in the removal of chips
    D. All of the above.
    E. None of the above.
15. What coolant should be used when drilling cast iron?
16. List the three factors that must be considered when repointing a drill.
17. What occurs when the cutting lips of a drill are not sharpened to the same lengths?
18. The _____ should be used frequently when sharpening to ensure a correctly sharpened drill.
19. The included angle of a drill point sharpened for general drilling is _____ degrees.
20. Large drills require a considerable amount of power and pressure to get started. They also have a tendency to drift off center. These conditions can be minimized by first drilling a(n) _____ hole.
21. When drilling a pilot hole, the hole should be as large as, or slightly larger than, the width of the _____ of the drill point.
22. What is a blind hole?
23. How is the depth of a drilled hole measured?
24. What is the name of the operation used to cut a chamfer in a hole to receive a flat-head screw?
25. The operation used to prepare a hole for a fillister or socket head screw is called _____.
26. _____ is the operation that machines a circular spot on a rough surface for the head of a bolt or nut.
27. The _____ is almost identical to the hand reamer except that the shank has been designed for machine use.
28. A(n) _____ expansion reamer provides rigidity and accuracy not possible with conventional expansion reamers.
28. The cutting speed for a high-speed reamer is approximately _____ that for a similar-sized drill.
30. How should a reamer be removed from a finished hole?

CHAPTER 13

Offhand Grinding

Learning Objectives

After studying this chapter, you will be able to:

- Identify the various types of offhand grinders.
- Dress and true a grinding wheel.
- Prepare a grinder for safe operation.
- Use an offhand grinder safely.
- List safety rules for offhand grinding.

Technical Terms

glazing
grinding
loading
offhand grinding
temper
tool rest
wheel dresser

Grinding is an operation that removes material by rotating an abrasive wheel or belt against the work, **Figure 13-1**. Grinding is used for the following tasks:

- Sharpening tools.
- Removing material too hard to be machined by other techniques.
- Cleaning the parting lines from castings and forgings.
- Finishing and polishing molds used in die casting of metals and injection molding of plastics.

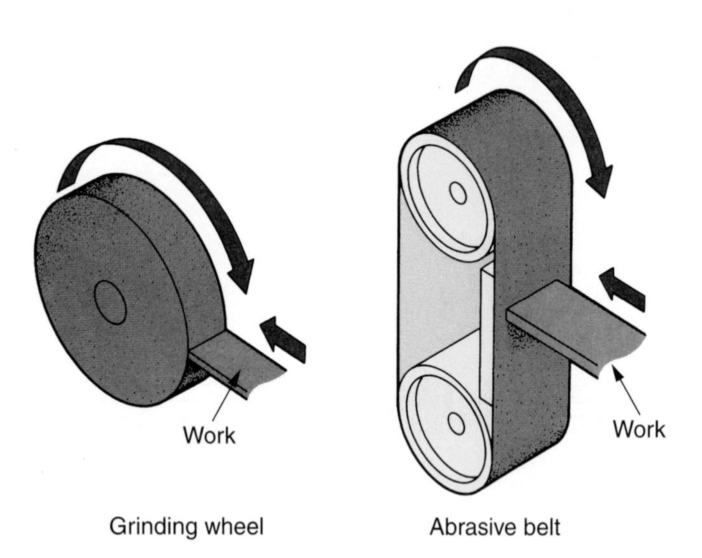

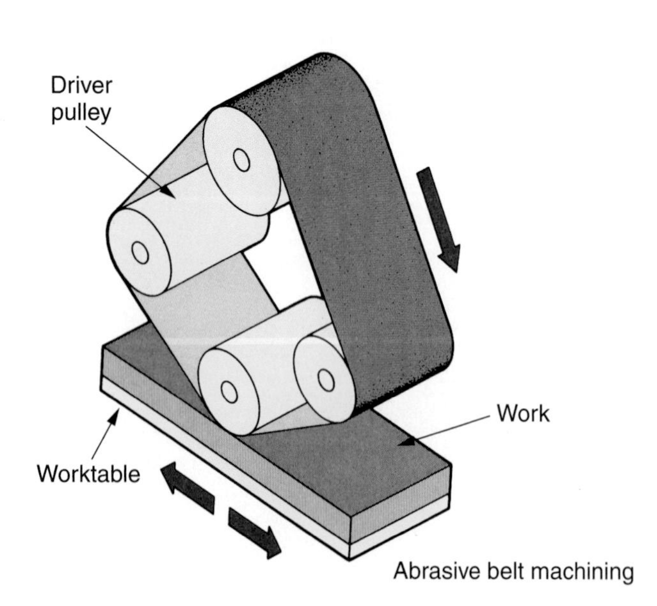

Goodheart-Willcox Publisher

Figure 13-1. Principles of how typical grinding machines work.

13.1 Abrasive Belt Grinders

Abrasive belt grinders are heavy-duty versions of the belt and disc sanders found in woodworking, **Figure 13-2**. A wide variety of abrasive belts permits these machine tools to be used for grinding to a line, finishing cast and forged parts, deburring, contouring, and sharpening. See **Figure 13-3**.

FEIN Power Tools, Inc.

Figure 13-2. An abrasive belt grinder. These machines can be used for offhand grinding or combined with other machine tools to perform more precise machining operations.

American Foundrymen's Society

Figure 13-3. Abrasive belt being used to clean up an aluminum casting of a base for an office chair.

13.2 Bench and Pedestal Grinders

The bench grinder and pedestal grinder are the simplest and most widely used grinding machines. Grinding done on a bench, pedestal, or belt grinder is called ***offhand grinding***. This type of work does not require great accuracy. The part is held in your hands and manipulated until it is ground to the desired shape.

The bench grinder is one that has been fitted to a bench or table, **Figure 13-4**. The grinding wheels mount directly onto the motor shaft. Normally, one wheel is coarse, for roughing, and the other is fine, for finish grinding.

A pedestal grinder is usually larger than the bench grinder and is equipped with a pedestal (base) fastened to the floor. See **Figure 13-5**. The dry-type pedestal grinder has no provisions for cooling the work during grinding other than a water container. The part is dipped into the water. A wet-type pedestal grinder, **Figure 13-6**, has a coolant system built into the grinder. This system keeps the wheels constantly flooded with fluid. The coolant washes away particles of loose abrasive material and metal and cools the work. Cooling prevents localized heat buildup, which can ruin tools and "burn" areas of other types of work.

SAFETY NOTE

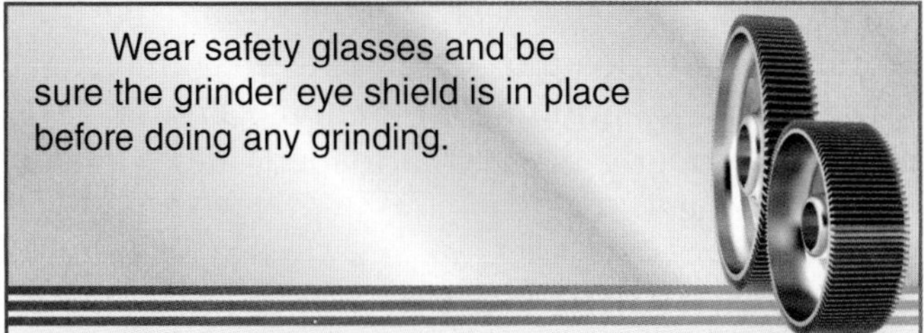

Wear safety glasses and be sure the grinder eye shield is in place before doing any grinding.

Baldor

Figure 13-4. A bench grinder can be used for many tasks. Never operate a bench or pedestal grinder unless all safety devices are in place and in sound condition. The tool rest must be properly spaced and eye protection must be worn, even though the grinder has eye shields.

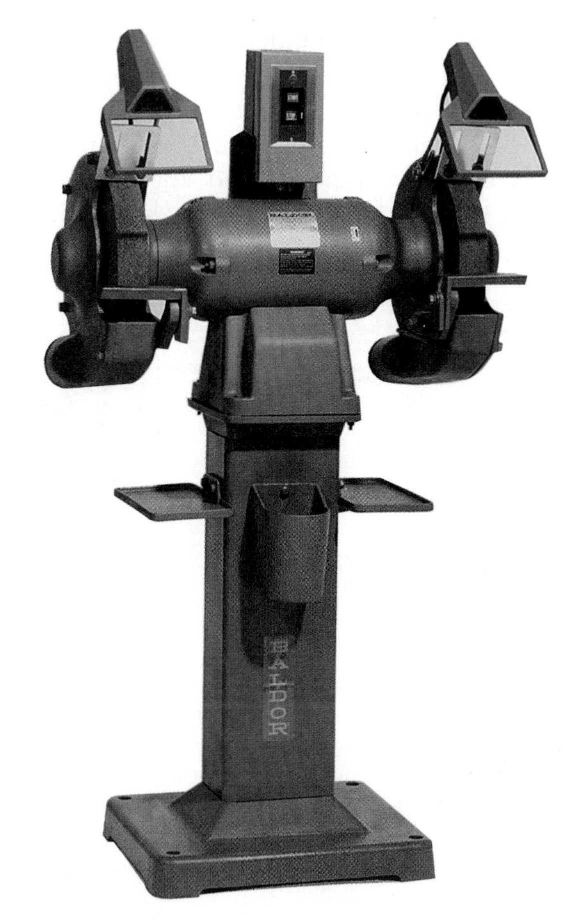

Baldor

Figure 13-5. This pedestal grinder has a roughing wheel on the left and a finishing wheel on the right.

Baldor

Figure 13-6. This grinder has a coolant attachment that keeps coolant dripping on the tool being sharpened.

The ***tool rest*** is provided to support the work being ground. It is recommended that the rest be adjusted to within 1/16″ (1.5 mm) of the wheel, **Figure 13-7**. This will prevent the work from being wedged between the rest and the wheel. After adjusting the rest, turn the wheel by hand to be sure there is sufficient clearance.

SAFETY NOTE

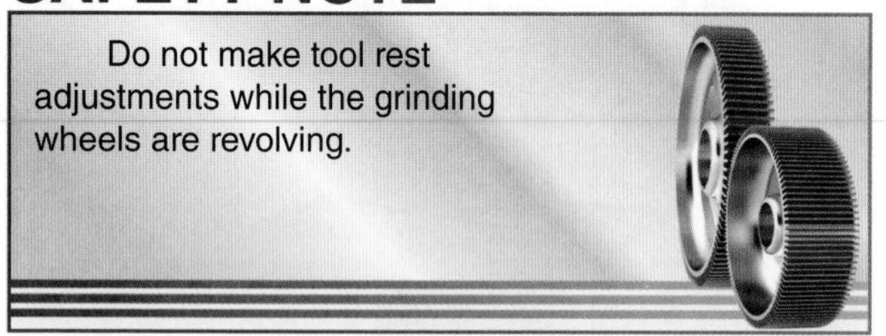

Do not make tool rest adjustments while the grinding wheels are revolving.

13.3 Grinding Wheels

Grinding wheels are made of coarse abrasive grains bonded together into circular shapes. As the wheel is used, the dulled abrasive particles wear away from the wheel to reveal new, sharp particles underneath.

Grinding wheels can be a source of danger and should be examined frequently for concentricity (running true), roundness, and cracks. A new wheel can be tested by suspending it on a string or wire, or holding it lightly with one finger, and tapping the side lightly with a metal rod or screwdriver handle. A solid wheel will give off a clear ringing sound. A wheel that does not give off a clear sound should be assumed to have a fault.

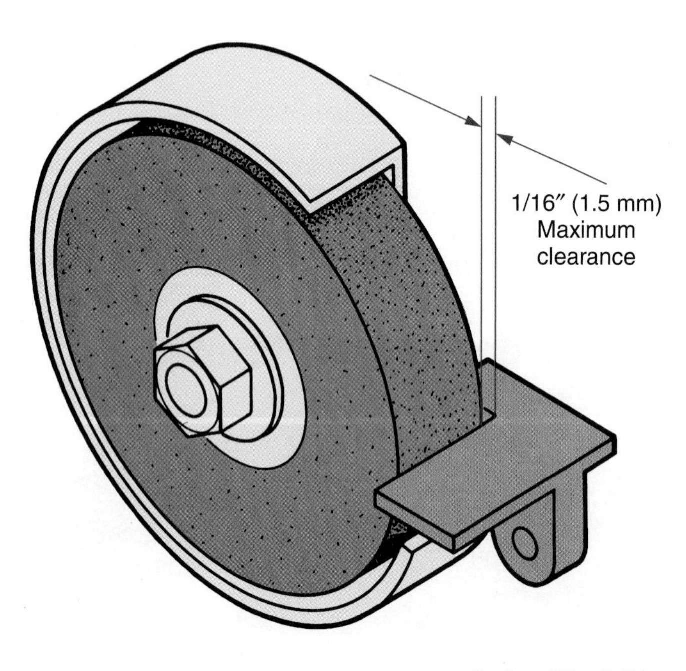

Goodheart-Willcox Publisher

Figure 13-7. The tool rest must be spaced properly for safety. Maximum safe clearance is 1/16″ (1.5 mm).

13.3.1 Wheel Dresser

The grinding wheels must run true and be balanced on the shaft. A ***wheel dresser***, **Figure 13-8**, is used to true the wheel and provide a better grinding surface.

The wheel dresser is supported on the tool rest and is held firmly against the wheel with both hands. It is moved back and forth across the wheel face to remove a thin layer of stone. See **Figure 13-9**.

During normal grinding operations, the chips and dulled grains of the grinding wheel wear away to reveal new, sharp grains. ***Loading*** of the wheel can occur when grinding soft metals. Small pieces of the material clog the wheel instead of falling away. ***Glazing*** can occur when grinding very hard metals. The grains of the wheel tend to dull before they can fall away to reveal the sharp grains underneath. The shiny appearance of the grinding wheel surface indicates the amount of glaze.

Both loading and glazing cause the wheel to cut inefficiently. A wheel dresser should be used to remove any loading or glazing that may have formed during grinding operations. **Figure 13-10** shows a properly dressed, a loaded, and a glazed grinding wheel.

13.3.2 Grinding Rules

To obtain maximum efficiency from a grinder, the following recommendations should be observed:

- Grind using the face of a wheel, not the sides.
- Move the work back and forth across the wheel face. This will wear the wheel evenly and prevent grooves from forming.

Photo courtesy of Grizzly Industrial, Inc. www.grizzly.com

Figure 13-8. A mechanical wheel dresser will true stone face.

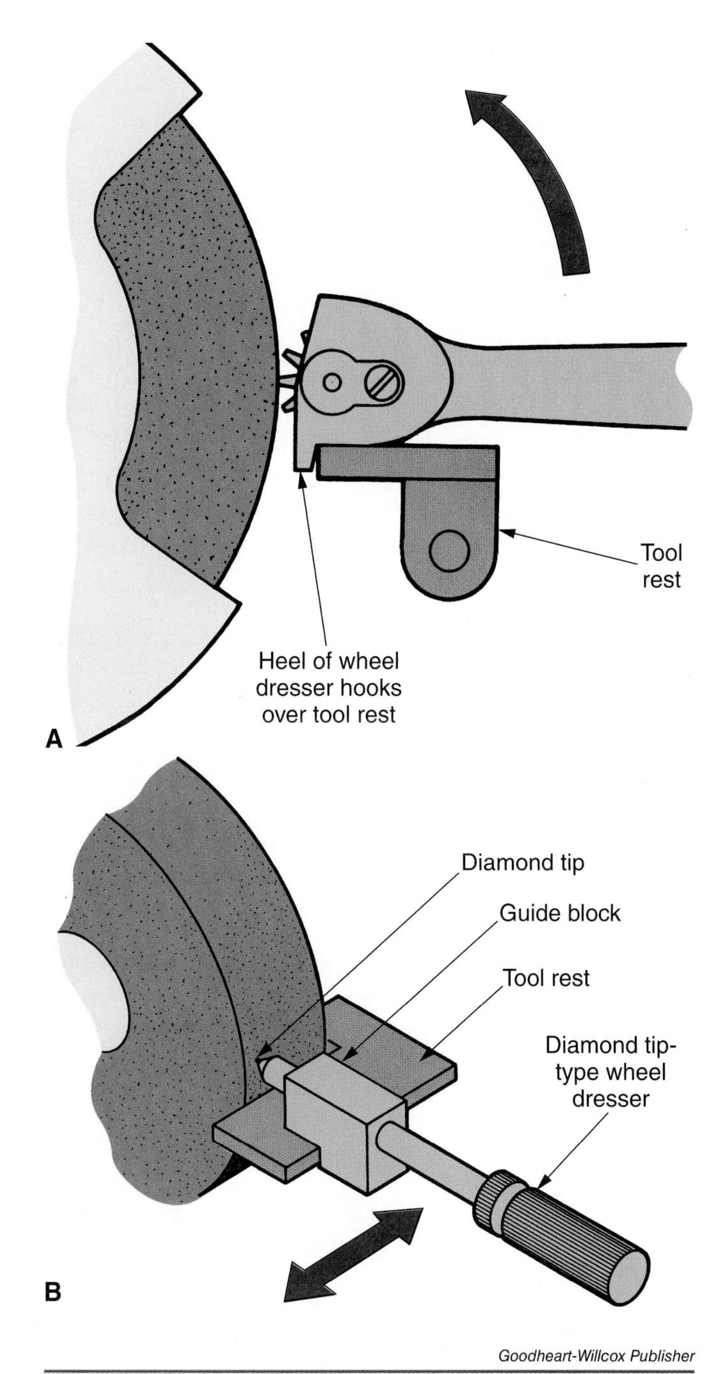

Figure 13-9. The proper way to use a mechanical wheel dresser. A—Move the tool back and forth over the face of the stone. Wear a dust mask, eye protection, and an apron when dressing wheels on grinders. B—Industrial diamonds are also used to dress and true grinding wheels. The guide block is used for grinders with slotted tool rests.

- Keep the wheel dressed and the tool rests properly adjusted.
- Soft metals (aluminum, brass, and copper) tend to load grinding wheels. When possible, these metals should be ground on an abrasive belt grinder.

Figure 13-10. Grinding wheels in various conditions.

13.4 Abrasive Belt and Wheel Grinder Safety

- Make sure the tool rest is properly adjusted.
- Wear goggles or a face shield when performing grinding operations, even though the machines are fitted with eye shields.
- Never attempt to operate an offhand grinding machine while your senses are impaired by medication or other substances.
- Check the machine thoroughly before using it. Lubricate the machine only as recommended by the manufacturer.
- Check a grinding wheel for soundness before putting it on the grinder. Destroy wheels that are not sound or that have a worn center hole.
- Do not use a wheel that is glazed or loaded with metal.
- Be sure all wheel guards and safety devices are in place before attempting to use a grinder or abrasive belt machine.
- If the grinding operation is to be performed dry, be sure to hook up all exhaust attachments before starting.
- Stand to one side of the machine during operation. Do not stand directly in front of the wheel.
- Hold small work in a clamp or hand vise. Under no condition should work be held with a cloth.

- Avoid work pressure on the side of the grinding wheel.
- Keep your hands clear of the rotating wheel.
- Never operate a grinding wheel at speeds higher than those recommended by the manufacturer.
- Have injuries caused by rotating grinding wheels treated immediately.
- Allow the wheels or belt to stop completely before attempting to make any machine adjustments.

13.5 Using a Dry-Type Grinder

After examining the grinder and making the necessary adjustments, turn on the machine. Be sure that you wear safety glasses whenever you are in the shop. Stand to one side until the grinder has reached operating speed.

Place the work on the tool rest and slowly push it against the grinding wheel. If too much pressure is applied, the work will begin to "burn" or discolor. Overheating can be minimized by dipping the work into the water container from time to time. Care must be taken when grinding edge tools because excessive heat will "draw" (remove) the ***temper*** (hardness) and ruin the tool.

Keep the work moving across the wheel face to prevent the formation of grooves or ridges. Dress and retrue the wheel as necessary for maximum efficiency.

Pieces of cloth should never be used to hold work while it is being ground. Serious injuries can result if the cloth is pulled into the wheel. Hold the work, especially small lathe cutter bits, in hand vises specially designed for that purpose, **Figure 13-11**.

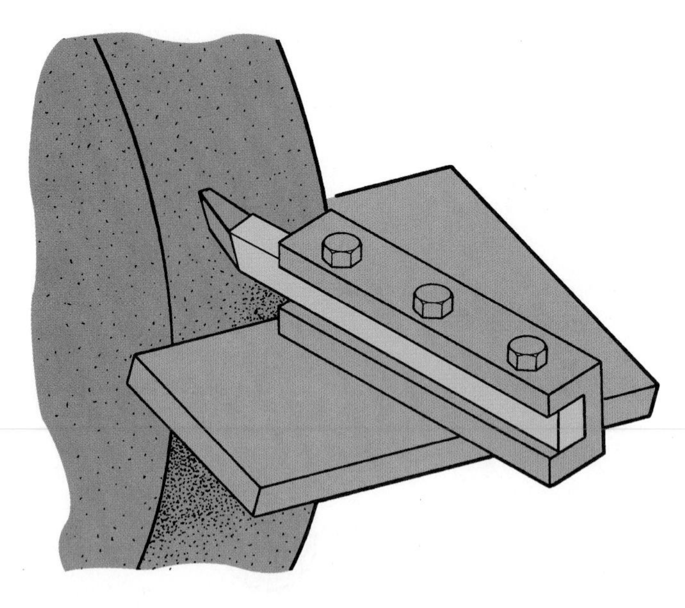

Goodheart-Willcox Publisher

Figure 13-11. A hand vise is used to hold cutter bits while they are sharpened.

13.6 Using a Wet-Type Grinder

The wet-type grinder is primarily used to grind carbide-tipped tools. Since carbide tools are often brazed onto a steel shank, both steel and carbide must be ground away when these tools are sharpened. Aluminum oxide wheels should be used to grind the steel shank. Silicon carbide or diamond-impregnated wheels should be used to grind the carbide tip.

Wet-type grinders should normally have a flat face, but a slightly crowned face should be used when grinding carbide-tipped tools, **Figure 13-12**.

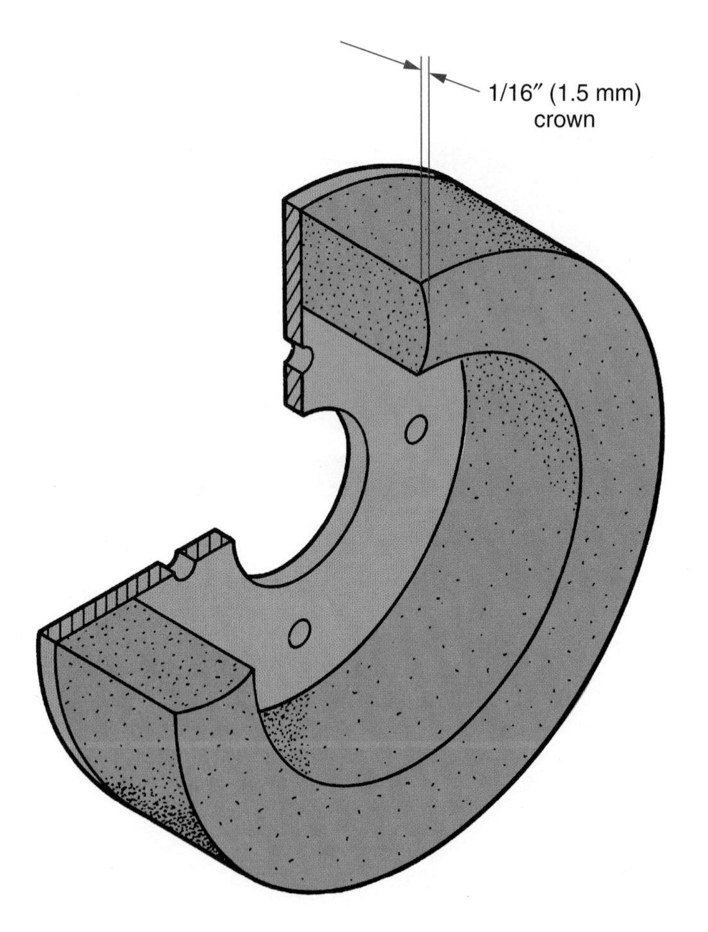

Goodheart-Willcox Publisher

Figure 13-12. A slight crown on the wheel face minimizes the amount of contact between the work and the wheel. A crown reduces the possibility of heat destroying the carbide tip of a tool.

The crown minimizes the contact between the wheel and the work. This reduces the possibility of the tip being damaged by excessive heat.

The coolant attachment must be adjusted to keep a full flow of liquid directed on the tool at all times. Adjust the tool table rest to obtain the correct clearance angle, **Figure 13-13**. A protractor guide is helpful when compound clearance angles are required.

Use the entire face of the wheel. Keep the tool in continuous motion to minimize wheel wear. Dress and retrue the wheel as needed.

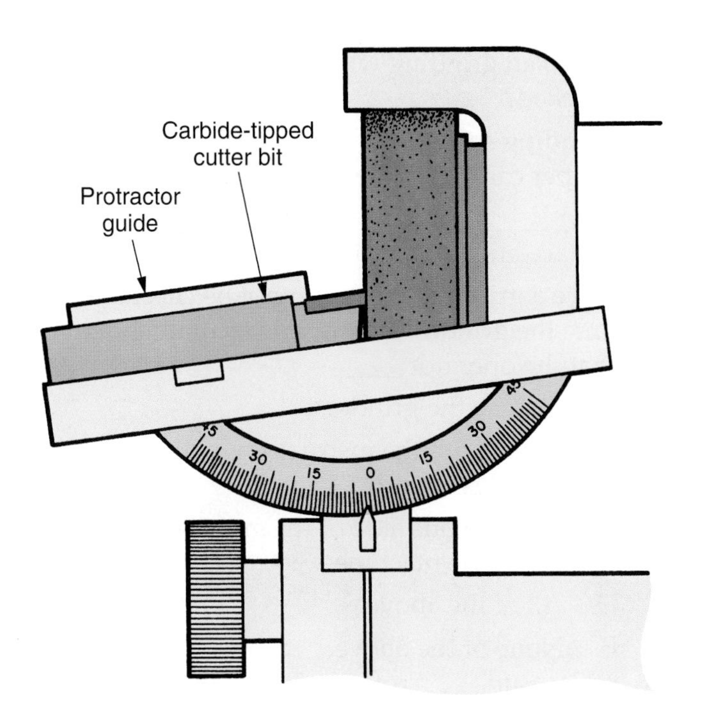

Goodheart-Willcox Publisher

Figure 13-13. The tool rest can be adjusted to any desired angle using a table protractor and protractor guide.

13.7 Portable Hand Grinders

Many grinding jobs, from light deburring to die-polishing operations, are done with small portable hand grinders. Flexible shaft grinders and precision microgrinders, **Figure 13-14**, are used to perform a variety of toolroom and production jobs. They can be powered by electricity or air.

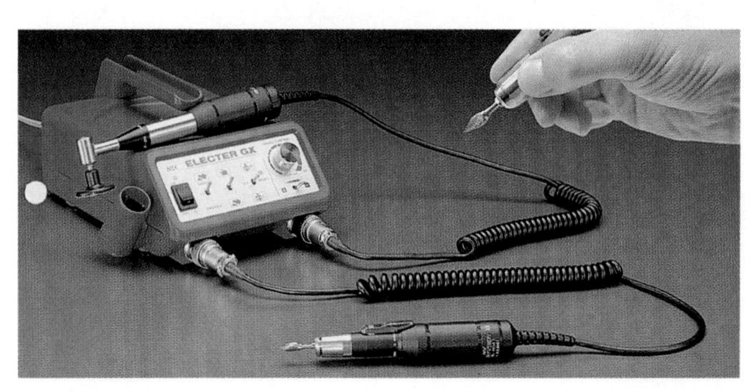

NSK America

Figure 13-14. A precision electric microgrinder is helpful for working in recessed areas on parts.

Chapter Review

Summary

- Offhand grinding can be performed using an abrasive belt, bench, pedestal, or portable hand grinder.
- It is important to properly dress and true a grinding wheel.
- Loading and glazing will cause the grinding wheel to cut inefficiently.
- It is important to check grinding wheels for soundness.
- Dry-type grinders can overheat the workpiece and can ruin the tool or cause discoloration.
- Wet-type grinders are used primarily to grind carbide cutting tools.
- Grinding can be a dangerous operation if not performed safely.

Review Questions

Answer the following questions using the information provided in this chapter.

1. Describe the grinding operation.
2. Bench and pedestal grinders are used to do _____ grinding.
3. The grinding technique referred to in the preceding statement is so named because _____.
   A. it can only do external work
   B. work is too hard to be machined by other methods
   C. work is manipulated with fingers until desired shape is obtained
   D. All of the above.
   E. None of the above.
4. Name the two types of pedestal grinders. How do they differ?
5. How far away should the tool rest be positioned from the grinding wheel? Why is this done?
6. How can grinding wheel soundness be checked?
7. Grinding soft metals like aluminum, brass, or copper can _____ the grinding wheel.
8. A(n) _____ grinding wheel will have a shiny surface and cut inefficiently.
9. Since a grinding wheel cannot be checked each time the grinder is used, it is recommended that the operator _____.
   A. not use the grinder
   B. check with the instructor whether the wheel is sound
   C. stand to one side of the grinder when using the machine
   D. All of the above.
   E. None of the above.
10. Work will _____ if it is forced against the wheel with too much pressure.
11. Carbide-tipped tools are usually sharpened on a(n) _____ grinder.
12. When grinding carbide-tipped tools, the face of the wheel on a wet-type grinder should be slightly _____.
13. Never mount a grinding wheel on a grinder without checking it for _____.
14. List four safety precautions to be observed when operating a grinder.

CHAPTER 14
The Lathe

Learning Objectives

After studying this chapter, you will be able to:

- Describe how a lathe operates.
- Identify the various parts of a lathe.
- Safely set up and operate a lathe using various work-holding devices.
- Calculate correct cutting speeds and feeds for lathe operations.
- Perform basic machining operations on a lathe.
- Sharpen lathe cutting tools.

Technical Terms

chipbreaker
compound rest
cross-slide
cutting speed
depth of cut
facing
feed
grooving
headstock
indexable insert cutting tool
lathe center
lathe dog
parting
plain turning
roughing cut
single-point cutting tool
spindle
swing
tailstock
tool post
ways

The lathe operates on the principle of the work being rotated against the edge of a cutting tool, **Figure 14-1**. It is one of the oldest and most important machine tools. The cutting tool is controllable and can be moved lengthwise on the lathe bed and across the revolving work at any desired angle. See **Figure 14-2**.

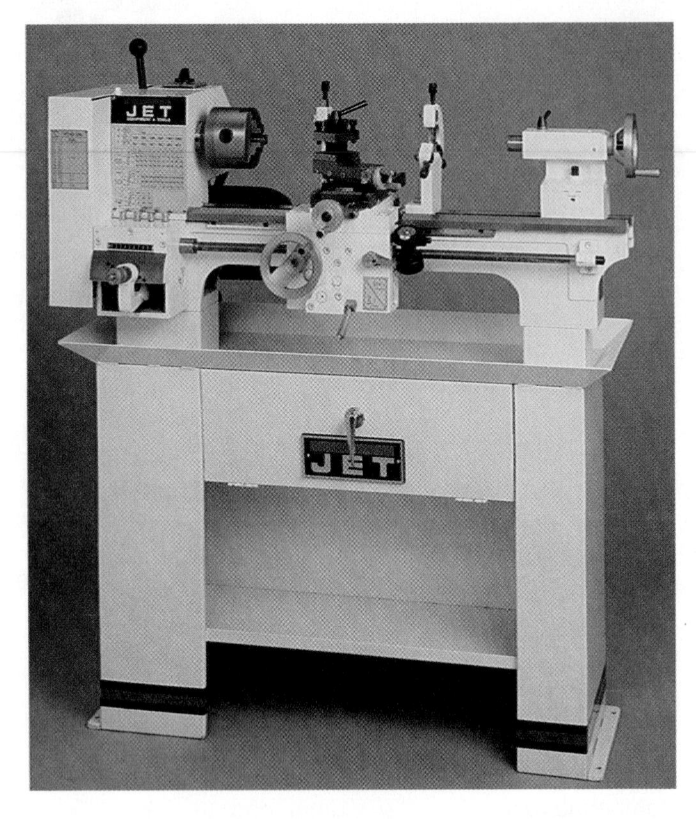

Jet Equipment & Tools

Figure 14-1. A basic metal-cutting lathe. All controls are operated manually. Most machinists begin their training on this type of lathe.

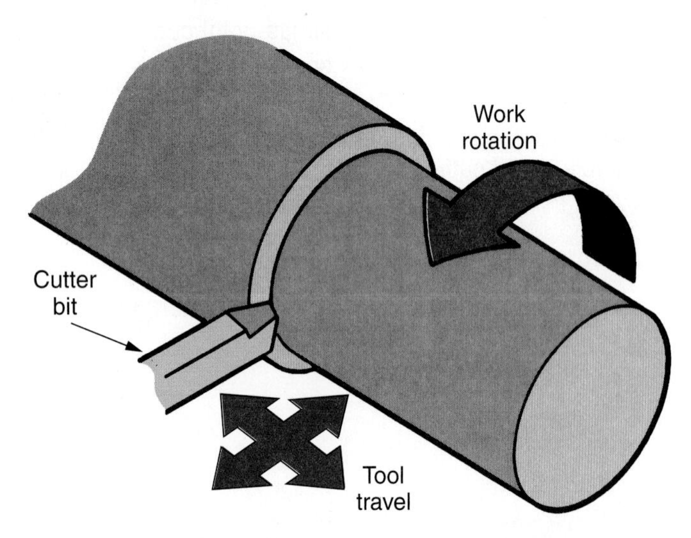

Goodheart-Willcox Publisher

Figure 14-2. Operating principle of the lathe. The cutting tool is fed into the revolving work.

14.1 Lathe Size

Lathe size is determined by the ***swing*** and the length of the bed, **Figure 14-3**. The swing indicates the largest diameter that can be turned over the ***ways*** (the flat or V-shaped bearing surface that aligns and guides the movable part of the machine). Bed length is the entire length of the ways.

Bed length must not be mistaken for the maximum length of the work that can be turned between centers. The longest piece that can be turned is equal to the length of the bed minus the distance taken up by the headstock and tailstock. Refer to measurement B in **Figure 14-3**.

As an example, consider the capacity and clearance of a modern 13″ × 6′ (325 mm × 1800 mm) lathe:

Swing over bed:	13″ (325 mm)
Swing over cross-slide:	8 3/4″ (218 mm)
Bed length:	72″ (1800 mm)
Distance between centers:	50″ (1240 mm)

14.2 Major Parts of a Lathe

The chief function of any lathe, no matter how complex it may appear to be, is to rotate the work against a controllable cutting tool. Each of the lathe parts in **Figure 14-4** can be assigned to one of the following three categories:

- Driving the lathe.
- Holding and rotating the work.
- Holding, moving, and guiding the cutting tool.

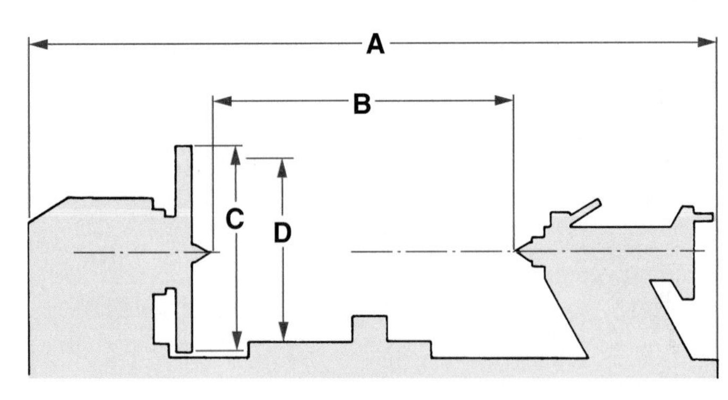

Goodheart-Willcox Publisher

Figure 14-3. Lathe measurements. A—Length of bed. B—Distance between centers. C—Diameter of work that can be turned over the ways. D—Diameter of work that can be turned over the cross-slide.

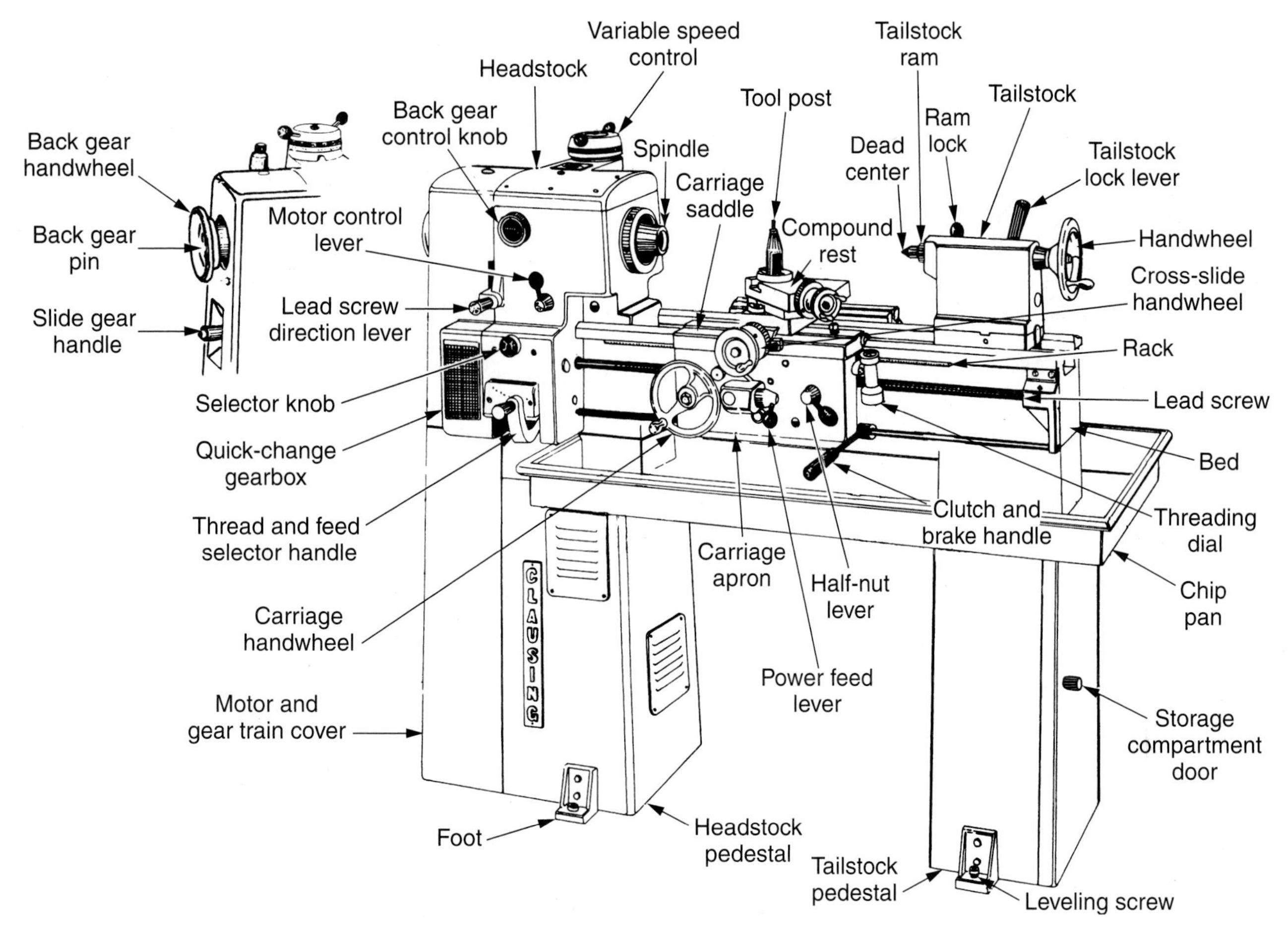

Clausing Industrial, Inc.

Figure 14-4. The engine lathe and its major parts.

14.2.1 Driving the Lathe

Power is transmitted to the drive mechanisms by a belt drive or gear train. Spindle speed can be varied by any of the following:

- Shifting to a different gear ratio, **Figure 14-5.**
- Adjusting a split pulley to another position, **Figure 14-6.**
- Moving the drive belt to another pulley ratio (seldom used today).
- Controlling the speed hydraulically.

Slower speeds with greater power are obtained on some machines by engaging a back gear. To avoid damaging the lathe's drive system, do not engage the back gear while the spindle is rotating.

14.2.2 Holding and Rotating the Work

The ***headstock*** contains the spindle to which the various work-holding attachments are fitted, **Figure 14-7.**

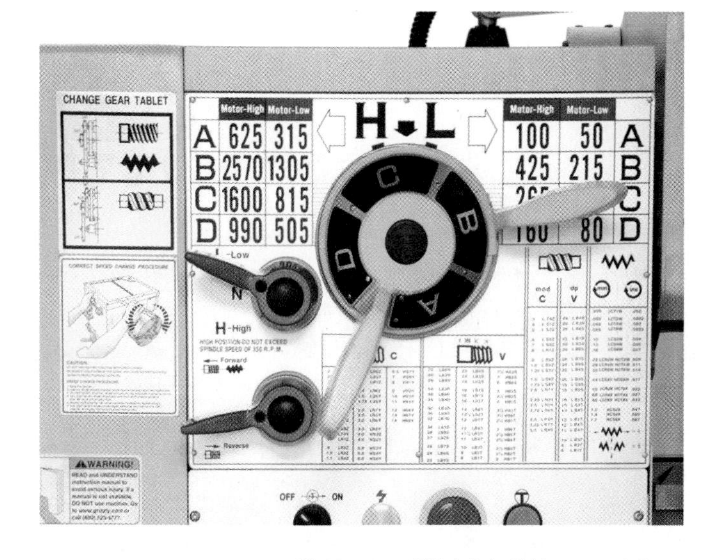

Photo courtesy of Grizzly Industrial, Inc. www.grizzly.com

Figure 14-5. Spindle speed control. Speed is increased or decreased by shifting to different gear ratios. On this machine, the desired speed is dialed in.

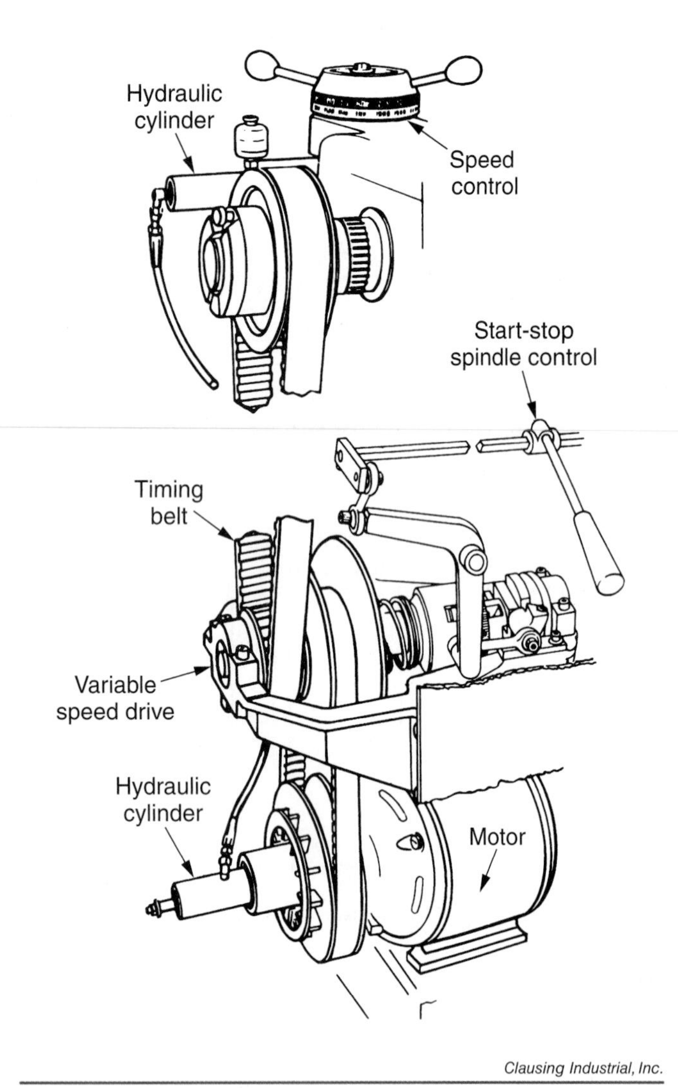

Clausing Industrial, Inc.

Figure 14-6. This split pulley is hydraulically actuated from the top of the machine by a speed control. A split pulley is used to control spindle speeds on many lathes.

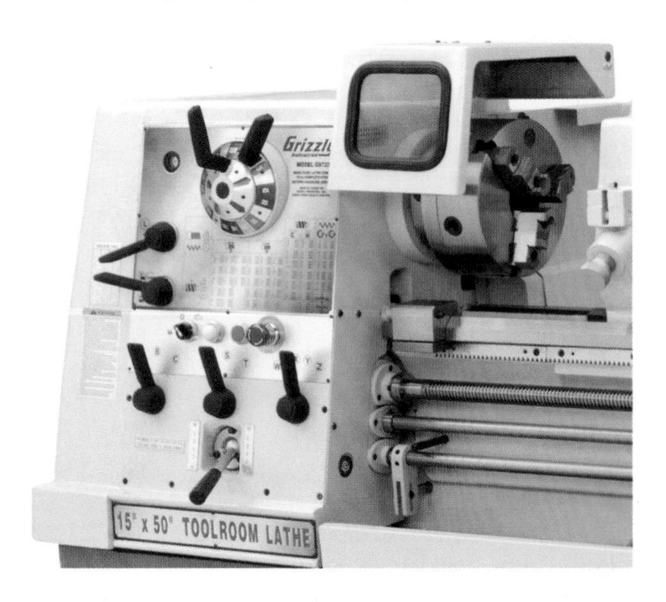

Photo courtesy of Grizzly Industrial, Inc. www.grizzly.com

Figure 14-7. The headstock is the driving end of the lathe.

The ***spindle*** revolves in heavy-duty bearings and is rotated by belts, gears, or a combination of the two. The front of the hollow spindle is tapered internally to receive tools and attachments with taper shanks, **Figure 14-8.** The hole through the spindle permits long stock to be turned without dangerous overhang. It also allows use of a knockout bar to remove taper-shank tools.

On the front end, a spindle may be threaded externally or fitted with one of two types of tapered spindle noses to receive work-holding attachments. See **Figure 14-9**. A threaded spindle nose is seldom used on modern lathes. It permits mounting an attachment by screwing it directly on the threads until it seats on the spindle flange.

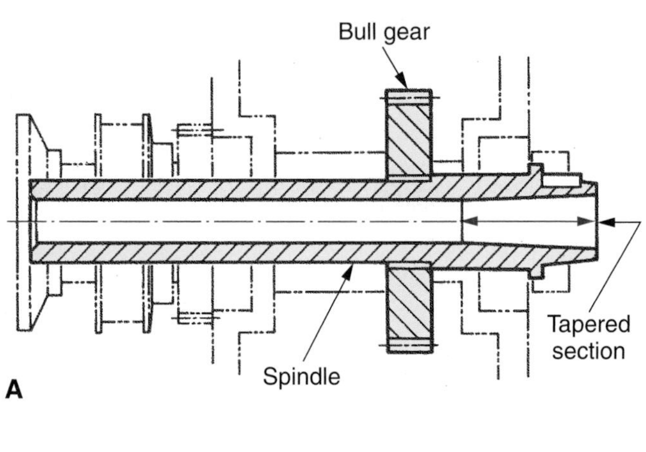

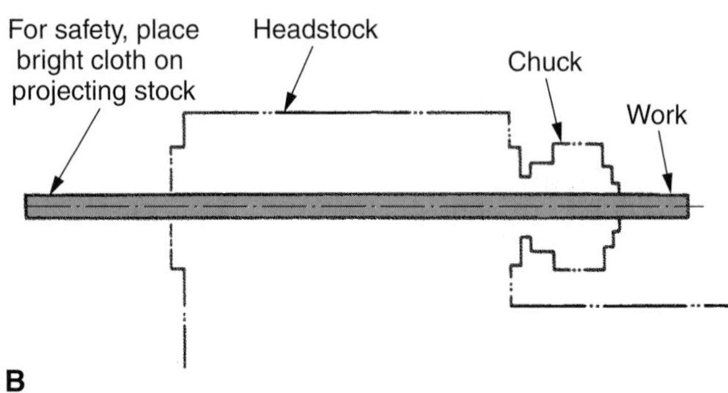

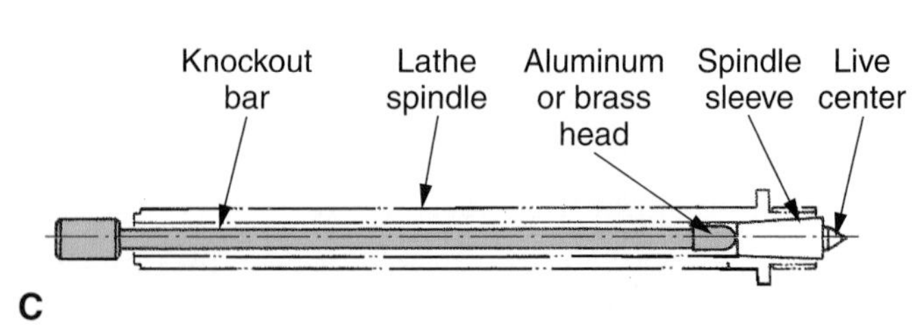

Goodheart-Willcox Publisher

Figure 14-8. Lathe spindle. A—Hollow construction of the spindle allows long stock to be turned without dangerous overhang. B—To prevent accidents that could cause injury, some sort of flag should be tied to the portion of stock that projects from the rear of the spindle. C—A knockout bar is used to tap tapered shank lathe accessories out of the spindle.

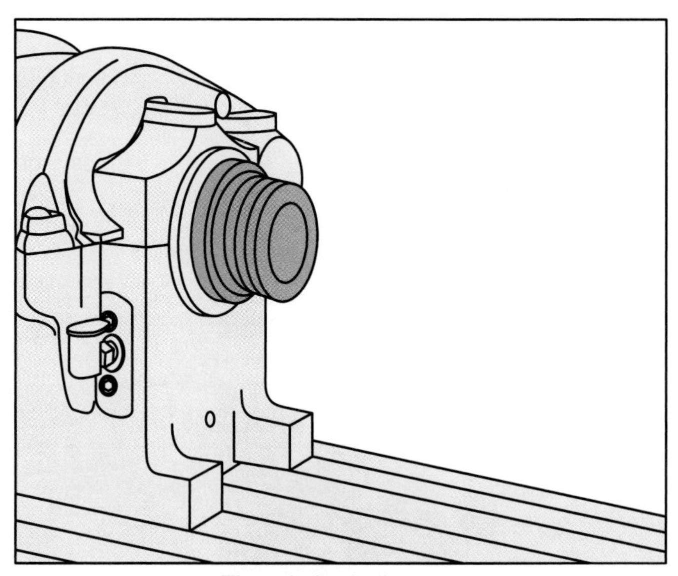

Threaded spindle nose

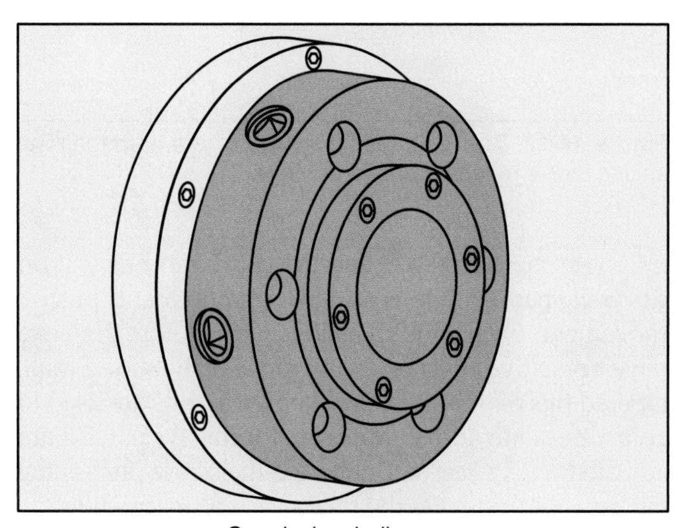

Cam-lock spindle nose

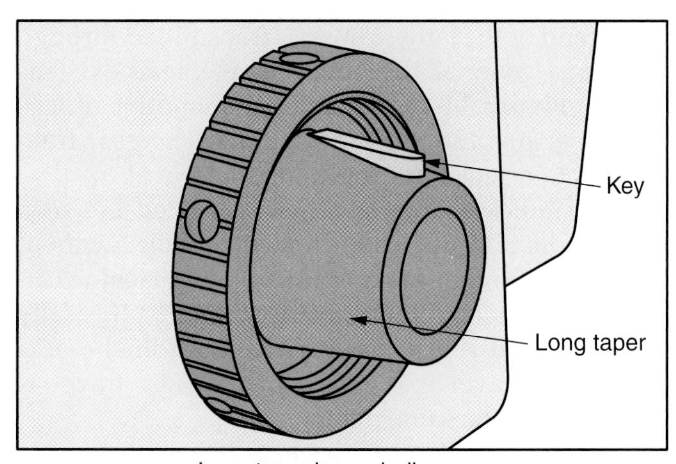

Long taper key spindle nose

Goodheart-Willcox Publisher

Figure 14-9. Spindle noses. The threaded spindle nose is seldom used today.

The cam-lock spindle nose has a short taper that fits into a tapered recess on the back of the work-holding attachment. A series of cam locking studs, located on the back of the attachment, are inserted into holes in the spindle nose. The studs are locked by tightening the cams located around the spindle nose.

A long taper key spindle has a protruding long taper and key that fits into a corresponding taper and keyway in the back of the work-holding device. To mount a work-holding device (a chuck or face-plate), the spindle is rotated until the key is on top. The keyway in the back of the work-holding device is slid over the key to support the device until the threaded spindle collar can be engaged with the threaded section of the device, then tightened.

SAFETY NOTE

Attachment points on the spindle nose and work-holding attachment must be cleaned carefully before mounting the device.

Work is held in the lathe by a chuck, faceplate, or collet, or by mounting it between centers. These attachments will be described in detail later in this chapter.

The outer end of the work is often supported by the lathe's ***tailstock***, **Figure 14-10**. The tailstock can be adjusted along the ways to accommodate different lengths of work.

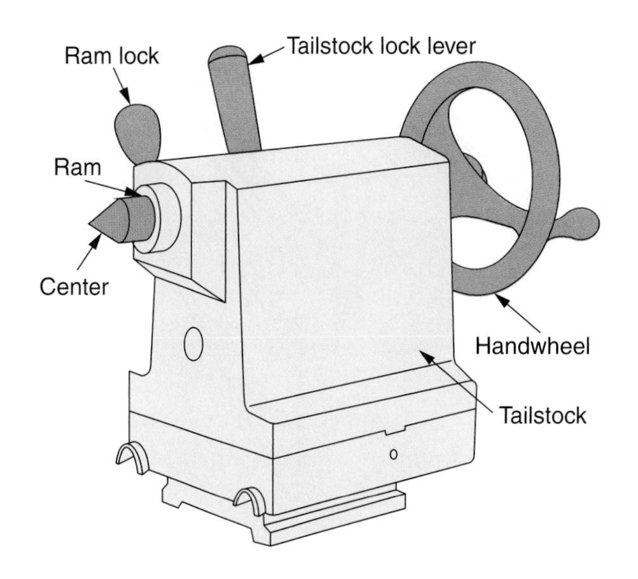

Goodheart-Willcox Publisher

Figure 14-10. Parts of the tailstock.

The tailstock is used to mount the lathe center, or can be fitted with tools for drilling, reaming, and threading. It can also be offset for taper turning.

The tailstock is locked onto the ways by tightening a clamp bolt nut or binding lever. The tailstock spindle is positioned by rotating the handwheel and can be locked in position by tightening a binding lever.

14.2.3 Holding, Moving, and Guiding the Cutting Tool

The bed, **Figure 14-11**, is the foundation of a lathe. All other parts are fitted to it. Ways are integral with the bed. The V-shaped rails maintain precise alignment of the headstock and tailstock, and guide the travel of the carriage.

The carriage, **Figure 14-12**, controls and supports the cutting tool. It is composed of the following parts:

- The saddle is fitted to the ways and slides along them.
- The apron contains a drive mechanism to move the carriage along the ways, using hand or power feed.
- The ***cross-slide*** permits transverse tool movement (movement toward or away from the operator, at a right angle to the axis of the lathe).
- The ***compound rest*** permits angular tool movement.
- The ***tool post*** is used to mount the cutting tool.

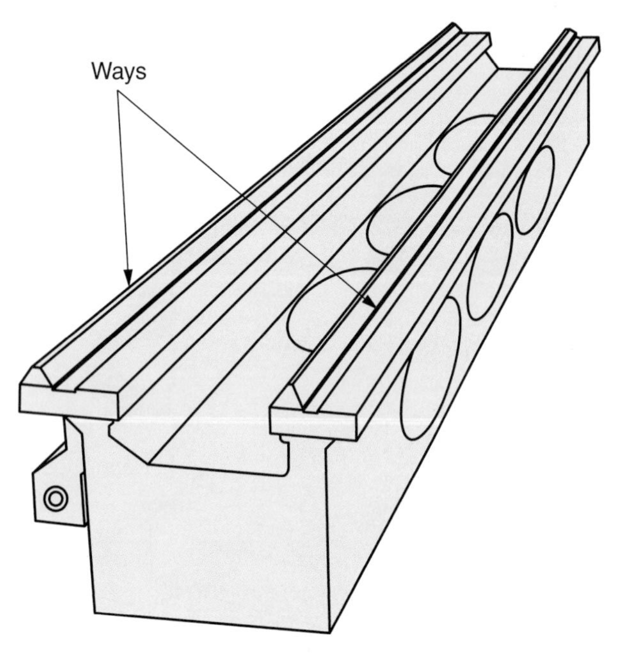

Goodheart-Willcox Publisher

Figure 14-11. The bed is the foundation of the lathe. The ways maintain alignment of the headstock and tailstock.

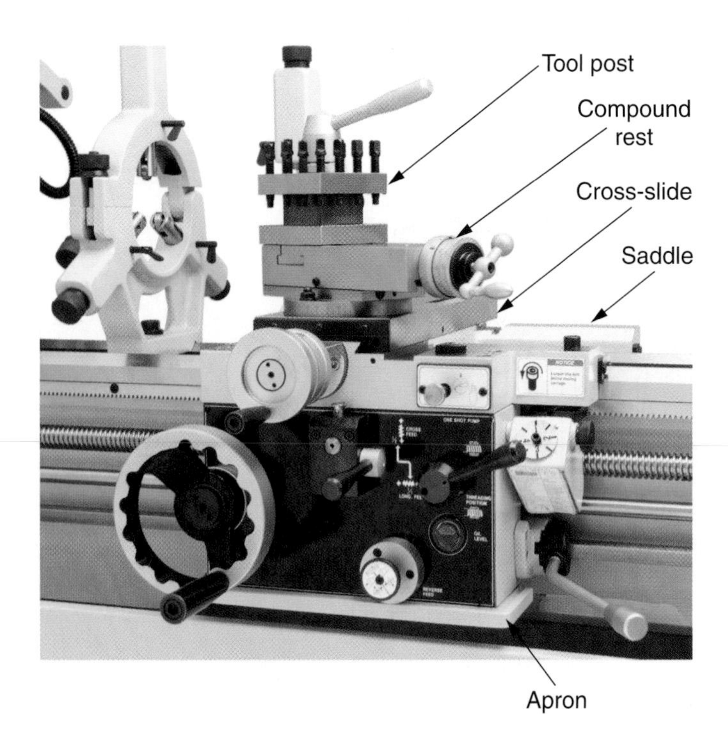

Photo courtesy of Grizzly Industrial, Inc. www.grizzly.com

Figure 14-12. The V-shaped ways guide the carriage. The cutting tool is mounted on the carriage.

Both the cross-slide and the compound rest sit on dovetail shaped slide bearings for smooth and precise movement. Over time the slides wear down. This can have a negative effect on the accuracy of the slide. Small, tapered pieces of iron or steel known as gibs are used to adjust the slide and compensate for this wear. Gibs are adjusted with a screw. Tightening the screw pushes the gib forward, reducing the clearance within the slide.

Power is transmitted to the carriage through the feed mechanism, which is located at the left (headstock) end of the lathe. Power is transmitted through a train of gears to the quick-change gearbox. This device, **Figure 14-13**, regulates the amount of tool travel per revolution of the spindle. The gear train also contains gears for reversing tool travel.

The quick-change gearbox is located between the spindle and the lead screw. It contains gears of various ratios that make it possible to machine different pitches of screw threads without physically removing and replacing gears. Longitudinal (back-and-forth) travel and cross (in-and-out) travel is controlled in the same manner.

An index plate provides instructions on how to set the lathe shift levers for various thread cutting and feed combinations. It is located on the face of the gearbox. Index plates often show the lever positions for both inch and metric feeds and thread pitches.

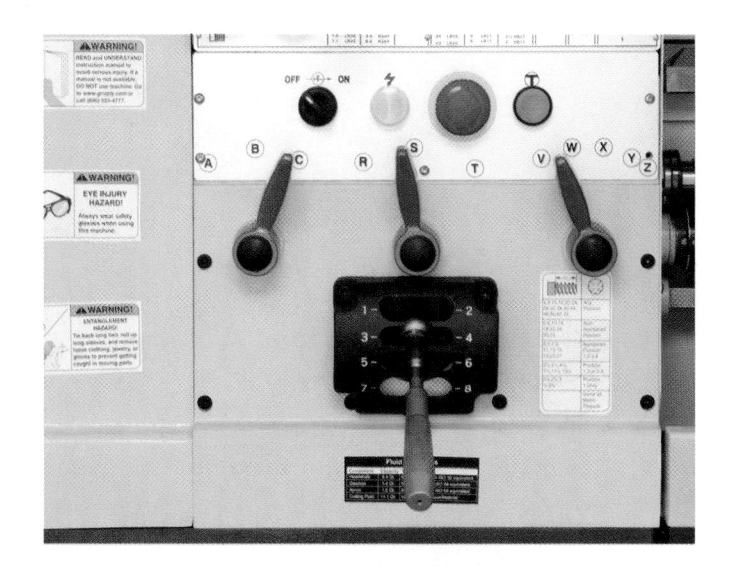

Photo courtesy of Grizzly Industrial, Inc. www.grizzly.com

Figure 14-13. A quick-change gearbox for cutting both inch and metric size threads.

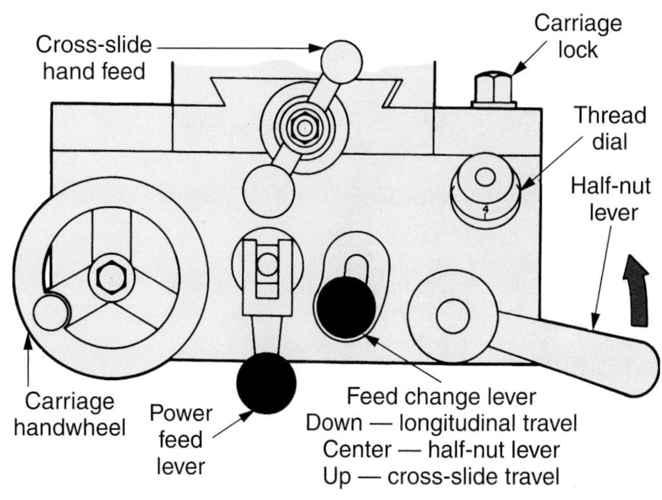

Goodheart-Willcox Publisher

Figure 14-15. Power feed functions are controlled by the feed change levers on the apron. The half-nut lever is engaged only for thread cutting, sometimes called "thread chasing."

The lead screw transmits power to the carriage through a gearing and clutch arrangement in the carriage apron, **Figure 14-14**. Feed change levers on the apron control the operation of power longitudinal feed and power cross-feed, **Figure 14-15**.

When the feed change lever is placed in neutral, the half-nuts may be engaged for thread cutting. The gear arrangement makes it possible to engage power feed and half-nuts simultaneously. The half-nuts are engaged only for thread cutting and are not used as an "automatic feed" for regular turning.

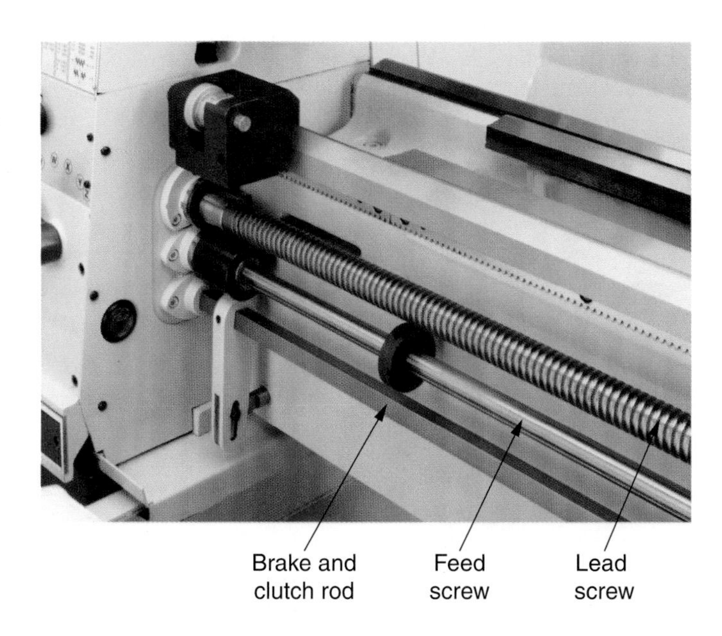

Photo courtesy of Grizzly Industrial, Inc. www.grizzly.com

Figure 14-14. The lead screw.

14.3 Work-Holding Attachments

One of the reasons the lathe is such a versatile machine tool is the great variety of ways that work may be mounted in or on it. The most common way is to mount the work so that it revolves, permitting the cutting tool to move across the work's surface. Large or odd-shaped pieces are sometimes mounted on the carriage and machined with a cutting tool that is mounted in the rotating spindle.

Most work is machined while supported by one of the following methods, as shown in **Figure 14-16**:

- Between centers using a faceplate and dog.
- Held in one of the three types of chucks.
  - 3-jaw universal chuck.
  - 4-jaw independent chuck.
  - Jacobs chuck.
- Held in a collet.
- Bolted to the faceplate.

14.4 Work-Holding between Centers

Considerable lathe work is done with the workpiece supported between centers. A ***lathe center***, or center, is a pointed work-holding device used to accurately align the workpiece along an axis, **Figure 14-17**. Heavy-duty ball bearings allow live centers to rotate freely with the workpiece. Dead centers do not rotate.

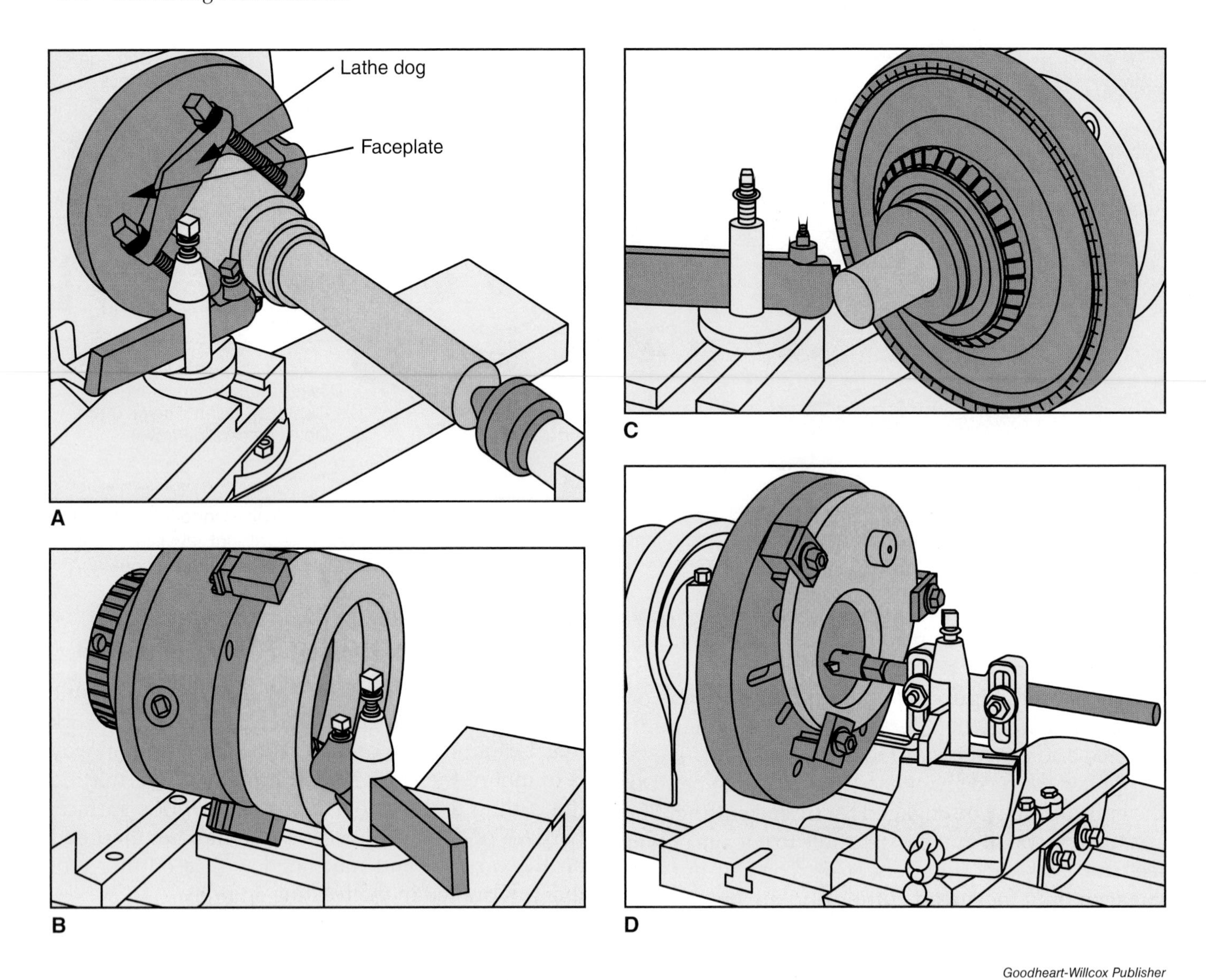

Goodheart-Willcox Publisher

Figure 14-16. Work-holding methods. A—Work being machined between centers. B—Work held in a chuck for machining. C—Work being machined while held in a collet. D—Work bolted to a faceplate for machining.

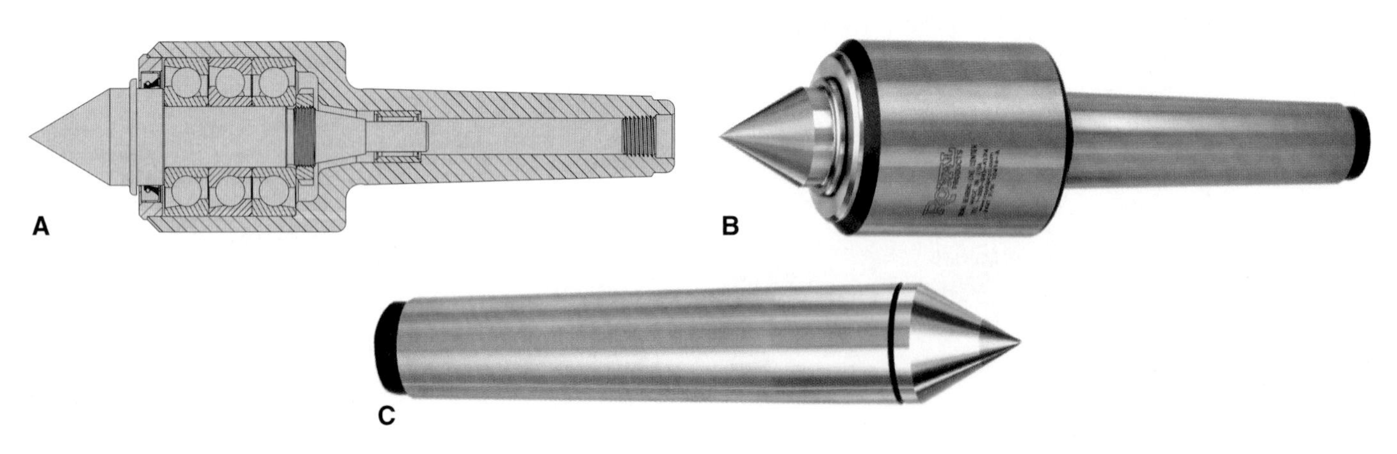

Royal Products, Division of Curran Manufacturing Corporation

Figure 14-17. Lathe centers. A—Sectional view shows construction of a heavy-duty ball bearing live center. B—High precision quad-bearing live center. C—A dead center does not rotate. The carbide tip provides great wear resistance.

When supporting the workpiece between centers, a faceplate, **Figure 14-18**, is often attached to the spindle nose. A sleeve and dead center are inserted into the spindle opening, **Figure 14-19**.

Either a live or dead center is fitted into the tailstock spindle to support one end of the work. The ends of the stock are drilled to fit over the center points.

A ***lathe dog*** is a device clamped to one end of the material to drive the workpiece. The three types of lathe dogs are as follows, as shown in **Figure 14-20**:

- The bent-tail standard dog has the setscrew exposed.
- The bent-tail safety dog has the setscrew recessed. This type dog is usually preferred over the standard lathe dog.
- The clamp-type dog is used for turning square or rectangular work.

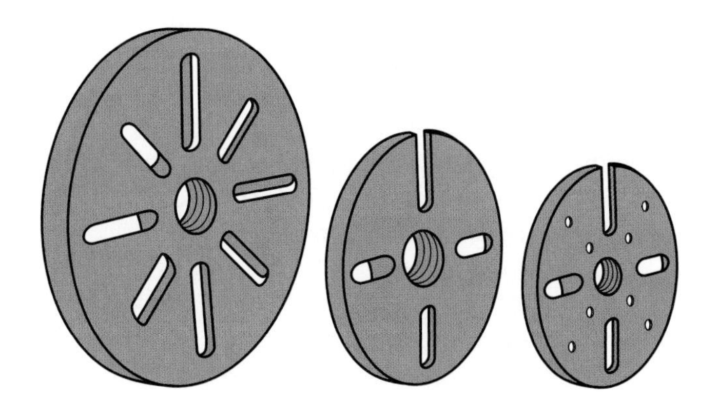

Goodheart-Willcox Publisher

Figure 14-18. Lathe faceplates come in various sizes.

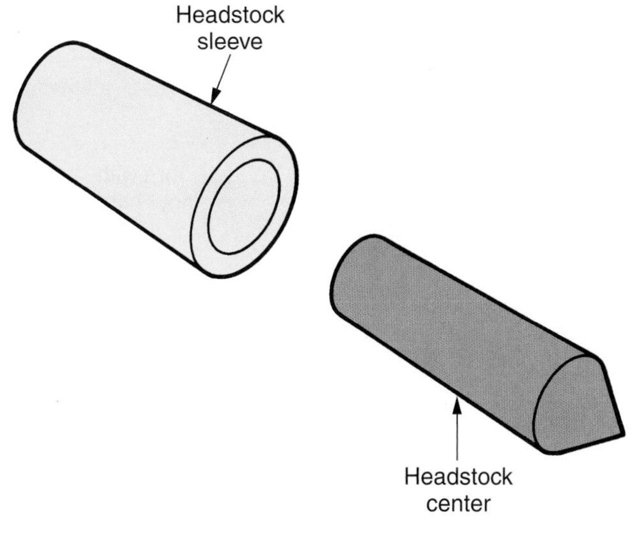

Goodheart-Willcox Publisher

Figure 14-19. Sleeve and headstock center.

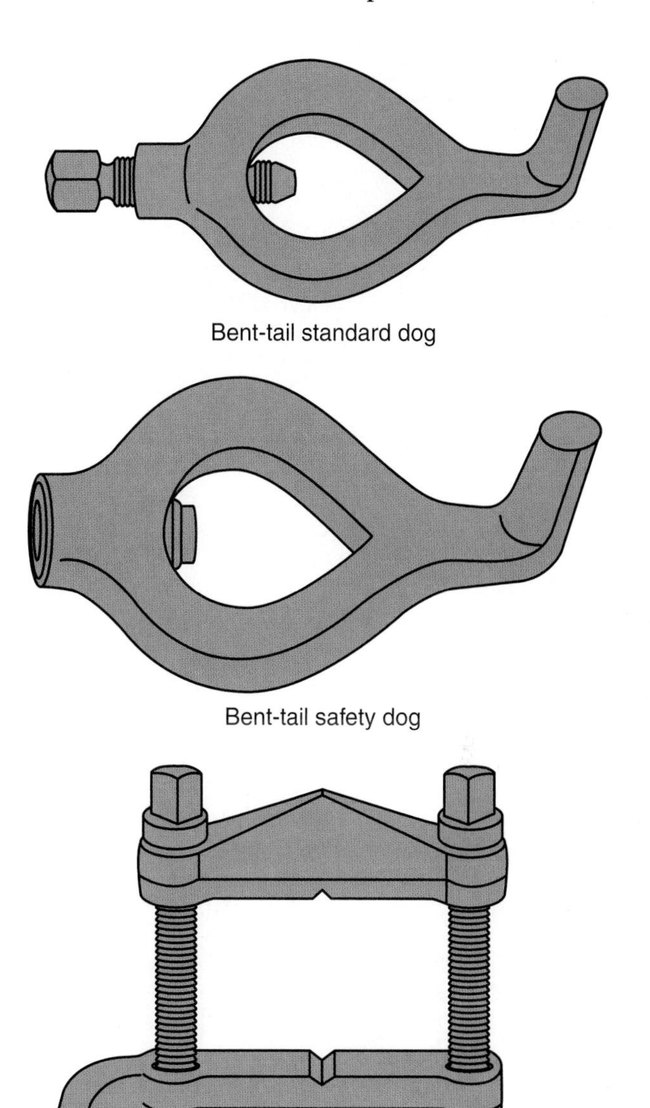

Goodheart-Willcox Publisher

Figure 14-20. Lathe dogs.

Because work-holding between centers requires the work to be supported at each end, only external machining operations can be performed.

14.4.1 Drilling Center Holes

Before work can be mounted between centers, it is necessary to locate and drill center holes in each end of the stock, **Figure 14-21**. Several methods for locating the center of round stock are shown in **Figure 14-22**.

Center holes are usually drilled with a combination drill and countersink, **Figure 14-23**. The drill angle is identical to that of the center point.

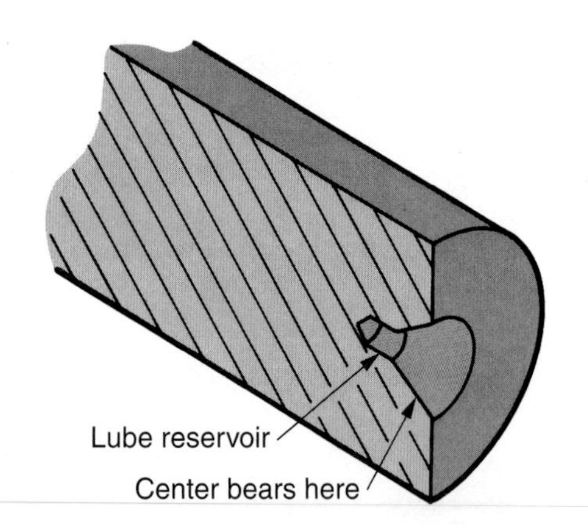

Goodheart-Willcox Publisher

Figure 14-21. The tailstock center rides in the drilled and countersunk center hole. If a dead center is used, a supply of lubricant is placed in the reservoir. The lubricant will expand and lubricate the center as the metals heat up.

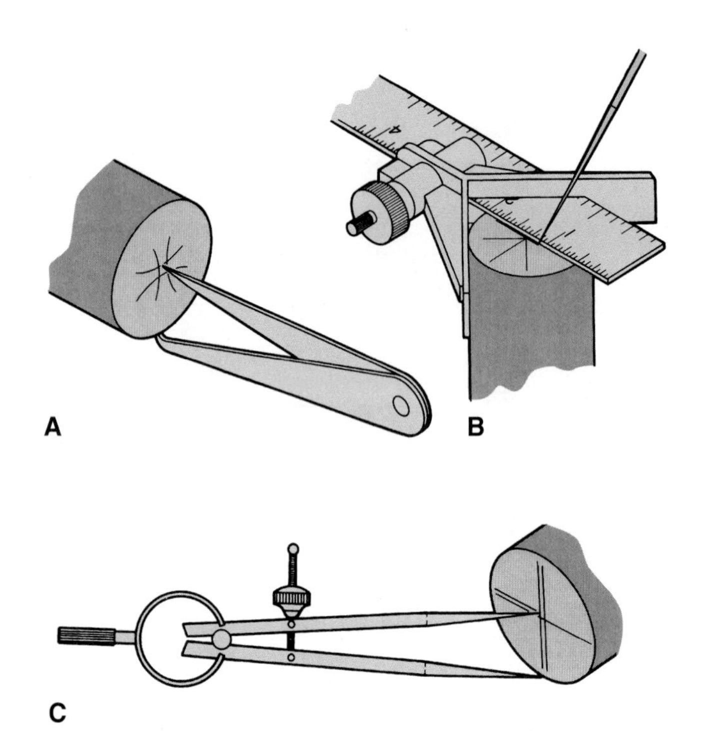

Goodheart-Willcox Publisher

Figure 14-22. Several ways to locate the center of round stock. A—With a hermaphrodite caliper. B—With center head and rule of a combination set (recommended method). C—With dividers.

The straight drill provides clearance for the center point and serves as a reservoir for a lubricant. The chart provides information needed to select the correct size center drill.

The center holes can be drilled on the lathe with the work centered in the chuck, on the lathe with the center drill held in the headstock, or on a drill press.

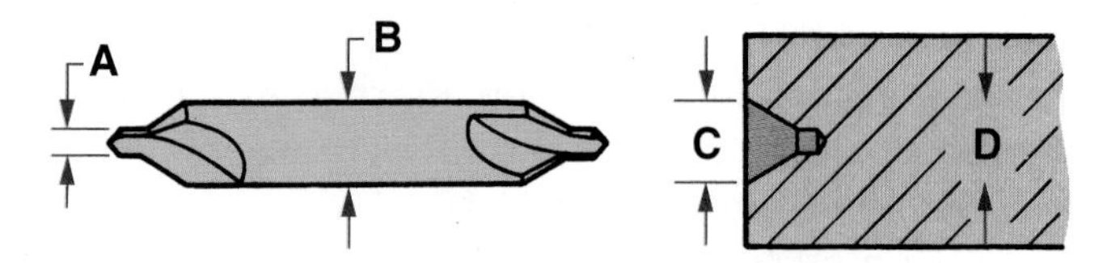

Combination drill and countersink no.	A	B	C	D
1	1/16	13/64	1/8	3/16 to 5/16
2	3/32	3/16	3/16	3/8 to 1
3	1/8	1/4	1/4	1 1/4 to 2
4	5/32	7/16	5/16	2 1/4 to 4

Goodheart-Willcox Publisher

Figure 14-23. This size chart contains information needed to select correct-size center drill. Combination drill and countersink makes the hole and countersinks it in one operation.

Some work can be held in a lathe chuck for center drilling. The work is centered in a lathe chuck mounted in the headstock. The center drill is fitted in a Jacobs chuck mounted to the tailstock.

Center holes can be drilled in large stock by mounting a Jacobs chuck in the headstock. Locate the center point of each end and center punch. Support one end on tail center and feed the other end into the center drill mounted in the Jacobs chuck. Repeat the operation on second end. See **Figure 14-24**.

When using a drill press, mount the work in a V-block for support. The center holes should be drilled deep enough to provide adequate support, **Figure 14-25**.

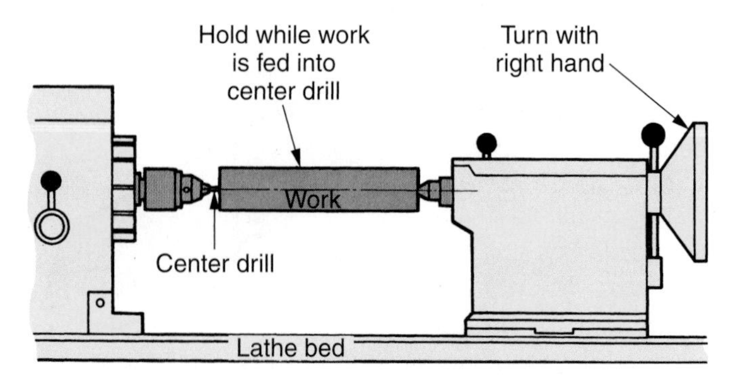

Goodheart-Willcox Publisher

Figure 14-24. Center holes can be drilled in large stock by mounting a Jacobs chuck in the headstock.

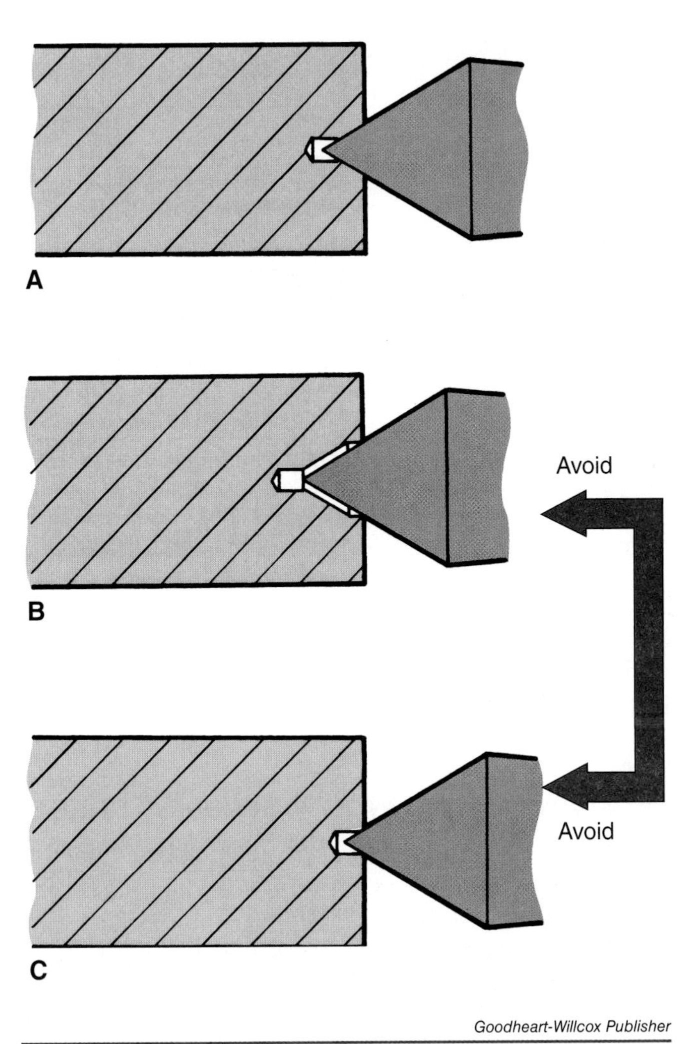

Figure 14-25. Correctly and incorrectly drilled center holes. A—Properly drilled center hole. B—Hole drilled too deep. C—Hole not drilled deep enough. Does not provide enough support. If used with a dead center, the center point will burn off.

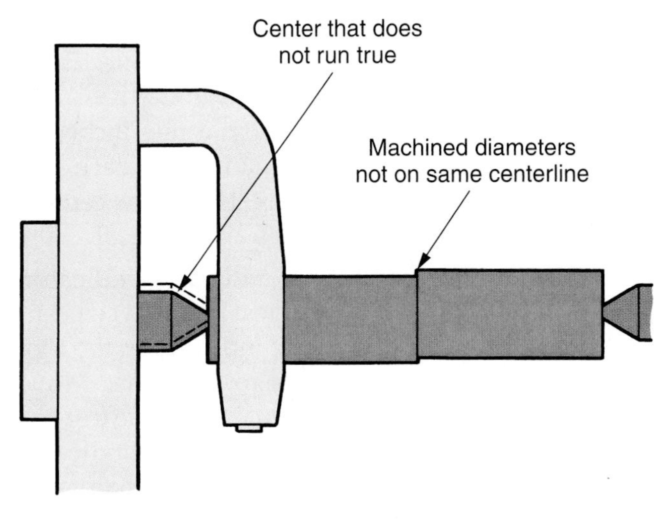

Figure 14-26. The workpiece must be reversed in the lathe dog so it can be machined for its entire length. If the live center does not run true, eccentric diameters will result.

14.4.2 Checking Center Alignment

Accurate turning between centers requires centers that run true and are in precise alignment. Since the work must be reversed to machine its entire length, care must be taken to make the live center run true. If the live center does not run true, the diameters will be eccentric (not aligned on the same center line), **Figure 14-26**. This can be prevented by truing the live center. If the center is not hardened, a light truing cut can be made. A tool post grinder will be needed to true a hardened center.

Approximate alignment can be determined by checking centers visually by bringing their points together, or by checking the witness lines on the base of the tailstock for alignment. See **Figure 14-27**.

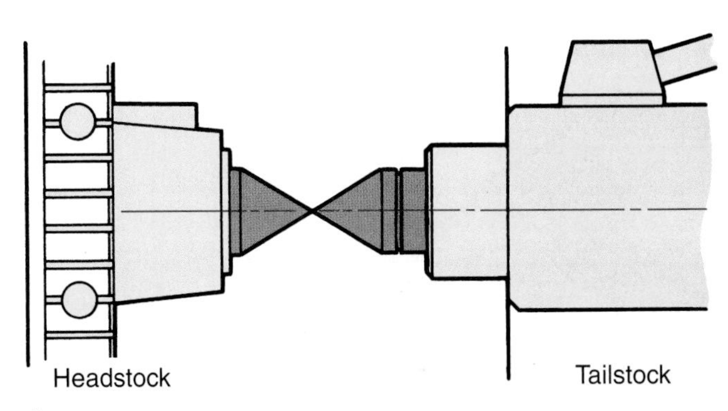

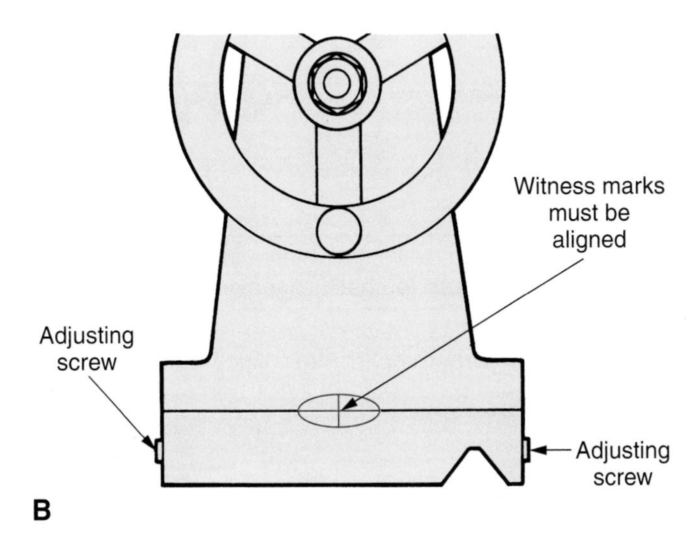

Figure 14-27. Checking center alignment. A—Checking alignment by bringing center points together. View is looking down on top of centers. B—Alignment can be determined by checking witness lines on base of tailstock.

A more precise method for checking alignment is needed if close tolerance work is to be done. Three such methods are as follows:

- Make a light trial cut across a few inches of the material. Check the diameter at each end with a micrometer, **Figure 14-28**. The centers are aligned if the readings are identical.
- Use a steel test bar and dial indicator, **Figure 14-29**. Mount the test bar between centers and position the dial indicator in the tool post and at right angles to the work. Move the indicator contact point against the test bar until a reading is shown. Move the indicator along the test bar. If the readings remain constant, the centers are aligned.
- Machine a section of scrap, **Figure 14-30**. Set the cross-feed screw to make a light cut at the right end of the piece. With the same tool setting, move it to the left and continue the cut. Identical micrometer readings indicate center alignment.

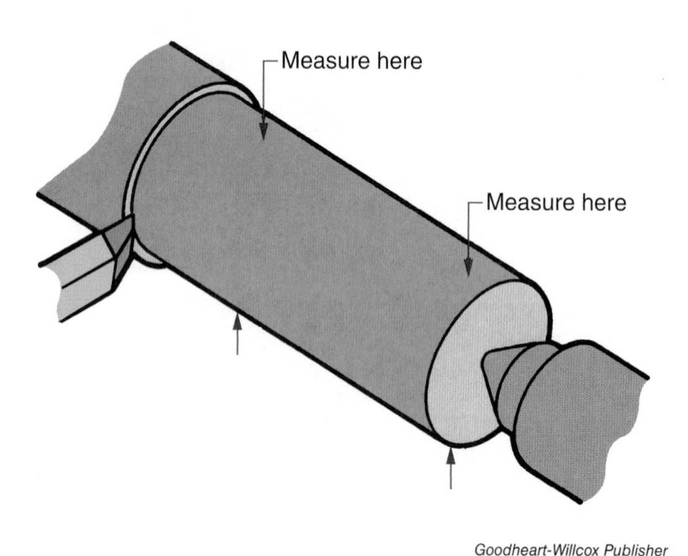

Goodheart-Willcox Publisher

Figure 14-28. Make a light cut on stock and measure diameter at two points to check alignment. Measurements must be equal.

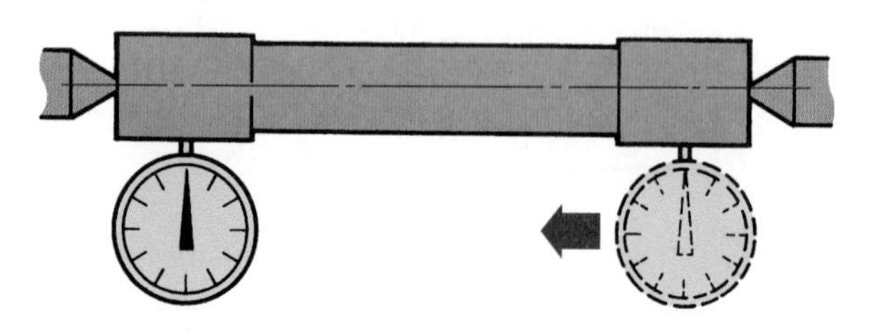

Goodheart-Willcox Publisher

Figure 14-29. Using a test bar and dial indicator to check center alignment.

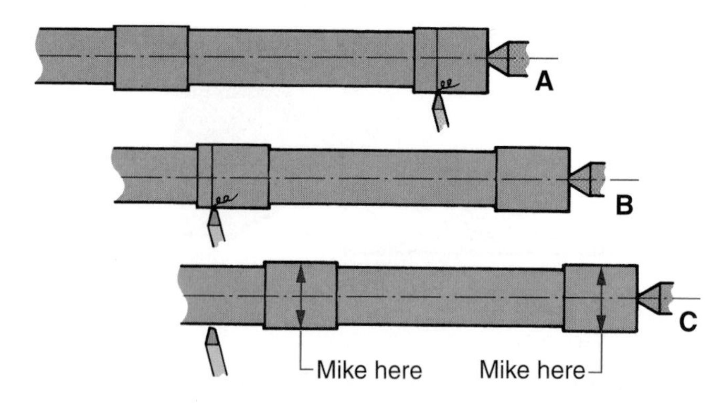

Goodheart-Willcox Publisher

Figure 14-30. Checking for center alignment. A—Machine two shoulders on a test piece. B—Keep same tool setting and make a cut on both shoulders. C—Measure resulting diameters.

Adjusting screws (one on each side at the base of the tailstock) are used to align the centers if checks indicate this is needed. Make adjustments gradually. See **Figure 14-31**.

14.4.3 Mounting Work between Centers

Clamp a dog to one end of the work. Place a lubricant (graphite and oil or a commercial center lubricant) in the center hole on the other end. Mount the piece on the centers and adjust the tailstock spindle until the work is snug. If the work is too loose, it will "clatter." If adjusted too tightly, it will score or burn the center point.

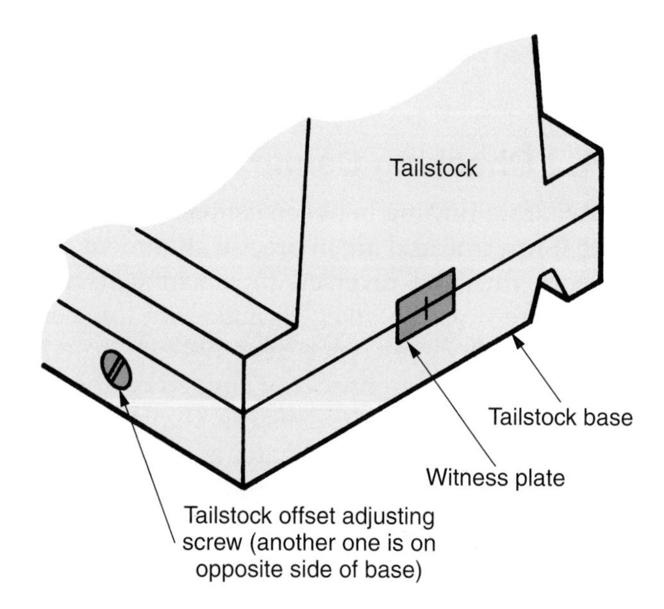

Goodheart-Willcox Publisher

Figure 14-31. Adjusting screws are located on both sides of tailstock base. Their primary use is to set over or shift the tailstock for taper turning.

Check the adjustment from time to time, since heat generated by the machining process will cause the work to expand. Using a live center, instead of a dead center, will reduce or eliminate many of the problems involved in working between centers.

Check to see if the dog tail binds on the faceplate slot, **Figure 14-32**. This can cause the work to be pulled off center. When machined, this will produce a surface that is not concentric with the center hole. If binding is occurring, use a different faceplate.

14.5 Using Lathe Chucks

The chuck is another device for holding work in a lathe. Chucking is the most rapid method of mounting work, and is widely preferred for that reason. Other operations, such as drilling, boring, reaming, and internal threading, can be done while the work is held in a chuck. Additional support can be obtained for the piece by supporting the free end with the tailstock center.

14.5.1 Three-Jaw Universal Chuck

The 3-jaw universal chuck is designed so that all jaws operate at the same time, **Figure 14-33**. It will automatically center round or hexagonal shaped stock.

Two sets of jaws are supplied with each universal chuck. One set is used to hold large-diameter work. The other set is used for small-diameter work, **Figure 14-34**.

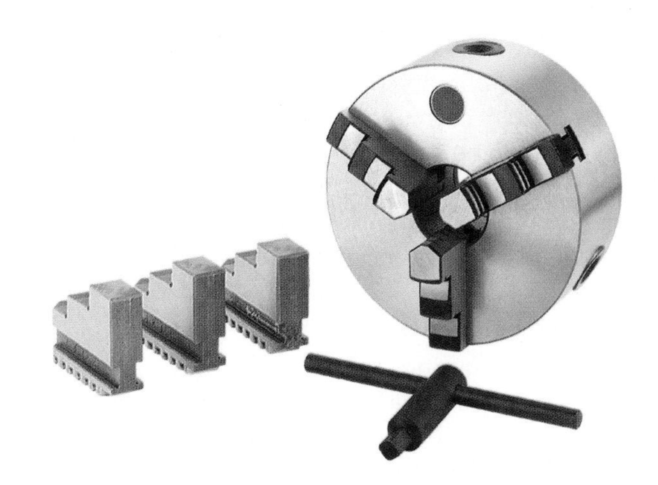

Photo courtesy of Grizzly Industrial, Inc. www.grizzly.com

Figure 14-33. The 3-jaw universal chuck automatically centers round or hexagonal stock.

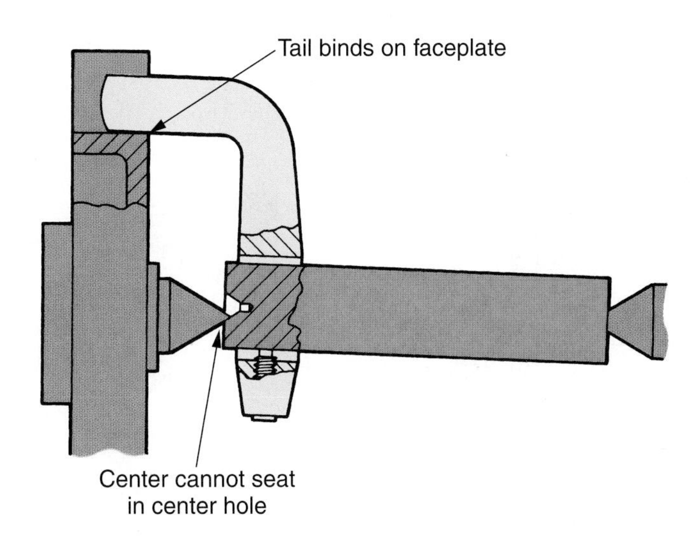

Goodheart-Willcox Publisher

Figure 14-32. Diameter of the turned surface will not be concentric with the center holes if center hole is not seated properly on the live center. A binding lathe dog is a common cause of this problem.

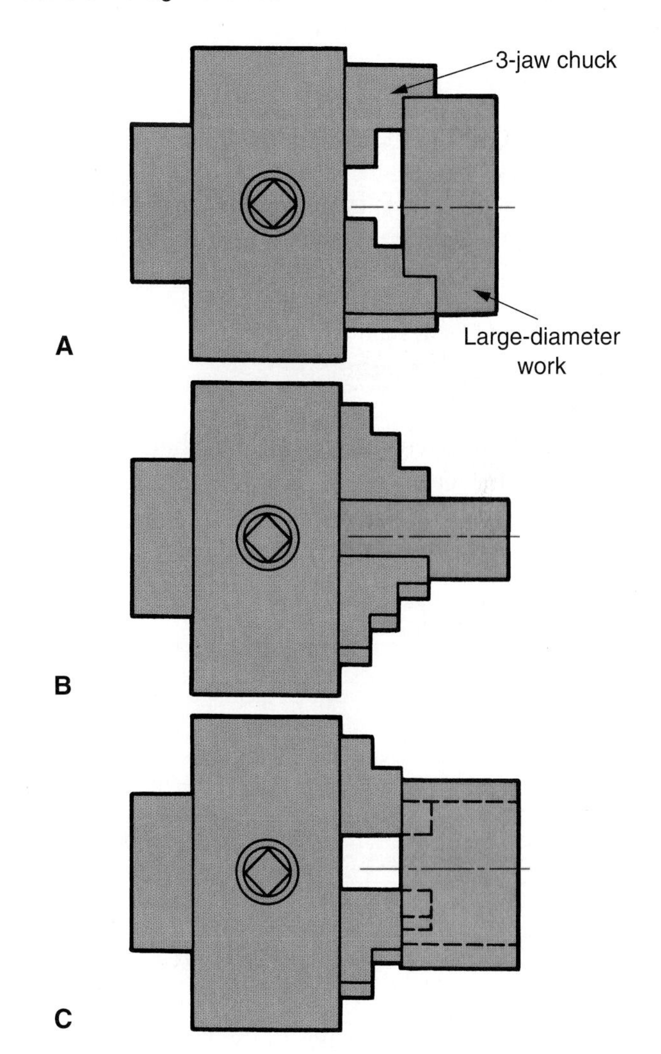

Goodheart-Willcox Publisher

Figure 14-34. Chuck jaws. A—One of the sets of jaws supplied with a 3-jaw universal chuck is used to mount large-diameter work. B—Holding work using the set of jaws supplied for smaller-size workpieces. C—Another method of mounting work in the chuck.

The jaws in each set are numbered 1, 2, and 3, as are the slots in which they are fitted. The jaw number must correspond with the slot number if the work is to be centered. Sets of jaws are made for a specific chuck and are not interchangeable with other chucks. Make sure the chuck and jaws have the same serial number!

14.5.2 Installing Chuck Jaws

Before installing jaws, clean the jaws, jaw slots, and scroll (spiral thread seen in the jaw slots). Turn the scroll until the first thread does not quite show in jaw slot 1. Slide the matching jaw into the slot as far as it will go. Now, turn the scroll until the spiral engages with the first tooth on the bottom of the jaw. Repeat the operation at slots 2 and 3, making sure the proper jaws are inserted.

SAFETY NOTE

Remove the chuck key when you finish using it. If the key is left in the chuck when the lathe is turned on, it could become a dangerous missile. Make it a habit to never let go of a lathe chuck key unless you are placing it on the tool tray or lathe board.

Jaws of a universal chuck lose their centering accuracy as the scroll wears. Accuracy is also affected when too much pressure is used to mount the work, or when work is gripped too near the front of the jaws.

SAFETY NOTE

Avoid gripping work near the front of the jaws. It can fly out and cause injuries.

14.5.3 Four-Jaw Independent Chuck

Each of the jaws of 4-jaw independent chuck, **Figure 14-35**, operates individually, instead of being coupled with the other jaws (as in the 3-jaw universal chuck). This permits square, rectangular, and odd-shaped work to be centered. Unlike those of the 3-jaw chuck, the jaws of a 4-jaw chuck can be removed from their slots and reversed. This reversing feature permits the jaws to be used to hold large-diameter work in one position and smaller-diameter work when reversed, **Figure 14-36**.

AMT—The Association for Manufacturing Technology

Figure 14-35. A 4-jaw independent chuck. The jaws on this type chuck are reversible.

Unlike the 3-jaw chuck, the 4-jaw type is not self-centering. The most accurate way to center round work in this type chuck is to use a dial indicator. The piece is first centered approximately, using the concentric rings on the chuck face as a guide. A dial indicator is then mounted in the tool post, **Figure 14-37**. The jaws are adjusted until the indicator needle does not fluctuate (move back and forth) when the work is rotated by hand. After the piece has been centered, all jaws must be tightened securely.

Another centering method uses chalk. Rotate the work while bringing the chalk into contact with it. The chalk mark indicates the "high point." Slightly loosen the jaws opposite the chalk mark.

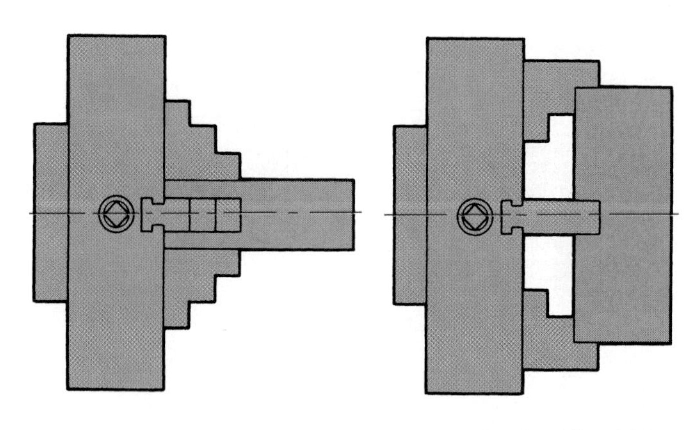

Goodheart-Willcox Publisher

Figure 14-36. Reversing feature of jaws in a 4-jaw independent chuck makes it possible to turn work having extreme differences in diameter without difficulty.

Goodheart-Willcox Publisher

Figure 14-37. Centering work in a 4-jaw chuck using a dial indicator. The machine shown is a small model maker's lathe.

Then tighten the jaws on the side where the chalk mark appears. Continue this operation until the work is centered. If the work is oversize enough, a cutting tool may be used instead of chalk.

Avoid trying to center stock in one or two adjustments, but rather, work in increments (very small steps). When making the final small adjustment, it may be necessary to loosen the jaw on the low side and retighten it, after which the high side is given a final tightening. This last method for making final adjustment applies, in particular, when centering work with a dial indicator.

14.5.4 Jacobs Chuck

When turning small-diameter work, such as screws or pins, the Jacobs chuck can be utilized. This chuck, **Figure 14-38**, is better suited for such work than the larger universal or independent chuck.

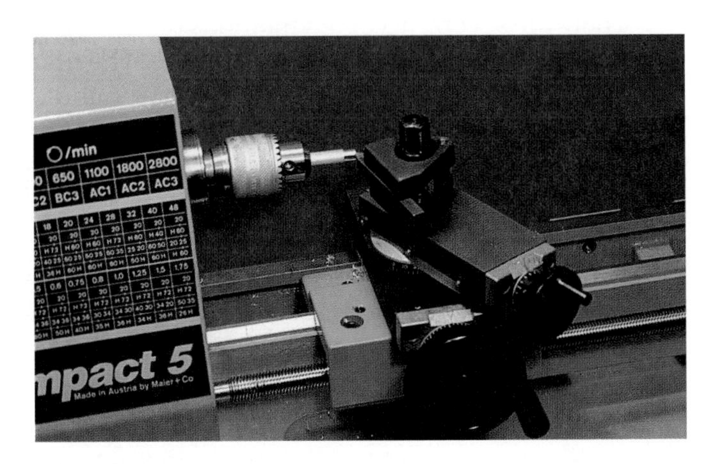

Goodheart-Willcox Publisher

Figure 14-38. Turning small-diameter work in a Jacobs chuck.

A standard Jacobs chuck is normally fitted in the tailstock for drilling. However, it also can be mounted by fitting it in a sleeve and then placing the unit in the headstock spindle. Wipe the chuck shank, sleeve, and spindle hole with a clean soft cloth before they are fitted together.

A headstock spindle Jacobs chuck is similar to the standard Jacobs chuck, but is designed to fit directly onto a threaded spindle nose, **Figure 14-39**. The chuck has the advantage of not interfering with the compound rest, making it possible to work very close to the chuck.

14.5.5 Draw-In Collet Chuck

The draw-in collet chuck is a work-holding device for securing work small enough to pass through the lathe spindle, **Figure 14-40**.

Royal Products, Division of Curran Manufacturing Corporation

Figure 14-39. This Jacobs chuck can be mounted on a lathe with a threaded spindle nose.

Royal Products, Division of Curran Manufacturing Corporation

Figure 14-40. A draw-in collet chuck.

Collets are accurately made sleeves with one end threaded and the other split into three even sections. The slots are cut slightly more than half the length of the collet and permit the jaws to spring in and clamp the work, **Figure 14-41**.

The standard collet has a circular hole for round stock, but collets for holding square, hexagonal, and octagonal material are available.

The chief advantages of collets are their ability to center work automatically and to maintain accuracy over long periods of hard usage. They have the disadvantage of being expensive, since a separate collet is needed for each different size or stock shape.

A collet chuck using steel segments bonded to rubber is also available. An advantage of this chuck is that each collet has a range of 0.100″ (2.5 mm), rather than being a single size, like steel collets. However, these collets are available only for round work.

14.5.6 Mounting and Removing Chucks

If a chuck is not installed on the spindle nose correctly, its accuracy will be affected. To install a chuck, remove the center and sleeve, if they are in place. Hold the center and sleeve with one hand and tap them loose with a knockout bar. Carefully wipe the spindle end clean of chips and dirt. Apply a few drops of spindle oil. Clean the portion of the chuck that fits on the spindle.

On a chuck that is fitted to a threaded spindle nose, clean the threads with a spring cleaner. See **Figure 14-42**.

Maswerks, Inc.

Figure 14-41. A variety of collets is necessary to clamp different stock sizes.

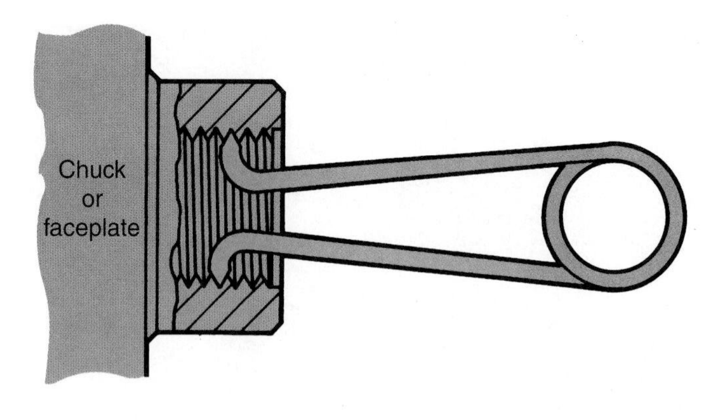

Goodheart-Willcox Publisher

Figure 14-42. A spring cleaner is used to clean the threads in a chuck before mounting it on a threaded spindle nose.

With the tapered key spindle nose, rotate the spindle until the key is in the up position. Slide the chuck into place and tighten the threaded ring. Pins on the cam-lock spindle nose are fitted into place and locked.

Fitting a chuck onto a threaded spindle nose requires a different technique. Hold the chuck against the spindle nose with the right hand and turn the spindle with the left hand. Screw the chuck on until it fits firmly against the shoulder. To avoid possible injury, do not spin the chuck on rapidly or use power. Release belt tension if possible, to eliminate any chance of power being transferred to the spindle.

During installation, place a board on the ways under the chuck to protect your hands and prevent damage to the machine ways if the chuck is dropped.

14.5.7 Removing a Chuck from Threaded Spindle

There are several accepted methods of removing chucks from a threaded headstock spindle. The first step in any method, regardless of the type of spindle nose, is to place a wooden lathe board across the ways beneath the chuck for support, **Figure 14-43A**. Then use one of the following techniques:

- Lock the spindle in back gear and use a chuck key to apply leverage.
- Place a suitable size adjustable wrench on one jaw and apply pressure to the wrench.
- If neither of the preceding methods works, place a block of hardwood between the rear lathe ways and one of the chuck jaws. Engage the back gear and give the drive pulley a quick rearward turn, **Figure 14-43B**.

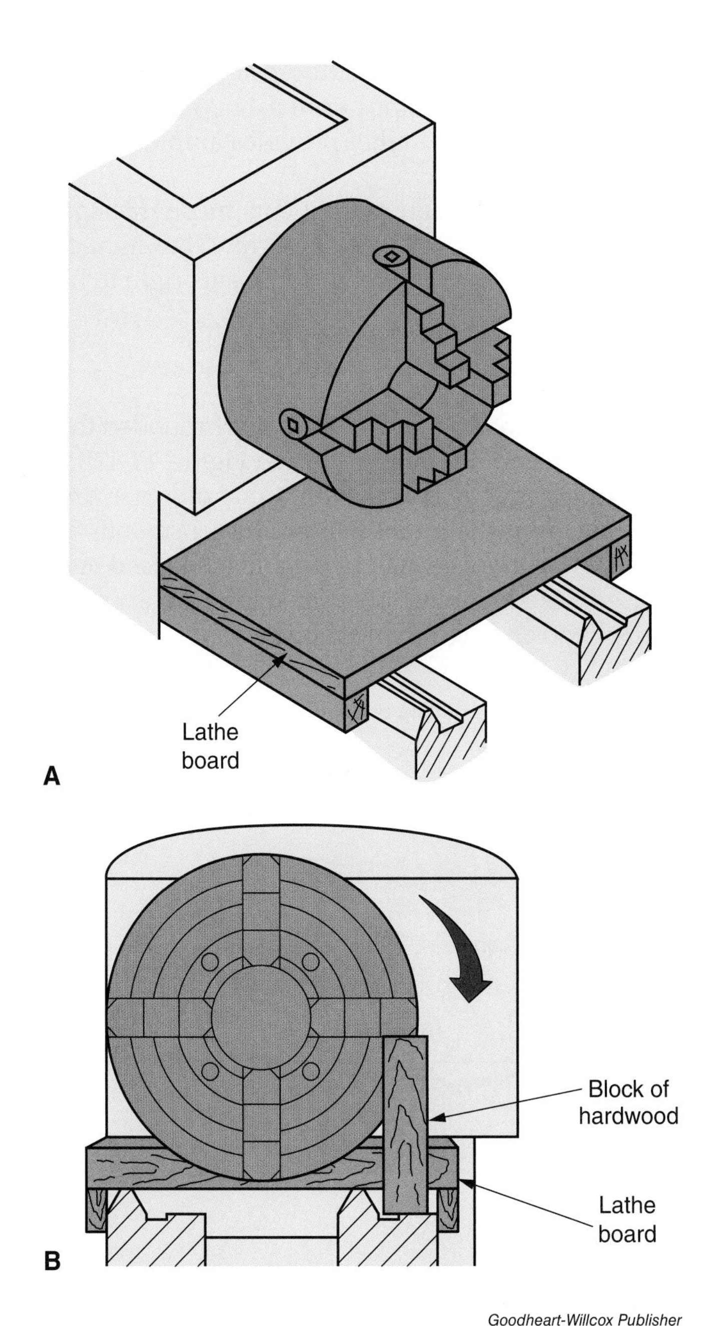

Goodheart-Willcox Publisher

Figure 14-43. Removing a chuck. A—A lathe board placed under a chuck when mounting or removing it will protect your hand should you drop the chuck. B—In truly stubborn cases, reversing the chuck against a block of hardwood is often used.

14.5.8 Removing a Chuck from Other Spindle Noses

Little difficulty should be encountered when removing a chuck from tapered and cam-lock spindle noses. For tapered spindle noses, first lock the spindle in back gear, then place the appropriate spanner wrench in the locking ring. Give it a tap or two with a leather or plastic mallet. Turn the ring until the chuck is released.

SAFETY NOTE

Place a wooden cradle under the chuck before attempting to remove it from the spindle. Removal will be easier and hand injuries will be avoided.

14.6 Cutting Tools and Toolholders

To operate a lathe efficiently, the machinist must have a thorough knowledge of cutting tools and know how they must be shaped to machine various materials. The cutting tool is held in contact with the revolving work to remove material from the work. Before the advent of indexable insert cutting tools, most applications were performed using a ***single-point cutting tool*** of high-speed steel (HSS).

The square cutter bit body is inserted in a lathe toolholder. Toolholders are made in straight, right-hand, and left-hand models. To tell the difference between right-hand and left-hand toolholders, hold the head of the tool in your hand and note the direction the shank points. The shank of the right-hand holder points to the right, the left-hand toolholders points to the left. A turret holder may also be utilized, **Figure 14-44**. Turret holders typically have four cutter bits. A bit can be changed by loosening the lock (handle) and pivoting the holder so the new bit is in cutting position, then locking it in place.

Garsya/Shutterstock.com

Figure 14-44. A turret holder with four cutting tools.

14.6.1 High-Speed Steel Cutting Tool Shapes

Figure 14-45 shows the parts of the cutter bit, and the correct terminology for those parts. To get best performance, the bit must have a keen, properly shaped cutting edge. The shape depends on the type of work, roughing or finishing, and on the metal to be machined.

Most cutter bits are ground to cut in one only direction (left or right). The exception is the round-nose tool, which can cut in either direction. Some cutting tools used for general purpose turning are shown in **Figure 14-46**.

14.6.2 Roughing Tools

The deep cuts made to remove considerable material from a workpiece are called ***roughing cuts***. Roughing tools have a tool shape (shape of cutting tip) that consists of a straight cutting edge with a small rounded nose. This shape permits deep cuts at heavy feeds. The slight side relief provides ample support to the cutting edge.

The left-cut roughing tool cuts most efficiently when it travels from left to right. The right-cut roughing tool operates just the opposite, right to left. See **Figure 14-47A**.

14.6.3 Finishing Tools

The nose of a finishing tool is more rounded than the nose of the roughing tool. See **Figure 14-47B**. If the cutting edge is honed with a fine oil stone after grinding, a finishing tool will produce a smooth finish on the workpiece. A light cut and a fine feed must be used. Like roughing tools, finishing tools are made in left-hand and right-hand models.

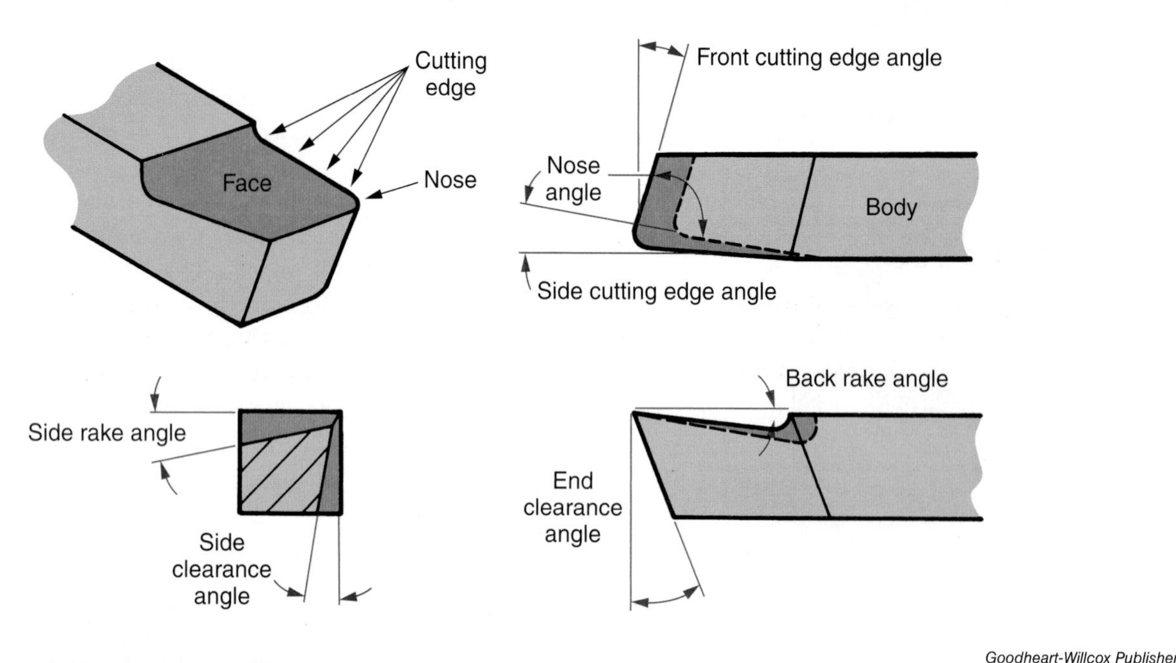

Goodheart-Willcox Publisher

Figure 14-45. Cutter bit nomenclature.

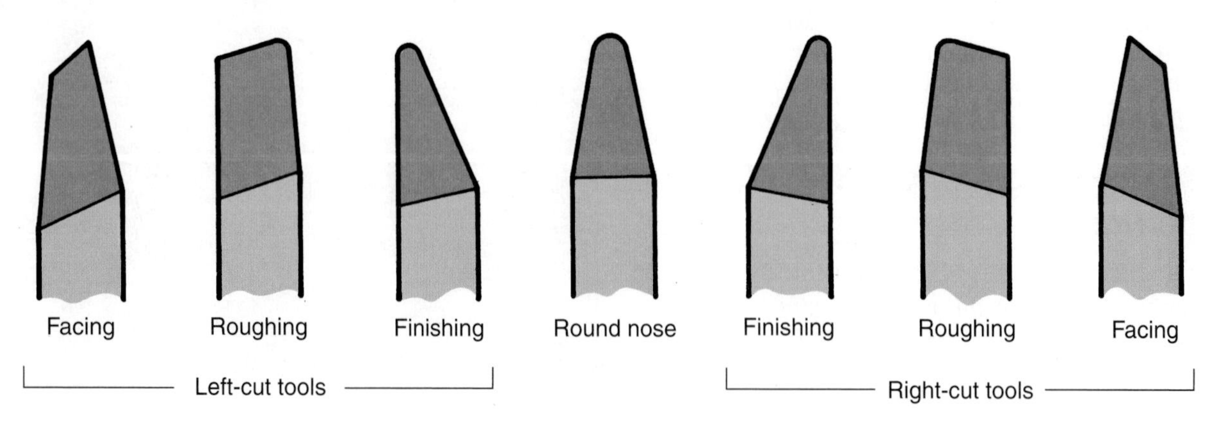

Goodheart-Willcox Publisher

Figure 14-46. Standard HSS cutting tool shapes.

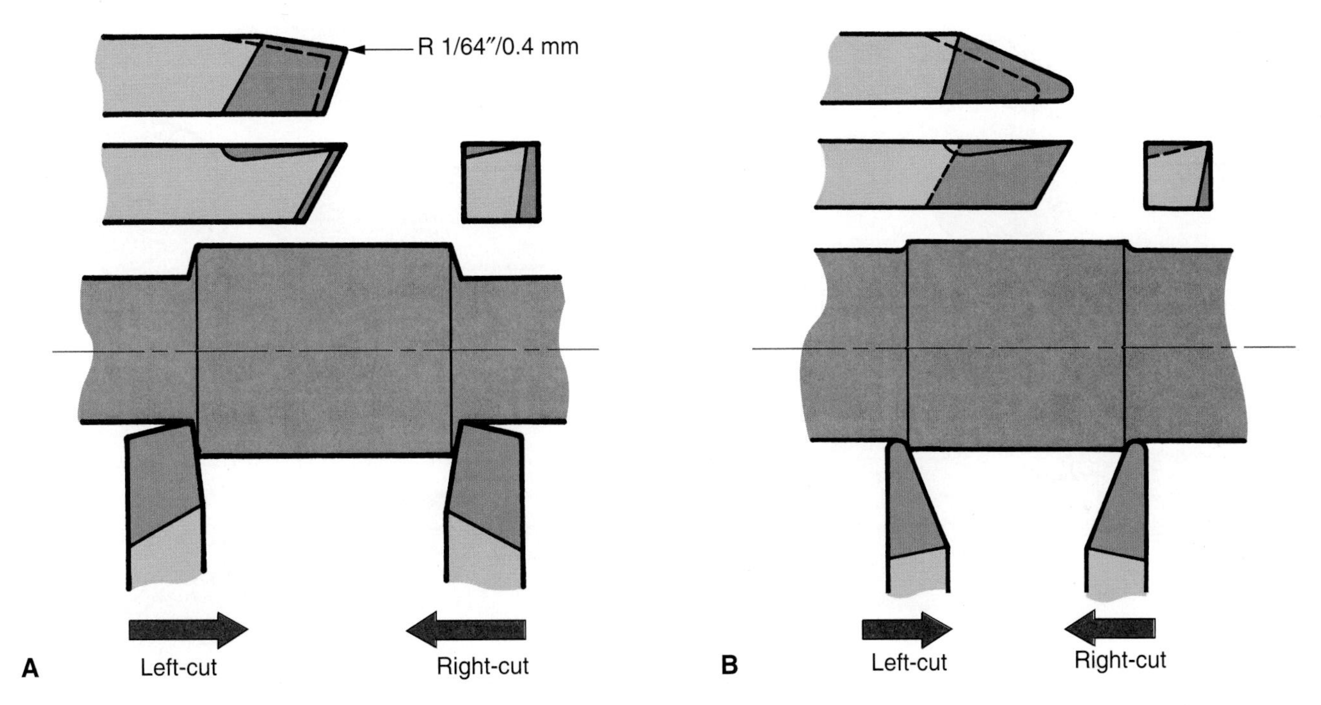

Figure 14-47. Lathe tools. A—The roughing tool is used for rapid material removal. B—The finishing tool will produce a smooth surface.

14.6.4 Facing Tool

The facing tool is ground to prevent interference with the tailstock center. The tool point is set at a slight angle to the work face with the point leading slightly. See **Figure 14-48A**.

14.6.5 Round-Nose Tool

A round-nose tool is designed for lighter turning and is ground flat on the face (without back or side rake) to permit cutting in either direction. See **Figure 14-48B**. A slight variation of the round-nose tool, with a negative rake ground on the face, is excellent for machining brass, **Figure 14-49**.

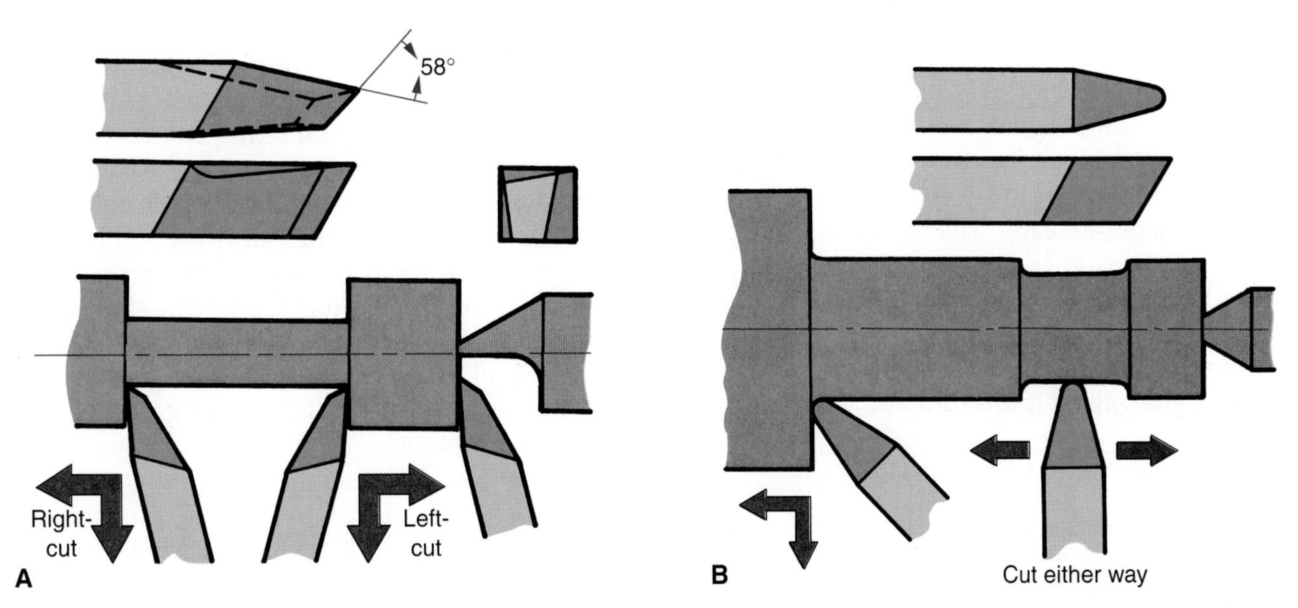

Figure 14-48. Lathe tools. A—A facing tool is used to machine surfaces perpendicular to the spindle centerline. B—A round-nose tool will produce fillets. Its shape permits it to cut either left or right.

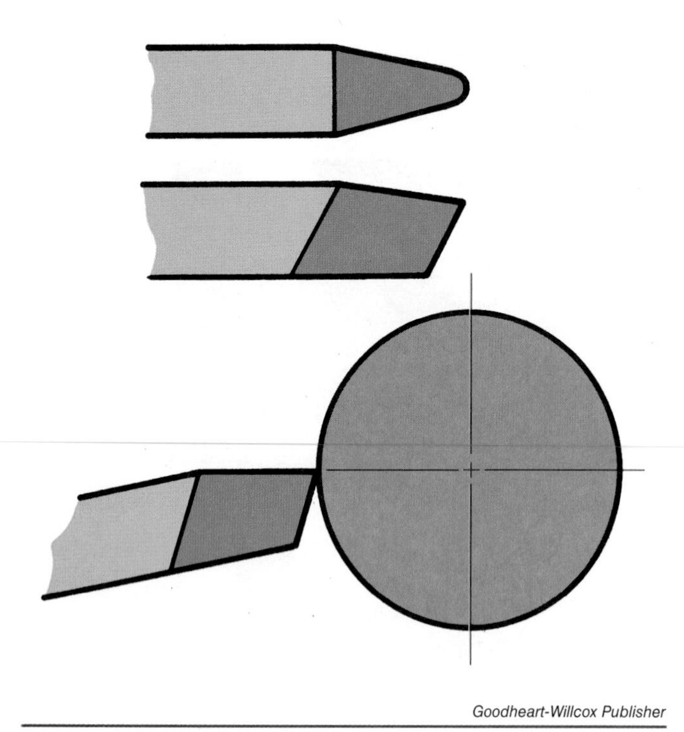

Goodheart-Willcox Publisher

Figure 14-49. Cutter nose shape with negative rake for machining brass.

Machining aluminum requires a tool with a considerably different shape from those previously described. As shown in **Figure 14-50**, the tool is set slightly above center to reduce any tendency to chatter (vibrate rapidly). The tool designs illustrated are typical of cutting tools used to machine aluminum alloys.

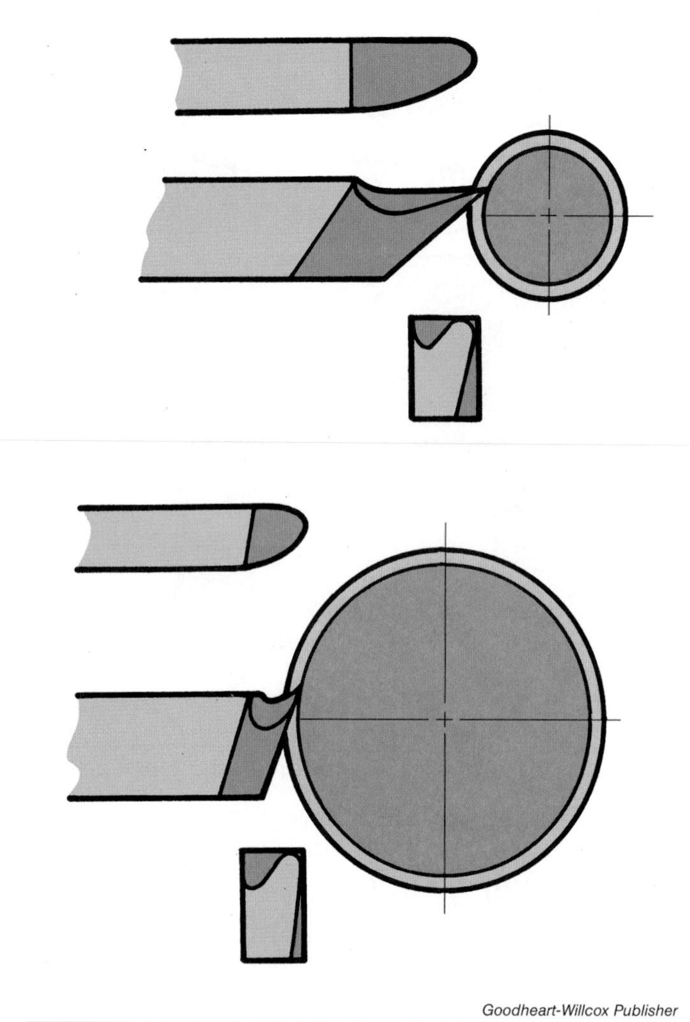

Goodheart-Willcox Publisher

Figure 14-50. For machining aluminum, cutter bit shapes different from those used for other metals are necessary. The tool is set slightly above center for a smoother cut.

14.6.6 Grinding High-Speed Cutter Bits

When first attempting to grind a cutter bit, it may be best if you first practice on sections of cold finished steel square stock. You may also want to use chalk or bluing and draw the desired tool shape on the front portion of the blank. The lines will serve as guides for grinding.

Figure 14-51 depicts the recommended grinding sequence for a cutter bit. Side clearance, top clearance, and end relief may be checked with a clearance and cutting angle gage, **Figure 14-52**.

14.6.7 Brazed-Tip Single-Point Cutting Tools

Brazed-tip single-point cutting tools are made by brazing a carbide cutting tip onto a shank made from less costly material, **Figure 14-53**. Many tip shapes (tool blanks) are available.

Cutting speeds can be increased by 300% to 400% when using carbide cutting tools. Powders of tungsten, carbon, and cobalt are molded into tool blanks and heated to extremely high temperatures. The hardness and strength of the blank can be controlled by varying the amount of cobalt that is used to cement (bind together) the tungsten and carbon particles.

For best results, these tools should be sharpened on a special silicon carbide or diamond-charged grinding wheel in which diamond dust particles or chips are embedded. A special type of grinder must be used, **Figure 14-54**.

Cutting tools designed for machining steel are chamfered 0.002″ to 0.003″ (0.050 to 0.075 mm) by honing them lightly with a silicon carbide or diamond hone. If the tools are not honed, the irregular edge produced by grinding will crumble when used. Honing, if done properly, does not interfere with the cutting action.

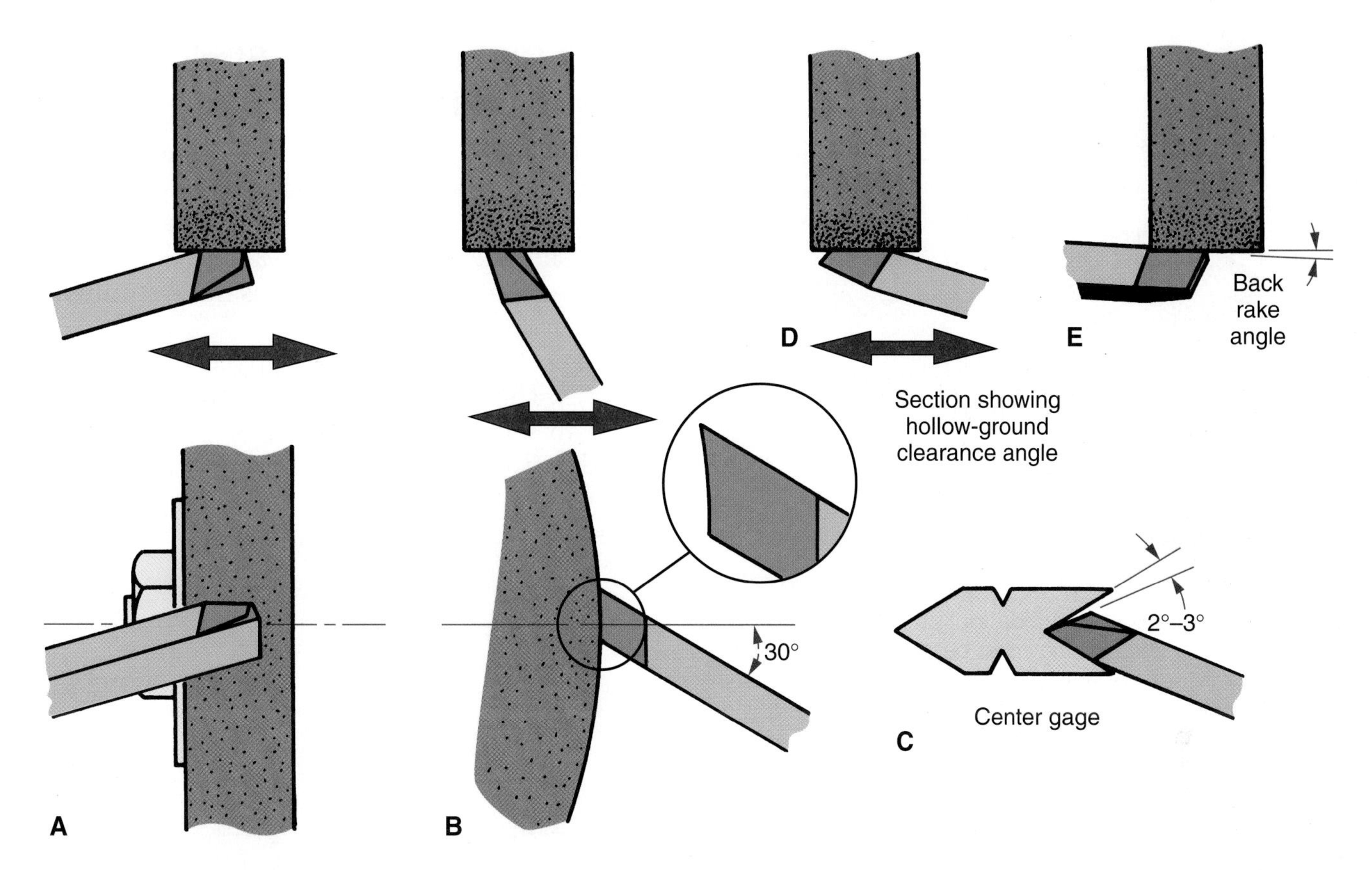

Goodheart-Willcox Publisher

Figure 14-51. Grinding sequence for a cutter bit. A—Two views showing how to position a cutter bit blank on the grinding wheel to shape side clearance angle and side cutting edge angle. B—Shaping end clearance angle and front cutting edge angle. C—Center gage being used to check nose angle. D—Grinding other side clearance angle, when required. E—Grinding back/side rake angles. Accuracy of clearance angles can be checked with cutter bit gage.

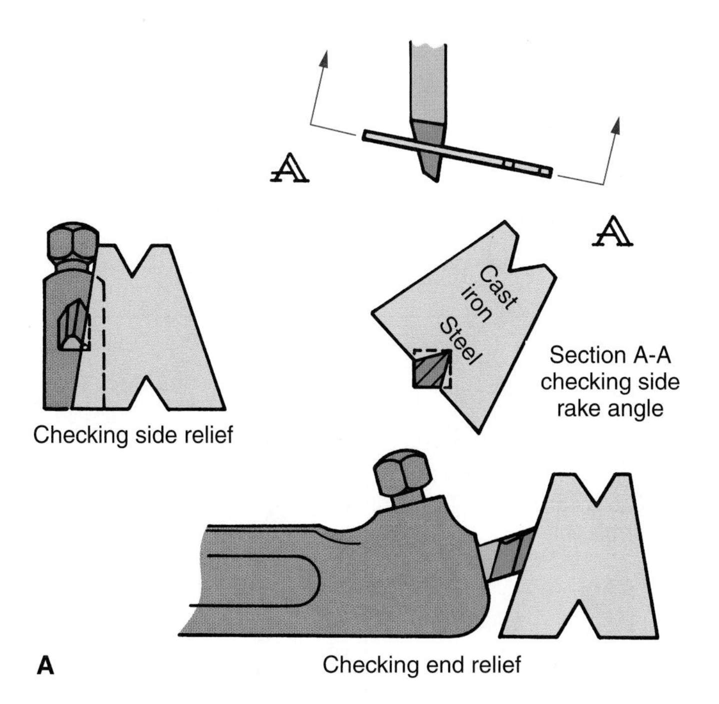

Rake and Clearance Angle for Lathe Tools (High-speed steel)

Cast iron		Low-carbon steel	High-carbon steel
Back rake	6–8°	8–12°	4–6°
Side rake	10–12°	14–18°	8–10°
Clearance*	6–9°	8–10°	6–8°
Alloy steels		**Soft brass**	**Aluminum**
Back rake	5–8°	0–2°	25–50°
Side rake	10–15°	0–2°	10–20°
Clearance*	6–8°	10–15°	7–10°
Copper			
Back rake	10–12°		
Side rake	20–25°		
Clearance*	6–8°		

*The end and side clearance angles are usually the same.

B

Goodheart-Willcox Publisher

Figure 14-52. Cutter bit gage. A—Bit gage being used to check accuracy after grinding cutter tip. B—This table provides rake and clearance angles for lathe tools to machine different metals.

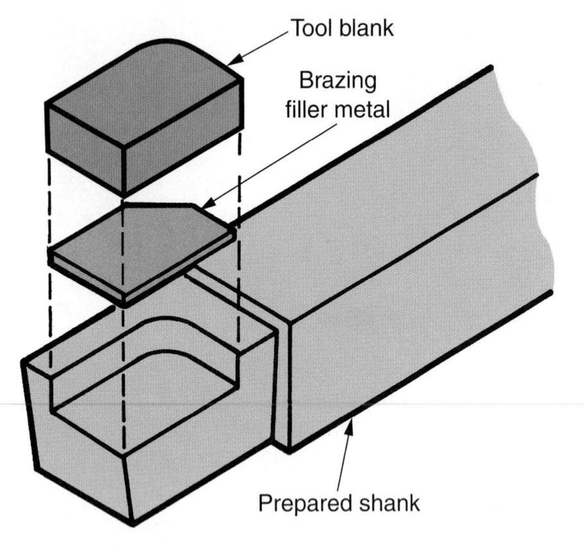

Goodheart-Willcox Publisher

Figure 14-53. Brazing a carbide tool blank into place on a prepared shank. The brazing must be done properly or the tool blank will not be solidly attached, causing it to wear rapidly. Tungsten carbide tool blanks are available in a wide selection of shapes, sizes, and degrees of hardness.

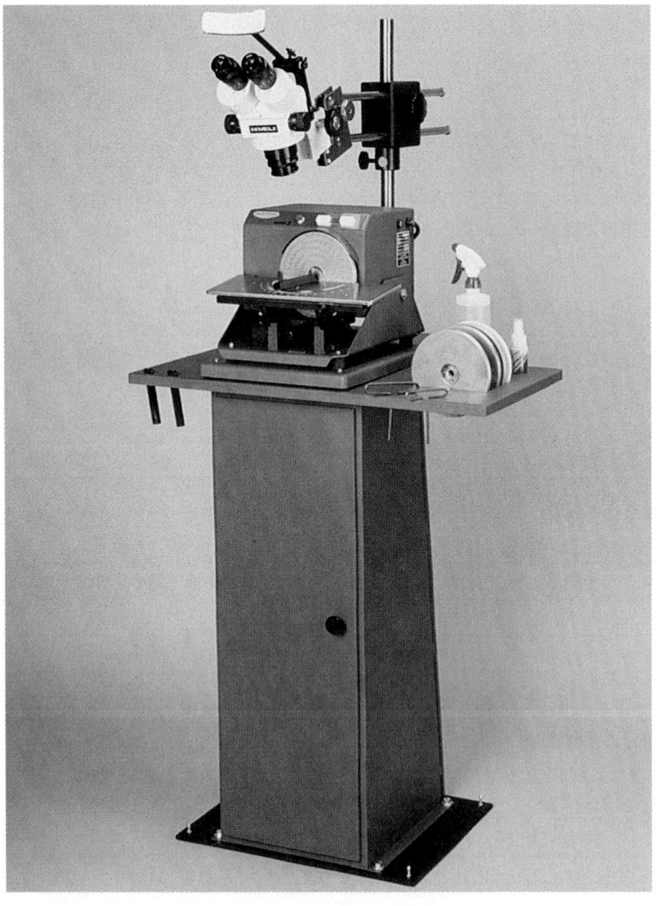

Goodheart-Willcox Publisher

Figure 14-54. A grinder designed for sharpening carbide, cermet, cubic boron nitride, and polycrystalline diamond (PCD) cutting tools. It uses special diamond charged wheels and is fitted with a microscope inspection system.

14.6.8 Carbide-Tipped Straight Turning Tools

The cutting tools shown in **Figure 14-55** are general purpose tools for facing, turning, and boring. The square nose shape permits machining to a square shoulder. The clearance angles of the carbide tools described are not as great as those required for high-speed steel cutting tools.

Also shown is a carbide-tipped threading tool (Style E). This tool has a 60° included angle that conforms to the Unified National 60° included-angle thread. It is used for V-grooving and chamfering.

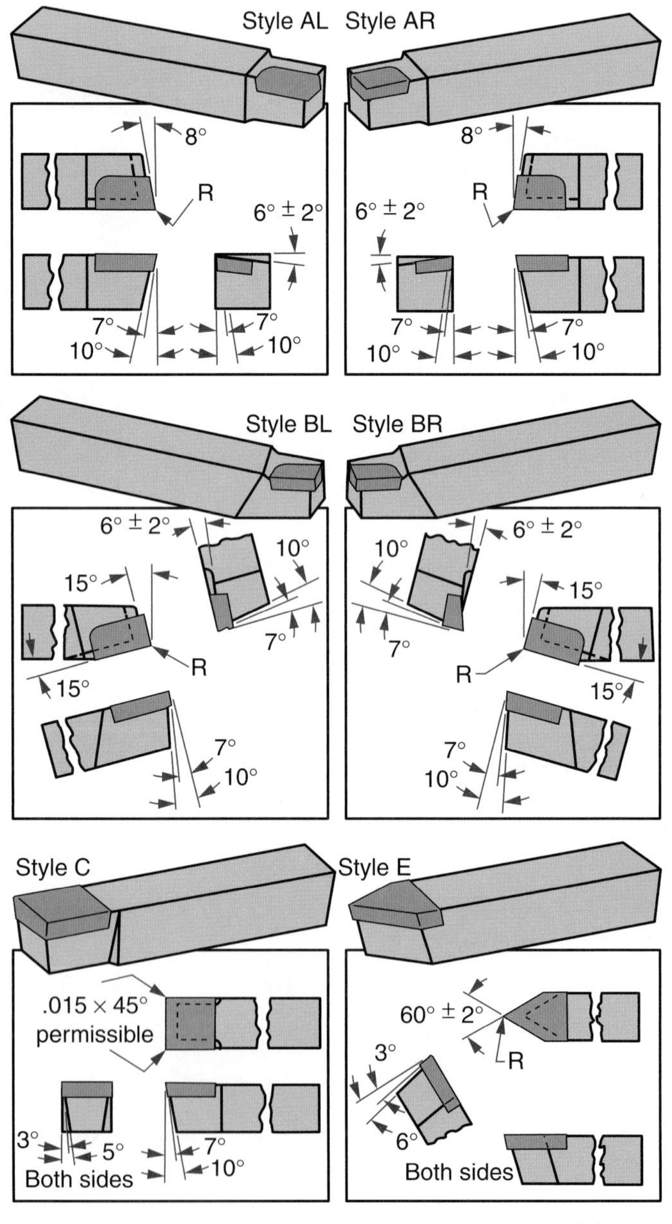

Carboloy, Inc.

Figure 14-55. Typical standard cemented-carbide single-point tools. Style E is a carbide-tipped threading tool.

14.6.9 Indexable Insert Cutting Tools

Brazed-tip cutting tools have almost completely been replaced by mechanically clamped ***indexable insert cutting tools***, **Figure 14-56**. Indexable insert cutting tools are widely used for both turning and milling operations. The inserts are manufactured in a number of shapes and sizes, **Figure 14-57**, for different turning geometries. Six of the most commonly used standard shapes are shown in order of increasing and decreasing strengths. See **Figure 14-58**.

Indexable inserts clamp to special toolholders, **Figure 14-59**. As an edge dulls, the next edge is rotated into position until all edges are dulled. Since it is less costly to replace inserts made from some materials than to resharpen them, they are usually discarded after use.

Inserts are manufactured from a number of materials, with each designed for a different metal requirement. See **Figure 14-60**. Carbide inserts are given increased versatility (higher abrasion resistance, chemical stability, and lubricity) when coated with various combinations of titanium carbide (TiC), titanium nitride (TiN), and alumina.

Sandvik Coromant Co.

Figure 14-56. Indexable insert cutting tools of carbide or sintered oxides (often referred to as cermets) are mechanically clamped into toolholders to perform cutting tasks. This insert is being used to machine stainless steel.

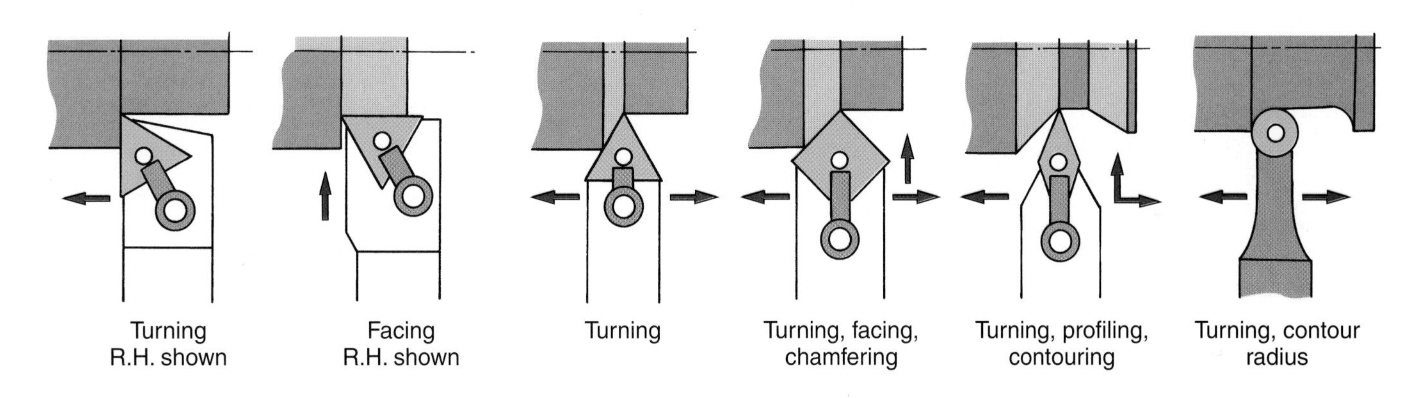

Goodheart-Willcox Publisher

Figure 14-57. Indexable inserts are manufactured in a number of different shapes and sizes for different turning operations.

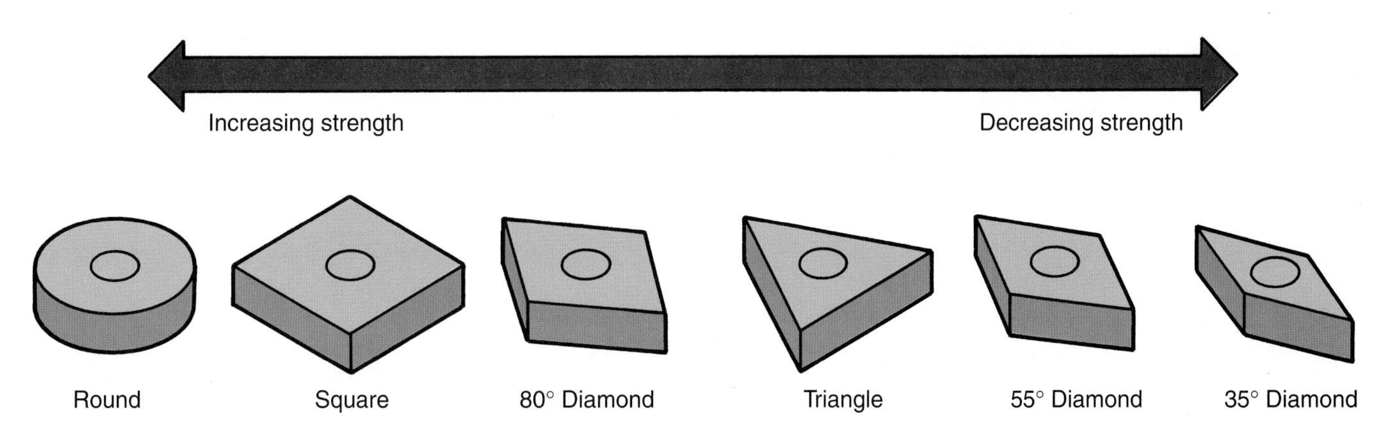

Goodheart-Willcox Publisher

Figure 14-58. Most commonly used indexable insert shapes are shown in order of increasing and decreasing strengths.

Carboloy, Inc.

Figure 14-59. A selection of typical holders and replaceable carbide insert cutting tools for the lathe. Each insert has three or four cutting tips. The inserts are clamped in place on the holder, and can be indexed (rotated into position) to present a new tip when the one in use becomes dull.

14.6.10 Chipbreakers

When some metals are machined, long continuous chips will be created, unless some method is used to break the chips into smaller pieces. This is accomplished by a small step or groove, called a ***chipbreaker***, which is located on the top of the cutter at the cutting edge. Most inserts manufactured today have molded-in chipbreakers. Other single-point cutting tools must have a chipbreaker ground into the top face of the tool, **Figure 14-61**.

14.6.11 Other Types of Cutting Tools

Diamonds, both natural and manufactured, are used as single-point cutting tools on materials whose hardness or abrasive qualities make them difficult to machine with other types of cutting tools. These diamonds are known as industrial diamonds.

14.7 Cutting Speeds and Feeds

The matter of cutting speed and feed is most important, since these factors govern the length of time required to machine the work and the quality of the surface finish.

Material	Strengths	Weaknesses	Typical Applications
HSS	Superior resistance Versatility	Poor speed capabilities Poor wear resistance	Screw machine and other low-speed operations, interrupted cuts, low-horsepower machining.
Carbide	Most versatile cutting material High shock resistance	Limited speed capabilities	Finishing to heavy roughing of most materials, including irons, steels, exotics, and plastics.
Coated Carbide	High versatility High shock resistance Good performance at moderate speeds	Limited to moderate speeds	Same as carbide, except with higher speed capabilities.
Cermet	High versatility Good performance at moderate speeds	Low shock resistance Limited to moderate speeds	Finishing operations on irons, steels, stainless steels, and aluminum alloys.
Ceramic- Hot/ Cold Pressed	High abrasion resistance High-speed capabilities Versatility	Low mechanical shock resistance Low thermal shock resistance	Steel mill-roll resurfacing, finishing operations on cast irons and steels.
Ceramic- Silicon Nitride	High shock resistance Good abrasion resistance	Very limited applications	Roughing and finishing operations on cast irons.
Ceramic- Whisker	High shock resistance High thermal shock resistance	Limited versatility	High-speed roughing and finishing of hardened steels, chilled cast iron, high-nickel superalloys.
Cubic Boron Nitride	High hot hardness High strength High thermal shock resistance	Limited performance on materials below 38Rc Limited applications High cost	Hardened work materials in 45-70 Rockwell C range.
Poly-Crystalline Diamond	High abrasion resistance High-speed capabilities	Limited applications Low mechanical shock resistance	Roughing and finishing operations on abrasive nonferrous or nonmetallic materials.

Valenite, Inc.

Figure 14-60. The nine basic categories of cutting tool materials.

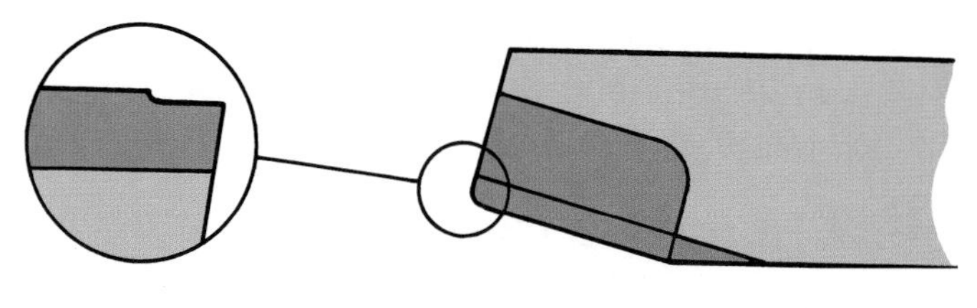

Goodheart-Willcox Publisher

Figure 14-61. Typical chipbreaker on a single-point tool.

Cutting speed indicates the distance the work moves past the cutting tool, expressed in feet per minute (fpm) or meters per minute (mpm). Measuring is done on the circumference of the work.

To explain this differently: if a lathe were to cut one long chip, the length of the chip cut in one minute (measured in either feet or meters) would be the cutting speed of the lathe. Cutting speed is not the revolutions per minute (rpm) of the lathe.

Feed is the distance that the cutter moves lengthwise along the lathe bed during a single revolution of the work.

There are a number of factors that must be considered when determining the correct cutting speed and amount of feed, including the following:

- Material used for the cutting tool.
- Kind of material being machined.
- Desired finish.
- Condition of the lathe.
- Rigidity of the workpiece.
- Kind of coolant being used (if any).
- Shape of the material being machined.
- Depth of cut.

If the machining is done with a cutting speed that is too slow, extra time will be needed to complete the job. If speed is too high, the cutting tool will dull rapidly and the finish will be substandard.

A speed and feed chart takes into consideration the many factors listed earlier. **Figure 14-62** is a chart for use with high-speed steel cutter bits. Cutting speeds and feeds on the chart can be increased by 50% if a coolant is used and by 300% to 400% if a cemented carbide cutting tool is used.

14.7.1 Calculating Cutting Speeds

The cutting speeds shown in **Figure 14-62** should be considered as only a starting point. Depending on the condition of the machine, they may have to be increased or decreased until optimum cutting conditions are obtained.

Material to be cut	Roughing cut 0.01″ to 0.020″ 0.25 mm to 0.50 mm feed		Finishing cut 0.001″ to 0.010″ 0.025 mm to 0.25 mm feed	
	fpm	mpm	fpm	mpm
Cast iron	70	20	120	36
Steel				
Low carbon	130	40	160	56
Med carbon	90	27	100	30
High carbon	50	15	65	20
Tool steel (annealed)	50	15	65	20
Brass—yellow	160	56	220	67
Bronze	90	27	100	30
Aluminum*	600	183	1000	300

The speeds for rough turning are offered as a starting point. It should be all the machine and work will withstand. The finishing feed depends on the finish quality desired.

* The speeds for turning aluminum will vary greatly according to the alloy being machined. The softer alloys can be turned at speeds upward of 1600 fpm (488 mpm) roughing to 3500 fpm (106 mpm) finishing. High silicon alloys require a lower cutting speed.

Goodheart-Willcox Publisher

Figure 14-62. Cutting speeds and feeds suggested for turning various metals with high-speed steel tools.

Cutting speed (CS) is given in feet per minute (fpm) or meters per minute (mpm). Speed of the work (spindle speed) is given in revolutions per minute (rpm). Thus, the peripheral speed (speed at the circumference or outside edge of the work) must be converted to rpm to determine the required spindle speed. The following formulas are used:

Inch-based.

$$rpm = \frac{CS \times 4}{D}$$

Where:

rpm = Revolutions per minute.

CS = Cutting speed recommended for the material being machined (steel, aluminum, etc.) in feet per minute.

D = Diameter of the work in inches. Convert all fractions to decimals.

Cutting speed problem. What spindle speed is required to finish-turn 4″ diameter aluminum alloy?

CS = Table recommends a cutting speed of 1000 fpm for finish-turning aluminum alloy.

D = 4″

$$rpm = \frac{CS \times 4}{D}$$

$$= \frac{1000 \times 4}{4}$$

$$= 1000 \text{ rpm}$$

Adjust the spindle speed to as close to this speed (1000 rpm) as possible. Increase or decrease speed as needed to obtain desired surface finish.

Metric-based.

$$rpm = \frac{CS \times 1000}{D \times \pi}$$

Where:

rpm = Revolutions per minute.

CS = Cutting speed recommended for material being machined (steel, aluminum, etc.) in meters per minute (mpm).

D = Diameter of work in millimeters (mm).

π = 3 (Since cutting speeds are approximate, π has been rounded off to 3 from 3.1416 to simplify calculations.)

Cutting speed problem. What spindle speed is required to finish-turn 100 mm diameter aluminum alloy?

CS = Table recommends a cutting speed of 300 mpm for finish-turning aluminum alloy.

D = 100 mm

π = 3

$$rpm = \frac{CS \times 1000}{D \times 3}$$

$$rpm = \frac{300 \times 1000}{100 \times 3}$$

$$= 1000 \text{ rpm}$$

Adjust the spindle to as close to this speed (1000 rpm) as possible. Increase or decrease speed as needed to obtain desired surface finish.

14.7.2 Roughing Cuts

Roughing cuts are taken to reduce work diameter to approximate size. The work is left 1/32″ (0.08 mm) oversize for finish turning. Since the finish obtained on the roughing cut is of little importance, use the highest speed and coarsest feed consistent with safety and accuracy.

14.7.3 Finishing Cuts

The finishing cut brings the work to the required diameter and surface finish. A high-spindle speed, sharp cutting tool, and fine feed are used.

14.7.4 Depth of Cut

The ***depth of cut*** refers to the distance the cutter is fed into the work surface. The depth of cut, like feed, varies greatly with lathe condition, material hardness, speed, feed, amount of material to be removed, and whether it is to be a roughing or finishing cut.

Depth of the cut can be set accurately with the micrometer dials on the cross-slide and compound rest, **Figure 14-63**.

The micrometer dial is usually graduated in 0.001″ or 0.02 mm increments. This means that a movement of one graduation feeds the cutting tool into the piece 0.001″ or 0.02 mm. However, material is removed around the periphery (outside edge) of

Ed Phillips/Shutterstock.com

Figure 14-63. Graduated micrometer dials on the cross-slide and compound rest handwheels of a lathe are used to set the depth of cut.

the rotating work at double the depth adjustment. For each 0.001″ (0.02 mm) of infeed, for example, the workpiece diameter is reduced by 0.002″ or 0.04 mm. See **Figure 14-64**. This must not be forgotten or twice as much material as specified will be removed.

Some lathes, however, have a micrometer dial set up so that the number of graduations the cutter is fed into the work will equal the amount that the work diameter will be reduced. That is, if the cutter is fed in 0.005″ (0.10 mm) or 5 graduations, the work diameter will be reduced 0.005″ (0.10 mm). Check the lathe you will be using to be sure which system it uses.

A common mistake when using a lathe is to remove too little material at too slow a speed. Cuts as deep as 0.125″ (3 mm) can be handled by light lathes. Cuts of 0.250″ (6 mm) and deeper can be made by heavier machines without overtaxing the lathe.

14.8 Preparing Lathe for Operation

Before an aircraft is permitted to take off, the pilot and crew must go through a checkout procedure to determine whether the engines, controls, and safety features are in first-class operating condition. The same applies to the operation of a machine tool such as a lathe. The operator should inspect the machine for safe and proper operation. The "checkout procedure" for the lathe should include the following actions:

- Clean and lubricate the machine. Use lubricant types and grades specified by the manufacturer. Many recommend a specific lubricating sequence to reduce any possibility of missing a vital lubrication point.
- Be sure all guards are in position and locked in place.
- Turn the spindle over by hand to be sure it is not locked nor engaged in back gear (unless you intend to use back gear).
- Move the carriage along the ways. There should be no binding.
- Check cross-slide movement. If there is too much play, adjust the gibs.
- Mount the desired work-holding attachment. Clean the spindle nose with a soft brush. A threaded nose spindle should have a drop of lubricating oil applied before the chuck or faceplate is attached.
- Adjust the drive mechanism for the desired speed and feed.
- If the tailstock is used, check it for proper alignment, **Figure 14-65**.
- Clamp the cutter bit into an appropriate toolholder and mount it in the tool post. Do not permit excessive compound rest overhang, since this often causes tool "chatter" and results in a poorly machined surface, **Figure 14-66**.
- Mount the work. Check for adequate clearance between the work and the various machine parts.

SAFETY NOTE

Sleeves should be rolled up and all jewelry removed before beginning to use the lathe.

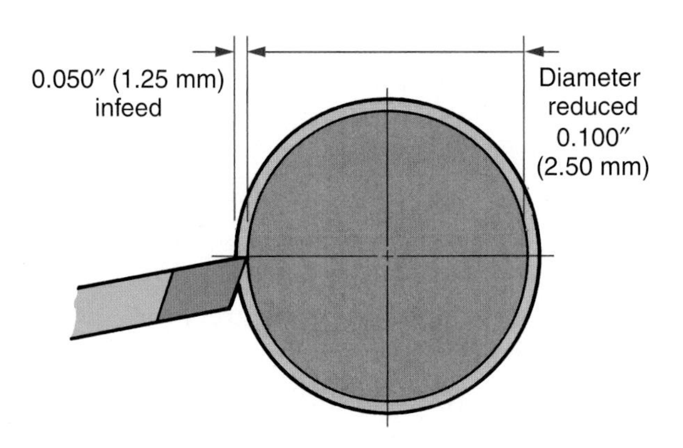

Goodheart-Willcox Publisher

Figure 14-64. Material is removed from the work on each cut at two times the infeed distance.

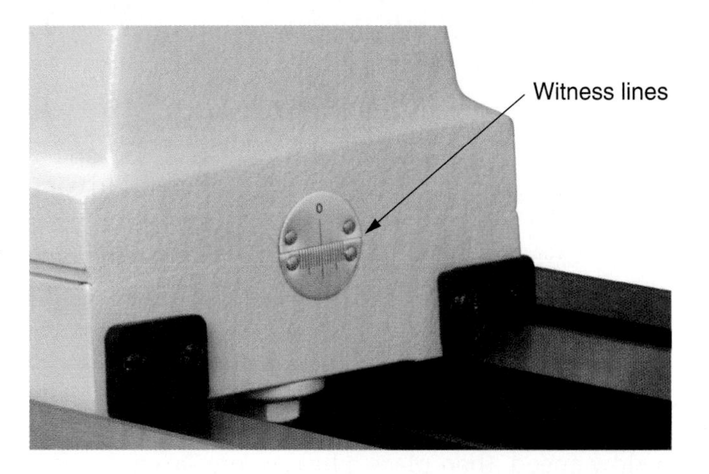

Photo courtesy of Grizzly Industrial, Inc. www.grizzly.com

Figure 14-65. Witness lines on tailstock indicate whether the tailstock is aligned properly with the headstock.

Goodheart-Willcox Publisher

Figure 14-66. Excessive overhang of the compound rest usually causes tool "chatter," resulting in a surface that is poorly machined.

Loose tools must never be placed on the lathe ways or carriage. A lathe board will aid in organizing and holding the tools and measuring instruments needed for the job. Some lathes come pre-equipped with a lathe tray built into the headstock.

14.9 Cleaning the Lathe

To maintain the accuracy built into a lathe, it must be thoroughly cleaned after each work period. Use a 2″ paintbrush (not a dust brush) to remove the accumulated chips.

SAFETY NOTE

Lathe chips are sharp. Do not remove them with your hands. Never use an air hose to remove chips. The flying particles could injure you or others.

Wipe all painted surfaces with a soft cloth. To complete the job, move the tailstock to the extreme right end of the ways. Use a soft cloth to remove any remaining chips, oil, and dirt from the machined surfaces.

To prevent rust until the next time the machine is used, apply a light coating of machine oil to all machined surfaces. The lead screw occasionally needs cleaning. To do so, adjust the screw to rotate at a slow speed, then place a heavy cord around it and start the machine. With the lead screw revolving, permit the cord to feed along the thread. Hold the cord just tightly enough to remove the accumulated dirt.

SAFETY NOTE

Never wrap the cord around your hand. The cord could catch and cause serious injury.

14.10 Lathe Safety

- Do not attempt to operate a lathe until you know the proper procedures and have been checked out on its safe operation by your instructor.
- Never attempt to operate a lathe while your senses are impaired by medication or other substances.
- Dress appropriately! Remove any necklaces or other dangling jewelry, wristwatch, or rings. Secure any loose-fitting clothing and roll up long sleeves. Wear an apron or a properly fitted shop coat. Safety glasses are a must!
- Clamp all work solidly. Use the correct size tool and work-holding device for the job. Get help when handling large sections of metal and heavy chucks and attachments.
- Check work frequently when it is being machined between centers. A workpiece expands as it heats up from friction and could damage the tailstock center.
- Be sure all guards are in place before attempting to operate the machine. Never attempt to defeat or bypass a safety switch.
- Turn the faceplate or chuck by hand to be sure there is no binding or danger of the work striking any part of the lathe.
- Keep the machine clear of tools, and always stop the machine before making measurements and adjustments.
- Metal chips are sharp and can cause severe cuts. Do not try to remove them with your hands when they become "stringy" and build up on the tool post, **Figure 14-67**. Stop the machine and remove them with pliers.
- Do not permit small-diameter work to project too far from the chuck without support from the tailstock. Without support, the work will be tapered, or worse, spring up over the cutting tool and could break. See **Figure 14-68**.

movit/Shutterstock.com

Figure 14-67. To avoid injury, always remove stringy chips with pliers. Never use your hands.

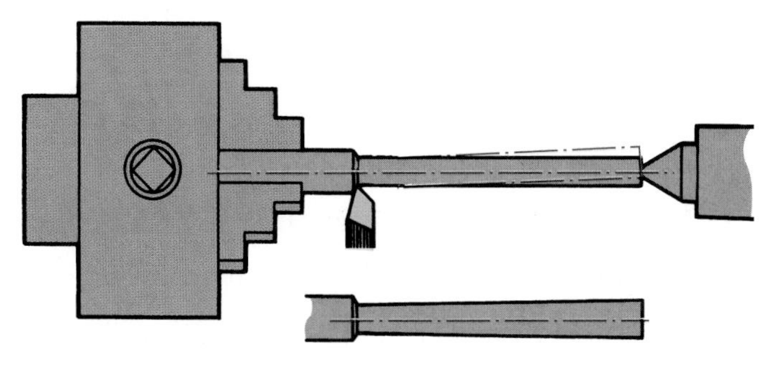

Goodheart-Willcox Publisher

Figure 14-68. If a small-diameter workpiece is not properly supported by a tailstock center, it will spring away from cutting tool and be machined on a slight taper.

- Do not run the cutting tool into the chuck or dog. Check any readjustment of the work or tool to make sure there is ample clearance when the cutter has been moved leftward to the farthest point that will be machined.
- Stop the machine before attempting to wipe down its surface, so the cloth does not become caught on rotating parts. When knurling, keep the coolant brush clear of the work.
- Before repositioning or removing work from the lathe, move the cutting tool clear of the work area. This will prevent accidental cuts on your hands and arms from the cutter bit.
- Avoid talking to anyone while running a lathe! Do not permit anyone to fool around with the machine while you are operating it. You are the only one who should turn the machine on or off, or make any adjustments.
- If the lathe has a threaded spindle nose, never attempt to run the chuck on or off the spindle using power. It is also dangerous practice to stop such a lathe by reversing the direction of rotation. The chuck could spin off and cause serious injury to you.
- Before engaging the half-nuts or automatic feed, you should always be aware of the direction of travel and speed of the carriage.
- Always remove the key from the chuck. Make it a habit to never let go of the key until it is out of the chuck and clear of the work area.
- Tools must not be placed on the lathe ways. Use a tool board or place them on the lathe tray.
- When doing filing on a lathe, make sure the file has a securely fitting handle.
- If any odd sounding noise or vibration develops during lathe operation, stop the machine immediately. If you cannot locate the trouble, get help from your instructor. Do not operate the machine until the trouble has been corrected.
- Remove sharp edges and burrs from the workpiece before dismounting it from the machine. Burrs and sharp edges can cause painful cuts.
- Use care when cleaning the lathe. Chips sometimes stick in recesses. Remove them with a paintbrush or wooden stick, not a dust brush. Never clean a machine tool with compressed air.

14.11 Facing Operations

Facing is an operation that machines the end of the work square and reduces it to a specific length. Facing is often the first operation performed in order to clean up the face of the workpiece before other machining operations are performed.

It is standard machining practice to cut stock slightly longer than needed. A steel rule may be used if the dimension is not critical. For more accuracy, a vernier caliper or large micrometer may be used. The difference between the rough length and the required length is the amount of material that must be removed.

14.11.1 Facing Work Held between Centers

Facing can be performed with the work mounted between centers. At times, considerable material must be removed. In this situation, it is best to leave the work longer than finished size and drill deeper center holes for better support during the roughing operation.

Face the work to length before starting the finish cut. A right-cut facing tool will be needed. The 58° point on this tool provides a slight clearance between the center point and the work face, **Figure 14-69**. Be careful not to damage the cutting tool point by running it into the center. A half center makes the operation easier, but is used only for facing. A half center does not provide an adequate bearing surface for general work and will not hold lubricant.

Set the compound rest at 30°, **Figure 14-70**. Bring the cutting tool up until it just touches the surface to be machined, and then lock the carriage. Remove material from each end of the stock until the specified length is attained.

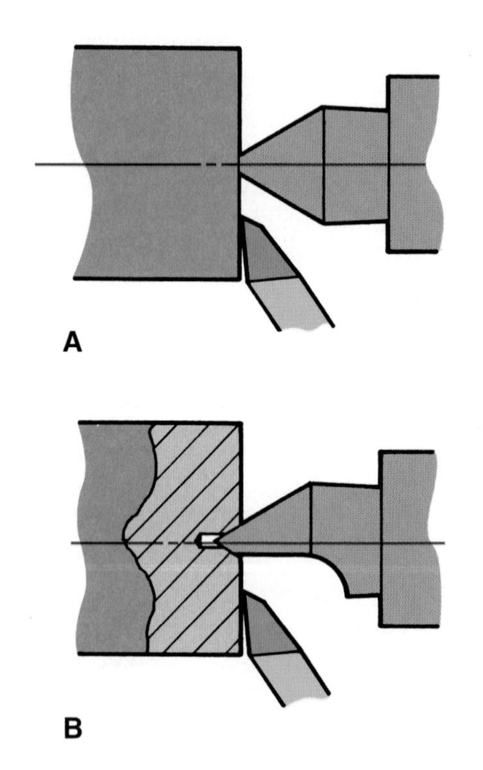

Goodheart-Willcox Publisher

Figure 14-69. Facing stock. A—Relationship of cutter bit to work face when making a facing cut. B—Using a half center will give more clearance when facing end of stock.

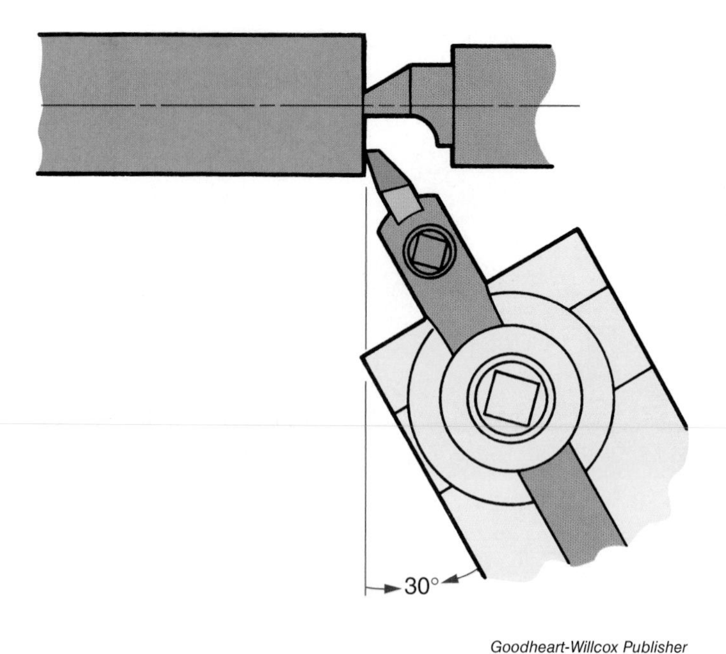

Goodheart-Willcox Publisher

Figure 14-70. Recommended compound rest setting when facing stock to length.

14.11.2 Facing Stock Held in a Chuck

A round-nose cutting tool, held in a straight toolholder, is used to face stock held in a chuck. The compound rest is pivoted 30° to the right. The toolholder is set to less than 90° to face the work, and the cutter bit is exactly on center. The carriage is then moved into position and locked to the way. See **Figure 14-71A**.

A facing cut can be made in either direction. The tool may be started in the center and fed out, or the reverse may be done. The usual practice is to start from the center and feed outward. If the material is over 1 1/2″ (38 mm) in diameter, automatic feed may be used.

With the cutting tool on center, a smooth face will result from the cut. A rounded "nubbin" (remaining piece of unmachined face material) will result if the tool is slightly above center, **Figure 14-71B**. A square-shoulder "nubbin" indicates that the cutter is below center, **Figure 14-71C**. Reposition the tool and repeat the operation if either condition is seen.

14.12 Turning Operations

Turning is a machining operation that reduces the outside diameter of the work. Turning is one of the most common machining operations performed on a lathe. Turning operations can be performed with the work held between centers or in a chuck.

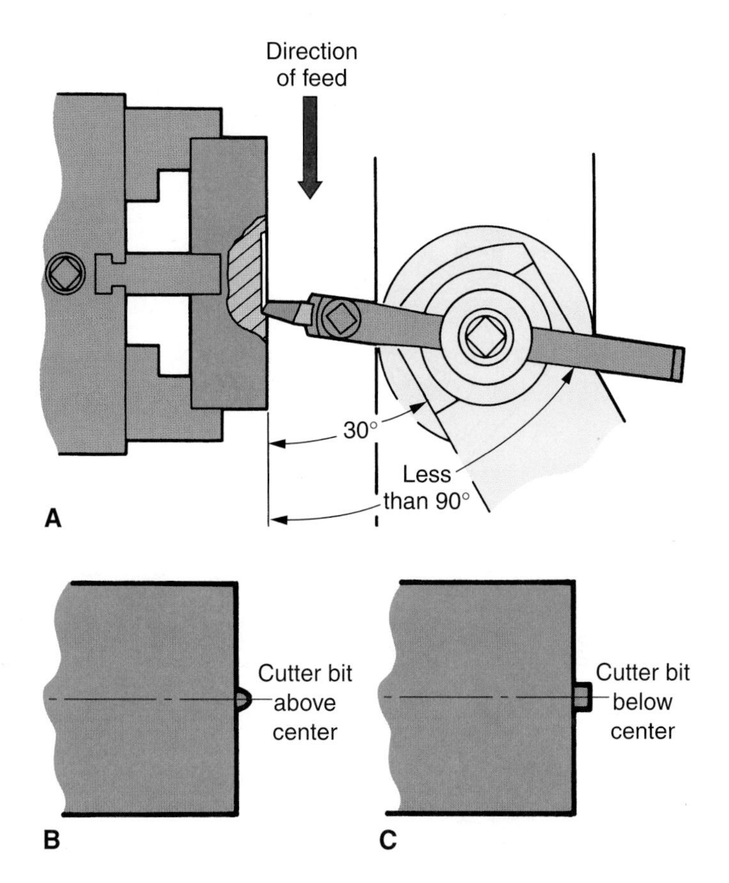

Goodheart-Willcox Publisher

Figure 14-71. Facing in a chuck. A—Correct tool and toolholder positions for facing. B—Rounded nubbin left by above-center cutter. C—Square-shoulder nubbin left by below-center cutter.

14.12.1 Rough Turning between Centers

Rough turning is an operation in which excess material is cut away rapidly with little regard for the quality of surface finish. The diameter is reduced to within 1/32″ (0.8 mm) of required size by using deep cuts and coarse feeds.

Set the compound rest at 30° from a right angle to the work, **Figure 14-72**. This will permit the tool to cut as close as possible to the left end of the work without the dog striking the compound rest.

SAFETY NOTE

Caution: Always check the maximum distance that compound rest can be fed toward the dog or chuck without striking them before you start the lathe.

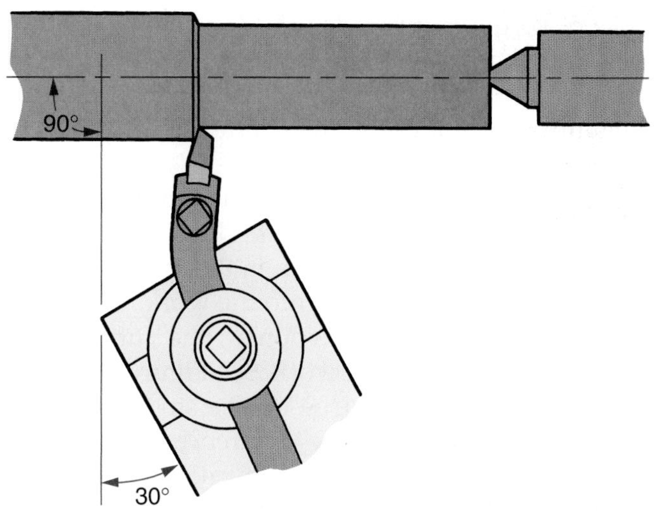

Goodheart-Willcox Publisher

Figure 14-72. Compound rest setting used for rough turning.

Use a left-hand toolholder. Position the tool post as far to the left as possible in the compound rest T-slot. Avoid excessive tool overhang, **Figure 14-73**.

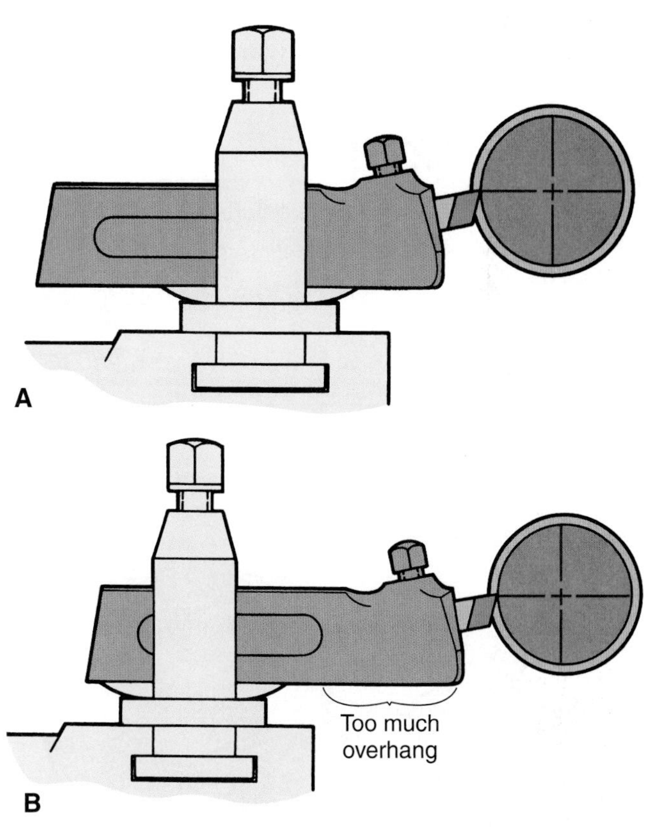

Goodheart-Willcox Publisher

Figure 14-73. Correct and incorrect mounting of toolholder. A—Toolholder and cutter bit in proper position. B—Too much overhang will cause the tool to "chatter" and produce a rough machined surface.

Locate the cutting edge of the tool about 1/16″ (1.5 mm) above work center for each inch of diameter, **Figure 14-74**. It can be set by comparing it with the tail center point or with an index line scribed on the tailstock ram of some lathes.

The toolholder must be positioned correctly. If it is not, the heavy side pressure developed during machining will cause it to turn in the tool post, forcing the cutting tool deeper into the work. When the toolholder is correctly positioned, the cutting tool will pivot away from the work. See **Figure 14-75**.

Make a trial cut to true up the stock. Measure the resulting diameter. The difference between the diameter and the required rough diameter is twice the distance the tool must be fed into the work. If the piece is greatly oversize, it will be necessary to make two or more cuts to bring it to size.

When depth of cut has been determined, engage the power feed. Observe the condition of the chips. They should be in small sections and slightly blue in color. Long, stringy chips indicate a cutting tool that is not properly sharpened. Stop the machine and remove stringy chips with pliers. Replace tool with one that is properly sharpened.

After each cut, measure work diameter to prevent excess metal removal.

SAFETY NOTE

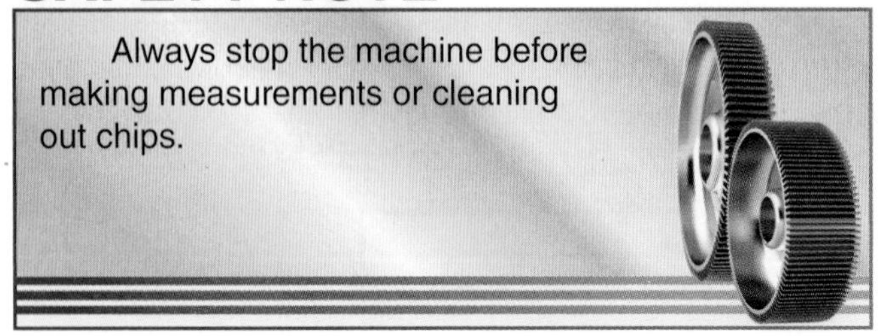

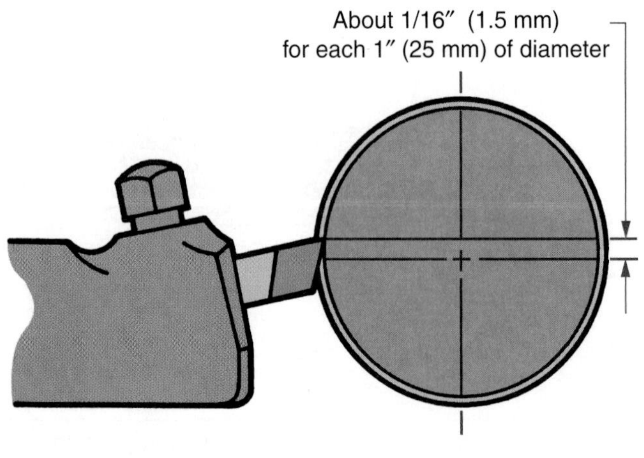

Goodheart-Willcox Publisher

Figure 14-74. With a modern tool post, set the tool on center or slightly above center when rough turning.

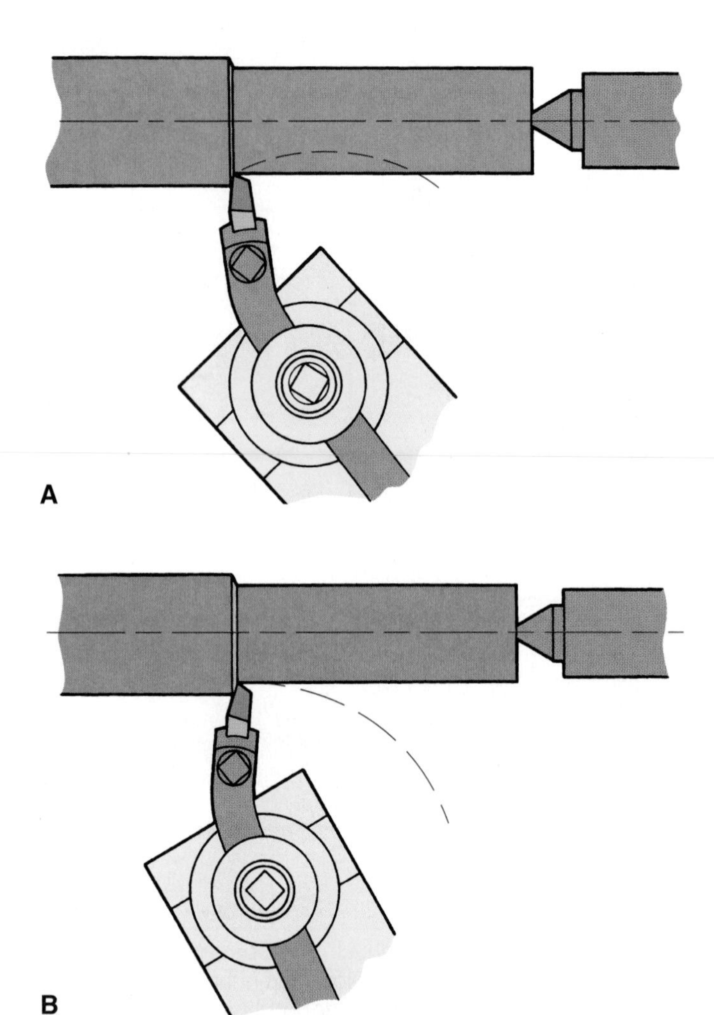

Goodheart-Willcox Publisher

Figure 14-75. Toolholder positioning. A—An incorrectly positioned tool will cut deeper into work if toolholder slips in the tool post. B—A correctly positioned tool will swing clear of work if toolholder slips.

If a dead center is used in the tailstock, lubricate the center frequently. Stop the machine immediately if the center heats up and starts to smoke or "squeal."

14.12.2 Finish Turning

After rough turning, the work is still oversize. It must be machined to the specified diameter and to a smooth surface finish by finish turning, **Figure 14-76**.

Fit a right-cut finishing tool into the toolholder. All rough and finish machining should be done toward the headstock (right to left) because the headstock offers a more solid base than the tailstock. Position the tool on center and check for adequate clearance between the compound rest and the revolving lathe dog.

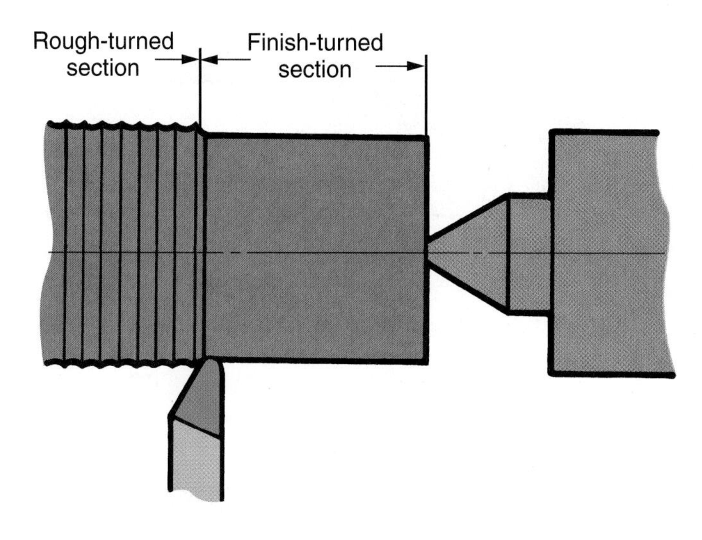

Goodheart-Willcox Publisher

Figure 14-76. After roughing work to approximate size, turn it to required size with a finishing tool. Cut should be made from right to left.

Adjust the lathe for a faster spindle speed and a fine feed. Run the cutting tool into the work until a light cut is being made, then engage the power feed. After a sufficient distance has been machined, disengage the power feed and stop the lathe.

SAFETY NOTE

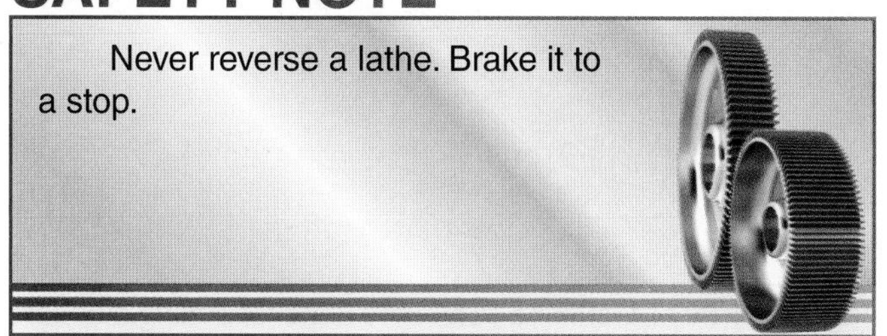

Never reverse a lathe. Brake it to a stop.

Do not interfere with the cross-slide setting. "Mike" the diameter of the machined area. The difference between the measurement and the specified diameter is the amount of material that must be removed. Move the cutting tool clear of the work and feed it in one-half the amount that must be removed. For example, if the diameter is 0.008″ (0.20 mm) oversize, tool infeed should be 0.004″ (0.10 mm). Make another cut about 1/2″ (13 mm) in width at the new depth setting. Measure again to make sure the correct diameter will be machined.

When reversing the work to permit machining its entire length, avoid marring of the finished surface by the lathe dog setscrew. Insert a small piece of soft aluminum or copper sheet between the setscrew and the workpiece.

14.12.3 Turning to a Shoulder

Up to this point, only ***plain turning*** has been described. This is turning in which the entire length of the piece is machined to a specified diameter. However, it is frequently necessary to machine a piece to several different diameters.

Locate the points to which the different diameters are to be cut. Scribe lines with a hermaphrodite caliper which has been set to the required length, **Figure 14-77**.

Machining is done as previously described, with the exception of cutting the shoulder, the point where the diameters change. The four types of shoulders, as shown in **Figure 14-78**, are as follows:

- Square.
- Angular.
- Filleted.
- Undercut.

A right-cut tool is used to make the square and angular type shoulders. See **Figure 14-79** and **Figure 14-80**. For machining a filleted shoulder, a round-nose tool is ground to the required radius using a fillet or radius gage to check radius accuracy. See **Figure 14-81**.

14.12.4 Turning Work Held in a Chuck

Work mounted in a chuck is machined in the same manner as if it were between centers. To prevent "springing" (flexing) while it is being machined, long work should be center drilled and supported with a tailstock center, **Figure 14-82**.

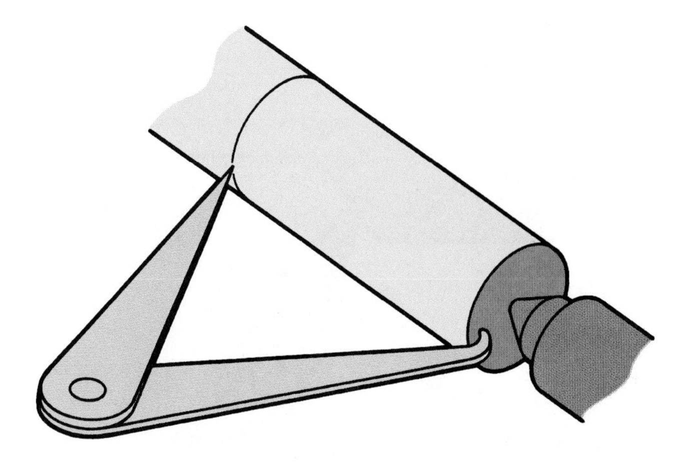

Goodheart-Willcox Publisher

Figure 14-77. Scribing reference lines on a workpiece with a hermaphrodite caliper.

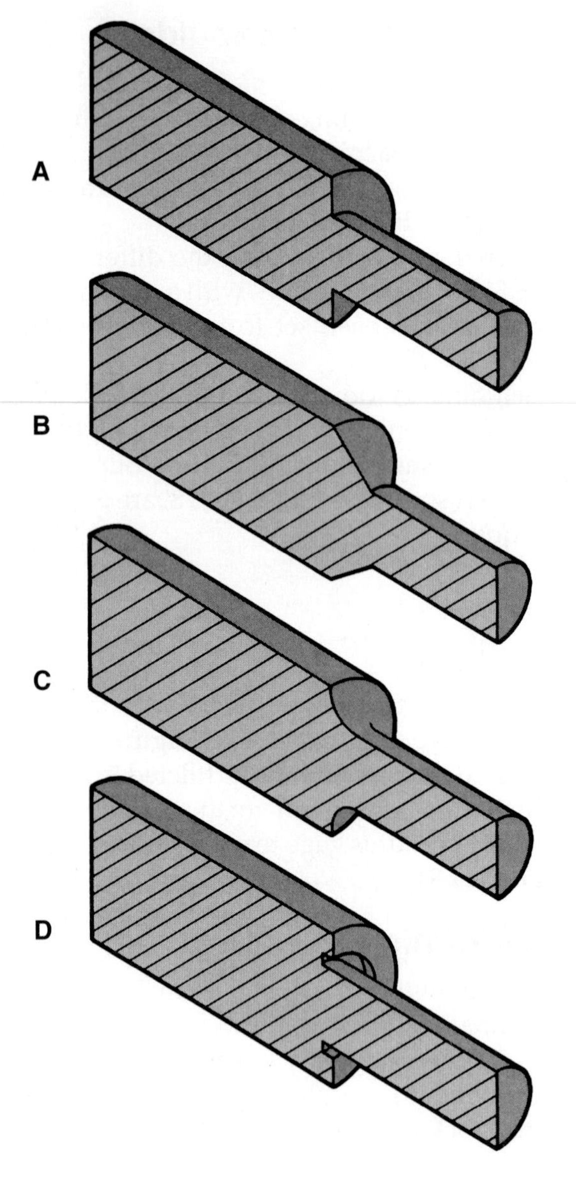

Goodheart-Willcox Publisher

Figure 14-78. The four types of shoulders. A—Square. B—Angular. C—Filleted. D—Undercut.

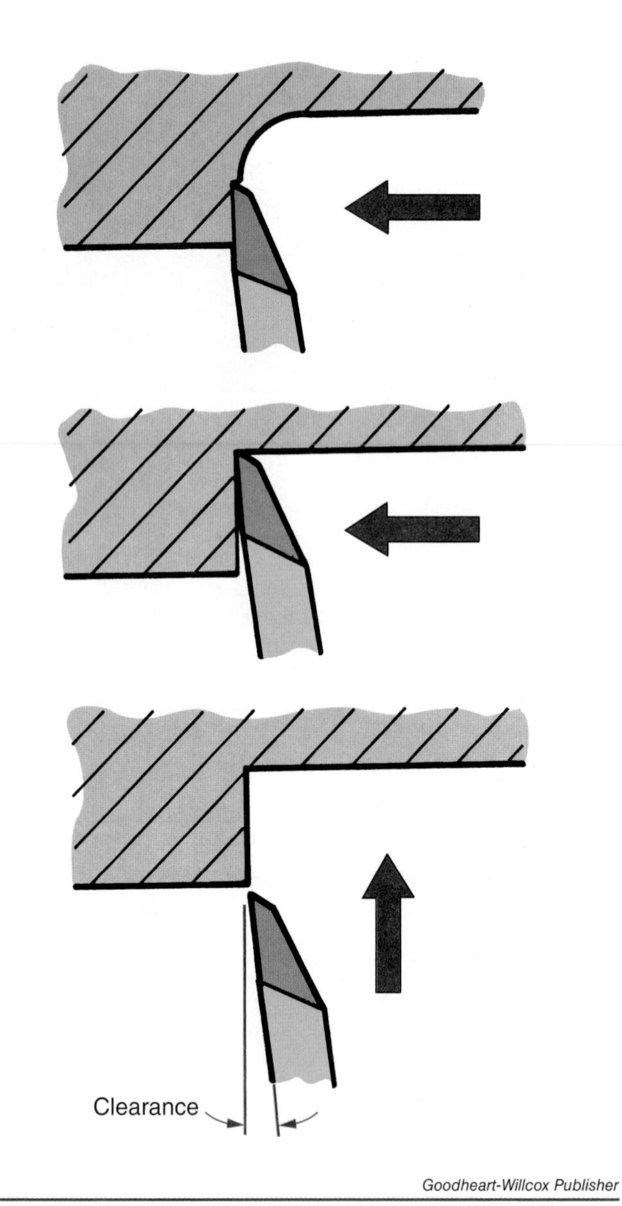

Goodheart-Willcox Publisher

Figure 14-80. Machining sequence used for cutting a square shoulder. A—First cut. B—Second cut. C—Facing cut.

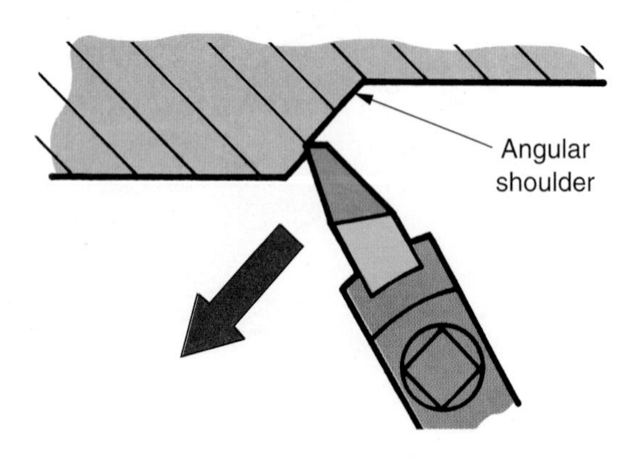

Goodheart-Willcox Publisher

Figure 14-79. To machine an angular shoulder, cut is made from smaller diameter to larger diameter.

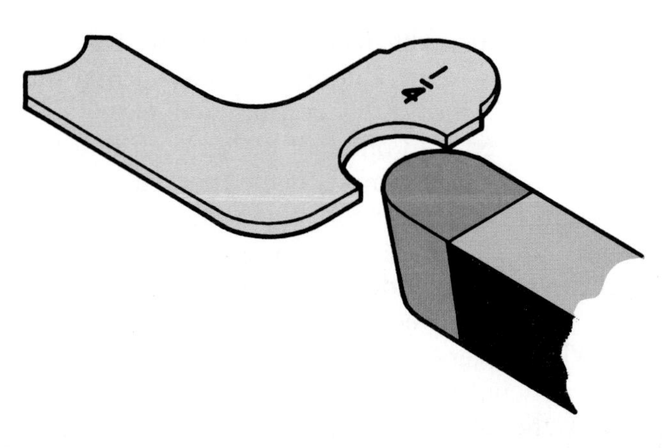

Goodheart-Willcox Publisher

Figure 14-81. The radius on a cutter bit can be checked with a fillet gage.

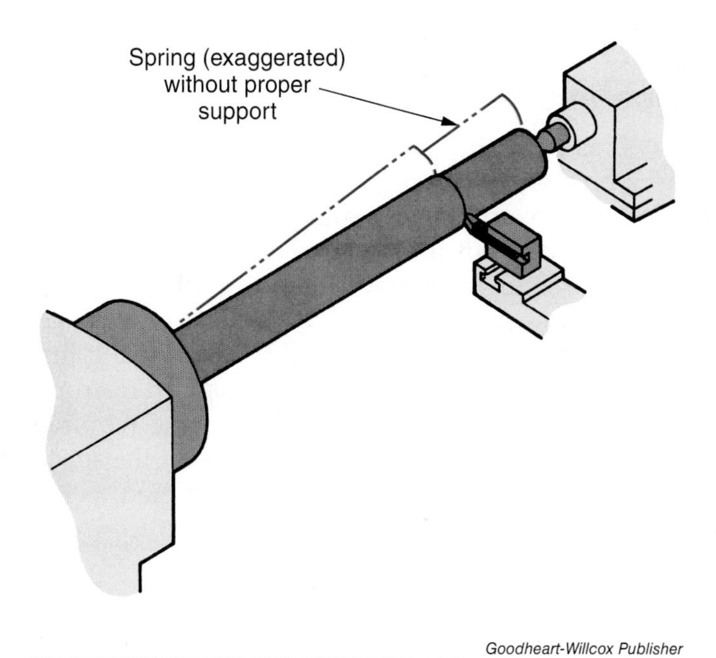

Goodheart-Willcox Publisher

Figure 14-82. For accurate turning, long work must be supported with a tailstock center.

Iscar Metals, Inc.

Figure 14-83. Parting is one of the more difficult jobs performed on the lathe. This illustration shows parting of thick-wall tubing. The replaceable tool has a helical twisted geometry to prevent binding during parting operations.

14.13 Parting and Grooving Operations

Parting and grooving are two common machining operations that can be performed on a lathe. ***Parting*** is the operation of cutting off material after it has been machined, **Figure 14-83**. ***Grooving*** is used to cut recesses or grooves into the surface of the workpiece.

Cutting tools for parting or grooving are held in a straight or offset toolholder. They must be ground with the correct clearance (front, side, and end). A concave rake is ground on top of the cutter to reduce chip width, and prevent it from seizing (binding) in the groove.

Keep the tool sharp. This will permit easy penetration into the work. If the tool is not kept sharp, it may slip and as pressure builds up, dig in suddenly and break.

14.13.1 Parting Operations

Parting is one of the more difficult operations performed on a lathe. The cutoff blade is set at exactly 90° to the work surface, **Figure 14-84**. The cutting edge should be set on center when parting stock 1″ (25.0 mm) in diameter. For larger pieces, the cutting edge should be positioned 1/16″ (1.5 mm) above center for each 1″ (25.0 mm) of diameter. The tool must be lowered as work diameter is reduced, unless the center of the piece has been drilled out.

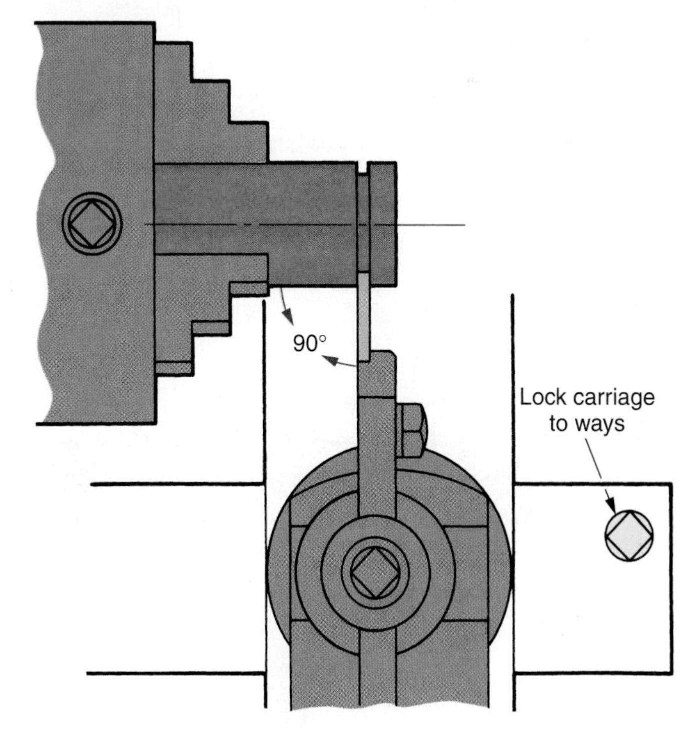

Goodheart-Willcox Publisher

Figure 14-84. Work is held close in chuck for the parting operation. Parting tool blade is set at a 90° angle to cut, and carriage is locked to the ways.

Spindle speed is about one-third that used for conventional turning. The compound rest and cross-slide must be tightened to prevent play.

SAFETY NOTE

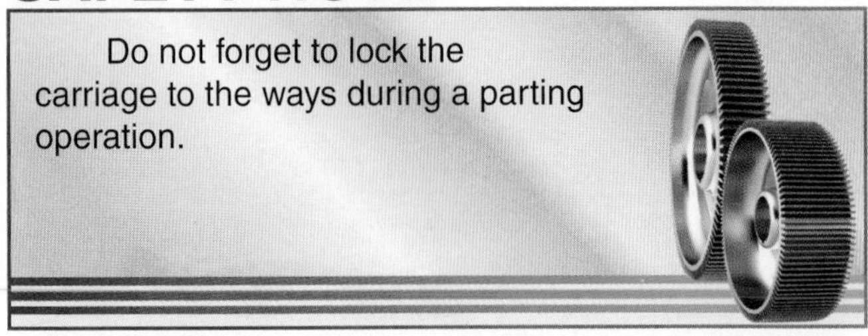

Feed should be ample to provide a continuous chip. If feed is too slow, "hogging" (the cutter digging in and taking a very heavy cut) can result. The tool will not cut continuously, but will ride on surface of the metal for a revolution or two, then bite suddenly. If the machine is in good condition, automatic cross-feed may be used.

When parting, apply ample quantities of cutting fluid. Whenever possible, hold the work "close" in the chuck and, if necessary, use an offset toolholder.

SAFETY NOTE

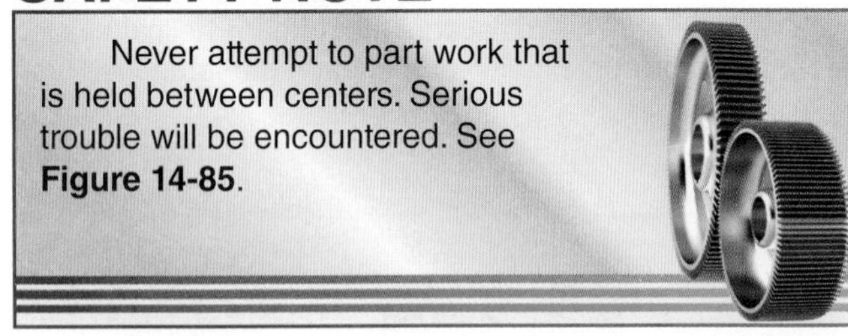

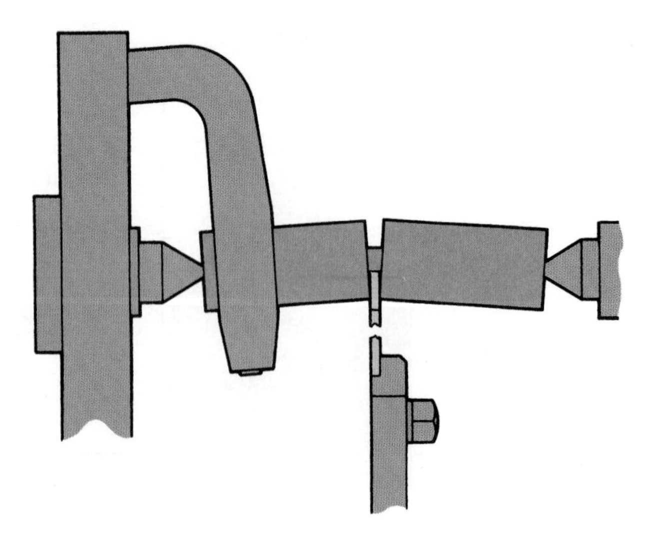

Goodheart-Willcox Publisher

Figure 14-85. Work cannot be parted safely while being held between centers.

14.13.2 Grooving or Necking Operations

It is sometimes necessary to cut a groove or neck on a shaft to terminate a thread or to provide adequate clearance for mating parts, **Figure 14-86**. As any recess cut into a surface has a tendency to weaken a shaft, it is better to make the groove round, rather than square.

The tool is set on center and fed in until it just touches the work surface. Set the cross-feed micrometer dial to zero and feed the tool in the required number of thousandths or millimeters for the specified depth. Square grooves can be machined with a parting tool or HSS tool ground to the correct width.

14.13.3 Cutting Grooves with a Parting Tool

When tolerances and surface finish requirements permit, grooves of various widths can often be machined with a parting tool. **Figure 14-87** illustrates two techniques that can be used. The parting tool should be mounted in the toolholder to reduce chatter and tool flexing.

Carboloy

Figure 14-86. A groove or neck can be cut into a shaft with a grooving or parting tool. Toolholders for grooving and parting can be straight or offset. The shape of the groove may be square, angular (square with sloping sides), or round.

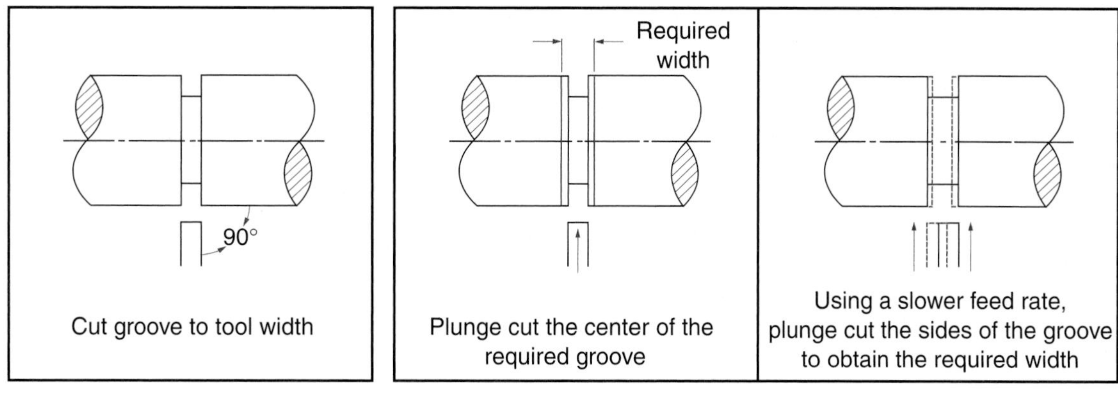

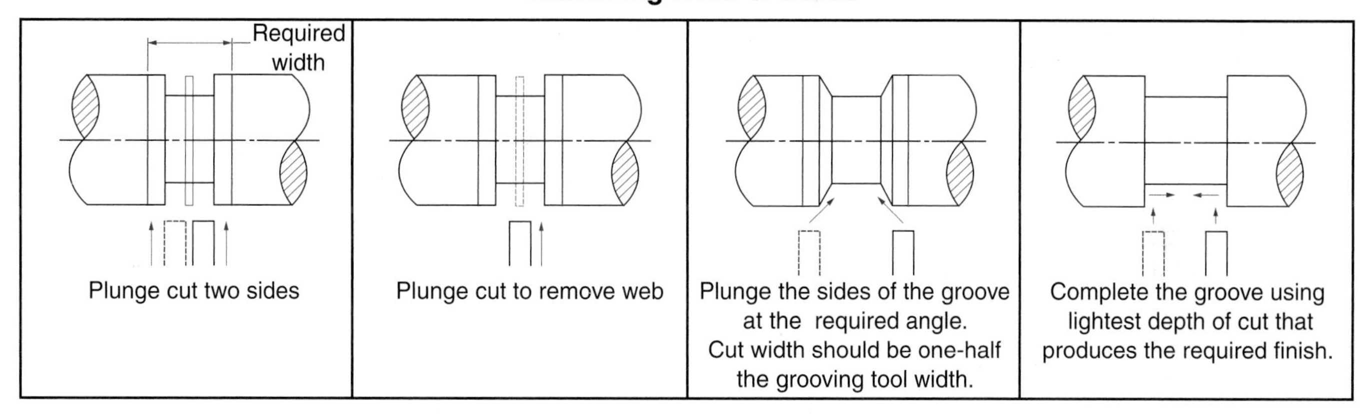

Goodheart-Willcox Publisher

Figure 14-87. These two techniques can be used to machine grooves of various widths using a parting tool.

Chapter Review

Summary

- Lathes operate by rotating the workpiece against the edge of a cutting tool.
- The major parts of a lathe can be broken down into the parts that drive the lathe, the work-holding and rotating parts, and the parts used to hold, move, and guide the cutting tools.
- Work can be held on a lathe between centers, using a chuck, in a collet, or bolted to a faceplate.
- A variety of cutting tools and toolholders can be used while turning on a lathe.
- Proper cutting speeds and rates of feed are very important to successful turning operations.
- It is important to properly maintain the lathe and follow all lathe safety precautions to ensure safe and efficient turning operations.
- Facing, facing to length, rough turning, finish turning, turning to a shoulder, parting, grooving, and necking are all operations that can be performed on a lathe.

Review Questions

Answer the following questions using information provided in this chapter.

1. The lathe operates on the principle of _____.
   A. the cutter revolving against the work
   B. the cutting tool, being controllable, can be moved vertically across the work
   C. the work rotating against the cutting tool, which is controllable
   D. All of the above.
   E. None of the above.
2. The size of a lathe is determined by the _____ and the length of the bed.
3. The largest piece that can be turned between centers is equal to _____.
   A. the length of the bed minus the space taken up by the headstock
   B. the length of the bed minus the space taken up by the tailstock
   C. the length of the bed minus the space taken up by the headstock and the tailstock
   D. All of the above.
   E. None of the above.
4. Into which of the following categories do the various parts of the lathe fall?
   A. Driving the lathe.
   B. Holding and rotating the work.
   C. Holding, moving, and guiding the cutting tool.
   D. All of the above.
   E. None of the above.
5. Some lathes come equipped with a(n) _____, allowing for slower speeds with greater power.
6. Explain the purpose of ways on the lathe bed.

The carriage supports and controls the cutting tool, and is composed of a number of parts. For Questions 7–10, match each description in the left column with the correct word in the right column. Place the appropriate letter in the blank.

7. _____ Fitted to the ways and slides along them.
8. _____ Permits transverse tool movement.
9. _____ Permits angular tool movement.
10. _____ Used to mount the cutting tool.

A. Compound rest.
B. Saddle.
C. Tool post.
D. Cross-slide.

11. Most work is machined while supported by one of four methods. List them.
12. Before work can be mounted between centers, a(n) _____ must be drilled at each end of the stock.
13. A tapered piece will result, when the work is turned between centers, if the centers are not aligned. Approximate alignment can be determined by two methods. What are they?
14. Briefly describe the three methods for checking center alignment if close tolerance work is to be done between centers.
15. What are the four most commonly used types of lathe chucks?
16. In most lathe operations, you will be using a single-point cutting tool made of _____.
17. Cutting speeds can be increased 300% to 400% by using _____ cutting tools.
18. What does cutting speed indicate?
19. _____ is used to indicate the distance that the cutter moves in one revolution of the work.

For Questions 20 and 21, calculate the cutting speeds for the following metals using the information provided below.

a. Formula: $\text{rpm} = \dfrac{\text{CS} \times 4}{\text{D}}$
b. CS = Cutting speed recommended for material being machined.
c. D = Diameter of work in inches.

20. What is the spindle speed (rpm) required to finish-turn 2 1/2″ diameter aluminum alloy? A rate of 1000 fpm is the recommended speed for finish-turning the material.
21. What is the spindle speed (rpm) required to rough-turn 1″ diameter tool steel? The recommended rate for rough-turning the material is 50 fpm.

For Question 22, calculate the cutting speed for the metric-sized material using the information provided.

a. Formula: $\text{rpm} = \dfrac{\text{CS} \times 1000}{\text{D} \times 3}$
b. CS = Cutting speed recommended for particular material being machined (steel, aluminum, etc.) in meters per minute (mpm).
c. D = Diameter of work in millimeters (mm).

22. What spindle speed is required to finish-turn 200 mm diameter aluminum alloy? Recommended cutting speed for the material is 300 mpm.

23. Accumulated metal chips and dirt are cleaned from the lathe with a(n) _____, *never* with your hands.

24. Which of the following actions are considered dangerous when operating a lathe?
    A. Wearing loose clothing and jewelry.
    B. Measuring with work rotating.
    C. Operating lathe with most guards in place.
    D Using compressed air to clean machine.
    E. All of the above.
25. List the four types of shoulders.
26. Why is a concave rake ground on top of the cutter when used for parting operations?
27. When using the parting tool, the spindle speed of the machine is about _____ the speed used for conventional turning.

CHAPTER 15

Other Lathe Operations

Learning Objectives

After studying this chapter, you will be able to:

- Perform boring and knurling operations on a lathe.
- Describe how drilling, reaming, filing, polishing, grinding, and milling operations can be performed on a lathe.
- Properly set up steady and follower rests.
- Safely set up and operate a lathe using various work-holding devices.
- Demonstrate familiarity with industrial applications of the lathe.

Technical Terms

automatic screw machine
boring
boring mill
follower rest
knurling
mandrel
steady rest
turret lathe

15.1 Boring on a Lathe

Boring is an internal machining operation in which a single-point cutting tool is used to enlarge a hole, **Figure 15-1**. Boring may be used to enlarge a hole to a specified size where a drill or reamer will not do the job. When properly set up, it produces a hole that is concentric with the outside diameter of the work.

15.1.1 Boring Difficulties

While the machining technique remains essentially the same as for external turning, several conditions will be encountered that could cause you difficulty. When boring on a lathe, you must make allowances for the following:

- Movement of the cross-slide screw is reversed.
- The machinist must work by "feel," since the cutting action cannot always be observed.
- Additional front clearance must be ground on the cutting tool to avoid rubbing, **Figure 15-2**. Otherwise, the shape of the cutting tool is identical to that used for external turning.
- Boring a deep or small-diameter hole requires a long, slender boring bar. The overhang makes the tool more likely to spring away from the surface being machined. It is also necessary to take several light cuts, instead of one heavy cut, to remove the same amount of material.

Kennametal

Figure 15-1. Boring or machining internal surfaces is sometimes done on a lathe.

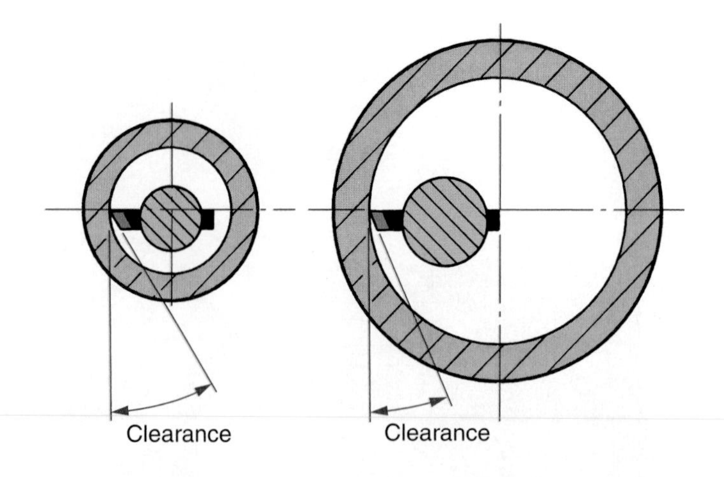

Goodheart-Willcox Publisher

Figure 15-2. Tool used to bore small diameter holes requires greater front clearance to prevent rubbing.

- Some people find that internal measuring tools are more difficult to use than those for making external measurements.

15.1.2 Boring the Hole

The hole size to be bored determines the type and size boring bar required, **Figure 15-3**. Always use the largest bar possible to give maximum tool support. The bar should extend from the holder only far enough to permit the tool to cut to the required hole depth, **Figure 15-4**.

The boring bar is set on center or slightly below center, with the bar parallel to tool travel, **Figure 15-5**. Check for adequate clearance when the tool is at maximum depth in the hole.

Begin by making a light cut in the same manner as you would for external machining. When the cut is completed, stop the machine. Set the cross-slide micrometer dial to zero and back the tool away from the work. Remove the boring bar from the hole.

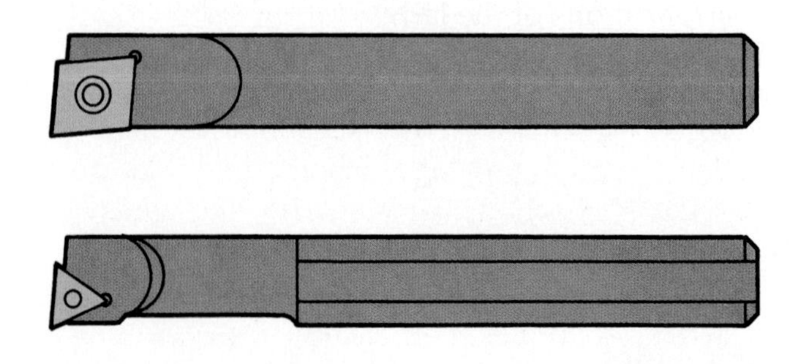

Goodheart-Willcox Publisher

Figure 15-3. Boring bars with indexable cutting tool inserts permit the machinist to use the most rigid bar for the job.

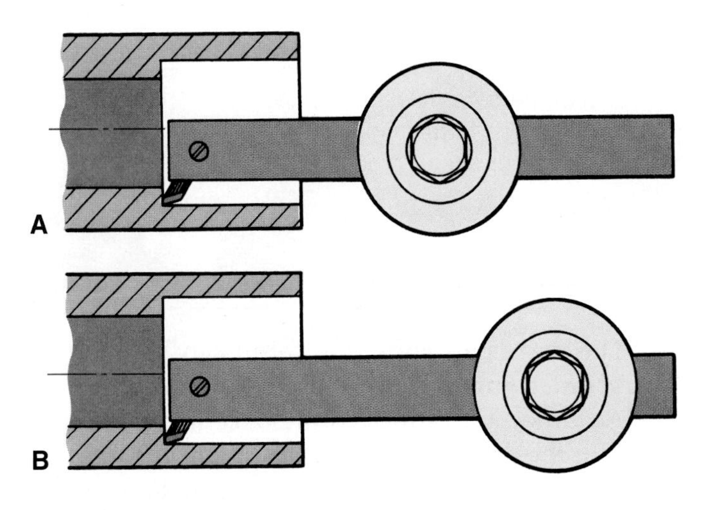

Goodheart-Willcox Publisher

Figure 15-4. Keep the cutting tool as close to tool post as possible for maximum tool support. A—Properly positioned boring bar. B—Boring bar projecting too far from tool post. The resulting vibration and "chatter" could produce a rough machined surface.

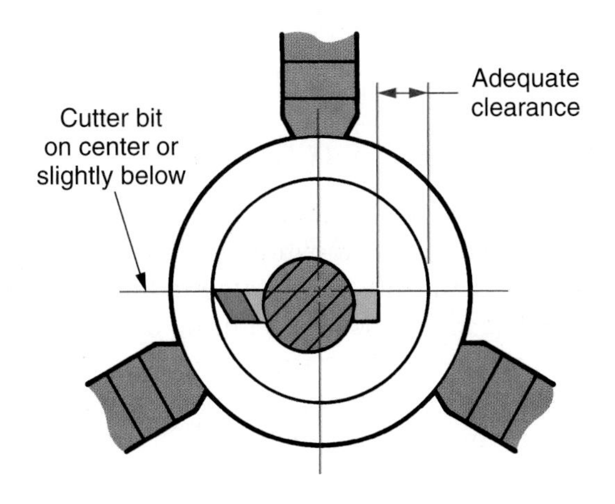

Goodheart-Willcox Publisher

Figure 15-5. The tool is set on, or slightly below, center when boring. Be sure to check for adequate clearance between boring bar and hole.

Check the hole diameter with an inside micrometer or with a telescoping gage and micrometer. After checking hole accuracy, bring the cross-slide back to zero, and advance the tool to make another cut. The amount of infeed will be determined by the boring bar in use and the material being bored. Make additional cuts, checking the hole size frequently, until the desired diameter is attained.

When making the final cut, it may be necessary to reverse tool travel after reaching the desired depth. Reverse the carriage feed, not the spindle rotation. Let the tool feed out of the hole without changing the tool setting. This will compensate for any tool spring.

When boring holes with long, slender boring bars, it may be necessary to run the tool into the hole without changing its setting after every second or third cut to compensate for tool spring.

With such long, slender boring bars, "chatter" is more likely to occur than when doing external work. Chatter can usually be eliminated by using one of the following methods:

- Use the largest boring bar practical for the hole size.
- Using a slower spindle speed.
- Reducing tool overhang.
- Grinding a smaller radius on the cutting tool nose.
- Placing a weight on the back overhang of the boring bar.
- Placing the tool slightly below center.
- Use a dampened boring bar.

Dampened boring bars are specifically designed to reduce vibration. The center of the bar is hollow and filled with lead shot or another material designed to absorb vibration.

15.2 Drilling on a Lathe

When a hole is to be cut in solid stock, the usual practice is to hold it in a suitable chuck and mount the drill in the tailstock. Drilling is accomplished on a lathe by feeding the stationary drill into the rotating workpiece, **Figure 15-6**.

Kennametal

Figure 15-6. When drilling on a lathe, the workpiece is held in the headstock and rotated while the stationary drill, held in the tailstock, is advanced into the work. The cutting fluid helps to reduce heat and flush away chips.

When using standard twist drills to drill holes that are 1/2″ (12.5 mm) or less in diameter, a straight shank drill is placed in a Jacobs chuck. The Jacobs chuck is then fitted into the tailstock spindle. Holes larger than 1/2″ (12.5 mm) in diameter are made with taper shank drills, **Figure 15-7**.

Drills that have taper shanks too large to be fitted in the tailstock can be used, if mounted as shown in **Figure 15-8**. A dog is fitted to the neck of the drill. The tool is set up to permit the tailstock center to press into the center hole in the drill tang.

The drill's cutting point bears against the rotating work. The drill is prevented from revolving by the dog bearing against the compound rest. The tailstock center keeps the drill aligned and enables it to be fed into the material by the tailstock handwheel.

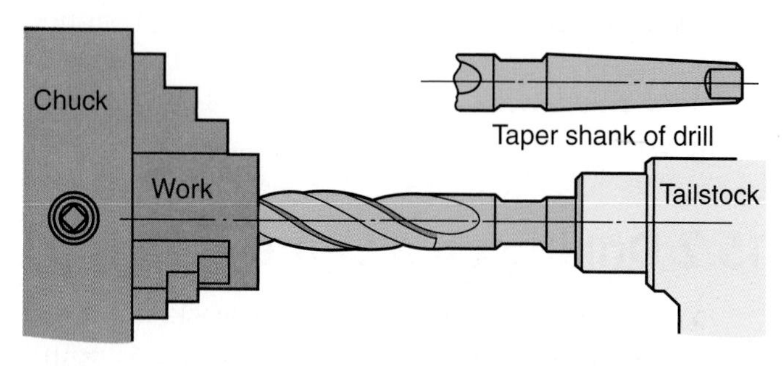

Goodheart-Willcox Publisher

Figure 15-7. Drills larger than 1/2″ (12.5 mm) in diameter are usually fitted with a self-holding taper that fits into the tailstock spindle of the lathe.

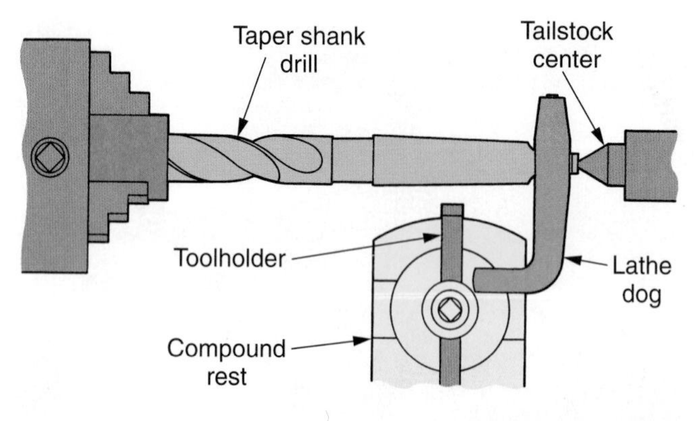

Goodheart-Willcox Publisher

Figure 15-8. When a drill shank is too large to be fitted into the tailstock, a lathe dog can be used to keep it from revolving. The tail of the lathe dog is supported by the compound rest. This type of drilling requires care to prevent the drill from slipping off the tailstock center when the full drill diameter breaks through the work.

SAFETY NOTE

Extreme care must be used to prevent the drill from slipping off the tailstock center after it breaks through the work.

The makeshift lathe dog setup can be avoided by use of a commercial drill holder, **Figure 15-9**.

Accuracy in drilling requires a centered starting point for the drill. A starting point made with a combination drill and countersink is adequate for most jobs. Holes over 1/2″ (12.5 mm) in diameter require a pilot hole. The pilot hole should have a diameter equal to the width of larger drill's dead center. See **Figure 15-10**.

Ample clearance must be provided in back of the work so that the drill will not strike the chuck or headstock spindle when it breaks through, **Figure 15-11**.

15.3 Reaming on a Lathe

Reaming is an operation used to make a hole accurate in diameter and finish, **Figure 15-12**. The hole is drilled slightly undersized to allow stock for reaming. The allowance for reaming depends on the hole size, as follows:

- With hole sizes ranging up to 1/4″ (6.5 mm) in diameter, allow 0.010″ (0.25 mm) of material for reaming.
- With a hole size from 1/4″ (6.5 mm) to 1/2″ (12.5 mm) in diameter, allow 0.015″ (0.4 mm).

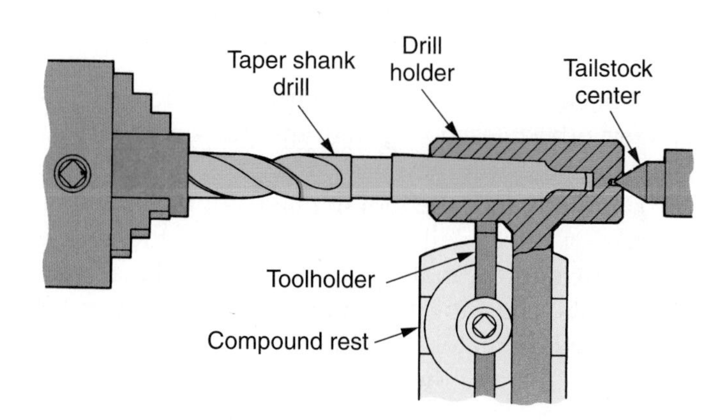

Goodheart-Willcox Publisher

Figure 15-9. Large taper shank drills can be used on the lathe by fitting them in a commercial drill holder.

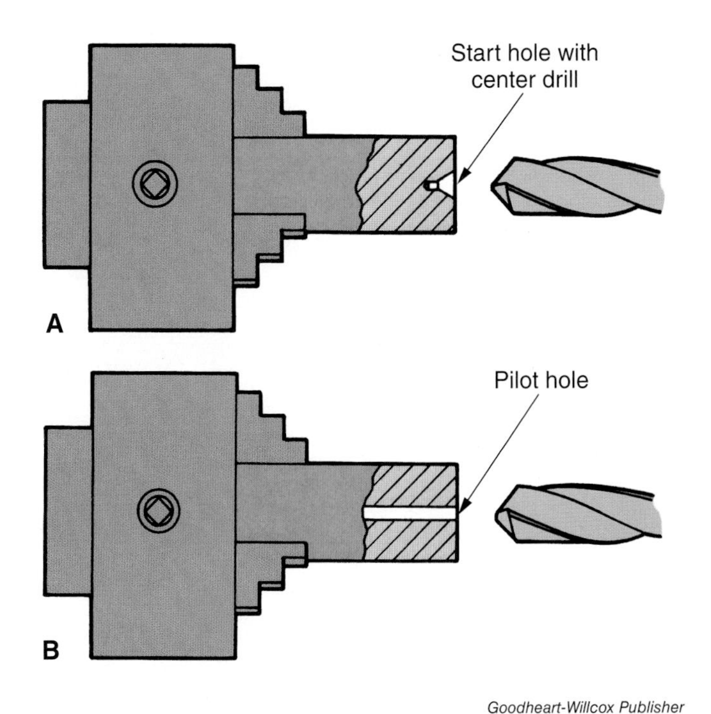

Goodheart-Willcox Publisher

Figure 15-10. Centering drill. A—The drill will cut exactly on center if the hole is started with a center drill. B—Holes larger than 1/2″ (12.5 mm) in diameter require drilling of a pilot hole.

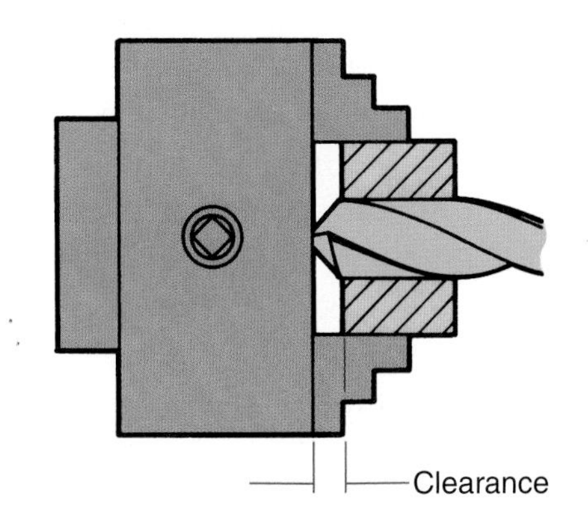

Goodheart-Willcox Publisher

Figure 15-11. There must be enough clearance between the back of the work and the chuck face to permit the drill to break through the work without damaging the chuck.

Goodheart-Willcox Publisher

Figure 15-12. A straight flute reamer mounted in an extension holder being used to finish a hole in a plastic workpiece.

- With a hole size from 1/2″ (12.5 mm) to 1.0″ (25.0 mm) in diameter, allow 0.020″ (0.5 mm).
- With a hole size from 1.0″ (25.0 mm) to 1.5″ (37.5 mm) in diameter, allow 0.025″ (0.6 mm).
- With a hole size above 1.5″ (37.5 mm) in diameter, allow 0.030″ (0.8 mm) for reaming.

When reaming, use a cutting speed about two-thirds the speed you would use for a similar size drill with the material being machined. Also, use a slow, steady feed with an adequate supply of cutting fluid. Remove the reamer from the hole before stopping the machine.

If a hand reamer is to be used, do not apply power to the workpiece mounted in the chuck. Fit the reamer into the hole, supporting the shank end with the tailstock center. Use an adjustable wrench to turn the reamer in a clockwise direction, **Figure 15-13.**

When removing a reamer from the hole, continue to rotate the tool clockwise. Avoid turning it counterclockwise, since that would ruin the tool's cutting edges.

15.4 Knurling on a Lathe

Knurling is the process of forming horizontal or diamond-shaped serrations (raised grooves or teeth) on the circumference of the work, **Figure 15-14**. Knurling is used to provide a gripping surface, change the appearance of the work, or increase the work's diameter. It is done with a knurling tool mounted in the tool post, **Figure 15-15.**

Goodheart-Willcox Publisher

Figure 15-13. Using a hand reamer on the lathe. Never turn on the power when performing hand reaming operations.

Photo courtesy of Grizzly Industrial, Inc. www.grizzly.com

Figure 15-14. Knurling rollers are being used to form serrations on a part.

Photo courtesy of Grizzly Industrial, Inc. www.grizzly.com

Figure 15-15. One type of knurling tool.

The knurled pattern is raised by rolling the knurls against the metal. This displaces the metal into the required pattern.

Angular knurls raise a diamond pattern, while a straight knurl produces a straight pattern along the length of the work. The patterns can be produced in coarse, medium, and fine pitch. See **Figure 15-16**.

15.4.1 Knurling Procedure

If a knurling tool setup is not made properly, the knurls will not track and will quickly dull. The following procedure is recommended:

1. Mark off section to be knurled.
2. Adjust the lathe to a slow back-geared speed and a fairly rapid feed.

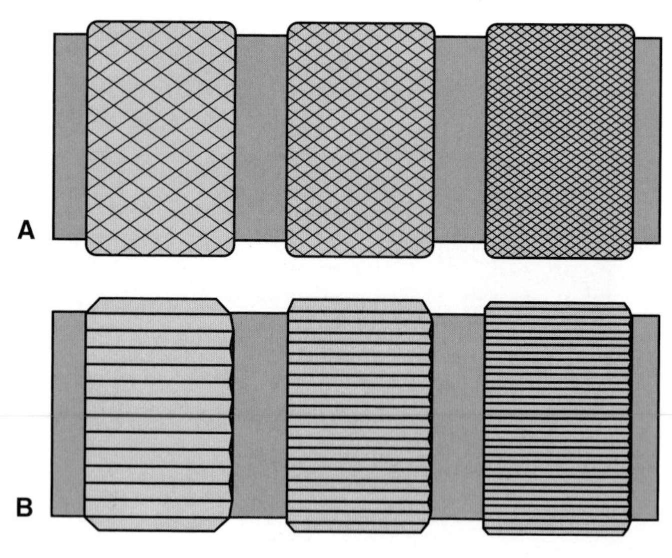

Goodheart-Willcox Publisher

Figure 15-16. Knurling patterns. A—Diamond knurl in coarse, medium, and fine pitch. B—Straight knurl in coarse, medium, and fine pitch.

3. Place the knurling tool in the tool post. Bring it up to the work. Both wheels must bear evenly on the work with their faces parallel with the centerline of the piece, **Figure 15-17**.
4. Start the lathe and slowly force the knurls into the work surface until a pattern begins to form. Tool travel should be toward the headstock whenever possible. Engage the automatic feed and let the tool travel across the work. Flood the work with cutting fluid.
5. When the knurling tool reaches the proper position, reverse spindle rotation and allow the tool to move back across the work to the starting point. Apply additional pressure to force the knurls deeper into the work.
6. Repeat the operation until a satisfactory knurl is formed.

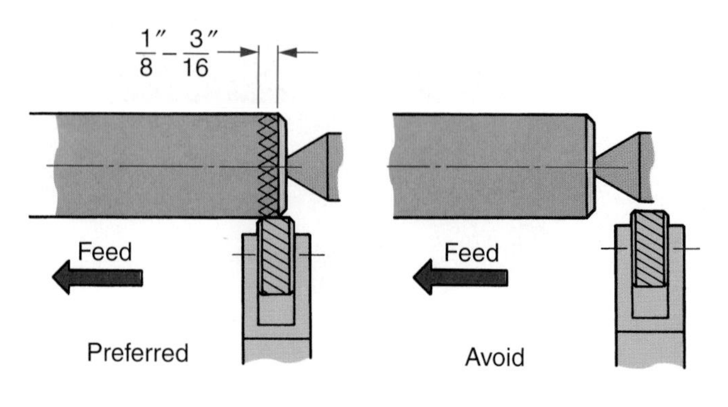

Goodheart-Willcox Publisher

Figure 15-17. Always start the knurl on the work.

15.4.2 Knurling Difficulties

If knurling is not performed properly, problems can arise and destroy the work. A common problem is the double-cut knurl, **Figure 15-18**. It occurs when one wheel of the knurling tool makes twice as many ridges as the other.

A double-cut knurl is usually caused by one wheel being dull. Raising or lowering the knurling tool to put more pressure on the dull wheel will frequently eliminate the trouble. Pivoting the tool slightly to allow the right side of the wheels to apply more pressure may also help.

Considerable side pressures are developed during knurling operations. Watch the tool carefully. Do not permit the work to slip into the chuck or loosen on the tailstock center. If a ball-bearing center is not used, keep the tailstock center well-lubricated.

Perform knurling before turning a shaft to a smaller diameter. If knurling is done after the smaller diameter has been machined, the work will spring away from the tool, giving the surface a superficial (light, nonpenetrating) knurl. It may also cause a permanent bend in the workpiece. See **Figure 15-19**.

Avoid applying too much pressure to the knurling tool. The work surface becomes hardened during the operation and the knurled section could "flake off." High pressure also tends to bend the shaft.

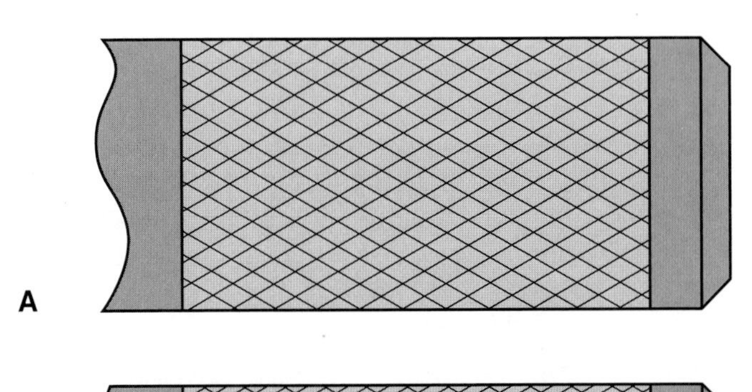

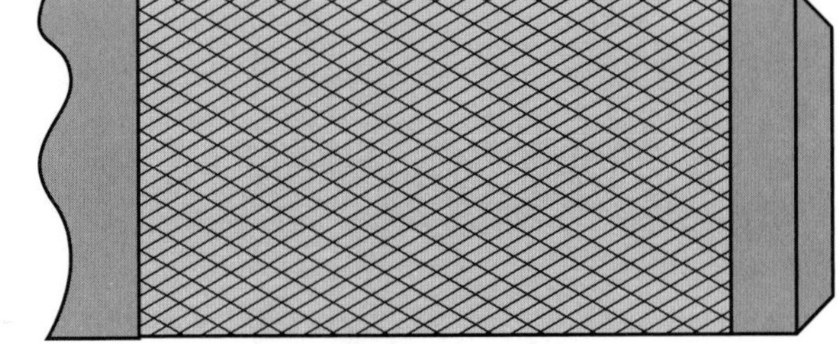

Goodheart-Willcox Publisher

Figure 15-18. Double-cut knurl. A—This is a correctly made diamond knurl pattern. B—A double-cut diamond knurl. It results when one knurl wheel is slightly above or below center.

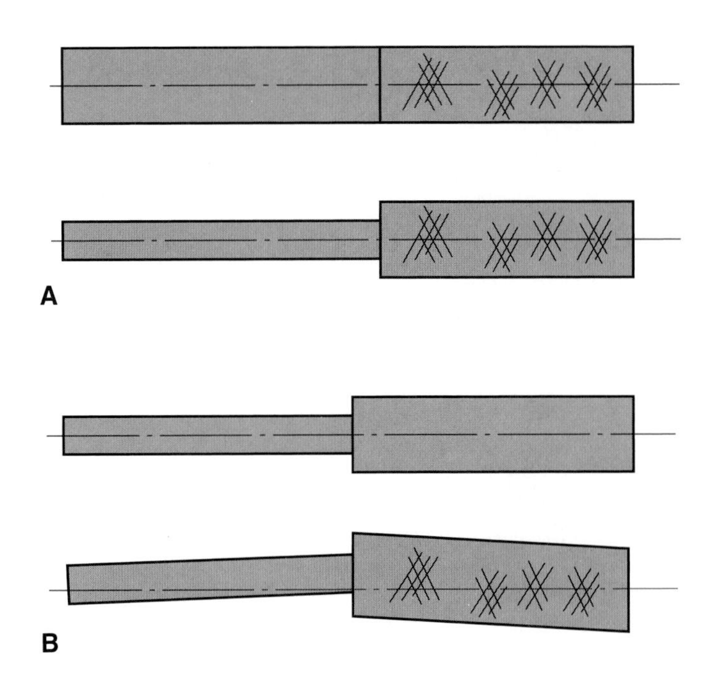

Goodheart-Willcox Publisher

Figure 15-19. Knurling problems. A—Do knurling before turning a shaft to a smaller diameter. B—If knurled after being turned to smaller diameter, shaft may take on a permanent bend and receive only a superficial (very light) pattern.

SAFETY NOTE

Never stop the lathe with the knurls engaged in the work. The piece will take on a permanent bend.

15.5 Filing and Polishing on a Lathe

Lathe filing is done to remove burrs, to round off sharp edges, and blend-in form cut outlines. A file is not intended to replace a properly sharpened cutting tool and should not be used to improve the surface finish on a turned section.

15.5.1 Filing on a Lathe

When filing on a lathe, avoid holding the tool stationary against the work. Keep the tool moving across the area being filed. If the file is held in one position, it will "load" with metal particles and score the surface of the work. An ordinary mill file will produce satisfactory results. However, a long-angle lathe file produces a superior cutting action. See **Figure 15-20**.

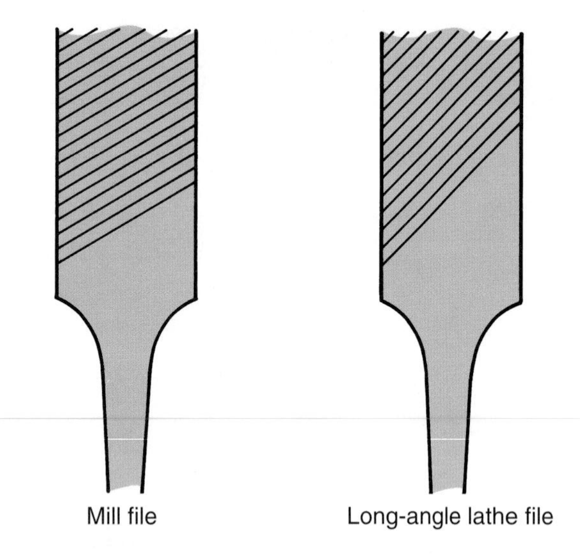

Goodheart-Willcox Publisher

Figure 15-20. How a standard mill file differs from the long-angle lathe file.

Operate the lathe at high spindle speed and apply long, even strokes. Release pressure on the return stroke. If uneven pressure is applied, out-of-round work will result. Clean the file often.

As simple as filing may appear, it can be quite dangerous if the following precautions are not observed:

- Move the carriage out of the way and remove the tool post.
- Use the left-hand method of filing, **Figure 15-21**. It involves holding the file handle in the left hand. The right hand is then clear of the revolving chuck or face plate.

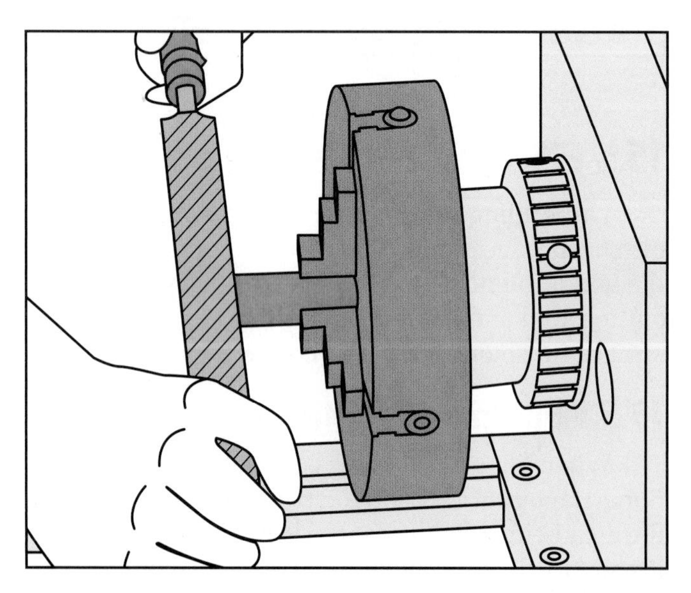

Goodheart-Willcox Publisher

Figure 15-21. The left-hand method of filing on the lathe is preferred.

SAFETY NOTE

Avoid the right-hand method of filing. The right-hand method places your left arm over the rotating chuck or faceplate. This method has the potential for injury.

15.5.2 Polishing on a Lathe

Polishing is sometimes done on a lathe using a strip of abrasive cloth suitable for the material to be polished. The strip is grasped between your fingers and held across the work, **Figure 15-22**. If more pressure is required, mount the abrasive cloth on a strip of wood or on a file, **Figure 15-23**. A high spindle speed is used for polishing.

The finer the abrasive used, the finer the resulting finish. A few drops of machine oil on the abrasive will improve the finish. For the final polish, reverse the cloth so the cloth backing, rather than the abrasive, is in contact with the work surface. Like filing, polishing is not a substitute for a properly sharpened tool bit.

Carefully and thoroughly clean the lathe after polishing operations. If not removed, abrasive particles from the cloth will cause rapid wear of the machine's moving parts.

15.6 Steady and Follower Rests

The ***steady rest*** and the ***follower rest*** are needed to provide additional support when the workpiece is long and thin. This additional support keeps the work from springing or bending away from the cutting tool. The support is also needed to reduce "chattering" when long shafts are machined. See **Figure 15-24**.

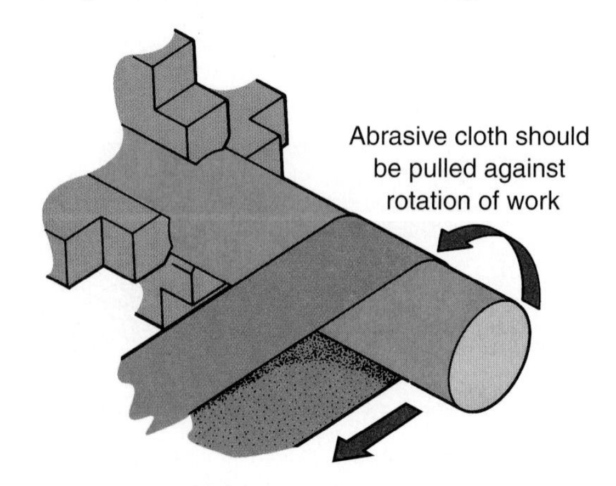

Goodheart-Willcox Publisher

Figure 15-22. Polishing with abrasive cloth held in hands. Keep hands away from revolving chuck or dog.

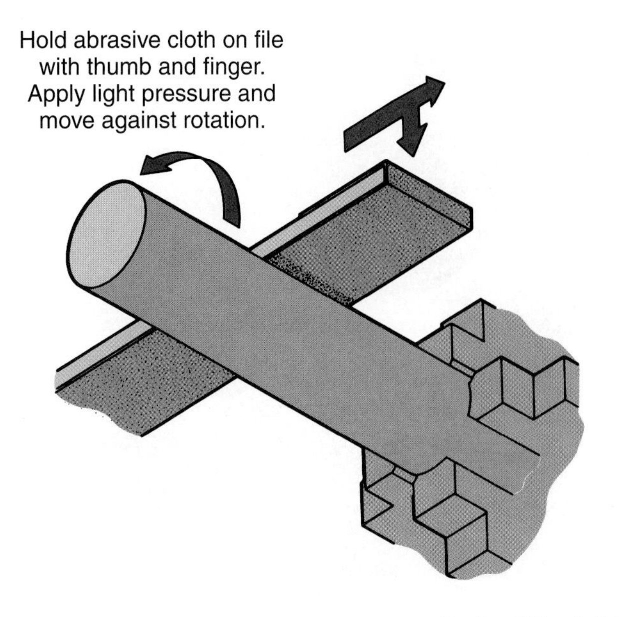

Goodheart-Willcox Publisher

Figure 15-23. More pressure can be applied if abrasive cloth is supported by a file or block of wood.

Photo courtesy of Grizzly Industrial, Inc. www.grizzly.com

Figure 15-24. When the nature or shape of work prevents it from being mounted between centers, special devices called "rests" are used to provide support. This lathe is equipped with a steady rest.

The steady rest, sometimes called a center rest, is bolted directly to the ways. It is provided with three adjustable jaws, each with individual locking screws. The upper portion of the attachment is fitted with a single jaw. It can be opened to permit the work to be placed in position, **Figure 15-25**.

15.6.1 Steady Rest Setup

To set up a steady rest, use the following directions:

1. Bolt the attachment to the ways at the desired position.
2. Back off all jaws and open the upper section.
3. Mount the work between centers or in a chuck. Support the free end with the tailstock center.
4. Lower and lock the upper segment in place.
5. Adjust the jaws up to the work and lock them into position. The jaws act as bearing surfaces where they contact the work. They must be well lubricated.

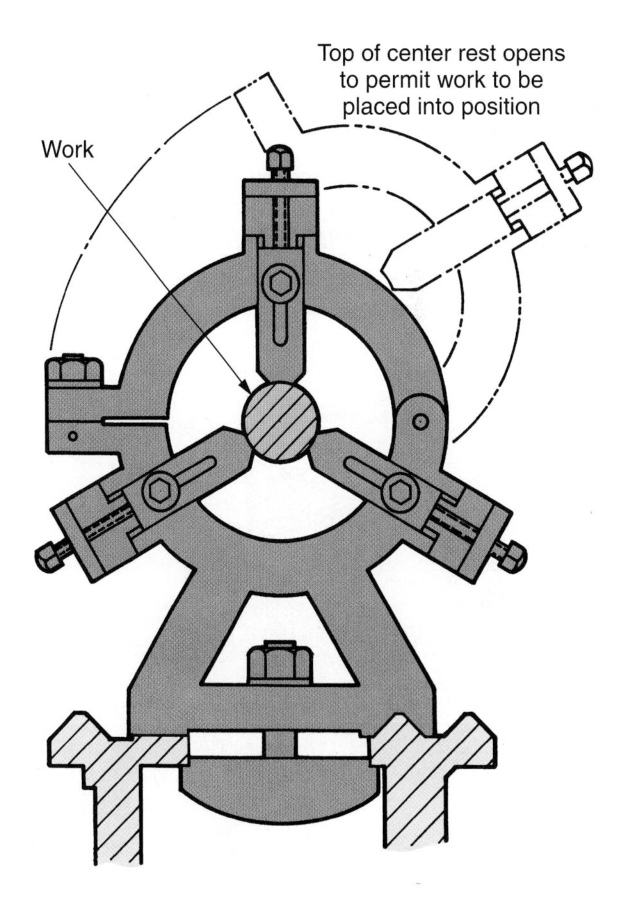

Goodheart-Willcox Publisher

Figure 15-25. To permit easy installation of work, the top of the steady or center rest swings open. Care must be taken to accurately center the work.

6. If the shaft being machined is unsuitable as a bearing surface (rough surface, out-of-round, square), a cat head is used. Care must be taken to center the shaft within the cat head. See **Figure 15-26**.
7. When machining at the end of a long shaft which cannot be supported with the tailstock, center the work in the chuck. Adjust the center rest to the work as close to the chuck as possible, then move it to a point where the support is needed. The same technique is used when performing drilling, reaming, tapping, or other operations on the end of a long shaft.

15.6.2 Follower Rest Setup

The follower rest operates on the same principle as the steady rest and is used in a similar manner. The follower rest differs slightly in that it provides support directly in back of the cutting tool and follows along during the cut. See **Figure 15-27**.

The follower rest bolts directly to the carriage and the jaws adjust in the same way as on the steady rest. The jaws must be readjusted after each cut.

Photo courtesy of Grizzly Industrial, Inc. www.grizzly.com

Figure 15-27. A follower rest is used to support long slender workpieces during machining operations.

15.7 Mandrels

At times, it is necessary to machine the outside diameter of a piece concentric with a hole that has been previously bored or reamed. This can be a simple operation if the material can be held in the lathe by conventional means. There are, however, times when the material cannot be gripped solidly to permit accurate machining. In such cases, the work is mounted on a ***mandrel***, a slightly tapered, hardened steel shaft, and turned between centers, **Figure 15-28**.

A solid mandrel is made from a section of hardened steel that has been machined with a slight taper (0.0005″ per inch), **Figure 15-29A**. These mandrels are made in standard sizes starting at 1/8″ in diameter. The size is stamped on the large end. The other end is slightly smaller than the specified size to permit easy installation in the work.

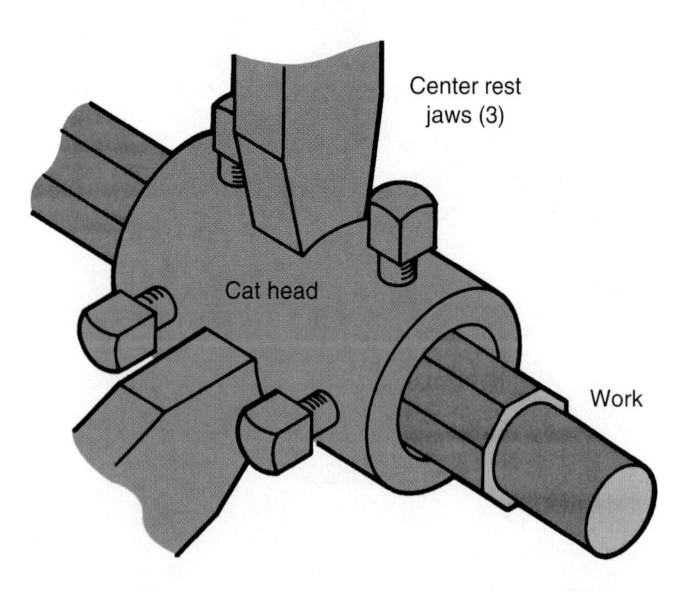

Goodheart-Willcox Publisher

Figure 15-26. A cat head will provide a bearing surface when needed.

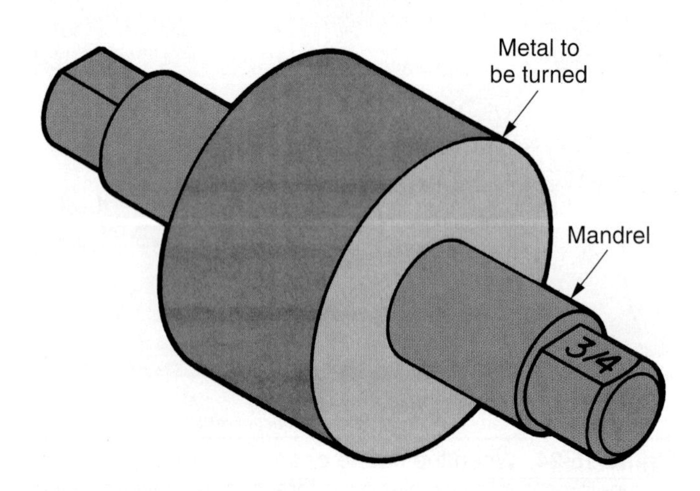

Goodheart-Willcox Publisher

Figure 15-28. This lathe mandrel has work mounted on it.

An expansion mandrel permits work with openings that vary from standard sizes to be turned. See **Figure 15-29B**. The shaft and sleeve have corresponding tapers and are machined from hardened steel. The sleeve is slotted so it can expand when forced onto the tapered shaft.

A gang mandrel, **Figure 15-29C**, is helpful when many pieces of the same configuration must be turned. Several pieces are mounted on the mandrel and separated with spacing collars. They are locked in place by tightening a nut.

Work is pressed on a mandrel with a mechanical arbor press. The work must first be checked for burrs and cleaned. Lubricate the work with a light oil to prevent it from "freezing" on the mandrel.

The mandrel is mounted between centers and driven by a lathe dog. Use care so the tool does not come into contact with the mandrel during the machining operation. In an emergency, a mandrel can be machined from a section of mild steel.

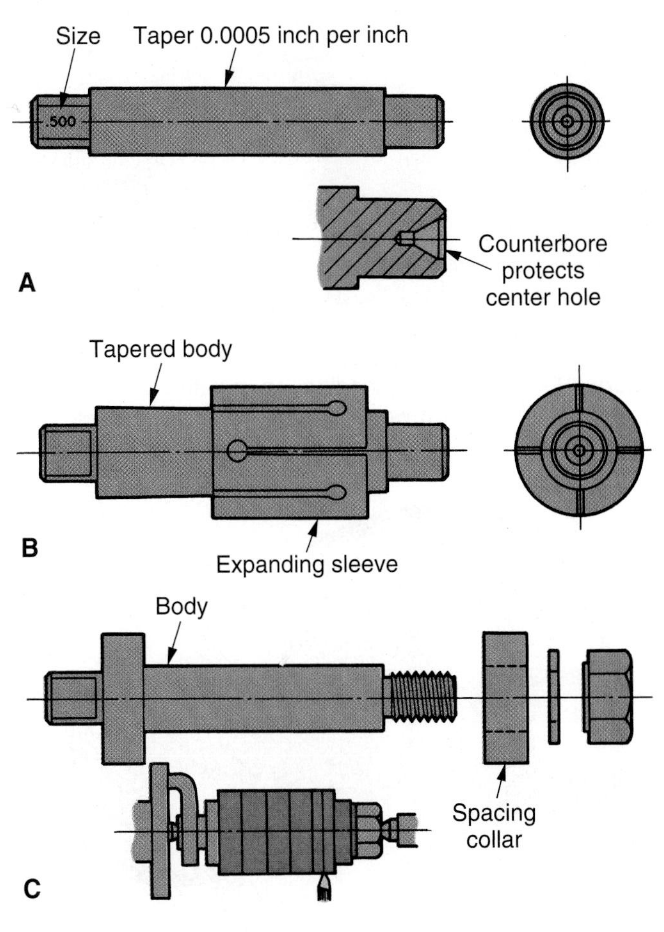

Goodheart-Willcox Publisher

Figure 15-29. Different kinds of mandrels. A—Solid type. B—Expansion type. C—Gang mandrel that allows mounting of several identical pieces.

15.8 Grinding on the Lathe

The tool post grinder permits the lathe to be used for internal and external grinding, **Figure 15-30**. With a few simple attachments, it is possible to sharpen reamers and milling cutters on the lathe. You can also grind shafts and true lathe centers.

Since steel parts sometimes warp during heat treatment, it is common to machine the piece to within 0.010″ to 0.015″ (0.2 mm to 0.3 mm) of finished size. After heat treatment, the metal is mounted on the lathe for grinding to finished size. A light grinding cut is made on each pass. When done properly, grinding produces a very smooth finish.

15.8.1 Preparing a Lathe for Grinding

Particles of the grinding wheel wear away during the grinding operation. Abrasive particles can cause excessive wear should they get into moving parts, so it is important to protect the lathe from them. When preparing to grind, cover the lathe bed, cross-slide, and other parts with canvas or heavy kraft paper to protect them from abrasive dust and grit. It is also good practice to place a small tray of water or oil just below the grinding wheel to collect as much grit and dust as possible.

SAFETY NOTE

When placing protective covering on the lathe, be sure the covering material cannot become entangled in the lead screw or other moving parts.

AMT—The Association for Manufacturing Technology

Figure 15-30. A tool post grinder used for internal grinding on a lathe.

15.8.2 Preparing the Grinder

Select the grinding wheel best suited for the job. It must be balanced and run true if a smooth, accurately sized job is to be obtained.

A diamond wheel dresser, consisting of an industrial diamond tip mounted on a steel shank, is used for the truing operation, **Figure 15-31**. It is mounted solidly to the lathe, on center or slightly below the center of the grinding wheel. The rotating wheel is moved back and forth across the diamond, removing about 0.001″ (0.02 mm) on each pass. Remove only enough material to true the wheel.

15.8.3 External Grinding

External grinding, **Figure 15-32**, is done to finish the exterior surface of the piece. The following steps are recommended to complete the job with the least amount of difficulty:

1. Mount the work solidly in the lathe. Provide adequate clearance.
2. Adjust lathe spindle speed for 80–100 rpm, and set a feed of 0.005″–0.007″ (0.12 mm–0.17 mm).
3. Turn on power for lathe and grinder. The work turns into the grinding wheel, **Figure 15-33**.
4. Feed the grinding wheel into the work until it just begins to "spark."
5. Engage the automatic longitudinal feed.
6. Check work diameter frequently with a micrometer. Use light cuts or the piece might overheat and warp.

NSK America

Figure 15-32. External grinding on a lathe. A—Grinding the circumference of a workpiece. B—Grinding the end of a workpiece.

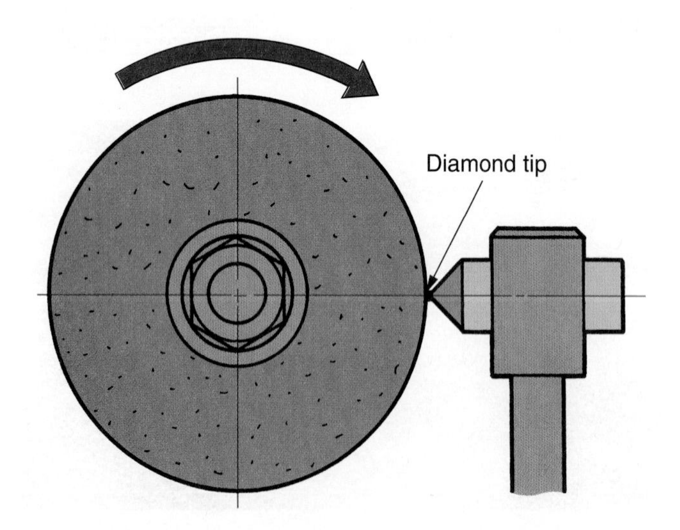

Goodheart-Willcox Publisher

Figure 15-31. Using a diamond wheel dresser to true the wheel on a tool post grinder before grinding on a lathe.

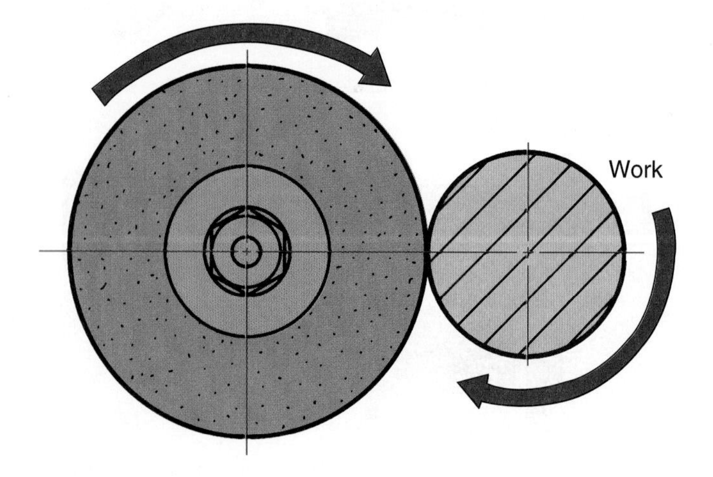

Goodheart-Willcox Publisher

Figure 15-33. With external grinding, the work turns into the grinding wheel.

7. Dress the grinding wheel again before making the final pass over the work. Allow the work to "spark out" (reach the point where the grinding wheel no longer cuts).

15.8.4 Internal Grinding

Internal grinding is done in much the same manner as external grinding but on the inside of the work, **Figure 15-34**. The work and grinding wheel must rotate in opposite directions, **Figure 15-35**.

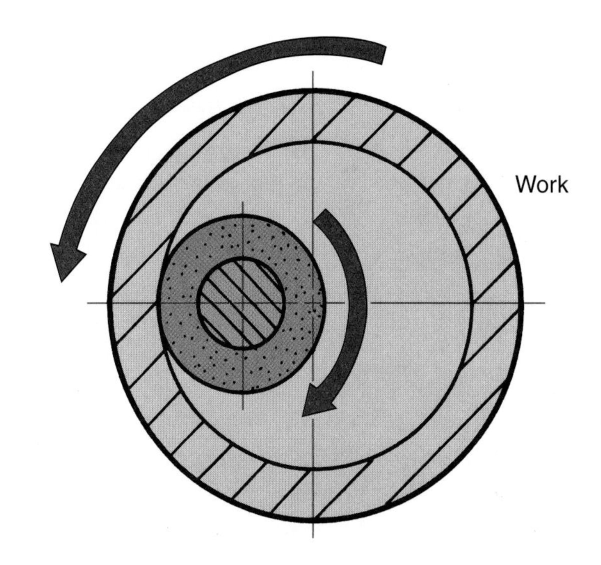

Goodheart-Willcox Publisher

Figure 15-35. With internal grinding, the work and grinding wheel turn in opposite directions.

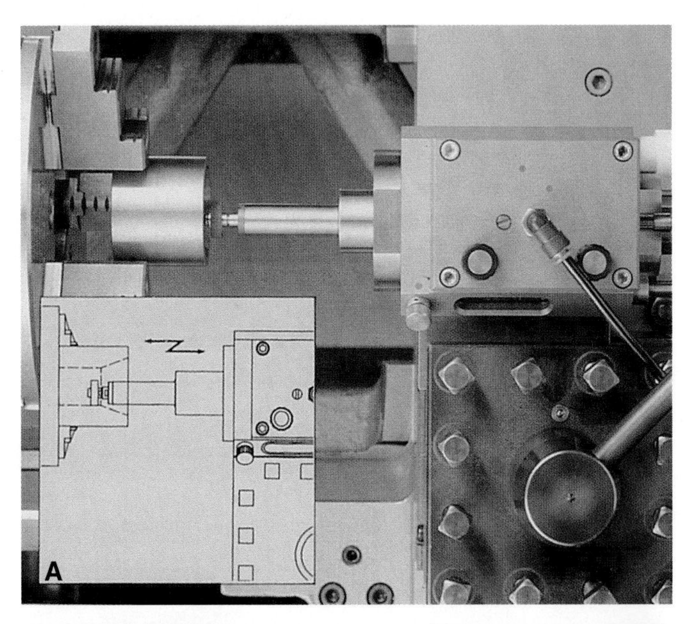

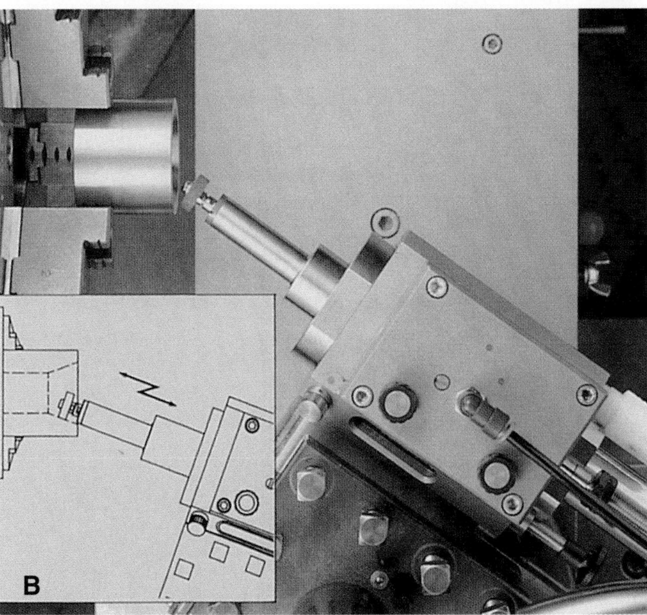

NSK America

Figure 15-34. Internal grinding operations on the lathe. A—Internal grinding. B—Taper grinding.

Because the quill (shaft for mounting the grinding wheel for internal work) is quite slender, use very light cuts and slow feeds to prevent the hole from "bell mouthing," as shown in **Figure 15-36**. For the same reason, it is suggested that the grinding wheel be allowed to "spark out" on the last cut.

SAFETY NOTE

When cleaning the lathe after grinding, use a brush to sweep particles off the lathe. Never use your hand or compressed air. Using compressed air can force particles into the joints of the machine and cause premature wear.

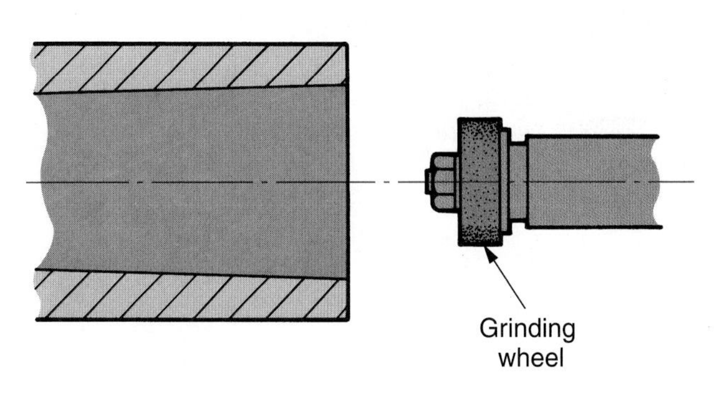

Goodheart-Willcox Publisher

Figure 15-36. Bell mouthing (grinding a hole larger at its mouth) is caused by taking too deep a cut with the grinder, or grinding with a feed that is too rapid.

15.9 Milling on a Lathe

Some lathes can be fitted with a vertical milling attachment, **Figure 15-37**. Such machines are primarily designed for home workshops, but are often used in model and experimental shops. A vise is mounted to the cross-slide which also provides traverse (in and out) movement while longitudinal (back and forth) feed is furnished by the carriage.

A special horizontal milling attachment is available for some lathes. It permits limited milling operations to be performed. The cutter is mounted on an arbor or fitted into the headstock. Cutter depth is controlled by the adjusting screw on the device. Cutter movement is controlled by carriage and cross-slide movements.

15.10 Special Lathe Attachments

A tracing or duplicating unit can be used when several identical pieces must be produced. Several types of duplicating or tracing units are available. One type makes use of flat templates. Another type uses a three-dimensional template or pattern. Most units are hydraulically operated.

The duplicating unit improves quality because each part is an exact duplicate of the master template. However, computer-controlled lathes now do the work formerly done by these units.

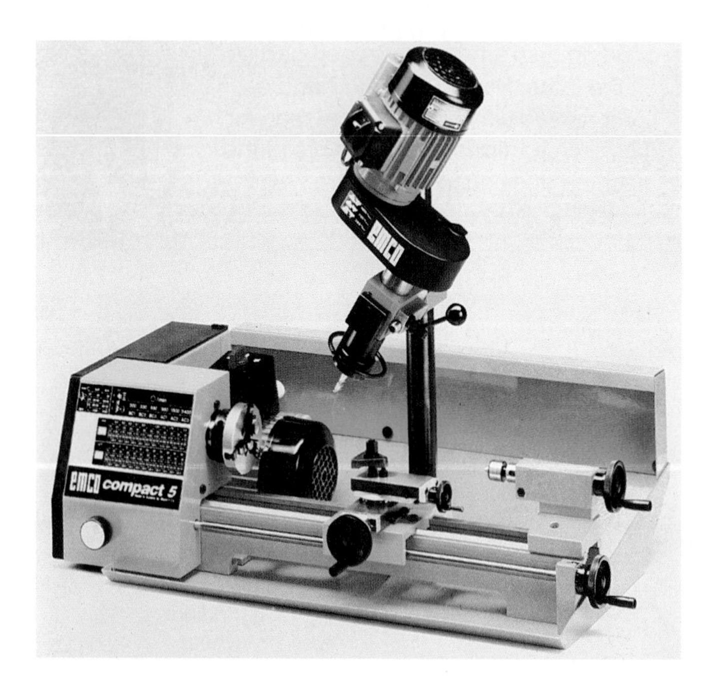

Emco-Maier Corp.

Figure 15-37. A light lathe fitted with a milling attachment. Work-holding devices for milling operations are fitted on the cross-slide after removing the toolholder.

15.11 Industrial Applications of the Lathe

Industry makes wide use of variations of the basic lathe. Conventional metalworking lathes are manufactured in a large range of sizes from the tiny jeweler's lathe to large machines that turn forming rolls for the steel industry.

Heavy-duty engine lathes are used to machine very large workpieces, **Figure 15-38**. The super-precision tool-room lathe is required to meet the close tolerances and fine surface finish specifications of toolrooms, model shops, and research and development laboratories. See **Figure 15-39**.

ID1974/Shutterstock.com

Figure 15-38. A heavy-duty engine lathe on display at a military technologies exposition.

Hardinge Super Precision HLV-DR is a registered trademark of Hardinge, Inc.

Figure 15-39. The Hardinge Super Precision HLV-DR toolroom lathe.

Limited production runs (usually less than 250 pieces) are sometimes produced on a manually operated ***turret lathe***. This is a conventional lathe equipped with a four- to six-sided tool holder called a turret. **Figure 15-40** illustrates how a number of different cutting tools are fitted to the turret. Stops control the length of tool travel and rotate the turret to bring the next cutting tool into position automatically. A cross-slide unit is fitted for turning, facing, forming, and cutoff operations, **Figure 15-41**.

The ***automatic screw machine*** is a variation of the lathe that was developed for high speed production of large numbers of small parts. The machine performs a maximum number of operations, either simultaneously, or in a very rapid sequence.

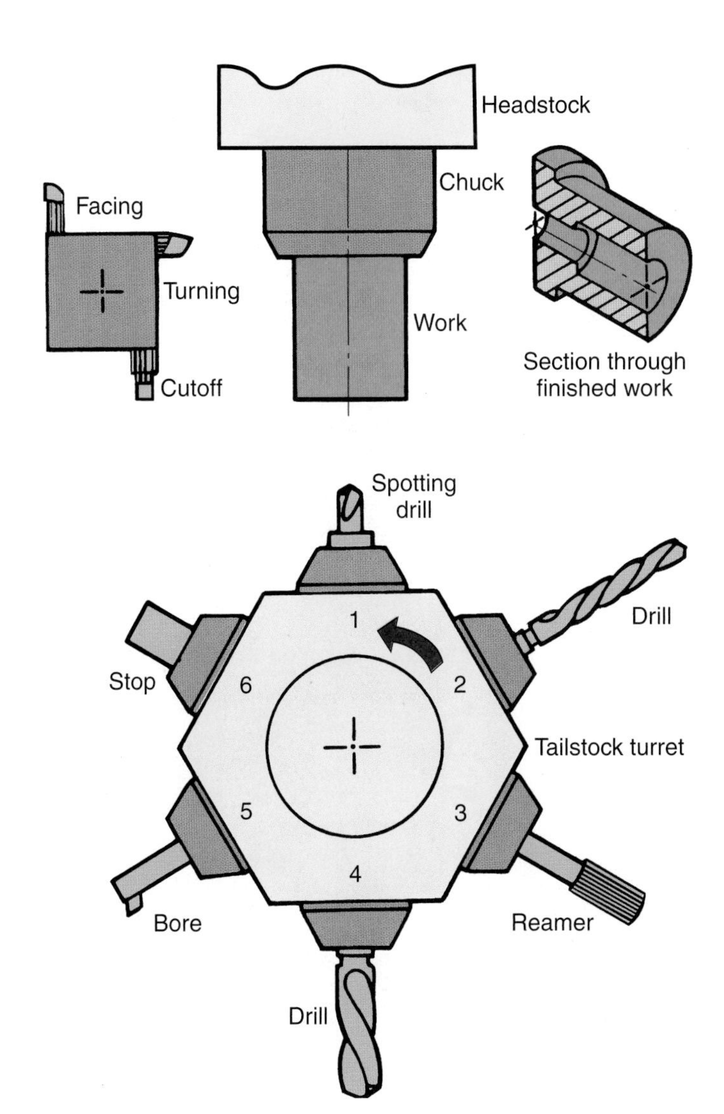

Goodheart-Willcox Publisher

Figure 15-40. Turret in relation to other parts of a lathe. The turret rotates to bring cutting tool into position. Stops control depth of tool cuts.

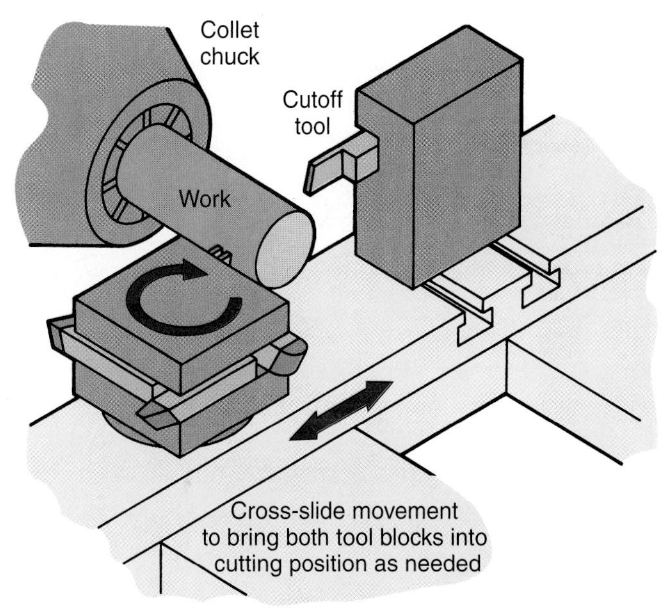

Goodheart-Willcox Publisher

Figure 15-41. The cross-slide on a turret lathe is similar to the cross-slide on a conventional lathe. However, it is fitted with several cutting tools that can be brought into position as needed.

Increasingly, industry is coming to rely on automatic turning centers to produce tiny precision parts in quantity. These centers, referred to as "Swiss-type" machines because they were originally used in the Swiss watchmaking industry, use computer control to perform a number of operations in sequence, producing a finished part. See **Figure 15-42**.

Work that is too large or too heavy to be turned in a horizontal position is machined on a vertical lathe, **Figure 15-43**. These huge machines, also known as ***boring mills***, are capable of turning and boring work with diameters up to 40′ (12 m).

Portable turning equipment is available for work in the field, such as chamfering the ends of large pipe prior to welding. See **Figure 15-44**.

Computer numerically controlled (CNC) lathes and turning machines are widely used for industrial production. With proper programming, these machine tools are capable of producing complex work with great accuracy and repeatability. A more detailed description of CNC machine tools and automated manufacturing operations will be discussed later in this book.

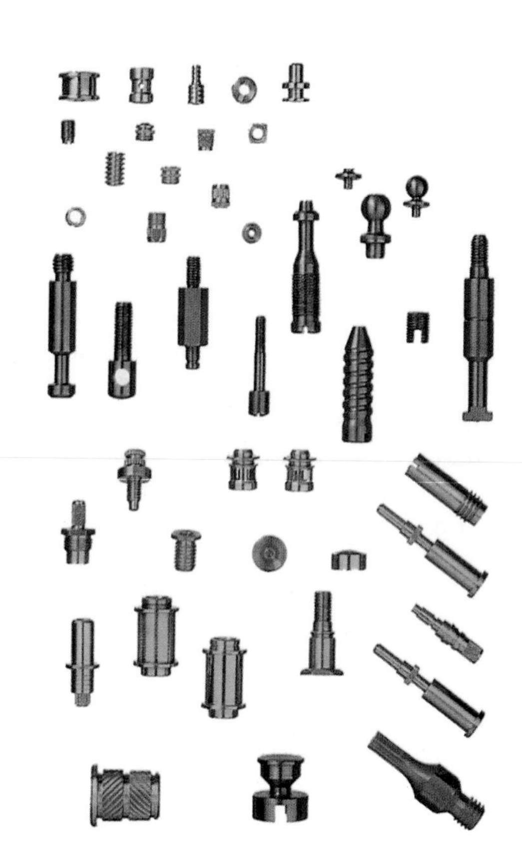

Tornos-Bechler S.A.

Figure 15-42. Swiss-type automatic turning center. A—This high-precision turning center machines small parts from bar stock ranging from 1 mm to 10 mm in diameter at high production rates. B—These tiny components, produced by a Swiss-type automatic turning center, are used in precision instruments.

AMT—The Association for Manufacturing Technology

Figure 15-43. A large boring mill. Workpieces are mounted on the large turntable and rotated into position for machining.

Tri-Tool, Inc.

Figure 15-44. A portable lathe that can be taken into the field. This worker is shown turning the end of a high-pressure gas pipeline to prepare it for being welded.

Chapter Review

Summary

- Boring is a machining operation performed on a lathe to enlarge a hole.
- Knurling forms horizontal or diamond shaped serrations on the circumference of the work.
- Drilling, reaming, filing, and polishing operations can be performed on a lathe with the proper tooling.
- Steady rests and follower rests are all work-holding devices that offer additional support to securely hold long, thin work during lathe operations.
- Mandrels allow an outside diameter to be machined concentric with an inside diameter and multiple workpieces to be machined at once.
- Grinding and milling operations can be performed on a lathe with the proper tooling and workholding attachments.
- Industrial applications of the lathe make use of special lathe attachments and variations of the basic lathe machine.

Review Questions

Answer the following questions using the information provided in this chapter.

1. Boring is a(n) _____.
   A. drilling operation
   B. internal machining operation in which a single-point cutting tool is used to enlarge a hole
   C. external machining operation in which a single-point cutting tool is used to reduce the diameter of a hole
   D. All of the above.
   E. None of the above.
2. Drills that are used on the lathe are fitted with _____ shanks.
   A. straight
   B. taper
   C. fluted
   D. Both A and B.
   E. Both B and C.
3. When is reaming done?
4. The process of forming horizontal or diamond-shaped serrations on the circumference of the work is called _____.
5. Knurling is commonly done to _____.
   A. provide a gripping surface
   B. change the appearance of the work
   C. increase the work's diameter
   D. All of the above.
   E. None of the above.
6. When is filing on the lathe usually done?

7. Polishing is an operation used to produce a(n) _____ on the work.
8. When is a steady rest used?
9. What is the difference between a steady rest and a follower rest?
10. There are times when a shaft is unsuitable as a bearing surface and cannot be used with a steady rest. When this occurs, a(n) _____ can be used so the shaft can be supported with the steady rest.
11. What is a mandrel and when is it used?
12. A mandrel is usually pressed into the work with a(n) _____.
13. Internal and external grinding can be done on a lathe with a(n) _____.
14. What should be done to protect the lathe from the abrasive dust and grit created during grinding operations?
    A. Use a nonabrasive grinding wheel.
    B. Cover the bed and moving parts with a heavy cloth.
    C. Use a soft abrasive grinding wheel.
    D. All of the above.
    E. None of the above.
15. Automatic turning centers, also known as _____ machines, can produce tiny precision parts in quantity.
16. Work that is too large or too heavy to be turned in the horizontal position is turned on huge machines called vertical lathes or _____.

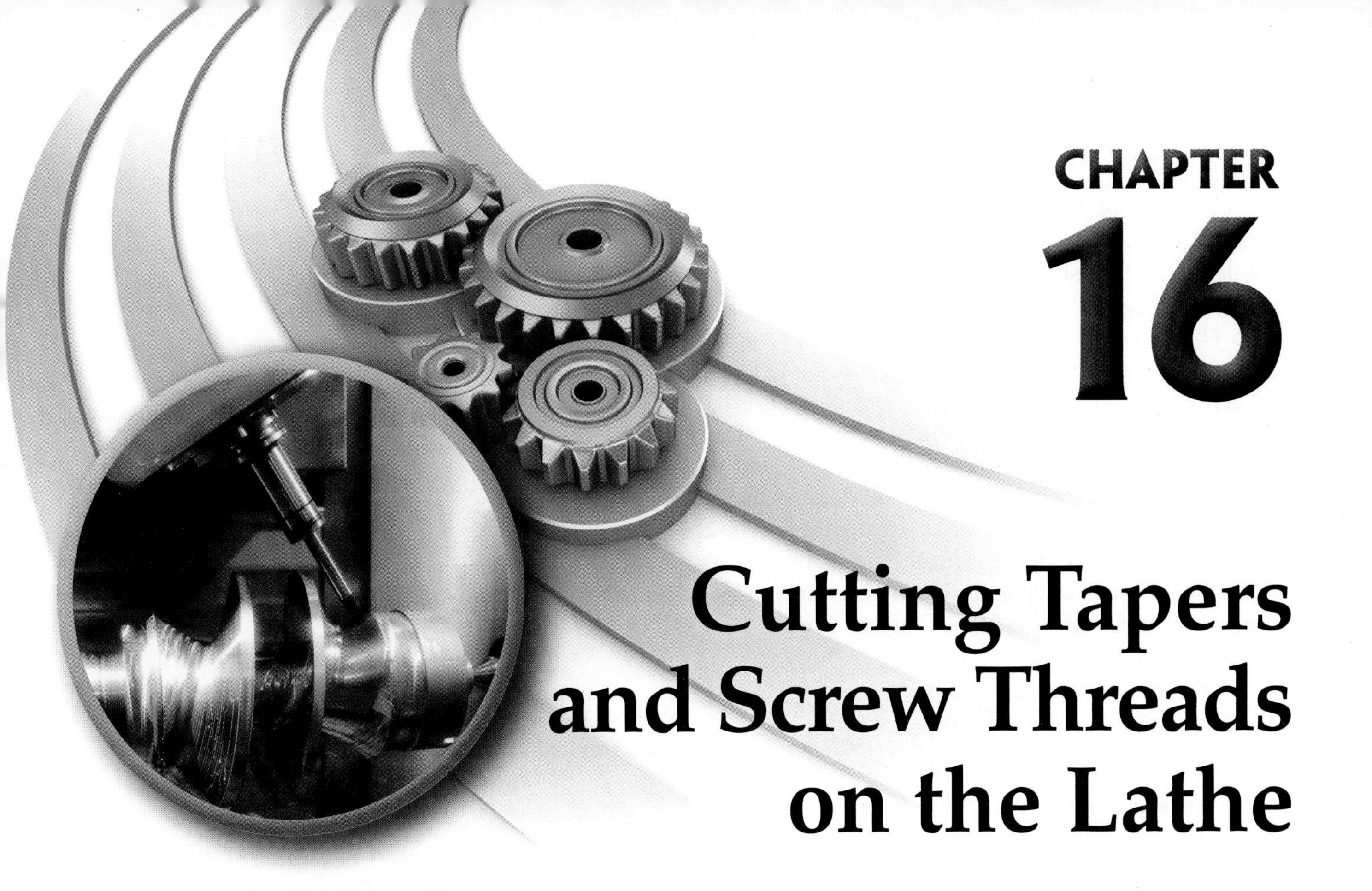

CHAPTER 16

Cutting Tapers and Screw Threads on the Lathe

Learning Objectives

After studying this chapter, you will be able to:

- Describe how a taper is turned on a lathe.
- Calculate tailstock setover for turning a taper.
- Safely set up and operate a lathe for taper turning.
- Describe the various forms of screw threads.
- Cut screw threads on a lathe.

Technical Terms

external thread
internal thread
lead
major diameter
minor diameter
offset tailstock method
pitch
pitch diameter
setover
taper
taper attachment
thread cutting stop
thread dial
three-wire method

16.1 Taper Turning

A section of material is considered to have a ***taper*** when it increases or decreases in diameter at a uniform rate. Bell-shaped pieces are not considered tapers. See **Figure 16-1**. A cone is an example of a taper. The "wedging" action of a taper makes it ideal as a means for driving drills, milling arbors, end mills, and centers. In addition, it can be assembled and disassembled easily, and will automatically align itself in a similarly tapered hole each time. Taper can be stated in taper per inch, taper per foot, degrees, millimeters per 25 mm of length, or as a ratio, **Figure 16-2**.

There are five principal methods of machining tapers on a lathe. Each has its advantages and disadvantages. The five methods, as listed in **Figure 16-3**, are as follows:

- The compound rest method.
- The offset tailstock method.
- Using a taper attachment.
- Using a square-nose tool.
- Using a reamer.

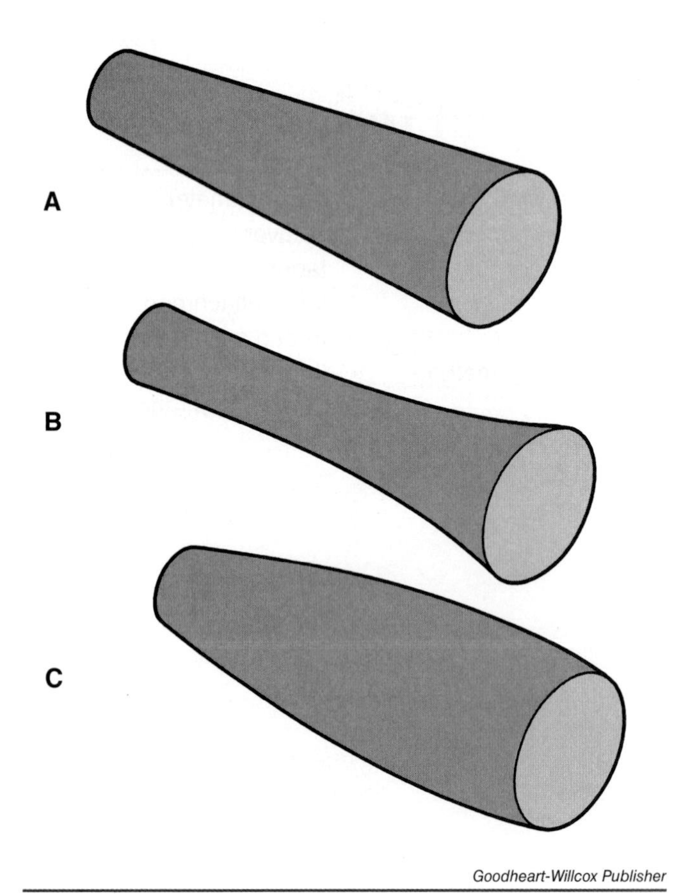

Figure 16-1. Taper. A—The diameter of a taper increases or decreases at a uniform rate. B and C—These pieces are "bell shaped," rather than tapered.

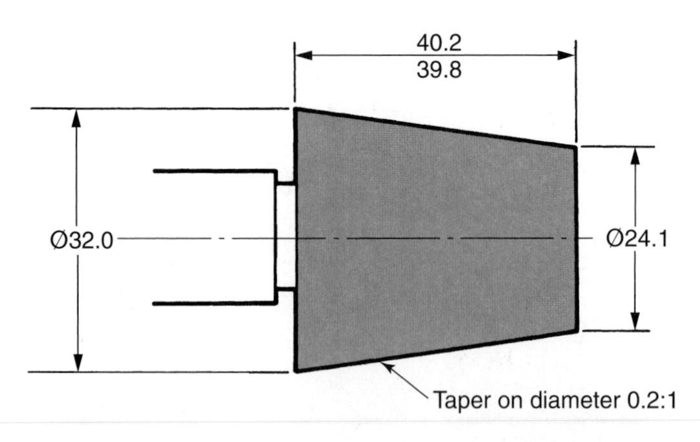

Figure 16-2. Taper may be stated as a ratio (0.2:1, in the example above), taper per inch, taper per foot, degrees, or in millimeters per 25 mm.

16.1.1 Taper Turning with Compound Rest

The compound rest method of turning a taper is the easiest. Either internal or external tapers can be machined, as shown in **Figure 16-4**.

Taper length is limited, however, by the movement of the compound rest. Because the compound rest base is graduated in degrees, **Figure 16-5**, the taper must be converted to degrees. A conversion table may be used. See **Figure 16-6**.

A careful study of the print will show whether the angle given is from center or is the included angle. **Figure 16-7** shows the difference in methods of measuring angles. If an included angle is given, it must be divided by two to obtain the angle from the centerline.

Ways of Machining Tapers

Method	Advantages and disadvantages	Information needed
1. Compound	Length of taper limited. Will cut external and internal taper.	Must know the taper angle.
2. Offset tailstock	External taper only. Must work between centers.	Taper per inch or taper per foot.
3. Taper attachment	Best method to use.	Angle or taper per inch or foot.
4. Tool bit	Very short taper.	Taper angle.
5. Reamer	Internal only.	Taper number.

Goodheart-Willcox Publisher

Figure 16-3. Methods by which tapers can be turned on a lathe.

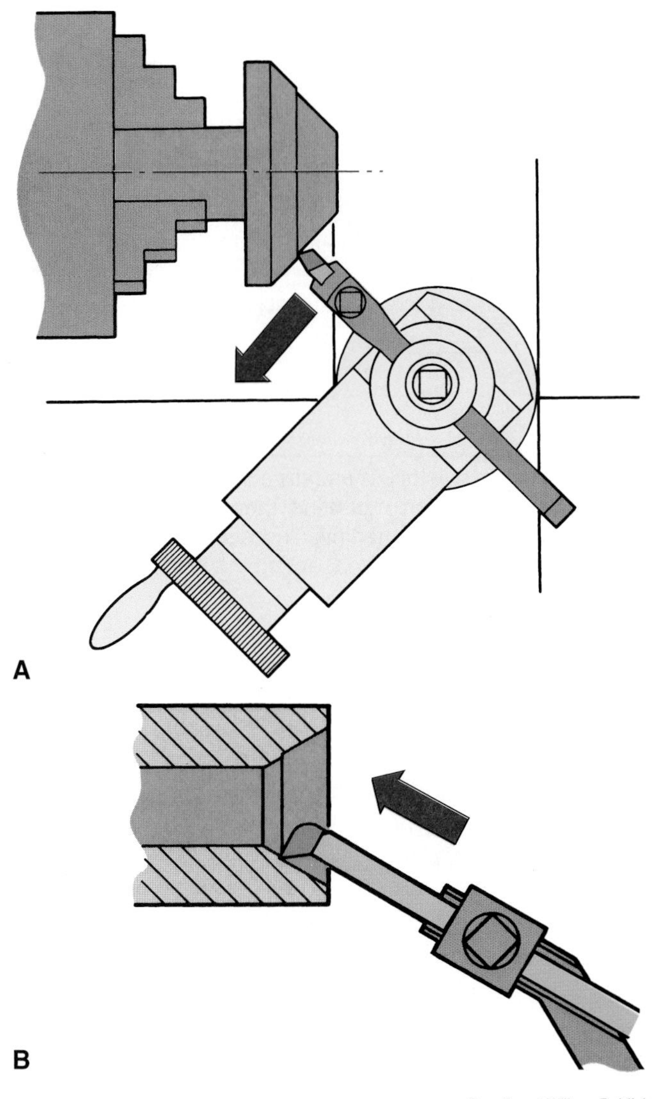

Goodheart-Willcox Publisher

Figure 16-4. Cutting tapers using the compound rest. A—External taper. The cut is being made from small diameter to large diameter. B—Internal taper being turned with the compound rest.

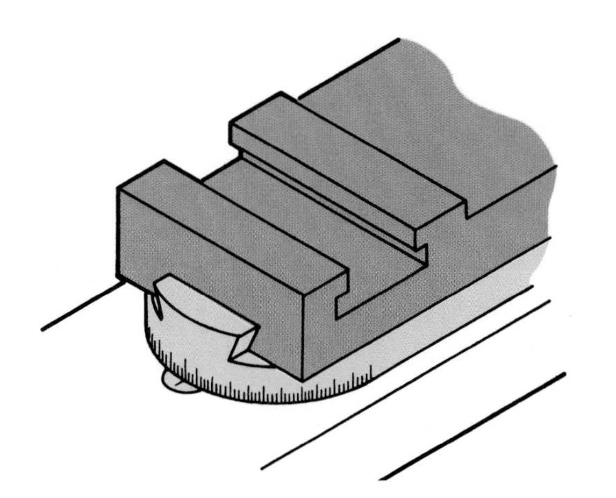

Goodheart-Willcox Publisher

Figure 16-5. Base of the compound rest is marked in degrees to aid in precise positioning.

Taper per Foot with Corresponding Angles

Taper per foot	Included angle	Angle with centerline
1/16	0° 17′ 53″	0° 8′ 57″
1/8	0° 35′ 47″	0° 17′ 54″
3/16	0° 53′ 44″	0° 26′ 52″
1/4	1° 11′ 38″	0° 35′ 49″
5/16	1° 29′ 31″	0° 44′ 46″
3/8	1° 47′ 25″	0° 53′ 42″
7/16	2° 5′ 18″	1° 2′ 39″
1/2	2° 23′ 12″	1° 11′ 36″
9/16	2° 41′ 7″	1° 20′ 34″
5/8	2° 58′ 3″	1° 29′ 31″
11/16	3° 16′ 56″	1° 38′ 28″
3/4	3° 34′ 48″	1° 47′ 24″
13/16	3° 52′ 42″	1° 56′ 21″
7/8	4° 10′ 32″	2° 5′ 16″
15/16	4° 28′ 26″	2° 14′ 13″
1	4° 46′ 19″	2° 23′ 10″

Goodheart-Willcox Publisher

Figure 16-6. You can use this table to convert taper per foot into corresponding angles for adjustment of the compound rest.

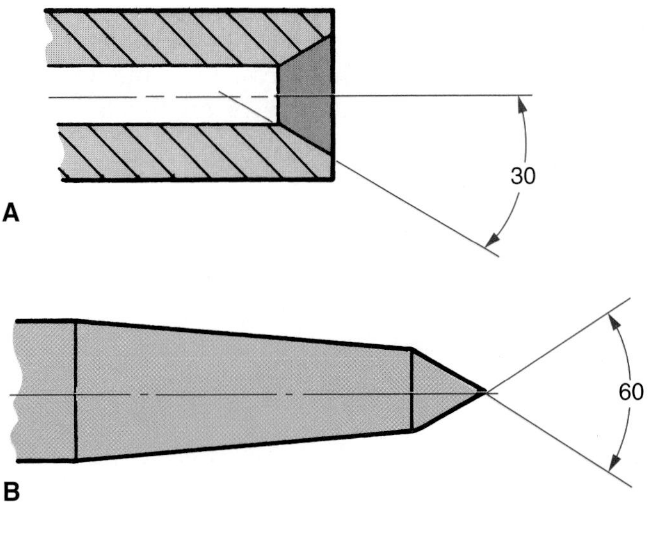

Goodheart-Willcox Publisher

Figure 16-7. The two methods used to measure angles. A—Angle measured from the centerline of the workpiece. B—Measurement of the included angle.

With the lathe's centerline representing 0°, pivot the compound rest to the desired angle and lock it in position. The usual practice is to turn a taper from the smaller diameter to the larger diameter. Refer to **Figure 16-4A**.

As will be the case when turning all tapers, the cutting tool must be set on exact center. Select a toolholder that will provide ample clearance.

To machine a taper, bring the cutting tool into position with the work and lock the carriage to prevent it from shifting during the turning operation.

Since there is no power feed for the compound rest, the cutting tool must be fed evenly with both hands to achieve a smooth finish. The entire cut must be made without stopping the cutting tool. The compound rest is moved back to the starting point and positioned with the cross-slide for the next cut.

When tapers are cut with a compound rest, the work can be mounted between centers or held in a chuck. A suitable boring bar is needed when machining internal tapers. Some internal tapers are finished to size with a taper reamer.

16.1.2 Taper Turning by Offset Tailstock Method

The ***offset tailstock method***, also known as the tailstock setover method, is also used for taper turning, **Figure 16-8**. Jobs that can be turned between centers may be taper turned by this technique. Only external tapers can be machined in this way, however.

Most lathe tailstocks consist of two parts, which permits the upper portion to be shifted off center, **Figure 16-9**. This movement, referred to as ***setover***, is accomplished by loosening the anchor bolt that locks the tailstock to the ways, then making the proper adjustments with screws on the tailstock.

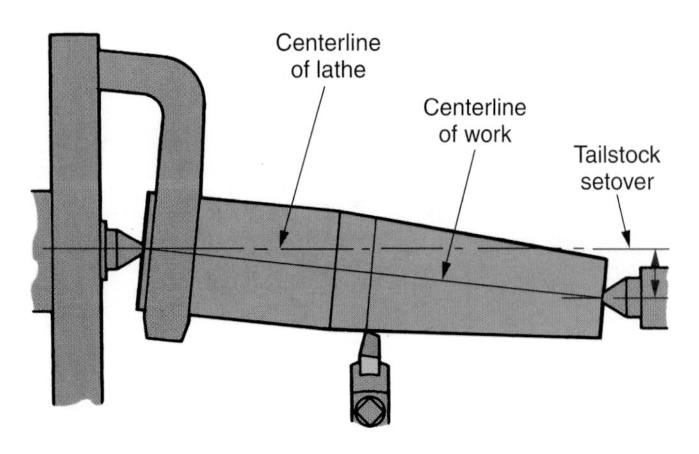

Goodheart-Willcox Publisher

Figure 16-8. Machining a taper using the offset tailstock method.

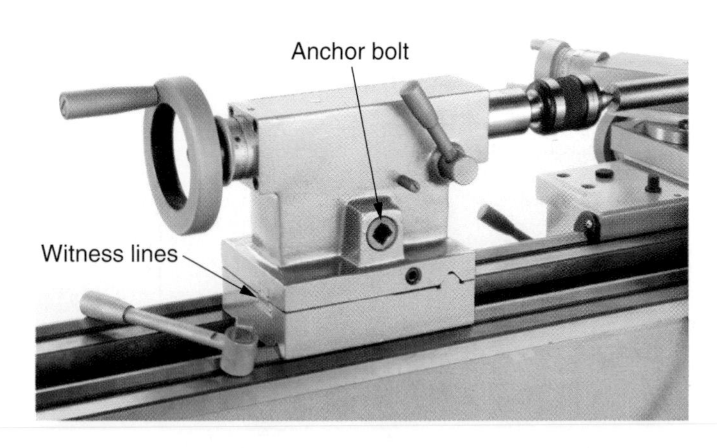

Photo courtesy of Grizzly Industrial, Inc. www.grizzly.com

Figure 16-9. The tailstock is usually constructed in two parts. This allows the section mounting the center to be shifted relative to the lathe's centerline. The distance off center, or setover, can be checked by observing the witness lines.

After the setover has been made, the screws are drawn up snug, but not tight.

Calculating Tailstock Setover

Taper turning by this technique is not a precise method and requires some "trial and error" adjustments to produce an accurate tapered section. The approximate setover can be calculated when certain basic information is known.

Offset must be calculated for each job, because the length of the piece plays an important part in the calculations. When lengths of the pieces vary, different tapers will be produced with the same tailstock offset, **Figure 16-10**.

The following terms are used with calculating tailstock setover. See **Figure 16-11**.

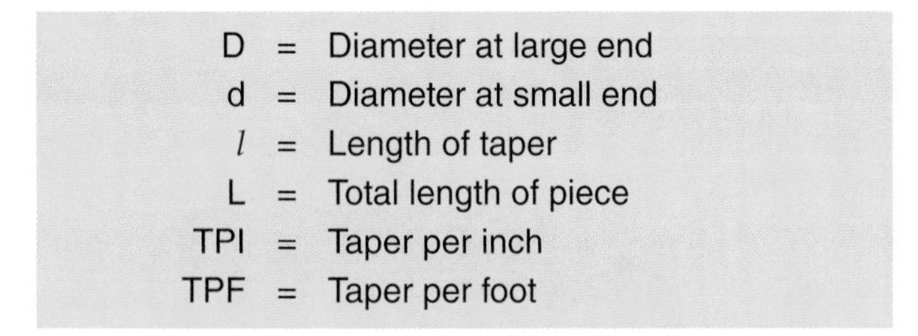

D	=	Diameter at large end
d	=	Diameter at small end
l	=	Length of taper
L	=	Total length of piece
TPI	=	Taper per inch
TPF	=	Taper per foot

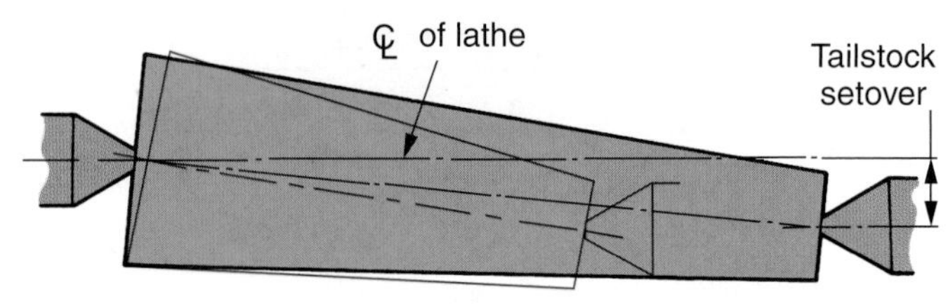

Goodheart-Willcox Publisher

Figure 16-10. Length of work causes taper to vary even though tailstock offset remains the same.

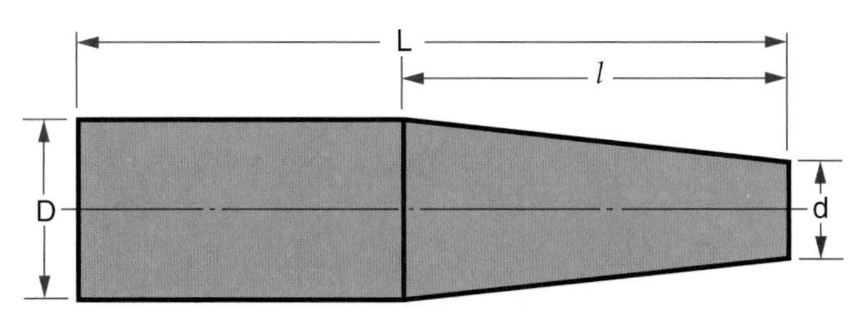

Goodheart-Willcox Publisher

Figure 16-11. Basic taper information. D = diameter at large end of taper; d = diameter at small end of taper; *l* = length of taper; L = total length of piece.

Calculating Setover When Taper per Inch Is Known

Information needed:

TPI = Taper per inch

L = Total length of piece

Formula used:

$$\text{Offset} = \frac{L \times TPI}{2}$$

Example: What will be the tailstock setover for the following job?

Taper per inch = 0.125

Total length of piece = 8.000

$$\text{Offset} = \frac{L \times TPI}{2} = \frac{8.000 \times 0.125}{2} = 0.500''$$

Note: The same procedure would be followed when using metric units. However, all dimensions would be in millimeters.

Calculating Setover When Taper per Foot Is Known

When taper per foot (TPF) is known, it must be converted to taper per inch (TPI). The following formula takes this into account:

$$\text{Offset} = \frac{TPF \times L}{24}$$

Calculating Setover When Dimensions of Tapered Sections Are Known but TPI or TPF Is Not Given

Plans often do not specify TPI or TPF, but do give other pertinent information. Calculations will be easier if all fractions are converted to decimals. All dimensions must either be in inches or in millimeters.

Information needed:

D = Diameter at large end

d = Diameter at small end

l = Length of taper

L = Total length of piece

Formula used:

$$\text{Offset} = \frac{L \times (D-d)}{2 \times l}$$

Example: Calculate the tailstock setover for the following job.

D = 1.250″

d = 0.875″

l = 3.000″

L = 9.000″

$$\text{Offset} = \frac{L \times (D-d)}{2 \times l} = \frac{9.000 \times (1.250-0.875)}{2 \times 3.000} = \frac{9.000 \times 0.375}{6} = 0.562''$$

Calculating Setover When Taper Is Given In Degrees

The space available in this text does not permit the introduction of basic trigonometry, which is necessary to make these calculations. However, any good machinist's handbook will provide this information. At least one such book should be part of every machinist's toolbox.

Measuring Tailstock Setover

When an ample tolerance (±0.015″ or 0.05 mm) is allowed, the setover can be measured with a steel rule. There are two ways to measure, as follows:

- Place a rule that has graduations on both edges between the center points, **Figure 16-12A**. Measure the distance between the center points.
- Measure the distance between the two witness lines on the tailstock base, **Figure 16-12B**.

Accurate work requires care in making the tailstock setover. An additional factor enters into the calculations—the distance that the center point enters the piece. Typically, 1/4″ (6.5 mm) is an ample allowance. It must be subtracted from the total length of the piece.

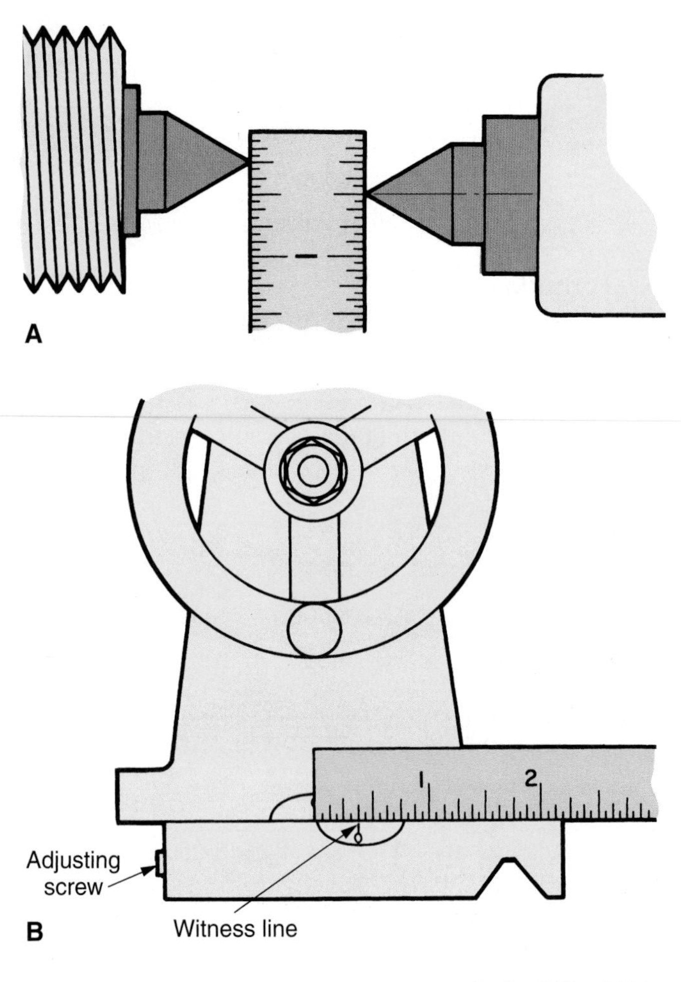

Goodheart-Willcox Publisher

Figure 16-12. Measuring setover. A—Approximate tailstock setover can be determined by measuring distance between center points. B—Approximate setover can also be determined by measuring distance between witness lines on the tailstock.

Use the appropriate method to calculate the offset. A precise setover may be made using the micrometer collar on the lathe cross-slide. See **Figure 16-13**.

1. Clamp the toolholder in a reverse position in the tool post.
2. Turn the cross-slide screw back to remove all play.
3. Turn in the compound rest until the toolholder can be felt with a piece of paper between the toolholder and tailstock spindle.
4. Use the micrometer collar and turn out the cross-slide screw the distance the tailstock is to be set over.
5. Move the tailstock over until the spindle touches the paper in same manner described in Step 3.
6. Check the setting again after "snugging up" the adjusting screws.

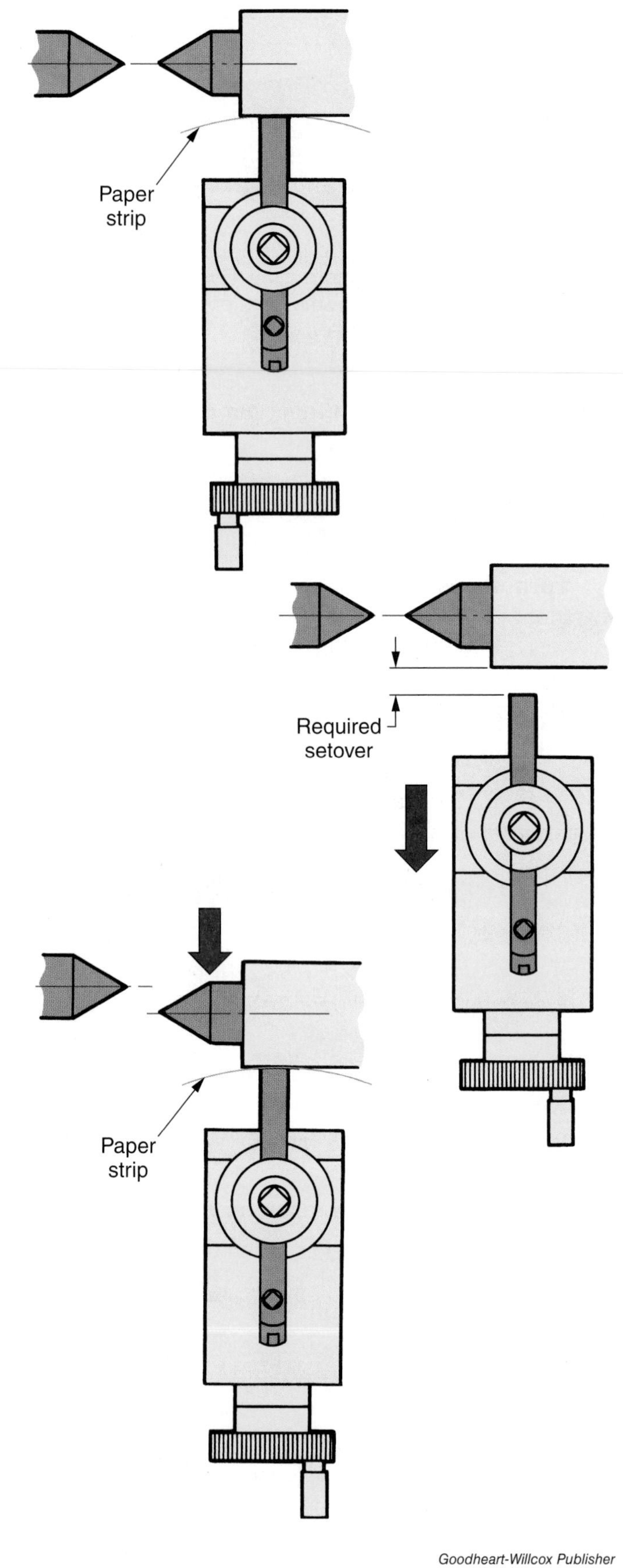

Goodheart-Willcox Publisher

Figure 16-13. Using the micrometer collar of the compound rest to make the setover measurement.

In place of the toolholder and paper strip, a dial indicator can be used to establish the offset. See **Figure 16-14**.

1. Mount the dial indicator in the tool post.
2. Position it with the cross-slide until the indicator reads zero when in contact with the tailstock spindle. There should be no "play" in the cross-slide.
3. Set the tailstock over the required distance using the dial indicator to make the measurement.
4. Recheck the reading after "snugging up" the adjusting screws. Make additional adjustments if any deviation in the indicator reading occurs.

Cutting a Taper

When cutting a taper, additional strain is imposed on the centers because they are out-of-line and do not bear true in the center holes. Because the pressures imposed are uneven, the work is more apt to heat up than when doing conventional turning between centers. It must be checked frequently for binding. A bell-type center drill offers some advantage in reducing strain. Some machinists prefer a center with a ball tip to produce an improved bearing surface. See **Figure 16-15**.

Make the cuts as in conventional turning. However, cutting should start at the small end of the taper.

16.1.3 Turning a Taper with a Taper Attachment

A ***taper attachment*** is a guide that can be attached to most lathes. It is an accurate way to cut tapers and offers advantages over other methods of machining tapers.

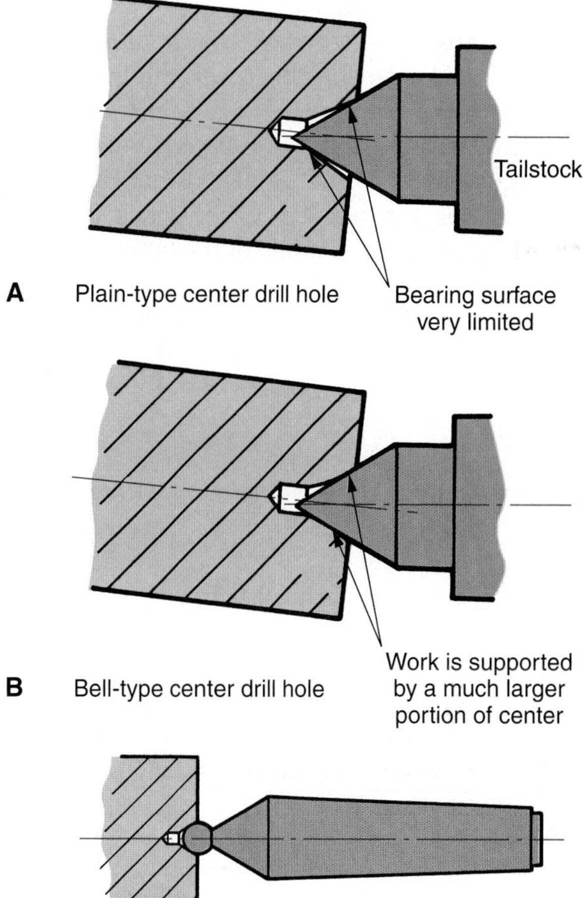

Goodheart-Willcox Publisher

Figure 16-15. Taper turning done by the offset tailstock method is hard on the tailstock center. A—Center point does not bear evenly in conventional center hole. B—A center hole drilled with a bell-type center drill reduces the problem by providing more bearing surface. C—A ball-tipped center lessens pressure on tail center when turning tapers.

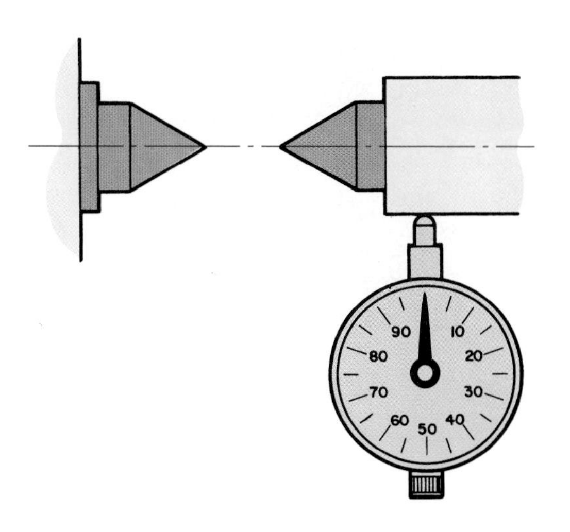

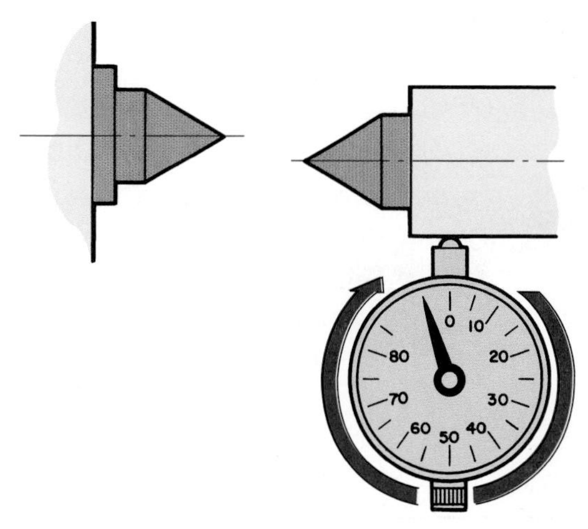

Goodheart-Willcox Publisher

Figure 16-14. A dial indicator can also be used to measure amount of setover.

Both internal and external tapers can be cut. This helps ensure an accurate fit for mating parts. Once the attachment has been set, the taper can be machined on material of various lengths. Work can be held by any conventional means. One end of the taper attachment swivel bar is graduated in total taper in inches per foot. The other end is graduated to indicate the included angle of the taper in degrees.

The lathe does not have to be altered. The machine can be used for straight turning by locking out the taper attachment. No realignment of the lathe is necessary.

Types of Taper Attachments

There are two types of taper attachments, plain and telescopic. See **Figure 16-16**. The plain taper attachment requires disengaging the cross-slide screw from the cross-slide feed nut. The cutting tool is advanced by using the compound rest feed screw.

The telescopic taper attachment is made in such a way that it is not necessary to disconnect the cross-slide feed nut. The tool can be advanced into the work with the cross-slide screw in the usual manner.

Setting a Taper Attachment

1. Study the plans and, if necessary, calculate the taper. Set the swivel bar as specified from the calculations.
2. Mount the work in the machine.
3. Slide the taper attachment unit to a position that will permit the cutting tool to travel the full length of the taper. Lock it to the ways.
4. Move the carriage to the right until the cutting tool is about 1″ (25 mm) away from the end of the work. This will permit any play to be taken up before the tool starts to cut.
5. If the machine is fitted with a plain taper attachment, tighten the binding screw that engages the cross-slide feed to the attachment.
6. Oil the bearing surfaces of the taper attachment and make a trial cut. If necessary, readjust until the taper is being cut to specifications. Complete the cutting operation.

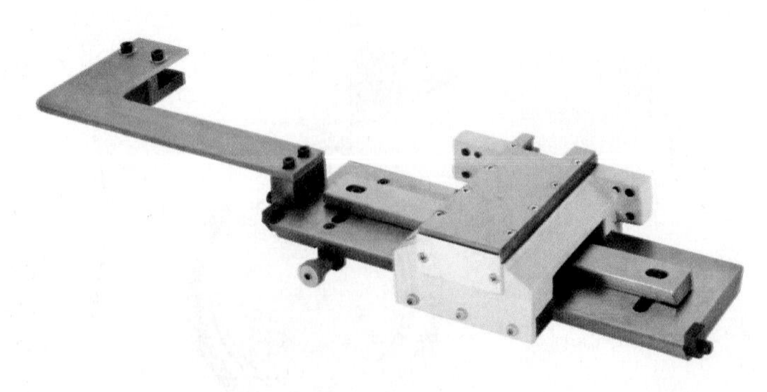

Photo courtesy of Grizzly Industrial, Inc. www.grizzly.com

Figure 16-16. This telescopic taper attachment is fitted to the rear of the carriage and permits taper turning without disengaging the cross-slide feed nut.

16.1.4 Turning a Taper with a Square-Nose Tool

Using a square-nose tool is a taper technique limited to the production of short tapers, **Figure 16-17**. The cutter bit is ground with a square nose and set to the correct angle with the protractor head and blade of a combination set.

The tool is positioned on center and fed into the revolving work. "Chatter" can be minimized by running the work at a slow spindle speed. The carriage must be locked to the ways.

Before using any of the taper-turning techniques on work mounted between centers, it is very important that centers be "zeroed in" (put in perfect alignment). Then the necessary adjustments (tailstock setover, taper attachment adjustment) can be made.

16.2 Measuring Tapers

There are two basic methods of testing the accuracy of machined tapers: the comparison method or direct measurement of the taper.

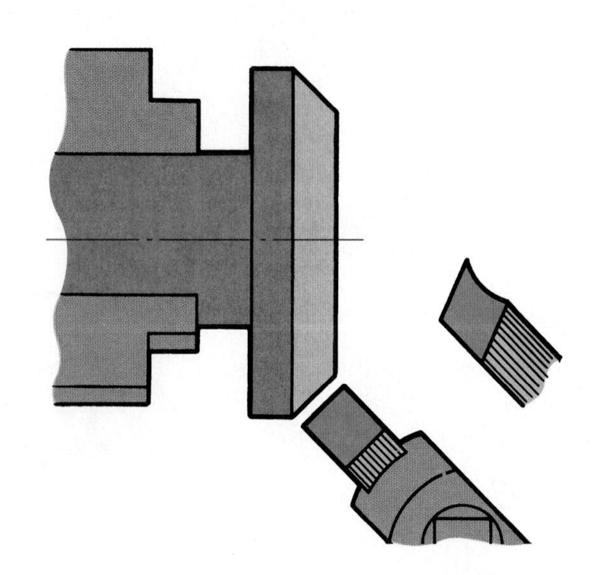

Goodheart-Willcox Publisher

Figure 16-17. A short taper can be turned with a square-nose tool.

16.2.1 Measuring Tapers by Comparison

Taper plug gages and taper ring gages serve two purposes, **Figure 16-18**. They measure the basic diameter of the taper as well as the angle of slope. The angle is checked by applying layout dye to the machined surface or plug gage. The blued section is inserted into the mating part and slowly rotated. If the layout dye rubs off evenly, it indicates that the taper is correct. If the layout dye rubs off unevenly, **Figure 16-19**, the remaining material will show where the taper is incorrect and indicate what machine adjustments are needed.

Gages are also provided with notches to indicate the specified tolerance in taper diameter. The indentations show the go and no-go limits, **Figure 16-20**.

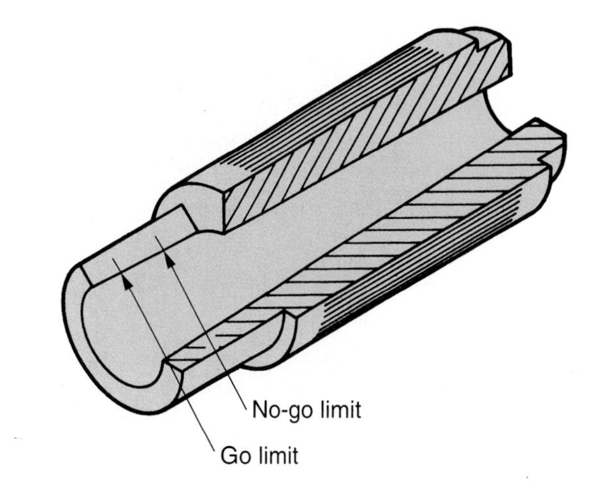

Goodheart-Willcox Publisher

Figure 16-20. Typical go and no-go ring gage for measuring tapers.

16.2.2 Direct Measurement of Tapers

A taper test gage is sometimes used to check taper accuracy, **Figure 16-21**. It consists of a base with two adjustable straight edges. Slots in the straight edges permit adapting the gage to check different tapers. The taper test gage is set by using two discs of known size which are located the correct distance apart.

Another technique for checking or measuring tapers is to set the tapered section on a surface plate.

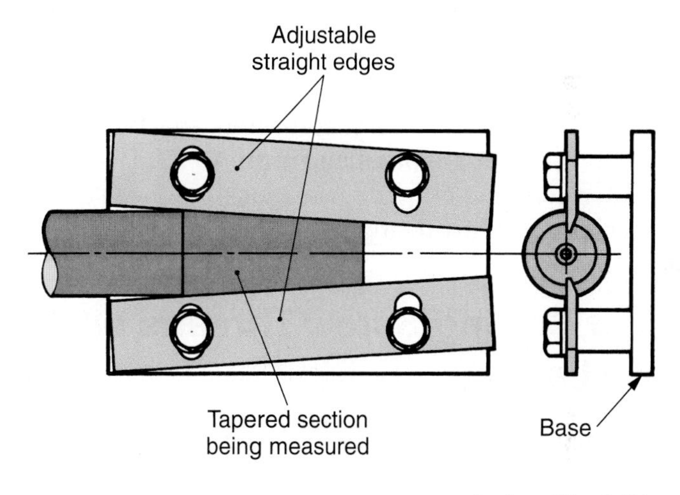

Goodheart-Willcox Publisher

Figure 16-21. A taper test gage can be set for different tapers.

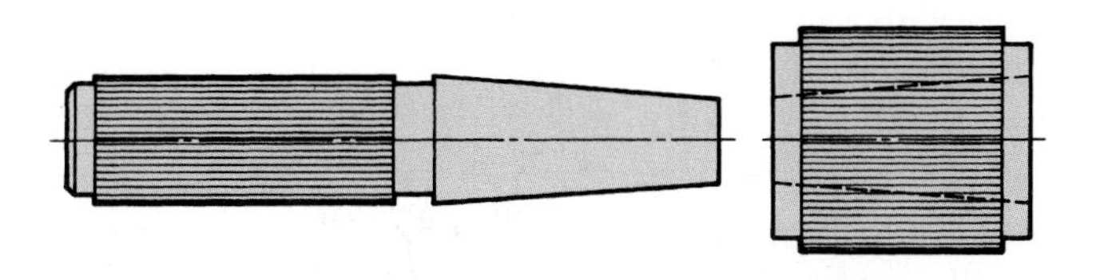

Goodheart-Willcox Publisher

Figure 16-18. Left—Plug gage. Right—Ring gage.

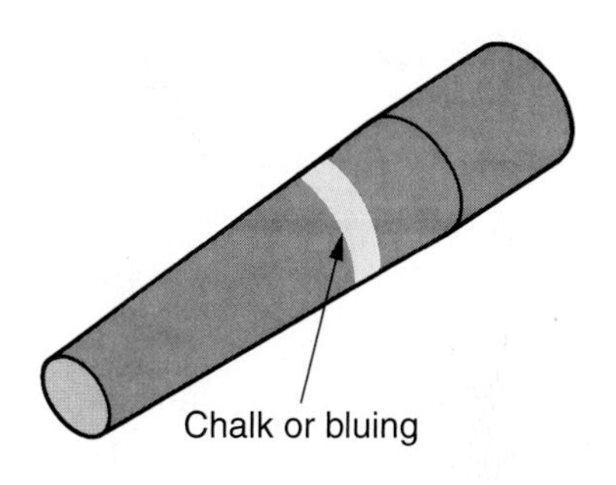

Goodheart-Willcox Publisher

Figure 16-19. When chalk or bluing does not rub off evenly, it indicates that taper does not fit properly and additional machine adjustments will have to be made.

Two gage blocks or ground parallels of the same height are placed on opposite sides of the taper. Two cylindrical rods (sections of drill rod are satisfactory) of the same diameter are placed on the blocks. See **Figure 16-22**. The distance across the rods is then measured with a micrometer.

Blocks 1″, 3″, or 6″ (25, 75, or 150 mm) taller than those used for the first reading are substituted. The rods are the same diameter as those used to make the first reading. A second reading is made, **Figure 16-22B**. The taper per foot then can be determined. First, subtract to find the difference between the two measurements. Then multiply it by twelve (if the readings were made 1″ apart), by four (if they were made 3″ apart), or by two (if they were made 6″ apart).

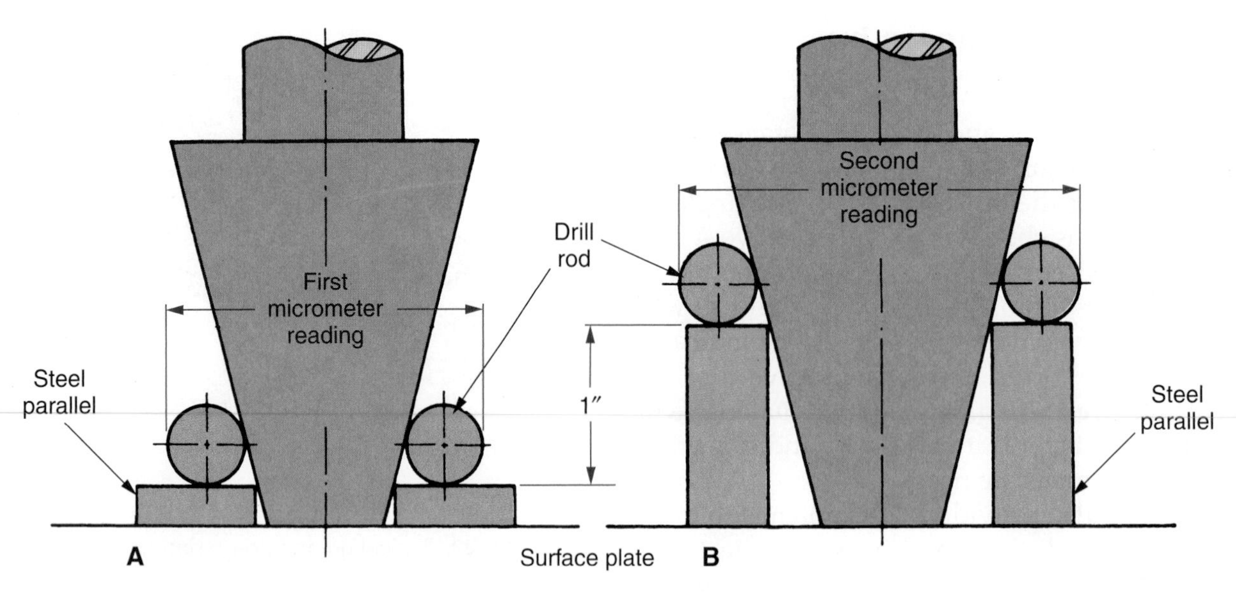

Figure 16-22. Measuring a taper using parallels, drill rod, micrometer, and a surface plate. A—Setup for first measurement. B—Setup for second measurement.

A sine bar is a very accurately machined bar with edges that are parallel, **Figure 16-23**. The bar is used in conjunction with gage blocks and sine tables to precisely measure angles.

16.3 Cutting Screw Threads on the Lathe

Screw threads are utilized for many applications. The more important are as follows:

- Making adjustments (cross-feed on a lathe).
- Assembling parts (nuts, bolts, and screws).

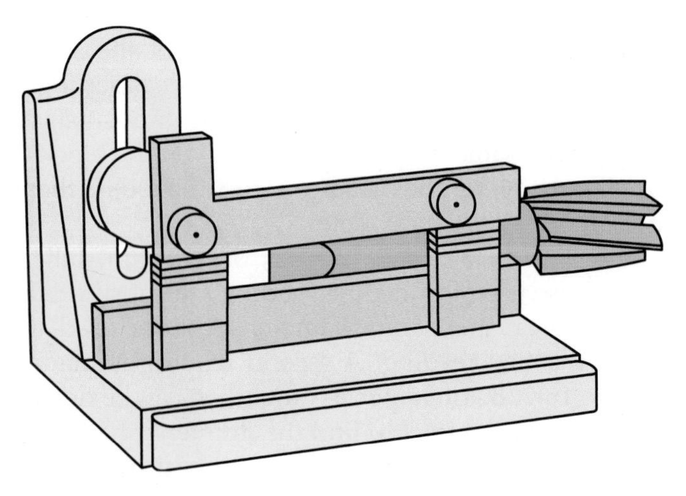

Figure 16-23. A sine bar and precision gage blocks can also be used to measure a taper.

- Transmitting motion (lead screw on a lathe).
- Applying pressure (clamps).
- Making measurements (micrometer).

16.3.1 Screw Thread Forms

The first screw threads cut by machine were square in cross-section. Since that time, many different thread forms have been developed, including American National, Unified, Sharp V, Acme, worm threads, and others. Each thread form has a specific use and a formula for calculating its shape and size. See **Figure 16-24**. More than 75% of all threads cut in the United States are of the Unified (UN) 60° type.

The following terms relate to screw threads, as shown in **Figure 16-25**:

- ***External threads*** are cut on the outside surface of the piece.
- ***Internal threads*** are cut on the inside surface of the piece.
- ***Major diameter*** is the largest diameter of the thread.
- ***Minor diameter*** is the smallest diameter of the thread.
- ***Pitch diameter*** is the diameter of an imaginary cylinder that would pass through threads at such points to make the width of the thread and width of the thread groove equal.

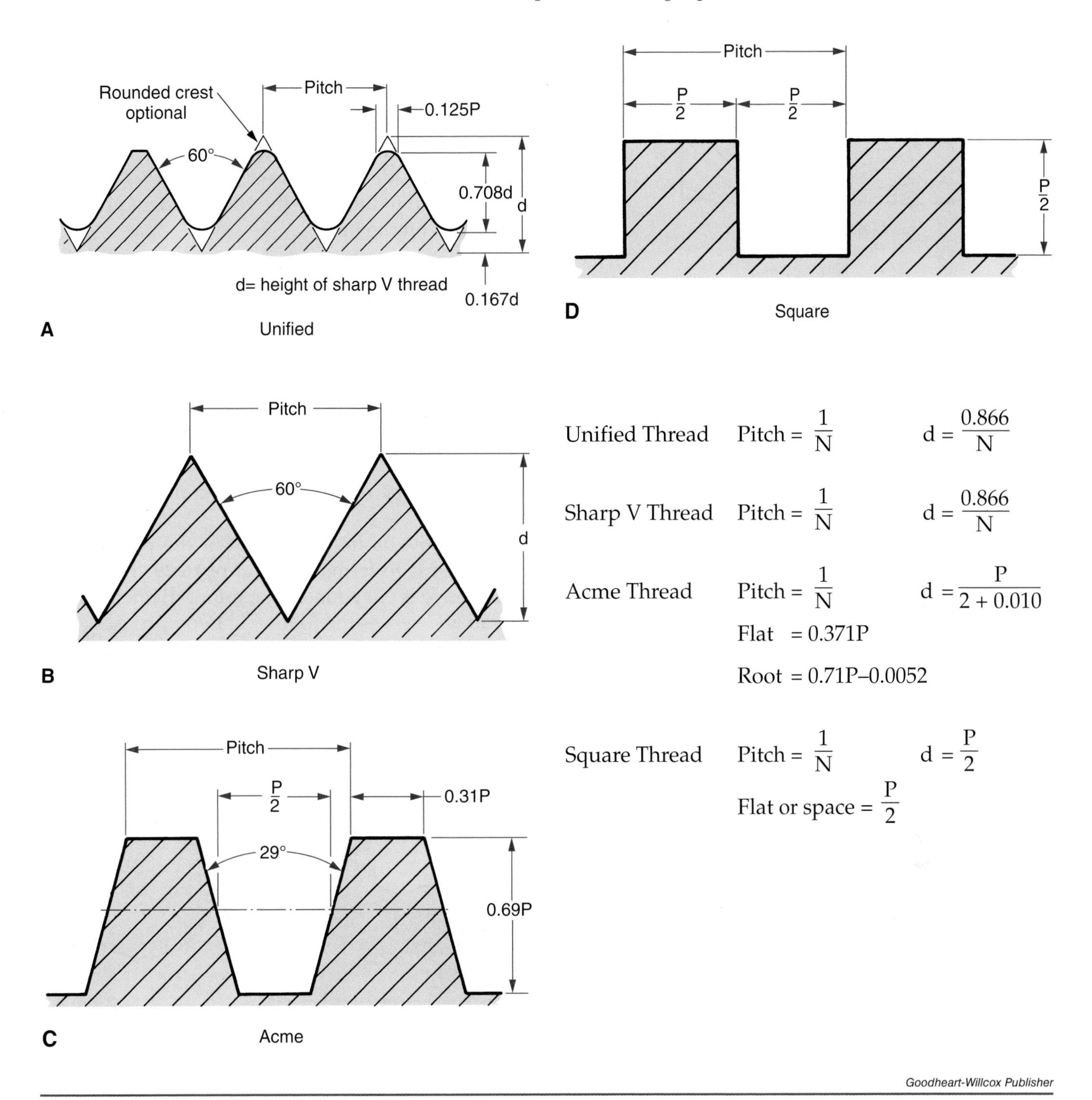

Goodheart-Willcox Publisher

Figure 16-24. Common thread forms. A—Unified thread form, interchangeable with American National thread. B—Sharp V thread form. C—Acme thread form. D—Square thread form. Note: In formulas above, N = Number of threads per inch; P = Pitch; d = depth of thread.

- ***Pitch*** is the distance from one thread point to the next thread point, measured parallel to the thread axis. Pitch of inch-based threads is equal to 1 divided by the number of threads per inch.
- ***Lead*** is the distance that a nut will travel in one complete revolution of the screw. On a single thread, the lead and pitch are the same. Multiple thread screws have been developed to secure an increase in lead without weakening the thread. See **Figure 16-26**.

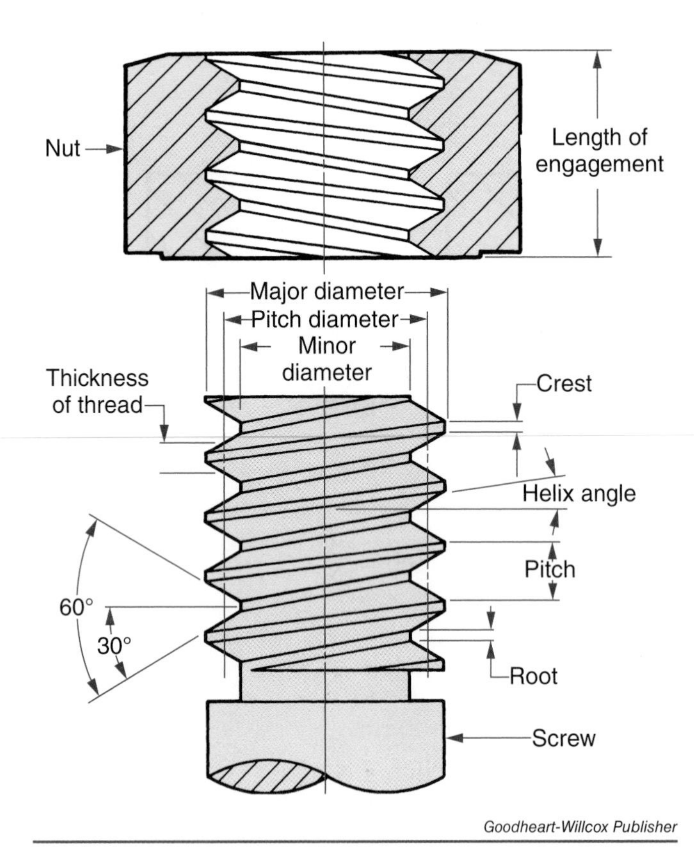

Figure 16-25. Nomenclature of a thread.

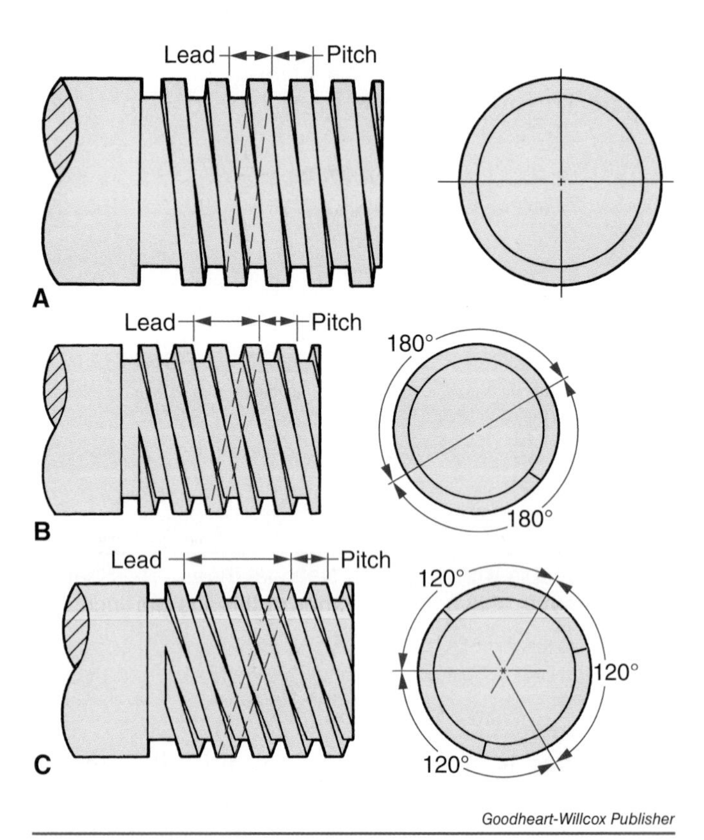

Figure 16-26. The difference between lead and pitch. A—Single thread screw, the pitch and lead are equal. B—Double thread screw, the lead is twice the pitch. C—Triple thread screw, the lead is three times the pitch.

16.3.2 Preparing to Cut 60° Threads on a Lathe

Sharpen the cutting tool to the correct shape, including the proper clearance. The top is ground flat with no side or back rake, **Figure 16-27**. An oilstone is used to touch up the cutting edges and form the radius on the tip.

A center gage is used for grinding and setting the tool bit in position, **Figure 16-28**. The gage is often referred to as a "fishtail."

The work is set up in the same manner as for straight turning. If mounted between centers, the centers must be precisely aligned or a tapered thread will be produced. If this occurs, the thread will not be usable unless it is cut excessively deep at one end. The work must also run true with no "wobble." The tail of the lathe dog must have no play in the face plate slot.

A groove is frequently cut at the point where the thread is to terminate, **Figure 16-29**. The thread end groove is cut equal to the minor diameter of the thread and serves the following two purposes:

- It provides a place to stop the threading tool at the end of its cut.
- It permits a nut to be run up to the end of the thread.

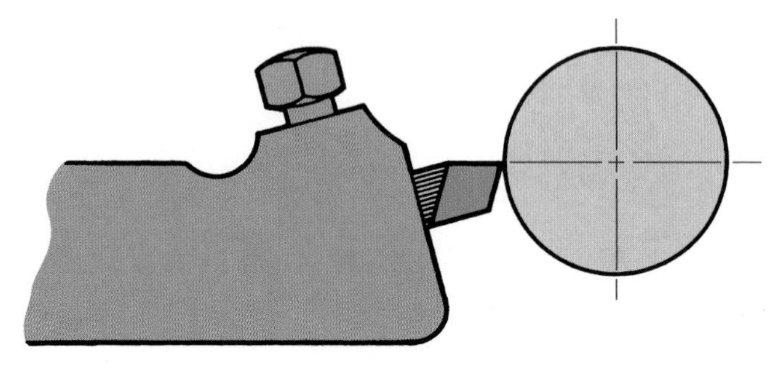

Figure 16-27. Cutting tool positioned for cutting 60° threads. The tool is set on center as shown.

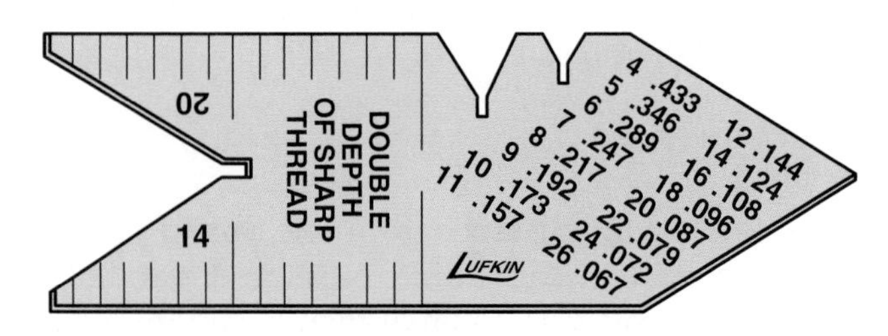

Figure 16-28. A center gage or "fishtail."

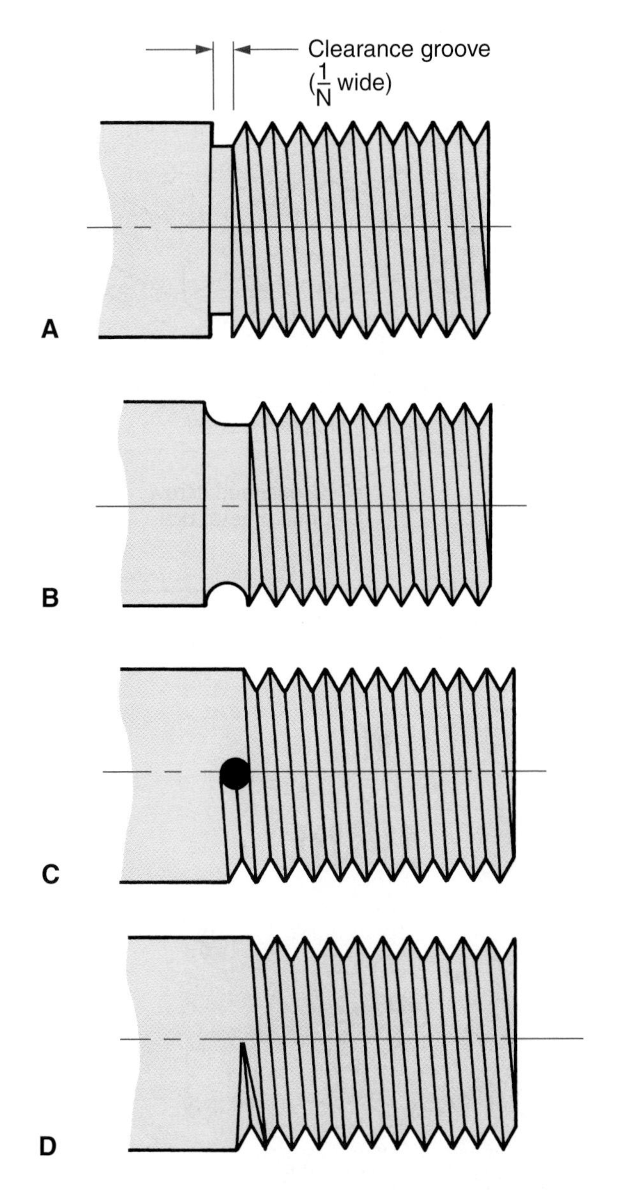

Goodheart-Willcox Publisher

Figure 16-29. Techniques for terminating a screw thread. A—Square groove. B—Round groove. C—Small shallow hole. D—Tool withdrawn from thread at end of cut.

Several methods may be used to terminate a thread, as shown in **Figure 16-29**. Ordinarily, the beginner should use a groove until sufficient experience has been gained. However, the design of some parts does not permit a groove to be used. In such a case, the threads must be terminated by another method. They require perfect coordination and very rapid operation of the cross-slide to get the tool out of position at the end of the cut.

The gearbox is adjusted to cut the correct number of threads. Make apron adjustments to permit the half-nuts to be engaged. After the proper apron and gear adjustments have been made, pivot the compound rest to 29° to the right, **Figure 16-30**. Then set the threading tool in place.

It is essential that the tool be set on center with the tool axis at 90° to the centerline of the work. This is done with the aid of a center gage. Place the gage against the work while the tool is set into a V, **Figure 16-31**. Tool height can be set by using the centerline scribed on the tailstock spindle or with the center point.

The compound rest is set at 29° to permit the tool to shear the chip better than if it were fed straight into the work, **Figure 16-32**. Since the angle of the tool is 30° and it is fed in at an angle of 29°, the slight shaving action that results will produce a smooth finish on the right side of the thread.

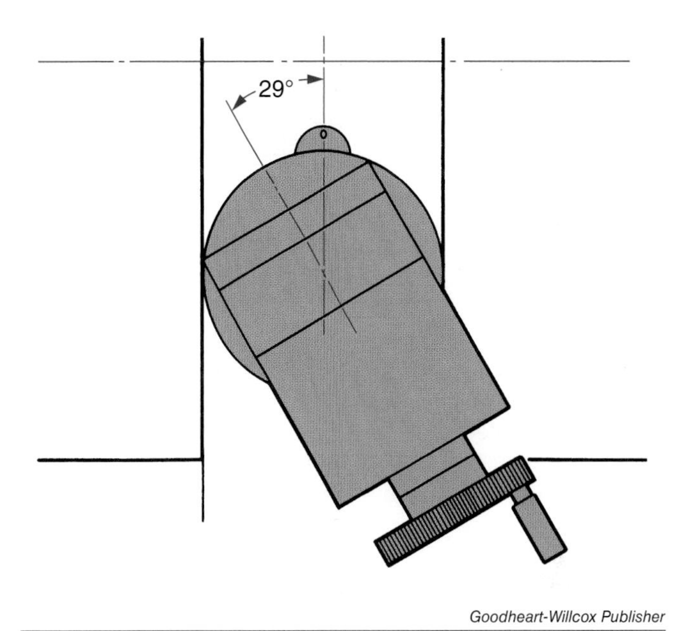

Goodheart-Willcox Publisher

Figure 16-30. The compound rest is set up for machining right-hand external threads.

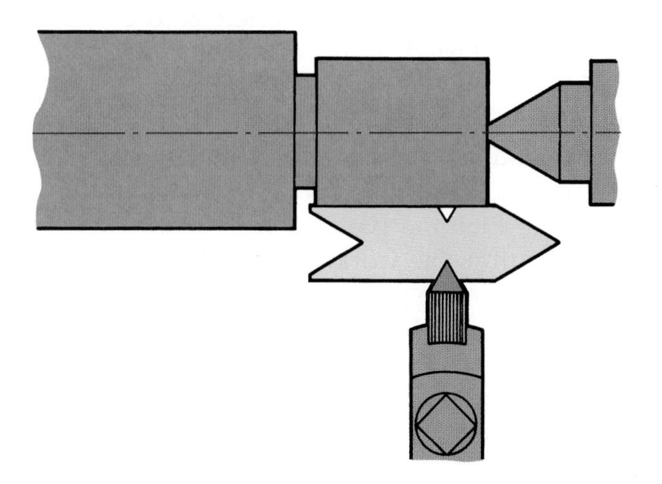

Goodheart-Willcox Publisher

Figure 16-31. Positioning a cutting tool for machining threads, using a center gage.

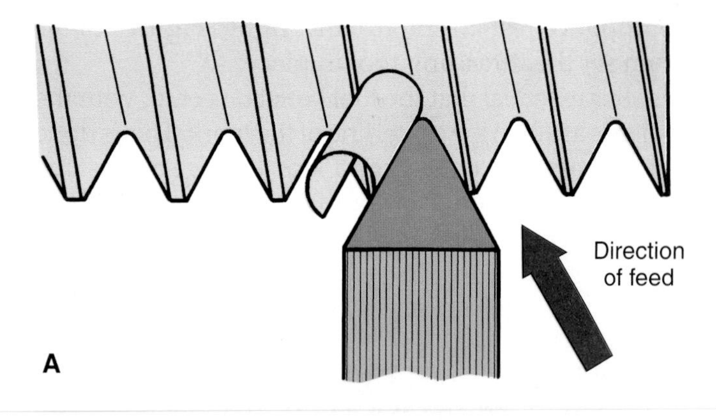

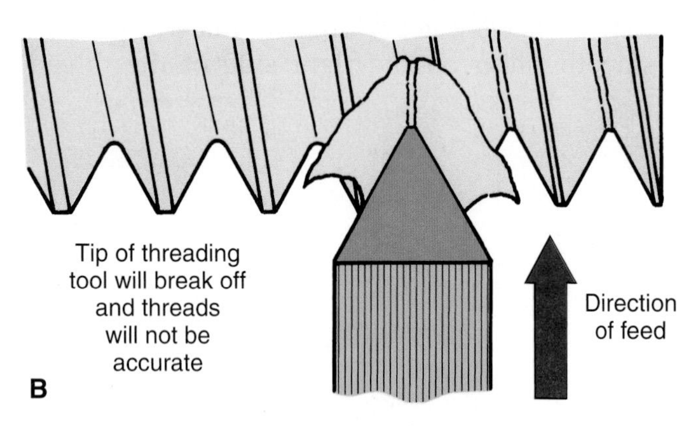

Goodheart-Willcox Publisher

Figure 16-32. Cutting action of tool. A—When the tool is fed in at 29° angle, only one edge is cutting, and that the cutting load is distributed evenly across the edge. B—When fed straight in, both edges are cutting and weakest part of tool, the point, is doing hardest work.

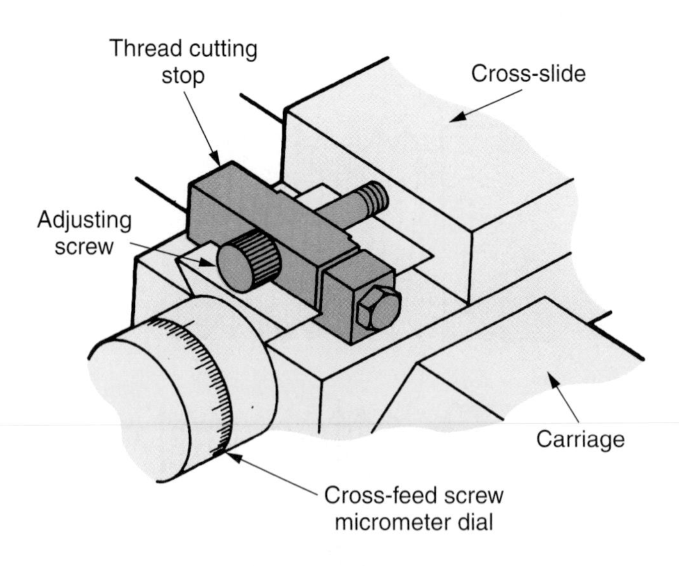

Goodheart-Willcox Publisher

Figure 16-33. After being properly adjusted, the thread cutting stop will let you start next cut in same location.

At the same time, not enough metal is removed to interfere with the main chip that is removed by the left edge of the tool.

Since the tool must be removed from the work after each cut and repositioned before the next cut can be started, a ***thread cutting stop*** may be used. After the point of the tool is set to just touch the work, lock the stop to the saddle dovetail with the adjusting screw just bearing on the stop, **Figure 16-33**.

After a cutting pass has been made, move the tool back from the work with the cross-slide screw. Move the carriage back to start another cut. Feed the tool into the work until the adjusting screw again bears against the thread cutting stop. By turning the compound rest in a distance of 0.002″ to 0.005″ (0.05 mm to 0.12 mm), the tool will be positioned for the next cut.

A ***thread dial*** that meshes with the lead screw is fitted to the carriage of most lathes, **Figure 16-34**. The thread dial is used to indicate when to engage

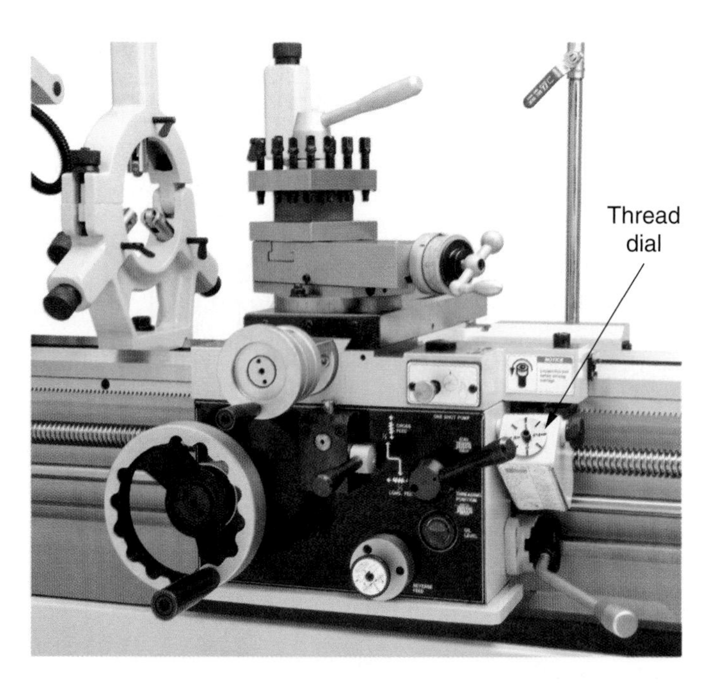

Photo courtesy of Grizzly Industrial, Inc. www.grizzly.com

Figure 16-34. A thread dial used for cutting either inch-based or metric-based threads. The housing contains a series of gears, with gear selection depending on the threads being cut.

the half-nuts, which permit the tool to follow exactly in the original cut. The thread dial eliminates the need to reverse spindle rotation after each cut to bring the tool back to the starting point.

The face of the thread dial, **Figure 16-35**, rotates when the half-nuts are not engaged. When the desired graduation moves into alignment with the index line, the half-nuts can be engaged.

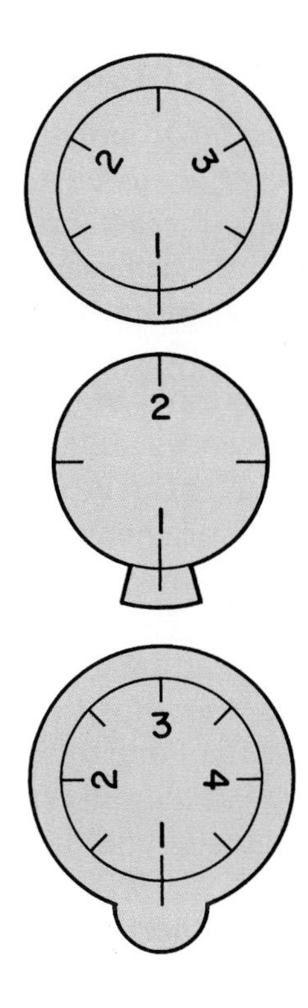

Goodheart-Willcox Publisher

Figure 16-35. Typical thread dial faces.

The thread dial is used as follows for all inch-based threads:

- For all even-numbered threads, close the half-nuts at any line on the dial.
- For all odd-numbered threads, close the half-nuts at any numbered line on the dial.
- For all threads involving one-half of a thread in each inch (such as 11 1/2), close the half-nuts at any odd numbered line.
- For all threads involving one-fourth of a thread in each inch (such as 4 3/4), return to the original starting line before closing the half-nuts.

On lathes that have been converted to metric threading capability, the thread dial cannot be used. When thread cutting with such a lathe, the half-nuts (once closed) must not be opened until the thread is completely cut. The spindle rotation must be reversed after each cut to return the tool to its starting position.

However, the thread dial can be used on lathes with full metric capabilities. The thread dial will vary with the lathe manufacturer and must be considered individually. To be sure of correct thread dial procedure, consult the manufacturer's handbook for the machine.

16.3.3 Making the Cut

Set the spindle speed to about one-fourth the speed that is used for conventional turning. Feed in the tool until it just touches the work. Then, move the tool beyond the right end of the work and adjust it to take a 0.002″ (0.05 mm) cut.

Turn on the power and engage the half-nuts when indicated by the thread dial. This cut is made to check whether the lathe is producing the correct threads. Thread pitch can be checked with a rule or with a screw pitch gage, **Figure 16-36**. When everything checks, make additional cuts, working in 0.005″ (0.12 mm) increments, until the thread is almost to size. The last few cuts should be no more than 0.002″ (0.05 mm) deep. All advances of the cutting tool are made with the compound rest feed screw.

A liberal application of cutting oil, before each cut, will help to obtain a smooth finish.

16.3.4 Resetting Tool in Thread

It is sometimes necessary to replace a broken cutting tool, or to resharpen it for the finish cuts. After replacing the tool, you must realign it with the portion of the thread already cut. This can be done as follows:

1. Set the tool on center and position it with a center gage.
2. Engage the half-nuts at the proper thread dial graduation.
3. Move the tool back from the work and rotate the spindle until the tool reaches a position about halfway down the threaded section.
4. Using the compound rest screw and the cross-slide screw, align the tool in the existing thread. Reset the thread cutting stop after the tool has been aligned.

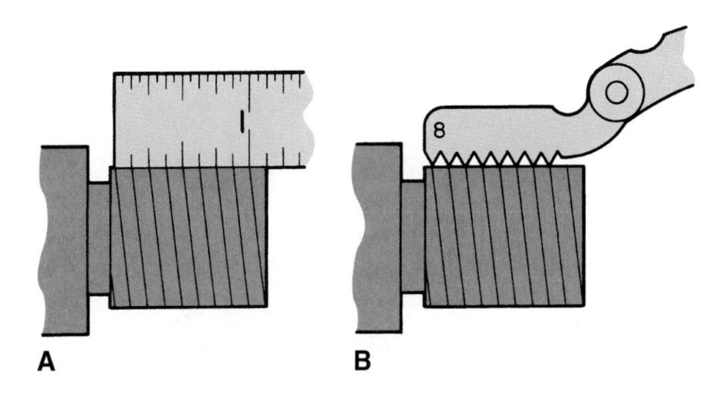

Goodheart-Willcox Publisher

Figure 16-36. Always check thread pitch after first light cut has been made. A—Checking with a rule. B—Checking with a screw pitch gage.

16.3.5 Cutting Threads with Insert-Type Cutting Tools

There are two basic types of 60° threading inserts, the partial-profile insert and the full-profile insert.

Partial-profile inserts, **Figure 16-37**, are most commonly used because they can cut a range of thread pitches. However, the major diameter (OD) of the thread must be cut to size prior to threading. Deburring may be required when cutting threads on most metals.

Full-profile inserts, **Figure 16-38**, produce the best thread form and finish. The tool cuts the leading flank, the root, and the trailing flank simultaneously. The machinist needs only to check the pitch diameter to determine if the major and minor diameters of the thread are to size. No deburring is necessary since the insert trims the thread crest. The disadvantage of the full-profile insert is that a separate insert is required for each thread pitch.

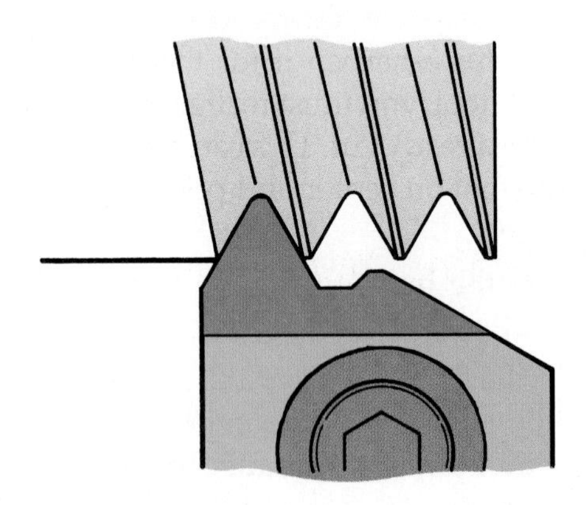

Goodheart-Willcox Publisher

Figure 16-37. Cutting threads with a partial-profile insert. The major (outside) diameter of the thread must be cut to size before using this type insert.

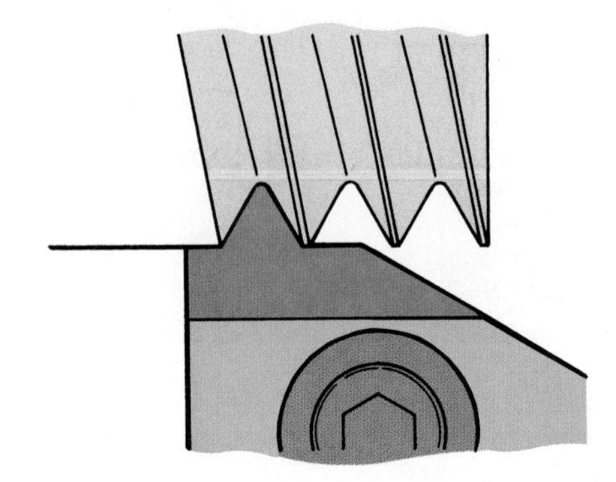

Goodheart-Willcox Publisher

Figure 16-38. Using a full-profile insert to cut a thread. A separate insert is required for each thread pitch.

16.3.6 Measuring Threads

Measure threads at frequent intervals during the machining operation to ensure accuracy. The easiest way to check thread size is to try fitting the threaded piece into a threaded hole or nut of the proper size. If the piece does not fit, it is too large and further machining is necessary. This technique is not very accurate, but is usually satisfactory when close tolerances are not specified.

A thread micrometer can be used to make quick, accurate thread measurements. It has a pointed spindle and a double-V anvil to engage the thread. See **Figure 16-39**.

The micrometer reading given is the true pitch diameter. It equals the outside diameter of the screw minus the depth of one thread. Each micrometer is designed to read a limited number of thread pitches and is available in both inch and millimeter graduations.

The ***three-wire method*** of measuring threads has proven to be quite satisfactory. As shown in **Figure 16-40A**, three wires of a specific diameter are fitted into the threads and a micrometer measurement is made over the wires. The formula in **Figure 16-40B** will provide the information necessary to calculate the correct measurement over the wires.

A three-wire thread measuring system has been developed to simplify and speed up the measuring process. It consists of a digital micrometer mounted in a special fixture that holds the threaded workpiece and the three wires. See **Figure 16-41**.

16.3.7 Cutting Left-Hand Threads

Left-hand threads are cut in basically the same manner as right-hand threads. The major differences involve pivoting the compound to the left and changing the lead screw rotation so the carriage travels toward the tailstock (left to right), **Figure 16-42**.

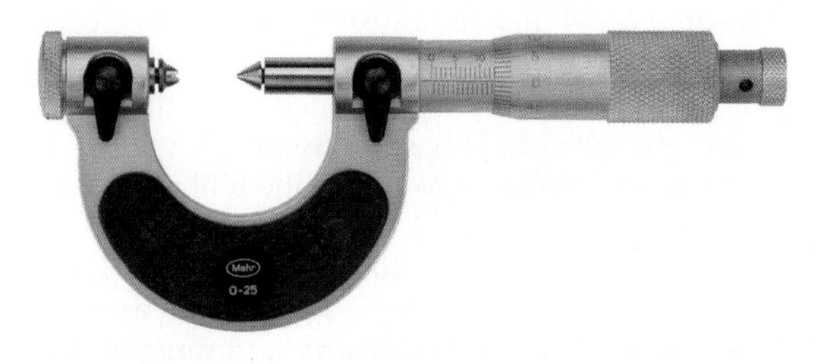

Mahr Federal Inc.

Figure 16-39. A thread micrometer can be used to check cut threads precisely.

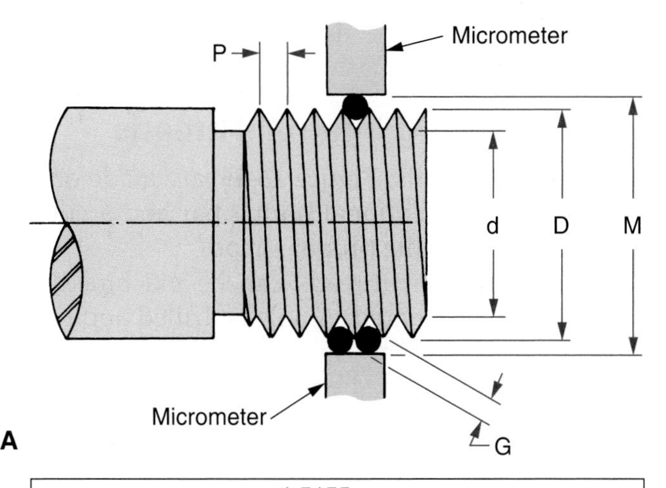

$$M = D + 3G - \frac{1.5155}{N}$$

Where: M = Measurement over the wires
D = Major diameter of thread
d = Minor diameter of thread
G = Diameter of wires
$P = \text{Pitch} = \frac{1}{N}$
N = Number of threads per inch

The smallest wire size that may be used for a given thread:

$$G = \frac{0.560}{N}$$

The largest wire size that can be used for a given thread:

$$G = \frac{0.900}{N}$$

The three-wire formula will work only if "G" is no larger or smaller than the sizes determined above. Any wire diameter between the two extremes may be used. All wires must be the same diameter.

B

Goodheart-Willcox Publisher

Figure 16-40. Three-wire method of measuring 60° screw threads. A—Arrangement of the workpiece, wires, and micrometer. B—The three-wire thread measuring formula.

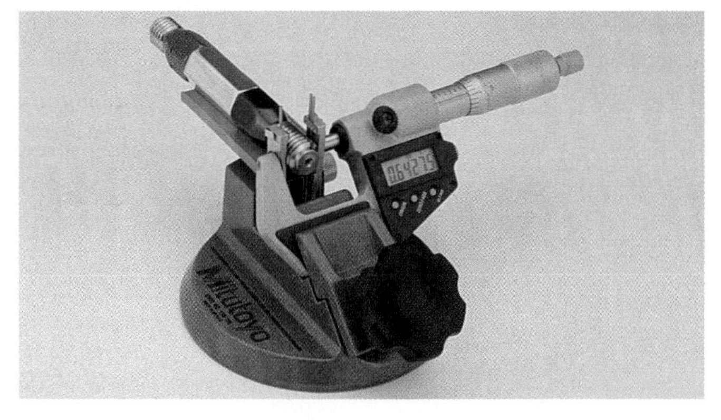

Mitutoyo/MTI Corp.

Figure 16-41. Thread measurements can be made in a fraction of the time normally needed with this three-wire measuring system. Wires are mounted in individual holders that fit into the clamping fixture.

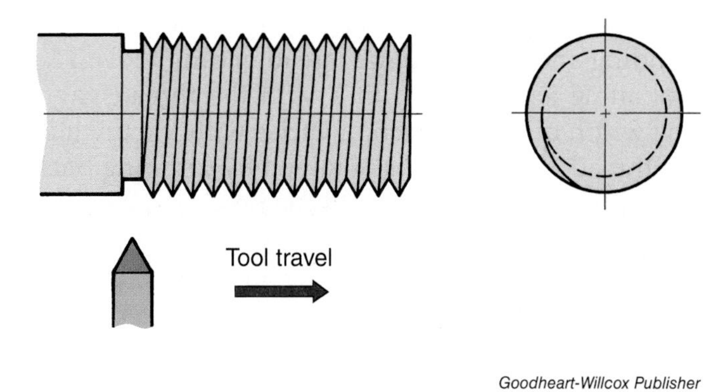

Goodheart-Willcox Publisher

Figure 16-42. Direction of tool travel for cutting left-hand threads.

16.3.8 Cutting Square Threads

Square threads are used to transmit motion. They are more difficult to cut than 60° threads.

To cut a square thread, first calculate the width of the required tool bit (0.5 × thread pitch). If the square thread is fairly coarse, a roughing tool is ground 0.010″ to 0.015″ (0.2 mm to 0.4 mm) smaller than the thread groove width. The cutting point of the finishing tool is ground 0.002″ to 0.003″ (0.05 mm to 0.08 mm) wider than the calculated groove width. Be sure adequate clearance is ground on the cutting tool, **Figure 16-43**.

16.3.9 Cutting Acme Threads

On the Acme thread, the top and bottom are flat, but the sides have a 29° included angle. It was originally developed to replace the square thread. Its advantages are the strength and ease with which it can be cut, compared to the square thread. The thread form is used in machine tools for precise control of component movement.

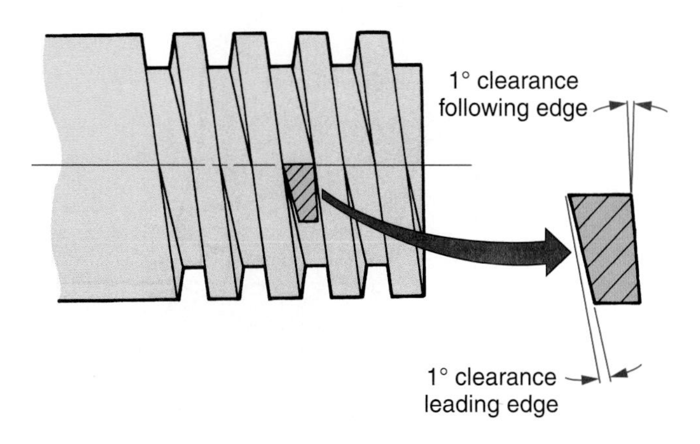

Goodheart-Willcox Publisher

Figure 16-43. Allow adequate side clearance when sharpening a tool to cut square threads.

The Acme screw thread gage is the standard for grinding and setting Acme thread cutting tools. The tool angle is ground to fit a V in the thread gage. The width of the flat section varies with the pitch of the thread. This width is obtained by grinding back the tool point until it fits into the notch appropriate for the thread being cut. See **Figure 16-44**.

In cutting the threads, the groove is usually roughed out with a square-nosed tool to approximate depth, and then finished with an Acme-shaped tool. The compound rest is set to 14° and the tool is positioned using the thread gage, **Figure 16-45**. Other than this, Acme threads are cut in the same manner as the Sharp V thread.

16.3.10 Cutting Internal Threads

Internal threads, **Figure 16-46**, are made on the lathe with a conventional boring bar and a cutting tool sharpened to the proper shape.

Before internal threads can be machined, the work must be prepared. A hole is drilled and bored to correct size for the thread's minor diameter. A recess is then machined with a square-nosed tool at the point where the thread terminates, **Figure 16-47**.

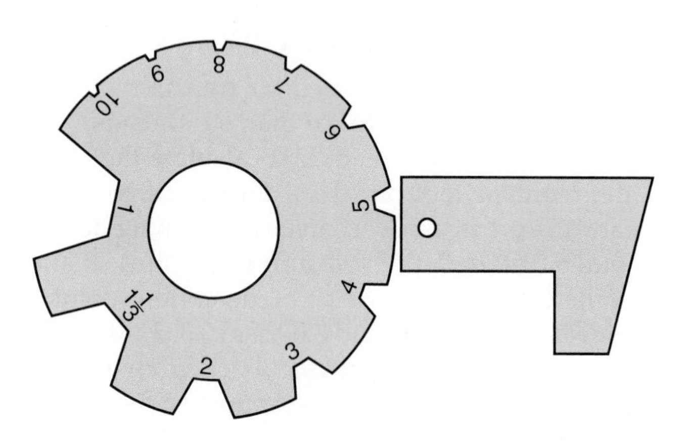

Goodheart-Willcox Publisher

Figure 16-44. The Acme screw thread gage and tool setup gage will allow you to check lathe settings.

Goodheart-Willcox Publisher

Figure 16-46. Internal threads.

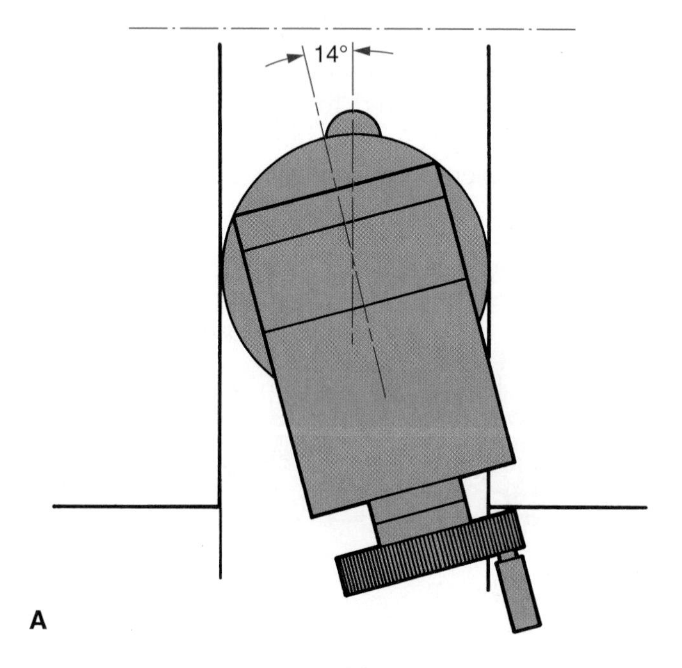

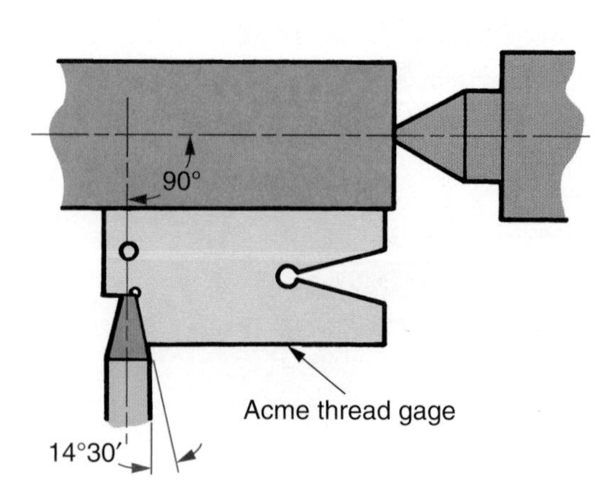

Goodheart-Willcox Publisher

Figure 16-45. Cutting Acme threads. A—Compound setting for cutting Acme threads. B—Cutting tool is positioned with an Acme thread gage.

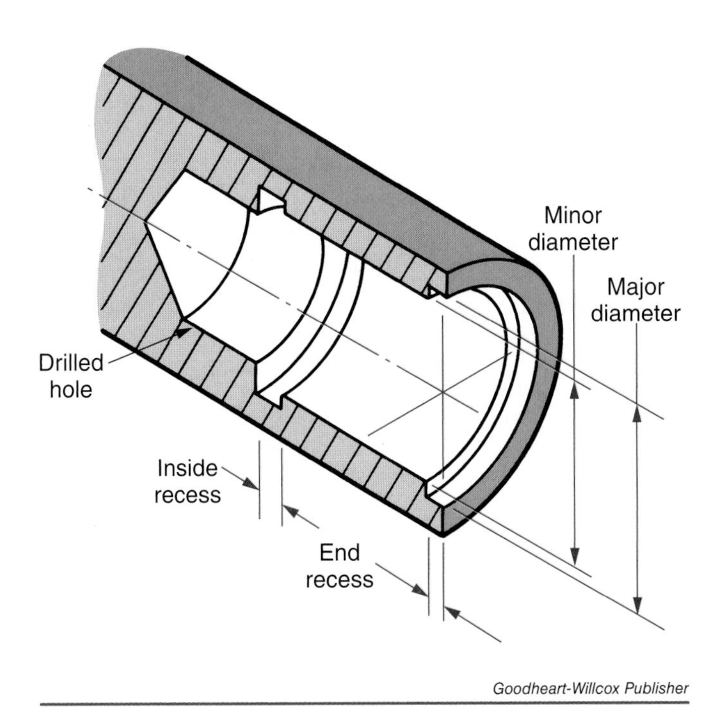

Figure 16-47. Opening for internal screw threads has been drilled and grooves machined to the major diameter.

The diameter of the recess is equal to the major diameter of the thread.

To cut right-hand internal threads, pivot the compound rest 29° to the left. See **Figure 16-48.** Mount the tool on center and align it, using a center gage, **Figure 16-49**.

Bring the tool up until it just touches the work surface. Adjust the micrometer collar on the cross-slide to zero with the tool in position. Using the compound rest screw, adjust the cutter to make a cut of 0.002" (0.05 mm).

When cutting internal threads, tool infeed and removal from the cut are the reverse of those used when cutting external threads.

A problem may arise in trying to determine when the tool has traveled far enough into the hole so the half-nuts can be disengaged. One method makes use of a line that has been lightly scribed in a blued area on the flat way of the lathe bed. The tool will have advanced far enough when the carriage reaches this point.

Another technique allows you to start at the back of the hole when cutting internal threads. Pivot the compound rest 29° to the right. Place the threading tool to the rear of the boring bar with the cutting edge up. See **Figure 16-50**.

The lathe spindle is run in reverse. To prevent the tool from being placed too far into the hole to start the cut, mount a micrometer carriage stop on the ways. See **Figure 16-51**. The carriage is returned until it touches the stop. For cutting the threads, follow the same general procedure previously described.

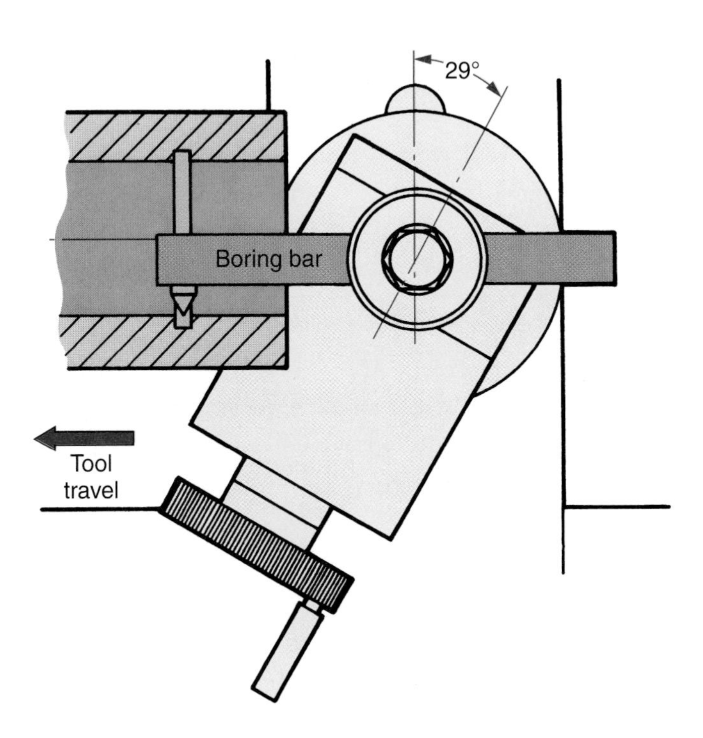

Figure 16-48. Compound setting for cutting internal right-hand screw threads.

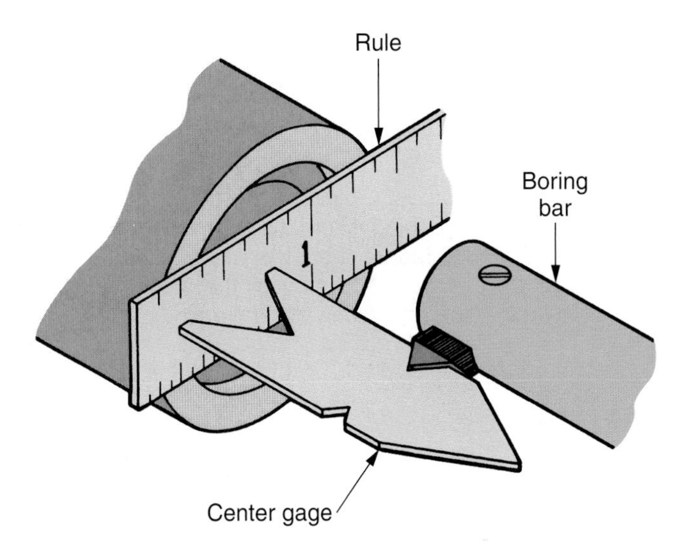

Figure 16-49. How to position cutting tool for machining internal screw threads.

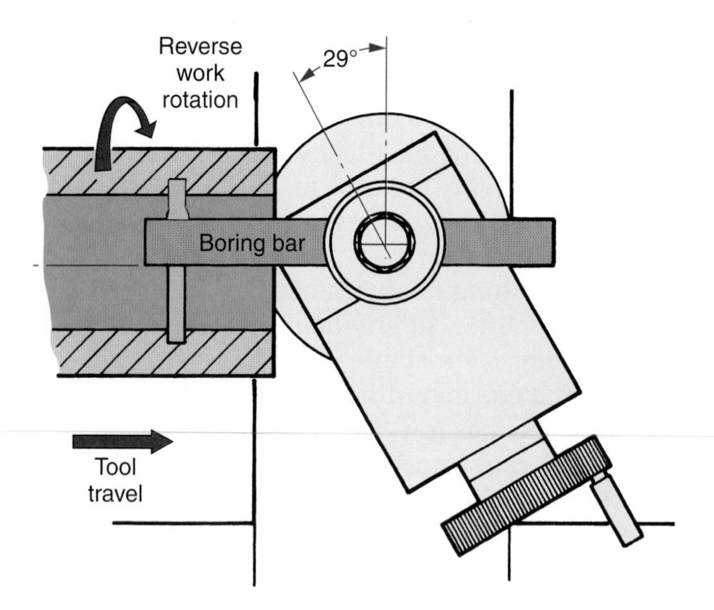

Goodheart-Willcox Publisher

Figure 16-50. An alternative setup for cutting internal right-hand threads. The work rotates in a direction opposite that of normal turning operations.

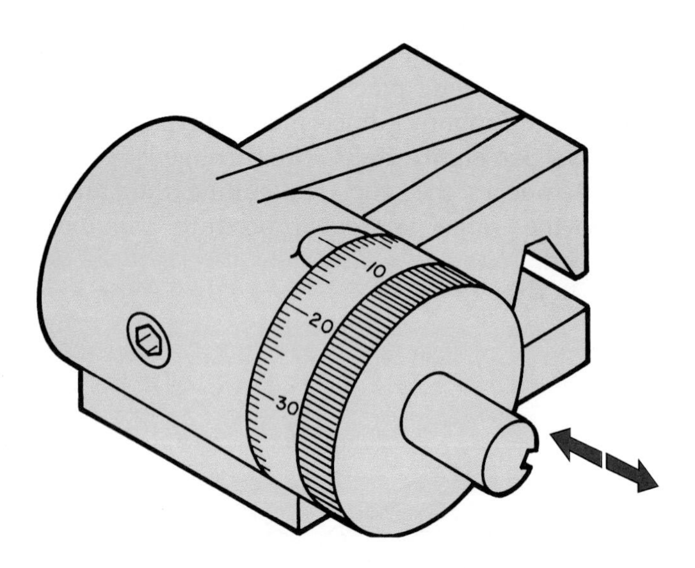

Goodheart-Willcox Publisher

Figure 16-51. When using alternate technique for cutting internal right-hand threads, mount a micrometer carriage stop on the ways. Adjust it to prevent the tool from being placed too far into the hole when starting each cut.

Continue making additional cuts until the threads are finished. Because the toolholder is not as rigid, lighter cuts must be taken when cutting internal threads than when machining external threads. Keep the work flooded with cutting fluid.

16.3.11 Cutting Threads on a Tapered Surface

Tapered threads must be cut, at times, to obtain a fluid- or gas-tight joint. When this situation arises, the threading tool must be positioned in relation to the centerline of the taper rather than to the taper itself, **Figure 16-52**.

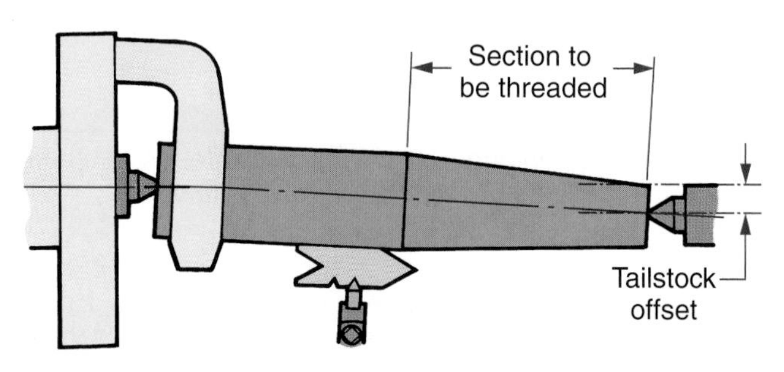

Goodheart-Willcox Publisher

Figure 16-52. Tool setup for machining screw threads on a taper. The tool is not positioned on the taper.

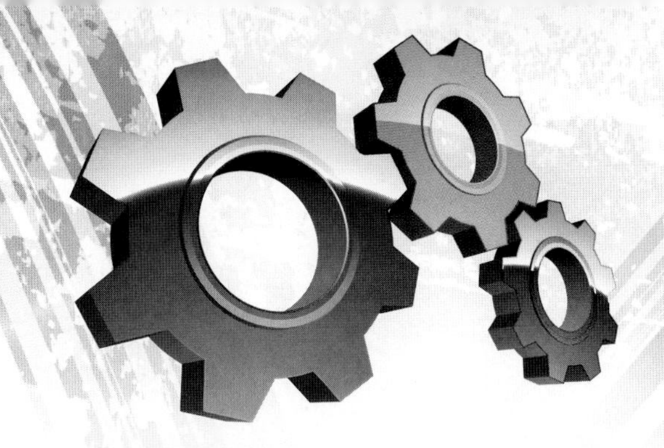

Chapter Review

Summary

- There are five methods for turning a taper on a lathe. Each has its own advantages and disadvantages.
- The tailstock setover can be calculated using a variety of measurements: taper per inch (TPI), taper per foot (TPF), diameter of the long end (D), diameter of the small end (d), length of the taper (l), and total length of the piece (L).
- When cutting a taper, additional strain is placed on the centers. The workpiece must be checked frequently for binding.
- Cutting a taper with a taper attachment is more accurate than other methods.
- The Unified 60° thread form is the most common type of thread. Other thread forms include the left-hand, square, and Acme threads.
- A variety of methods and tools exist for cutting threads.
- Using a threaded hole or nut, a thread micrometer, or the three-wire method are three methods for measuring threads.

Review Questions

Answer the following questions using the information provided in this chapter.

1. Explain what is considered a taper.
2. There are five ways of machining tapers on a lathe. List them, with their advantages and disadvantages.

Machine adjustments must be calculated for each tapering job. The information given below will enable you to calculate the necessary tailstock setover for questions 3–5.

Formula: When taper per inch is known,

$$\text{Offset} = \frac{L \times TPI}{2}$$

When taper per foot is known,

$$\text{Offset} = \frac{L \times TPF}{24}$$

When dimensions of tapered section are known but TPI or TPF is not given,

$$\text{Offset} = \frac{L \times (D-d)}{2 \times l}$$

Where:

TPI = Taper per inch
TPF = Taper per foot
D = Diameter at large end of taper
d = Diameter at small end of taper
l = Length of taper
L = Total length of piece

Note: These formulas, except for the TPF formula, can be used when dimensions are in mm.

3. What will the tailstock setover be for the following job?

   Taper per inch = 0.125″
   Total length of piece = 4.000″

4. What will the tailstock setover be for the following job?

   D = 2.50″
   d = 1.75″
   l = 6.00″
   L = 9.00″

5. What will the tailstock setover be for the following job?

   D = 45.0 mm
   d = 25.0 mm
   l = 175.0 mm
   L = 275.0 mm

6. List the five most important uses of screw threads.

For questions 7–13, match the following terms and identifying phrases.

7. _____ External thread.
8. _____ Internal thread.
9. _____ Major diameter.
10. _____ Minor diameter.
11. _____ Pitch diameter.
12. _____ Pitch.
13. _____ Lead.

A. Smallest diameter of thread.
B. Largest diameter of thread.
C. Distance from one point on a thread to a corresponding point on next thread.
D. Cut on outside surface of piece.
E. Diameter of imaginary cylinder that would pass through threads at such points as to make width of thread and width of space at these points equal.
F. Cut on inside surface of piece.
G. Distance a nut will travel in one complete revolution of screw.

14. The tip of a cutting tool to cut a Sharp V thread is sharpened using a(n) _____ to check that it is the correct shape.
15. A groove is cut at the point where a thread is to terminate. It is cut to the depth of the thread and serves to:
    A. Provide a place to stop the threading tool after it makes a cut.
    B. Permits a nut to be run up to the end of the thread.
    C. Terminate the thread.
    D. All of the above.
    E. None of the above.
16. The compound rest is set at _____ when cutting threads to permit the cutting tool to shear the material better than if it were fed straight into the work.
17. The _____ is fitted to many lathe carriages. It meshes with the lead screw and is used to indicate when to engage the half-nuts to permit the thread cutting tool to follow exactly in the original cut.

The three-wire thread measuring formula for inch-based threads is:

$$M = D + 3G - \frac{1.5155}{N}$$

Where:

G = Wire diameter

D = Major diameter of thread (Convert to decimal size)

M = Measurement over the wires

N = Number of threads per inch

For questions 18–21, calculate the correct measurement over the wires for the following threads. Use the wire size given in the problem.

18. 1/2-20 UNF
    (wire size 0.032″)
19. 1/4-20 UNC
    (wire size 0.032″)
20. 3/8-16 UNC
    (wire size 0.045″)
21. 7/16-14 UNC
    (wire size 0.060″)

CHAPTER 17

Broaching Operations

Learning Objectives

After studying this chapter, you will be able to:

- Describe the broaching operation.
- Explain the advantages of broaching.
- Set up and cut a keyway using a keyway broach and an arbor press.

Technical Terms

arbor press
broach
broaching
burnishing
keyway

Broaching is a manufacturing process for machining flat, round, and contoured surfaces. Both internal and external surfaces can be shaped by this process. Broaching is ideal for producing keyways, splines, and irregularly shaped openings.

17.1 Broaches and Broaching Machines

With broaching, a multitoothed cutting tool is pushed or pulled across the work, **Figure 17-1**. Each tooth on the ***broach*** (cutting tool) removes only a small portion of the material being machined, **Figure 17-2**. Broaching is usually performed with a cutting fluid.

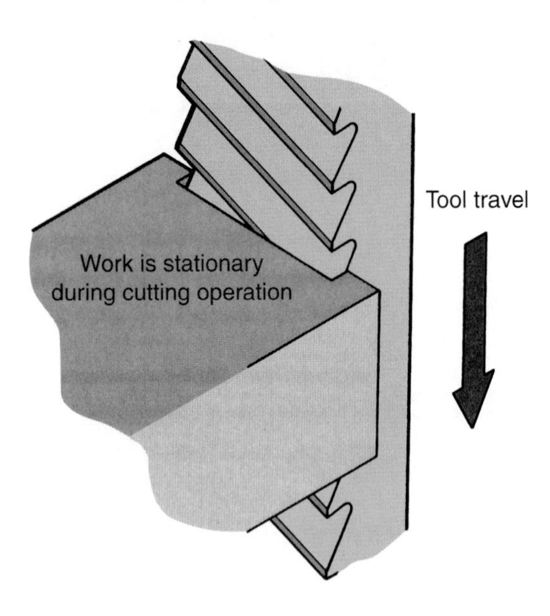

Figure 17-1. Broaching involves the use of a multitooth cutting tool (the broach) that moves against the stationary work. The operation may be on a vertical or horizontal plane, and may involve making internal or external cuts.

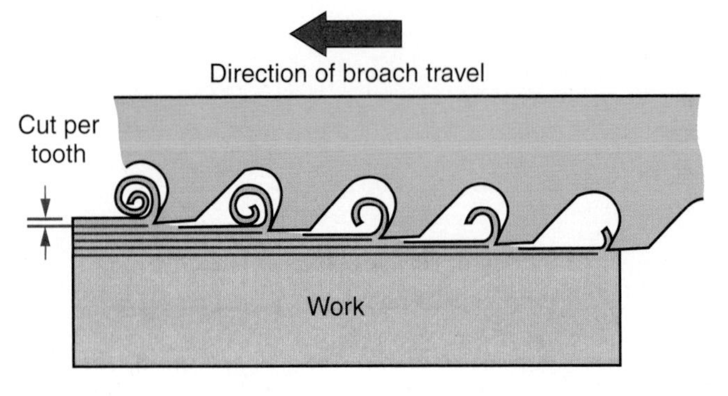

Figure 17-2. Each tooth on a broaching tool removes only a small portion of the material being machined.

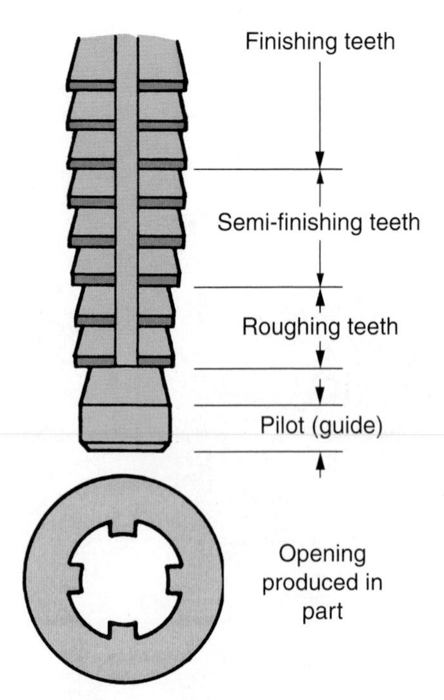

Figure 17-3. This drawing shows a greatly shortened section of an internal broaching tool and the splines it cuts into a part. The pilot guides the cutter into a cut or hole previously made in the work. Each tooth of the broach increases slightly in size until the specified size is attained.

A broach has three kinds of teeth: roughing teeth, semifinishing teeth, and finishing teeth. The roughing teeth remove the most material while the semifinishing and finishing teeth produce a finer finish. See **Figure 17-3**. In many industrial applications, the broach is assembled from units with different tooth sizes. The tooth units are stacked on a mandrel to be pulled through the work.

There are three basic types of broaching operations: internal broaching, external broaching, and pot broaching. Each type of broaching operation makes use of a different type of broach.

Internal broaching makes use of a pull broach. Internal broaching requires a starting hole so that the cutting tool can be inserted. The pull broach is tapered. The first teeth are the smallest. Each tooth is only a few thousandths of an inch larger than the previous tooth. The broaching machine pulls the broach by the pull end, which extends below the first teeth and the pilot guide, **Figure 17-4**.

External broaching uses the slab broach, a flat toothed strip that is usually held (singly or in groups) in a slotted fixture. Slab broaches can be several inches wide and just as thick. External broaching is most often used to machine flat surfaces, but can also be used to cut slots or splines on the outside of a workpiece.

National Broach & Machine Co.

Figure 17-4. Two large pull broaches positioned above the clamping fixtures used to hold workpieces in this broaching machine. The pull end, pilot, and first set of roughing teeth are all visible.

In pot broaching, the tool is stationary and the work is pushed or pulled through the broach, **Figure 17-5**. Slab broaches or ring-shaped broaches are held in a fixture called a pot. The pot is designed to hold multiple cutting tools concentrically. Pot broaching is often used to machine precision gears, **Figure 17-6**.

A variety of broaching equipment is available. The machines range in size from large hydraulically powered broaching machines, **Figure 17-7**, to small ***arbor press*** units (manually operated presses used on smaller work that can have a leverage of several tons) to cut keyways in gears, pulleys, and similar components.

National Broach & Machine Co.

Figure 17-5. Slab broaches, shown to the lower left, are mounted to the pot and used to cut slots on the outside of the workpiece as it is pushed through the stationary fixture.

National Broach & Machine Co.

Figure 17-6. The internal teeth of these precision gears were machined using the pot broaching process.

National Broach & Machine Co.

Figure 17-7. Two broaches, similar to the ones shown in **Figure 17-4**, are mounted on this hydraulically powered broaching machine. Parts are typically completed in a single pass of the broach.

17.2 Advantages of Broaching

The broaching process offers several manufacturing advantages, including the following:

- High productivity.
- Capability of maintaining close tolerances.
- Production of good surface finishes.
- Economy (even though initial tooling costs can be high unless standard tooling is used).
- Long tool life, since only a small amount of material is removed by each tooth.
- Capability of using semiskilled workers, since equipment is automated.

Almost any material that can be machined by other techniques can be broached. Consistently close tolerances can be maintained by the broaching process. While the surface finishes produced by broaching are smooth compared to many other machining processes, they can be further improved by adding ***burnishing*** (noncutting) elements to the finishing end of the broach.

The machining operation can usually be completed in a single pass of the cutting tool. When properly employed, broaching can remove metal faster than almost any other machining technique. Small parts can be stacked and shaped in a single pass, **Figure 17-8**, though larger units may require several passes to machine all surfaces.

TheFinalMiracle/Shutterstock.com

Figure 17-8. Typical small parts machined by broaching. Small parts can be stacked and machined in a single pass.

17.3 Keyway Broaching

Cutting a ***keyway*** in a gear, pulley, or similar component is a simple broaching operation that can be done in the average machine shop. See **Figure 17-9**. A typical keyway broach set, **Figure 17-10**, contains an assortment of precision broaches, slotted bushings, the necessary shims, instructions, and a lubrication guide.

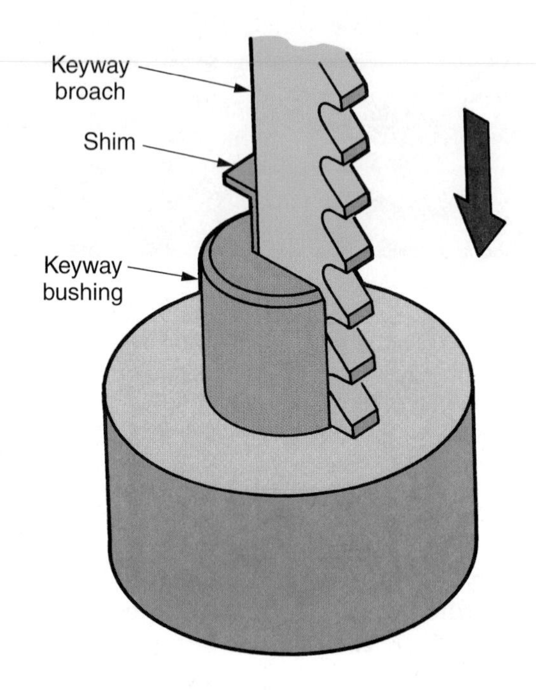

Goodheart-Willcox Publisher

Figure 17-9. Cutting a keyway using an arbor press to push a broach through the work. Several passes may be required. Different size shims move the cutting tool into the work.

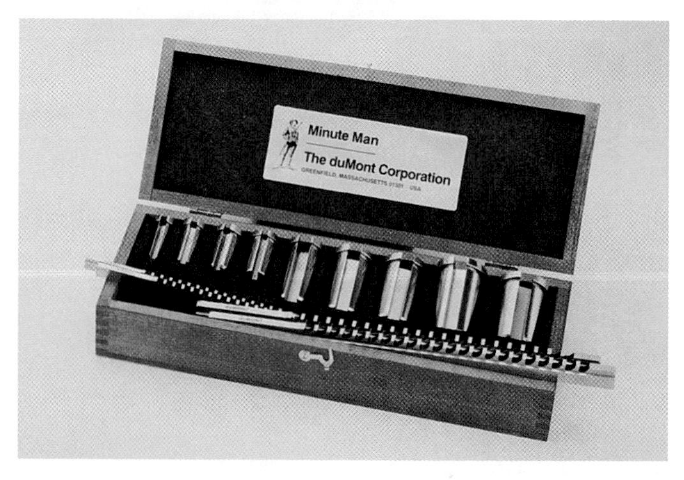

duMont Corp.

Figure 17-10. A typical keyway broach set. It contains precision broaches, slotted bushings, and necessary shims. Ample lubrication is necessary.

SAFETY NOTE

Handle the broach with care. Its sharp teeth can cause a serious injury to the hand.

First, measure the bore into which the keyway is to be cut. Then complete the following steps:

1. Select the bushing that fits the hole and the required broach.
2. Place the bushing into the hole and insert the broach.
3. Set the assembly into position on the arbor press, making sure there is ample clearance for the broach to pass through the work. Be sure that the broach is centered on the arbor press ram.

SAFETY NOTE

If the broach is not centered, it could be damaged by being pushed to one side. A loose or worn arbor press ram can also damage a broach by pushing it to one side.

4. Lubricate the broach as instructed by the broach manufacturer.
5. Push the broach through the workpiece.

SAFETY NOTE

Do not allow the tool to fall to the floor as it is pushed through the workpiece. Dropping the broach could damage the tool.

6. Clean the broach and insert the second pass shim.
7. Lubricate the broach again, and push the broach through.
8. Repeat the sequence until the keyway is the correct depth. Take care to properly align the broach with each previous path.
9. Use a clean cloth to wipe the broach, bushing, and shims clean. Apply a thin coating of oil to prevent rusting and return them to storage.
10. Remove any burrs from the keyway.

Chapter Review

Summary

- The three basic types of broaching are pull broaching, slab broaching, and pot broaching.
- Broaching has many advantages over other machining processes.
- Broaching can be used on both large and small pieces.
- Cutting a keyway can be performed with a keyway broach set on an arbor press in a single pass.

Review Questions

Answer the following questions using the information provided in this chapter.

1. Broaching is a manufacturing process for machining _____ surfaces.
   A. flat
   B. round
   C. contoured
   D. All of the above.
   E. None of the above.
2. What are the three types of teeth on the cutting tool of a broaching machine?
3. What does internal broaching require that external broaching does not?
4. How does pot broaching differ from internal and external broaching?
5. List three advantages offered by broaching.
6. With broaching, the machined surface can be further improved by adding _____ elements to the finishing end of the broach.

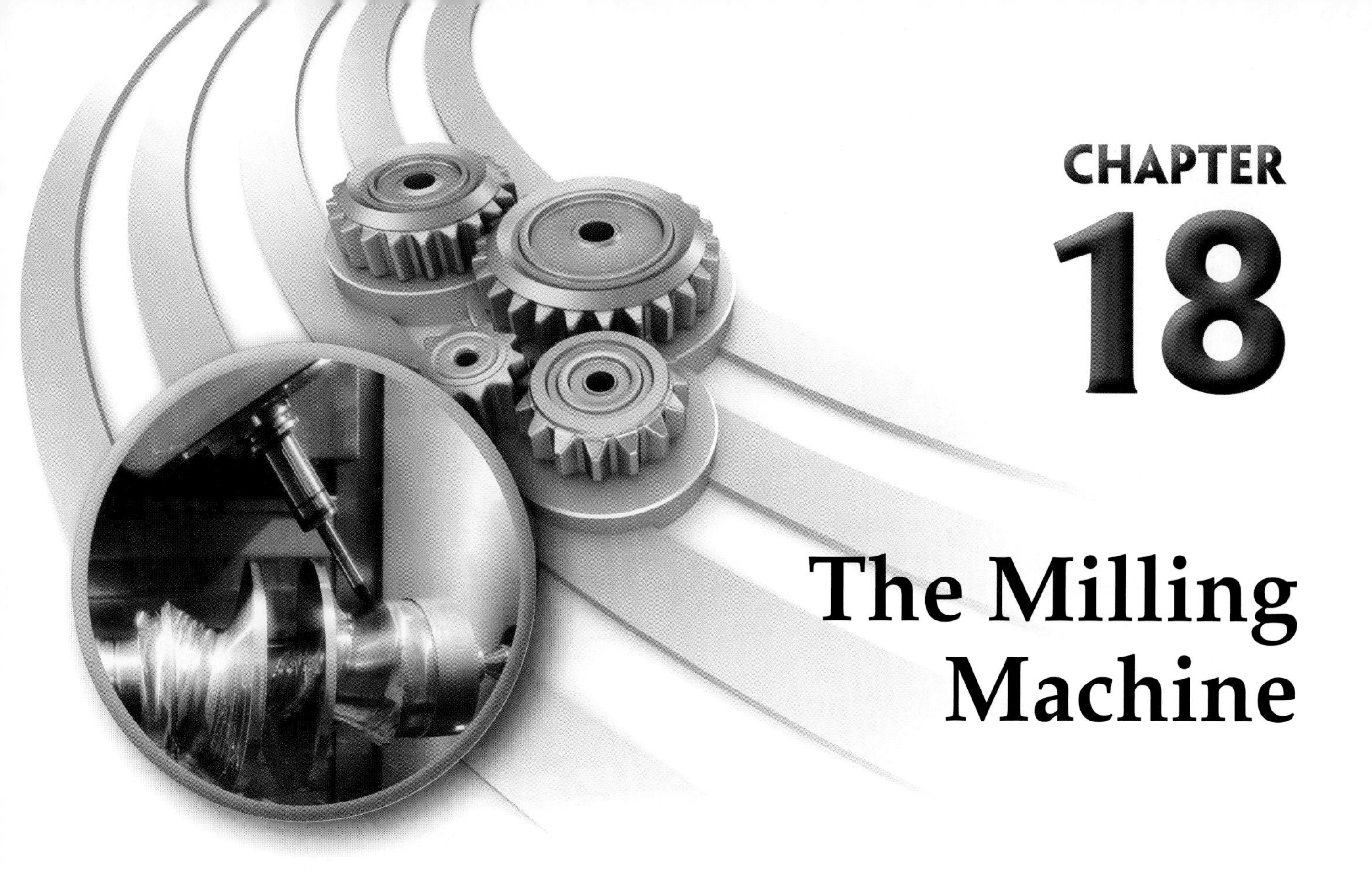

CHAPTER 18

The Milling Machine

Learning Objectives

After studying this chapter, you will be able to:

- Describe how milling machines operate.
- Identify the various types of milling machines.
- Select the proper cutter for the job to be done.
- Calculate cutting speeds and feeds.

Technical Terms

arbor
climb milling
column and knee milling machine
conventional milling
cutting speed
face milling
feed
fixed-bed milling machine
horizontal milling machine
peripheral milling
side milling cutter
vertical milling machine

A milling machine rotates a multitoothed cutter into the workpiece to remove material, **Figure 18-1**. Each tooth of the cutter removes a small individual chip of material. A wide variety of cutting operations can be performed on a milling machine. The milling machine is capable of machining flat or contoured surfaces, slots, grooves, recesses, threads, gears, spirals, and other configurations.

Milling machines are available in more variations than any other family of machine tools, **Figure 18-2**. These machines are well suited to computer-controlled operation. Work may be clamped directly to the machine table, held in a fixture, or mounted in or on one of the numerous work-holding devices available for milling machines.

18.1 Types of Milling Machines

It is difficult to classify the various categories of milling machines, because their designs tend to merge with one another. For practical purposes, however, milling machines may be grouped into two large families, as follows:

- Fixed-bed milling machines.
- Column and knee milling machines.

Both groups are made with horizontal or vertical spindles, **Figure 18-3**. On a ***horizontal milling machine***, the cutter is fitted onto an arbor mounted in the machine on an axis parallel with the worktable. Multiple cutters may be mounted on the spindle for some operations.

The cutter on a ***vertical milling machine*** is normally perpendicular (at a right angle) to the worktable. However, on many vertical milling machines the spindle can be tilted to perform angular cutting operations.

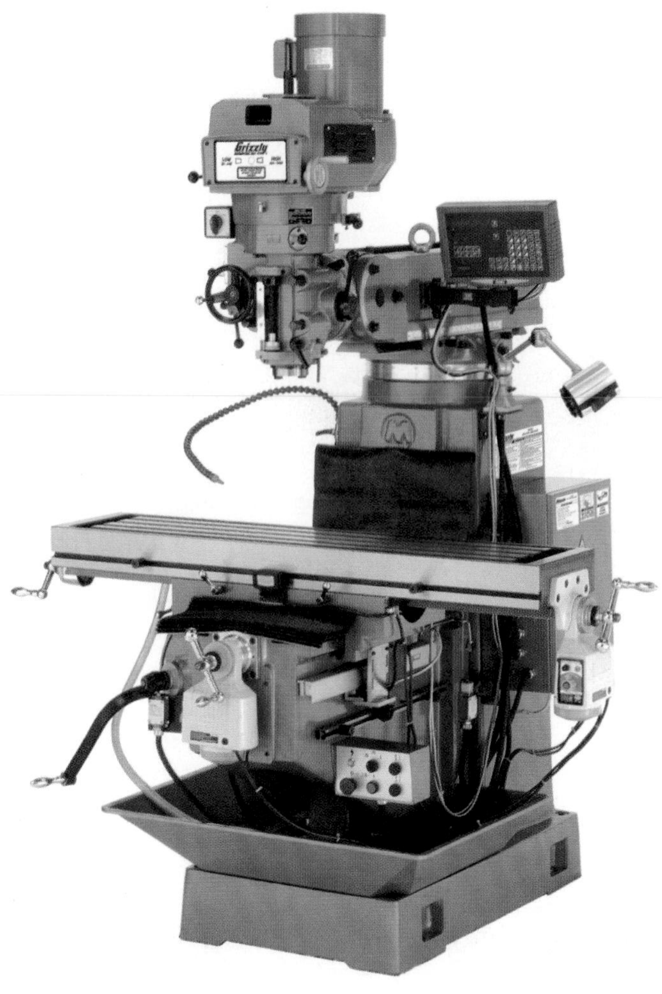

Photo courtesy of Grizzly Industrial, Inc. www.grizzly.com

Figure 18-2. A knee-type vertical milling machine that incorporates longitudinal and traverse power feeds, digital readout, and a built-in coolant system.

BIG Kaiser Precision Tooling Inc.

Figure 18-1. A milling machine rotates a multitoothed cutter into the workpiece.

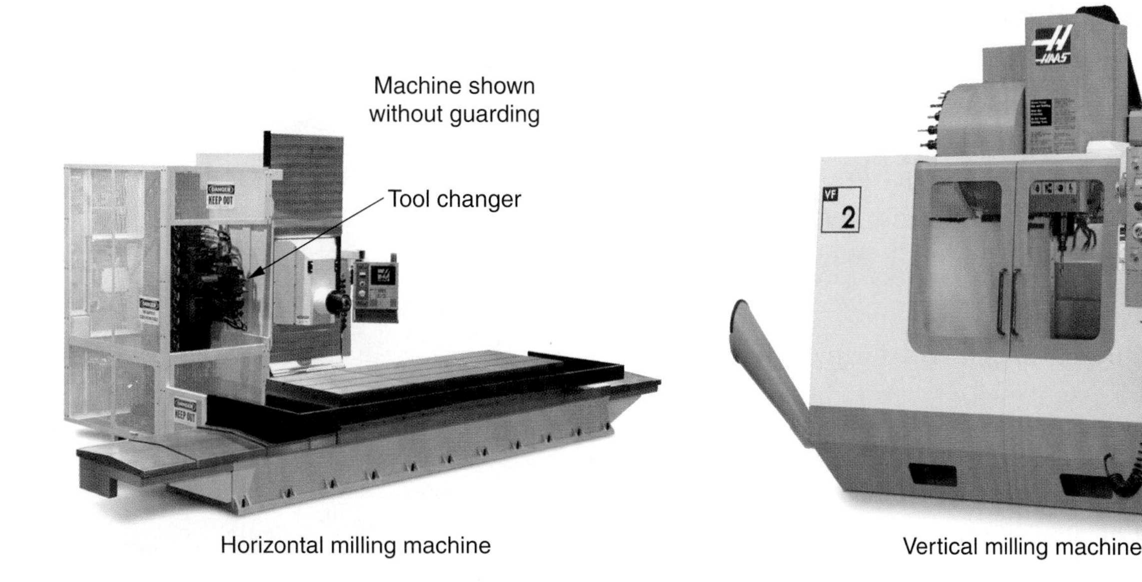

Figure 18-3. Horizontal and vertical CNC milling machines.

18.1.1 Fixed-Bed Milling Machines

Fixed-bed milling machines are characterized by very rigid worktable construction and support, **Figure 18-4**. The worktable moves only in a longitudinal (back and forth/X-axis) direction, and can vary in length from 3′ to 30′ (0.9 to 9.0 m). Vertical (up and down/Z-axis) and cross (in and out/Y-axis) movements are obtained by moving the cutter head.

Bed-type milling machines can be further classified as horizontal-, vertical-, or planer-type machines. The type of bed permits heavy cutting on large workpieces, **Figure 18-5**.

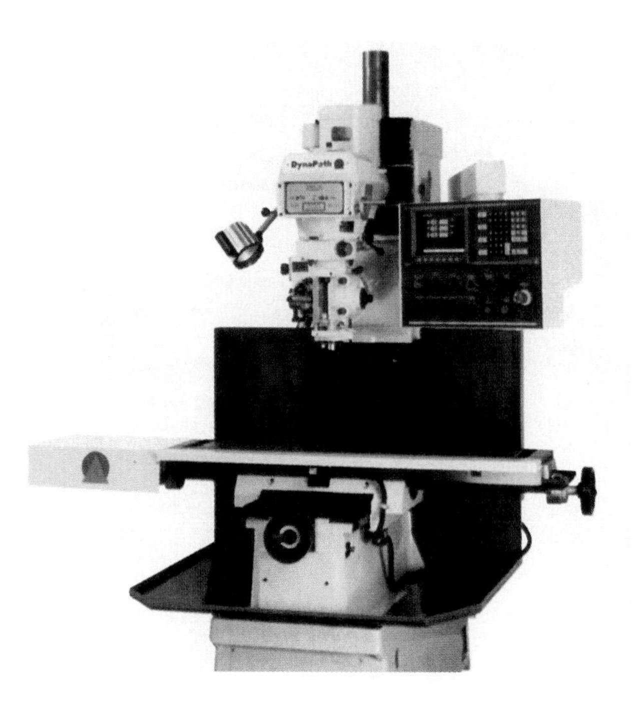

Autocon Technologies, Inc.

Figure 18-4. Fixed-bed or bed mills have a very rigid worktable that moves only in a longitudinal direction. The machine shown can be manually or CNC operated.

MAG IAS, LLC

Figure 18-5. This large, 5-axis fixed-bed vertical profiler milling machine is cutting a contoured section of an aircraft wing skin.

18.1.2 Column and Knee Milling Machines

The ***column and knee milling machine*** is so named because of the parts that provide movement to the workpiece. They consist of a column that supports and guides the knee in vertical (up and down/Z-axis) movement, and a knee that supports the mechanism for obtaining table movements. These movements are traverse (in and out/Y-axis) and longitudinal (back and forth/X-axis). See **Figure 18-6**.

These machines are commonly referred to as "knee-type milling machines." The three basic categories of knee-type milling machines are plain (horizontal) milling machines, universal milling machines, and vertical milling machines.

Plain Milling Machine

On the plain milling machine, the cutter spindle projects horizontally from the column, **Figure 18-7**. The worktable has three movements: vertical, cross, and longitudinal (X-, Y-, and Z-axes), **Figure 18-8**.

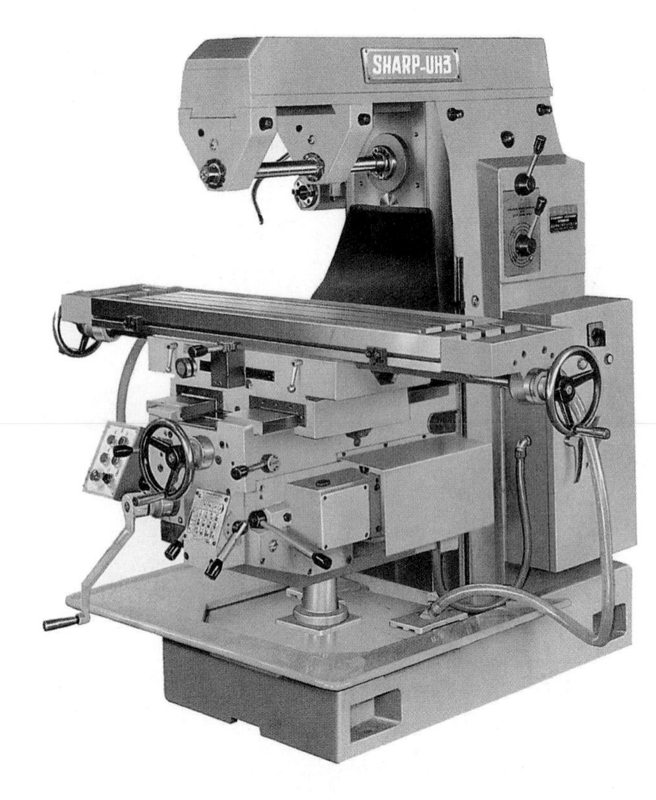

Sharp Industries, Inc.

Figure 18-7. Plain-type milling horizontal machine.

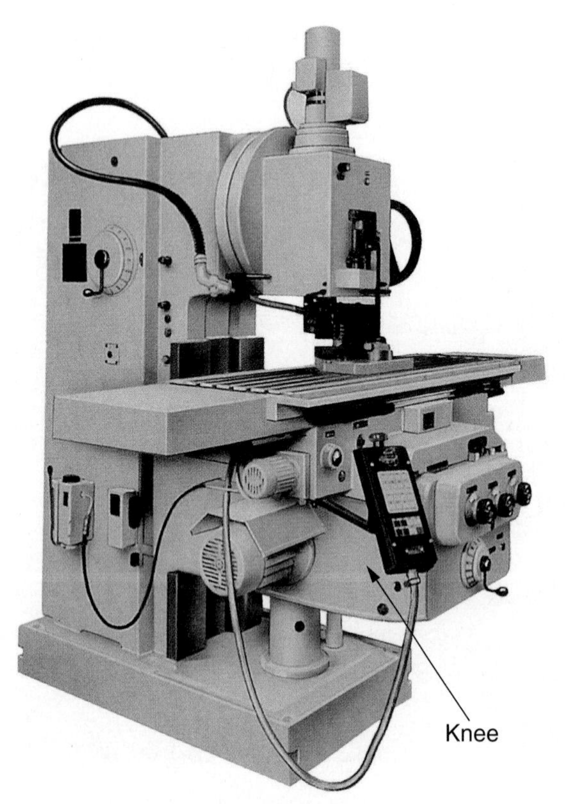

Vertical column and knee milling machine

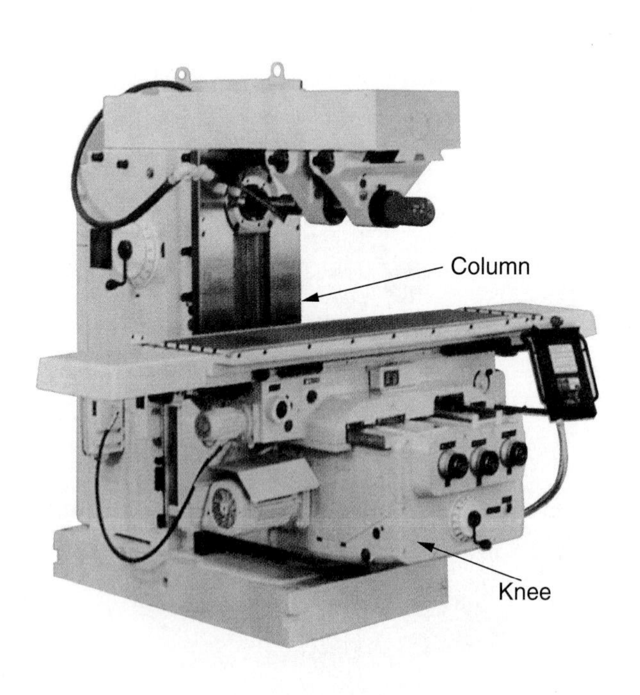

Horizontal column and knee milling machine

WMW Machinery Company, Inc.

Figure 18-6. Column and knee milling machines.

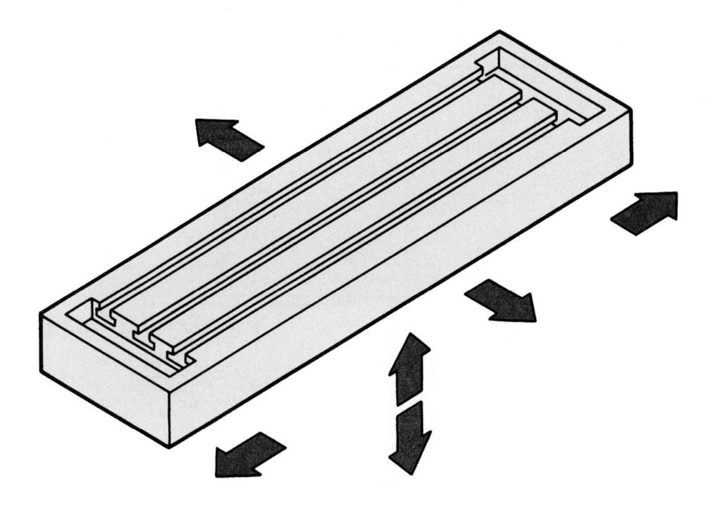

Goodheart-Willcox Publisher

Figure 18-8. Table movements of plain-type horizontal milling machine.

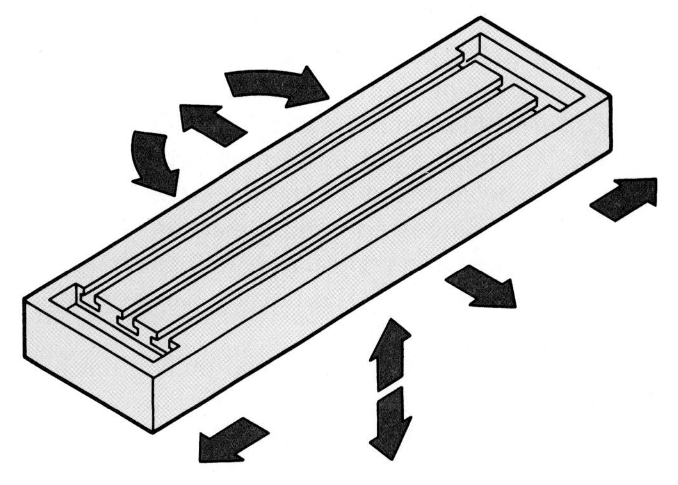

Goodheart-Willcox Publisher

Figure 18-10. Table movements of the universal-type milling machine.

Universal Milling Machine

A universal milling machine, **Figure 18-9**, is similar to the plain milling machine, but the table has a fourth axis of movement. On this type of machine, the table can be swiveled on the saddle through an angle of 45° or more, **Figure 18-10**. This makes it possible to produce spiral gears, spiral splines, and similar workpieces.

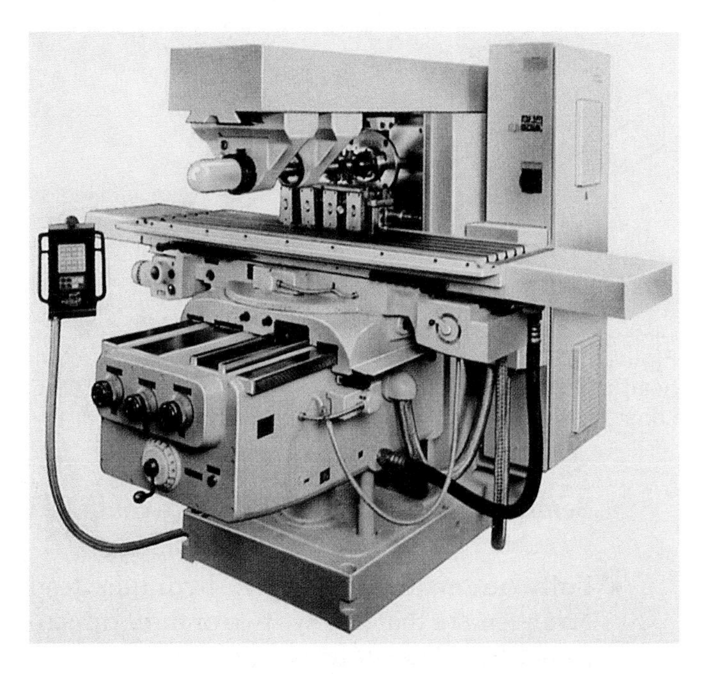

WMW Machinery Company, Inc.

Figure 18-9. On a universal-type horizontal milling machine, the table can be swiveled 45° or more.

Vertical Milling Machine

A vertical milling machine differs from the plain and universal machines by having the cutter spindle in a vertical position, at a right angle to the top of the worktable. See **Figure 18-11**. The cutter head can be raised and lowered by hand or by power feed.

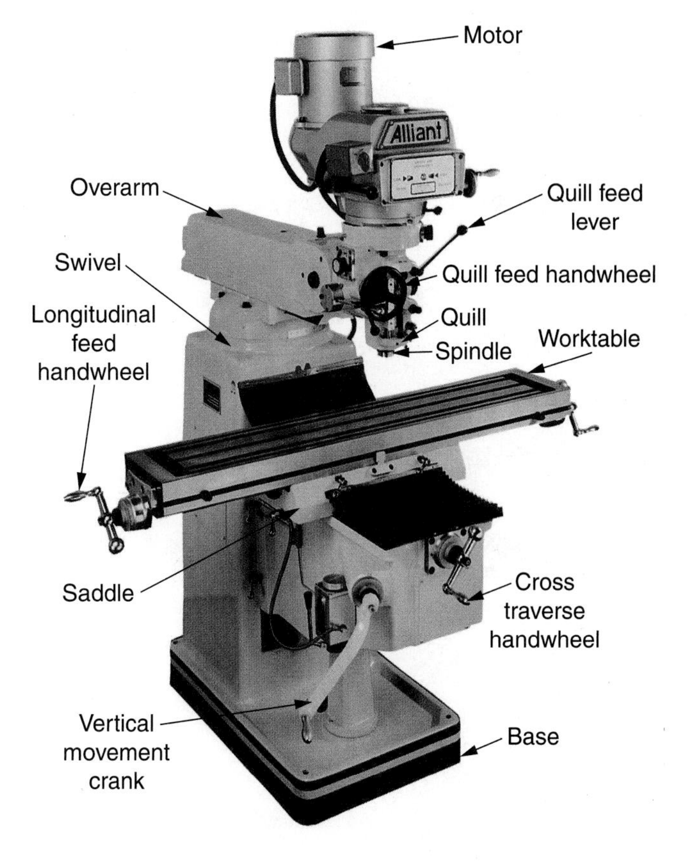

Rem Sales, Inc.

Figure 18-11. The vertical milling machine.

This type of milling machine is best suited for use with an end mill or face mill cutter. Vertical mills include swivel-head, sliding-head, and rotary-head types.

A swivel-head milling machine, **Figure 18-12**, is the type often found in training programs. The spindle can be swiveled for angular cuts.

On the sliding-head milling machine, the spindle head is fixed in a vertical position. The head can be moved in a vertical direction by hand or under power, **Figure 18-13**.

The spindle on the rotary-head milling machine can be moved vertically and in circular arcs of adjustable radii about a vertical center line. It can be adjusted manually or under power feed, **Figure 18-14**.

18.1.3 Methods of Milling Machine Control

The method employed to control table movement is another way of classifying milling machines, and all machine tools in general. The four methods of control are as follows:

- **Manual.** All movements are made by hand lever control.

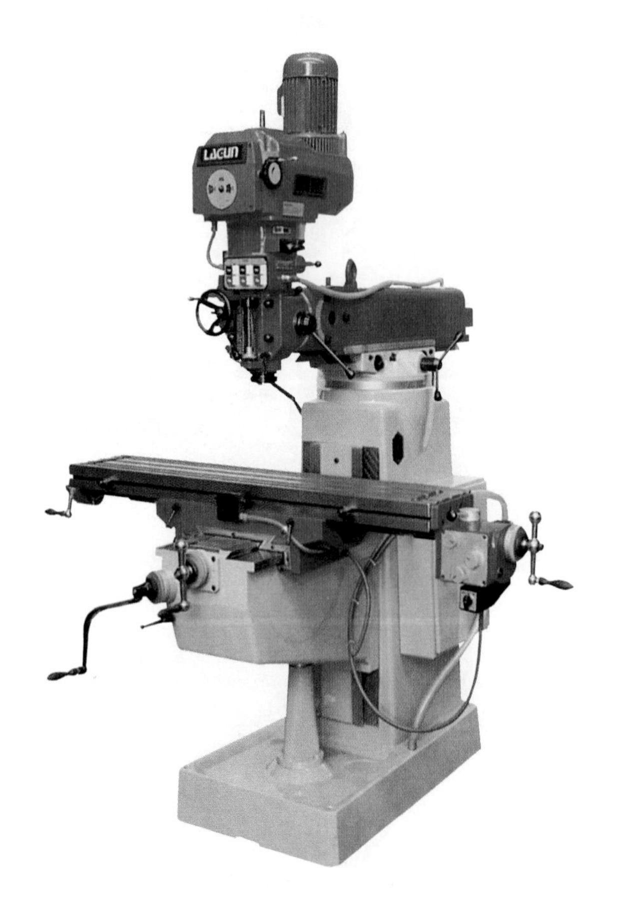

Republic-Lagun Machine Tool Co.

Figure 18-12. A typical swivel-head milling machine.

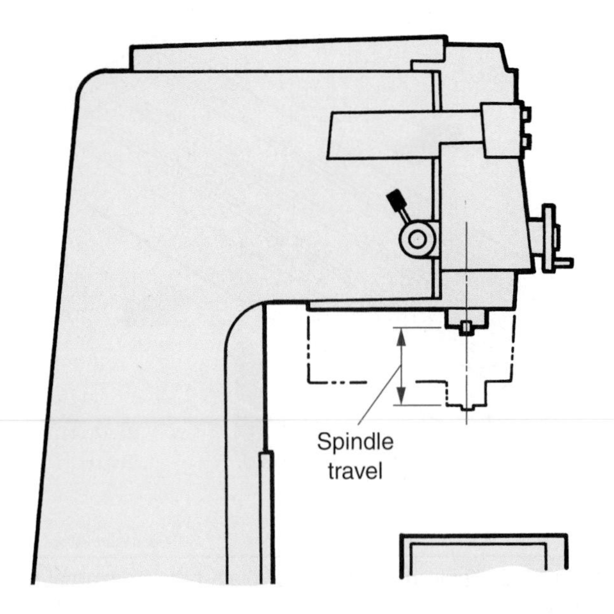

Goodheart-Willcox Publisher

Figure 18-13. Spindle head is fixed in a vertical position on sliding head milling machine. The entire head is moved to make cutting adjustments.

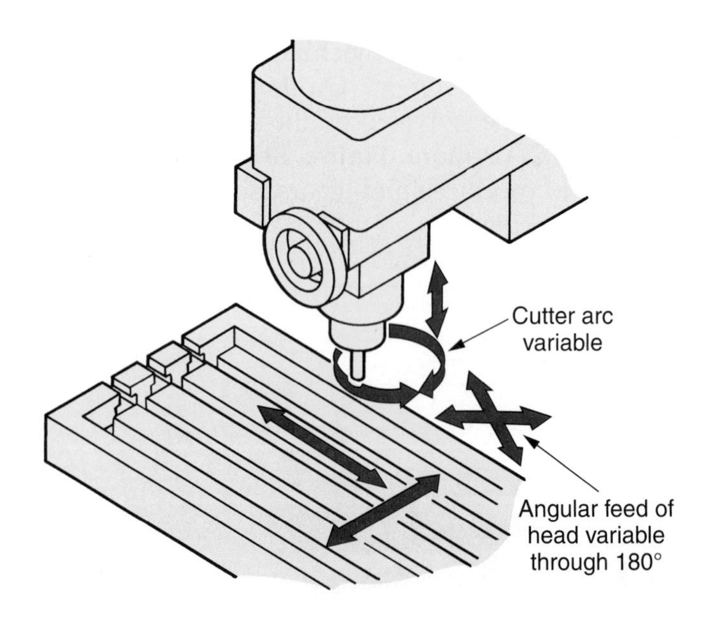

Goodheart-Willcox Publisher

Figure 18-14. Spindle movements possible on a rotary head vertical milling machine. Milling machines with CNC capabilities are replacing this type machine.

- **Semi-automatic.** Movements can be controlled by either hand or power feed.
- **Fully automatic.** A complex hydraulic feed arrangement that follows two or three dimensional templates to automatically guide one or more cutters. Specifications can also be programmed to guide the cutters and table through the required machining operations.

- **Computerized (CNC).** Machining coordinates are entered into a computer, using a special programming language. Instructions from the computer operate actuators (electric, hydraulic, or pneumatic devices) that move the table and cutter or cutters through the required machining sequence. Manually operated milling machines can be retrofitted with computerized control systems, **Figure 18-15.**

Small milling machines may only have power feed available for longitudinal table movement, **Figure 18-16.** On larger machines, automatic feed or power feed is used for all table movements.

Table movement (feed) can be engaged at cutting speed. However, there is a rapid traverse feed that allows fast power movement in any direction of feed engagement. This permits work to be positioned at several times the fastest rate indicated on the feed chart. It operates by positioning the automatic power feed control lever to give the desired directional movement and activating the rapid traverse lever.

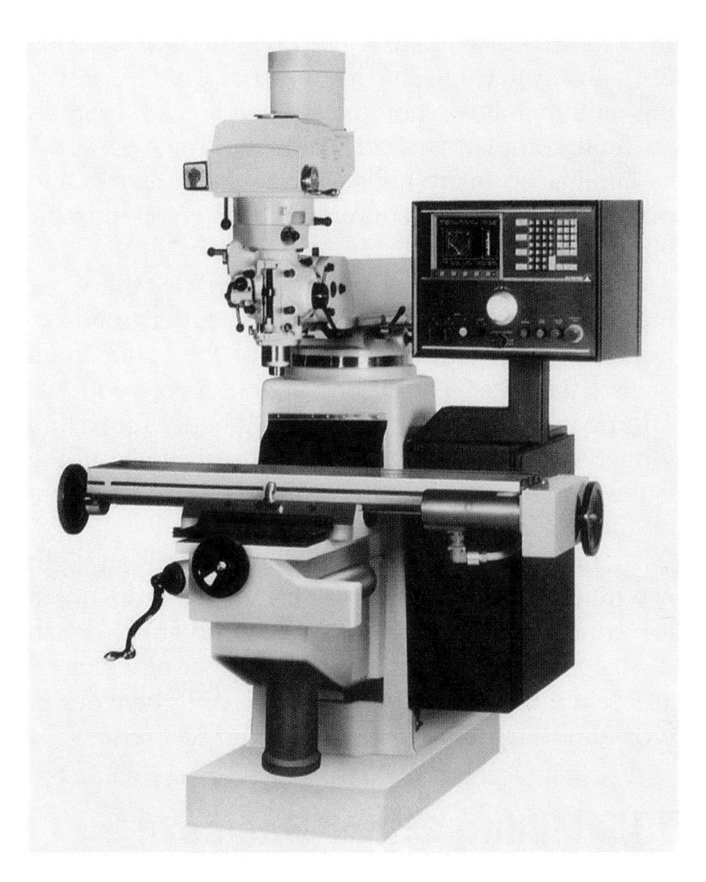

Autocon Technologies, Inc.

Figure 18-15. This manual vertical milling machine was retrofitted with a CNC control unit. The retrofit provides two-axis (X and Y) machine control capabilities and depth (Z axis) information displayed on the monitor.

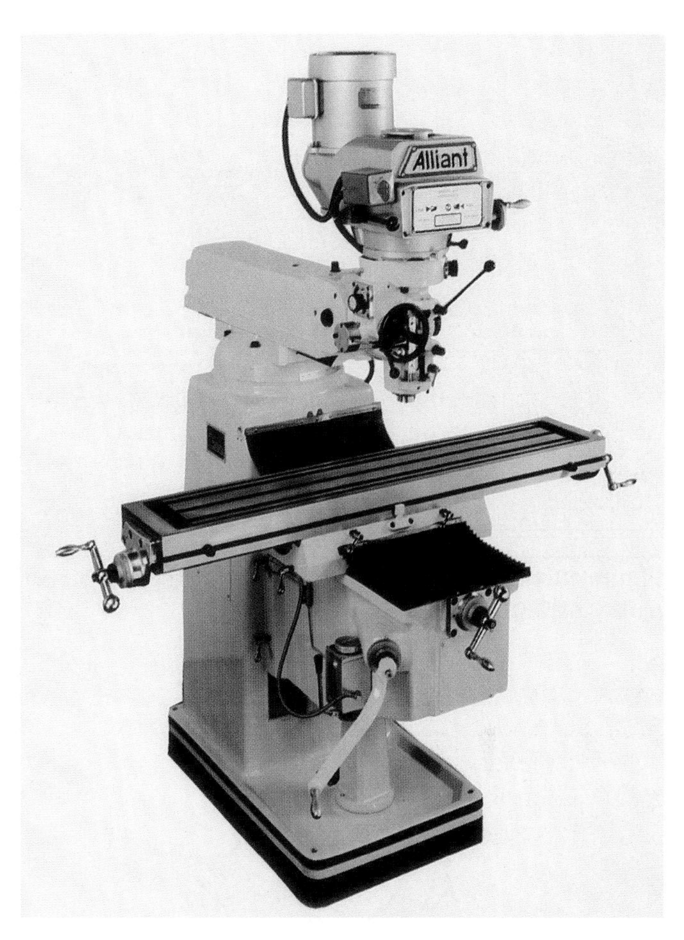

Rem Sales, Inc.

Figure 18-16. Manually controlled vertical milling machine. It has power feed for longitudinal table movement.

SAFETY NOTE

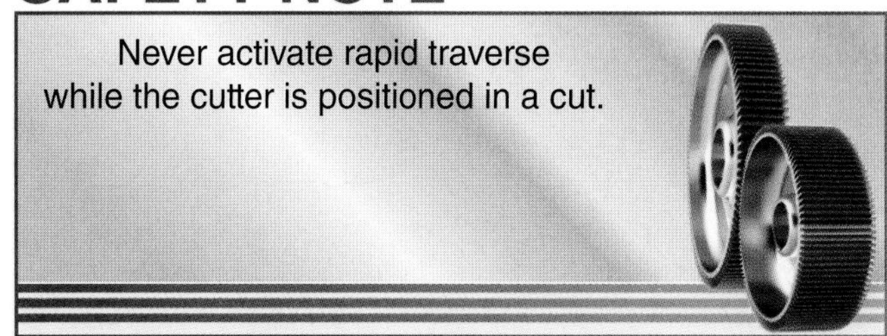

18.2 Milling Operations

There are two main categories of milling operations. ***Face milling*** is done when the surface being machined is parallel with the cutter face, **Figure 18-17.** Large, flat surfaces are machined with this technique. ***Peripheral milling*** is done when the surface being machined is parallel with the periphery of the cutter, **Figure 18-18.** There are also two distinct methods of milling, conventional milling and climb milling.

Sandvik Coromant Co.

Figure 18-17. Face milling. A—With face milling, the surface being machined is parallel with the cutter face. B—An example of face milling.

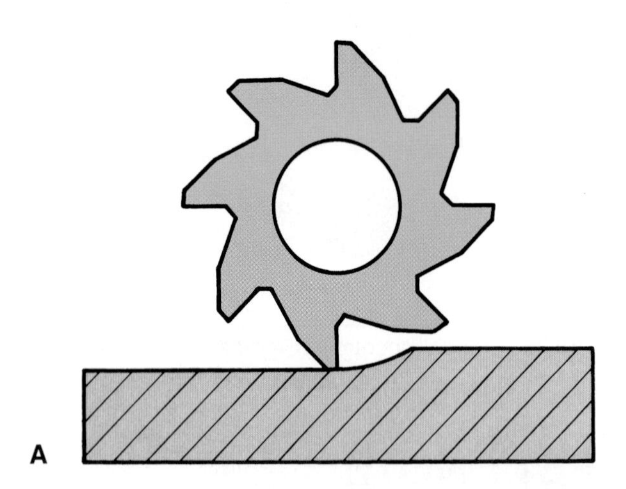

Goodheart-Willcox Publisher; BIG Kaiser Precision Tooling Inc.

Figure 18-18. Peripheral milling. A—In this milling method, the surface being machined is parallel with the periphery of the cutter. B—An example of peripheral milling.

With ***conventional milling*** or up-milling, the work is fed into the rotation of the cutter, **Figure 18-19A**. The chip is at minimum thickness at the start of the cut. The cut is so light that the cutter has a tendency to slide over the work until sufficient pressure is built up to cause the teeth to bite into the material. This initial sliding motion, followed by the sudden breakthrough as the tooth completes the cut, leaves the "milling marks" so familiar on many milled surfaces. The marks and ridges can be kept to a minimum by keeping the table gibs properly adjusted.

With ***climb milling*** or down-milling, the work moves in the same direction as cutter rotation, **Figure 18-19B**. Full engagement of the cutter tooth is instantaneous. The sliding action of conventional milling is eliminated, resulting in a better finish and longer tool life. The main advantage of climb milling is the tendency of the cutter to press the work down on the worktable or holding device.

Climb milling is not recommended on light machines, nor on large older machines that are not in top condition or are not fitted with an antibacklash device to take up play. There is danger of a serious accident if there is play in the table, or if the work or work-holding device is not mounted securely.

18.3 Milling Cutter Basics

The typical milling cutter is circular in shape with a number of cutting edges (teeth) located around its circumference. Milling cutters are manufactured in a large number of stock shapes, sizes, and kinds. See **Figure 18-20**.

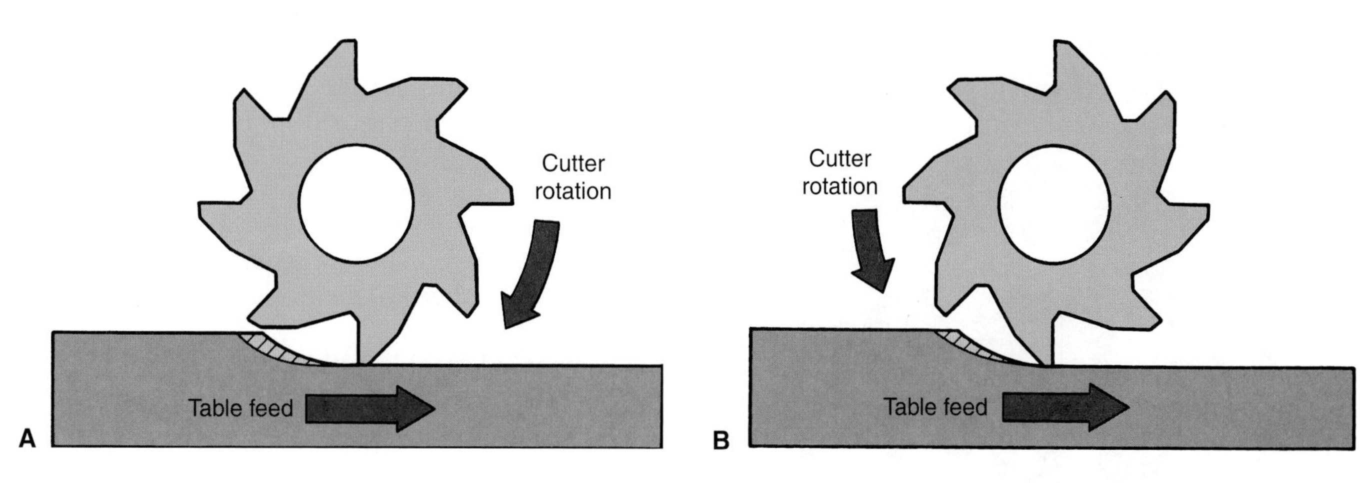

Goodheart-Willcox Publisher

Figure 18-19. Milling methods. A—Cutter and work movement with conventional or up-milling. B—Cutter and work movement with climb or down-milling.

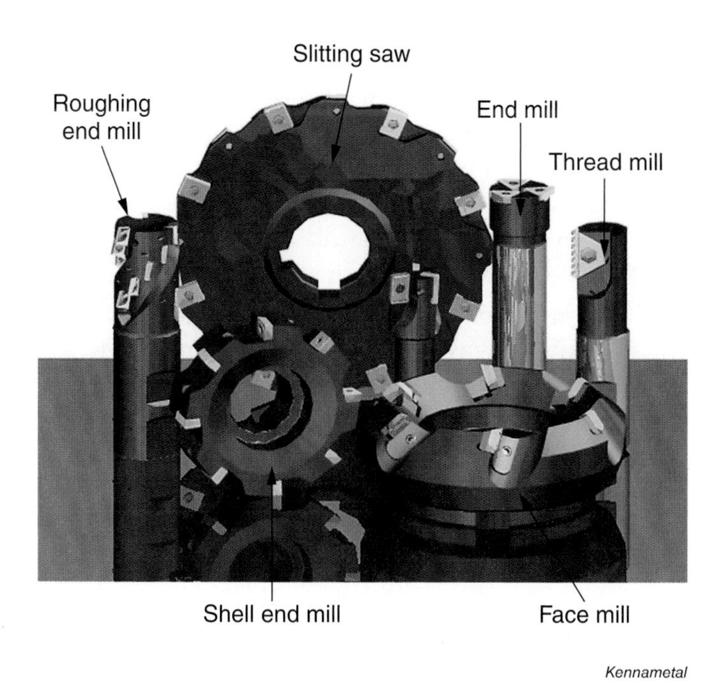

Kennametal

Figure 18-20. A selection of indexable insert milling cutters.

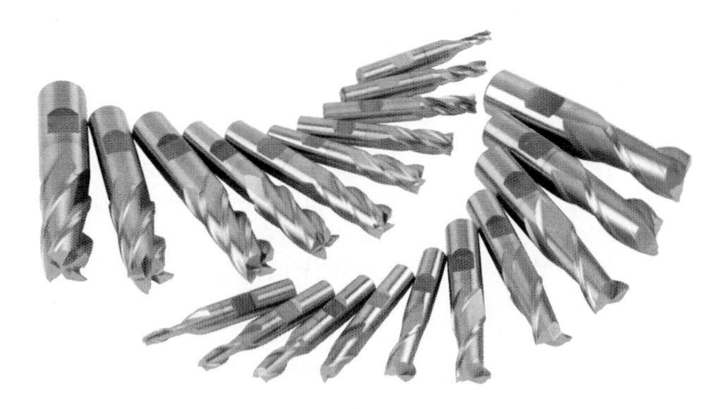

Photo courtesy of Grizzly Industrial, Inc. www.grizzly.com

Figure 18-21. A selection of two and four flute HSS (high speed steel) solid milling cutters. The gold color on the cutters is a titanium nitride (TiN) coating that increases wear resistance and the lifetime of the cutter significantly.

18.3.1 Milling Cutter Classification

Milling cutters are often classified by the method used to mount them on the machine. Milling cutters can be mounted directly by their shanks, directly to the spindle nose, or with an arbor. Shank cutters are fitted with either a straight or taper shank that is an integral part of the cutter. They are held in the machine by collets or special sleeves. Arbor cutters have a suitable hole for mounting to an arbor

There are two general types of milling cutters. Solid cutters are made with the shank, body, and cutting edges all in one piece, **Figure 18-21**. The inserted-tooth cutter has teeth made of special cutting material which are brazed or clamped in place, **Figure 18-22**. Worn and broken teeth can be replaced easily instead of discarding the entire cutter.

18.3.2 Milling Cutter Materials

Considering the wide range of materials that must be machined, the ideal milling cutter should have the following attributes:

- **High abrasion resistance.** The cutting edges should not wear away rapidly due to the abrasive nature of some materials.
- **Red hardness.** The cutting edges should not be affected by the terrific heat generated by many machining operations.
- **Edge toughness.** The cutting edges should not readily break down due to the loads imposed upon them by the cutting operation.

Dapra Corporation

Figure 18-22. A variety of cutters with indexable carbide inserts. When the cutting edges dull, new cutting edges are rotated into position. The inserts are available in different grades depending upon the material being machined.

Since no single material can meet these requirements in all situations, cutters are made from materials that are, by necessity, a compromise.

High-speed steels (HSS) are the most versatile of the cutter materials. Cutters made from high-speed steels are excellent for general purpose work or where vibration and chatter are problems. They are preferred for use on machines of low power.

HSS milling cutters can be improved by the application of surface lubricating treatments, surface hardening treatments, or by the application of coatings (such as chromium, tungsten, or tungsten carbide) to the cutting surfaces. While the treated tools cost 2 to 6 times as much as conventional HSS tools, they may last 5% to 10% longer or provide 50% to 100% percent higher metal removal rates with the same tool life.

Cemented tungsten carbides include a broad family of hard metals. They are produced by powder metallurgy techniques and have qualities that make them suitable for metal cutting tools. Cemented carbides can, in general, be operated at speeds 3 to 10 times faster than conventional HSS cutting tools.

Normally, only the cutting tips are made of cemented carbides rather than the entire cutter. They are brazed or clamped to the cutter body. See **Figure 18-23**.

Most inserted-tooth cutters make use of indexable inserts. Each insert has several cutting edges at various corners. When they become dull, the inserts are indexed (turned) so a new cutting edge contacts the metal.

Cemented carbide cutters are excellent for long production runs and for milling materials with a scale-like surface like cast iron, cast steel, or bronze.

18.4 Types and Uses of Milling Cutters

Milling cutters are commonly grouped into categories based on their shape and function. Selection of the proper cutting tool is very important to efficient milling operations. The following are the most commonly used milling cutters, with a summary of the work to which they are best suited.

18.4.1 End Mills

End milling cutters are designed for machining slots, keyways, pockets, and similar work, **Figure 18-24**. The cutting edges are on the circumference and end. End mills may have straight or helical flutes, **Figure 18-25**, and have straight or taper shanks, **Figure 18-26**. Straight shank end mills are available in single and double end styles, **Figure 18-27**.

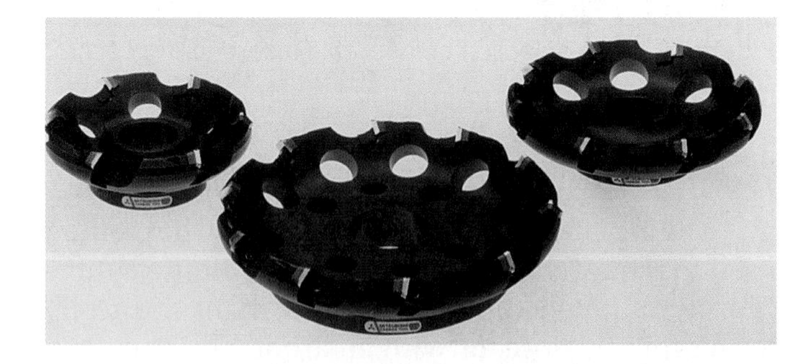

Mitsubishi Materials USA Corporation

Figure 18-23. These inserted-tooth cutters use teeth that are clamped to the cutter body. The cutter teeth (colored gold in this photo) can be indexed four times to present a fresh cutting edge as they wear.

Kennametal Inc.

Figure 18-24. An indexable-insert end mill cutting clearance slots on a face milling cutter.

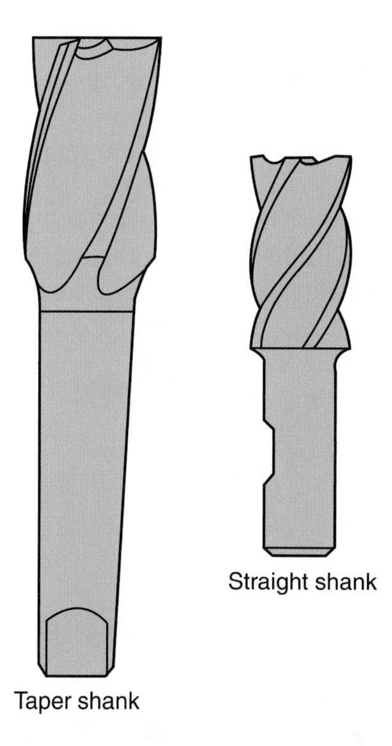

Goodheart-Willcox Publisher

Figure 18-26. Straight shank and taper shank end mills.

Mitsubishi Materials USA Corporation

Figure 18-25. End mill with multiple indexable inserts are made in both helical-fluted and straight-fluted types. Two face-type cutters are also shown.

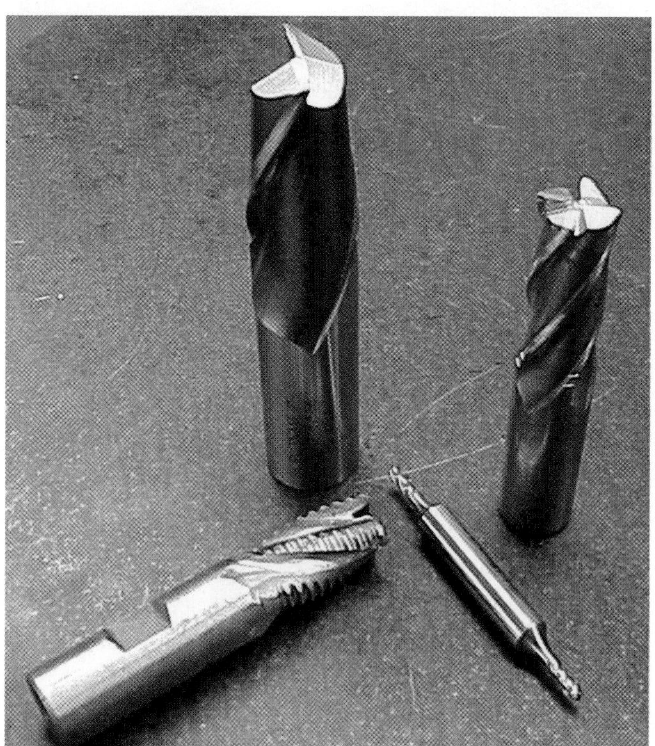

Goodheart-Willcox Publisher

Figure 18-27. Three large single-end-type end mills and a smaller double-end-type end mill.

The term hand is used to describe the direction of cutter rotation and the helix of the flutes, **Figure 18-28**. When viewed from the cutting end, a right-hand cutter rotates counterclockwise, and a left-hand cutter rotates clockwise.

Ball-nose end mills, **Figure 18-29**, are utilized for tracer milling, computer-controlled contour milling, die-sinking, fillet milling, and other radius work. A cut with a depth equal to one-half the end mill diameter can generally be taken in solid stock. See **Figure 18-30**.

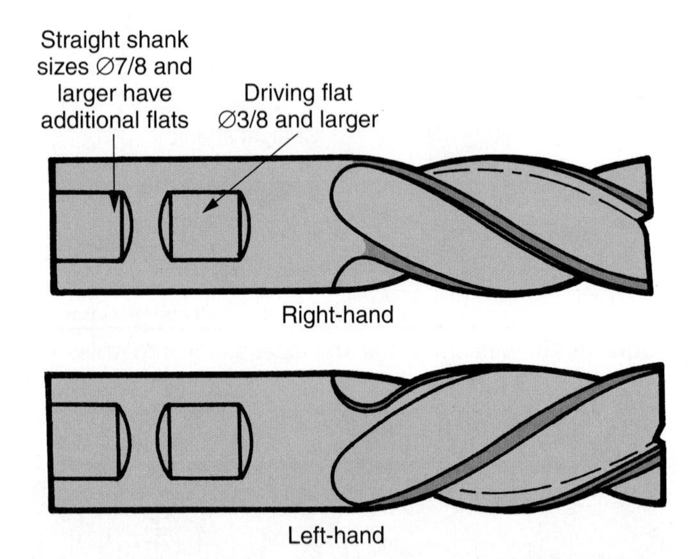

Goodheart-Willcox Publisher

Figure 18-28. A right-hand cutter rotates counterclockwise when viewed from cutting end. A left-hand cutter rotates clockwise.

Mitsubishi Materials USA Corporation

Figure 18-29. A ball-nose end mill has a rounded tip. These mills have two replaceable cutting inserts.

Delcam International

Figure 18-30. Three-dimensional milling is being done with a ball-nose end mill.

Several end mill styles are available, including the following:

- A two-flute end mill can be fed into the work like a drill. There are two cutting edges on the circumference, with the end teeth cut to the center, **Figure 18-31**.
- The multiflute end mill can be run at the same speed and feed as a comparable two-flute end mill, but it has a longer cutting life and will produce a better finish. It is recommended for conventional milling where plunge cutting (feeding into the work like a twist drill) is not necessary. See **Figure 18-32**.
- A shell end mill, **Figure 18-33**, has teeth similar to the multiflute end mill but is mounted on a stub arbor. The cutter is designed for both face and end milling. Shell end mills are made with right-hand cut, right-hand helix, or with left-hand cut, left-hand helix.

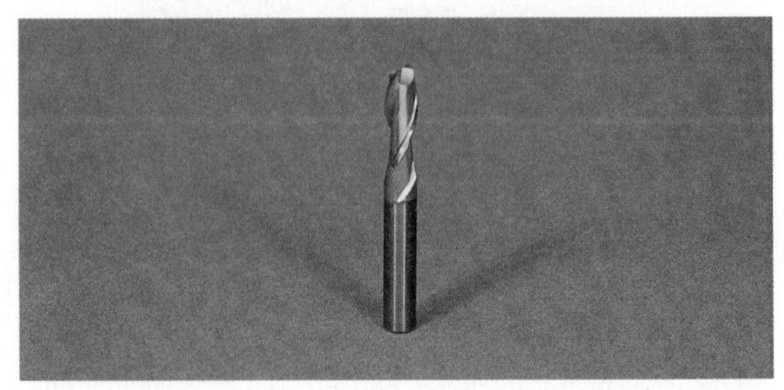

Goodheart-Willcox Publisher

Figure 18-31. The two-flute end mill can be fed into work like a drill.

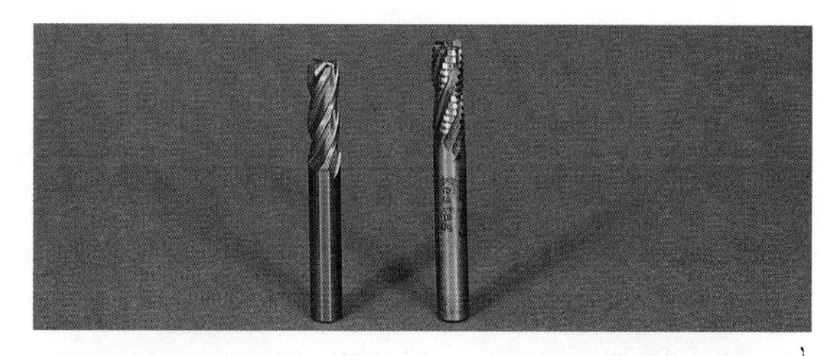

Goodheart-Willcox Publisher

Figure 18-32. Multifluted end mills. The peripheral grooves in the cutter at right reduce chip size, lowering cutting forces. Most modern cutters have a nitride coating that improves resistance to abrasive wear and corrosion.

Goodheart-Willcox Publisher

Figure 18-33. Three indexable insert shell end mills.

18.4.2 Face Milling Cutters

Face milling cutters are intended for machining large, flat surfaces parallel to the face of the cutter, **Figure 18-34**. The teeth are designed to make the roughing and finishing cuts in one operation. Because of their size and cost, most face milling cutters have inserted cutting edges. Facing cutters can be mounted directly to a machine's spindle nose or on a stub arbor. See **Figure 18-35**.

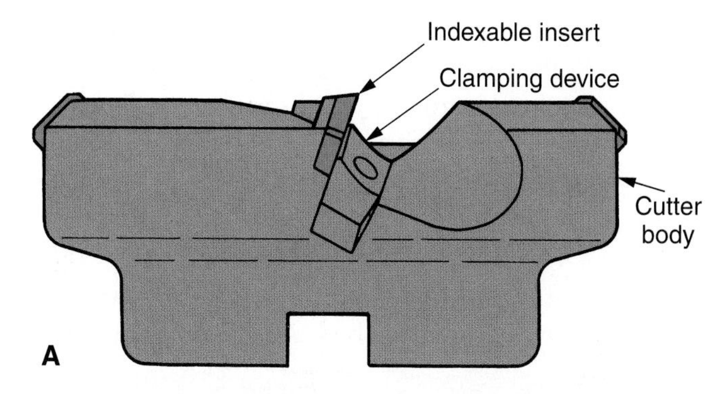

Mitsubishi Materials USA Corporation

Figure 18-34. Face mill with indexable inserts. A—The replaceable inserts are mechanically clamped in place on the cutter body. As cutting edges wear, the inserts can be turned (indexed) to present a fresh cutting edge. B—An example of a face mill with indexable inserts. Insert selection is based on material to be machined.

Valenite, Inc.

Figure 18-35. This face-type milling cutter has an unusual design, making use of bearing-mounted inserts that rotate as they cut. The manufacturer claims better heat dissipation provided by the rotating inserts will increase cutter life and permit higher cutting speeds.

A fly cutter is a single-point cutting tool used as a face mill. An example is shown in **Figure 18-36**. Fly cutters are capable of machining very fine surface finishes. Fly cutters should not be used to make heavy roughing cuts.

18.4.3 Arbor Milling Cutters

The more common arbor milling cutters and the work for which they are best adapted include the following:

- Plain milling cutter
- Side milling cutter
- Angle cutters
- Metal slitting saws
- Formed milling cutters

Plain Milling Cutter

Plain milling cutters are cylindrical, with teeth located around the circumference. Plain milling cutters less than 3/4″ (20 mm) are made with straight teeth. Wider plain cutters, called slab cutters, are made with helical teeth designed to cut with a shearing action. See **Figure 18-37**. This reduces the tendency for the cutter to chatter.

The different designs of plain milling cutters serve different purposes, as follows:

- The light-duty plain milling cutter is used chiefly for light slabbing cuts and shallow slots.

Goodheart-Willcox Publisher

Figure 18-36. The fly cutter is a single-point cutting tool used for face milling.

Lovejoy Tool Company, Inc.

Figure 18-37. A helical indexable insert plain milling cutter.

- The heavy-duty plain milling cutter is recommended for heavy cuts where considerable material must be removed. It has fewer teeth than a comparable light-duty cutter. The cutting edges are better supported and chip spaces are ample to handle the larger volume of chips.
- A helical plain milling cutter has fewer teeth than either of the two previously mentioned cutters. The helical cutter can be run at high speeds and produces exceptionally smooth finishes.

Side Milling Cutter

Cutting edges are located on the circumference and on one or both sides of ***side milling cutters***. They are made in solid form or with inserted teeth. See **Figure 18-38**.

- Plain side milling cutter teeth are on the circumference and on both sides of the cutter. It is recommended for side cutting, straddle milling, and slotting. Plain side milling cutters are available in diameters ranging from 2″ (50 mm) to 8″ (200 mm), and in widths from 3/16″ (5 mm) to 1″ (25 mm).
- A staggered-tooth side milling cutter has alternating right-hand and left-hand helical teeth. They aid in reducing chatter while providing adequate chip clearance for higher operating speeds and feeds than are possible with the plain side milling cutter. This type of cutter is especially good for machining deep slots.

Lovejoy Tool Company, Inc.

Figure 18-38. Staggered-tooth indexable insert side milling cutter.

- The half side milling cutter has helical teeth on the circumference but side teeth on only one side. It is made as a right-hand or left-hand cutter, and is recommended for heavy straddle milling and milling to a shoulder.
- The interlocking side milling cutter is ideally suited for milling slots or bosses, and for making other types of cuts that must be held to extremely close tolerances. The unit is made as two cutters with interlocking teeth that can be adjusted to the required width by using spacers or collars, **Figure 18-39**. The alternating right and left shearing action eliminates side pressures, producing a good surface finish.

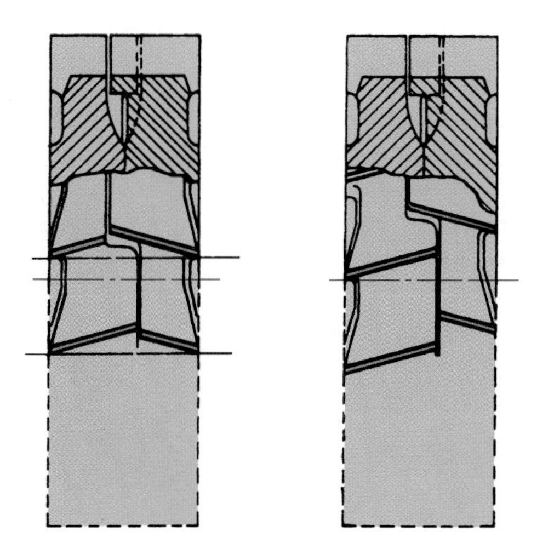

Morse Tool Co.

Figure 18-39. Tooth pattern on interlocking side milling cutters.

Angle Cutters

Angle cutters differ from other cutters in that the cutting edges are neither parallel, nor at right angles to the cutter axis.

- On a single-angle milling cutter, the teeth are on the angular face and on the side adjacent to the large diameter. Single-angle cutters are made in both right-hand and left-hand cut, with included angles of 45° and 60°.
- The double-angle milling cutter is used for milling threads, notches, serrations, and similar work. Double-angle cutters are manufactured with included angles of 45°, 60°, and 90°. Other angles can be special ordered.

Metal Slitting Saws

Metal slitting saws are thin milling cutters that resemble circular saw blades, **Figure 18-40**. They are employed for narrow slotting and cutoff operations. Slitting saws are available in diameters as small as 2 1/2″ (60 mm) and as large as 8″ (200 mm).

- The plain metal slitting saw is essentially a thin plain milling cutter. It is used for ordinary slotting and cutoff operations. Both sides are ground concave for clearance. The hub is the same thickness as the cutting edge. It is stocked in thicknesses ranging from 1/32″ (0.8 mm) to 3/16″ (5 mm).
- A side chip clearance slitting saw is similar to the plain side milling cutter. It is especially suitable for deep slotting and sawing applications because of its ample chip clearance.

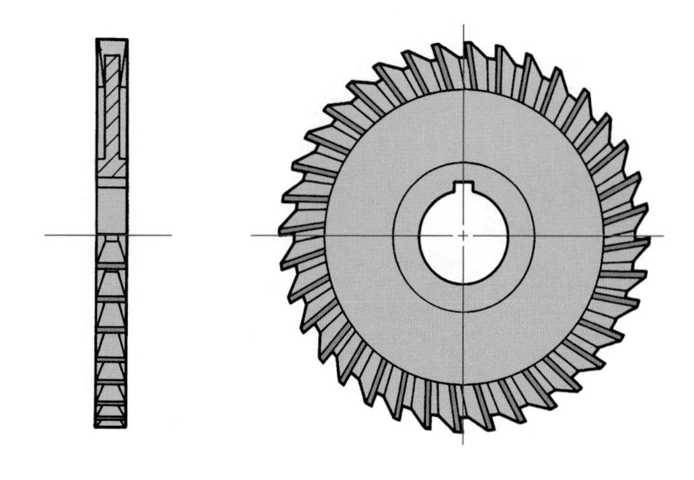

Goodheart-Willcox Publisher

Figure 18-40. Metal slitting saws are used for slotting and cutoff operations. The construction of the side chip clearance slitting saw allows it to be used for deep slotting applications.

Formed Milling Cutters

Formed milling cutters are employed to accurately duplicate a required contour. A wide range of shapes can be machined with standard cutters available. See **Figure 18-41**. Included in this cutter classification are the concave cutter, convex cutter, corner-rounding cutter, and gear cutter.

18.4.4 Miscellaneous Milling Cutters

Included in this category are cutters that do not fit into any of the previously mentioned groups.

- The T-slot milling cutter has cutting edges for milling the bottoms of T-slots after cutting with an end mill or side cutter, **Figure 18-42**.

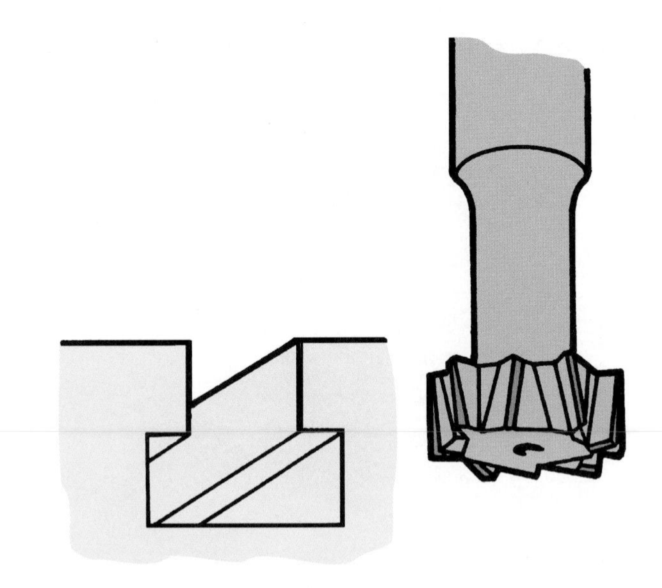

Goodheart-Willcox Publisher

Figure 18-42. A T-slot milling cutter and the cut it produces.

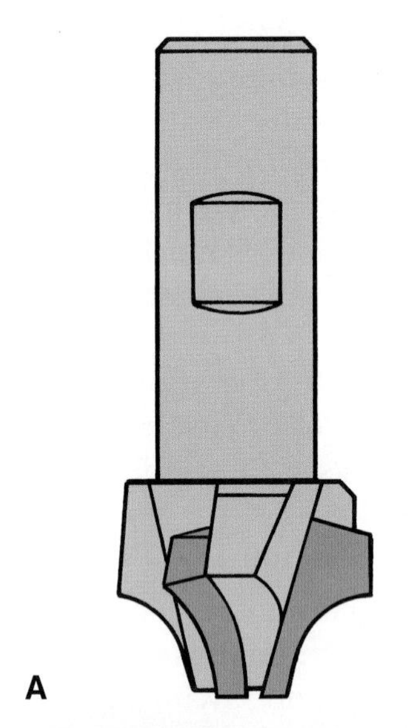

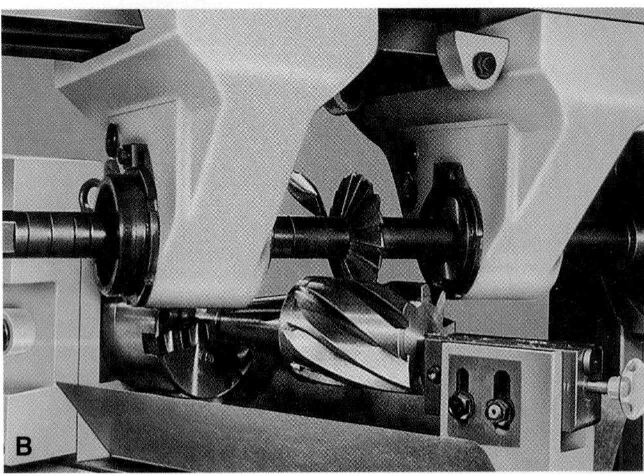

Goodheart-Willcox Publisher; WMW Machinery Company, Inc.

Figure 18-41. Formed milling cutters. A—Corner-rounding end mill with replaceable tungsten carbide cutting edges. B—Gear cutter. Note how table is swiveled at an angle to cutting tool. This feature makes it possible to machine helical-flute workpieces like the one being cut.

- A Woodruff keyseat cutter is used to mill the semicircular keyseat for a Woodruff key.
- A dovetail cutter mills dovetail-type ways and it is used in much the same manner as the T-slot cutter. See **Figure 18-43**.

18.4.5 Care of Milling Cutters

Milling cutters are expensive and easily damaged if care is not taken in their use and storage. The following recommendations will help extend cutter life:

- Use sharp cutting tools. Machining with dull tools results in low-quality work, and eventually damages the cutting edges beyond the point where they can be salvaged by grinding.

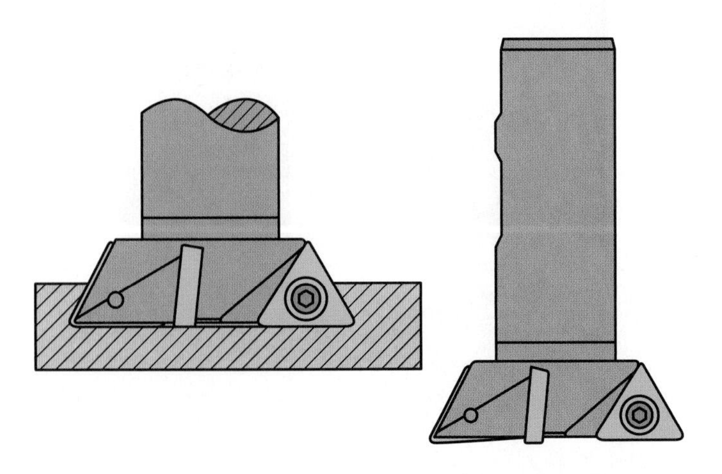

Goodheart-Willcox Publisher

Figure 18-43. A dovetail cutter with indexable insert cutting edges.

- Properly support tools and make sure the work is held rigidly.
- Use the correct cutting speed and feed for the material being machined.
- Ensure an ample supply of cutting fluid, **Figure 18-44**.
- Employ the correct cutter for the job.
- Store cutters in individual compartments or on wooden pegs. They should never come in contact with other cutters or tools.
- Clean cutters before storing them. If they are to be stored for any length of time, it is best to give them a light protective coating of oil.
- Never hammer a cutter onto an arbor. Examine the arbor for nicks or burrs if the cutter does not slip onto it easily. Do not forget to key the cutter to the arbor.
- Place a wooden board under an end mill when removing it from a vertical milling machine. This will prevent cutter damage if it is dropped accidentally. Protect your hand with a heavy cloth or gloves.

18.5 Holding and Driving Cutters

The ***arbor*** is the most common method employed to hold and drive cutters. It is made in a number of sizes and styles. Arbors with self-holding tapers were used on some small hand milling machines and older models of larger millers, but are seldom used today. Today, there are three basic arbor styles in general use, **Figure 18-45**.

- Style A is fitted with a small pilot end that runs in a bronze bearing in the arbor support. This style is best used when maximum arbor support clearance is required.
- Style B is characterized by the large bearing collar that can be positioned on any part of the arbor. This feature makes it possible to mount the bearing support as close to the cutter as possible for maximum cutter support. This permits heavy cuts.
- Style C is used to hold smaller sizes of shell end and face milling cutters that cannot be mounted directly to the spindle nose.

Sharnoa Corp.

Figure 18-44. Rigidly supported cutters will permit heavier cuts and prolong cutter life. An ample supply of cutting fluid is essential.

A

B

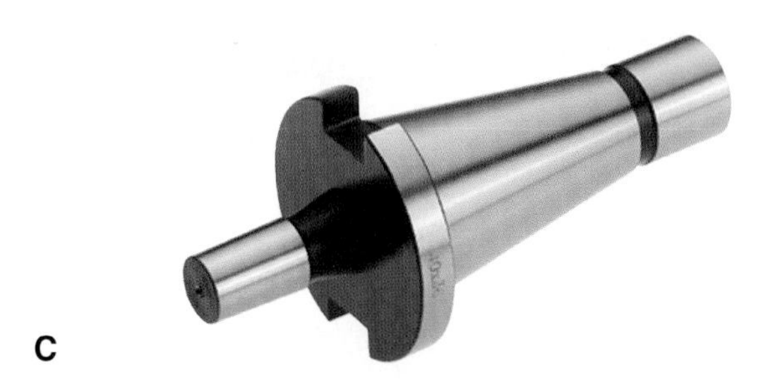

C

Photo courtesy of Grizzly Industrial, Inc. www.grizzly.com; WMW Machinery Company, Inc.

Figure 18-45. Basic arbor styles. A—Style A arbor. B—Style B arbor. C—Style C arbor.

In general, use the shortest arbor possible that will permit adequate clearance between the arbor support and the work.

Both style A and style B arbors have a keyway milled their entire length, allowing a key to be employed to prevent the cutter from revolving on the arbor. See **Figure 18-46**.

Spacing collars allow the cutter to be positioned on the arbor, **Figure 18-47**. Accurately made in a number of widths, collars permit two or more cutters to be precisely spaced for gang and saddle milling. A left-hand threaded nut tightens the cutter and collars on the arbor. The nut should not be tightened directly against the bearing collar on the style B arbor, because the bearing may be damaged.

A draw-in bar is used on most vertical and horizontal milling machines, **Figure 18-48**. It fits through the spindle and screws into the arbor or collet to hold it firmly on the spindle. Drive keys on the nose of the spindle fit into corresponding slots on the arbor, collet, or collet holder to provide positive (nonslip) drive.

End mills may be mounted in spring collets, adapters, shell end mill holders, or stub arbors, depending upon the type of work to be done. See **Figure 18-49**.

Spring collets accommodate straight shank end mills and drills. Some collets must be fitted in a collet chuck. Adapters are used for taper shank end mills and drills. Shell end mill holders enable shell end mills,

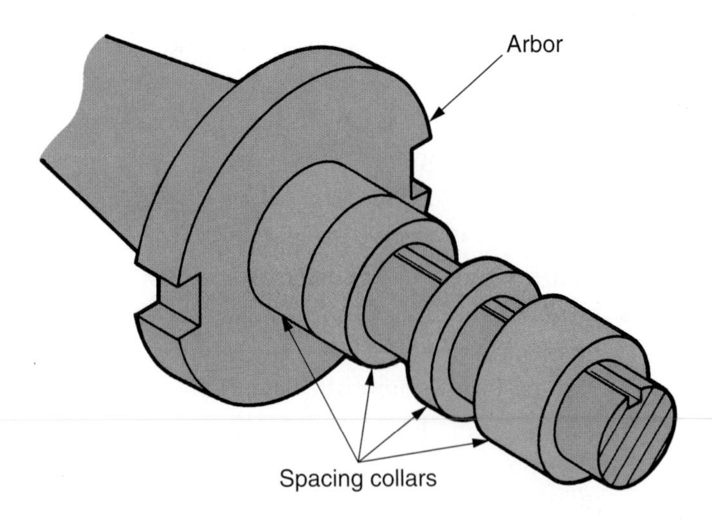

Goodheart-Willcox Publisher

Figure 18-47. Spacing collars are manufactured in many different widths. They are used to position one or more cutters on the arbor.

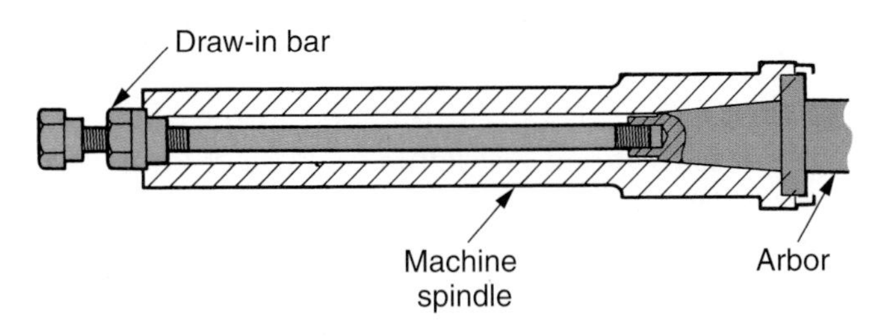

Goodheart-Willcox Publisher

Figure 18-48. The draw-in bar holds the arbor onto the spindle. Avoid operating a milling machine if the arbor is not held in place with a draw-in bar.

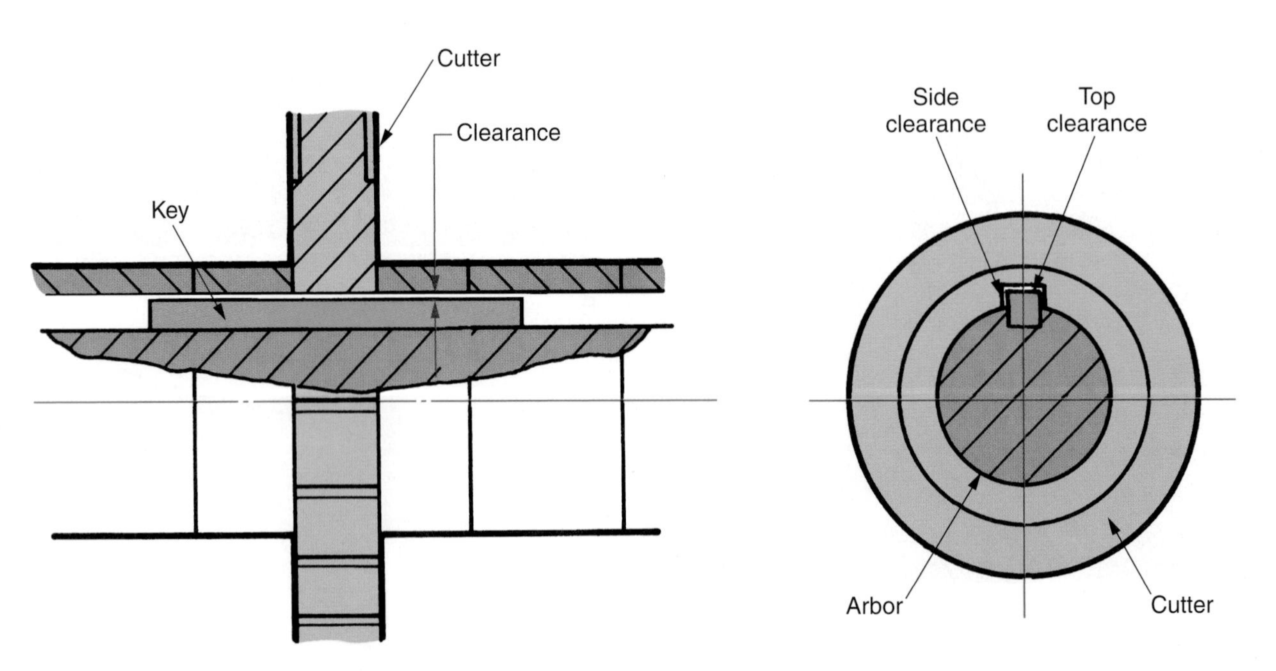

Goodheart-Willcox Publisher

Figure 18-46. Keying the cutter to the arbor prevents it from slipping during the cutting operation.

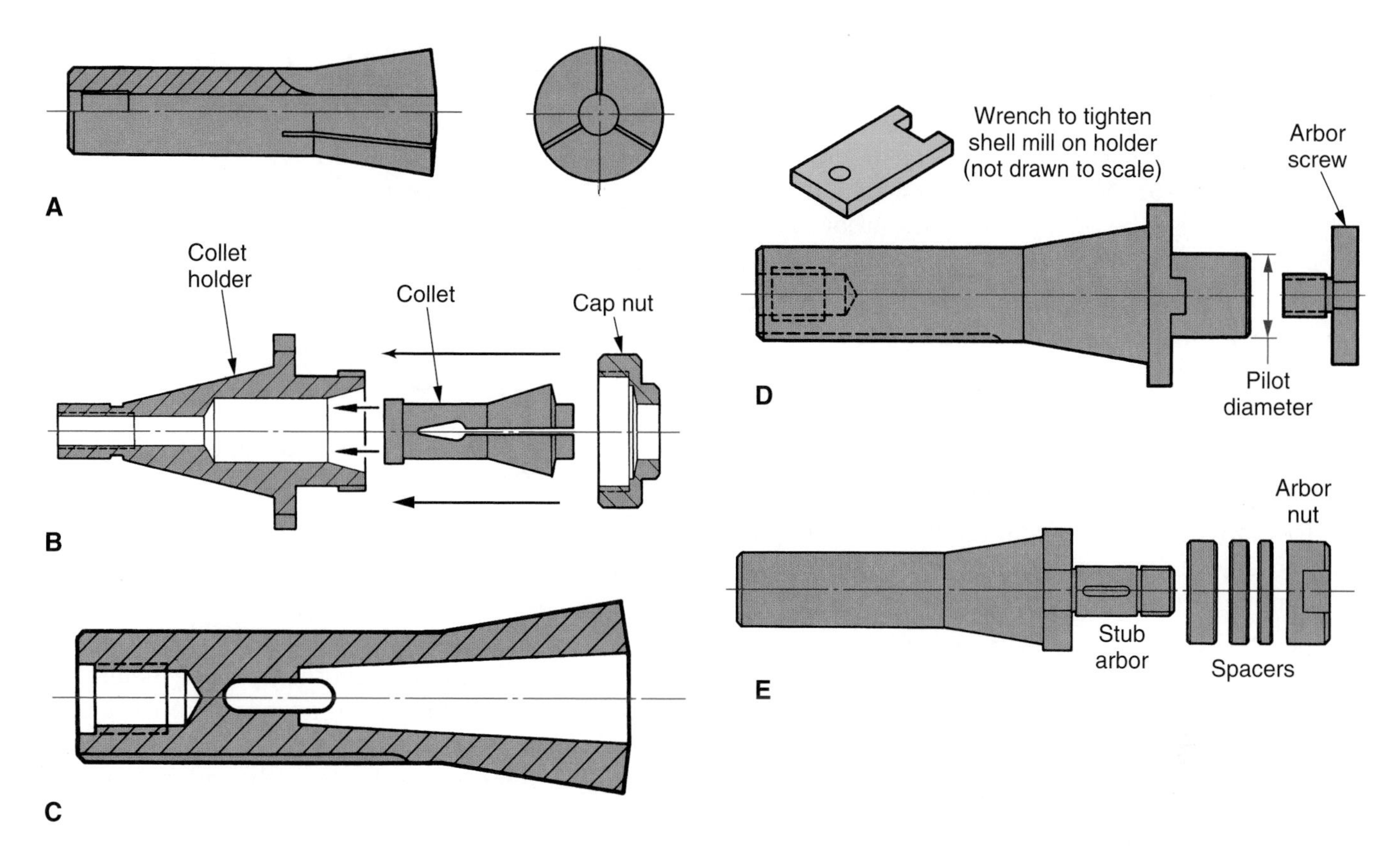

Goodheart-Willcox Publisher

Figure 18-49. Mounting devices. A—Spring collet (R-8 taper type). B—Collet chuck and collet. C—Adapter used with taper shank cutting tools. D—Shell end mill holder (R-8 taper). E—Stub arbor (R-8 taper).

used for face and side milling, to be fitted to a vertical milling machine. Stub arbors are short arbors that permit various side cutters, slitting saws, formed cutters, and angle cutters to be used on a vertical mill.

18.5.1 Care of Cutter Holding and Driving Devices

To maintain precision and accuracy during a milling operation, care must be taken to prevent damage to the cutter holding and driving devices.

- Keep the taper end of the arbor clean and free of nicks. The same applies to the spindle taper.
- Clean and lubricate the bearing sleeve before placing the arbor support on it and make sure the bearing sleeve fits snugly.
- Clean the spacing collars before slipping them onto an arbor, **Figure 18-50**. Otherwise, cutter runout will occur, making it difficult to make an accurate cut.
- Store arbors separately and in a vertical position.

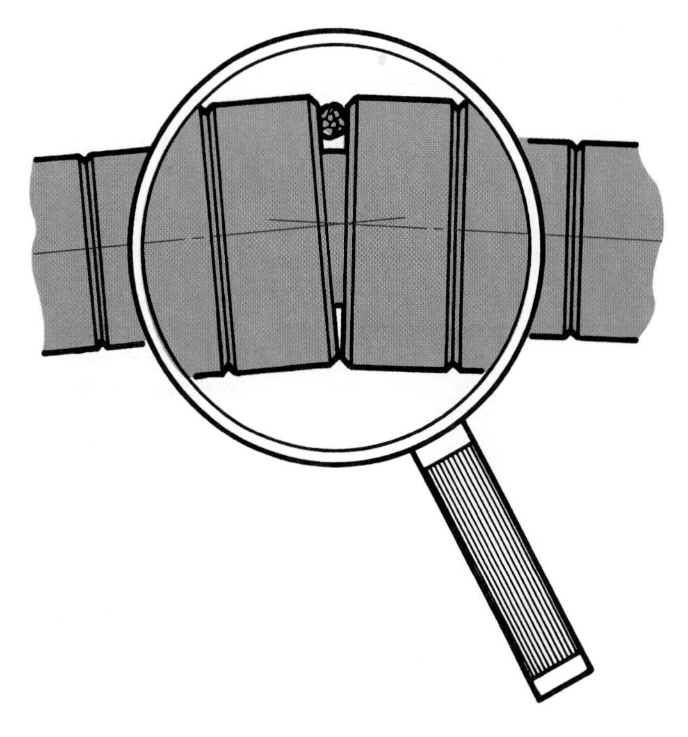

Goodheart-Willcox Publisher

Figure 18-50. A chip between spacing collars will cause an arbor to be sprung out of true. The end in the arbor is greatly exaggerated in this drawing.

- Never loosen or tighten an arbor nut unless the arbor support is locked in place, because this could spring the arbor so that it will not run true.
- Use a wrench of the correct size on the arbor nut, **Figure 18-51**. Make sure at least four threads are engaged before tightening the arbor nut.
- Avoid tightening an arbor nut by striking the wrench with a hammer or mallet. This can crack the nut and distort the threads.
- Do not force a cutter onto an arbor. Check to see what is making it difficult to slide on. Correct any problem.
- Key all cutters to the arbor.

To remove an arbor or adapter from the machine:

1. Loosen the draw-in bar nut a few turns. Do not remove it from the arbor completely.
2. Tap the draw-in bar with a lead hammer to loosen the arbor in the spindle.
3. Hold the loosened arbor with one hand and unscrew the draw-in bar with the other.
4. Remove the arbor from the spindle. Clean and store it properly.

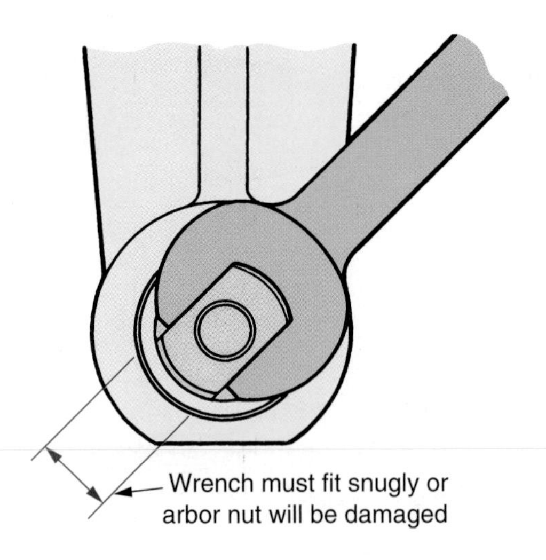

Goodheart-Willcox Publisher

Figure 18-51. Use a wrench of correct type and size to loosen an arbor nut.

18.6 Milling Cutting Speeds and Feeds

The time required to complete a milling operation and the quality of the finish is almost completely governed by the cutting speed and feed rate of the cutter.

Milling ***cutting speed*** refers to the distance, measured in feet or meters, that a point (tooth) on the cutter's circumference will move in one minute. It is expressed in feet per minute (fpm) or meters per minute (mpm). Milling cutting speed is directly dependent on the revolutions per minute (rpm) of the cutter.

Milling ***feed*** is the rate at which work moves into the cutter. It is given in feed per tooth per revolution (ftr). Proper feed rate is probably the most difficult setting for a machinist to determine. In view of the many variables (width of cut, depth of cut, machine condition, cutter sharpness), feed should be as coarse as possible, consistent with the desired finish.

18.6.1 Calculating Cutting Speeds and Feeds

Considering the previously mentioned variables, the speeds listed in **Figure 18-52**, and the feeds listed in **Figure 18-53**, are suggested. The usual procedure

Material	High-speed steel cutter		Carbide cutter	
	Feet per minute	Meters per minute*	Feet per minute	Meters per minute*
Aluminum	550–1000	170–300	2200–4000	670–1200
Brass	250–650	75–200	1000–2600	300–800
Low carbon steel	100–325	30–100	400–1300	120–400
Free cutting steel	150–250	45–75	600–1000	180–300
Alloy steel	70–175	20–50	280–700	85–210
Cast iron	45–60	15–20	180–240	55–75

Reduce speeds for hard materials, abrasive materials, deep cuts, and high alloy materials. Increase speeds for soft materials, better finishes, light cuts, frail work, and setups. Start at midpoint on the range and increase or decrease speed until best results are obtained.
*Figures rounded off.

Goodheart-Willcox Publisher

Figure 18-52. Recommended cutting speeds for milling. Speed is given in surface feet per minute (fpm) and in surface meters per minute (mpm).

Type of cutter	Material: Aluminum	Brass	Cast iron	Free cutting steel	Alloy steel
End mill	0.009 (0.22) 0.022 (0.55)	0.007 (0.18) 0.015 (0.38)	0.004 (0.10) 0.009 (0.22)	0.005 (0.13) 0.010 (0.25)	0.003 (0.08) 0.007 (0.18)
Face mill	0.016 (0.40) 0.040 (1.02)	0.012 (0.30) 0.030 (0.75)	0.007 (0.18) 0.018 (0.45)	0.008 (0.20) 0.020 (0.50)	0.005 (0.13) 0.012 (0.30)
Shell end mill	0.012 (0.30) 0.030 (0.75)	0.010 (0.25) 0.022 (0.55)	0.005 (0.13) 0.013 (0.33)	0.007 (0.18) 0.015 (0.38)	0.004 (0.10) 0.009 (0.22)
Slab mill	0.008 (0.20) 0.017 (0.43)	0.006 (0.15) 0.012 (0.30)	0.003 (0.08) 0.007 (0.18)	0.004 (0.10) 0.008 (0.20)	0.001 (0.03) 0.004 (0.10)
Side cutter	0.010 (0.25) 0.020 (0.50)	0.008 (0.20) 0.016 (0.40)	0.004 (0.10) 0.010 (0.25)	0.005 (0.13) 0.011 (0.28)	0.003 (0.08) 0.007 (0.18)
Saw	0.006 (0.15) 0.010 (0.25)	0.004 (0.10) 0.007 (0.18)	0.001 (0.03) 0.003 (0.08)	0.003 (0.08) 0.005 (0.13)	0.001 (0.03) 0.003 (0.08)

US Customary value expressed in inches per tooth. Metric value (shown in parentheses) expressed in millimeters per tooth.
Increase or decrease feed until the desired surface finish is obtained.
Feeds may be increased 100 percent or more depending upon the rigidity of the machine and the power available, if carbide tipped cutters are used.

Goodheart-Willcox Publisher

Figure 18-53. Recommended feed rates in inches per tooth and millimeters per tooth for high speed steel (HSS) milling cutters.

is to start with the midrange figure and increase or decrease speeds until the most satisfactory combination is obtained, consistent with cutter life and surface quality.

In general, speed is reduced for hard or abrasive materials, deep cuts, or high alloy content metals. Speed is increased for soft materials, better finishes, and light cuts. Refer to **Figure 18-54** to calculate the cutting speed and feed for a specific material.

Example Problem: Determine the approximate cutting speed and feed for a 6″ (152 mm) diameter side cutter (HSS) with 16 teeth, when milling free cutting steel.

Information Available:

Recommended cutting speed for free cutting steel (midpoint in range) = 200 fpm

Recommended feed per tooth (midpoint in range) = 0.008″

Cutter diameter = 6″

Number of teeth on cutter = 16

To determine speed setting (cutter rpm), the following is given in **Figure 18-54**.

Rule: Divide the feet per (fpm) by the circumference of the cutter, expressed in feet.

Formula:
$$\text{rpm} = \frac{\text{fpm} \times 12}{\pi D}$$
$$= \frac{200 \times 12}{3.14 \times 6}$$
$$= \frac{2400}{18.84}$$
$$= 127.39 \text{ rpm}$$

To determine feed setting (feed in inches per minute or F), the following is given in **Figure 18-54**.

Rule: Multiply feed per tooth per revolution by the number of teeth on the cutter and by speed (rpm).

Formula:
$$F = \text{ftr} \times T \times \text{rpm}$$
$$= 0.008 \times 16 \times 127$$
$$= 16.25$$

Rules for determining speed and feed			
To find	**Having**	**Rule**	**Formula**
Speed of cutter in feet per minute (fpm)	Diameter of cutter and revolutions per minute	Diameter of cutter (in inches) multiplied by 3.1416 (π) multiplied by revolutions per minute, divided by 12	$fpm = \frac{\pi D \times rpm}{12}$
Speed of cutter in meters per minute	Diameter of cutter and revolutions per minute	Diameter of cutter multiplied by 3.1416 (π) multiplied by revolutions per minute, divided by 1000	$mpm = \frac{D(mm) \times \pi \times rpm}{1000}$
Revolutions per minute (rpm)	Feet per minute and diameter of cutter	Feet per minute, multiplied by 12, divided by circumference of cutter (πD)	$rpm = \frac{fpm \times 12}{\pi D}$
Revolutions per minute (rpm)	Meters per minute and diameter of cutter in millimeters (mm)	Meters per minute multiplied by 1000, divided by the circumference of cutter (D)	$rpm = \frac{mpm \times 1000}{\pi D}$
Feed per revolution (FR)	Feed per minute and revolutions per minute	Feed per minute, divided by revolutions per minute	$FR = \frac{F}{rpm}$
Feed per tooth per revolution (ftr)	Feed per minute and number of teeth in cutter	Feed per minute (in inches or millimeters) divided by number of teeth in cutter × revolutions per minute	$ftr = \frac{F}{T \times rpm}$
Feed per minute (F)	Feed per tooth per revolution, number of teeth in cutter, and rpm	Feed per tooth per revolution multiplied by number of teeth in cutter, multiplied by revolutions per minute	$F = ftr \times T \times rpm$
Feed per minute (F)	Feed per revolution and revolutions per minute	Feed per revolution multiplied by revolutions per minute	$F = FR \times rpm$
Number of teeth per minute (TM)	Number of teeth in cutter and revolutions per minute	Number of teeth in cutter multiplied by revolutions per minute	$TM = T \times rpm$

rpm = Revolutions per minute
T = Teeth in cutter
D = Diameter of cutter
π = 3.1416 (pi)
frm = Speed of cutter in feet per minute
TM = Teeth per minute
F = Feed per minute
FR = Feed per revolution
ftr = Feed per tooth per revolution
mpm = Speed of cutter in meters per minute

Goodheart-Willcox Publisher

Figure 18-54. Rules for determining cutting speed and feed.

18.6.2 Adjusting Cutting Speed and Feed for Milling

Depending upon the make of the milling machine, cutting speed (cutter rpm) and rate of feed (table movement speed) may be changed by any of the following:

- Shifting V-belts.
- Adjusting variable speed pulleys.
- Utilizing a quick change gear box and shifting or dialing to the required speed and feed setting, **Figure 18-55**.

On some machines, speed and feed changes are made hydraulically or electronically. Always make speed changes as specified in the instruction manual for the specific machine being used.

SAFETY NOTE

Make sure the machine has come to a complete stop before attempting to adjust V-belts.

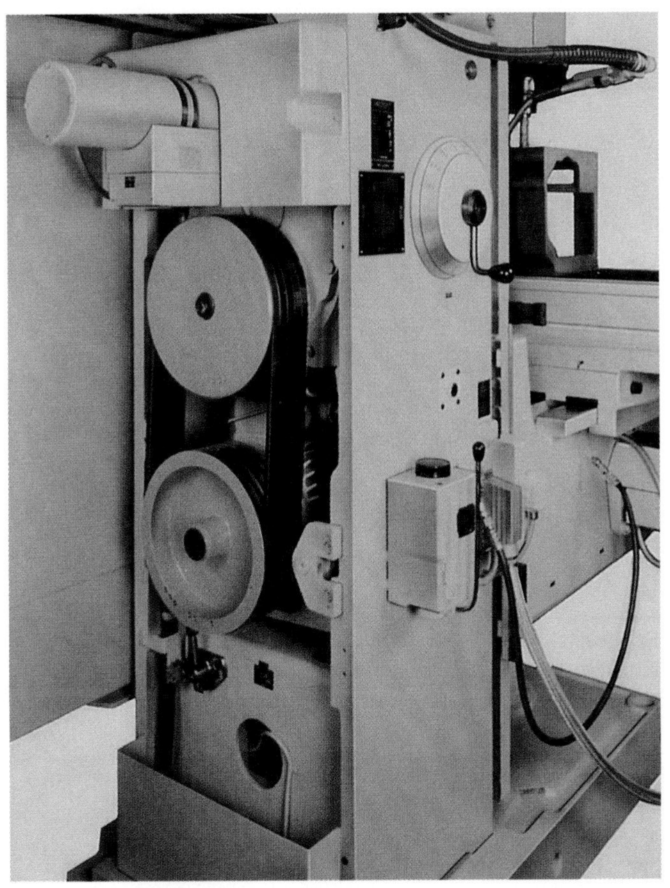

A

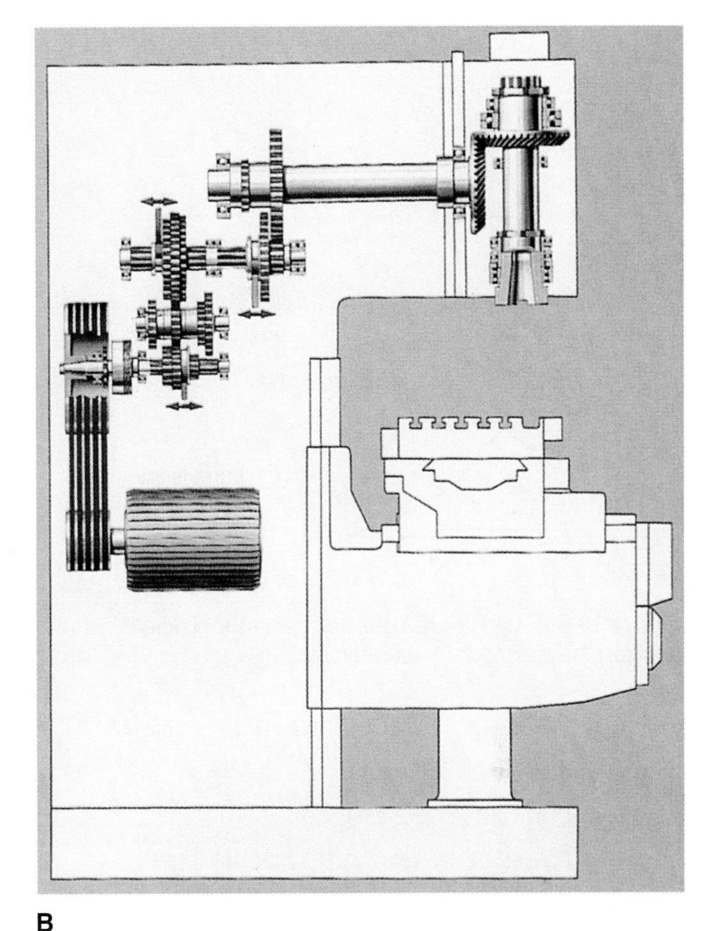

B

WMW Machinery Company, Inc.

Figure 18-55. Spindle speed control. A—Main belt drive and spindle speed control dial on a vertical milling machine. B—Cutaway showing sliding gear mechanism that permits dialing any of 18 speed choices.

The speed and feed are only approximate. Set machine to closest setting, either higher or lower. Increase or reduce speed until satisfactory cutting conditions are achieved.

18.7 Cutting Fluids

Cutting fluids serve several purposes, including the following:

- They carry away the heat generated during the machining operation.
- They act as a lubricant.
- They prevent the chips from sticking to or fusing with the cutter teeth.
- They flush away chips.
- They influence the finish quality of the machined surface.

It is important that the correct cutting fluid be used for the material being machined.

18.8 Milling Work-Holding Attachments

One of the more important features of the milling machine is its adaptability to a large number of work-holding attachments. Each of these attachments increases the usefulness of the milling machine.

The vise is probably the most widely employed method of holding work for milling. The jaws are hardened to resist wear, and are ground for accuracy. A milling vise, like other work-holding attachments, is keyed to the table slot with lugs, **Figure 18-56**.

A flanged vise has slotted flanges for fastening the vise to the table, **Figure 18-57**. The slots permit the vise to be mounted on a horizontal milling machine either parallel to, or at right angles to, the spindle.

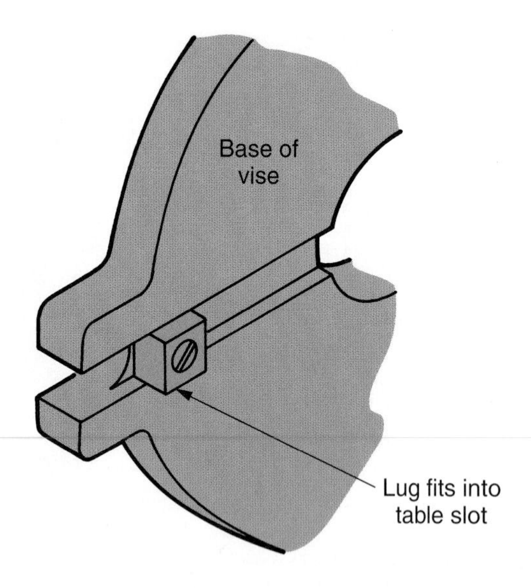

Goodheart-Willcox Publisher

Figure 18-56. Lugs on the base of a work-holding attachment position the device on the worktable.

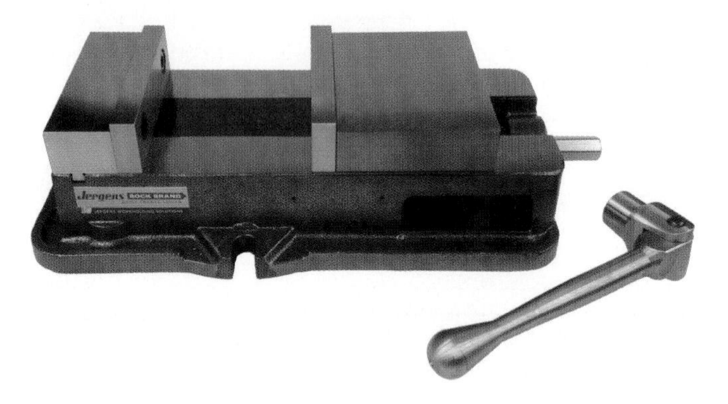

© Copyright Jergens, Inc. 2013

Figure 18-57. A typical flanged vise.

The body of a swivel vise is similar to a flange vise but is fitted with a circular base, graduated in degrees. This permits it to be locked at any angle to the spindle. See **Figure 18-58**.

The toolmaker's universal vise permits compound or double angles to be machined without complex or multiple setups. See **Figure 18-59**.

A magnetic chuck, shown in **Figure 18-60**, is ideally suited for many milling operations. The magnet eliminates the need for time-consuming hold-down clamps to mount the work to the table. The magnetic chuck can be used only with ferrous metals.

A rotary table, **Figure 18-61**, can perform a variety of operations, such as: cutting segments of circles, circular slots, cutting irregular-shaped slots, and similar operations. A dividing attachment can be fitted to many rotary tables in place of the handwheel.

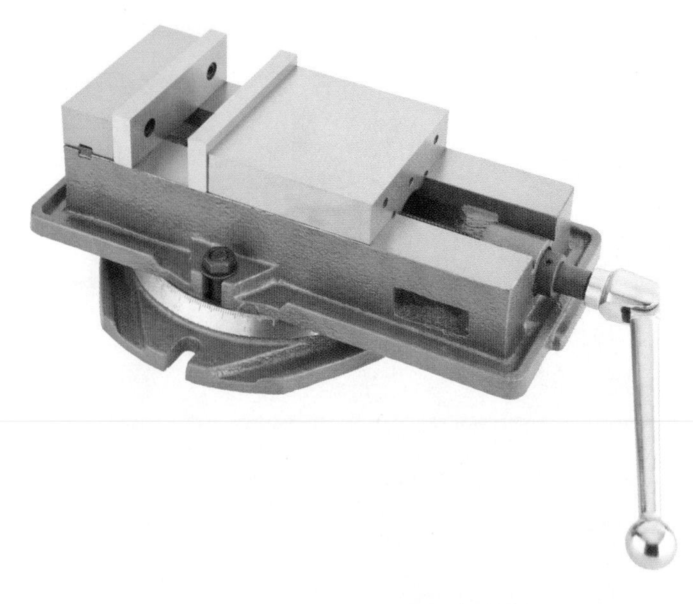

Photo courtesy of Grizzly Industrial, Inc. www.grizzly.com

Figure 18-58. The swivel vise will rotate to easily align work. Note that the base of the vise is graduated in degrees. It can be positioned and locked at an angle to the machine spindle.

Photo courtesy of Grizzly Industrial, Inc. www.grizzly.com

Figure 18-59. The toolmaker's universal vise can be pivoted on several planes.

O.S. Walker Co.

Figure 18-60. Milling a workpiece held with a magnetic chuck.

WMW Machinery Company, Inc.

Figure 18-61. A rotary table can be used to machine parts with round and curved shapes.

The table is graduated in degrees around its circumference and adjustments can be made accurately with the handwheel to 1/30 of a degree (2 minutes).

An index table permits the rapid positioning of work, **Figure 18-62**. Indexing is usually in 15° increments. However, a clamping device allows the table to be locked at any setting.

A dividing head will divide the circumference of circular work into equally spaced units. This feature makes a dividing head indispensable when

Yukiwa Seiko USA, Inc.

Figure 18-62. An index table permits work to be rapidly positioned. The 3-jaw chuck is used for round workpieces. Note that the unit can be mounted on the milling machine vertically or horizontally.

milling gear teeth, cutting splines, and spacing holes on a circle. It also makes possible the milling of squares, hexagons, and various other regular shapes. A precision device, a dividing head has an indexing accuracy of about one minute of arc. This is the equivalent of 1/21,600 part of a circle.

The dividing head consists of two parts, the dividing unit and the foot stock. Work may be mounted between centers, in a chuck, or in a collet. See **Figure 18-63**. An index plate, identified by circles of holes on its face, and the index crank, which revolves on the index plate, are fundamental in the dividing operation.

Rotating the index crank causes the dividing head spindle (to which work is mounted) to rotate. The standard ratios for the dividing head are five turns of the index crank for one complete revolution of the spindle (5:1) or 40 turns of the index crank for one revolution of the spindle (40:1).

The ratio between index crank turns and spindle revolution, plus the index plate with its series of equally spaced hole circles, makes it possible to divide the circumference of the work into the required number of equal spaces. For example, if 10 teeth were to be cut on a gear, it would require 1/10 of 5 turns (assuming dividing head has a 5:1 ratio), or one-half turn of the index crank for each tooth.

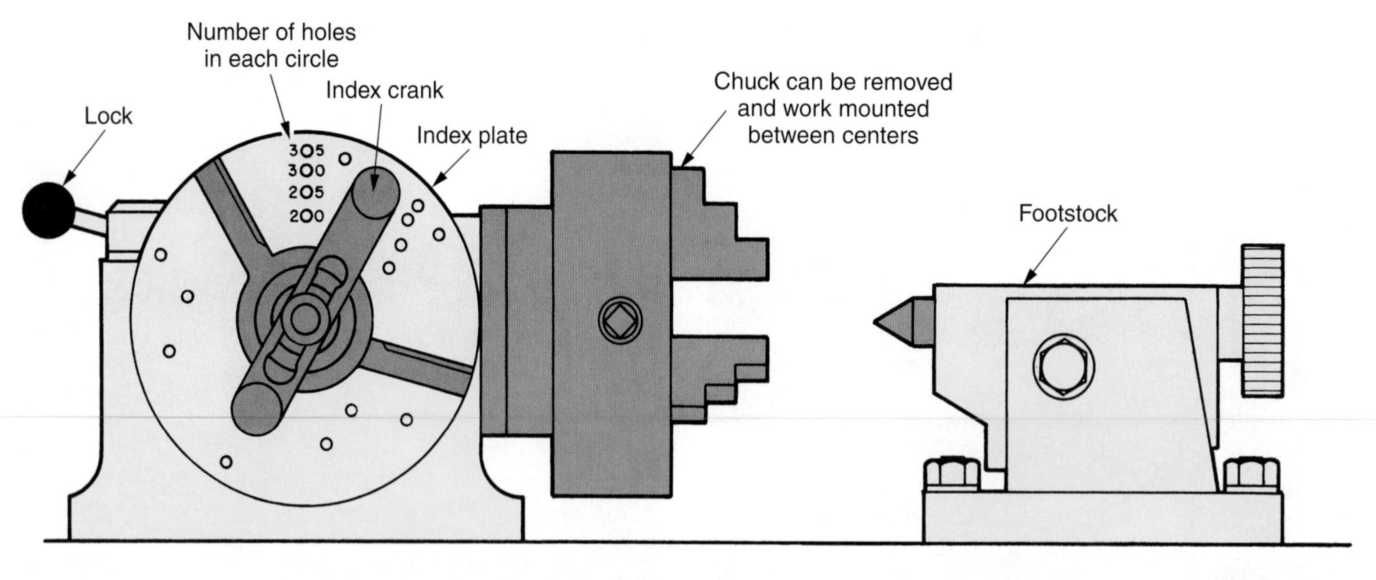

Goodheart-Willcox Publisher

Figure 18-63. Dividing head and foot stock. Work can be mounted between centers, in a collet, or in a chuck, as shown. The index plate and index head are fundamental to the dividing operation.

For 25 teeth, the number of crank turns would be 1/25 of 5, or dividing 5 by 25, or 5/25. By further reduction, this becomes 1/5 of a turn of the crank for each tooth. This is where the holes in the index plate come into use. The holes allow fractional turns to be made accurately.

Select an index plate with a series of holes divisible by 5. On one such plate, the circles have 47, 49, 51, 53, 54, 57, and 60 holes. In this situation, 60 is divisible by 5. Thus, indexing would be through 12 holes on the 60-hole circle for each tooth.

When indexing, it is not necessary to count 12 holes each time the work is repositioned after a tooth has been cut. Two arms, called sector arms or index fingers, are loosened and positioned. One is located against the pin on the index crank. The other is moved clockwise until the arms are 12 holes apart, not including the hole that the pin is in. See **Figure 18-64**.

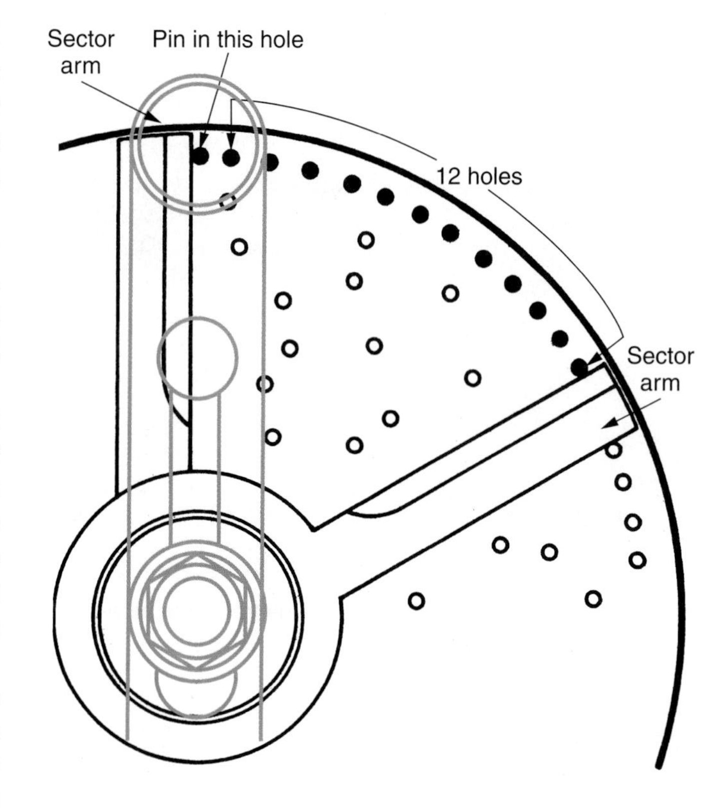

Goodheart-Willcox Publisher

Figure 18-64. Positioning sector arms for proper movement of the spindle. When positioning sector arms, do not count the hole that the pin on index crank is in.

To index, first move the workpiece clear of the cutter. Disengage the crank by withdrawing the pin from the index plate and rotating it clockwise through the section marked by the sector arms. Drop the pin into the hole at the position of the second sector arm and lock the dividing head mechanism. Next, move the sector arms in the same direction as crank rotation to catch up with the pin in the index crank. For each cut, repeat the operation. The dividing head spindle typically can be moved through an arc of 100°. It swivels 5° below horizontal, and 5° beyond perpendicular. See **Figure 18-65**.

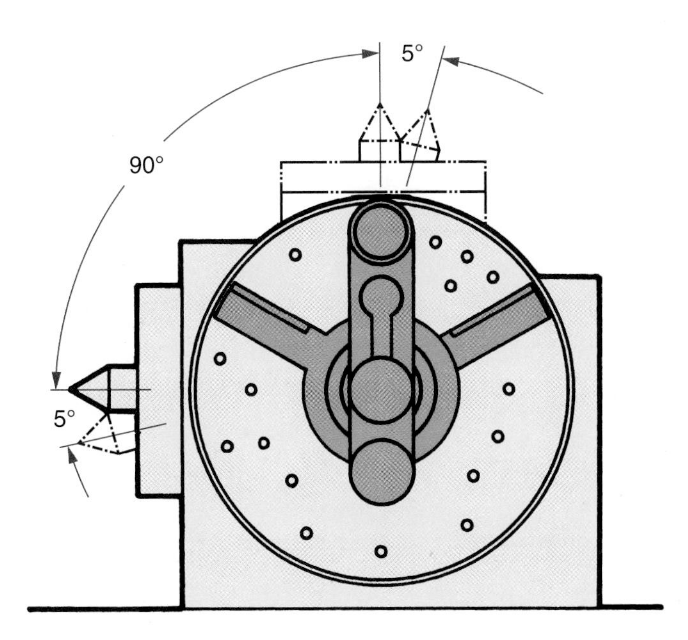

Goodheart-Willcox Publisher

Figure 18-65. Most dividing heads can pivot the spindle through an arc of 100°.

Photo courtesy of Weiler Corporation

Figure 18-66. Use a brush to remove metal chips. Never use your hand, since sharp chips can cause serious cuts.

18.9 Milling Safety Practices

Milling machines, like all machine tools, should be cleaned after each work session. A brush should be used to remove accumulated chips, **Figure 18-66.** Chips are razor-sharp. Never use your hand to remove them.

Never remove chips with compressed air. The flying chips may injure you or a nearby person. Also, if cutting oil was used in the machining operation, the compressed air will create a highly flammable oily mist. If ignited by an open flame, the mist can explode. Finish by wiping down the machine with a soft cloth.

The following procedures are suggested for the safe operation of a milling machine.

- Become thoroughly familiar with the milling machine before attempting to operate it. When in doubt, obtain additional instructions.
- Never attempt to operate a milling machine while your senses are impaired by medication or other substances.
- Wear appropriate clothing and approved safety glasses.
- Stop the machine before making adjustments, measurements, or trying to remove accumulated chips.
- Be sure all power to the machine is turned off before opening or removing guards and covers.
- Use a piece of heavy cloth or gloves for protection when handling milling cutters. Avoid using your bare hands.
- Get help to move any heavy machine attachment, such as a vise, dividing head, rotary table, or large work.
- Be sure the work-holding device is mounted solidly to the table, and the work is held firmly. Spring or vibration in the work can cause thin cutters to jam and shatter.
- Never reach over or near a rotating cutter.
- Avoid talking with anyone while operating a machine tool, do not allow anyone to turn on your machine for you.
- Never "fool around" when operating a milling machine. Keep your mind on the job and be ready for any emergency.
- Keep the floor around your machine clear of chips and wipe up spilled cutting fluid immediately. Place sawdust or special oil-absorbing compound on slippery floors. Place all oily rags in an approved metal container that can be closed tightly.
- Be thoroughly familiar with the placement of the machine's emergency stop switch, lever or button.

- Treat any small cuts and skin punctures as potential infections. Clean them thoroughly. Apply antiseptic and cover injury with a bandage. Report any injury, no matter how minor, to your instructor or supervisor.
- Launder work clothes frequently. Greasy clothing is a fire hazard.

Chapter Review

Summary

- Milling machines can be grouped into two general families: fixed-bed and column and knee.
- Cutting speed and rate of feed can be changed in several different ways, depending on the type of milling machine.
- The two main categories of milling operations are face milling and peripheral milling.
- Milling cutters can be classified as either a solid cutter or an inserted-tooth cutter, and can be made of a variety of materials to suit the cutting needs.
- Great care should be taken of milling tools to ensure precision and accuracy.
- The material being milled should be taken into account when choosing a milling cutter, milling method, and cutting fluid.
- Work-holding attachments like vises, magnetic chucks, rotary tables, index tables, and dividing heads provide the milling machine with much of its versatility.

Review Questions

Answer the following questions using the information provided in this chapter.

1. The two broad classifications of milling machines are fixed-bed and _____ milling machines.
2. A(n) _____-type milling machine has a horizontal spindle and the worktable has three movements.
3. A(n) _____-type milling machine is similar to the above machine but a fourth movement has been added to the worktable to permit cutting helical shapes.
4. A(n) _____-type milling machine has the spindle perpendicular or at right angles to the worktable.
5. List the four methods of machine control.
6. The surface being machined is parallel with the cutter face in _____ milling.
7. The surface being machined is parallel with the periphery of the cutter in _____ milling.
8. In climb or down-milling, _____.
   A. the work is fed into the rotation of the cutter
   B. the work moves in the same direction as the rotation of the cutter
   C. Neither of the above.
9. In or up-conventional milling, _____.
   A. the work is fed into the rotation of the cutter
   B. the work moves in the same direction as the rotation of the cutter
   C. Neither of the above.
10. What are two general types of milling cutters?
11. What is the term "hand" used to describe, in reference to an end mill?

For questions 12–21, match each term with the correct sentence below.

12. _____ Two-flute end mill.
13. _____ Multiflute end mill.
14. _____ Fly cutter.
15. _____ Shell end mill.
16. _____ Face milling cutter.
17. _____ Plain milling cutter.
18. _____ Slab cutter.
19. _____ Side milling cutter.
20. _____ Staggered-tooth side cutter.
21. _____ Metal slitting saw.

A. Has cutting teeth on the circumference and on one or both sides.
B. Cutter with helical teeth designed to cut with a shearing action.
C. A facing mill with a single-point cutting tool.
D. Can be fed into work like a drill.
E. Cutter with teeth located around the circumference.
F. Recommended for conventional milling where plunge cutting (going into work like a drill) is *not* required.
G. Intended for machining large flat surfaces parallel to the cutter face.
H. Thin milling cutter designed for machining narrow slots and for cutoff operations.
I. Mounts on a stub arbor.
J. Has alternate right-hand and left-hand helical teeth.

22. What is a draw-in bar, and how is it used?
23. The distance, measured in feet or meters, that a point (tooth) on the circumference of a cutter moves in one minute, is known as the _____.
24. The rate at which the work moves into the cutter is known as the _____.
25. Calculate machine speed (rpm) and feed (F) for a 1.5″ diameter tungsten carbide 5 tooth (T) end mill when machining cast iron. Recommended cutting speed is 190 fpm. Feed per tooth (ftr) is 0.004″. Use the following formulas:

$$\text{rpm} = \frac{\text{fpm} \times 12}{\pi D}$$

and

$$F = \text{ftr} \times T \times \text{rpm}$$

26. Determine machine speed (rpm) and feed (F) for a 2.5″ diameter HSS shell end mill with 8 teeth (T), machining aluminum. Recommended cutting speed is 550 fpm. Feed per tooth (ftr) is 0.010″.
27. Calculate machine speed (rpm) for machining aluminum with a 6″ diameter HSS side milling cutter. Recommended cutting speed is 550 fpm.
28. Determine machine speed (rpm) and feed (F) for a 4″ diameter HSS side milling cutter with 16 teeth (T) milling free cutting steel. Recommended cutting speed is 200 fpm. Feed per tooth (ftr) is 0.005″.
29. Calculate machine speed (rpm) and feed (F) for a 2.5″ diameter HSS slab milling cutter with 8 teeth (T) machining brass. Recommended cutting speed is 250 fpm. Feed per tooth (ftr) is 0.006″.
30. Cutting fluids serve several purposes. List three of them.

For questions 31–38, match each term with the correct sentence below.

31. _____ Flanged vise.
32. _____ Swivel vise.
33. _____ Universal vise.
34. _____ Rotary table.
35. _____ Dividing head.
36. _____ Indexing table.
37. _____ Magnetic chuck.
38. _____ Vise lug.

A. Keys vise to a slot in worktable.
B. Can only be mounted parallel to or at right angles on worktable.
C. Needed when cutting segments of circles, circular slots, and irregular-shaped slots.
D. Can only be used with ferrous metals.
E. Has a circular base graduated in degrees.
F. Permits rapid positioning of circular work in 15° increments and can be locked at any angular setting.
G. Used to divide circumference of round work into equally spaced divisions.
H. Permits compound angles (angles on two planes) to be machined without complex or multiple setups.

39. Stop the machine before making adjustments and _____.
40. Before removing a machine's guards or covers, be sure that all power to the machine is _____.
41. Milling cutters are sharp. Protect your hands with a cloth or _____ when handling them.
42. Treat all small cuts and skin punctures as potential sources of infection. Which of the following should be done?
    A. Clean them thoroughly.
    B. Apply antiseptic and cover with a bandage.
    C. Promptly report the injury to your instructor.
    D. All of the above.
    E. None of the above.

CHAPTER 19

Milling Machine Operations

Learning Objectives

After studying this chapter, you will be able to:

- Describe how milling machines operate.
- Set up and safely operate horizontal and vertical milling machines.
- Perform various cutting, drilling, and boring operations on a milling machine.
- Make the needed calculations and cut spur gears.
- Make the needed calculations and cut a bevel gear.
- Point out safety precautions that must be observed when operating a milling machine.

Technical Terms

addendum
bevel gear
circular pitch
dedendum
diametral pitch
gang milling
pitch circle
slitting
slotting
spur gear
straddle milling

The versatility of the milling machine family permits many different machining operations to be performed. Because of the number of varied operations that can be performed, it is not possible to cover all of them in a book of this type. Only basic operations will be described.

19.1 Vertical Milling Machine

The vertical milling machine is capable of performing milling, drilling, boring, and reaming operations. It differs from the horizontal mill in that the spindle is mounted in a vertical position.

The spindle head swivels 90° left or right for machining at any angle, **Figure 19-1**. The ram, on which it is mounted, can be adjusted in and out. On many vertical mills, it also revolves 180° on a horizontal plane. Both swivels are graduated in degrees with a vernier scale to ensure accurate angular settings.

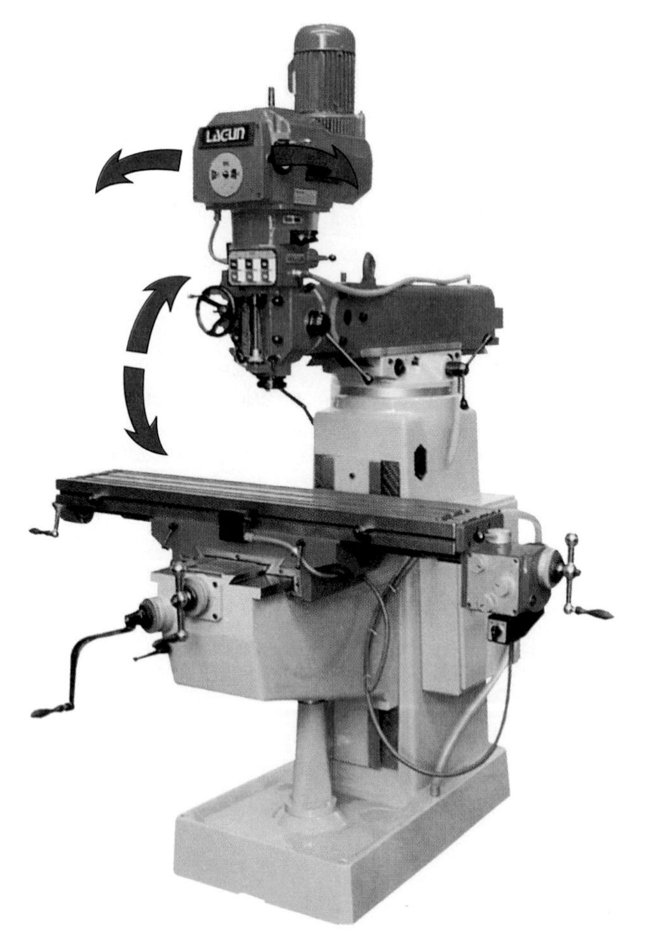

Republic-Lagun Machine Tool Co.

Figure 19-1. Angular head adjustments are possible on many vertical milling machines.

19.1.1 Cutters for Vertical Milling Machine

Although adapters are available that permit the use of side and angle cutters, **Figure 19-2**, face mills and end mills are the cutters normally used in vertical milling machines.

Taper shank end mills and drills are fitted in an adapter, **Figure 19-3A**. Some machine spindles have a Brown & Sharpe taper. When taper shanks are large enough, they are mounted directly, **Figure 19-3B**. When a taper is too small to fit directly into the spindle, a sleeve must be used. Straight shank end mills are held in a spring collet, **Figure 19-3C**, or in an end mill adapter, **Figure 19-3D**. Small drills, reamers, and similar tools are held in a standard Jacobs chuck fitted to the spindle by one of the above methods.

19.2 Vertical Milling Machine Operations

In addition to the usual precautions that must be observed when getting a machine tool ready for a job, the spindle head alignment must be checked. Make sure that the spindle head is at an exact right angle (perpendicular) to the worktable. If the spindle is not perpendicular, it is not possible to machine a flat surface. The surface will be irregular or "dished."

Parlec, Inc.

Figure 19-2. Adapters permit arbor and other type cutters to be mounted on a vertical milling machine.

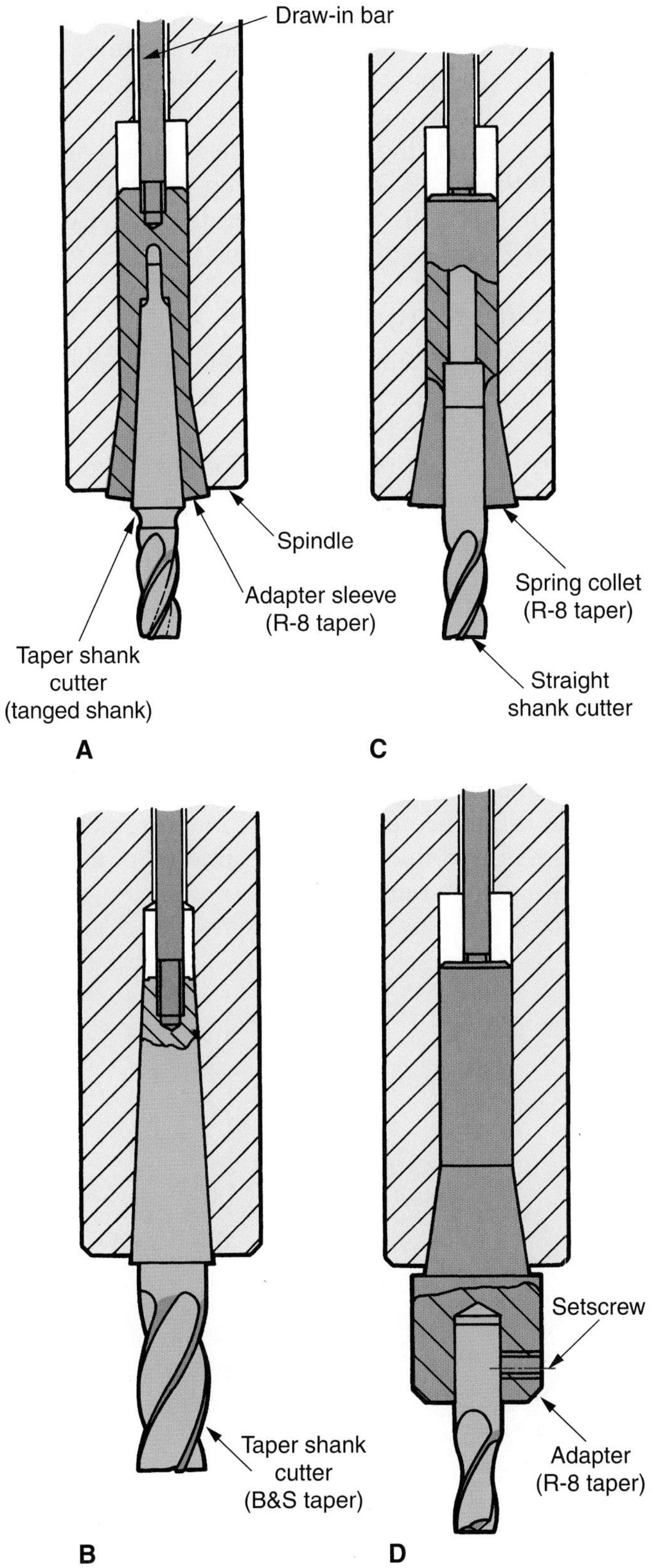

Draw-in bar
Spindle
Adapter sleeve (R-8 taper)
Taper shank cutter (tanged shank)
A
Spring collet (R-8 taper)
Straight shank cutter
C
Taper shank cutter (B&S taper)
B
Setscrew
Adapter (R-8 taper)
D

Goodheart-Willcox Publisher

Figure 19-3. Four of the most common methods used to mount end mills in a vertical milling machine. A—Adapter sleeve with taper shank cutter. B—B&S taper mounted directly in the spindle. C—Spring collet with straight shank cutter. D—Adapter with setscrew on straight shank cutter.

Milling head perpendicularity can be checked with the use of a dial indicator, as shown in **Figure 19-4**. The device holding the indicator may be shop made or purchased.

If a vise is used to mount the work, wipe the vise base and worktable clean. Inspect for burrs and nicks. They prevent the vise from seating properly on the table. Bolt it firmly to the machine.

If extreme accuracy is required, the next step is to align the vise with a dial indicator, **Figure 19-5**. However, for many jobs the vise can be aligned with a square, as shown in **Figure 19-6**. Angular settings can be made with a protractor, **Figure 19-7**, or by using the degree divisions on the base of a swivel vise.

If you have not already done so, wipe the vise jaws and bottom clean of chips and dirt. Place clean parallels in the vise and place the work on them.

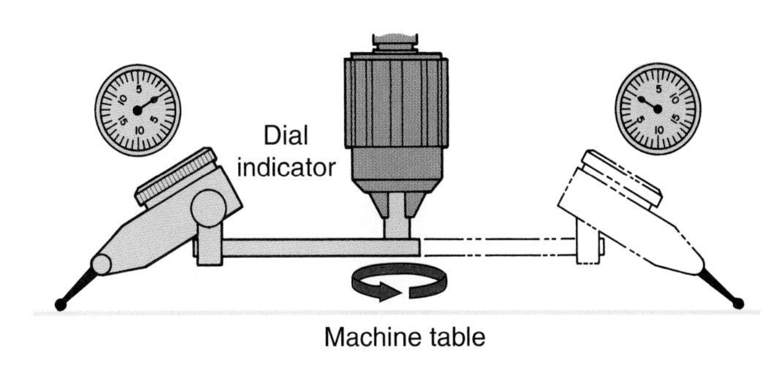

Goodheart-Willcox Publisher

Figure 19-4. A dial indicator can be used, as shown, to check whether the milling head is perpendicular to the table.

Goodheart-Willcox Publisher

Figure 19-5. Use of a dial indicator permits extreme accuracy in aligning a solid vise jaw.

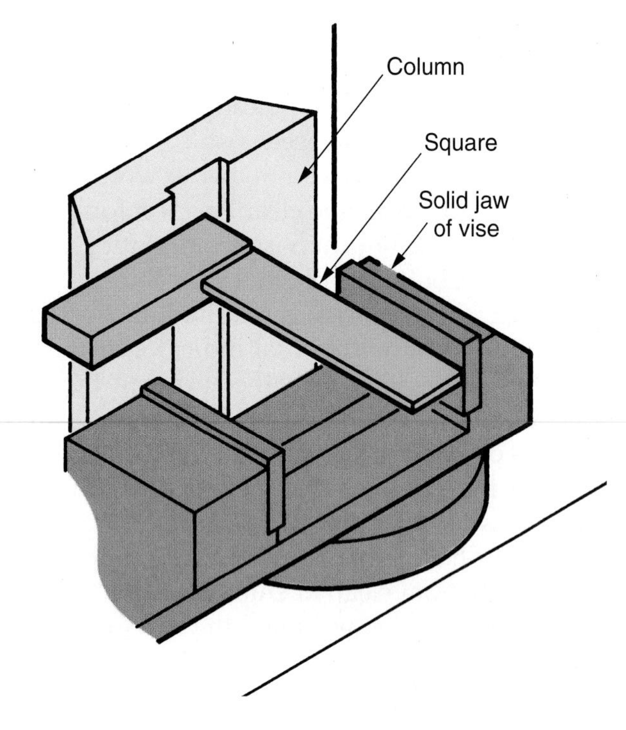

Goodheart-Willcox Publisher

Figure 19-6. Procedure for squaring a solid vise jaw using a machinist's steel square.

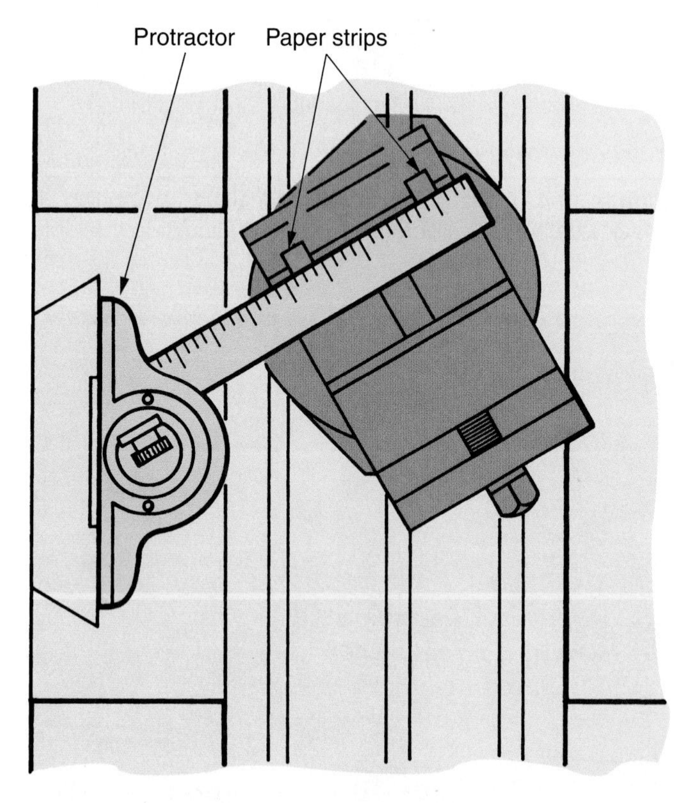

Goodheart-Willcox Publisher

Figure 19-7. Angular settings can be made with protractor head and steel rule of a combination set. Paper strips aid in determining whether setting is accurate.

Tighten the jaws and tap the work onto the parallels with a mallet or soft-faced hammer. Thin paper strips can be used to check whether the work is firmly on the parallels, **Figure 19-8.**

Never strike the vise handle with a hammer or mallet to put additional holding pressure on the jaws. If the workpiece is rough, protect the vise jaws and parallels with soft metal strips.

19.2.1 Squaring Stock

A specific sequence must be followed to machine several surfaces of a piece square with one another, as shown in **Figure 19-9.**

1. Machine the first surface. Remove the burrs and place the first machined surface against the fixed vise jaw. Insert a piece of soft metal rod between the work and movable jaw if that portion of the work is rough or not square.
2. Machine the second surface.
3. Remove the burrs and reposition the work in the vise to machine the third side. This side must be machined to dimension. Take a light cut and use a micrometer to measure for size. The difference between this measurement and the required thickness is the amount of material that must be removed.

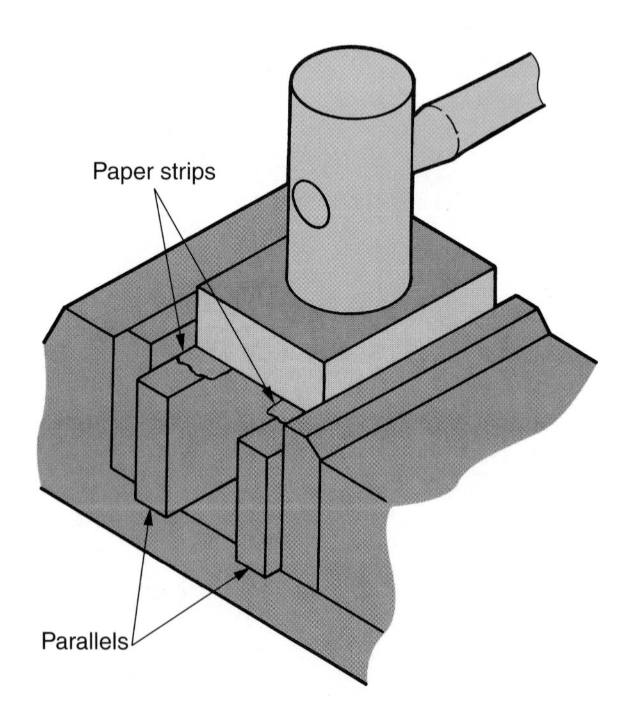

Goodheart-Willcox Publisher

Figure 19-8. Setting work on parallels with a soft-faced hammer. Thin paper strips are used to check whether the work is firmly on the parallels.

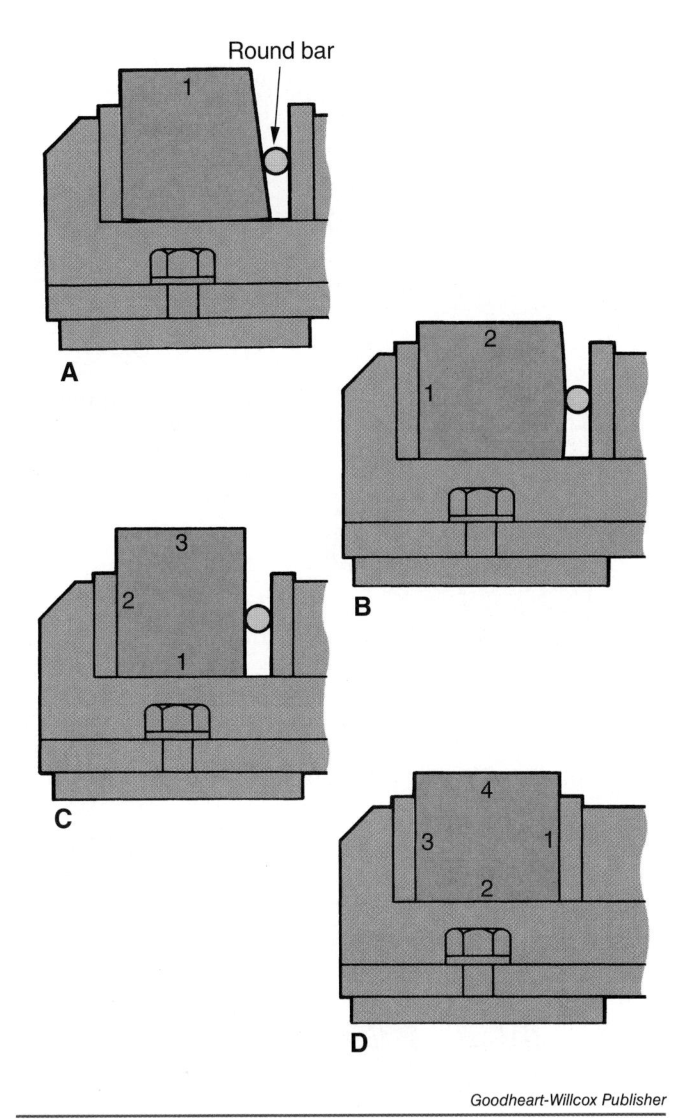

Goodheart-Willcox Publisher

Figure 19-9. Sequence for squaring work on a milling machine. A—Square top. B—Square side. C—Square bottom. D—Square other side.

4. Repeat the above operation to machine the fourth side.
5. If the piece is short enough, the ends may be machined by placing it in a vertical position with the aid of a square, **Figure 19-10A**. Otherwise, it may be machined as shown in **Figure 19-10B**.

SAFETY NOTE

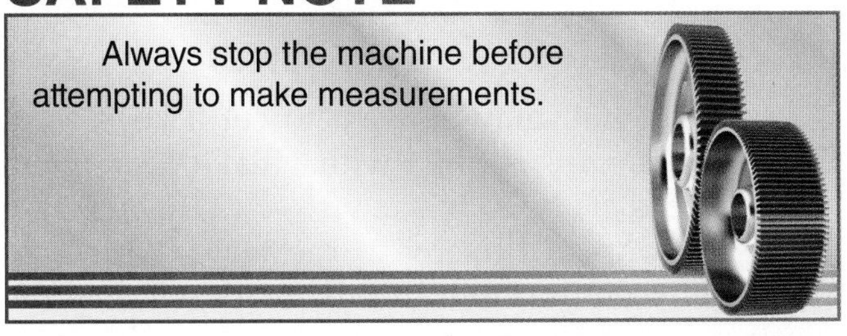

Always stop the machine before attempting to make measurements.

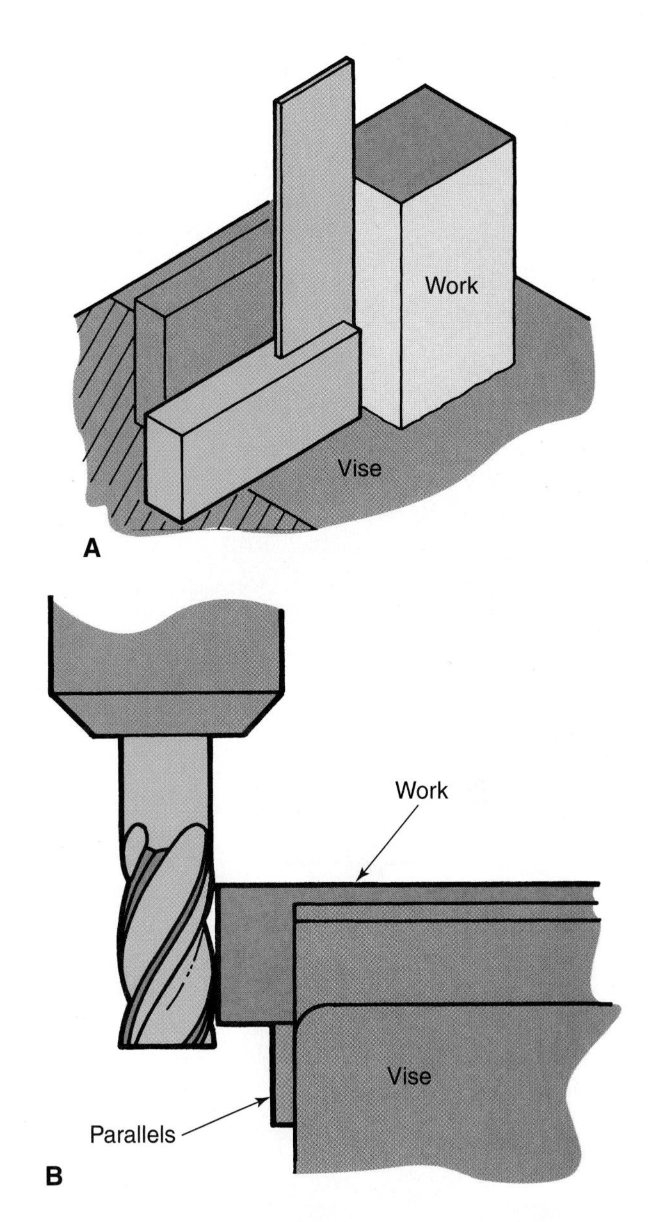

Goodheart-Willcox Publisher

Figure 19-10. Squaring ends. A—Using a square to position short pieces for machining ends. Movable jaw is not shown for clarity. B—Another technique for squaring ends of work. Jaw must be checked with a dial indicator to be sure it is at a right angle to the column.

19.2.2 Machining Angular Surfaces

Angular surfaces (bevels, chamfers, and tapers) may be milled by tilting the spindle head assembly to the required angle. They may also be made by setting the work at the specified angle in a vise. See **Figure 19-11**.

A fixture, **Figure 19-12**, is often used when many similar pieces must be milled. Compound angles (angles on two planes) are made in a universal vise.

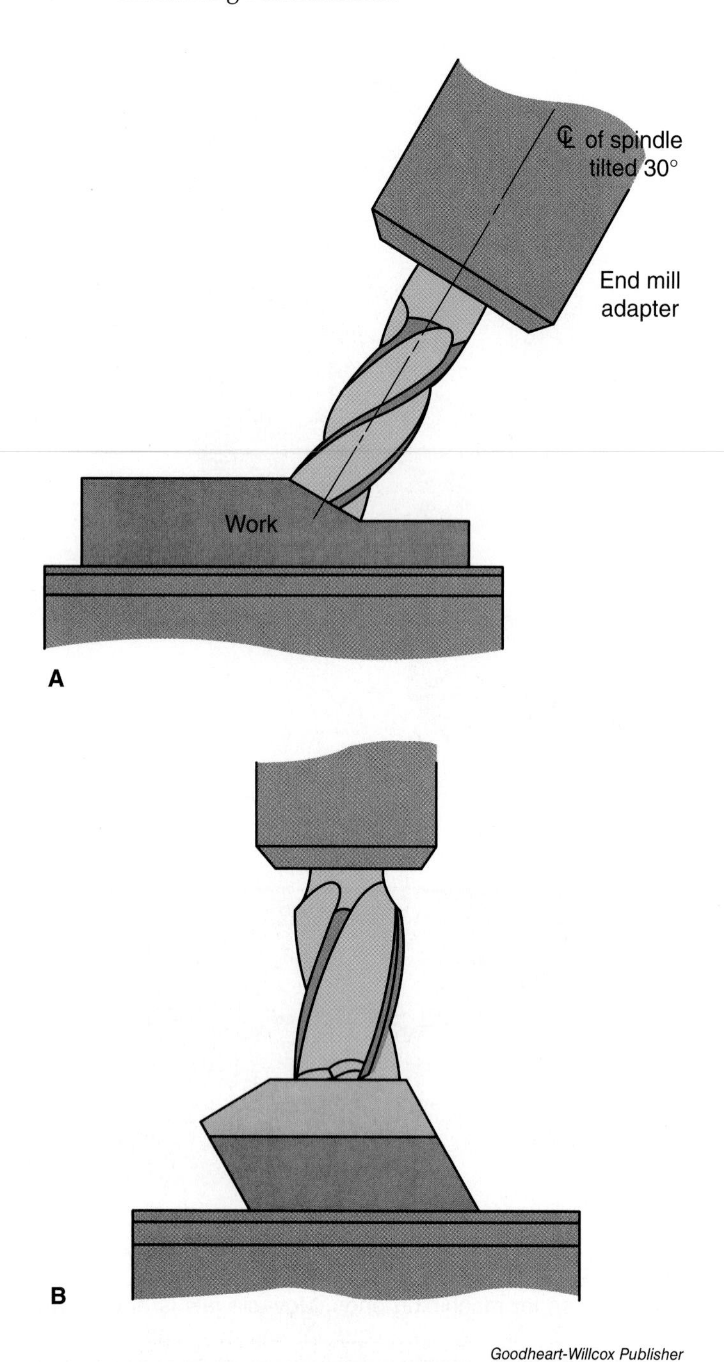

Goodheart-Willcox Publisher

Figure 19-11. Cutting an angular surface. A—Cutting an angular surface with the spindle head pivoted to the desired angle. B—Cutting an angular surface by positioning the work at the desired angle in a vise.

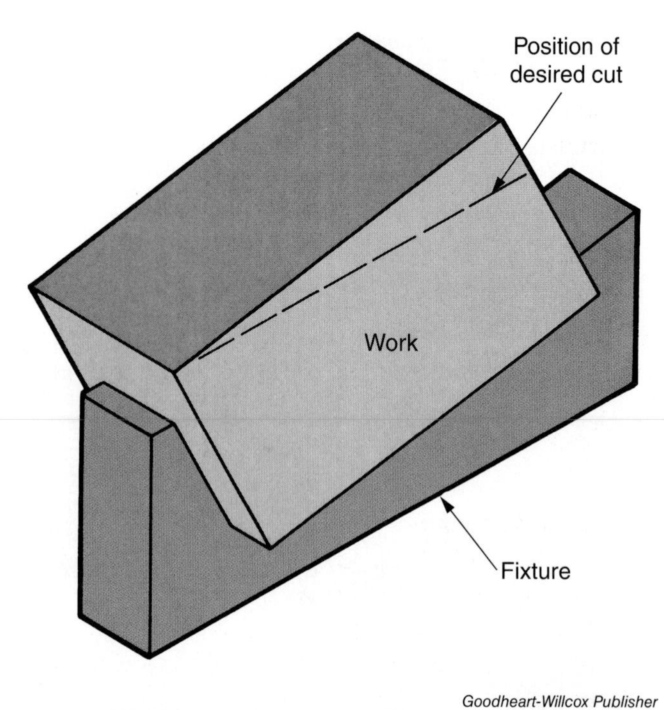

Goodheart-Willcox Publisher

Figure 19-12. If many pieces are to have angular surfaces machined, considerable time can be saved by using a fixture.

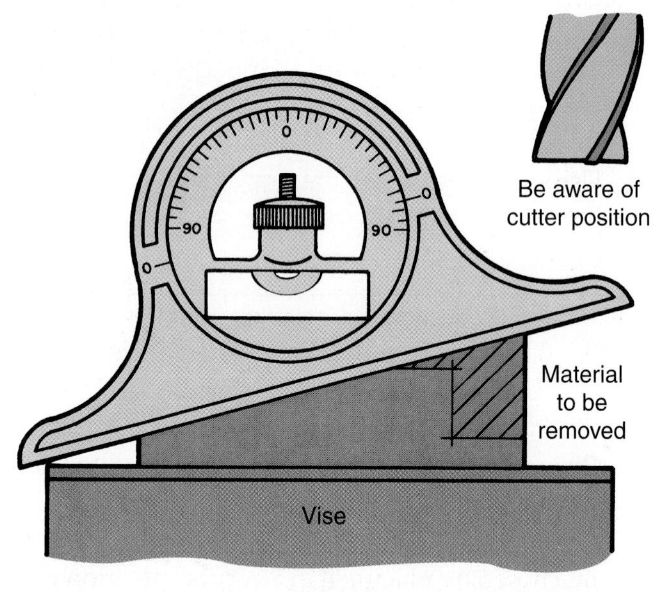

Goodheart-Willcox Publisher

Figure 19-13. Work can be quickly set at the desired angle with aid of a protractor head.

When the pivoted spindle head is used for angular cutting, it is essential that the vise be aligned with a dial indicator. Make a layout of the desired angle on the work and clamp it in the vise. Position the cutter and machine to the line.

Work mounted at an angle in the vise for machining must be set up carefully. Alignment may be made with a protractor head fitted with a spirit level, **Figure 19-13**, or with a surface gage.

19.2.3 Milling a Keyseat or Slot

An end mill may be used to cut a keyseat or slot. After aligning the vise with a dial indicator, the workpiece is clamped in the vise or to the machine table. If mounted directly to the table, a piece of paper between the table and the work will seat the work more solidly and prevent slippage.

A sharp cutter, equal in diameter to the keyseat or slot, must be used. A two-flute end mill is used when a blind keyseat or slot is to be machined, **Figure 19-14**. Otherwise, a multiflute end mill is used.

19.2.4 Locating End Mill to Cut Keyseat or Slot on Round Work

After the milling machine has been set up, the work is secured in a vise, between centers, in V-blocks, or in a fixture. Before machining begins, precise centering of the end mill must be done as follows:

1. Lock the knee to the column.
2. Lay out the slot length, as shown in **Figure 19-15**.
3. Hold the end of a long, narrow strip of paper between the cutter and the work. Carefully move the cutter toward the work until the paper strip is pulled lightly from your fingers by the rotating cutter. Pay close attention when using this alignment technique. Use a paper strip long enough to keep your fingers well clear of the cutter. Release the paper as soon as you feel the cutter "grabbing" it.
4. Unlock the knee from the column and lower the table until the cutter is slightly above the work. Move the cutter inward half the work diameter, plus half the cutter diameter, plus the paper thickness.
5. Using another long, narrow strip of paper, use the same technique to get the required depth.

Correct keyseat depth may be obtained from tables in a machinist's handbook.

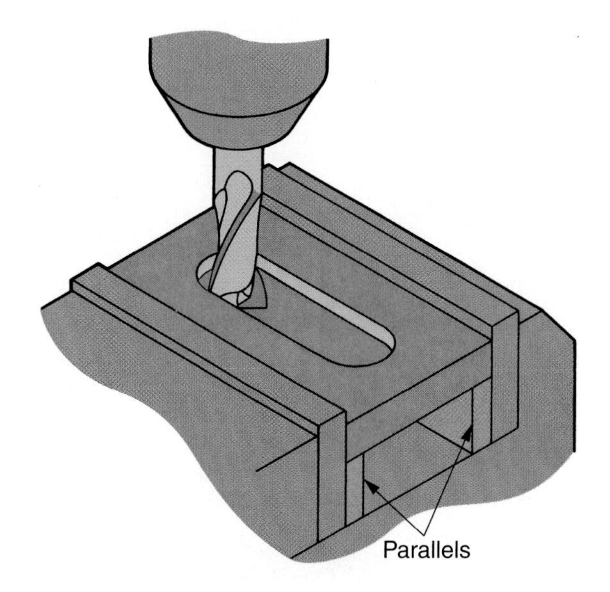

Goodheart-Willcox Publisher

Figure 19-14. Blind slot being machined with a two-flute end mill.

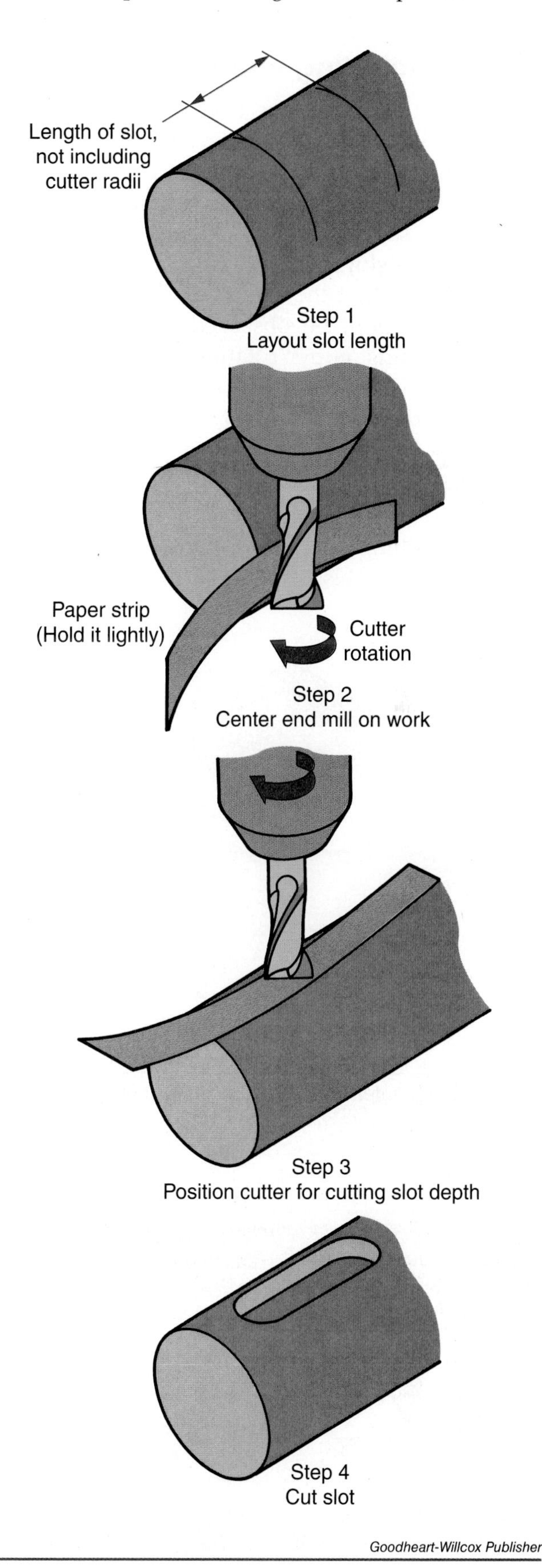

Goodheart-Willcox Publisher

Figure 19-15. Procedure for using a paper strip to position cutter on the exact center of round stock for cutting a keyseat. Keep your fingers clear of the rotating cutter!

19.2.5 Machining Internal Openings

Internal openings are easily machined with a vertical milling machine, **Figure 19-16**. A two-flute end mill must be used if the cutter is to make the initial opening. It can be fed directly into the material in much the same manner as a drill.

When the slot is wider than the cutter diameter, it is important that the direction of feed, in relation to cutter rotation, be observed. Feed direction is normally against cutter rotation, **Figure 19-17**. This applies only when the cutter is removing metal from one side of the opening.

19.2.6 Machining Multilevel Surfaces

Milling a multilevel surface is probably the easiest of the milling operations, **Figure 19-18**. A layout of the various levels is made on the work's surface.

Goodheart-Willcox Publisher

Figure 19-18. A stepped (multilevel) surface being milled.

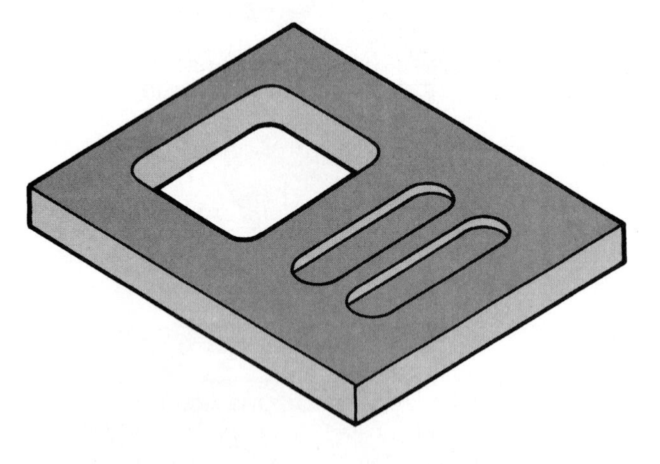

Goodheart-Willcox Publisher

Figure 19-16. Internal openings have been milled in this aircraft part.

Cuts are made until the lines are reached. For accuracy, the cut must be checked with a depth micrometer (remove any burrs before making the measurement), **Figure 19-19**. Make necessary table adjustments accordingly.

19.2.7 Drilling, Reaming, and Boring

Holes may be located to very close tolerances for drilling, reaming, or boring on a vertical milling machine. The first hole can be located with a wiggler, **Figure 19-20**, a centering scope, **Figure 19-21**,

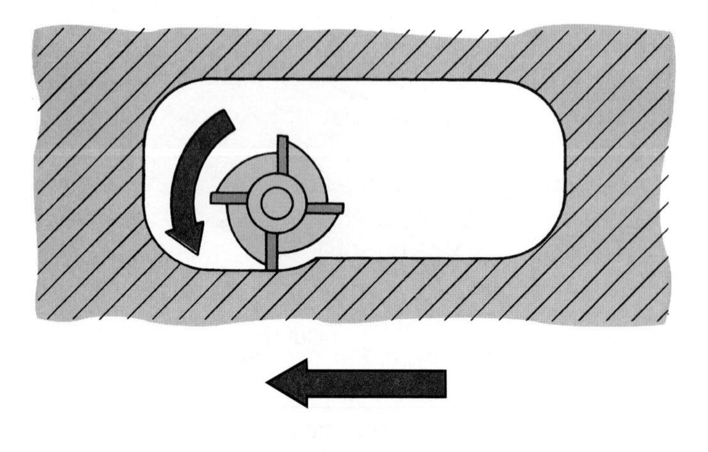

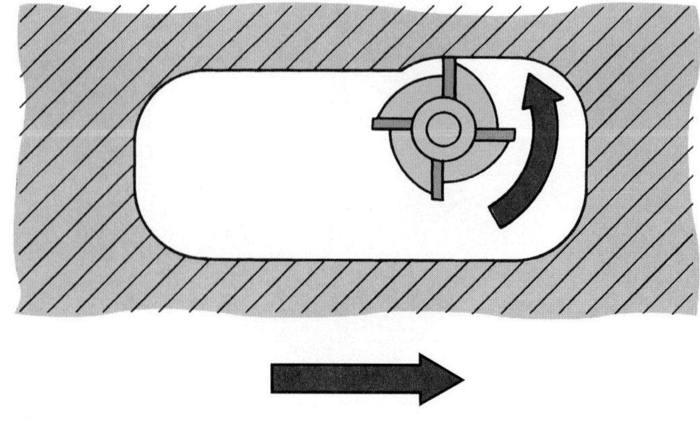

Goodheart-Willcox Publisher

Figure 19-17. When making this type cut, remember that feed direction is always against cutter rotation.

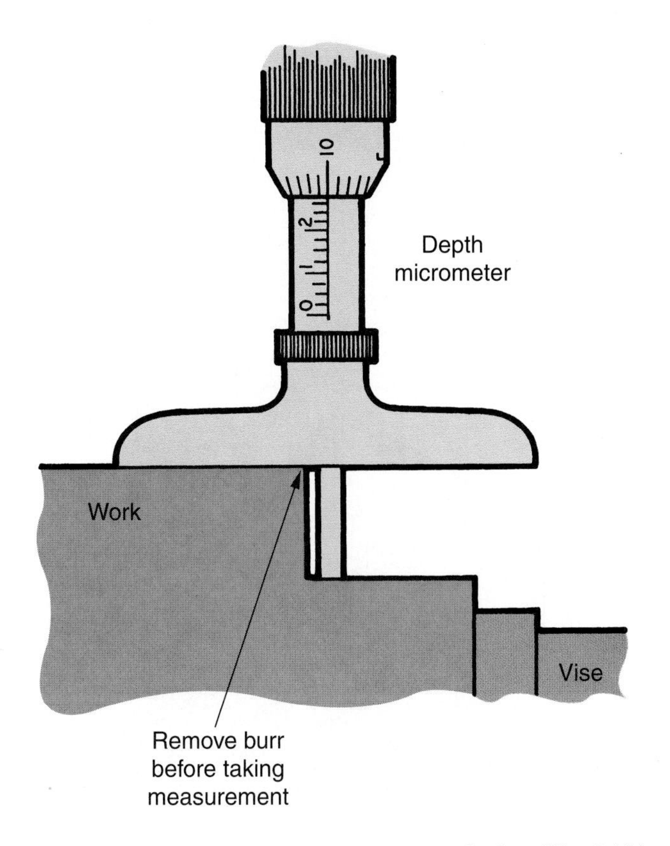

Figure 19-19. Check depth of cut with a depth micrometer before making final cutter adjustments. Remove any burrs before making the measurement. Always stop the machine before attempting to remove burrs and making a measurement.

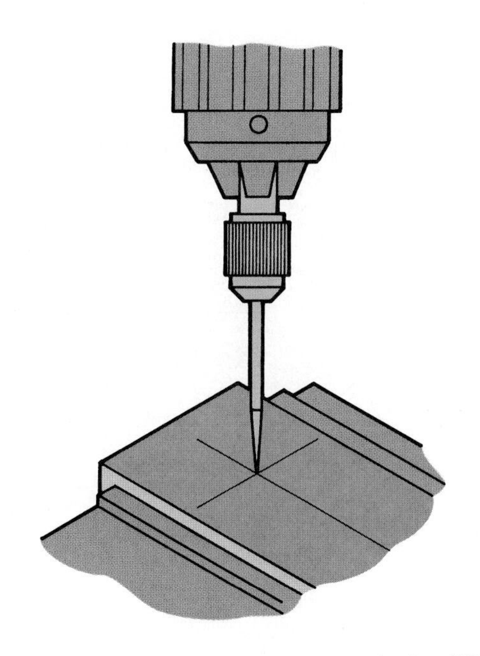

Figure 19-20. Accuracy can be ensured when locating holes on a milling machine by aligning the first hole with a "wiggler."

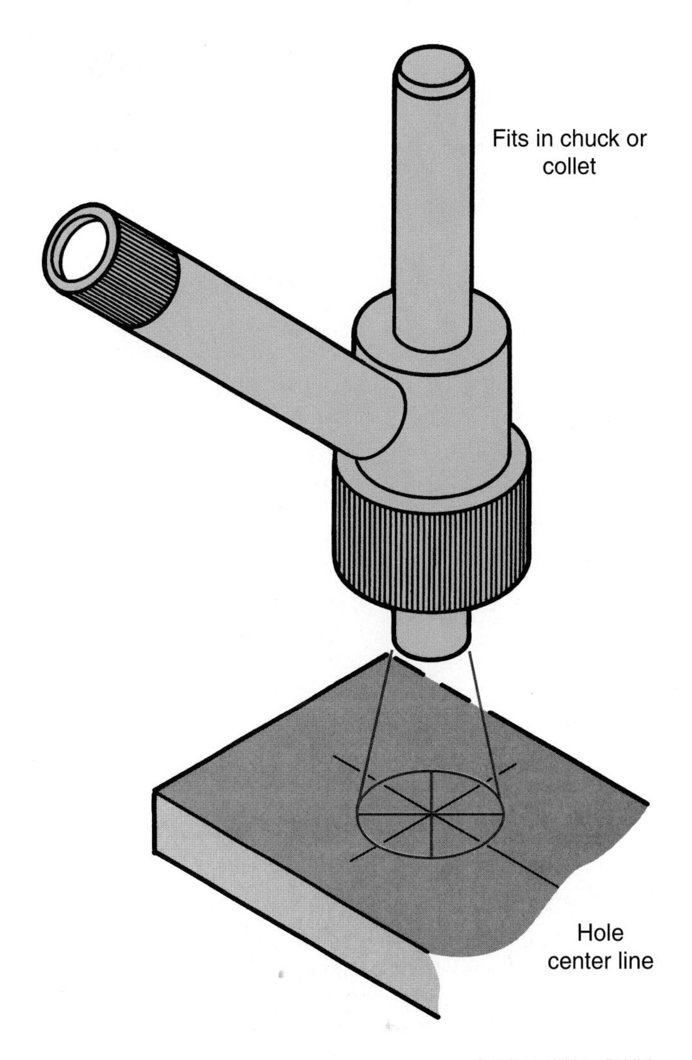

Figure 19-21. A centering scope is another tool that aids in centering work for drilling or boring on vertical milling machines. It is fitted in the chuck (the chuck must run true) or collet. The machinist sights through the optical system and positions work using the crosshairs in the scope. It is capable of locating true position within ±0.0001″ (0.0025 mm).

or an edge finder, **Figure 19-22**. Then, it is possible to locate any remaining holes within 0.001″ (0.025 mm) using the micrometer feed dials. Tolerances of 0.0001″ (0.0025 mm) are possible with a measuring rod and dial indicator attachment or a digital readout gaging system fitted to the machine.

Boring permits holes to be machined accurately with fine surface finishes. A hole must first be drilled in the work before boring can be started. A single-point tool is fitted to a boring head, which in turn is mounted in the spindle, **Figure 19-23**. Hole diameter is obtained by offsetting the tool point from center. The adjustment is graduated for a direct reading. Some boring head dials indicate actual tool movement, while others indicate actual material removal.

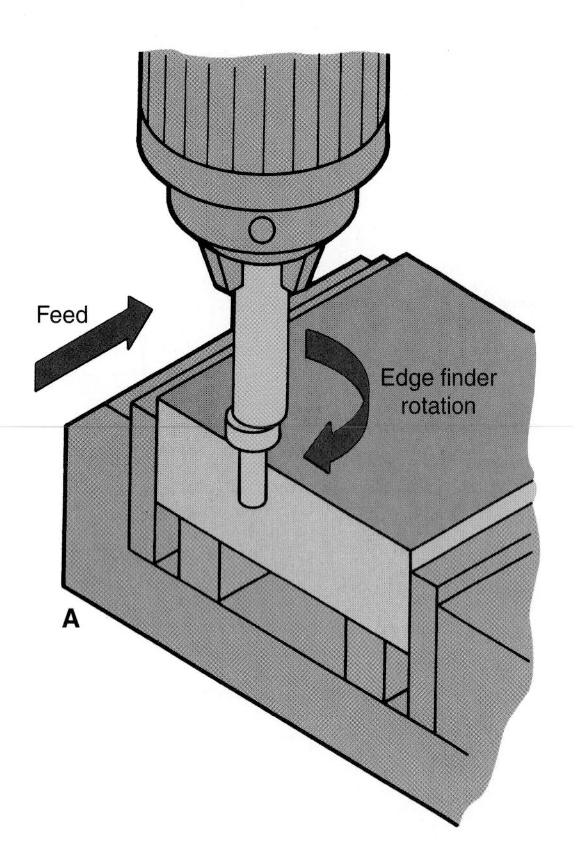

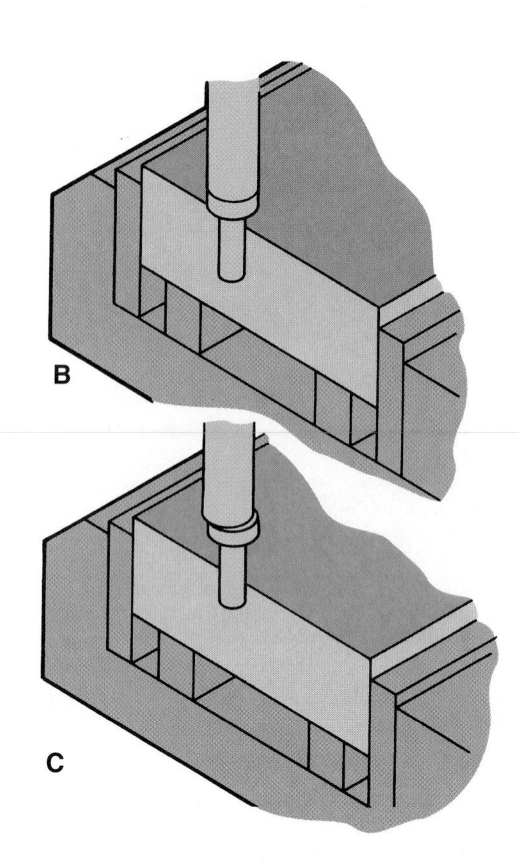

Goodheart-Willcox Publisher

Figure 19-22. The edge finder is a precision positioning tool that will locate the edge of the work in relation to center of the spindle with 0.0002″ (0.005 mm) accuracy. A—With spindle rotating at moderate speed, and with edge finder tip as shown, slowly feed tip of tool against work. B—Edge finder tip will gradually become centered with its shank. C—When the tip becomes exactly centered, it will abruptly jump sideways about 1/32″ (0.8 mm). When this occurs, stop table movement immediately. Center of the spindle will be exactly one-half tip diameter away from edge of work. Set the micrometer dial to "0" and, with edge finder clear of work, move table longitudinally the required distance plus one-half the tip diameter. Follow the same procedure to get traverse measurement.

Marposs Corp.; Komet of America, Inc.

Figure 19-23. Boring heads. A—Micro-adjustable boring head with digital display. This unit has an accuracy of 0.00004″ (0.01 mm), and has a decimal or metric readout display. B—Boring head with a programmable electronic controller interfaced to the machine control. It relates measurement data to preset tolerance limits, and signals tool to make size adjustments to compensate for tool wear. The boring head incorporates a small servomotor that adjusts the boring slide in extremely fine limits. Communication to, and powering of, the tool are entirely wireless.

19.3 Milling Machine Care

- Check and lubricate the machine with the recommended lubricants.
- Clean the machine thoroughly after each job. Use a brush to remove chips. Never attempt to clean the machine while it is running. Never use your hands.
- Keep the machine clear of tools.
- Check each setup for adequate clearance between the work and the various parts of the machine.
- Never force a cutter into a collet or holder. Check to see why it does not fit properly.
- Use a sharp cutter. Protect your hands when mounting it.
- Check the machine to determine whether it is level. Checking should be done at regular intervals.
- Have all guards in place before attempting to operate a milling machine.
- Check coolant level and condition if the reservoir is built into the milling machine. Change it if it becomes contaminated.
- Start the machining operation only after you are sure that everything is in satisfactory working condition. It may be necessary to make special fixtures to hold odd-shaped or difficult-to-mount work.
- Use attachments designed for the machine.

19.4 Horizontal Milling Machine Operations

Like the vertical milling machine, the horizontal mill is a versatile machine tool. Many different machining operations can be performed on it.

19.4.1 Milling Flat Surfaces

A careful study of the part drawing will let you determine what operation is to be performed, what cutter is best suited for the job, and the best way to hold the workpiece. Flat surfaces may be milled with a plain cutter or slab cutter mounted on an arbor (peripheral milling), or with an inserted tooth face or shell milling cutter (face milling). The method used will be determined by the size and shape of the work.

After the milling method and cutter have been selected, the following sequence of operations is recommended:

1. Check and lubricate the machine. Wipe the worktable clean and examine it for nicks and burrs. Nicks or burrs will prevent the workpiece or holding attachments from seating properly on the table.
2. Mount the work directly on the table, if possible. **Figure 19-24** illustrates one method for mounting long work. If the work cannot be mounted to the table, use a vise. Clean its base and bolt it firmly in place. Locate the vise as close to the machine column as workpiece shape and the arbor support will permit. When possible, pivot the vise so that the solid jaw supports the work against cutting rotation, **Figure 19-25**.

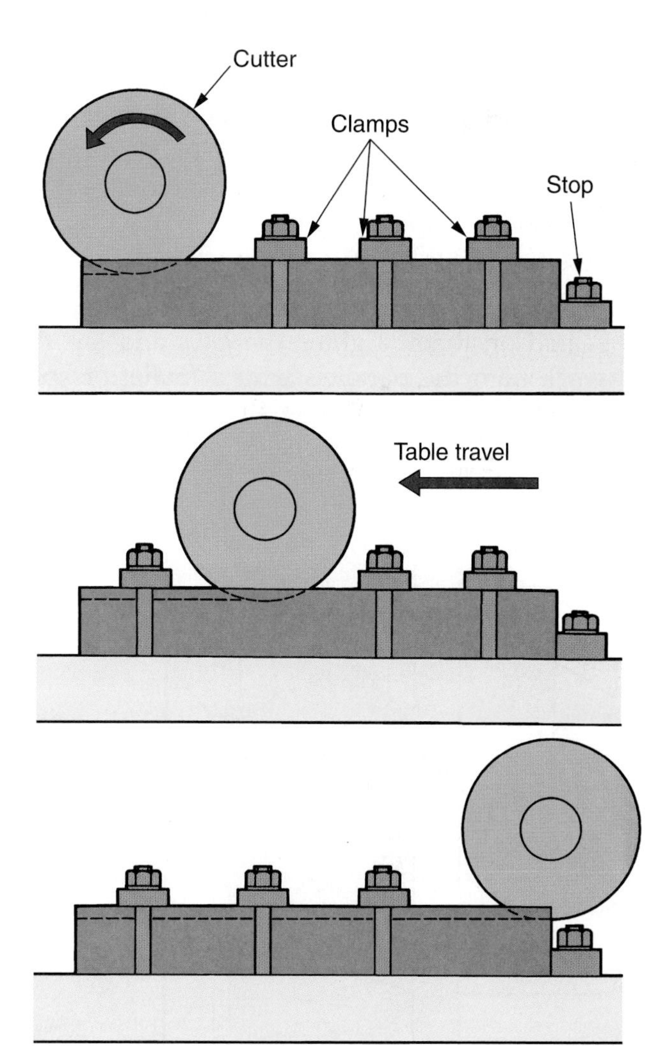

Figure 19-24. A method used to mount long work. Reposition clamps as the cut progresses across the workpiece.

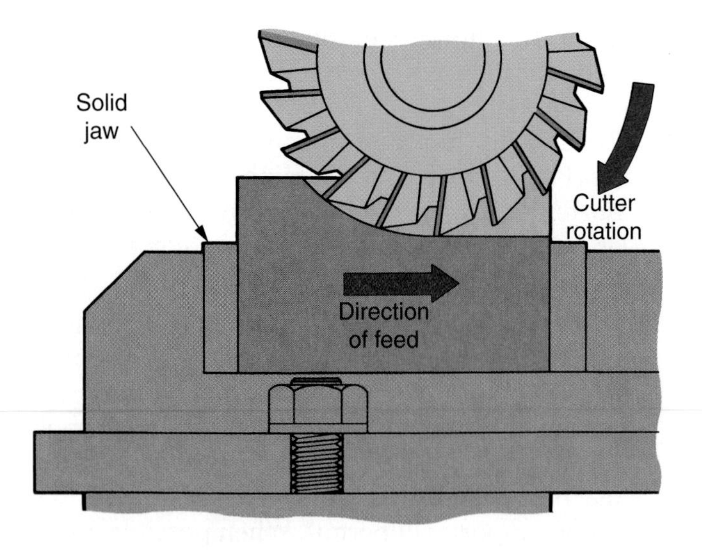

Figure 19-25. Whenever the setup permits, the solid jaw of the vise should be in this position.

3. If extreme accuracy is required, align the vise using a dial indicator. Otherwise, a square or machine arbor will do, **Figure 19-26**. Angular vise settings can be made using the vise base graduations or a protractor.
4. Wipe the vise jaws and bottom clean.
5. Place clean parallels in the vise with the work seated on them. Tighten the jaws and tap the work onto the parallels with a mallet or soft-faced hammer. Thin paper strips can be used to check that the work is seated tightly on top of the parallels.

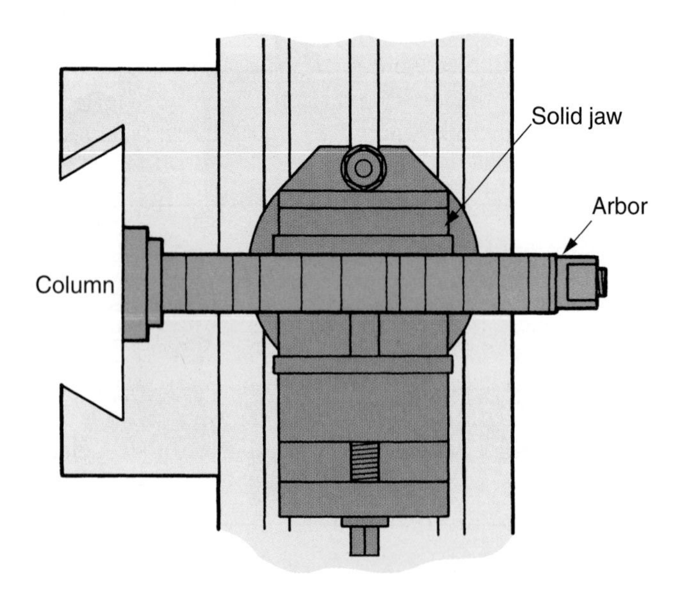

Figure 19-26. An arbor can also be used to align a vise.

6. Select an arbor that is as short as the job will permit. Wipe the taper section of the arbor and the spindle opening with a dry cloth. Insert the arbor and draw it tightly into place with the draw-in bar.
7. When possible, the cutter should be wide enough to machine the area in one pass. It should have the smallest diameter possible, while being large enough to provide adequate clearance. See **Figure 19-27**.
8. Key the cutter to the arbor to prevent it from slipping during machining. Position it as close to the column as the work will permit. To protect your hands when mounting a cutter on the arbor, use a piece of cloth or gloves. If a helical slab mill is used, mount it so the cutting pressure forces it toward the column.
9. Position and lock the arbor support into place, **Figure 19-28**. Then, tighten the arbor nut.

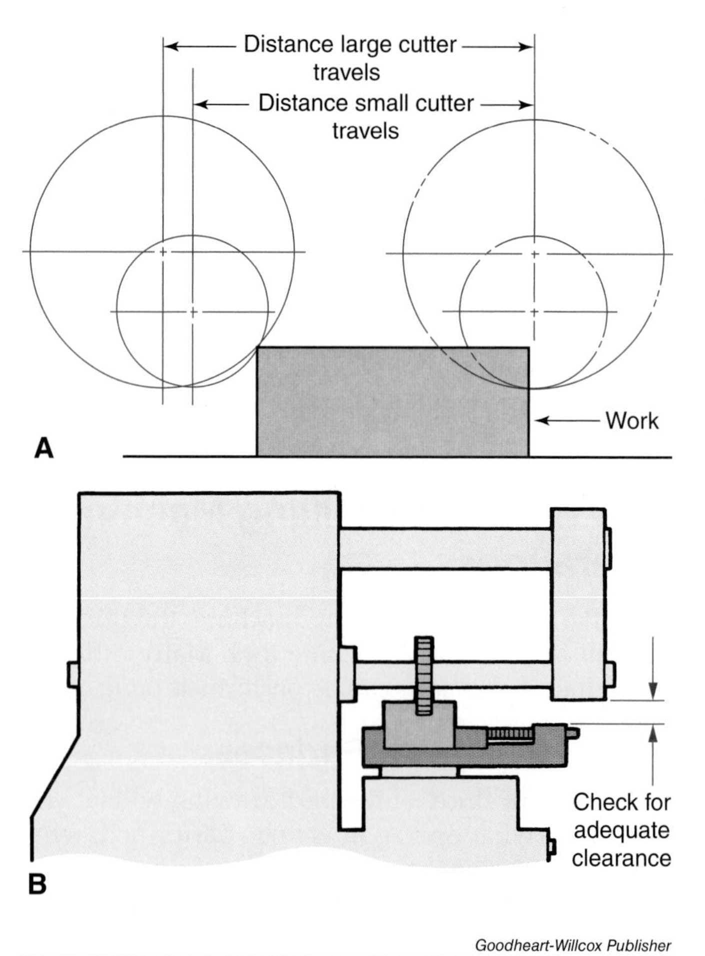

Figure 19-27. Cutter diameter. A—A small-diameter cutter is more efficient than a large-diameter cutter because it travels less distance while doing the same amount of work. B—Use the smallest cutter diameter possible, but be certain it is large enough for adequate clearance.

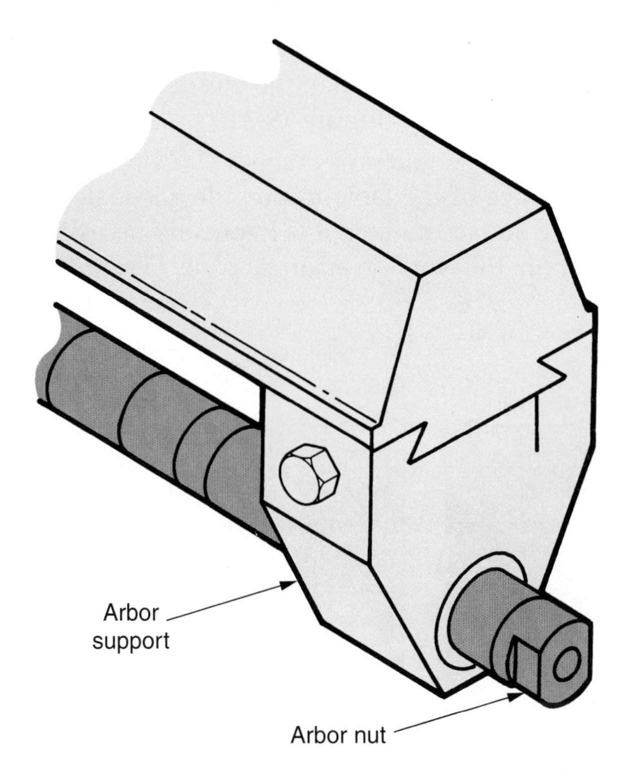

Goodheart-Willcox Publisher

Figure 19-28. The arbor nut must not be tightened until after the arbor support has been positioned and locked in place.

10. Adjust the machine to the proper cutting speed and feed.
11. Turn on the machine and check cutter rotation and direction of power feed. If satisfactory, loosen all worktable and knee locks. Position the workpiece under the rotating cutter until the cutter just touches its surface. Set the micrometer dial to 0. Back the work away from the rotating cutter. Make a light cut with ample cutting fluid flooding the surface. Make a measurement, then raise the table the required distance.

Should a previously machined surface require additional machining, it will be best to position the cutter in the following manner. Hold a long, narrow strip of paper with its loose end between the work and the cutter. Raise the table until the paper is pulled lightly from your fingers.

SAFETY NOTE

Pay close attention when positioning a cutter by the paper strip method. Use a long strip of paper, hold it lightly, and keep your fingers well clear of the rotating cutter.

Once the cutter has been positioned, it is only necessary to move the cutter clear of the work and raise the table the required distance, plus the thickness of the paper. Tighten all locks (except longitudinal) and feed the work into the cutter. As soon as cutting starts, turn on the coolant and power feed.

Unless there is an emergency, do not stop the work during the machining operation. This will cause a slight depression to be cut in the machined surface. Do not bring the rotating cutter back over the machined surface. It will create ridges on the surface of the work.

Complete the cut and stop the cutter. Return the work to the starting position. Avoid feeding the work back to the starting position while the cutter is rotating. This will cause a series of depressions to be made on the newly machined surface.

Repeat the above operations if additional metal must be removed to bring the work to size.

SAFETY NOTE

Do not attempt to feel the machined surface while the cut is in progress or while the cutter is rotating.

19.4.2 Squaring Stock

The sequence for squaring stock on a horizontal milling machine is the same as that used on a vertical milling machine. Whenever possible, the cutter should be wide enough to make a full-width cut on the material in one pass. If the material is short enough, the ends can be machined by placing it in a vertical position with the aid of a square, as shown in **Figure 19-10A**. If the work is too long for this technique, the ends can be squared as shown in **Figure 19-29**.

19.4.3 Face Milling

Face milling makes use of a cutter that machines a surface at right angles to the spindle axis and parallel to the face of the tool, **Figure 19-30**. Face milling cutters over 6″ (150 mm) in diameter are usually of the inserted tooth type and mount directly to the spindle nose. They are used to mill large, flat surfaces. Face mills smaller than 6″ (150 mm) are called shell end mills and are held on a Style C arbor.

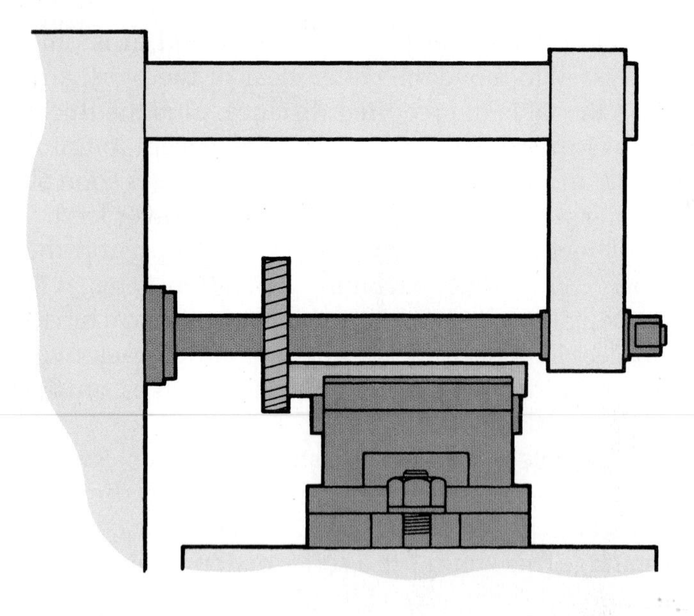

Goodheart-Willcox Publisher

Figure 19-29. Another technique for squaring work ends. The solid vise jaw must be checked with a dial indicator to ensure that it is properly aligned. Be sure there is adequate clearance between the work and the arbor.

Mazak Corp.

Figure 19-30. In face milling, both the cutter and the workpiece face being machined are at a right angle to the axis of the spindle.

1. Select a cutter that is 3/4″ (20 mm) to 1″ (25 mm) larger in diameter than the width of the surface to be machined, **Figure 19-31.**
2. The work should project about 1″ (25 mm) beyond the edge of the table to provide adequate clearance. In face milling, it is frequently necessary to mount the work on an angle plate, **Figure 19-32.**

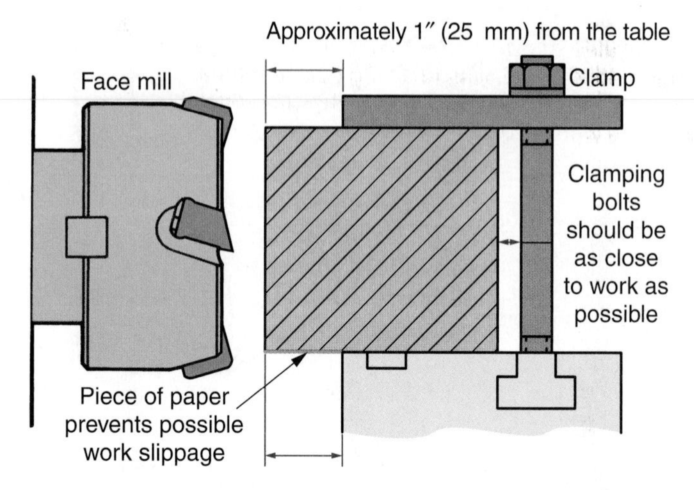

Goodheart-Willcox Publisher

Figure 19-31. Select a cutter that is 3/4″ (20 mm) to 1″ (25 mm) larger in diameter than width of surface to be machined. The work should project approximately 1″ (25 mm) beyond the table edge to provide adequate clearance.

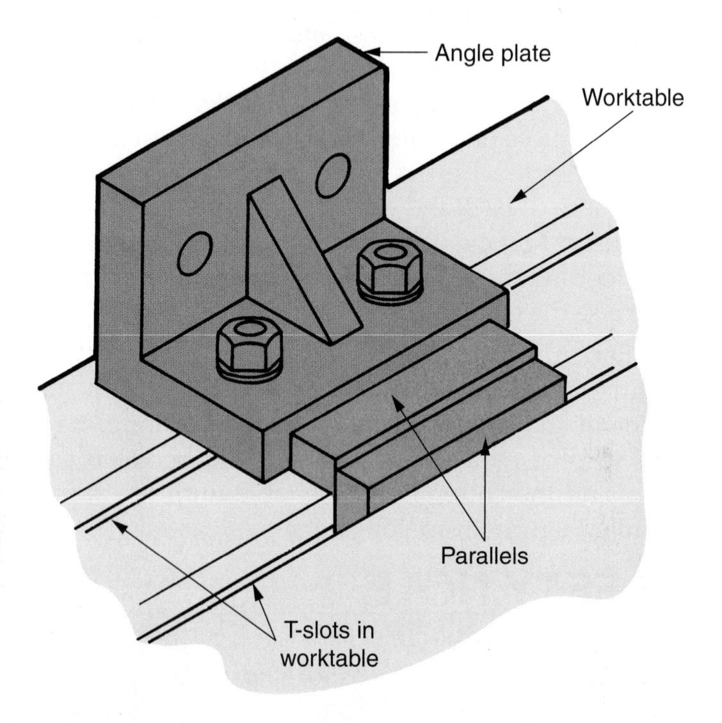

Goodheart-Willcox Publisher

Figure 19-32. An angle plate is often used to mount work for face milling. Check that mounting clamps clear the cutter. Note the use of parallels to align the angle plate.

3. Adjust the machine for correct speed and feed.
4. Slowly feed the work into the cutter until it starts to remove material. Roughing cuts up to 1/4" (6 mm) may be taken. Use adequate cutting fluid.
5. When the cut is complete, stop the cutter. Return the work to the starting position for additional machining, if needed.
6. Make the finishing cut and tear down the setup. Use a brush to remove chips. Clean and store the cutter.

19.4.4 Side Milling

Side milling refers to any milling operation that involves the use of half side and side milling cutters. When cutters are used in pairs to machine opposite sides of a piece at the same time, the setup is called ***straddle milling***, **Figure 19-33**.

Cutters used for this operation should be kept in matched pairs. That is, they should be sharpened at the same time to maintain equal diameters. Shoulder width of the machined surface is determined by the thicknesses of the spacers between the cutters, **Figure 19-34**.

Spacers are available in a large selection of sizes. Special sizes can be made by surface grinding standard sizes to needed dimensions. Shim stock spacers can be used to build up standard size spacers to desired dimension.

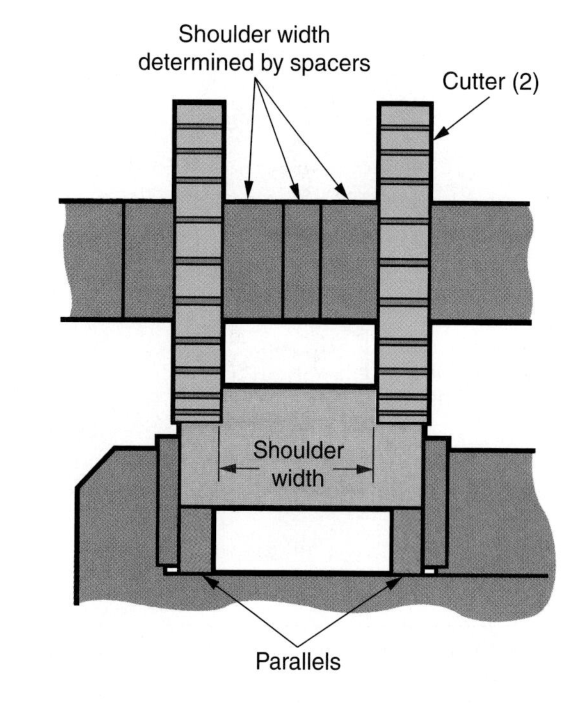

Goodheart-Willcox Publisher

Figure 19-34. Arbor spacers are used to set distance between cutters.

Gang milling involves mounting several cutters on an arbor to machine several surfaces in a single pass, **Figure 19-35**. It is a variation of straddle milling. Gang milling is used when many identical pieces must be made.

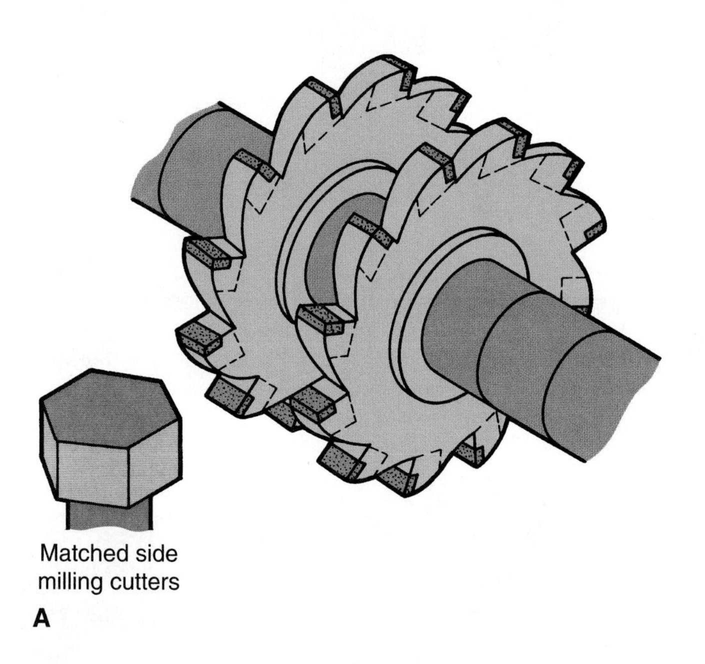

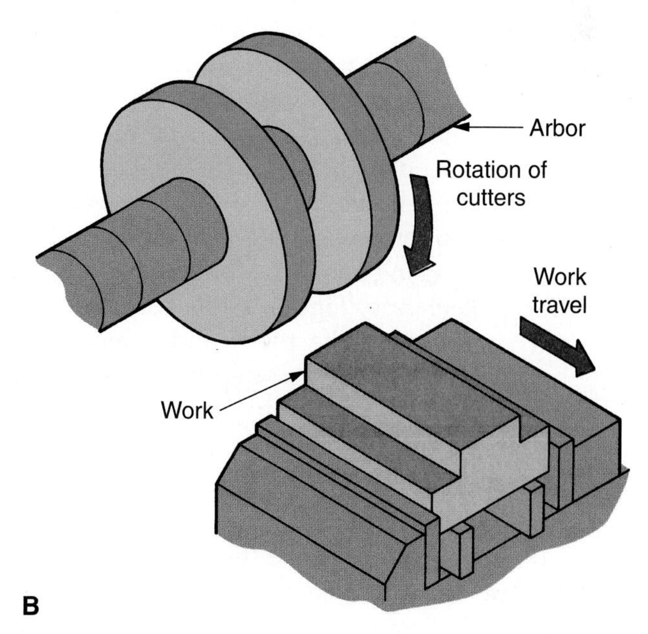

Goodheart-Willcox Publisher

Figure 19-33. Straddle milling. A—This machining method uses a spacer between cutters. These two cutters are machining a hexagon on a machine part. B—An example of straddle milling on flatwork.

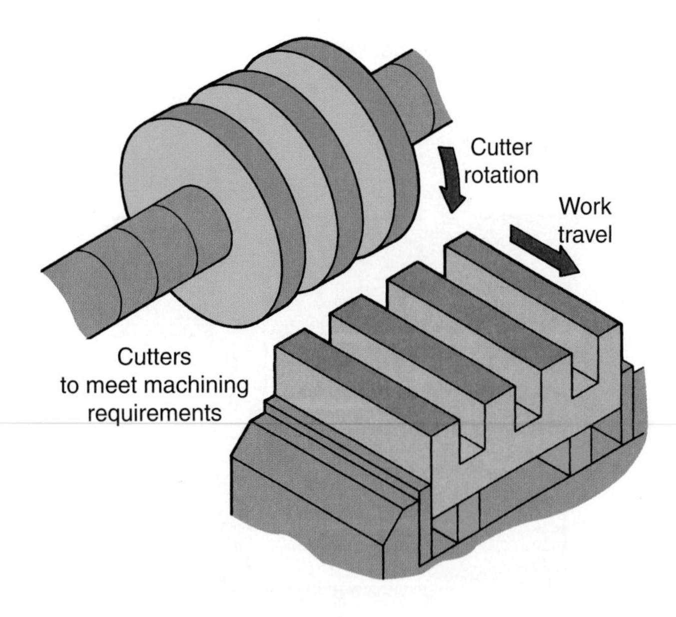

Goodheart-Willcox Publisher

Figure 19-35. Gang milling involves the use of two or more milling cutters mounted on a single arbor.

A side milling cutter can also be utilized to machine grooves, keyseats, and when used with a dividing head or rotary table, squares, hexagons, or other shapes on round stock.

19.4.5 Locating Side Cutter for Milling a Slot in Square or Rectangular Work

When milling a slot, the machine is set up in much the same manner as it was for milling flat surfaces. Use a dial indicator to check the vise to ensure accuracy. Exercise the same care in placing the side cutter on the arbor as was followed with the slab cutter. A plain side milling cutter may be used if the slot is not too deep. Otherwise, a staggered-tooth side milling cutter should be used.

Make a layout on the end of the work, then position the cutter by using one of the following methods:

- Use a steel rule, **Figure 19-36**. Make a light cut partway up the piece and remove the burrs. Measure the cut depth with a depth micrometer. The difference between this measurement and the required depth equals the amount of material that must be removed.
- Use the paper strip technique to bring the side of the cutter against the side of the work, **Figure 19-37**. Move the cutter inward the required distance, plus the thickness of the paper. The paper strip or depth micrometer positioning technique may be used to set the cutter to the desired depth.

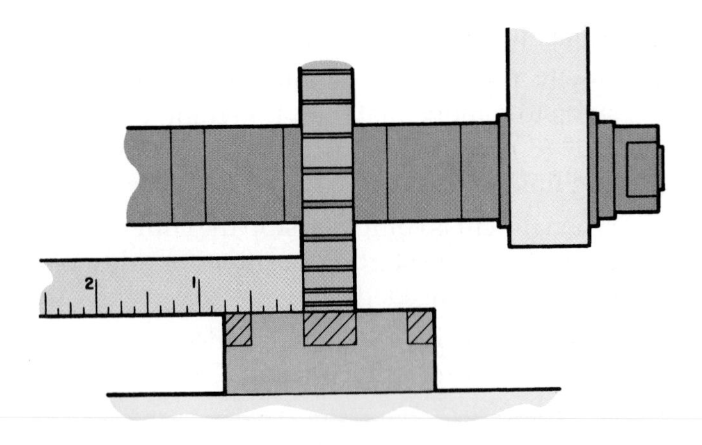

Goodheart-Willcox Publisher

Figure 19-36. A cutter being positioned with aid of a steel rule. The layout lines are visible on the front of the workpiece.

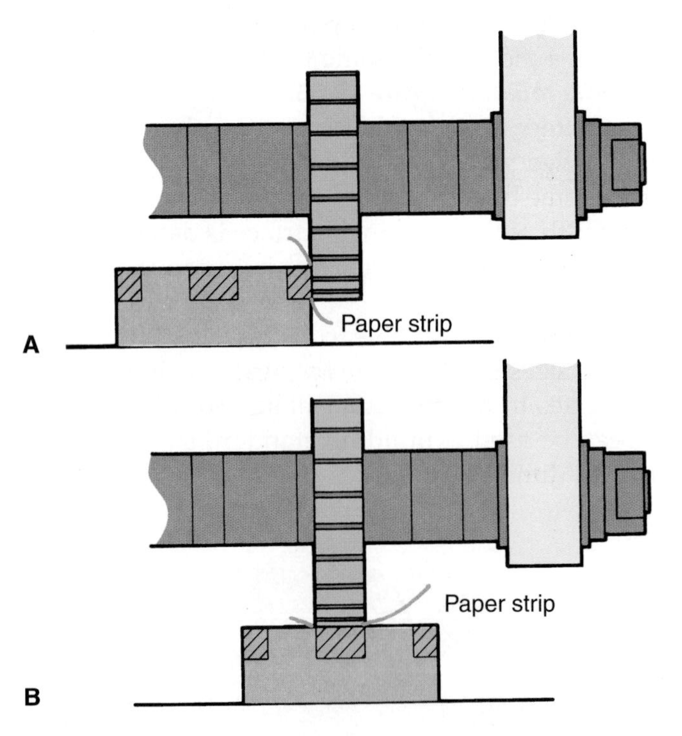

Goodheart-Willcox Publisher

Figure 19-37. Paper strip method of positioning the cutter. A—Using a paper strip to secure internal dimension. Read micrometer to move cutter the correct distance over the work. B—Using a paper strip to position the cutter for depth. Read micrometer dial as cutter is lowered.

SAFETY NOTE

When using the paper strip technique to position a cutter, remember to use a long paper strip and keep your fingers clear of the revolving cutter.

19.4.6 Locating Side Cutter for Milling a Slot or Keyseat in Round Stock

There are many situations that require keyseats for the standard square key to be cut in round stock. The keyseat must be kept precisely on center if it is to be in alignment with the keyway in the mating piece.

After the milling machine has been set up and work positioned in a vise, between centers, in V-blocks, or in a fixture, you must center the cutter. Precise centering of the cutter may be accomplished by one of the following methods:

- Center the cutter visually on the work. With the aid of a steel square and rule, adjust the table until both sides measure the same, **Figure 19-38**. Due to the difficulty of obtaining precise measurements with a rule, most machinists prefer to use a depth micrometer in place of the rule.
- Short pieces cannot always be centered by the above method. For situations of this type, the work is positioned under the rotating cutter and brought lightly into contact with it. Traverse (in/out) feed is used to pass the work under the cutter. Because the work is circular in shape, an oval-shaped cut will result, and the oval will be perfectly centered. To center the cutter, position it on the oval, **Figure 19-39**.

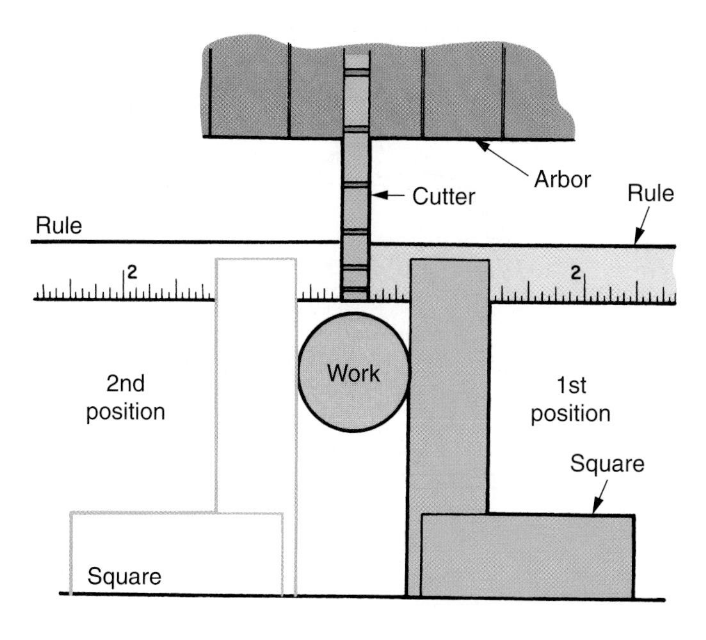

Goodheart-Willcox Publisher

Figure 19-38. How to center cutter on round stock with a steel rule and machinist's square.

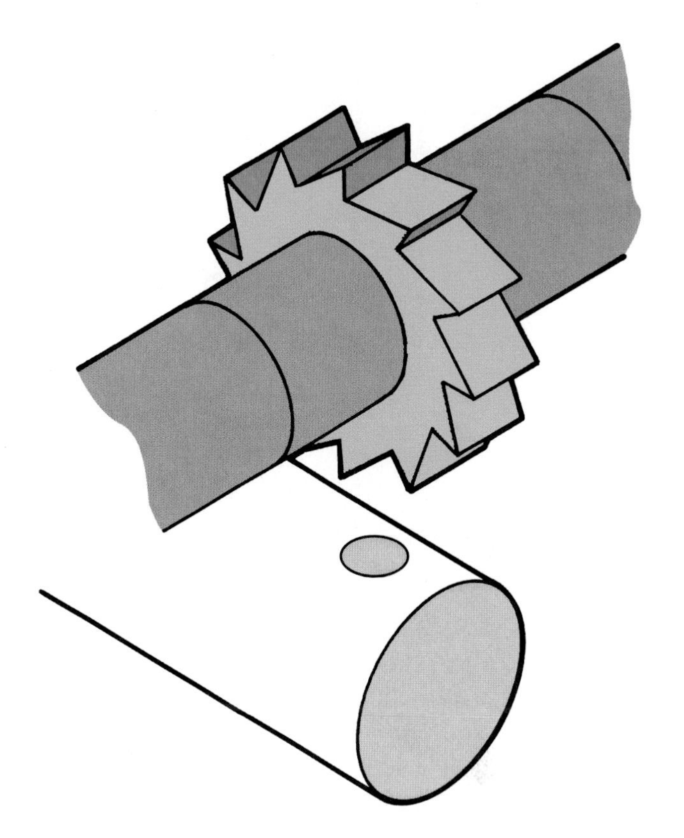

Goodheart-Willcox Publisher

Figure 19-39. Cutter is being positioned on center using an oval made in the work with cutter as a guide.

- The previously mentioned narrow paper strip technique may be used to center the cutter. Hold the strip between the work and the cutter. Carefully move the work toward the cutter until it causes the paper to be lightly pulled from between your fingers. Lower the table until the cutter is slightly above the work. Move the cutter inward half the diameter of the work, plus half the cutter thickness, plus the paper thickness, **Figure 19-40**. The same technique may be used to center a Woodruff keyseat cutter.

Lock the saddle to prevent traverse table movement after the cutter has been centered.

Correct keyseat depth can be obtained from tables in one of the many machinist's handbooks. The paper strip technique is used to set the cutter to the required depth. Tighten the knee locks after the depth setting has been made. Cutting fluid should be applied liberally during the cutting operation.

When the Woodruff keyseat cutter is used, it must also be positioned longitudinally on the work. Slowly feed into the piece until the required depth is attained. This can be checked by placing a key in the cut and using a micrometer to measure the section.

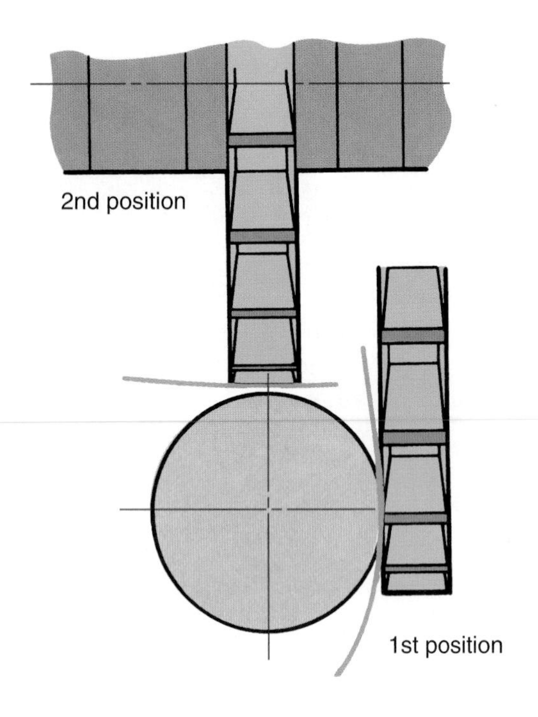

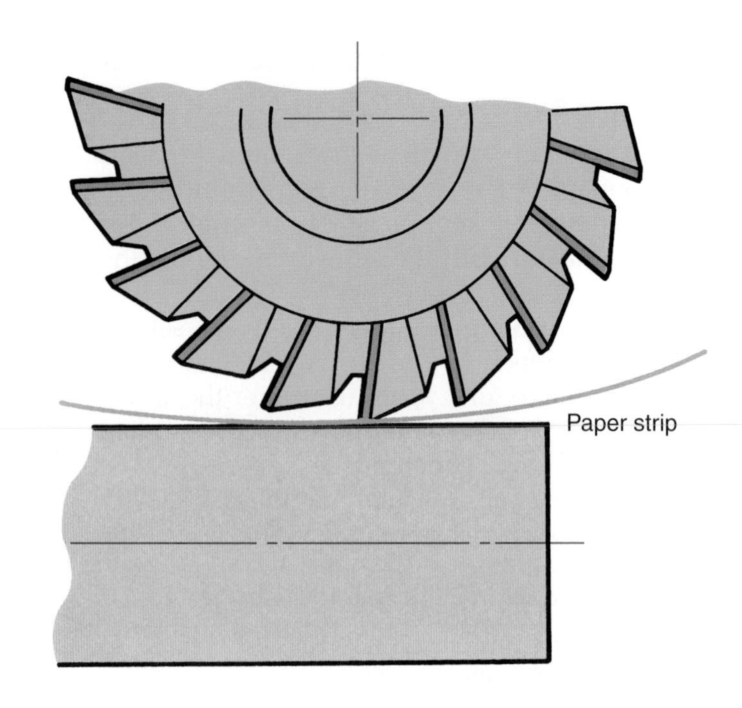

Goodheart-Willcox Publisher

Figure 19-40. Using the paper strip technique to position round stock. Use a long strip of paper. Hold lightly between your fingers and keep them well clear of the revolving cutter.

19.4.7 Slitting

Slitting thin stock into various widths for the production of flat gages or templates is a fairly common milling operation, **Figure 19-41**. It is performed with a slitting saw and is likely to give considerable trouble if extreme care is not exercised.

A slitting saw of the smallest diameter permitting adequate clearance is used. It must be keyed to the arbor (the key should also fit into spacers on either side of cutter). Best results can be obtained if the cutter is mounted for climb milling. That is, the work and cutter move in the same direction at the point of contact, **Figure 19-42**. Cutting pressure is downward and will tend to press the work onto the table or holding device.

Goodheart-Willcox Publisher

Figure 19-41. Typical slitting or sawing operation setup. Work must be positioned over a table slot and clamped securely. Be sure the clamping bolts clear the arbor.

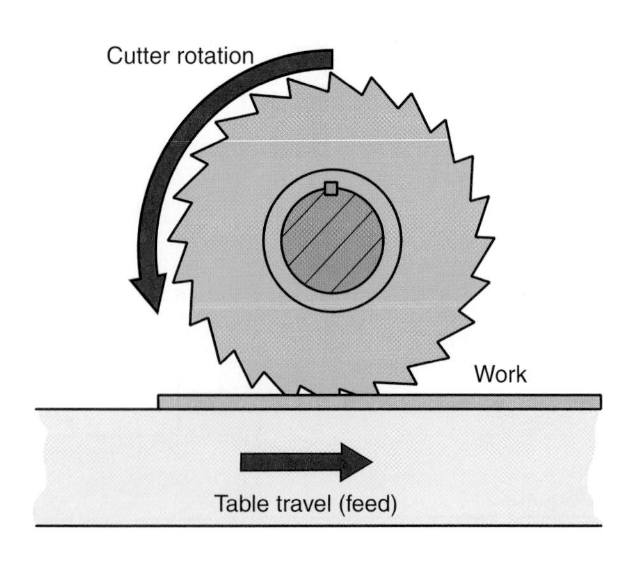

Goodheart-Willcox Publisher

Figure 19-42. Cutter rotation and feed direction for best slitting results. Use a slow feed.

Adjust the table gibs until there is heavy drag felt when the table is moved by hand. This will remove table "play" and prevent the cutter from jumping in the cut.

If the section is narrow enough, the piece may be clamped in a vise, **Figure 19-43**. It should be well supported on parallels. Do not permit the parallels to project out into the cutter path.

Long strips must be clamped to the worktable. The shop-made angle iron clamp shown in **Figure 19-44** is recommended. The work is aligned with the column face and must be positioned to permit the saw to make the cut over the center of a table T-slot.

A sheet of paper between the work and table will prevent the metal from slipping during the slitting operation. The cutter is set to a depth equal to the work thickness plus 1/16″ (1.5 mm). Always use a sharp cutter.

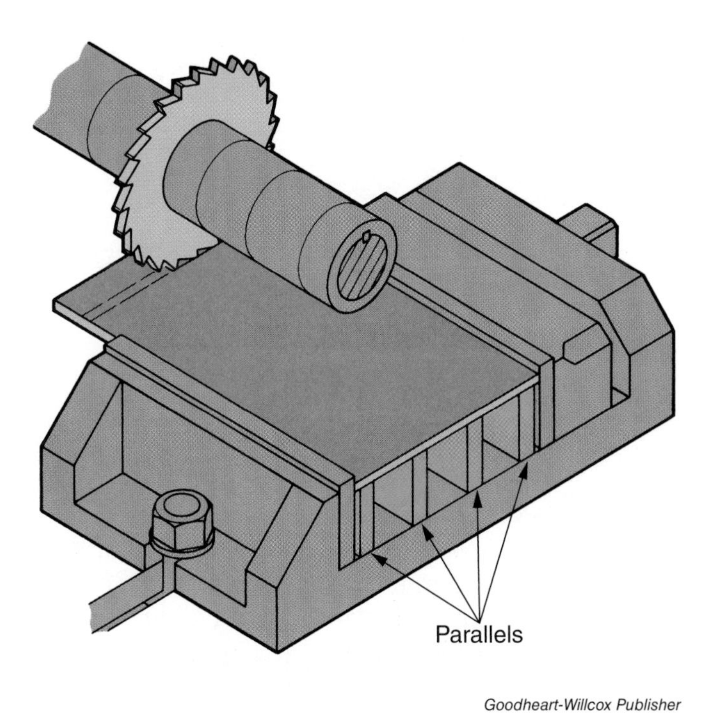

Figure 19-43. Slitting work is held in vise. Be sure parallels do not project into cutter path!

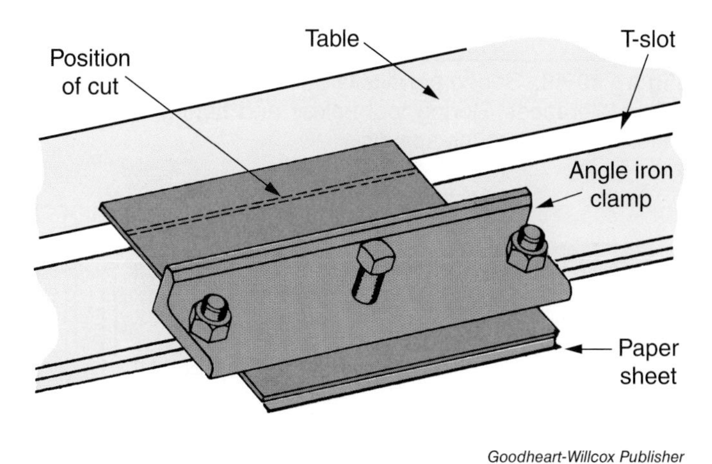

Figure 19-44. A worktable clamp made from angle iron. Paper sheet prevents work movement. Position the work so that the cut is made over a T-slot.

19.4.8 Slotting

Slotting is similar to slitting, except that the cut is made only partway through the work, **Figure 19-45**. The slot in a screw head is an example of slotting.

1. Mount the cutter as for conventional milling. Use a sharp cutter of a width suitable for the job. Note the difference between a slotting cutter and a slitting cutter, **Figure 19-46**.
2. Set the machine for the correct cutting speed. Use the slowest feed possible, increasing feed rate if conditions warrant.
3. Align the vise and mount the work.

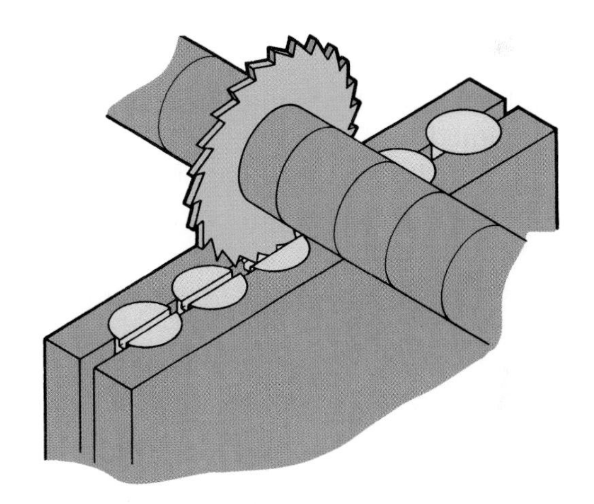

Figure 19-45. Screw heads being slotted with a slitting saw. Special slotting saw is available which has many more teeth than a slitting saw of a similar diameter.

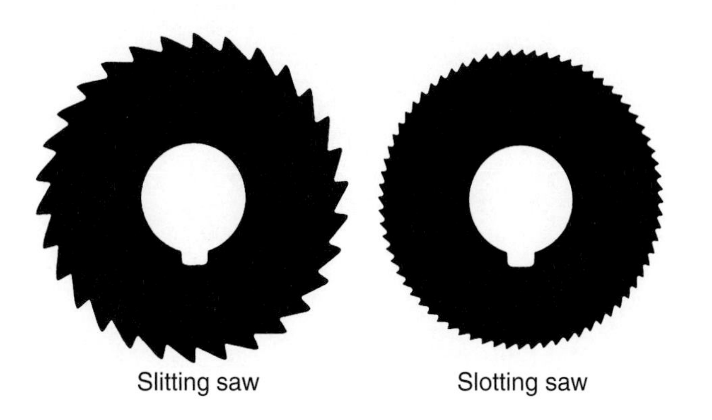

Figure 19-46. How slitting and slotting cutters differ.

4. Position the cutter and make a light cut. Check the trial cut and make adjustments if necessary. Stop the machine before making measurements or adjustments.
5. Adjust the work for proper cut depth.
6. Apply cutting fluid and make the cut.

SAFETY NOTE

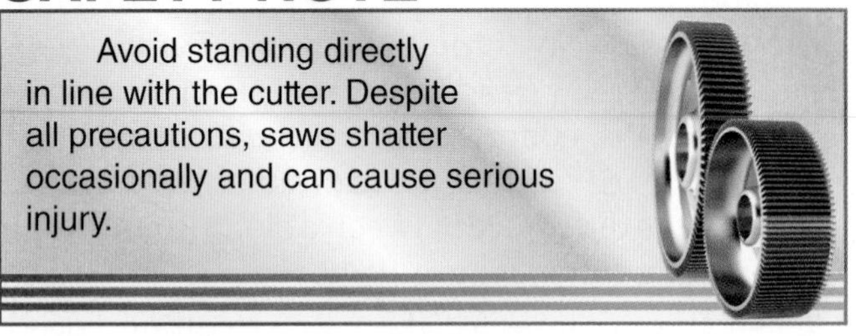

Avoid standing directly in line with the cutter. Despite all precautions, saws shatter occasionally and can cause serious injury.

19.4.9 Drilling and Boring

The machinist often finds it necessary to produce accurately spaced and drilled holes in work. The milling machine offers a convenient way to make these holes in a specified and precise alignment.

Small drills are held in a standard Jacobs-type chuck mounted in the machine spindle or in collet chucks, **Figure 19-47**. Larger, taper shank drills are fitted into an adapter sleeve.

Parlec, Inc.

Figure 19-47. Drills can also be mounted in collets. Collet holders are available in a range of sizes.

Boring is done with a single point cutting tool fitted in a boring head. The boring head may be equipped with a taper shank and mounted directly in the spindle, **Figure 19-48**, or a straight shank and held in a collet or adapter.

A wiggler, **Figure 19-49**, will aid in aligning the machine for drilling the hole prior to boring, or when holes are small enough to be drilled and reamed. A dial indicator must be used to realign previously made holes for boring to final size, **Figure 19-50**.

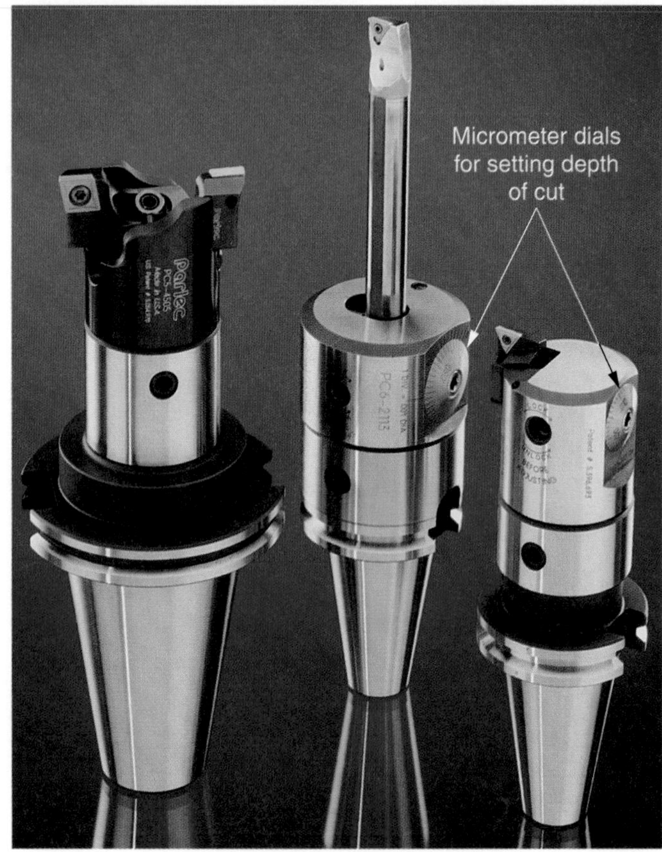

Parlec, Inc.

Figure 19-48. Boring permits large holes to be machined to close tolerances. Boring tool holder and boring heads that mount directly into the spindle.

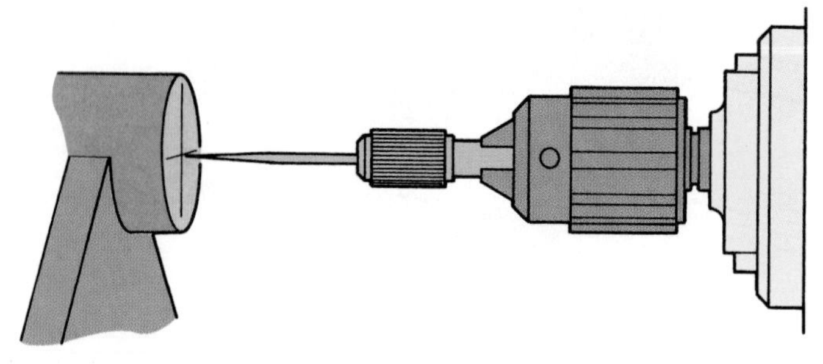

Goodheart-Willcox Publisher

Figure 19-49. Holes can be located with a wiggler.

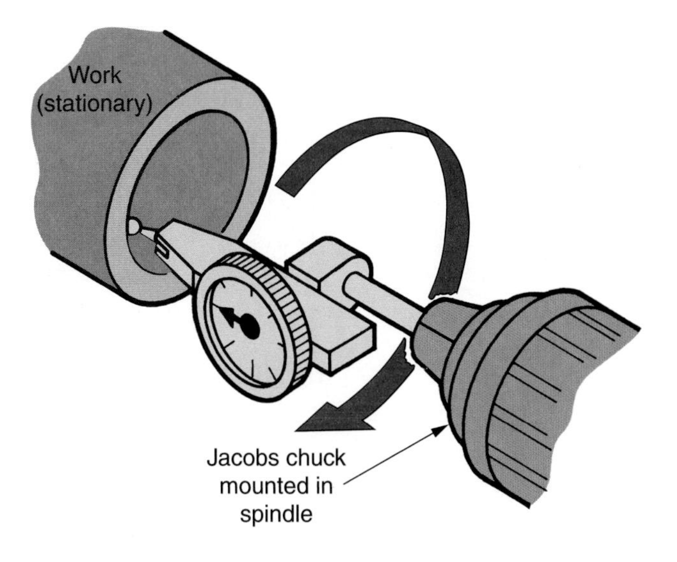

Goodheart-Willcox Publisher

Figure 19-50. Existing holes can be realigned on a horizontal milling machine with the aid of a dial indicator.

19.5 Cutting a Spur Gear

A gear, **Figure 19-51**, is a toothed wheel, usually fitted to a shaft. It typically engages a similar toothed wheel to smoothly transmit power or motion at a definite ratio between the shafts. The teeth are shaped so that contact between the mating gears is continually maintained while they are in operation.

The ***spur gear*** has teeth that run straight across the face and are perpendicular to the sides. It is the simplest gear and is widely used. Knowledge of gear nomenclature (terminology) is necessary to calculate the data needed to machine a simple inch-based spur gear. Inch-based gears and metric-based gears are not interchangeable.

Eric Milos/Shutterstock.com

Figure 19-51. A few of the many types of gears available. A large spur gear, a pinion gear, and two bevel gears.

19.5.1 Gear Nomenclature

Gear cutting requires knowledge of gear nomenclature to aid in determining the proper gear cutter to use, the depth of the teeth, and the dividing head setup. The various gear parts and formulas are shown in **Figure 19-52**.

The ***pitch circle*** is an imaginary circle located approximately half the distance from the roots and tops of gear teeth. It is tangent to the pitch circle of the mating gear. The pitch diameter (D) is the diameter of the pitch circle.

Diametral pitch (P) is the number of teeth per inch of pitch diameter. ***Circular pitch*** (p) is the distance measured on the pitch circle between similar points on adjacent teeth.

Tooth thickness (t) is the thickness of the tooth at the pitch circle. This dimension is used when measuring the tooth thickness with a vernier gear tooth caliper. The ***addendum*** (a) is the distance the tooth extends above the pitch circle. The ***dedendum*** (b) is the distance the tooth extends below the pitch circle.

The working depth (h_k) is the sum of the addendums of the two mating gears. The whole depth (h_t) is the total depth of a tooth space, equal to the addendum plus dedendum, or to the depth to which each tooth is cut.

Clearance (c) is the difference between the working depth and the whole depth of a gear tooth. The clearance is the amount by which the dedendum of one gear exceeds the addendum of the mating gear.

19.5.2 Gear Cutters

No one gear cutter can be used to cut all gears. Gear cutters are made with eight different forms depending on the number of teeth for which the cutter is to be used. **Figure 19-53** illustrates the comparative sizes for gear teeth. The cutter range is as follows:

No. of Cutter	Range of Teeth
1	135 to a rack
2	55 to 134
3	35 to 54
4	26 to 34
5	21 to 25
6	17 to 20
7	14 to 16
8	12 to 13

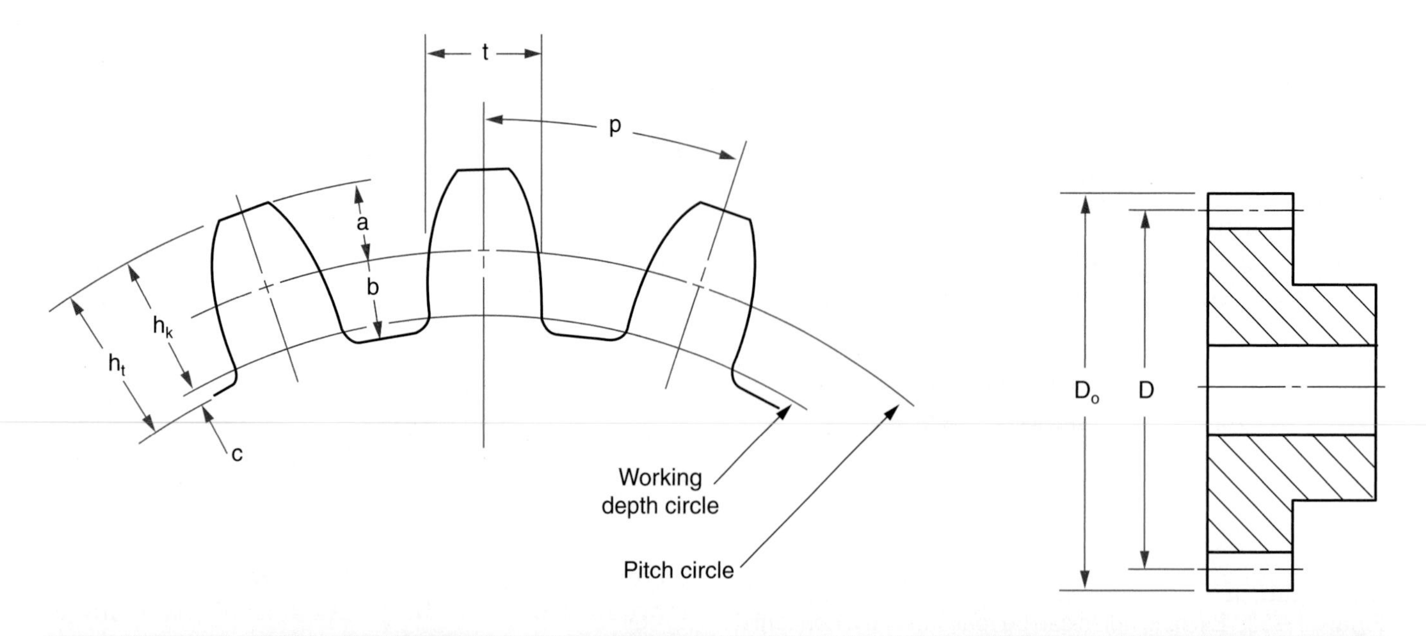

Spur Gear Parts and Formulas		
Gear Part	**Symbol**	**Definition**
Number of teeth on the gear	N	
Pitch circle		An imaginary circle located approximately half the distance from the roots and tops of gear teeth. It is tangent to the pitch circle of the mating gear.
Pitch diameter	D	The diameter of the pitch circle.
Diametral pitch	P	The number of teeth per inch of pitch diameter.
Circular pitch	p	The distance measured on the pitch circle between similar points on adjacent teeth.
Tooth thickness	t	Thickness of the tooth at the pitch circle. The dimension used in measuring tooth thickness with a vernier gear tooth caliper.
Addendum	a	The distance the tooth extends above the pitch circle.
Dedendum	b	The distance the tooth extends below the pitch circle.
Working depth	h_k	The sum of the addendums of the two mating gears.
Whole depth	h_t	Total depth of a tooth space, equal to the addendum (a) plus dedendum (b), or to the depth to which each tooth is cut.
Clearance	c	The difference between the working depth and the whole depth of a gear tooth. The amount by which the dedendum on a given gear exceeds the addendum of the mating gear.
Center distance	C	Distance between the centers of two mating gears.
Outside diameter	D_o	Diameter or size of gear blank.
Pressure angle	θ	The angle of pressure between contacting teeth of mating gears. It represents the angle at which the forces from the teeth on one gear are transmitted to the mating teeth of another gear. Pressure angles of 14½°, 20°, and 25° are standard. However, the 20° is replacing the older 14½°.

Goodheart-Willcox Publisher

Figure 19-52. Chart showing gear nomenclature, including gear parts and formulas used for gear calculations.

Formulas for Spur Gear Calculations		
To Find	**Rule**	**Formula**
Diametral pitch	Divide 3.1416 by the circular pitch.	$P = 3.1416 / p$
Circular pitch	Divide 3.1416 by the diametral pitch.	$p = 3.1416 / P$
Pitch diameter	Divide the number of teeth by the diametral pitch.	$D = N / P$
Outside diameter	Add 2 to the number of teeth and divide the sum by the diametral pitch.	$D_o = \frac{N + 2}{P}$
Number of teeth	Multiply the pitch diameter by the diametral pitch.	$N = D \times P$
Tooth thickness	Divide 1.5708 by the diametral pitch.	$T = 1.5708 / P$
Addendum	Divide 1.0 by the diametral pitch.	$a = 1.0 / P$
Dedendum	Divide 1.157 by the diametral pitch.	$b = 1.157 / P$
Working depth	Divide 2 by the diametral pitch.	$h_k = 2 / P$
Whole depth	Divide 2.157 by the diametral pitch.	$h_t = 2.157 / P$
Clearance	Divide 0.157 by the diametral pitch.	$c = 0.157 / P$
Center distance	Add the number of teeth in both gears and divide the sum by two times the diametral pitch.	$C = \frac{N_1 + N_2}{2P}$
Length of rack	Multiply the number of teeth in the rack by the circular pitch.	$L = N \times p$

Goodheart-Willcox Publisher

Figure 19-52. Continued.

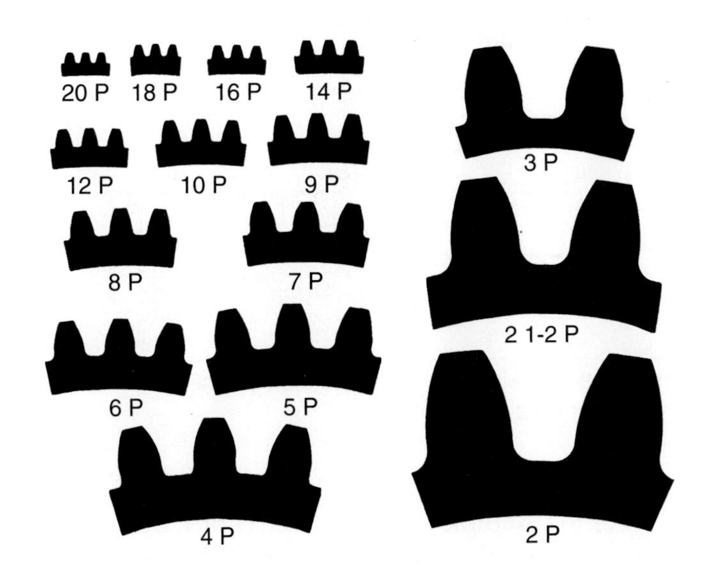

Goodheart-Willcox Publisher

Figure 19-53. Comparative sizes of gear teeth. Diametral pitch is shown.

With the information furnished, it is possible to calculate the data needed to cut a simple inch-based spur gear.

Example Problem: Calculate the data needed to cut a 40 tooth, 10 diametral pitch gear.

1. Diameter (D_o) of gear blank needed.
   Having:

   Diametral pitch (P) = 10
   Number of teeth (N) = 40

   Formula:

$$D_o = \frac{N + 2}{P} = \frac{40 + 2}{10} = \frac{42}{10}$$

$$= 4.200$$

$$= 4.200'' \text{ diameter}$$

2. Whole depth of tooth (h_t) needed. This will be the depth of the cut.
   Having:

   Diametral pitch (P) = 10

   Formula:

$$h_t = \frac{2.157}{P} = \frac{2.157}{10} = 0.216$$

$$= 0.216''$$

3. The dimension of the addendum (a) is needed to measure the gear tooth for determining whether it is being machined to specifications.
   Having:

   Diametral pitch (P) = 10

   Formula:

$$a = \frac{1}{P} = \frac{1}{10} = 0.100$$
$$= 0.100''$$

4. Tooth thickness (t) is needed to determine whether the gear is being machined to specifications.
   Having:

   Diametral pitch (P) = 10

   Formula:

$$t = \frac{1.5708}{P} = \frac{1.5708}{10} = 0.157$$
$$= 0.157''$$

5. Reference to the gear cutter chart indicates that a No. 3 cutter, with a range of 35 to 54 teeth, must be used to cut 40 teeth.
6. Using a 40:1 ratio dividing head for this job means that the index crank must be turned through one complete revolution to position the gear blank for each cut. A 5:1 ratio dividing head would require the use of an index plate that would permit a setting of one-eighth turn for each cut.

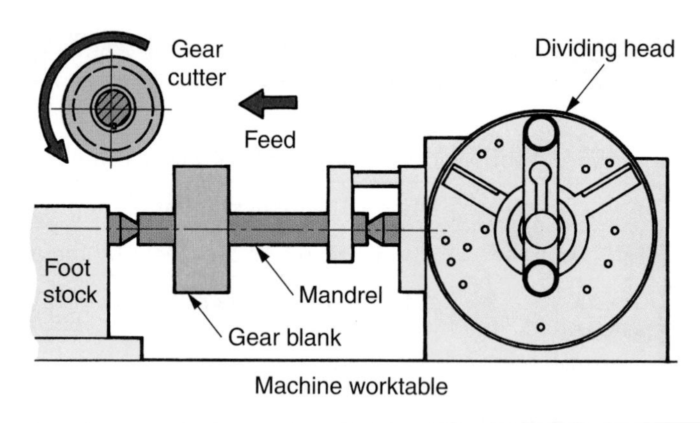

Figure 19-54. Cutting is done toward dividing head.

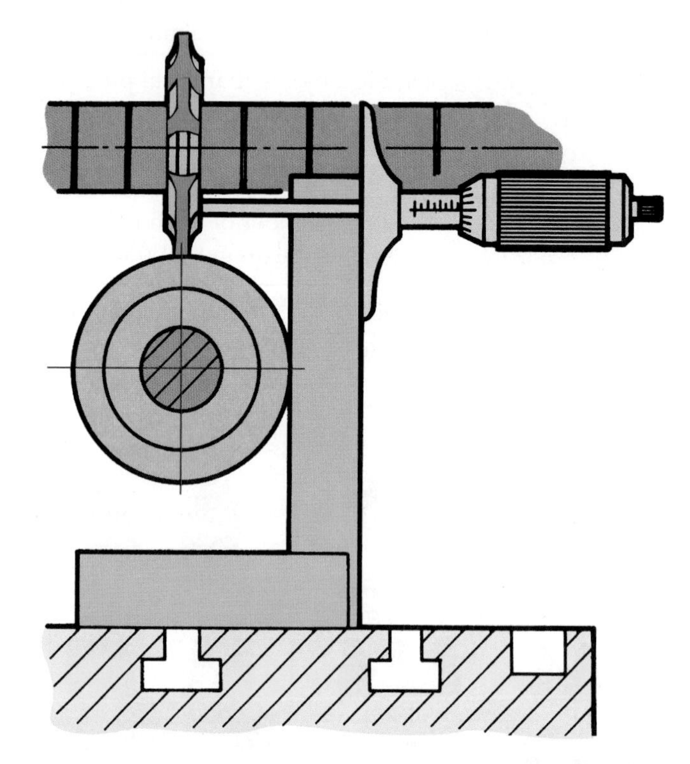

Figure 19-55. A depth micrometer can be used to center the gear cutter on a gear blank.

19.5.3 Cutting the Gear

Following a few simple precautions will greatly reduce the possibility of an inaccurately machined gear.

1. Set up the milling machine as previously described. Check center alignment of the dividing head and foot stock.
2. Press the gear blank onto a mandrel and mount the unit to the dividing head. Cutting is done toward the dividing head, **Figure 19-54**.
3. Use a dial indicator on the gear blank longitudinally and turn the blank through one complete revolution. Make any adjustments that are necessary.
4. Center the cutter on the gear blank. Use a depth micrometer and a steel square, **Figure 19-55**. Position the cutter for depth and use the paper strip technique. Set the micrometer dial to 0, and then raise the table to within 0.040″ of finished depth. For the example problem given earlier, it would be raised to make a cut of 0.216″ – 0.040″, or 0.176″.
5. Move the work until the cutter just begins removing metal. Back it away from the cutter. Using the dividing head, bring the next cut into position. Repeat this sequence around the gear blank until you are back to the original cutting position. If there is exact alignment with the first cut, you are ready to cut the gear.
6. Make the roughing cuts. Use liberal quantities of cutting fluid.

7. The finish cut requires more care. Set the vertical scale on a gear tooth vernier caliper to the distance calculated for the addendum (a), **Figure 19-56**. Raise the work to within a few thousandths of the calculated whole depth of the tooth (h_t) and make cuts at two positions. Make your measurement and adjust until the reading equals the distance calculated for tooth thickness (t). Make the finish cuts. Press the completed gear from the mandrel. Remove all burrs and cut the keyway, if required.

Spur gears can also be measured using Van Keuren wires, **Figure 19-57**. A table furnished with the wire set specifies the wire diameter to use according to the diametral pitch of the gear. The table also includes dimensions for checking external spur gear measurement (M) over the wires. Measurement is made with outside micrometers.

Spiral gears and helical gears are cut on a universal type milling machine utilizing a universal dividing head geared to the table lead screw.

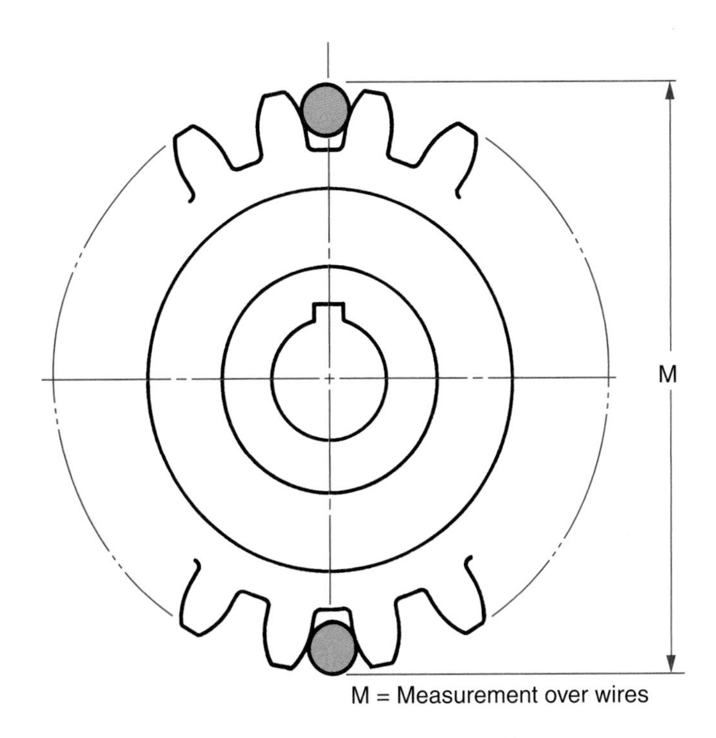

Figure 19-57. Gears can also be measured using Van Keuren wires. Tables furnished with the wire set provide information needed when measuring gears with even number and odd number of teeth.

19.6 Cutting a Bevel Gear

Bevel gears, **Figure 19-58**, are used to change the angular direction of power between shafts. The teeth are either straight or curved. The procedure for cutting a straight tooth bevel gear is illustrated and explained in this section.

Figure 19-56. Machinist is measuring gear tooth with gear tooth vernier caliper. Tool is read in same manner as a vernier caliper and vernier height gage.

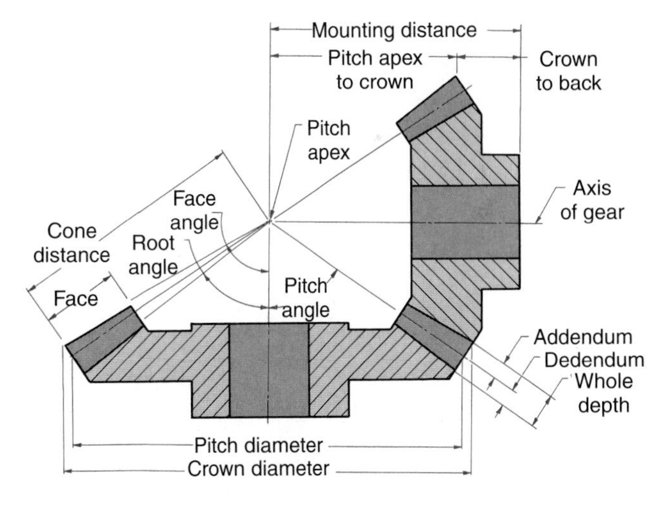

Figure 19-58. Nomenclature of bevel gears. The smaller gear is called the pinion.

Since tooth space at the pitch diameter is narrower at the small end than at the large end, special form relieved cutters have been designed to cut bevel gears. To achieve the required gear tooth dimensions, additional material must be removed from the flanks of the teeth after the preliminary roughing operation.

Measurements of the finished gear are made of the blank size and shape, tooth thickness, and depth. However, there is no simple method for checking tooth surfaces. Final inspection is made by running the mating gears and checking for quietness and shape of the tooth contact.

19.6.1 How to Mill a Bevel Gear

The following information was obtained using the dimensions shown in **Figure 19-59**, and the formulas given in this chapter.

- Cone distance = 3.535″
- Face width = 1.000″
- Pitch diameter at large end = 5.000″
- Pitch diameter at small end = 3.585″
- Circular pitch at large end = 0.5236″
- Circular pitch at small end = 0.3756″
- Addendum at large end of gear = 0.166″
- Addendum at small end of gear = 0.1195″
- Dedendum at large end of gear = 0.193″
- Dedendum at small end of gear = 0.139″
- Whole depth of tooth at large end = 0.3595″
- Whole depth of tooth at small end = 0.2585″

The tooth parts at the small end of the gear are in exact proportion to those at the large end. Dimensions at the small end can be found by multiplying the dimensions at the large end by the ratio of the respective cone distances.

Determining the ratio:

$$\frac{C_s}{C_r}$$

Where:

$C_s = C_r - F$

C_r = Cone distance

F = Face width

Example Problem: Calculate the tooth thickness and tooth space at the small end of the bevel gear in **Figure 19-59**.

1. Bevel gear ratio needed.

   Having:

   Cone distance (C_r) = 3.535″

   Face width (F) = 1.000″

   Formula:

   $$C_s = C_r - F_s = 3.535'' - 1.000'' = 2.535''$$

   $$\frac{C_s}{C_r} = \frac{2.535''}{3.535''} = 0.717$$

   $$\text{The ratio } \frac{C_s}{C_r} = 0.717$$

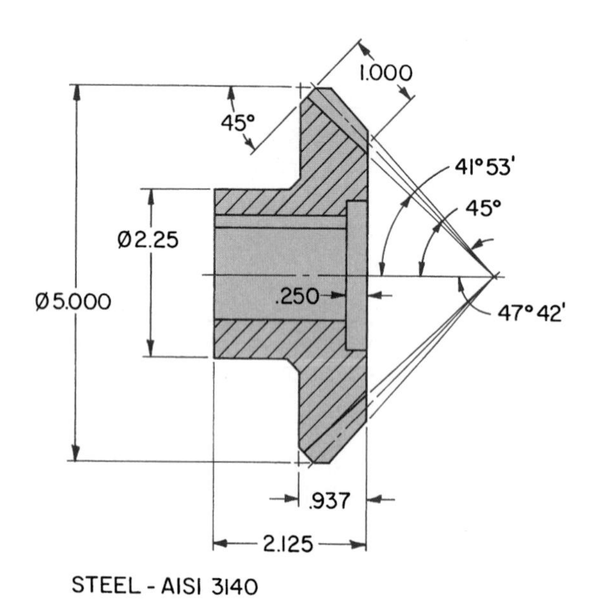

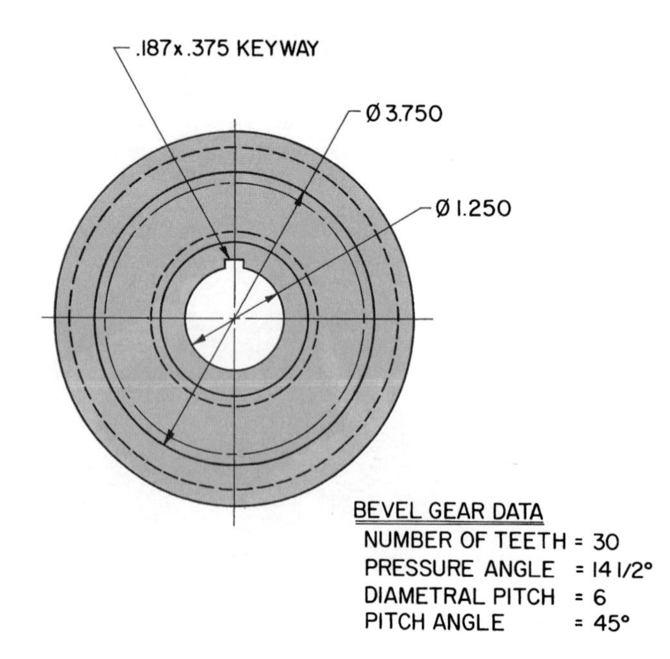

Goodheart-Willcox Publisher

Figure 19-59. Dimensions of bevel gear to be cut.

2. Tooth thickness and tooth space at the small end (t_s) needed.
   Having:

Tooth thickness and tooth space at the large end (t_L) = 0.2618″

The ratio $C_s / C_r = 0.717$

Formula:

$$t_s = t_L \times \frac{C_s}{C_r} = 0.2618'' \times 0.717 = 0.1877$$

$$= 0.1877''$$

19.6.2 Preparing to Cut a Bevel Gear

1. Mount the dividing head and tilt it to 45° 53′. See **Figure 19-60**.
2. Calculate the correct index plate to cut 30 teeth.
3. Secure the gear blank on an arbor.
4. Mount the gear blank and arbor in the dividing head with the large end of the gear toward the dividing head.

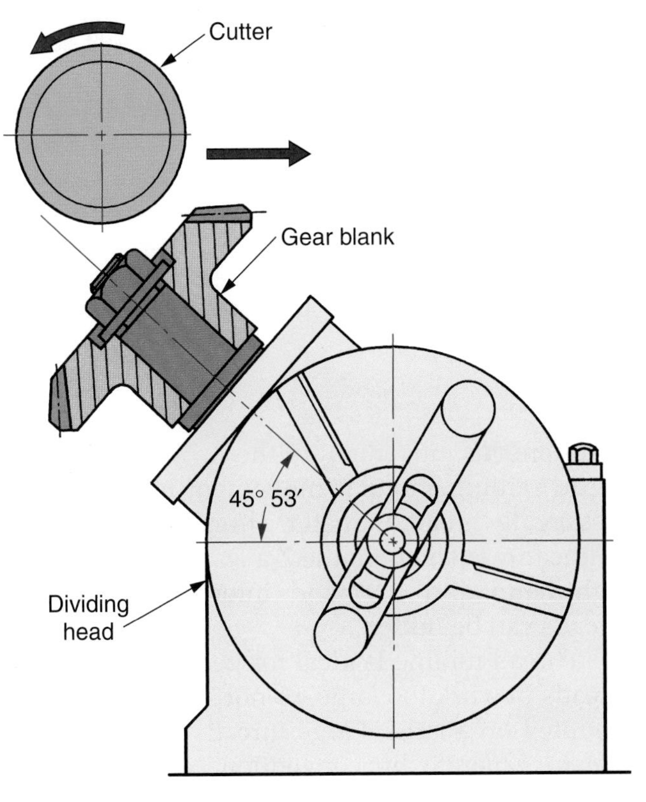

Figure 19-60. Tilt dividing head to the required angle (45° 53′) and mount the gear blank.

5. Select the proper bevel cutter and mount it on the arbor. Cutting is toward the dividing head. Position the arbor bearing as close to the cutter as possible, allowing for adequate clearance.
6. Center the cutter on the gear blank. Lock the cross slide. Set the graduated collar to zero.
7. Position the cutter for depth, using the paper strip technique. Set the knee graduated collar to zero.
8. Clear the cutter and raise the table to the whole tooth depth at the large end (0.3595″).
9. Set the machine for the proper cutting speed and feed. Many machinists prefer to make the cut in two runs before making the final cut.
10. Cut all teeth by plain indexing. Use adequate coolant.
11. The correct dimensions for the finished gear teeth are 0.2618″ and 0.1878″. To obtain these dimensions, an additional amount of material must be removed on each side of the teeth, as shown in **Figure 19-61**.
12. To remove this additional material, the rough finished blank must be rotated slightly (2° for this gear) and the table set over (0.044″). The following formulas were used to make the calculations.

Determining angle of roll:

$$A = \frac{57.3}{D}\left[\frac{p}{2} - \frac{C_r}{W}(T_L - T_s)\right]$$

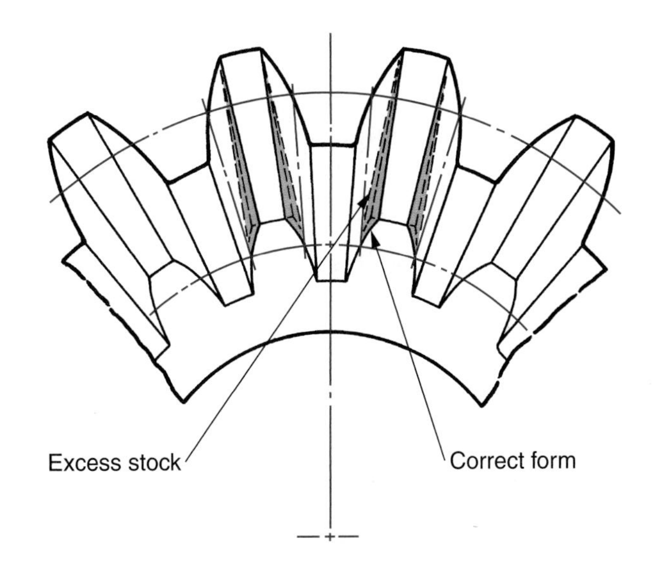

Figure 19-61. Excess material that must be removed when using the described technique to cut a bevel gear.

Where:

A = Angle of roll in degrees

D = Pitch diameter at large end of gear

p = Circular pitch at large end of gear

C_r = Pitch cone distance at large end of gear

T_s, T_L = Chordal thickness of gear cutter corresponding to pitch line at small and large ends of gear, respectively

57.3 = Degrees per radian

W = Width of gear tooth face

Determining table setover:

$$n = \frac{T_L}{2} - T_L - \frac{T_s}{2} \times \frac{C_r}{W}$$

Where:

n = Table setover

The direction of roll and setover must always be made in opposite directions. See **Figure 19-62**. Remove all backlash before making work movements.

13. Finish machining the gear. Remove all burrs. Measure tooth thickness at small and large ends. It may be necessary to remove a slight amount of material at the small end to get proper meshing. Refer to **Figure 19-61**.

This method of making bevel gears does not produce a tooth form that is accurate throughout the length of the tooth face. This is especially true at the small end of the gear even though the tooth form is correct at the large end. Remove this small amount of excess material by rotating the blank through a small angle with the dividing head and taking light cuts until proper meshing is attained.

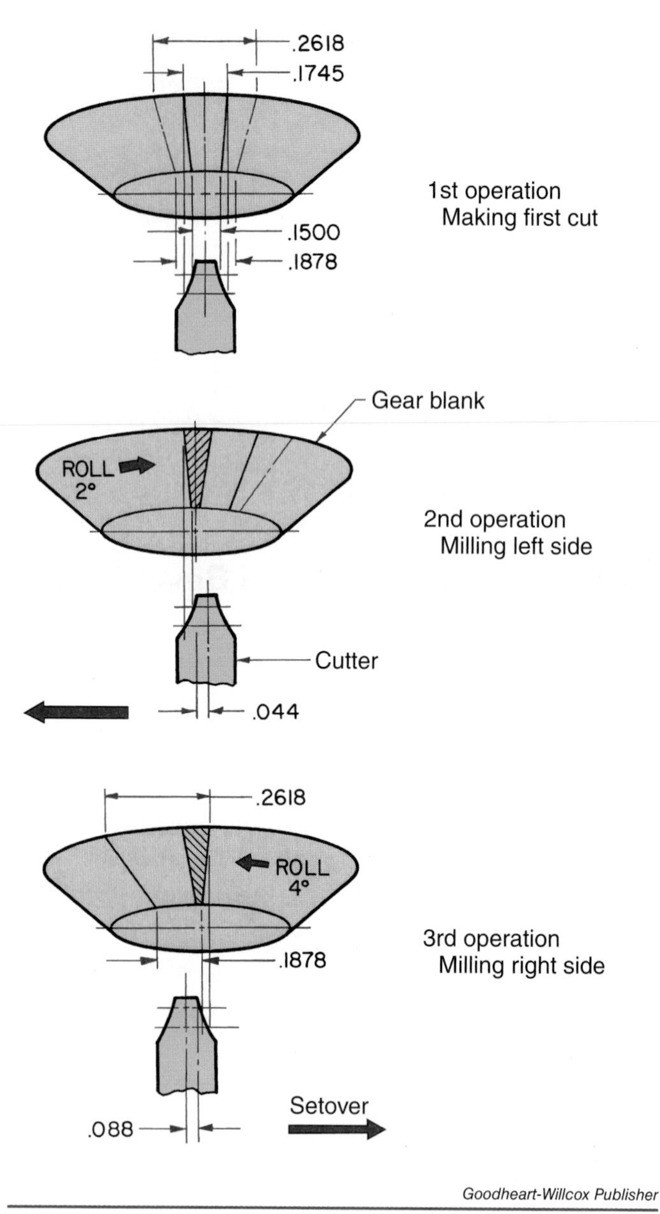

Figure 19-62. Sequence to be followed when milling the teeth of a bevel gear.

19.7 Thread Milling

For years, threads have been milled on the lathe and special thread milling machines. The technique was used to produce coarse pitch threads that required the removal of relatively large quantities of metal.

Today, CNC machining centers with helical interpolation can mill high-quality threads in holes ranging from 3/8″ (9.5 mm) diameter to diameters of almost any size. Helical interpolation creates tool movement in a helical path and requires simultaneous movement in the X, Y, and Z axes. In thread milling, circular movement in the X and Y axes create thread diameter. Thread pitch is created by linear movement in the Z axes. See **Figure 19-63**. Right-hand and left-hand internal and external threads can be milled.

Thread milling is ideal for generating precision threads in work too large, or not symmetrical, to be mounted on a lathe. Large threads are very expensive to generate by conventional methods. Since most threads can be milled in a single pass of the cutting tool, the process is a cost saving solution by greatly reducing production time.

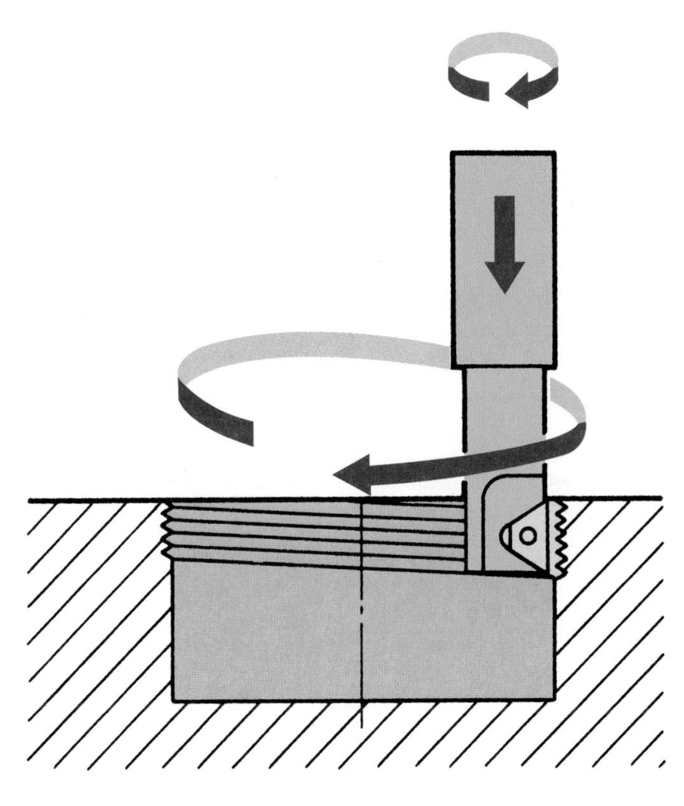

Goodheart-Willcox Publisher

Figure 19-63. In thread milling, helical interpolation creates tool movement in a helical path. This involves simultaneous movement on the X, Y, and Z axes. Thread diameter is created by the circular movement in the X and Y axes. The simultaneous linear movement in the Z axis creates the pitch of the thread.

Thread milling cutters are available in a wide range of sizes. See **Figure 19-64**. Most cutters have carbide inserts. The type of tool used is determined by thread size and depth, class of fit, type of material machined, and its relative hardness. Internal threads and external threads can be milled using the same tool.

19.8 Milling Machine Safety

- Avoid performing any machining operation on the milling machine until you are thoroughly familiar with how it should be done.
- Some materials that are machined can produce chips, dust, and fumes that are dangerous to your health. Never machine materials that contain asbestos, fiberglass, beryllium, or beryllium copper unless you are fully aware of the precautions that must be taken.

Greenfield Industries

Figure 19-64. Typical thread milling cutters. Pitch of cutter teeth must match thread pitch.

- Make sure there is adequate ventilation when performing jobs where dust and fumes are a hazard.
- If the area where you work is extremely noisy, wear hearing protectors. Take no chances. Protect your hearing and sight at all times in the shop.
- Maintain cutting fluids properly. Discard them when they become rancid or contaminated. Never pour used coolants or solvents down the drain.
- Carefully read instructions when using synthetic oils, solvents, and adhesives. Many are dangerous if not handled correctly. Return all oils and solvents to proper storage. Always wipe up spills.
- Never start a cut until you are sure there is adequate clearance on all moving parts!
- Be sure the cutter rotates in the proper direction. Expensive cutters can be quickly ruined.
- Carefully store milling cutters, arbors, collets, adapters, and other tools after use. They can be damaged if not stored properly.
- Exercise care when handling long pieces of metal. Accidentally contacting a light fixture or busbar can cause severe electrical burns and even electrocution!

19.9 Industrial Applications

The milling machines found in industry operate on the same basic principles as those found in training programs. In many cases, the same equipment is utilized, **Figure 19-65.**

There are many types of milling machines in use. They range in size from small tabletop models to machines capable of handling work that weighs many tons. Most milling machines in use have CNC capability.

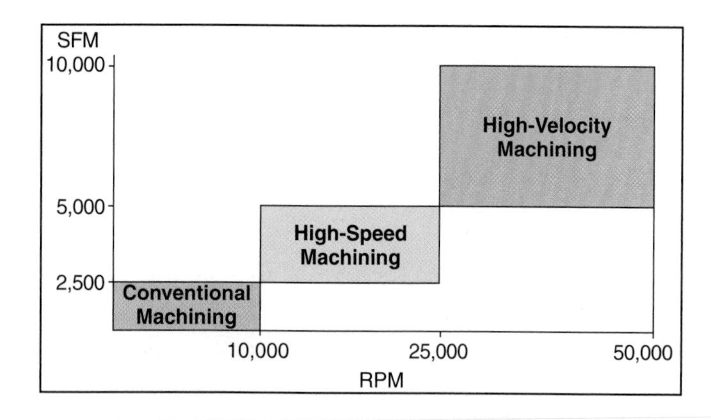

Goodheart-Willcox Publisher

Figure 19-66. This graph illustrates how the three machining techniques differ. The high-velocity machining range is expected to increase as new equipment is developed.

19.10 High-Velocity Machining

Often referred to as high-speed machining, the term high-velocity machining is preferred because it implies more than high spindle speeds. See **Figure 19-66.**

In addition to high spindle speeds (25,000 rpm and higher), high-velocity machining also involves high feed rates (600 ipm and up), high rapid traverse rates to reduce noncutting time, high acceleration and deceleration rates, and fast tool changes. To meet these requirements, special bearings, motors, highly responsive servo systems, spindle cooling techniques, and rapid chip removal methods had to be developed. Cutting tools and tool holders must also be carefully balanced. CNC programs for high-velocity machining also require large amounts of computer data.

High-velocity machining is used primarily by the aerospace industry. The one-piece aluminum structural components used in aerospace craft are "hogged," carved from large sections of metal. See **Figure 19-67.** Most weigh less than 5% as much as

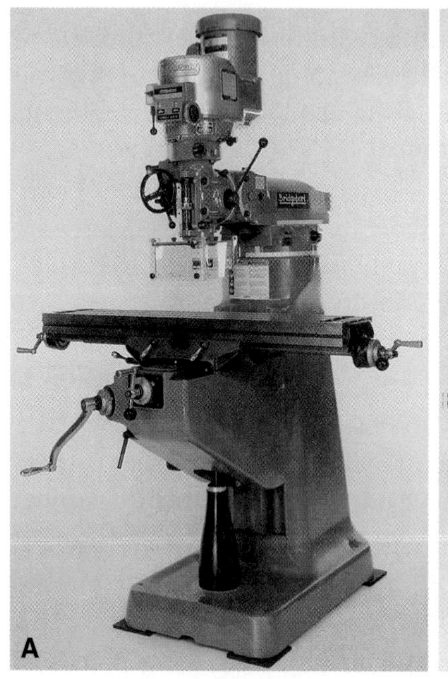

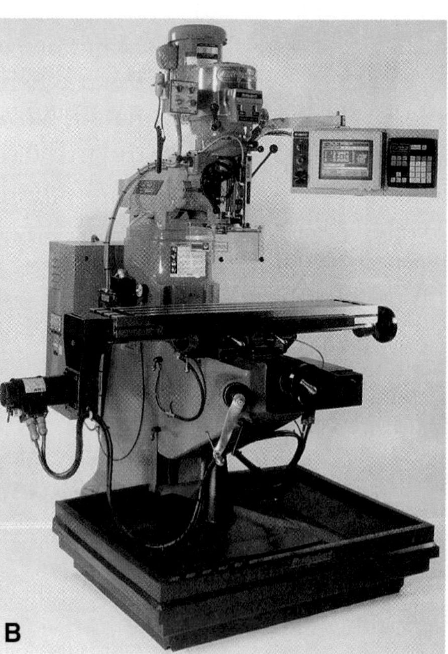

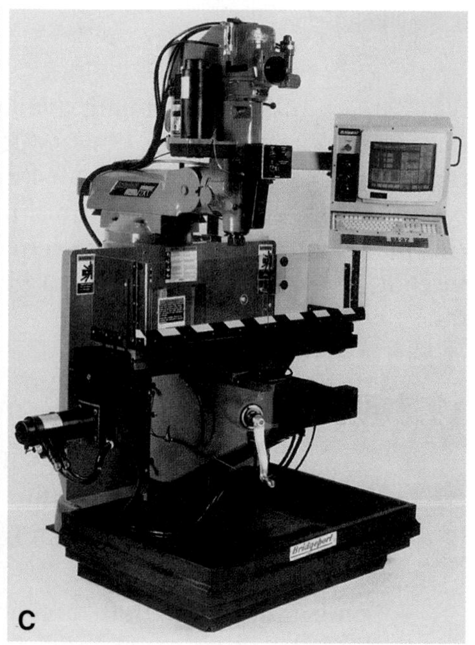

Bridgeport Machines, Inc.

Figure 19-65. Three types of vertical milling machines found in both industry and training programs. A—Manually operated vertical milling machine. B—CNC 2-axis vertical milling machine. C—CNC 3-axis vertical milling machine.

MAG IAS, LLC

Figure 19-67. Typical aircraft structural element carved from a solid aluminum section with a 5-axis horizontal profiler. It is much stronger than a similar part fabricated from aluminum sheet. Many of these sections weigh less than 5% as much as the aluminum plate from which they are cut.

the aluminum plate from which they are machined. These parts are stronger, lighter, and easier to join to other sections. They are less expensive to produce than similar parts fabricated from many smaller sheet metal pieces by conventional techniques. Contoured surfaces can be finished so smoothly that little or no additional finishing work is needed. Dies and other expensive tooling are eliminated. The safety factor is also increased because no riveting is required—each rivet hole is a source of potential structural weakness.

Chapter Review

Summary

- Squaring, machining flat and angular surfaces, machining keyseats and slots, and face and side milling are all common milling operations.
- Vertical milling machines are capable of milling, drilling, boring, and reaming operations.
- Horizontal milling machines are capable of milling, slitting, slotting, drilling, and boring operations.
- Cutting spur and bevel gears requires knowledge of gear nomenclature, gear cutters, and specific calculations.
- Chips, dust, fumes, cutting fluids, and noise are all potential hazards when milling.
- Many milling machines found in industry are similar to training machines.
- Most milling machines in use have CNC capabilities.

Review Questions

Answer the following questions using the information provided in this chapter.

1. A(n) _____ scale on the spindle head of a vertical milling machine ensures accurate angular settings.
2. _____ mills are the cutters normally used on a vertical milling machine.
   A. Face
   B. Spur
   C. End
   D. Both A and B.
   E. Both A and C.
3. The most accurate way to align a vise on a milling machine is with a(n) _____.
4. List three methods for machining chamfers, bevels, and tapered sections on a vertical milling machine.
5. Blind keyseats or slots are made with a(n) _____ end mill.
6. Explain how to center an end mill on round stock for the purpose of machining a keyseat using the paper strip technique.
7. A(n) _____ end mill is used when the cutter must be fed into the work like a drill.
8. A(n) _____ can be used to locate the first hole of a series to be drilled on a vertical milling machine.
   A. edge finder
   B. wiggler
   C. centering scope
   D. All of the above.
   E. None of the above.
9. In general, use the _____ arbor possible that permits adequate clearance between the arbor support and the work.
10. Describe how to safely mount a milling cutter on an arbor.
11. Why should a milling cutter be keyed to the arbor?
12. Face milling cutters over 6″ (150 mm) in diameter are usually of the _____ type.
13. What does "gang milling" mean?
    A. Several cutters being used at the same time to machine a job.
    B. Two or more cutters straddling the job.
    C. Several side cutters being used at the same time to machine a job.
    D. All of the above.
    E. None of the above.
14. When sawing (slitting) thin stock, the _____ diameter cutter that provides adequate clearance should be used.
15. What is a spur gear?
16. How does a rack differ from a spur gear?
17. How would a dividing head be set up to cut a 100-tooth gear? The dividing head has a 40:1 ratio and the index plate has the following series of holes: 33, 37, 41, 45, 49, 53, 57.
    Number of full turns. _____
    Hole series used. _____
    Number of holes in sector arm spacing. _____
18. List five precautions to be observed when operating a milling machine.

CHAPTER 20

Precision Grinding

Learning Objectives

After studying this chapter, you will be able to:

- Explain how precision grinders operate.
- Identify the various types of precision grinding machines.
- Select, dress, and true grinding wheels.
- Safely operate a surface grinder using various work-holding devices.
- Solve common surface grinding problems.
- List safety rules related to precision grinding.
- Identify other types of precision grinding operations.

Technical Terms

centerless grinding
creep grinding
dwell
electrolytic grinding
form grinding
internal grinding
platens
plunge grinding
precision grinding
tooth rest
traverse grinding
universal tool and cutter grinder

Grinding, like milling, drilling, sawing, and turning, is a cutting operation. However, instead of using one, two, or several cutting edges, grinding makes use of an abrasive tool composed of thousands of cutting edges. See **Figure 20-1**. Since each of the abrasive particles is actually a separate cutting edge, the grinding wheel might be compared to a many-toothed milling cutter.

In ***precision grinding***, each abrasive grain removes a relatively small amount of material, permitting a smooth, accurate surface to be generated. It is also one of the few machining operations that can produce a smooth, accurate surface on material regardless of its hardness. Grinding is frequently used as a finishing operation.

20.1 Types of Surface Grinders

While all grinding operations might be called surface grinding because all grinding is done on the surface of the material, industry classifies surface grinding as the grinding of flat surfaces. There are two basic types of surface grinding machines:

- Planer-type surface grinders make use of a reciprocating motion to move the worktable back-and-forth under the grinding wheel. Three variations of planer-type surface grinding are illustrated in **Figure 20-2**.
- Rotary-type surface grinders have circular worktables that revolve under the rotating grinding wheel, **Figure 20-3**. Two variations of the technique are shown in **Figure 20-4**.

Goodheart-Willcox Publisher

Figure 20-1. A grinding wheel removes material in the same manner as a milling cutter, but the chips of metal removed are much smaller.

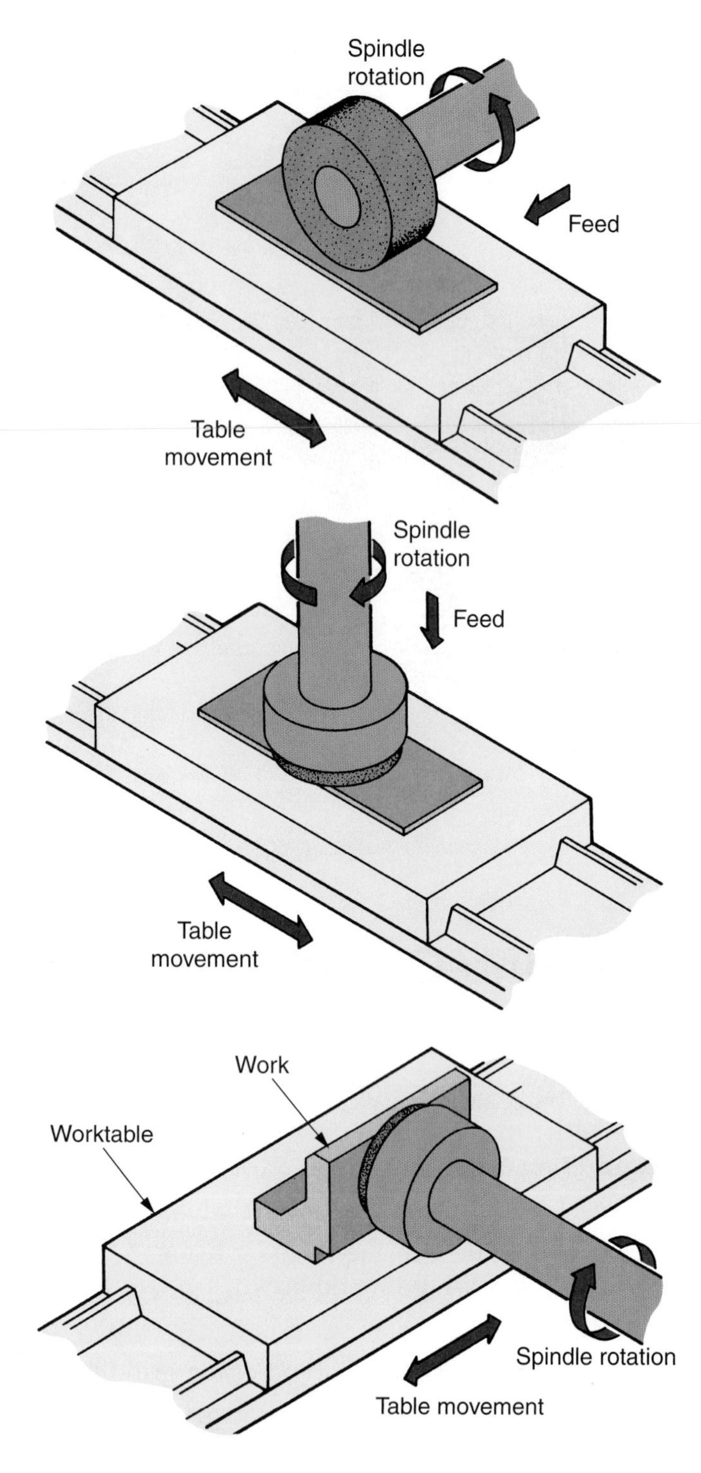

Goodheart-Willcox Publisher

Figure 20-2. Three variations of the planer-type surface grinder.

The planer-type surface grinder is frequently found in training situations. It slides the work back-and-forth under the edge of the grinding wheel. Table movement can be controlled manually or by means of a mechanical or hydraulic drive mechanism.

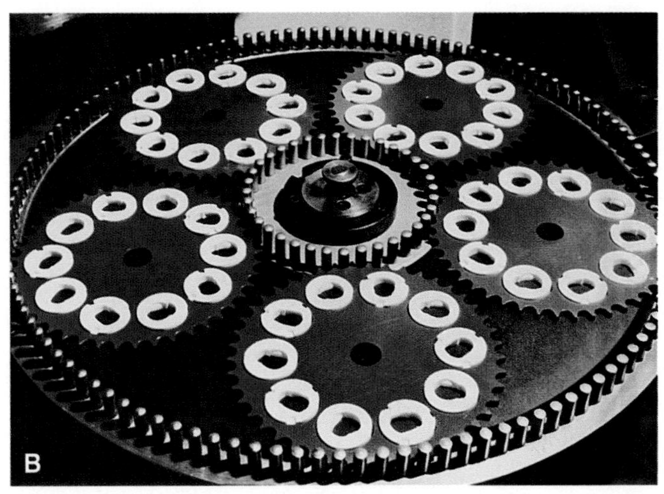

Peter Wolters of America, Inc.

Figure 20-3. Rotary surface grinder. A—This rotary fine grinding system can handle work from 5/32″ to 3 5/8″ (0.4 mm to 90 mm) thick and 13/32″ to 13 5/8″ (10 mm to 340 mm) in length. It is a rapid, safe, clean, and economical way to finish material to close tolerances. B—Ceramic pieces in place for grinding to required thickness. Not only does the grinding head rotate, but the work-holders also rotate to provide a superior finish.

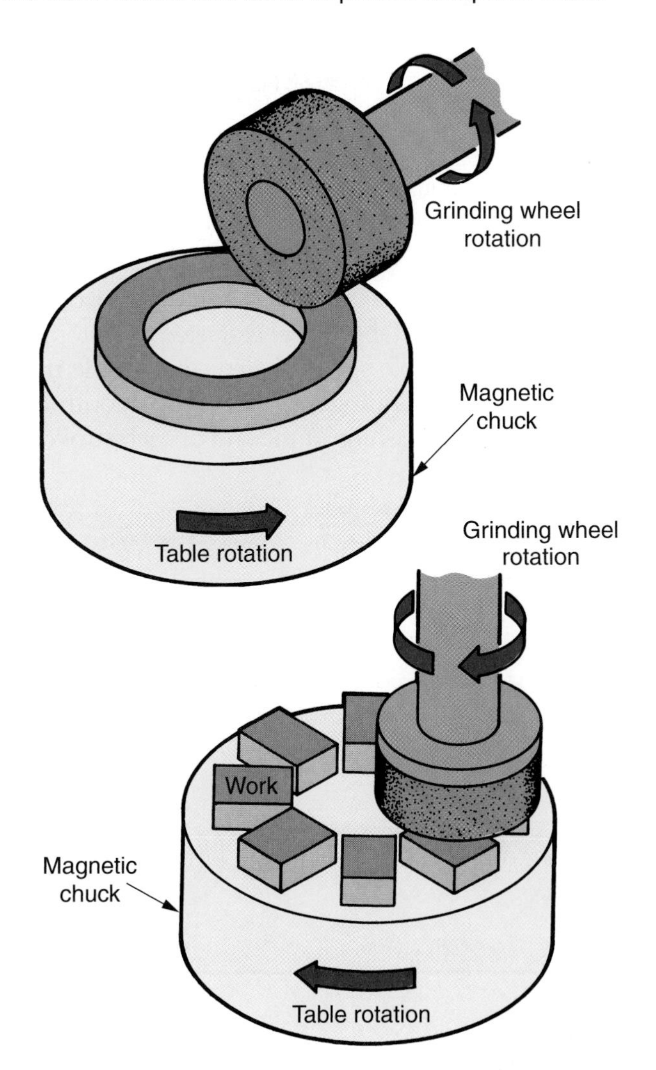

Goodheart-Willcox Publisher

Figure 20-4. Two variations of the rotary-type surface grinder.

A manually operated machine is shown in **Figure 20-5A**. All work and grinding wheel movements are made by hand. The large traverse handwheel controls the left-and-right movement of the table. Cross-feed (in-and-out motion) is controlled by the smaller cross-feed handwheel. The down-feed handwheel controls the up-and-down adjustment of the grinding wheel. This handwheel is located on the top of the vertical column.

A variation of this type of surface grinder can be run manually or automatically, **Figure 20-5B**. To prepare the machine for automatic operation, the operator simply fills in the blanks when requested by the menu prompts. No computer or CNC experience is needed. As the operator creates the part manually, pressing a button after each move, the machine "memorizes" how the part is made. The machine can run subsequent parts automatically.

Both manual and automatic machines operate in much the same manner. However, the person using a manually operated machine must develop a rhythm to get a smooth, even cutting stroke. Spring stops act as cushions at the end of maximum table travel, **Figure 20-6**.

Adjustable table stops on the hydraulically activated traverse feed permit the operator to establish exact table positioning, **Figure 20-7**. At the end of the stroke, table direction is reversed automatically. Automatic cross-feed moves the work in or out a predetermined distance at the completion of each cutting cycle.

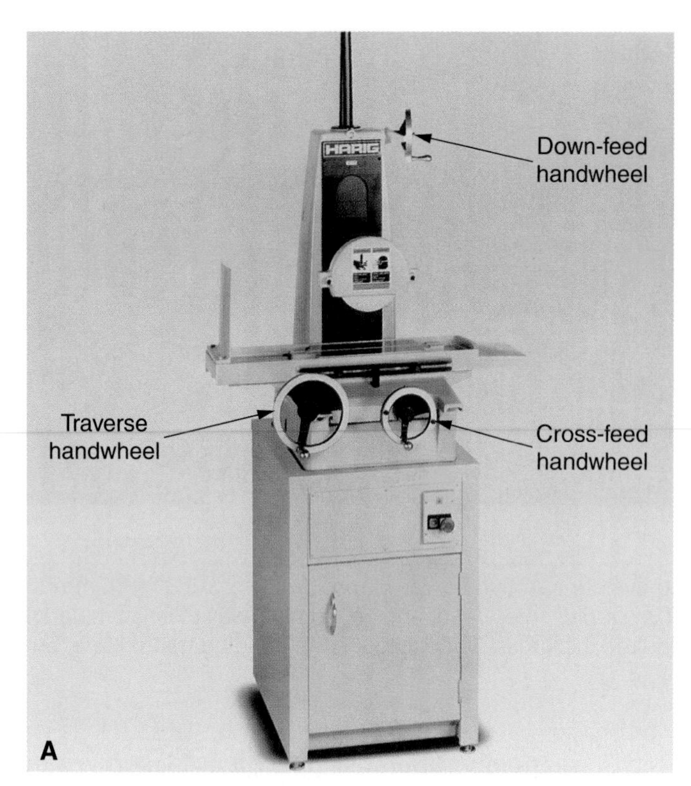

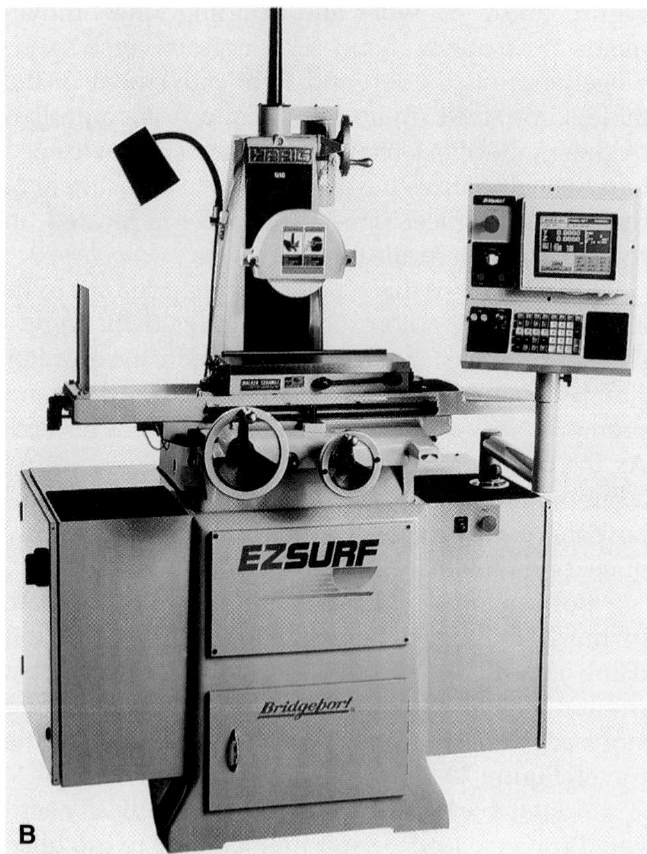

Harig Div. of Bridgeport Machines Inc.

Figure 20-5. Planer-type surface grinders. A—A manually-operated surface grinder. B—A surface grinder that operates either manually or automatically. It permits an operator in training to step up to automatic grinding when ready to do so.

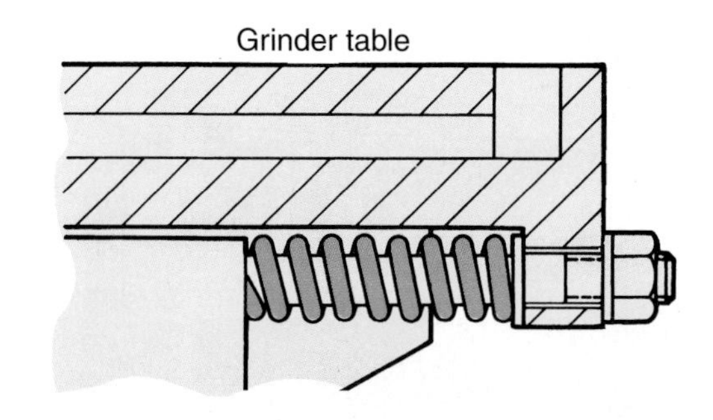

Goodheart-Willcox Publisher

Figure 20-6. Springs on worktable guide are often used to cushion the end of stroke on manually operated surface grinders.

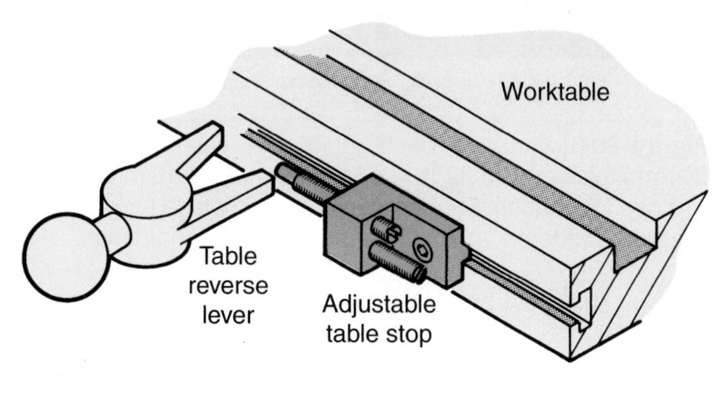

Goodheart-Willcox Publisher

Figure 20-7. An adjustable table stop is used to regulate length of the worktable stroke.

A control console, **Figure 20-8**, is located on the front of the machine. Table travel is started and stopped from this station. Table speed is also controllable from this location. Some grinding machines have a control for ***dwell***—a hydraulic cushion at the end of each stroke.

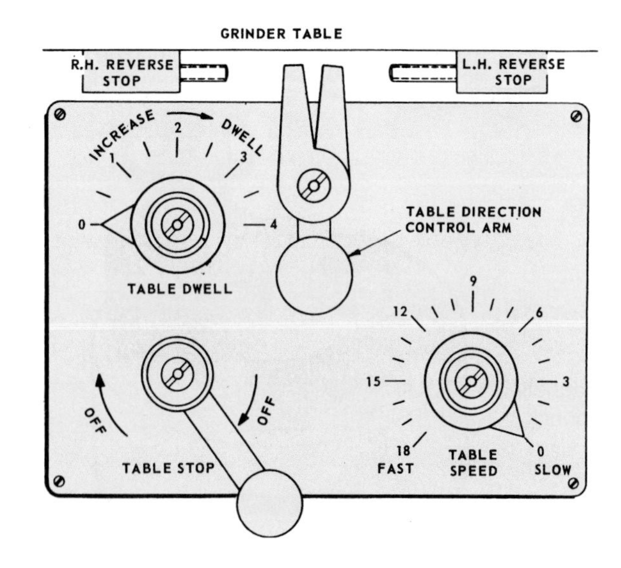

Goodheart-Willcox Publisher

Figure 20-8. Typical surface grinder control console. Table dwell sets up a cushioning action at the end of each table stroke.

20.2 Work-Holding Devices

Much of the work done on a surface grinder is held in position by a magnetic chuck. This holds the work by exerting a magnetic force. Nonmagnetic materials (such as aluminum and brass) can be ground by bracing with steel blocks or parallels to prevent movement.

An electromagnetic chuck uses an electric current to create a strong magnetic field. Another type of magnetic chuck makes use of a permanent magnet. This eliminates cords needed for electromagnets and the danger of the work flying off the chuck if the electrical connection is accidentally broken. See **Figure 20-9**.

Frequently, work mounted on a magnetic chuck becomes magnetized and must be demagnetized before it can be used. A demagnetizer can be used to neutralize the magnetic field.

Other ways to mount work on a surface grinder are:

- A universal vise, **Figure 20-10A**.
- An indexing head with centers, **Figure 20-10B**.
- Clamps to hold the work directly on worktable.
- A precision vise.
- Double-faced masking tape can be used to hold thin sections of nonmagnetic materials. Refer to **Figure 20-11**.

A

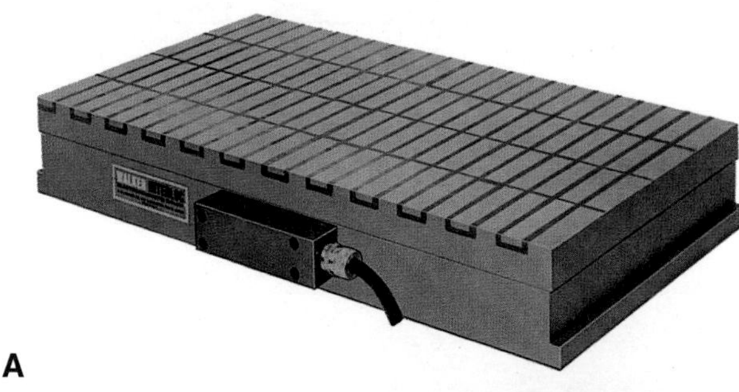

B

O.S. Walker Co., Inc.

Figure 20-9. Magnetic chucks. A—An electromagnetic chuck makes use of an electric current to create a magnetic field. B—Magnetic chuck with a permanent magnet.

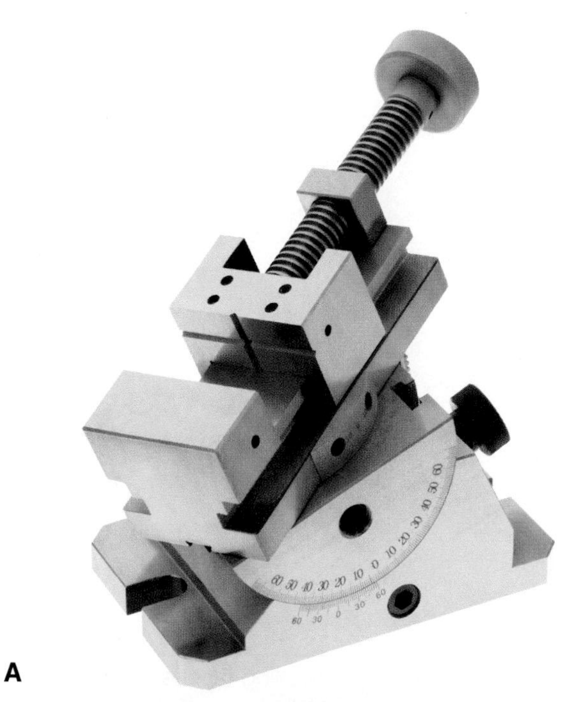

A

B

Photo Courtesy of Grizzly Industrial, Inc. www.grizzly.com; Courtesy of Vermont Machine Tool and the Bryant Grinder Division

Figure 20-10. Work-holding devices. A—A universal vise can be used for grinding operations. B—Centers and an indexing head are used for grinding tasks when shape of work permits. An indexing head is used in much the same manner as the dividing head in milling.

20.3 Grinding Wheels

As mentioned at the opening of this chapter, each abrasive particle in a grinding wheel is a cutting tooth. As the wheel cuts, metal chips dull the abrasive grains and wear away the bonding material (medium that holds abrasive particles together). The ideal grinding wheel, of course, would be one in which the bonding material wears away slowly enough to get maximum use from the individual abrasive grains.

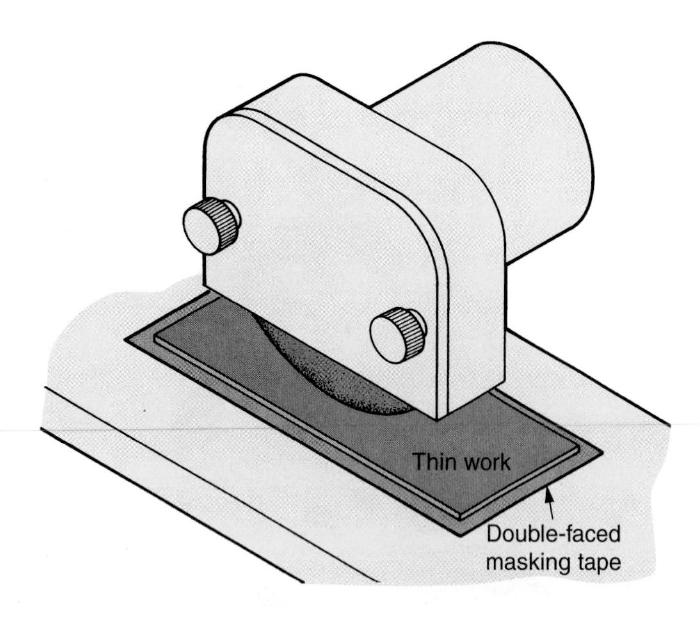

Goodheart-Willcox Publisher

Figure 20-11. Double-faced masking tape is used to mount thin nonferrous material for grinding.

However, it would also wear rapidly enough to permit dulled abrasive particles to drop off and expose new particles.

Because so many factors affect grinding wheel efficiency, the wheel eventually dulls and must be dressed with a diamond dressing tool. See **Figure 20-12**. Failure to dress the wheel of a precision grinding machine will, in time, result in the wheel face becoming loaded or glazed and unable to cut freely, **Figure 20-13**.

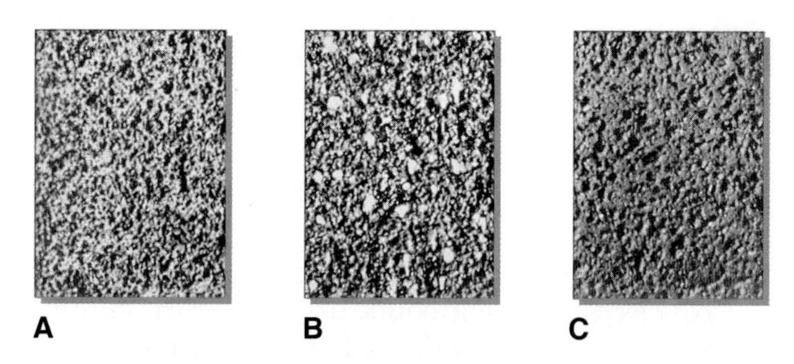

Norton Co.

Figure 20-13. Grinding wheels in various conditions: A—Properly dressed. B—Loaded. C—Glazed.

Only manufactured abrasives are suitable for high-speed grinding wheels. The properties and spacing of abrasive particles and composition of the bonding medium can be controlled to get the desired grinding performance, **Figure 20-14**.

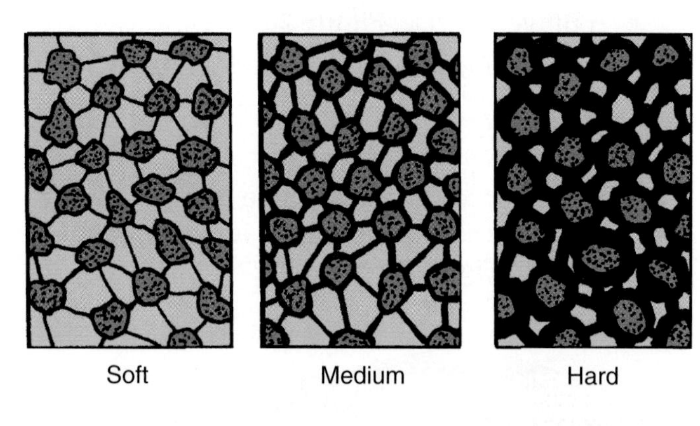

Goodheart-Willcox Publisher

Figure 20-14. Wheel hardness is determined by the type and percentage of bond and grain spacing.

CITCO Div., Western Atlas, Inc.

Figure 20-12. Diamond wheel dressing tools. A—Close-up of a natural diamond chip on a grinding wheel dresser. The diamond should be rotated a partial turn each time it is used. This will put a new edge of the diamond into position. B—Dressing tools manufactured from manmade diamonds eliminate the irregularities of natural diamonds, resulting in consistent diamond exposure and longer wear.

20.3.1 Grinding Wheel Marking System

To aid in achieving consistent grinding performance, a standard system of marking grinding wheels has been defined by American National Standards Institute (ANSI). Called ANSI Standard B74.13, it is used by all grinding wheel manufacturers. Five factors were considered:

- Abrasive-type classifies the abrasive material in the grinding wheel. Manufactured abrasives fall into two main groups identified by letter symbols:
    - A = aluminum oxide
    - C = silicon carbide

  An optional prefix number may be used to designate a particular type of aluminum oxide or silicon carbide abrasive.
- Grain size is indicated by a number, usually from 8 (coarse) to 600 (very fine).
- Grade is the strength of the bond holding the wheel together, ranging from A (soft) to Z (hard).
- Structure refers to grain spacing or the manner in which the abrasive grains are distributed throughout the wheel. It is numbered 1 to 16. The higher the number, the more "open" the structure (wider the grain spacing). The use of this number is optional.
- Bond indicates the type of material that holds the abrasive grains (wheel) together. Eight types are used:
    - B = Resinoid
    - BF = Resinoid reinforced
    - E = Shellac
    - O = Oxychloride
    - R = Rubber
    - RF = Rubber reinforced
    - S = Silicate
    - V = Vitrified

An additional number or one or more letters may be used as the manufacturer's private marking to identify the grinding wheel. Its application is optional.

The adoption of a standardized grinding wheel marking system has guaranteed, to a reasonable degree, duplication of grinding performance. The wheel marking system is shown in **Figure 20-15**.

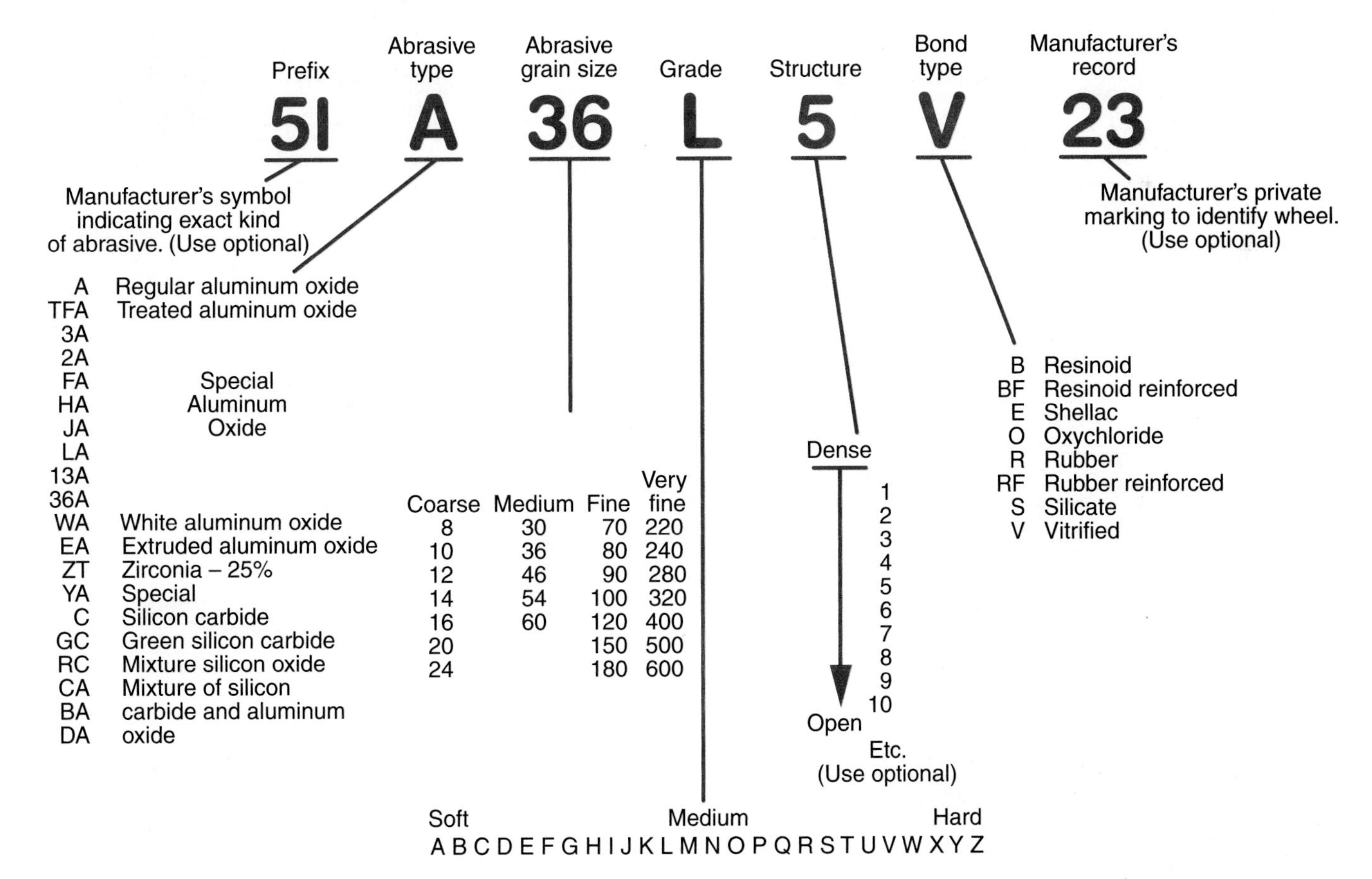

Figure 20-15. Standard system for marking grinding wheels.

20.3.2 Grinding Wheel Shapes

Grinding wheels are made in many standard shapes, **Figure 20-16**. While twelve basic face shapes are generally available, the face may be changed to suit specific job requirements, **Figure 20-17**. Wheels used for internal grinding are manufactured in a large selection of shapes and sizes, **Figure 20-18**.

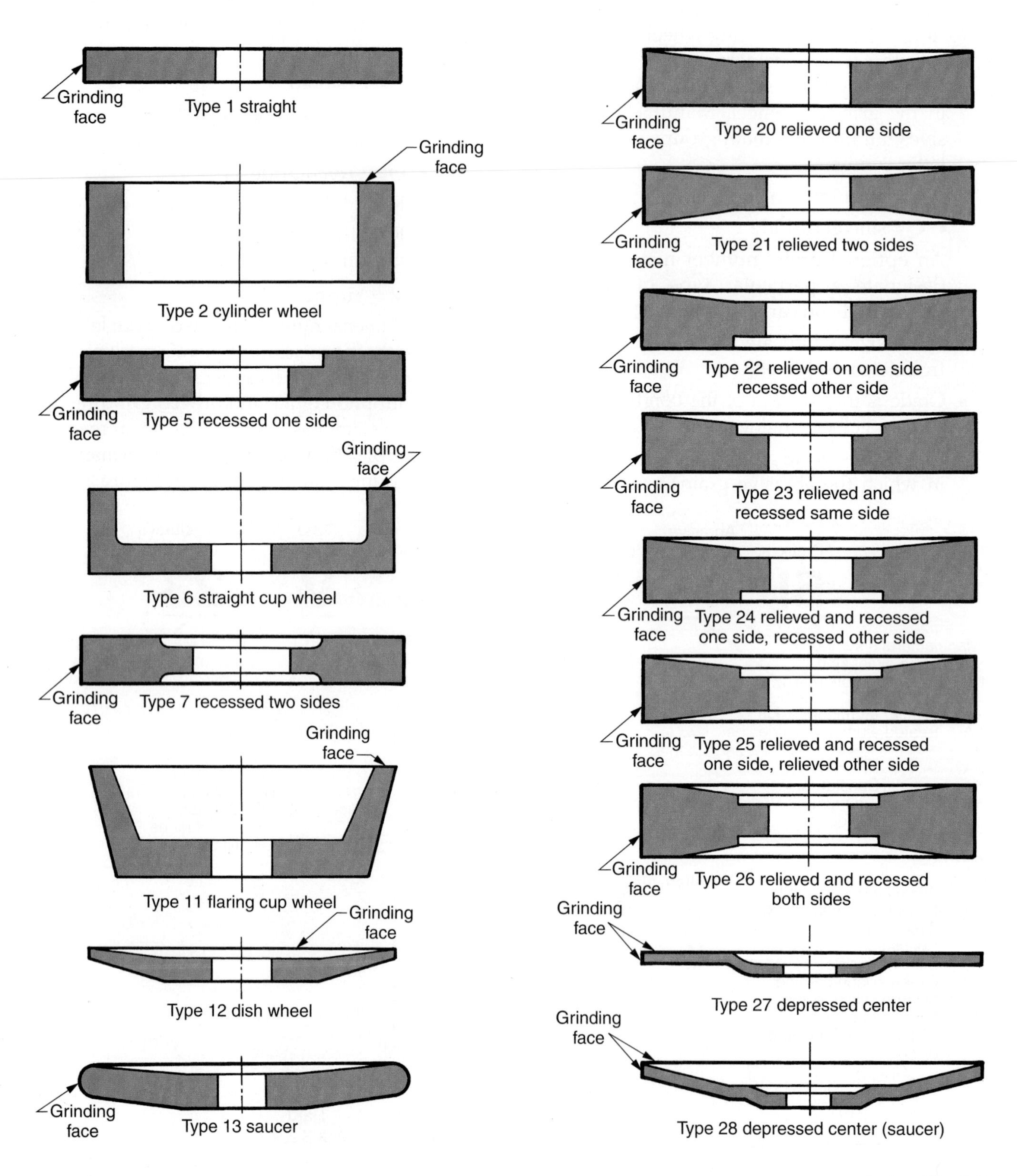

Figure 20-16. Standard grinding wheel shapes.

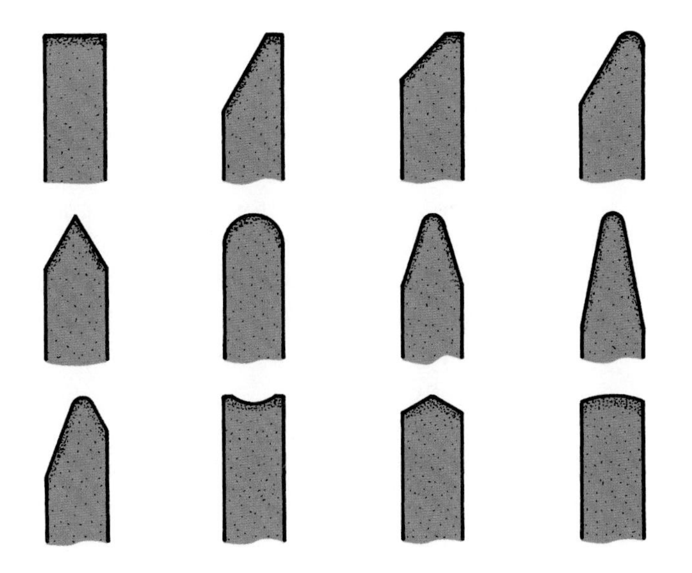

Goodheart-Willcox Publisher

Figure 20-17. The twelve basic face shapes that are generally available.

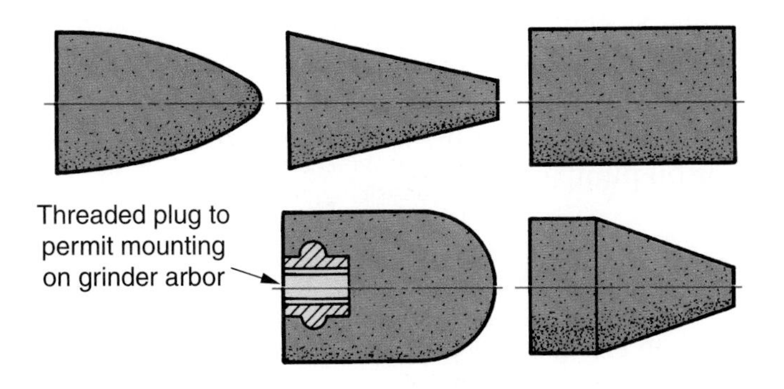

Goodheart-Willcox Publisher

Figure 20-18. A few of the many grinding wheel shapes available for internal grinding.

20.3.3 Mounting Grinding Wheels

Select a grinding wheel recommended for the job. Check its soundness by lightly tapping the wheel with a plastic or wooden screwdriver handle. A good wheel will produce a clear metallic ring. If the wheel is cracked, the tone will be flat or dull, rather than a clear ringing sound.

SAFETY NOTE

Always discard cracked grinding wheels. If possible, break them into several pieces to ensure they are not used.

Unbalanced wheels will cause irregularities on the finished ground surface. They should be statically balanced as shown in **Figure 20-19A**. On CNC grinders, automatic wheel balancing systems can lengthen the life of the wheel and provide improved surface finishes. Automatic systems like the one shown in **Figure 20-19B** use vibration sensors and

Revolution Tool Company; Marposs Corp.

Figure 20-19. Grinding wheel balancing. A—Static balance method. The wheel nut shown features a series of threaded holes on a bolt circle. By adding or removing setscrews of different lengths opposite the wheel's heavy and light sections, the wheel can be statically balanced. B—Automatic grinding wheel balancing system uses a flange- or spindle-type balancing head, vibration and ultrasound sensors, and a microprocessor-based controller to keep the wheel in balance.

ultrasound wheel contact sensors to monitor operation of the wheel. A microprocessor-based controller signals a flange or spindle-mounted balancing head to make necessary adjustments.

Mount the wheel on the spindle. It should fit snugly. Never force a grinding wheel on a shaft. The blotter rings or compressible washers should be large enough to extend beyond the wheel flanges, **Figure 19-20**. It is essential that the wheel be mounted properly. If it is not, excessive strains will develop during the grinding operation and the wheel could shatter.

SAFETY NOTE

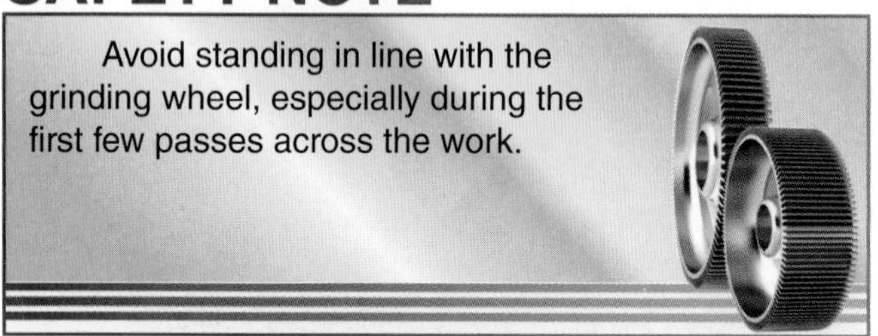

20.4 Cutting Fluids

Cutting fluids are an important factor in reducing wear on the grinding wheel. They help to maintain accurate dimensions, and are important to the quality of the surface finish produced. As a coolant, the cutting fluid must remove the heat generated during the grinding operation. Heat must be removed as fast as it is generated.

Several types of cutting fluids are used in grinding operations:

- Water-soluble chemical fluids are solutions that take advantage of the excellent cooling ability of water. They are usually transparent and include a rust inhibitor, water softeners, detergents to improve the cleaning ability of water, and bacteriostats (substances that regulate and control the growth of bacteria).
- Polymers are added to water to improve lubricating qualities.
- Water-soluble oil fluids are coolants that are usually "milky white," since they consist of a mixture of oil and water. They are also less expensive than most chemical-type fluids. Bacteriostats are added to control bacteria growth.

Coolant can be applied by flooding the grinding area, **Figure 20-21**. The fluid recirculates by means of a pump and holding tank built into the machine. A mist system forces the coolant over the wheel or applies it to the work surface under pressure (air). It cools by evaporation. A coolant can also be applied manually by pumping the fluid from a pressure-type oil pump can.

SAFETY NOTE

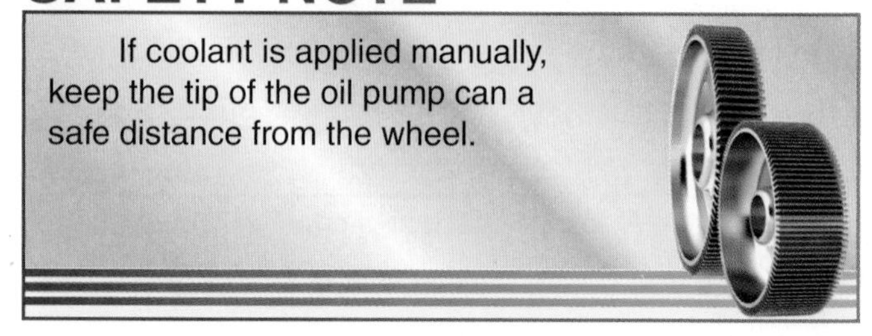

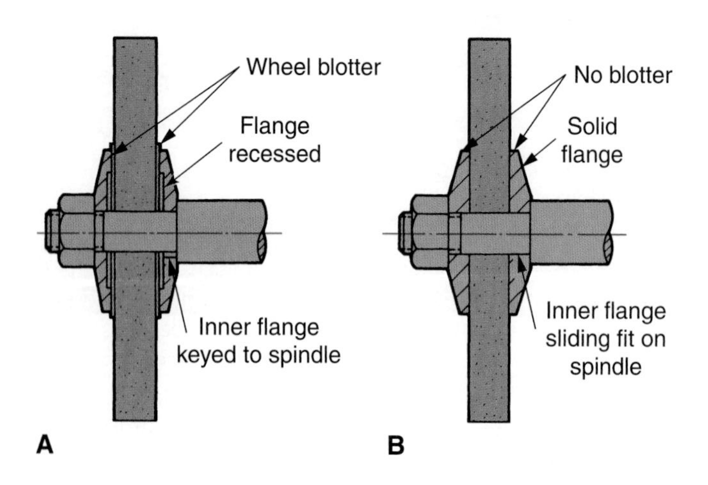

Goodheart-Willcox Publisher

Figure 20-20. Do not operate a grinder unless the wheel is properly mounted. A—Correctly mounted wheel. B—Wheel incorrectly and dangerously mounted.

Stana/Shutterstock.com

Figure 20-21. Coolant must flood area being ground.

For safety, long equipment life, and quality control, a coolant system should be cleaned at regular intervals. To clean the coolant system, remove all dirt and sludge from the holding tank. Chips and grinding wheel residue in the coolant can mar the ground surface of the workpiece. Discard the fluid when it becomes contaminated.

20.5 Grinding Applications

The following procedure is recommended to produce a surface that is flat or free of waviness:

1. Select and mount a suitable grinding wheel.
2. True and dress the wheel with a diamond dressing tool, **Figure 20-22**.
3. Mount the work-holding device. If a magnetic chuck is used, it should be "ground-in" to ensure a surface that is true and parallel to table travel. Grind off as little material as possible from the chuck in order to true the surface. Also, be sure to flood the surface with coolant during the procedure. For high-precision work, the magnetic chuck should be ground-in each time it is remounted on the machine.
4. Check the coolant system to be sure it is operating satisfactorily.
5. Position the work, and energize the chuck. If the work is already ground on one surface, protect it and the chuck surface by fitting a piece of oiled paper between the machined surface and the chuck before energizing the chuck, **Figure 20-23**.
6. Adjust the table stops, **Figure 20-24**.
7. Check the holding power of the magnetic chuck by trying to move the work.

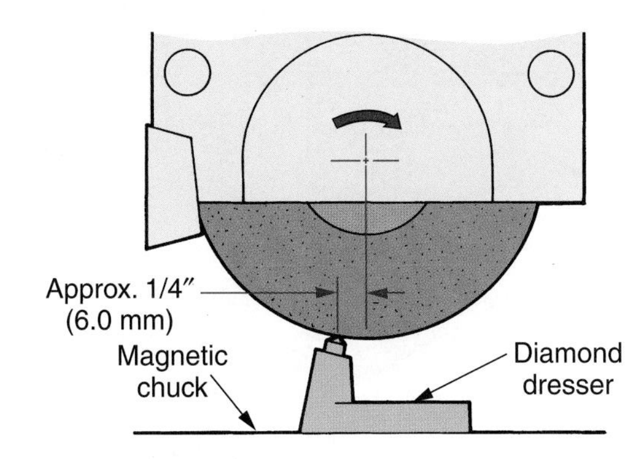

Goodheart-Willcox Publisher

Figure 20-22. For best results when cleaning or truing a grinding wheel, position the diamond wheel dresser as shown.

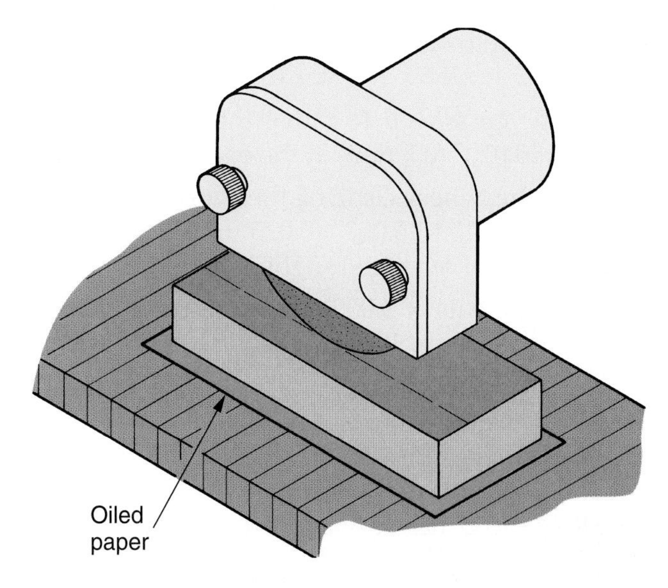

Goodheart-Willcox Publisher

Figure 20-23. A piece of oiled paper placed between the magnetic chuck and a newly ground surface will protect the finish of the work and surface of the chuck.

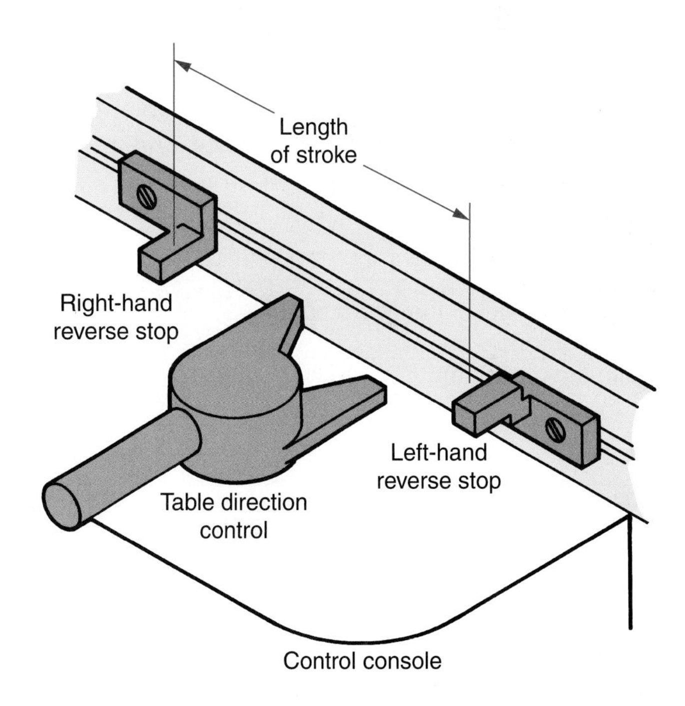

Goodheart-Willcox Publisher

Figure 20-24. Adjustable stops regulate the length of the table stroke. Care must be taken to ensure the stop adjustment permits the entire work surface to be ground.

8. Down-feed the grinding wheel until it just touches the highest point on the work surface. The grinding wheel can be set to the approximate position by down-feeding until it just touches a sheet of paper placed between the wheel and the work surface.

9. Turn on the coolant, spindle, and hydraulic pump motors.
10. Set the cross-feed to move the table in or out about 0.020″ (0.5 mm) at the end of each cycle.
11. With the wheel clear of the work, down-feed about 0.001″ to 0.003″ (0.025 mm to 0.075 mm) per pass for average roughing cuts.
12. Use light cuts of 0.0001″ (0.0025 mm) for finishing the surface. It is wise to redress the wheel for finishing cuts.

When the work surface has been ground to the required dimension and finish, use the following procedure to turn off the machine:

1. Move the grinding wheel clear of the work.
2. Turn off table travel.
3. Turn off coolant.
4. Let the grinding wheel run for a few moments after the coolant has been turned off. This will permit the wheel to free itself of all traces of fluid. Otherwise, the wheel can absorb some of the coolant and become out of balance.
5. Use a squeegee to remove excess coolant from the work. De-energize the chuck and remove the work.

SAFETY NOTE

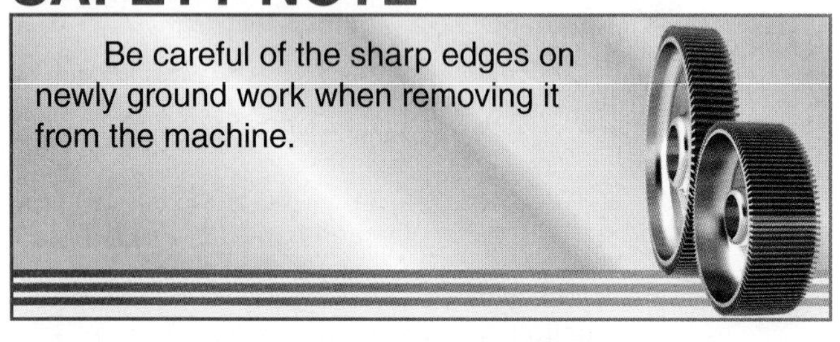

6. Clean the machine. Apply a light coating of oil on the chuck's work surface to prevent possible rusting.
7. Place all tools in proper storage.

20.5.1 Grinding Edges Square and Parallel with Face Sides

Most rectangular work requires the edges to be parallel to each other and square with the finished face sides. In one commonly used technique, **Figure 20-25**, edge 1 is ground while being held in a precision vise. After burrs are removed, the adjacent edge 2 is ground. Its squareness is checked vertically with a dial indicator.

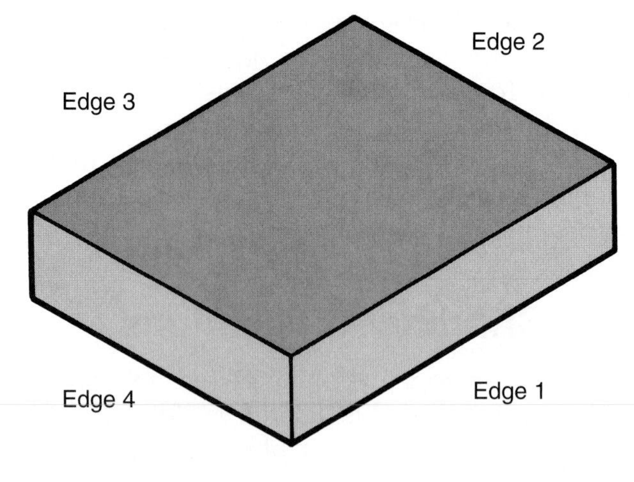

Goodheart-Willcox Publisher

Figure 20-25. The sequence for squaring edges of a rectangular workpiece after the faces have been ground flat and parallel.

After edges 1 and 2 are ground square with each other, they will serve as reference planes to grind edges 3 and 4 to required dimensions. Use oiled paper in the vise to protect the ground faces and edges from stray metal and abrasive particles. The vise must be carefully wiped clean and burrs removed after each edge is ground.

An angle plate can also be used when grinding edges square and parallel with the finished faces. See **Figure 20-26**. A parallel may be used to set the work in approximate position.

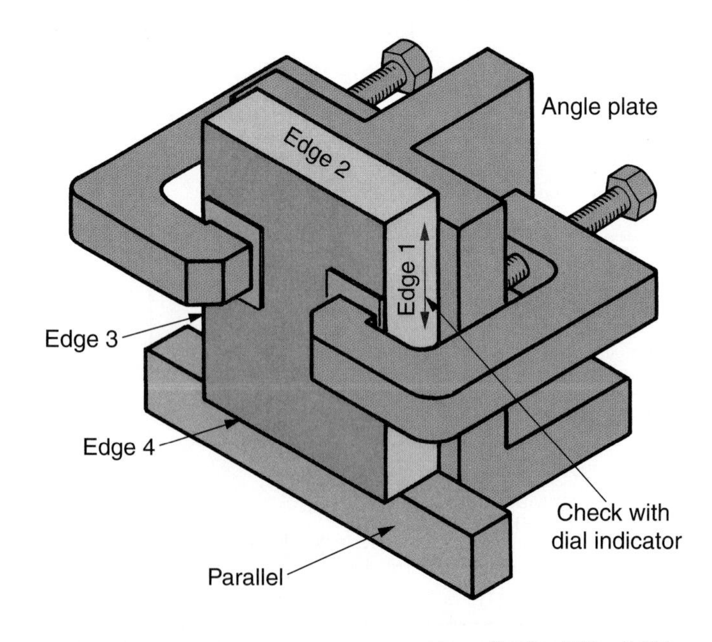

Goodheart-Willcox Publisher

Figure 20-26. An angle plate can also be used to hold work for grinding the edges square and parallel.

The same positioning and grinding sequences are followed as previously described.

20.5.2 Creep Grinding

Creep grinding is a production-type machining technique that makes a deep cut into the work. It is also sometimes known as deep grinding. Special grinding machines are required for this type of work.

Creep grinding is a surface grinding operation that is often performed in a single pass with an unusually large depth of cut, **Figure 20-27**.

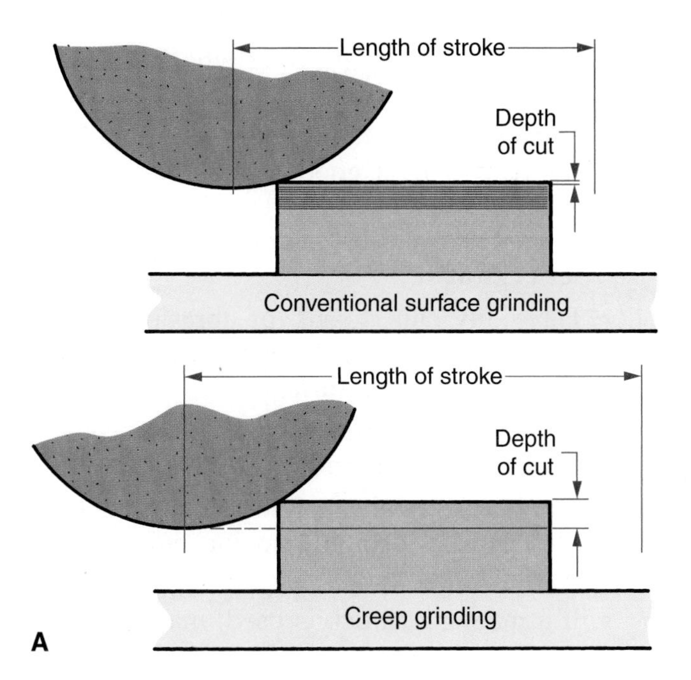

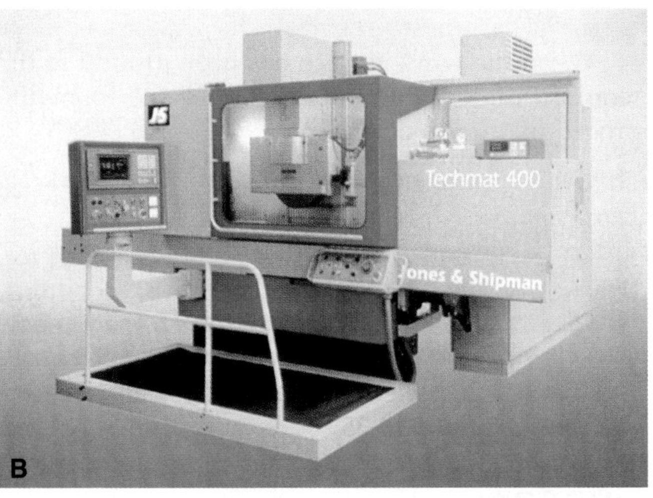

Goodheart-Willcox Publisher; Jones & Shipman, Inc.

Figure 20-27. Creep grinding. A—The difference between conventional and creep grinding. Creep grinding equipment must be specially designed for this heavy-duty work. B—A heavy-duty CNC creep grinding machine.

In comparison to conventional surface grinding, the depth of cut is increased 1000–10,000 times and the work speed is reduced in the same proportion. Machining time can be reduced by 50% to 80%. The tools (such as grinding wheels and work-holding devices) must be designed for this heavy-duty work.

20.6 Grinding Problems

There are many problems peculiar to precision surface grinding. The following paragraphs will address a few of the more common difficulties and provide suggestions for their solution.

Irregular table movement or no table movement (on hydraulic-type machines) can be caused by clogged hydraulic lines, insufficient hydraulic fluid, a hydraulic pump that is not functioning properly, or inadequate table lubrication. A cold hydraulic system can also cause these symptoms. Let the machine warm up for at least 15 minutes before use. Air in hydraulic lines can cause erratic table movement. Make corrections as recommended by the manufacturer of the machine.

Irregular scratches, of no identifiable pattern, are frequently caused by a dirty coolant system, or by particles becoming loosened in the wheel guard. This could also mean that the grinding wheel is too soft, and the abrasive particles are carried to the wheel by the coolant system.

Work surface waviness can be caused by a wheel being out of round. It can be corrected by truing the wheel.

Chatter or vibration marks can be caused by a glazed or loaded grinding wheel. There is a slipping action between the wheel and the work. The wheel cuts until the glazed section comes into position and slides over the work, rather than cutting. Correct the problem by redressing the grinding wheel. The same effect can also be caused by a grinding machine that is not mounted solidly, or by a wheel that is loose on the spindle. Check for these conditions and make corrections if necessary.

Burning or work surface checking may be the result of too little coolant reaching the work surface, a wheel that is too hard, or a wheel with grain that is too fine. Make needed corrections as indicated by inspection.

Wheel glazing or loading often indicates that the wrong coolant is being used. A dull diamond on the wheel dresser can also cause this problem.

Deep, irregular marks on the work surface may be caused by a loose grinding wheel.

Work that is not flat may be caused by insufficient coolant, a nicked or dirty chuck surface, or a wheel that is too hard. Check and make any necessary corrections.

Work that is not parallel is frequently caused by a chuck that has not been "ground in" since the last time it was mounted on the machine. A nicked or dirty chuck can also cause the same problem. Insufficient coolant can allow the work to heat up and expand in the center of the cut, permitting more material to be removed in that area than at each end. As the piece cools, the center will become depressed, **Figure 20-28**. Correct the problem by directing more fluid to the cutting area.

20.7 Grinding Safety

- Never attempt to operate a grinder until you have been instructed in its proper and safe operation. When in doubt, consult your instructor.
- Do not use a grinder unless all guards and safety devices are in place and securely attached.

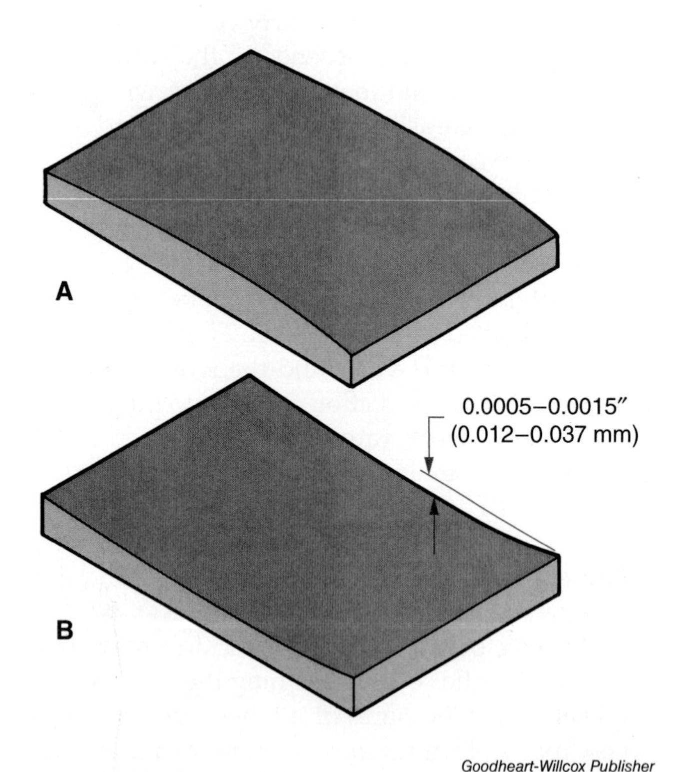

Figure 20-28. Effect of insufficient coolant. A—If not enough coolant is used, frictional heat can cause the work to expand in the center of cut, causing more material to be removed in the center than at each end. B—As the piece cools, its center becomes depressed.

- Never try to operate grinding machines while your senses are impaired by medication or other substances.
- Always wear proper eye protection when performing any grinding operation.
- Never place a wheel on a grinder before checking it for soundness. Destroy faulty wheels!
- Check the wheel often to prevent it from becoming glazed or loaded. Dress the wheel when required.
- Make sure the grinding wheel is clear of the work before starting the machine.
- Never operate a grinding wheel at a speed higher than specified by the manufacturer.
- Change coolant fluid before it becomes contaminated. It is good practice to set up a schedule for replacing the fluid at regular intervals or adding chemicals to control bacterial growth.
- Have any cuts, burns, or abrasions treated promptly—major infections can result from untended minor injuries.
- Immediately wipe up any spilled coolants from the floor around the machine.
- Stop the machine before making measurements or performing major machine and work adjustments.
- If a magnetic chuck is used, make sure it is holding the work solidly before starting to grind.
- Remove any jewelry, including watch and rings, before using a magnetic chuck to prevent them from becoming magnetized or being pulled toward the machine.
- If automatic feed is used, run the work through one cycle by hand to be sure there is adequate clearance and that the dogs are adjusted properly.
- Keep all tools clear of the worktable.

20.8 Universal Tool and Cutter Grinder

The ***universal tool and cutter grinder*** is a grinding machine designed to sharpen cutters (primarily milling cutters) to specified tolerances.

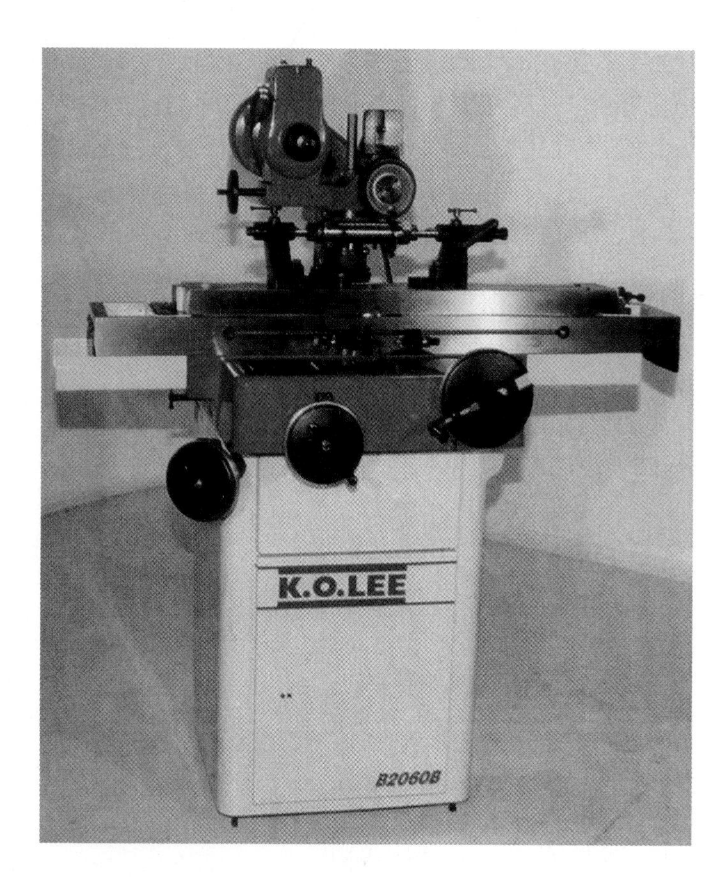

K.O. Lee Co.

Figure 20-29. Universal tool and cutter grinder.

See **Figure 20-29**. Special attachments permit straight, spiral, and helical cutters to be sharpened accurately. Other attachments enable the machine to be adapted to all types of internal and external cylindrical grinding. See **Figure 20-30**.

The wheel shapes most frequently used for tool and cutter grinding are shown in **Figure 20-31**. Charts prepared by the various grinding wheel manufacturers are used to determine the correct wheel composition (abrasive, grain size, and bond) for the job at hand.

Keep the grinding wheel clean and sharp by frequent dressing with a diamond tool. Use light cuts to avoid drawing the temper out of the tooth cutting edge.

Crowding the wheel into the cutter is a common mistake when grinding cutters. The cutters are made from materials like HSS and cemented carbides, which do not give off a brilliant shower of sparks when in contact with a grinding wheel. This creates the illusion that the cut being made is too light.

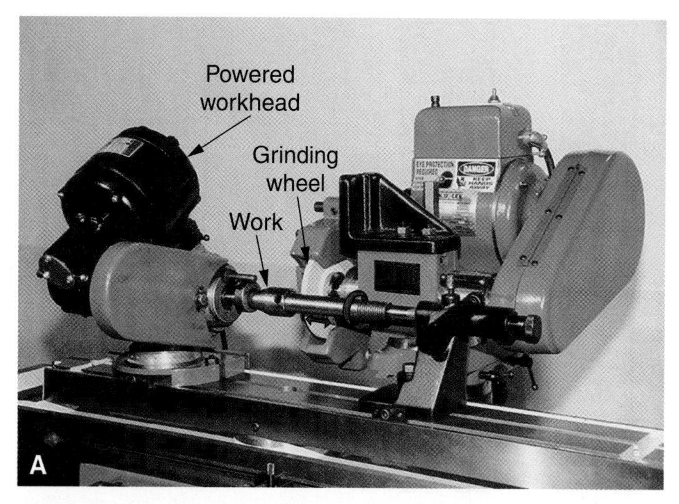

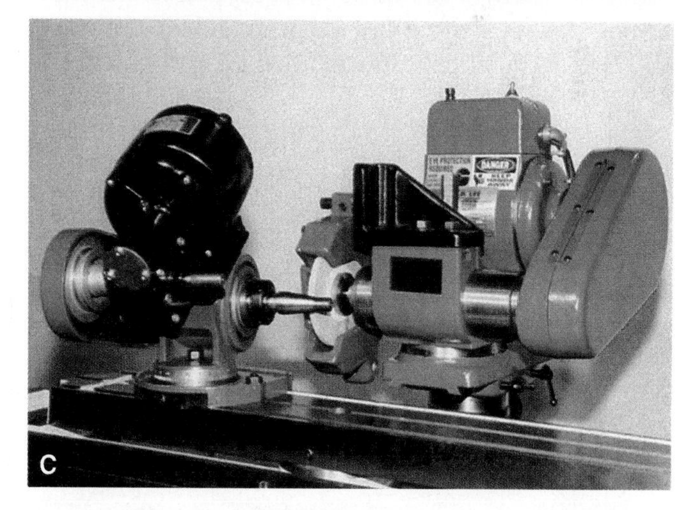

K.O. Lee Co.

Figure 20-30. Tool and cutter grinder applications. A—Limited cylindrical grinding can be done on a universal tool and cutter grinder. B—A universal tool and cutter grinder can also be used for internal grinding. C—The center of a spindle is being trued using a universal tool and cutter grinder. The work is rotated by a powered workhead.

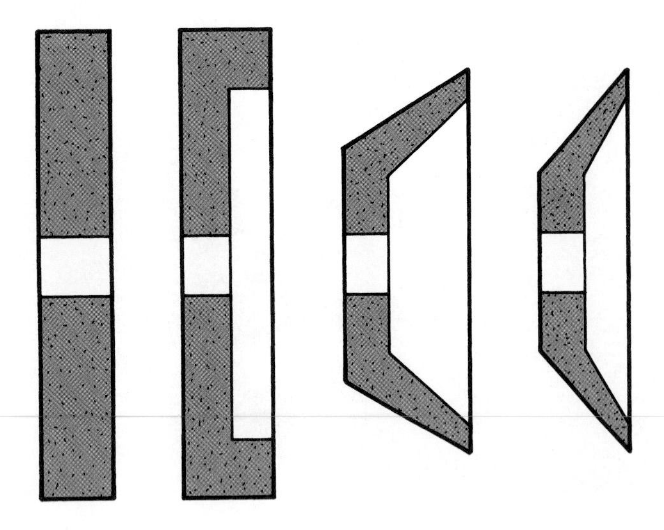

Goodheart-Willcox Publisher

Figure 20-31. Typical wheel shapes used for grinding cutters.

20.9 Sharpening Cutters

A ***tooth rest*** locates each tooth quickly and accurately into position. Several types are used to permit different cutter types to be sharpened. The bracket that supports the tool rest can be mounted to the worktable or on the grinding wheel housing. See **Figure 20-32.**

20.9.1 Grinding Plain Milling Cutters

For this procedure, proper eye protection is a must. Also, if the machine is not equipped with a vacuuming system, a respirator mask should also be worn.

1. Select the correct wheel for the job. True it with a diamond tool.
2. Mount the cutter on a suitable arbor and place the unit between centers.
3. Mount the tooth rest to the wheel head. Position the edge about 1/4″ (6.0 mm) above the center line of the grinding wheel, **Figure 20-33.** This will produce a 5° to 6° clearance angle on the tooth cutting edge of a 6″ (150 mm) diameter cutter. Adjust to suit the cutter being ground.

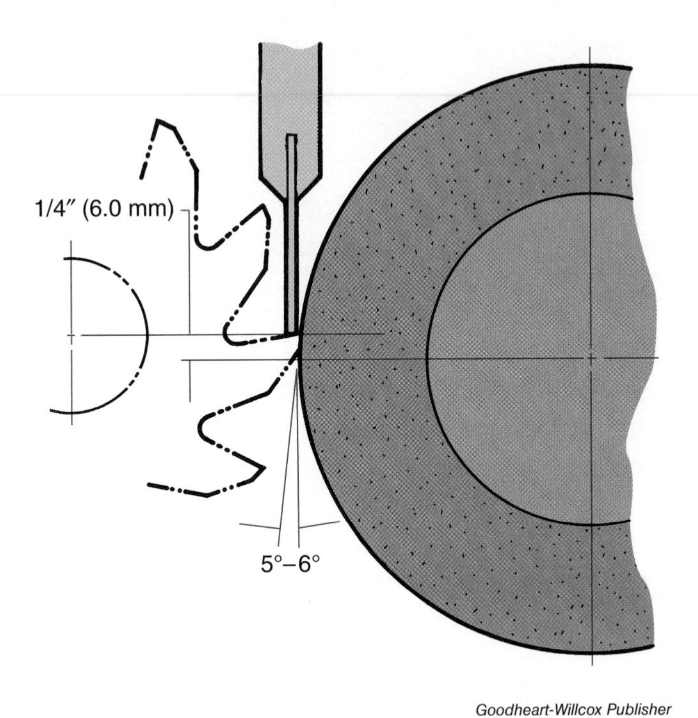

Goodheart-Willcox Publisher

Figure 20-33. Setup is for grinding a 6″ (150 mm) diameter cutter.

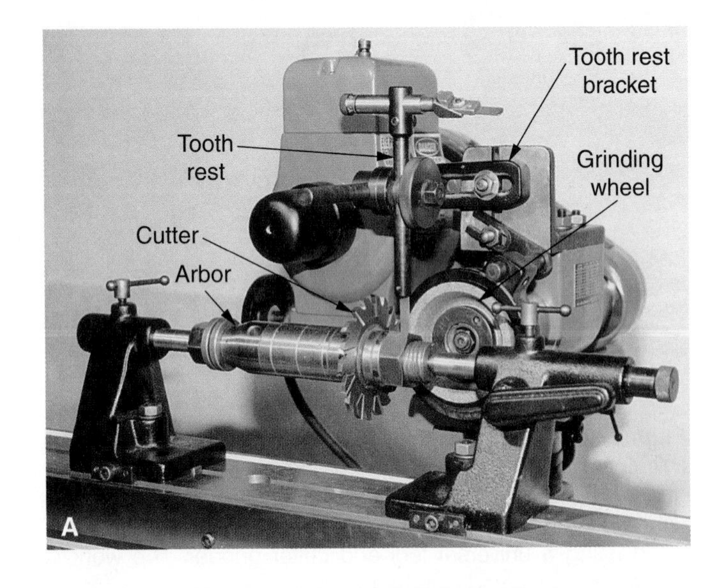

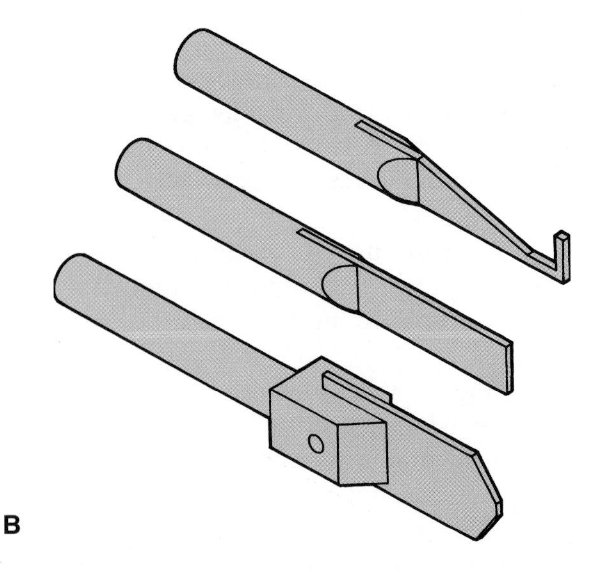

K.O. Lee Co.; Goodheart-Willcox Publisher

Figure 20-32. Tooth rest. A—A tooth rest is used to position the teeth of a cutter. Note the support bracket and the cup-shaped wheel that does the grinding. B—Several-types of tooth rests.

4. The setup should permit the wheel to grind away from the tooth cutting edge. While requiring more machining care than grinding into the cutting edge of the tooth, there is less chance of drawing the temper. Also, it prevents the formation of a burr, which would need to be taken off with an oilstone in order to secure a sharp edge. See **Figure 20-34**.

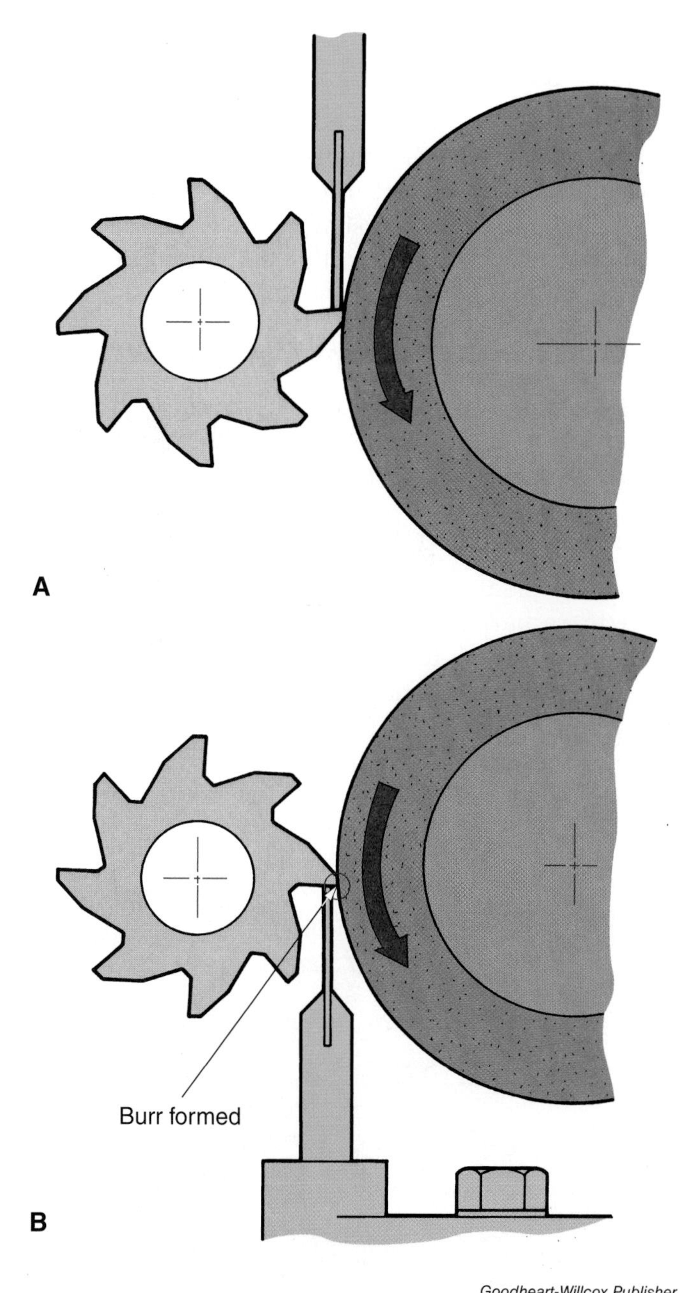

Figure 20-34. Tooth grinding setup. A—The wheel grinds away from the cutting edge of the tooth. With this technique, there is less chance of drawing the temper out of the tooth and no burr is formed. B—If the wheel grinds into the cutting edge of the tooth, there is some danger that a burr will be formed or the temper drawn.

5. Flare cup wheels are also used for cutter and tool grinding. They are set up as shown in **Figure 20-35**. Since there is a greater area of contact when using a flare cup wheel, lighter cuts should be taken than with straight grinding wheels.
6. Start the machine and feed the cutter into the wheel. Take a light cut. A bit of thinned layout bluing should be applied to the back of the tooth. This will allow a visual check of how the grinding operation is progressing and whether the setup is producing the proper clearance angle.
7. When satisfied with the setup, bring the next tooth into position on the tooth rest and grind that tooth.
8. Repeat the operation until all of the teeth are sharpened. Make necessary adjustments to ensure tooth concentricity. The cutting surfaces of all teeth must be the same distance from arbor hole center line.

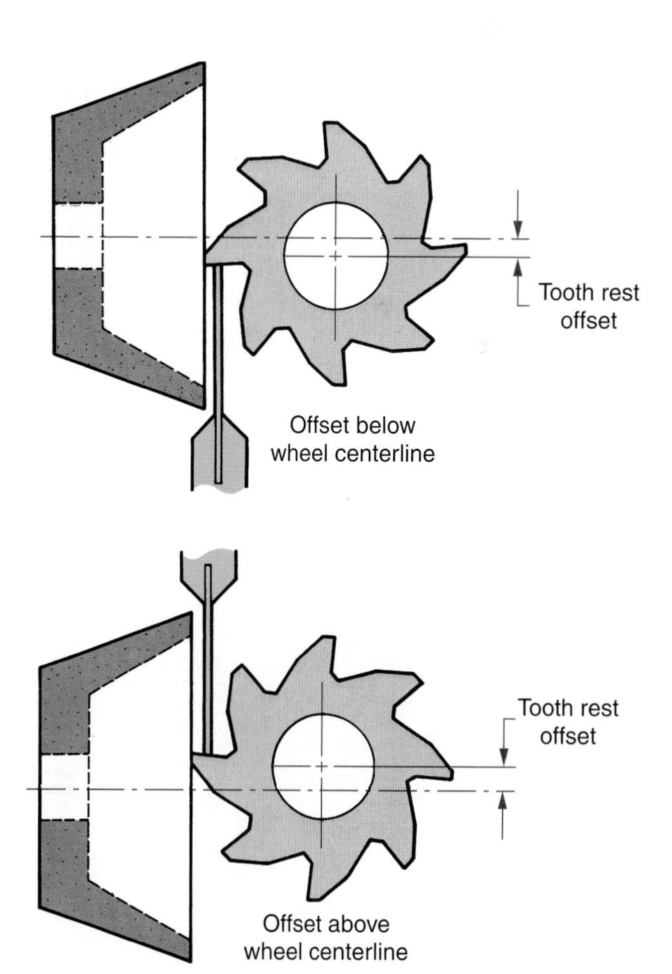

Figure 20-35. Milling cutter being ground with a flare cup wheel.

After a cutter has been sharpened several times, the clearance angle flat (land) will become too wide. Then, it becomes necessary to grind in a secondary clearance angle.

If it becomes apparent that more material is being removed from some teeth than others, a quick check must be made to determine the cause:

- The grinding wheel may be too soft and wearing down too rapidly. As the wheel wears, less material is removed from the cutter tooth.
- The tooth rest may not be mounted solidly, allowing it to move during the grinding operation.
- The arbor may not be running true on the centers. Check the arbor runout with a dial indicator as the arbor is rotated.

When the cause of the problem has been identified, make the necessary corrections and continue the operation.

An indexing disc may also be used to position each tooth for sharpening, **Figure 20-36**. It is mounted on the arbor. The divisions are normal to each other, plus or minus 4 minutes (1/15°). They are available in a range of graduations.

Teeth on a side-milling cutter must also be sharpened. This is done by mounting the cutter on a stub arbor and fitting the unit into a workhead, rather than positioning it between centers. Facing mills are sharpened in the same manner.

20.9.2 Grinding Cutters with Helical Teeth

Slabbing cutters and other cutters that have helical teeth are sharpened in much the same manner as plain milling cutters, **Figure 20-37**. However, these cutters must be held against the tooth rest as the table is traversed. This will impart a twisting motion to keep the tooth correctly located against the grinding wheel.

20.9.3 Grinding End Mills

End mills are sharpened in basically the same way as helical teeth cutters, with the end mill mounted in a workhead rather than between centers. The end teeth are sharpened with the same technique used to sharpen the side teeth on a side milling cutter.

20.9.4 Grinding Form Cutters

Form tooth cutters must be ground radially to preserve the tooth shape, **Figure 20-38**. An index disc or a special form cutter grinder can be used.

20.9.5 Grinding Taps

A universal tool and cutter grinder can also be used to resharpen taps. Normally, a tap becomes dull when the leading edges of the starting chamfer become worn. The chamfer can be reground by mounting the tap in a workhead, **Figure 20-39**. Flutes are reground using a straight wheel with an edge that has been shaped to fit the flutes.

K.O. Lee Co.

Figure 20-36. An indexing disc is often used to position each tooth for sharpening.

K.O. Lee Co.

Figure 20-37. Setup for sharpening a cutter with helical teeth.

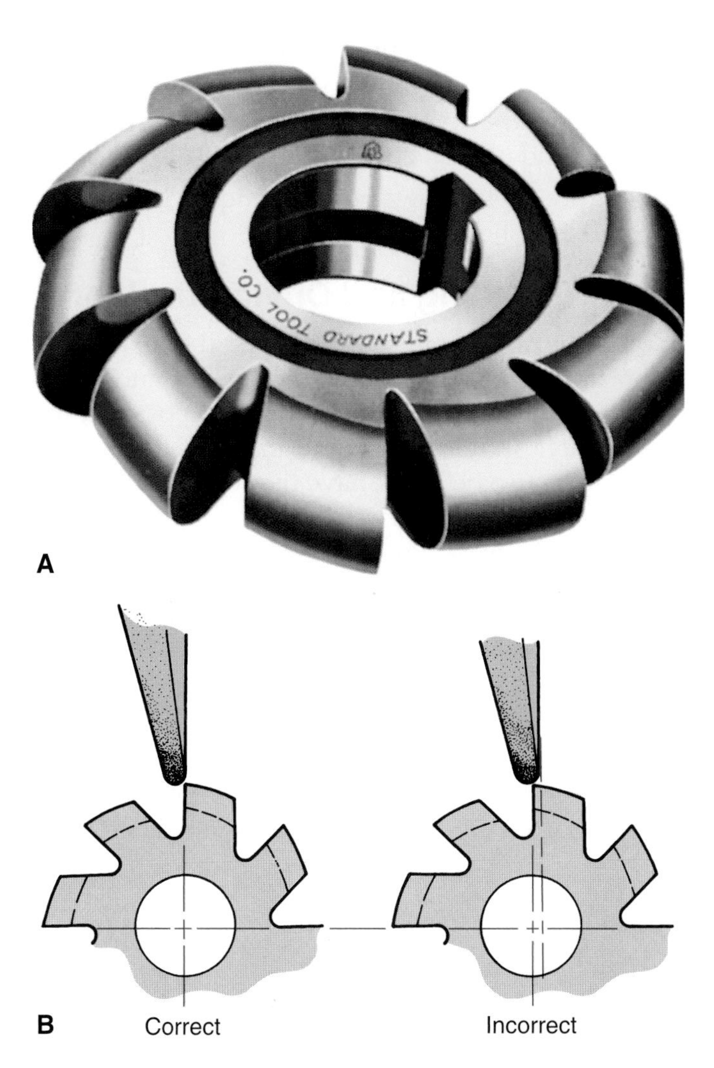

Standard Tool Co.; Goodheart-Willcox Publisher

Figure 20-38. Form tooth cutters. A—This convex cutter is typical of form tooth cutters. B—Form tooth cutters must be ground radially. Otherwise, the form or shape that the cutter was designed to machine will be altered.

20.9.6 Grinding Reamers

The cutting action of a machine reamer takes place at the front end of the teeth, **Figure 20-40**. Sharpen the reamer in the same manner used to sharpen a face milling cutter. The worktable is pivoted at a 45° angle. Using a cup wheel, adjust the tooth rest and grinding head to give the correct clearance.

20.10 Cylindrical Grinding

With a cylindrical grinder, it is economically feasible to machine hardened steel to tolerances of 0.00001″ (0.0002 mm) with extremely fine surface finishes. Work is mounted between centers and

K.O. Lee Co.

Figure 20-39. Grinding taps. A—Setup for regrinding a chamfer on a tap. B—Flutes are being reground on a tap to renew the cutting edges of the teeth.

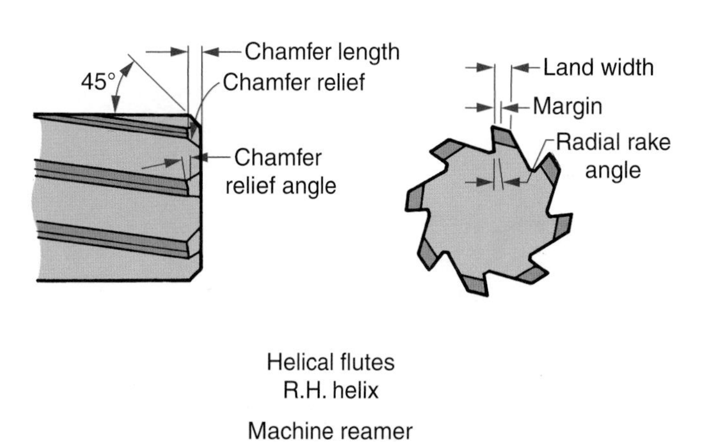

Goodheart-Willcox Publisher

Figure 20-40. Cutting edges of a machine reamer.

rotates while in contact with the grinding wheel, **Figure 20-41**. Straight, taper, and form grinding operations are possible with this technique. Two variations of cylindrical grinding are traverse grinding and plunge grinding.

- In ***traverse grinding***, a fixed amount of material is removed from the rotating workpiece as it moves past the revolving grinding wheel. Work wider than the face of the grinding wheel can be ground. See **Figure 20-42**.
- In ***plunge grinding***, the work still rotates. However, it is not necessary to move the grinding wheel across the work surface. The area being ground is no wider than the wheel face. Grinding wheel infeed is continuous rather than incremental (minute changes at end of each cut), **Figure 20-43**.

Landis Div. of Western Atlas

Figure 20-41. Close-up of a cylindrical grinding operation.

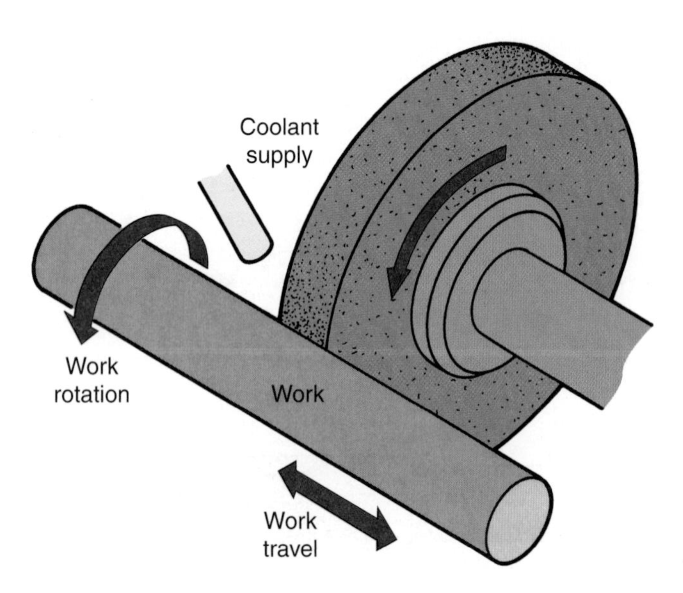

Goodheart-Willcox Publisher

Figure 20-42. The principle of traverse grinding. The rotating work moves past the rotating grinding wheel.

Figure 20-44 shows both techniques being used to grind to a shoulder.

20.10.1 Holding and Driving the Work

As the work rotates on centers, it is extremely important that the centers be free of dirt and nicks. They must also run absolutely true. If possible, the head center should be ground in place. The center holes must also be clean, have the correct shape and depth, and be well-lubricated.

Long work is best supported by work rests, **Figure 20-45**. These support the workpiece from the back and bottom and are adjustable to compensate for material removed in the grinding operation.

Work rotation is accomplished through the use of a drive plate that revolves around the headstock center and an adjustable drive pin and dog, **Figure 20-46**. Work can also be mounted in a chuck.

20.10.2 Machine Operation

To ensure a good finish and size accuracy, it is vital that work rotation and traverse table movement (back-and-forth in front of the grinding wheel) be smooth and steady.

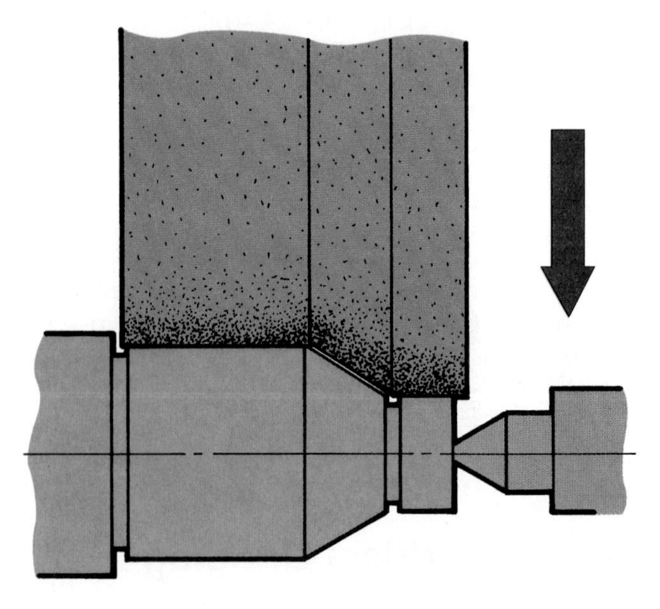

Goodheart-Willcox Publisher

Figure 20-43. With plunge grinding, the grinding wheel is fed into the rotating work. Since the work is no wider than the grinding wheel, reciprocating motion is not needed.

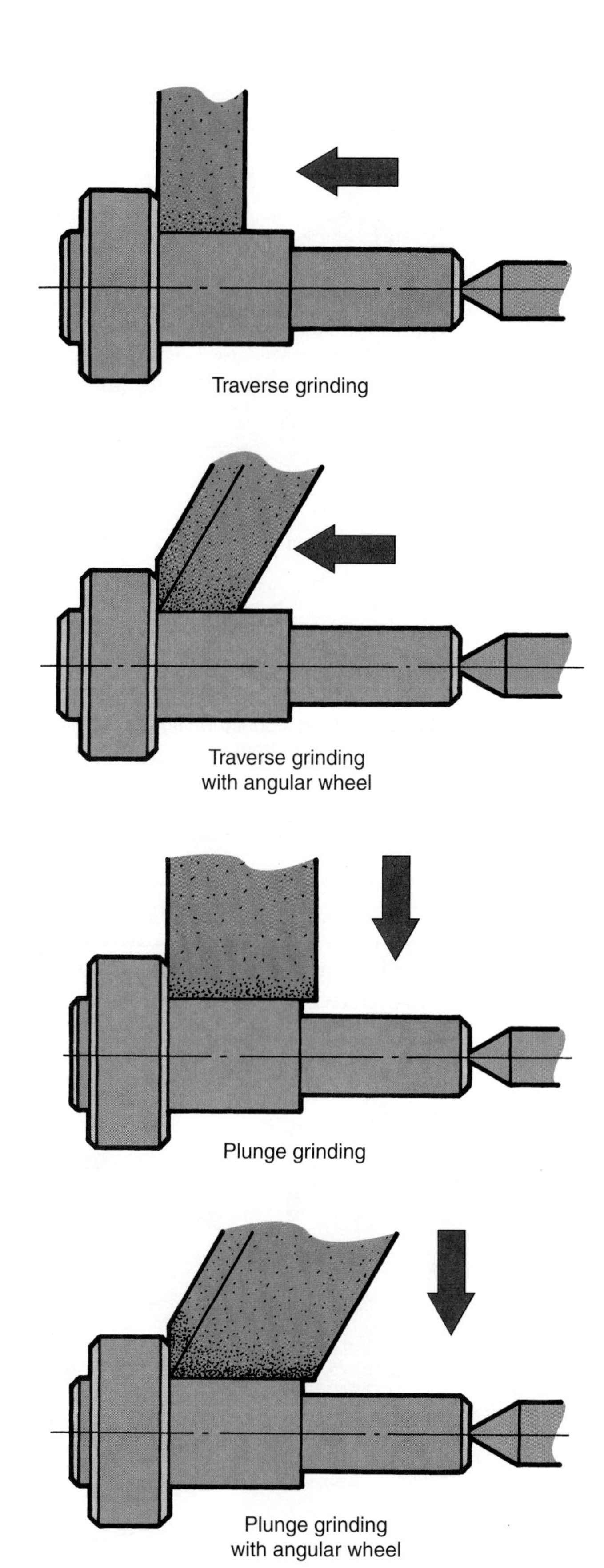

Goodheart-Willcox Publisher

Figure 20-44. Grinding to a shoulder by traverse and plunge grinding techniques.

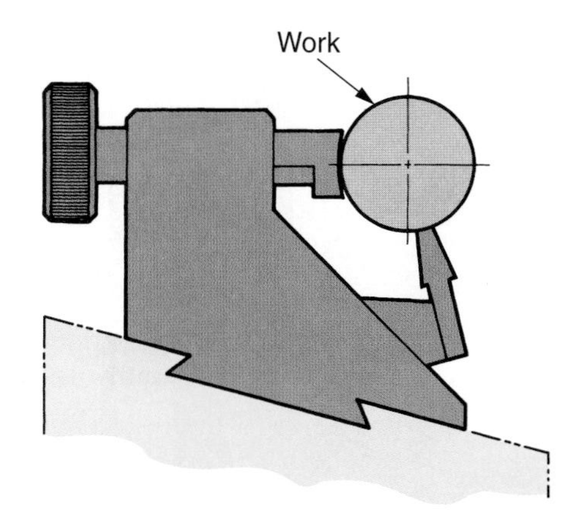

Goodheart-Willcox Publisher

Figure 20-45. Work rests should be placed every four or five diameters along the work for support. They must be adjusted after each grinding pass.

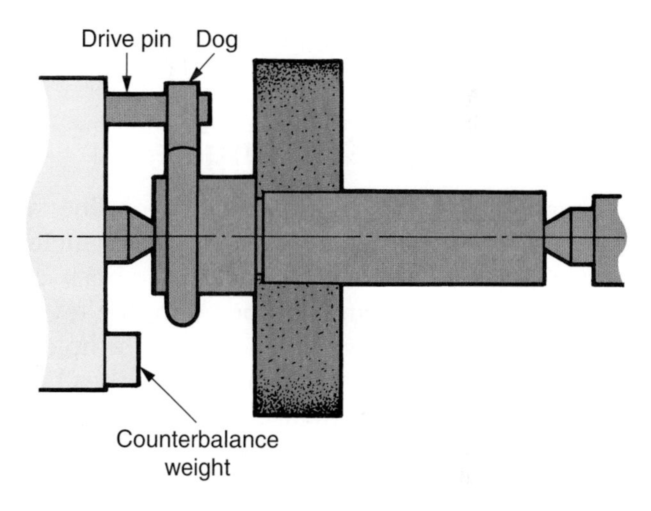

Goodheart-Willcox Publisher

Figure 20-46. One method used to rotate work on a cylindrical grinder.

Table movement should be adjusted so that the wheel will overrun the work end by about one-third the width of the wheel face, **Figure 20-47**. This permits the grinding wheel to do a more accurate grinding job. Insufficient runoff will result in work that is oversize. Complete runoff of the grinding wheel will cause the piece to be undersize.

The grinding wheel must be trued and balanced. Otherwise, vibration will cause chatter marks on the work and may cause it to be out-of-round.

Cutting speeds and feeds can be determined from information available on charts furnished by the grinding machine and grinding wheel manufacturers. They will also specify what coolant will give the best results.

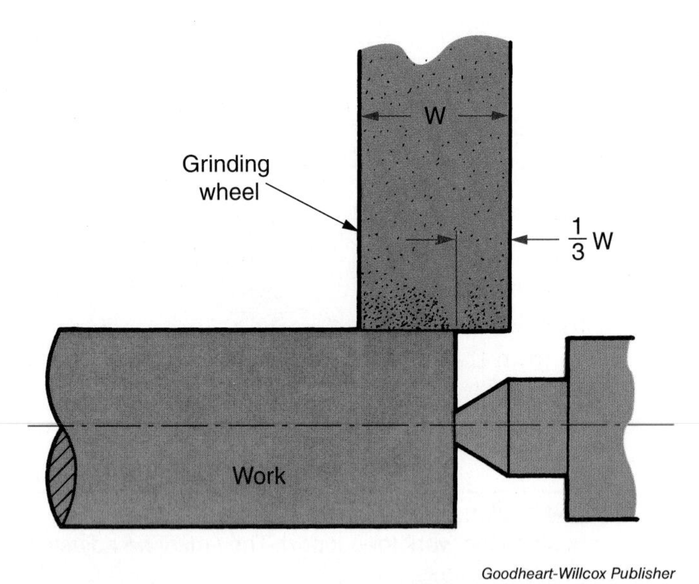

Goodheart-Willcox Publisher

Figure 20-47. Table movement should permit to run the wheel extend about one-third of its width beyond the end of the work. This ensures that the wheel removes the proper amount of metal, improving the accuracy of the job.

20.11 Internal Grinding

Internal grinding is done to secure a fine surface finish and accuracy on inside diameters, **Figure 20-48**. Work is mounted in a chuck and rotates. During the grinding operation, the revolving grinding wheel moves in and out of the hole.

A special grinding machine that finishes holes in pieces too large to be rotated by the conventional machine is shown in **Figure 20-49**. Hole diameter is controlled by regulating the diameter of the circle in which the grinding head moves.

sspopov/Shutterstock.com

Figure 20-48. Internal grinding operation.

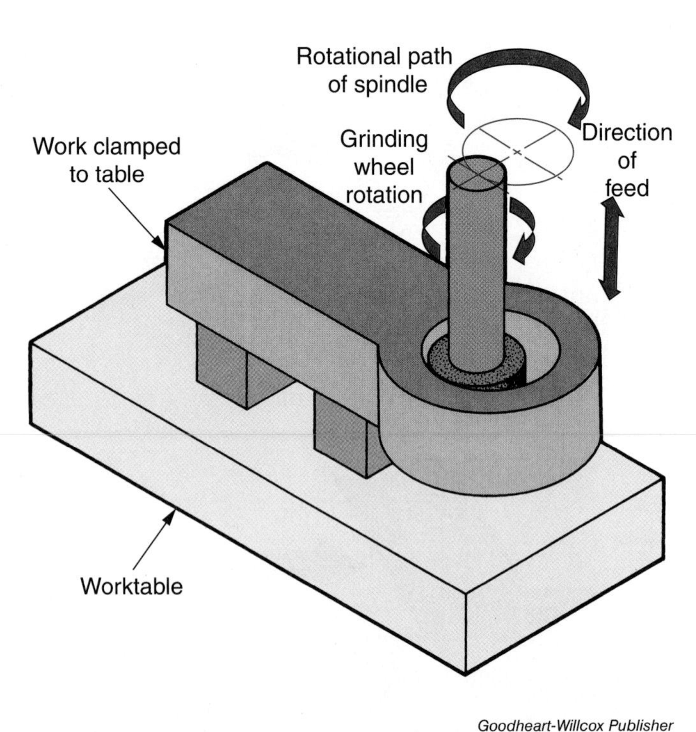

Goodheart-Willcox Publisher

Figure 20-49. Internal grinding technique used when work is too large or odd-shaped to be rotated.

20.12 Centerless Grinding

In ***centerless grinding***, the work does not have to be supported between centers because it is rotated against the grinding wheel. Instead, the piece is positioned on a work support blade, and fed automatically between a regulating or feed wheel and a grinding wheel. See **Figure 20-50**.

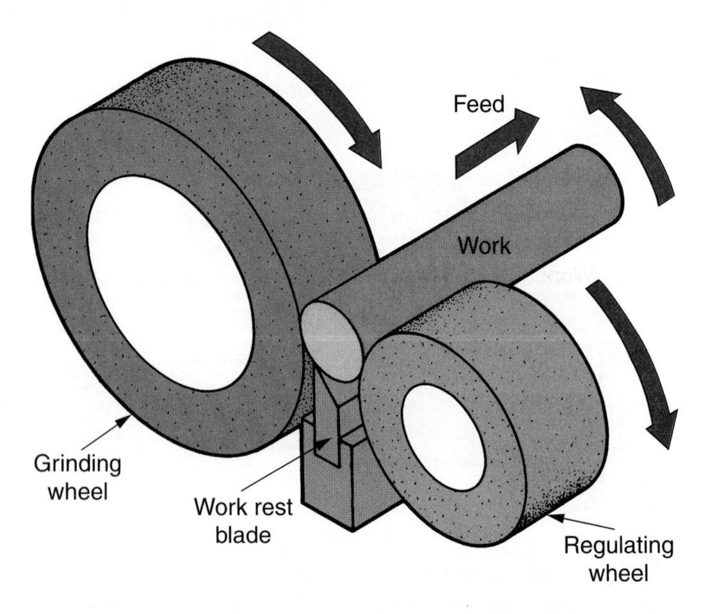

Goodheart-Willcox Publisher

Figure 20-50. Basic arrangement for centerless grinding.

Basically, the regulating wheel causes the piece to rotate, and the grinding wheel does the cutting. Feed through the wheels is obtained by setting the regulating wheel at a slight angle.

There are four variations of centerless grinding, **Figure 20-51**:

- **Through feed grinding.** This method can be used only to produce simple cylindrical shapes. Work is fed continuously by hand, or from a feed hopper, into the gap between the grinding wheel and the regulating wheel. The finished pieces drop off the work support blade.
- **Infeed grinding.** This is a centerless grinding technique that feeds the work into the wheel gap until it reaches a stop. The piece is ejected at the completion of the grinding operation. Work diameter is controlled by adjusting the width of the gap between the regulating wheel and the grinding wheel. Work with a shoulder can be ground using this technique.
- **End feed grinding.** This form of centerless grinding is ideally suited for grinding short tapers and spherical shapes. Both wheels are dressed to the required shape and work is fed in from the side of the wheel to an end stop. The finished piece is ejected automatically.
- **Internal centerless grinding.** This method minimizes distortion in finishing thin-wall work and eliminates reproduction of hole-size errors and waviness in the finish.

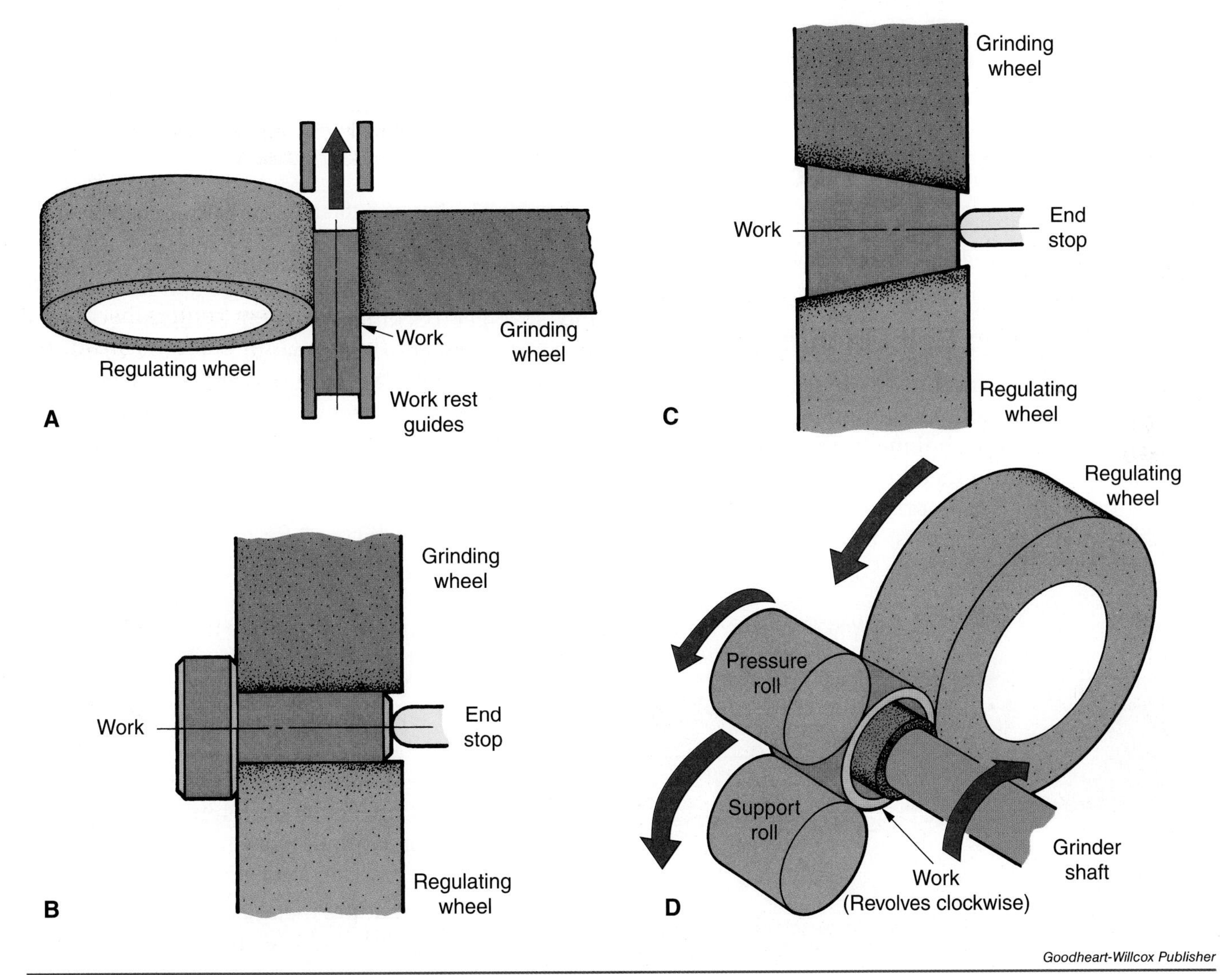

Figure 20-51. Centerless grinding variations. A—Through feed centerless grinding. The angle of the regulating wheel pulls the work over the grinding wheel. B—Infeed centerless grinding. The work is fed into the wheel gap until it reaches a stop. The piece is ejected at completion of the grinding operation. C—End feed centerless grinding is best suited for grinding short tapers and spherical shapes. D—Setup for internal centerless grinding.

Centerless grinding is used when large quantities of the same part are required. Production is high and costs are relatively low, because there is no need to drill center holes or to mount work in a holding device. Almost any material can be ground by this technique.

20.13 Form Grinding

In ***form grinding***, the grinding wheel is shaped to produce the required contour on the work. **Figure 20-52** shows this principle.

Thread grinding is an example of form grinding. A form or template guides a diamond dressing wheel that shapes the grinding wheel used to grind the required thread shape. The pattern formed in the grinding wheel is a mirror image of the pattern in the template, **Figure 20-53**. There is automatic compensation on the grinding machine for the material removed from the grinding wheel when it is dressed.

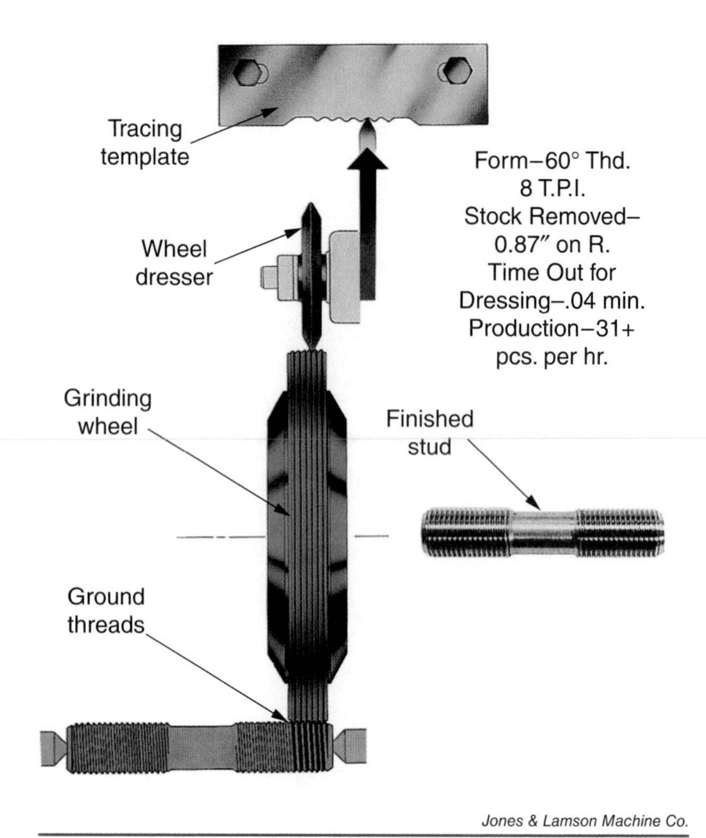

Figure 20-53. Precision threads are being form-ground on a special stud.

20.14 Other Grinding Techniques

In addition to the grinding techniques already described, industry makes considerable use of other abrasive-type processes.

20.14.1 Abrasive Belt Machining

Abrasive belt grinding was first used for light stock removal and polishing operations. However, the capability of the technique has advanced to the stage where high-rate metal removal to close tolerances is possible. This is primarily due to tougher and sharper abrasive grains, improved adhesives, and stronger backings. Several abrasive grinding machine applications are shown in **Figure 20-54**.

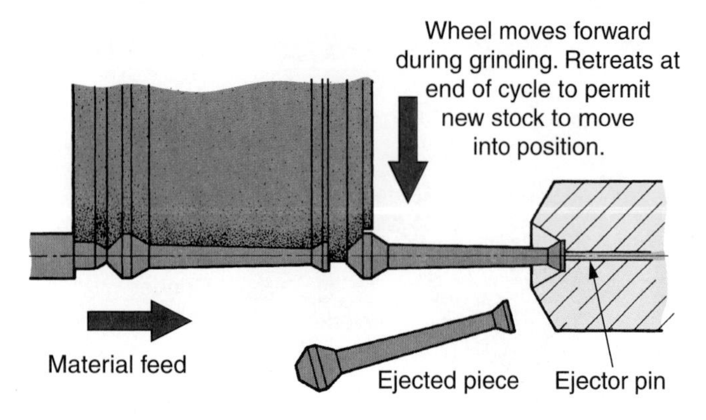

Figure 20-52. Form grinding of this engine part is done at rate of 200 pieces an hour. The material was heat-treated before grinding.

Abrasive belts, because of their length, run cool and require light contact pressure, thus reducing the possibility of metal distortion caused by heat. Soft contact wheels and flexible belts conform to irregular shapes. Belts may be used dry or with a coolant. The most satisfactory belt speed for grinding ferrous and nonferrous metals is between 5000–9000 sfm (surface feet per minute). Slower speeds of 1500–3000 sfm are required for tougher materials like titanium.

Abrasive belt grinding usually requires support behind the belt, **Figure 20-55**. This may be in the form of contact wheels or platens.

- Contact wheels are usually made of cloth or rubber. The hardness and density of the contact wheel affects stock removal and finish. Serrated or slotted wheels improve cutting action and prolong abrasive belt life.
- ***Platens*** are made of metal (some have cemented carbide inserts) and are usually not as effective as contact wheels. They are flat, but can be shaped to conform to the contour required on the work. Jets of air or water may be applied between the belt and platen to reduce friction.

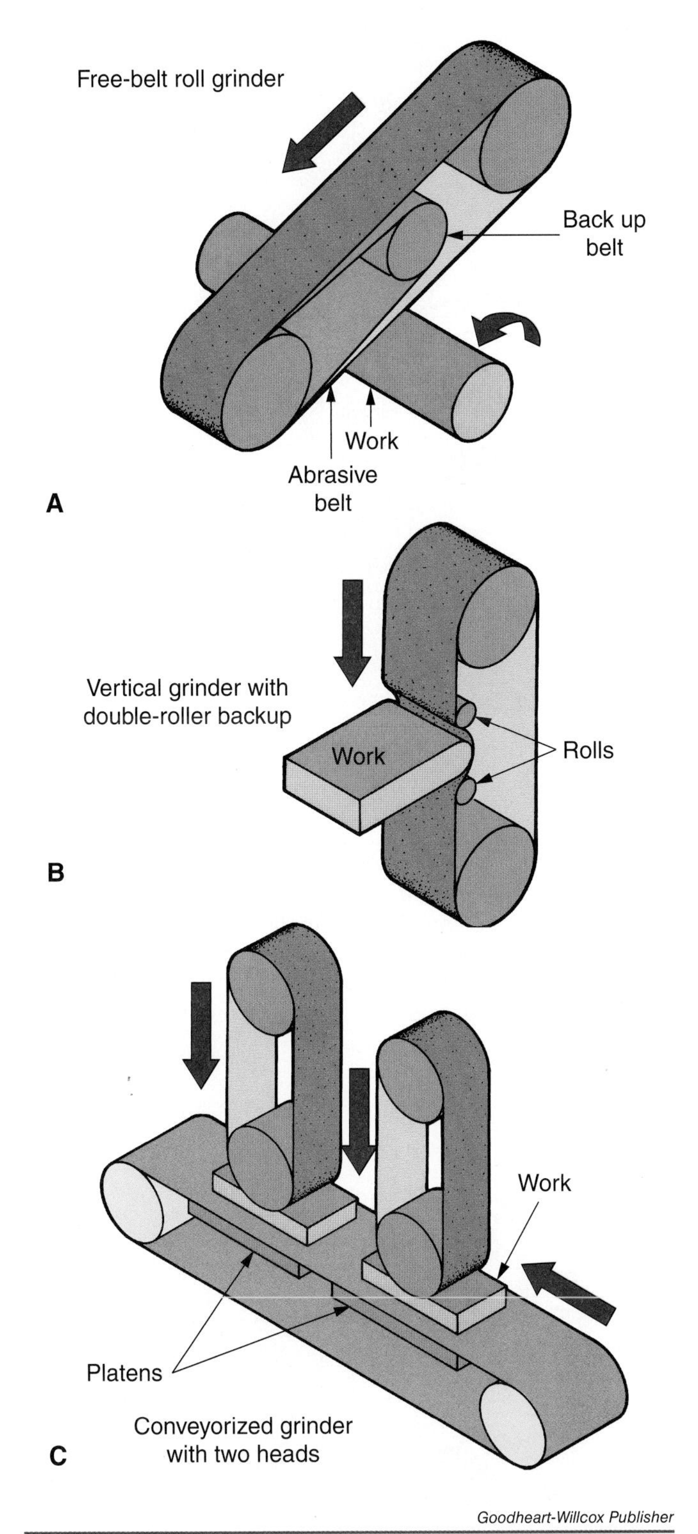

Figure 20-54. A few of the many abrasive belt grinding techniques.

A major advantage of abrasive belt grinding is its versatility. A machine can be converted quickly from heavy stock removal to finishing operations, or for grinding a different material, by simply changing the abrasive belt.

Figure 20-55. The contact wheel supports the abrasive belt to prevent it from deflecting when it contacts the work.

20.14.2 Electrolytic Grinding

Electrolytic grinding is actually a form of electrochemical machining. Applications of the technique include rapid removal of stock from alloy steel parts, sharpening carbide tools, and machining heat-sensitive work.

An electric current is passed between a metal-bonded grinding wheel (cathode) and the work (anode) through a conductive electrolyte, **Figure 20-56**. The surface of the work is attacked electrochemically and dissolved in a process similar to electroplating, but in reverse.

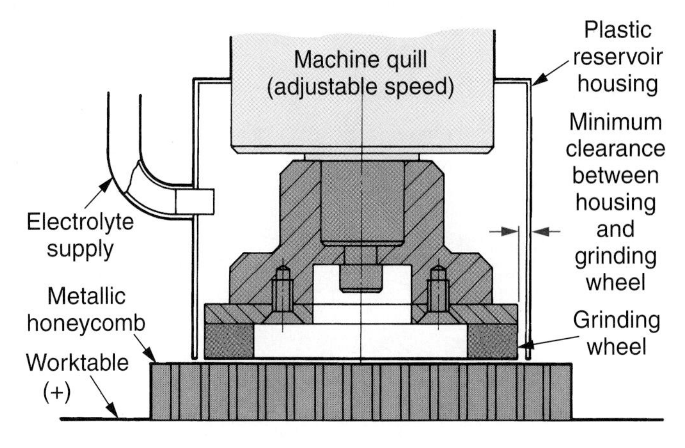

Figure 20-56. Electrolytic grinding process. It is actually an electrochemical machining process.

The dissolved material is removed by the wheel. No burr is developed, making it possible to machine materials like stainless steel and exotic metal honeycomb sections. No heat is generated, and there is no metallurgical change in the metal.

20.14.3 Computer Controlled (CNC) Grinders

Many types of computer numerical control (CNC) grinders are available, **Figure 20-57**. They are designed to operate automatically. Such functions as positioning, spindle start and stop, vertical feed motion, and linear feed rates are programmed into the machine's computer. The operator is then relieved from the responsibility of controlling and monitoring the numerous coordinate settings and related machine functions.

The grinding wheel path representing the contours of the part (at a selected distance from the part edge) is programmed directly to its dimensional specifications. Grinding wheel wear (wheel diameter decreases each time it is dressed) is compensated for automatically.

The part's contour is generated in a continuous motion by the X- and Y-axes slides of the machine. See **Figure 20-58**.

Goodheart-Willcox Publisher

Figure 20-58. Contour grinding using CNC to shape a cam. In this illustration, it appears that the grinding wheel is moving around the cam edge. In reality, the part moves along its programmed path around the vertical machine spindle holding the rotating wheel.

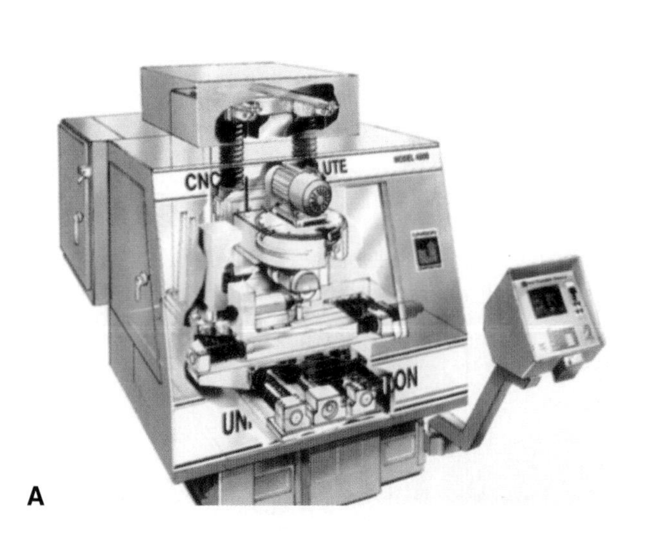

A

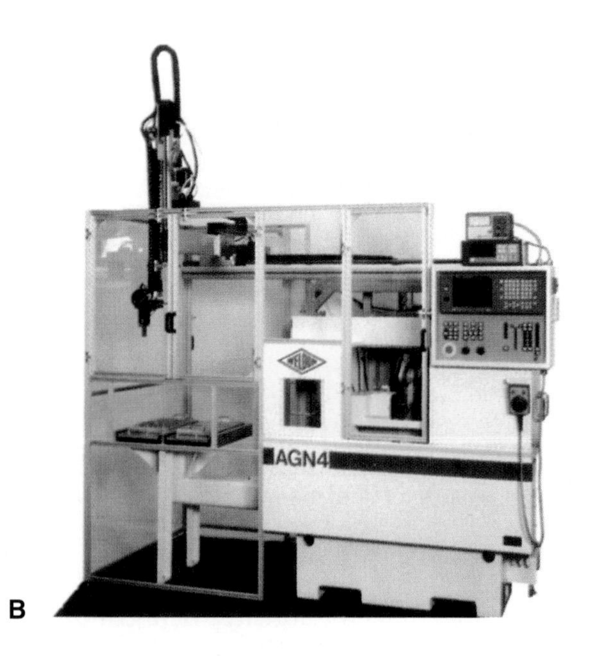

B

Unison Corp.; Weldon Machine Tool, Inc.

Figure 20-57. CNC grinders. A—Cutaway of CNC grinder for remanufacturing and sharpening fluted cutting tools such as drills, reamers, and end mills. B—CNC cylindrical grinder with automatic loading system. The machine is capable of plunge, contour, and traverse operations.

Chapter Review

Summary

- Precision grinding operates by removing a small amount of material through abrasive action to generate a smooth, accurate surface.
- The two basic types of precision grinders are planer- and rotary-type surface grinders.
- Selecting the correct abrasive type, grain size, grade, structure, and bond, as well as the correct shape grinding wheel are all important aspects of proper grinding operations.
- Universal tool and cutter grinders can be used to support and sharpen tools and cutters.
- Other types of precision grinding operations include cylindrical, internal, centerless, and form grinding.
- Industry also makes use of abrasive belt, electrolytic, and CNC grinding.

Review Questions

Answer the following questions using the information provided in this chapter.

1. Industry classifies surface grinding as the grinding of _____ surfaces.
2. Name the two categories of surface grinding machines.
3. Name three work-holding devices used to hold work for surface grinding.
4. The ideal grinding wheel will wear away _____.
   - A. as the abrasive particles become dull
   - B. at a predetermined rate
   - C. slowly to save money
   - D. All of the above.
   - E. None of the above.
5. A(n) _____ dressing tool is usually used to true and dress wheels for precision grinding.
6. List the two conditions that commonly prevent a grinding wheel from cutting efficiently.
7. List the five distinguishing characteristics of a grinding wheel.
8. A solid grinding wheel will give off a(n) _____ when struck lightly with a metal rod.
9. Why are cutting fluids or coolants necessary for grinding operations?
10. What is the difference between conventional grinding and creep grinding?
11. Irregular scratches on the work are usually caused by _____.
    - A. rancid coolant
    - B. using the wrong coolant
    - C. too little coolant
    - D. too much coolant
    - E. a dirty coolant system

12. How can the problem in the above question be corrected?
13. Chatter and vibration marks on the work are caused when the grinding wheel is _____.
    A. too hard
    B. too soft
    C. glazed or loaded
    D. out-of-round
    E. loose
14. The problems in the above question can be corrected by _____ the grinding wheel.
15. A(n) _____ is a grinding machine designed to support cutters (usually milling cutters) while they are being sharpened.
16. List the two variations of cylindrical grinding.
17. With _____ grinding, it is not necessary to support work between centers or mount work in a chuck while it is being rotated against the grinding wheel.
18. The grinding technique that uses a belt on which abrasive particles are bonded for stock removal, finishing, and polishing operations is known as _____.
19. _____ grinding is actually an electrochemical machining process.
20. Describe how the process in the above question is done.

CHAPTER 21

Band Machining

Learning Objectives

After studying this chapter, you will be able to:

- Describe how a band machine operates.
- Explain the advantages of band machining.
- Select the proper blade for the job to be done.
- Weld a blade and mount it on a band machine.
- Safely operate a band machine.

Technical Terms

band machining
diamond-edge band
file band
knife-edge blade
mist coolant
raker set
straight set
tooth form
wavy set

Band machining is a widely used machining technique that makes use of a continuous saw blade. Each tooth is a precision cutting tool, so accuracy can be held to close tolerances. This eliminates or minimizes many secondary machining operations.

21.1 Band Machining Advantages

Band machining offers several major advantages over other machining techniques, **Figure 21-1**.

- Band machining maintains sharpness. Wear is distributed over many teeth. Chip load is uniform and constant on each tooth, minimizing tool wear.
- Band machining provides unrestricted cutting geometry. Cutting can be done at any angle, in any direction, and the length of the cut is unlimited.
- Band machining provides a built-in workholder. Cutting action is downward, so cutting forces hold the workpiece to the table. In most situations, work need not be clamped.
- Band machining is efficient. Excess chip production wastes power. Band machining produces the desired shape with a minimum of chips. There is little waste, since band machining cuts directly to shape, and unwanted material is removed in solid sections.

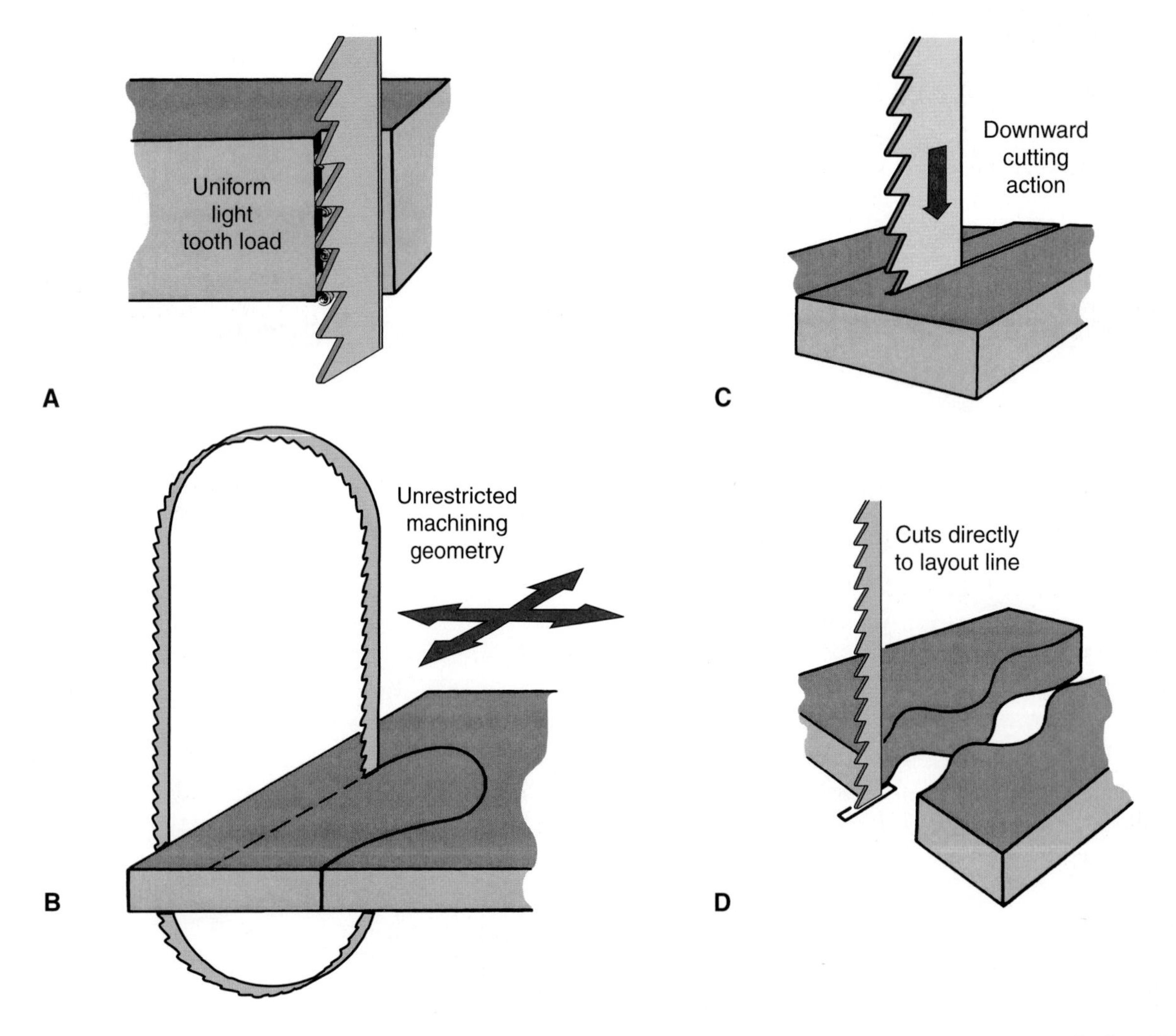

Figure 21-1. Band machining advantages. A—Wear is distributed over many cutting edges (teeth) with band machining. B—Band machining permits machining at any angle or direction. Cut length is almost unlimited. C—The downward force of the cutting action helps hold work on table. D—Band machining is very efficient and produces little waste. Unwanted material is removed in solid sections.

21.2 Band Blade Selection

Some blade manufacturers list more than 500 different band saw blades. Points that must be considered by the machinist when selecting the correct blade for a specific job are the blade type and blade characteristics.

There are six basic band saw blade types. **Figure 21-2** lists them, along with their applications:

- Tungsten carbide.
- Bimetal (high-speed steel cutting edge with a flexible carbon steel back).
- High-speed steel.
- Shock resistant high-speed steel.
- Hard edge with spring-tempered back.
- Carbon steel with a flexible back.

Blade characteristics include width, pitch, set, gage, and tooth form.

- **Width.** The wider the blade, the greater its strength and the more accurately it will cut, **Figure 21-3**. For making straight cuts, use the widest blade the machine will accommodate. Contour cutting should use the widest blade that will cut the required radius, **Figure 21-4**. Widths from 1/16″ to 2″ (1.5 mm to 50 mm) are available.

Match the Tools to the Job		
Type of blade	**Applications**	**Band machine**
T/C Inserted tungsten carbide teeth on fatigue-resistant blade.	Heavy production and slabbing operations in tough materials.	Horizontal cutoff machines over 5 hp with positive feed. Vertical contour machines over 5 hp with positive feed.
Imperial bimetal HSS cutting edge with flex-resistant carbon-alloy back.	Mild to tough production and cutoff applications.	Horizontal cutoff machines over 1 1/2 hp with controlled feed, generally with variable-speed drives and with coolant system. Vertical contour machines over 1 1/2 hp with coolant system.
Demon M–2 HSS blade.	Heavy-duty toolroom and maintenance shop work. Full-time production applications.	Horizontal cutoff machines over 1 1/2 hp with controlled feed, generally with variable-speed drives and with coolant system. Vertical contour machines over 1 1/2 hp with coolant system.
Demon shock-resistant M–2 HSS blade specially processed for greater shock resistance.	Structurals, tubing, materials of varying cross section.	Horizontal cutoff machines over 1 1/2 hp with controlled feed, generally with variable-speed drives and with coolant system. Vertical contour machines with 1 1/2 hp with coolant system.
Dart Carbon-alloy, hard-edge, spring-tempered back blade.	Superior accuracy for light toolroom and maintenance shop applications as well as light manufacturing.	Horizontal cutoff machines under 1 1/2 hp with coolant system. Vertical contour machines under 1 1/2 hp with coolant system.
Standard carbon All-purpose, hard-edge, flexible back blade.	Light toolroom and maintenance shop applications.	Horizontal cutoff machines under 1 1/2 hp with weight feed and without coolant. Generally step speeds. Vertical contour machines under 1 1/2 hp without coolant.

DoALL Co.

Figure 21-2. Recommendations for using basic blade types.

Goodheart-Willcox Publisher

Figure 21-3. Blade width is measured from the tooth tip to the back of the blade.

Width of blade	Smallest radius	The width of the blade is determined by the smallest radius to be cut
1/16	1/16	
3/32	1/8	
1/8	7/32	
3/16	3/8	
1/4	5/8	
5/16	7/8	
3/8	1 1/4	
1/2	3	

Goodheart-Willcox Publisher

Figure 21-4. Blade width affects the smallest radius that can be cut.

- **Pitch.** Refers to the number of teeth per inch or the distance between teeth measured in millimeters. The thickness of the material to be cut determines the proper pitch blade to use, **Figure 21-5**. At least three teeth should be in contact with the work for best performance. Blades are available with pitches from 2 to 32 teeth per inch.

Material thickness	Band pitch
Less than 1″ (25 mm)	10 or 14
1 to 3″ (25 to 75 mm)	6 or 8
3 to 6″ (75 to 150 mm)	4 to 6
6 to 12″ (150 to 300 mm)	2 or 3

Goodheart-Willcox Publisher

Figure 21-5. Recommended band pitches to saw various thicknesses of material.

- **Set.** Provides clearance for the blade back, **Figure 21-6**. There are three basic forms, each with its own specific application:
  - ***Raker set*** is a three-tooth saw set in which one tooth is angled toward the left, another one straight, and the next one angled toward the right, alternating continuously along the length of the blade. Raker set is recommended for cutting large solids or thick plate and bar stock.
  - ***Wavy set*** is a saw tooth set in which several teeth are angled to the right and several to the left, alternating continuously along the blade. Wavy set should be used for work with varying thicknesses, such as pipe, tubing, and structural materials.
  - ***Straight set*** is a two-tooth saw set in which one tooth is angled to the side and the other one straight, alternating continuously along the length of the blade. Straight set is specified for free cutting materials, such as aluminum and magnesium.
- **Gage.** Refers to blade thickness, **Figure 21-7**. Heavier gage blades are stronger than thin gauge blades.
- **Tooth form.** The term ***tooth form*** refers to the shape of the tooth, **Figure 21-8**. There are three basic tooth forms. Each has its specific application:
  - Standard tooth blades, with well-rounded gullets, are usually best for most ferrous metals, hard bronze, and brass.

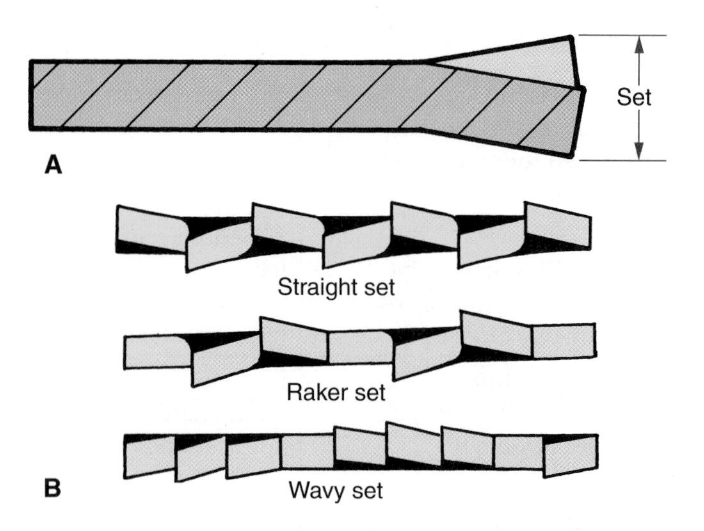

Goodheart-Willcox Publisher

Figure 21-6. Blade set. A—The term "blade set" refers to the side angle of the teeth. B—The different types of blade set.

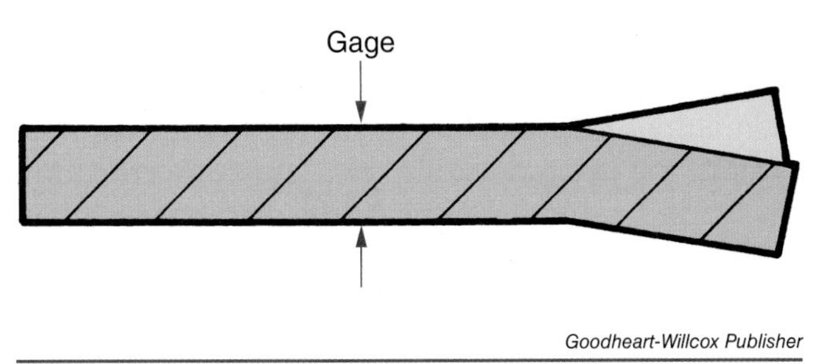

Figure 21-7. Blade thickness is referred to as gage.

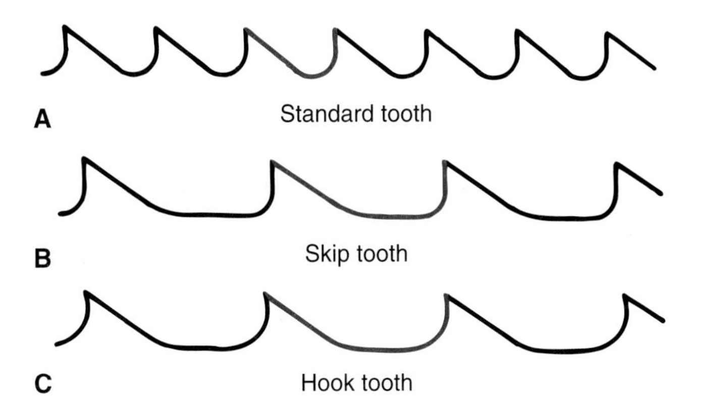

Figure 21-8. Tooth forms. A gullet from each type has been colored red to emphasize its geometry. A—A standard tooth blade has shallow, rounded gullets. B—A skip tooth blades has long, shallow gullets. The back of the gullet rises straight up to the tip of the next tooth. C—A hook tooth blade has extended gullets like a skip tooth blade. However, the back of gullet curves up and back to the tip of the next tooth. The shape of the gullet resembles a fish hook.

- Skip tooth blades provide more gullet and better chip clearance without weakening the blade body. They are recommended for most aluminum, magnesium, and brass alloys.
- Hook tooth blades offer two advantages over the skip tooth blade. The blade's design helps the blade feed easier and prevents the blade from gumming up.

The parts of a blade are shown in **Figure 21-9**. Many band saws have built-in blade selection devices. With them, it is a simple matter to dial in the various bits of information necessary to determine the best blade for the job. Following this recommendation will result in the job being done faster and with a better finish.

21.3 Welding Blades

Band saw blade stock can be purchased as ready-to-use welded bands. However, it is more economical

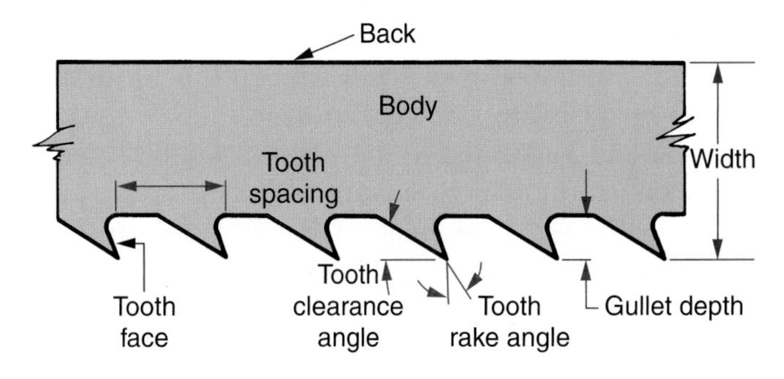

Figure 21-9. Saw blade terminology.

to buy it in 100′, 300′, or 500′ (30 m, 90 m, or 150 m) strip-out containers. The desired length of blade material is withdrawn from the container, the ends squared, and the blade welded. Extreme care must be taken to make a good weld (one that is as strong as the blade).

SAFETY NOTE

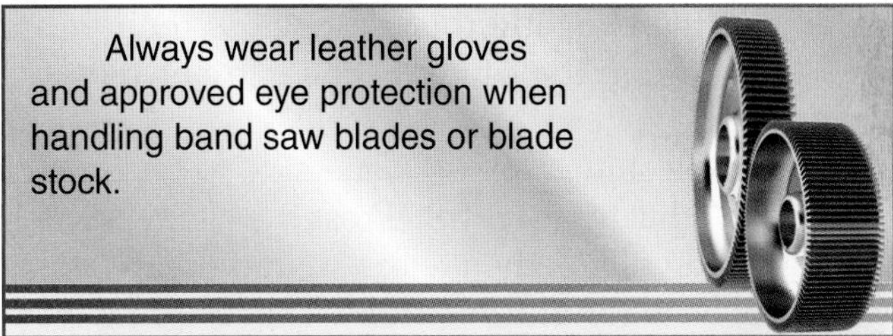

Always wear leather gloves and approved eye protection when handling band saw blades or blade stock.

21.3.1 Preparing Blade for Welding

Use snips or blade cutoff shears to trim the blade to length. Cuts in blade stock must be square or the problems shown in **Figure 21-10** will result.

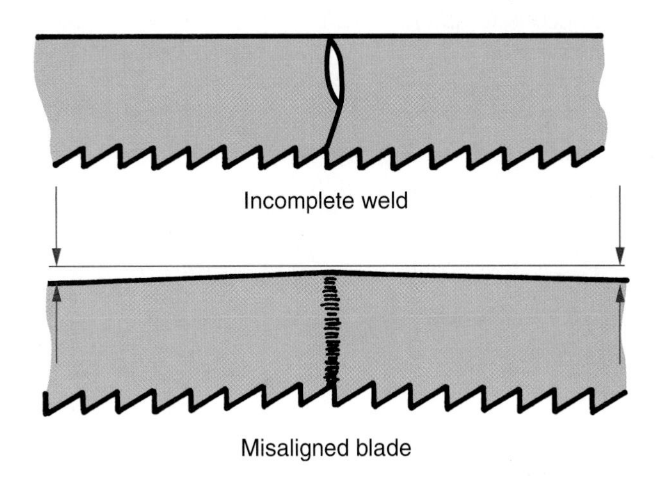

Figure 21-10. Common problems encountered when welding band saw blades. They are usually the result of poorly squared blade ends or dirty blade material.

After squaring the blade ends, it will be necessary to remove several teeth (depending on blade pitch) by grinding. Since about 1/4″ (6.5 mm) of the blade is consumed in the welding process, teeth must be ground off to ensure uniform tooth spacing after the weld has been made. Remove only the teeth. Do not grind into the back of the blade.

21.3.2 Making the Blade Weld

For convenience, most band machines have a welder built-in. This is a resistance butt welder. Blades up to 1/2″ (12.5 mm) wide can be welded in a light-duty welder. A heavy-duty resistance, or "flash," butt welder is required for heavier blades. Clean the welder jaws before and after making a weld.

The blade ends are butted together and clamped in the jaws of the welder. The saw teeth should be placed against the aligning plates. Pressure (determined by the width and thickness of the blade being joined) is applied to the band ends. Check the blade to be sure the ends are touching across their entire width and are in the center of the gap between the welder jaws.

After checking, stand to one side to be clear of any "flash" that might result. Press the welder switch as far in as it will go, then release it immediately. The weld will be made automatically. The resulting weld should look like the one shown in **Figure 21-11**.

SAFETY NOTE

Wear approved eye protection with tinted filter lenses when welding band saw blades. The bright flash can cause eye injury. Refer to the manufacturer's instructions to determine the proper filter lens number.

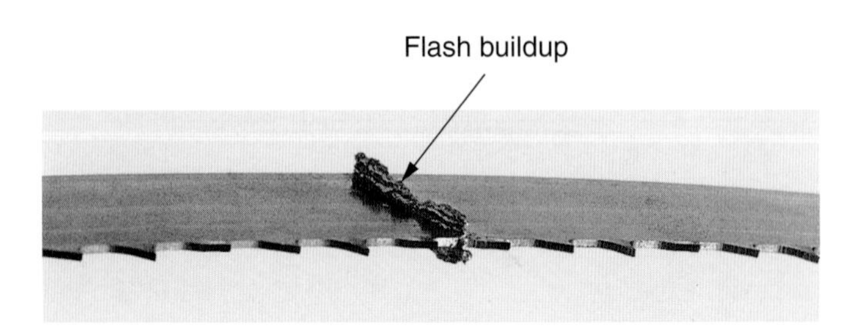

DoALL Co.

Figure 21-11. The flash buildup at the point of the weld should be uniform across the blade. Do not flex the blade at the weld until the welded section has been annealed. It is very brittle and will break if flexed.

The weld, at this point, is brittle and must be annealed before it can be used. Follow the recommendations for annealing furnished by the manufacturer of the machine being used. Avoid overheating the blade, or it will remain brittle. Let it cool slowly after heating.

Remove the "flash" formed during welding on the grinder built into the welder. A finished weld should look like **Figure 21-12**. Use care when grinding to avoid dulling the teeth. The blade will also be weakened if it becomes "dished" during the grinding, **Figure 21-13**.

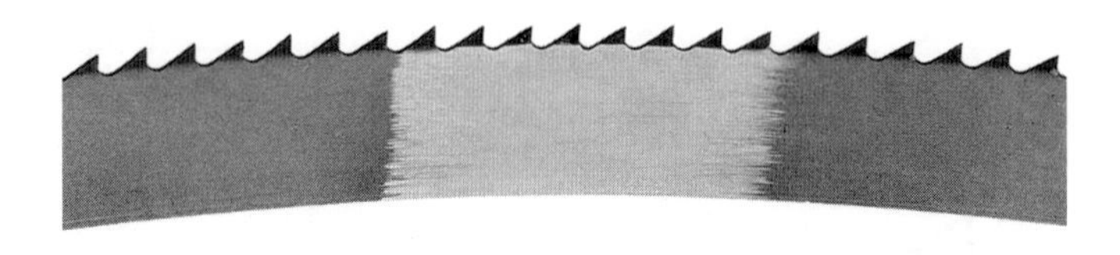

Goodheart-Willcox Publisher

Figure 21-12. A properly made and cleaned weld. If done properly, the weld will be as strong as the blade itself.

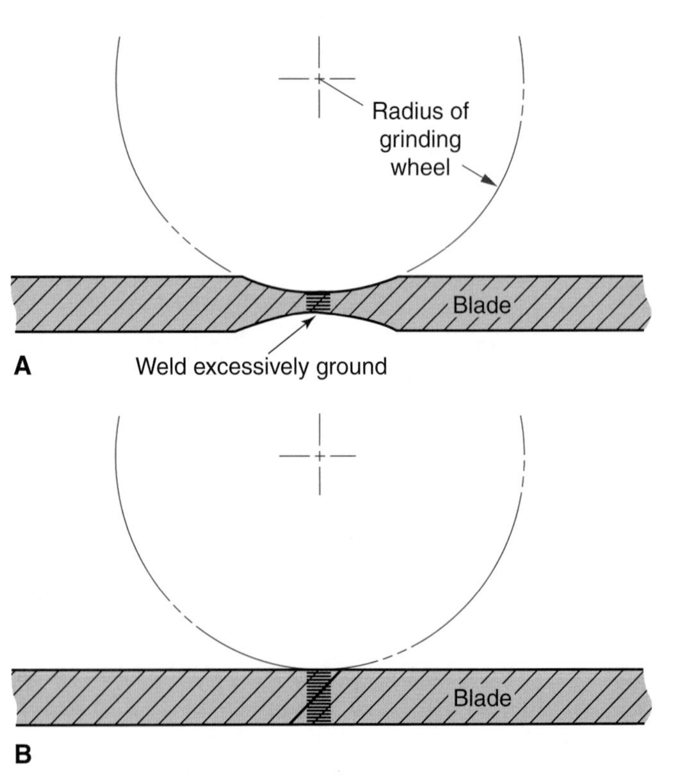

Goodheart-Willcox Publisher

Figure 21-13. A blade is seriously weakened if it becomes "dished" during the grinding operation that removes weld flash. A—Incorrect grinding. Too much material is being removed from the saw blade, causing it to become dished. B—Correct grinding. Only the flash buildup and surface irregularities around the weld are removed.

21.4 Band Machine Preparation

As with all other machine tools, a band machine must be made ready with care if the tool is to operate at maximum efficiency.

21.4.1 Band Machine Lubrication

Use the grades of lubricants specified in the manufacturer's manual for the machine. Develop and use a specific lubrication sequence. It will reduce the possibility of missing a vital point.

21.4.2 Blade Guides

Select and install blade guides suitable for the job at hand, **Figure 21-14**. Use blade guide inserts for light sawing. Roller guides are recommended for continuous high-speed sawing.

Guides must be the proper width, **Figure 21-15**. If they are too wide, the saw teeth will be damaged. If they are too narrow, the blade will tend to twist in the work, making it difficult to follow the desired cutting path.

Before sawing, inspect and clean the upper and lower blade guides. Make sure the backup bearings are not clogged with chips. Typically, there should be 0.001″ to 0.002″ (0.025 mm to 0.050 mm) clearance between the guide and blade, **Figure 21-16**. For best results, follow the manufacturer's recommendations.

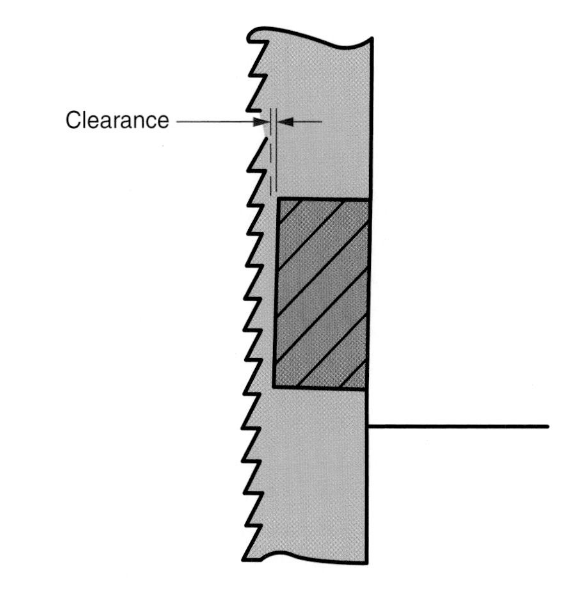

Goodheart-Willcox Publisher

Figure 21-15. Blade guides must be wide enough to prevent the blade from twisting, but narrow enough so they will not damage the teeth.

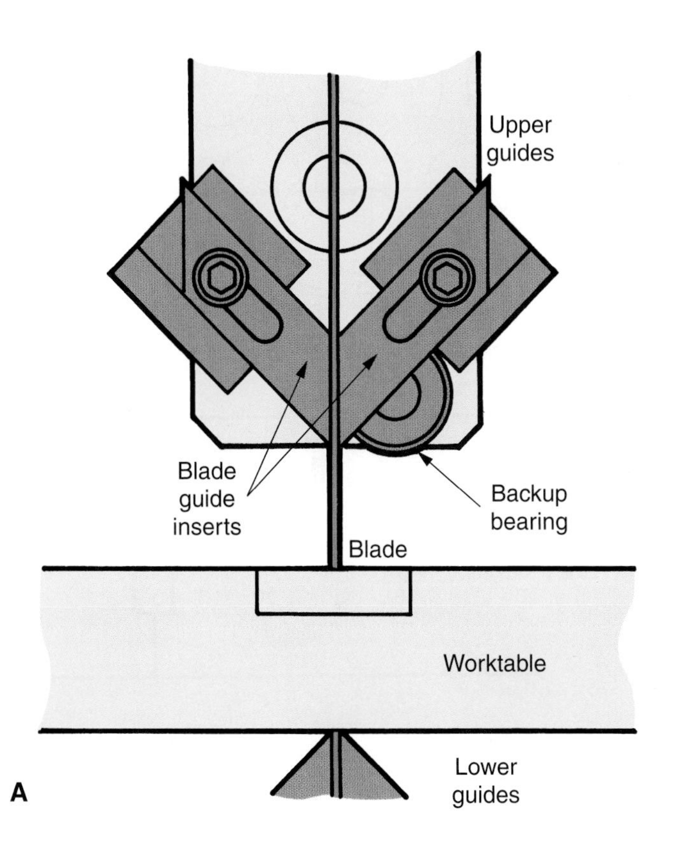

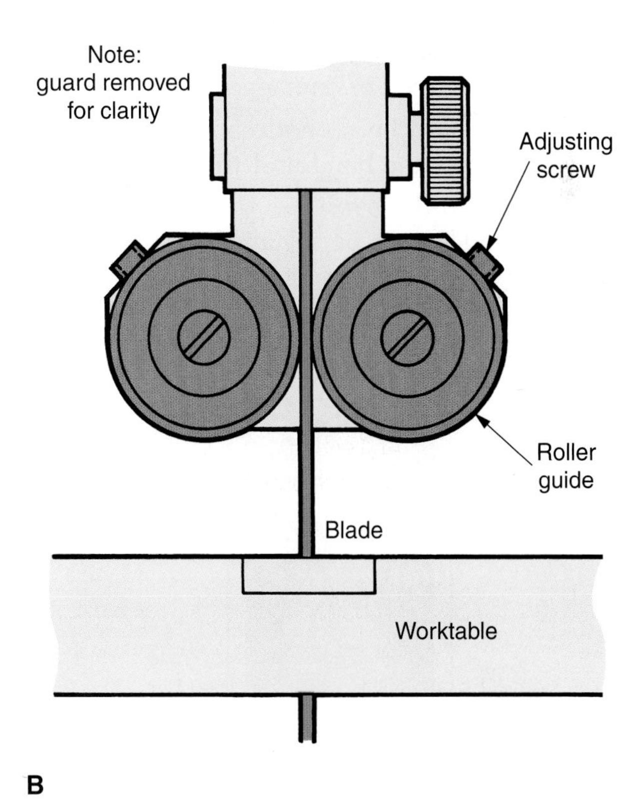

Goodheart-Willcox Publisher

Figure 21-14. Blade guides. A—Blade guide inserts. B—Roller guides.

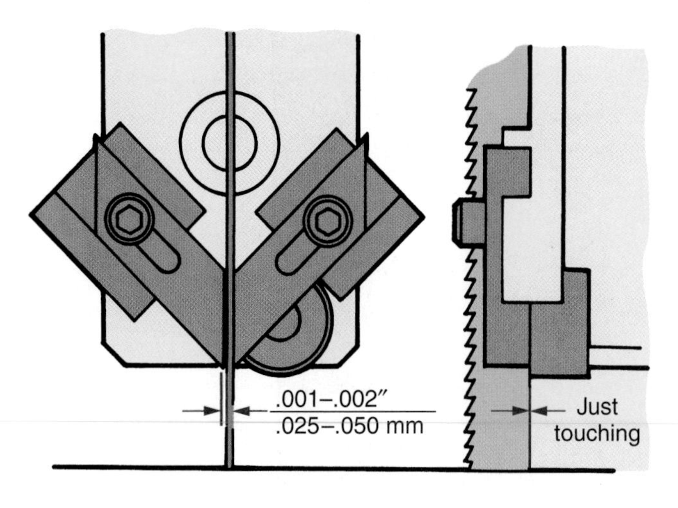

Goodheart-Willcox Publisher

Figure 21-16. Proper blade guide adjustment.

21.4.3 Blade Tracking

Adjust the band carrier wheels so the blade will track correctly. Again, it is important that the manufacturer's recommendations be carefully followed. If the manufacturer's information is not available, observing the following points will usually permit satisfactory operation:

- The center of the band should ride directly over the center of the wheel crown on the rubber tire, **Figure 21-17**. Replace the tire if it becomes frayed or damaged.
- There should be no noticeable gap between the back of the band and the back-up bearings of the saw guides.
- The blade must be installed with the teeth downward.

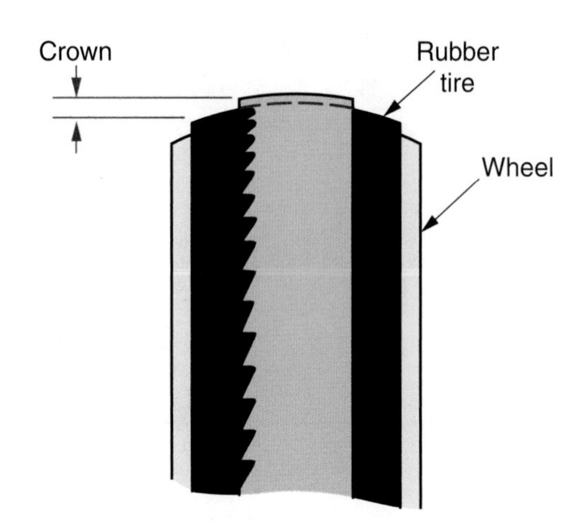

Goodheart-Willcox Publisher

Figure 21-17. The blade should ride directly on the center line of the wheel's crown.

SAFETY NOTE

Exercise extreme caution when handling saw blades. They are sharp and can cause serious injury. Be sure that power to the band machine has been turned off at the master switch before attempting to install a blade.

21.4.4 Band Blade Tension

Blade tension refers to the pressure put on the saw band to keep it taut and tracking properly. On smaller machines, tension is usually applied by means of a hand crank, **Figure 21-18**. On large heavy-duty machines, it is applied hydraulically. The amount of blade tension is determined by the width and pitch of the blade. Use the band tension chart furnished with the machine.

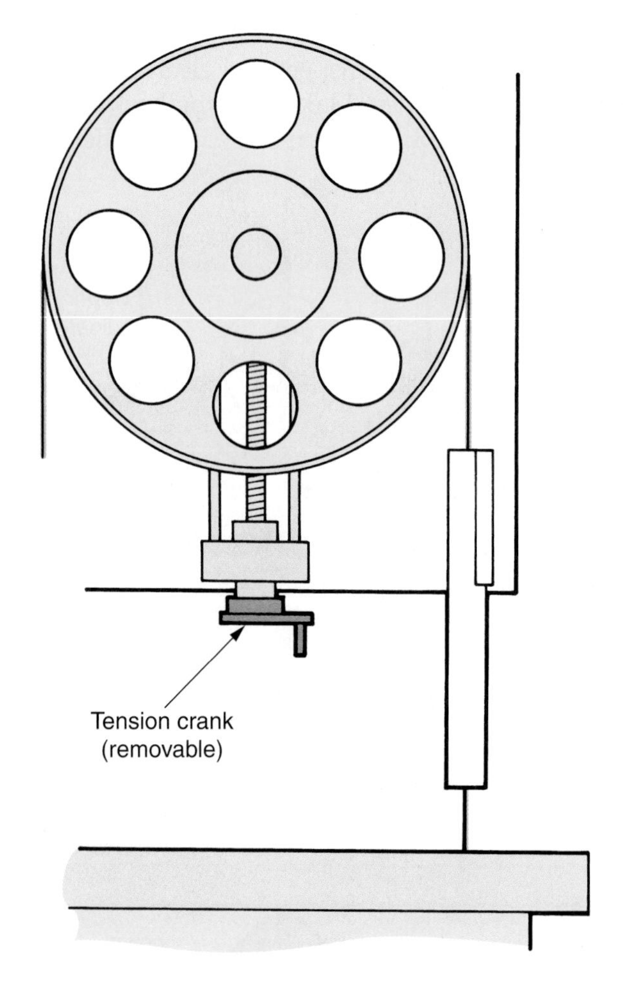

Goodheart-Willcox Publisher

Figure 21-18. Tension-adjusting hand crank.

Many band machines have built-in tension meters that make it easy to adjust and maintain proper blade tension. This is especially important when a new blade is installed. A new blade has a tendency to stretch slightly when first used. This can create a safety problem if tension is not readjusted before it falls off too far.

21.4.5 Band Cutting Speed

As with other machining operations, best results will be obtained if recommended cutting speeds are maintained. See **Figure 21-19** for band speeds of a few selected materials.

21.4.6 Band Cutting Fluids

Cutting fluid can be applied on a band machine by flooding, in the form of a mist, or as a solid-type lubricant.

- Flooding is recommended for heavy-duty band machining.
- ***Mist coolant*** is used for high-speed sawing of free machining nonferrous metals. It is also used when tough, hard-to-machine materials are cut.
- Solid lubricants are applied when the machine does not have a built-in coolant system.

If requested, coolant manufacturers will furnish a coolant chart with recommendations for band machining operations.

21.5 Band Machining Operations

Vertical band machines, like the one shown in **Figure 21-20**, are designed to perform the following sawing operations.

21.5.1 Straight Sawing

Straight, two-dimensional sawing is band machining in its simplest form, **Figure 21-21**. The operator just follows a straight layout line. Slitting, slotting, and notching, can also be done rapidly on a band machine.

Material	Thickness Inches (millimeters)	Band speed Surface feet per minute (meters per minute)
Aluminum alloys	—	360+ sfm (110+) mpm
Low-to-medium carbon steels	Under 1″ (25 mm)	345–360 sfm (105–110) mpm
	1″–6″ (25 mm–150 mm)	295–345 sfm (90–105) mpm
Medium-to-high carbon steels	Under 1″ (25 mm)	225–250 sfm (70–75) mpm
	1″–6″ (25 mm–150 mm)	200–225 sfm (60–70) mpm
Free machining steels	Under 1″ (25 mm)	260–395 sfm (80–120) mpm
	1″–6″ (25 mm–150 mm)	260–345 sfm (80–105) mpm
Titanium, pure and alloys	Under 1″ (25 mm)	100–115 sfm (30–35) mpm
	1″–6″ (25 mm–150 mm)	90–110 sfm (30–35) mpm

Goodheart-Willcox Publisher

Figure 21-19. Recommended cutting speeds for selected metals and alloys.

DoALL Co.

Figure 21-20. Vertical band machine designed for metal machining.

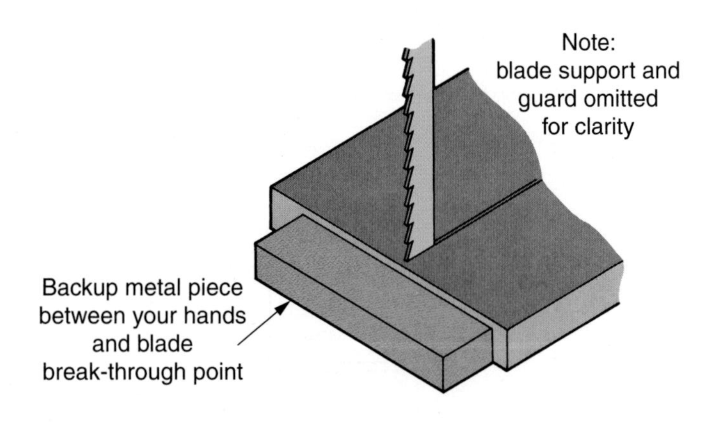

Goodheart-Willcox Publisher

Figure 21-21. An example of two-dimensional band machining. To protect your hands, be sure that a piece of backup metal is in place at the point where the saw blade breaks through.

SAFETY NOTE

Exercise extreme care when the blade breaks through the work at the completion of a cut. If possible, a piece of backup metal, called a "push block," should be between your hand and the point where the band will break through the work. The sharp moving blade can cause serious injury!

21.5.2 Contour Sawing

Contour sawing is possible on a vertical band machine. Machine size is the limiting factor on the work dimensions that can be cut.

21.5.3 Angular Sawing

The table on a vertical band machine is usually mounted on trunnions, permitting it to be tilted. This makes it possible to machine compound angular cuts.

21.5.4 Internal Cuts

Precision internal cuts can be made on a vertical band machine. The band is threaded through a hole drilled in the piece, it is then welded, and the work is maneuvered along the prescribed lines. Additional holes must be drilled if sharp corners are to be made, **Figure 21-22**. It may be necessary to use the blade as a file to get the work into cutting position.

After completing the internal cut, cut the band close to the weld so that the entire weld can be cut away prior to rewelding. It is recommended that there be no more than one weld in the band.

21.6 Band Machine Power Feed

Power feed or mechanical pressure attachments are available for band machines. The simplest attachment makes use of weights to pull work into the blade, **Figure 21-23**. Both hands of the operator are free to guide the work.

Several types of hydraulic power feed attachments have been devised. On some vertical band machines, the worktable is hydraulically actuated and feeds the work into the blade at a constant rate. Accidental overfeeding is eliminated, greatly extending band life.

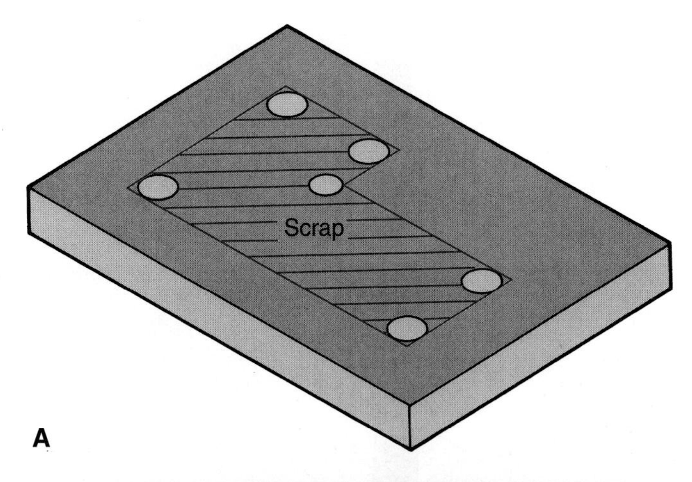

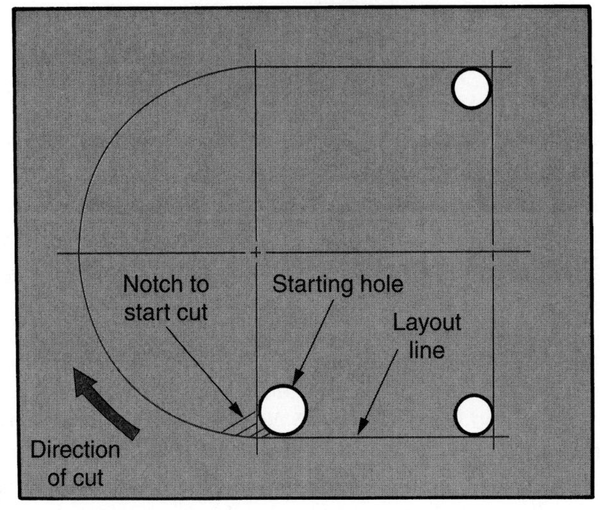

Goodheart-Willcox Publisher

Figure 21-22. Making internal cuts. A—Drilled holes are needed where sharp corners are to be made in an internal cut. B—Start internal cuts by using the blade as a file, and notch the work until the blade can be positioned to start the cut.

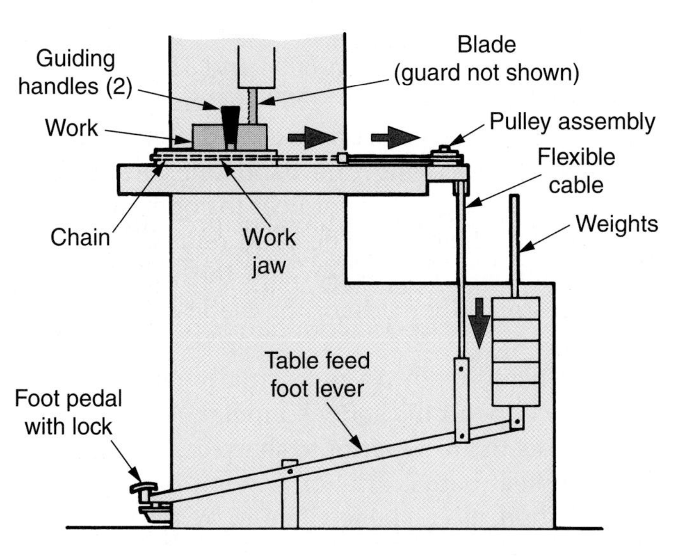

Goodheart-Willcox Publisher

Figure 21-23. The simplest type of power feed system uses weights to pull the work into the blade.

A hand wheel, connected to the work by a sprocket and chain, guides work along the layout line. A servomechanism on the lower blade guide senses changes in feed force on the work and automatically counteracts them. To maintain a constant feeding force on the work, the device advances, slows, stops, or reverses the worktable movement. To leave the operator's second hand free, a foot switch permits moving the table by remote control.

Another type of feed mechanism is a self-contained unit attached to the machine. A hydraulic cylinder applies and maintains constant pressure on the work through a sprocket and chain system. By turning a handwheel one direction or another, the operator can move the sprocket left or right, which in turn rotates the work. **Figure 21-24** shows the parts of this type of system.

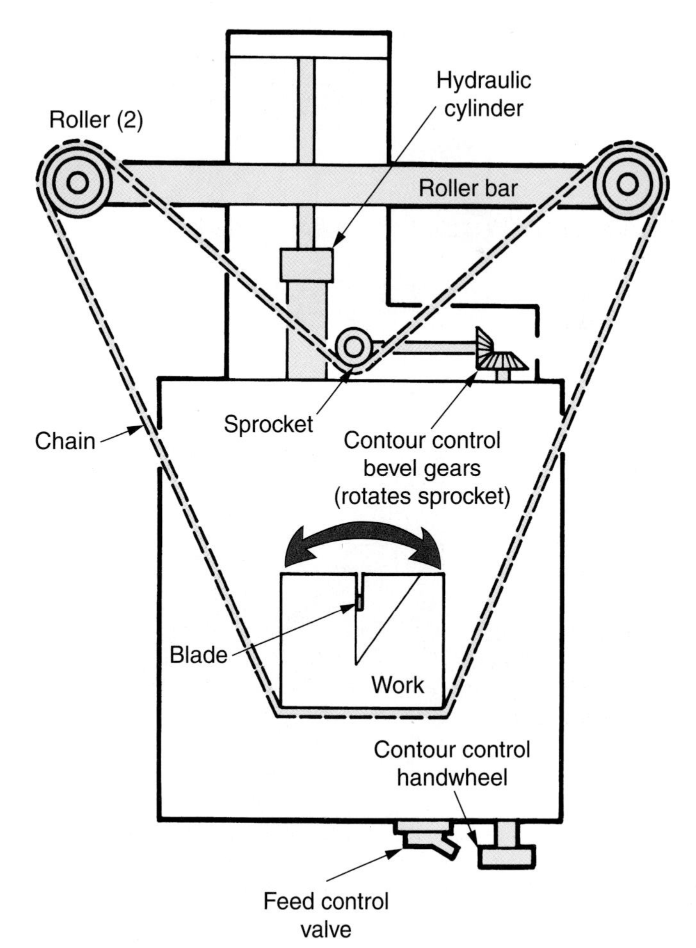

Goodheart-Willcox Publisher

Figure 21-24. A hydraulically actuated power feed unit.

21.7 Other Band Machining Applications

The great versatility of the band machine is further enhanced by the addition of accessories or minor tool modifications.

21.7.1 Band Filing

A smooth, uniformly finished surface may be obtained rapidly and with considerable accuracy on a band machine fitted for filing, **Figure 21-25**. A series of small file segments make up the ***file band***. The individual units interlock and form a continuous file. The segments are fitted to a flexible back. File guides replace the regular saw guides when a band file is used. A variety of file shapes and cuts are available.

21.7.2 Band Polishing

Parts can be polished on a band machine with a polishing attachment, **Figure 21-26**. A continuous band abrasive cloth replaces the saw blade.

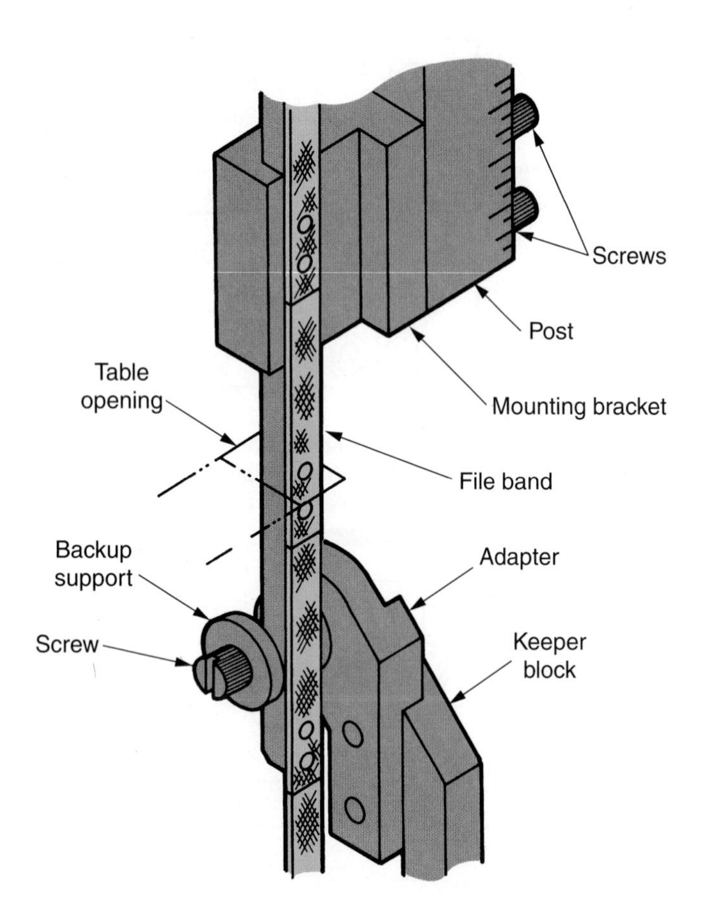

Goodheart-Willcox Publisher

Figure 21-25. Band filing requires that special guides be used. In this illustration, the worktable has been removed for clarity.

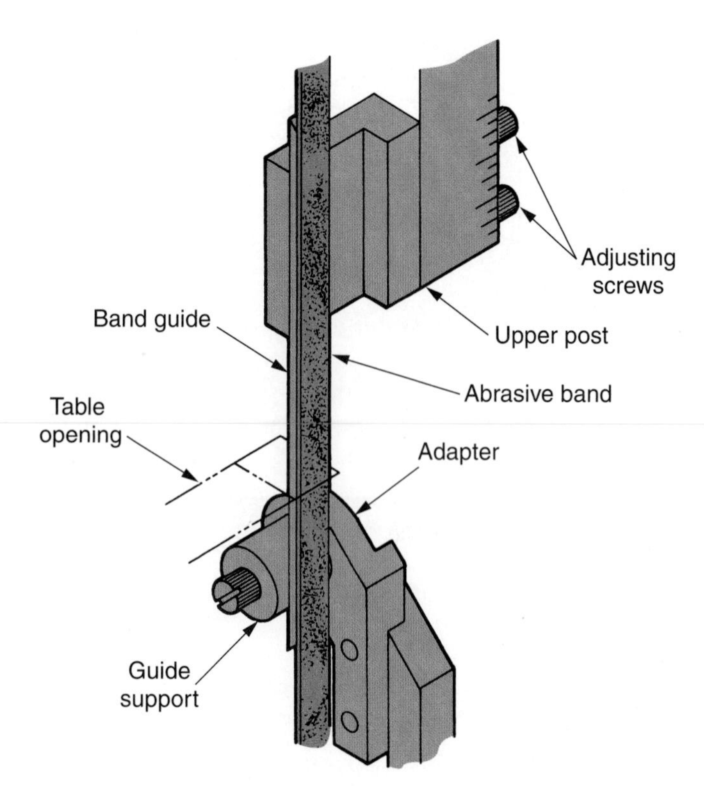

Goodheart-Willcox Publisher

Figure 21-26. Polishing can be done on a band machine by replacing the saw blade with an abrasive band, and using the guide and support shown. Worktable has been removed for clarity.

Best results can be obtained if the back of the abrasive band is lubricated with graphite powder. Abrasive band life will also be greatly extended.

21.7.3 Friction Sawing

Friction sawing makes use of extremely high cutting speeds of between 6000 and 15,000 feet per minute (fpm) or 1800 to 4500 meters per minute (mpm) and heavy pressure to cut ferrous metals.

In friction sawing, the band operates at high speed, generating sufficient heat to soften the metal just ahead of the blade. The band removes the softened metal as it moves through the work. Only a small area on either side of the blade is affected by the heat.

The blade teeth do not actually cut—they are used to scoop out the softened metal. As a matter of fact, dull teeth are superior to sharp teeth, since they generate heat better. Friction sawing is a spectacular operation, as can be testified to by the shower of sparks produced.

The technique is a rapid way to cut ferrous metals under 1″ (25 mm) in thickness. Thicker stock can be cut if a rocking movement is used. The hardness of the metal being cut does not affect the effectiveness of this cutting method.

The band drive wheels on machines used for friction sawing are usually large in diameter and are carefully balanced for smooth, vibration-free operation. Large wheels are needed to reduce fatigue on the band due to the high operating speed. A larger diameter keeps the band from bending as sharply as it loops around the wheel, reducing stress on the band.

Band machines designed for friction sawing cannot be used for conventional band machining unless they are equipped with variable speed drives. Most band machines are not adaptable for friction sawing.

SAFETY NOTE

Friction sawing requires a full face shield, leather gloves, and a transparent shield fitted around the cutting area.

21.7.4 Other Band Tools

Tooth-type bands are most commonly used on the vertical band machine. However, other types of blades have been developed for special work, **Figure 21-27**.

The ***knife-edge blade*** is used to cut material that would tear or fray when machined by a conventional blade. For example, sponge rubber, cork, cloth, corrugated cardboard, and rubber would tear easily.

The ***diamond-edge band*** is specially designed to cut material that is difficult or impossible to cut with a conventional toothed blade. The diamonds are only on the front edge of the band where the cutting is accomplished. On a wire band, diamonds are fused around the circumference on the band permitting it to cut in any direction.

Friable materials (those that are easily crumbled or reduced to powder) of extreme hardness or with abrasive qualities can be cut economically with special bands.

In addition to the bands mentioned, blades with unusual characteristics are available to meet almost any band machining requirement.

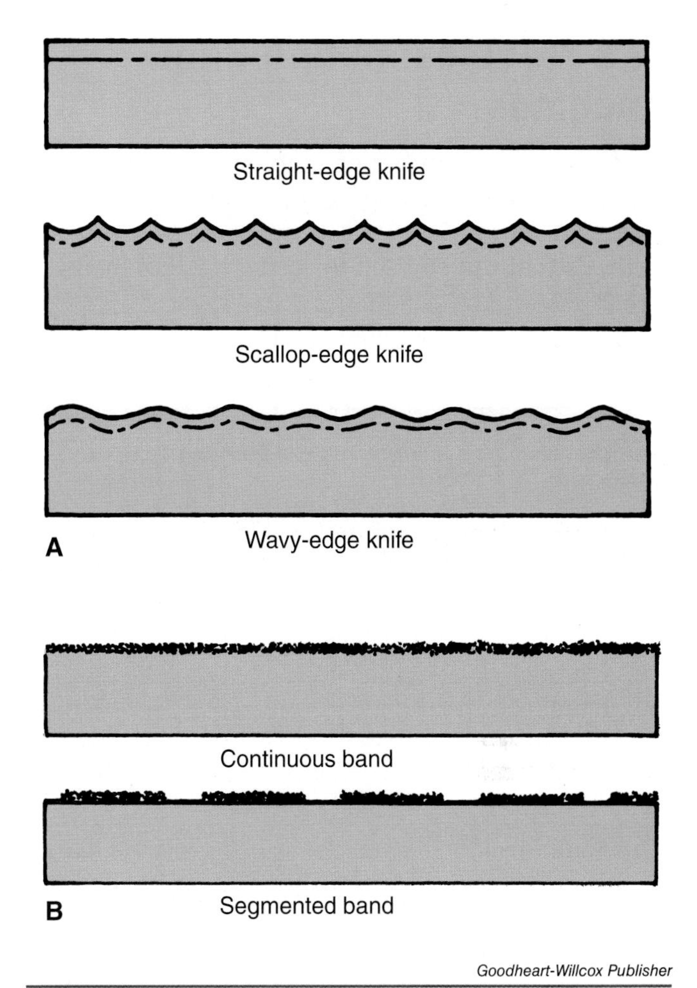

Figure 21-27. Other band tools. A—Types of knife edge blades. B—The cutting edge of a diamond-edge band is impregnated with diamond dust.

21.7.5 Specialized Vertical Band Machines

The vertical band machine is manufactured in a wide range of sizes, and has been adapted to do many kinds of band machining. Some large machines have been fitted with closed circuit TV and a remote control console to permit the operator to contour-machine large sections of material or to perform hazardous or dangerous work. For example, this type of machine is used to machine toxic and radioactive materials.

Band machines with computer numerical control (CNC) are also available. They have "X" and "Y" table movement capability plus circular interpolation (allowing them to cut circular and elliptical contours).

21.8 Troubleshooting Band Machines

There are a number of problems that can occur during band machining. Use the table shown in **Figure 21-28** to identify various problems and the methods that can be used to correct each of them.

21.9 Band Machining Safety

- Do not attempt to operate a band machine until you have received instructions on its safe operation.
- Never attempt to operate the machine while your senses are impaired by medication or other substances. Be sure all guards are in place before starting to operate a band machine.

Troubleshooting Band Machines	
Problem	**Correction**
1. Teeth dull prematurely.	a. Use slower cutting speed. b. Replace blade with a finer pitch band. c. Be sure proper type cutting fluid is used. d. Increase feed pressure. e. Check to be sure band is installed with teeth pointing down.
2. Band teeth breaking out.	a. Reduce feed pressure. b. Use finer pitch band if thin material is being cut. c. Be sure work is held solidly as it is fed into band. d. Use a heavier-duty cutting fluid.
3. Band breaks.	a. Change to a heavier band. b. Reduce cutting speed. c. Check wheels for damage. d. If blade breaks at weld, use longer annealing time. Reduce heat gradually. e. Use finer pitch blade. f. Reduce feed pressure. g. Decrease band tension. h. Check blade guides for proper adjustment. i. Use cutting fluid.
4. Cutting rate too slow.	a. Increase band speed. b. Use coarser pitch blade. c. Increase feed presssure. d. Use cutting fluid.
5. Band makes "belly-shaped" cut.	a. Increase blade tension. b. Adjust guides close to work. c. Use coarser pitch band. d. Increase feed pressure.
6. Band does not run true against saw guide backup bearing.	a. Remove burr on back of band where joined. b. If hunting back and forth against backup bearing on guide, reweld blade with back of band in true alignment. c. Check alignment of wheels. d. Check backup bearing. Replace if worn.
7. Premature loss of set.	a. Use narrower band. b. Reduce cutting speed. c. Apply cutting fluid.

Goodheart-Willcox Publisher

Figure 21-28. Common band machine problems and their possible solutions.

- Do not start a cut until the guides have been set properly. Guides should be positioned as close to the work as the job will permit.
- Other than changing blade speeds (some machines require the band to be running when speed changes are made), make no adjustments until the blade has come to a complete stop.
- Wear eye protection and leather gloves when handling band blades or blade material.
- Get help when handling heavy material.
- Remove burrs and sharp edges from the work as soon as possible. They can cause serious cuts. Have cuts and bruises treated immediately. Report all injuries.
- Do not clean chips from the machine with your hands. Use a brush. Stop the machine before attempting to clean it.
- Keep your hands away from the moving blade. Use a push block for additional safety. Never have your hands in line with the cutting edge of the band.

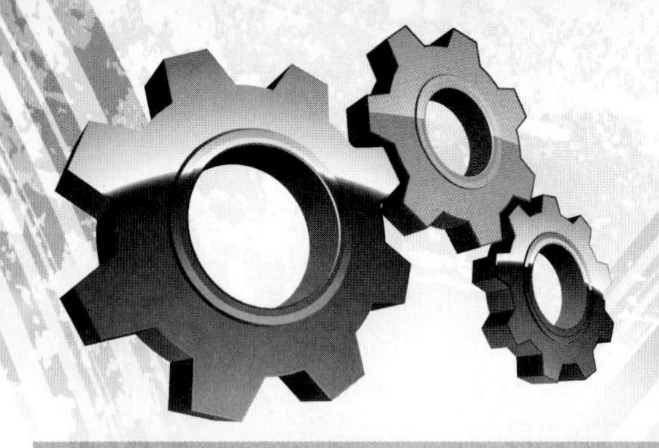

Chapter Review

Summary

- Band machining offers a variety of advantages over other machining operations including efficient material removal and downward cutting action.
- Blade type and blade characteristics like the width, pitch, set, gage, and tooth form are all important factors when choosing the proper blade.
- For a blade to be properly welded, the ends must be squared, several teeth are removed, the ends are welded together, the flash is ground away, and the blade is annealed.
- Proper band machine and blade maintenance are important for safe band machining operation.
- Special blades and attachments for band filing, polishing, friction sawing, and other operations make the band machine very versatile.

Review Questions

Answer the following questions using the information provided in this chapter.

1. Band machining makes use of a(n) _____ saw blade.
2. List three advantages band machining has over other machining techniques.
3. What two points must be considered when selecting a blade for a specific job?
4. When making straight cuts, use the _____ blade the machine can accommodate.
5. What does blade pitch refer to?
   A. Width of the blade in inches or millimeters.
   B. Thickness of the blade in inches or millimeters.
   C. Number of teeth per inch of blade or tooth spacing in millimeters.
   D. All of the above.
   E. None of the above.
6. Tooth form is the _____.
   A. shape of the tooth
   B. thickness of the blade
   C. number of teeth on the blade
   D. All of the above.
   E. None of the above.
7. A resistance _____ welder is used to weld band machine blades.
8. The blade must be _____ after welding because the joint is extremely brittle and cannot be used in this condition.
9. The cutting tool must be installed with the teeth facing _____.
10. Blade tension is the pressure put on the saw band to _____.
    A. cut the metal more rapidly
    B. keep it taut and tracking properly
    C. reduce the power needed to do the cutting
    D. All of the above.
    E. None of the above.

For questions 11–13, match each description with the correct term.

11. _____ is recommended for heavy-duty sawing.
12. _____ is used for high-speed sawing of free machining nonferrous metals.
13. _____ are applied when the machine is not fitted with a built-in coolant system.
    A. Mist coolant
    B. Flooding
    C. Solid lubricants

14. What is the simplest form of band machining?
15. The worktable on many vertical band machines can be tilted to make _____ cuts.
16. How can internal cuts be made on a band machine?
17. Smooth, uniformly finished surfaces are possible when the machine is fitted for _____.
18. Describe friction sawing.
19. Of what use is a knife-edge blade on a band machine?
20. When are diamond-edge bands used?
21. What is unique about a diamond impregnated wire band?

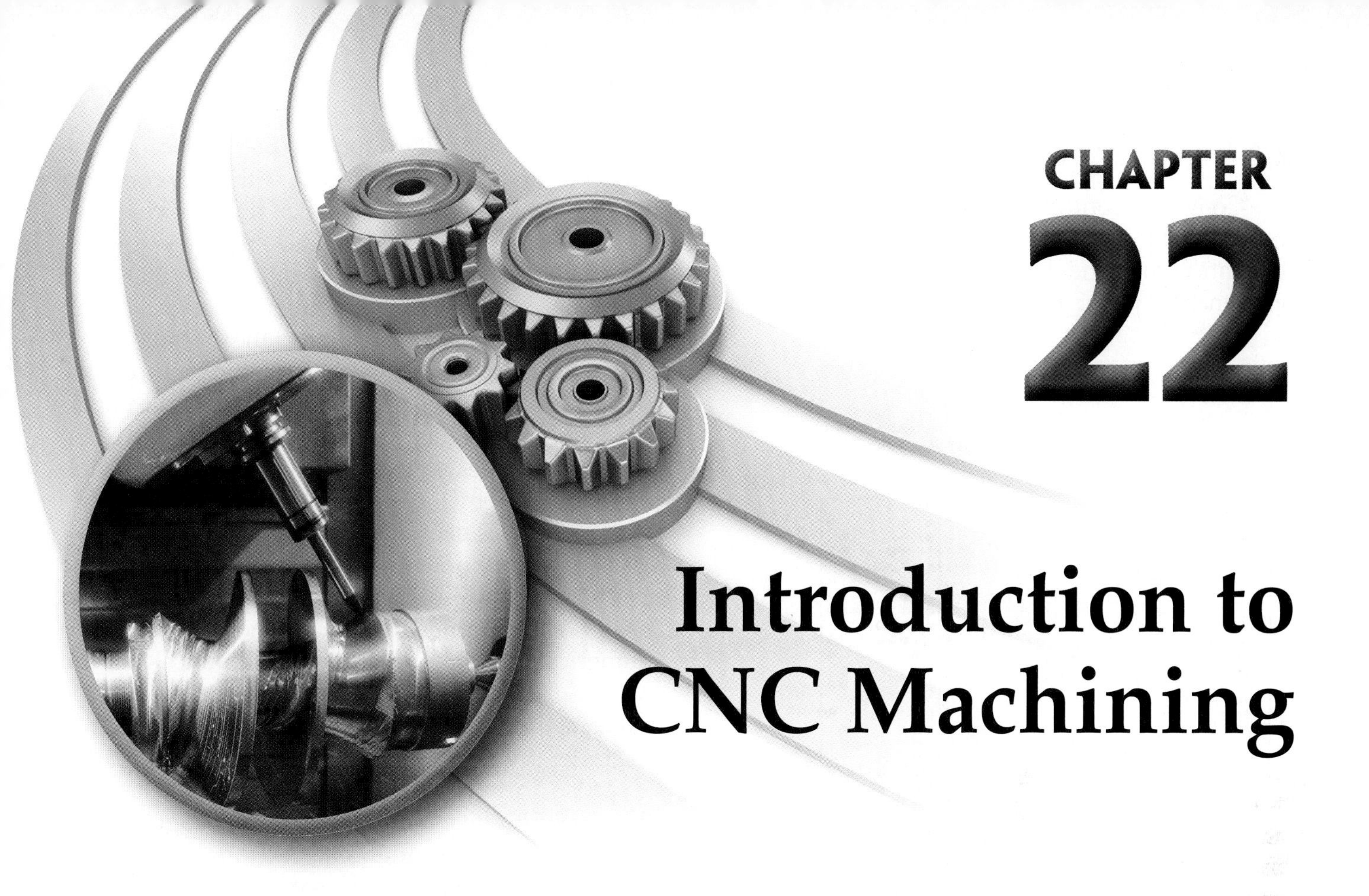

CHAPTER 22

Introduction to CNC Machining

Learning Objectives

After studying this chapter, you will be able to:

- Describe the development of CNC technology.
- List advantages and disadvantages of using CNC technology.
- Describe the features of CNC milling machines.
- Compare the characteristics of various types of CNC turning machines.
- Identify safety guidelines for CNC machining processes.
- Summarize the use of the Cartesian and polar coordinate systems in CNC technology.
- Contrast the two types of motors that are commonly used to drive CNC machines.

Technical Terms

automatic tool changer (ATC)
backlash
Cartesian coordinate system
closed-loop system
conversational language
coordinate system
dry cycle
encoder
horizontal machining center (HMC)
machining center
open-loop system
polar coordinate system
servomotor
stepper motor
Swiss-type turning center
turning center
vertical machining center (VMC)

22.1 History of CNC

Computer-controlled machining got its start during the 1950s as numerical control (NC). Programs to control machine tools were created and stored on punched paper tape, **Figure 22-1**. Programmers punched holes in the tape at specific locations to produce multiple combinations, with each combination representing a separate letter, number, character, or command. Paper tape readers read the rows of punched holes on the tape at a rate of up to 1,000 lines per second and transferred the information to the computer that controlled the motion of the machine tool.

When these machines were first introduced, the cost of the machines and the skills needed to operate them outweighed the potential cost savings, so many companies were slow to develop the technology. However, by the end of the 1950s, the use of NC equipment was beginning to increase.

The development of the minicomputer in the 1960s dramatically increased the acceptance and use of NC equipment. This technology converted numerical control to computer numerical control (CNC). Computer control lowered the cost of the equipment and made it more cost-effective due to the faster processing capability of the minicomputer.

Also during the 1960s, programming codes were developed to help standardize CNC programming languages, as up to this time each machine manufacturer had implemented its own language. These codes, which are still in use today, are described in Chapter 23, *Basic CNC Programming*.

As computer and machine drive technologies continued to develop from the 1970s through the 1990s, equipment became faster and more capable of producing complex shapes, making CNC machining a more cost-effective method, **Figure 22-2**. Also, the use of computer memory to store programs began to displace the use of punched paper tape. Today, the use of punched paper tape is all but nonexistent.

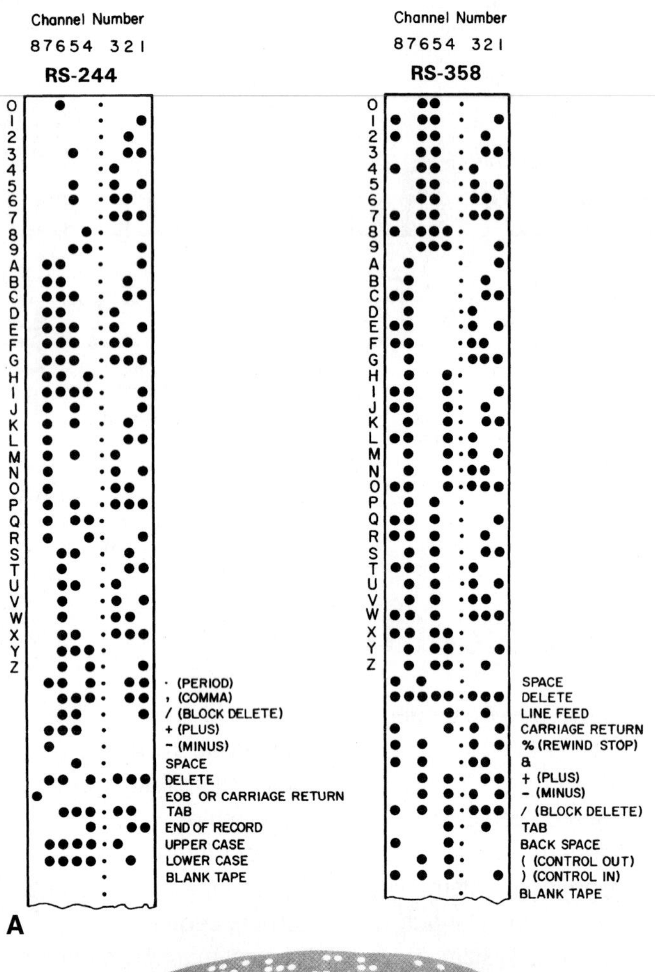

Goodheart-Willcox Publisher; Ailisa/Shutterstock.com

Figure 22-1. Punched tape. A—Tape code. Note that every level of RS-244 tape has an odd number of punches for parity check. RS-358 tape has an even number. Parity check is a method of automatically checking to reduce possibility of tape errors caused by malfunctioning tape punch. B—A roll of punched tape.

22.2 Advantages and Disadvantages of Using CNC

The primary advantage of using CNC equipment is that it removes much of the variation caused by human interaction during the manual machining process. Even the best machinists cannot perfectly duplicate their actions for every part. A CNC machine is not perfect, but its movements are better controlled than those of a human, which reduces variation in the parts that are produced. Another general advantage is the rate at which parts are produced. The faster parts can be machined, the less they cost per unit to manufacture. This allows manufacturers to offer lower prices to customers, making them more competitive in the global marketplace.

Figure 22-2. A helical impeller machined on a vertical machining center. This part can only be machined thanks to the additional axes of rotation provided by the tilting trunnion table and rotary table.

The biggest disadvantage of using CNC is the cost of the machine. As costs continue to decrease, CNC technology is becoming more affordable. However, some small companies still lack the resources to purchase the necessary equipment.

22.3 CNC Milling Machines

A basic CNC milling machine is a traditional milling machine with built-in CNC capabilities. ***Machining centers*** are CNC milling machines equipped with an ***automatic tool changer (ATC)***, a device that automatically changes and stores the tools, **Figure 22-3**. The machining cycle pauses for the automated tool change process, but the automated tooling change takes much less time than a manual tooling change. Most machining centers are enclosed.

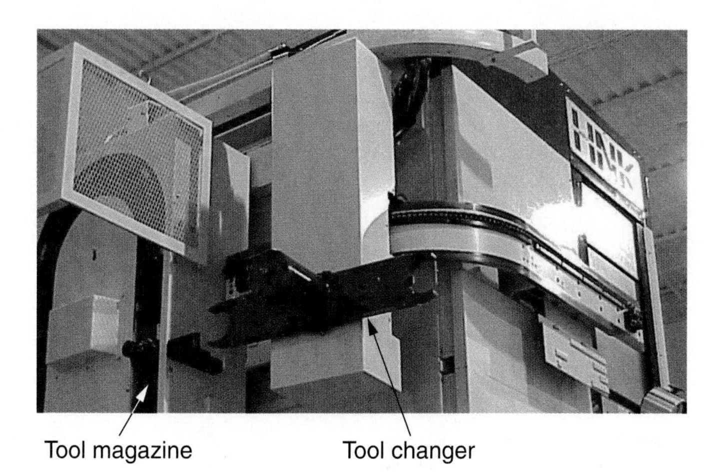

Goodheart-Willcox Publisher

Figure 22-3. This automatic tool changer (ATC) holds the tools in magazine and uses a pivot insertion system to quickly and automatically change tools on a horizontal machining center.

The advantages of CNC milling machines is that they increase part output through faster operation compared to manual machining, and their reduced variation yields a higher quality product. Options such as additional rotating axes can be added, which increase the versatility of the machine, but can also significantly increase the cost.

The greatest disadvantage of CNC milling machines is their high cost. Various types of CNC milling machines are available, with prices ranging from moderately expensive to hundreds of thousands of dollars or more.

22.3.1 Basic CNC Milling Machines

The basic CNC milling machine, like its manual counterpart, has three axes of movement and requires tooling to be changed manually. The machine shown in **Figure 22-4** is an example of a

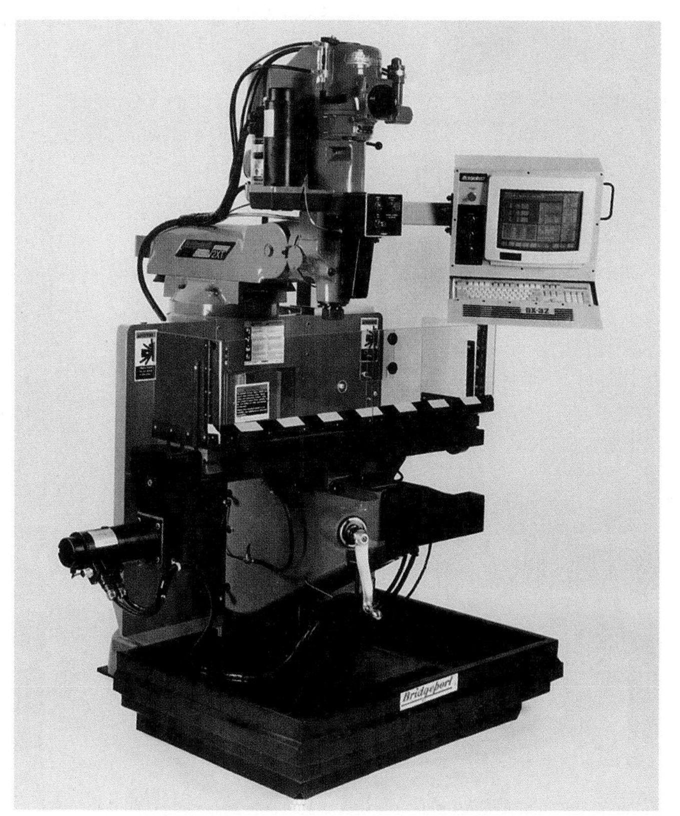

Bridgeport Machines, Inc.

Figure 22-4. A 3-axis CNC vertical milling machine. This basic CNC milling machine can be controlled both manually and in full CNC mode. These milling machines are often called Bridgeport mills after the manufacturer.

basic CNC milling machine. It can be operated both manually and in full CNC mode.

Traditional machinists can operate a basic CNC milling machine manually, or they can be trained to create programs using a method known as ***conversational language***. In this method, the control is equipped with software that allows the operator to select operations from menus. The software converts the operator's input into standard programming code for the control, which then drives the equipment to machine the parts. Conversational language is described in Chapter 23, *CNC Programming Basics*.

Basic CNC milling machines are usually equipped with quick-change collets to allow the operator to change tools during the programmed cycle, **Figure 22-5**. A CNC program can be written with pauses to allow an operator to change tools manually. Quick-change collets are less rigid than normal collets and can only be used for light- to medium-duty operations.

Royal Products, Division of Curran Manufacturing Corporation

Figure 22-5. Inserting an endmill mounted in a quick-change collet. Quick-change tool systems provide a fast method for changing tools on basic CNC milling machines.

The biggest advantage of the basic CNC milling machine is its cost. It is well suited to the toolroom environment and allows traditional machinists to operate it manually or in full CNC mode to create small batches of parts. However, the basic CNC milling machine is not suited for high-volume production because it lacks the speed, precision, and repeatability necessary for such applications.

22.3.2 Machining Centers

Machining centers come in two varieties based on the orientation of the spindle, either horizontal or vertical. ***Horizontal machining centers (HMCs)*** have the same features as manual horizontal milling machines. HMCs offer rigid, horizontally mounted spindles and stable of workholding methods, **Figure 22-6**.

Vertical machining centers (VMCs) have vertically mounted spindles and feature the same capabilities of manual vertical milling machines with the benefit of computer control and automatic tool changing, **Figure 22-7**. VMCs can be equipped with additional axes of rotation for machining parts with complex features that cannot be machined in a single CNC program cycle within the cubic space provided by a basic 3-axis system, **Figure 22-8**. This option allows for more flexibility in the types and complexities of parts that can be produced.

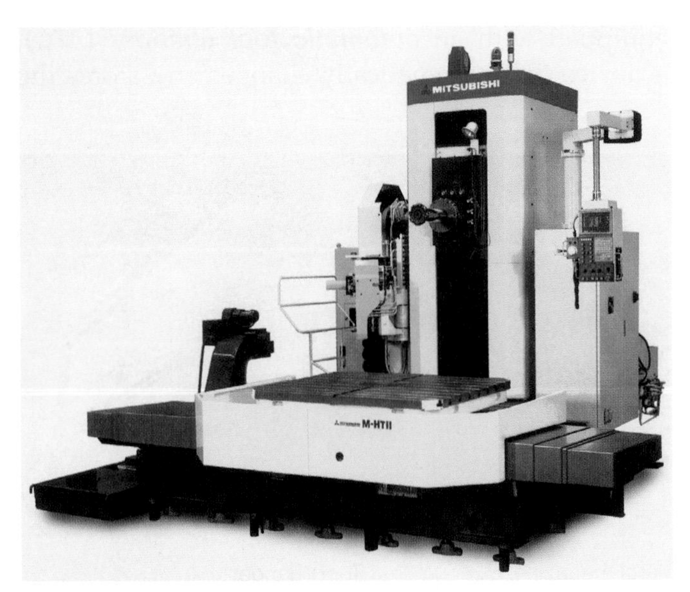

MHI Machine Tool U.S.A., Inc.

Figure 22-6. A fixed-bed horizontal machining center. This CNC milling machine has a 40-ton automatic tool changer and can handle work weighing up to 17,600 lb (8000 kg).

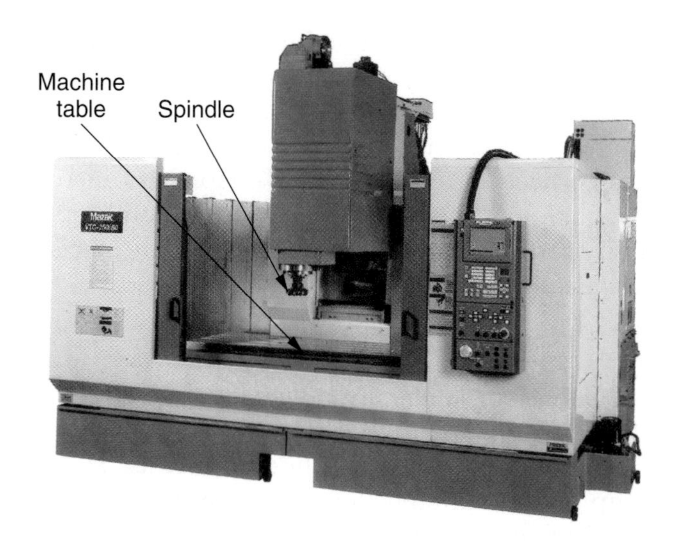

Mazak

Figure 22-7. A vertical machining center (VMC) with twenty-four tool capacity.

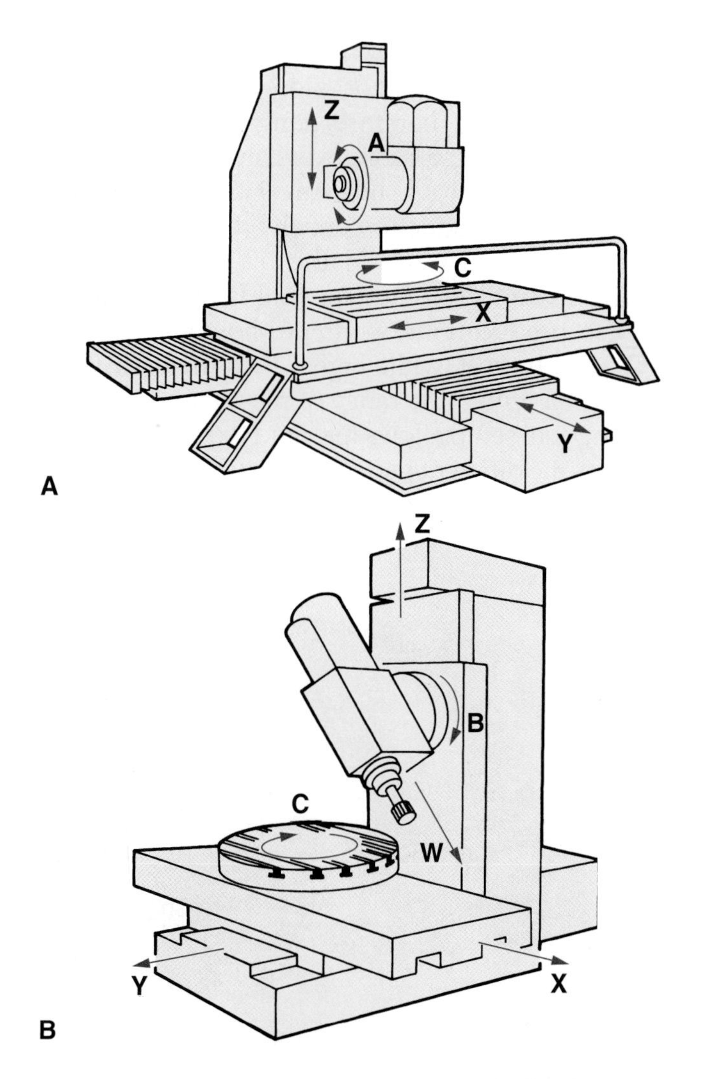

Goodheart-Willcox Publisher

Figure 22-8. Multiaxis machining. A—A 5-axis CNC milling machine. B—A 6-axis vertical machining center (VMC).

The advantages of the VMC are tied to its machining capabilities. VMC machines vary in accuracy, precision, speed, and power, along with tool change capabilities and the ability to add rotational axes. Their biggest disadvantage is cost. As capabilities increase, so do the costs of the machines. Costs can be controlled by selecting equipment with fewer options, then tailoring the machine setups in-house to meet the requirements. However, such practices limit the flexibility of the machines.

22.4 CNC Turning Machines

CNC turning machines are designed to machine cylindrical parts about an axis of rotation. A CNC turning machine is basically a computer-controlled lathe, with the same features that a manual lathe would have. A CNC turning machine equipped with an automatic tool changer is known as a ***turning center***, **Figure 22-9.**

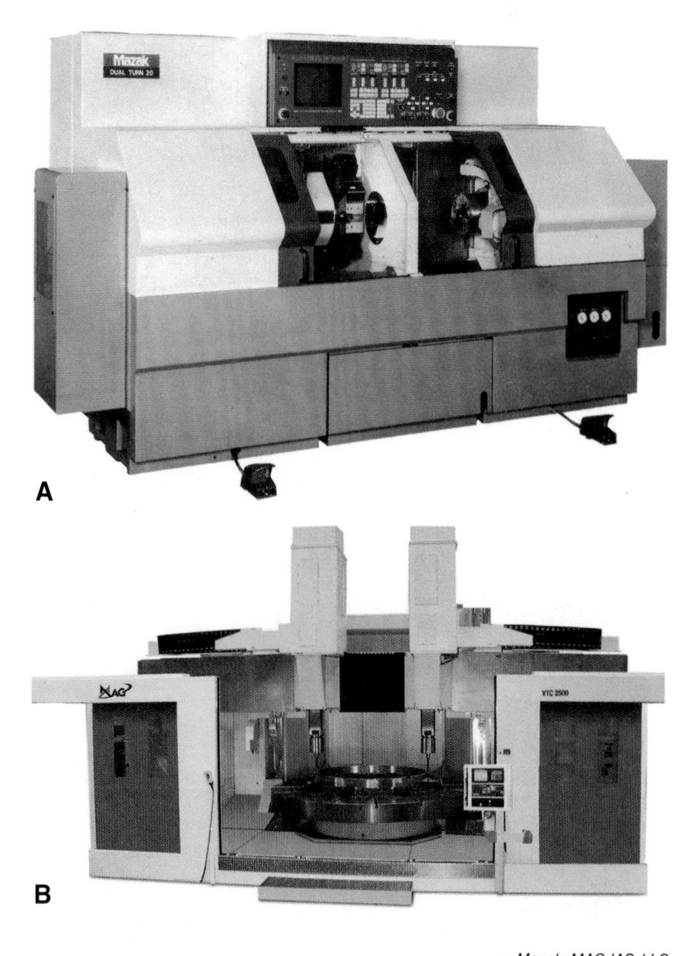

Mazak; MAG IAS, LLC

Figure 22-9. Turning centers. A—This 4-axis turning center permits operations to be performed on both sides of the workpiece, reducing machining time. B—A 4-axis vertical turning center.

There are many types of CNC turning machines, each with their own advantages and disadvantages. The four types of CNC turning machines described in this chapter are presented in order from least expensive to most expensive because one of the first things a company studies when considering the purchase of CNC equipment is the cost of the machinery.

22.4.1 Basic CNC Lathes

Basic CNC lathes are common lathes with CNC capabilities included, **Figure 22-10**. Some CNC lathes can be operated in manual mode, just like a traditional engine lathe, as well as in full CNC mode. The same type of programming and control system is built into this machine as in more expensive models. Its computer capabilities differ only in the amount of storage memory available. As memory capacity increases, so does the cost of the machine.

Tooling changes are done by the operator. The biggest advantage of the basic CNC lathe is its cost. Since tool changes are not automated on a basic machine, the cost of the hardware required for such automation is avoided.

CNC lathes work well in a toolroom environment. Traditional machinists can operate them manually or, if a special run of multiple pieces is required, operators can be trained to create programs using conversational language. This allows operators to program machines without an intricate knowledge of CNC programming.

The disadvantage of CNC lathes is that they are not suited for high-volume production. Their speed, precision, and repeatability are better than manual machining, but they cannot match the capabilities of more advanced CNC machines. CNC lathes can be equipped with gang-tool setups or automated turrets, described later in this chapter, to increase speed. However, their lack of precision and repeatability can create quality problems in high-precision manufacturing.

22.4.2 Gang-Tool Lathes

Gang-tool lathes are a step up from the basic CNC lathe. Gang-tool lathes have multiple tools mounted along the cross-slide, **Figure 22-11**. This setup allows different tools to be used in a single program cycle. The tool to be used is controlled by the position of the cross-slide relative to the workpiece mounted in the spindle and collet.

Although this type of setup can be added to a basic CNC lathe, the cross-slide travel on a basic machine is usually limited to a maximum of 12″ to 18″, which does not allow room for several tools to be set up. Therefore, gang-tool setups typically require a larger, more expensive machine. The need for a larger cross-slide travel and the need for higher-volume production usually drive the need to upgrade from a basic CNC lathe to a gang-tool lathe.

The advantages of a gang-tool lathe include the ability to produce more parts per hour than a basic CNC lathe. Its cost, while higher than the basic CNC lathe, is still within the range of most manufacturers. The gang-tool setup allows for rapid tool changes and, as long as the different parts being machined are of the same or similar diameter, low changeover times.

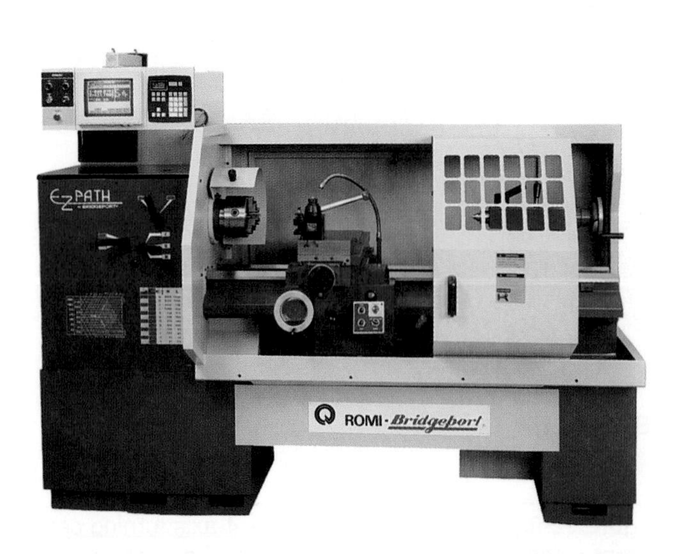

Bridgeport Machines, Inc.

Figure 22-10. A basic CNC lathe with 2-axis CNC control.

Goodheart-Willcox Publisher

Figure 22-11. Multiple tool setup on a CNC gang-tool lathe. The workpiece is held in a collet chuck.

The disadvantages of the gang-tool lathe are its limited flexibility and the space required between the tools in the setup. If a changeover is required between parts whose diameters are drastically different, the entire toolholder setup on the cross-slide must be adjusted, adding to the time required to execute the changeover. This is especially true when there are several tools set up along the cross-slide.

Gang-tool setups work well for machining small-diameter parts. Because the part turns between the tools on the cross-slide, the smaller diameters allow for more tools to be set up in gang-type fashion. Larger-diameter parts do not work well with a gang-tool setup, because the space needed between tools for the larger-diameter workpiece restricts the number of tools that can be set up on the cross-slide.

22.4.3 Turret Lathes

CNC lathes that have a rotating turret attached to the cross-slide of the lathe are called turret lathes. The computer controls the rotation of the turret and the movement of the cross-slide based on the information written into the CNC program. The part is held in a collet or chuck on the spindle. As the spindle is rotated, the turret rotates the appropriate tool into place, and the cross-slide feeds the tool into the workpiece, **Figure 22-12**. The apron feeds the cross-slide along the axis of the workpiece as it rotates. When this machining step is complete, the cross-slide withdraws from the workpiece, the turret rotates the next tool into place, and the next machining step begins. This process repeats until the machining process is complete.

The advantages of a turret-type CNC lathe are its accuracy and speed. With the proper toolholding technology, the machine can be changed over from one part to another in a minimal amount of time.

The disadvantages of turret-type machines include their cost and the skill set needed by operators to operate the equipment effectively. Tooling changes and part setups and changeovers are part of the daily operation of this machine, so the operator must be familiar with the machine control. Often, the operator has to make frequent adjustments to the program to modify tooling offsets. The level of operator training required is therefore higher than for basic and gang-tool CNC lathes.

22.4.4 Swiss-Type Turning Centers

Complex details require a high level of precision in their machining. When a part to be machined includes elaborate detail, ***Swiss-type turning centers*** are used, **Figure 22-13**. These machines use a guide bushing to hold the workpiece tightly at the point of machining. With Swiss-type turning centers, the tools are not fed along the cylindrical axis of the part being machined. Rather, the main spindle behind the guide bushing moves the part along the cylindrical axis. The tools are fed along a direction perpendicular to the cylindrical axis of the part, as with cross-slide motions on other lathe types, **Figure 22-14**. The tools on a Swiss-type turning center can be set up gang-tool style, or a turret can be used.

Figure 22-12. A large turret lathe machining a cam shaft. The workpiece is held in a 3-jaw universal chuck while the cutting tools are held in a 12-station turret and rotated into place. The tool presetter measures the offset lengths of the tools held in the turret for locating purposes.

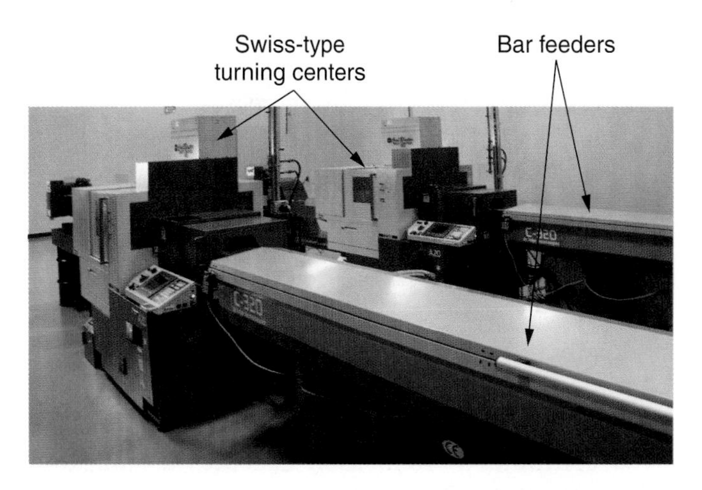

Mack Molding Co., Arlington, Vermont

Figure 22-13. These 7-axis Swiss-type turning centers have twenty-one tools each and are coupled with bar feeders to automatically load precision ground stock into the machine.

Mack Molding Co., Arlington, Vermont

Figure 22-14. This Swiss-type turning center comes equipped with an additional axis on the sub spindle. This allows the center to machine both ends of the workpiece simultaneously. This feature, along with a rapid feed rate, significantly reduces cycle times.

The advantage of a Swiss-type machine is its ability to accurately machine small, complex parts with intricate details, **Figure 22-15**. The biggest disadvantage is cost. The machines are extremely expensive, and the raw materials needed to take advantage of the capabilities of this type of equipment are also expensive. The guide bushing diameter is not adjustable. To ensure that the part does not deflect during the machining operation, the stock must be ground to a precise diameter before machining with the Swiss-type turning center can begin. This increases the cost of the raw materials significantly. Swiss-type turning centers should be considered when the parts being produced require a high level of accuracy and precision.

22.5 CNC Safety

The only difference between CNC machines and their manually operated counterparts is that the CNC machines are automated. They are controlled by a computer system as it executes a computer program written by a human. Therefore, the safety warnings and cautions related to manual machines are applicable to CNC equipment as well. Review Chapter 3, *Shop Safety*, for more information.

Figure 22-15. A small intricate component made of brass being machined on a Swiss-type turning center with a gang-tool style setup.

The following are additional safety concerns related solely to CNC equipment:

- Before operating a machine for the first time, obtain training on the machine. You must have a working knowledge of the machine and safe operating practices. Different machines may have different operating mechanisms. Read the operation and maintenance manuals provided with each machine.
- Do not remove or bypass guards, doors, and other safety devices on any CNC machine so that they do not work as they were designed.
- Know the sequence of operations executed by any program run by the machine you are operating. This allows you to recognize problems with the machine when it does something out of the ordinary.
- When changing tools manually, wait for the machine axes and the spindle to come to a complete stop.
- Know where the emergency stop buttons are located. One will always be located on the control panel and others may be located in other areas around the machine. The emergency stop brings the program to an immediate halt.
- Know how to manually jog the machine to a point clear of the workpiece.

- Know how to restart the program from its beginning and, if multiple tools are part of the program, how to manually change the tools. This will have to be done in the event of an emergency stop.
- Be familiar with all error codes and cautions that may appear on the control screen. Know what steps to take to correct each error code.
- Obtain training in the use of all auxiliary devices on the machine. Many machines have coolant systems, automated lubrication systems for spindles and other moving parts, and automated chip handlers. Know how to maintain these.
- Never try to remove chips while the machine is running, even if the chips appear to be clear of the work area.
- Cycle all programs once without a part in place during setup (this is called a ***dry cycle***). It is best to run the program in single-step mode, with the feed rate overridden to its lowest range. This can prevent an accidental machine crash.
- CNC programs can also be verified by machining a sample part from plastic, wax, or a similar inexpensive material, **Figure 22-16.**
- Be careful around automated tool changers. They can move without warning and trap an operator against a part of the machine. Automated tool changers are usually pneumatically powered, so care should be taken around compressed air systems as well.
- Keep all cooling fans, intakes, and exhausts in the computer control panel clean and free of obstructions. This can prevent equipment overheating and malfunction and possibly even equipment fires.
- If the machine crashes or does anything out of the ordinary, stop the machine, observe lockout/tagout practices, and contact the area supervisor.

22.6 CNC Coordinate Systems

The CNC program drives the motors that position all of the moving axes. The program uses a ***coordinate system*** to communicate the direction and distance the workpiece or tool must move to the motors and the control. The most common systems in use today are the Cartesian coordinate system and the polar coordinate system.

Figure 22-16. CNC programs can be verified by producing a part in an inexpensive material, such as plastic or wax.

22.6.1 Cartesian Coordinate System

The ***Cartesian coordinate system***, shown in **Figure 22-17**, forms the basis of CNC programming. Programs, written in either inch or metric units, specify the destination of a particular movement.

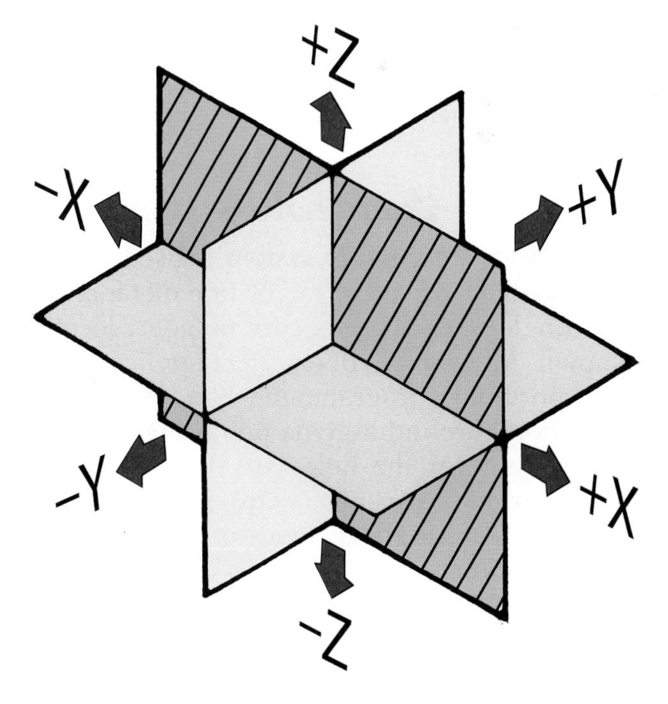

Goodheart-Willcox Publisher

Figure 22-17. The Cartesian coordinate system is the basis of all CNC programming. Each of the three major axes (X, Y, and Z) is perpendicular (at a 90° angle) to the other two. The arrows indicate the direction of travel, positive or negative.

With the destination established, the axis of movement (X, Y, or Z) and the direction of movement (+ or –) can be identified. To determine whether the movement is positive (+) or negative (–), the program is written as though the tool, rather than the work, is doing the moving.

Spindle motion is assigned the Z axis. This means that for a drill press or vertical milling machine, the Z axis is vertical. For machines such as a lathe or horizontal milling machine, however, the Z axis is horizontal. See **Figure 22-18**. The system of coordinates used for machine axis designation is specified according to the right-hand rule of Cartesian coordinates, as shown in **Figure 22-19**.

Machining instructions can be programmed directly into computer memory by entering the information through the keypad on the control. Programs can also be created and stored separately from the machine. These programs can then be uploaded using portable storage devices such as USB memory devices. Sometimes machines are networked into a company's computer systems. In these cases, programs can be uploaded into the control through the computer network.

Most programs are written using the Cartesian coordinate system because defining the location of a given point relative to another point is easier in Cartesian coordinates. This is because distances between important points are usually provided on prints using dimensions that are based on the Cartesian coordinate system.

22.6.2 Polar Coordinate System

The ***polar coordinate system*** is used in CNC programs only when a straight-line distance and travel angle from a tool's current point location is known. An example of a perfect application for polar coordinate programming is a circular pattern of holes around a given point, **Figure 22-20**. The locations for the holes are defined using a distance from the center point of the circular pattern and an angle relative to an axis that is parallel to either the X or Y axis of the machine. Cartesian coordinates are not used in these cases because doing so would require the programmer to use trigonometry to convert the polar coordinate information into Cartesian coordinates. This would involve rounding multiple decimal values to the third or fourth place, which would cause errors in programming.

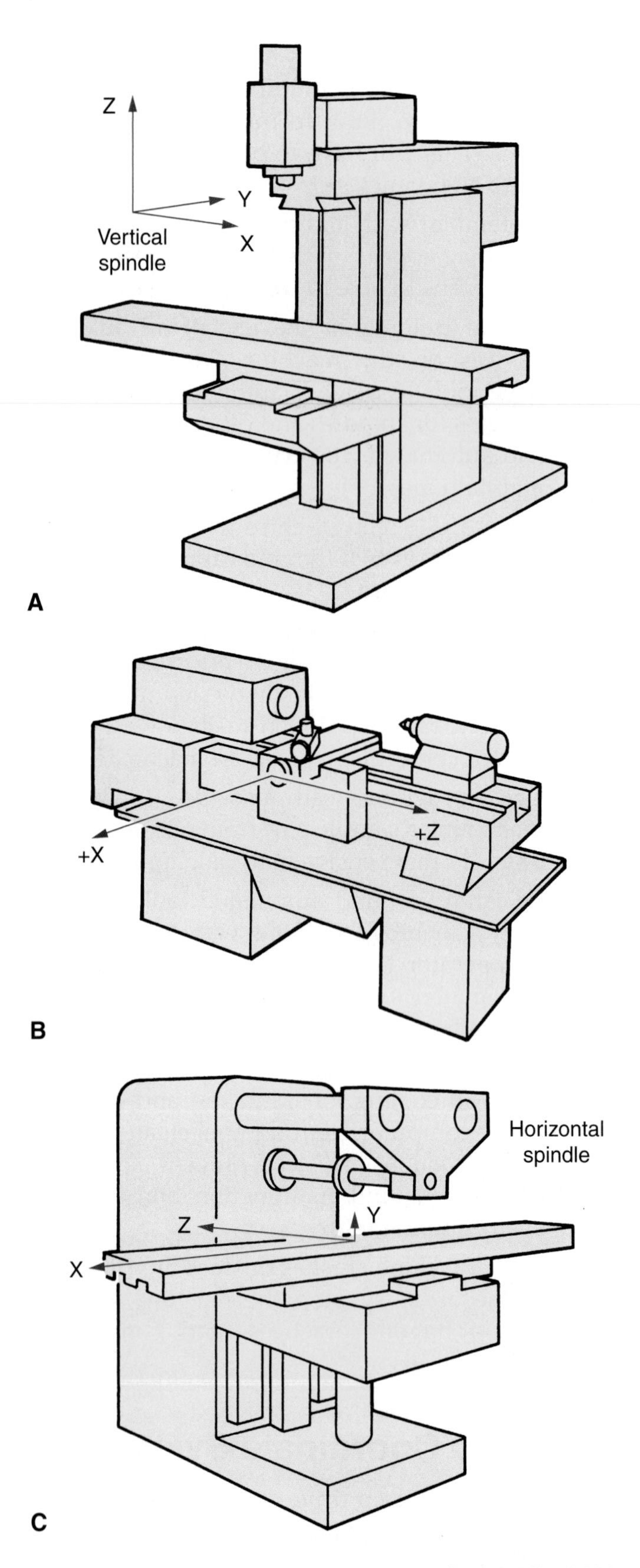

Figure 22-18. Axes of machine tool movements. A—Vertical milling machine. B—Lathe. C—Horizontal milling machine. Spindle motion is assigned to the Z axis. The Z axis differs between vertical spindle and horizontal spindle machines.

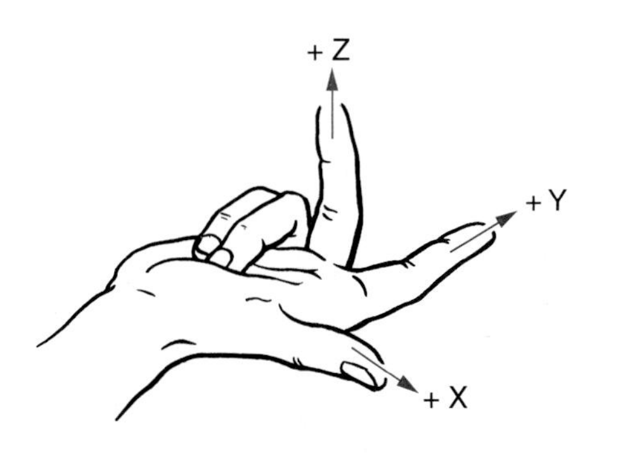

Figure 22-19. Machine tool axes are specified according to right-hand system of Cartesian coordinates. When right hand is held as shown, the thumb, forefinger, and next finger point in positive (+) directions of X, Y, and Z axes.

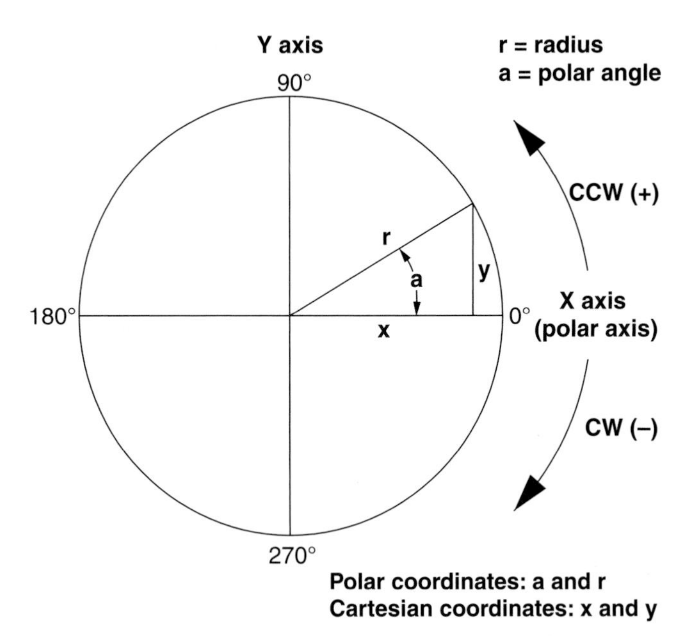

Figure 22-20. Polar coordinates describe a new location from the present location (usually the center of a circle) by showing the value of the radius (r) and the polar angle (a). Counterclockwise rotation is positive, clockwise rotation is negative.

The polar coordinate system is not often used. However, when a part is defined with dimensions using the polar coordinate system, having the capability to create programs using this system reduces time and errors associated with the mathematical conversions that would otherwise be required.

22.7 CNC Movement Systems

CNC axes are positioned using electric motors. The two basic types of motors used to drive CNC axes are stepper motors and servomotors.

22.7.1 Stepper Motors

Stepper motors are often used in less expensive CNC machines, especially machines that are used for woodworking purposes, **Figure 22-21**. They do not require encoders to keep track of the motor's revolutions relative to the position of the axis they are driving. An ***encoder*** is a transducer that measures the position of moving axes and provides electronic feedback to the control.

Stepper motors are typically used in ***open-loop systems***. An open-loop system has no means to provide feedback to the control relative to the position of the axis. This requires the machine to be reset to its zero position, often called machine homing, more frequently than a machine using servomotors. Another major drawback of stepper motors is that they lose torque at higher speeds, which means their machine cycle times must be reduced when cutting forces require higher torque.

22.7.2 Servomotors

Servomotors do not lose as much torque at high speeds as stepper motors, so machines equipped with servomotors can remove more material faster, **Figure 22-22**. However, because of their design, they require encoders to provide feedback to the control relative to the position of the axis they are driving. This is known as a ***closed-loop system***. The use of encoders is critical when precision machining is needed.

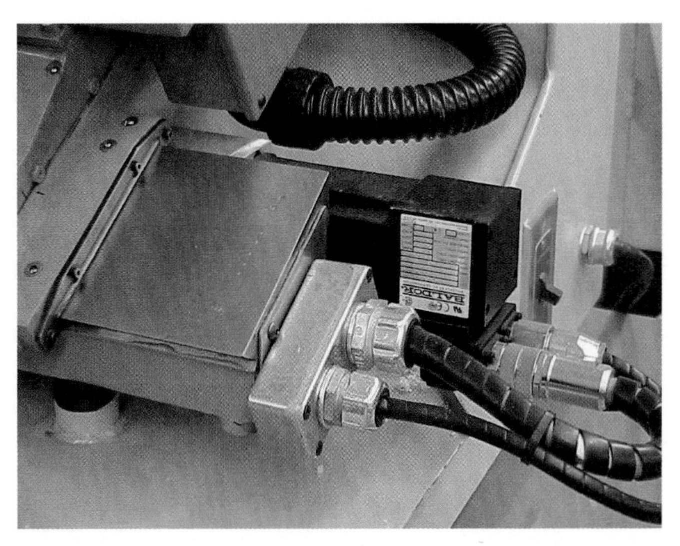

Figure 22-21. This stepper motor controls the Z axis movement on a small benchtop turning center. Stepper motors are commonly used in open-loop systems.

Goodheart-Willcox Publisher

Figure 22-22. This servomotor controls the Y axis movement on a horizontal machining center (HMC). Servomotors use a closed-loop system, requiring encoders to provide feedback to the control.

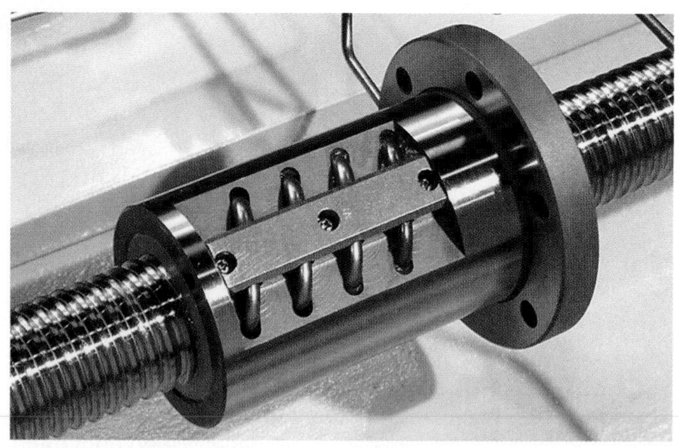

Haas Automation, Inc.

Figure 22-23. A lead screw assembly consisting of a lead screw and a ball bearing nut. The lead screw translates the rotary motion of the motor into linear motion.

22.7.3 Lead Screws and Lead Nuts

Both stepper and servomotors drive the axes of a machine using lead screws. Lead screws are designed to translate the rotating motion of an electric motor into straight-line, or linear, motion. As lead screws rotate, they cause a lead nut to move along the threads of the screw, **Figure 22-23**. Lead nuts are attached to the moving table of a milling machine or the cross-slide of a lathe.

Threaded rods and nuts do not fit tightly together. If they did, you would not be able to turn one while holding the other. The clearance designed into the mating of the rods and nuts is called ***backlash***. Backlash is present in all threaded applications.

Minimizing backlash is critical in CNC machines to ensure precise positioning as the axis changes direction. Lead screws and lead nuts are precision-ground to exacting tolerances to minimize backlash. Encoders help filter out minor backlash but, as these components wear, they often need to be replaced. Worn lead screws and lead nuts result in too much clearance for the encoder to compensate.

Chapter Review

Summary

- Computer numerical control (CNC) is a computer-controlled version of the numerical control machines developed in the 1950s.
- CNC technology results in faster, more precise machining, but CNC machines typically cost much more than their manual counterparts.
- CNC milling machines are similar to traditional milling machines but have built-in CNC capabilities.
- Machining centers are CNC milling machines equipped with automatic tool changers.
- CNC turning machines consist of computer-controlled lathes, including basic CNC lathes, gang-tool lathes, turret lathes, and Swiss-type turning centers.
- All of the safety guidelines for manual machines should be followed for their CNC counterparts.
- CNC machines have additional safety guidelines related to the computer and the automation of the equipment.
- CNC programs use the Cartesian coordinate system and the polar coordinate system to communicate the direction and distance a workpiece or tool must move.
- CNC axes are driven by stepper motors or servomotors, depending on the speed and cost of the equipment.

Review Questions

Answer the following questions using the information provided in this chapter.

1. What medium was used to store the programs that controlled machine tools in the original numerical control machines?
2. What technological advance in the 1960s was responsible for dramatically increasing the use of numerical control technology?
3. Name two advantages of using CNC equipment instead of manual methods.
4. What is the biggest disadvantage of using CNC?
5. How many axes of movement does a basic milling center have?
6. Basic CNC milling centers use a method called _____ that allow operators to select operations from menus without having to know a programming language.
7. Why might a milling center need more axes of rotation than are provided on a basic model?
8. Describe the difference between a basic CNC lathe and a gang-tool lathe.
9. Why might you avoid using a gang-tool lathe for parts that have a relatively large diameter?
10. What is the purpose of the turret on a turret lathe?
11. Why do operators of turret lathes require more training than those who operate basic CNC lathes?
12. Which type of CNC turning center would be used to machine small parts that have a high degree of elaborate detail?
13. When changing tools manually, wait for the machine axes and the _____ to come to a complete stop.

14. What is the purpose of a dry cycle?
15. In the _____ coordinate system, movement is specified in a positive or negative direction along the X, Y, or Z axis.
16. The _____ coordinate system specifies movement by giving a straight-line distance and travel angle from a tool's current point location.
17. A(n) _____ is a transducer that provides electronic feedback about the position of the moving axes to the control in a CNC machine.
18. What is the difference between an open-loop system and a closed-loop system?
19. Why must the lead screws and lead nuts in the motors that drive the CNC machine axes be checked and replaced at regular intervals?

CHAPTER 23

CNC Programming Basics

Learning Objectives

After studying this chapter, you will be able to:

- Explain the process of planning and developing a CNC program.
- Describe three methods of generating CNC code.
- Summarize the use of CAD and CAM software in CNC programming.
- Classify the types of codes used in the ANSI/EIA 274D code format.
- Identify commonly used CNC modal commands.

Technical Terms

absolute positioning
assigned code
block
circular interpolation
G-codes
incremental positioning
linear interpolation
machine control unit (MCU)
manual data input (MDI)
M-codes
modal command
offline programming
post-processing
unassigned code
word address format

CNC programming methods have changed greatly since the days of paper tape storage. However, the programming language that commands the motion of the machine tools has remained virtually the same, **Figure 23-1**. Even so, great strides have been taken to simplify the task of CNC program creation and uploading. Advances in computing power and software have made it easier for operators to create CNC programs without requiring an in-depth knowledge of control codes. These advances have greatly benefitted manufacturing, as it enables machine shops to quickly and accurately utilize CNC technologies.

However, for safe machine operations, maximum efficiency, and effective troubleshooting, it is essential that programmers and operators understand codes and their functions. As most programs will have to be modified at least once during their run cycle, programmers and operators must be able to read and write a CNC program. This chapter covers the basics of CNC program creation, with an emphasis on language and code format.

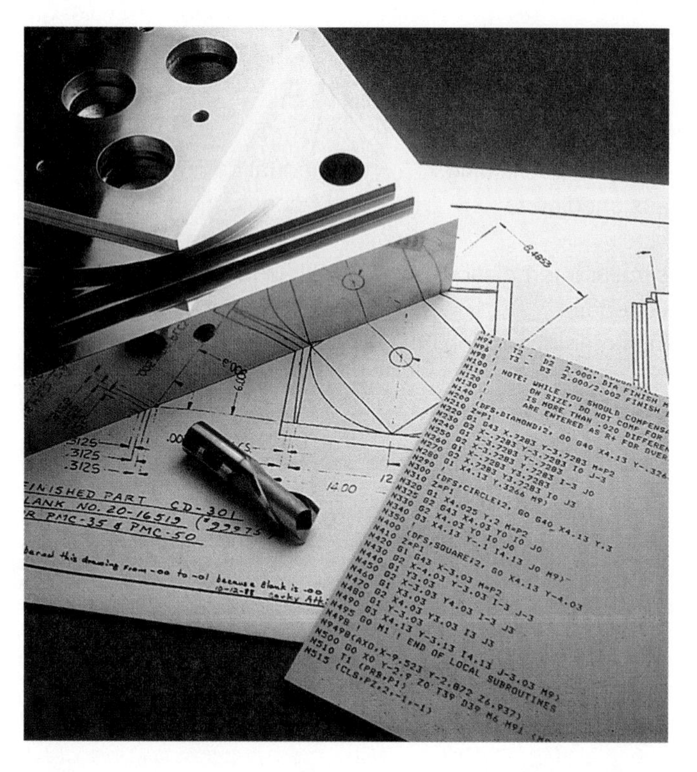

Giddings & Lewis, Inc.

Figure 23-1. A series of coded instructions, called a program, is used to control CNC machine tool movements and operations. Shown, from left to right, is a part machined on a VMC, an endmill, the print of the part, and a print-out of the CNC program.

23.1 Developing CNC Programs

In order to develop a valid CNC program to machine a part, the programmer needs a copy of a print of the part to use as a planning tool. In most cases, the company's quality control department can provide a copy that can be marked up as needed for planning purposes. Making unofficial photocopies of the print is unacceptable. All copies of the print should be officially controlled by the quality control department or its equivalent. Always ask permission before copying a print.

Figure 23-2 shows a part drawing that can be used to demonstrate how CNC programs are planned. After examining the print, the programmer determines the condition and dimensions of the stock to be used in the process. In this example, the plate to be machined is 6061 aluminum that has already been machined to the proper length, width, and thickness of the finished part.

The necessary machining processes are then determined based on the print. In this case, the machining processes will consist of drilling four holes and milling the pocket in the middle of the plate. After the machining processes and tooling have been determined, the programmer calculates the feed rates and spindle speeds as described in earlier chapters of this textbook.

For this part, the two features (holes and pocket) will be machined using different processes and tools. Notice in **Figure 23-3** that the holes and the pocket of the part have been color-coded. The color codes can be made with colored pencils or markers or even highlighters. Color-coding allows the programmer to readily associate tooling and processes with the colors of the features on the drawing.

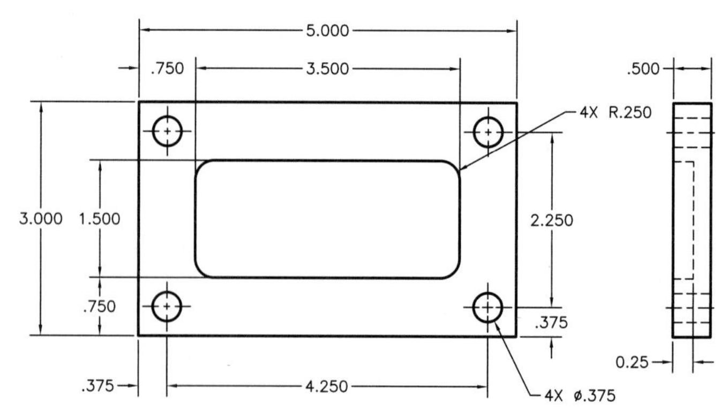

Goodheart-Willcox Publisher

Figure 23-2. A simple part drawing. This plate will be made of 6061 aluminum, and will be machined from stock that has already been machined to the proper length, width, and thickness.

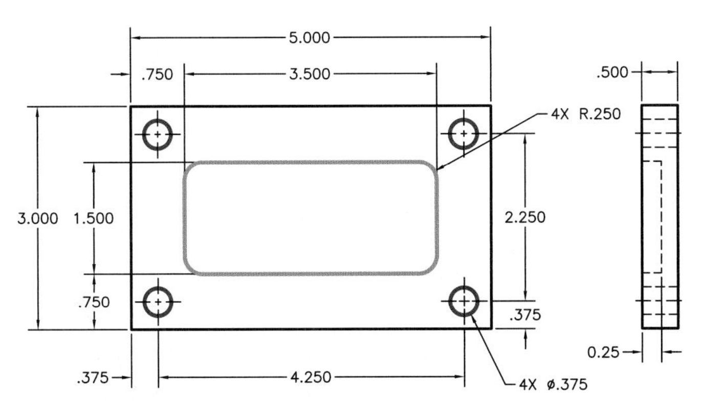

Goodheart-Willcox Publisher

Figure 23-3. The same part drawing from **Figure 23-2** with the holes color-coded red and the pocket color-coded green. Color-coding the print can assist the programmer in associating the tooling and processes with the features on the part while writing the CNC program.

In this example, the holes will be drilled using a center drill and a twist drill. The pocket will be milled using a larger end mill to rough out the pocket and a 0.500″ diameter end mill to create the corners of the pocket, which are filleted at a radius of 0.250″.

At this point, the programmer can begin planning how the drilling and pocket milling processes should be executed. For this uncomplicated part, a prewritten subroutine can be used to machine the holes. The tool path for milling the pocket will have to be written manually, unless a CAM software package is used. Chapter 24, *CNC Milling*, describes the programming techniques for milling processes.

23.2 Programming Methods

There are many methods available to generate CNC machine code today. All of these methods can be grouped into three distinct categories: manual programming, conversational language programming, and CAM programming.

23.2.1 Manual Programming

Manual data input (MDI) consists of entering the CNC program codes at the ***machine control unit (MCU)***, also simply called a control, **Figure 23-4**. The MCU reads the program from memory and translates it into the electronic signals needed for machine operation. MDI programming is primarily used for small programs with few lines or for entering a small number of lines into an existing program. The first format for CNC coded language was the ANSI/EIA 274-D format, which is described later in this chapter.

Dmitry Kalinovsky/Shutterstock.com

Figure 23-4. A CNC machine operator using the manual data input (MDI) method of programming to enter a program into the machine control unit (MCU) of this machining center.

SAFETY NOTE

Great caution must be taken when programming using the MDI method to ensure that the code has been entered 100% correctly. Incorrect code could cause a machine crash or worse. An incorrect spindle speed could cause the tool to break, fly off, and injure or kill the operator.

Manual data input can be done two ways. The first method of MDI available on CNC machines was "MDI execute." The code is entered and executed one line at a time. This is a slow and tedious process that has to be repeated for each part. The improved method of manual data input is "MDI store." The entire code is entered, reviewed, and saved to the machine's on-board memory, and then the whole sequence is executed. Controls with machine graphics can verify the code graphically before any machine movements take place.

The disadvantage of MDI is that it may interfere with the processing of products scheduled for shipment. Some machines allow code for another process to be entered and stored while production parts are being machined. However, this may mean that two people are needed at the machine—one to enter the code and one to manage the production run. An alternative would be to have the operator enter code between machine cycles. However, this typically slows both programming and machining cycle times, and can introduce errors into the code, negatively impacting safety.

Offline programming, **Figure 23-5**, is a better method for manually creating CNC code. In this method, a plain text editor, such as Notepad in computers driven by the Microsoft® Windows operating system, is used to type the code on an offline computer. The term offline computer refers to a computer that is not currently in use controlling the operation of a CNC machine. The program is then stored electronically until needed. The primary advantage of offline programming is it can be done without slowing production machining operations.

Dmitry Kalinovsky/Shutterstock.com

Figure 23-5. A CNC machine operator using the offline programming method at a computer station to write a CNC program.

23.2.2 Conversational Language Programming

Conversational language programming has become a popular option on CNC machine controls over the last several years. Conversational language is a series of software codes that allow operators to enter information into the MCU without the tedious practice of typing the program using programming codes, **Figure 23-6**. Fewer mathematical calculations are needed, and information can be entered directly from the print. As the operator enters the information, the software converts the information into the format needed by the CNC control to execute the program.

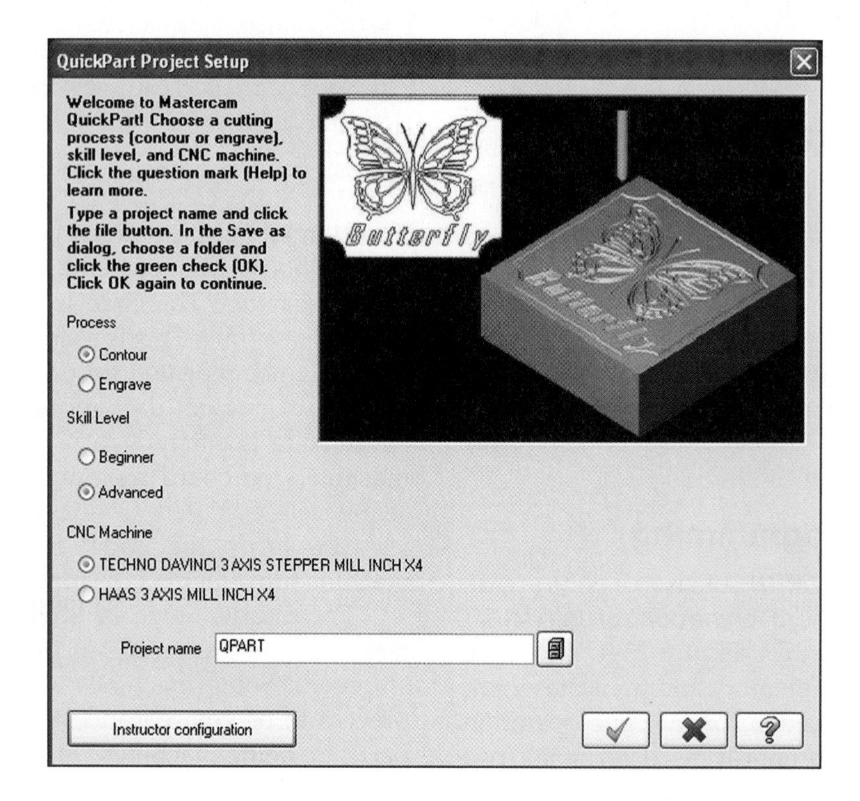

Figure 23-6. Conversational programming allows operators to machine a part by answering a series of questions and prompts instead of writing machine code. Mastercam QuickPart gives students a fast, easy way to design and cut parts while learning the principles of CAD/CAM.

The advantages of this method of programming include the ease of inputting multiple programs in a job shop environment where several production runs of small quantities do not warrant the setup needed to generate code in an offline setting. Also, in-depth knowledge of programming is not required in conversational language programming. Unfortunately, the fact that the machine does the bulk of code generation has eroded some of the programming skills of CNC operators to the point that many now find it difficult to read code and make changes when they are required.

23.2.3 CAM Programming

In computer-aided manufacturing (CAM) programming, computer software is used to generate the CNC program. In CAM, the part can be created graphically in either a two-dimensional (2D) or three-dimensional (3D) format. After the part and all of its features have been defined, the programmer defines the various cutting tools and processes needed using the CAM software, **Figure 23-7**.

23.3 CAD and CAM Software

Two types of software are commonly used to support CNC programming. Computer-aided manufacturing (CAM) software is used to define the part and the tools and machining processes needed to manufacture it. However, information from computer-aided drafting software (CAD) is often used by CNC programmers as a basis for the CAM program.

23.3.1 CAD Software

Computer-aided drafting (CAD) software allows drafters to create precise, accurate working drawings of objects such as mechanical parts to show how they should be built or constructed. Both 2D drawings and 3D parametric models created using CAD software can be used with CAM technology.

Two-dimensional software packages allow the designer to replace the traditional drafting board with a computerized system, **Figure 23-8**. However, 2D CAD software offers much more potential to the manufacturing process than just a faster method

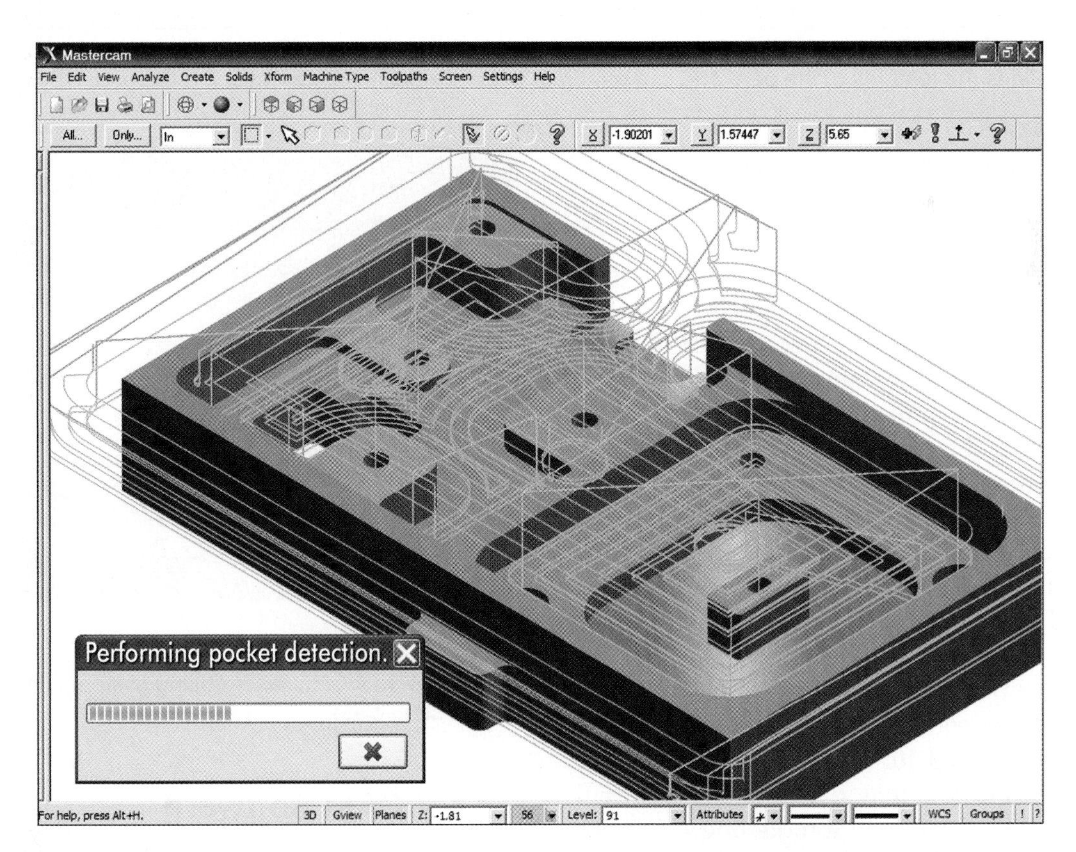

Figure 23-7. Three-dimensional CAD drawings can be imported into a CAM software package to generate a CNC program. Mastercam Feature Based Machining eliminates the manual processes involved in identifying features for programming milling and drilling operations on solid parts.

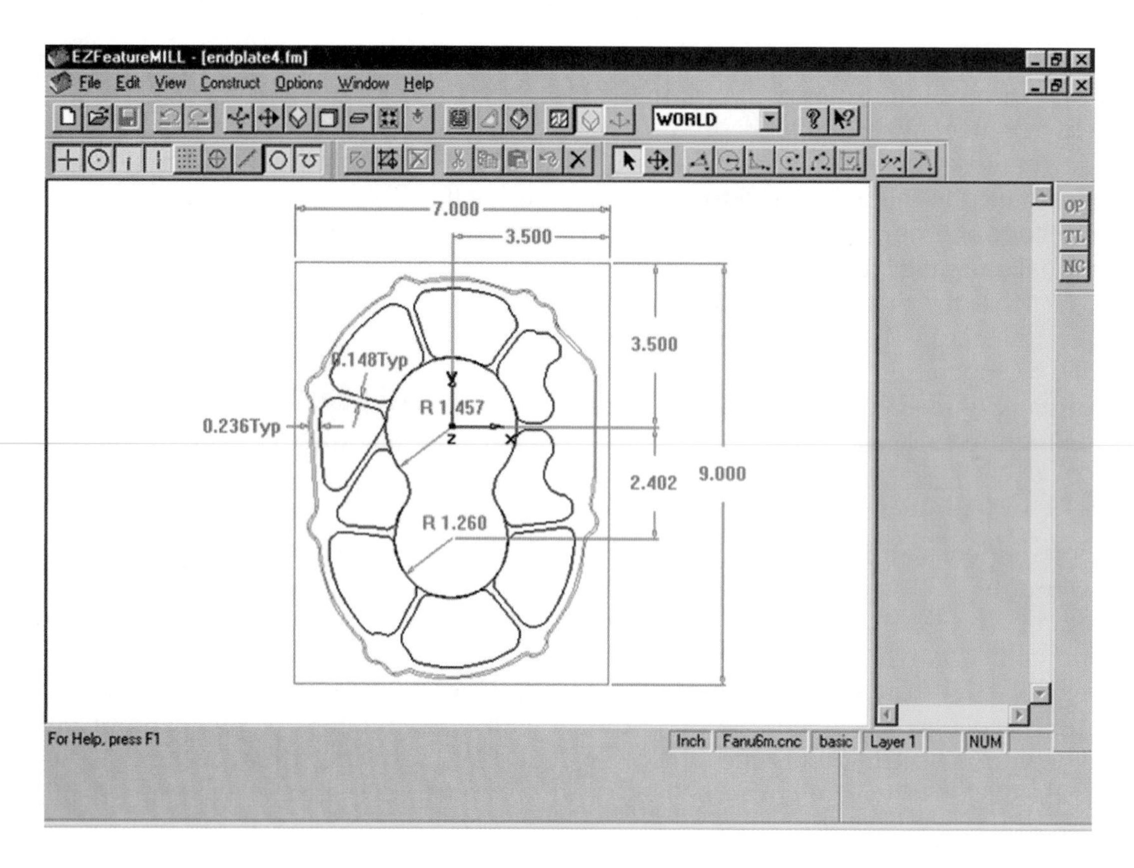

EZFeatureMILL—Engineering Geometry Systems

Figure 23-8. With some CAD software, the 2D geometry of the part can be created at the machine tool or imported from files. The software defines the features of the part, and can generate the code necessary to machine the part.

of drawing part designs. Programmers can use the electronic data from a 2D part file to facilitate the creation of CNC programs.

Some CNC machines, such as water jet cutters and some laser cutting systems, machine profiles from a piece of stock that has a constant thickness, **Figure 23-9**. For these applications, 2D CAD files can be imported directly into the CNC machine's control. To complete the program, the operator needs only to program information into the control regarding material type and thickness, along with a few other commands needed to operate the machine. However, if the design is more complex and cannot be machined from a piece of flat stock, 3D models are needed.

Created using solid modeling CAD software, 3D parametric models are truly three-dimensional in their design, **Figure 23-10**. Using 3D CAD software, the operator can define materials and features with precise detail. Most parametric modeling CAD programs are also capable of mass properties calculations and various mechanical analyses of parts. These models can be imported into the CAM software and used directly as a basis for the CNC programming.

hxdbzxy/Shutterstock.com

Figure 23-9. A CNC laser cutting system cutting 2D shapes from a steel plate.

23.3.2 CAM Software

CAM software uses drawings or models to define the shape and features of a part. The programmer can create a 3D model of the part to be machined directly within the CAM software. Alternatively, the

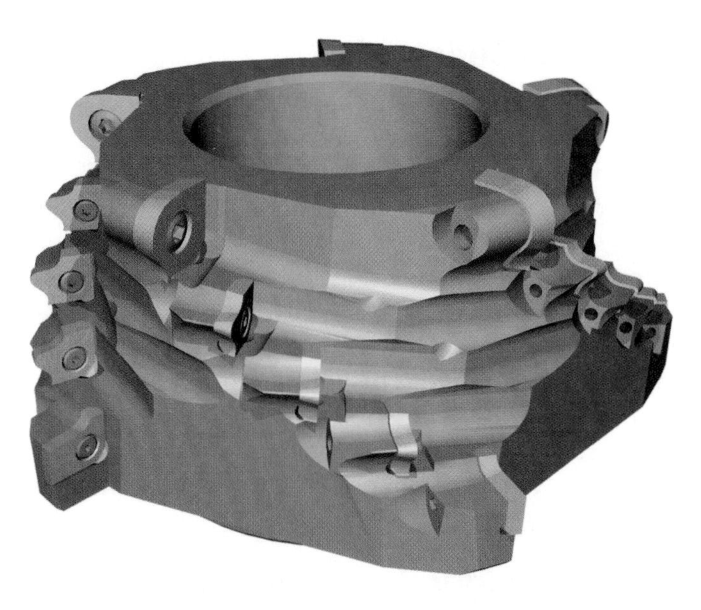

Lovejoy Tool Company, Inc.

Figure 23-10. A 3D CAD model of an indexable-insert slab mill. The carbide inserts and bolts have also been modeled. Before this model is imported into the CAM software to create a CNC program, the inserts and bolts will first be removed.

Figure 23-11. CAM software calculates the tool paths for the machining operations, and can include specialty applications for specific machining operations. The Mastercam Blade Expert add-on is a powerful and easy-to-use custom application, designed to generate efficient, smooth, and gouge-free tool paths for complex multi-bladed parts.

programmer can import 2D or 3D CAD files to form the basis for the programming. If 2D CAD files are used, the programmer adds depth to the part after it has been imported into the CAM software.

CAM software has three functions. First, it allows the programmer to define the tooling and process steps to generate the paths the cutters will take during the machining process, **Figure 23-11**. The programmer supplies the tooling and process data, and the CAM software reads the geometries needed for tool path generation from the 3D model of the part.

When tool path generation is complete, the second function of the CAM software is to simulate the program graphically, **Figure 23-12**. The programmer can include work-holding devices in the program along with the part being machined so that clearance moves around clamps and fixture details can be included in the program.

Finally, the CAM software converts the CAM program into a language that the CNC machine control unit can understand. This process is known as ***post-processing***. CAM systems come with post-processing programs for most of the machines in use today. However, on occasion, the post-processor needed for a specific machine may not be included with the CAM software. When this occurs, post-processors may be available from the machine manufacturer. If not, a programmer knowledgeable in post-processor modification and in the code format of the specific machine will need to write the post-processing program or edit an existing program to work with the specific machine.

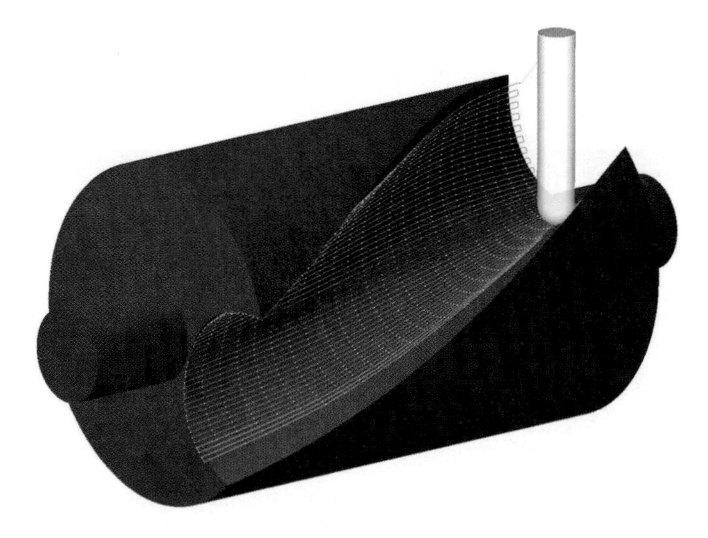

Figure 23-12. CAM software can simulate the CNC program graphically. Cut patterns guide the tool along specified paths. Mastercam's multiaxis machining has many tool path types to govern the cut pattern.

23.4 CNC Programming Codes

The ANSI/EIA 274D code format was the first standard CNC code format. Commonly called G-code or G- and M-code because of the frequent use of these two letters, codes in this format have been used in numerical control code generation since the 1960s.

G-code uses a ***word address format*** style of programming. The word address format uses combinations of letters (A to Z, except E and V) with numerical data. These combinations are called ***words***. The letters precede the numbers and define the purpose of the numerical data. Each word gives the machine a single instruction. Words that are to be executed together are written on a single line of code called a ***block***.

23.4.1 G-Codes

G-codes are preparatory codes. They control the positioning of the machining cutter relative to:

- The path the cutter takes.
- Offsets from a defined path.
- Temporary reference coordinates.

G-codes also include other information used to control the relationship of the cutter to the workpiece. **Figure 23-13** contains a list of the various G-codes used in CNC programming.

23.4.2 M-Codes

The miscellaneous codes that control machine functions such as starting and stopping spindle rotation and turning coolant flow on and off are known as ***M-codes***. The numbers associated with many G- and M-codes are standardized. They perform the same function for all machines that use the ANSI/EIA 274D format. However, there are several numeric values that are not assigned to any particular task or command. This gives individual machine manufacturers the flexibility to use these codes for specialized functions that are specific to their machines. **Figure 23-14** contains lists of the various M-codes used in CNC programming.

23.4.3 Assigned and Unassigned Codes

Most of the G- and M-codes are assigned codes. ***Assigned codes*** are codes that have the same application no matter what machine control is being used and no matter what type of machining process is being performed. For example, G00 is rapid travel, no matter whether you are programming a milling machine or a lathe. Likewise, M00 is a program stop for both applications.

G-Codes—Preparatory Functions

Code	Function
G00	Rapid traverse.
G01	Linear interpolation.
G02	CW (clockwise) circular interpolation.
G03	CCW (counterclockwise) circular interpolation.
G04	Dwell (timed delay of established duration). Length is expressed in X or F word.
G20	Inch programming.
G21	Metric programming.
G33	Constant pitch thread cutting.
G81	Simple drilling cycle.
G90	Absolute coordinates.
G91	Incremental coordinates.

Goodheart-Willcox Publisher

Figure 23-13. Examples of CNC programming G-codes. G-codes are preparatory codes and are standardized.

M-Codes—Miscellaneous Functions

Code	Function
M00	Stop machine until operator restart.
M02	End of program.
M03	Start spindle—CW.
M04	Start spindle—CCW.
M05	Stop spindle.
M06	Tool change.
M07	Coolant on (mist).
M08	Coolant on (flood).
M09	Coolant off.
M30	End program and return to program top.
M52	Advance spindle.
M53	Retract spindle.
M56	Tool inhibit.

Goodheart-Willcox Publisher

Figure 23-14. Examples of CNC programming M-codes. M-codes are miscellaneous codes and are standardized.

Unassigned codes are codes that can be used by different types of machine controls to perform tasks for specialty purposes. Every machine's manual includes a table showing how these codes are used for the particular machine.

23.4.4 Other Letter Codes

Other letters are used in generating CNC codes as well. For instance, when the control reads the letter "F," it uses the numbers immediately after the letter to set the programmed feed rate for the movement of the machine axes. The letter "S" signals the control that the numbers immediately after the letter determine the rpm of the spindle. **Figure 23-15** contains a list of these letters and describes their purpose in programming.

23.5 CNC Modal Commands

Modal commands are those G-code commands that, once activated, remain activated until canceled or another modal command is activated. Many of the G-code commands are modal. This section describes the modal commands most commonly used in CNC programming.

Other Letter Codes	
Code	**Function**
A	Designates rotation about the X axis.
B	Designates rotation about the Y axis.
C	Designates rotation about the Z axis.
D	Specifies cutter radius compensation offset.
F	Specifies a feed rate.
H	Specifies tool length compensation offset.
I	Specifies the location of the arc center for a circular move. The I refers to the distance and direction in the X axis. Also specifies the amount of tool "back away" at the bottom of a fine boring cycle.
J	Specifies the location of the arc center for a circular move. The J refers to the distance and direction in the Y axis.
K	Specifies the location of the arc center for a circular move. The K refers to the distance and direction in the Z axis.
L	Used with subroutines to specify the number of times the subroutine should be repeated.
N	Designates a sequence number, line number, or block number in a program.
O	Identifies a program.
P	Specifies the length of time for dwell commands.
Q	Specifies the peck depth for each pass in a peck drilling canned cycle.
R	Designates the radius value in a circular movement. Also specifies the rapid plane height for a canned cycle command.
S	Used to program a spindle speed.
T	Used to select a tool station.
U	Replaces the letter X when doing incremental moves on a turning center.
W	Replaces the letter Y when doing incremental moves on a turning center.
X	Identifies a coordinate position along the X axis.
Y	Identifies a coordinate position along the Y axis.
Z	Identifies a coordinate position along the Z axis.

Goodheart-Willcox Publisher

Figure 23-15. Other letter codes used in CNC programming.

23.5.1 G00—Rapid Travel

The G00 command activates the rapid travel setting for all of the positioning axes on the machine. When this command is activated, all axes, when called by the program to move, will move at the fastest rate the machine is capable of moving. This command is for positioning moves, quickly moving the cutter to or away from the areas to be machined. The G00 command enables the CNC program to cycle faster, allowing more parts to be made per time period.

SAFETY NOTE

Great care must be taken with the G00 command. Because it is modal, it remains active until another G-code requiring a different programmed feed rate is read by the control. Also, rapid travel commands do not take the shortest path to the next position, instead moving at the maximum feed rate for each axis independently. This may result in unexpected paths of travel. Mistakes with this command can cause severe damage to the tool, the workpiece, the CNC machine, and possibly the operator.

23.5.2 G01—Linear Interpolation

In CNC programming, ***linear interpolation*** means straight-line movement. The G01 command specifies a linear move to a specified location, **Figure 23-16.** It is one of the positioning codes that requires a programmed feed rate. A sample line of code containing a linear interpolation command is shown below:

```
N0010 G01 X0.500 Y0.500 F30
```

Where:

- N0010 addresses the line of code.
- G01 is the call for a linear interpolation move.
- X0.500 is the X axis coordinate for the target location.
- Y0.500 is the Y axis coordinate for the target location.
- F30 is the feed rate at which the machine will move the cutter to the target point.

When this line of code is executed by the control, the machine axes move from their current position to the 0.500, 0.500 coordinate in a straight-line motion, at a rate of 30 units per minute. The units may be either inches or millimeters, depending on which unit was specified earlier in the program. The codes for specifying units are described later in this chapter.

Some machines require that all lines of code be addressed in order. Other machines ignore the N sequence number and do not require their use. However, N numbers assist the operator and programmer with troubleshooting and are recommended for most applications.

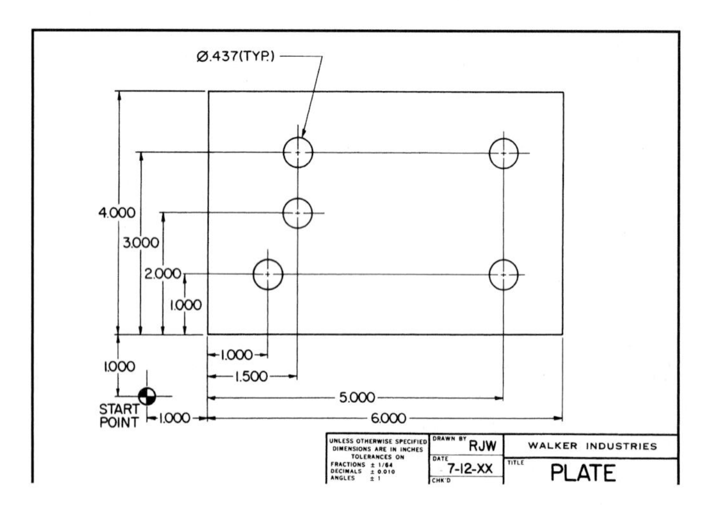

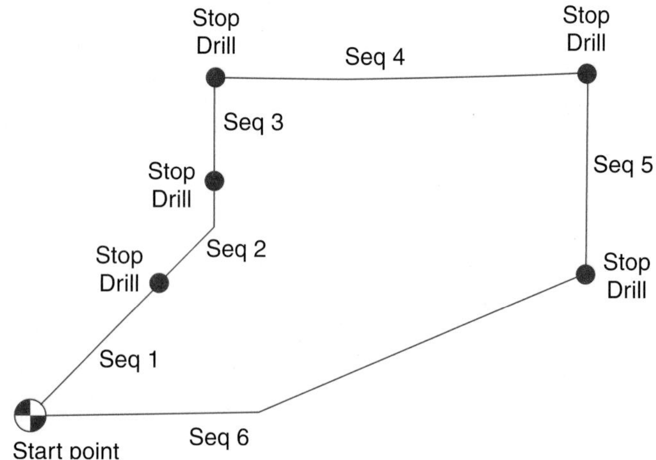

Goodheart-Willcox Publisher

Figure 23-16. Linear interpolation is straight-line movement. A—A simple part drawing requiring linear interpolation commands. B—The planned drilling sequence for the above part. Linear interpolation is required to move the drill to the proper location.

If the line of code specifying linear interpolation does not have a programmed feed rate, one of two things will happen. The machine will move at a previously programmed feed rate, or it will not move at all and the control screen will signal that a programming error has occurred. Most controls have a registry where information for parameters, such as feed rate, is stored. The feed rate is pulled from this registry if no reference is found on the line of code.

If a line of code fails to provide a feed rate and the control uses a previous feed rate, the rate may have been programmed for a different tool performing a different machining step. That feed rate may not be suitable for the tool and machining step currently being performed. The controls of some CNC machines hold parameters in their registries from previous programs for other parts. This too could result in a feed rate that is not appropriate for the task at hand. Therefore, it is a good practice to include a feed rate specification on every line of code that calls for a linear or circular interpolation move. This ensures that no incorrect feed rates will be pulled from the control's registry.

23.5.3 G02 and G03—Circular Interpolation

Just as linear interpolation refers to straight-line movement, ***circular interpolation*** refers to movement in a circular or radial pattern, **Figure 23-17**. The G02 command signals the control to drive the axes to move the cutter in a clockwise direction, and the G03 command specifies a counterclockwise direction.

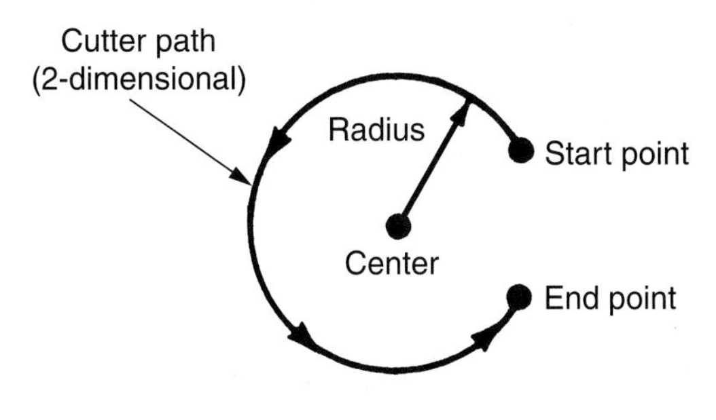

Goodheart-Willcox Publisher

Figure 23-17. When using circular interpolation, only the coordinate location of the start and end points of the arc, the radius of the circle, the coordinate location of the circle's center, and the direction of the cut need to be programmed. The intermediate points required to describe the circular movement are computed by the MCU. On some machines the circle is divided into 90° segments and must be programmed accordingly.

When visualizing the direction of movement, imagine looking down on the machine from the ceiling. Whether the machine moves the workpiece around a stationary tool, the tool around a stationary workpiece, or any other combination of axis movements, you must visualize the cutter moving around a stationary workpiece. This is how the machine will interpret your program.

A sample line of code containing a G02 circular interpolation command is shown below:

```
N0010 G02 X0.500 Y0.500 R+0.500 F30
```

Where:

- N0010 addresses the line of code.
- G02 is the call for a clockwise circular interpolation move.
- X0.500 is the X axis coordinate for the target location.
- Y0.500 is the Y axis coordinate for the target location.
- R+0.500 is the radius of the circular movement.
- F30 is the feed rate the machine will take to move the cutter to the target point.

Note the plus sign between the letter R and the 0.500 value. This "R address" specifies the location of the center point of the arc. In a circular movement from one point to another point, the center point of the arc can be in one of two places, depending on whether the arc motion is greater than or less than 180°. If the arc to be machined is less than 180°, the plus sign is used. If the arc to be machined is greater than 180°, a minus sign is used. If no sign is placed after the letter R, most controls interpret the arc angle to be less than 180°.

Another method of specifying the center point for the arc to be machined is to use I and J codes instead of the R address. An example of this method is shown below:

```
N0010 G03 X0.500 Y0.500 I0.000 J0.500 F30
```

In this example, the G02 from the previous example has been replaced with a G03 command, so this arc will be machined in a counterclockwise rotation. The R+0.500 address has been replaced with the code I0.00 J0.500. The I and J values provide the location of the center point of the arc. The letter I specifies distance along the X axis, and the letter J specifies distance along the Y axis. The numeric values are incremental distances from the point where the machine is currently positioned to the center point of the arc. In this example, the center point of the arc is 0.500 units away from the start point in the Y axis direction. Since the number associated with I is 0.000, there is no distance along the X axis between the start point and the center point. If either I or J is not written into the line of code, most controls assume the value for the omitted address is zero.

As with the G01 command, the G02 and G03 commands require a programmed feed rate. If the rate is not included on the line of code with the circular interpolation callout, the control pulls the last feed rate in the registry. If the registry is empty, the control stops machine movement and displays an error on the control screen.

Both G02 and G03 are modal commands. Once they have been activated, all movements following the activation will be circular motions until they are cancelled.

23.5.4 G20 and G21—Units

Another pair of modal commands are the G20 and G21 commands, which set the working units in the control. The standard units recognized by most CNC machines are decimal inches and millimeters. G20 sets the working units in the control to inches, and G21 sets the units to millimeters. If this command is not present in the program before the first command that causes axis movement, the machine will do one of two things when it makes its first move. It will either use the default units specified in the machine control, or it will pull the units used for the previous program. Either of these actions may result in incorrect machining. It is therefore a good practice to include the appropriate code in one of the first lines of a program.

23.5.5 G90 and G91—Absolute and Incremental Positioning

The G90 and G91 modal commands determine the relationship between points in a program based on their coordinate values. Recall from Chapter 22, *Introduction to CNC Machining*, that most programs are written using the X, Y, and Z axes of the Cartesian coordinate system. The X axis specifies table movement to the right and left. The Y axis moves the table forward and backward, or toward and away from an operator standing in front of the machine. Finally, Z axis movement raises and lowers the machining spindle.

The G90 command indicates to the control that absolute positioning is active. In ***absolute positioning***, the coordinate values for any point are interpreted in relation to the X0,Y0 (X = 0, Y = 0) position, **Figure 23-18**. For instance, in **Figure 23-19**, the starting point for the R60 arc is 200 units along the X axis and 40 units along the Y axis, relative to the X0,Y0 position. The ending point for the arc is at X=140 units and Y=100 units. Assuming that the cutter is currently positioned at the starting point of the arc, the line of code to move the cutter to the endpoint of the arc is:

```
N0010 G03 G90 X140 Y100 R+60 F100
```

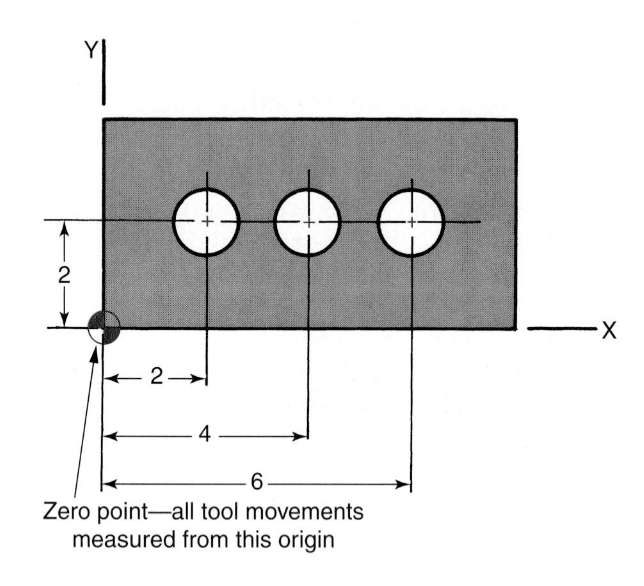

Goodheart-Willcox Publisher

Figure 23-18. In the absolute positioning system all coordinates are measured from a fixed point of origin. The coordinate values for any point are interpreted in relation to the X0,Y0 (X = 0, Y = 0) position.

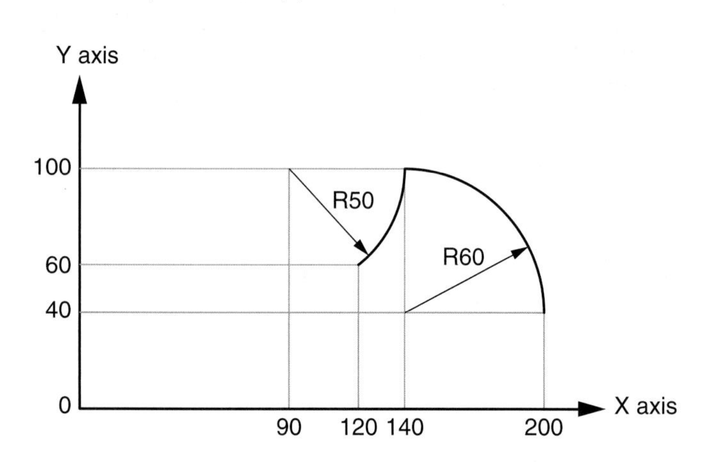

Goodheart-Willcox Publisher

Figure 23-19. An example of two circular movements executed by a CNC machine. Arc R60 travels counterclockwise and arc R50 travels clockwise.

This command can be read as follows: "Moving in a counterclockwise circular direction using absolute positioning, move to point X140, Y100 along an arc path of radius 60 at a feed rate of 100 millimeters per minute." Notice the use of the plus sign after the letter R because the arc is less than 180°.

Incremental positioning, specified by the G91 command, is not related back to the X0,Y0 position. Rather, it is based on the location of the previous point, **Figure 23-20**. For example, for the same R60 arc shown in **Figure 23-19**, with incremental positioning, the line of code might be:

```
N0010 G03 G91 X–60 Y60 R+60 F100
```

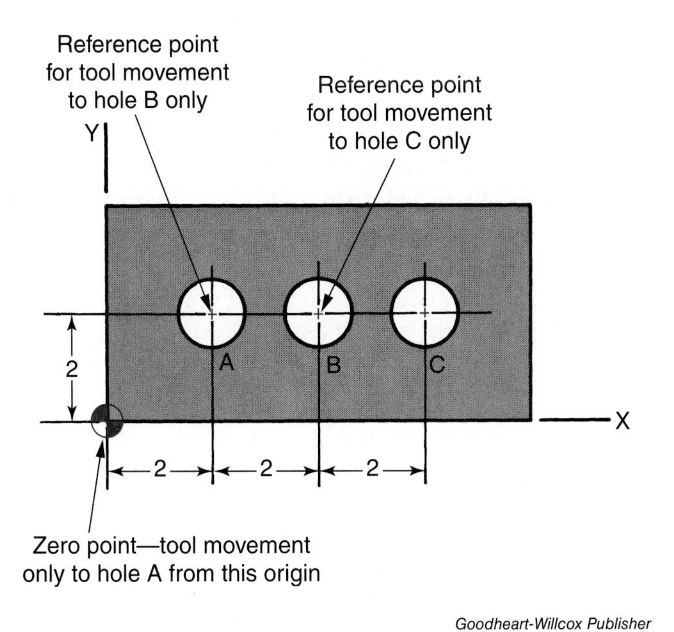

Figure 23-20. The incremental positioning system. Each set of coordinates is based on the location of the previous point.

Note the changes in the X and Y coordinate values. The difference between the X200 coordinate and X140 coordinate is 60 units. The minus sign indicates that movement is in the negative X direction. Similarly, the difference between the Y100 and Y40 coordinates is 60 units. Because the movement is in the positive Y direction, the value is unsigned.

Programs are not commonly written using incremental positioning because doing so increases the chance that damage could occur as a result of accidentally leaving out negative signs or making mistakes on coordinate values. Another disadvantage of incremental positioning relates to program modification. In the event that a point coordinate defined by incremental positioning must be changed, the coordinates of every point after the changed point must be changed as well, or they will be incorrect. However, incremental positioning is an ideal method in some situations. The best example of a good use for incremental positioning is within subroutines, which will be described in Chapter 24, *CNC Milling*.

23.5.6 Setting the Modal Commands

Although it is not strictly necessary, it is a good CNC programming practice to establish the modal command setting in the first line of CNC code. If you have a control that stores parameters for units and positioning from previous programs in its registries, writing a simple line of code such as the one below clears those registries and prepares the new program for operation.

```
N0010 G00 G20 G90
```

This line of code overrides any previous settings and sets the machine to operate at rapid travel, sets the units to inches, and sets the positioning to absolute.

Chapter Review

Summary

- A programmer plans and develops CNC programs by reviewing the print for the part to be created, determining the tools and processes needed to create the part, and planning how the processes should be executed.
- CNC code can be generated manually, using conversational language, or by using computer-aided manufacturing software.
- Computer-aided manufacturing software allows programmers to create (or input from a CAD file) a 2D or 3D model of a part to be machined and specify the tools and processes needed to machine it.
- Sets of codes in the ANSI/EIA 274D code format, commonly called G-code, are prefaced by letters such as "G" and "M" that allow the machine control unit to identify the type of command to be executed.
- Commonly used modal commands include those that specify rapid travel, linear and circular movement, units of measurement, and the type of positioning to be used to machine the part.

Review Questions

Answer the following questions using the information provided in this chapter.

1. Why should all CNC programmers and operators have at least a basic understanding of CNC code formats?
2. What is the first thing a programmer needs in order to develop a CNC program?
3. After a programmer has determined the machining processes and _____ to be used to machine a part, the feed rates and spindle speeds can be calculated.
4. In _____ CNC programming, the programmer types the coded language in a format that can be read directly by the machine control unit.
5. Why is offline programming the preferred method of manual programming?
6. Using a series of software codes to enter information into the CNC control without using programming codes is commonly known as _____ programming.
7. In which type of programming does the programmer use computer software to define the part and the tools and processes needed to machine it?
8. What two types of software are commonly used to support CNC programming?
9. Which type of software allows drafters to create precise, accurate working drawings of objects such as mechanical parts to show how they should be built or constructed?
10. List the three basic functions of CAM software.
11. Converting a CAM program into a format that the CNC machine control unit can understand is known as _____.
12. The first standard CNC code format was the _____ code format, commonly called G-code.
13. In CNC programming, a group of codes called _____ control the positioning of the CNC machining cutter.
14. The miscellaneous codes that control machine functions are known as _____.
15. What do the letters "F" and "S" signify in CNC programming codes?
16. In what way are modal commands different from other CNC commands?
17. Which modal command activates the rapid travel setting for the positioning axes?
18. What will happen if a line of code that specifies linear interpolation does not have a programmed feed rate?
19. Explain the difference between the G02 and the G03 commands.
20. Which command would you use to set the units for a machining process to decimal inches?
21. In _____ positioning, the coordinate values for any point are interpreted in relation to the X0,Y0 position of the Cartesian coordinate system.
22. Which command would you use to specify that the coordinates provided should be read from the location of the previous point, rather than the X0,Y0 position?
23. What line of code would you use near the beginning of a CNC program to clear the control's registries and set up the machine for rapid travel, units in millimeters, and incremental positioning? For this question, assume that the address for the line of code is N0010.

CHAPTER 24

CNC Milling

Learning Objectives

After studying this chapter, you will be able to:

- Identify the purpose of common miscellaneous function codes.
- Describe work-holding devices that are commonly used in CNC machining processes.
- List information the programmer needs to gather before beginning to write a CNC program.
- Carry out the procedure for writing a CNC milling program.

Technical Terms

canned cycle
comment code
machine home
peck drilling
program zero position
subroutine

In ANSI/EIA 274D code format, both G-codes and M-codes are required to complete a CNC program. Chapter 23, *CNC Programming Basics*, explained G-codes and described the functions of several modal G-codes. This chapter builds on that information by introducing M-code functions and describing work-holding devices for CNC machining. The chapter also includes a step-by-step example of the process of creating a CNC program.

24.1 Miscellaneous Function Codes

As you may recall, M-codes are miscellaneous codes that control specific machine functions. This section defines some of the more frequently used M-codes.

24.1.1 M00—Program Stop

The M00 command causes the machine cycle to stop. When the control encounters this command, all movement, including spindle rotation, table feed rate movement, and coolant flow immediately come to a halt.

This command can be used for many applications. If the machine is not equipped with an automatic tool changer, then the operator is responsible for changing tools. The M00 command can be used to stop the machine to allow the operator to change tools. After the tool has been changed, the operator resumes the program by pressing the cycle start button on the control. Other tasks, such as a part or tool inspection or a chip clearing operation, can be accomplished after stopping the machine with the M00 command, **Figure 24-1**.

Note that when the cycle is restarted, the machine does not automatically resume the operations that were stopped by the M00 command. A line of code is necessary to restart the spindle and coolant flow before the machine axes move to begin the next step in the machining processes.

24.1.2 M01—Optional Program Stop

The M01 command allows the programmer to include an optional stop in the machine cycle. Most CNC machines are equipped with an optional stop button or toggle switch. If the button or toggle switch is set to the "on" position, the machine stops when the control reads an M01 command just as it would when reading an M00 command. If the button or toggle switch is set to the "off" position, the machine ignores the M01 command and continues to the next line of code.

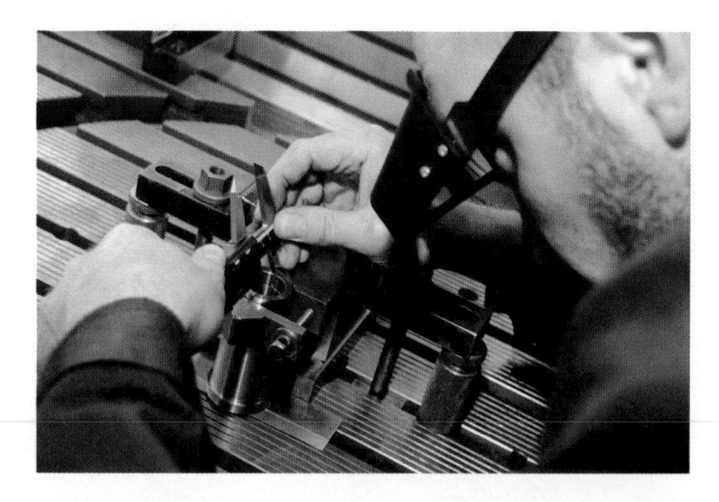

Dmitry Kalinovsky/Shutterstock.com

Figure 24-1. The M00 and M01 codes can be used to stop the CNC machine, allowing the operator to make measurements, clear chips, change tools, or inspect the part, tools, or work-holding device. The CNC operation is resumed by pressing the cycle start button.

The M01 is useful during the intial setup of an operation or for periodic inspection. The command can be used after each tool has finished cutting, allowing the operator to inspect the part after each operation. If everything is running satisfactorily, the toggle switch or button can be set to the "off" position to ignore the M01 code in future cycles.

24.1.3 M02 and M30—Program End

The M02 code is one of two options to signal the machine that the end of the program has been reached. On reading this command, the control stops all machine functions and the machine signals the operator, either through a program end light or a message on the control screen, that the program has finished its cycle. The M02 command does not cause the cycle to repeat automatically.

Like the M02 code, the M30 code signals the end of the program. Unlike M02, however, M30 also resets the program back to the first line of code so the program can be repeated. For this reason, most programmers use M30 instead of M02. Machines that were equipped with paper tape readers required this code to physically rewind the tape to the beginning of the program. Both codes have their uses. The M02 code may be used in a machine shop where only one or a few parts need to be machined. The M30 code is better suited to the needs of mass production applications because the operator does not need to manually reset the program to the beginning after each cycle as with the M02 command.

24.1.4 M03, M04, and M05—Spindle Operation

The M03 code turns the spindle on to rotate in a clockwise direction as viewed from above the machine. The M04 code also turns the spindle on, but specifies counterclockwise rotation. The M05 code is the spindle stop code. It cancels both the M03 code and the M04 code.

As with the programmed feed rates discussed in Chapter 23, *CNC Programming Basics*, the spindle's speed must be specified in the line of code, as follows:

```
N0030 M03 S2500
```

The designation for spindle speed definition is the letter "S," and the number following it defines the spindle rpm. Therefore, the line of code above starts the spindle rotation in a clockwise direction (M03) at 2500 rpm (S2500).

It is considered best practice to include the spindle speed code on every line of code that contains the M03 or M04 command. If the code is not present, the machine function may stop and display an error code, or it may pull a previously programmed spindle speed from the registry. As with feed rates, spindle speeds pulled from the registry are not likely to be correct. Using an improper spindle speed can cause damage to the tool, part, machine, or operator.

Once the spindle has been turned on, it continues to rotate until it is cancelled. In addition to M05, codes that cancel spindle rotation include M00, M01, M02, and M30. The M03 and M04 codes cancel each other out, because they are the same code with the exception of the direction of spindle rotation. However, it is unwise to use one of these codes to stop the other. Many spindles, such as high-speed pneumatic spindles, need to come to a complete stop before the command to change direction is activated. Activating the command to change the direction of spindle rotation before the rotation completely stops can damage the spindle.

24.1.5 M06—Tool Change

Another miscellaneous code that is commonly used with CNC machines equipped with automatic tool changers is the M06 code, which initiates a tool change. The M06 code requires an indication of the location of the next tool to be used. The letter "T" signals the control that the following number designates a position in the automatic tool change system, **Figure 24-2**. In the example below, the T2 defines the tool that is to be loaded based on the tool's location in the turret or magazine:

```
N0060 M06 T2
```

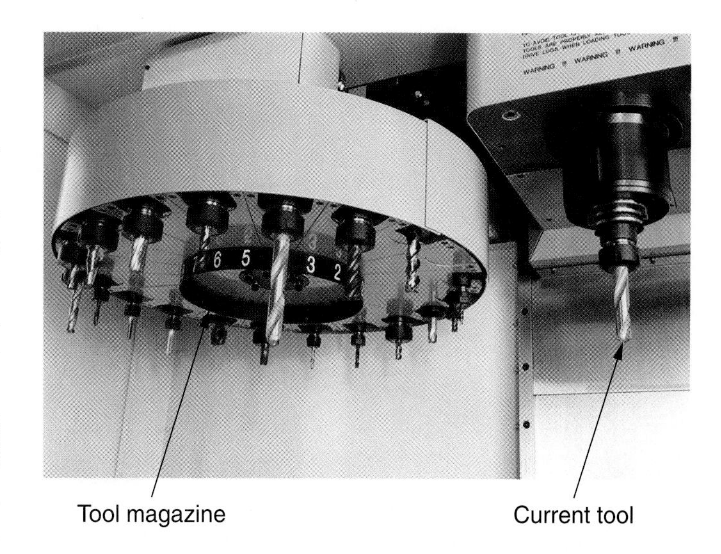

Haas Automation, Inc.

Figure 24-2. This CNC machine has a tool magazine capable of holding twenty tools. The numbers of each position in the magazine are clearly visible. An M06 code would signal a tool change designated by the letter "T" and the position number of the tool in the magazine.

SAFETY NOTE

The machine has no way of knowing if the proper tool is in this location. The operator must assume responsibility for proper machine and tool setup. Failing to ensure that tools are in their proper locations may cause damage to the machine, the part, or the operator.

24.1.6 M07, M08, and M09—Coolant

The M07, M08, and M09 codes control coolant flow through the CNC machine, **Figure 24-3**. Both M07 and M08 start coolant flow. M07 starts the flow as a mist, and M08 starts the flow as a flood. M09 stops the flow of coolant through the machine. Some machines are equipped with through-spindle coolant systems, which are activated by different, manufacturer specific, M-codes.

SAFETY NOTE

Care must be taken when using through-spindle coolant systems. Activating through-spindle coolant when a solid tool holder is in the spindle could damage the spindle bearings.

Goodheart-Willcox Publisher

Figure 24-3. Sufficient coolant is needed to flush away chips, prevent damage to the tool, reduce workpiece deformation from excessive heat, and improve surface finish.

24.2 Work-Holding Devices

When creating CNC programs, the programmer must consider the work-holding system. During a manual machining process, the operator can see what is being machined and can avoid moving the work-holding device into the cutter. However, the CNC machine moves from programmed point to programmed point. If part of the work-holding device is in the path of the cutter, they will collide. Knowledge of the work-holding devices and methods allows the programmer to write the program in a way that avoids accidental contact between the cutter and any part of the device. Common work-holding devices for CNC milling processes include the machining vise and specialty jigs and fixtures.

SAFETY NOTE

Check the program carefully to avoid interference between the cutter and work-holding device. A collision can result in serious damage to the cutter, work-holding device, the part, or the machine. It is also possible that the operator can be injured by flying debris when a collision occurs.

24.2.1 Machining Vise

In a machine shop, the machining vise is a common work-holding device. It is versatile and can be set up to hold a variety of parts and configurations. The disadvantage of the machining vise is that it takes more time to remove finished parts and load unmachined parts for the next cycle. The operator has to manually open the vise, remove the part, locate the next part, and retighten the vise.

The machining vise is a good option for production runs that involve only a few parts. However, as the volume of parts increases, it becomes more cost-effective to build specialty jigs and fixtures.

24.2.2 Specialty Jigs and Fixtures

Specialty jigs and fixtures are devices that allow the operator to unload, load, and locate parts faster, saving machining time. In these devices, the clamping mechanisms vary. Commonly used clamping mechanisms for CNC milling include the following:

- Threaded clamps, similar to the movable jaw on a vise.
- Cam locks in which a 1/4 turn locks the part into place.
- Toggle clamps.
- Pneumatic or hydraulic vises, **Figure 24-4**.
- Vacuum systems, **Figure 24-5**.
- Magnetic tables.

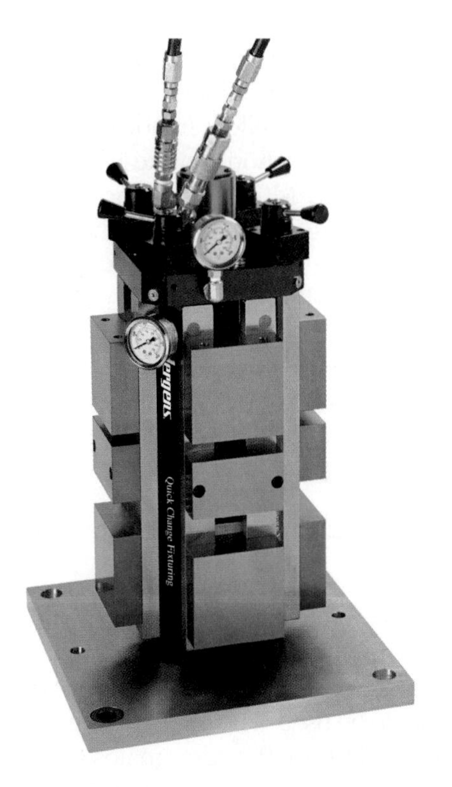

© Copyright Jergens, Inc. 2013

Figure 24-4. A 4-sided hydraulic vise column. The column has eight stations for holding workpieces. The jaws of the vises are made of machinable aluminum. Pockets can be machined into the jaws to grip oddly shaped workpieces.

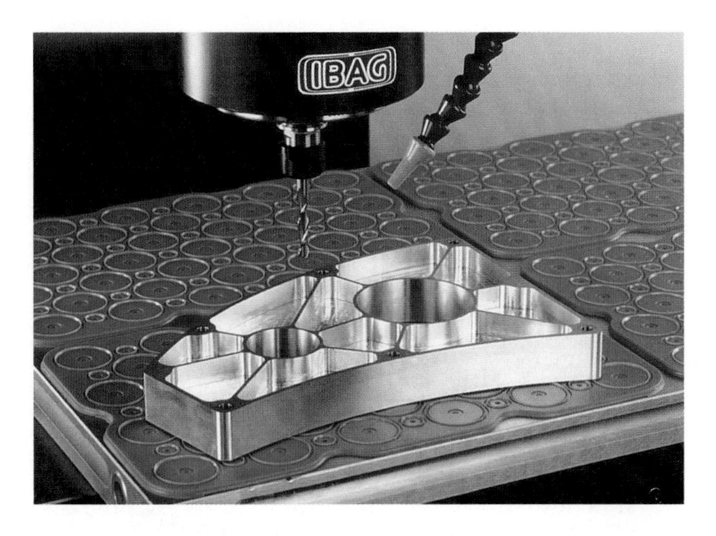

IBAG North America

Figure 24-5. A vacuum work-holding system, here being used to hold a part on a VMC.

24.3 Planning the Program

Figure 24-6 shows the part to be used in this example of CNC programming. This part is .500″ thick and is made from a low carbon steel. Nine holes will need to be drilled and the outside edges machined. All machined features are through features.

24.3.1 Identifying the Work-Holder

As explained earlier in this chapter, the first consideration in writing a CNC program is how the part is going to be held. For this example, a specialty fixture will be used to hold and locate the part. The part sits in the fixture in such a way that the outside edges and all the holes are clear of the nesting system. The fixture is equipped with a magnetic system that can be turned on and off by the operator.

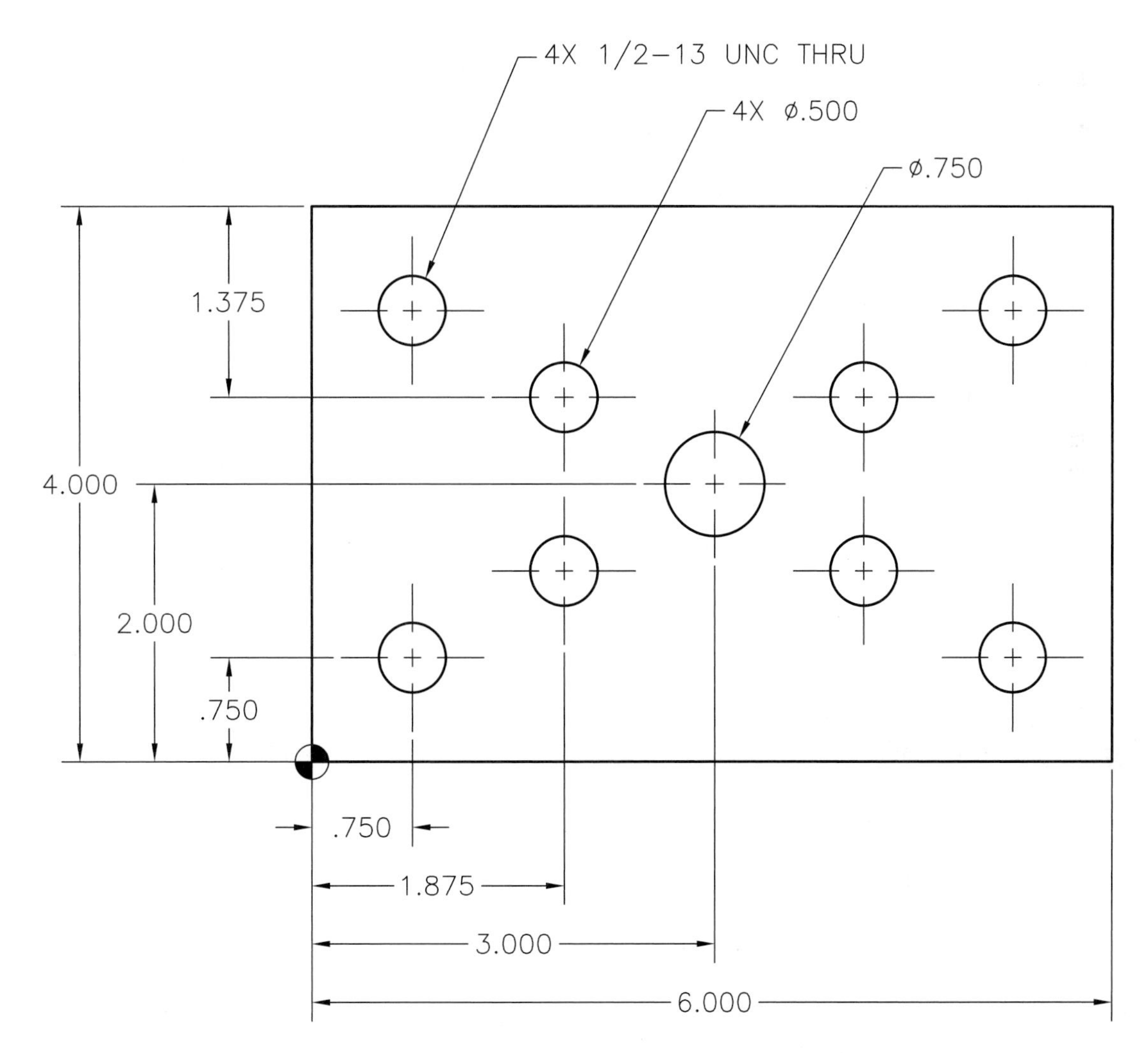

Goodheart-Willcox Publisher

Figure 24-6. This part will be used for all of the example CNC programming in the rest of the chapter. The part will be made of .500″ thick low carbon steel, and will require nine holes to be drilled (four of which will be tapped) and the outside edges to be machined.

It also has a fail-safe device that will not allow the machine to start unless the magnetic device is activated.

24.3.2 Identifying the Required Tools

Next, the programmer determines the tools that will be needed for the process. The example part has the following features:

- The outside edges of a 6.000″ × 4.000″ mild steel block.
- Four 1/2-13 UNC threaded holes.
- One hole that is .750″ in diameter.
- Four holes that are .500″ in diameter.

The table in **Figure 24-7** shows the suggested tools for each feature. In addition to these tools, a spot drill, or center drill, will be used to locate the centers of the holes. The center drill helps keep the twist drills from "walking" when they come in contact with the base metal. For this part, a #2 center drill will be used to spot the nine holes.

24.3.3 Determining Feed Rates and Spindle Speeds

Next, the programmer determines the feed rates and spindle speeds for each of the identified tools. The information in Chapter 12, *Drills and Drilling Machines*, and Chapter 18, *The Milling Machine*, can be used to specify suitable speeds and feeds for the program. These can always be adjusted as needed to optimize the program. For this example, feeds and speeds for high-speed steel cutters will be used. The table in **Figure 24-8** shows the feed rates and spindle rpms to be used for each of the tools needed to machine the example part.

Tools, Feeds, and Speeds for Sample Part

Tool #	Tool	Feed rate (in/min)	Spindle speed (rpm)
1	.750″ four-fluted end mill	9	1050
2	#2 center drill	3	4000
3	27/64″ drill	6	950
4	.500″ drill	6	800
5	.750″ drill	6	530
6	1/2-13 tap	20	260

Goodheart-Willcox Publisher

Figure 24-8. Suggested feed rates (in/min) and spindle speeds (rpm) for each tool.

24.4 Initial Programming and Preparing the Machine

As described in the previous chapter, it is a good practice to begin the program with a line of code that sets the machine up properly to run the rest of the program. The units for this program will be decimal inches, and the coordinates will be absolute. Therefore, the first line of code will look like this:

```
N0010 G00 G53 G20 G90
```

Recall that the first group of numbers specifies the line number in the program. N0010 is the first line in the program. It sets the machine movement to rapid travel (G00), the units to inches (G20), and the coordinate system to absolute (G90). The G53 code changes the coordinate system to the machine's

Tools Required for Sample Part

Feature	Tools Required
Outside edges	A .750″ four-flute end mill is suggested because it has a large diameter that will allow the edges to be machined faster.
1/2-13 UNC threaded holes	Tap drill using a 27/64″ drill bit, then cut the threads with a 1/2-13 UNC tap.
.750″-diameter hole	A .500″ diameter drill will be needed to drill a pilot hole for the larger drill, because the dead center of the .750″ diameter drill is too large to fit inside the center drilled hole.
.500″-diameter holes	.500″ diameter drill.

Goodheart-Willcox Publisher

Figure 24-7. Suggested tool for each feature.

preset coordinate system. Refer to Chapter 23, *CNC Programming Basics*, for more information about these specifications.

24.4.1 Specifying Machine Offsets (G53–G59)

The next task is to set the machine offsets so the machine knows where the workpiece is in relation to the table and spindle. When programs are written offline, the location of the part relative to the machine home position is not considered. ***Machine home*** is the location of the table and spindle, given in Cartesian coordinates. The location of the spindle relative to the table is considered to be X0.000, Y0.000, and Z0.000. A reference point on the part or perhaps a point on the work-holding device is then used to establish a ***program zero position***. Then, when the work-holding device is set up on the table, the program zero position point can be located relative to the machine home position.

The X, Y, and Z locations can be recorded in a machine register under any of six G codes: G54 through G59. These six codes can all be different. Using multiple zero positions, several parts can be machined using a single program, **Figure 24-9**. Alternatively, some complex machining operations may require multiple zero positions for a single part. When the control encounters one of these codes, it offsets the zero position from the machine home position to the location identified in the machine register. This is now the program zero.

The program zero function is cancelled by the G53 code. This code changes the machine parameters to position the axes based on the machine home being the zero point. G53 is non-modal. All subsequent movement commands will use the program zero defined by the G54 code.

For this example, the axes will be positioned so that machine home is the zero point. The program X and Y zero points are directly over the lower-left corner of the block, and the Z zero point is .100″ above the top of the block. The line of code looks like this:

```
N0020 G54 X0.000 Y0.000 Z0.100
```

All of the information in lines N0010 and N0020 could have been placed on one line. Several years ago, it would have been best practice to do so. Machines were limited in the amount of memory available to store programs, and the size of the program was determined by the number of lines of code it contained. Modern machines are equipped with more memory capacity, so limiting the lines of code is not as critical as it once was. The program is easier to edit or troubleshoot later if each function has a separate line of code, so in this program, the second line specifies the program zero position.

24.4.2 Defining Tool Changes

The next lines in the program define the first tool to be used in the machining process. This will be the .750″ end mill that will be used to machine the outside edges of the part. The third and fourth lines of code might look like the following:

```
N0030 M05
N0040 M06 T1 (Change to .750″ End Mill)
```

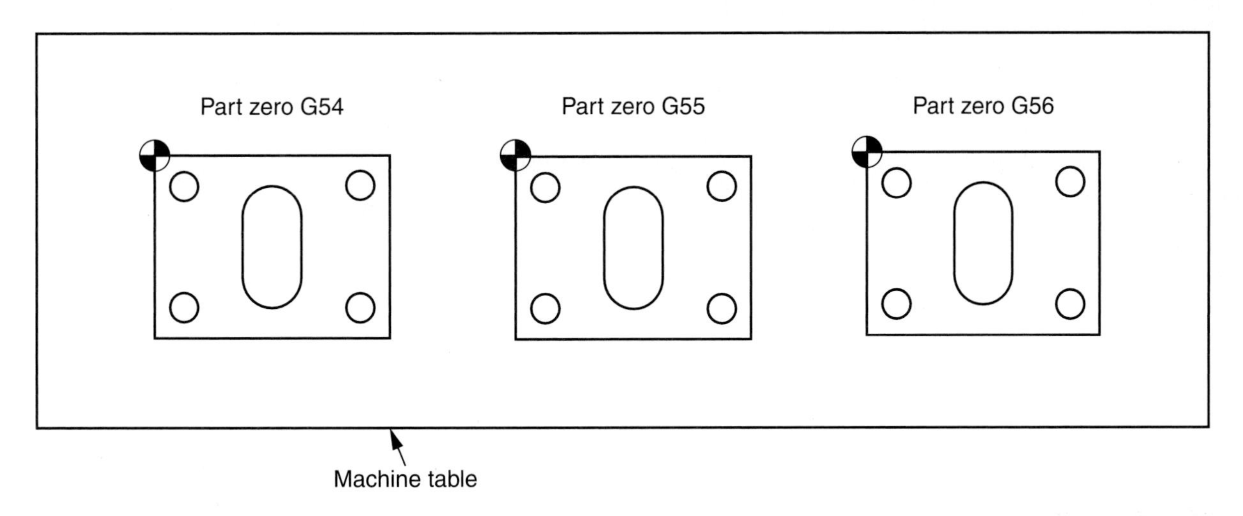

Goodheart-Willcox Publisher

Figure 24-9. Multiple parts can be machined using a single program by defining multiple zero positions with the G54–G59 codes. The zero positions of three parts, relative to the machine home, are defined using three G-codes (G54, G55, and G56).

The third line of code is a fail-safe operation to stop the spindle rotation. Some programmers may consider this unnecessary. However, it never hurts to think about the many ways a program can malfunction. A crash could cause damage to the machine, the tool, the workpiece, or the operator. A few extra lines of code are well worth the small amount of machine memory to ensure that such malfunctions do not occur.

The fourth line of code uses the M06 function to begin the automated tool change cycle, **Figure 24-10**. The T1 callout specifies the tool in the turret location designated as location 1. Remember that neither the machine nor the program knows which tool is stored in that location. If this program is being run for the first time after a setup or after a long idle period, it is a good idea to manually remove any toolholders from the spindle and make sure all tools are in their proper locations.

Note the text inside the parentheses in line 4. This is a ***comment code***. The machine ignores any text inside the parentheses, but the text will appear on the control screen. Comment codes make it easier for the operator to follow the program. In this case, the comment code serves as a reminder to the operator of which tool should be located in the T1 position. Comment codes also make it easier to locate specific lines of code during editing and troubleshooting applications.

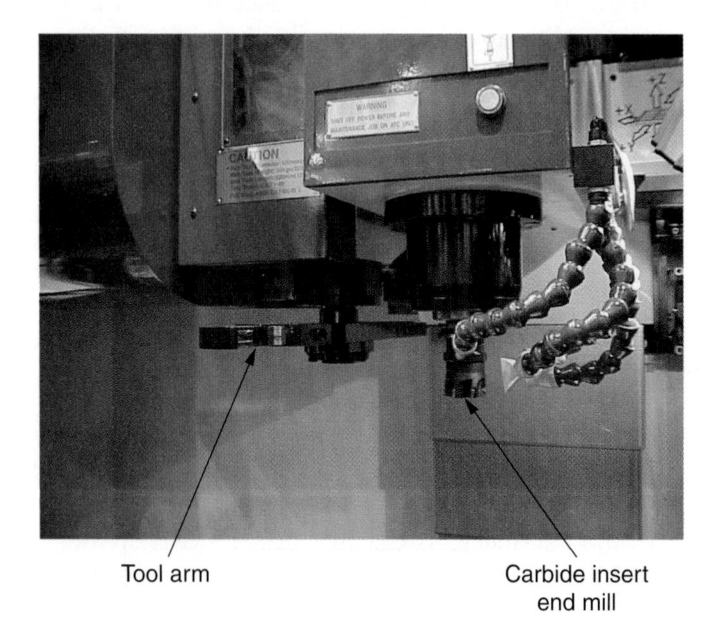

Goodheart-Willcox Publisher

Figure 24-10. A 180° rotation automatic tool changer on a VMC. The automated tool change cycle is called by the M06 code. A letter "T" and a number specifies the tool in the corresponding turret location. Here a carbide insert end mill is being placed back into the tool magazine.

Some machines also allow the use of a semicolon to separate comment codes from actual program codes on a program line. However, it is usually easier to locate the comment codes when parentheses are used.

24.4.3 Specifying Cutter Compensation (G40–G42)

If the lower-left corner was used as the zero point of the workpiece and the other three corners were defined using the dimensions of the block, the machine would place the center of the cutter over the zero point. This would result in the dimensions of the block being too small by the diameter of the cutter in both the X and Y directions. To avoid this, the machine must be programmed to compensate for the diameter of the cutter being used.

The codes G40, G41, and G42 are used to control cutter compensation, **Figure 24-11**. G41 specifies cutter compensation to the left, or counterclockwise. This means that, when envisioning the direction of the cutter moving away from the operator, it will be moving with the edge being machined to the left. G42 is similar to G41, but specifies compensation to the right, or in a clockwise direction. Code G40 cancels both G41 and G42.

For this example, the block will be machined in a counterclockwise direction, so the G41 command is used:

```
N0050 G41 X-.100 Y-.100 Z-.650 D1
```

The letter D defines the diameter of the cutter being used, corresponding to tool T1, and the X and Y coordinates place the edge of the cutter .100″ in front of and to the left of the corner of the block. The Z coordinate places the bottom of the cutter .050″ below the bottom of the part.

24.5 Programming the Machining Operations

Now that the initial specifications have been programmed and the cutter has been positioned, the next step is to program the actual machining operations. The operations will be performed in the following order:

- Machining the edges.
- Spot drilling the centers of all of the holes.
- Peck drilling all of the holes.
- Tapping the 1/2-13 UNC threaded holes.

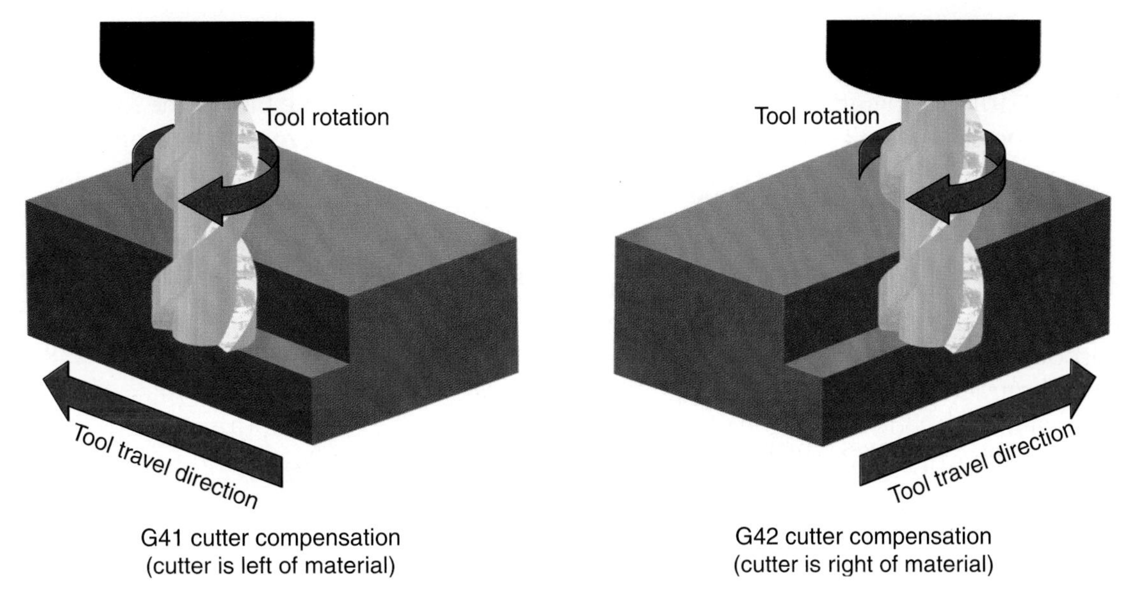

Goodheart-Willcox Publisher

Figure 24-11. Envision the cutter moving away from the operator. Use G41 if the edge being machined is on the left. Use G42 if the edge being machined is on the right. Use G40 to cancel both G41 and G42.

24.5.1 Machining the Outside Edges

The first operation will be to machine the outside edges of the part. The lines of code are:

```
N0060 M03 S1050
N0070 M08
N0080 G01 X0.000 Y0.000 F9.0
```

Line N0060 turns the spindle on in a clockwise rotation, and the S1050 specifies an rpm of 1050. Line N0070 starts the coolant flow. Line N0080 brings the edge of the cutter into contact with the part and changes the feed from rapid to a programmed feed rate of 9.0 inches per minute.

The next lines of code move the table so that the cutter will machine the outside edges of the part:

```
N0090 X6.000
N0100 Y4.000
N0110 X0.000
N0120 Y0.000
N0130 G00 X-.100 Y-.100 Z.100 M05
N0140 M09
```

The last two lines, N0130 and N0140, moves the edge of the cutter away from the part after the machining is complete using rapid positioning (G00), stops the spindle rotation (M05), and turns the coolant off (M09).

24.5.2 Spot Drilling

The next machining step is to change the tool to the #2 center drill in preparation for the spot drilling operation. As shown in **Figure 24-8**, the center drill has been defined as tool #2, or T2. The lines of code include:

```
N0150 Z.500
N0160 M06 T2 (Change to #2 Center Drill)
N0170 X.750 Y.750 Z.200
N0180 M03 S4000
N0190 M08
```

Line N0160 accomplishes the tool change, and line N0170 moves the cutter to a point .100″ above the top of the part and just over the location for the 1/2-13 threaded hole at the lower-left corner of the part. In line N0180, the spindle is again started in the clockwise direction, this time at 4000 rpm, and Line N0190 starts the coolant flow. All nine locations for the holes will be spot drilled in this step.

The drilling operation will be perfomed using a code called a ***subroutine*** or ***canned cycle***, which is a set of commands that follows a prescribed sequence, almost like a miniature program. Based on the information the programmer includes in the line of code that activates the canned cycle, the program follows a prescribed series of moves. A canned cycle remains active and repeats the same sequence until it is cancelled using the G80 code.

There are several canned cycles related to the drilling operation: G73, G74, and G76, along with

G81 through G89. G81 is the code for a simple drilling operation. It will be used in this case because it is best suited for the spot drilling operation. The line of code that begins this process will look something like this:

```
N0200 G99 G81 R.100 Z-.075 F100
```

The G99 code directs the machine to retract the Z axis to the location defined with the letter R. The R is usually used to define a programmed radius for circular interpolation but, in this case, it defines the position for tool retraction. There is a G98 code that can be used as well. This code directs the machine to retract the tool to the original Z position (which in this case would be Z.200).

The G81 code starts the canned drill cycle. This cycle advances the drill at the programmed feed rate (F) to the Z depth defined. The execution of line N0190 spot drills the first hole. The following lines of code define the locations for the rest of the holes.

```
N0210 Y3.250
N0220 X1.875 Y2.625
N0230 Y1.375
N0240 X3.000 Y2.000
N0250 X4.125 Y1.375
N0260 Y2.625
N0270 X5.250 Y3.250
N0280 Y.750
N0290 G80
N0300 G00 X-.100 Y-.100 Z.100 M05
N0310 M09
```

Each of the points defined in lines N0210 through N0280 places the center drill above the part in the proper location. The G81 command is still active, so it executes the canned drilling cycle after each of the linear moves defined in these lines of the code. Line N0290 contains the G80 code that cancels the G81 canned drilling cycle. Lines N0300 moves the cutter to a clearance position away from the part and stops the spindle rotation. Line N0310 stops the coolant flood.

24.5.3 Peck Drilling

The next machining step is to change the tool to the 27/64″ drill, in preparation for the drilling operation for the pilot holes needed for the threading operation. Recall that the 27/64″ drill has been specified as tool #3, or T3. The lines of code to prepare for drilling the pilot code include:

```
N0320 M06 T3 (Change to #3 27/64″ Drill)
N0330 X.750 Y.750 Z.100
N0340 M03 S950
N0350 M08
```

The tool change is accomplished in line N0320. Line N0330 places the drill .200″ above the part and at the position for the first of the four threaded holes at the lower-left corner of the block. Lines N0340 and N0350 start the spindle and the coolant flow.

The pilot holes will be drilled using the G83 code, which calls a canned cycle for peck drilling. ***Peck drilling*** is used in order to clear the chips away at intervals while each hole is drilled. This command requires more information to define the cycle. The correct way to write a line of code for the G83 command is:

```
G83 X# Y# Z# R# Q# F#
```

Where:

- G83 is the command to activate the peck drilling canned cycle.
- X# and Y# are the XY coordinates for the hole position.
- Z# is the Z coordinate for the bottom of the part, with additional clearance to allow for the included angle at the tip of the drill.
- R# is the position of the clearance plane to which the drill retracts at the end of each cutting feed.
- Q# is the depth of cut for each cutting feed.
- F# specifies the feed rate in the drilling operation.

Figure 24-12 illustrates how this canned cycle works. The drill begins a programmed feed rate from the clearance position above the part (R position)

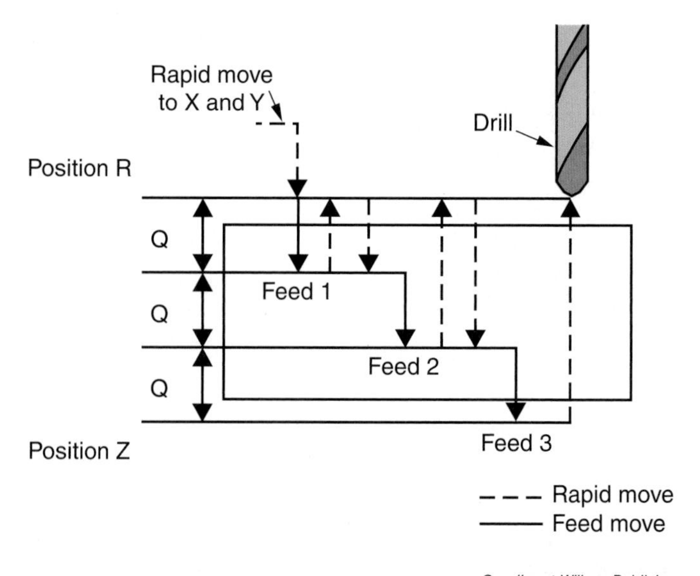

Figure 24-12. A representation of the G83 peck drill canned cycle.

downward a distance equal to the value assigned to the letter Q. It then retracts to the clearance plane to allow any buildup of chips to break away and clear the hole. It then feeds down to the second level, which is a depth of two times the value for the letter Q. Again, it retracts to the R position. The cycle continues until it reaches the Z depth specified in the line of code. The tool then retracts to either the Z axis of the location defined with the letter R or to the original Z axis position, depending on whether G98 or G99 was used.

In the example program, the tip of the drill is already positioned at a point directly above the location of the hole at a height of .100″ above the top surface. To determine the value of Q, the programmer adds the .100 value for the clearance location, the .500″ part thickness, and .060″ for clearance below the part so that the included angle of the drill extends beyond the bottom surface. A full .060″ clearance is not necessary but, when added to the other two numbers it equals .660″, which is easily divisible by three. This program will use three cutting feeds, or pecks. Each peck will be equal to .220″. This is the value for Q. Therefore, the next line of code will read:

```
N0360 G99 G83 Z-.560 R.100 Q.220 F6.0
```

Since the tool was already positioned in the proper XY location, the coordinates for those two axes do not need to be specified. Line N0360 drills the first hole. The following lines of code define the other three hole positions:

```
N0370 Y3.25
N0380 X5.25
N0390 Y.750
N0400 G80
N0410 G00 X-.100 Y-.100 Z.100 M05
N0420 M09
```

The code to drill the four .500″ holes and the pilot hole for the .750″ diameter hole in the middle is similar to the code used for these four holes. Including code for the tool changes, the remainder of the peck drilling cycles will look like the following:

```
N0430 M06 T4 (Change to #4 1/2″ Drill)
N0440 X1.875 Y1.375 Z.100
N0450 M03 S800
N0460 M08
N0470 G99 G83 Z-.560 R.100 Q.220 F6.0
N0480 Y3.25
N0490 X5.25
N0500 Y.750
N0510 X3.000 Y2.000
N0520 G80
N0530 G00 X-.100 Y-.100 Z.100 M05
N0540 M09
N0550 M06 T5 (Change to #5 3/4″ Drill)
N0560 X3.000 Y2.000 Z.100
N0570 M03 S530
N0580 M08
N0590 G99 G83 Z-.560 R.100 Q.220 F6.0
N0600 G80
N0610 G00 X-.100 Y-.100 Z.100 M05
N0620 M09
```

24.5.4 Tapping the Holes

The requirements for tapping holes are different from other types of drilling, **Figure 24-13**. First, the feed rate and spindle rpm are much slower. Second, the feed and the spindle rpm must be synchronized so that

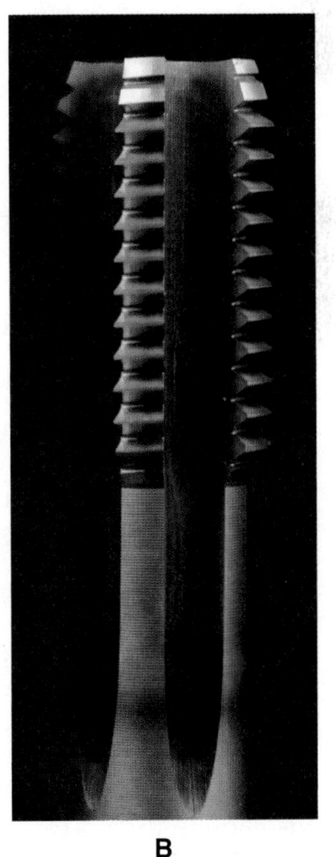

A B

OSG Tap & Die, Inc.

Figure 24-13. Threading tools. A—A thread forming tap for soft metals where chip formation is a problem. B—A thread cutting tap.The flutes in the thread cutting tap allow chips to fall away from the cutter.

the feed rate equals the lead of the thread multiplied by the rpm of the spindle. The lead is the distance from the crest of one thread to the crest of the thread next to it. Finally, when the depth of the thread is reached, both the spindle rotation and the feed must stop, then reverse direction to back the tap out of the newly threaded hole. The G84 code accomplishes these tasks based on the information provided. The next few lines of code needed to start the tapping process include:

```
N0630 M06 T4 (Change to #5 1/2-13 Rigid Tap)
N0640 X.750 Y.750 Z.100
N0650 M03 S260
N0660 M08
```

Notice that the spindle rpm has been programmed at 260 rpm. This was done for two reasons. First, this rpm is on the low end of recommended tapping spindle speeds. If the trial runs indicate that the spindle speed and feed rates can be increased, this can be done before the program is released for production runs. Second, the lead of the thread must be considered. In a 1/2-13 thread, there are 13 threads per every inch of length on the tap. Since the tapping feed rate must be an even ratio of the thread lead and the spindle rpm, the formula for the tapping feed rate is:

$$F = \text{Lead of tap} \times \text{rpm}$$

Where:

F	=	tapping feed rate
Lead of tap	=	1/thread per inch
rpm	=	spindle rpm

The tapping feed rate is as follows:

$$F = 1/13 \times 260$$

$$F = 20 \text{ inches per minute}$$

A spindle rpm of 390 could have also been used. When plugged into the formula above, this would have yielded a feed rate of 30 inches per minute. However, it is best to start slow. Now that the spindle rpm and feed rate have been determined, the code to thread the four holes can be written as follows:

```
N0670 G99 G84 Z-.560 R.100 F20.0
```

This line of code threads the hole, then reverses the feed and spindle rotation to retract the tap from the hole. The next step is to complete the threading cycle by specifying the XY locations of the other three holes.

```
N0680 Y3.25
N0690 X5.25
N0700 Y.750
N0710 G80
N0720 G00 X-.100 Y-.100 Z.100 M05
N0730 M09
N0740 M30
```

As with the other canned cycles, the G80 cancels the tapping cycle. Line N0720 retracts the tool and stops the spindle. Line N0730 stops the coolant flow. The last line of the program contains the M30 code, which ends the program and resets it to the beginning. The complete code is shown below:

```
N0010 G00 G53 G20 G90
N0020 G54 X0.000 Y0.000 Z0.000
N0030 M05
N0040 M06 T1 (Change to .750" End Mill)
N0050 G41 X-.100 Y-.100 Z-.650 D.750
N0060 M03 S1050
N0070 M08
N0080 G01 X0.000 Y0.000 F9.0
N0090 X6.000
N0100 Y4.000
N0110 X0.000
N0120 Y0.000
N0130 G00 X-.100 Y-.100 Z.100 M05
N0140 M09
N0150 Z.500
N0160 M06 T2 (Change to #2 Center Drill)
N0170 X.750 Y.750 Z.200
N0180 M03 S4000
N0190 M08
N0200 G99 G81 R.100 Z-.075 F3.0
N0210 Y3.250
N0220 X1.875 Y2.625
N0230 Y1.375
N0240 X3.000 Y2.000
N0250 X4.125 Y1.375
N0260 Y2.625
N0270 X5.250 Y3.250
N0280 Y.750
N0290 G80
N0300 G00 X-.100 Y-.100 Z.100 M05
N0310 M09
```

```
N0320 M06 T3 (Change to #3 27/64″ Drill)
N0330 X.750 Y.750 Z.100
N0340 M03 S950
N0350 M08
N0360 G99 G83 Z-.560 R.100 Q.220 F6.0
N0370 Y3.25
N0380 X5.25
N0390 Y.750
N0400 G80
N0410 G00 X-.100 Y-.100 Z.100 M05
N0420 M09
N0430 M06 T4 (Change to #4 1/2″ Drill)
N0440 X1.875 Y1.375 Z.100
N0450 M03 S800
N0460 M08
N0470 G99 G83 Z-.560 R.100 Q.220 F6.0
N0480 Y3.25
N0490 X5.25
N0500 Y.750
N0510 X3.000 Y2.000
N0520 G80
N0530 G00 X-.100 Y-.100 Z.100 M05
N0540 M09
N0550 M06 T5 (Change to #5 3/4″ Drill)
N0560 X3.000 Y2.000 Z.100
N0570 M03 S530
N0580 M08
N0590 G99 G83 Z-.560 R.100 Q.220 F6.0
N0600 G80
N0610 G00 X-.100 Y-.100 Z.100 M05
N0620 M09
N0630 M06 T4 (Change to #5 1/2″-13 Rigid Tap)
N0640 X.750 Y.750 Z.100
N0650 M03 S260
N0660 M08
N0670 G99 G84 Z-.560 R.100 F20.0
N0680 Y3.25
N0690 X5.25
N0700 Y.750
N0710 G80
N0720 G00 X-.100 Y-.100 Z.100 M05
N0730 M09
N0740 M30
```

Chapter Review

Summary

- Miscellaneous function codes, or M-codes, include machine functions such as stopping the program execution, ending the program, turning the spindle and coolant flow on and off, and performing automatic tool changes.
- Work-holding devices for CNC machining include machining vises and specialty jigs and fixtures that are designed to hold specific workpieces efficiently.
- Before a CNC program can be written, the programmer needs to know how the workpiece will be held in place, what tools will be needed, and the feed rates and spindle speeds required for the tools.
- A CNC program consists of lines of code that guide the CNC machine to select the proper tools and perform the required processes in a logical, efficient order.

Review Questions

Answer the following questions using the information provided in this chapter.

1. The code most likely to be used by a programmer to stop the machine for a required manual tool change is _____.
2. What is the difference between the M00 code and the M01 code?
3. The _____ code is used to stop the machine at the end of the program and to reset the program to the first line of code, ready for re-use.
4. To turn on the spindle and specify clockwise rotation, the programmer uses the _____ code.
5. What is the meaning of the following line of code: N0030 M04 S3000?
6. Briefly explain how an automatic tool change can be written into a program.
7. Why is it important to identify work-holder to be used before beginning to write a CNC program?
8. Under what circumstances might a machining vise be preferred instead of a specialty jig or fixture as the work-holding device for a machining operation?
9. Explain why it is good practice to include a spot drill step to locate the centers of holes before the holes are drilled.
10. The location of the spindle, given in Cartesian coordinates, is known as the _____ home.
11. Which codes provide the X, Y, and Z coordinates of the program zero location?
12. Text enclosed in parentheses that is not read as part of the program is known as a(n) _____.
13. Which code would a programmer use to specify cutter compensation to the right (clockwise)?
14. Suppose you have programmed a canned drilling cycle to drill six holes at specific locations on the workpiece. What line of code would you write to move the cutter away from the part to a location of X=–0.200, Y=–0.200, Z=0.100 using rapid positioning; stop the spindle rotation, and turn off the coolant? Assume that the line number is N0540.
15. A canned cycle, or _____, is a set of commands that follows a prescribed sequence.
16. In the code that specifies a peck drilling operation, what does the value of Q determine?
17. How is the value of Q calculated?
18. In what three ways are the requirements for tapping holes different from other types of drilling?
19. What would be the feed rate for a tapping operation for a 1/4-20 thread with a spindle speed of 400?
20. Which code is used at the end of a tapping cycle to end the cycle?

CHAPTER 25

CNC Turning

Learning Objectives

After studying this chapter, you will be able to:

- Identify work-holding devices that are commonly used with CNC lathes and turning centers.
- List the information needed to prepare for writing a program for a CNC turning center.
- Explain how to set the machine offsets and prepare the CNC machine for a turning operation.
- Carry out the procedure for writing a program to machine a part using a two-axis CNC lathe.

Technical Terms

bar puller
part catcher
repetitive cycle

This chapter introduces the creation of CNC programs for turning centers. The machining process is different than for a CNC machining center, but there are many similarities. Once you understand programming for a CNC machining center, as described in Chapter 24, *CNC Milling*, it is fairly easy to learn how to program a CNC turning center.

25.1 Work-Holding Devices for CNC Turning Centers

Work-holding devices used with CNC turning centers do not differ significantly from those used with manual lathes. Differences are often dictated by the application of the technology. For example, in a toolroom setting, parts are typically made in small quantities. Sometimes only a single part may be needed. In this application, three-jaw and four-jaw chucks are often used.

It is not unusual to find collets being used in a toolroom setting as well, **Figure 25-1**. Because collets tend to reduce loading and unloading times, the use of collets best suits applications in which large volumes of parts are being manufactured. In a production setting, chucks are seldom used because of the extra time it takes to load and unload parts. Also, with collets, automated systems can be used to load and unload parts, as well as activate the collets.

A turning center can be equipped with a bar feeder, bar puller, and part catcher to automate the machining process. Bar feeders feed bar stock into the turning center. ***Bar pullers*** are mounted in the turret with the cutting tools, **Figure 25-2**. Bar pullers clamp and pull bar stock through the spindle for machining.

Collet chuck
Main spindle
Sub spindle

Royal Products, Division of Curran Manufacturing Corporation

Figure 25-1. A turning center with collet chucks in both the main and sub spindle. The part can be held by the collet in the sub spindle so the rear end of the part can be machined.

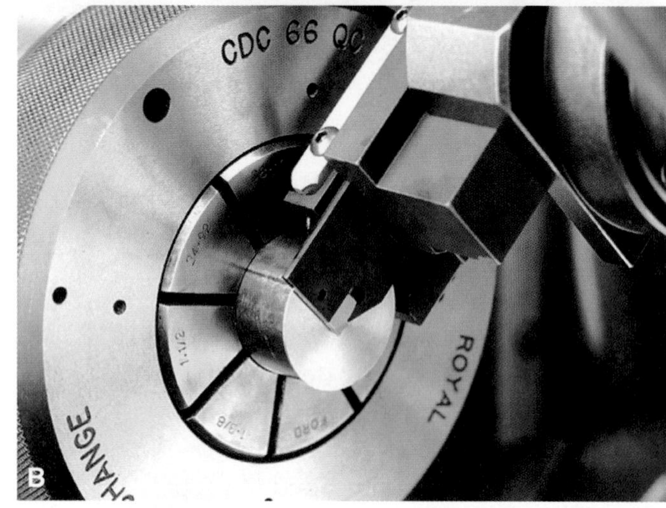

Royal Products, Division of Curran Manufacturing Corporation

Figure 25-2. Bar pullers. A—Heavy-duty CNC bar puller. B—Combination bar puller and parting tool. Reduces cycle time by combining the cutoff and bar pulling operations. C—Coolant-actuated bar puller. This bar puller is actuated by the through-spindle coolant system.

Part catchers are programmable devices used to catch finished parts as they are cut off the bar stock. Combined with an automatic accumulator to collect finished parts, **Figure 25-3**, turning centers can be completely automated.

25.2 Planning for a CNC Turning Program

This chapter describes how to create a program to machine the part shown in **Figure 25-4**. The part is a cylindrical piece with three different diameters. The base material is bar stock 3.00″ in diameter.

25.2.1 Identifying the Processes and Tools

As when developing a milling program, the first step in developing a turning program is to determine the processes to be performed and the tools that will be needed. Because the processes determine the tools to be used, begin by identifying the processes. For the part to be machined in this chapter, the following four processes are required:

1. Facing the end of the part.
2. Roughing out the basic shape, **Figure 25-5**.
3. Making a finish cut to smooth the surfaces.
4. Performing a cutoff operation to separate the finished part from the bar stock.

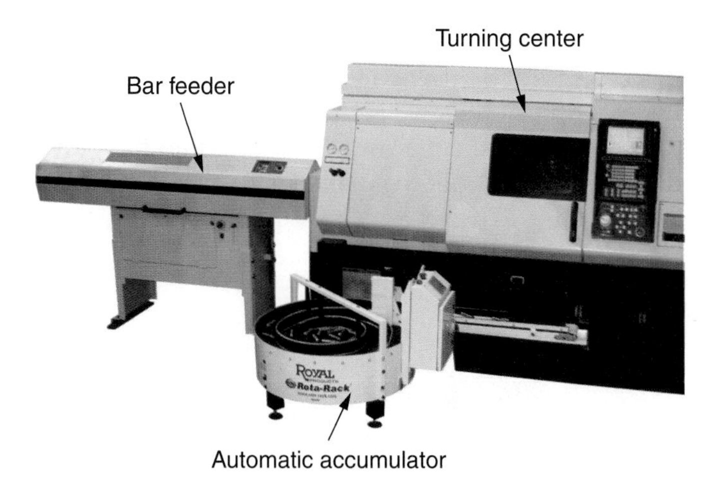

Royal Products, Division of Curran Manufacturing Corporation

Figure 25-3. Combined with a bar feeder and automatic accumulator, this turning center is completely automated. The spiral-shaped Rota-Rack® automatic accumulator collects finished parts as they are ejected from the turning center and indexes the parts toward its center.

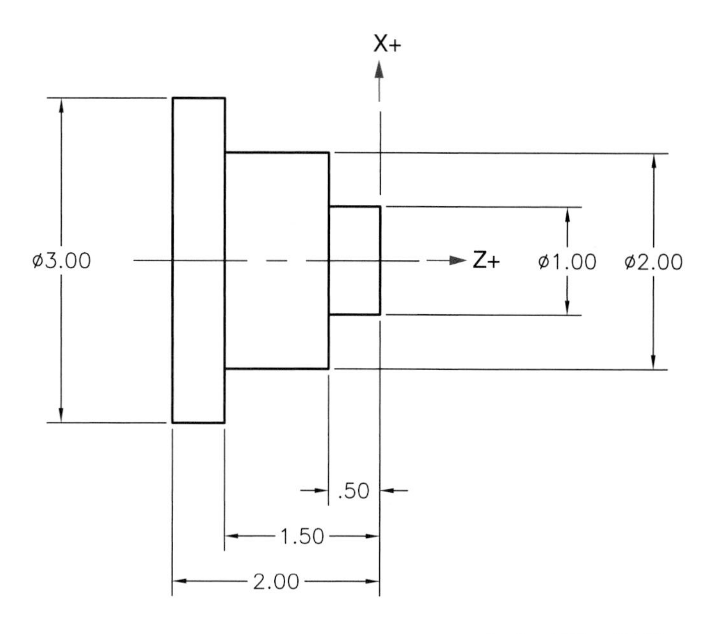

Goodheart-Willcox Publisher

Figure 25-4. This chapter will describe how to create the program to machine this part on a two-axis turning center.

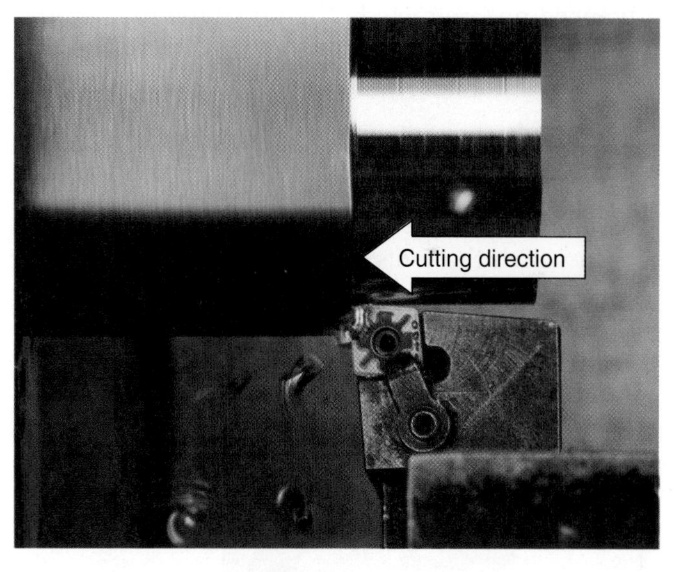

Carboloy

Figure 25-5. Using a rough pass to machine a basic profile with an indexable insert cutting tool.

To perform these steps, the following three tools will be needed:

- A roughing tool to remove the majority of the material (Steps 1 and 2).
- A finishing tool to remove a small amount of material to create a smoother finish (Step 3).
- A parting tool to cut the finished part away from the bar stock (Step 4).

Many machinists use the same tool for the roughing and finishing operations. However, the roughing operation dulls the cutter faster than the finish operation because of the amount of material being removed. In a high-volume production operation, using a second tool for finishing allows the tool for the roughing operation to be used longer. This may result in fewer cutters needed per production run.

25.2.2 Coordinate System Orientation

Before writing a program for a CNC lathe or turning center, it is useful to visualize the positioning of the part along the coordinate system axes. The Z axis is the cylindrical axis that passes through the part, as shown in **Figure 25-4**. The positive direction of the Z axis runs from the headstock, which turns the part, toward the tailstock end of the machine, if a tailstock is being used. The X axis runs perpendicular to the Z axis.

The tool post should be set up so that the tip of the tool travels along the X axis as the cross-slide is advanced and retracted. The tip of the tool should intersect the Z axis. The program created in this chapter is for a two-axis lathe like the one shown in **Figure 25-6**.

The Y axis typically is not used with a standard two-axis CNC lathe equipped with a cross-slide. It is more commonly used with CNC turning centers in which the tools are held in a tool turret. See **Figure 25-7**. The Y axis moves in a vertical motion, with the positive direction upward and the negative direction downward.

Hardinge

Figure 25-6. A slant bed two-axis turning center.

Sandovik Coromant

Figure 25-7. The tool turret on this three axis turning center allows vertical movement along the Y axis. The milling head attachment is being used to machine slots in this part.

25.2.3 Radius versus Diameter

In a turning operation, the part turns around the Z axis. As the cutter advances along the X axis, the tool reduces the radius of the part.

When programming, remember that the diameter of a cylindrical part is two times its radius, **Figure 25-8**. For example, if the goal is to reduce the diameter of a part by .100″, the cutter must advance into the part a distance equal to half the diameter, or .050″.

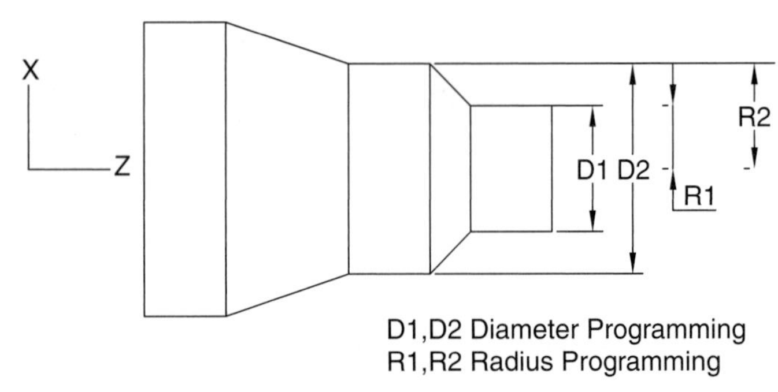

Goodheart-Willcox Publisher

Figure 25-8. The difference between diameter and radius programming.

Older machines required programmers to compensate mathematically for the difference between the diameter and radius. Modern equipment can be programmed using either diameter or radius input data. This is controlled in the machine's registers and cannot be adjusted from within individual programs. These settings are usually protected so that only advanced programmers have access to them. It is up to the programmer to verify these settings when creating a program for a specific machine. For the example program developed in this chapter, the equipment is assumed to be set to take diameter input.

25.2.4 Diameter for Spindle Speed Formula

The spindle speed formula in a program for a lathe or turning center requires a cutting speed (CS, based on the material being machined) and a diameter (D) component:

$$\text{rpm} = \frac{\text{CS} \times 4}{\text{D}}$$

In the milling operation, the diameter used in the formula is the cutter diameter. However, in turning, the diameter of the workpiece is used. As the part is machined and the diameter is reduced, the formula may need to be changed. However, if the changes in diameter are small, the changes in rpm may be negligible. It is a good idea to calculate the changes as material is removed on each pass.

25.3 Initial Programming

The first few lines in a program for a CNC lathe or turning center are similar to those for a milling operation. They set up the machine and prepare it for the current operation.

25.3.1 Setting the Machine Offsets

Typically, the machine zero position is set to a point where the tip of the tool is in line with the cylindrical axis and flush with the end of the workpiece. During the initial setup, the tool is manually positioned at this point and the value from the digital readout is recorded in the machine register under G-codes 54 through 59. Recall from Chapter 24 that these codes are used to specify machine offsets.

25.3.2 Preparing the Machine

The first lines of code set the machine to a point where the cutter is away from the part, and also set the appropriate program defaults:

```
N0010 G00 G53 G20 G90
N0020 G54 X-3.25 Z.25
N0030 M05
N0040 M00 (Change to a roughing tool)
```

Line N0010 sets the travel speed to rapid traverse (G00), clears any machine offset position that may be loaded in the register (G53), sets the units to inches (G20), and sets positioning to absolute (G90). This clears any information that may be active from a previous program or from a previous cycle in the current program that did not run to completion.

Line N0020 sets the machine zero position. Notice that the coordinates in this case are stored under code G54. Line N0030 is a fail-safe operation to stop the spindle rotation.

SAFETY NOTE

Some programmers may think it is unnecessary to stop spindle rotation at this point. However, it never hurts to think about the many ways a program can malfunction and cause damage to the machine, the tool, or the workpiece. A few extra lines of code are well worth the small amount of machine memory to help ensure that such malfunctions do not occur.

Line N0040 introduces the miscellaneous function M00. Recall from Chapter 24 that M00 is a program stop. It is used to pause a program when needed to allow an operator to clear chips away from a part, inspect the tool for wear, measure a feature, or change tools manually if the machine is not equipped with an automatic tool changer, **Figure 25-9**. In the example in this chapter, the machine requires the operator to manually change the tools. The text inside the parentheses is a note prompting the operator to change the tool to a roughing tool for the first machining operation.

Marcelo Costa/Shutterstock.com

Figure 25-9. An M00 code can be used to stop the program to inspect a machined feature.

Sandvik Coromant

Figure 25-10. Facing removes material from the end of the workpiece. Typically, facing is one of the first operations performed when turning a part.

25.4 Programming the Machine Operations

After the correct tool has been loaded, the machine is ready to begin the first machining step, which is facing the end of the part. At this point, the machine is still paused from the execution of the M00 command. To restart the machine, the operator is required to press the cycle start button on the control panel. Pressing cycle start returns the machine to the state it was in before the M00 command. For example, if the spindle was rotating and the coolant was flowing, they will restart using the same settings as before the M00 command was read by the control.

25.4.1 Facing the End of the Part

The following lines of code direct the machine to perform the facing operation, **Figure 25-10**:

```
N0050 M03 S600
N0060 G00 Z-.050
N0070 G01 X0.0 F.019
N0080 G00 X-3.05 Z.05 (End of Facing
                     Operation)
```

Line N0050 starts the spindle in a clockwise rotation at 600 rpm. The next line moves the tip of the cutting tool to a position just off the outside diameter of the workpiece, .050″ away from the tip of the part. Line N0070 feeds the tip of the cutter across the end of the part to the part's center at a feed rate of .019 inches per revolution. This removes .050″ stock from the end of the part. Finally, line N0080 moves the cutter away from the part at a rapid travel rate, ending the facing operation.

25.4.2 Stock Removal Cycles

The next step in the machining process is to remove the majority of the material using a rough cutting operation. CNC lathes have codes that perform operations similar to the canned cycles on CNC machining centers. These cycles are called ***repetitive cycles***. The G71 and G72 codes remove large amounts of material. They program the tool to start at a certain point and then to machine down by the specified amount using a series of passes. The G71 command removes material by feeding the tool along the Z axis during the cutting feed. G72 removes material by feeding the tool along the X axis during the cutting feed. Since the maximum travel for this part is .500″ along the X axis and 1.500″ along the Z axis, using the G71 code is the more efficient code in this case. See **Figure 25-11**.

The G71 code activates the repetitive cycle. There are two ways to activate the repetitive cycle: using a single line of code or using two lines of code. For simplicity, this example uses the two-line option. The first line is:

```
N0090 G71 U.125 R.063
```

On this line, the letter U defines the amount of stock to be removed on each roughing pass, and the letter R defines the retracting distance for the tool after each pass.

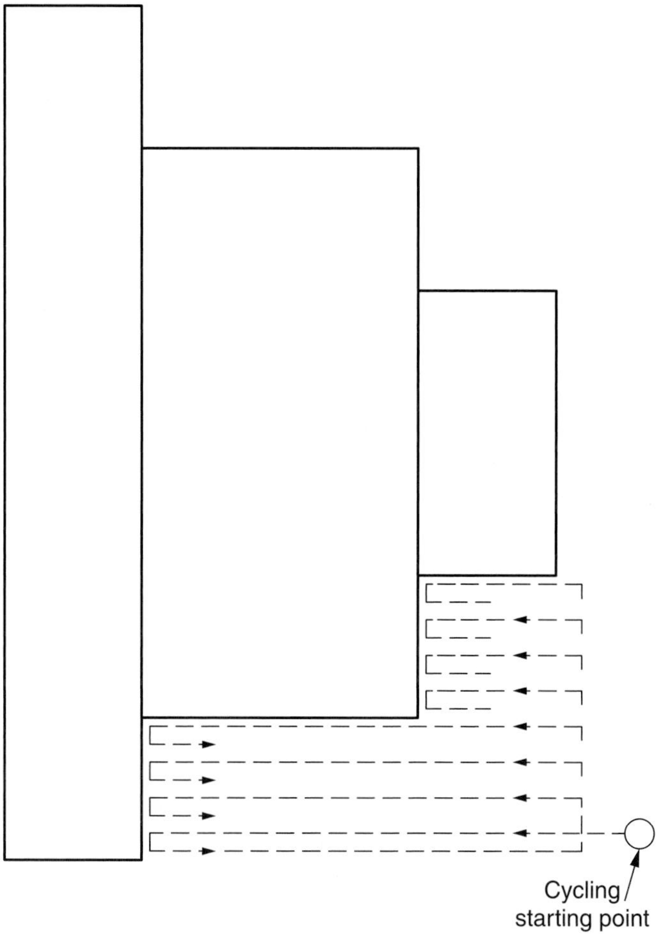

Cycling starting point

Goodheart-Willcox Publisher

Figure 25-11. The tool path taken while performing a rough turning operation with the G71 repetitive cycle.

SAFETY NOTE

The letter U will be used again in the next line of code, and it will mean something different. Do not get these two confused! The result could crash the machine.

The next line defines the rest of the repetitive cycle:

```
N0100 G71 P### Q### U.030 W.030 F.020
```

Line N0100 contains two unknown variables. The letter P defines the line of code that begins the cycle, and the letter Q defines the line of code that ends the cycle. Number (pound) signs can be inserted following the letters P and Q to fill in space until the lines of code for the beginning and end of the cycle are known.

The letter U on line N0100 defines the amount of stock to be left for the finishing pass. The control references the lines of code containing the finishing pass using the offset value prefaced by the letter U for the X direction and the letter W for the Z direction. The control automatically calculates the .030 offset defined to leave the material for the finishing cycle. The letter F defines the feed rate in terms of inches of feed per revolution of the part.

The process of using the G71 code requires the tool to be sent to a start point away from the part. This cycle will start from the position defined on line N0080. Next, the finish profile must be defined. **Figure 25-12** shows the start point and where the final cut will begin. This position is at the end of the part at the point where the Z axis intersects the part. First, position the tip of the cutter at the proper X axis dimension and just off the part in the Z direction:

```
N0110 G00 X0.0 Z0.0
```

During the facing operation, .050″ material was removed from the end of the part, so the tip of the cutter will be .050″ from the end of the part at the Z = 0.00″ position. The programmer also plans to leave .030″ of stock for the finishing cycle. The next few lines of code specify the path of the cutter to define the final shape of the part.

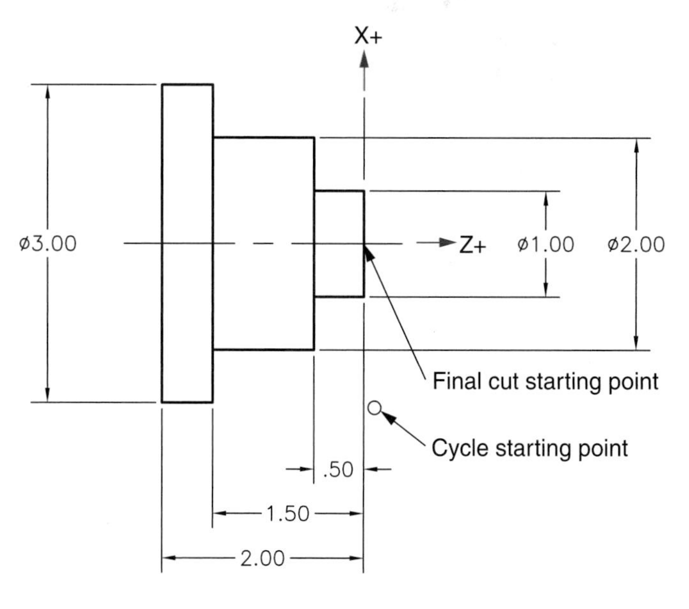

Goodheart-Willcox Publisher

Figure 25-12. The cycle starting point defined by line N0080 and the starting point of the final cutting pass. The tool will retreat to the starting point when the cycle is complete. The starting point of the final cutting pass helps define the entire repetitive cycle.

```
N0120 G01 Z-.08 F.015
N0130 X-1.00
N0140 Z-.580
N0150 X-2.00
N0160 Z-1.580
N0170 X-3.01
N0180 G00 X-3.25 Z.25
```

The dashed line in **Figure 25-13** shows the cutter path produced by these lines of code. The space between the dashed line and the finished shape represents the .030″ of stock left from the stock removal cycle to be removed by the finishing cycle. In line N0170, the movement extends beyond the diameter of the part to ensure that the entire surface is machined. The cycle ends by returning the tool to the starting point.

Now that the finish profile has been defined and the lines have been numbers, go back and fill in the values for variables P and Q in line N0100:

```
N0100 G71 P0110 Q0170 U.030 W.030 F.020
```

Notice that the value for Q is 0170, not 0180. N0180 is not part of the final pass. It merely moves the tool to a clearance position. Also notice that the "N" in the line numbers is not used in the P and Q values.

25.4.3 Finish Turning Cycle

A G70 code is used after the G71 code. The G70 code is a finish turning cycle, **Figure 25-14**. It uses the same start point, cycle start line number, cycle stop line number, and finish point as the roughing cycle. After the roughing cycle, it is necessary to change to a finishing tool. The next two lines of code are needed for the tool change:

```
N0190 M00 (Change to a finishing tool)
```

Line N0190 uses the M00 code to stop the tool for a manual change to the finishing tool. When the cycle start button is pressed, the spindle resumes rotation.

After the tool change, the finishing pass can be activated.

```
N0200 G70 P0110 Q0170
N0210 G00 X-3.25 Z.25
```

Line N0200 starts the finishing cycle. The line numbers for the start and end of the cycle are the same numbers used in line N0100 for the roughing cycle. Line N0210 moves the cutter to a clearance position.

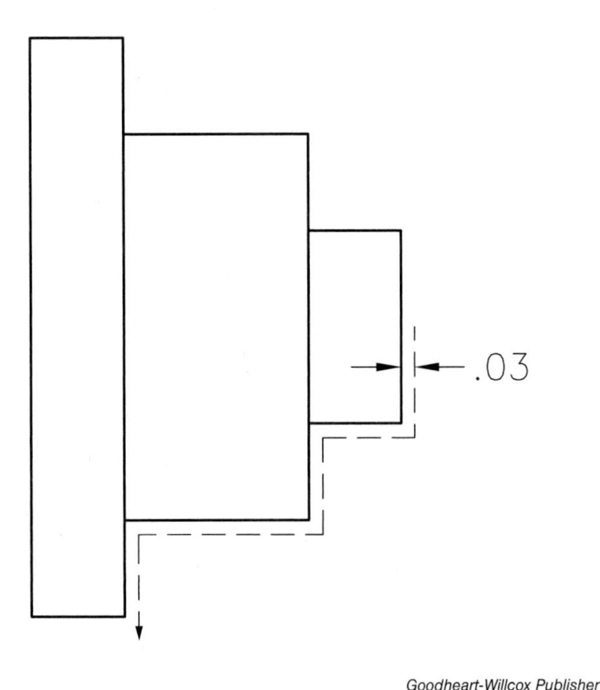

Goodheart-Willcox Publisher

Figure 25-13. The dashed line shows the cutter path defined by the finish profile. The .030″ gap between the part and the line represents the stock left to be removed by the finishing cycle.

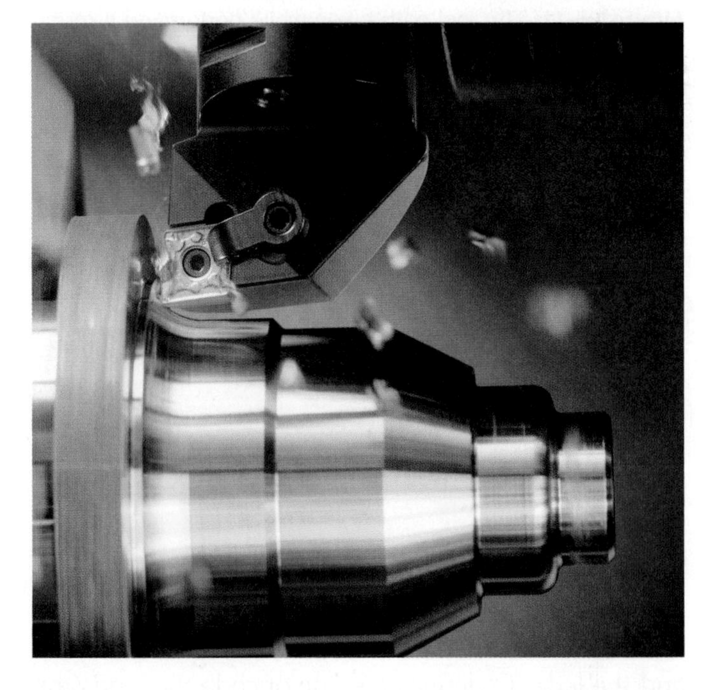

Sandvik Coromant

Figure 25-14. Finishing cuts use fine feed and a shallow depth of cut to machine the workpiece to size and provide a good surface finish.

25.4.4 Cutoff Operation

The cutoff operation is the final step for many procedures performed on a CNC lathe or turning center, **Figure 25-15**. For the final machining step, the program is paused one last time so the tool can be changed to a parting tool. Then the finished part is cut away from the bar stock:

```
N0220 M00 (Change to a parting tool)
N0230 G00 X-3.050 Z-2.08
N0240 G01 X0.0 F.019
```

As noted in the program, line N0220 pauses the program for the tool change. Line N0230 uses rapid positioning to position the cutter for the final cut, and line N0240 performs the final cutting operation.

The G75 code could also be used to perform a cutoff operation. The G75 code calls for a pecking cutoff or grooving cycle used for both inside and outside diameters. The G75 code requires an X axis coordinate (X), the peck distance (I), and the feed rate (F).

Figure 25-15. Cutting off the finished part is a commonly performed operation on CNC lathes and turning centers. The parting tool should be as short as possible when performing a cutoff operation.

25.4.5 Ending the Program

The last two lines finish up the program:

```
N0250 G00 X-3.25 Z.25 M05
N0260 M30
```

Line N0250 uses rapid positioning to move the cutter out of the way and stops the spindle. The M30 code in line N0260 ends the program and resets it to the beginning. The complete code is shown below:

```
N0010 G00 G53 G20 G90
N0020 G54 X-3.25 Z.25
N0030 M05
N0040 M00 (Change to a roughing tool)
N0050 M03 S600
N0060 G00 Z-.050
N0070 G01 X0.0 F.019
N0080 G00 X-3.05 Z.05 (End of Facing
      Operation)
N0090 G71 U.125 R.063
N0100 G71 P0110 Q0170 U.030 W.030 F.020
N0110 G00 X0.0 Z0.0
N0120 G01 Z-.080 F.015
N0130 X-1.00
N0140 Z-.58
N0150 X-2.00
N0160 Z-1.58
N0170 X-3.01
N0180 G00 X-3.25 Z.25
N0190 M00 (Change to a finishing tool)
N0200 G70 P0110 Q0170
N0210 G00 X-3.25 Z.25
N0220 M00 (Change to a parting tool)
N0230 G00 X-3.050 Z-2.08
N0240 G01 X0.0 F.019
N0250 G00 X-3.25 Z.25 M05
N0260 M30
```

Chapter Review

Summary

- Work-holding devices for CNC turning centers are similar to those used for manual lathes and may include chucks and collets, depending on the application.
- The first step in creating a CNC turning program is to identify the processes and tools needed for the operation.
- CNC lathes and turning centers may use two axes with a cross-slide, or three axes on turning centers in which tools are held in a turret.
- Before writing the program, the programmer must determine whether the machine is set up to accept radius or diameter input data.
- The initial lines of a program for a turning center set the machine offsets, position the cutter, and perform any other necessary tasks, such as stopping the program for a manual tool change.
- Stock removal is performed in multiple steps using repetitive cycles.
- The final step in a turning procedure is to cut the finished part away from the bar stock.

Review Questions

Answer the following questions using the information provided in this chapter.

1. What type of work-holding device is most likely to be used in a production setting in which large volumes of parts are being manufactured?
2. What two things must be determined before a programmer can begin writing a program for a CNC turning center?
3. The process of machining a part so that it is perpendicular to the center axis passing through the part is known as _____.
4. Rough cuts and finish cuts can be performed by the same tool. Why do many machinists prefer to use two separate tools?
5. Along which axis of the Cartesian coordinate system is the cylindrical axis of the part aligned?
6. Which axis of the Cartesian coordinate system is typically not needed for two-axis CNC lathes equipped with a cross-slide?
7. Why is it important to know whether a machine is set up to take radius or diameter input before programming?
8. The diameter of the _____ is used in the spindle speed formula for a lathe or turning center.
9. Where is the machine zero position typically set for a CNC lathe?
10. Why do many programmers automatically insert the M05 code before pausing the program?
11. Which code would you use to pause the program so that a tool can be changed manually?
12. What action is required to restart the machine after a pause and return it to the state it was in when the program paused?
13. Stock removal cycles are a sequence of steps called _____ that make several passes to cut away the defined amount of material.
14. Which code removes material by feeding the tool along the Z axis during a cutting feed?
15. Which code removes material by feeding the tool along the X axis during a cutting feed?
16. In the following line of code, how much stock is to be removed in each pass, assuming the program is set up to use inches?

    N0130 G71 U.130 R.087
17. When a G71 code is used, what do the letters P and Q specify?
18. In the following line of code, what does the number after the U specify?

    N0320 G71 P0240 Q0310 U.030 W.025 F.015
19. What is commonly the final step in the procedure to machine a part using a CNC lathe or turning center?

CHAPTER 26

Automated Manufacturing

Learning Objectives

After studying this chapter, you will be able to:

- Define the term "automation."
- Summarize the traits of a flexible manufacturing system.
- Define the term "industrial robot."
- Identify the safety issues associated with automated manufacturing.
- Explain the similarities and differences between different rapid prototyping techniques.

Technical Terms

computer integrated manufacturing (CIM)
Direct Shell Production Casting (DSPC)
flexible manufacturing system (FMS)
Fused Deposition Modeling (FDM)™
just-in-time (JIT) inventory system
Laminated Object Manufacturing (LOM)™
robot
smart tooling
stereolithography
work envelope

Automation is a term coined in 1947 by a Ford Motor Company engineer. It is not a revolutionary new form of manufacturing, but a method of manufacturing that has evolved over many years. Automation is a system for the continuous automatic production of a product, **Figure 26-1**. It relies on a machine or group of machines activated electronically, hydraulically, mechanically, or pneumatically (or a combination of these means) to automatically perform one or more of the five basic manufacturing processes:

- Making.
- Inspecting.
- Assembling.
- Testing.
- Packaging.

The principles of automation have been known for many years. An automated flour mill was in operation in the late 1700s near Philadelphia. It was able to continuously mill grain into flour. The mill used many of the elements found in modern automated operations.

Hirata Corporation of America

Figure 26-1. A robotic system used to assemble the components of automobile engines.

The integration of the computer, **Figure 26-2**, with specially designed machine tools and equipment has revolutionized production technology, resulting in increased productivity, improved product quality, and reduced manufacturing costs. Automation has also reduced human involvement to an absolute minimum in many phases of the manufacturing process.

26.1 Flexible Manufacturing Systems

Due to its extensive use of computer-controlled machinery and adaptive tooling, a ***flexible manufacturing system (FMS)*** is able to be quickly adapted to changes in the product or the manufacturing process. This general category of machining/manufacturing technology is also widely referred to as ***computer integrated manufacturing (CIM)***. A flexible manufacturing system is made up of one or more groupings of machines that are used to perform multiple operations automatically. Such a grouping of machines is often called a flexible manufacturing cell (FMC).

A flexible manufacturing cell brings together workstations (machine tools), automated handling and transfer systems, and computer control in an integrated manner, **Figure 26-3**. It is capable of performing multiple manufacturing actions simultaneously. Work is

Manufacturing Technology, Inc.

Figure 26-2. The integration of computers and specially designed machine tools and equipment has revolutionized automation technology. Computerized systems improve product quality and reduce manufacturing costs.

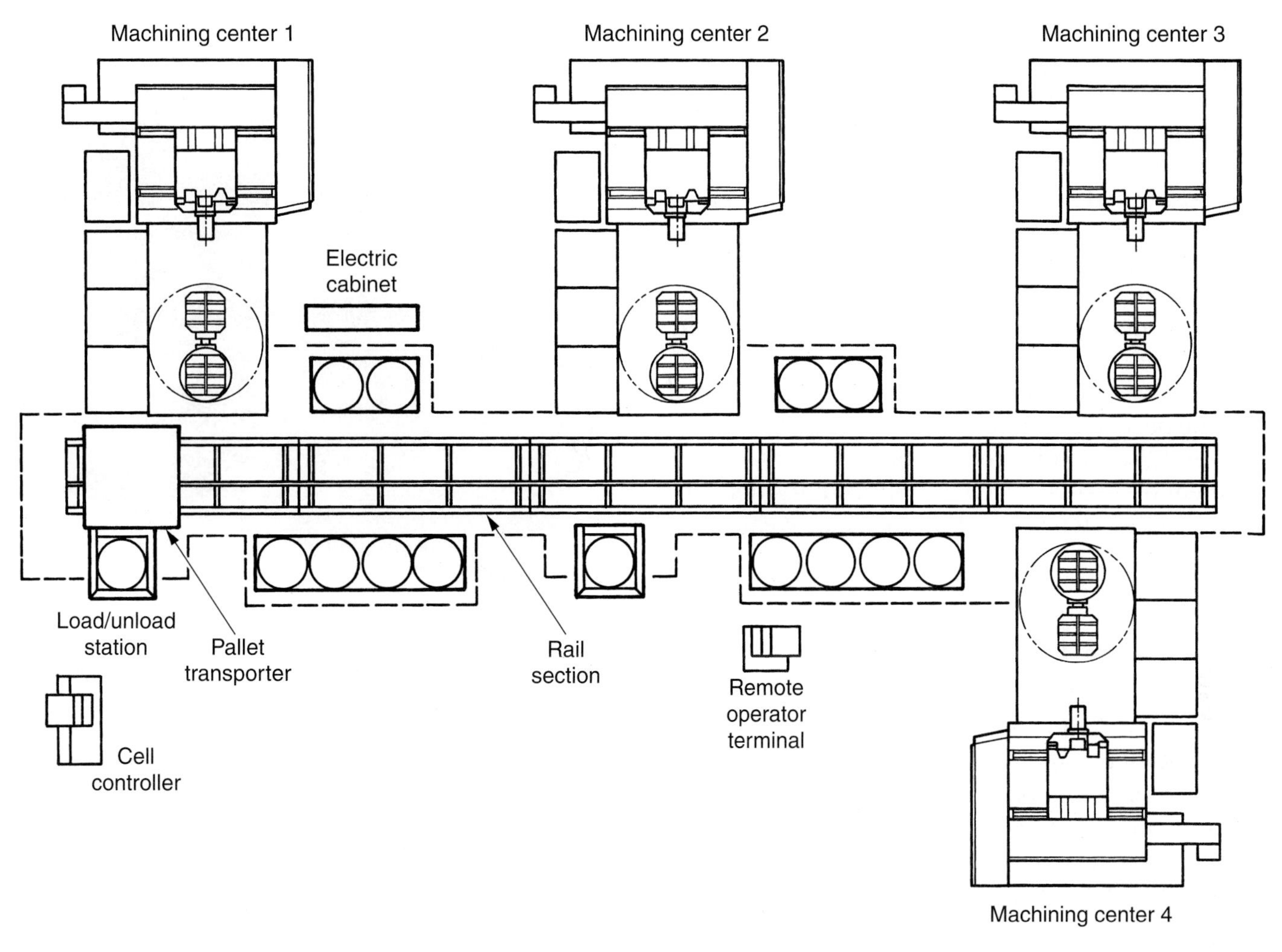

Cincinnati Milacron

Figure 26-3. Flexible manufacturing cell that uses a pallet transporter to link the machines. A cell controller automatically queues work for immediate delivery to the next machine available.

transferred to and from the individual machines in the flexible manufacturing cell by automated fixture carts, conveyors, or specially designed loaders, **Figure 26-4**. Robots may also be employed in some operations.

The machine tools typically used in flexible manufacturing systems are CNC (computer-aided numerical control) vertical and horizontal machining centers. Other CNC machine tools, such as grinders, and automatic gaging equipment may also be included in flexible manufacturing systems.

Each step of the manufacturing process is computer-controlled and linked with the succeeding one. Measuring sensors, often utilizing lasers, are sensitive enough to detect tool wear as it occurs and automatically compensate for it. Every part is inspected. Problems such as tool malfunction or tool breakage are immediately identified so corrections can be made before additional parts are produced that do not meet specifications.

Westech Products Group/Gantrex Machine Tool Loaders

Figure 26-4. A conveyor (foreground) transfers workpieces to an unattended horizontal machining center in this manufacturing cell.

Flexible manufacturing systems can also make use of smart tooling; just-in-time (JIT) inventory of materials, parts, subassemblies, and robots.

Smart tooling consists of cutting tools and workholding devices that can be easily reconfigured to produce a variety of shapes and sizes within a given part family. This makes it economically feasible to manufacture products in smaller lot sizes.

A ***just-in-time (JIT) inventory system*** is an inventory management system that eliminates the need to store large quantities of materials and parts. The parts and materials are scheduled for arrival at the time needed and not before. While JIT inventory systems lessen (and often eliminate) storage costs, production can be reduced or stopped if the delivery of the needed items is delayed by bad weather or strikes.

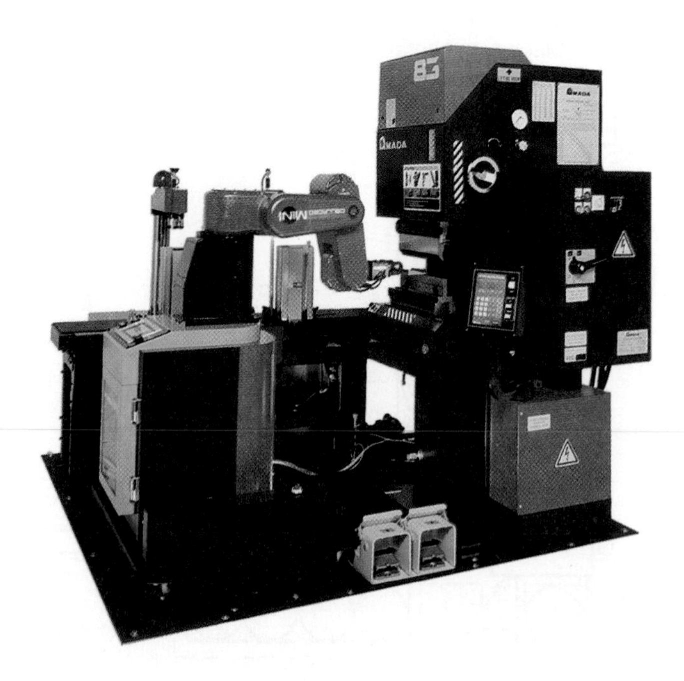

U.S. Amada, Ltd.

Figure 26-5. Robots can be programmed to do many types of jobs that are hazardous or tedious for human workers. Spot welding, paint spraying, and loading parts onto machines are among the jobs performed by robots. This robot is paired with a press brake to load and remove sheet metal parts.

26.2 Robotics

The Robotic Industries Association (RIA) has adopted the following definition for industrial robots, like the one shown in **Figure 26-5**.

"A ***robot*** is a reprogrammable, multifunctional manipulator designed to move material, parts, tools, or specialized devices through variable programmed motions for the performance of a variety of tasks."

26.2.1 Robotics in Automation

There are two implied requirements in the RIA definition of an industrial robot:

- The robot must be a multiaxis machine (three or more axes) that is capable of moving parts or tools to any given location within its work envelope with a high degree of repeatability and precision. The ***work envelope*** of a robot is the volume of space defined by the reach of the robot arm in three-dimensional space. See **Figure 26-6**.

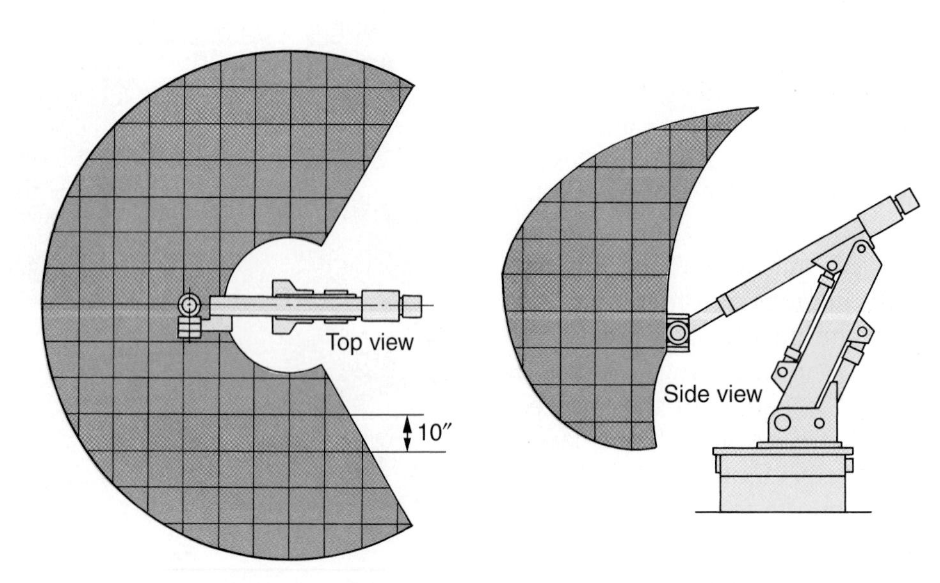

Goodheart-Willcox Publisher

Figure 26-6. The work envelope of a robot is the volume of space defined by the reach of the robot arm in three dimensions.

- There must be a control system that can be programmed to drive the robot through a series of specified motions and be capable of interacting with other machines and equipment.

26.2.2 Industrial Robot Design

As shown in **Figure 26-7**, an industrial robot consists of four basic components:

- **Controller.** Performs computations for controlling the movement of the waist, arm, and wrist to the proper location.
- **Power supply.** Supplies the power to operate the robot. May be hydraulic, pneumatic, or electric. Most modern robots have electric drive.
- **Manipulator.** The articulated "arm" of the robot. The end of the arm is fitted with a "wrist" capable of rotational motion around one or more axes.
- **End-of-arm tooling.** A device attached to the robot wrist for specific applications, such as a gripper, welding head, spray gun, etc.

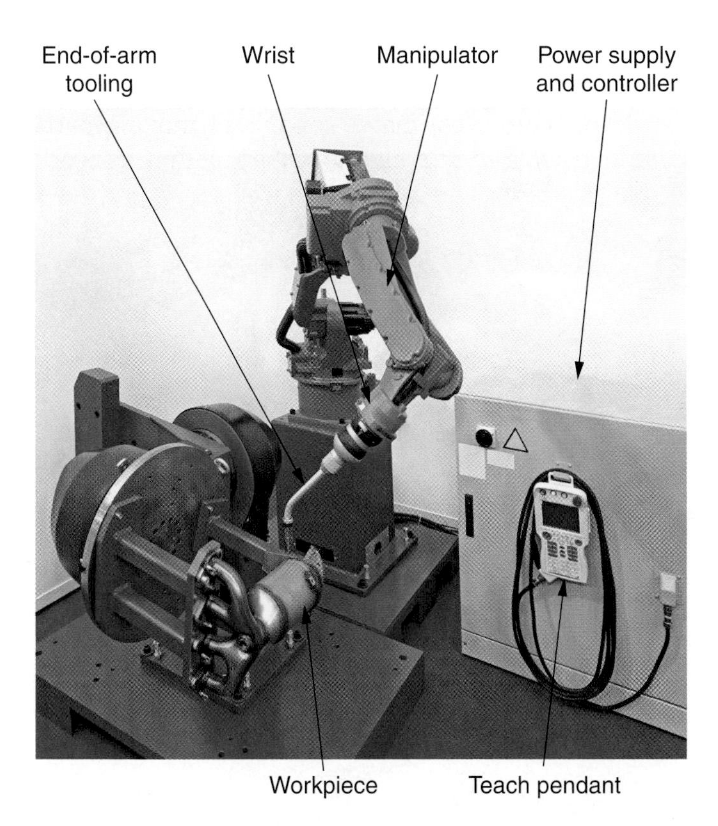

Baloncici/Shutterstock.com

Figure 26-7. An industrial robot consists of four basic parts: a controller, a power supply, a manipulator, and end-of-arm tooling. The end-of-arm tooling on this robot is a gas metal arc welding gun.

26.2.3 Industrial Robot Applications

Industrial robots are widely used in a number of different applications:

- **Hazardous and harsh environments.** Fumes produced by spray painting and welding, heat in foundry or forging operations, or feeding material into punch presses. See **Figure 26-8**.
- **Tedious operations.** Repetitive operations such as feeding, loading, and unloading of parts and materials, and some assembly operations. See **Figure 26-9**.

Fanuc Robotics North America, Inc.

Figure 26-8. A robot ladling molten aluminum into a wheel mold.

Fanuc Robotics North America, Inc.

Figure 26-9. This robot is being used to load and unload castings from a drilling machine.

- **Precision operations.** Precisely repeated positioning operations, maintaining consistent tool speed, following complex welding and cutting paths without patterns, and quality control using lasers.
- **Handling heavy or unwieldy materials.** Lifting material onto or from a stack, moving material beyond the normal reach of a human, or mounting heavy or bulky workpieces on machine tools, **Figure 26-10**.

Many types of robots have been developed, but almost all can be classified into one of the four basic geometric configurations shown in **Figure 26-11**. Advanced technology robots can be programmed to serve a wide range of automated manufacturing applications. They can be interfaced with testing devices to perform various types of measurements.

Some robots are capable of selecting and positioning complex shaped parts for machining or storage. A laser is utilized to "see" and define the part outline so the correct item will be selected and positioned for machining.

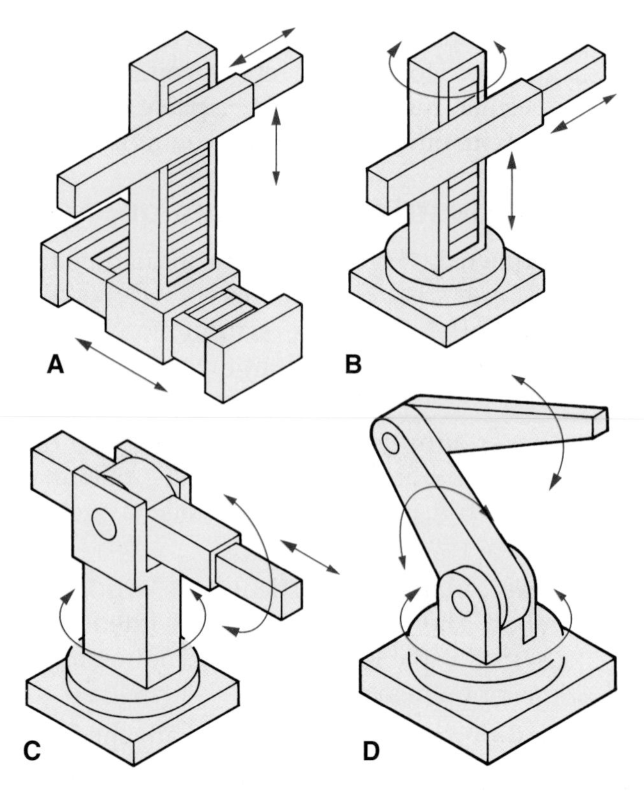

Goodheart-Willcox Publisher

Figure 26-11. Basic geometric configurations of robots. All provide three articulations (specific arm movements). A—Cartesian coordinates. B—Cylindrical coordinates. C—Polar coordinates. D—Revolute coordinates.

Ford Motor Co.

Figure 26-10. This robot lifts an entire vehicle chassis so it can be transferred from one assembly line to another.

26.3 Safety in Automated Manufacturing

- Observe the same safe operating procedures that are used with traditional machine tools and machining.
- Wear approved eye protection and a snug-fitting apron or shop coat.
- Be sure work-holding devices are positioned correctly and are securely fastened to the worktable.
- Carefully check tool clearance to be sure the cutter will clear the workpiece, work-holding devices, clamps, etc., when manually positioning the work and during rapid traverse. Make a "dry run" for a safety check of tool positioning for each machining operation.
- Never enter the work envelope of a robot until you are sure power is turned off.
- Establish that the machine tool and control unit are functioning correctly, and continually monitor machining operation to be sure each tool is cutting properly. Know how to safely stop machining operations in case of an emergency.
- To avoid injury, remove burrs and sharp edges before inspecting finished parts for accuracy or finish of machined surfaces.

26.4 Rapid Prototyping Techniques

A number of new techniques have been developed to allow designers to quickly generate three-dimensional models or prototypes of parts in relatively inexpensive materials. These techniques help to improve design and identify potential machining problems.

26.4.1 Laminated Object Manufacturing (LOM)™

Laminated Object Manufacturing (LOM)™ is a process used to rapidly produce models, prototypes, accurate patterns, and molds. See **Figure 26-12**. This three-dimensional modeling method was developed in 1991 by Helisys, Inc. Since then, many automotive, aerospace, appliance, and medical product manufacturers have adopted the process. The equipment is compact and can operate 24 hours a day.

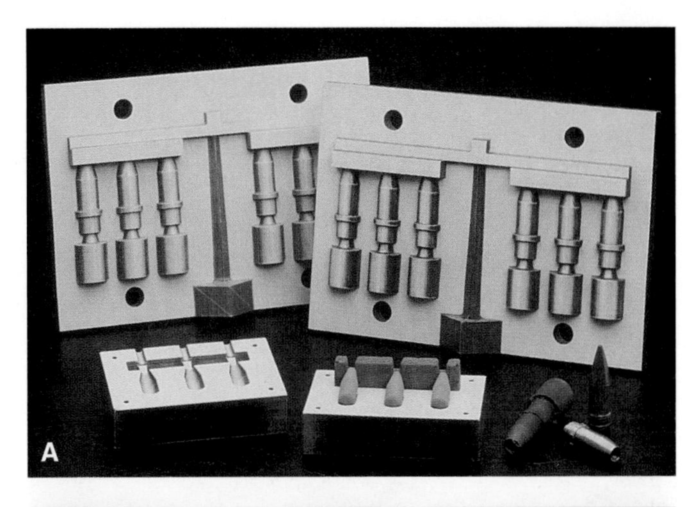

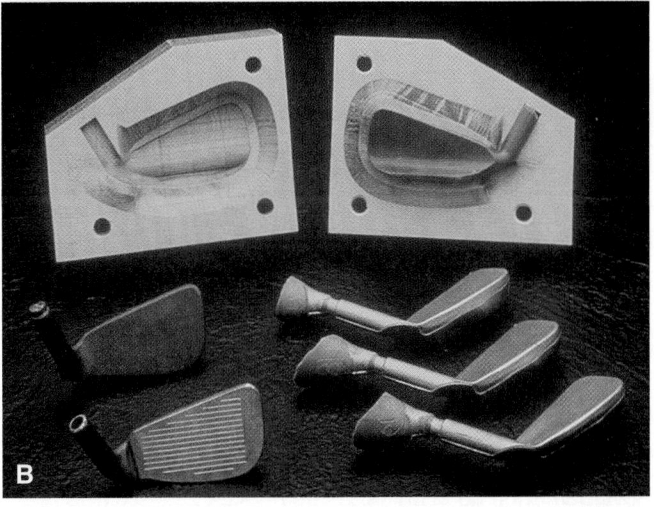

Helisys, Inc.

Figure 26-12. Laminated Object Manufacturing. A—LOM was used to produce these extremely accurate, full-size matchplates (patterns) and core box to be used in casting prototype metal projectiles. B—Rapid prototyping of the molds to make the wax patterns for the investment casting of oversized golf club irons was done with LOM. Design changes could be carried out before making the expensive molds necessary for quantity production.

LOM uses inexpensive solid sheet material, such as paper, plastic, and composites, to form the desired designs. Molds made from the composite materials can be used in direct tooling applications for such processes as injection molding, vacuum forming, and sand casting. See **Figure 26-13**. LOM parts are accurate enough for form-and-fit verification applications. The finished part has a composition similar to wood and can be easily machined or modified to obtain the exact fit required. Modifications and corrections can be incorporated into the CAD design before manufacturing.

Helisys, Inc.

Figure 26-13. The CAD drawing onscreen was used to produce the LOM model (being held). That model was then used to produce a mold. The mold was then used to produce an aluminum sand casting of the prototype rear-engine power takeoff housing for a diesel truck engine. This technique cut development time in half.

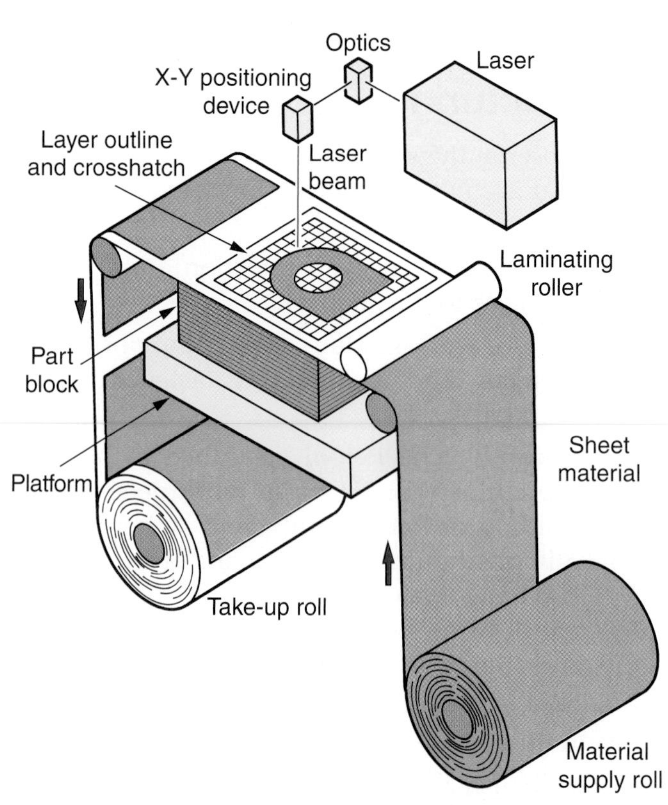

Goodheart-Willcox Publisher

Figure 26-14. The Laminated Object Manufacturing process.

Operation of the LOM System

CAD data is programmed into the LOM's process controller. A cross-sectional slice is generated and a laser cuts the outline of the cross section, then crosshatches the excess material for later removal, **Figure 26-14**. A new layer of material is bonded to the top of the previously cut layer. The next cross section is prepared and cut. This automatic process continues until all layers are laminated and cut. Excess material is removed to expose the finished part. The completed object's surface can then be sanded, polished, or painted as desired.

26.4.2 Stereolithography

Another rapid prototyping technology, called ***stereolithography***, can also produce complex design prototypes of castings and other objects in hours, instead of days or weeks, **Figure 26-15**. The three-dimensional hard plastic models produced by the stereolithography process can be studied to determine whether they are the best solution to a design problem.

3D Systems

Figure 26-15. Stereolithography equipment is used to produce durable, finely detailed patterns extensively used in modeling, functional prototyping, and tooling applications.

Since new models can be made quickly, design changes and modifications can be evaluated without the expense of making new patterns or molds.

The stereolithography process uses a computer-guided low-power laser beam to harden a liquid photocurable polymer plastic into the programmed shape. The process starts by creating the required design on a CAD system and orienting it in three-dimensional space. A support structure is added to hold the various elements in place while the model is built up.

The design data is downloaded into the stereolithography machine, which operates in a way similar to a CNC machine tool. A program within the stereolithography machine's controller slices the design into cross sections of 0.005″ to 0.020″ (0.12 mm to 0.50 mm) in thickness. The machine's control unit guides a fine laser beam onto the surface of a vat containing the liquid photocurable polymer. The liquid solidifies wherever it is struck by the laser beam, **Figure 26-16**.

The model is created from the bottom up, on a platform located just below the surface of the liquid plastic. After each "slice" is formed, the platform drops a programmed distance. The sequence is repeated until the entire model is formed, **Figure 26-17**. The model is removed from the platform and rinsed in a chemical bath to remove excess resin. The model is then placed in an ultraviolet oven for curing. When the model has cured, the support structure is clipped away. The model can then be finished by filing, sanding, and polishing. Paint or dye can be applied. The entire stereolithography process is shown in **Figure 26-18**.

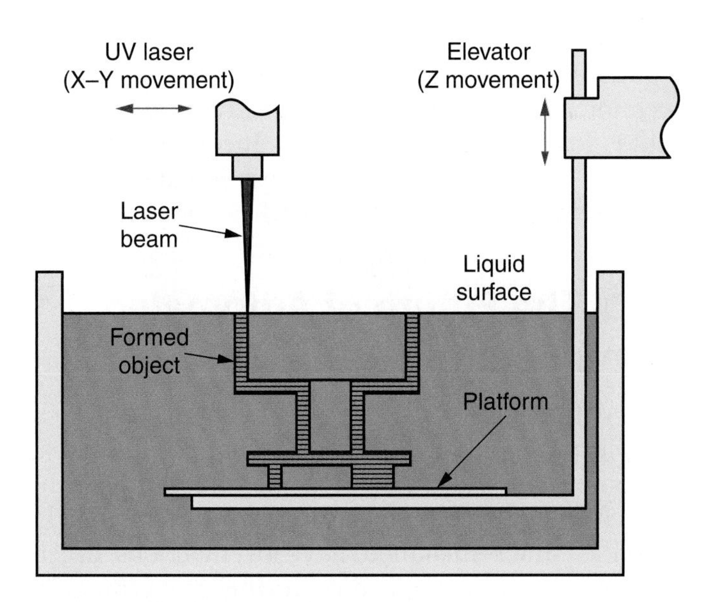

Goodheart-Willcox Publisher

Figure 26-16. Basic diagram of the stereolithography process.

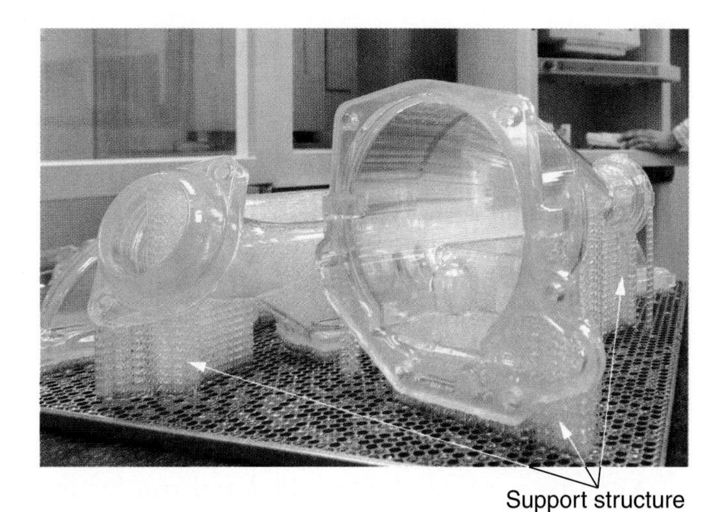

Ford Motor Co.

Figure 26-17. A model created using the stereolithography process. Note the support structure on the model. Once the model is cured and the support structure is removed, this model will be used to train assembly robots. By training the robots with accurate prototypes rather than waiting for production tooling, the plant can greatly reduce setup time.

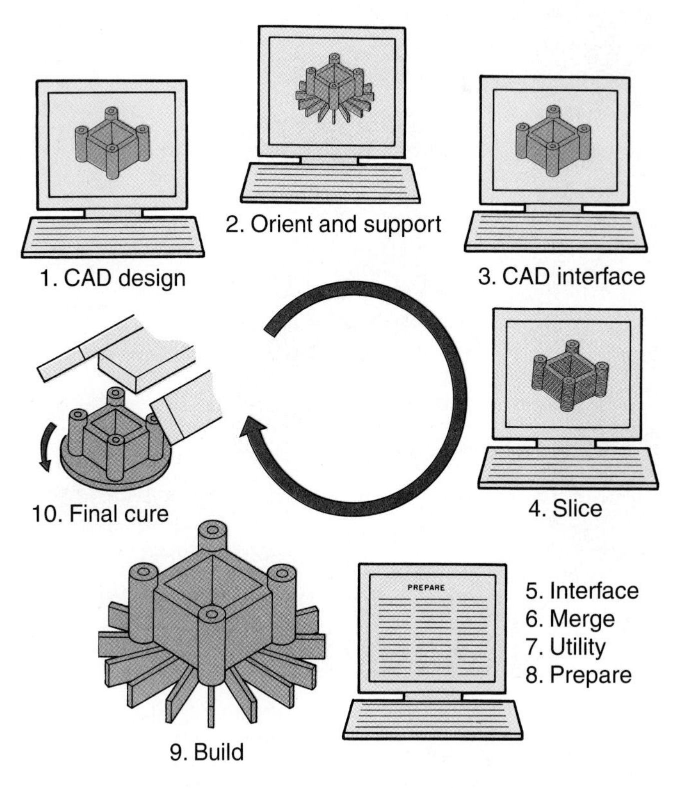

Goodheart-Willcox Publisher

Figure 26-18. The stereolithography process.

26.4.3 Other Rapid Prototyping Techniques

Rapid prototyping techniques that operate on a system similar to stereolithography can produce fully functional prototypes made of ABS (acrylonitrile butadiene styrene), medical ABS, or investment casting wax. When made of investment casting wax, the three-dimensional models can be used as the master for a cast part.

A rapid prototyping process called ***Fused Deposition Modeling (FDM)***™ can produce fully functional prototypes, **Figure 26-19**. When made of ABS, the parts can be installed and run for the best proof that a design works. The ABS parts can also be made in color, **Figure 26-20**.

The FDM process can produce three-dimensional objects based on CAD-generated solid or surface models. A temperature-controlled head extrudes thermoplastic material layer by layer. The designed object emerges as a three-dimensional part without tooling.

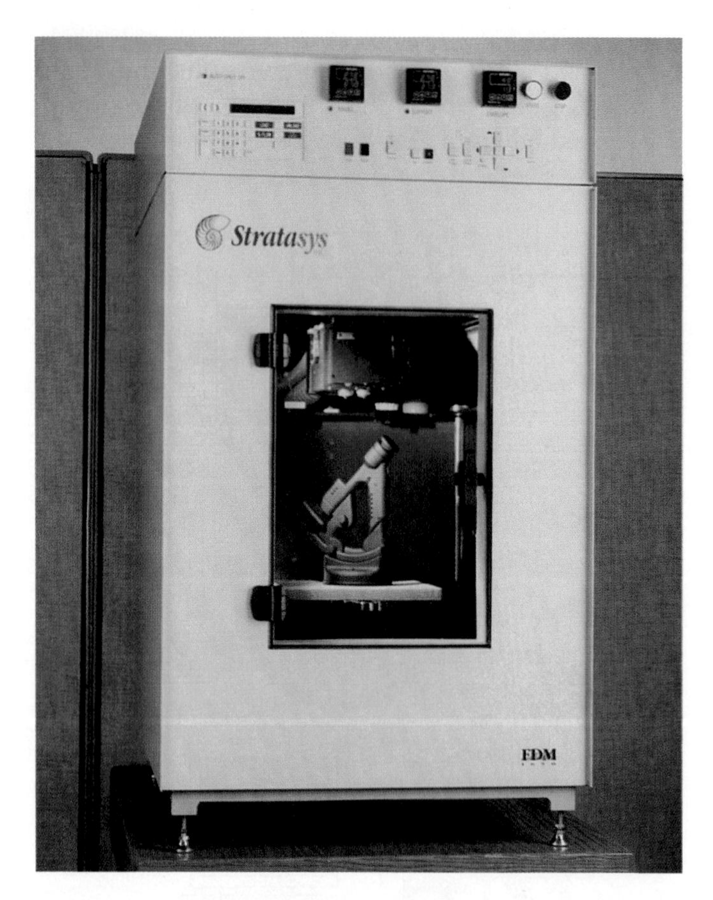

Stratasys, Inc.

Figure 26-19. Fused Deposition Modeling equipment can produce prototypes from ABS, medical ABS, or investment casting wax. When made from ABS, the parts can be installed and run to proof the design.

Stratasys, Inc.

Figure 26-20. ABS parts can be made in color.

A variation of the technology has been developed that produces parts made of sand or ceramic material, instead of plastic. Layers of special material are built up, using an inkjet-style printer to spray a quick-hardening binder to solidify each layer. The technology has been used to quickly produce shell molds for casting such metals as Inconel™ and aluminum. The moldmaking process is called ***Direct Shell Production Casting (DSPC)***. See **Figure 26-21**.

26.5 The Future of Automated Manufacturing

With technological advances being made at such a rapid pace, only time will tell what the full effect of computers and automated manufacturing techniques will be on our society. Some workers, mostly unskilled and semiskilled, have lost their jobs, much like the home artisans did during the Industrial Revolution. The same thing happened to carriage makers, blacksmiths, buggy whip manufacturers,

Ford Motor Co.

Figure 26-21. 3D sand printing equipment being used to produce sand-cast molds for production purposes.

and feed dealers when Henry Ford started to mass-produce the automobile, **Figure 26-22.**

Advances in manufacturing technology displaced some skilled workers then, just as automated manufacturing and robots have displaced certain workers in the modern economy. However, machines eventually helped to employ, directly or indirectly, many more people than the number they originally displaced. Better jobs, with higher pay and improved working conditions, were created. These changes also demanded that workers and technicians become better educated. The key to the future, then, will be men and women who are well versed in the various areas of industrial technology.

A

B

Ford Motor Co.

Figure 26-22. Then and now in the manufacture of motor cars. A—When Henry Ford opened the first production line, he paid the workers $5 per day, a very good salary for the time. B—On a modern production line, robots perform many of the jobs. As a result, both productivity and quality have increased dramatically while production costs have decreased.

Chapter Review

Summary

- Automation is a system for the continuous automatic production of a product.
- Flexible manufacturing systems are adaptable, allowing a single production line to produce different products simultaneously.
- A robot is a programmable manipulator, capable of moving parts or tools to any location within its work envelop with a high degree of precision and repeatability.
- Safety concerns in an automated manufacturing environment include all the safety concerns encountered in traditional machining, plus the safety concerns directly related to the automated equipment.
- Rapid prototyping techniques can quickly and inexpensively produce accurate models that can be used to improve design and identify potential machining problems.
- Common rapid prototyping techniques include Laminated Object Manufacturing, stereolithography, and Fused Deposition Modeling.

Review Questions

Answer the following questions using the information provided in this chapter.

1. What is automation?
2. How are automated machines activated?
3. List the five basic manufacturing processes that automated machines can perform.
4. What do the following automation-related acronyms stand for? (An acronym is a word formed using the initial letters of words in a phrase. For example: RPM stands for revolutions per minute.)
   A. FMS.
   B. CIM.
   C. FMC.
   D. CNC.
   E. JIT.
5. List three means of transferring work to and from the individual machines in a flexible manufacturing cell.
6. Explain the following terms:
   A. Smart tooling.
   B. JIT inventory.
   C. Robot.
7. In robotics, what is meant by the term "work envelope"?
8. List three general uses for robots.
9. List three techniques that are available for the rapid prototyping of a CAD design.
10. In stereolithography, a(n) _____ is used to harden a thin layer of material as the object is built up.

CHAPTER 27

Quality Control

Learning Objectives

After studying this chapter, you will be able to:

- Explain the need for quality control.
- Summarize the difference between destructive and nondestructive testing.
- Describe the various nondestructive testing methods.

Technical Terms

coordinate measuring machine (CMM)
destructive testing
dye penetrant inspection
eddy-current inspection
magnetic particle inspection
megahertz (MHz)
nondestructive testing
optical comparator
quality control (QC)
radiographic inspection
statistical process control (SPC)
ultrasonic testing

Quality control (QC) is a process that identifies and prevents potential product defects in the manufacturing process before they can cause injuries or damage and substandard products. The ultimate goal of quality control is *not* to detect imperfect products after manufacture, but rather to *prevent* them from ever being made.

Quality control is an essential component of industry. It plays a vital role in improving the competitive position of a manufacturer.

Bill Hannan

Figure 27-2. This Fokker DRI triplane, used by the Germans in the First World War, was constructed more-or-less to specifications. Some of the materials were inspected and tested before use. However, some aircraft components had tolerances of ±25 mm (1″).

27.1 The History of Quality Control

The development of quality control closely parallels the development of the airplane. Early aircraft were made with little concern for quality control, **Figure 27-1.** A full-size drawing was usually drawn in chalk on the shop floor. The plane was built over the chalk drawing. If a wood strip was straight with no knots and looked strong enough, it was used. The design was adjusted as needed to make parts fit. As a result, there were a lot of variations from plane to plane.

As the theory of flight became more refined, more care was taken in selecting materials that went into the planes, **Figure 27-2.** Tests were made to determine the strength of materials. Engines were made to specific standards and were checked many times during their manufacture. As a result, aircraft dependability increased greatly. In the automotive field, Henry Ford began the mass production of the Model-T using similar quality control methods.

The large-scale production of all-metal aircraft in the early 1930s introduced many new quality control techniques, **Figure 27-3.** Jigs and fixtures, which held the parts while they were machined or fabricated, were aligned with optical tools. The use of iron filings and a magnetic current, called magnetic particle inspection, helped to find defects and flaws in ferrous metals. Tolerances measured in ten-thousandths of an inch became common for engine components. Inspectors made up a larger percentage of the workforce than ever before. Because of the success of quality control programs in aeronautics, many other industries established quality control programs as well.

Goodheart-Willcox Publisher

Figure 27-1. There was very little quality control in the manufacture of early aircraft, like this one on display at an aeronautical museum.

John Winter

Figure 27-3. The manufacturing of all-metal aircraft during the 1930s brought about the development of quality control techniques that are still in use. This PBY Catalina amphibious aircraft was built in 1935 and is still in flying condition.

The need for thousands of high-performance aircraft during World War II, **Figure 27-4**, led to the introduction of many of the quality control techniques in use today. The inspector became a vital part of the manufacturing team.

Advanced vehicles, such as supersonic jets and spacecraft, **Figure 27-5**, require quality control programs of great scope and magnitude. Because these vehicles are subjected to extreme pressures, stresses, and temperatures, a breakdown or malfunction of any of the thousands of critical parts would be disastrous.

Bob Walker

Figure 27-4. The B-17 Flying Fortress was the backbone of the Air Force during the 1940s. Mass production was made possible only because of modern quality control procedures.

Following the lead of the aerospace industry, other industries began placing increasing emphasis on quality control techniques. As products became more complex and demand for reliability continued to grow, an increasing percentage of industry's budget had to be spent on quality control.

27.2 Types of Quality Control

Quality control techniques fall into two basic classifications:

- ***Destructive testing*** results in the part being destroyed during the quality control testing program.
- ***Nondestructive testing*** is done in such a manner that the usefulness of the product is not impaired.

27.2.1 Destructive Testing

Destructive tests range from the simple "break-it-and-look" test to costly and time-consuming quality control procedures involving various techniques, chemicals, and equipment. A specimen to use for testing is selected either at random or intervals from a predetermined number of pieces, **Figure 27-6**. At a statistical level, destructive testing indicates the characteristics of the untested (and undestroyed) remaining pieces. However, it cannot ensure that 100% of the parts will be quality compliant. In fact, any number of the untested parts could be defective and unwittingly be used in the manufacture of a product.

Christopher Parypa/Shutterstock.com

Figure 27-5. F-22 Raptor in a steep climb. Even the smallest defect in such a high-performance plane could be catastrophic. A rigorous quality control program must be followed during its construction and maintenance.

Ford Motor Company

Figure 27-6. In this destructive test, a known weight is thrown against the vehicle door at a known velocity. The test is used to check the sensitivity of the side-impact air bag sensors.

27.2.2 Nondestructive Testing

Nondestructive testing is a basic tool of industry. It is well-suited for testing electronic and aerospace products where the performance of each part is critical. Each piece can be tested individually and as part of a completed assembly.

27.3 Nondestructive Testing Techniques

You are familiar with several methods of nondestructive testing: measuring, weighing, and visual inspection. While satisfactory for many products, these methods leave much to be desired for other products. To ensure effective quality control in all areas of manufacturing, industry has developed more sophisticated testing techniques.

27.3.1 Measuring

The use of micrometers, vernier tools, dial indicators, gages, and similar devices falls into the dimensional measuring category of quality control testing. See **Figure 27-7**. To ensure accuracy, these tools must be used properly and frequently calibrated using known standards, such as those established by the National Institute of Standards and Technology (NIST). The calibration is done in a precision tool calibration laboratory, **Figure 27-8**.

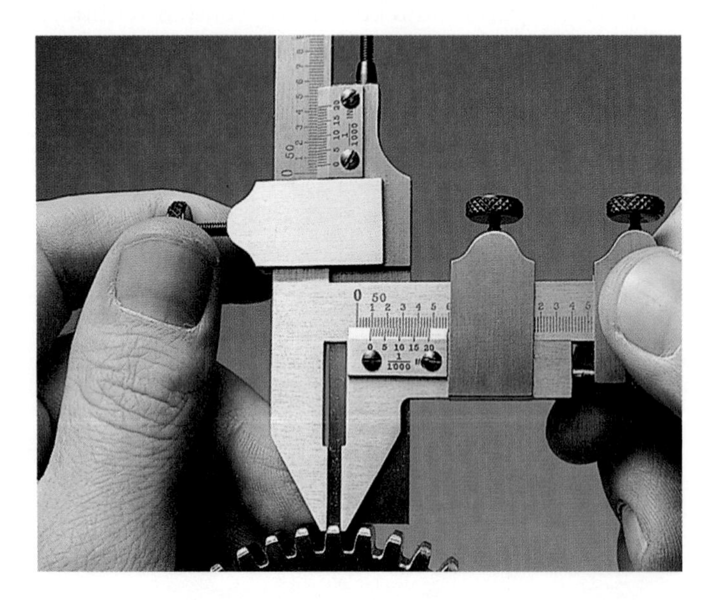

L.S. Starrett Co.

Figure 27-7. Some quality control techniques are based on linear measurement. To guarantee accuracy of measuring tools, they must be carefully checked against known standards at frequent intervals.

Master Lock Co.

Figure 27-8. Personnel in the precision tool calibration laboratory test and keep an accurate set of records on all production measuring tools to ensure their accuracy. This technician is using a dial indicator to check the accuracy of a plate lamination gage used by automatic press operators at a production site.

The shape of some products, **Figure 27-9**, prevents accurate measurements from being made by conventional measuring tools. A ***coordinate measuring machine (CMM)***, **Figure 27-10**, is an instrument that makes precise measurements electronically. A CMM can be used manually or programmed to check any number of individual reference points on the object against specifications.

Measuring can be done visually on an ***optical comparator***, which is a gaging system for inspection and precise measurement of small parts and sections of larger parts. See **Figure 27-11**.

Mark Yuill/Shutterstock.com

Figure 27-9. Complex parts, such as these turbine wheels, cannot be checked for accuracy with conventional measuring tools. Why do you think those measuring tools cannot be used?

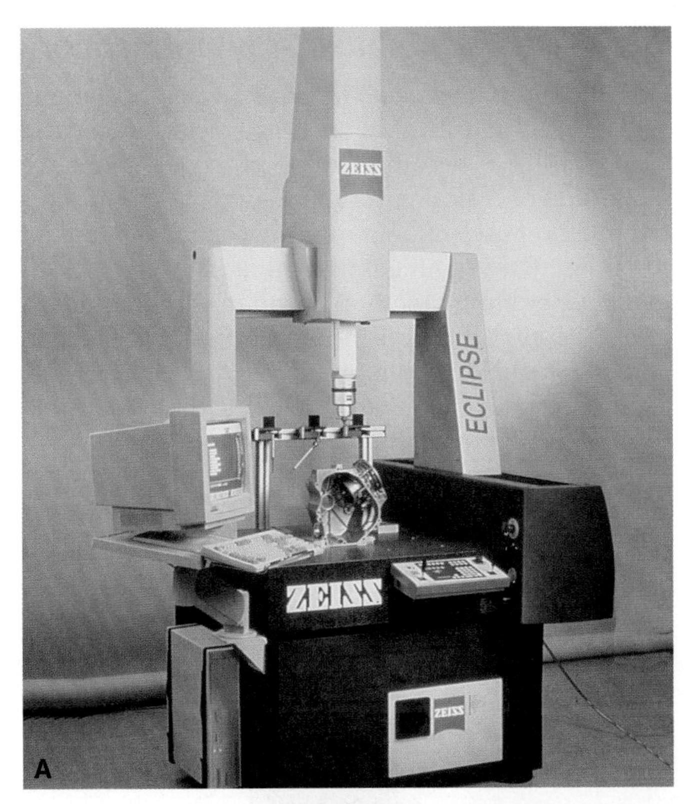

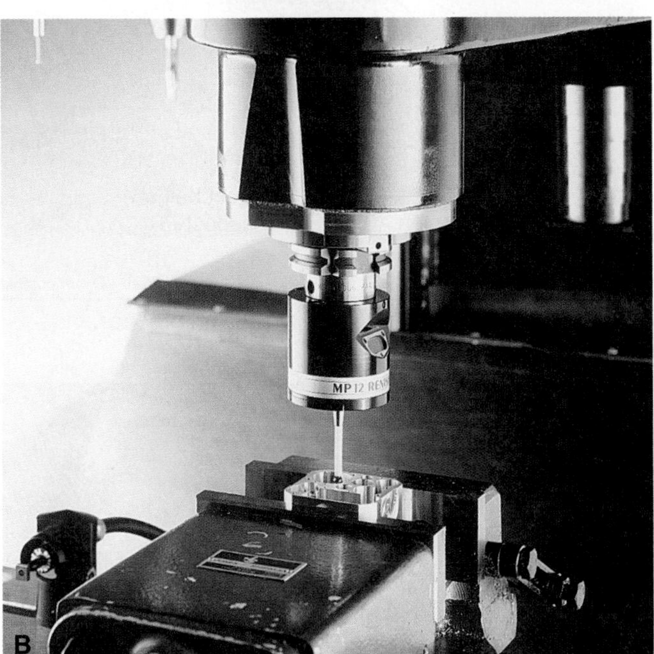

Carl Zeiss, Inc.; Renishaw, Inc.

Figure 27-10. Coordinate measuring machines. A—This CNC 3-axis coordinate measuring machine, with a weight capacity of 1000 lb, can measure parts with an accuracy of 0.0002″ (0.005 mm). Because of ceramic material used in parts of its construction, accuracy is guaranteed between 64°F and 78°F. The probe system can scan the work at a velocity of 17″ (425 mm) per second. B—This probe is being used to verify the accuracy of the holes machined into a complex part. A CMM is more accurate than any manual measuring method for measuring complex contours and surfaces.

Bridgeport Machines, Inc.

Figure 27-11. Parts can be magnified up to 500× on an optical comparator.

A part can be magnified up to 500× without distortion to permit accurate measurement. An enlarged image of the part being inspected is projected onto a screen, where it is superimposed on a grid or an accurate drawing overlay of the part. At maximum magnification, variations as small as 0.00001″ (0.00025 mm) can be noted by a skilled operator, **Figure 27-12.**

Master Lock Co.

Figure 27-12. A quality control engineer is checking hole location on a plate lamination of an automatic lock body assembly, using an optical comparator with a part overlay transparency.

An *optical gaging system*, automates the process of optical comparator inspection by incorporating a computer. The computer's software builds a graphical model of the measurements and remembers each step as the user measures a part. The resulting procedure is saved to the computer's memory. When measuring additional samples, the software prompts the user by highlighting each measurement in the model. The user simply follows these onscreen instructions to finish the inspection.

Manufacturers can also take a statistical approach to quality control. This is carefully worked out in a mathematical approach to quality measurement known as ***statistical process control (SPC)***. In statistical process control, a certain percentage of parts made in a production run are measured. If a wide variance is found among the inspected parts, adjustments are made and the inspection rate is increased to include more or all of the units manufactured. Through statistical analysis of the variations found in the manufacturing process, the manufacturer can predict factors such as tool and equipment wear and correct the problems before they result in unacceptable products.

Special gaging and inspection tools can be designed for almost any application, **Figure 27-13**. Combining precision tools with electronic devices permits accurate inspection to be made by semi-skilled workers.

Kennametal, Inc.

Figure 27-13. A technician using a radial measuring system to ensure roundness of the taper at the cutting unit gageline of an element used in a quick-change tooling system.

27.3.2 Radiographic (X-ray) Inspection

Radiographic inspection uses X rays and gamma radiation to detect flaws such as cracks or voids in an object. It has become routine in the acceptance or inspection of critical parts and materials. This inspection method uses radiographic inspection equipment to project X rays and gamma radiation (highly energetic, penetrating radiation found in certain radioactive elements) through the object under inspection and onto a section of photographic film, **Figure 27-14**. The developed film has an image of the internal structure of the part or assembly, **Figure 27-15**.

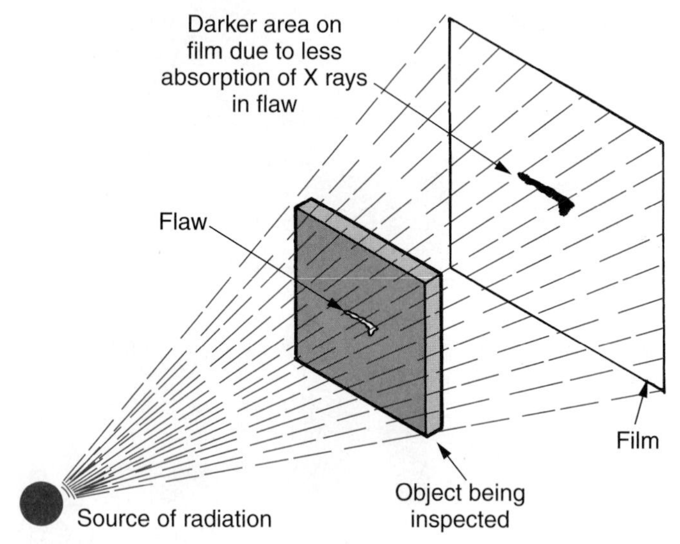

Goodheart-Willcox Publisher

Figure 27-14. The radiographic inspection process.

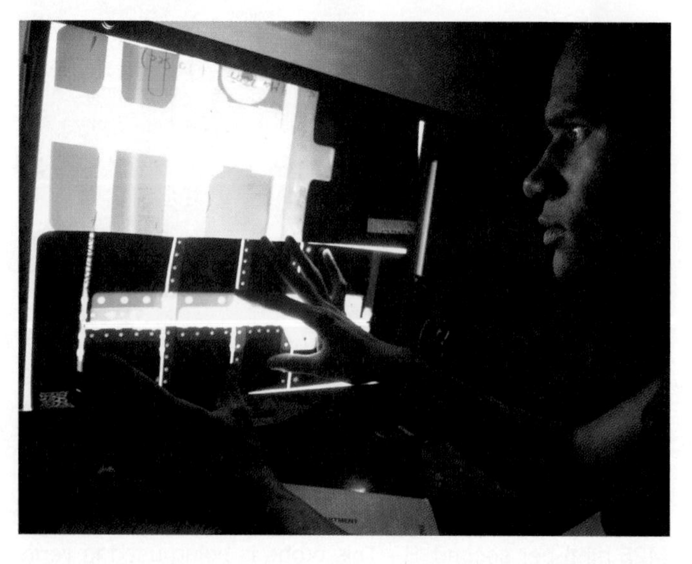

US Air Force

Figure 27-15. Radiographic examination being used to look for cracks and other flaws in an aircraft landing gear door.

Many kinds of *peripheral* (outside circumference) inspections can be made because of the *omnidirectional* (all directions) characteristics of the X rays, **Figure 27-16**.

The radiographic inspection process offers many advantages:

- Inspection sensitivity is high.
- The projected image is geometrically correct.
- A permanent record is produced.
- Image interpretation is highly accurate.

In addition, the internal and hidden parts of complex assemblies can be inspected without fear of damage. Objects can also be inspected without being taken out of use. For example, wheels and axles on locomotives and railway cars are inspected at regular intervals without having to remove them and take them to special facilities.

SAFETY NOTE

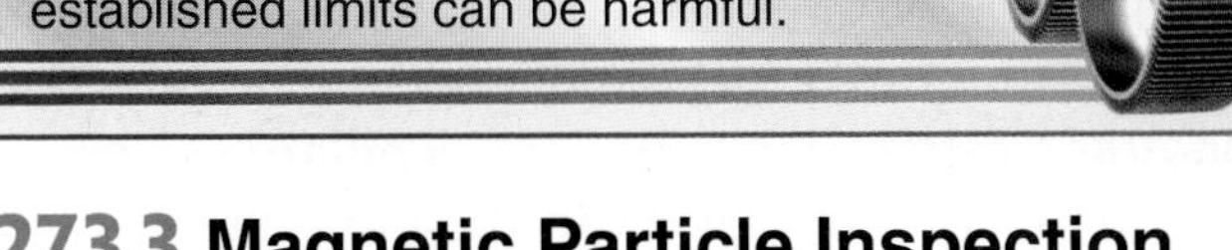

If you perform radiographic inspections, you must observe strict safety precautions when handling radioactive materials. You must also avoid exposure to X-ray radiation during testing. Exposure to X rays beyond established limits can be harmful.

27.3.3 Magnetic Particle Inspection

Magnetic particle inspection, commonly known as "magnafluxing" (a trademark of Magnaflux Corp.) is a nondestructive testing technique that uses iron particles and induced magnetic fields to detect flaws on or near the surface of ferromagnetic (iron-based) materials. The technique is rapid, but shows only serious defects. It does not show scratches or minor visual defects.

The magnetic particle inspection technique is based on the theory that every conductor of electricity is surrounded by a circular magnetic field. If the part is made of ferromagnetic material, these lines of force will, to a large extent, be contained within the piece. A circular magnetic field, if not interrupted, has no poles. However, because a flaw or other imperfection cuts through these magnetic lines of force, poles will be formed at each edge of the flaw. These poles will hold the finely divided iron particles, thus outlining the flaw. The limitations of this technique are apparent when the flaw is parallel to the line of magnetic force. The flaw will not interrupt the force, so no indication of it will appear when iron particles are applied, **Figure 27-17**.

In practice, a magnetic field is induced into the part and fine particles of iron are blown (dry method) or poured in liquid suspension (wet method) over it, **Figure 27-18**. As noted, a flaw will disturb or distort the magnetic field, giving it magnetic properties different from the surrounding metal. Many of the iron particles will be attracted to the area and form a definite indication of the flaw (its exact location, shape, and extent), **Figure 27-19**.

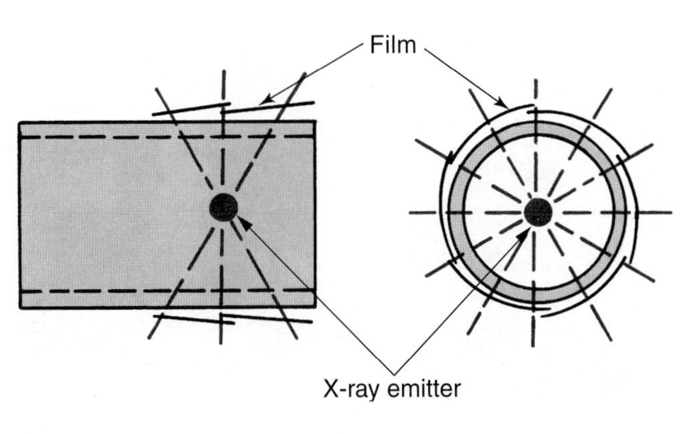

Goodheart-Willcox Publisher

Figure 27-16. Technique used to perform radiographic inspection on cylindrical objects. The X-ray emitter is placed in the center of the object and X-ray film is placed all around the circumference of the object. A flaw causes more exposure of the film, so an image of the flaw is shown on the film when it is developed.

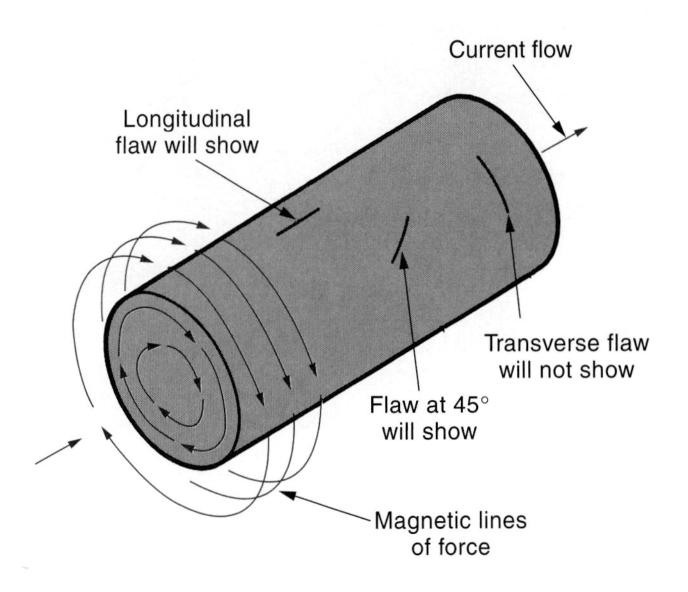

Magnaflux Corp.

Figure 27-17. Theory, scope, and limitations of magnetic particle inspection technique.

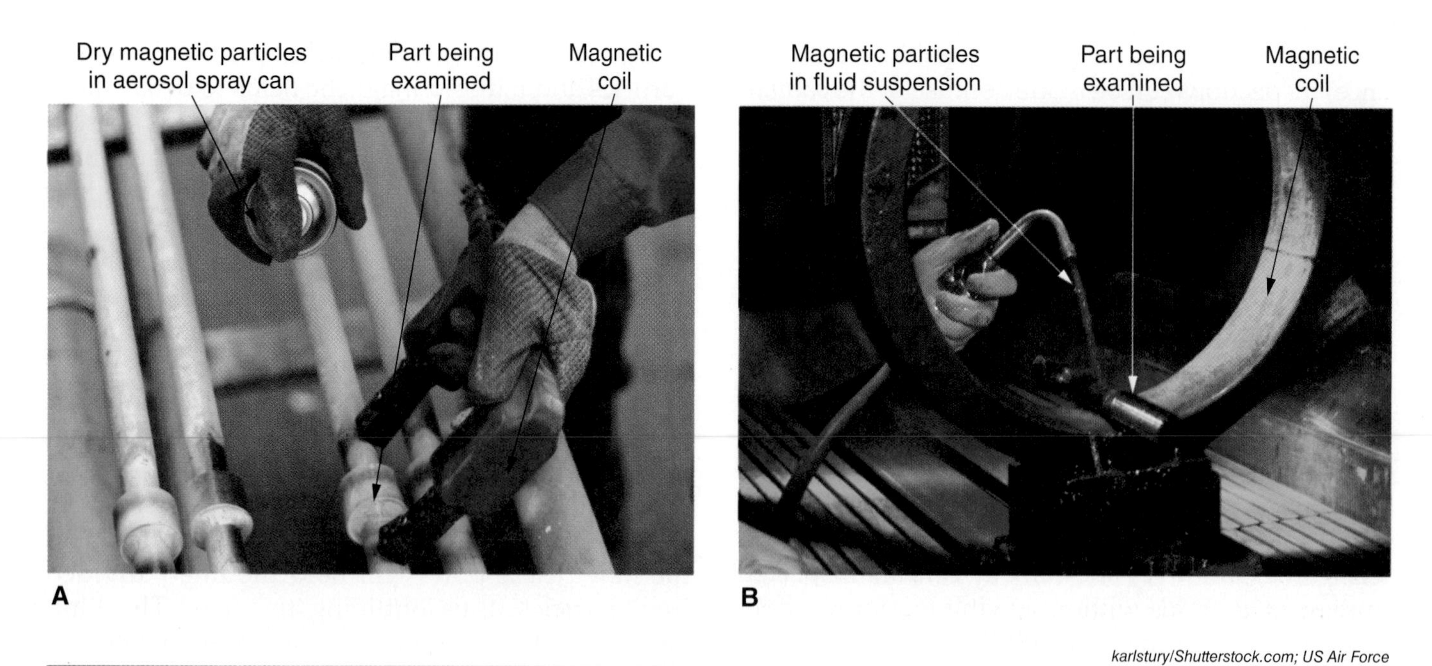

karlstury/Shutterstock.com; US Air Force

Figure 27-18. Variations of magnetic particle inspection. A—Dry method. In this test, fine iron particles are sprayed over the test area using an aerosol spray can. A handheld magnetic coil is placed on the test site. B—In the wet method, iron particles suspended in a fluid are poured over the part being inspected. The iron particles have been colored with fluorescent dye to improve their visibility under a black light.

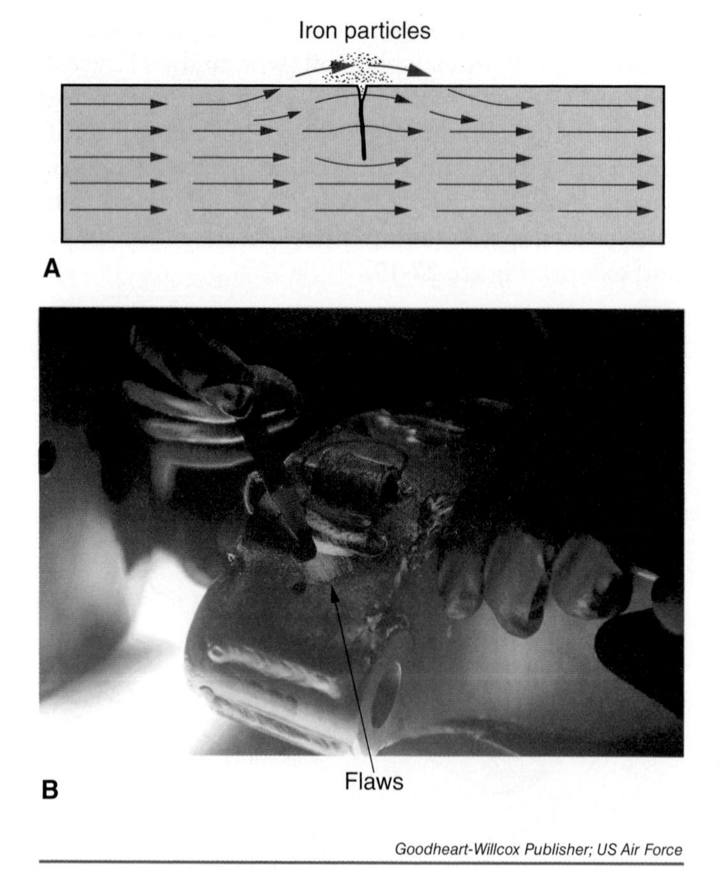

Goodheart-Willcox Publisher; US Air Force

Figure 27-19. Magnetic particles reveal cracks in the assembly. A—Cracks in the part generate a magnetic field outside the surface of the part, which holds the iron particles in place. B—A buildup of iron particles makes even tiny flaws visible.

27.3.4 Dye Penetrant Inspection

Because magnetic particle inspection is only useful on steel and iron, a different inspection method is required for nonmagnetic materials. ***Dye penetrant inspection*** relies on capillary action to show flaws in parts, and can therefore be used on nonmagnetic metals. In dye penetrant inspection, a fluorescent dye solution is applied to the part's surface by dipping, spraying, or brushing, **Figure 27-20.**

US Air Force

Figure 27-20. Parts being dipped in fluorescent dye penetrant. The excess dye will be rinsed off and a developer will be applied to the parts before they are inspected.

Capillary action draws the dye solution into the defect. The surface is rinsed clean, and a developer is applied. The developer acts as a blotter, drawing the dye penetrant back to the surface.

When the part is inspected under ultraviolet light (black light), any defects will glow with fluorescent brilliance, making them easily visible, **Figure 27-21**. This is because dye penetrant has flowed into and remains in the flaw.

Visible dye penetrant inspections use a dye that makes flaws visible under normal light. This type of dye penetrant test is easy to use, accurate, economical, and does not require a black light. Application is similar to that described for the fluorescent dye penetrant. The specimen is coated with a red liquid dye which soaks into the surface crack or flaw. The liquid is then washed off and the part is dried. A developer is dusted or sprayed on the part, and any flaws or cracks show up red against the white background of the developer.

27.3.5 Ultrasonic Inspection

Ultrasonic testing techniques make use of sound waves above the audible range. They can be used to detect cracks and flaws in almost any kind of material that is capable of conducting sound. Sound waves can also be used to measure the thickness of the same materials from one side. Ultrasonic test equipment is shown in **Figure 27-22**.

The human ear can hear sound waves whose frequencies range from about 20 to 20,000 hertz (cycles per second, abbreviated Hz). Sound waves that oscillate (vibrate) with a frequency greater than 20,000 Hz are inaudible (cannot be heard) and are known as ultrasound. Ultrasonic testing equipment produces sound waves whose frequency is measured in millions of cycles per second, or ***megahertz (MHz)***.

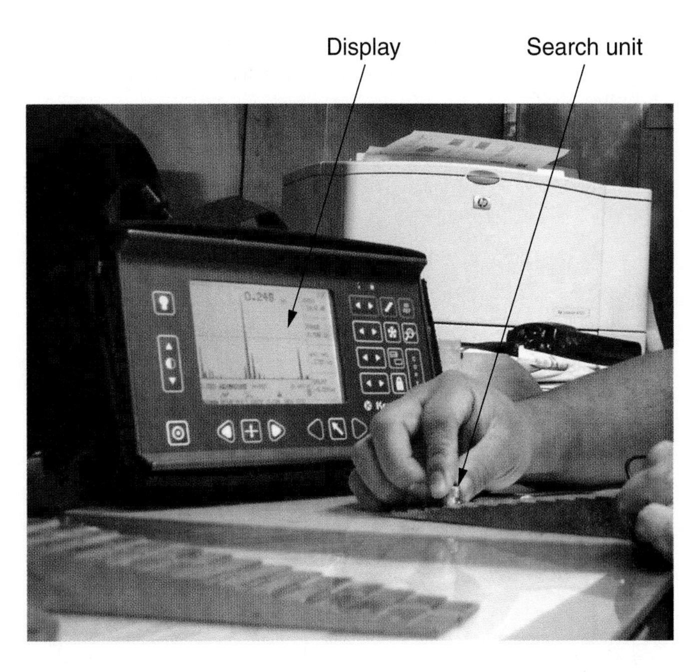

US Navy

Figure 27-22. An ultrasonic inspection is being performed with portable, contact-type ultrasonic testing equipment.

US Air Force

Figure 27-21. A part treated with dye penetrant is being examined under ultraviolet light. The areas of the part that are glowing brightly are potential flaws.

Sound waves are used to obtain information about the internal structure of a material. By observing the echoes that are reflected from within the material, it is possible to judge distances by the length of time required to receive an echo from an obstruction (flaw). See **Figure 27-23**. The ultrasonic testing equipment includes a timing device for measuring the relative length of elapsed time between the sending of the sound waves and the return of the various echoes.

The search unit contains a piezoelectric transducer (crystal) that is electrically pulsed and then vibrates at its own natural frequency to produce the ultrasonic waves, **Figure 27-24**. A liquid coupling is needed in order to transmit the sound waves from the search unit to the metal and to return the echoes to the search unit. The liquid coupling is usually a film of oil, glycerin, or water between the transducer and the test piece. The same results can be achieved by immersing both the test piece and search unit in water, **Figure 27-25**. Immersion-type testing is ideal for production testing, since there is no contact between the search unit and the work. The lack of contact prevents the transducer from wearing out.

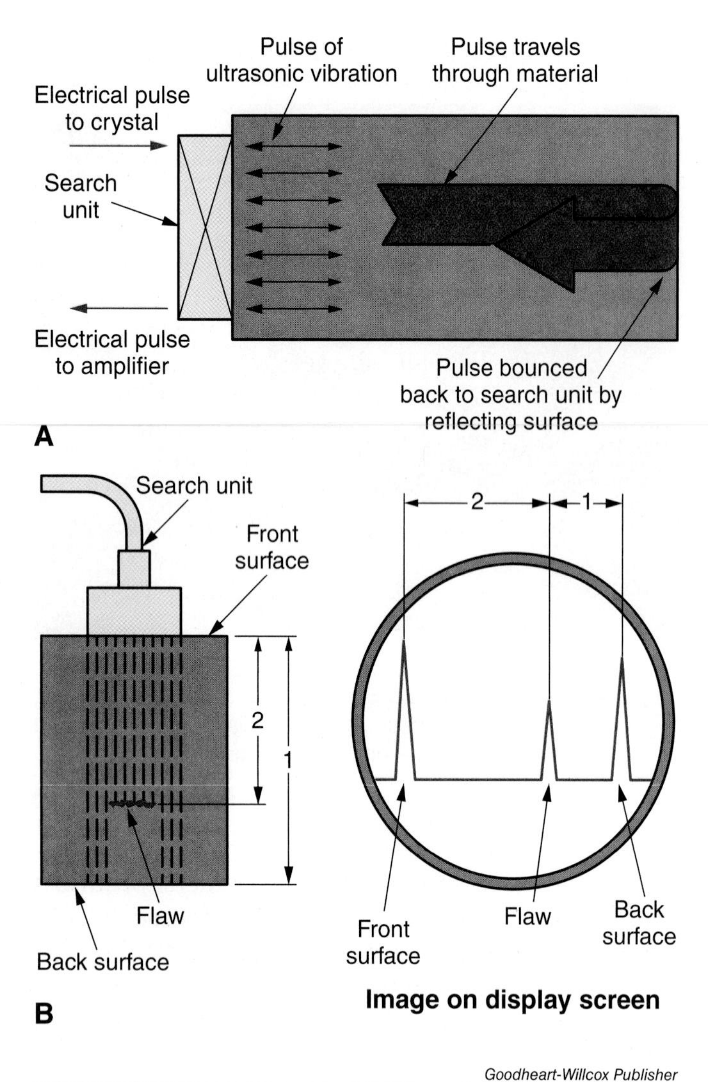

Goodheart-Willcox Publisher

Figure 27-23. Ultrasonic testing. A—Sound waves are produced by the search unit. They pass through the material being inspected. When they strike the back surface of the material, they are reflected back to the search unit. B—When the search unit passes over a flaw, the length of time that it takes the ultrasonic sound waves to be reflected back to the search unit is reduced.

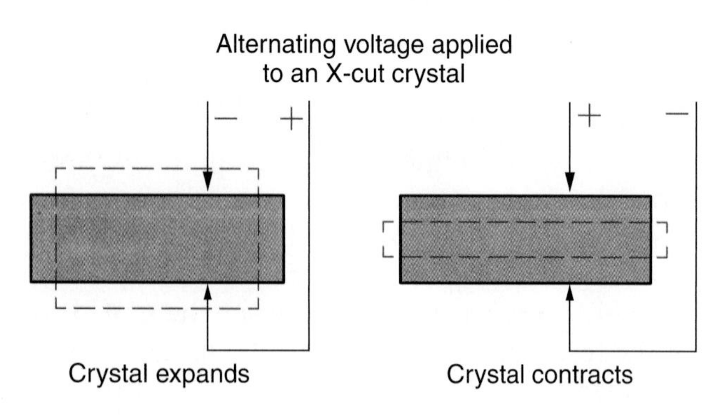

Goodheart-Willcox Publisher

Figure 27-24. A search unit's piezoelectric transducer is a device containing a crystal that changes shape based on electric polarity. As it is exposed to alternating polarity, the crystal's shape fluctuates, producing ultrasonic sound waves.

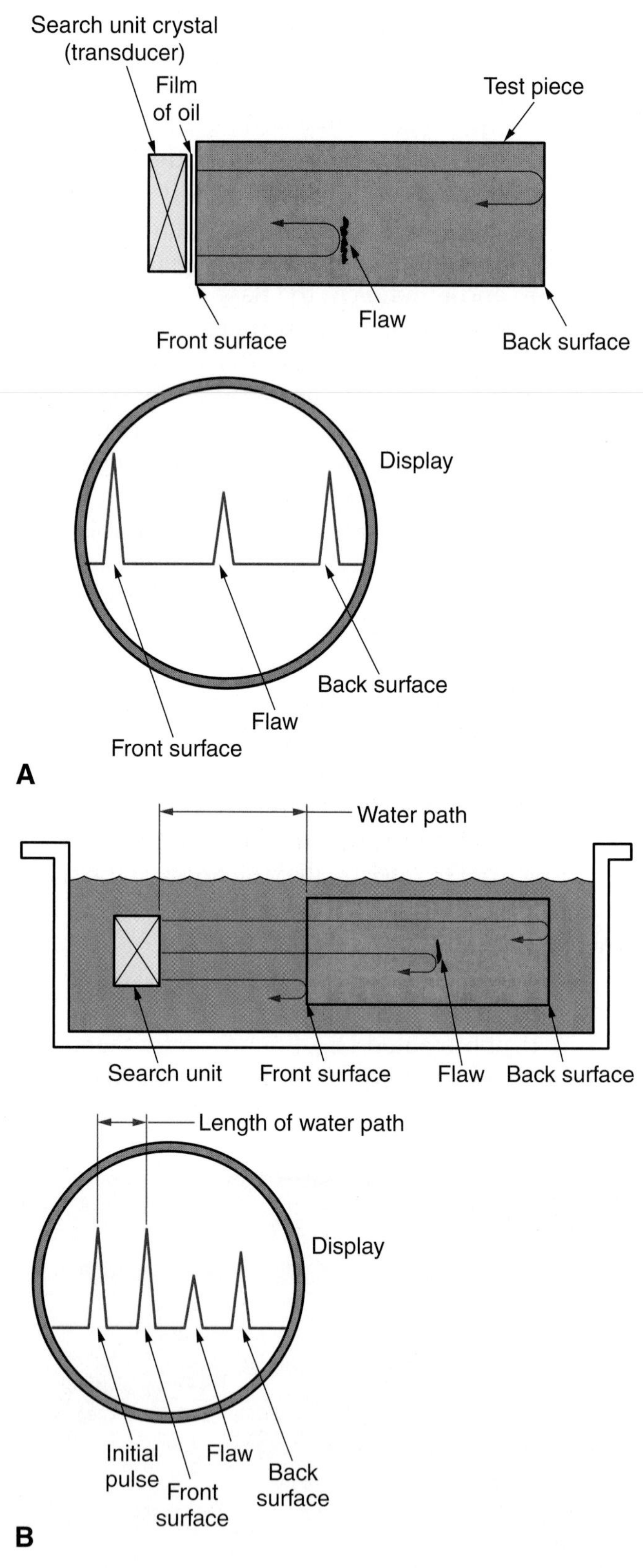

Goodheart-Willcox Publisher

Figure 27-25. Liquid coupling. A—Action of a contact-type ultrasonic inspection device. A film of oil, water, or glycerin is used to make a positive contact between the transducer and the test piece. B—Immersion-type ultrasonic testing. Note the extra spike on the display, indicating the path through water.

The search unit's transducer vibrates for about two-millionths of a second. The result is a very short burst of sound waves that travel through the liquid to the surface of the test material. A portion of the sound is immediately reflected from the surface of the metal as a very large echo. Part of the sound will not be reflected and will continue into the test material. If this portion of the sound encounters no interference, such as a discontinuity (flaw) in the material, it will continue until it is partially reflected from the back surface as a second echo or "back reflection." If there is a flaw in the interior, a portion of the sound will be reflected from the flaw and will return to the transducer as a separate echo between the echoes received from the front and back surfaces.

After the transducer has given off its short burst of sound waves, it stops vibrating and "listens" for the returning echoes. When the echoes are received, they cause the transducer to vibrate and to generate an electric current which can be displayed as a graph, **Figure 27-26**. The graph displayed onscreen can be expanded or condensed to improve its readability.

There are two basic categories of ultrasonic testing, as shown in **Figure 27-27**:

- **Pulse echo.** Uses sound waves that travel through the test piece until they are reflected by flaws or the back of the test piece. The relative positions of the echoes on a graph identify the location of flaws. The same search unit is used to both transmit the sound and receive the echoes.
- **Through inspection.** Uses one search unit to transmit ultrasonic waves through the piece, and another search unit to pick up the signal on the opposite end of the piece. If the beam is partially blocked by a flaw, the reduced intensity of the signal activates a warning device (such as a flashing light or ringing bell) to alert the operator to the flaw. The through inspection technique is frequently used to check the integrity of helicopter rotor blades, for example.

27.3.6 Laser Inspection

Lasers have been adapted for quality control, **Figure 27-28**. In addition to being able to check the fit of components on an assembly line, the computer-controlled laser can be made sensitive enough to detect tool wear and automatically compensate for it, **Figure 27-29**. This helps prevent imperfect parts from being made. Laser measuring devices can also be used to evaluate the condition of products that are in use.

27.3.7 Eddy-Current Inspection

Eddy-current inspection is based on the fact that flaws in a metal product will cause impedance (resistance) changes in a coil brought near to the part. Different eddy-currents will result when test coils are placed next to metal parts with and without flaws.

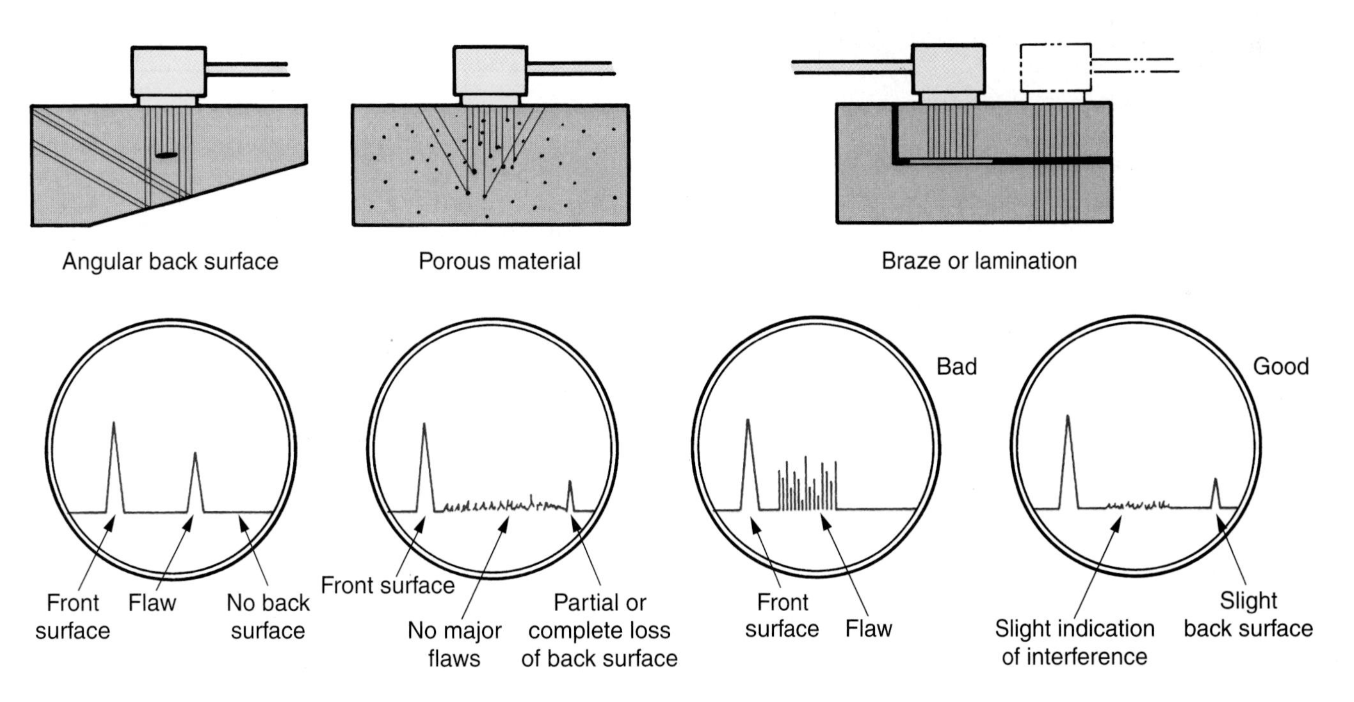

Figure 27-26. Various flaws as they appear on a video display.

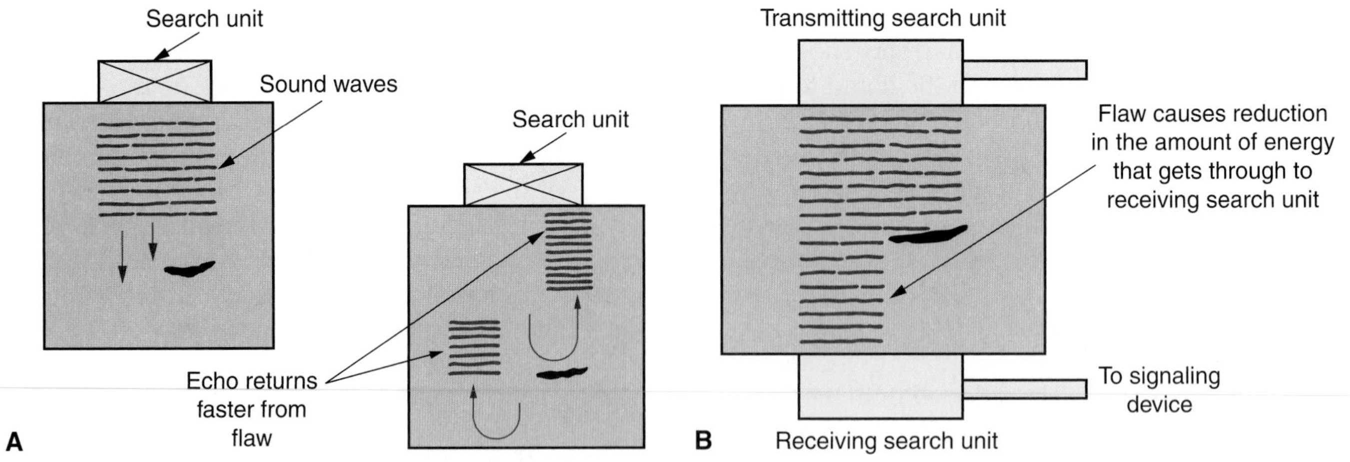

Search unit
Sound waves
Search unit
Echo returns faster from flaw
A
Transmitting search unit
Flaw causes reduction in the amount of energy that gets through to receiving search unit
To signaling device
B
Receiving search unit

Goodheart-Willcox Publisher

Figure 27-27. Types of ultrasonic inspection. A—With pulse echo ultrasonic inspection, reflected wave will return sooner when it bounces off a flaw. B—Through-type ultrasonic inspection uses one search unit to generate ultrasonic waves and a separate search unit to detect the ultrasonic waves when they reach the opposite side of the piece.

Ford Motor Company

Figure 27-28. Robotic laser inspection being used to check the fit of a door as it is installed on a vehicle.

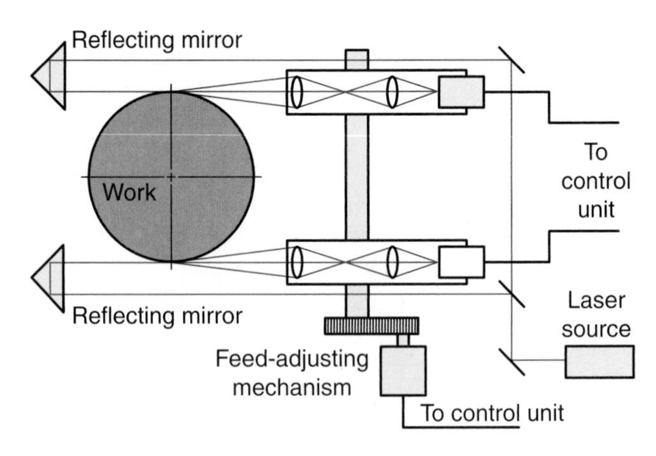

Goodheart-Willcox Publisher

Figure 27-29. One type of laser inspection device. With it, tool wear can be continuously monitored. Note how laser beams can detect work that is oversize or undersize.

The differences between the eddy currents determine which parts pass inspection and which parts fail.

Eddy-current inspection methods can be divided into two general categories:

- **Eddy-current differential system.** Used to detect cracks, seams, holes, or other flaws in metal parts (such as wire, tubing, and bar stock) as they move off the production line. The test equipment must be sensitive to rapid change.
- **Eddy-current absolute system.** Used to detect variations in dimension, composition, and other physical properties of a metal product. Instrumentation for this type operation must be responsive to relatively small changes.

Eddy-current inspection equipment is designed to detect changes in impedance and convert them into a form that can be monitored by the operator. See **Figure 27-30**. The material being tested can be passed through encircling detection coils at speeds up to 500 fpm (150 mpm). When a flaw in the test piece is detected, the eddy-current test equipment does one or more of the following:

- Flashes a warning light to alert the operator.
- Sounds a tone alarm.
- Activates a rejection device to eject the part that does not meet standards.
- Marks the section that contains the flaw so it can be removed.
- Provides an electronic record of the test.

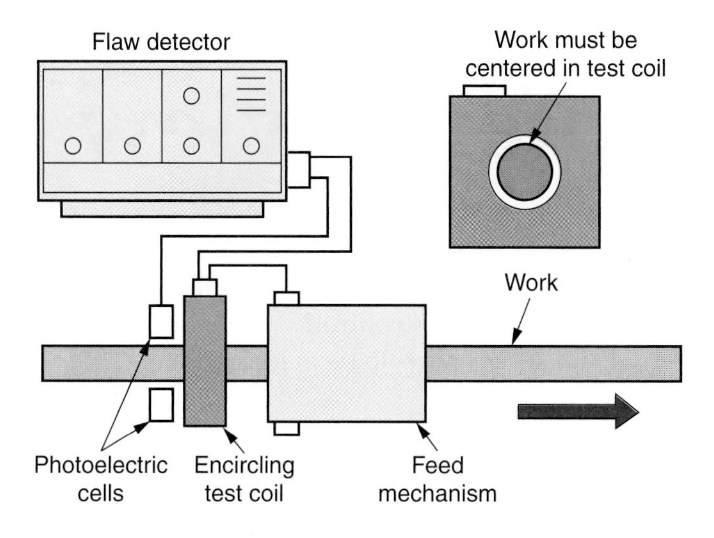

Goodheart-Willcox Publisher

Figure 27-30. Eddy-current inspection system. Photoelectric cells turn off the alarm system when the end of a test piece passes into the test coil. Any flaw would cause small current changes in the test coil.

27.4 Other Quality Control Techniques

In addition to the quality control techniques described, industry makes extensive use of other specialized testing devices. For example, the quality of a machined surface may be critical. The surface finish can be inspected visually with a microfinish surface comparator or electronically with a profilometer. See **Figure 27-31**.

Because of the versatility of the computer, many new quality control techniques are being developed as the need arises. The use of computers makes complete inspection of all parts or products much faster, **Figure 27-32**.

Ford Motor Company

Figure 27-32. New quality control devices make 100% inspection possible. Lasers scan front end, doors, and other assemblies to make sure the vehicle's fit and finish are acceptable.

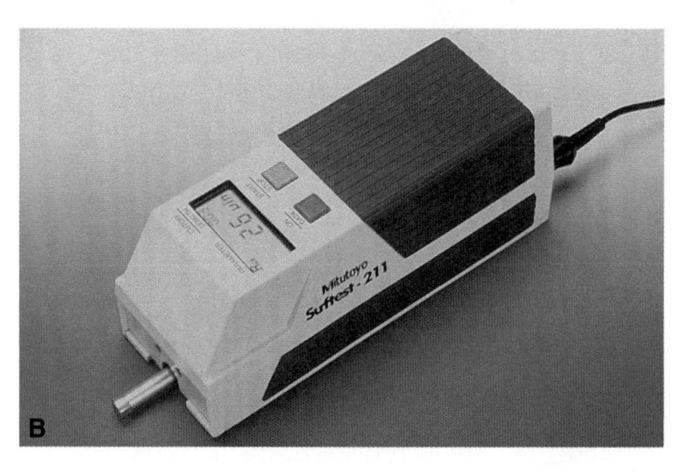

Mitutoyo/MTI Corp.

Figure 27-31. Methods of checking surface finishes. A—Machinist is using a microfinish surface comparator. B—Surface roughness can also be checked electronically with a profilometer.

Chapter Review

Summary

- The purpose of quality control (QC) is to seek out and prevent potential product defects in the manufacturing process before they can cause injuries or damage and substandard products.
- The preferred goal of quality control is not to detect and discard imperfect parts, but to prevent the defective parts from being manufactured in the first place.
- Quality control measures can be divided into two categories: destructive testing and nondestructive testing.
- In destructive testing, a sample part is selected from a lot and tested. The testing process destroys the specimen.
- In nondestructive testing, the testing process does not damage the specimen. As a result, every part can be tested during the manufacturing process.
- Nondestructive testing methods include measuring, radiographic (x-ray) inspection, magnetic particle inspection, dye penetrant inspection, ultrasonic inspection, laser inspection, and eddy-current inspection.

Review Questions

Answer the following questions using the information provided in this chapter.

1. Which of the following best describes the ultimate goal of a quality control program?
   A. Identify imperfect products at the end of the production process.
   B. Prevent imperfect products from being made.
   C. Reduce the cost of manufacturing.
   D. None of the above.
2. Quality control falls into two basic classifications. Name and explain each.
3. Precision measuring tools, such as micrometers, vernier tools, or dial indicators, must be inspected and _____ frequently to ensure their accuracy.
4. The _____ is an instrument that projects the magnified image of a small part on a screen so it can be measured.
5. Which of the following is an advantage of statistical process control?
   A. It allows parts to be measured more accurately.
   B. It predicts problems so they can be corrected before they result in unsatisfactory products.
   C. It relies entirely on visual inspection and does not require any measuring or testing instruments.
   D. All of the above.
6. Radiographic inspection uses _____ and gamma radiation to detect flaws in an object.
7. List four advantages of the radiographic inspection process.
8. Magnetic particle inspection is commonly known as _____.
9. Describe the magnetic particle inspection process.
10. Only _____ metals can be inspected by the magnetic particle technique.
11. Describe the dye penetrant inspection process.
12. Which of the following explains why dye penetrant inspection is more appropriate than magnetic particle inspection in some situations?
    A. Magnetic particle inspection requires much more training.
    B. Magnetic particle inspection is more expensive than dye penetrant inspection.
    C. Dye penetrant inspection can be used on nonmagnetic materials.
    D. Dye penetrant inspection causes less damage to the material than magnetic particle inspection.
13. Ultrasonic inspection makes use of _____.
    A. accurately made measuring fixtures
    B. high-frequency sound waves
    C. X rays
    D. All of the above.
14. List the two basic categories of ultrasonic testing. Briefly describe each.
15. Eddy-current inspection relies on changes in _____ in a test coil to identify flaws in a part.

CHAPTER 28

Metal Characteristics

Learning Objectives

After studying this chapter, you will be able to:

- Explain how metals are classified.
- Describe the characteristics of different metals.
- Recognize the hazards that are posed when certain metals are machined.
- Explain the characteristics of some reinforced composite materials.

Technical Terms

alloy
Aluminum Association Designation System
base metal
carbon content
cold finished steel
ductility
ferrous
high-carbon steels
high-speed steels (HSS)
honeycomb sandwich panels
hot rolled steel
low-carbon steels
medium-carbon steels
mild steel
nonferrous
red hardness

More than a thousand different metals and alloys are used by the metalworking industry. Most of them are worked in the machine shop. However, it must be noted that there is a "new materials revolution" taking place, especially in the aerospace and medical equipment industries. Many nonmetals and composites of metals and nonmetals are rapidly emerging in applications that were once the exclusive domain of metals.

Can you tell by looking at a piece of metal whether it is ***ferrous*** (containing iron), or ***nonferrous*** (containing no iron)? Is it an ***alloy*** (a mixture of two or more metals)? Could it be a ***base metal*** (a pure, nonferrous, nonprecious metal), like tin, copper, or zinc? From a practical point of view, it is almost impossible to find out much about a piece of metal by just looking at it.

A machinist is not expected to know everything there is to know about all metals. However, a working knowledge of common materials, and the terms associated with them, is essential. In addition to the conventional metals worked in a modern machine shop, several newer materials will be described in this chapter.

28.1 Classifying Metals

Modern industrial metals can be classified as:

- Ferrous metals.
- Nonferrous metals.
- High-temperature metals.
- Rare metals.

Because of space exploration and military research, great strides have been made in the development of the last two groups in recent years. Refer to the *Reference Section* of this text for tables that show physical properties of metals, dimensional tolerances, and feeds and speeds for machining.

28.2 Ferrous Metals

Irons, steels, and their alloys, make up the family of ferrous metals. For simplicity and easier understanding, ferrous metals can be further subdivided into various categories.

28.2.1 Cast Irons

Cast irons are iron alloys that contain 2.0% to 5.0% carbon with small quantities of silicon and manganese. There may also be traces of other elements in the alloy.

Of the cast irons, gray iron and malleable iron are most widely used in industry. They can be found in great quantities in automotive, railroad, farm equipment, and machine tool bodies. See **Figure 28-1**.

Malleable iron can be hammered into shape without fracturing. Most cast irons can be readily machined once the hard surface scale has been penetrated. Carbide cutting tools are recommended because of the abrasive nature of the iron scale. No cutting fluids should be used on cast iron. Compressed air is recommended if a coolant is needed.

SAFETY NOTE

Use extreme care when using compressed air as a coolant. To prevent injury, chips and dust must be contained.

28.2.2 Steels

Steel is often considered the "backbone" of the metalworking industry. Carbon steel is very common. It is an alloy of iron and various alloying elements, with carbon as the major alloying agent. The alloying elements give the iron the desired characteristics needed to perform a specific job.

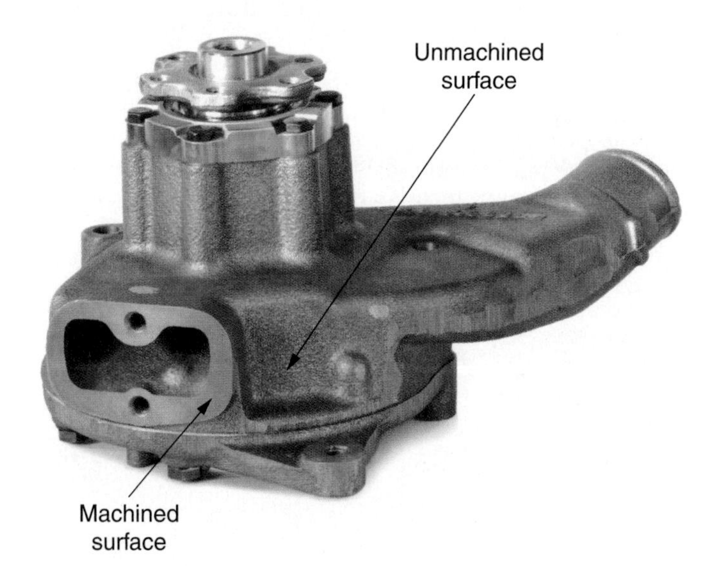

Shell114/Shutterstock.com

Figure 28-1. Cast water pump. Note that the machined surfaces on the part are shiny silver, while the unmachined surfaces are dull grey with a pebbled texture.

The physical properties of steel are unique. Steel can be made soft enough to be easily machined. By careful heat treatment, the soft steel can be transformed into a very hard material. By slightly varying the heat treatment procedure, steel can be given a hard, wear-resistant surface while retaining a soft, tough core to resist breaking. The magnetic and electrical qualities of steel also make it ideal for many electrical applications.

Carbon steels are classified according to their ***carbon content*** (amount of carbon they contain). The carbon content is measured in percentage or in points (100 points equal 1%). They are available in all standard mill forms, **Figure 28-2**.

Low-carbon steels do not contain enough carbon (less than 0.30%, or 30 points) to be hardened. They are easy to work and can be case-hardened by heating them while exposing their surface to an external source of carbon. Low-carbon steel is often called ***mild steel*** or machine steel. It is used for nuts, bolts, screws, gun parts, precision shafting, tie rods, tool cylinders, and similar applications.

Hot rolled steel is steel that has been rolled to finished size while hot, **Figure 28-3**. It is easily identified by its black oxide surface scale. ***Cold finished steel*** is steel that has been "pickled" or treated with a dilute acid solution to remove the oxide coating. After pickling, the steel is drawn or rolled to finished size and shape while cold. Cold finished steel is characterized by a smooth, bright finish, **Figure 28-4**. Cold finishing improves the machinability of the steel.

Medium-carbon steels contain 0.30% to 0.60% (30 to 60 points) carbon. The carbon content is sufficient to allow partial hardening with proper heat treatment. The heat treatment process also improves the strength of the steel. Available in all standard mill forms, medium-carbon steels are used for items such as machine parts, automotive gears, camshafts, crankshafts, cap screws, and precision shafting.

Goodheart-Willcox Publisher

Figure 28-2. A small portion of the hundreds of shapes in which steel stock is available.

Mircea BEZERGHEANU/Shutterstock.com

Figure 28-3. Hot rolled steel is formed into the desired shape and thickness while it is still hot.

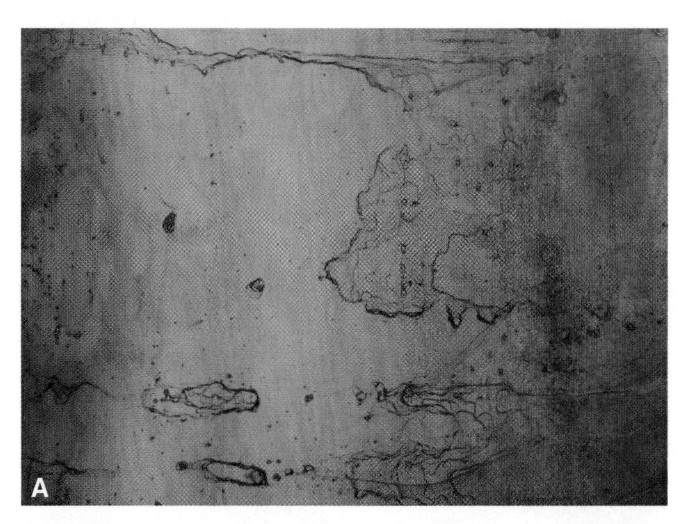

upixa/Shutterstock.com; R-studio/Shutterstock.com

Figure 28-4. Comparison of hot rolled steel and cold finished steel appearances. A—Hot rolled steel has a black oxide coating. B—Cold finished steel has a smooth shiny surface.

High-carbon steels contain 0.60% to 1.50% (60 to 150 points) carbon. They are available in hot rolled form. However, some high-carbon steel shapes may be purchased with ground surfaces. Drill rod and ground flat stock are examples.

High-carbon steels are used in products that must be heat-treated. Applications include heavy machinery parts, control rods, tools, springs, and a large variety of agricultural equipment.

Adding controlled amounts of sulfur or lead to carbon steels will result in improved machinability without greatly affecting the metal's mechanical properties. Usually, machining involves the removal of considerable metal. The machine tool must be capable of increased cutting speeds before the machining of high-carbon steels will prove economical.

28.2.3 Alloy Steels

Alloy steels have other metal elements added to change their characteristics. Alloy steels are more costly to produce than carbon steels because of the increased number of special operations that must be performed in their manufacture, **Figure 28-5**.

Elements such as nickel, chromium, molybdenum, vanadium, manganese, and tungsten are used to make alloy steels harder, stronger, or tougher. A combination of two or more of the above elements usually imparts some of the characteristic properties of each.

NASA

Figure 28-5. Many alloy and high-temperature steels were used in the manufacture of this large rocket engine (for size comparison, note the technician at bottom center). The engine has been designed to safely handle the super cold of liquid oxygen at one end and the blast furnace temperature of exhaust gases at the other, while operating in the cold of space.

Chromium-nickel steels, for example, develop good hardening properties with good ***ductility*** (a property of metal that permits permanent deformation by hammering, rolling, and drawing without breaking or fracturing). Chromium-molybdenum combinations develop excellent hardenability with satisfactory ductility and a certain amount of heat resistance.

28.2.4 Metallic Alloying Elements

Metallic elements that may be added to alloy steels, and properties they impart to the steel, include:

- **Nickel.** Imparts toughness and strength, particularly at low temperatures. Nickel steels permit more economical heat treatment and have improved resistance to corrosion. They are especially suitable for the case-hardening process and are used for applications such as armor plate, roller bearings, and aircraft engine parts.
- **Chromium.** Added when toughness, hardness, and wear resistance are desired. It is the primary alloying element in stainless steel. Chromium steel is found extensively in automotive and aircraft parts, **Figure 28-6**.

US Air Force Thunderbirds

Figure 28-6. Chromium steel is used extensively in the landing gear assembly of this aircraft. It has ability to withstand the repeated shocks of landing the aircraft.

- **Molybdenum.** Used as an alloying agent when the steel must remain tough at high temperatures.
- **Vanadium.** An alloying element that gives steel a fine grain structure and increased toughness at high temperatures.
- **Manganese.** Purifies steel and adds strength and toughness. Manganese steel is used for parts that must withstand shock and resist wear, **Figure 28-7**.
- **Tungsten.** Gives steel a fine, dense structure and improved heat treatment qualities. It is one of the principal alloying agents in many tool steels. Tools made with these steels retain their strength and hardness at high temperatures.
- **Cobalt.** The chief alloying element in high-speed steels because it improves the ***red hardness*** (quality of remaining hard when red hot) of cutting tool materials. Wear resistance is also improved.

28.2.5 Tool Steel

Tool steel is the term usually applied to the steels used to make tools that cut, shear, or form materials. They may be either carbon or alloy steels. Tool steels in the lower carbon content range (0.70% to 0.90% or 70 to 90 points) are used in tools that are subjected to shock. The low carbon content gives the steel the toughness it needs to withstand impact. Higher-carbon-content tool steels (1.10% to 1.30% or 110 to 130 points) are used to make tools with sharp cutting edges. The hardness of these steels prevents the cutting edge from dulling with use.

kaband/Shutterstock.com

Figure 28-7. Many components of earthmoving vehicles are made from manganese steel. This type of steel is strong and tough enough to withstand constant abrasive wear.

Drills, reamers, milling cutters, punches, and dies are made from alloy tool steels. Although several tool steels can be hardened using water as a quenching medium, most must be hardened in oil or in air. These are known as oil-hardened and air-hardened steels.

Some tool steels are also classified as ***high-speed steels (HSS)***, because they are capable of making deeper cuts at higher cutting speeds than regular tool steels. They have red hardness, or the ability to retain their hardness at high temperatures, **Figure 28-8**. They also possess high abrasion resistance. Despite the development and wide spread use of cemented carbides and ceramics, high-speed steel remains a major cutting tool material.

28.2.6 Tungsten Carbide

Tungsten carbide is the hardest known metal. It is almost as hard as a diamond. The metal is shaped by molding tungsten, carbon, and cobalt powders under heat and pressure in a process known as sintering. The metals fuse together without melting.

Tungsten carbide, while not a true steel, is usually classified with the steels. Tools made from this family of materials can cut many times faster than cutting tools made from high-speed steel, **Figure 28-9**.

Applying a 0.0001″ (0.002 mm) thick coating of titanium nitride (TiN) and titanium carbide (TiC) to the surface of tungsten carbide tools extends their life 3 to 8 times longer than uncoated tools. Material builds up on the cutting edge of uncoated tools, resulting in a ragged surface finish on the work. Coated tools resist buildup from the chips. They also run cooler, last longer, hold tolerances better, and produce a better surface finish. See **Figure 28-10**.

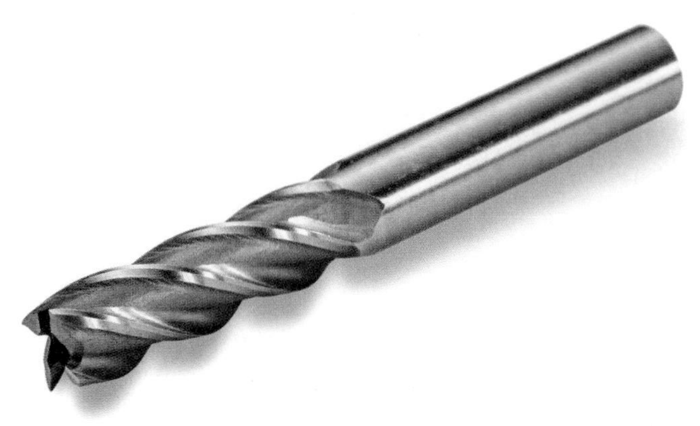

Matee Nuserm/Shutterstock.com

Figure 28-8. End mill cutter made from high-speed steel.

Valenite, Inc.

Figure 28-9. Cutting tools made from tungsten carbide can cut many times faster than high-speed steel cutters.

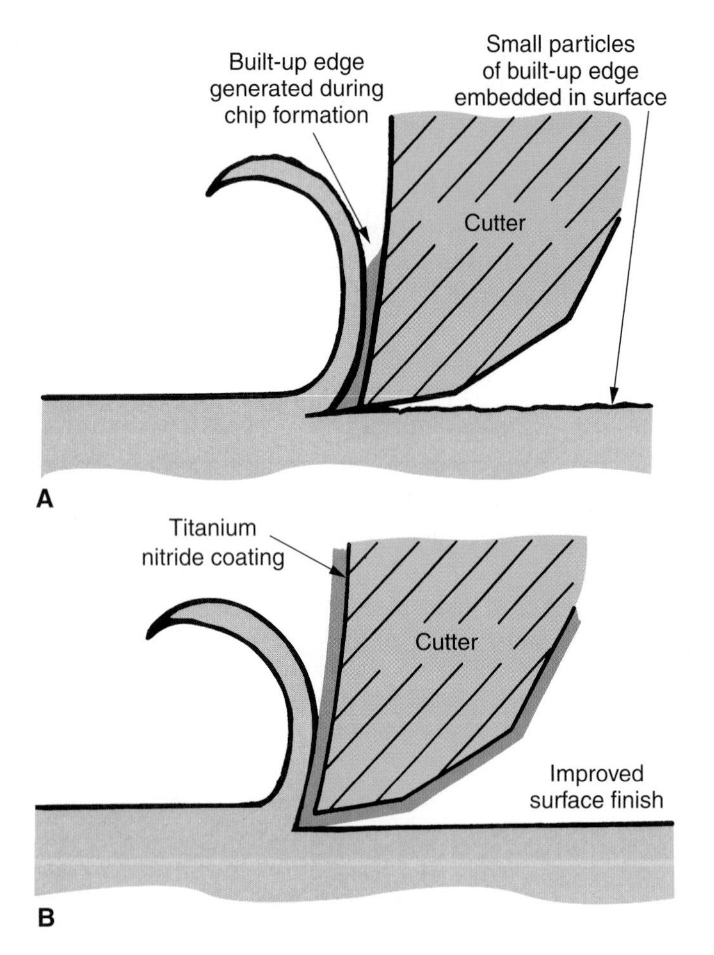

Goodheart-Willcox Publisher

Figure 28-10. Tool coating benefits. A—Uncoated cutting tools acquire a buildup of material on the cutting edge, which produces a "ragged" surface finish as parts of the buildup flake off. B—Tools coated with titanium nitride (TiN) and titanium carbide (TiC) resist the buildup from the chips. The coated cutting tools run cooler, stay sharp longer, and produce a better surface finish.

Recommended tungsten carbide grade applications and how the various manufacturers list their corresponding products are shown in chart form in **Figure 29-11**.

28.2.7 Stainless Steels

There are more than a hundred different stainless steels. However, one characteristic common to all of them is that they contain enough chromium to render them corrosion resistant. Stainless steels can be divided into three basic groups:

- **Austenitic.** This classification includes the chromium-nickel and chromium-nickel-manganese stainless steels. Generally, they are hardenable only by cold working and are nonmagnetic. The American Iron and Steel Institute (AISI) 300 series of stainless steels are in this category.
- **Martensitic.** These stainless steel alloys of iron, carbon, and chromium are characteristically magnetic in nature and obtain their hardness through normal heat-treating processes.
- **Ferritic.** These stainless steels have more than 18% chromium. They are nonhardenable and magnetic.

Stainless steels can be machined using the same techniques used for mild steels. However, some precautions must be observed with stainless steel:

- Feeds must be high enough to ensure that the cutting edge(s) get under the previous cuts and thus avoid the hardened portions.
- Tools must be as large as possible, because the life of the cutting edge(s) depends on good heat dissipation into the body of the cutting tool.
- Finishing cuts should be used when working to close tolerances.
- The machine should be adjusted so there is minimum play. Otherwise, the cutting tool may "ride" the work and glaze and/or harden the surface.

28.2.8 Identifying Steels

Because different kinds of steels look alike, several methods of identification have been devised. They include identification by chemical composition, mechanical properties, the ability to meet a standard specification or industry accepted practice, or the ability to be fabricated. The shape and intended use for the metal can also determine the method of identification.

Recommended Tungsten Carbide Insert Grade Applications	
Grade C2 uncoated	For cast iron, nonferrous materials, and general-purpose use.
Grade C4 uncoated	Light finishing cast iron, nonferrous, and general-purpose.
Grade C6 uncoated	For steel, steel casting, malleable cast iron, stainless steels, and free cutting steels.
Grade C5–C6 Titanium nitride coated (TiN)	For carbon steels, tool steels, alloy steels, steel castings, malleable cast iron, austenitic and martensitic stainless steels, and free cutting steels.
Grade C6–C7 Titanium carbide coated (TiC)	For steel, steel castings, malleable cast iron, nodular iron, and martensitic stainless steels.
Grade C2–C4–C6–C8 Aluminum oxide, ceramic coated	For carbon steels, tool steels, stainless steels, alloy steels, steel castings, gray cast iron, malleable cast iron, and nodular iron.

A

Tungsten Carbide Comparison Chart

Class	RTC	Newcomer	Carboloy	Iscar	Kennametal	Mitsubishi	Sandvik	Seco	Sumitomo	Valenite	V.R. Wesson	RTW
C-2	RTC2	N21	833	IC20	K-68	UTi20T HTi10	CG-20	HX	G10E	VC-1 VC-2/VC-28	2A5	CQ-2
C-6	RTC5	N60	395 78	IC54	K420	STi20 UTi20T	S-2 S-35	S-2 S-4	ST20E	VC-6	VR-75	Cy-5
TiN	RTC052	NN60	516	IC656	KC810 KC950	–	GC-425 GC-315 GC-225	TP15	AC720 AC815 AC15	VN-5 VC-7 VC-88 VO1	663 650 660	TRW-755
TiN & AlO_2 & TiN	RTC054	1000	550 560	IC635	KC850	U610	GC4025 GC235	TP10	–	VN2	653	718

B

Goodheart-Willcox Publisher

Figure 28-11. Recommendations for various tungsten carbide insert-type cutting tools. A—Recommended applications. B—Tungsten carbide insert comparisons.

AISI/SAE Codes

The American Iron and Steel Institute (AISI) and the Society of Automotive Engineers (SAE) have devised almost identical standards that are widely used for identifying steel. Both systems use an identical four-number code (some steels require a fifth digit) that describes the physical characteristics of the steels. The AISI system also makes use of a prefix letter (A, B, C, etc.) that indicates the steel manufacturing process used.

The four-numeral code works as follows: The first digit classifies the steel. The second digit indicates the approximate percentage of the alloying element in the steel. The last two digits show the approximate carbon content of the steel in points (hundredths of one percent). For example, a steel designated SAE 1020 is a carbon steel with approximately 20 points or 0.20% carbon. The AISI and SAE four-digit code applies primarily to bar, rod, and wire products.

Color Coding

Color coding is another method of identifying the many kinds of steel. Each commonly used steel is designated by a specific color. The color coding is painted on the ends of the stock. In cases where the color code may not be visible on the ends of the stock, it can be placed on an attached tag. See **Figure 28-12**.

Spark Test

A spark test is also used at times to determine grades of steel. The metal is touched to the grinding wheel lightly and the resulting sparks are carefully observed. See **Figure 28-13**. The patterns and characteristics of the sparks created differ from alloy to alloy. Because of this, it is possible to identify an unknown steel by observing the sparks it makes and comparing them to the spark characteristics of the different steels.

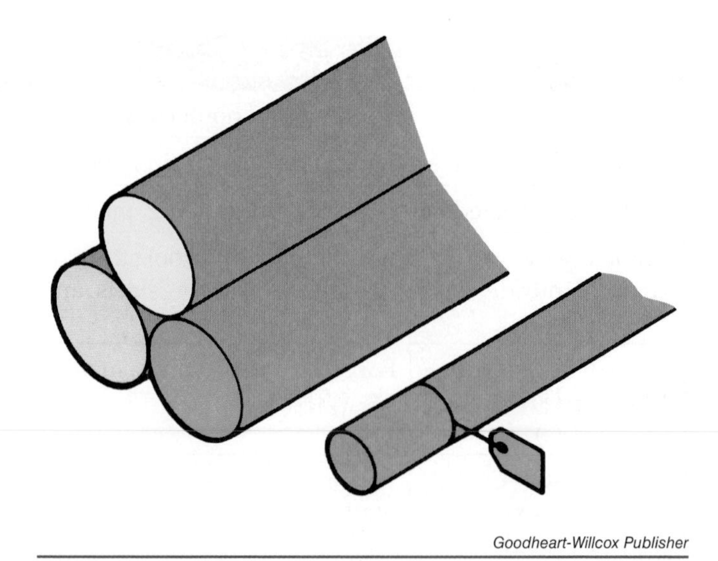

Goodheart-Willcox Publisher

Figure 28-12. Note color coding of steel bar stock. The color code identifies the type of steel.

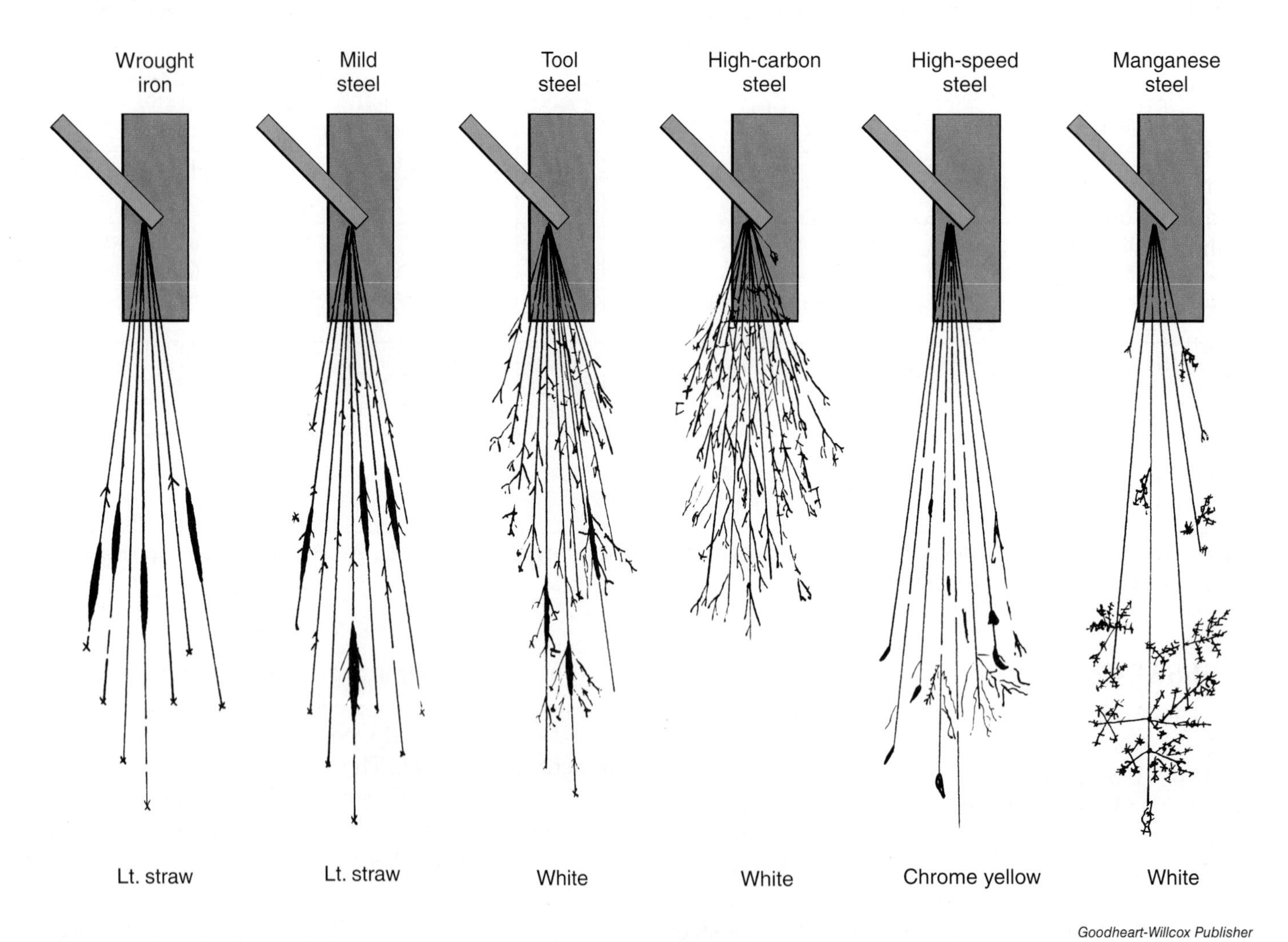

Goodheart-Willcox Publisher

Figure 28-13. A spark test is sometimes used to determine the grade of steel. To perform the test, touch the steel to the grinding wheel lightly and observe the color and form of the resulting sparks.

SAFETY NOTE

When performing or observing a spark test, wear approved eye protection. The grinder's eye shield should be clean and in position. The tool rest must also be properly adjusted.

28.3 Nonferrous Metals

There are many metals that do not have iron as their basic ingredient. These metals are known as nonferrous metals. Nonferrous metals have specific properties that make them ideal for tasks where ferrous metals are not suitable. See **Figure 28-14**.

28.3.1 Aluminum

Aluminum has come to mean a large family of aluminum alloys, not just a single metal. As first produced, aluminum is 99.5% to 99.76% pure. It is somewhat soft and not very strong.

The strength of aluminum can be greatly increased by adding small amounts of alloying elements, by heat-treating, or by cold working. A combination of the three techniques has produced aluminum alloys that, pound for pound, are stronger than structural steel. In addition to increasing strength, alloying elements can be selected to improve welding characteristics, corrosion resistance, machinability, and other desirable traits.

There are two main classes of aluminum alloys: wrought alloys and cast alloys. The shape of wrought alloys is changed by mechanically working them by forging, rolling, extruding, hammering, or other techniques. Cast alloys are shaped by pouring metal into a mold and allowing it to solidify, **Figure 28-15**.

Each alloy is given an identifying number that contains a four digit code plus a temper designation. This method of identifying aluminum alloys is known as the ***Aluminum Association Designation System***. The temper designation indicates the degree of hardness of the alloy. It follows the alloy identification number and is separated from it by a dash.

Aluminum alloys possess many desirable qualities. They are extremely strong and corrosion-resistant under most conditions. Aluminum alloys are lighter than most other commercially available metals. They can be shaped and formed easily, and are readily available in a multitude of sizes, shapes, and alloys.

Alexandar Iotzov/Shutterstock.com

Figure 28-14. Aluminum is used in many airframes because of its high strength to weight ratio.

Chuck Rausin/Shutterstock.com

Figure 28-15. Aluminum casting. Note that the molten aluminum is the same silver color as solid aluminum.

Machining Aluminum

Most of the wrought aluminum alloys possess excellent machining characteristics. They are capable of being machined to intricate shapes at high cutting speeds. However, the makeup of an aluminum alloy is a factor that can affect machinability. Some aluminum alloys of a nonabrasive nature (those containing copper, magnesium, or zinc), have improved machinability. Other alloys with abrasive constituents (such as silicon) reduce tool life and machined surfaces may have a slightly gray finish with little luster.

Most aluminum alloys are easier to machine to a good finish when in full hard temper than when in an annealed state. Machining characteristics of more commonly used aluminum alloys are:

- **Number 1100 and 3003 alloys.** Have good machinability but are gummy in nature. Turnings are long and stringy, causing difficulty in chip disposal. Good results can be obtained if the cutting tools have large top and side rake angles, with keen, smooth cutting edges.
- **Number 5052 alloy.** Has turnings that are long and stringy, and the machined surface is not as good as on 3003. Machinability is good, however.
- **Number 5056 alloy.** Has good machinability with the advantage of fairly easy chip disposal.
- **Numbers 2017-T4 and 2014-T6 alloys.** Machine to an excellent finish. Of the two, 2014-T6 has better machinability because of the heat-treating method used. It causes greater tool wear.
- **Number 2024-T3 alloy.** Has good machining characteristics with properly sharpened and honed tools. Surface finishes are excellent.
- **Number 6061-T6 alloy.** Contains silicon and magnesium. It is more difficult to machine than the 2000 series alloys. Properly sharpened cutting tools and coolants with good lubricating qualities are essential. Fine finishes are obtainable with moderately heavy cuts.
- **Number 7075-T6 alloy.** The highest strength aluminum alloy commercially available. Machining qualities are good.

High-speed cutting tools will produce satisfactory results when machining most aluminum alloys. However, the gumminess of the material can cause it to stick to the tool. As a result, tungsten carbide or ceramic tools are needed to produce the best results. The recommended lathe tool geometry for aluminum is shown in **Figure 28-16**.

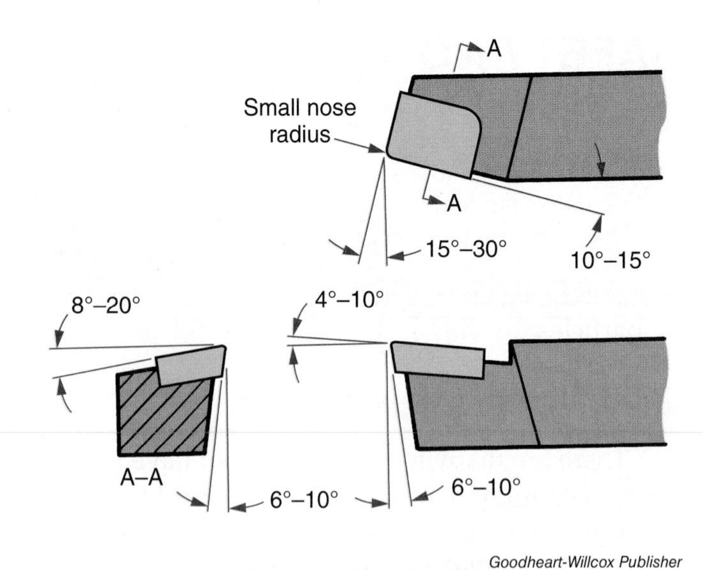

Figure 28-16. Configuration of a tungsten carbide lathe tool for machining aluminum.

28.3.2 Magnesium

Magnesium alloys are the lightest of the structural metals. They have a high strength-to-weight ratio. These alloys have excellent machining properties, and can be machined by all common metalworking techniques. The recommended lathe tool geometry for magnesium is shown in **Figure 28-17**.

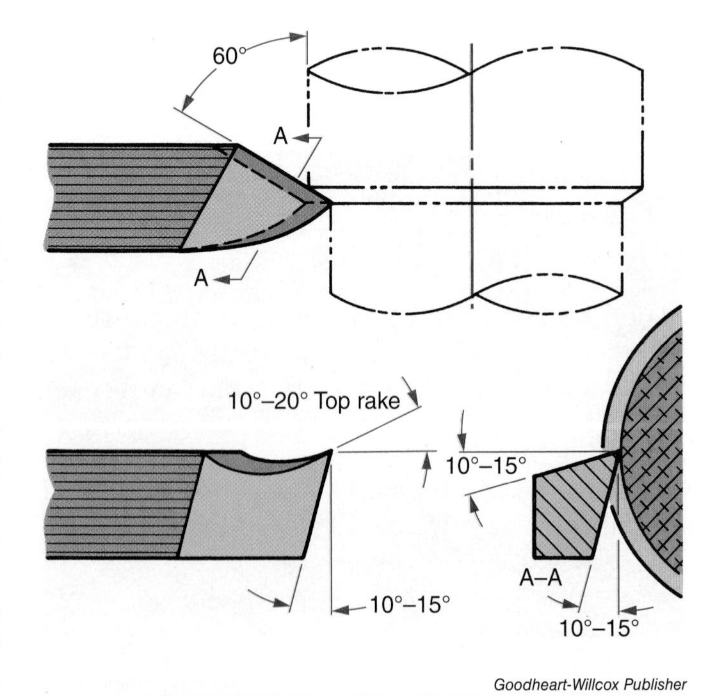

Figure 28-17. Configuration of HSS lathe tool recommended for machining magnesium.

Despite their many advantages, magnesium alloys must be worked with extreme care. Several of the alloys developed for use at elevated temperatures in aerospace vehicles contain thorium, a low-level radioactive material. They must be handled according to strict safety precautions for radioactive materials.

Another concern when machining magnesium is that extreme care must be taken because the chips or particles are highly flammable. (Because of the relatively high thermal conductivity of magnesium, there is normally no fire hazard when a solid section of the metal is exposed to fire, however.) Burning magnesium chips are intensely hot (5610°F, or 3100°C). In addition, due to the reactive nature of the metal, magnesium fires cannot be extinguished by conventional firefighting techniques. Water or commercial extinguishing agents will actually intensify the fire. A special (Class D) extinguishing agent is made for flammable metals fires.

To guard against magnesium fires, do not allow magnesium chips to accumulate on or around the machine, and use a straight mineral oil cutting fluid in adequate quantities to flood the work.

SAFETY NOTE

When machining magnesium and its alloys, do not use water-based coolants. Water will react with the magnesium chips and actually intensify a magnesium fire once it gets started.

28.3.3 Titanium

Titanium is a metal as strong as steel but only half as heavy. It bridges the gap between aluminum and steel. It is silvery in appearance, and extremely resistant to corrosion. Most titanium alloys are capable of continuous use at temperatures up to about 800°F (427°C). These characteristics make titanium ideal for use in high-speed aircraft components.

Machining Titanium

Titanium can be machined with conventional tools if the following practices are observed:

- Tool and work setups must be rigid.
- Tools must be kept sharp.
- Good coolants must be applied in adequate quantities.
- Cutting speeds must be slower, with heavier feeds than those used for steel.

Turning titanium that is commercially pure is very similar to turning 18-8 stainless steel. The alloys are somewhat more difficult to machine. Tungsten carbide and some types of ceramic tools produce the best results.

Milling titanium is more difficult than turning because the chips tend to weld to the cutter teeth. Climb milling may alleviate the problem to a great extent. Cast alloy tools often prove more economical to use than tungsten carbide tools. A water-based coolant is recommended.

Drilling titanium with conventional high-speed steel drills will produce satisfactory work. Drills should be no longer than necessary to produce the required depth hole and still allow the chips to flow unhampered.

Tapping titanium is one of the more difficult machining operations. The tap has a tendency to freeze or bind in the hole. Careful selection of cutting fluid will minimize this problem. Black oxide coated taps are recommended.

Sawing titanium requires a slow speed of about 50 fpm (15 mpm) with heavy, constant pressure. The tooth geometry of the blade must be designed for sawing titanium.

28.3.4 Copper

Copper can be shaped easily, but becomes hard when worked and must be annealed or softened. It is difficult to machine because of its toughness and softness. Copper also has good thermal and electrical conductivity.

With copper, keep tools honed sharp and make as deep a cut as possible. Coolant should be used, as copper heats up quickly. Contaminated coolant can cause corrosion (discoloration of surfaces).

Brass and bronze are the most familiar of the copper-based alloys. However, lesser-known heat-treatable alloys are available. The newer alloys combine copper and exotic metals like zirconium and beryllium. Most copper-based alloys are available in rod, bar, tube, wire, strip, and sheet forms.

Brass

Brass is an alloy of copper and zinc. It ranges in color from reddish yellow to a silvery yellow, with the color determined by the percentage of zinc it contains. Most brasses can be readily machined.

Bronze

Bronze, **Figure 28-18**, is an alloy of copper and tin. It is harder than brass and is much more expensive.

Bethlehem Steel Co.; Bird-Johnson Company

Figure 28-18. Bronze is an excellent metal for ship propellers. A—The propeller shown is 31 feet in diameter. B—Ship propellers are machined on large CNC multiaxis machine tools.

Many special bronze alloys include additional alloying elements, such as aluminum, nickel, silicon, and phosphorous. Most bronzes are relatively easy to machine with sharp tools. The recommended lathe tool geometry for brass and bronze is shown in **Figure 28-19**.

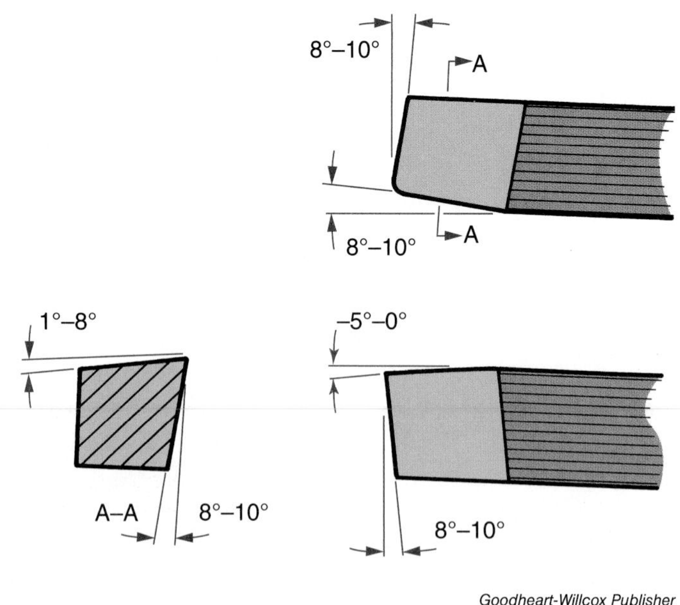

Goodheart-Willcox Publisher

Figure 28-19. Configuration of HSS lathe tool for turning brass and bronze.

Beryllium Copper

Beryllium copper is one of the newer copper-based alloys. Its machining qualities are similar to those of copper.

SAFETY NOTE

Machining of beryllium copper can pose a definite health hazard if precautions are not observed. Beryllium is toxic. The fine dust generated by machining and filing beryllium copper can cause severe respiratory damage. A respirator-type face mask must be worn. Special procedures must also be followed when cleaning machines used to machine beryllium copper.

A vacuum system should be used to remove the beryllium copper dust, or the work should be liberally flooded with cutting fluid. Do not permit cutting tools to become dull; since dull tools generate more dust than sharp tools.

Beryllium copper can be heat-treated. It should not be machined in the annealed state. The recommended lathe tool geometry for beryllium copper is shown in **Figure 28-20**. When machining beryllium, use a mineral oil-based cutting fluid. The fluid should be selected for its cooling properties, rather than for lubrication.

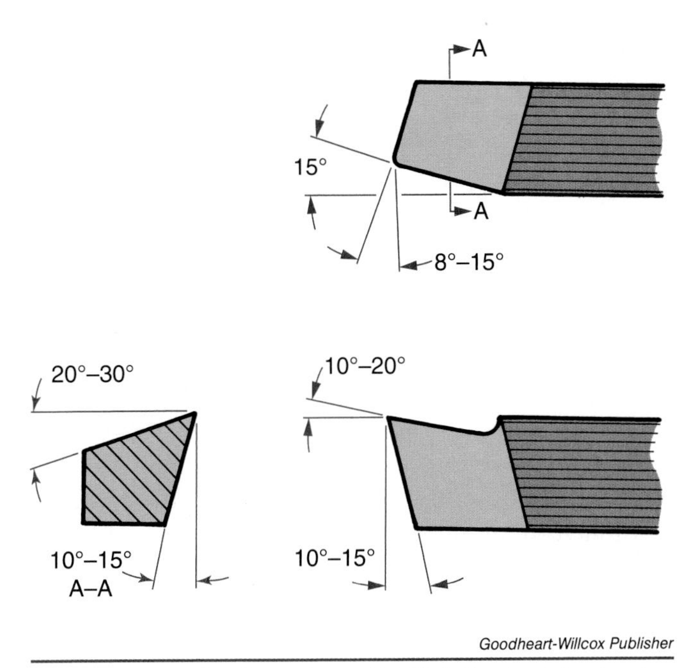

Figure 28-20. Configuration of HSS lathe tool for turning beryllium copper.

28.4 High-Temperature Metals

The nuclear and aerospace industries are chiefly responsible for the development of a number of high-temperature metals. These metals maintain their high strength during extended periods at elevated temperatures. They are sometimes called superalloys.

28.4.1 Nickel-Based Alloys

Nickel-based alloys are known commercially as Iconel X, Hastelloy X, Rene 41, etc. They have many uses in jet engines, rocket engines, and electric heat-treating furnaces. These applications require metals that can operate at temperatures of 1200°F to 1900°F (649°C to 1038°C). These metals are not easy to machine by conventional methods.

28.4.2 Molybdenum

Molybdenum has excellent strength at elevated temperatures in the 1900°F to 2500°F (1038°C to 1372°C) range, and has found many applications in modern technology. It also has great resistance to corrosion by acids, molten glass, and metals.

If both the work and cutting tool are mounted rigidly, molybdenum machines similar to cast iron. For most work, tungsten carbide and ceramic tools are preferred over high-speed steel tools.

28.4.3 Tantalum

Tantalum alloys are specified where dependability at temperatures above 2000°F (1094°C) is required. Tantalum is used for rocket nozzles, heat exchangers in nuclear reactors, and in some space structures.

Tantalum is not an easy metal to machine. It is gummy and has a tendency to tear. High-speed steel tools are usually recommended. Extreme cutting angles are used to keep the tool and chip clear of the work. The tool should be well supported, with little overhang.

28.4.4 Tungsten

Tungsten melts 6200°F (3429°C), a higher temperature than any other metal. However, tungsten is not resistant to oxidation at high temperatures (above 930°F or 499°C) and must be protected with a suitable coating, such as one of the silicides. It has many uses in rocket engines, welding electrodes, and high-temperature furnaces. Tungsten is an ideal metal for breaker points in electrical devices.

Machining is quite difficult, but can be done with carbide and ceramic tools if the work is preheated to about 400°F (204°C). The final shaping of a tungsten part is frequently done by grinding. Adequate cooling of the grinding wheel with an oil-based compound is recommended.

28.5 Rare Metals

Name just about any rare metal, and the odds are that someone is experimenting with its use in aerospace applications. Most metals in this category are available only in small quantities for experimental purposes. Many of them cost considerably more than gold.

Included in the rare metals group are such elements as scandium, yttrium, cerium, europium, lanthanum, and holmium. While they may seem strange and almost unknown at the present time, it was not too long ago that uranium, titanium, and beryllium were in the same category. In fact, until the 1880s, aluminum was considered a rare metal worth many times more than gold!

28.6 Other Materials

In addition to conventional metals and plastics, the modern machine shop is also expected to work other types of materials.

28.6.1 Honeycomb

Many ways have been devised to give existing metals greater strength and rigidity while reducing weight. ***Honeycomb sandwich panels*** are an example. Sections of thin material (aluminum, stainless steel, titanium, and nonmetals like Nomex® fabric) are bonded together to form a structure that is similar in appearance to the wax comb that bees create to store honey, **Figure 28-21.**

A strong structural panel is formed by rigidly bonding the honeycomb between two metal or composite sheets. The resulting material has a very high strength-to-weight ratio and rigidity-to-weight ratio. The bonding is done with an adhesive, or the materials fused by brazing or resistance welding.

Because the honeycomb material is fragile (before being shaped and bonded into rigid units), it can be difficult to machine. In addition to special tools that literally pare the material off, electrolytic grinding is the most rapid method for machining honeycomb. It does not leave a burr that would create problems in its removal.

Many advanced aerospace vehicles use large quantities of aluminum, stainless steel, and titanium honeycomb in their structures.

28.6.2 Composites

Composites are a relatively new class of material composed of fibers that are bonded together in a special plastic matrix (binding substance, such as an epoxy) under heat and pressure. The fibers in a composite can be made of many different materials, including various metals, graphite, boron, or fiberglass. Composite materials are generally lighter, stronger, and more rigid than many conventional metals. Some are capable of withstanding temperatures of 600°F (315°C) for long periods of time, and temperatures of 1000°F (540°C) for short periods of time.

Present uses for composites are concentrated in the aerospace and automotive industries, **Figure 28-22.** However, composites are used in many other products, including fishing poles, skis, golf clubs, tennis rackets, and bicycle frames, **Figure 28-23.** Much research is being done to reduce the cost of composites. As their cost comes down, the role of composite materials in manufacturing will continue to increase.

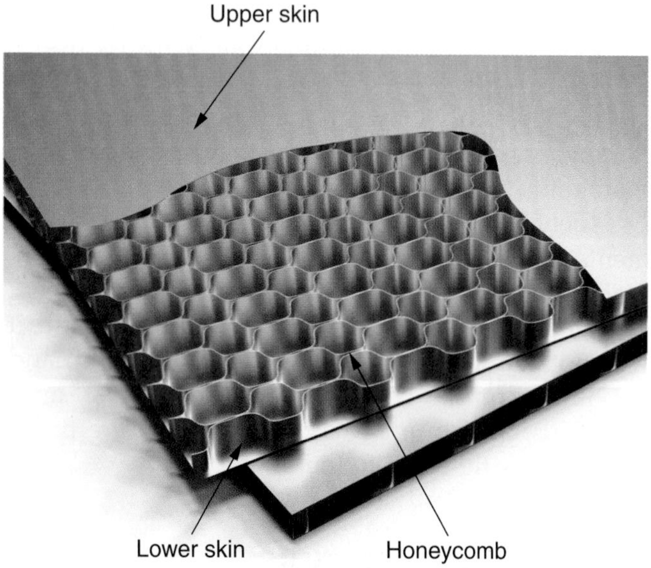

Upper skin

Lower skin

Honeycomb

Peter Sobolev/Shutterstock.com

Figure 28-21. Honeycomb sandwich panels have great strength and rigidity for their weight. They have many applications in aerospace industries.

miroungaShutterstock.com

Figure 28-22. The Boeing 787 Dreamliner was the first commercial airliner to be built primarily from composite materials. By weight, the aircraft is composed of 50% composite materials, 20% aluminum, 15% titanium, 10% steel, and 5% of other materials.

Compositek Corporation

Figure 28-23. One of the many types of composite materials is glass or carbon fiber bound together with a resin. Here, glass fiber is being wound around a large metal mandrel before resin is applied. Once the resin cures (hardens), grinding and other machining techniques can be used to achieve final dimensions. Note the dust mask being worn by the technician to prevent inhaling tiny airborne fibers.

Chapter Review

Summary

- Ferrous metals contain iron; nonferrous metals do not. An alloy is a mixture of two or more metals. A base metal is a pure, nonferrous, non-precious metal.
- Ferrous metals include cast irons (2%–5% carbon), low carbon steels (< 0.3% carbon), medium carbon steels (0.3%–0.6% carbon), high-carbon steel (0.6%–1.5% carbon), and various alloy steels. In alloy steel, various metallic elements are mixed with the steel to give it specific properties.
- The term tool steel refers to steels used to make tools that cut, shear, or form materials. Tool steel can be carbon steel or alloy steel. Tool steels are heat-treated to give them the hardness or toughness they need to perform well. High speed steel (HSS) is a special type of tool steel with high red hardness and high abrasion resistance.
- Tungsten carbide is a nonferrous metal that is commonly grouped with ferrous metals. It is the hardest known metal and is commonly used to make cutting tools.
- Stainless steels are steels that contain enough chromium to make them corrosion resistant. Austenitic stainless steels are magnetic and hardenable only by cold working. Martensitic stainless steels are nonmagnetic and hardenable through heat treatment. Ferritic stainless steels are nonmagnetic and nonhardenable.
- Steels are identified by their AISI/ASE codes, by color coding, or by a spark test. Spark test identify different steels based on the characteristics of the sparks they produce.
- Aluminum is a soft, light, nonferrous metal. It can be strengthened by adding alloying elements. Most aluminum alloys have high corrosion resistance. When machining aluminum, tungsten carbide or ceramic tools produce better results than high-speed steel tools.
- Magnesium is a nonferrous metal with a high strength-to-weight ratio and excellent machining properties. Because some magnesium alloys contain thorium and because magnesium chips and dust are highly flammable, strict safety precautions must be observed when working with magnesium.
- Titanium is a nonferrous metal that is strong, lightweight, and corrosion resistant. Titanium's light weight and ability to withstand high temperatures without losing strength makes it ideally suited for aerospace applications.
- Copper is a soft, tough nonferrous metal. It is easily shaped but becomes hard when worked. The two most familiar copper-based alloys are bronze and brass.
- High-temperature metals are those capable of maintaining high strength under extended exposure to high temperatures. High-temperature metals include nickel-based alloys, molybdenum, tantalum, and tungsten.
- In addition to the traditional metals, rare metals (such as scandium, yttrium, cerium, europium, lanthanum, and holmium), honeycomb panels, and composite materials are seeing increased use in manufacturing.

Review Questions

Answer the following questions using the information provided in this chapter.

1. What are the four main classifications of modern industrial metals?
2. Iron and steel are classified as _____ metals.
3. Carbide cutting tools are recommended for machining cast iron because _____.
   A. cast iron is soft
   B. cast iron has an abrasive surface scale
   C. high speed steel is incompatible with the recommended cutting fluid
   D. All of the above.

4. Carbon steel is an alloy of _____ and carbon.
5. How are carbon steels classified?
6. Hot rolled steel is characterized by the _____ on its surface.
7. The machinability of carbon steel is improved if sulfur or _____ is added as an alloying element.
8. The primary alloying element in stainless steel is _____.
9. Drills, reamers, and some milling cutters are usually made from _____ steel.
10. The chief characteristic of stainless steel is its resistance to _____.
11. List the three basic groups of stainless steel.
12. Aluminum, magnesium, and titanium are _____ metals.
13. Magnesium is the _____ of the structural metals.
14. _____ is a metal that is as strong as steel but only half as heavy.
15. Brass is an alloy of copper and _____.
16. Bronze is an alloy of copper and _____.
17. Why must a machinist take special precautions when working beryllium copper?
18. Nickel-based alloys, molybdenum, tantalum, and tungsten are classified as _____ metals.
19. What is the structural material known as honeycomb?
20. What are composites?

CHAPTER 29

Heat Treatment of Metals

Learning Objectives

After studying this chapter, you will be able to:

- Explain why some metals are heat-treated.
- List some of the metals that can be heat-treated.
- Describe some types of heat treatment techniques and how they are performed.
- Explain how to case harden low-carbon steel.
- Summarize the process for hardening and tempering some carbon steels.
- Compare hardness testing techniques.
- List the safety precautions that must be observed when heat-treating metals.

Technical Terms

annealing
box annealing
Brinell hardness test
case hardening
critical temperature
hardening
heat treatment
normalizing
process annealing
quenched
Rockwell hardness test
scleroscope
stress-relieving
surface hardening
tempering
Webster hardness tester

Since many parts produced in the machine shop must be heat-treated before use, it is important that the machinist be familiar with the basic science of heat-treating metals. ***Heat treatment*** is the controlled heating and cooling of a metal or alloy to obtain certain desirable changes in its physical characteristics. These changes include improving resistance to shock, developing toughness, and increasing wear resistance and hardness, **Figure 29-1**.

In a heat treatment process, the metal is heated to a predetermined temperature and then ***quenched*** (cooled rapidly) in water, brine, oil, blasts of cold air, or liquid nitrogen. The desired qualities are not always achieved after quenching. Stresses can develop that, under certain conditions, may cause some steels to shatter. Therefore, the metal may need to be reheated to a lower temperature, followed by another cooling cycle to develop the proper degree of hardness and toughness.

Heat treatment involves a number of processes, which are described in later sections of this chapter. Similar techniques can be used to anneal (soften) metals to make them easier to machine, or to case harden (produce a hard exterior surface) steel for better resistance to wear.

29.1 Heat-Treatable Metals

Steel and most of its alloys are hardenable. However, when carbon steel is heat-treated, the carbon content of the metal is an important consideration. Carbon steels are classified by the percentage of carbon they contain in "points" or hundredths of a percent. For example: 60 point carbon steel contains 0.60% carbon. Steel with less than 50 points carbon cannot be hardened.

Magnesium, copper, beryllium, titanium, and many aluminum alloys are also capable of being heat-treated.

MARCELODLT/Shutterstock.com

Figure 29-1. Many parts of this huge ore-carrying truck are heat-treated. Without heat-treated parts (such as wheels, drive shaft, gears, and axles), the vehicle would not be able to maintain its grueling workload for long without part wear and failure.

29.2 Types of Heat Treatment

The heat treatment of metals can be divided into two major categories. One deals with ferrous metals, the other with nonferrous metals. Because each area is so broad, it is beyond the scope of this text to include more than basic information on the heat treatment of metals. Heat treatment of both ferrous and nonferrous metals can be further divided into different processes. The following sections explain basic heat treatment processes and explain how they affect steel. Details regarding the heat treatment of nonferrous metals will be presented later in the chapter.

29.2.1 Stress-Relieving

Stress-relieving is done to remove internal stresses that have developed in parts that have been cold worked, machined, or welded. Stress in steel parts is relieved by heating the parts to 1000°F to 1200°F (547°C to 660°C). The parts are held at this temperature one hour or more per inch of thickness, and then slowly air- or furnace-cooled. The technique is sometimes called ***process annealing***.

29.2.2 Annealing

Annealing is a process that reduces the hardness of a metal to make it easier to machine or work. It involves heating the metal to slightly above its ***critical temperature*** (temperature at which the metal's crystal structure changes), but never more than 50°F to 75°F (28°C to 40°C) above this point, **Figure 29-2**. The time that it is held at this temperature depends on the shape and thickness of the part. After the holding period, the piece is allowed to cool slowly in the furnace or other insulated enclosure.

For some steels, it may be necessary to use the box annealing method or a controlled atmosphere furnace to prevent the work from scaling or decarbonizing (losing carbon at the surface). When annealing is done in a controlled atmosphere, oxidation does

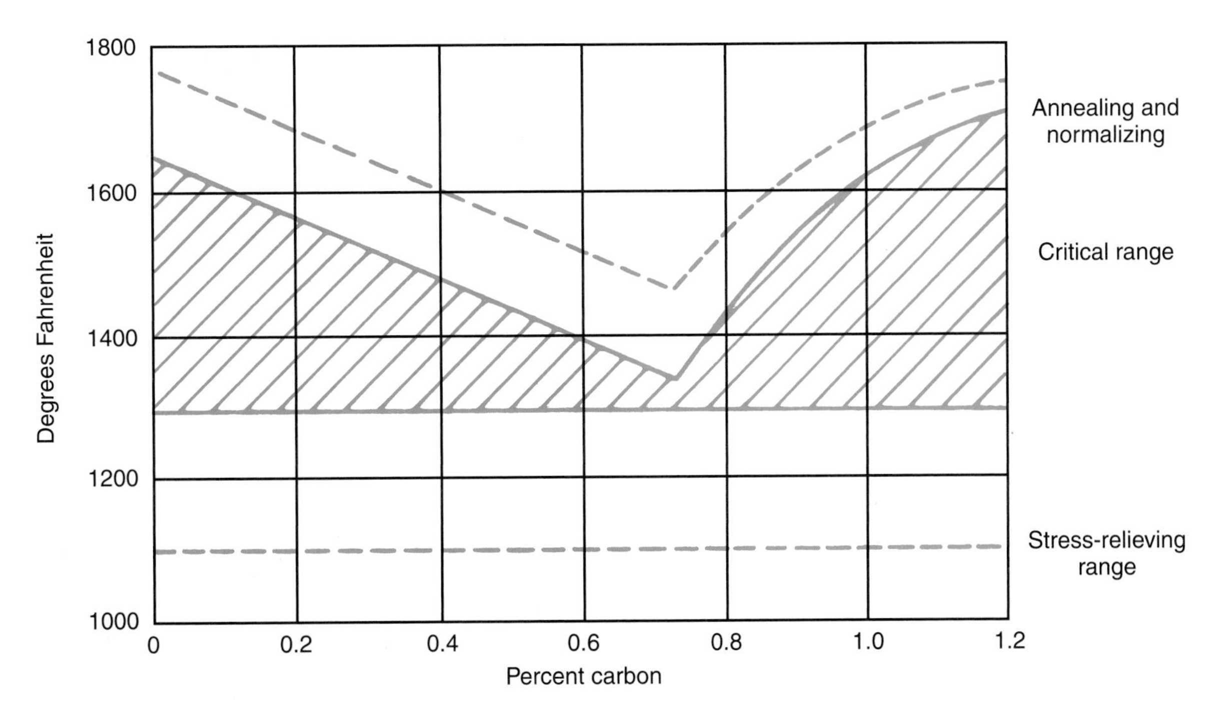

Figure 29-2. Critical range diagram for plain carbon steel.

not take place and the part remains bright. With ***box annealing***, the part is placed in a metal box and the entire unit is heated and then allowed to cool slowly in the sealed furnace.

In reducing the hardness of the metal, its machinability is improved. Many nonferrous metals can also be softened by annealing.

29.2.3 Normalizing

Normalizing is a process that refines the grain structure of some metals and thereby improves their machinability. In the normalizing process, the metal is heated to slightly above its upper critical temperature and allowed to cool to room temperature. It is a process closely related to annealing.

29.2.4 Hardening

Hardening is a heat treatment that reduces the ductility of steel. This is accomplished by heating the metal to a predetermined temperature for a specified period of time. The temperature at which steel will harden is called its critical temperature, and ranges from 1400°F to 2400°F (760°C to 1316°C), depending on the alloy and carbon content. For a "rule-of-thumb" range of hardening temperatures for carbon steel, see **Figure 29-3**.

After being heated, the part is quenched in water, brine, oil, liquid nitrogen, or blasts of cold air. Water

Carbon Content	Hardening Temperature Range
0.65% to 0.80%	1450F to 1550F (788C to 843C)
0.80% to 0.95%	1410F to 1460F (766C to 793C)
0.95% to 1.10%	1400F to 1450F (760C to 788C)
Over 1.10%	1380F to 1430F (749C to 777C)

Goodheart-Willcox Publisher

Figure 29-3. This chart lists approximate ranges of hardening temperatures for carbon steel. The heated metal must be quenched in water, brine, light oil, or with blasts of cold air.

or brine is used to quench plain carbon steel. Oil is usually used to quench alloy steels. Blasts of cold air or liquid nitrogen are used for high-alloy steels.

Quenching leaves the steel hard and brittle. It may fracture if exposed to sudden changes in temperature. For most purposes, this brittleness and hardness must be reduced by a tempering or drawing operation.

29.2.5 Surface Hardening

Surface hardening is often used when only a medium-hard surface is required on high-carbon or alloy steels, **Figure 29-4**. The internal structure of the metal is not affected. Flame hardening, induction hardening, and laser hardening are used to achieve this.

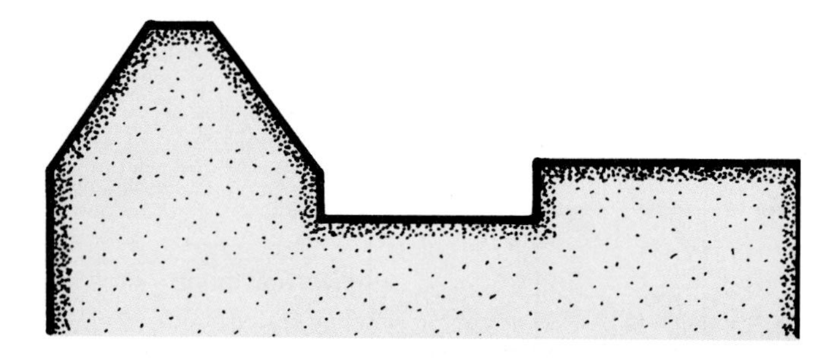

Goodheart-Willcox Publisher

Figure 29-4. Lathe ways are frequently surface-hardened for improved wear resistance. The denser dot pattern at the edges of this lathe way cross section show that only the outer surface is hardened.

Flame hardening is the rapid heating of the surface with an acetylene torch, and immediate quenching of the heated surface. The flame must be moved constantly to prevent burning or hardening the metal too deeply.

Induction hardening is the use of a high-frequency electrical induction current to heat the metal, **Figure 29-5**. The quenching medium follows the induction coil. The technique is quick and tends to minimize distortion in the piece being heat-treated. Induction hardening is ideal for production hardening operations.

Laser hardening relies on lasers to heat the metal being hardened. It works on the same principle as the previously described hardening techniques. A laser beam 1/8″ to 5/8″ (3.2 mm to 15.9 mm) wide is focused on the area to be hardened, **Figure 29-6**. The light energy emitted by the laser is converted into heat energy and is absorbed by the metal.

The surface heats rapidly, so the part must be moved under the beam or the area will be heated to the melting temperature. Because the laser is focused on a small area and the part is continuously moved, there is very little heat input into the part outside of the target area. As a result, the hardened area cools rapidly (self-quenches) enough to touch within a few seconds.

With the laser technique, the area being hardened can be carefully controlled. It may be as small as 1/4″ (6.5 mm) square. There is very little chance of part warping or distortion with laser hardening. The process produces a fine grain structure that has a tougher wearing surface than other techniques.

Radyne

Figure 29-5. A stand-alone vertical-scanning induction system for heat treatment. It is programmable and equipped with a touch screen. The furnace can be loaded and unloaded manually or by a robotic manipulator.

Light Beam Technology, Inc.

Figure 29-6. A laser beam being used to heat-treat the bearing area of a shaft. Black paint on the bearing area prevents the beam from being reflected back from the surface without heating the metal. The small flame rising from the spot being treated is caused by the black paint burning off.

29.2.6 Case Hardening

Low-carbon steel cannot be hardened to any great degree by conventional heat treatment. However, a hard shell can be put on the surface, while the inner portion remains relatively soft and tough. The process used to do this is called ***case hardening***, **Figure 29-7**.

Case hardening is accomplished by heating the piece to a red heat and introducing small quantities of carbon or nitrogen to the metal's surface. This can be done by one of the following methods: carburizing, cyaniding, or nitriding.

- During carburizing, sometimes termed the pack method, the steel is buried in a dry carbonaceous material (material rich in carbon) and heated to just above its transformation range, **Figure 29-8**. In the transformation range (1350°F–1650°F or 730°C–900°C), steels undergo internal structural changes that radically affect their properties. The part is held at this temperature for 15 minutes to one hour, until the desired case thickness is attained. The part is removed from the furnace and quenched. Deep (very thick) cases can be obtained by this method.
- During the liquid salt method of case hardening, also known as cyaniding, the part is heated in a molten cyanide salt bath, then quenched. The immersion period is usually less than one hour. A high hardness is imparted to the work, and the treated parts have good wear resistance.
- The nitriding method of case hardening develops high hardness without quenching, and distortion is virtually nonexistent. In the nitriding method, or gas method, of case hardening, parts are placed in a special airtight heating chamber where ammonia gas is introduced at high temperature. The ammonia decomposes into nitrogen and hydrogen. The nitrogen enters the steel to form nitrides which give an extreme hardness to the metal's surface. Wear-resistance and high-temperature hardness are greatly increased.

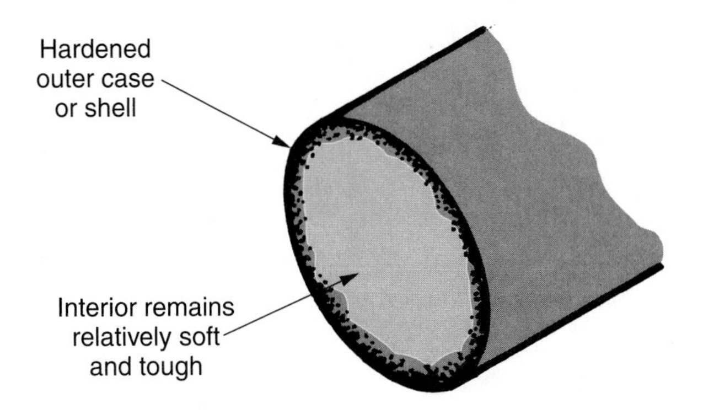

Goodheart-Willcox Publisher

Figure 29-7. Cross section of a case-hardened part shows that interior remains relatively soft and tough, while a hard "shell" is formed on the exterior.

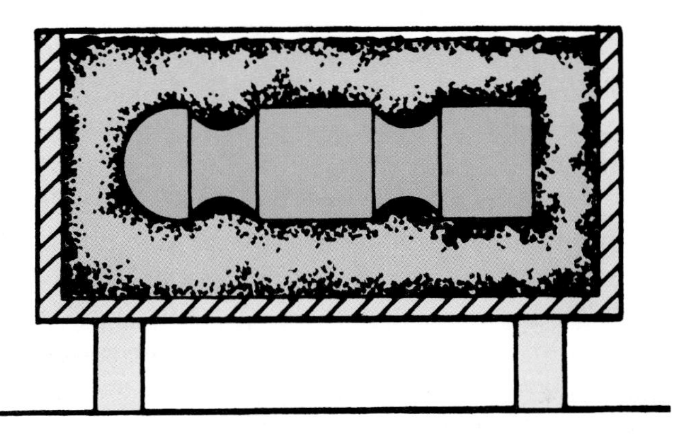

Goodheart-Willcox Publisher

Figure 29-8. A part packed in a container of carbonaceous material, ready to be case hardened.

29.2.7 Tempering

Tempering is a heat treatment process that reduces a metal's brittleness or hardness. In the tempering process, steel is heated to a temperature below its critical range. Refer to **Figure 29-2**. The exact temperature will depend on the type of steel being treated and its intended application. This information can be found in steelmakers' catalogs and the various machinists' handbooks. The temperature is held until complete penetration is achieved, and then the steel is quenched.

With the internal stresses released, the toughness and impact resistance increase. As the temperature is raised, ductility is improved, but there is a decrease in hardness and strength.

29.3 Heat Treatment of Other Metals

In addition to steel, many other metals and their alloys are potentially heat-treatable. You should have a basic understanding of these processes.

29.3.1 Heat Treatment of Aluminum

Aluminum is a general term applied to the base metal and its many alloys. When heat-treating aluminum, it is imperative that the exact alloy be known or the part being heat-treated may be ruined.

Aluminum alloys are heat-treatable in much the same manner as steel. That is, the metal is heated to a predetermined temperature, quenched, and then reheated to a lower temperature. However, some aluminum alloys age-harden at room temperature. These alloys must be kept refrigerated to remain soft and ductile (property of metal which permits it to be drawn out or hammered thin) while they are being worked.

Heat treatment temperatures for several of the more common aluminum alloys are shown in **Figure 29-9**. Information on other alloys, each of which requires special treatment to bring out optimum physical qualities, can be obtained from handbooks available from the various producers of aluminum.

29.3.2 Heat Treatment of Brass

Brass can be annealed after cold working by heating to 1100°F (593°C) and cooling. The rate of cooling has no appreciable effect on the metal.

29.3.3 Heat Treatment of Copper

Copper is annealed in much the same manner as brass. The metal may be quenched or allowed to cool slowly at room temperature.

29.3.4 Heat Treatment of Titanium

Most titanium alloys are heat-treatable. However, special facilities are necessary because titanium is a reactive metal: it readily absorbs oxygen, carbon, and nitrogen. These elements greatly affect the strength of titanium and its resistance to fatigue and corrosion.

29.4 Equipment for Heat Treatment

Heat treatment involves three distinct steps: high temperature heating, quenching (rapid cooling) to harden, and tempering for final hardness and physical properties. Each step has a significant effect on the final result.

29.4.1 Quenching Media

The main problem in heat treatment is to cool the metal at a uniform rate over its entire area. Water, oil, air, and liquid nitrogen are standard quenching media used to draw heat from the part being treated. Quench tanks for oil and water are available with temperature controlling systems, **Figure 29-10**.

Water cools metal very quickly. It is used mainly when plain carbon steel is being treated for maximum hardness. Water has the disadvantage of forming gases when the hot metal is immersed in it. The gas bubbles cling to the metal, slow cooling, and cause soft spots in the treated piece. The use of brine

Heat-Treatable Aluminum Alloys						
Solution				Precipitation (Aging)		
Aluminum alloy	Heat to F (C)	Quench	Resulting temper	Heat to F (C)	Hold for hours	Resulting temper
2024-0	910°F–930°F (488°F–499°C)	In cold water as quickly as possible after removal from furnace	—	Room	48–96	2024–T4
6061-0	960°F–980°F (515°C–527°C)		6061-W	315°F–325°F (157°C–162°C)	16–20	6061–T6
				345°F–355°F (174°C–179°C)	6–10	
7075-0	860°F–930°F (460°C–499°C)		7075-W	245°F–255°F (118°C–124°C)	20–26	7075–T6
Note: When heat-treating clad 2024 and clad 7075 aluminum, hold temperature for shortest possible time.						

Goodheart-Willcox Publisher

Figure 29-9. Heat treatment temperatures for several common aluminum alloys.

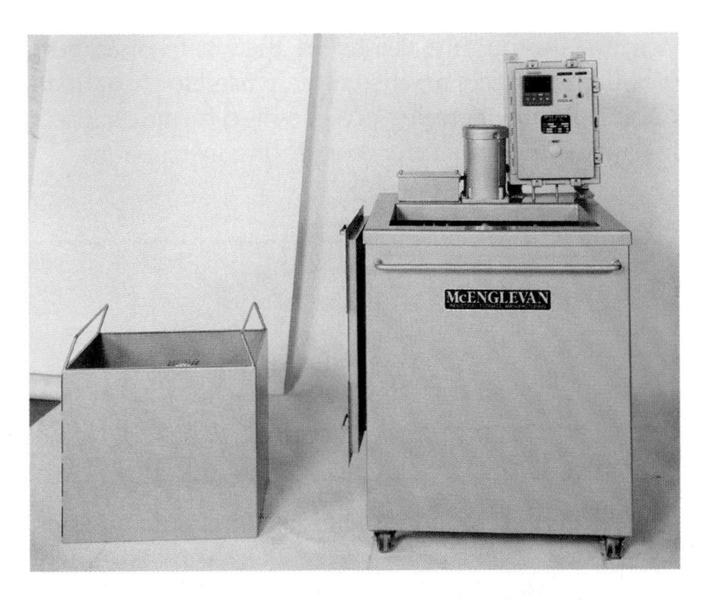

MIFCO McEnglevan Industrial Furnace Co., Inc.

Figure 29-10. A circulating oil-quench system. The shop rule for a circulating quench tank is one gallon of quench medium for each pound of steel at 1500°F quenched per hour. Standard quench oils give best results when heated to 120°F to 140°F (50°C to 60°C).

(5% to 10% salt in the quench water) prevents the formation of gases and gives better cooling results.

Mineral oils cool more slowly and produce less distortion in the treated part than water. Special quenching oils have been developed. They have a high flash point (the lowest temperature at which the oil vapors will ignite in air) and do not have a disagreeable odor. The mineral oil quenching baths used for production must be filtered and cooled down to room temperature. Oils are used to harden alloy steels.

SAFETY NOTE

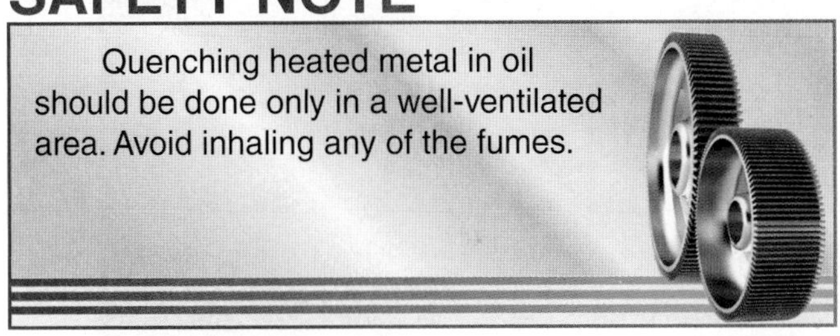
Quenching heated metal in oil should be done only in a well-ventilated area. Avoid inhaling any of the fumes.

Freely circulated air is used to cool some highly alloyed steels. The air used as a cooling medium must be dry. Any moisture in the air could cause the steel to fracture.

Liquid nitrogen is a cryogenic medium used for quenching several aluminum alloys and space-age metals. Because liquid nitrogen has a temperature of about –320°F (196°C), its use as a quenching medium requires special facilities.

29.4.2 Furnaces

The heat treatment furnace must be capable of reaching and maintaining the temperatures needed for heat treatment. They are heated by electricity, gas, or oil.

Most small furnaces are heated by electricity. They are safe to operate, quiet, require no elaborate venting systems, reach temperature quickly, and can be controlled with accuracy. If the furnace is equipped with a microprocessor-based controller, it is possible to program the furnace for precise time/temperature cycles, **Figure 29-11**.

Some models are equipped with two chambers, **Figure 29-12**. The upper high-temperature chamber can provide atmospheric control for heat-treating with inert gases. The lower chamber is used for tempering and drawing.

To provide atmospheric control, the furnace is sealed and a vacuum is drawn to remove atmospheric gases that might contaminate the metal being heat-treated. The chamber is then flooded with an inert gas (one that will not oxidize or be absorbed by the metal's surface) during the heat treatment operation.

Industry uses many types of heat treatment furnaces. Most are automated and continuous in operation. Modern electric furnaces are fitted with numerous safety devices. Avoid using a furnace until

NEYTECH

Figure 29-11. Bench-top muffle type furnace with a programmable controller for precise time/temperature management. This electric-powered furnace has a temperature range of 90°F to 2012°F (32°C to 1100°C).

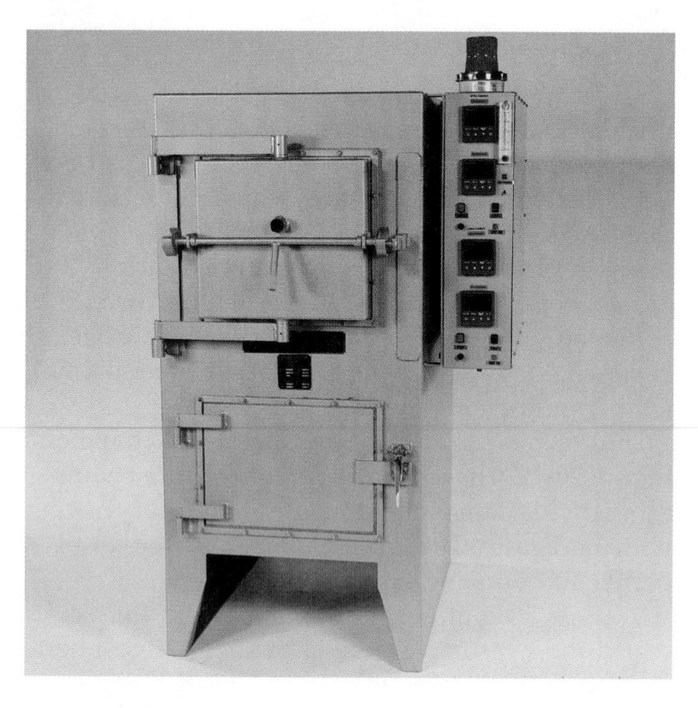

MIFCO McEnglevan Industrial Furnace Co., Inc.

Figure 29-12. Dual electric furnace with digital readout and microprocessor-based controller. Top unit features atmospheric control for hardening with inert gases; the lower unit is used for drawing and tempering.

you are thoroughly versed in their safe operation. Gas-fired furnaces are also widely used for heat treatment. For safe operation of a gas-fired furnace, closely follow the manufacturer's operating instructions.

SAFETY NOTE

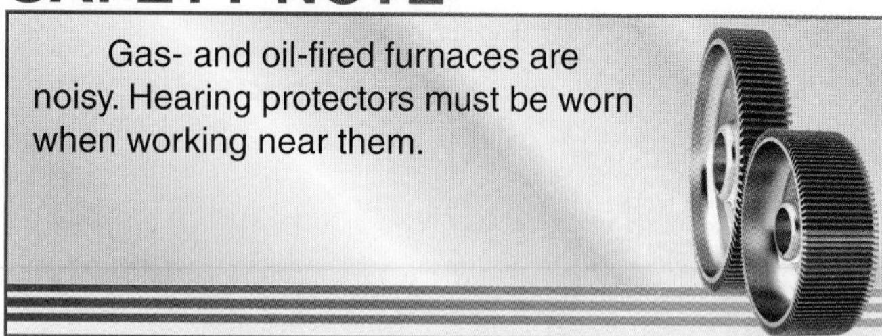

Gas- and oil-fired furnaces are noisy. Hearing protectors must be worn when working near them.

Heat-treating any metal requires maintaining accurate temperatures. A pyrometer is an instrument that accurately measures furnace temperature. In some furnaces, a pyrometer is integrated into a thermostat that can be set to the desired temperature and used to maintain that temperature once it is reached.

If the furnace is not equipped with a pyrometer, it will be necessary to judge the temperature by the color of the metal as it heats. Color charts are available from the metal suppliers. See **Figure 29-13**.

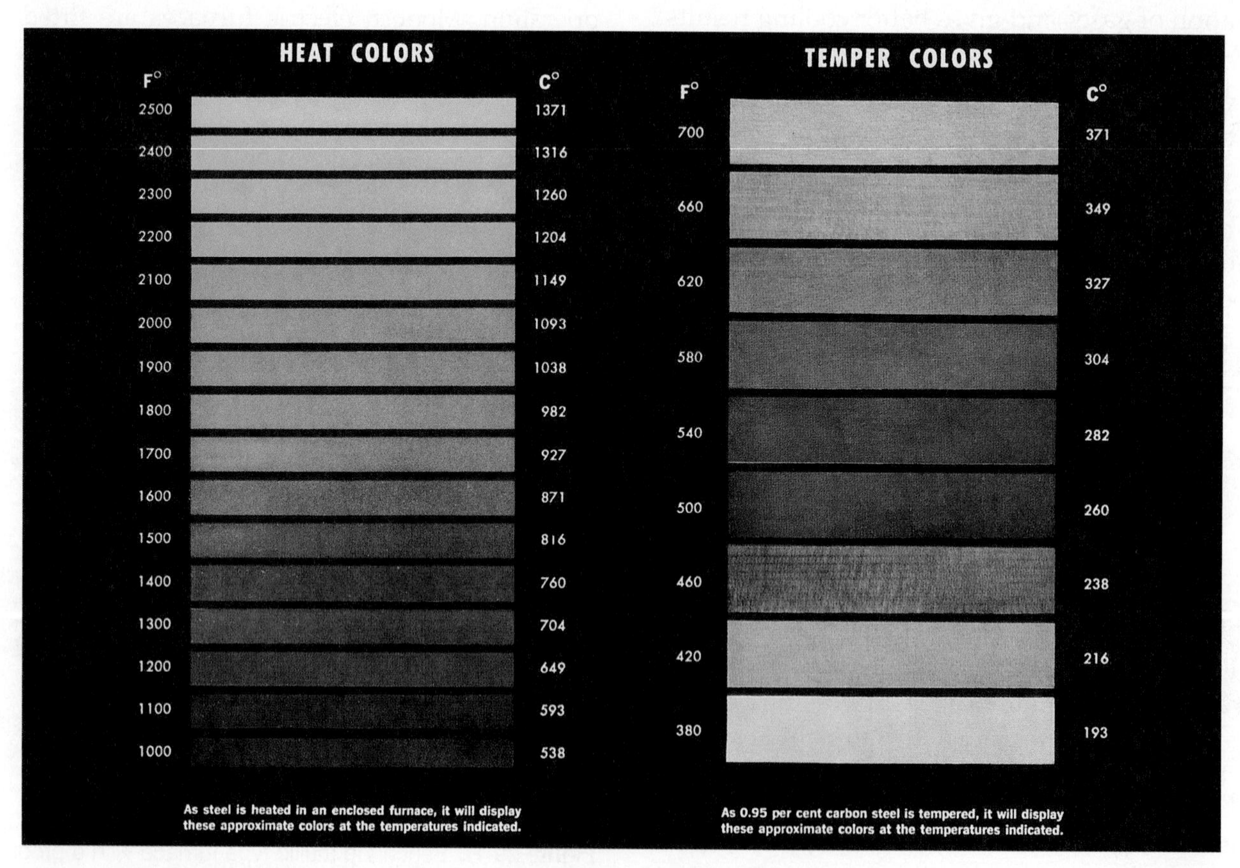

Bethlehem Steel Co.

Figure 29-13. As steel is heated, it changes to the colors listed in the left column of this table. The colors in the right column are useful for tempering steel if no pyrometer is fitted to the furnace.

29.5 Hardening Carbon Steel

The following procedure is recommended for hardening carbon steels. Oxidation of the metal during heat treatment can be avoided by wrapping the part in stainless steel foil, **Figure 29-14.**

1. Place the metal in the furnace and set controls (if applicable) for the desired temperature.
2. Heat the metal to its critical temperature (1300°F to 1600°F, or 705°C to 870°C). Avoid placing the part being heat-treated directly in the gas flames. If it has not been wrapped in stainless steel foil, position it in a piece of iron pipe, as shown in **Figure 29-15.** Allow the part to "soak" in the furnace until it is heated evenly throughout.
3. Preheat the tong jaws and then remove the piece from the furnace.

SAFETY NOTE

Dress properly for any operation involving the furnace, **Figure 29-16.** The metal is very hot and serious burns can result from relatively minor accidents.

4. Quench the piece in water, brine, or oil (depending on the type of steel being treated). Except for some alloy steels, steels are usually classified as water-hardening or oil-hardening types, based on the quenching medium to be used on them.

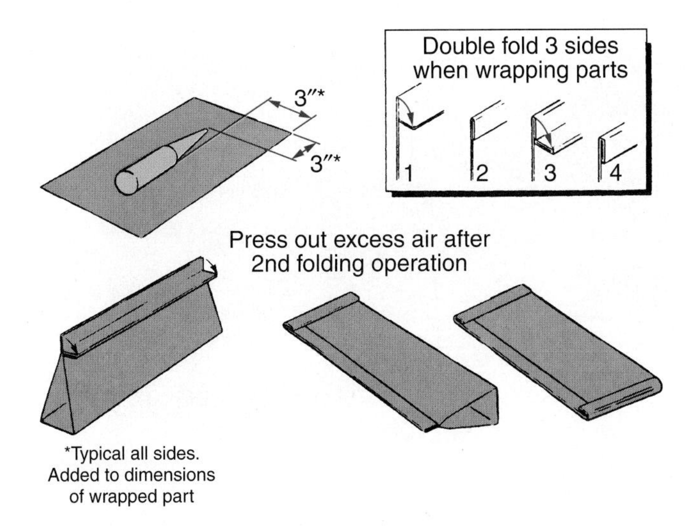

Figure 29-14. Procedure for wrapping parts in stainless steel foil to protect the parts against oxidation and decarburization during heat treatment.

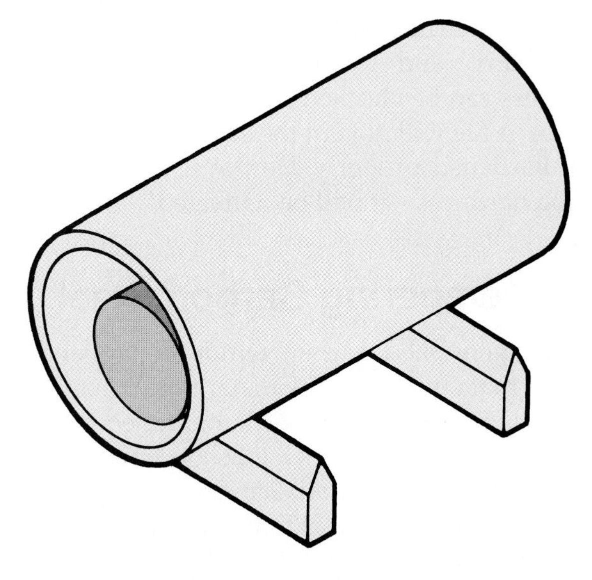

Figure 29-15. Work being heat-treated must be protected from direct flames in a gas furnace by inserting it in a section of pipe. Keep the pipe raised off the furnace floor to permit uniform heating.

Figure 29-16. Heat treatment temperatures are very hot. Dress properly for job and keep the area around furnace clean so there is no danger of slipping or stumbling. Also, preheat tongs before grasping the heated part.

The quenching technique is critical. To ensure an even hardness throughout the piece, dip long slender sections straight down into the quenching fluid with an up-and-down motion. Avoid a circular motion, since this may warp the piece. Parts with other shapes should be moved around in a manner that will permit them to cool quickly and evenly.

Steel that has been hardened properly will be "glass-hard" and too brittle for most purposes. Hardness can be checked by trying to file the work surface. A file will not cut the surface if the piece has been hardened properly. Do not use a new file for testing hardness—it will be damaged!

Hot liquid baths of oil, molten salts, or lead are often used in place of a furnace for heating parts to their proper tempering temperature. The pieces are held in the bath until the heat permeates them. They are then removed from the bath and allowed to cool in still air.

29.6 Tempering Carbon Steel

As mentioned earlier, tempering relieves the stresses that develop in the metal during hardening. Until it is tempered, the brittle hardened steel may crack or shatter from shock (such as being dropped or struck) or from sudden changes in temperature.

Tempering is done as follows:

1. Polish the hardened piece with abrasive cloth.
2. Heat the piece to the correct tempering temperature. The correct temperature is based on the type of steel and the job the finished part is to do. Proper tempering temperatures range from about 380°F or 193°C (the metal will turn a silvery yellow or straw color) to 700°F or 371°C (the metal will turn light gray or blue color). Color charts are available from steel companies and can be used as a guide for determining when the desired temperature is reached. Refer to **Figure 29-13**. Quench the metal as soon as it reaches the required temperature.
3. Small tools are best tempered by placing them on a steel plate that has been heated red-hot. Have the point of the tool extend beyond the edge of the plate, as shown in **Figure 29-17**. Watch the temper color as the metal heats up. Quench the metal as soon as the tool point turns the proper color.

Goodheart-Willcox Publisher

Figure 29-17. Use a heated steel plate when tempering small tools. Have point of tool extend beyond edge of plate. Rotate the tool as it heats up to ensure that it is heated evenly throughout.

29.7 Case Hardening Low-Carbon Steel

Of the several case hardening techniques, the simplest is carburizing, which requires a minimum of equipment. It uses a nonpoisonous commercial case hardening compound.

SAFETY NOTE

The case hardening technique known as cyaniding is not recommended. Cyanide is a deadly poison and is very dangerous to use under any but ideal conditions.

There are two methods recommended for using case hardening compound to harden low-carbon steel. The first method is as follows:

1. Bring the furnace to temperature.
2. Bring the workpiece to a bright red (1650°F–1700°F, or 900°C–930°C). Use a pyrometer or thermocouple to monitor the temperature, **Figure 29-18**.

Honeywell, Inc., Industrial Automation and Control

Figure 29-18. A microprocessor-based digital controller for electric heat treatment furnaces. It can accept ten different thermocouple types.

3. Dip, roll, or sprinkle the case hardening compound on the piece, **Figure 29-19**. The powder will melt and adhere to the surface, forming a shell.
4. Reheat the workpiece to a bright red and hold at that temperature for a few minutes.
5. Quench in cold water.

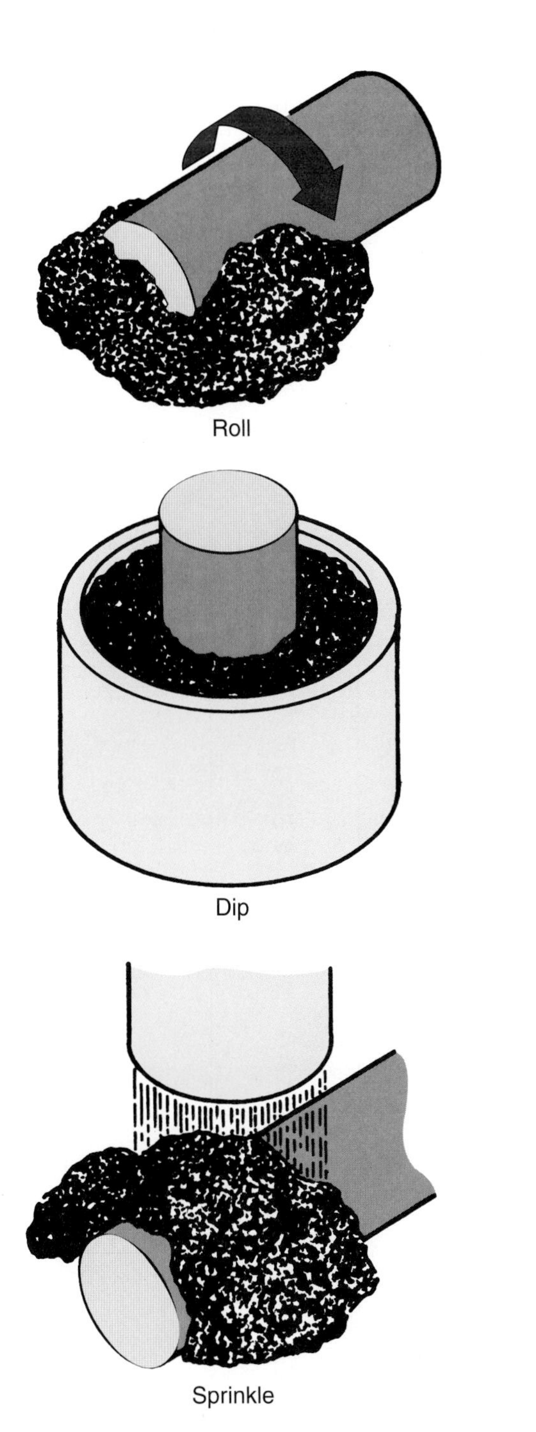

Figure 29-19. Dip, roll, or sprinkle case hardening compound on the work until a shell of compound has been formed.

The second method for case hardening low-carbon steel is to:

1. Find a steel container large enough to hold the work.
2. Put the part to be hardened into the container and completely cover it with case hardening compound. Refer to **Figure 29-8**.
3. Place the container into the furnace and heat it to a red heat. Hold the temperature for 5 to 30 minutes, depending on the depth of case required.
4. Use dry tongs to remove the piece from the container and then quench the piece in clean, cool water.

SAFETY NOTE

With either method of using case hardening compound, work in a well-ventilated area and wear full face protection, leather apron, and heat-resistant gloves.

29.8 Hardness Testing

Hardness testing ensures that the metal has been given the proper degree of hardness for its intended use. With hardness testing, it is also possible to establish standards for hardness that can be cited on drawings and specifications.

The most commonly used technique is to press a steel ball or diamond probe into the metal under a specific load to see how deep it penetrates. This type of test is performed using Brinell and Rockwell testing machines, which are known as indention hardness testers. The testing machines display the results as a hardness number, which indicates the degree of the metal's hardness. See **Figure 29-20**.

29.8.1 Brinell Hardness Test

The ***Brinell hardness test*** is used extensively in both laboratory and production situations and measures a metal's resistance to deformation. It is an excellent index of such factors as machinability, uniformity of grade, temper after heat treatment, and body hardness of the metal.

The Brinell hardness test is performed by applying a known load to the surface of the metal through a hardened steel ball of known diameter.

Hardness Conversion Table
(Approximate)

Values vary depending on grades and conditions of material involved. Rockwell "B" Scale should not be used over B-100. The "C" Scale should not be used under C-20.

Brinell	Rockwell		Shore scleroscope	Tensile lb sq. in.	Brinell	Rockwell	Shore scleroscope	Tensile lb sq. in.
Hardness No.	B Scale	C Scale	Hardness No.	In 1000 lb	Hardness No.	B Scale	Hardness No.	In 1000 lb
782	...	72	107	383	163	84	25	84
744	...	69	100	365	159	83	25	82
713	...	67	96	350	156	82	24	80
683	...	65	92	334	153	81	24	79
652	...	63	88	318	149	80	23	78
627	...	61	85	307	146	78	23	77
600	...	59	81	294	143	77	22	76
578	...	58	78	284	140	76	..	74
555	...	56	75	271	137	75	..	73
532	...	54	72	260	134	74	..	71
512	...	52	70	251	131	72	..	70
495	...	51	68	242	128	71	..	69
477	...	49	66	233	126	70	..	67
460	...	48	64	226	124	69	..	66
444	...	47	61	217	121	67	..	65
430	...	45	59	210	118	66	..	63
418	...	44	57	205	116	65	..	62
402	...	43	55	197	114	64	..	61
387	...	41	53	189	112	62	..	60
375	...	40	52	183	109	61	..	59
364	...	39	50	178	107	59	..	58
351	*(110)*	38	49	172	105	58	..	57
340	*(109)*	37	47	167	103	57	..	56
332	*(108.5)*	36	46	162	101	56	..	55
321	*(108*	35	45	157	99	54	..	54
311	*(107.5)*	34	44	152	97	53	..	53
302	*(107)*	33	42	148	96	52	..	53
293	*(106*	31	41	144	95	51	..	52
286	*(105.5)*	30	40	140	93	50	..	52
277	*(104.5)*	29	39	136	92	49	..	51
269	*(104)*	28	38	132	90	48	..	50
262	*(103)*	27	37	128	88	47	..	49
255	*(102)*	26	36	125	87	46	..	48
248	*(101)*	25	36	121	86	45	..	48
241	100	24	35	118	85	44	..	47
235	99	*(22)*	34	115	83	43	..	47
228	98	*(21)*	33	113	82	42	..	46
223	97	*(20)*	33	109	81	41	..	46
217	96	*(19)*	32	106	80	40	..	45
212	95	*(18)*	31	104	79	39	..	45
207	94	*(17)*	30	101	78	38	..	44
202	93	*(15)*	30	99	77	37	..	44
196	92	*(13)*	29	96	76	36	..	43
192	91	*(12)*	29	94	75	35	..	43
187	90	*(10)*	28	91	74	33	..	42
183	89	*(9)*	28	90	73	31	..	42
179	88	..	27	89	72	30	..	41
174	87	..	27	88	71	29	..	41
170	86	..	26	86	70	27	..	40
166	85	..	26	85	69	26	..	40

Goodheart-Willcox Publisher

Figure 29-20. Hardness conversion table.

The diameter of the resulting impression is measured. The Brinell hardness number is calculated using the applied load, the diameter of the steel ball, and the diameter of the impression, as follows:

$$BHN = \frac{P}{\frac{\pi D}{2}(D - \sqrt{D^2 - d^2})}$$

Where:

BHN = Brinell hardness number in kilograms per mm^2.
P = Applied load in kilograms.
D = Diameter of steel ball in millimeters.
d = Diameter of impression in millimeters.

To perform a Brinell hardness test using a compressed air-type tester, follow these instructions:

1. Adjust the air regulator until desired load is indicated on the load gauge. Check the load reading when making the initial test of a series, and adjust the air regulator valve, if necessary, so that the desired load is indicated on the dial when a specimen is actually under load.
2. Place the test specimen on the anvil, and adjust the gap between the surface of the specimen and the steel ball to a minimum. The test specimen must be thick enough to prevent a bulge or other marking from appearing on the side opposite the impression. If necessary, the surface on which the impression is to be made should be filed, ground, machined, or polished with abrasive material. Preparing the surface this way will help make the indentation clear enough that it can be measured with the necessary accuracy.
3. Make sure the test specimen is in place and then apply and release the load. Do not apply a load without a specimen in place. Doing so will damage the machine's anvil.
4. Read the impression made in the test specimen with the special microscope and obtain the Brinell hardness number from the hardness table.

29.8.2 Rockwell Hardness Test

The most widely used of all hardness testing methods is the ***Rockwell hardness test***, **Figure 29-21.** Either a steel ball or a specially designed diamond cone penetrator is used in the Rockwell hardness tester, depending on the material being tested. See **Figure 29-22.**

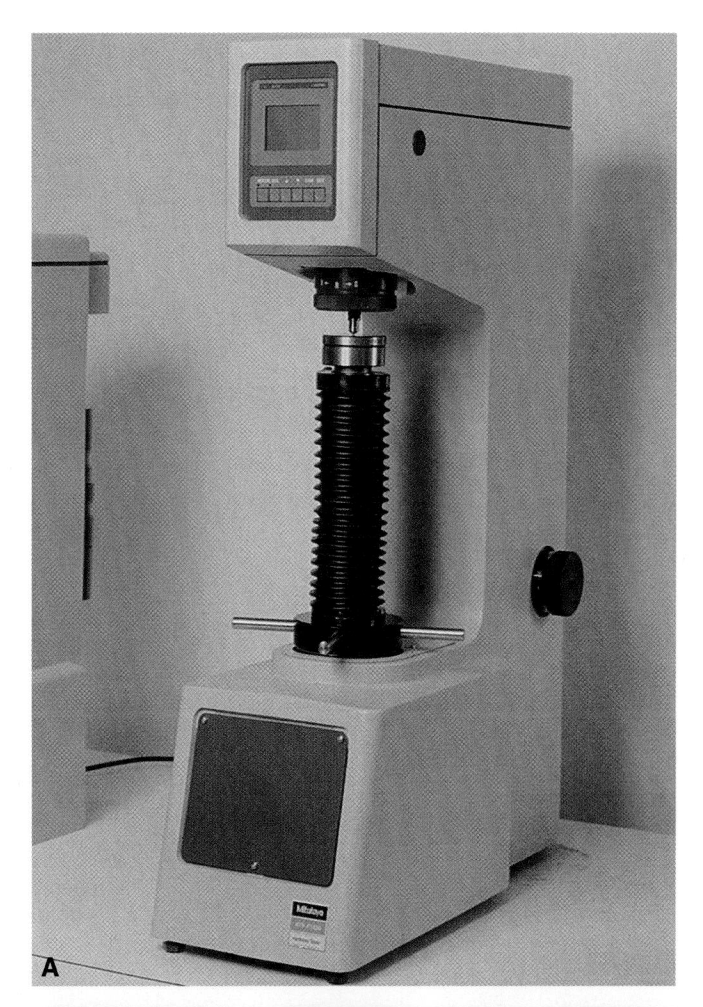

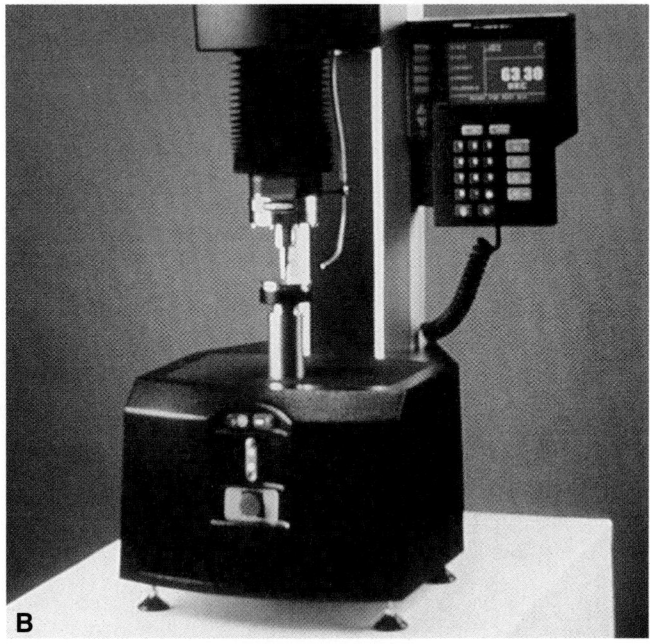

Mitutoyo/MTI Corp.; Wilson Instruments/Instron Corporation

Figure 29-21. Rockwell hardness testers. A—The basic-type tester. B—This tester has pushbutton controls for all functions. It digitally processes test results, statistical calculations, hardness scale conversions, and corrections for round parts.

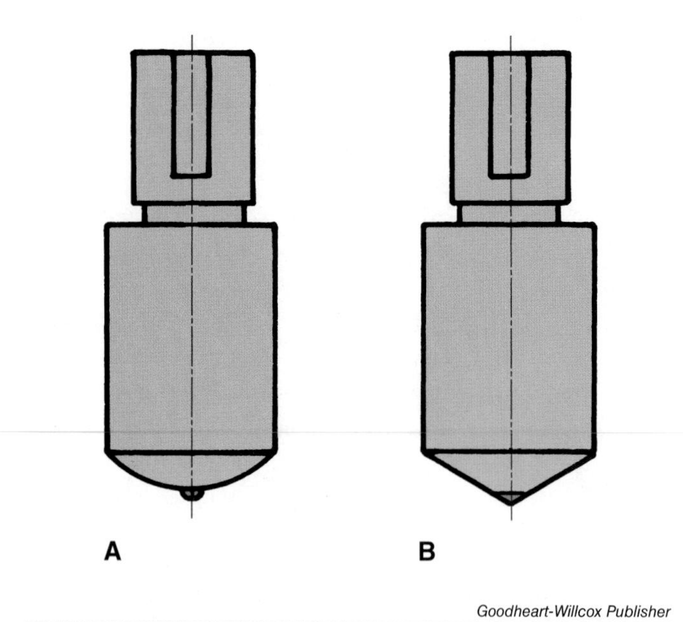

Figure 29-22. Rockwell penetrators. A—Ball-type. B—Conical diamond, or brale, type.

A minor load of 10 kg is first applied, and the dial gage is set to zero. The major load is then added and removed, **Figure 29-23**. The hardness number is based on the additional depth to which the test ball or conical diamond penetrator is driven by the major load beyond the depth of the previously applied light load. The hardness number is automatically indicated on the gage.

Typically, a 1/16″ steel ball is used in conjunction with a 100 kg load for testing such metals as brass, bronze, and soft steel. All readings made with the 1/16″ ball and 100 kg load are *Rockwell B readings*; the letter *B* must be placed before the number. There is no Rockwell hardness designated by a number alone. It must always be prefixed by the proper scale letter.

The conical diamond test point, known as a brale penetrator, is used with a 150 kg load for testing hardened steel or any hard metals, **Figure 29-24**. All readings with the brale penetrator and 150 kg load are *Rockwell C readings*; the letter *C* must precede the hardness number.

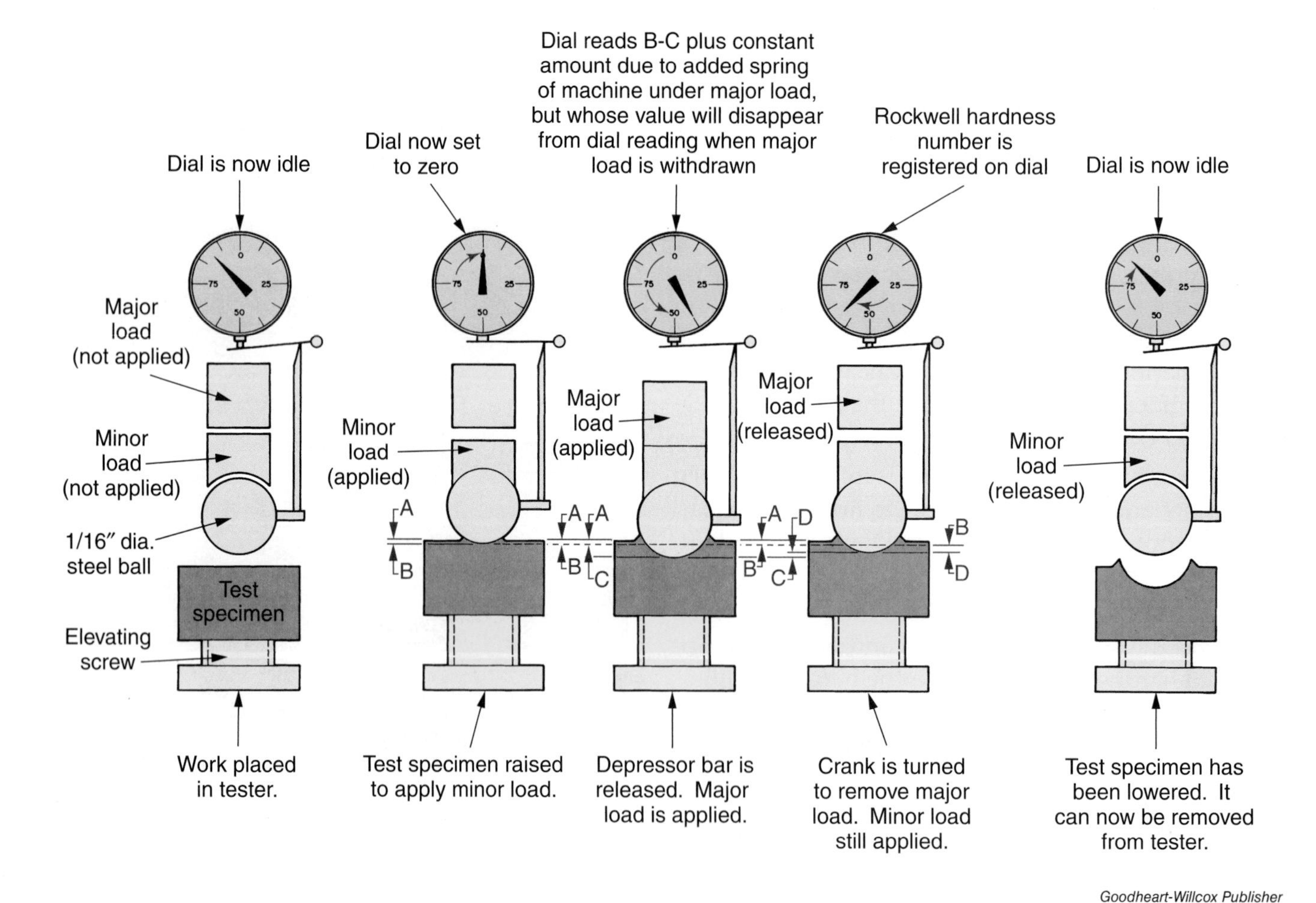

Figure 29-23. This diagram shows how 1/16″ steel ball penetrator is used to make a Rockwell B hardness reading. Size of the ball has been greatly exaggerated for clarity.

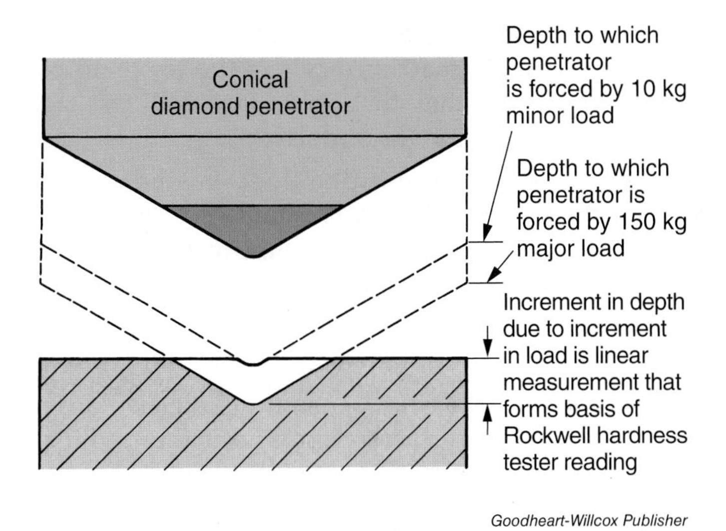

Figure 29-24. How conical diamond penetrator (brale) is used to determine Rockwell C hardness.

Special penetrators are available for testing soft materials. The scale designation is based on the ball size used to make the test. See **Figure 29-25.**

Two weights are normally supplied with a Rockwell hardness tester. One of them has a red marking and the other black. However, three different loads—60 kg, 100 kg, and 150 kg—can be applied. The weight arm, together with the link and weight pan, will apply a load of 60 kg. The weight with the red marking is placed on the weight pan for the 100 kg load. When the weight with the black marking is added, a 150 kg load is applied. The black weight is never used alone. When tests are made with the 100 kg load, the Rockwell B scale is used. The Rockwell C scale is used with the 150 kg load.

The 60 kg load (weight arm and weight pan alone) is used with the brale penetrator for testing extremely hard metals, such as tungsten carbide alloys. The 1/16″ ball is extensively used with the 60 kg load for testing sheet brass.

To operate the Rockwell hardness tester, select and mount the proper penetrator point, then check that the weight for the desired test load is in position. Place the correct anvil in the elevating screw with extreme care, or the penetrator might be damaged. Inspect the test specimen and remove any scale or burr that would flatten under the test and give a false reading.

The following procedures are considered to give the most precise hardness readings.

1. Place the test specimen on the anvil.
2. Gently raise the specimen until it comes into contact with the penetrator. Continue to raise the specimen until the minor load of 10 kg has been applied.
3. Set the machine to zero.

Scale symbol	Penetrator	Major load (kg)	Dial figures	Typical applications of scales
B	1/16″ ball	100	Red	Copper alloys, soft steels, aluminum alloys, malleable iron, etc.
C	Diamond cone	150	Black	Steel, hard cast iron, titanium, deep case hardened steel, etc.
A	Diamond cone	60	Black	Cemented carbides, thin steel, and shallow case hardened steel.
D	Diamond cone	100	Black	Thin steel, medium case hardened steel, and pearlite malleable iron.
E	1/8″ ball	100	Red	Cast iron, aluminum and magnesium alloys, and bearing materials.
F	1/16″ ball	60	Red	Annealed copper alloys, thin soft sheet metals.
G	1/16″ ball	150	Red	Phosphor bronze, beryllium copper, malleable iron, etc.
H	1/8″ ball	60	Red	Aluminum, zinc, lead.

Goodheart-Willcox Publisher

Figure 29-25. A letter is used as a prefix to the hardness value read from the Rockwell tester dial. The letter depends on load, type of penetrator, and scale from which dial readings are taken.

4. Carefully apply the major load. The penetrator is forced into the work. The depth to which it penetrates depends on the metal's hardness.
5. Wait until the reading stops changing.
6. Lift the major load, but leave the minor load still applied.
7. Read the Rockwell hardness number. If the test has been made with the 1/16″ ball and the load is 100 kg, the reading is made with the Rockwell B scale, and the letter B is prefixed to the number to signify the condition of the test. The Rockwell C scale is used and the letter C is prefixed to the number if the brale penetrator and a 150 kg load were used.

After completing the test, lower the specimen away from the penetrator and remove the specimen from the testing machine.

Like other precision tools, a Rockwell hardness tester must be handled with care if it is to maintain its accuracy. There are a few precautions that must be observed:

- When moving the tester, grasp it only by the cast iron base, never by any parts that are attached to the base.
- The tester should be leveled on a solid bench, in a location free from grit and vibration. Keep the machine covered when not in use!
- False readings will result if the shoulder on the brale penetrator is not kept clean, **Figure 29-26**. Carefully wipe the penetrator with a clean, soft cloth before installing it in the tester.
- Use only lubricants specified by the manufacturer.
- Support long work properly.
- Clean the work with an abrasive cloth to remove scale and roughness; a smooth surface is needed to ensure accurate readings. Castings and forgings should have a spot ground or machined where the test is to be made, so that the penetrator will test the metal beneath the surface.
- The specimen must be thick enough so that the underside of the specimen does not show the slightest indication of the test.
- Corrections must be added to readings made on round stock, if it is not possible to file or grind a flat spot in the test area.
- Never force the penetrator and the anvil together when a test piece is not in the machine. Doing so will damage the equipment.
- If the tester is used on case hardened steel, accurate readings cannot be made unless the "case" is several times thicker than the indentation depth.

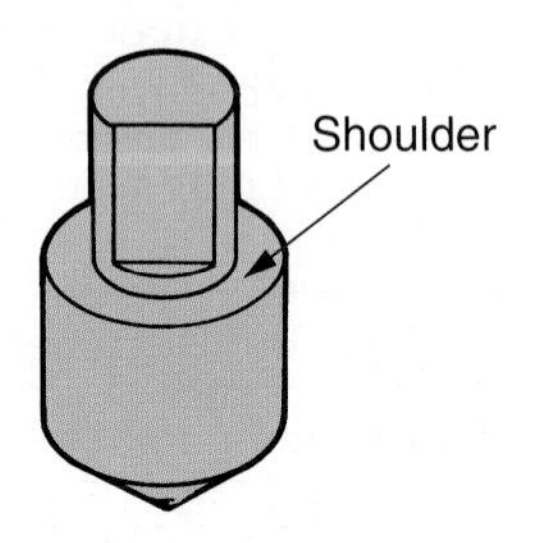

Goodheart-Willcox Publisher

Figure 29-26. The shoulder of the brale penetrator must be free of dirt and burrs before it is mounted in tester. Otherwise, incorrect readings will result.

29.8.3 Webster Hardness Tester

The ***Webster hardness tester***, **Figure 29-27**, is a portable testing device for checking the hardness of materials such as aluminum, brass, copper, and mild steel. It can be used on assemblies that cannot be brought into the laboratory, or to test a variety of shapes that other testers cannot check, such as extrusions, tubing, or flat stock.

The tester's dial reading is converted to the Rockwell hardness scale by referring to a conversion chart furnished with the tool.

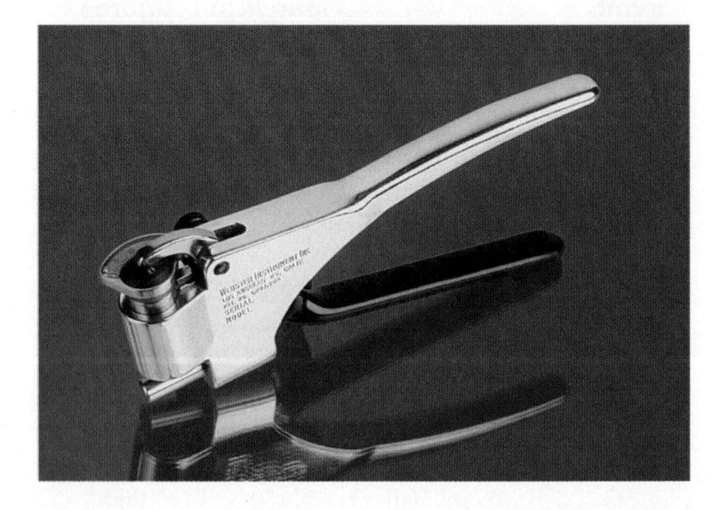

Webster Instrument, Inc.

Figure 29-27. The Webster hardness tester is a portable device used with materials such as aluminum, brass, copper, or mild steel. By referring to a conversion chart, the tester's dial indicator reading can be converted to the appropriate Rockwell scale.

29.8.4 Scleroscope Hardness Testing Machine

A ***scleroscope*** is a testing device that drops a hammer onto the test piece. The resulting rebound of the hammer determines the hardness.

Two styles of scleroscope are in use. One is fitted with a vertical scale, the other with a dial that records the test results. Scleroscopes can be used to test the hardness of all metals, ferrous and nonferrous, polished or unpolished, with virtually no limitation in size or shape. Hardness testing with the scleroscope is essentially a nonmarring test. No craters are produced that would require refinishing of the test area.

In a scleroscope hardness test, a diamond hammer is dropped from a fixed height and makes a tiny indentation in the metal. The hammer rebounds, but not to its original height. The rebound height is always lower than the initial height because some of the energy in the falling hammer is dissipated in producing an indentation. The rebound of the hammer varies in proportion to the hardness of the metal—the harder the metal, the smaller the indentation and the higher the rebound.

The tester scale consists of units determined by dividing into 100 parts the average rebound of the hammer from quenched tool steel of ultimate hardness. These rebounds will range from 95 to 105. The scale is carried higher than 100 to cover super-hard metals. See **Figure 29-28**. A scleroscope is capable of making accurate hardness readings on the softest or the hardest metals without changing the scale or diamond hammer.

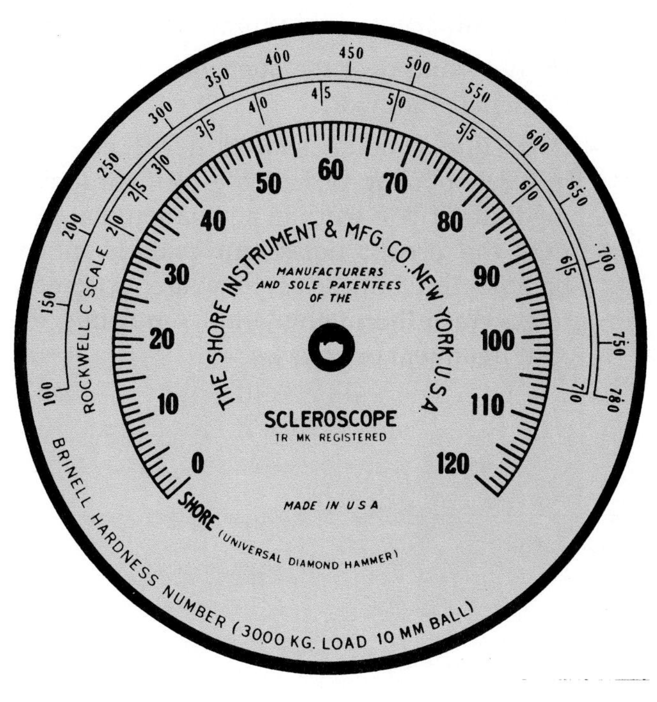

Goodheart-Willcox Publisher

Figure 29-28. Scales on a scleroscope dial. Note that the dial includes scales for the equivalent Brinell and Rockwell C hardness values.

For hardness testing on a scleroscope, the specimen must be mounted on the anvil. The unit should be leveled, which is done by turning the leveling screws while observing the built-in spirit level. The instrument is operated pneumatically by means of a rubber squeeze bulb.

To perform a test, bring the barrel cap firmly into contact with the test specimen. It is essential that you maintain a firm pressure on the specimen during the test. Squeeze and release the rubber bulb to draw the hammer to the up position. While maintaining torque on the knob, again squeeze and release the rubber bulb to release the hammer. Observe the reading on the scale or dial.

The height to which the hammer rebounds on the first bounce indicates the hardness of the specimen. For greatest accuracy, several tests should be averaged. Do not make more than one test at a given spot or false readings will result. While the scleroscope hardness testing method may sound unorthodox, the results are very close to those obtained with Brinell and Rockwell testers.

29.9 Heat Treatment Safety

- Never attempt to heat-treat metals while your senses are impaired by medication or other substances.
- Make sure that the furnace is in good operating condition before attempting to use it. Avoid lighting a furnace until you have been instructed in its safe operation. Never stand in front of a gas furnace while igniting it.
- Never look into the furnaces unless you are wearing tinted goggles or glasses under your face shield.
- Heat treatment involves raising metal to very high temperatures. Handle the hot metal with appropriate tools, and always wear an approved full face safety shield and the proper protective clothing. Wear heat resistant gloves and a leather apron, never a cloth apron.

- Work only in areas that are well ventilated.
- Do not stand over the quenching bath when immersing hot work.
- Potassium cyanide should never be used in a school shop or lab as a case hardening medium. If you work in a situation that permits the use of potassium cyanide, never breathe the fumes, as they are extremely toxic. Wash thoroughly after completing the heat treatment operations.

Chapter Review

Summary

- Steel and most of its alloys are hardenable. When carbon steel is heat-treated, the carbon content of the metal is an important consideration.
- Stress-relieving is done to remove internal stresses that have developed in parts that have been cold worked, machined, or welded.
- Annealing is a process that reduces the hardness of a metal to make it easier to machine or work.
- Normalizing is a process that refines the grain structure of some metals and thereby improves their machinability.
- Hardening is a heat treatment that reduces the ductility of steel.
- Surface hardening is heat treatment technique used to create a medium-hard surface on high-carbon or alloy steels without affecting the internal structure of the metal.
- Case hardening is heat treatment technique that creates a hard outer shell on low-carbon steel.
- Tempering is a heat treatment technique that reduces a metal's brittleness or hardness.
- Water, oil, air, and liquid nitrogen are standard quenching media used to draw heat from the part being treated.
- Furnaces are used to heat the metal for most heat treatments. Furnaces can be heated by electricity, gas, or oil. Some furnaces are equipped with a special chamber from which air can be removed and replaced with an inert gas.
- The temperature of steel can be determined using a pyrometer or by comparing the color of the heated steel to a color chart supplied by the steel company.
- In addition to the traditional metals, rare metals (such as scandium, yttrium, cerium, europium, lanthanum, and holmium), honeycomb panels, and composite materials are seeing increased use in manufacturing.
- The simplest method of case hardening is carburizing, in which the low-carbon steel is covered with a case hardening compound and heated.
- Hardness testing ensures that the metal has been given the proper degree of hardness for its intended use. Brinell, Rockwell, Webster, and scleroscope hardness testers can be used to measure hardness. Each type of hardness test has its own scale, but can be converted to other scales.

Review Questions

Answer the following questions using the information provided in this chapter.

1. Heat treatment can be used to _____.
   A. change a ferrous metal to nonferrous metal
   B. increase or decrease the hardness of a metal
   C. change the thermal conductivity of a metal
   D. All of the above.
   E. None of the above.
2. What are the general steps in heat-treating metal?
3. Water, brine, oil, blasts of cold air, or liquid _____ can be used as a quenching medium.
4. Carbon steels are classified by the percentage of carbon they contain, expressed in _____ or hundredths of a percent.

5. List four metals other than steel that can be heat-treated.
6. The _____ heat treatment process reduces the stress that has developed in parts that have been welded, machined, or cold worked during processing.
7. The _____ heat treatment process reduces the hardness of metal.
8. The _____ heat treatment process refines the grain structure of steel to improve its machinability.
9. The _____ heat treatment process is used to create a medium-hard surface on high-carbon or alloy steels.
10. The _____ heat treatment process is used to create a hard outer surface on low-carbon steel while keeping the inner portion relatively soft and tough.
11. Tempering a piece of hardened steel makes it _____.
    A. harder
    B. tougher
    C. more brittle
    D. All of the above.
    E. None of the above.
12. Some _____ alloys age-harden at room temperature and must be refrigerated to remain soft and ductile while they are being worked.
13. Although _____ can be heat-treated, special facilities are required because it is a reactive metal.
14. Water is mainly used as the quenching medium when _____.
    A. moderate hardness is desired in plain carbon steel
    B. moderate hardness is desired in alloy steel
    C. maximum hardness is desired in plain carbon steel
    D. All of the above.
    E. None of the above.
15. List three characteristics of electric heat treatment furnaces.
16. A(n) _____ is used to measure and monitor the high temperatures needed in heat treatment.
17. The most commonly used technique for testing the hardness of metal is to press a steel ball or _____ probe into the metal under a specific load.
18. List three types of commonly used hardness testers.
19. A _____ determines the hardness of a metal based on the distance that a diamond hammer rebounds after it is dropped on the metal.
    A. Rockwell hardness tester
    B. Brinell hardness tester
    C. Webster hardness tester
    D. scleroscope
20. List three furnace-related safety precautions.

CHAPTER 30

Metal Finishing

Learning Objectives

After studying this chapter, you will be able to:

- Describe how the quality of a machined surface is determined.
- Explain why the quality of a machined surface has a direct bearing on production costs.
- Identify organic coatings and describe how they are applied.
- List common inorganic coatings and summarize how they are applied.
- Describe different methods of applying metal coatings.
- Explain various methods of deburring.

Technical Terms

anodizing
chemical blackening
electroplating
lay
metal finish
metal spraying
microinches
micrometers
roller burnishing
surface roughness standards
vitreous enamel
waviness

The term ***metal finish*** refers to the degree of smoothness or roughness remaining on the surface of a part after it has been machined. A machined surface has geometric irregularities that are produced by the cutting action of the tool, **Figure 30-1**.

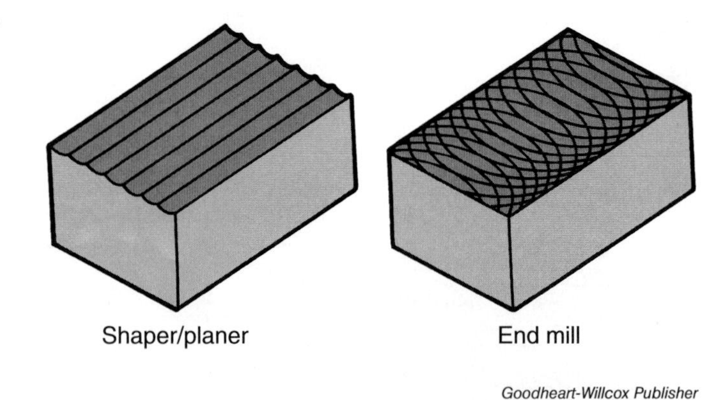

Goodheart-Willcox Publisher

Figure 30-1. Each type of cutting tool leaves characteristic markings.

30.1 Quality of Machined Surfaces

At one time, the quality of a machined surface was noted by the symbol "*f*". This was not based on specific standards. Therefore, the engineer or drafter included explanatory notes on the drawing. These notes indicated the desired quality of general surface finish, such as rough grind, smooth turn, or surface grind.

The technique left much to be desired, since each machinist interpreted the specifications differently. Often, the piece was better finished than it had to be, increasing its production cost. The problem reached such proportions that in the early 1940s, the standards associations of Canada, Great Britain, and the United States developed tentative surface roughness or texture standards.

The terms and ratings of ***surface roughness standards***, or texture standards, detail how surfaces produced by machining, grinding, casting, molding, forging, or similar processes are to be measured and communicated. These standards are not concerned with luster, appearance, color, corrosion resistance, wear resistance, hardness, and the many other characteristics that may affect a part's suitability for specific applications.

The standards also do not define the different degrees of surface roughness and waviness that are suitable for specific purposes, nor do they address how the irregularities are to be made. The standards deal only with the height, width, and direction of surface irregularities, since these are of practical importance in specifications.

The current surface roughness standards arrive at roughness values by mathematically averaging the size of irregularities on a surface. The values are given in ***microinches*** (millionths of an inch, shown as XX μin.) or ***micrometers*** (millionths of a meter, shown as XX μm). With established standards, a universal set of numbers and symbols indicating surface roughness or texture are now used on drawings and in specifications. See **Figure 30-2** and **Figure 30-3**.

Symbol	Description
	Basic surface roughness/texture symbol. Surface may be produced by any method.
	Material removal by machining required. Horizontal bar indicates that material removal by machining is required. Material must be provided for that purpose.
.187	*Material removal allowance.* Number indicates amount of material that must be removed in inches/millimeters. Tolerances may be added.
	Material removal prohibited. Circle in vee indicates that surface must be produced by processes such as casting, forging, hot finishing, cold finishing, powder metallurgy, or injection molding without subsequent removal of material.
	Surface texture symbol. To be used when any surface characteristics are specified above horizontal line or to right of symbol. Surface may be produced by any method.

Goodheart-Willcox Publisher

Figure 30-2. Surface roughness or texture symbols.

Roughness Height Rating		Surface Description	Process
Microinch	Micrometer		
1000	25.2	Very rough	Saw and torch cutting, forging, or sand casting
500	12.5	Rough machining	Heavy cuts, coarse feeds in turning, milling, and boring
250	6.3	Coarse	Very coarse surface grind, rapid feeds in turning, planing, milling, boring, and filing
125	3.2	Medium	Machining operations with sharp tools, high speeds, fine feeds, and light cuts
63	1.6	Good machine finish	Sharp tools, high speeds, extra fine feeds, and cuts
32	0.8	High grade machine finish	Extremely fine feeds and cuts on lathe, mill, and shaper required. Easily produced by centerless, cylindrical, and surface grinder
16	0.4	High quality machine finish	Very smooth reaming or fine cylindrical or surface grinding, or coarse hone or lapping of surface
8	0.2	Very fine machine finish	Fine honing and lapping of surface
2–4	0.05 0.1	Extremely smooth machine finish	Extra fine honing and lapping of surface

Goodheart-Willcox Publisher

Figure 30-3. Roughness values. When used on a drawing, a number indicates the roughest surface in microinches or micrometers that is acceptable for that specific application.

The same standards that address surface roughness also address other surface conditions and give values for them. ***Waviness*** describes the presence of smoothly rounded peaks and valleys caused by tool and machine vibration and chatter, **Figure 30-4**. Waviness is of greater magnitude than roughness. It is measured with reference to a nominal or geometrically perfect surface.

Waviness is specified by the maximum allowable peak-to-valley height in inches or millimeters. It is measured using a sensitive dial indicator with a ball contact 0.06″ (1.5 mm) in diameter. **Figure 30-5** shows how acceptable waviness tolerances are specified on drawings, in reports, and as specifications.

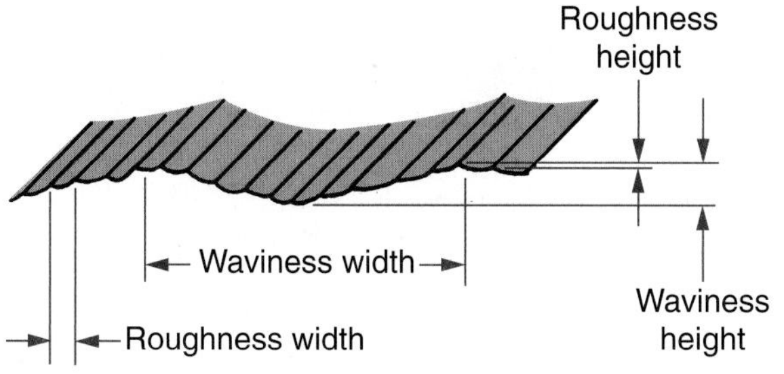

Roughness height

Waviness width

Waviness height

Roughness width

Goodheart-Willcox Publisher

Figure 30-4. How surface waviness is measured. Note the difference in magnitude between waviness and roughness.

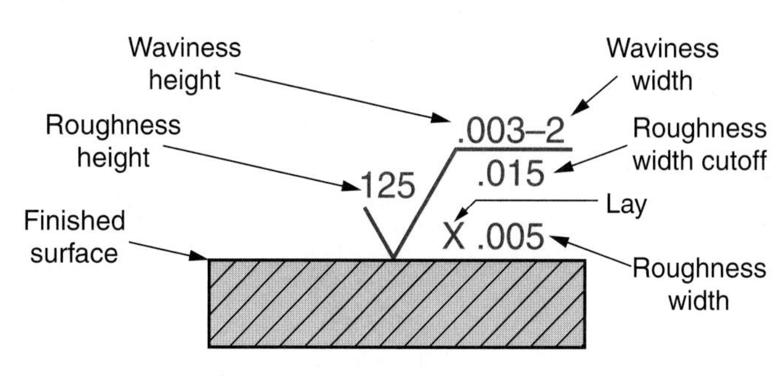

Goodheart-Willcox Publisher

Figure 30-5. On drawings, symbols and numbers show roughness, waviness, and lay. They specify finishes required on a surface.

Lay is the term that describes the direction of the predominant tool marks, grain, or pattern of surface roughness. See **Figure 30-6**.

30.1.1 Degrees of Surface Roughness

Milling and turning can produce surface finishes in the order of 125 μin. to 8 μin. (3.2 μm to 0.2 μm). Grinding, depending on the coarseness of the wheel and feed rate, has a range of 63 μin. to 4 μin. (1.6 μm to 0.1 μm).

Lapping produces the smoothest finish on a production basis. It is used by automotive manufacturers to produce mating surfaces that are flat and smooth enough to form gasketless oil-tight seals in automatic transmissions and other applications. Surfaces are as fine as 2 μin. to 3 μin. (0.05 μm to 0.07 μm). See **Figure 30-7** and **Figure 30-8**.

Several tools have been developed to measure surface quality. The most accurate is the profilometer, which measures and amplifies surface roughness electronically, **Figure 30-9**. Surface roughness is usually read on a meter. A printout of the results can be provided by some models.

Sunnen Products Company

Figure 30-7. Computerized vertical honing machine. It is designed for precision honing of bores in such applications as small engine blocks, air compressor cylinders, valve bodies, and aircraft cylinders. The unit has a microprocessor-based control system that enables the operator to control all aspects of the production cycle.

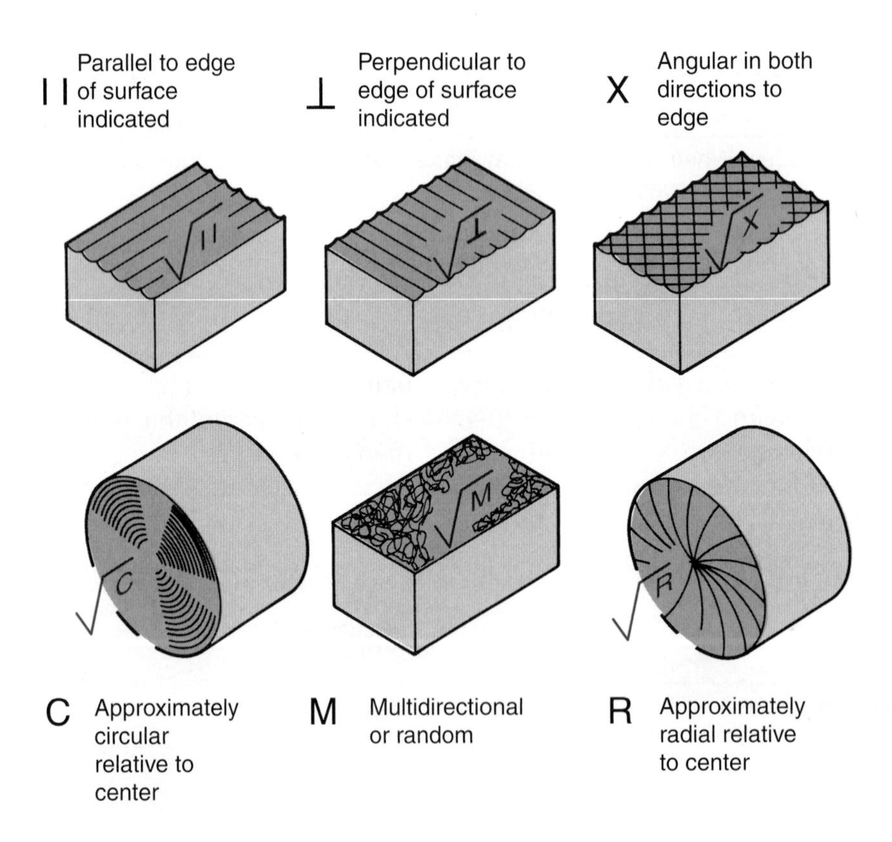

Goodheart-Willcox Publisher

Figure 30-6. The lay symbols and their meanings. Lay symbols are located beneath the horizontal bar on a surface texture symbol.

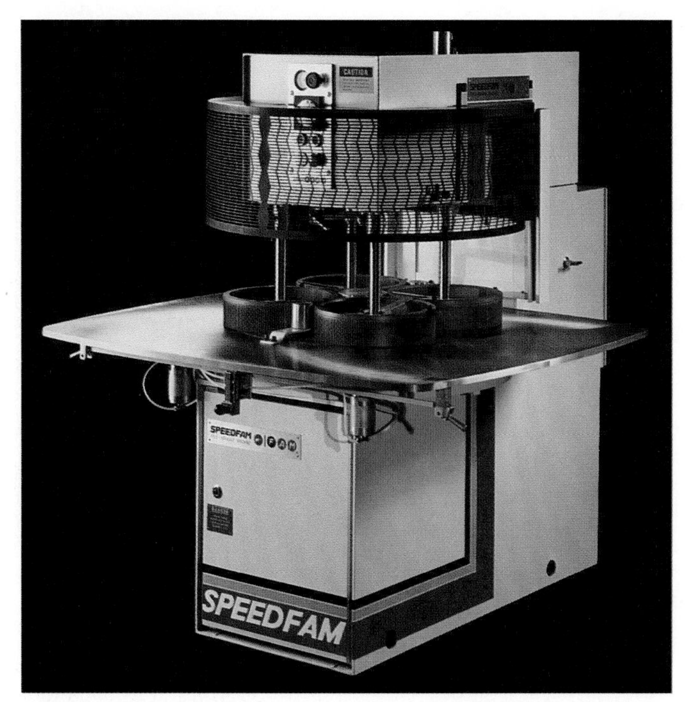

A

B

SpeedFam; Engis Corporation

Figure 30-8. Lapping machines. A—Free abrasive lapping machines provide exceptional accuracy with part thickness accuracies of ±0.00015″, flatness of 0.00001″, and a fine surface finish. B—A twin-plate diamond lapping system. It laps or polishes both sides of a workpiece simultaneously and produces parallel parts with uniform edge-to-edge flatness. The unit uses a menu-driven microprocessor controller.

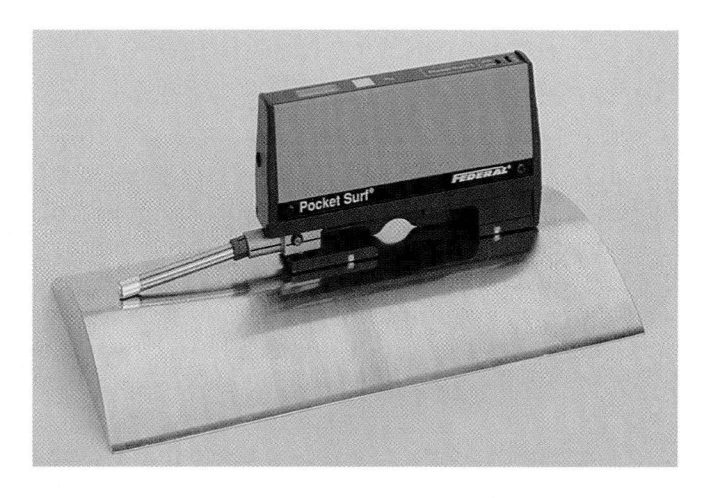

Federal Products Co.

Figure 30-9. Handheld surface roughness gage (profilometer) with a digital display that displays the measured value in microinches or micrometers. It has a variety of interchangeable probes for different applications.

A microfinish surface comparator, or surface roughness gage, is a visual comparison tool. It contains sample specimens of the various degrees of surface finishes that conform to values established by the American National Standards Institute (ANSI Y14.36M). The part being inspected is placed next to the gage, **Figure 30-10**. The finish on the part is determined by finding the corresponding finish on the gage.

Goodheart-Willcox Publisher

Figure 30-10. A surface roughness gage is being used to visually compare surface of a machined block. Since it is often difficult to check machined surfaces visually, a test by feel is also used.

30.1.2 Economics of Machined Surfaces

The quality of a machined surface has a direct bearing on production costs: the finer the finish, the higher the cost. If costs are to be kept within acceptable limits, care must be taken to meet, but not exceed, the required specifications.

The chart in **Figure 30-11** illustrates the range of surface finishes that can be achieved by the various machining processes. Values are relative and will vary depending on the condition of the machine, sharpness of the cutting tool, and the material being machined.

30.2 Other Metal Finishing Techniques

While the quality of the machined surface is of paramount importance in the machining of metal, other finishing techniques may be used for one or more of the following reasons:

- **Appearance.** A proper finish affects a product's salability and is more important than often realized. A product is much more attractive with a proper finish than when left unfinished. Finishes can also be applied to make the product blend into its surroundings.

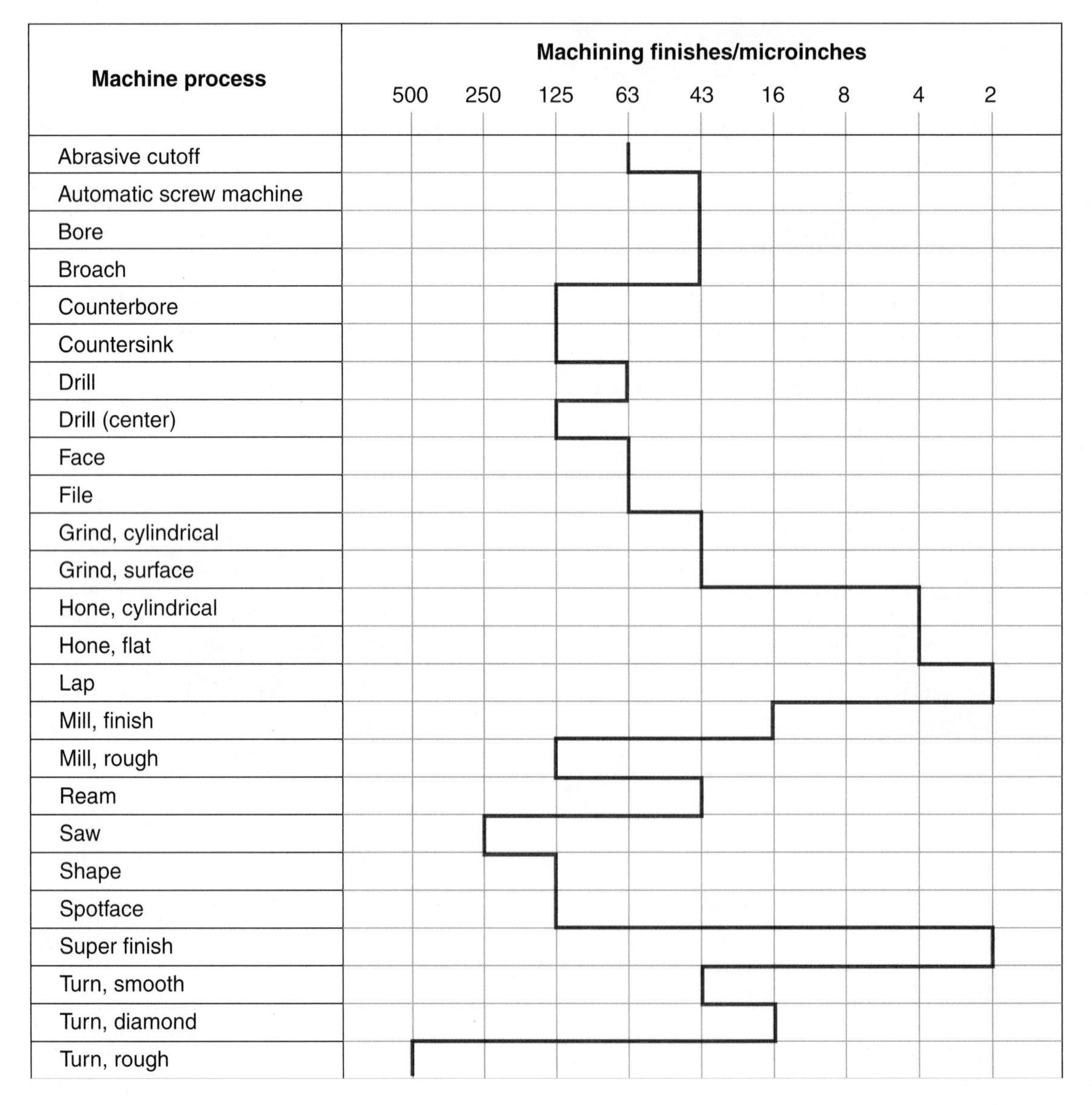

Goodheart-Willcox Publisher

Figure 30-11. Typical surface finishes that result from various machining processes. The finer the finish (the lower the roughness value in microinches), the higher the cost of achieving it.

- **Protection.** All metals are affected to some degree by contaminants in the atmosphere and by abrasion. A proper finish helps protect materials from exposure to contaminants, reducing corrosion and wear.
- **Identification.** Some manufacturers use distinctive finishes on their products that make their products stand out from their competition.
- **Cost reduction.** An expensive metal can be coated onto a less costly metal, or other material. The finished part will be less costly than if it were made from the more costly metal. The part will retain the properties of the higher priced metal. Silver and gold, for example, are often applied to steel to improve electrical conduction, heat distribution properties, solderability, and appearance.

Figure 30-12 shows the many finishes that can be applied to aluminum. Regardless of the finish that is to be used, the surfaces must be cleaned of all contaminants before the finish can be applied. Oxidation can be removed mechanically (sandblasting or burnishing), or chemically (etching with an acid). Solvents are often used to remove oil and grease.

Special methods have been devised to clean complex machined castings. They are needed to remove loose casting sand, metal chips, and other foreign matter trapped in recesses and passages during manufacturing operations. One method is to dip the part in a solvent tank. The flow and agitation of the part will remove foreign matter from deep cavities, holes, and recesses of the casting. The final operation of the cleaning cycle is the application of rust-inhibiting chemicals.

30.2.1 Organic Coatings

A wide range of finishes fall into the category of organic coatings: paints, varnishes, lacquers, enamels, various plastic-base materials, and epoxies in both clear and pigmented (color) formulas. With the exception of the epoxies, they are set by the evaporation of their solvents. This may be accomplished by air drying or baking. Epoxies require the addition of a catalyst or hardener to set. Organic finishes are seldom applied to machined surfaces.

A primer is often required to ensure satisfactory bonding between the metal and the finish. Some types of castings may need a filler to smooth out the rough cast surface.

Organic coatings are applied in the following ways:

- **Brushing.** With this method, a brush applies the finish by direct contact with the part. At one time, brushing was the only way to apply finishes with any degree of control.
- **Spraying.** This method atomizes the finishing material and carries it to the work surface by air pressure. Spraying is easily adapted to mass-production techniques, **Figure 30-13**. For small jobs, small pressurized spray cans, offered in a wide range of colors, are available.
- **Roller coating.** A roller applies the finish by direct contact with the part. This method can be used only on flat surfaces. It is a low-cost technique that can be mechanized.
- **Dipping.** The part is submersed into the finish, removed, and allowed to dry. Dipping is widely used today by the automotive industry to apply body primer and rustproofing, **Figure 30-14**. The coating is dried by warm air.

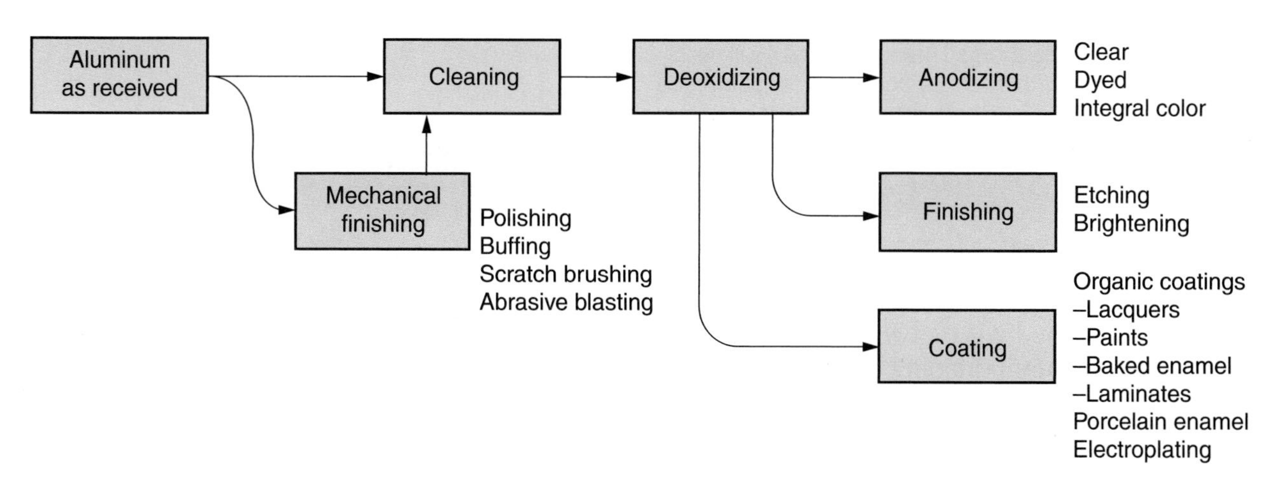

Figure 30-12. Various types of finishes that can be applied to aluminum.

A

B

Corepics VOF/Shutterstock.com; Ford Motor Co.

Figure 30-13. With spraying, the finishing material is atomized and carried to the work surface by air pressure. A—Many finishing materials and solvents pose a serious health hazard when atomized. This worker is wearing the proper protective gear, including rubber gloves, air-supplied respirator, and a protective suit. B—Because of the health hazards and repetitive nature of the task, mass-production spray painting is often automated.

Ford Motor Co.

Figure 30-14. Electrically charged particles of paint will evenly coat all bare surfaces of this vehicle body as it is submerged into a tank of primer. Because the paint will be drawn into every crevice, this technique provides better corrosion prevention than spraying.

- **Flow coating.** The part is flooded with the finish and allowed to drain while held in an atmosphere saturated with solvent vapor. Drying is thus delayed until draining is complete.

30.2.2 Inorganic Coatings

Several well-known finishing techniques, such as anodizing, glass coating, chemical blackening, fall into the inorganic coatings category. ***Anodizing*** is the best known of this type finish. This process forms a protective layer of aluminum oxide on aluminum parts, **Figure 30-15**.

There are three classes of anodizing: ordinary anodizing, hard coat anodizing, and electrobrightening. The basic procedure for producing all three is the same, **Figure 30-16**.

The anodized coating of aluminum oxide forms by reaction of the aluminum with an electrolyte when the aluminum is used as the anode. Oxygen is liberated at the surface of the aluminum and an oxide forms. Ordinary anodizing leaves an oxide layer of 0.0001″ to 0.0006″ (0.003 mm to 0.015 mm) thick on the surface of the aluminum. Hard coat anodizing produces an oxide layer about 10 times thicker than that produced by ordinary anodizing.

Evgany Korshenkov/Shutterstock.com

Figure 30-15. Anodized aluminum fitting.

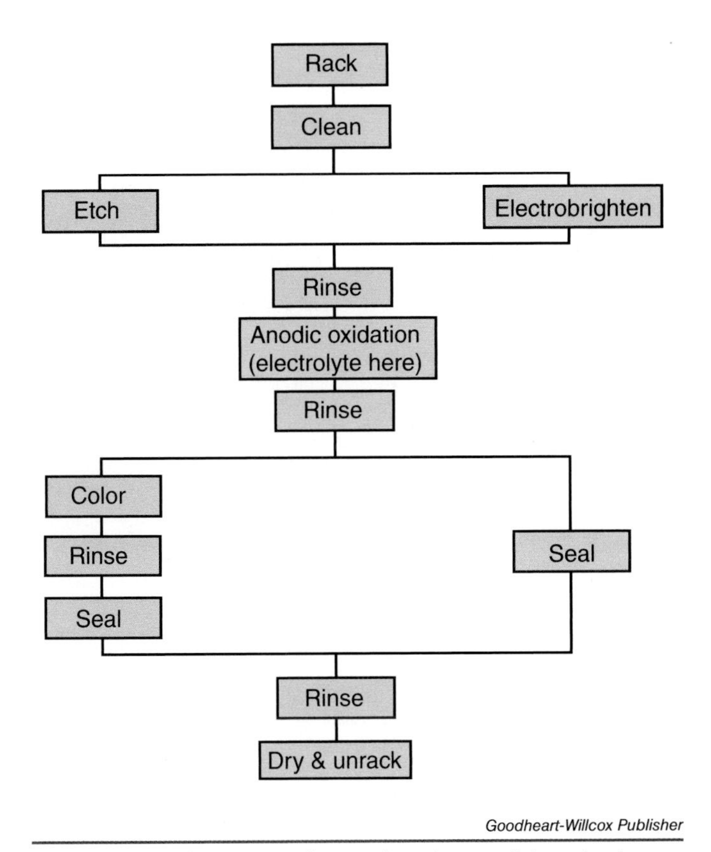

Figure 30-16. Sequence of operations for anodizing aluminum.

This gives a hard coat anodized finish superior resistance to abrasion, erosion, and corrosion. However, the strength of the material is slightly reduced.

Electrobrightening is an anodizing process in which the oxide film is dissolved by electrolyte at about the same rate that it is formed. The process leaves a smooth, bright, mirror-like finish. The anodized coating can be dyed in a wide range of colors. The color becomes part of the surface of the metal.

Vitreous enamel, or porcelain, is a glass coating that can be fused to most metals, including steel sheet and cast iron surfaces. It forms an extremely hard coating that is smooth and easy to clean, **Figure 30-17**. Vitreous enamel is available in many colors. It can be applied to most metals that remain solid at the firing (melting) temperatures of the coating material. The finish is applied as a powder (frit) or as a thin slurry (slip). After drying, the material is fired at about 1500°F (815°C) until it fuses to the metal surface.

Chemical blackening is a finishing process that produces a black oxide coating on the surfaces of ferrous metals, **Figure 30-18**. It chemically bonds with the metal surface, instead of merely adhering to it like paint or other applied finishes.

J. Waldron/Shutterstock.com

Figure 30-17. Vitreous enamel, or porcelain, is a glass coating that is commonly fused to steel or cast surfaces. It is an extremely hard coating that is smooth, easy to clean, and heat resistant. These qualities make it a popular finish for a wide range of products, including the stove and teapot shown here.

mrHanson/Shutterstock.com

Figure 30-18. Chemical blackening is a surface finishing technique that can be applied to steel parts and products. It bonds with the metal's surface, rather than coating it; there is no surface buildup. The bottom of the drill chuck and the chuck key have been chemically blackened.

Blackening is used as a finish on precision parts that cannot tolerate the added dimensional thickness of paint or plating (the blackening is 0.00003″ or 0.0008 mm thick). It will not chip or peel.

The blackening process can be carried out at room temperature and takes about 15 minutes to complete. Operators must wear eye protection and gloves that provide chemical protection and a plastic or rubber apron to protect clothing.

Chemical blackening offers several advantages:

- The black finish enhances the appearance of many items. Because the cost is so low, it is possible to finish parts that were previously left unfinished.
- It aids in protecting the machined surface against humidity and corrosion.
- It reduces glare from the material. Using blackened tools and other machine parts results in less eye fatigue in the operator, improving safety.
- It improves abrasion resistance.
- It improves the adhesion qualities of the treated metal. Paint and other finishes applied over blackening take hold faster, adhere more strongly, and last longer.

There are two drawbacks to the process, both involving environmental treatment and disposal laws:

- The water used in the process must be filtered and treated before it can be discharged as wastewater.
- Chemical residue from the process must be handled as toxic waste.

Baloncici/Shutterstock.com

Figure 30-19. The cutting edges of several of these cutting tools are coated with titanium nitride. The titanium nitride has a golden appearance. It is extremely hard and improves a tool's cutting ability and wear resistance.

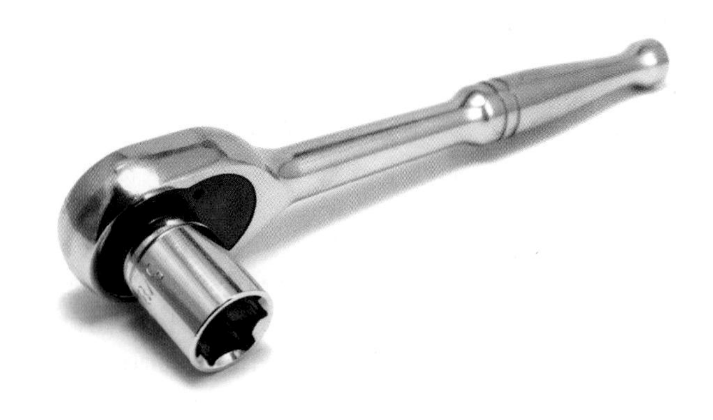

Allison Herreid/Shutterstock.com

Figure 30-20. Many hand tools are chrome plated. Chrome plating improves the wear- and corrosion-resistance of the tool, and also gives the tool an attractive luster.

30.2.3 Metal Coatings

With the exception of electroplating, which can be used on a number of materials, metal coatings are applied primarily to steel. Metal coatings adhere tightly to the surface and provide protection from corrosion, improved wear resistance, enhanced appearance, or a combination of these qualities, **Figure 30-19**. The most commonly used metal coating processes include electroplating, metal spraying, and detonation gun coating.

Electroplating is a process in which a layer of one metal is deposited on the surface of another metal by the use of an electrical current. Practically any metal can be used as a coating. Chrome plating is one common example of electroplating. See **Figure 30-20**.

With electroplating, coating thickness can be closely controlled. Unlike many other metal coating processes, electroplating can deposit coatings that improve both wear-resistance and appearance. When a product is machined, allowance must be made for the additional metal thickness that will be deposited by the plating process.

In ***metal spraying***, a metal wire or powder is heated to its melting point and sprayed by air pressure to produce the desired coating on the work surface. See **Figure 30-21**.

Most inorganic materials that can be melted without decomposition can be applied by spraying.

Photo:Sulzer

Figure 30-21. The plasma spray coating is one of several metal spraying techniques used in industry. In this photo, a hard metallic surface is being applied to the stator case of an aviation turbine.

Flame-sprayed coatings can be applied to build up worn or scored surfaces so they can be remachined to the required size. Superhard coatings can be sprayed when abrasion-resistant surfaces are needed, **Figure 30-22.**

The detonation gun coating, or D-Gun®, process is another technique for depositing a metallic coating on a workpiece. The process was invented and developed by the Union Carbide Corporation. The detonation gun is essentially a water-cooled barrel several feet long and about 1″ (25 mm) in diameter. It has valves for introducing gases and material to be sprayed. See **Figure 30-23.**

A carefully measured mixture of gases (usually oxygen and acetylene) is fed into the barrel, along with a charge of coating material in powder form. The powder has a particle size of less than 100 microns. A spark ignites the gases. This heats and melts the powder and creates a high-temperature, high-velocity gas stream. The stream of gas carries the molten material to the surface to be coated. A pulse of nitrogen purges the barrel after each detonation; the process is repeated many times a second.

Each detonation, called a "pop," results in a circle of coating material a few microns thick and about an inch in diameter. The molten coating droplets quickly solidify on the workpiece surface. The complete coating consists of many overlapping "pops" that build up to the required thickness.

The process can be used to apply coatings with very high melting points to fully heat-treated parts without danger of changing the metallurgical properties or strength of the part and without danger of thermal distortion. The application process can also be fully automated.

Almost any material that can be melted without decomposing can be sprayed. Coatings include pure metals and metallic alloys such as nickel and nichrome, tungsten carbide, and ceramics. These coatings are used in many applications, especially where there is need to combat wear (abrasive, erosive, or adhesive), often in very corrosive environments. Refer to **Figure 30-24.**

Ford Motor Co.

Figure 30-22. This plasma transferred wire arc gun applies a hard, low-friction coating to the inside of engine cylinders. The gun is inserted into each cylinder, rotated to evenly apply the material to the cylinder walls, and then withdrawn. After the cylinders are coated, they are machined to the proper size.

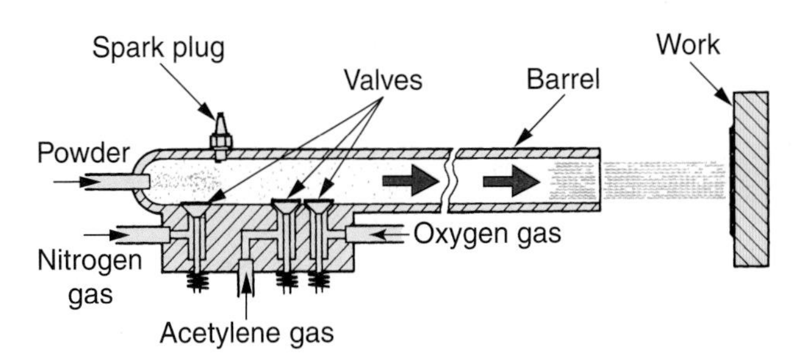

Union Carbide Corp.

Figure 30-23. Cross section of a detonation gun used to apply metal coatings. The oxygen and acetylene valves open to allow explosive gases into the barrel and then seal the gas passages during the detonation phase. After detonation, the nitrogen valve opens briefly to allow pulse of nitrogen gas to clear the barrel before the process repeats.

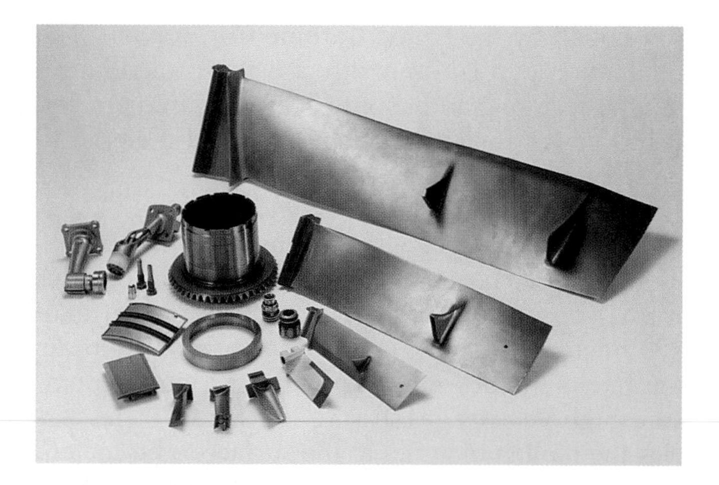

Union Carbide Corp.

Figure 30-24. Some examples of metal and ceramic coating that have been applied to gas turbine engine parts by the detonation gun coating process.

30.2.4 Mechanical Finishes

Buffing is a power polishing operation. Buffing wheels are attached to a buffing lathe or bench grinder. The wheels are charged with different grades of abrasives that remove scratches and polish the metal's surface to a high luster.

SAFETY NOTE

Unless the buffing lathe is equipped with a high efficiency dust-collection system, an approved filter mask must be worn during any buffing operation.

Diamond dust and air-powered, handheld polishing units are used to polish the hardened steel dies used for die casting or the injection molding of plastics. See **Figure 30-25**.

Roller burnishing, **Figure 30-26**, is a cold-working operation. It removes no metal, but rather uses rollers to compress or flatten the peaks of a metal's surface into the valleys. With the process, no honing or grinding is necessary.

Both ID and OD burnishing is possible and can be done at speeds and feeds comparable to cutting tools. A roller burnishing tool for finishing the inside surfaces or bores is shown in **Figure 30-27**. The tool can size, finish, and work-harden metal parts to tolerances as close as 0.0002″ (0.005 mm), and surface finishes as fine as two to four microinches.

NSK America

Figure 30-25. Polishing a steel die with a diamond dust to produce a mirror-like finish.

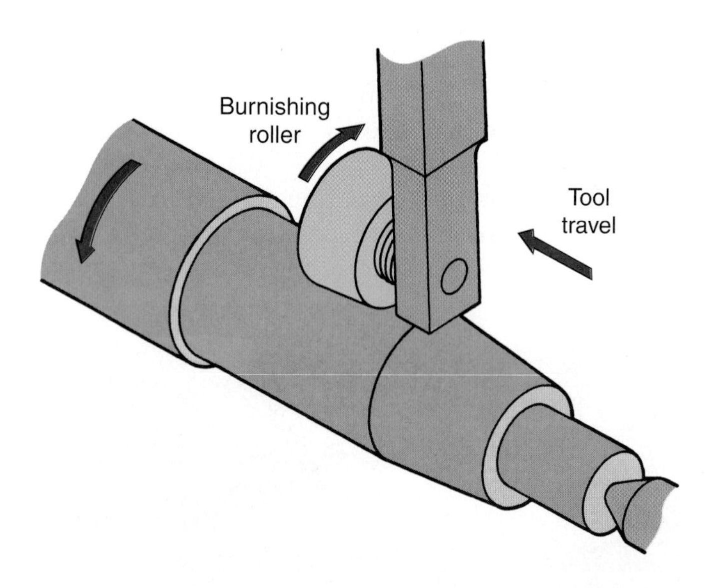

Goodheart-Willcox Publisher

Figure 30-26. Roller burnishing is a cold-working operation that gives a high quality finish to the metal. It flattens the peaks of a metal's surface into the valleys.

30.2.5 Deburring Techniques

Power brushing is frequently used to remove burrs from machined surfaces. Wire or fiber brushes replace the buffing wheels on the buffing lathe for flat surfaces or installed in a drill press to deburr bores, **Figure 30-28**. In removing burrs, the wheels produce a satin sheen on the brushed surface, **Figure 30-29**.

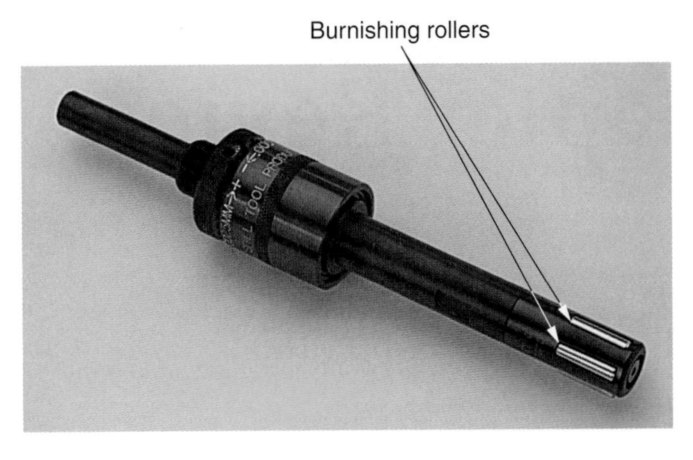

Cogsdill Tool Products, Inc.

Figure 30-27. Roller burnishing tool for finishing holes. It sizes, finishes, and work-hardens the metal. The process removes no metal. Each hole size requires a different roller burnishing tool.

Weiler Corporation

Figure 30-28. Drill press equipped with a wire brush to remove burrs from bores. Always wear a dust mask when performing a deburring, brushing, or buffing operation.

Weiler Corporation

Figure 30-29. A wire brush installed in a drill press is being used to deburr the edges of bores through a metal plate.

Hand deburring of small holes and intricate parts is tedious and expensive so other methods are preferred for high-volume work. One technique known as abrasive flow machining, or extrude-honing, makes use of silicon putty permeated with finely divided abrasive particles. The silicon putty is forced into and through the part to be deburred. As the putty flows through passages in the part, the abrasive grains remove any burrs. Abrasive flow machining produces a very uniform finish.

Chapter Review

Summary

- Surface roughness standards detail the method that the height, width, and direction of surface irregularities are determined and how they are to be communicated in drawings.
- Different machining processes create different degrees of roughness in the machined surface. Generally, the finer the finish, the greater the production cost to achieve that finish.
- Finishes can perform any of the following four functions, or any combination of the four: enhance appearance, provide protection against wear and corrosion, make the part more identifiable, and reduce production cost of the part.
- Organic coatings include paints, varnishes, lacquers, enamels, plastic-based materials, and epoxies. They are applied by brushing, spraying, roller coating, dipping, or flow coating.
- Inorganic coatings include finishes produced by anodizing, applying vitreous enamel, and chemical blackening.
- Common processes for applying metal coatings include electroplating, metal spraying, and detonation gun coating.
- Other finishing processes include buffing and roller burnishing.
- Methods of deburring include power brushing with wire or fiber brushes and abrasive flow machining (extrude honing).

Review Questions

Answer the following questions using the information provided in this chapter.

1. At one time, the "*f*" symbol was used on drawings to designate a(n) _____ surface.
2. Why was the use of the "*f*" symbol discontinued?
3. Surface roughness is now measured in _____ or _____.
4. The term "waviness" describes _____.
   A. very rough surfaces
   B. smoothly rounded undulations caused by tool and machine vibration and chatter
   C. scratches on the machined surface
   D. All of the above.
   E. None of the above.
5. Lay is another surface finish condition. What does it mean?
6. While the quality of a machined surface is of paramount importance in the machining of metal, other finishing methods are used in the machine shop. They are used for one or more of the following reasons. Explain each.
   A. Appearance.
   B. Protection.
   C. Identification.
   D. Cost reduction.
7. Regardless of the type of finish applied to aluminum, the surface must be thoroughly _____ of all contaminants.
8. Paints, lacquers, and enamels are in the family of _____ coatings.
9. List the five ways used to apply the finishes in Question 8.
10. List three types of anodizing.
11. What is electroplating?

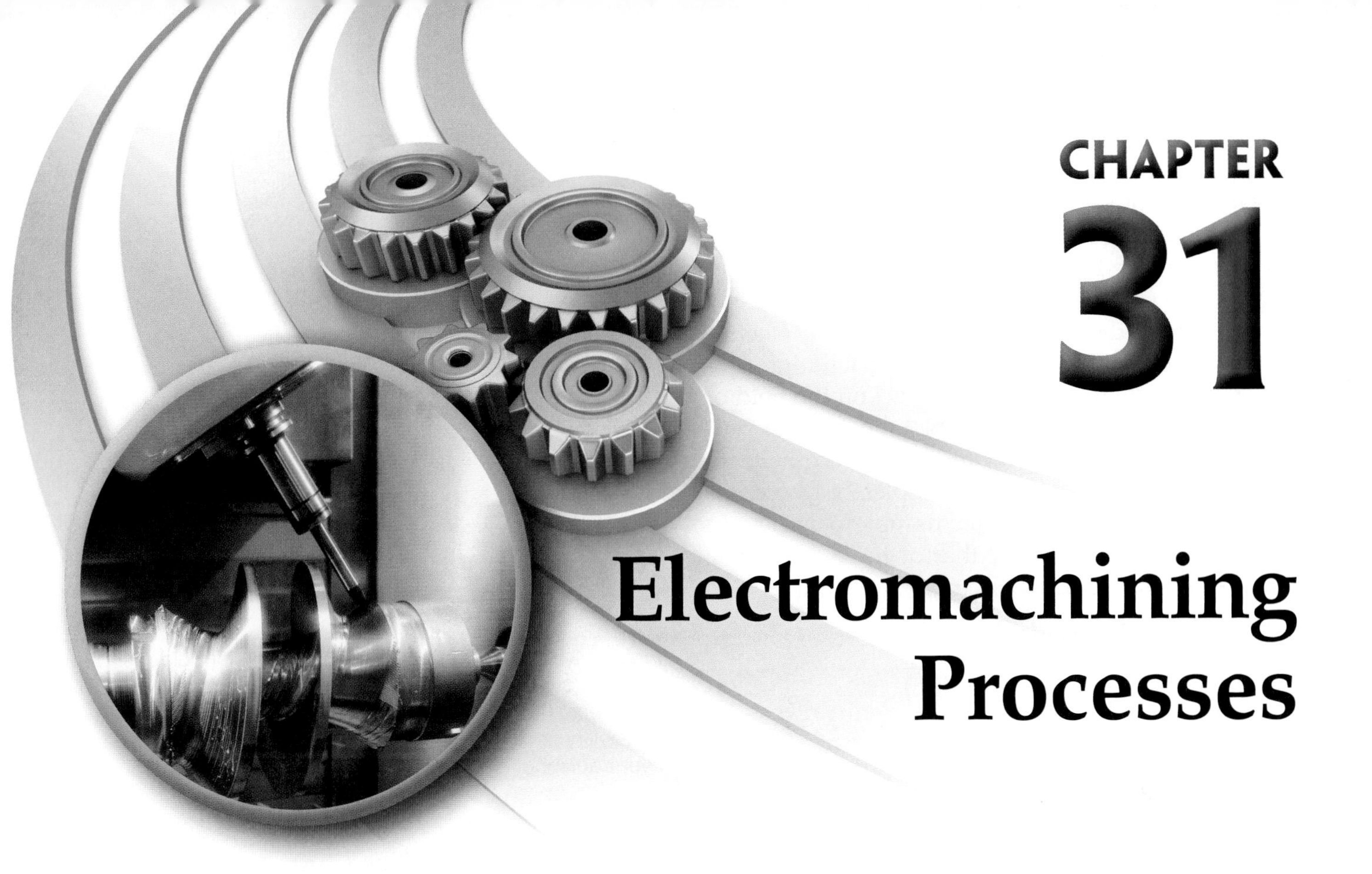

CHAPTER 31

Electromachining Processes

Learning Objectives

After studying this chapter, you will be able to:

- Explain the advantages and disadvantages of the electromachining processes.
- Describe electrical discharge machining.
- Explain electrical discharge wire cutting.
- Summarize the small hole EDM drilling process.
- Describe electrochemical machining.

Technical Terms

dielectric fluid
electrical discharge machining (EDM)
electrical discharge wire cutting (EDWC)
electrochemical machining (ECM)
electrode
electromachining
servomechanism
small hole EDM drilling

The most notable advantage of ***electromachining*** is that mechanical forces have no influence on the processes. The tool is not in direct physical contact with the material being removed, as is the case with conventional machining techniques. Rather, electrical energy is applied to remove metal through erosion.

Electrical discharge machining and electrochemical machining are two electromachining processes that have made a great impact on the field of metalworking and machining. Notably, neither process produces a chip as metal is removed. Instead, the particles are disposed of completely by vaporization or are reduced to microscopic particles. In order to be machined by either of these processes, the metals must conduct electricity.

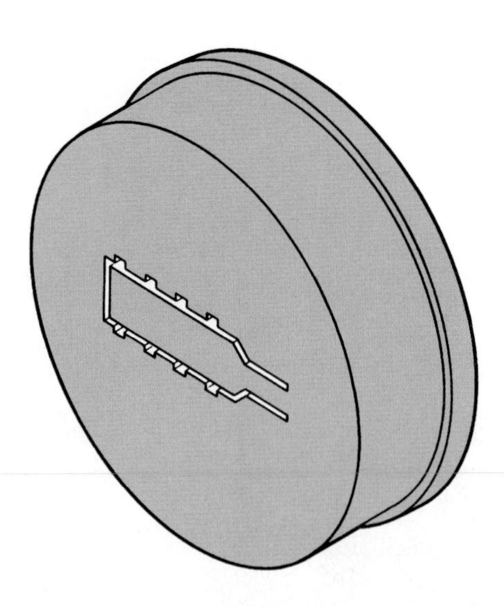

Goodheart-Willcox Publisher

Figure 31-1. Die blanks, like this extrusion die used to produce storm window frames, can be machined by EDM after being heat-treated. The job can be done quickly, with no possibility of die warping, and at a considerable cost savings over traditional machining techniques.

SAFETY NOTE

Never attempt to operate any type of machining equipment while your senses are impaired by medication or other substances.

31.1 Electrical Discharge Machining (EDM)

Electrical discharge machining (EDM) is a process that can work tough, hard, fragile, or heat-sensitive metals to close tolerances. The same qualities would make the metal difficult to machine by conventional "chip-making" techniques.

Die blanks, for example, can be worked after heat-treating, **Figure 31-1**. This eliminates the warping and distortion that frequently occur when a finished die is heat-treated. Superhard metals are easily worked to tolerances as close as 0.0002″ (0.005 mm). Surface finishes can be varied from very rough to almost mirror-smooth. "Washed-out" (worn) dies can also be reworked in the heat-treated state.

31.1.1 EDM Principle

In a gasoline engine, sparking (or arcing) takes place at the spark plug gap when the ignition coil fires to ignite the fuel mixture. The spark plug's electrodes are gradually eroded by the action of the electric arcs. This is the basis of EDM. See **Figure 31-2**.

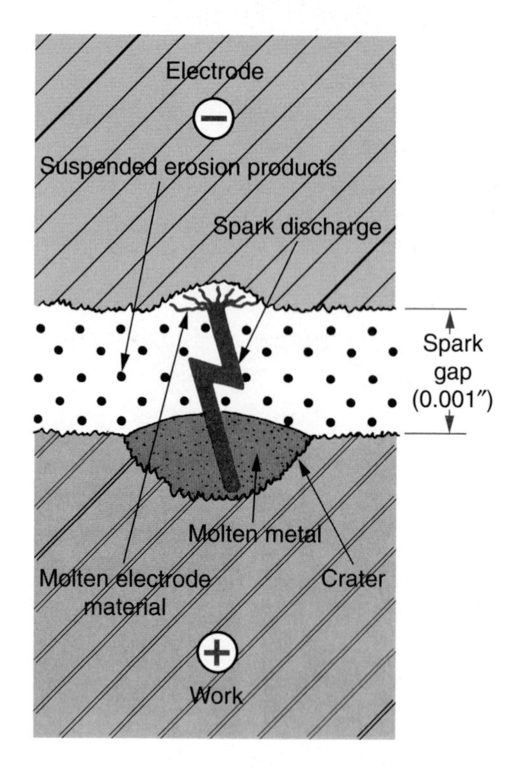

Goodheart-Willcox Publisher

Figure 31-2. How the EDM process works. A spark (or arc) from an electrode causes the work to erode.

An electrical discharge machine, **Figure 31-3**, is composed of the following parts.

- The power supply is needed to provide direct current and to control voltage and frequency.

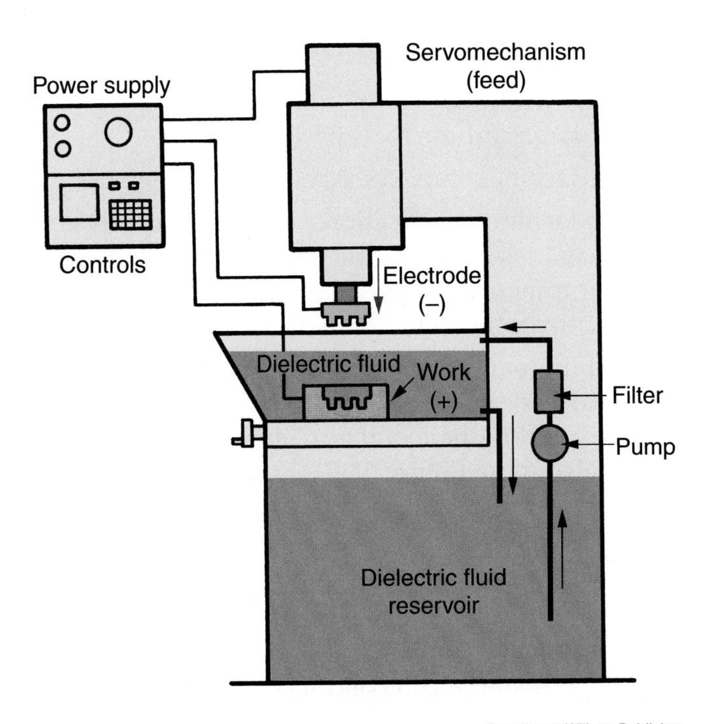

Goodheart-Willcox Publisher

Figure 31-3. Electrical discharge machine.

- The ***electrode*****, Figure 31-4**, can be compared to the cutting tool on a conventional machine tool. Graphite has become the material most commonly used for EDM electrodes. Because of the dust produced when machining graphite electrodes, special machine tools with enclosed cutting zones and powerful suction systems have been developed, **Figure 31-5**.
- The ***servomechanism*** (drive unit) is used to accurately control electrode movement and to maintain the correct distance between the work and the electrode as machining progresses.
- The ***dielectric fluid***, usually a light mineral oil, is used to form a nonconductive barrier between the electrode and the work at the arc gap.

The servomechanism maintains a thin gap of about 0.001″ (0.025 mm) between the electrode and the work. The electrode and the work are submerged in the dielectric fluid, **Figure 31-6**. When the voltage across the gap reaches a point sufficient to cause the dielectric fluid to break down, a spark occurs. Each spark erodes only a tiny particle of metal, but since the sparking occurs 20,000 to 30,000 times per second, appreciable quantities of metal are removed.

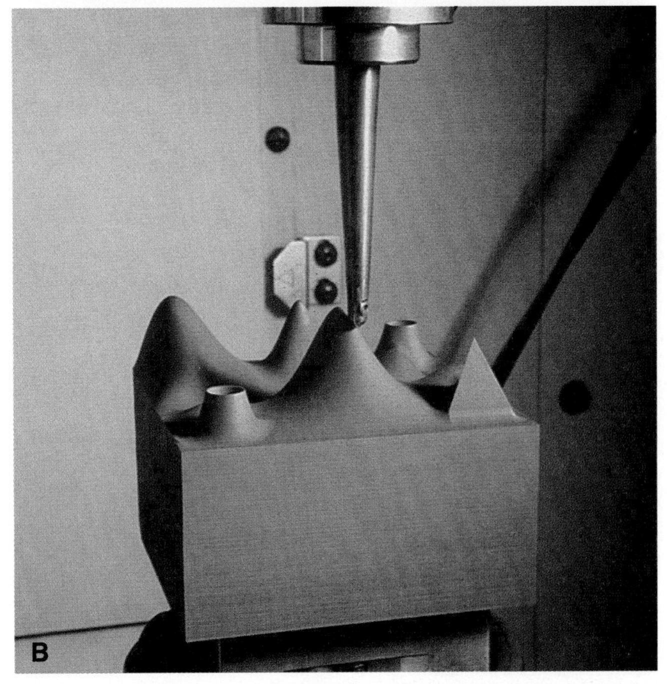

LeBlond Makino Machine Tool Co.

Figure 31-4. Graphite electrodes. A—This graphite electrode has thin cutting areas. It would be difficult to machine them by conventional means. B—A more complex shaped graphite electrode. A probe is being used to check whether the electrode's shape meets specifications.

Besides providing a nonconductive barrier, the dielectric fluid also flushes particles from the gap, keeps the electrode and the work cool, and prevents fusion of the electrode with the workpiece. A filter removes particles from the fluid as it is recirculated through the machine.

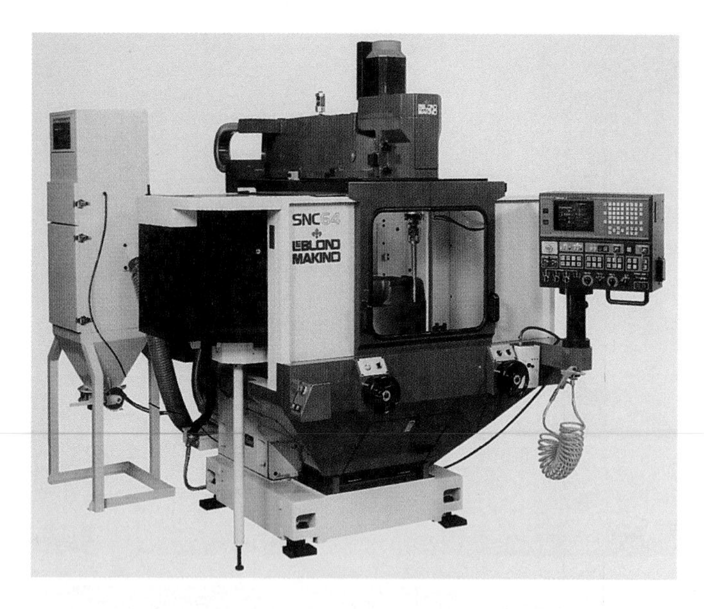

LeBlond Makino Machine Tool Co.

Figure 31-5. Machines for milling graphite electrodes must be environmentally clean. This CNC graphite mill has a totally enclosed cutting zone and a powerful suction system that contains, captures, and removes chips and dust.

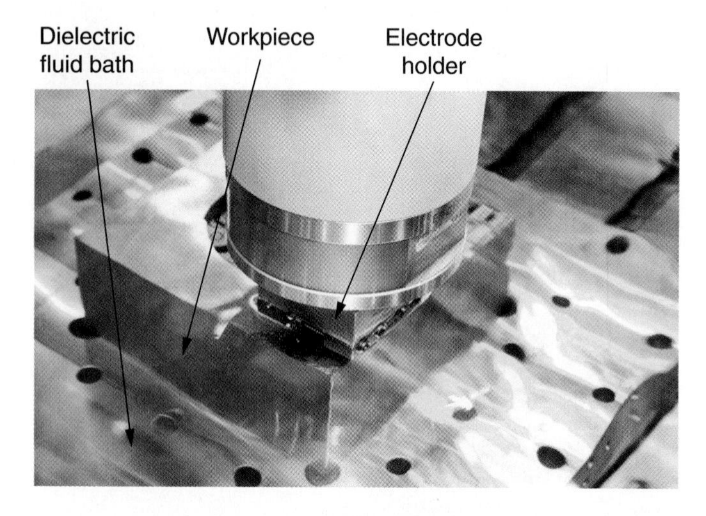

AMT—Association for Manufacturing Technology

Figure 31-6. A thin gap separates the electrode and the work. The dielectric fluid fills the gap and insulates the workpiece from the electrode.

Roughing cuts are made at low voltage and low frequency, with high amperage and high capacitance (opposition to any change in voltage). Finishing cuts require high voltage and high frequency, with low amperage and low capacitance.

Hard metals erode at a much slower rate than soft metals. Since the electrode is also consumed, but at a much slower rate than the work, considerable savings can be achieved by making interchangeable electrodes for roughing, sizing, and finishing. Long runs may require several sets of electrodes.

31.1.2 EDM Applications

The EDM process is used to:

- Shape carbide tools and dies.
- Machine complex shapes in hard, tough metals.
- Machine applications where the physical characteristics of the metal or its use make it impractical or very expensive to machine by conventional methods.
- Eliminate tedious and expensive handwork in die making. With EDM, the cavity produced in the metal is a "mirror image" of the electrode, **Figure 31-7**.
- Drill holes ranging in size from 0.0012″ to 0.120″ (0.3 mm to 3 mm) in diameter. They can be square, rectangular, triangular, or round. Multiple holes can be produced at the same time. Hardness is not a factor as long as the material can conduct electricity.

31.1.3 Electrical Discharge Wire Cutting (EDWC)

Electrical discharge wire cutting (EDWC) is a cutting process similar to band machining, but uses a small-diameter wire electrode in place of a saw. Like EDM, this process removes material by spark erosion as the wire electrode is fed through the workpiece, **Figure 31-8**. The wire electrode is fed from a spool over sapphire or diamond guides.

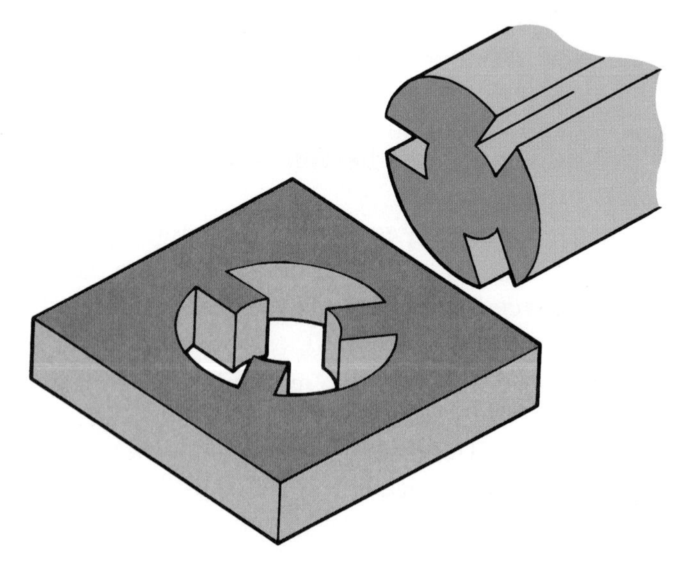

Goodheart-Willcox Publisher

Figure 31-7. An EDM electrode must be an exact reversal or mirror image of the cut to be made. Many electrodes are made from graphite.

A starter or threading hole is required. Some machines start their own holes and thread the wire electrode automatically. A steady stream of dielectric fluid cools the electrode and the work. The wire electrode is used only once because it becomes warped or distorted after just one pass through the work.

EDWC is well-suited for CNC applications and can produce dies, shaped carbide cutting tools, and punches in less than one-third the time required by conventional methods. With EDWC, layers of sheet metal stacked up to 15″ (150 mm) thick can be gang-cut to produce a number of parts in one pass. EDWC is also useful for cutting intricate shapes that would be difficult to machine by convention methods, **Figure 31-9**.

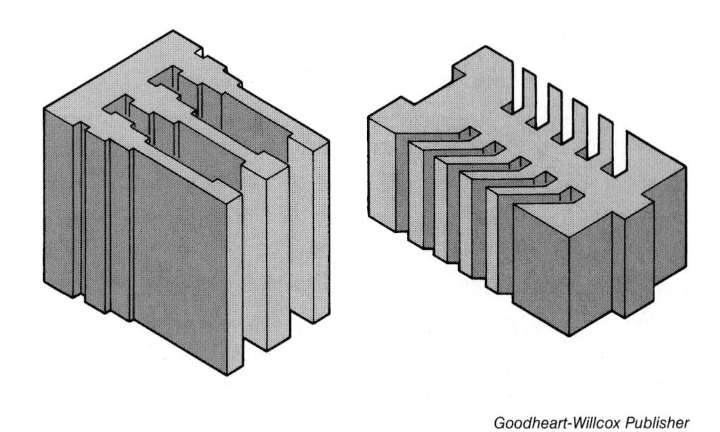

Goodheart-Willcox Publisher

Figure 31-9. Examples of work produced using EDWC.

31.1.4 Small Hole EDM Drilling

The ***small hole EDM drilling*** process operates on the same basic principle as the EDM and EDWC processes. In this process, a spinning, hollow electrode is brought close to the workpiece. High-pressure dielectric fluid is pumped through the center of the electrode and floods the small gap between the workpiece and the electrode. When sufficient voltage is built up, an arc jumps the gap, eroding metal from the workpiece. In addition to the high-pressure dielectric fluid provided through the electrode, additional dielectric fluid can be sprayed on the cut through nozzles. The additional fluid helps to clear and cool the cut area. See **Figure 31-10**. Small hole EDM drilling is often used to create the starter hole for EDWC.

Small hole EDM drilling has certain advantages over other drilling processes. It is able to drill holes in hardened metals. Because the electrode never comes

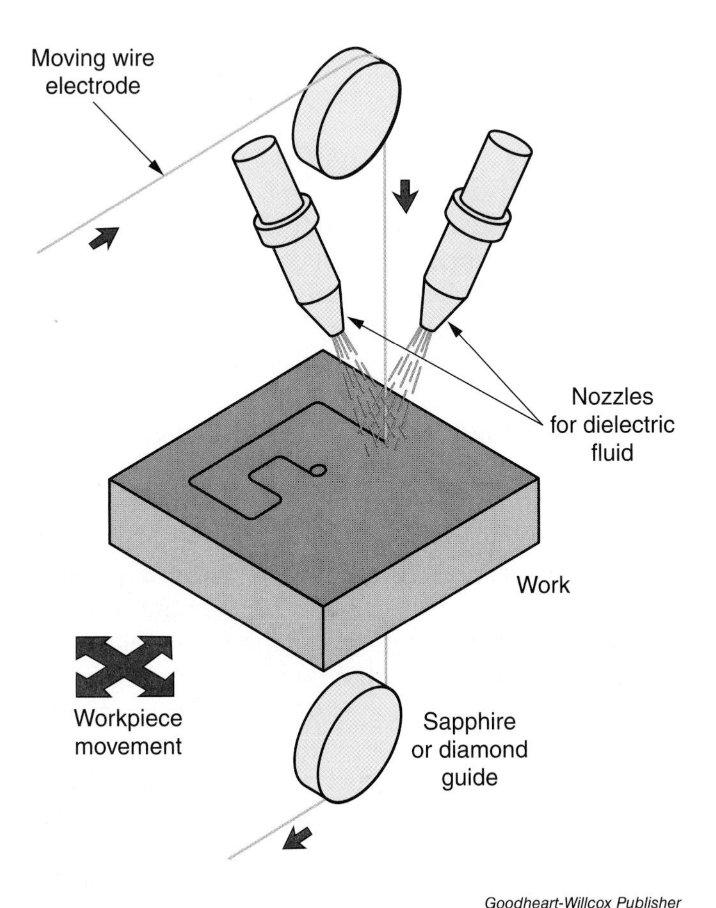

Goodheart-Willcox Publisher

Figure 31-8. Electrical discharge wire cutting (EDWC) differs from EDM in that cuts are made with a fine, moving wire electrode instead of a solid electrode. This technique is ideal for CNC operations.

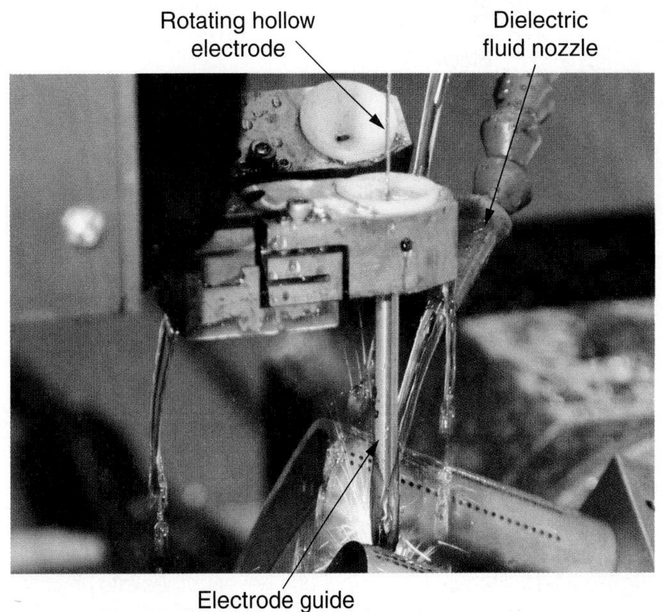

AMT—Association for Manufacturing Technology

Figure 31-10. Small hole EDM drilling in progress.

into physical contact with the workpiece, small hole EDM drilling can be used to accurately drill holes on highly curved or angled surfaces that would be difficult to drill by mechanical means, **Figure 31-11**. Small hole EDM drilling is also able to produce deep holes without the drift associated with mechanical drilling. Also, there is no risk of breaking off a hardened drill bit inside the hole. Lastly, the holes created by the EDM drilling process do not require deburring.

The small hole EDM drill process also has some drawbacks. The electrodes are consumed at nearly the same rate that the hole is drilled. For larger diameter holes, the drilling rate is slower than can be achieved by mechanical drilling. A new or freshly dressed electrode must be used at the bottom of a blind hole, or the bottom of the hole will not be flat.

31.2 Electrochemical Machining (ECM)

Electrochemical machining, more commonly known as ECM, might be classified as electroplating in reverse. Like electroplating, the process requires DC electricity and a suitable electrolyte (an electrically conductive fluid). However, with ECM, the metal is removed from the work rather than deposited onto it, **Figure 31-12**.

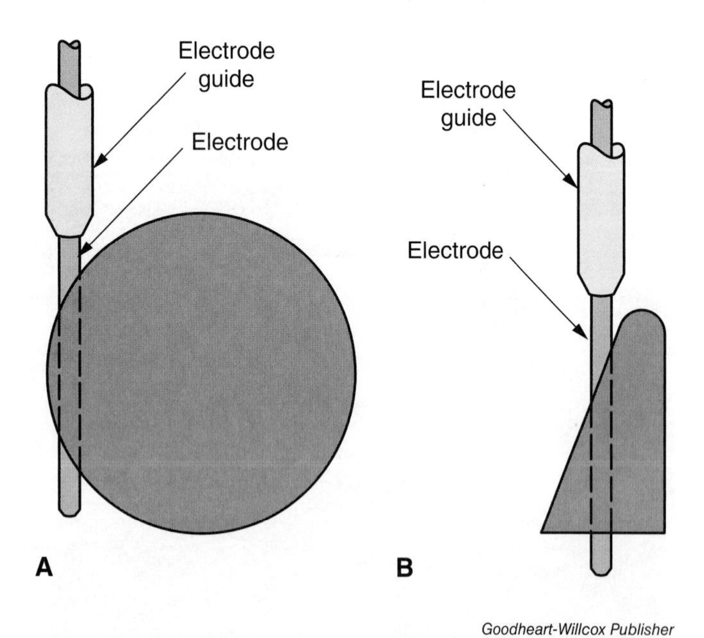

Goodheart-Willcox Publisher

Figure 31-11. Because the electrode never comes into physical contact with the workpiece, small hole EDM drilling can be used to drill holes in areas that would be problematic for mechanical drilling. A—Drilling a hole through the outer extents of a sphere. B—Drilling a hole through an angled surface.

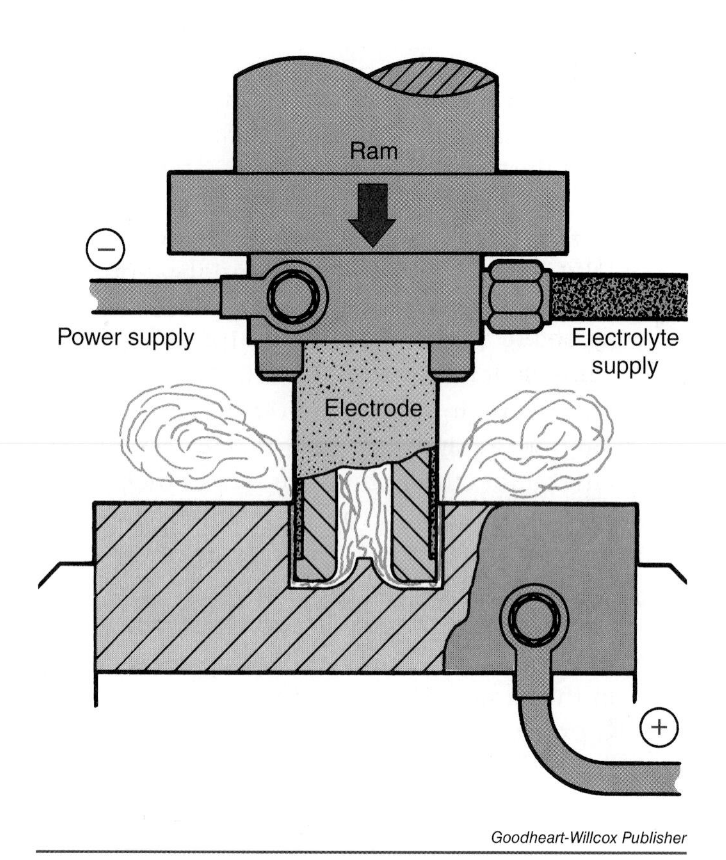

Goodheart-Willcox Publisher

Figure 31-12. How electrochemical machining (ECM) works.

ECM is also similar to EDM, but has some significant differences. First, the fluid between the electrode and the workpiece is a conductive electrolyte rather than a dielectric fluid. As a result, there is no spark between the workpiece and the electrode. Metal is removed through electrolysis rather than spark erosion.

The electrolyte for ECM is usually common salt (NaCl) mixed with water. A stream of electrolyte is pumped at high pressure through a gap between the positively (+) charged work and the negatively (–) charged tool (electrode). The current passing through the gap removes material from the work by electrolysis, duplicating the shape of the electrode tool as it advances into the metal. In some applications, tolerances as close as 0.0004″ (0.010 mm) can be maintained.

The work is not touched by the tool; therefore, no friction, heat, sparking, or tool wear occur. The machined surface is free of burrs and, in some instances, is highly polished. The operation of the machine is unique because the only sound heard is the rush of liquid.

31.2.1 Advantages of ECM

ECM offers many advantages:

- Metal is removed rapidly—up to 1 cubic inch per minute for each 10,000 amps of machining current.
- The kind of metal or its hardness does not affect the speed of material removal. Cast iron is about the only metal that presents problems, so it is machined by other processes.
- ECM is accurate. Difficult shapes can be machined easily.
- The machined metal is stress-free and will not warp or spring out of shape when removed from the machine.
- There is no tool wear.
- Several operations (milling, grinding, deburring, and polishing) often can be eliminated with ECM.
- Advances continually are being made in ECM. The ability to produce highly complex shapes with simple tooling widens the range of application for this machining process.

31.2.2 Disadvantages of ECM

ECM also has some significant disadvantages:

- Equipment and tooling costs are high.
- The electrolyte is corrosive.
- Extremely high voltage must be applied to the workpiece.

Chapter Review

Summary

- Electrical discharge machining (EDM can work metals that are difficult to work with conventional machining processes. It is capable of working tough, hard, fragile, or heat-sensitive metals to close tolerances.
- In the EDM process, a dielectric fluid fills the gap between the electrode and the workpiece. When voltage builds to a sufficient level, it creates an arc between the electrode and the workpiece. This arc erodes some metal from the workpiece.
- In electrical discharge wire cutting (EDWC), a small-diameter wire electrode removes material by spark erosion as it is fed through the workpiece.
- In small hole EDM drilling, a spark is created between the end of a hollow, rotating electrode and the workpiece. As the electrode is fed into the workpiece, it bores a hole.
- Electrochemical machining is similar to electroplating, but in reverse. Unlike EDM processes, electrochemical machining removes material by electrolysis rather than spark erosion.

Review Questions

Answer the following questions using the information provided in this chapter.

1. EDM stands for _____.
2. EDM can be used to work _____.
   A. metals that are too hard to be machined by conventional means
   B. nonconductive metals
   C. conductive and insulating materials
   D. All of the above.
   E. None of the above.
3. The EDM machine _____ maintains a very thin gap of about 0.001″ (0.025 mm) between the electrode and the work.
4. Unlike conventional machining processes, EDM does not produce a(n) _____ as metal is removed.
5. Explain the functions of the dielectric fluid used in EDM.
6. EDWC stands for _____.
7. How does EDWC differ from EDM?
8. Small hole EDM drilling uses a _____, rotating electrode.
   A. nonconductive
   B. hollow
   C. flexible
   D. case hardened
9. Which of the following statements regarding small hole EDM drilling is true?
   A. A highly conductive electrolyte is used in the process.
   B. The electrode is in direct physical contact with the workpiece.
   C. The hole is created by arc erosion of the workpiece.
   D. All of the above.
   E. None of the above.
10. ECM stands for _____.
11. Metals to be machined by EDM and ECM must conduct _____.
12. ECM is similar to _____ in reverse.
13. In ECM, _____.
    A. the work is not touched by the tool
    B. there is no friction or heat generated
    C. there is no tool wear
    D. All of the above.
    E. None of the above.
14. In ECM, metal is removed by _____.
15. What are five advantages that ECM offers?

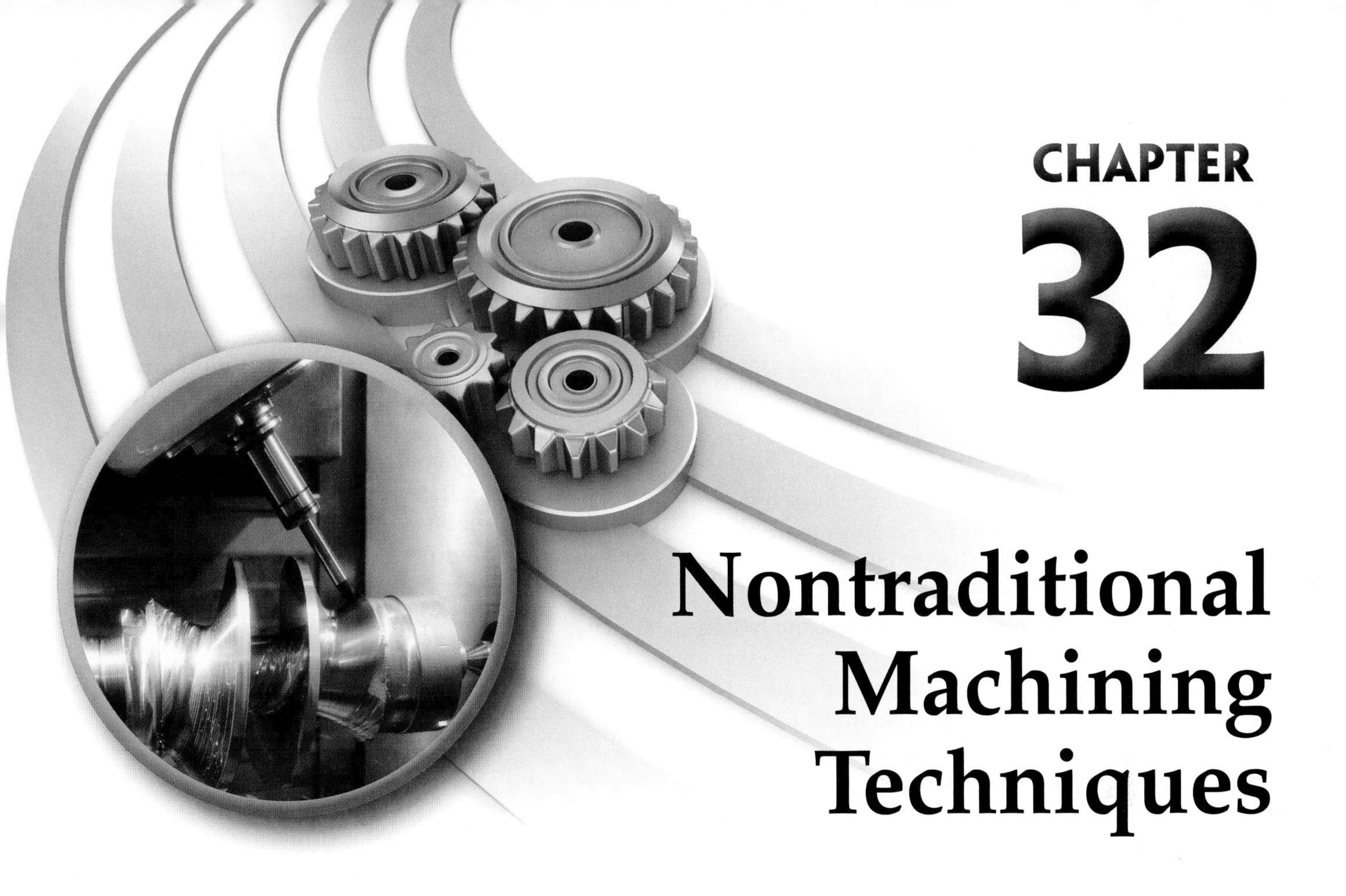

CHAPTER 32

Nontraditional Machining Techniques

Learning Objectives

After studying this chapter, you will be able to:

- Describe the chemical milling process and its advantages and disadvantages.
- Explain water jet cutting and water jet abrasive milling.
- Summarize the various ultrasonic machining processes.
- Explain electron beam machining.
- Describe the laser beam machining process.

Technical Terms

chemical blanking
chemical machining
chemical milling
electron beam machining (EBM)
etchant
hydrodynamic machining (HDM)
impact machining
laser
ultrasonic-assist machining
water jet cutting

The metalworking industry is responsible for cutting, shaping, and fabricating both metals and nonmetals. As the use of existing materials has evolved and new materials have been developed, new machining techniques have been devised to keep pace. This chapter will describe several of the new techniques that differ from the traditional chip-removal method of the lathe, drill press, milling machine, grinder, and saw.

32.1 Chemical Machining

Chemical machining is a refinement of the process once used by photoengravers to prepare printing cuts and plates. Chemicals, usually in an aqueous (with water) solution, are used to etch away selected portions of the metal to produce an accurately contoured part.

In general, chemical machining falls into two categories: chemical milling and chemical blanking.

32.1.1 Chemical Milling

Chemical milling, also called chem-milling or contour etching, is a recognized and accepted technique for machining metal to exacting tolerances through chemical action. The process makes it possible to remove metal selectively from relatively large surface areas, **Figure 32-1**. For example, chem-milling can be used to reduce the weight of sheet metal parts, which is critical to aerospace vehicle performance.

In the chem-milling process, masks (special coating materials) are applied to the areas of the part that should not have metal removed. The prepared part is then immersed in an ***etchant*** (usually a strong alkaline solution). Where bare metal is exposed on the part, the resulting chemical action gradually erodes the metal. The areas that are masked do not react to the etching solution. The immersion time must be carefully controlled.

Chem-milling and conventional milling are complementary processes. Refinements in chem-milling make it possible to remove metal to form shapes or microscopic parts that would be difficult or impossible to do by conventional machining techniques, **Figure 32-2**.

Tapers and multiple-depth cuts can also be produced. Multiple-depth cuts are possible with chem-milling by masking shallower sections after they have been eroded to the desired depth. When the part is reimmersed in the etchant, additional metal is removed from only the unmasked areas. Tapers are produced by withdrawing metal from an etchant at a predetermined rate. See **Figure 32-3**.

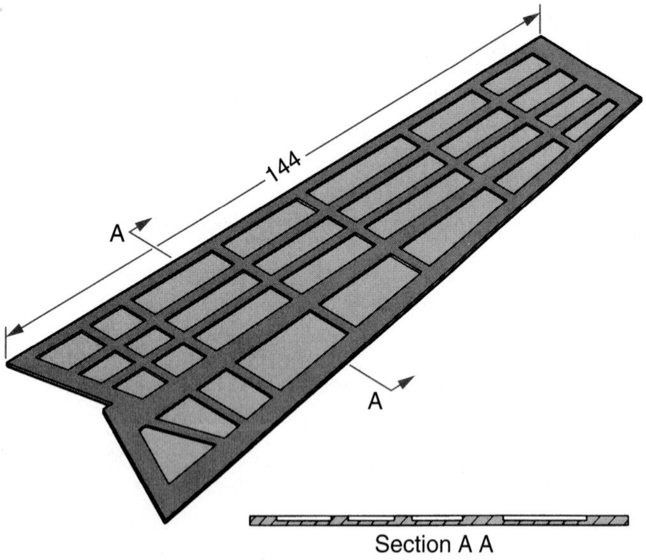

Goodheart-Willcox Publisher

Figure 32-1. Chemical milling is used to remove metal to close tolerances. This aircraft wing panel has been chemically reduced in sections where spars are not attached. Considerable weight reduction is accomplished with no sacrifice of structural strength.

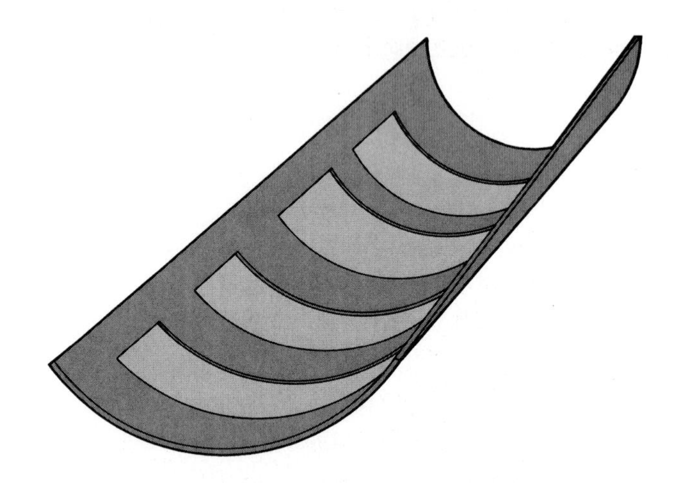

Goodheart-Willcox Publisher

Figure 32-2. With chem-milling, parts like the outer skin of an aircraft engine housing can be milled after they are formed.

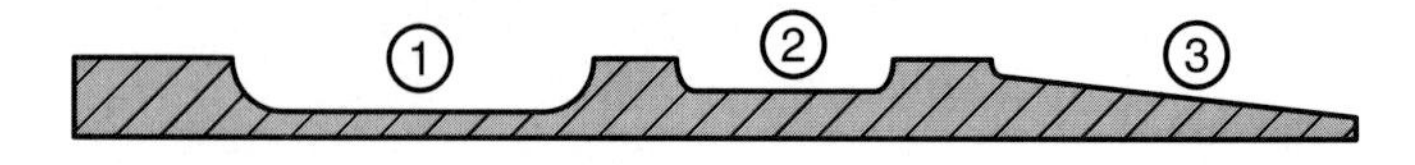

Goodheart-Willcox Publisher

Figure 32-3. Surface 2 on this part is less deep than surface 1 because it was masked partway through the chem-milling operation. As a result, surface 1 was etched for a longer period of time. Surface 3 was created by dipping that end of the part in the etchant and withdrawing it gradually.

Steps in Chemical Milling

There are six major steps in the chemical milling process:

1. **Cleaning.** Removes all grease and dirt that might affect the etching process.
2. **Masking.** The entire part is coated with a masking material, applied by brushing, dipping, spraying, or roller coating. The masked metal sheet is then baked to remove all solvents.
3. **Scribing and stripping.** A template is placed over the entire part. Then, the areas to be exposed are circumscribed, and the masking material is stripped away.
4. **Etching.** Parts are racked and lowered into an etchant for milling operation. See **Figure 32-4**.
5. **Rinsing and solvent stripping.** After being rinsed, the parts are lowered into a solvent tank, which releases the maskant bond. Maskant residue is stripped from the part.
6. **Inspection.** The accuracy of the chemical milling etch is measured with the aid of an ultrasonic thickness gage.

Advantages of Chemical Milling

Chemical milling offers many advantages:

- Tooling costs are low.
- Tolerances of ±0.003″ (0.07 mm) are possible on cuts up to 0.50″ (12.5 mm) deep.
- The size of the workpiece size is limited only by the size of the immersion tank.
- Warping and distortion of formed sections is negligible.
- Contoured or shaped parts can be chem-milled after they are formed, **Figure 32-5**.
- Many parts can be produced simultaneously.
- Unsupported pieces that are as thin as 0.015″ (0.4 mm) can be machined without danger of buckling.
- Both sides of the metal can be milled at the same time.
- Any metal, regardless of its state of heat treatment, can be machined chemically.
- No burrs are produced in the machined area.

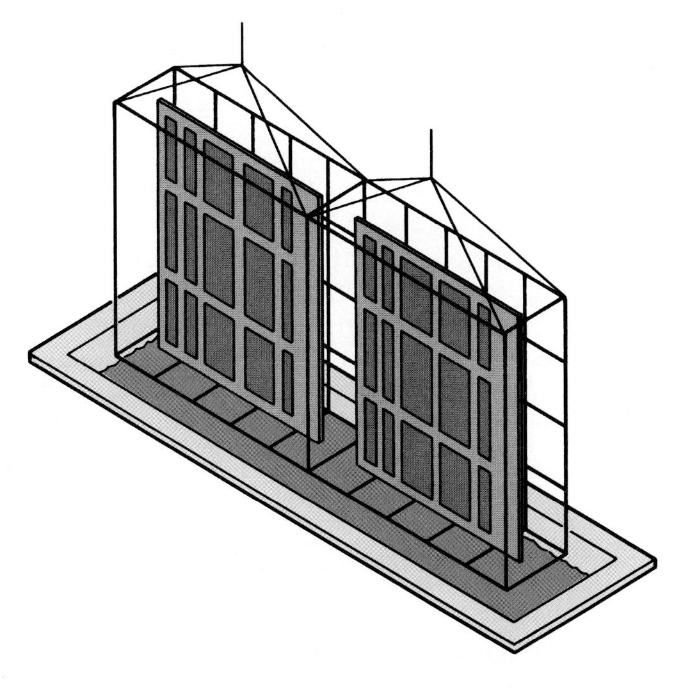

Goodheart-Willcox Publisher

Figure 32-4. Masked parts on a rack, ready to be lowered into a tank of etchant. The dark gray areas are unmasked.

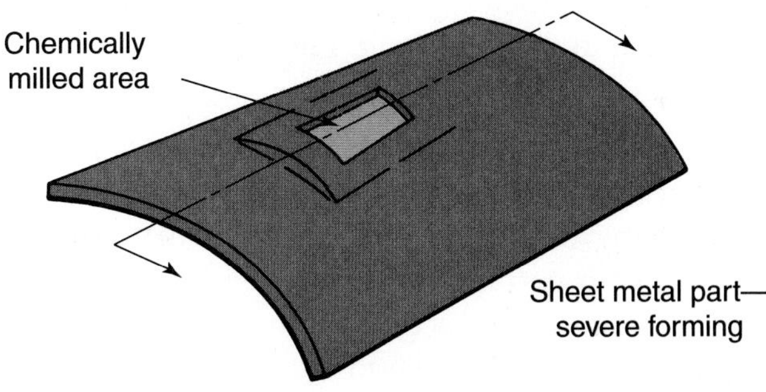

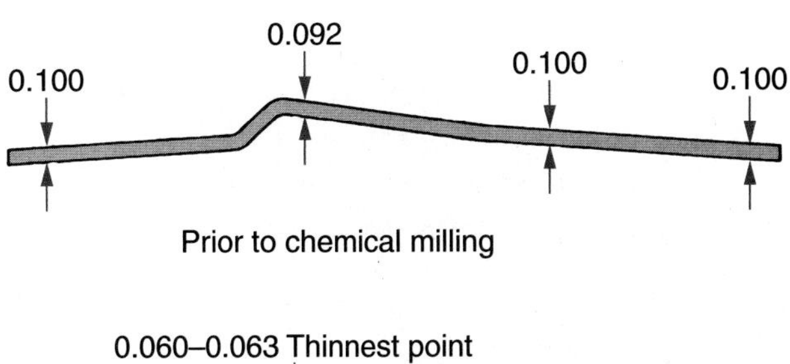

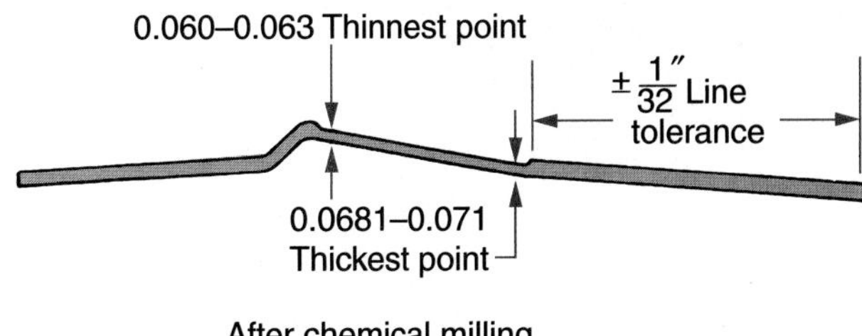

Northrop-Grumman Corp.

Figure 32-5. This aircraft part was chem-milled after it was formed. Pieces even more severely formed than this part can be chem-milled economically.

Disadvantages of Chemical Milling

The chemical milling process does have some disadvantages:

- The process is slow. It takes considerable time to remove large quantities of metal.
- All surface imperfections must be removed before etching. Otherwise, these areas would etch at a faster rate and be amplified on the finished surface.
- Chem-milling is not recommended for etching holes.
- Surface finishes on deep etches are not as fine as conventionally machined surfaces.
- Lateral dimensions are difficult to hold because the etchant works sideways as well as in depth. Typical lateral tolerances work on an etch factor of 3:1. This means that for every 0.003″ (0.08 mm) of etched depth, 0.001″ (0.03 mm) of undercut will occur. The top edges of the cavity will be sharp; however, the inside edges and corners will have a radius approximately equal to the cut depth, **Figure 32-6**.

32.1.2 Chemical Blanking

Chemical blanking, also called chem-blanking, photoforming, or photoetching, involves total removal of metal from certain areas by chemical action. It is a variation of chemical milling. Chem-blanking is used by the aerospace and electronics industries to produce small, intricate, ultrathin parts. See **Figure 32-7**.

Metal foil as thin as 0.00008″ (0.002 mm) can be worked by chem-blanking. (By comparison, a sheet of copier paper is about 0.004″ or 0.1 mm thick.) This ultrathin metal is laminated to a plastic backing to protect it from damage during shipping.

The process is not recommended for metals thicker than 0.09″ (2.3 mm). However, almost any metal can be chem-blanked.

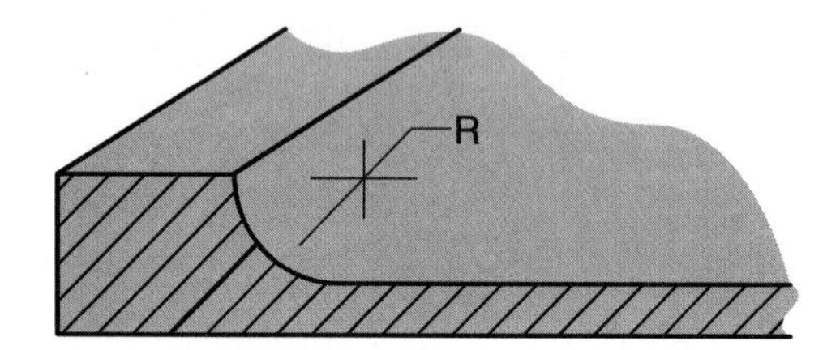

Goodheart-Willcox Publisher

Figure 32-6. The inside edges of a chem-milled section will have a radius equal to the depth of the etch.

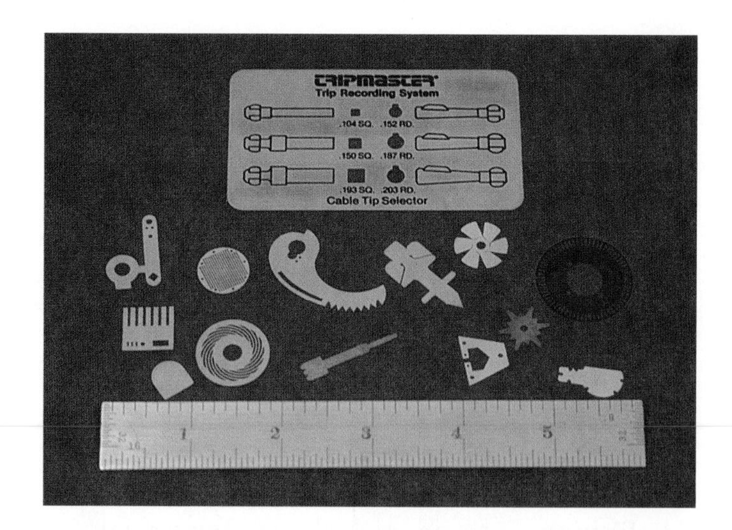

Microphoto, Inc.

Figure 32-7. Made by chemical blanking, these electronic parts range in thickness from 0.0001″ to 0.002″ (0.0025 mm to 0.05 mm).

32.2 Hydrodynamic Machining (HDM)

Hydrodynamic machining (HDM), or ***water jet cutting***, is a popular cutting process, **Figure 32-8**. It uses a high-velocity, high-pressure stream of water to cut through materials. It was developed to shape composites, which are layers of a tough fabric-like material bonded together into three-dimensional shapes called layups. A nontraditional cutting method was needed because the texture of composites quickly dulls conventional cutting tools. Water jet cutting can be accomplished with or without abrasives. The addition of abrasives to the water jet is preferred for shaping metals and harder nonmetallic materials, **Figure 32-9**.

Depending on the material being cut, tolerances can be held to ±0.004″ (0.1 mm). No heat is generated that could damage the material being cut, nor is airborne dust produced. Instead of going into the air, particulates (fragments) are carried away by the water jet.

SAFETY NOTE

Since the speed of the water jet leaving the nozzle is approximately three times the speed of sound, it can cause serious injury. Hands must be kept clear of the work area.

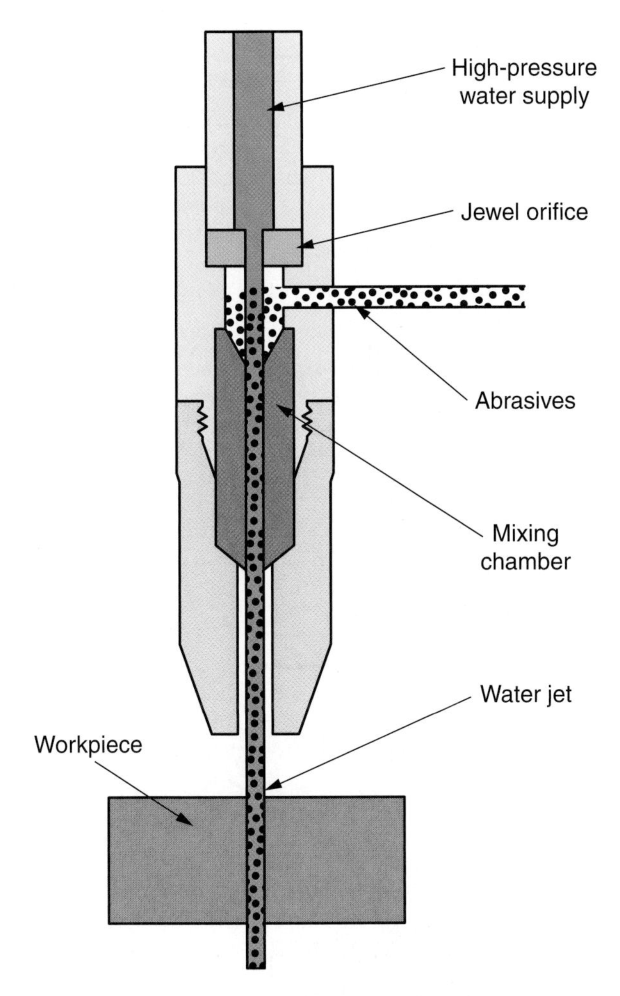

Goodheart-Willcox Publisher

Figure 32-8. Water jet cutting head. Water under extremely high pressure is forced through an orifice made of a very hard material like diamond, ruby, or sapphire. Abrasives are added to the high-pressure water stream by induction.

xtrekx/Shutterstock.com

Figure 32-9. Abrasives are added to cut hard materials, like this stainless steel plate.

32.2.1 Water Jet Abrasive Milling

Originally, water jet machining was used only in applications where the jet passed all of the way through the material being cut. An abrasive was added when metals and other hard materials were to be cut. Cut depth was not a consideration. Techniques and processes have been developed to control cut depth. The process generates no heat that could cause distortion to the finished part.

Water jet abrasive milling is similar to chem-milling. However, there is no expensive disposal of environmentally hostile by-products (etchants and material residue). A mask made from material more resistant to the abrasive action of the water jet is placed over the workpiece. The mask, rather than variation in a tool path, controls the geometry of the area being machined. See **Figure 32-10**.

Each pass of the workpiece under the abrasive water jet removes a small quantity of material. Workpiece speed determines the amount of material removed on each pass.

Many difficult-to-machine aerospace materials with complex shaped areas can be machined by this technique. Tolerances of ±0.002″ (0.05 mm) can be maintained.

32.3 Ultrasonic Machining

The term "ultrasonic" is used to describe sound waves at frequencies higher than the human ear can detect. The science of ultrasonics has found applications in many areas—machining, welding, quality control, and cleaning, to name a few. Considerable research is being done to develop new uses and to improve existing techniques.

The average person can hear sounds that vibrate between 20 to 20,000 hertz (cycles per second). Below 20 hertz, sound waves are called infrasonic. Above 20,000 hertz, sound waves are called ultrasonic. Industrial ultrasonic machining applications make use of ultrasonic sound waves with frequencies up to 100,000 hertz.

Ultrasonic waves are created by passing an electric current (usually 60 hertz ac) into a converter to produce the desired frequency. Then, current from the converter is passed to a transducer, which converts the electricity to mechanical motion, producing sound waves. The sound waves may be used in conjunction with a fluid (as in quality control and cleaning applications) or applied directly to the cutting tool or metal as it is being machined, welded, or formed. See **Figure 32-11**.

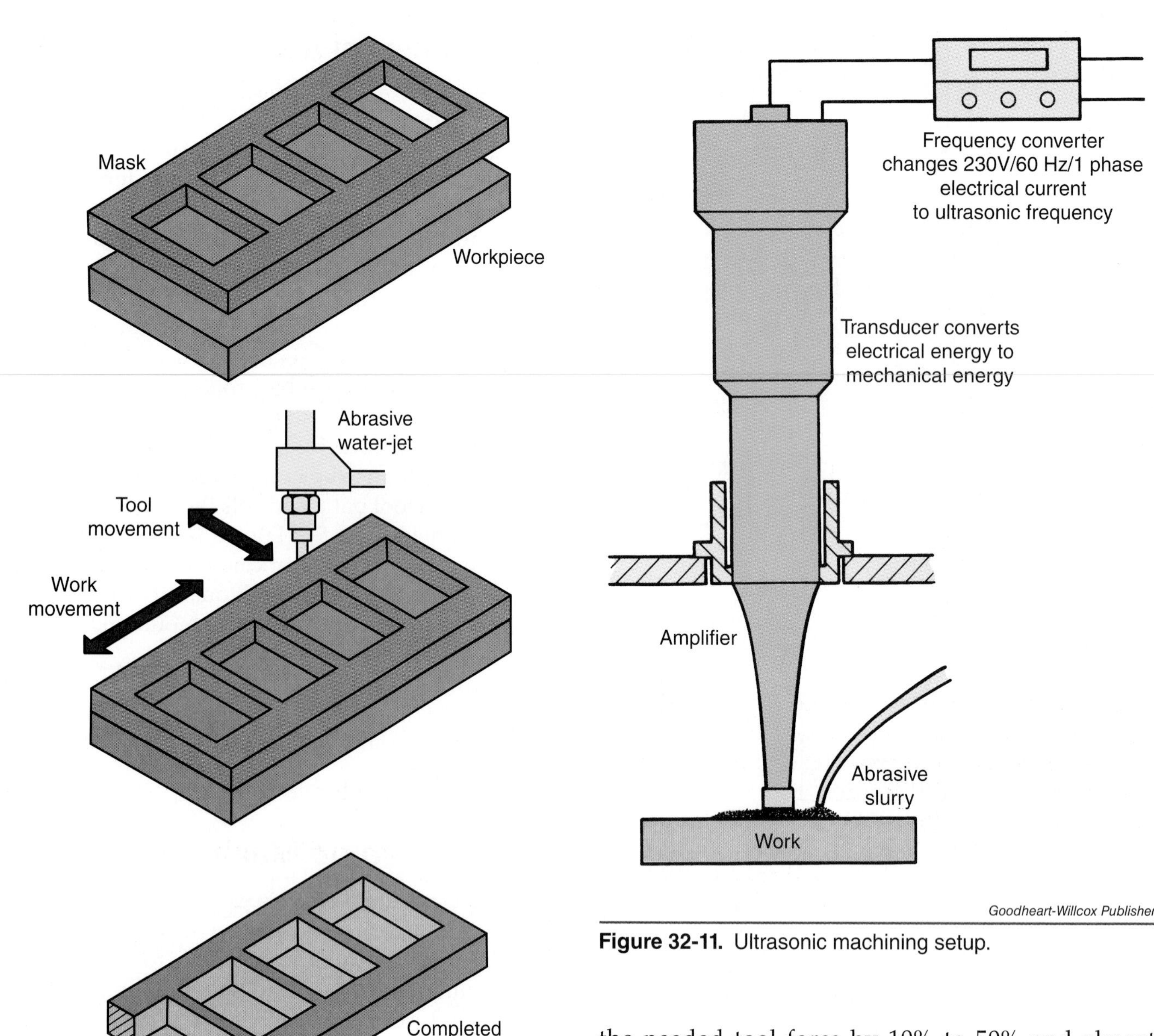

Figure 32-10. The water jet milling operation. A mask of an abrasive-resistant material can be used several times before it is worn away and must be replaced.

Figure 32-11. Ultrasonic machining setup.

32.3.1 Ultrasonic-Assist Machining

Ultrasonic-assist machining applies sound waves to the tool or metal as it is cutting or being cut. In this way, the ultrasonic sound waves assist the conventional metal cutting processes. In ultrasonic-assist machining, the transducer is fitted to a standard machine tool to make the tool vibrate at a high frequency. The ultrasonic assist can reduce the needed tool force by 10% to 50% and almost completely eliminate tool chatter. Tool wear is reduced and more cutting can be done between sharpenings. Surface finishes are also improved. Chatter reduction is a distinct advantage when the boring operations require the use of long, slender boring tools.

Ultrasonic assist is also useful in grinding operations. Ultrasonic waves are passed through the grinding wheel. Particles and chips that normally become embedded in the wheel are vibrated loose and washed away by the coolant. Grinding temperatures are reduced, and wheel life is extended. Material can be removed more rapidly and the surface finishes are improved without an increase in power consumption. Ultrasonic vibrations imparted to grinder coolant have been found to produce similar results.

Ultrasonic assist can be used in drilling, reaming, honing, milling, and EDM (electrical discharge machining) processes.

32.3.2 Impact Machining

Impact machining, also called slurry machining, is a machining process that removes metal by using ultrasonics and a special tool to force abrasives against the work. With the exception of diamond tools, it is the only commercially feasible way to machine extremely hard, brittle, and fragile materials. In fact, the technique works best on hard, brittle materials (glass, quartz, silicon, and carbides). It is ineffective on soft materials, such as aluminum and copper.

Machining is done by a shaped cutting tool oscillating about 25,000 times per second. This rapid oscillation pounds a slurry of fine, abrasive particles against the work. The tool stroke, at the vibrating end, is only 0.003″ (0.076 mm), **Figure 32-12**.

A solid funnel-shaped horn must be used to amplify and transmit the vibrations. Ultrasonic vibrations directly from the transducer do not produce enough motion to produce the required tool movement. The tool does not touch the work and no heat is generated, so there is no risk of the process causing heat-related distortion in the work. Microscopic portions of the work are chipped away to produce the desired shape. The machined section is a mirror image of the cutting tool, **Figure 32-13**.

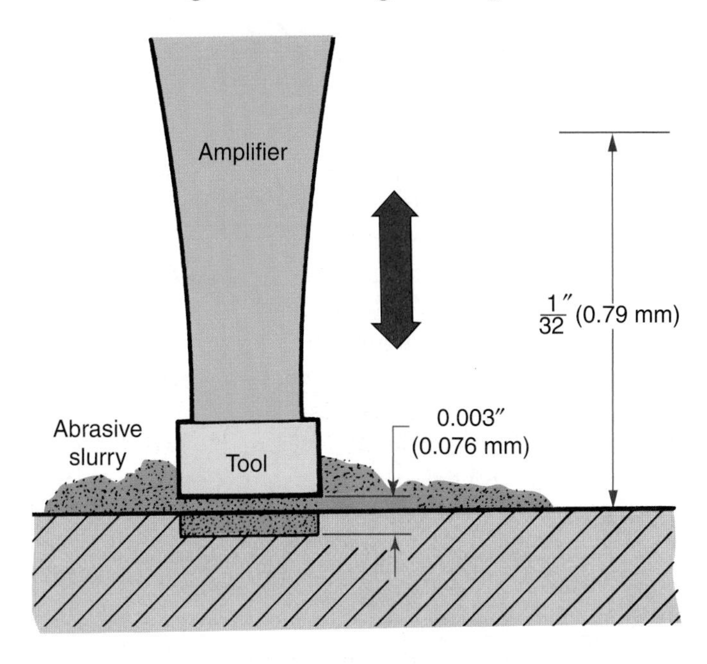

Goodheart-Willcox Publisher

Figure 32-12. Tool motion in ultrasonic (impact) machining is slight, only 0.003″ (0.076 mm). The 1/32″ (0.79 mm) measurement is used to indicate scale.

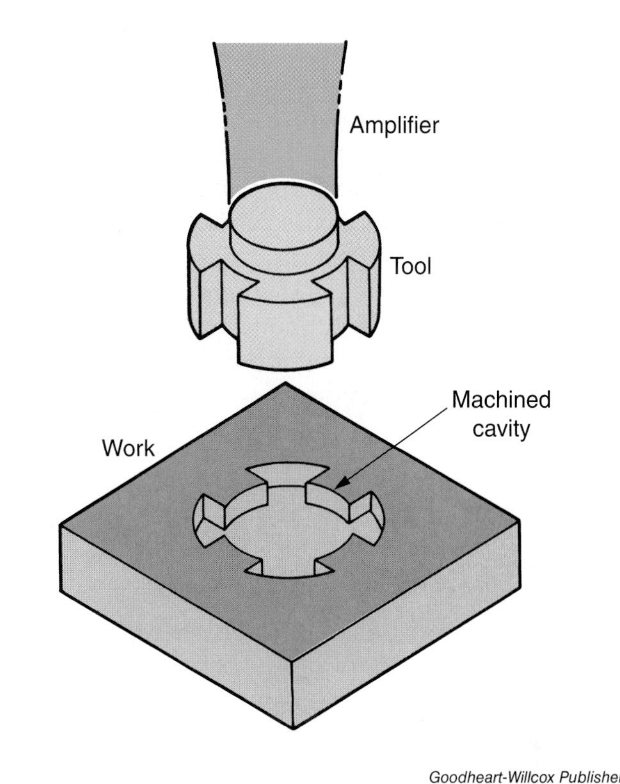

Goodheart-Willcox Publisher

Figure 32-13. The machined section is a mirror image of the tool.

Industrial applications of impact machining include:

- Slicing and cutting germanium and silicon wafers into tiny chips for transistors, diodes, and rectifiers for the electronics industry.
- Machining complex shapes in nonconductive and semiconductive materials that cannot be satisfactorily handled by EDM and ECM.
- Shaping virtually unmachinable space-age materials.

The disadvantages of impact machining are that it is slow, the surface finish is dependent on the size of the abrasive grit used, and the maximum cut depth is only about 1″ (25 mm). On the other hand, tolerances of 0.001″ (0.025 mm) can be maintained on hole size and geometry in most materials. Equipment cost is moderate. The training period for machine operators is short and does not require special skills.

32.3.3 Other Ultrasonic Applications

Industrial applications for ultrasonics vary. A common use is to improve the cleaning power of chemical solvents. As the ultrasonic waves pass through the cleaning solution, microscopic vapor pockets are created in the fluid. These "bubbles" (high-vacuum areas) form and collapse about 20,000 times per second. This creates local pressure as high as 10,000 psi, as well as heat. The bubbles smash against the work and literally tear away the dirt, oil, grease, chips, flux, and other contaminants on the surface. The cleaning action is so thorough that the technique is used to decontaminate work that has been exposed to radioactive solutions and gases.

Nondestructive testing using ultrasonics has been developed to the point where the testing procedure is fully automated. Sound waves are also used in the precision welding of dissimilar materials. For example, ultrasonic waves can weld aluminum wires to special glass on some electronic circuits.

32.4 Electron Beam Machining (EBM)

The electron beam microcutter-welder uses a high-energy, highly focused beam of electrons to weld or cut materials. ***Electron beam machining (EBM)*** is the direct result of the specific needs of the atomic energy, electronics, and aerospace industries. In some respects, it is the most precise and versatile of the nontraditional machining techniques.

Because the electron beam will cut any known metal or nonmetal that can exist in a high vacuum, its microcutting capabilities are almost unlimited. The development of refined focusing systems has made it possible to control the cutting action with a high degree of precision.

Electron beam cutting action can be controlled so precisely that it is possible to drill holes as small as 0.0002″ (0.005 mm) in diameter and mill slots having widths of 0.0005″ (0.0125 mm). The finish of the completed work is similar to a very fine machined edge. See **Figure 32-14**.

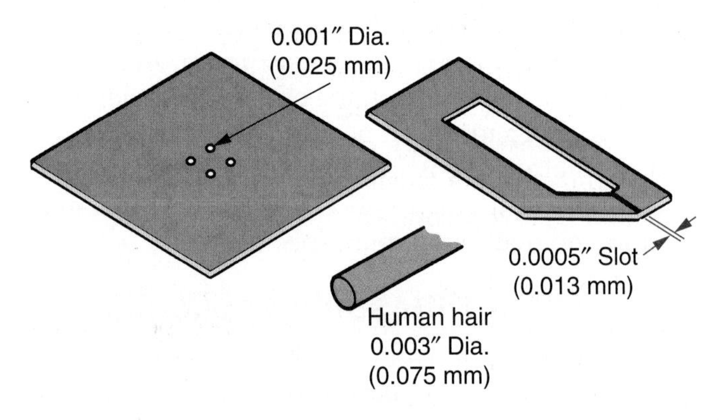

Figure 32-14. Note the size of the work done using the electron beam technique. The parts and the shaft of human hair are drawn to the same scale.

The electron beam machine is basically a source of thermal energy, **Figure 32-15**. The beam of electrons is focused to a very sharp point by the magnetic lens. The workpiece can be moved under the electron beam. The electron beam itself can be repositioned by changing the current input to the deflection coil.

Cutting is achieved by alternately heating and cooling the area to be cut. The heating and cooling must be controlled carefully so the material at the point of focus is heated to a temperature high enough to vaporize it, yet not enough to cause the surrounding area to melt. This is accomplished by pulsing the beam. The beam is on for only a few milliseconds and is off for a considerably longer period of time. The focal point of the beam can reach a temperature of 12,000°F (6649°C), but because the beam is on for only a short period of time, very little of the heat is transferred to the metal surrounding the cut.

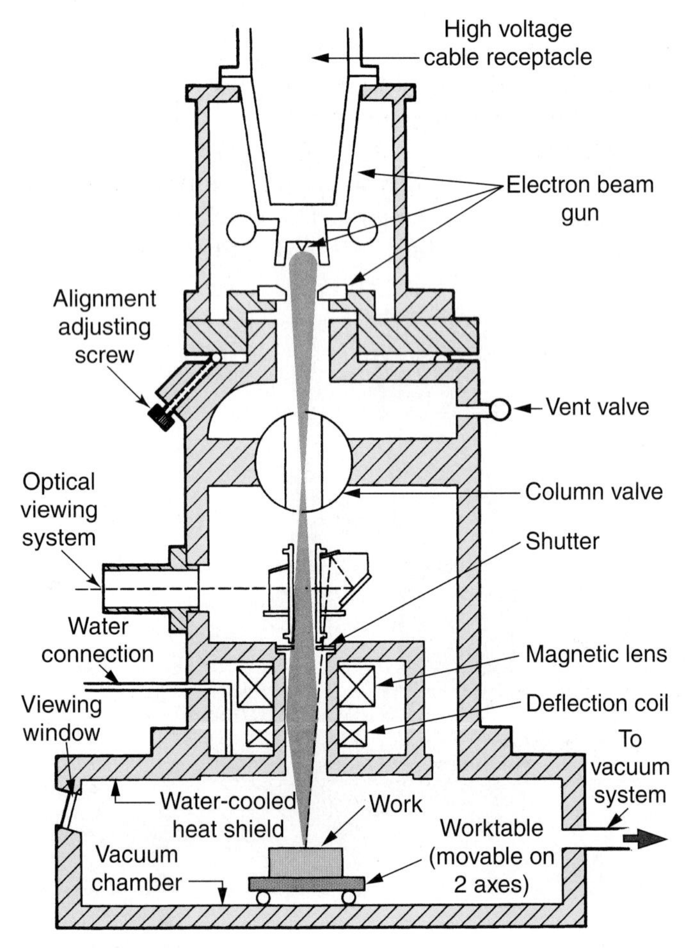

Figure 32-15. Cross-sectional view of an electron beam microcutter-welder.

Cut geometry (the shape of the cut) is controlled by movement of the worktable in the vacuum chamber and by using the deflection coil to bend the beam of electrons to the desired cutting path. Initial hole diameter or cut width is controlled by the amount of power applied and the duration of the cutting time. See **Figure 32-16**.

32.5 Laser Beam Machining

The term ***laser*** is an acronym for Light Amplification by Stimulated Emission of Radiation. The laser produces a narrow and intense beam of light that can be focused optically onto an area only a few microns in diameter. See **Figure 32-17**.

Depending upon the initial energy source used to activate the laser, it is possible to instantaneously create temperatures up to 75,000°F (41,650°C) at the point of focus. This is almost seven times the average temperature of the sun. No known material can withstand such heat.

There has been a dramatic increase in the use of lasers in parts manufacturing. Lasers are used for cutting, drilling, slotting, scribing, heat-treating, and welding. See **Figure 32-18**.

The energy output of a laser is usually not continuous. It lasts only a fraction of a second. When used for machining and cutting, it operates at 1 to 5 cycles per second. The cycle can be controlled manually or electronically. The laser operates on the principle described in **Figure 32-19**.

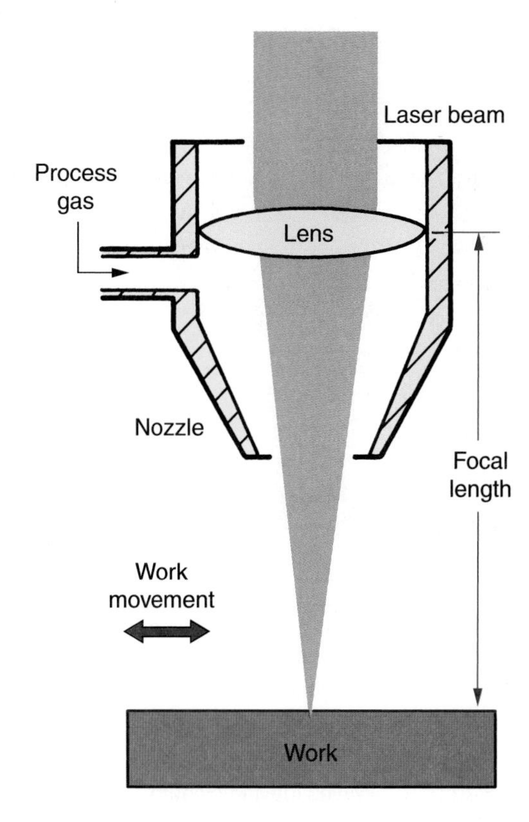

Goodheart-Willcox Publisher

Figure 32-17. A laser produces a narrow and intense beam of light. The beam can be concentrated to a point as small as 0.0002″ (0.05 mm) diameter.

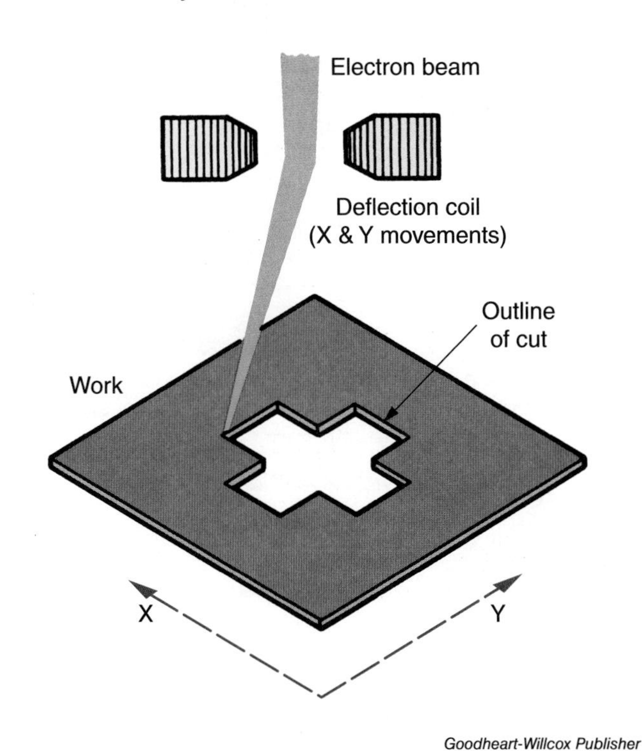

Goodheart-Willcox Publisher

Figure 32-16. The path of the cut is controlled by a deflecting electron beam or by movement of the worktable. Cutting must be done in a vacuum.

Some laser units can operate in either continuous or pulsed modes. With most materials, the edge quality of the cut is better with a continuous beam. Cutting speed is also much greater. The disadvantage of pulsing is that starting, stopping, and turning corners concentrates heat in localized areas on the work. This overheating causes the part to burn away from the programmed path, affecting cut quality and dimensional accuracy.

In the past, cutting aluminum, stainless steel, and titanium parts to shape by a laser required a time-consuming secondary operation to remove the oxide and dross that formed on the edge and surface of the cut. Titanium had a tendency to become embrittled by the oxygen that was absorbed during the cutting operation. These problems are solved by flooding the work surface with argon or nitrogen during the cutting operation. Since no oxygen can contaminate the metal, there is no oxide formation or embrittlement. Parts can be welded immediately after being cut because they require no cleaning.

Lumonics Corp.; Rofin-Sinar, Inc.

Figure 32-18. Laser cutting. A—When used with CNC, a laser is capable of cutting 3-D workpieces. B—A CNC CO_2 laser cutting specially shaped slots in a stainless steel plate. The high velocity gas jet aids in material removal by blowing out molten metal through the backside of the work. It also protects the lens from spatter ejected from the cut zone, especially during the piercing operation.

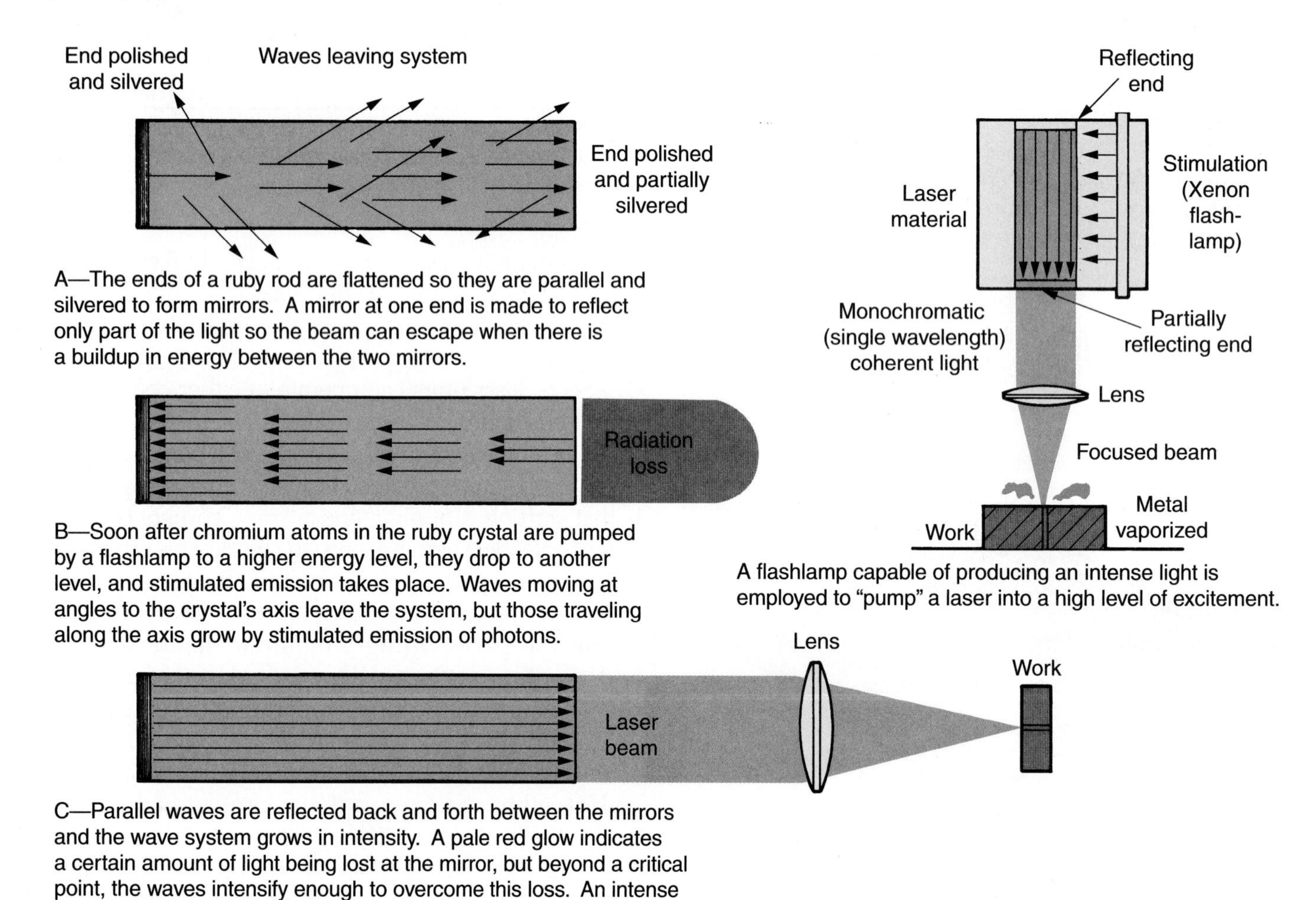

Goodheart-Willcox Publisher

Figure 32-19. Operation of a laser.

Chapter Review

Summary

- In chemical machining, chemicals are used to etch away selected portions of the metal to produce an accurately contoured part. Chemical machining falls into two categories: chemical milling and chemical blanking.
- Chemical milling is used to remove metal selectively from relatively large surface areas. Areas that are not to have metal removed are masked before the part is immersed in the etchant. Areas can be masked off in stages to produce multiple-depth cuts.
- The steps in chemical milling are cleaning the part, masking the entire part, scribing and stripping the mask away from the areas to be etched, etching, rinsing and solvent stripping, and inspection.
- Chemical blanking, or photoetching, is a process that uses chemicals to cut parts out of ultrathin materials.
- Hydrodynamic machining, or water jet cutting, is a process that uses a high-velocity stream of water to cut through materials. Abrasives can be added to the water stream in order to cut through very hard materials.
- Water jet abrasive milling works on the same principle as water jet cutting. However, as with chemical milling, a mask is applied to protect certain areas of the workpiece. The water jet and the workpiece are moved so that the jet does not cut all the way through the material.
- Ultrasonic waves can be used to enhance the conventional cutting process. This is known as ultrasonic-assist machining. Ultrasonic waves can also be used in conjunction with an abrasive slurry to wear away metal from the workpiece. This is known as impact machining.
- Electron beam machining uses a focused beam of electrons to cut through materials. Because the electron beam is pulsed on and off, very little heat transfers from the cut into the surrounding metal.
- Laser beam machining uses the intense heat of a tightly focused beam of light to cut through materials. The cut area can be flooded with an inert gas to protect the metal being cut from oxidation and oxygen embrittlement.

Review Questions

Answer the following questions using the information provided in this chapter.

1. Chemical machining falls into two categories. Briefly describe each of them.
2. Chemical milling is also known as chem-milling or _____.
3. List the six major steps in chemical milling, in order.
4. During chemical milling, a mask protects the portions of a workpiece that are _____.
   A. made from nonferrous metal
   B. to be etched
   C. to be cleaned
   D. All of the above.
   E. None of the above.
5. Briefly describe water jet machining.
6. Sound waves below 20 hertz are called _____.
7. Sound waves above 20,000 hertz are called _____.
8. Impact machining makes use of a special tool that forces _____ against the work to do the cutting.
9. Impact machining is one of the very few commercially feasible methods for machining which types of materials?
   A. Hard.
   B. Brittle.
   C. Fragile.
   D. All of the above.
   E. None of the above.

10. What are three disadvantages of impact machining?
11. With impact machining, tolerances of _____ can be maintained on hole size and geometry in most materials.
12. Holes as small as _____ in diameter can be drilled using electron beam machining.
13. The electron beam machine is basically a source of what type of energy?
    A. Thermal.
    B. Sonic.
    C. Fluid.
    D. All of the above.
    E. None of the above.
14. List two methods used to control the shape of the cut with EBM.
15. What does the term "laser" stand for?

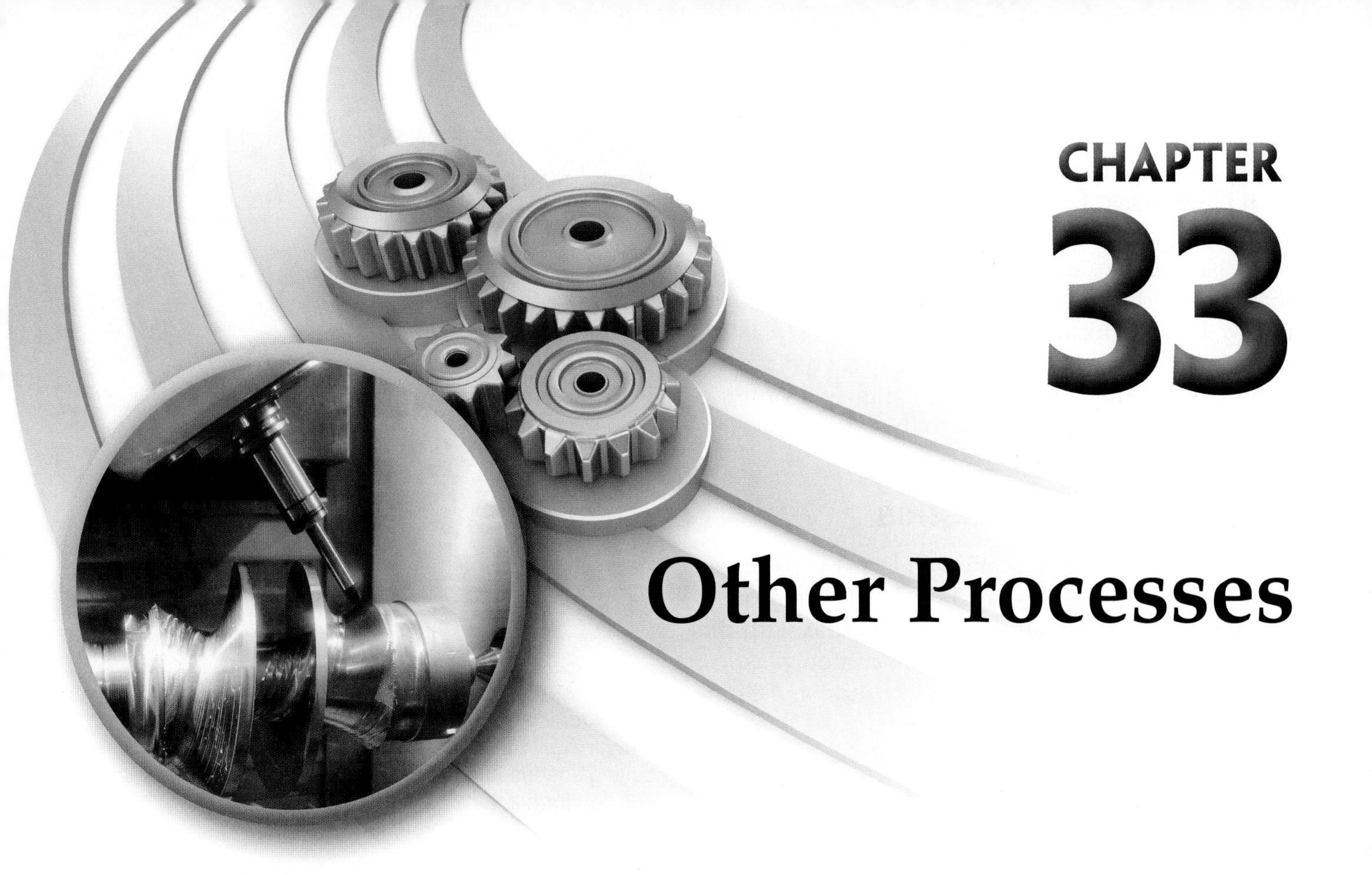

CHAPTER 33

Other Processes

Learning Objectives

After studying this chapter, you will be able to:

- Discuss the general machining characteristics of various plastics.
- Describe the hazards associated with machining plastics.
- Describe the five basic operations of chipless machining and their variations.
- Explain how the Intraform process differs from other chipless machining techniques.
- Describe how powder metallurgy parts are produced.
- Compare the advantages and disadvantages of various HERF techniques.
- Explain how the science of cryogenics is used in industry, and list some applications.

Technical Terms

briquetting
chipless machining
cold forming
cold heading
cryogenic
electrohydraulic forming
explosive forming
high-energy-rate forming (HERF)
magnetic forming
powder metallurgy (P/M)
sintering

New materials and new processes are continually being introduced to the field of machining and its related areas. In addition to the nontraditional machining processes described in Chapter 32, such processes as chipless machining, high-energy-rate forming, and cryogenic treatment are being used today. The machining of various types of plastics and the fabrication of parts from powdered metals have become important. This chapter will provide you with a basic introduction to these technologies.

33.1 Machining Plastics

Plastics are being machined in increasingly larger quantities each year, **Figure 33-1**. For this reason, it is important that the machinist be familiar with the machining characteristics of the common plastics, such as nylon, Teflon®, Delrin®, acrylics, and laminated plastics.

Special care must be taken when working with plastics. The dust and fumes given off by some plastics may be irritating to the skin, eyes, and respiratory system. Other plastics have fillers, such as glass fibers, that can be harmful to your health. Be sure you are aware of the safety precautions that must be observed before you attempt to machine any plastic.

33.1.1 Nylon

Nylon is the name for a group of polyamide resins. Nylon is tough and exhibits high tensile (pull), impact, and flexural (bending) strengths. This material is highly resistant to abrasion and is not affected by most common chemicals, greases, and solvents.

This plastic is excellent for bearings, gears, cams, and rollers. Another application for parts machined from nylon is in the preparation of experimental components. Research and development projects also make considerable use of machined nylon.

General Machining Precautions for Nylon

Most types of nylon can be machined with the same techniques used to machine soft brass. The use of coolants allows higher cutting speeds, but is not necessary to produce good-quality machined surfaces. When used, coolants should be of the soluble-oil type.

Since nylon is not as rigid as metal, it must be well supported during machining operations. Otherwise, deflection of the unsupported stock will result in inaccurate dimensions and distortions. To ensure accuracy, bring parts machined from nylon to room temperature before checking dimensions.

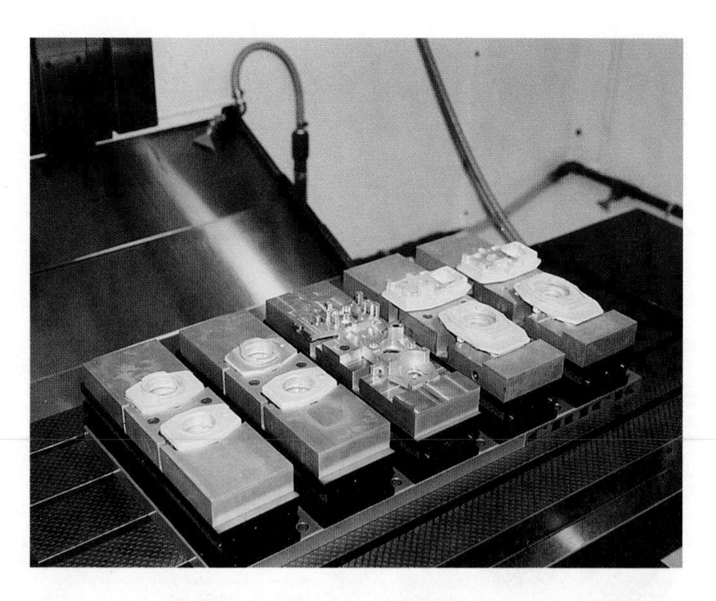

Chick Machine Tool, Inc.

Figure 33-1. More and more plastics are being machined. The fixtures shown here hold both plastic and metal parts for machining. With CNC, machine tools automatically change cutting speeds and feeds when different materials come into position for machining.

Turning Nylon

Nylon can be turned on a standard metalworking lathe. While tool bits sharpened to machine soft brass will prove satisfactory, best results are obtained using the tool bit shapes shown in **Figure 33-2**. Tools must be kept very sharp. The best finish on nylon is obtained with a high speed and fine feed.

Milling Nylon

Conventional milling cutters, providing they are kept very sharp, can be used for milling nylon. Climb milling will minimize the formation of burrs. Surface speeds in excess of 100 fpm (30 mpm) can be used.

Vertical milling is practical using fly cutters or two-lip end mills. Cutters must be kept very sharp to prevent plastic from melting or becoming gummy. Feeds of 10″ (25 mm) per minute or higher have proven satisfactory. Smoother surface finishes can be achieved using lighter feeds.

Drilling Nylon

Since drilling produces considerable heat, it requires extra care. Standard twist drills, when sharpened as shown in **Figure 33-3**, will produce acceptable results. However, best results are obtained by using drills designed specifically for plastics. Drills for plastics have flutes that are highly polished and have a long lead.

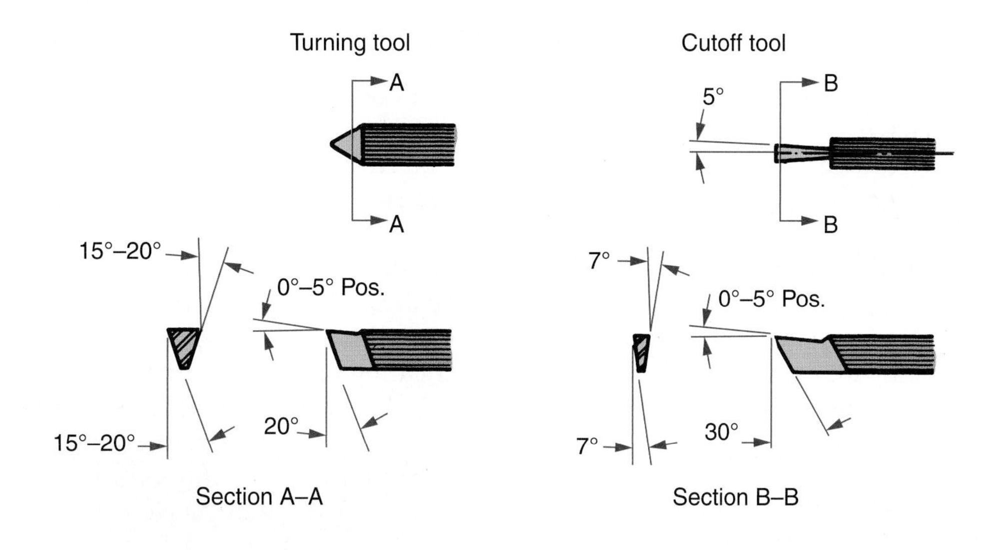

Figure 33-2. Typical cutting tools for machining nylon on a lathe.

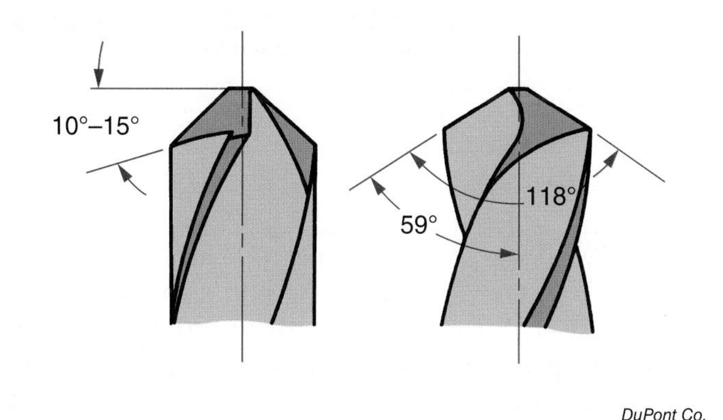

Figure 33-3. Drill point recommended for drilling nylon.

Use heavy feeds to prevent the drill from scraping the plastic rather than cutting it. Scraping results in excessive heat, which melts the plastic. If a coolant is not used, the drill must be withdrawn from the hole frequently to clean out chips and prevent overheating. If the drill is kept cool, holes will be drilled to size.

Nylon can be reamed using an expansion reamer that is adjusted to a few thousandths oversize. Holes that are finished with solid reamers tend to be undersize.

Threading and Tapping Nylon

Nylon can be threaded and tapped with conventional equipment. But before being tapped, the hole should be chamfered to reduce the chance of the first few threads tearing. Production tapping requires a tap that is 0.005″ (0.125 mm) oversize, unless a self-locking thread is desired. The tap can be made oversize by chrome plating.

Nylon can be threaded on a lathe using a single point cutting tool and the same procedures recommended for metal. However, the cutting tool must be kept very sharp and have plenty of clearance. Because of nylon's resilience, the finish cut should not be less than 0.005″ (0.125 mm). Support long work with a ball bearing follower rest.

Sawing Nylon

For sawing nylon and other plastics, good results can be obtained by using a band saw. A band saw blade quickly dissipates the heat. Dry cutting is best accomplished with a skip tooth metal-cutting blade that has 4 to 6 teeth per inch. However, the blade must be sharp to prevent gumming of the nylon. If gumming occurs, it will usually freeze the blade in the cut. Hollow-ground circular saw blades are also used extensively in cutting plastics. Blades with a slight set are recommended for cutting multiple layers of thin plastic or cutting thicknesses greater than 3/4″ (19 mm).

Annealing Nylon

Like metal, machined nylon parts require annealing to prevent dimensional changes. Annealing of plastics should be carried out in the absence of air, preferably by immersion in a suitable liquid. High-temperature boiling hydrocarbons, such as waxes and oils, are recommended for annealing nylon. A temperature of 300°F (149°C) is often used for general annealing.

An annealing time of 15 minutes per 0.125″ (3.175 mm) of thickness is normally required. Allow the part to cool slowly in a draft-free area. Placing the heated piece in a cardboard container is a simple way to ensure slow, even cooling.

33.1.2 Delrin

Parts manufactured from Delrin (the DuPont trade name for acetal resin) have an unusual combination of physical properties that bridge the gap between metals and plastics. These properties include excellent dimensional stability, high strength, and rigidity. The plastic is used extensively for parts in business machines (gears, cams, bearings, and printing wheels). Delrin is also replacing brass and zinc for many applications in the automotive and plumbing industries. The material has low friction, and thus requires a minimum use of lubricants. It is very quiet in operation.

In machining characteristics, Delrin is very similar to nylon. Recommended machining, cutting, and finishing operations are shown in **Figure 33-4**.

33.1.3 Teflon

Teflon (the DuPont trade name for its fluorocarbon resins) is filling a wide range of needs in the electronic, electrical, chemical, and processing industries. It has a very low friction coefficient: rubbing two flat pieces of Teflon together generates about the same friction as rubbing together two ice cubes. This makes Teflon ideal for use as bearings and seals in food processing equipment, where lubricating oil would contaminate the food.

Teflon works well in both high-temperature and ***cryogenic*** (very-low-temperature) applications. It is an expensive material, but will do things that no other material can do as well.

Machining, Cutting, and Finishing with Delrin Acetal Resin

Equipment			Cutting Speed			
Operation	Machines	Tools	RPM/FPM	Feed	Coolant use	Remarks
Sawing	Std.	Std. 14 T.P.I. Slight set	100–300 FPM (30.5–91.5 M/min)	Med.	—	Coolant improves finish of cut
Drilling	Std.	Std. twist drills 118°	1500 RPM for 0.500″ (1.27 cm) drill	Med.	At med. and high speeds	On-size holes drilled without coolant
Turning	Std.	Std.	690–840 FPM (210–256 M/min)	.002″–.005″ (0.051–0.13 mm)	At high speeds	Depth of cut—.016″–.200″ (0.41–5.08mm); support long lengths
Milling	Std.	Std. cutters; single fluted end mills	Similar to brass	Similar to brass	Not required	—
Shaping	Std.	Std.	Max.	Similar to brass	Not required	—
Reaming	Std.	Expansion type preferred	Similar to brass	Med.	At med. and high speeds	—
Threading and tapping	Std.	Std.	Similar to brass	Similar to brass	At med. and high speeds	Coolant facilitates cutting to dimensions
Blanking and punching	Std.	Std.	—	—	Not required	Primarily for 1/16″ (0.16 cm) thick stock
Filing and sanding	Std.	Std. file, std. abrasive paper and discs	—	—	Wet sanding	Finish will vary with type of file
Finishing	Std.	6″–12″ (15.2–30.4 cm) dia. muslin pumice and water polishing compound	1000–2000 RPM	—	Not required	Use light pressure and rotate part

DuPont Co.

Figure 33-4. Recommended machining, cutting, and finishing operations for Delrin acetal resins.

Teflon General Machining Characteristics

Teflon has a high thermal expansion rate. It expands at a rate about 10 times that of steel. When tolerances are critical, measurements should be made at room temperature, or (where applicable) at the temperature at which the part will be used. It is recommended that the plastic be stored at a temperature of 74°F (23°C) or higher for at least 48 hours before the machining operations.

Teflon has a tendency to pick up metal shavings and chips. For this reason, no machining should be attempted until the equipment has been thoroughly cleaned of all metal particles.

Turning Teflon

Teflon is more flexible than many other machinable plastics. To prevent deflection of the material away from the cutting tool, take care to support the work properly.

Tools must be sharp and have generous clearances so that the cutting edge of the tool will not rub, **Figure 33-5**. Chips must not be allowed to accumulate around the work because they prevent the heat from dissipating. See **Figure 33-6**.

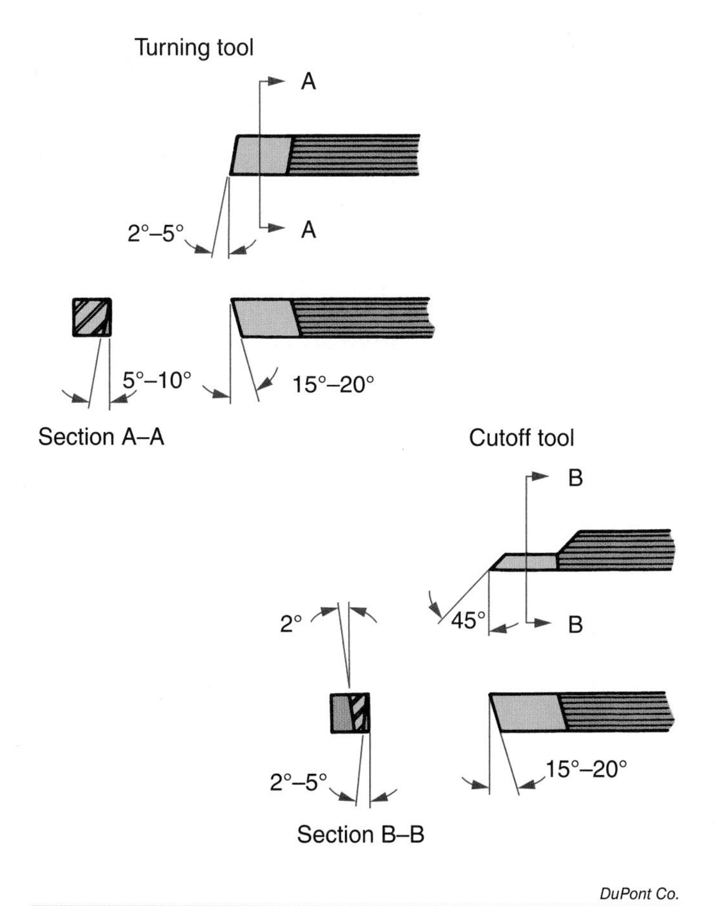

DuPont Co.

Figure 33-5. Typical lathe cutting tools for machining Teflon.

Goodheart-Willcox Publisher

Figure 33-6. When turning plastics, do not allow chips to accumulate around the work. Chips prevent heat from being dissipated, which may distort the work.

Cutting fluids are needed when tolerances are critical. Large amounts of water-based coolant will deter thermal expansion in Teflon. The best surface finishes are achieved at cutting speeds of 200 to 500 fpm (60 to 150 mpm) with feeds of 0.0002″ to 0.010″ (0.005 mm to 0.25 mm).

Drilling Teflon

Drills sharpened as shown in **Figure 33-7** will provide satisfactory cutting action in Teflon. Teflon tends to swell slightly during the drilling operation, which results in a hole smaller than the drill. To compensate for this swelling, the machinist must use a drill that is slightly oversize. You should test drill a piece of scrap material to determine the exact drill size needed. For close-tolerance drilling in Teflon, feeds of 0.004″ to 0.006″ (0.10 mm to 0.15 mm) are suggested.

Milling Teflon

Teflon is milled in much the same manner as the other plastics. Only newly sharpened and honed cutters should be used. The work must be solidly supported.

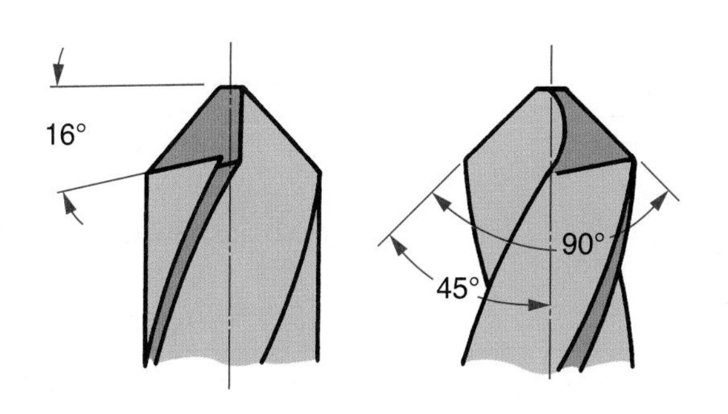

DuPont Co.

Figure 33-7. Drill point recommended for Teflon.

With shell mills, the milling head of the machine should be tilted slightly into the cut, as shown in **Figure 33-8**. This will eliminate cutter drag marks on the material.

Reaming Teflon

Reaming of Teflon is not advised. If close tolerances are specified, holes should be bored with a single-point tool.

Threading and Tapping Teflon

Teflon is threaded and tapped with the same general techniques as those suggested for nylon.

Sawing Teflon

If a square cut is to be made in Teflon, a rigid machine with first-class saw guides is essential. No coolant is needed. Maximum machine speeds can be used with a skip tooth blade having 4 to 6 teeth per inch.

Annealing Teflon

To maintain dimensional stability, Teflon should be annealed. Normally, the plastic is heated to a temperature above that to which the finished part will be exposed, but below 621°F (327°C). Above 621°F, Teflon turns to a gel. One hour of annealing for each 1″ (25 mm) of thickness is adequate. Allow the part to cool slowly. Heating is usually done in an oven.

The following oven annealing procedure for Teflon is recommended:

1. Anneal the rod or tubing from which the part is to be made.
2. Rough machine the part to within 0.06″ (1.5 mm) of finished size.
3. Anneal again, but at temperatures slightly lower than that of the initial annealing.
4. Finish machine after the Teflon has returned to room temperature.

33.1.4 Acrylics

The acrylic plastics Lucite® (the DuPont trade name) and Plexiglas® (the Rohm & Haas Co. trade name) have an unusual combination of desirable characteristics. They have good dimensional stability and high impact strength, even at temperatures as high as 200°F (93°C). Acrylics are also easy to machine, form, and polish.

Because of their unusual "light piping" ability (they transmit light as a hose carries water) and edge lighting qualities, acrylics are commonly used in light control and optical applications.

General Machining Characteristics for Acrylics

Acrylics are machined in much the same manner as other plastics. Generally, little difficulty will be encountered if sharp tools with adequate clearance are used. Drills should be sharpened as shown in **Figure 33-9**. Transparent acrylics present a unique

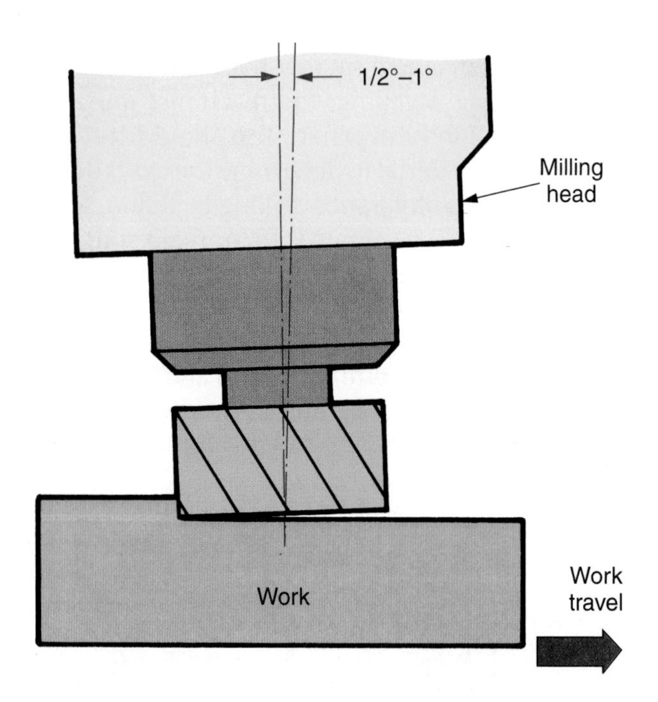

Goodheart-Willcox Publisher

Figure 33-8. For shell milling, it is recommended that the cutter be tilted slightly (1/2° to 1°) into the cut. This ensures the cut is made with the leading edge of the cutter, eliminating cutter marks on the machined surface.

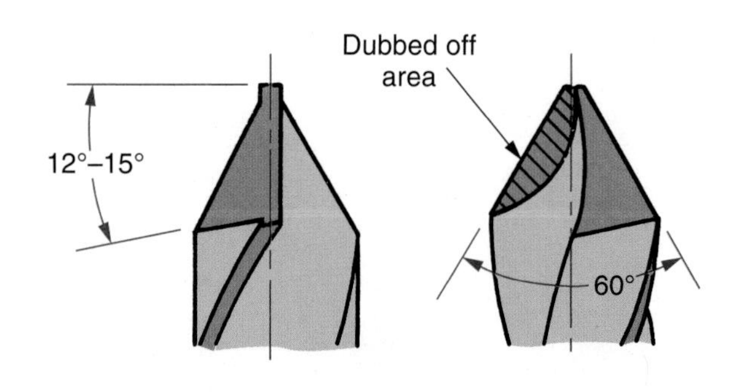

Rohm and Haas

Figure 33-9. Drill point recommended for making through holes in acrylics. The tip angle should be increased to 118° for blind holes. Smoothly finished holes can be produced by first drilling a pilot hole and filling it with wax. Wax will allow chips to move up flutes without sticking to drill.

machining experience. You can see the tool cutting inside the work. Poor cutting action can be spotted immediately and corrections made.

33.1.5 Laminated Plastics

Laminated plastics consist of layers of reinforcing materials (such as cotton fabric, paper, or glass fiber) that have been impregnated with synthetic resins. The resins are cured under heat and pressure.

Machining of laminated plastics can be accomplished with conventional machine tools. Some laminated plastics, like those containing glass fiber, are highly abrasive. Carbide tools are recommended.

SAFETY NOTE

A dust collector system and filtered dust mask or respirator must be used for operator safety when machining plastics that contain glass fiber.

Turning Laminated Plastics

A round-nose lathe tool produces the best surface finish. Speeds up to 4000 fpm (1220 mpm) can be used. Lathe work is usually done dry. However, internal threading may sometimes require use of a lubricant.

Drilling Laminated Plastics

Drilling operations are similar to those used with nylon. However, drilling parallel with the laminations should be avoided whenever possible, because the material may split along the laminations. See **Figure 33-10**.

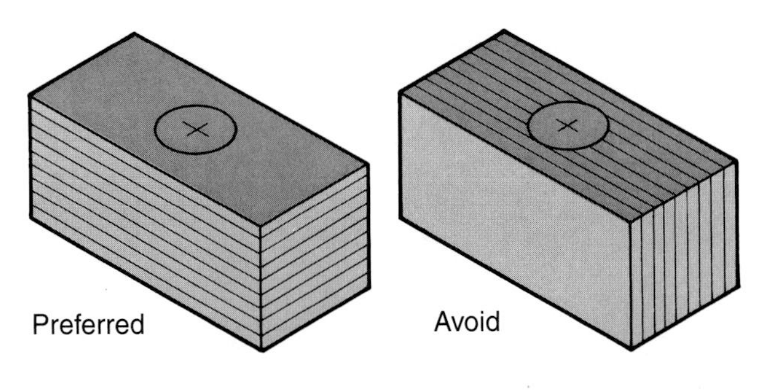

Goodheart-Willcox Publisher

Figure 33-10. Whenever possible, holes should be drilled at a right angle to the laminations. Holes drilled parallel to laminations tend to split some types of material.

Milling Laminated Plastics

Speeds up to 1000 fpm (305 mpm) are possible with good results. Feeds up to 20″ (500 mm) per minute have been used. For cotton fabric-base laminates, best results are obtained by using the highest spindle speed the cutter will stand, with the maximum feed that will produce an acceptable surface finish.

Climb milling is recommended to keep the work held tightly in the holding device and to prevent an edge from being raised.

Sawing Laminated Plastics

Band sawing is recommended for curved or straight cuts in laminated plastics when smooth edges and close tolerances are not specified. Blades with 5 to 8 teeth per inch and medium to high set can be operated at speeds up to 8000 fpm (2400 mpm). Feed the work as fast as it will cut without forcing the blade.

Threading and Tapping Laminated Plastics

Hand threading of laminated plastics can be done with standard taps and dies. It is normally done dry. High-speed steel taps that are oversize by 0.002″ to 0.005″ (0.05 to 0.13 mm) should be used if available. A slight chamfer on the hole to be tapped, or rod to be threaded, will improve the work quality by preventing the first few threads from tearing.

33.2 Chipless Machining

Chipless machining forms a metal wire or rod into the desired shape using a series of dies. This metalworking technique will not replace conventional machining, but it does result in substantial cost savings for some jobs. It can reduce waste and increase production speed. The process is sometimes called ***cold heading*** or ***cold forming***.

In chipless machining, a series of dies replace conventional machine tools like the lathe, drill press, and milling machine. Material used in chipless machining is usually in coil form and is referred to as wire. Work is transferred from station-to-station where the various dies are used to form the desired shapes, which can be quite complex.

Accuracy typically can be held to tolerances of 0.002″ (0.05 mm), and closer tolerances can be achieved, if required. However, cost increases in proportion to the precision wanted. In most cases, waste is totally eliminated.

There are five basic operations that can be performed by the chipless machining process, **Figure 33-11**.

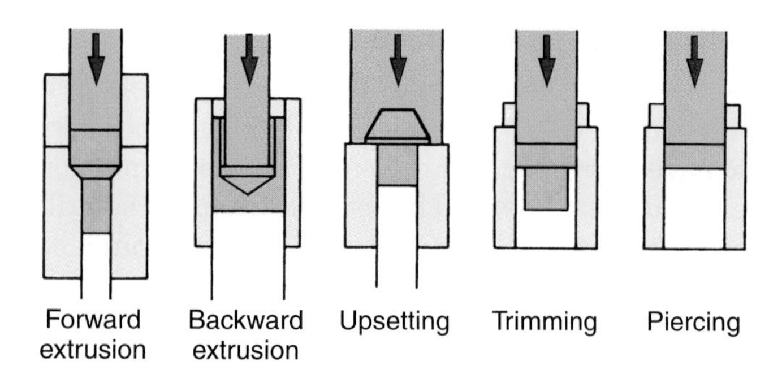

Figure 33-11. The five basic operations performed by machines designed for chipless machining.

Combinations and variations of these operations make possible a wide range of applications.

Chipless machining is an economical and efficient way to make bolts, nuts, screws, and other fasteners, **Figure 33-12**. It is also used to make the vast majority of spark plug bases, **Figure 33-13**.

Metals ranging from aluminum alloys to medium-carbon steel can be shaped using this technique. Stainless steel, copper, and nickel alloys can be cold formed, but the ease with which they are shaped depends upon the part design. Material up to 1.5″ (37.5 mm) in diameter can be formed on some machines.

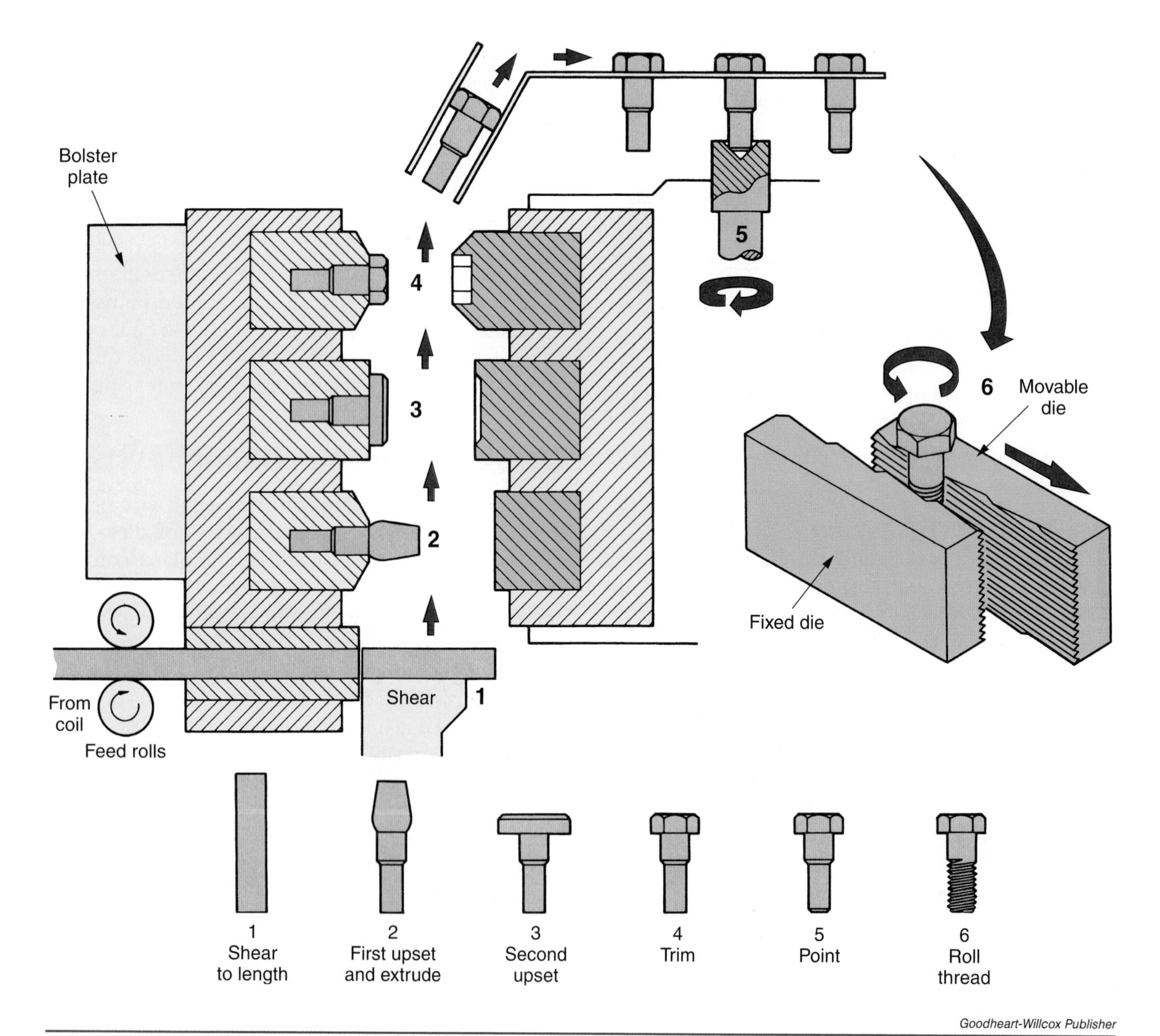

Figure 33-12. Arrows and numbers indicate sequence involved in producing bolts by chipless machining. Trace the part flow through the sequence of operations.

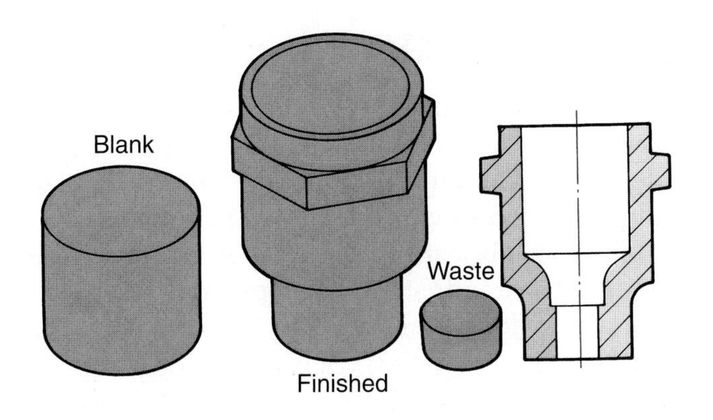

Goodheart-Willcox Publisher

Figure 33-13. Almost all spark plug bases are made by chipless machining. The resulting scrap is minimal.

33.2.1 Intraform Machining

Intraform® is a type of chipless machining used to form profiles on the inside diameter of a hollow cylindrical workpiece. Forming inside profiles would be extremely difficult and expensive to do by conventional machining techniques.

In this process, a section of hollow cylindrical stock is placed over a steel mandrel, **Figure 33-14.** It is then squeezed by rapidly pulsating dies, **Figure 33-15.** At the completion of the operation, the mandrel's profile is produced on the inside diameter of the part, **Figure 33-16.**

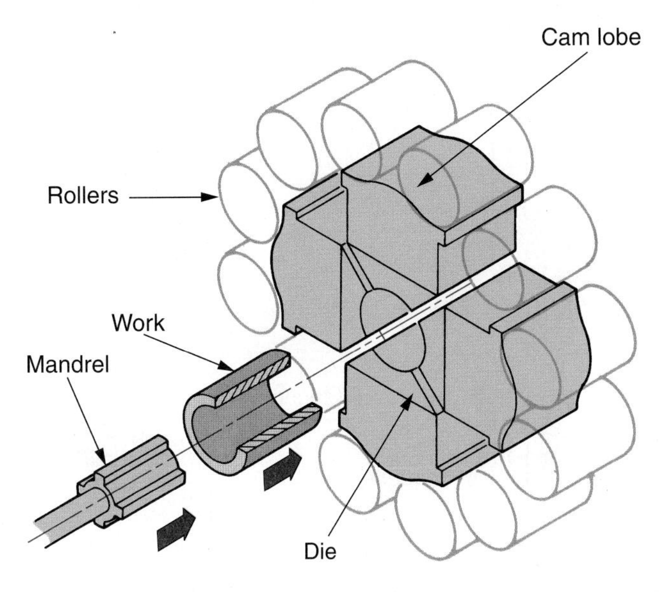

Goodheart-Willcox Publisher

Figure 33-14. A part ready to be shaped by the Intraform process. The mandrel is inserted into the work. The work and mandrel are then inserted into the dies. Note that the cam lobes are positioned in the spaces between the rollers, so the pressure on the dies is minimal.

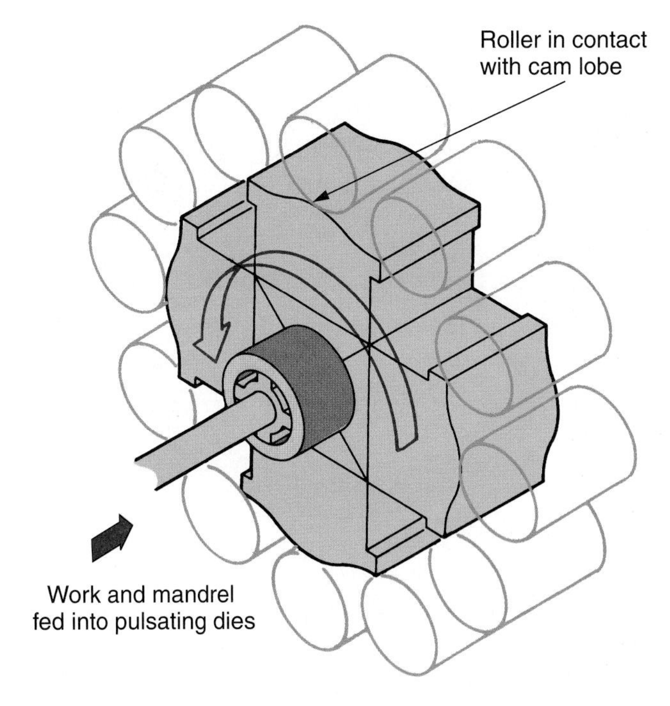

Goodheart-Willcox Publisher

Figure 33-15. As the dies rotate, the rollers apply increasing pressure as they ride up on the cam lobes. Contact with the rotating dies causes the work and mandrel to rotate at about 80% of the die rpm. The work feeds over the mandrel.

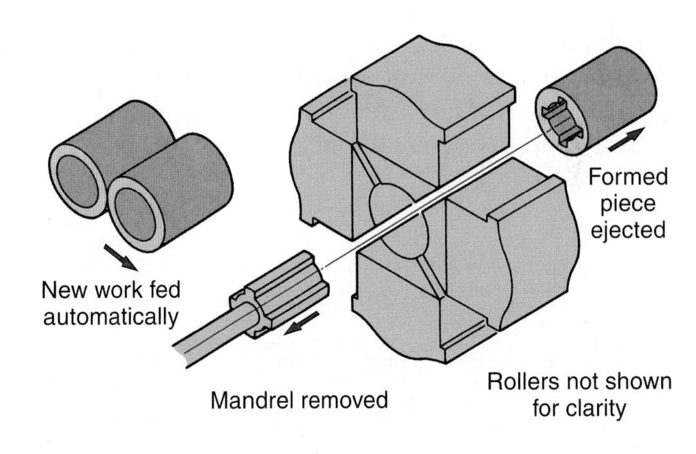

Goodheart-Willcox Publisher

Figure 33-16. When the operation is completed, the mandrel is retracted. The next piece feeds automatically into position while the formed part is ejected.

Figure 33-17 shows how fixed rollers cause the four dies to pulsate rapidly around the outside diameter of the work. The cam lobes are shaped to provide a smooth continuous squeezing action of the dies. Even though the work is being squeezed by the dies more than 1000 times per minute, there is no excessive noise or vibration.

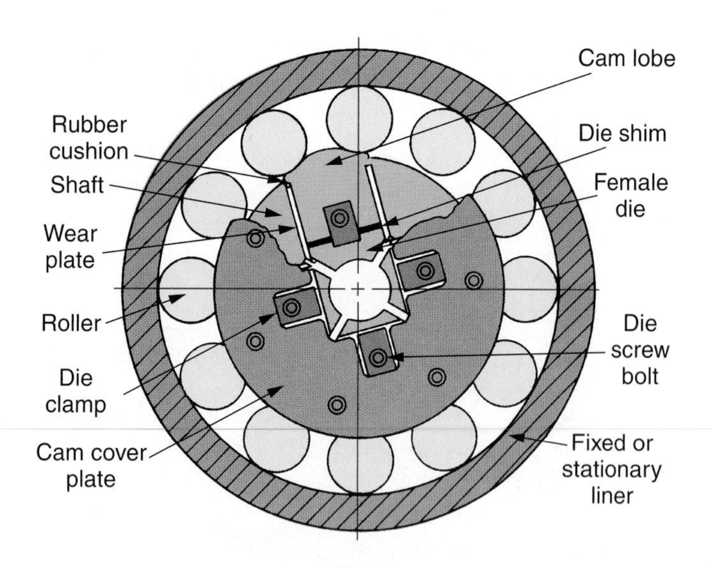

Goodheart-Willcox Publisher

Figure 33-17. An Intraform machine die head in the open position (cam lobes between rollers). Interaction of the cams and rollers squeezes each die more than 1000 times per minute.

The technique has proven to be a practical way to produce rifle barrels for example. Predrilled steel blanks are fed into the machine which forms the chamber and rifling. In addition to improving the surface finish of the bore, the operation also improves the physical characteristics of the metal. Approximately 80% of all barrels produced by the process are of target rifle quality, compared to only 10% shaped by conventional methods.

33.3 Powder Metallurgy

Powder metallurgy (P/M) is a technique used to shape parts from metal powders. These parts can be quite complex, **Figure 33-18**. Sometimes called sintering, the process was developed in the late 1920s to make self-lubricating electric motor bearings for the automotive industry. The steps involved in fabricating powder metallurgy products are shown in **Figure 33-19**.

Metal Powder Industries Federation

Figure 33-18. An assortment of products made by the powder metallurgy process.

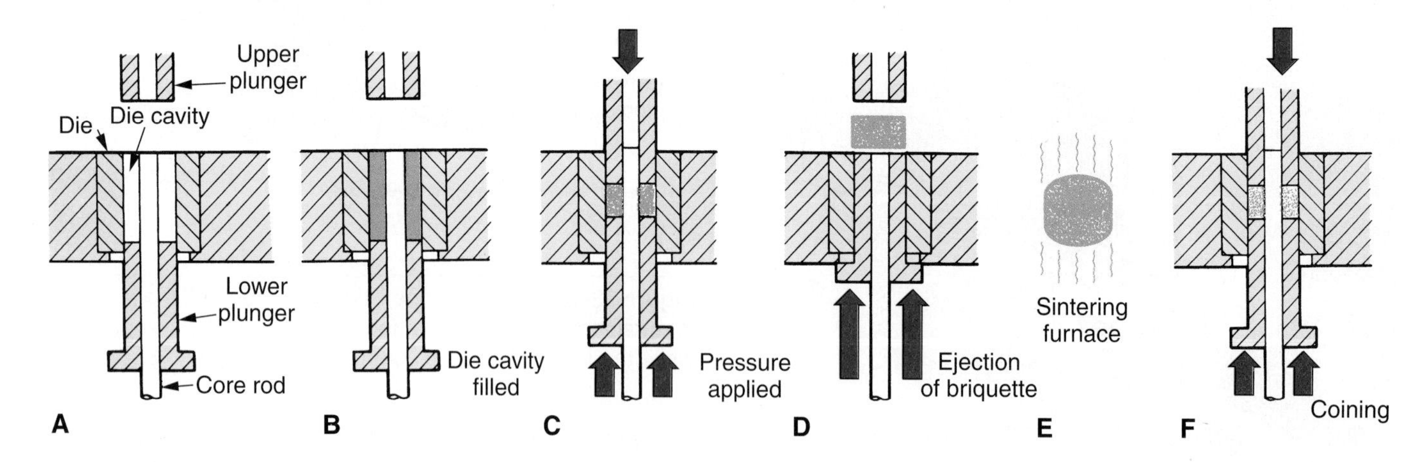

Goodheart-Willcox Publisher

Figure 33-19. Steps in fabricating a part, using the powder metallurgy process. A—Note the cross section of the die and the die cavity. The depth of the cavity is determined by the thickness of the required part, and the amount of pressure that will be applied. B—The die cavity is filled with the proper metal powder mixture. C—Pressure as high as 50 tons per square inch is applied. D—A briquette or "green compact" is pushed from die cavity. E—Pieces are then passed through a sintering furnace to convert them into a strong, useful product. F—Some pieces can be used as they come from the furnace. Others may require a coining or sizing operation to bring them to exact size and to improve their surface finish.

33.3.1 Powder Metallurgy Applications

The P/M process is widely used by industry for such applications as:

- Self-lubricating bearings and bearing materials.
- Precision finished machine parts, such as gears, cams, or ratchets, with tolerances as close as ±0.0005″ (0.0127 mm).
- Permanent metal filters (sintered bronze fuel filters, for example).
- Fabricating materials that are difficult to work.
  - Tough cutting tools (tungsten carbide), **Figure 33-20**.
  - Aluminum-nickel alloy (Alnico) magnets.
  - Mixtures of metals and ceramics (cermets) for jet and rocket applications that require the heat resistance of ceramics as well as the heat transfer qualities of metals. These materials are also used to make special cutting tools.
  - High-density counterweights for aerospace instruments that require maximum weight concentration in a minimum space.
  - Nickel-cadmium (NiCad) storage battery elements.

Kennametal, Inc.

Figure 33-20. A technician examines a tray of tungsten carbide cutting tool inserts following the application of a wear-resistant coating that extends tool life. The inserts were made using the powder metallurgy process.

- Wrought powder metallurgy tool steels are now available in mill forms (bars, rods, flats, wire, and plate). They offer improved machinability, wear resistance, increased toughness, and better dimensional stability than conventional cast or wrought tool steels. P/M tool steels are primarily used for metal-cutting and metal-forming operations.

33.3.2 Powder Metallurgy Process

The first phase in the manufacture of powder metal products is the careful mixing of high-purity metal powders. The powders are carefully weighed and thoroughly mixed into a blend of correct proportions. Many materials, including iron, steel, stainless steel, brass, bronze, nickel, and chromium can be used. Combinations of these metals and nonmetals can also be used.

Briquetting

The powder blend is then fed into a precision die. The die cavity has the shape of the desired part, but it is several times deeper than the thickness of the part. The powder is compressed by an upper and lower punch. Pressures applied range from 15 to 50 tons per square inch. This portion of the operation is known as ***briquetting***.

The piece, as ejected from the die, appears to be solid metal. However, this "green compact" is quite brittle and fragile. It will crumble if not handled carefully.

Sintering

To transform the green compact into a strong, useful unit, it must be heated to a temperature of 1500°F to 2300°F (816°C to 1260°C) for 30 minutes to 2 hours, depending on the powder mixture. This process, known as ***sintering***, is done in a controlled atmosphere furnace. The atmosphere within the sintering furnace is carefully controlled to prevent oxidation or contamination of the parts being processed.

Finishing

Many parts can be used as they come from the furnace. However, because of shrinkage and distortion caused by the heating operation, the pieces may have to go through a sizing, coining, or forging operation. The part must be reheated just prior to undergoing these operations, which press the sintered pieces into accurate size dies to obtain precise finished dimensions, higher densities, and smoother surface finishes. Powder metal parts can also be drilled, tapped, plated, heat treated, machined, or ground.

33.3.3 Powder Metallurgy Costs

The tool cost is moderate and the process is best suited to quantity production. If production quantities exceed several thousand units, the finished piece can be produced by the P/M method at less than the cost of rough sand castings.

33.4 High-Energy-Rate Forming (HERF)

The introduction of super-tough alloys for aerospace vehicles and the need for shaping thin, brittle metal has been responsible for the development of new ways to do the work. One of these new techniques is known as ***high-energy-rate forming (HERF)***. HERF uses extreme pressures to form a material around a die. There is little similarity between it and conventional metalworking processes like turning, drilling, and milling.

Problems develop when attempting to shape the super-tough alloys by conventional means. They exhibit "spring back," where the metal tends to try to regain its original shape, **Figure 33-21**.

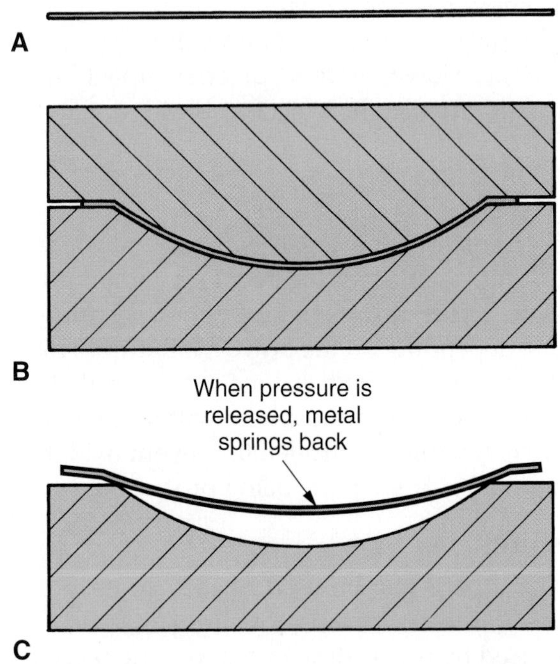

A

B

When pressure is released, metal springs back

C

Goodheart-Willcox Publisher

Figure 33-21. Many metals tend to spring back to near their original shape after being formed by conventional means. A—A flat sheet-metal blank ready for forming. B—The metal is formed between dies. C—The metal tries to return to its original shape when the male die is removed or pressure is released.

It is difficult and costly to shape these metals to acceptable tolerances by conventional means.

In HERF, the metal is shaped in microseconds with pressures generated by the sudden application of large amounts of pressure. The great pressures are generated by detonating explosives, releasing compressed gases, discharging powerful electrical sparks, or electromagnetic energy. The metal, in most cases, is slammed against the die and shaped so rapidly that there is no tendency for the material to try to return to its original shape.

HERF offers many advantages. Tool costs are reduced, and there is usually no need for expensive machinery. There appears to be no limitation to the size of the sections that can be formed.

33.4.1 Explosive Forming

Explosive forming uses the high-pressure wave of an explosive charge to form the metal. It is an older technique, originating in the late 1800s to shape ornate door knobs and similar products. In more recent years, it has been adapted to modern machining. See **Figure 33-22**.

The size of many aerospace and marine components makes them impossible to form them in existing presses, **Figure 33-23**. The presses are either too small or are not powerful enough to develop the pressure required to shape the high-strength alloys.

Alan Freed/Shutterstock.com

Figure 33-22. High-energy-rate forming (HERF) was used to shape end sections of the space shuttles' external fuel tanks.

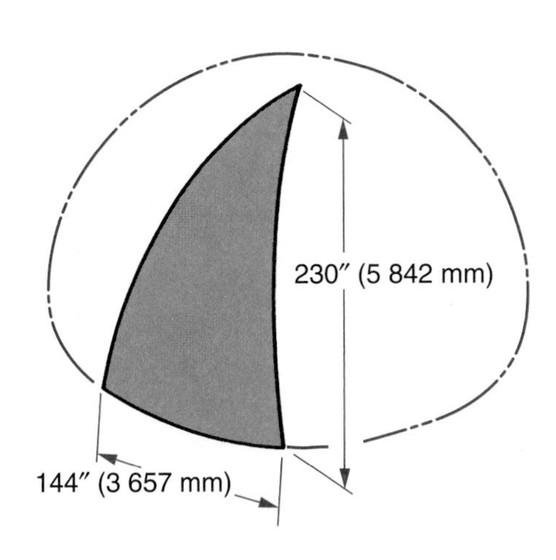

Goodheart-Willcox Publisher

Figure 33-23. Explosive forming was used to shape segments for the domes of the space shuttles' external fuel tanks. The segments were welded together after forming.

Explosive forming uses the pressure wave generated by an explosion in a fluid to force the material against the walls of the die, **Figure 33-24**. The fluid helps to distribute the pressure pulse generated by the detonation equally across the surface of the material.

In preparation for explosive forming, the metal is cut or fabricated to a shape determined by the contours of the finished part. This preform is placed in the die, the die is filled with water, and an explosive charge suspended in the water.

A large holding ring, clamped over the outer edge of the work, provides the necessary seal for drawing a vacuum in the die. A vacuum is necessary between the work and the die. Otherwise, the air trapped between the preform and die would become pressurized by the explosion, preventing the metal from seating in the die and assuming its proper shape.

When the explosive is detonated, the resulting pressure slams the material against the die walls. The forming process is accomplished in microseconds.

Placement and quantity of the high explosive is critical. The charge can range from a few ounces to form small parts, to the many pounds needed to form large sections of aluminum and steel up to 4″ (100 mm) thick. For example, the heavy steel missile hatches on the Navy's submarines, **Figure 33-25**, are formed by this technique. Many forms of explosives are used: rod, sheet, granules, liquid, stick, cord, and plastic.

U.S. Navy

Figure 33-25. Many parts used on nuclear submarines are shaped by HERF.

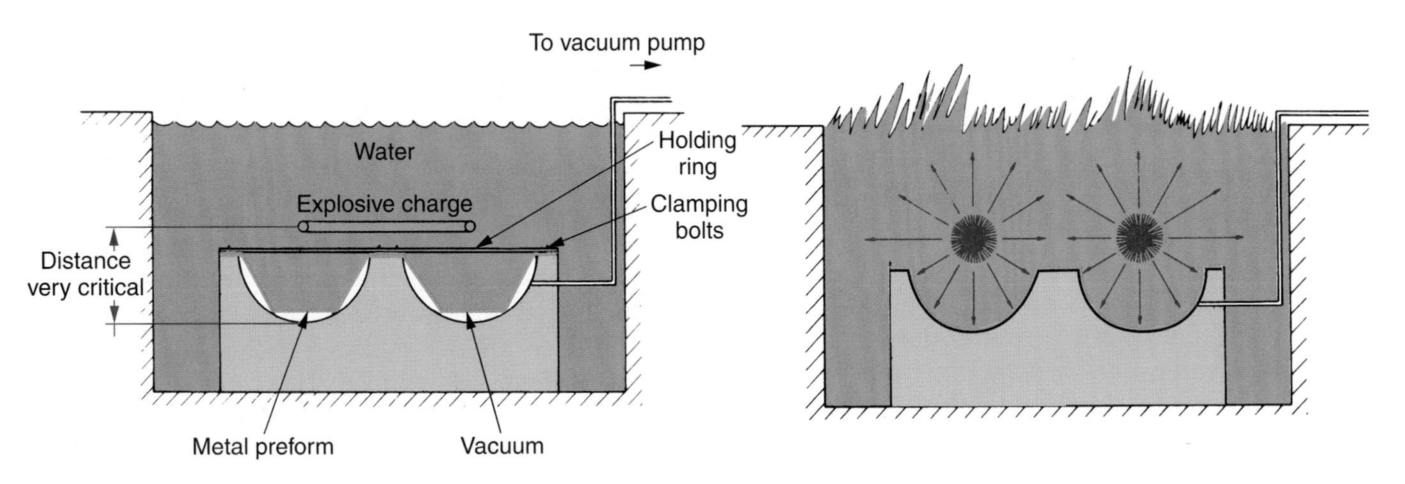

Goodheart-Willcox Publisher

Figure 33-24. This diagram shows the principle of the explosive forming process. In some applications, a few dollars worth of explosives will do the work of a press that may cost a million dollars or more.

Depending on the placement of the explosive, HERF operations fall into two categories: stand-off and contact.

Stand-Off Operations

In stand-off HERF, the charge is located some distance from the work. Its energy is transmitted through a fluid medium, such as water. This technique is used to form and size parts.

Contact Operations

In contact HERF, the charge is touching the work and the explosive energy acts directly on the metal. Welding, hardening, compacting powdered metals, and controlled cutting are done with this technique.

Disadvantages of Explosive Forming

While explosive forming offers many advantages, there are also some drawbacks associated with the process:

- The technique has not been developed to the stage where a part can always be formed properly on the first shot.
- Since the operation uses an explosion to do the forming, the noise can be a problem.
- Strict laws prohibit the use of explosives in populated areas. As a result, it is usually necessary to locate the facility in an isolated site. This increases transportation and handling costs.
- Personnel must be highly skilled in the safe handling of high explosives.
- Insurance rates are high.

33.4.2 Electrohydraulic Forming

Electrohydraulic forming, also called capacitor discharge forming or spark forming, is similar to explosive forming, but uses electricity rather than an explosive charge to generate the required pressure. See **Figure 33-26**. Many of the titanium parts on aerospace vehicles are formed using this technique. Titanium is tough and light metal, but it is difficult to work by conventional methods.

In electrohydraulic forming, the work is submerged in water, as is done with explosive forming. High-voltage electrical energy is discharged from a capacitor bank (an array of devices used to store electrical energy) into a thin wire or foil suspended between two electrodes, which are also submerged.

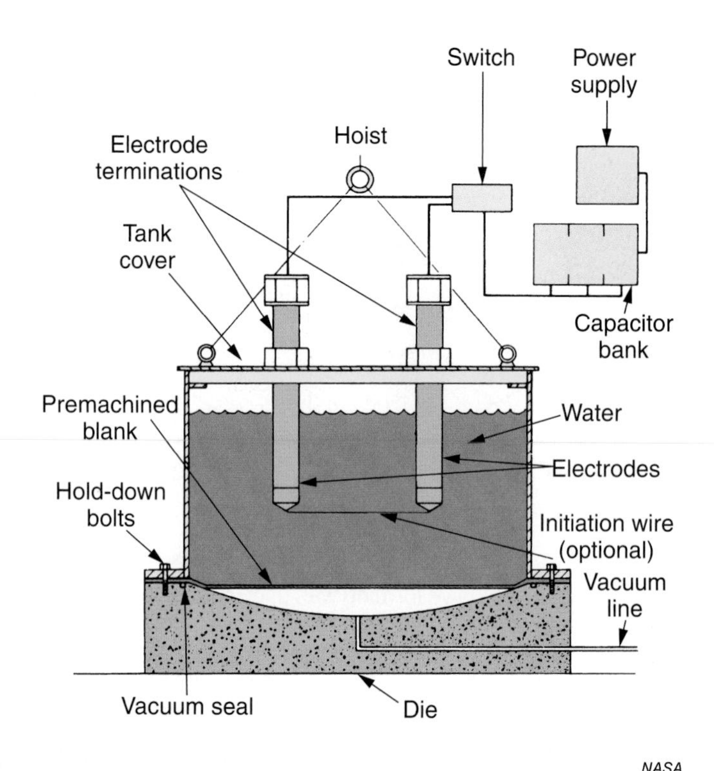

Figure 33-26. Diagram showing the setup for electrohydraulic forming, which uses electrical energy as a source of power for HERF operations.

As the wire or foil is vaporized by the electric current discharge, the vapor products expand, converting the electrical energy to hydraulic energy. The shock wave forms the metal against the die. Since the energy produced is less than that associated with explosives, it is usually necessary to repeat the operation several times to achieve the desired results.

A well-designed electrohydraulic forming facility can be adapted to automation. Because it generates high pressures without an explosion, the electrohydraulic forming process can be used in conventional industrial facilities.

33.4.3 Magnetic Forming

Magnetic forming, also termed electromagnetic forming or magnetic pulse forming, uses an insulated induction coil wrapped around or placed within the work, **Figure 33-27**. As very high momentary currents are passed through the coil, an intense magnetic field is developed. This causes the work to collapse, compress, shrink, or expand depending on the design of the coil. Coil location depends upon whether the metal is to be squeezed inward or bulge outward. The coil is shaped to produce the desired shape in the work. See **Figure 33-28**.

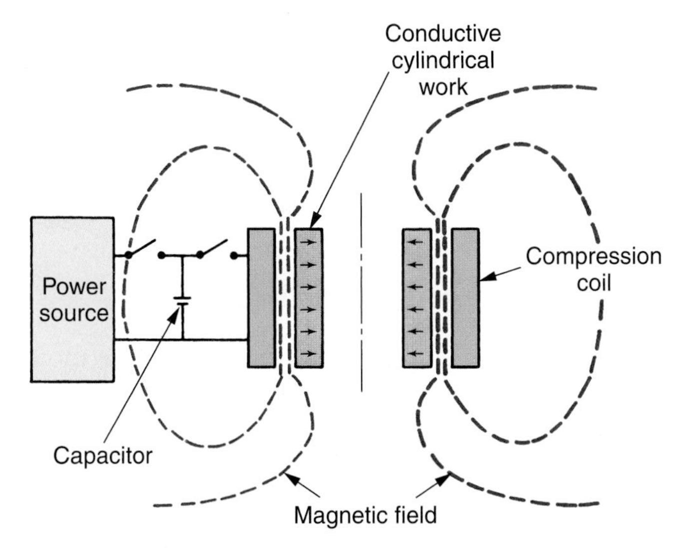

Figure 33-27. Magnetic pulse metal forming. A strong magnetic field is produced by discharging a capacitor through a coil. During the brief impulse, eddy currents in the work restrict the magnetic field to surface of the workpiece. This creates a uniform force to form the metal. The process can be used directly on highly conductive metals. Low-conductivity metals can be shaped by using a layer of highly conductive aluminum between work and coils.

The power source is basically the same as that used for electrohydraulic forming—a capacitor bank. The spark gap, however, is replaced by a coil. Energy is obtained by capacitor discharge through the coil. In fact, properly designed equipment can be used to perform either electrohydraulic or magnetic pulse operations.

The energy storage capacity and the ability to use that energy determine the size of the work that can be formed. Highly conductive metals can be formed easily by this process. Nonconductive or low-conductivity materials can be formed if they are wrapped or coated with a high-conductivity auxiliary material.

33.4.4 Pneumatic-Mechanical Forming

Pneumatic-mechanical forming uses a punch and die operated by high-pressure gas. It was the first of the HERF techniques to become a standard production tool. The operation has much in common with conventional forging, since a punch and die are used. However, the forces developed are many times more powerful and are sufficient to shape hard-to-work materials. The metal blank is heated prior to the forming operation. The machine requires less space than the conventional forging press.

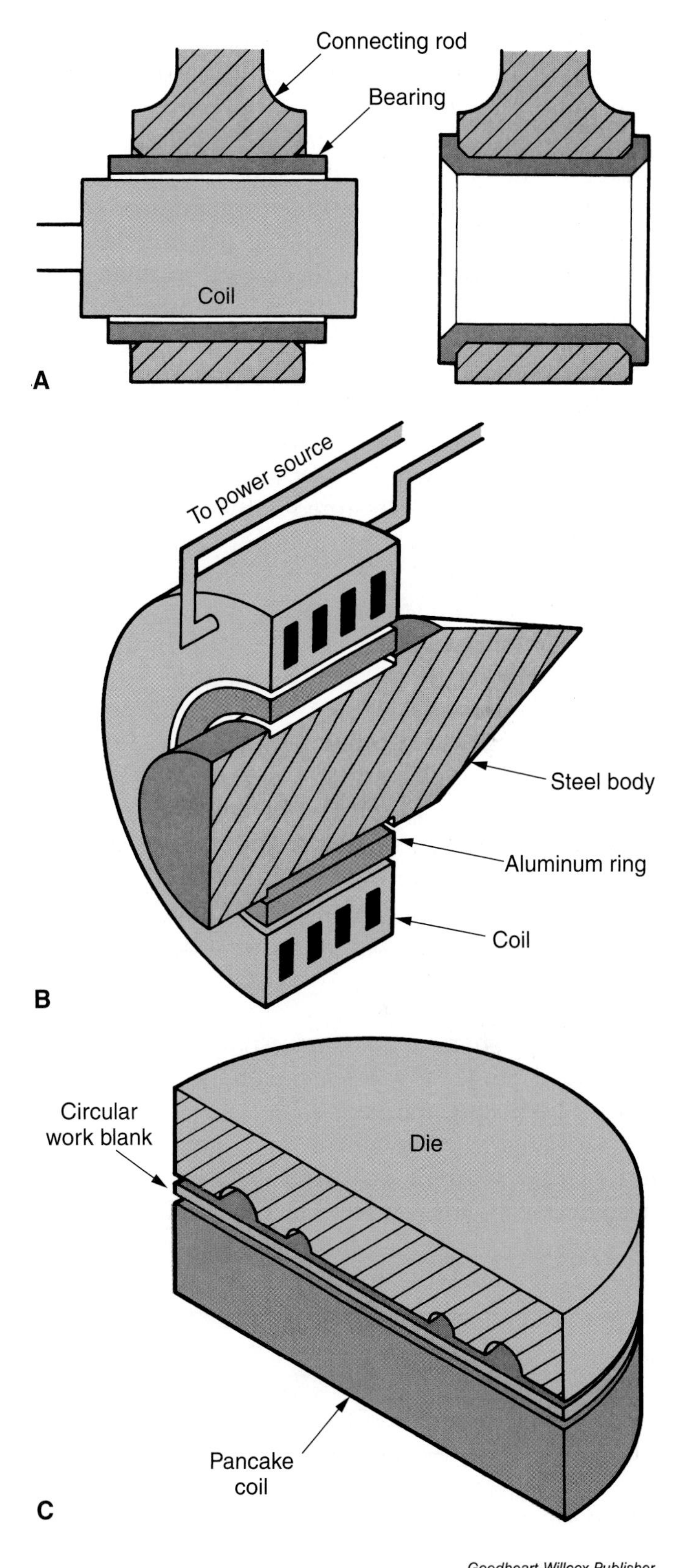

Figure 33-28. Magnetic pulse forming applications. A—Magnetic forming can be used to expand a bearing sleeve into a connecting rod. B—A magnetic field can be used to shrink or squeeze parts together. C—Flat forming applications, which involve forcing sheet metal into a die, require a pancake coil to provide uniform magnetic pressure.

In pneumatic-mechanical forming, a high-pressure gas is used to accelerate a punch into a die. The punch and die are mounted on opposed rams that meet with equal force, thus taking most of the strain off the frame of the machine. Some machines make use of a recoil mechanism similar to that used on an artillery piece. Because of their self-reacting designs, most machines impart no shock load to floor. This makes it possible to use them in close proximity to conventional chip-making machines.

Pneumatic-mechanical forming can be accomplished by several means. The most common method uses a two-part cylinder, **Figure 33-29**. Gas is stored in one part of the cylinder at approximately 2000 psi (13,800 kPa). Gas is also stored in the second section of the cylinder, but at a much lower pressure, about 200 psi (1380 kPa). A plate with an orifice separates the two sections of the cylinder. The piston (ram), with a special seal, closes off the orifice. In this way, the high-pressure gas acts only on a small section of the piston (area of the orifice) while the low-pressure acts on the entire area of the piston. This maintains a stable balance between the two sections.

To make the machine operate, the pressure is increased slightly in the high-pressure section. This upsets the balance and the piston starts to move. When the seal is disengaged, the high pressure instantly acts on the entire surface area of the piston driving it forward at tremendous speed, **Figure 33-30**.

Hard-to-shape metals are usually formed with one blow. Some advanced materials can best be shaped by this technique.

With the pneumatic-hydraulic method of forging, precise control is possible. Fewer operations are needed (most parts can be formed with one stroke of

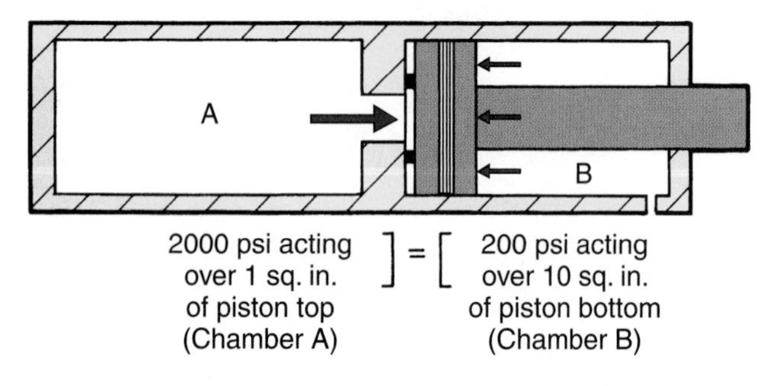

Figure 33-29. High-pressure gas acts on only a small area of the piston top (area of the orifice) while low-pressure gas acts on the entire area of the piston bottom. This maintains a balance between the two forces.

the press), so production is rapid. The parts are produced to close tolerances with smoothly finished surfaces that require a minimum of machining. Because less material is used, fewer operations must be performed to finish the part, and higher production rates are obtained, substantial savings are possible.

33.5 Cryogenic Applications

The science of cryogenics is one of the more recent additions to field of machining and metalworking. Cryogenics is not a new or different way to work metal; rather, it is used to improve or reinforce other metalworking techniques. The term "cryogenic" literally means "to make icy cold." It combines the Greek word *kryos*, meaning "icy cold," and the Latin *generatus*, which means "to make or create."

Cryogenics deals with temperatures beginning at the point where oxygen liquefies (approximately –300°F or –184°C), and going down to just about absolute zero (–460°F or –273°C). At this temperature, every element but helium freezes. Oxygen and nitrogen look something like white beach sand or table salt. Metal also acts strangely. Lead coils act like steel springs, some metals increase tremendously in strength, and others become superconductors of electricity.

Cryogenics is widely used in annealing and the heat treatment of metals. The characteristics of a number of aluminum alloys and some space-age metals are greatly improved by first heating them, and then quenching them in liquid nitrogen.

Cryogenics is also used to shrink-fit metal parts together. Since most metals shrink in size as they become cold, one part of the assembly is made slightly oversize and then immersed in liquid nitrogen. The exact amount of oversize is determined by part size and type of metal. The diameter is reduced (shrunk) by the extreme temperature drop until it fits easily into its mating part. As the cooled part returns to room temperature, it expands and is thus locked in place. The parts do not become distorted as they would if they were mechanically pressed together or heated (expanded) until they could be fitted together.

33.5.1 Cryogenic Treatment of Cutting Tools

Research has shown that cryogenic treatment of cutting tools can greatly extend their operating life. Cryogenic treatment produces carbide inserts that last two to eight times longer than untreated inserts.

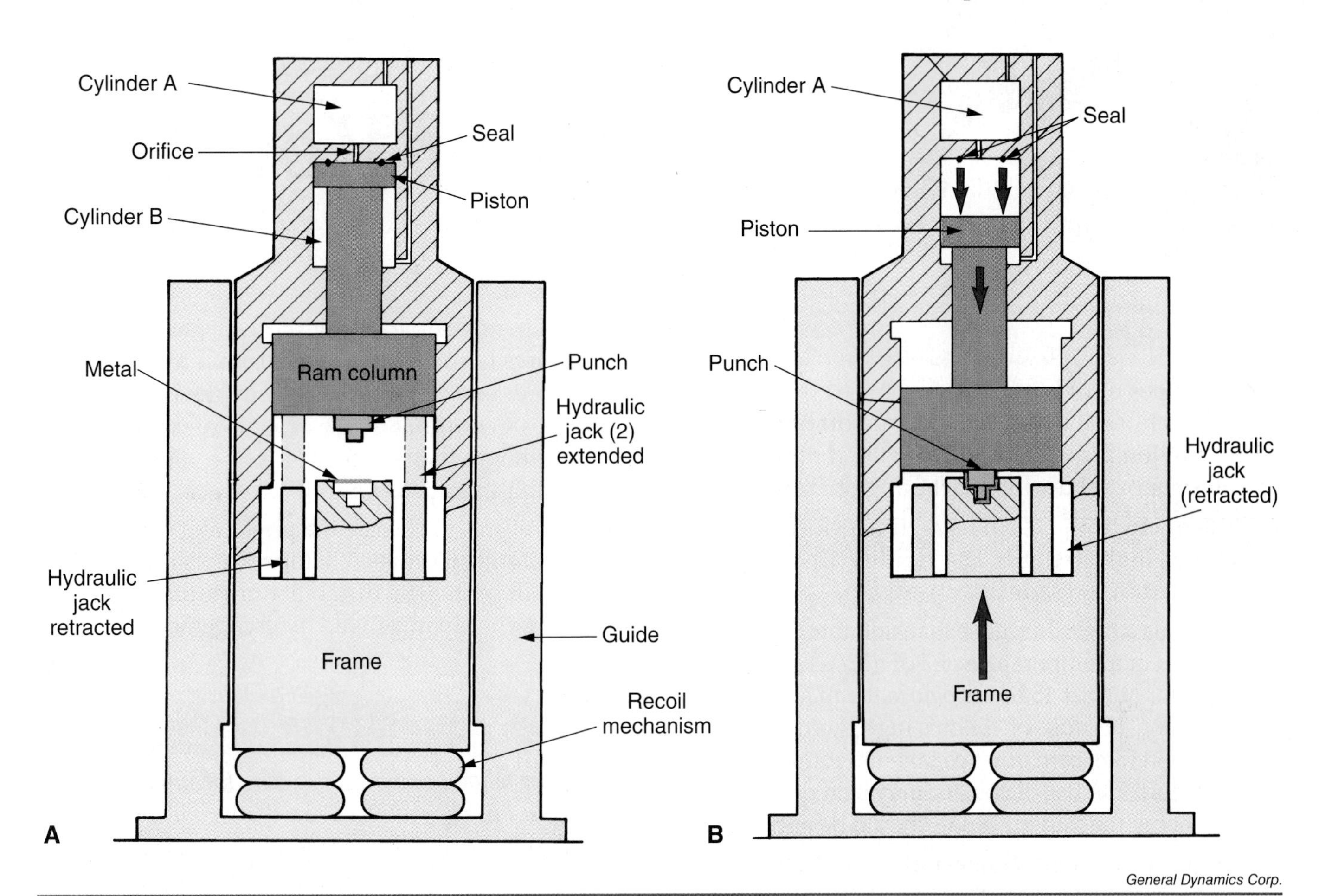

General Dynamics Corp.

Figure 33-30. Pneumatic-mechanical forming. A—Cross-sectional view of a pneumatic-mechanical forming press. The hydraulic jacks extend at the end of each operating cycle to lift the ram column back into position for the next cycle. B—The operation of the press is triggered when the pressure in Cylinder A is increased enough to break the seal. This slight movement allows high-pressure gas to act instantaneously over entire area of the piston. The ram is driven downward at great speed. At the same time, frame moves upward by reaction of gas pressure over a driven piston. The frame and ram are acted upon with equal thrust; so each has equal momentum but in opposite directions. To reset for the next cycle, the jacks lift ram column upward until it seats against the seal.

Similar treatment of copper spot welding electrodes increases their useful life by 300%. Other types of cutting tools have shown increases of up to 400% in tool life.

During cryogenic treatment, liquid nitrogen enters the cooling chamber as a gas, at a carefully controlled rate. The cutting tools are not immersed in liquid nitrogen. Cooling occurs at a slow rate to avoid a sudden change in temperature that would cause damaging thermal shock.

The full cooling cycle takes about 40 hours. Four to six hours are required to lower the temperature to about –320°F (–196°C). The tools "soak" at this temperature for 24 hours or more. Another four to six hours are required to return the tools to room temperature. See **Figure 33-31**. For larger or more massive items, the time required to lower and raise temperature in the cooling chamber must be increased.

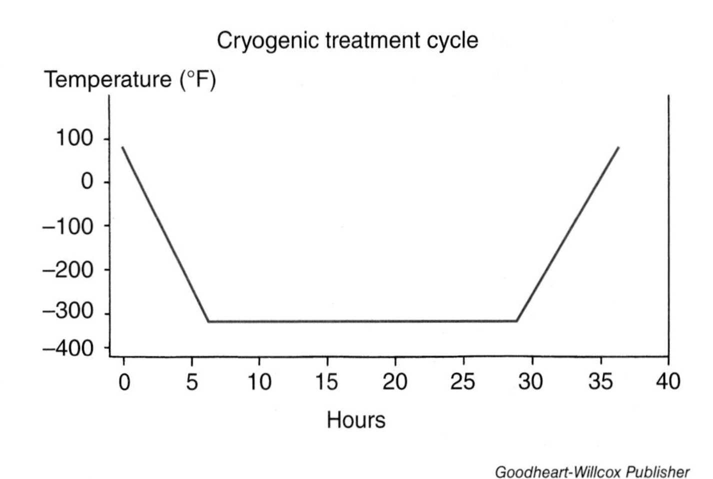

Goodheart-Willcox Publisher

Figure 33-31. To achieve the maximum benefit from the cryogenic treatment, cutting tools require a gradual lowering of temperature, an extended soak period, and a slow return to room temperature.

Chapter Review

Summary

- Most types of nylon can be machined with the same techniques used to machine soft brass. Since nylon is not as rigid as metal, it must be well supported during machining operations.
- Parts made from Delrin have dimensional stability, high strength, and rigidity. Delrin is machined in the same way as nylon.
- Teflon has a high thermal expansion rate. It should be stored at a temperature of 74°F (23°C) or higher for at least 48 hours before the machining operations. Because of Teflon's high thermal expansion rate, care must be taken to minimize expansion. The use of coolant may be required. After being machined, Teflon should be annealed.
- Acrylics have good dimensional stability and high impact strength, even at elevated temperatures. They are easy to machine, form, and polish. Sharp tools with adequate clearances should be used.
- Laminated plastics consist of layers of reinforcing materials that have been impregnated with synthetic resins. Carbide tools may be required for machining some laminated plastics. Drilling parallel with the laminations should be avoided whenever possible.
- Chipless machining is a process in which a metal wire or rod is forced into the desired shape using a series of dies. It is an economical and efficient way to make bolts, nuts, screws, and other fasteners.
- Intraform® is a type of chipless machining used to form profiles on the inside diameter of a hollow cylindrical workpiece.
- Powder metallurgy, or sintering, is a method of making parts from metal powders. Metal powders are pressed in a die to form a green compact. The green compact is fragile when it emerges from the die. The green compact is placed in a controlled-atmosphere furnace to strengthen. If necessary, the part can be sized, coined, or forged after it is taken out of the furnace.
- High-energy rate forming (HERF) uses extreme pressures to form a material around a die. The pressures can be generated by explosives, the release of compressed gas, electrical discharge, or electromagnetism.
- Cryogenics is the application of extremely cold temperatures to change the physical properties of a material. Cryogenic temperatures are used to shrink-fit parts. The durability of cutting tools is also commonly improved by cryogenic treatment.

Review Questions

Answer the following questions using the information provided in this chapter.

1. Give a common trade name for each of the following types of plastics:
   A. Polyamide resins.
   B. Acetal resins.
   C. Fluorocarbon resins.
   D. Acrylic resins.
2. If plastics are to be machined with any degree of accuracy, the cutting tools must be _____.
3. Like metal, machined nylon parts require _____ to ensure against dimensional changes.
4. When turning Teflon on a lathe, care must be taken to prevent _____ from accumulating around the work. If this is not done, heat will build up and cause the plastic to become distorted.
5. When drilling laminated plastics, avoid drilling holes _____.
   A. parallel to the laminations
   B. perpendicular to the laminations
   C. all the way through the material
   D. All of the above.
   E. None of the above.
6. Chipless machining is also known as _____ or cold forming.

7. In chipless machining, metal is formed into the desired shape using a series of _____.
8. List the five basic operations that can be performed by the chipless machining process.
9. Intraform is a chipless machining technique that can form profiles on the inside of hollow _____ pieces.
10. The powder metallurgy process is used to make _____.
    A. self-lubricating bearings
    B. precision machine parts
    C. permanent metal filters
    D. All of the above.
    E. None of the above.
11. List the steps, in proper order, for making a part by the powder metallurgy technique.
12. What is a briquette or "green compact?"
13. The abbreviation HERF means _____.
14. In HERF, metal is shaped _____.
    A. by the slow application of great pressure
    B. with pressure generated by the sudden application of large amounts of energy
    C. by conventional forging methods
    D. All of the above.
    E. None of the above.
15. The pressures needed in HERF are generated by _____.
    A. detonating explosives
    B. releasing compressed gases
    C. electromagnetic energy
    D. All of the above.
    E. None of the above.
16. _____ is a HERF technique that uses an induction coil to shape metal.
17. In pneumatic-mechanical forming, _____ is used to accelerate the punch into the die.
    A. pressurized oil
    B. pressurized gas
    C. explosives
    D. All of the above.
    E None of the above.
18. The science of cryogenics deals with temperatures ranging from –300°F (–184°C) to _____.
    A. 32°F (0°C)
    B. 0°F (–18°C)
    C. –100°F (–73°C)
    D. –460°F (–273°C)
19. Why is it better to use cryogenics, rather than heat, to shrink-fit parts together?
20. Why must the cooling of treated cutting tools be done at a slow rate?

Reference Section

The following pages contain a number of tables, charts, and other materials that will be useful as reference in a variety of machining-related areas. To make locating information easier, the material in this section is listed below, along with the page number.

Common Shapes of Metals

Shapes		Length	How Measured	How Purchased
	Sheet less than 1/4″ thick	Up to 144″	Thickness × width, widths to 72″	Weight, foot, or piece
	Plate more than 1/4″ thick	Up to 20′	Thickness × width	Weight, foot, or piece
	Band	Up to 20′	Thickness × width	Weight, or piece
	Rod	12′ to 20′	Diameter	Weight, foot, or piece
	Square	12′ to 20′	Width	Weight, foot, or piece
	Flats	Hot rolled 20′-22′ Cold finished	Thickness × width	Weight, foot, or piece
	Hexagon	12′ to 20′	Distance across flats	Weight, foot, or piece
	Octagon	12′ to 20′	Distance across flats	Weight, foot, or piece
	Angle	Up to 40′	Leg length × leg length × thickness of legs	Weight, foot, or piece
	Channel	Up to 60′	Depth × web thickness × flange width	Weight, foot, or piece
	I-beam	Up to 60′	Height × web thickness × flange width	Weight, foot, or piece

Color Codes for Marking Steels

S.A.E. Number	Color Code	S.A.E. Number	Color Code	S.A.E. Number	Color Code	S.A.E. Number	Color Code
	Carbon steels	2115	Red and bronze	T1340	Orange and green	3450	Black and bronze
1010	White	2315	Red and blue	T1345	Orange and red	4820	Green and purple
1015	White	2320	Red and blue	T1350	Orange and red		**Chromium steels**
X1015	White	2330	Red and white		**Nickel-chromium steels**	5120	Black
1020	Brown	2335	Red and white	3115	Blue and black	5140	Black and white
X1020	Brown	2340	Red and green	3120	Blue and black	5150	Black and white
1025	Red	2345	Red and green	3125	Pink	52100	Black and brown
X1025	Red	2350	Red and aluminum	3130	Blue and green		**Chromium-vanadium steels**
1030	Blue	2515	Red and black	3135	Blue and green		
1035	Blue		**Molybdenum steels**	3140	Blue and white	6115	White and brown
1040	Green	4130	Green and white	X3140	Blue and white	6120	White and brown
X1040	Green	X4130	Green and bronze	3145	Blue and white	6125	White and aluminum
1045	Orange	4135	Green and yellow	3150	Blue and brown	6130	White and yellow
X1045	Orange	4140	Green and brown	3215	Blue and purple	6135	White and yellow
1050	Bronze	4150	Green and brown	3220	Blue and purple	6140	White and bronze
1095	Aluminum	4340	Green and aluminum	3230	Blue and purple	6145	White and orange
	Free cutting steels	4345	Green and aluminum	3240	Blue and aluminum	6150	White and orange
1112	Yellow	4615	Green and black	3245	Blue and aluminum	6195	White and purple
X1112	Yellow	4620	Green and black	3250	Blue and bronze		**Tungsten steels**
1120	Yellow and brown	4640	Green and pink	3312	Orange and black	71360	Brown and orange
X1314	Yellow and blue	4815	Green and purple	3325	Orange and black	71660	Brown and bronze
X1315	Yellow and red	X1340	Yellow and black	3335	Blue and orange	7260	Brown and aluminum
X1335	Yellow and black		**Manganese steels**	3340	Blue and orange		**Silicon-manganese steels**
	Nickel steels	T1330	Orange and green	3415	Blue and pink	9255	Bronze and aluminum
2015	Red and brown	T1335	Orange and green	3435	Orange and aluminum	9260	Bronze and aluminum

Metal Sheet Materials Chart

Material (Sheet less than 1/4″ thick)	How Measured	How Purchased	Characteristics
Copper	Gage number (Brown & Sharpe and Amer. Std.)	24″ × 96″ sheet or 12″ or 18″ by lineal feet on roll	Pure metal
Brass	Gage number (Brown & Sharpe and Amer. Std.)	24″ × 76″ sheet or 12″ or 18″ by lineal feet on roll	Alloy of copper and zinc
Aluminum	Decimal	24″ × 72″ sheet or 12″ or 18″ by lineal feet on roll	Available as commercially pure metal or alloyed for strength, hardness, and ductility
Galvanized steel	Gage number (Amer. Std.)	24″ × 96″ sheet	Mild steel sheet with zinc plating, also available with zinc coating that is part of sheet
Black annealed steel sheet	Gage number (Amer. Std.)	24″ × 96″ sheet	Mild steel with oxide coating, hot-rolled
Cold-rolled steel sheet	Gage number (Amer. Std.)	24″ × 96″ sheet	Oxide removed and cold-rolled to final thickness
Tin plate	Gage number (Amer. Std.)	20″ × 28″ sheet 56 or 112 to pkg.	Mild steel with tin coating
Nickel silver	Gage number (Brown & Sharpe)	6″ or 12″ wide by lineal feet on roll	Copper 50%, zinc 30%, nickel 20%
Expanded	Gage number (Amer. Std.)	36″ × 96″ sheet	Metal is pierced and expanded (stretched) to diamond shape; also available rolled to thickness after it has been expanded
Perforated	Gage number (Amer. Std.)	30″ × 36″ sheet 36″ × 48″ sheet	Design is cut in sheet; many designs available

Order of Ductility of Metals

1. Gold
2. Platinum
3. Silver
4. Iron
5. Copper
6. Aluminum
7. Nickel
8. Zinc
9. Tin
10. Lead

Physical Properties of Metals

Metal	Symbol	Specific Gravity	Specific Heat	Melting Point*		Lbs. per Cubic Inch
				°C	°F	
Aluminum (cast)	Al	2.56	.2185	658	1217	.0924
Aluminum (rolled)	Al	2.71	–	658	1217	.0978
Antimony	Sb	6.71	.051	630	1166	.2424
Bismuth	Bi	9.80	.031	271	520	.3540
Boron	B	2.30	.3091	2300	4172	.0831
Brass	–	8.51	.094	–	–	.3075
Cadmium	Cd	8.60	.057	321	610	.3107
Calcium	Ca	1.57	.170	810	1490	.0567
Chromium	Cr	6.80	.120	1510	2750	.2457
Cobalt	Co	8.50	.110	1490	2714	.3071
Copper	Cu	8.89	.094	1083	1982	.3212
Columbium	Cb	8.57	–	1950	3542	.3096
Gold	Au	19.32	.032	1063	1945	.6979
Iridium	Ir	22.42	.033	2300	4170	.8099
Iron	Fe	7.86	.110	1520	2768	.2634
Iron (cast)	Fe	7.218	.1298	1375	2507	.2605
Iron (wrought)	Fe	7.70	.1138	1500-1600	2732-2912	.2779
Lead	Pb	11.37	.031	327	621	.4108
Lithium	Li	.057	.941	186	367	.0213
Magnesium	Mg	1.74	.250	651	1204	.0629
Manganese	Mn	8.00	.120	1225	2237	.2890
Mercury	Hg	13.59	.032	–39	–38	.4909
Molybdenum	Mo	10.2	.0647	2620	47.48	.368
Monel metal	–	8.87	.127	1360	2480	.320
Nickel	Ni	8.80	.130	1452	2646	.319
Phosphorus	P	1.82	.177	43	111.4	.0657
Platinum	Pt	21.50	.033	1755	3191	.7767
Potassium	K	0.87	.170	62	144	.0314
Selenium	Se	4.81	.084	220	428	.174
Silicon	Si	2.40	.1762	1427	2600	.087
Silver	Ag	10.53	.056	961	1761	.3805
Sodium	Na	0.97	.290	97	207	.0350
Steel	–	7.858	.1175	1330-1378	2372-2532	.2839
Strontium	Sr	2.54	.074	769	1416	.0918
Tantalum	Ta	10.80	–	2850	5160	.3902
Tin	Sn	7.29	.056	232	450	.2634
Titanium	Ti	5.3	.130	1900	3450	.1915
Tungsten	W	19.10	.033	3000	5432	.6900
Uranium	U	18.70	–	1132	2070	.6755
Vanadium	V	5.50	–	1730	3146	.1987
Zinc	Zn	7.19	.094	419	786	.2598

*Circular of the Bureau of Standards No.35, Department of Commerce and Labor

Wire Gages in Decimal Inches

Number of Wire Gage	American or Brown & Sharpe	Washburn & Moen Mfg. Co., A.S.& W. Roebling	Imperial Wire Gage	Stubs' Steel Wire	Birmingham or Stubs' Iron Wire
0000000	...	.4900	.5000	...	...
000000	.5800	.4615	.4640	...	...
00000	.5165	.4305	.4320	...	.500
0000	.460	.3938	.4000	...	.454
000	.40964	.3625	.3720	...	.425
00	.3648	.3310	.3480	...	.380
0	.32486	.3065	.3240	...	.340
1	.2893	.2830	.3000	.227	.300
2	.25763	.2625	.2760	.219	.284
3	.22942	.2437	.2520	.212	.259
4	.20431	.2253	.2320	.207	.238
5	.18194	.2070	.2120	.204	.220
6	.16202	.1920	.1920	.201	.203
7	.14428	.1770	.1760	.199	.180
8	.12849	.1620	.1600	.197	.165
9	.11443	.1483	.1440	.194	.148
10	.10189	.1350	.1280	.191	.134
11	.090742	.1205	.1160	.188	.120
12	.080808	.1055	.1040	.185	.109
13	.071961	.0915	.0920	.182	.095
14	.064084	.0800	.0800	.180	.083
15	.057068	.0720	.0720	.178	.072
16	.05082	.0625	.0640	.175	.065
17	.045257	.0540	.0560	.172	.058
18	.040303	.0475	.0480	.168	.049
19	.03589	.0410	.0400	.164	.042
20	.031961	.0348	.0360	.161	.035
21	.028462	.0317	.0320	.157	.032
22	.025347	.0286	.0280	.155	.028
23	.022571	.0258	.0240	.153	.025
24	.0201	.0230	.0220	.151	.022
25	.0179	.0204	.0200	.148	.020
26	.01594	.0181	.0180	.146	.018
27	.014195	.0173	.0164	.143	.016
28	.012641	.0162	.0148	.139	.014
29	.011257	.0150	.0136	.134	.013
30	.010025	.0140	.0124	.127	.012
31	.008928	.0132	.0116	.120	.010
32	.00795	.0128	.0108	.115	.009
33	.00708	.0118	.0100	.112	.008
34	.006304	.0104	.0092	.110	.007
35	.005614	.0095	.0084	.108	.005
36	.005	.0090	.0076	.106	.004
37	.004453	.0085	.0068	.103	...
38	.003965	.0080	.0060	.101	...
39	.003531	.0075	.0052	.099	...
40	.003144	.0070	.0048	.097	...

Cutting Speeds for Round Stock

Diameter	Material	Roughing Cut (rpm)	Finishing Cut (rpm)	Threading (rpm)	Diameter	Material	Roughing Cut (rpm)	Finishing Cut (rpm)	Threading (rpm)
1/8″	Machine steel/bronze	2880	3200	1020	1 1/2″	Machine steel/bronze	240	270	67
	Cast iron	1920	2560	800		Tool steel	134	200	53
	Tool steel (annealed)	1600	2400	640		Brass	400	534	134
	Brass	4800	6400	1600		Aluminum	534	800	134
	Aluminum	6400	9600	1600					
3/16″	Machine steel/bronze	2880	3200	1120	1 3/4″	Machine steel/bronze	205	230	80
	Tool steel	1600	2400	640		Tool steel	115	170	50
	Brass	4800	6400	1600		Brass	340	450	115
	Aluminum	6400	9600	1600		Aluminum	456	680	115
1/4″	Machine steel/bronze	1440	1600	560	2″	Machine steel	180	200	50
	Tool steel	800	1200	320		Tool steel	100	150	40
	Brass	2400	3200	800		Brass	300	400	100
	Aluminum	3200	4800	800		Aluminum	400	600	100
3/8″	Machine steel/bronze	960	1066	270	2 1/2″	Machine steel	141	160	56
	Tool steel	540	800	220		Tool steel	80	120	32
	Brass	1700	2100	530		Brass	240	320	80
	Aluminum	2130	3200	540		Aluminum	320	480	80
1/2″	Machine steel/bronze	720	800	280	3″	Machine steel	120	140	40
	Tool steel	400	600	160		Tool steel	65	100	40
	Brass	1200	1600	400		Brass	200	270	65
	Aluminum	1600	2400	400		Aluminum	270	400	65
5/8″	Machine steel/bronze	576	640	160	3 1/2″	Machine steel	103	115	40
	Tool steel	320	480	200		Tool steel	60	85	23
	Brass	960	1280	320		Brass	171	228	57
	Aluminum	1280	192	320		Aluminum	228	342	57
3/4″	Machine steel/bronze	500	550	176	4″	Machine steel	90	100	35
	Tool steel	266	400	106		Tool steel	50	75	20
	Brass	800	1066	266		Brass	150	200	50
	Aluminum	1066	1600	266		Aluminum	200	300	50
1″	Machine steel/bronze	360	400	140	4 1/2″	Machine steel	80	90	31
	Tool steel	200	300	80		Tool steel	45	67	18
	Brass	600	800	200		Brass	133	178	45
	Aluminum	800	1200	200		Aluminum	178	267	45
1 1/4″	Machine steel	288	320	112	5″	Machine steel	72	80	28
	Tool steel	160	240	64		Tool steel	40	58	16
	Brass	480	640	160		Brass	120	160	40
	Aluminum	640	960	160		Aluminum	160	240	40

Rules for Determining Speeds and Feeds

To Find	Having	Rule	Formula
Speed of cutter in feet per minute (FPM)	Diameter of cutter and revolutions per minute	Diameter of cutter (in inches) multiplied by 3.1416 (π) multiplied by revolutions per minute, divided by 12	$FPM = \frac{\pi D \times RPM}{12}$
Speed of cutter in meters per minute (MPM)	Diameter of cutter and revolutions per minute	Diameter of cutter multiplied by 3.1416 (π) multiplied by revolutions per minute, divided by 1000	$MPM = \frac{D(mm) \times \pi \times RPM}{1000}$
Revolutions per minute (RPM)	Feet per minute and diameter of cutter	Feet per minute, multiplied by 12, divided by circumference of cutter (πD)	$RPM = \frac{FPM \times 12}{\pi D}$
Revolutions per minute (RPM)	Meters per minute and diameter of cutter in millimeters (mm)	Meters per minute, multiplied by 1000, divided by the circumference of cutter (πD)	$RPM = \frac{MPM \times 1000}{\pi D}$
Feed per revolution (FR)	Feed per minute and revolutions per minute	Feed per minute, divided by revolutions per minute	$FR = \frac{F}{RPM}$
Feed per tooth per revolution (FTR)	Feed per minute and number of teeth in cutter	Feed per minute (in inches or millimeters) divided by number of teeth in cutter × revolutions per minute	$RTR = \frac{F}{T \times RPM}$
Feed per minute (F)	Feed per tooth per revolution, number of teeth in cutter, and RPM	Feed per tooth per revolutions multiplied by number of teeth in cutter, multiplied by revolutions per minute	$F = FTR \times T \times RPM$
Feed per minute (F)	Feed per revolution and revolutions per minute	Feed per revolution multiplied by revolutions per minute	$F = FR \times RPM$
Number of teeth per minute (TM)	Number of teeth in cutter and revolutions per minute	Number of teeth in cutter multiplied by revolutions per minute	$TM = T \times RPM$

RPM = Revolutions per minute
T = Teeth in cutter
D = Diameter of cutter
π = 3.1416 (pi)
FPM = Speed of cutter in feet per minute
TM = Teeth per minute
F = Feed per minute
FR = Feed per revolution
FTR = Feed per tooth per revolution
MPM = Speed of cutter in meters per minute

Recommended Turning Rates for Stainless Steels Using High-Speed Tools

Nature of Stock	Type No.	Speed (sfpm)	Feed (inches per revolution)
Free machining grades	430 F	100 – 140	0.003-0.005 for finish cuts and up to 0.015 for roughing cuts
	416	90 – 135	
	303	80 – 120	
High-carbon grades that are slowed down due to their abrasive action on tools	410	75 – 115	0.003–0.008
	430	75 – 115	0.003–0.008
	420	45 – 85	0.003–0.008
	431	45 – 85	0.003–0.008
	440	30 – 60	0.003–0.008
	302		
	304		
	316	45 – 80	0.004–0.008

Feeds and Speeds for HSS Drills, Reamers, and Taps

Material	Brinell	Drils			Reamers		Taps (sfm)			
							Threads per Inch			
		(sfm)	Point	Feed	(sfm)	Feed	3–7 1/2	8–15	16–24	25–up
Aluminum	99–101	200–250	118°	M	150–160	M	50	100	150	200
Aluminum bronze	170–187	60	118°	M	40–45	M	12	25	45	60
Bakelite	...	80	60-90°	M	50–60	M	50	100	150	200
Brass	192–202	200–250	118°	H	150–160	H	50	100	150	200
Bronze, common	166–183	200–250	118°	H	150–160	H	40	80	100	150
Bronze, phosphor, 1/2 hard	187–202	175–180	118°	M	130–140	M	25	40	50	80
Bronze, phosphor, soft	149–163	200–250	118°	H	150–160	H	40	80	100	150
Cast iron, soft	126	140–150	90°	H	100–110	H	30	60	90	140
Cast iron, medium soft	196	80–110	118°	M	50–65	M	25	40	50	80
Cast iron, hard	293–302	45–50	118°	L	67–75	L	10	20	30	40
Cast iron, chilled*	402	15	150°	L	8–10	L	5	5	10	10
Cast steel	286–302	40–50*	118°	L	70–75	L	20	30	40	50
Celluloid	...	100	90°	M	75–80	M	50	100	150	200
Copper	80–85	70	100°	L	45–55	L	40	80	100	150
Drop forgings (steel)	170–196	60	118°	M	40–45	M	12	25	45	60
Duralumin	90–104	200	118°	M	150–160	M	50	100	150	200
Everdur	179–207	60	118°	L	40–45	L	20	30	40	50
Machinery steel	170–196	110	118°	H	67–75	H	35	50	60	85
Magnet steel, soft	241–302	35–40	118°	M	20–25	M	20	40	50	75
Magnet steel, hard*	321–512	15	150°	L	10	L	5	10	15	25
Manganese steel, 7% – 13%	187–217	15	150°	L	10	L	15	20	25	30
Manganese copper, 30% Mn.*	134	15	150°	L	10–12	L	...	...	...	...
Malleable iron	112–126	85–90	118°	H	...	H	20	30	40	50
Mild steel, .20 –.30 C	170–202	110–120	118°	H	75–85	H	40	55	70	90
Molybdenum steel	196–235	55	125°	M	35–45	M	20	30	35	45
Monel metal	149–170	50	118°	M	35–38	M	8	10	15	20
Nickel, pure*	187–202	75	118°	L	40	L	25	40	50	80
Nickel steel, 3 1/2%	196–241	60	118°	L	40–45	L	8	10	15	20
Rubber, hard	...	100	60-90°	L	70–80	L	50	100	150	200
Screw stock, C.R.	170–196	110	118°	H	75	H	20	30	40	50
Spring steel	402	20	150°	L	12–15	L	10	10	15	15
Stainless steel	146–149	50	118°	M	30	M	8	10	15	20
Stainless steel, C.R.*	460–477	20	118°	L	15	L	8	10	15	20
Steel, .40 to .50 C	170–196	80	118°	M	8–10	M	20	30	40	50
Tool, S.A.E., and forging steel	149	75	118°	H	35–40	H	25	35	45	55
Tool, S.A.E., and forging steel	241	50	125°	M	12	M	15	15	25	25
Tool, S.A.E., and forging steel*	402	15	150°	L	10	L	8	10	15	20
Zinc alloy	112–126	200–250	118°	M	150–175	M	50	100	150	200

*Use specially constructed heavy-duty drills.
Note: Carbon steel tools should be run at speeds 40% to 50% of those recommended for high speed steel.
Spiral point taps may be run at speeds 15% to 20% faster than regular taps.

Feeds and Speeds for HSS Drills in Various Metals

Drill Diameter	Cast Iron		Bronze or Brass		Drop Forgings Alloy Steel Tool Steel Annealed		Drop Forgings Alloy Steel Heat-Treated		Steel Castings		Mild Steel	
	Feed	Speed	Feed	Speed	Feed	Speed	Feed	Speed	Feed	Speed	Feed	Speed
1/16″	.002	4550	.002	9150	.002	3650	.002	2750	.002	3650	.002	4250
	.004	6700	.004	12000	.003	4550	.003	3650	.003	4550	.003	5600
1/8″	.002	2550	.002	4550	.002	1800	.002	1225	.002	1800	.002	2100
	.004	3350	.004	5600	.003	2250	.003	1800	.003	2250	.003	2800
3/16″	.004	1500	.004	3100	.003	1200	.003	900	.003	1200	.003	1400
	.006	2200	.007	5600	.004	1500	.004	1200	.005	1500	.005	1900
1/4″	.004	1150	.004	2300	.003	925	.003	750	.003	925	.003	1050
	.006	1650	.007	2750	.004	1150	.004	925	.005	1150	.005	1500
5/16″	.006	925	.007	1825	.004	725	.004	500	.004	725	.005	850
	.009	1325	.010	2200	.006	925	.005	725	.006	925	.007	1200
3/8″	.006	750	.007	1525	.004	600	.004	400	.004	600	.005	700
	.009	1100	.010	1850	.006	750	.005	600	.006	750	.007	925
7/16″	.009	650	.010	1300	.006	525	.005	350	.006	525	.006	600
	.012	950	.014	1525	.009	650	.006	525	.010	650	.010	800
1/2″	.008	575	.010	1150	.006	375	.005	300	.006	375	.006	525
	.012	850	.014	1375	.009	575	.006	375	.010	575	.010	700
9/16″	.012	500	.014	1000	.008	350	.007	275	.010	350	.010	575
	.016	750	.018	1200	.012	500	.010	350	.014	500	.014	625
5/8″	.012	450	.014	900	.008	300	.007	250	.010	300	.010	425
	.016	675	.018	1100	.012	450	.010	300	.014	450	.014	565
11/16″	.012	410	.014	800	.008	275	.007	225	.010	275	.010	375
	.016	625	.018	1000	.012	410	.010	275	.014	410	.014	525
3/4″	.012	375	.014	750	.008	250	.007	200	.010	250	.010	350
	.016	550	.018	900	.012	375	.010	250	.014	375	.014	475
13/16″	.014	350	.016	700	.010	240	.009	190	.014	240	.014	325
	.020	525	.022	850	.014	350	.012	240	.016	350	.016	450
7/8″	.014	325	.016	650	.010	225	.009	175	.014	225	.014	300
	.020	475	.022	800	.014	325	.012	225	.016	325	.016	400
15/16″	.014	300	.016	625	.010	200	.009	160	.014	200	.014	275
	.020	450	.022	725	.014	300	.012	200	.016	300	.016	375
1″	.014	280	.016	575	.010	185	.009	150	.014	185	.014	265
	.020	425	.022	675	.014	280	.012	185	.016	280	.016	350

Chicago-Latrobe

Speeds and feeds shown apply to average working conditions and materials. They are recommended with regard to conserving drills and avoiding excessive machine tool wear. Under many conditions, these speeds and feeds may be considerably increased; under others they must be decreased. This is dependent on judgment of operator, and performance obtained. Excessive speeds and feeds will show up by action of machine and drill. Same applies to lower speeds and feeds. Operator will notice whether he/she is getting proper performance by experience, and will advance or retard as case may justify. Feeds and speeds should be changed in proper proportions and a liberal use of cooling compound will increase life of tools.

Never dip a drill into water to cool it while grinding. This will cause tiny checks, or cracks at the cutting edge, which will cause the drill to dull quickly.

Do not leave a drill in after it shows signs of dulling or laboring; then is the time to regrind. Proper grinding is essential.

To determine feed and speed according to the above chart, proceed as follows:

You are going to drill heat-treated drop forgings. We suppose you will use a 1/2″ drill. Follow column down to where the 1/2″ drill meets it; there you will find that a feed from .005 to .006 and a speed of from 300 to 375 rpm are recommended. Start by using .005 feed and 300 rpm. If drill and machine seem to turn smoothly without strain, then both feed and speed can be advanced. Operator will soon find which is best.

Formulas for Machining Bar Stock

Surface speed—feet/minute	
Round bars	$\frac{\text{Diameter} \times 3.1416 \times \text{rpm}}{12}$
Hexagon bars (distance across corners)	$\frac{\text{Size} \times 3.1416 \times \text{rpm} \times 1.155}{12}$
	$\frac{\text{Distance} \times 3.1416 \times \text{rpm}}{12}$
Square bars (distance across corners)	$\frac{\text{Size} \times 3.1416 \times \text{rpm} \times 1.414}{12}$
	$\frac{\text{Distance} \times 3.1416 \times \text{rpm}}{12}$
Revolutions—number/minute	
Round bars	$\frac{\text{SFM} \times 12}{\text{Distance} \times 3.1416}$
Hexagon bars	$\frac{\text{SFM} \times 12}{\text{Size} \times 3.1416 \times 1.155}$
	$\frac{\text{SFM} \times 12}{\text{Distance} \times 3.1416}$
Square bars	$\frac{\text{SFM} \times 12}{\text{Size} \times 3.1416 \times 1.414}$
	$\frac{\text{SFM} \times 12}{\text{Distance} \times 3.1416}$
Feed—inches/revolution	$\frac{\text{Feed inches per minute}}{\text{rpm}}$
	$\frac{\text{Diameter} \times 3.1416 \times \text{Feed}}{\text{SFM} \times 12}$
Feed—inches/tooth	$\frac{\text{Feed}}{\text{Number of teeth}}$
Time for actual machining—seconds	$\frac{\text{Revolutions required} \times 60 \text{ seconds}}{\text{rpm}}$
Machine time	Time for machining + idle time
Tapping or threading time—seconds	$\frac{\text{Number of threads} \times 60 \text{ seconds}}{\text{Actual threading speed in rpm}}$

Standard Dimensional Tolerances for Bar Stock

Hot-Rolled Bars: Rounds and Squares

Specified Size (Inches)	Variations from Size (Inches)		Out-of-Round (1) or Square (2) Inches
	Over	Under	
1/4 to 5/16 inclusive (3)	(4)	(4)	(4)
Over 5/16 to 7/16 inclusive (3)	0.006	0.006	0.009
Over 7/16 to 5/8 inclusive (3)	0.007	0.007	0.010
Over 5/8 to 7/8 inclusive	0.008	0.008	0.012
Over 7/8 to 1 inclusive	0.009	0.009	0.013
Over 1 to 1 1/8 inclusive	0.010	0.010	0.015
Over 1 1/8 to 1 1/4 inclusive	0.011	0.011	0.016
Over 1 1/4 to 1 3/8 inclusive	0.012	0.012	0.018
Over 1 3/8 to 1 1/2 inclusive	0.014	0.014	0.021
Over 1 1/2 to 2 inclusive	1/64	1/64	0.023
Over 2 to 2 1/2 inclusive	1/32	0	0.023
Over 2 1/2 to 3 1/2 inclusive	3/64	0	0.035
Over 3 1/2 to 4 1/2 inclusive	1/16	0	0.046
Over 4 1/2 to 5 1/2 inclusive	5/64	0	0.058
Over 5 1/2 to 6 1/2 inclusive	1/8	0	0.070
Over 6 1/2 to 8 inclusive	5/32	0	0.085

(1) Out-of-round is difference between maximum and minimum diameters of bar, measured at same cross section.
(2) Out-of-square is difference in two dimensions at same cross section of a square bar, each dimension being distance between opposite faces.
(3) Round sections in size range of 1/4″ to approximately 5/8″ diameter are commonly produced on rod mills in coils. Tolerances on product made this way have not been established; for such tolerances, producer should be consulted. Variations in size of coiled product made on rod mills are greater than size tolerances for product made on bar mills.
(4) Squares in this size are not commonly produced as hot-rolled product.

Hot-Rolled Bars: Hexagons and Octagons

Specified Sizes Between Opposite Sides (Inches)	Variations from Size (Inches)		Maximum Difference 3 Measurements for Hexagons Only (Inches)
	Over	Under	
1/4 to 1/2 inclusive	0.007	0.007	0.011
Over 1/2 to 1 inclusive	0.010	0.010	0.015
Over 1 to 1 1/2 inclusive	0.021	0.021	0.025
Over 1 1/2 to 2 inclusive	1/32	1/32	1/32
Over 2 to 2 1/2 inclusive	3/64	3/64	3/64
Over 2 1/2 to 3 1/2 inclusive	1/16	1/16	1/16

Cold-Finished Bars: Flats

Specified Width Size or Thickness (Inches)	Variations from Width Over or Under (Inches)*		Variations from Thicknesses Over or Under (Inch)
	For Thicknesses 1/4″ and Under	For Thicknesses Over 1/4″	
Over 1/8 to 1 inclusive	—	—	0.002
Over 3/8 to 1 inclusive	0.004	0.002	0.002
Over 1 to 2 inclusive	0.006	0.003	0.003
Over 2 to 3 inclusive	0.008	0.004	0.004
Over 3 to 4 1/2 inclusive	0.010	0.005	0.005

*When it is necessary to heat treat, or heat treat and pickle after cold finishing, because of special hardness or mechanical property requirements, tolerances are double those shown in the table.

Hot-Rolled Bars: Flats

Specified Widths (Inches)	Variations from Thickness for Thicknesses Given Over or Under (Inches)			Variations from Width (Inches)	
	1/8 to 1/2 Inclusive	Over 1/2 to 1 Inclusive	Over 1 to 2 Inclusive	Over	Under
To 1 inclusive	0.008	0.010	—	1/64	1/64
Over 1 to 2 inclusive	0.012	0.015	1/32	1/32	1/32
Over 2 to 4 inclusive	0.015	0.020	1/32	1/16	1/32
Over 4 to 6 inclusive	0.015	0.020	1/32	3/32	1/16
Over 6 to 8 inclusive	0.016	0.025	1/32	1/8	5/32
Over 8 to 10 inclusive	0.021	0.031	1/32	5/32	3/16

Cold-Finished Bars: Hexagons, Octagons, and Squares

Specified Size (Inches)	Variations from Size (Inches)*	
	Over	Under
Over 1/2 to 1 inclusive	0	0.004
Over 1 to 2 inclusive	0	0.006
Over 2 to 3 inclusive	0	0.008
Over 3	0	0.010

*When it is necessary to heat treat, or heat treat and pickle after cold finishing, because of special hardness or mechanical property requirements, tolerances are double those shown in the table.

Cold-Finished Bars: Rounds

Specified Size (Inches)	Variations from Size (Inches)*	
	Over	Under
Over 1/2 to 1 exclusive	0.002	0.002
1 to 1 1/2 exclusive	0.0025	0.0025
1 1/2 to 4 inclusive	0.003	0.003

*When it is necessary to heat treat, or heat treat and pickle after cold finishing, because of special hardness or mechanical property requirements, tolerances are double those shown in table.

Machine-Cut Bars (Cut after Machine Straightening)

Specified Sizes as They Apply to Rounds, Squares, Hexagons, Octagons, and Width of Flats (Inches)	Variations from Specified Lengths (Inches)			
	To 12′ Inclusive		Over 12′ to 25′ Inclusive	
	Over	Under	Over	Under
To 3 inclusive	1/8	0	3/16	0
Over 3 to 6 inclusive	3/16	0	1/4	0
Over 6 to 9 inclusive	1/4	0	5/16	0
Over 9 to 12 inclusive	1/2	0	1/2	0

Allowances for Turning Machine-Straightened Bars

When ordering bars that are to be turned, it is recommended that allowances be made for finishing from hot-rolled diameters not less than amounts shown in following table, and specify hot-rolled sizes accordingly:

Nominal Diameter of Hot-rolled Bar (Inches)	Minimum Allowance on Diameter for Turning (Inches)
1 1/2 to 3 inclusive	1/8
Over 3 to 6 inclusive	1/4
Over 6 to 8 inclusive	3/8

Hot or Cold Cutting Length Tolerances

Specified Sizes as They Apply to Rounds, Squares, Hexagons, Octagons, and Width of Flats (Inches)	Variations from Specified Lengths (Inches)			
	To 12′ Inclusive		Over 12′ to 25′ Inclusive	
	Over	Under	Over	Under
To 2 inclusive	1/2	0	3/4	0
Over 2 to 4 inclusive	3/4	0	1	0
Over 4 to 6 inclusive	1	0	1 1/4	0
Over 6 to 9 inclusive	1 1/4	0	1 1/2	0
Over 9 to 12 inclusive	1 1/2	0	2	0

Hot-Finished and Cold-Finished Bars for Machining

Camber is greatest deviation of a side from a straight line. Measurement is taken on concave side of bar with a straightedge. Unless otherwise specified, hot-finished and cold-finished bars for machining purposes are furnished machine-straightened to following tolerances:

Hot-finished: 1/8″ in any 5 feet; but may not exceed $1/8'' \times \frac{\text{No. of feet in length}}{5}$

Cold-finished: 1/16″ in any 5 feet; but may not exceed $1/16'' \times \frac{\text{No. of feet in length}}{5}$

Standard Dimensional Tolerances for Wire

Drawn, Centerless Ground, Centerless Ground and Polished Round and Square Wire

Specified Size (Inches)	Tolerance (Inches)	
	Over	Under
1/2	0.002	0.002
Under 1/2 to 5/16 inclusive	0.0015	0.0015
Under 5/16 to 0.050 inclusive	0.001	0.001

The maximum out-of-round tolerance for round wire is one-half of total size tolerance shown in above table.

Tolerance for Wire for Which the Final Operation Is a Surface Treatment for the Purpose of Removing Scale or Drawing Lubricant

Specified Size (Inches)	Tolerance (Inches)	
	Over	Under
1/2	0.004	0.004
Under 1/2 to 5/16 inclusive	0.003	0.003
Under 5/16 to 0.050 inclusive	0.002	0.002

Drawn Wire in Hexagons and Octagons

Specified Size* (Inches)	Tolerance (Inches)	
	Over	Under
1/2	0	0.004
Under 1/2 to 5/16 inclusive	0	0.003
Under 5/16 to 1/8 inclusive	0	0.002

*Distance across flats

(Carpenter Steel Co.)

Rules and Formulas for Bevel Gear Calculations

(Use with drawing on page 584)

To Find	Rule	Formula
Diametral pitch (P)	Divide the number of teeth by the pitch diameter.	$P = \frac{N}{D}$
Circular pitch (p)	Divide 3.1416 by the diametral pitch.	$p = \frac{3.1416}{P}$
Pitch diameter (D)	Divide the number of teeth by the diametral pitch.	$D = \frac{N}{P}$
Pitch angle of pinion tan (b_p)	Divide the number of teeth in the pinion by the number of teeth in the gear to obtain the tangent.	$\tan b_p = \frac{N_p}{N_g}$
Pitch angle of gear tan (b_g)	Divide the number of teeth in the gear by the number of teeth in the pinion to obtain the tangent.	$\tan b_g = \frac{N_g}{N_p}$
Pitch cone distance (C_r)	Divide the pitch diameter by twice the sine of the pitch angle.	$C_r = \frac{D}{2 (\sin b)}$
Addendum (a)	Divide 1.0 by the diametral pitch.	$a = \frac{1.0}{P}$
Addendum angle tan (A_1)	Divide the addendum by the pitch cone distance to obtain the tangent.	$\tan A_1 = \frac{a}{C_r}$
Angular addendum (A_a)	Multiply the addendum by the cosine of the pitch angle.	$A_a = a \cos b$
Outside diameter (D_o)	Add twice the angular addendum to the pitch diameter.	$D_o = D + 2A_a$
Dedendum angle tan (c_1)	Divide the dedendum by the pitch cone distance to obtain the tangent.	$\tan c_1 = \frac{a + c}{C_r}$
Addendum of small end of tooth (a_s)	Subtract the width of face from the pitch cone distance, divide the remainder by the pitch cone distance and multiply by the addendum.	$a_s = a \left(\frac{C_r - W}{C_r}\right)$
Thickness of tooth at pitch line (t_L)	Divide the circular pitch by 2.	$t_L = \frac{P_c}{2}$
Thickness of tooth at pitch line at small end of gear (t_s)	Subtract the width of face from the pitch cone distance, divide the remainder by the pitch cone distance and multiply by the thickness of the tooth at the pitch line.	$t_s = t_L \left(\frac{C_r - W}{C_r}\right)$
Face angle (F_a)	Face cone of blank turned parallel to root cone of mating gear.	$F_a = b + c_1$
Whole depth of tooth space (h_t)	Divide 2.157 by the diametral pitch.	$h_t = \frac{2.157}{P}$
Apex distance at large end of tooth (V)	Multiply one-half the outside diameter by the tangent of the face angle.	$V = \left(\frac{D_o}{2}\right) \tan F_a$
Apex distance at small end of tooth (v)	Subtract the width of face from the pitch cone distance, divide the remainder by the pitch cone distance and multiply by the apex distance.	$v = V \left(\frac{C_r - W}{C_r}\right)$
Gear ratio (m_g)	Divide the number of teeth in the gear by the number of teeth in the pinion.	$m_g = \frac{N_g}{N_p}$
Number of teeth in gear and/or pinion (N_g, N_p)	Multiply the pitch diameter by the diametral pitch.	$N_g = DP$ $N_p = DP$
Cutting angle (d)	Subtract the addendum plus clearance angle from the pitch angle.	$d = b - c_1$
Number of teeth of imaginary spur gear for which cutter is selected (N_o)	Divide the number of teeth in actual gear by the cosine of the pitch angle.	$N_o = \frac{N}{\cos b}$

Note: Use with table on page 583

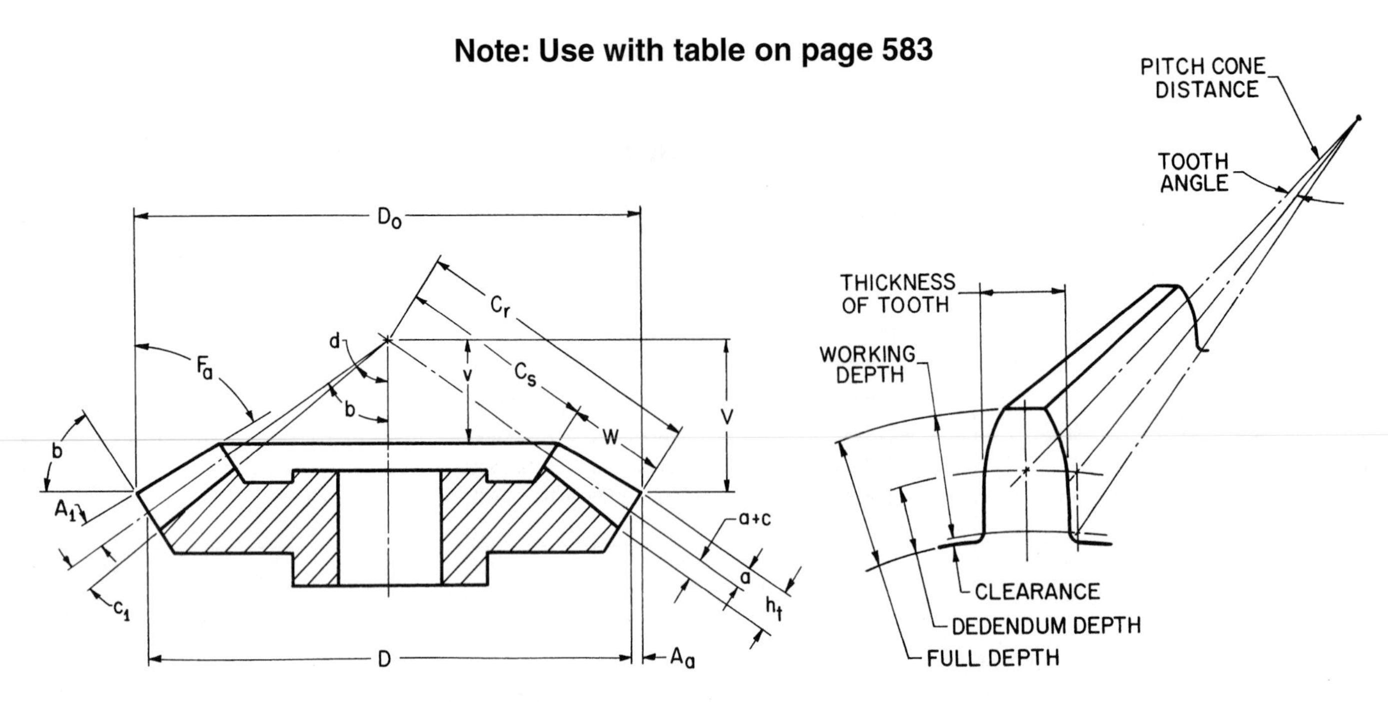

Decimal Equivalents: Number-Size Drills

Drill	Size of Drill in Inches	Drill	Size of Drill in Inches	Drill	Size of Drill in Inches	Drill	Size of Drill in Inches
1	.2280	21	.1590	41	.0960	61	.0390
2	.2210	22	.1570	42	.0935	62	.0380
3	.2130	23	.1540	43	.0890	63	.0370
4	.2090	24	.1520	44	.0860	64	.0360
5	.2055	25	.1495	45	.0820	65	.0350
6	.2040	26	.1470	46	.0810	66	.0330
7	.2010	27	.1440	47	.0785	67	.0320
8	.1990	28	.1405	48	.0760	68	.0310
9	.1960	29	.1360	49	.0730	69	.0292
10	.1935	30	.1285	50	.0700	70	.0280
11	.1910	31	.1200	51	.0670	71	.0260
12	.1890	32	.1160	52	.0635	72	.0250
13	.1850	33	.1130	53	.0595	73	.0240
14	.1820	34	.1110	54	.0550	74	.0225
15	.1800	35	.1100	55	.0520	75	.0210
16	.1770	36	.1065	56	.0465	76	.0200
17	.1730	37	.1040	57	.0430	77	.0180
18	.1695	38	.1015	58	.0420	78	.0160
19	.1660	39	.0995	59	.0410	79	.0145
20	.1610	40	.0980	60	.0400	80	.0135

Decimal Equivalents: Letter-Size Drills

Drill	Size of Drill in Inches	Drill	Size of Drill in Inches	Drill	Size of Drill in Inches	Drill	Size of Drill in Inches
A	0.234	H	0.266	O	0.316	V	0.377
B	0.238	I	0.272	P	0.323	W	0.386
C	0.242	J	0.277	Q	0.332	X	0.397
D	0.246	K	0.281	R	0.339	Y	0.404
E	0.250	L	0.290	S	0.348	Z	0.413
F	0.257	M	0.295	T	0.358		
G	0.261	N	0.302	U	0.368		

60° V-Type Thread Dimensions—Fractional Sizes

National Special Thread Series

Nominal Size	Threads per Inch	Major Diameter Inches	Minor Diameter Inches	Pitch Diameter Inches	Tap Drill for 75% Thread †	Clearance Drill Size*
1/16″	64	.0625	.0422	.0524	3/64″	51
5/64″	60	.0781	.0563	.0673	1/16″	45
3/32″	48	.0938	.0667	.0803	49	40
7/64″	48	.1094	.0823	.0959	43	32
1/8″	32	.1250	.0844	.1047	3/32″	29
9/64″	40	.1406	.1081	.1244	32	24
5/32″	32	.1563	.1157	.1360	1/8″	19
5/32″	36	.1563	.1202	.1382	30	19
11/64″	32	.1719	.1313	.1516	9/64″	14
3/16″	24	.1875	.1334	.1604	26	8
3/16″	32	.1875	.1469	.1672	22	8
13/64″	24	.2031	.1490	.1760	20	3
7/32″	24	.2188	.1646	.1917	16	1
7/32″	32	.2188	.1782	.1985	12	1
15/64″	24	.2344	.1806	.2073	10	1/4″
1/4″	24	.2500	.1959	.2229	4	17/64″
1/4″	27	.2500	.2019	.2260	3	17/64″
1/4″	32	.2500	.2094	.2297	7/32″	17/64″
5/16″	20	.3125	.2476	.2800	17/64″	21/64″
5/16″	27	.3125	.2644	.2884	J	21/64″
5/16″	32	.3125	.2719	.2922	9/32″	21/64″
3/8″	20	.3750	.3100	.3425	21/64″	25/64″
3/8″	27	.3750	.3269	.3509	R	25/64″
7/16″	24	.4375	.3834	.4104	X	29/64″
7/16″	27	.4375	.3894	.4134	Y	29/64″
1/2″	12	.5000	.3918	.4459	27/64″	33/64″
1/2″	24	.5000	.4459	.4729	29/64″	33/64″
1/2″	27	.5000	.4519	.4759	15/32″	33/64″
9/16″	27	.5625	.5144	.5384	17/32″	37/64″
5/8″	12	.6250	.5168	.5709	35/64″	41/64″
5/8″	27	.6250	.5769	.6009	19/32″	41/64″
11/16″	11	.6875	.5694	.6285	19/32″	45/64″
11/16″	16	.6875	.6063	.6469	5/8″	45/64″
3/4″	12	.7500	.6418	.6959	43/64″	49/64″
3/4″	27	.7500	.7019	.7259	23/32″	49/64″
13/16″	10	.8125	.6826	.7476	23/32″	53/64″
7/8″	12	.8750	.7668	.8209	51/64″	57/64″
7/8″	18**	.8750	.8028	.8389	53/64″	57/64″
7/8″	27	.8750	.8269	.8509	27/32″	57/64″
15/16″	9	.9375	.7932	.8654	53/64″	61/64″
1″	12	1.0000	.8918	.9459	59/64″	1 1/64″
1″	27	1.0000	.9519	.9759	31/32″	1 1/64″
1 5/8″	5 1/2	1.6250	1.3888	1.5069	1 29/64″	1 41/64″
1 7/8″	5	1.8750	1.6152	1.7451	1 11/16″	1 57/64″
2 1/8″	4 1/2	2.1250	1.8363	1.9807	1 29/32″	2 5/32″
2 3/8″	4	2.3750	2.0502	2.2126	2 1/8″	2 13/32″

† Refer to tables "Decimal Equivalents: Number-Size Drills" and "Decimal Equivalents: Letter-Size Drills."

* Clearance drill makes hole with standard clearance for diameter of nominal size.

** Standard spark plug size.

60° V-Type Thread Dimensions—Metric Sizes

International Standard

Major Diameter (mm)	Pitch (mm)	Minor Diameter (mm)	Pitch Diameter (mm)	Top Drill for 75% Thread (mm)	Tap Drill for 75% Thread † (No. or Inches)	Clearance Drill Size*
2.0	.40	1.48	1.740	1.6	1/16″	41
2.3	.40	1.78	2.040	1.9	48	36
2.6	.45	2.02	2.308	2.1	45	31
3.0	.50	2.35	2.675	2.5	40	29
3.5	.60	2.72	3.110	2.9	33	23
4.0	.70	3.09	3.545	3.3	30	16
4.5	.75	3.53	4.013	3.75	26	10
5.0	.80	3.96	4.480	4.2	19	3
5.5	.90	4.33	4.915	4.6	14	15/64″
6.0	1.00	4.70	5.350	5.0	9	1/4″
7.0	1.00	5.70	6.350	6.0	15/64″	19/64″
8.0	1.25	6.38	7.188	6.8	H	11/32″
9.0	1.25	7.38	8.188	7.8	5/16″	3/8″
10.0	1.50	8.05	9.026	8.6	R	27/64″
11.0	1.50	9.05	10.026	9.6	V	29/64″
12.0	1.75	9.73	10.863	10.5	Z	1/2″
14.0**	1.25	12.38	13.188	13.0	33/64″	9/16″
14.0	2.00	11.40	12.701	12.0	15/32″	9/16″
16.0	2.00	13.40	14.701	14.0	35/64″	21/32″
18.0	1.50	16.05	17.026	16.5	41/64″	47/64″
18.0	2.50	14.75	16.376	15.5	39/64″	47/64″
20.0	2.50	16.75	18.376	17.5	11/16″	13/16″
22.0	2.50	18.75	20.376	19.5	49/64″	57/64″
24.0	3.00	20.10	22.051	21.0	53/64″	31/32″
27.0	3.00	23.10	25.051	24.0	15/16″	13/32″
30.0	3.50	25.45	27.727	26.5	13/64″	113/64″
33.0	3.50	28.45	30.727	29.5	1 11/64″	121/64″
36.0	4.00	30.80	33.402	32.0	1 17/64″	17/16″
39.0	4.00	33.80	36.402	35.0	1 3/8″	19/16″
42.0	4.50	36.15	39.077	37.0	1 29/64″	143/64″
45.0	4.50	39.15	42.077	40.0	1 37/64″	113/16″
48.0	5.00	41.50	44.752	43.0	1 11/16″	129/32″

† Refer to tables "Decimal Equivalents: Number-Size Drills" and "Decimal Equivalents: Letter-Size Drills."
* Clearance drill makes hole with standard clearance for diameter of nominal size.
** Standard spark plug size.

Tap Drill Sizes

Probable Percentage of Full Thread Produced in Tapped Hole Using Stock Sizes of Drill

Tap	Tap Drill	Decimal Equivalent of Tap Drill	Theoretical % of Thread	Probable Oversize (Mean)	Probable Hole Size	Percentage of Thread	Tap	Tap Drill	Decimal Equivalent of Tap Drill	Theoretical % of Thread	Probable Oversize (Mean)	Probable Hole Size	Percentage of Thread
0–80	56	.0465	83	.0015	.0480	74	8–32	29	.1360	69	.0029	.1389	62
	3/64	.0469	81	.0015	.0484	71		28	.1405	58	.0029	.1434	51
1–64	54	.0550	89	.0015	.0565	81	8–36	29	.1360	78	.0029	.1389	70
	53	.0595	67	.0015	.0610	59		28	.1405	68	.0029	.1434	57
1–72	53	.0595	75	.0015	.0610	67		9/64	.1406	68	.0029	.1435	57
	1/16	.0625	58	.0015	.0640	50	10–24	27	.1440	85	.0032	.1472	79
2–56	51	.0670	82	.0017	.0687	74		26	.1470	79	.0032	.1502	74
	50	.0700	69	.0017	.0717	62		25	.1495	75	.0032	.1527	69
	49	.0730	56	.0017	.0747	49		24	.1520	70	.0032	.1552	64
2–64	50	.0700	79	.0017	.0717	70		23	.1540	67	.0032	.1572	61
	49	.0730	64	.0017	.0747	56		5/32	.1563	62	.0032	.1595	56
3–48	48	.0760	85	.0019	.0779	78		22	.1570	61	.0032	.1602	55
	5/64	.0781	77	.0019	.0800	70	10–32	5/32	.1563	83	.0032	.1595	75
	47	.0785	76	.0019	.0804	69		22	.1570	81	.0032	.1602	73
	46	.0810	67	.0019	.0829	60		21	.1590	76	.0032	.1622	68
	45	.0820	63	.0019	.0839	56		20	.1610	71	.0032	.1642	64
3–56	46	.0810	78	.0019	.0829	69		19	.1660	59	.0032	.1692	51
	45	.0820	73	.0019	.0839	65	12–24	11/64	.1719	82	.0035	.1754	75
	44	.0860	56	.0019	.0879	48		17	.1730	79	.0035	.1765	73
4–40	44	.0860	80	.0020	.0880	74		16	.1770	72	.0035	.1805	66
	43	.0890	71	.0020	.0910	65		15	.1800	67	.0035	.1835	60
	42	.0935	57	.0020	.0955	51		14	.1820	63	.0035	.1855	56
	3/32	.0938	56	.0020	.0958	50	12–28	16	.1770	84	.0035	.1805	77
4–48	42	.0935	68	.0020	.0955	61		15	.1800	78	.0035	.1835	70
	3/32	.0938	68	.0020	.0958	60		14	.1820	73	.0035	.1855	66
	41	.0960	59	.0020	.0980	52		13	.1850	67	.0035	.1885	59
5–40	40	.0980	83	.0023	.1003	76		3/16	.1875	61	.0035	.1910	54
	39	.0995	79	.0023	.1018	71	1/4–20	9	.1960	83	.0038	.1998	77
	38	.1015	72	.0023	.1038	65		8	.1990	79	.0038	.2028	73
	37	.1040	65	.0023	.1063	58		7	.2010	75	.0038	.2048	70
5–44	38	.1015	79	.0023	.1038	72		13/64	.2031	72	.0038	.2069	66
	37	.1040	71	.0023	.1063	63		6	.2040	71	.0038	.2078	65
	36	.1065	63	.0023	.1088	55		5	.2055	69	.0038	.2093	63
6–32	37	.1040	84	.0023	.1063	78		4	.2090	63	.0038	.2128	57
	36	.1065	78	.0026	.1091	71	1/4–28	3	.2130	80	.0038	.2168	72
	7/64	.1094	70	.0026	.1120	64		7/32	.2188	67	.0038	.2226	59
	35	.1100	69	.0026	.1126	63		2	.2210	63	.0038	.2248	55
	34	.1110	67	.0026	.1136	60	5/16–18	F	.2570	77	.0038	.2608	72
	33	.1130	62	.0026	.1156	55		G	.2610	71	.0041	.2651	66
6–40	34	.1110	83	.0026	.1136	75		17/64	.2656	65	.0041	.2697	59
	33	.1130	77	.0026	.1156	69		H	.2660	64	.0041	.2701	59
	32	.1160	68	.0026	.1186	60							

Tap Drill Sizes
(continued)

Tap	Tap Drill	Decimal Equivalent of Tap Drill	Theoretical % of Thread	Probable Oversize (Mean)	Probable Hole Size	Percentage of Thread
5/16–24	H	.2660	86	.0041	.2701	78
	I	.2720	75	.0041	.2761	67
	J	.2770	66	.0041	.2811	58
3/8–16	5/16	.3125	77	.0044	.3169	72
	O	.3160	73	.0044	.3204	68
	P	.3230	64	.0044	.3274	59
3/8–24	21/64	.3281	87	.0044	.3325	79
	Q	.3320	79	.0044	.3364	71
	R	.3390	67	.0044	.3434	58
7/16–14	T	.3580	86	.0046	.3626	81
	23/64	.3594	84	.0046	.3640	79
	U	.3680	75	.0046	.3726	70
	3/8	.3750	67	.0046	.3796	62
	V	.3770	65	.0046	.3816	60
7/16–20	W	.3860	79	.0046	.3906	72
	25/64	.3906	72	.0046	.3952	65
	X	.3970	62	.0046	.4016	55
1/2–13	27/64	.4219	78	.0047	.4266	73
	7/16	.4375	63	.0047	.4422	58
1/2–20	29/64	.4531	72	.0047	.4578	65
9/16–12	15/32	.4688	87	.0048	.4736	82
	31/64	.4844	72	.0048	.4892	68
9/16–18	1/2	.5000	87	.0048	.5048	80
	33/64	.5156	65	.0048	.5204	58
5/8–11	17/32	.5313	79	.0049	.5362	75
	35/64	.5469	66	.0049	.5518	62
5/8–18	9/16	.5625	87	.0049	.5674	80
	37/64	.5781	65	.0049	.5831	58
3/4–10	41/64	.6406	84	.0050	.6456	80
	21/32	.6563	72	.0050	.6613	68
3/4–16	11/16	.6875	77	.0050	.6925	71
7/8–9	49/64	.7656	76	.0052	.7708	72
	25/32	.7812	65	.0052	.7864	61
7/8–14	51/64	.7969	84	.0052	.8021	79
	13/16	.8125	67	.0052	.8177	62
1″–8	55/64	.8594	87	.0059	.8653	83
	7/8	.8750	77	.0059	.8809	73
	57/64	.8906	67	.0059	.8965	64
	29/32	.9063	58	.0059	.9122	54
1″–12	29/32	.9063	87	.0060	.9123	81
	59/64	.9219	72	.0060	.9279	67
	15/16	.9375	58	.0060	.9435	52

Tap	Tap Drill	Decimal Equivalent of Tap Drill	Theoretical % of Thread	Probable Oversize (Mean)	Probable Hole Size	Percentage of Thread
1″–14	59/64	.9219	84	.0060	.9279	78
	15/16	.9375	67	.0060	.9435	61
1 1/8–7	31/32	.9688	84	.0062	.9750	81
	63/64	.9844	76	.0067	.9911	72
	1″	1.0000	67	.0070	1.0070	64
	1 1/64	1.0156	59	.0070	1.0226	55
1 1/8–12	1 1/32	1.0313	87	.0071	1.0384	80
	1 3/64	1.0469	72	.0072	1.0541	66
1 1/4–7	1 3/32	1.0938	84	No test results available. Reaming recommended		
	1 7/64	1.1094	76			
	1 1/8	1.1250	67			
1 1/4–12	1 5/32	1.1563	87			
	1 11/64	1.1719	72			
1 3/8–6	1 3/16	1.1875	87			
	1 13/64	1.2031	79			
	1 7/32	1.2188	72			
	1 15/64	1.2344	65			
1 3/8–12	1 9/32	1.2813	87			
	1 19/64	1.2969	72			
1 1/2–6	1 5/16	1.3125	87			
	1 21/64	1.3281	79			
	1 11/32	1.3438	72			
	1 23/64	1.3594	65			
1 1/2–12	1 13/32	1.4063	87			
	1 27/64	1.4219	72			

Taper Pipe		Straight Pipe	
Thread	Drill	Thread	Drill
1/8–27	R	1/8–27	S
1/4–18	7/16	1/4–18	29/64
3/8–18	37/64	3/8–18	19/32
1/2–14	23/32	1/2–14	47/64
3/4–14	59/64	3/4–14	15/16
1–11 1/2	1 5/32	1–11 1/2	1 3/16
1 1/4–11 1/2	1 1/2	1 1/4–11 1/2	1 33/64
1 1/2–11 1/2	1 47/64	1 1/2–11 1/2	1 3/4
2–11 1/2	2 7/32	2–11 1/2	2 7/32
2 1/2–8	2 5/8	2 1/2–8	2 21/32
3–8	3 1/4	3–8	3 9/32
3 1/2–8	3 3/4	3 1/2–8	3 25/32
4–8	4 1/4	4–8	4 9/32

(Standard Tool Co.)

Taper Pin and Reamer Sizes

Length of Pin	Reamer															
	6/0	5/0	4/0	3/0	2/0	0	1	2	3	4	5	6	7	8	9	10
3/8	50	44	38	32	29											
1/2	51	45	39	33	30	27										
5/8	52	46	41	34	30	27	21									
3/4	1/16	47	42	7/64	1/8	9/64	5/32	16	13/64	15/64	I	P				
1		49	44	37	31	29	25	11/64	9	1	H	O	W			
1 1/4					32	30	26	19	10	2	G	5/16	V	29/64		
1 1/2						*1/8	*9/64	20	3/16	7/32	F	N	V	29/64	35/64	43/64
1 3/4						*31	*29	*5/32	14	3	E	N	U	29/64	35/64	21/32
2						*33	*30	*25	*16	4	D	M	U	7/16	35/64	21/32
2 1/4						*7/64	*1/8	*26	*11/64	*13/64	C	M	23/64	7/16	17/32	21/32
2 1/2						*37	*31	*28	*19	*8	*B	L	T	7/16	17/32	41/64
2 3/4						*40	*33	*29	*21	*11	*1	9/32	S	27/64	17/32	41/64
3						*42	*35	*30	*5/32	*3/16	*2	J	11/32	27/64	33/64	41/64
3 1/4									*24	*14	*2	*I	R	27/64	33/64	5/8
3 1/2									*26	*16	*3	*H	Q	Z	33/64	5/8
3 3/4										*17	*4	*G	*Q	Z	1/2	5/8
4										*19	*5	*F	*P	13/32	1/2	39/64
4 1/4												*1/4	*O	Y	1/2	39/64
4 1/2												*D	*O	X	31/32	39/64
4 3/4												*C	*5/16	*25/64	31/64	19/32
5												*B	*N	*W	31/64	19/32
5 1/4															15/32	19/32
5 1/2															*15/32	37/64
5 3/4															*15/32	37/64
6															*29/64	37/64

* Hole sizes too small to admit taper pin reamers of standard length. Special, extra-length reamers are required for these cases.

Hardness Conversions

Brinell Indentation Diameter, (mm)	Brinell Hardness Number		Rockwell Hardness Number		Rockwell Superficial Hardness Number (Superficial Diamond Penetrator)			Tensile Strength (Approximate) × 1000 psi
	Standard Ball	Tungsten-Carbide Ball	B Scale	C Scale	15 N Scale	30 N Scale	45 N Scale	
2.45	—	627	—	58.7	89.6	76.3	65.1	347
2.50	—	601	—	57.3	89.0	75.1	63.5	328
2.55	—	578	—	56.0	88.4	73.9	62.1	313
2.60	—	555	—	54.7	87.8	72.7	60.6	298
2.65	—	534	—	53.5	87.2	71.6	59.2	288
2.70	—	514	—	52.1	86.5	70.3	57.6	274
2.75	—	495	—	51.0	85.9	69.4	56.1	264
2.80	—	477	—	49.6	85.3	68.2	54.5	252
2.85	—	461	—	48.5	84.7	67.2	53.2	242
2.90	—	444	—	47.1	84.0	65.8	51.5	230
2.95	429	429	—	45.7	83.4	64.6	49.9	219
3.00	415	415	—	44.5	82.8	63.5	48.4	212
3.05	401	401	—	43.1	82.0	62.3	46.9	202
3.10	388	388	—	41.8	81.4	61.1	45.3	193
3.15	375	375	—	40.4	80.6	59.9	43.6	184
3.20	363	363	—	39.1	80.0	58.7	42.0	177
3.25	352	352	—	37.9	79.3	57.6	40.5	170
3.30	341	341	—	36.6	78.6	56.4	39.1	163
3.35	331	331	—	35.5	78.0	55.4	37.8	158
3.40	321	321	—	34.3	77.3	54.3	36.4	152
3.45	311	311	—	33.1	76.7	53.3	34.4	147
3.50	302	302	—	32.1	76.1	52.2	33.8	143
3.55	293	293	—	30.9	75.5	51.2	32.4	139
3.60	285	285	—	29.9	75.0	50.3	31.2	136
3.65	277	277	—	28.8	74.4	49.3	29.9	131
3.70	269	269	—	27.6	73.7	48.3	28.5	128
3.75	262	262	—	26.6	73.1	47.3	27.3	125
3.80	255	255	—	25.4	72.5	46.2	26.0	121
3.85	248	248	—	24.2	71.7	45.1	24.5	118
3.90	241	241	100.0	22.8	70.9	43.9	22.8	114
3.95	235	235	99.0	21.7	70.3	42.9	21.5	111
4.00	229	229	98.2	20.5	69.7	41.9	20.1	109
4.05	223	223	97.3	—	—	—	—	104
4.10	217	217	96.4	—	—	—	—	103
4.15	212	212	95.5	—	—	—	—	100
4.20	207	207	94.6	—	—	—	—	99
4.25	201	201	93.8	—	—	—	—	97
4.30	197	197	92.8	—	—	—	—	94
4.35	192	192	91.9	—	—	—	—	92
4.40	187	187	90.7	—	—	—	—	90
4.45	183	183	90.0	—	—	—	—	89
4.50	179	179	89.0	—	—	—	—	88
4.55	174	174	87.8	—	—	—	—	86
4.60	170	170	86.8	—	—	—	—	84
4.65	167	167	86.0	—	—	—	—	83
4.70	163	163	85.0	—	—	—	—	82
4.80	156	156	82.9	—	—	—	—	80
4.90	149	149	80.8	—	—	—	—	73
5.00	143	143	78.7	—	—	—	—	71
5.10	137	137	76.4	—	—	—	—	67
5.20	131	131	74.0	—	—	—	—	65
5.30	126	126	72.0	—	—	—	—	63
5.40	121	121	69.0	—	—	—	—	60
5.50	116	116	67.6	—	—	—	—	58
5.60	111	111	65.7	—	—	—	—	56

(Carpenter Steel Co.)

Conversion Table: US Customary to SI Metric

When You Know:	Multiply By:		To Find:
	Very accurate	Approximate	
Length			
inches	*25.4		millimeters
inches	* 2.54		centimeters
feet	* 0.3048		meters
feet	*30.48		centimeters
yards	* 0.9144	0.9	meters
miles	* 1.609344	1.6	kilometers
Weight			
grains	15.43236	15.4	grams
ounces	*28.349523125	28.0	grams
ounces	* 0.028349523125	0.028	kilograms
pounds	* 0.45359237	0.45	kilograms
short ton	* 0.90718474	0.9	tonnes
Volume			
teaspoons		5.0	milliliters
tablespoons		15.0	milliliters
fluid ounces	29.57353	30.0	milliliters
cups		0.24	liters
pints	* 0.473176473	0.47	liters
quarts	* 0.946352946	0.95	liters
gallons	* 3.785411784	3.8	liters
cubic inches	* 0.016387064	0.02	liters
cubic feet	* 0.028316846592	0.03	cubic meters
cubic yards	* 0.764554857984	0.76	cubic meters
Area			
square inches	* 6.4516	6.5	square centimeters
square feet	* 0.09290304	0.09	square meters
square yards	* 0.83612736	0.8	square meters
square miles		2.6	square kilometers
acres	* 0.40468564224	0.4	hectares
Temperature			
Fahrenheit	*5/9 (after subtracting 32)		Celsius

* = Exact

Conversion Table: SI Metric to US Customary

When You Know:	Multiply By:		To Find:
	Very accurate	Approximate	
Length			
millimeters	0.0393701	0.04	inches
centimeters	0.3937008	0.4	inches
meters	3.280840	3.3	feet
meters	1.093613	1.1	yards
kilometers	0.621371	0.6	miles
Weight			
grains	0.00228571	0.0023	ounces
grams	0.03527396	0.035	ounces
kilograms	2.204623	2.2	pounds
tonnes	1.1023113	1.1	short tons
Volume			
milliliters		0.2	teaspoons
milliliters	0.06667	0.067	tablespoons
milliliters	0.03381402	0.03	fluid ounces
liters	61.02374	61.024	cubic inches
liters	2.113376	2.1	pints
liters	1.056688	1.06	quarts
liters	0.26417205	0.26	gallons
liters	0.03531467	0.035	cubic feet
cubic meters	61023.74	61023.7	cubic inches
cubic meters	35.31467	35.0	cubic feet
cubic meters	1.3079506	1.3	cubic yards
cubic meters	264.17205	264.0	gallons
Area			
square centimeters	0.1550003	0.16	square inches
square centimeters	0.00107639	0.001	square feet
square meters	10.76391	10.8	square feet
square meters	1.195990	1.2	square yards
square kilometers		0.4	square miles
hectares	2.471054	2.5	acres
Temperature			
Celsius	*9/5 (then add 32)		Fahrenheit

* = Exact

Decimal Conversion Chart

Fraction	Inches	mm
1/64	.01563	.397
1/32	.03125	.794
3/64	.04688	1.191
1/16	.0625	1.588
5/64	.07813	1.984
3/32	.09375	2.381
7/64	.10938	2.778
1/8	.12500	3.175
9/64	.14063	3.572
5/32	.15625	3.969
11/64	.17188	4.366
3/16	.18750	4.763
13/64	.20313	5.159
7/32	.21875	5.556
15/64	.23438	5.953
1/4	.25000	6.350
17/64	.26563	6.747
9/32	.28125	7.144
19/64	.29688	7.541
5/16	.31250	7.938
21/64	.32813	8.334
11/32	.34375	8.731
23/64	.35938	9.128
3/8	.37500	9.525
25/64	.39063	9.922
13/32	.40625	10.319
27/64	.42188	10.716
7/16	.43750	11.113
29/64	.45313	11.509
15/32	.46875	11.906
31/64	.48438	12.303
1/2	.50000	12.700

Fraction	Inches	mm
33/64	.51563	13.097
17/32	.53125	13.494
35/64	.54688	13.891
9/16	.56250	14.288
37/64	.57813	14.684
19/32	.59375	15.081
39/64	.60938	15.478
5/8	.62500	15.875
41/64	.64063	16.272
21/32	.65625	16.669
43/64	.67188	17.066
11/16	.68750	17.463
45/64	.70313	17.859
23/32	.71875	18.256
47/64	.73438	18.653
3/4	.75000	19.050
49/64	.76563	19.447
25/32	.78125	19.844
51/64	.79688	20.241
13/16	.81250	20.638
53/64	.82813	21.034
27/32	.84375	21.431
55/64	.85938	21.828
7/8	.87500	22.225
57/64	.89063	22.622
29/32	.90625	23.019
59/64	.92188	23.416
15/16	.93750	23.813
61/64	.95313	24.209
31/32	.96875	24.606
63/64	.98438	25.003
1	1.00000	25.400

Grade Marking for Bolts

Bolt Head Marking	SAE = Society of Automotive Engineers ASTM = American Society for Testing and Materials	Bolt Material	Minimum Tensile Strength in Pounds per Square Inch (psi)
No Marks	SAE Grade 1 SAE Grade 2 Indeterminate quality	Low-carbon steel Low-carbon steel	65,000 psi
2 Marks	SAE Grade 3	Medium-carbon steel, cold worked	110,000 psi
3 Marks	SAE Grade 5 ASTM – A 325 Common commercial quality	Medium-carbon steel, quenched and tempered	120,000 psi
Letters BB (BB)	ASTM – A 354	Low-alloy steel or medium-carbon steel, quenched and tempered	105,000 psi
Letters BC (BC)	ASTM – A 354	Low-alloy steel or medium-carbon steel, quenched and tempered	125,000 psi
4 Marks	SAE Grade 6 Better commercial quality	Medium-carbon steel, quenched and tempered	140,000 psi
5 Marks	SAE Grade 7	Medium-carbon alloy steel, quenched and tempered, roll-threaded after heat treatment	133,000 psi
6 Marks	SAE Grade 8 ASTM – A 345 Best commercial quality	Medium-carbon alloy steel, quenched and tempered	150,000 psi

Machine Screw and Cap Screw Heads

	Size	A	B	C	D
Fillister Head	#8	.260	.141	.042	.060
	#10	.302	.164	.048	.072
	1/4	3/8	.205	.064	.087
	5/16	7/16	.242	.077	.102
	3/8	9/16	.300	.086	.125
	1/2	3/4	.394	.102	.168
	5/8	7/8	.500	.128	.215
	3/4	1	.590	.144	.258
	1	1 5/16	.774	.182	.352
Flat Head	#8	.320	.092	.043	.037
	#10	.372	.107	.048	.044
	1/4	1/2	.146	.064	.063
	5/16	5/8	.183	.072	.078
	3/8	3/4	.220	.081	.095
	1/2	7/8	.220	.102	.090
	5/8	1 1/8	.293	.128	.125
	3/4	1 3/8	.366	.144	.153
Round Head	#8	.297	.113	.044	.067
	#10	.346	.130	.048	.073
	1/4	7/16	.183	.064	.107
	5/16	9/16	.236	.072	.150
	3/8	5/8	.262	.081	.160
	1/2	13/16	.340	.102	.200
	5/8	1	.422	.128	.255
	3/4	1 1/4	.526	.144	.320
Hexagon Head	1/4	.494	.170	7/16	
	5/16	.564	.215	1/2	
	3/8	.635	.246	9/16	
	1/2	.846	.333	3/4	
	5/8	1.058	.411	15/16	
	3/4	1.270	.490	1 1/8	
	7/8	1.482	.566	1 5/16	
	1	1.693	.640	1 1/2	
Socket Head	#8	.265	.164	1/8	
	#10	5/16	.190	5/32	
	1/4	3/8	1/4	3/16	
	5/16	7/16	5/16	7/32	
	3/8	9/16	3/8	5/16	
	7/16	5/8	7/16	5/16	
	1/2	3/4	1/2	3/8	
	5/8	7/8	5/8	1/2	
	3/4	1	3/4	9/16	
	7/8	1 1/8	7/8	9/16	
	1	1 5/16	1	5/8	

Screw Thread Elements for Unified and National Thread Form

Threads per Inch (n)	Pitch (p) $p = \frac{1}{n}$	Single Height	Double Height	83 1/3% Double Height	Basic Width of Crest and Root Flat $\frac{P}{8}$	Constant for Best Size Wire Also Single Height of 60° V-Thread	Diameter of Best size Wire
		Subtract from basic major diameter to get basic pitch diameter	Subtract from basic major diameter to get basic minor diameter	Subtract from basic major diameter to get minor diameter of ring gage			
3	.333333	.216506	.43301	.36084	.0417	.28868	.19245
3 1/4	.307692	.199852	.39970	.33309	.0385	.26647	.17765
3 1/2	.285714	.185577	.37115	.30929	.0357	.24744	.16496
4	.250000	.162379	.32476	.27063	.0312	.21651	.14434
4 1/2	.222222	.144337	.28867	.24056	.0278	.19245	.12830
5	.200000	.129903	.25981	.21650	.0250	.17321	.11547
5 1/2	.181818	.118093	.23619	.19682	.0227	.15746	.10497
6	.166666	.108253	.21651	.18042	.0208	.14434	.09623
7	.142857	.092788	.18558	.15465	.0179	.12372	.08248
8	.125000	.081189	.16238	.13531	.0156	.10825	.07217
9	.111111	.072168	.14434	.12028	.0139	.09623	.06415
10	.100000	.064952	.12990	.10825	.0125	.08660	.05774
11	.090909	.059046	.11809	.09841	.0114	.07873	.05249
11 1/2	.086956	.056480	.11296	.09413	.0109	.07531	.05020
12	.083333	.054127	.10826	.09021	.0104	.07217	.04811
13	.076923	.049963	.09993	.08327	.0096	.06662	.04441
14	.071428	.046394	.09279	.07732	.0089	.06186	.04124
16	.062500	.040595	.08119	.06766	.0078	.05413	.03608
18	.055555	.036086	.07217	.06014	.0069	.04811	.03208
20	.050000	.032475	.06495	.05412	.0062	.04330	.02887
22	.045454	.029523	.05905	.04920	.0057	.03936	.02624
24	.041666	.027063	.05413	.04510	.0052	.03608	.02406
27	.037037	.024056	.04811	.04009	.0046	.03208	.02138
28	.035714	.023197	.04639	.03866	.0045	.03093	.02062
30	.033333	.021651	.04330	.03608	.0042	.02887	.01925
32	.031250	.020297	.04059	.03383	.0039	.02706	.01804
36	.027777	.018042	.03608	.03007	.0035	.02406	.01604
40	.025000	.016237	.03247	.02706	.0031	.02165	.01443
44	.022727	.014761	.02952	.02460	.0028	.01968	.01312
48	.020833	.013531	.02706	.02255	.0026	.01804	.01203
50	.020000	.012990	.02598	.02165	.0025	.01732	.01155
56	.017857	.011598	.02320	.01933	.0022	.01546	.01031
60	.016666	.010825	.02165	.01804	.0021	.01443	.00962
64	.015625	.010148	.02030	.01691	.0020	.01353	.00902
72	.013888	.009021	.01804	.01503	.0017	.01203	.00802
80	.012500	.008118	.01624	.01353	.0016	.01083	.00722
90	.011111	.007217	.01443	.01202	.0014	.00962	.00642
96	.010417	.006766	.01353	.01127	.0013	.00902	.00601
100	.010000	.006495	.01299	.01082	.0012	.00866	.00577
120	.008333	.005413	.01083	.00902	.0010	.00722	.00481

Note: Using the Best Size Wires, measurement over three wires minus Constant for Best Size Wire equals Pitch Diameter.

Bolt Torquing Chart

Metric Standard						
Grade of Bolt		5D	.8G	10K	12K	
Min. Ten. Strength		71,160 P.S.I.	113,800 P.S.I.	142,200 P.S.I.	170,679 P.S.I.	
Markings on Head		5D	.8G	10K	12K	Size of Socket or Wrench Opening
Metric		Foot Pounds				Metric
Bolt Dia.	U.S. Dec. Equiv.					Bolt Head
6mm	.2362	5	6	8	10	10mm
8mm	.3150	10	16	22	27	14mm
10mm	.3937	19	31	40	49	17mm
12mm	.4720	34	54	70	86	19mm
14mm	.5512	55	89	117	137	22mm
16mm	.6299	83	132	175	208	24mm
18mm	.7090	111	182	236	283	27mm
22mm	.8661	182	284	394	464	32mm

SAE Standard/Foot Pounds						
Grade of Bolt	SAE 1 & 2	SAE 5	SAE 6	SAE 8		
Min. Ten. Strength	64,000 P.S.I.	105,000 P.S.I.	133,000 P.S.I.	150,000 P.S.I.		
Markings on Head					Size of Socket or Wrench Opening	
U.S. Standard	Foot Pounds				U.S. Standard	
Bolt Dia.					Bolt Head	Nut
1/4	5	7	10	10.5	3/8	7/16
5/16	9	14	19	22	1/2	9/16
3/8	15	25	34	37	9/16	5/8
7/16	24	40	55	60	5/8	3/4
1/2	37	60	85	92	3/4	13/16
9/16	53	88	120	132	7/8	7/8
5/8	74	120	167	180	15/16	1.
3/4	120	200	280	296	1-1/8	1-1/8

Cutting Fluids for Various Metals

Metal	Cutting Fluid
Aluminum and its Alloys	Kerosene, kerosene and lard oil, soluble oil
Plastics	Dry
Brass, Soft	Dry, soluble oil, kerosene and lard oil
Bronze, High Tensile	Soluble oil, lard oil, mineral oil, dry
Cast Iron	Dry, air jet, soluble oil
Copper	Soluble oil, dry, mineral lard oil, kerosene
Magnesium	Low viscosity neutral oils
Malleable Iron	Dry, soda water
Monel Metal	Lard oil, soluble oil
Slate	Dry
Steel, Forging	Soluble oil, sulfurized oil, mineral lard oil
Steel, Manganese	Soluble oil, sulfurized oil, mineral lard oil
Steel, Soft	Soluble oil, mineral lard oil, sulfurized oil, lard oil
Steel, Stainless	Sulfurized mineral oil, soluble oil
Steel, Tool	Soluble oil, mineral lard oil, sulfurized oil
Wrought Iron	Soluble oil, mineral lard oil, sulfurized oil

EIA and AIA National Codes for CNC Programming

Preparatory (G) Functions	
G word	**Explanation**
G00	Denotes a rapid traverse rate for point-to-point positioning.
G01	Describes linear interpolation blocks; reserved for contouring.
G02, G03	Used with circular interpolation.
G04	Sets a calculated time delay during which there is no machine motion (dwell).
G05, G07	Unassigned by the EIA. May be used at the discretion of the machine tool or system builder. Could also be standardized at a future date.
G06	Used with parabolic interpolation.
G08	Acceleration code. Causes the machine, assuming it is capable, to accelerate at a smooth exponential rate.
G09	Deceleration code. Causes the machine, assuming it is capable, to decelerate at a smooth exponential rate.
G10-G12	Normally unassigned for CNC systems. Used with some hard-wired systems to express blocks of abnormal dimensions.
G13-G16	Direct the control system to operate on a particular set of axes.
G17-G19	Identify or select a coordinate plane for such functions as circular interpolation or cutter compensation.
G20-G32	Unassigned according to EIA standards. May be assigned by the control system or machine tool builder.
G33-G35	Selected for machines equipped with thread-cutting capabilities (generally referring to lathes). G33 is used when a constant lead is sought, G34 is used when a constantly increasing lead is required, and G35 is used to designate a constantly decreasing lead.
G36-G39	Unassigned.
G40	Terminates any cutter compensation.
G41	Activates cutter compensation in which the cutter is on the left side of the work surface (relative to the direction of the cutter motion).
G42	Activates cutter compensation in which the cutter is on the right side of the work surface.
G43, G44	Used with cutter offset to adjust for the difference between the actual and programmed cutter radii or diameters.
G43	refers to an inside corner, and G44 refers to an outside corner.
G45-G49	Unassigned.
G50-G59	Reserved for adaptive control.
G60-G69	Unassigned.
G70	Selects inch programming.
G71	Selects metric programming.
G72	Selects three-dimensional CW circular interpolation.
G73	Selects three-dimensional CCW circular interpolation.
G74	Cancels multiquadrant circular interpolation.
G75	Activates multiquadrant circular interpolation.
G76-G79	Unassigned.
G80	Cancel cycle.
G81	Activates drill, or spotdrill, cycle.
G82	Activates drill with a dwell.
G83	Activates intermittent, or deep-hole, drilling.
G84	Activates tapping cycle.
G85-G89	Activates boring cycles.
G90	Selects absolute input. Input data is to be in absolute dimensional form.

EIA and AIA National Codes for CNC Programming (continued)

Preparatory (G) Functions	
G word	**Explanation**
G91	Selects incremental input. Input data is to be in incremental form.
G92	Preloads registers to desired values (for example, preloads axis position registers).
G93	Sets inverse time feed rate.
G94	Sets inches (or millimeters) per minute feed rate.
G95	Sets inches (or millimeters) per revolution feed rate.
G97	Sets spindle speed in revolutions per minute.
G98, G99	Unassigned.

Miscellaneous (M) Functions	
M word	**Explanation**
M00	Program stop. Operator must cycle start in order to continue with the remainder of the program.
M01	Optional stop. Acted upon only when the operator has previously signaled for this command by pushing a button. When the control system senses the M01 code, machine will automatically stop.
M02	End of program. Stops the machine after completion of all commands in the block. May include rewinding of tape.
M03	Starts spindle rotation in a clockwise direction.
M04	Starts spindle rotation in a counterclockwise direction.
M05	Spindle stop.
M06	Executes the change of a tool (or tools) manually or automatically.
M07	Turns coolant on (flood).
M08	Turns coolant on (mist).
M09	Turns coolant off.
M10	Activates automatic clamping of the machine slides, workpiece, fixture, spindle, etc.
M11	Deactivates automatic clamping.
M12	Inhibiting code used to synchronize multiple set of axes, such as a four-axis lathe that has two independently operated heads or slides.
M13	Combines simultaneous clockwise spindle motion and coolant on.
M14	Combines simultaneous counterclockwise spindle motion and coolant on.
M15	Sets rapid traverse or feed motion in the + direction.
M16	Sets rapid traverse or feed motion in the – direction.
M17, M18	Unassigned.
M19	Oriented spindle stop. Stops spindle at a predetermined angular position.
M20-M29	Unassigned.
M30	End of data. Used to reset control and/or machine.
M31	Interlock bypass. Temporarily circumvents a normally provided interlock.
M32-M39	Unassigned.
M40-M46	Signals gear changes if required at the machine; otherwise, unassigned.
M47	Continues program execution from the start of the program, unless inhibited by an interlock signal.
M48	Cancels M49.
M49	Deactivates a manual spindle or feed-override and returns to the programmed value.

EIA and AIA National Codes for CNC Programming (continued)

Miscellaneous (M) Functions	
M word	**Explanation**
M50-M57	Unassigned.
M58	Cancels M59.
M59	Holds the rpm constant at its value.
M60-M99	Unassigned.

Other Address Characters	
Address character	**Explanation**
A	Angular dimension about the X-axis.
B	Angular dimension about the Y-axis.
C	Angular dimension about the Z-axis.
D	Can be used for an angular dimension around a special axis, for a third feed function, or for tool offset.
E	Used for angular dimension around a special axis or for a second feed function.
H	Fixture offset.
I, J, K	Centerpoint coordinates for circular interpolation.
L	Not used.
O	Used on some N/C controls in place of the customary sequence number word address N.
P	Third rapid traverse code—tertiary motion dimension parallel to the X-axis.
Q	Second rapid traverse code—tertiary motion dimension parallel to the Y-axis.
R	First rapid traverse code—tertiary motion dimension parallel to the Z-axis (or to the radius) for constant surface speed calculation.
U	Secondary motion dimension parallel to the X-axis.
V	Secondary motion dimension parallel to the Y-axis.
W	Secondary motion dimension parallel to the Z-axis.

Powder Metallurgy Processes

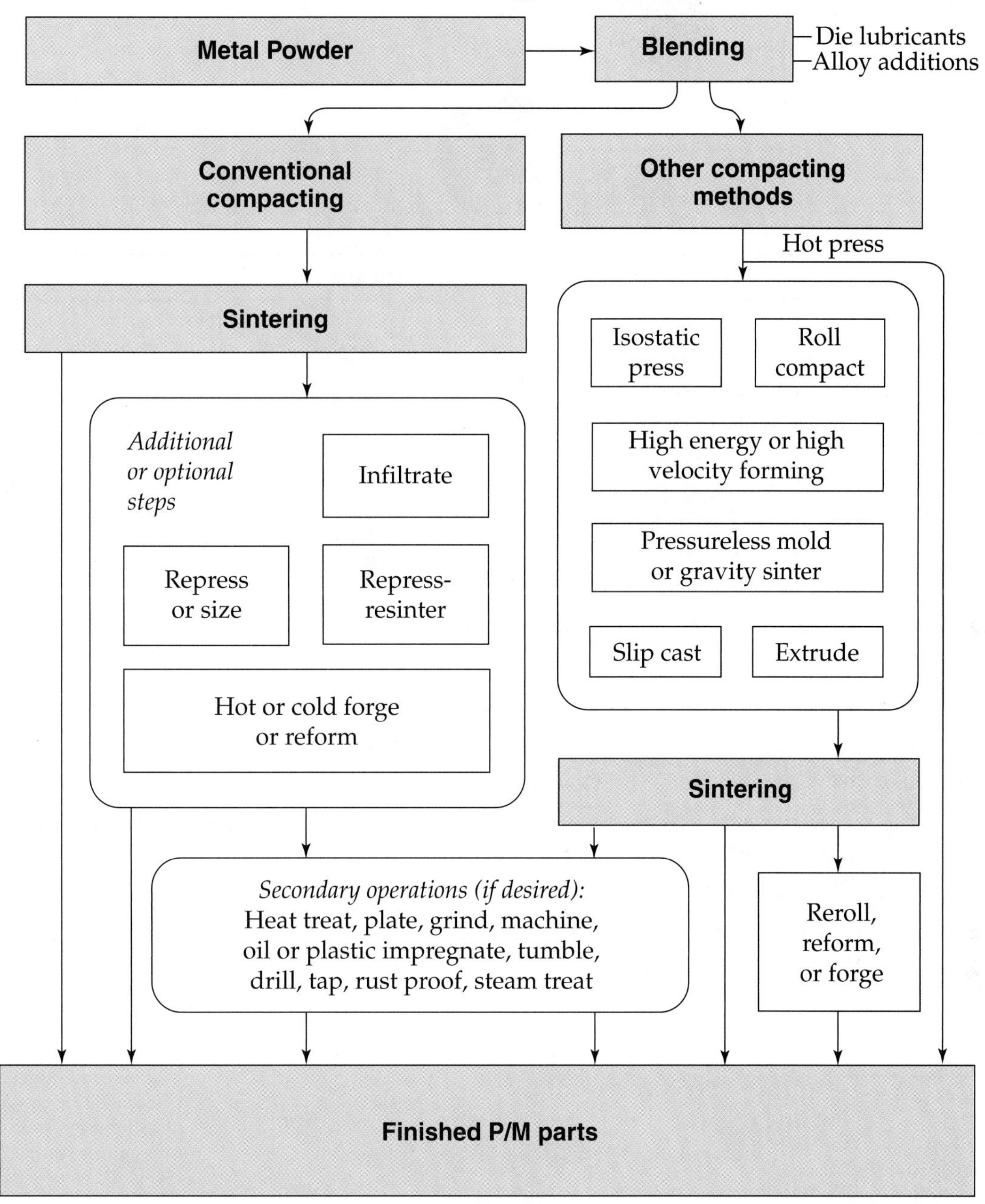

Sintering Sintering is a solid-state phenomenon in which powdered metal particles become metallurgically bonded below the melting point of the metal. No adhesives or cements are used.

Infiltration Pores of P/M parts are filled with a lower-melting-point metal such as copper-based alloy. When the part is sintered, the infiltrant material melts and penetrates into the P/M part by capillary action.

Coining and sizing Basically, this operation involves repressing sintered parts in a die similar to the original compacting die.

Impregnation The pores of the P/M part are filled with a lubricant or other nonmetallic such as plastic resin. This may be done by means of a vacuum or by soaking. The part then becomes self-lubricating (or pressure-tight if resin is used).

Master Chart of Welding and Allied Processes

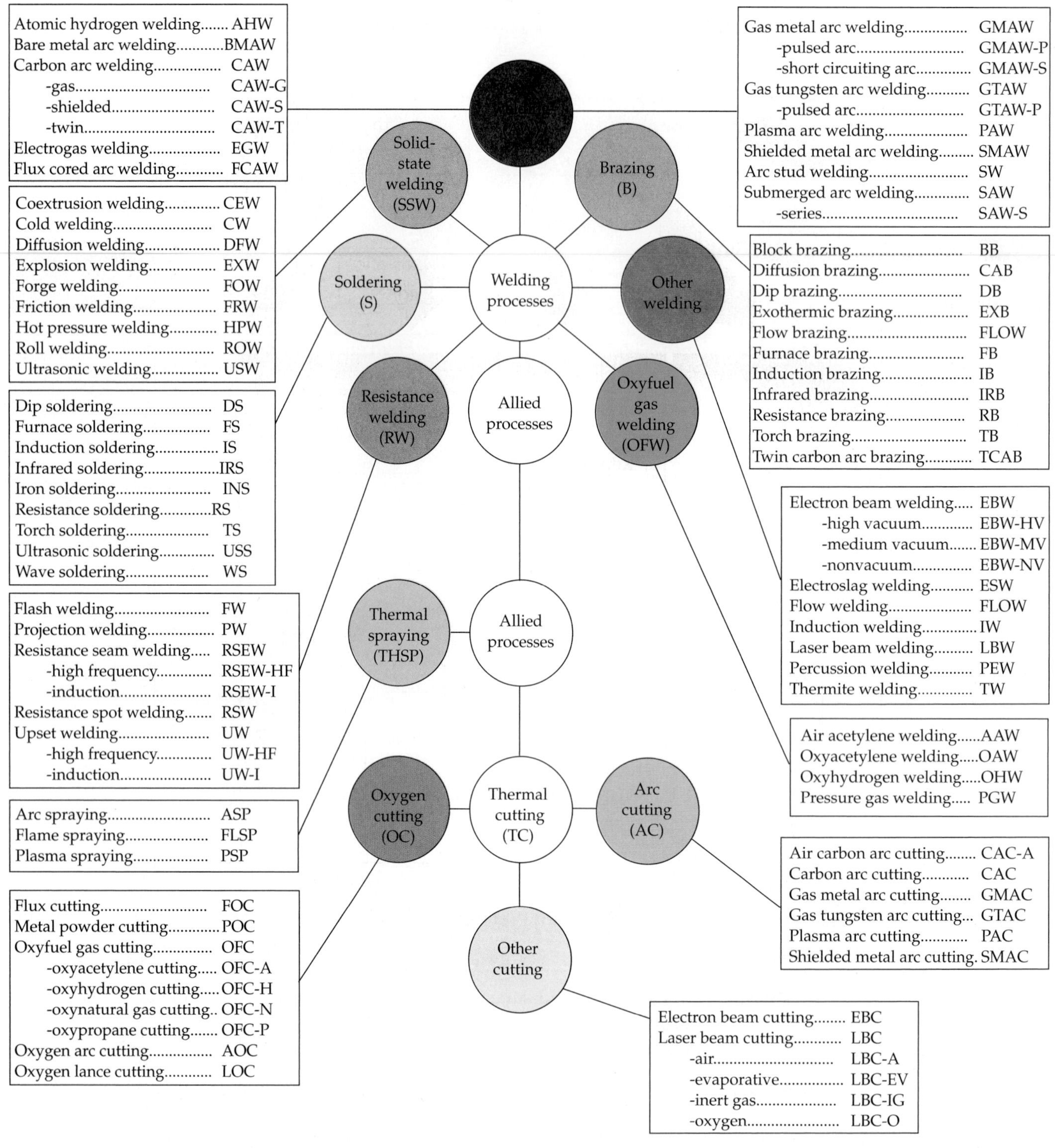

(American Welding Society)

Grinding Wheel Markings

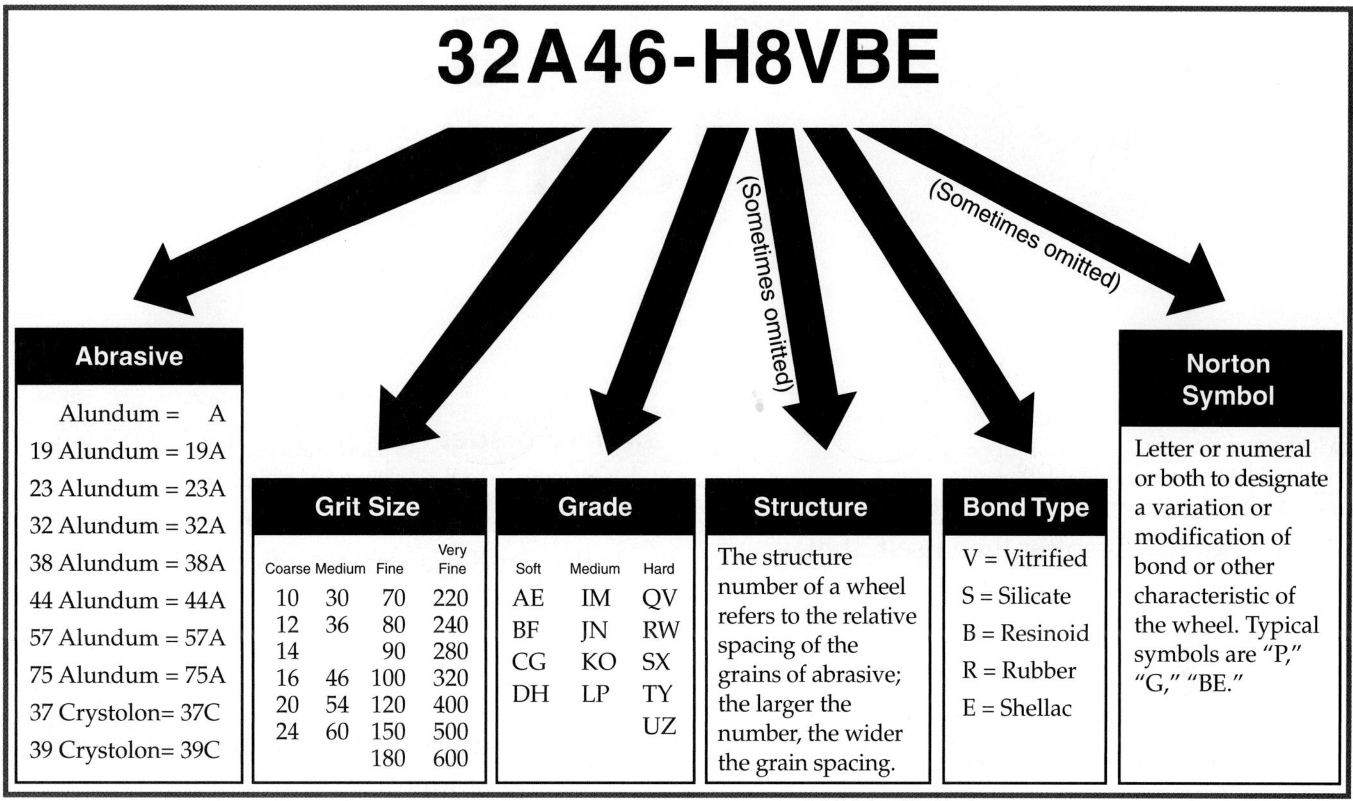

(Norton Co.)

Standard Symbols Used in Dimensioning

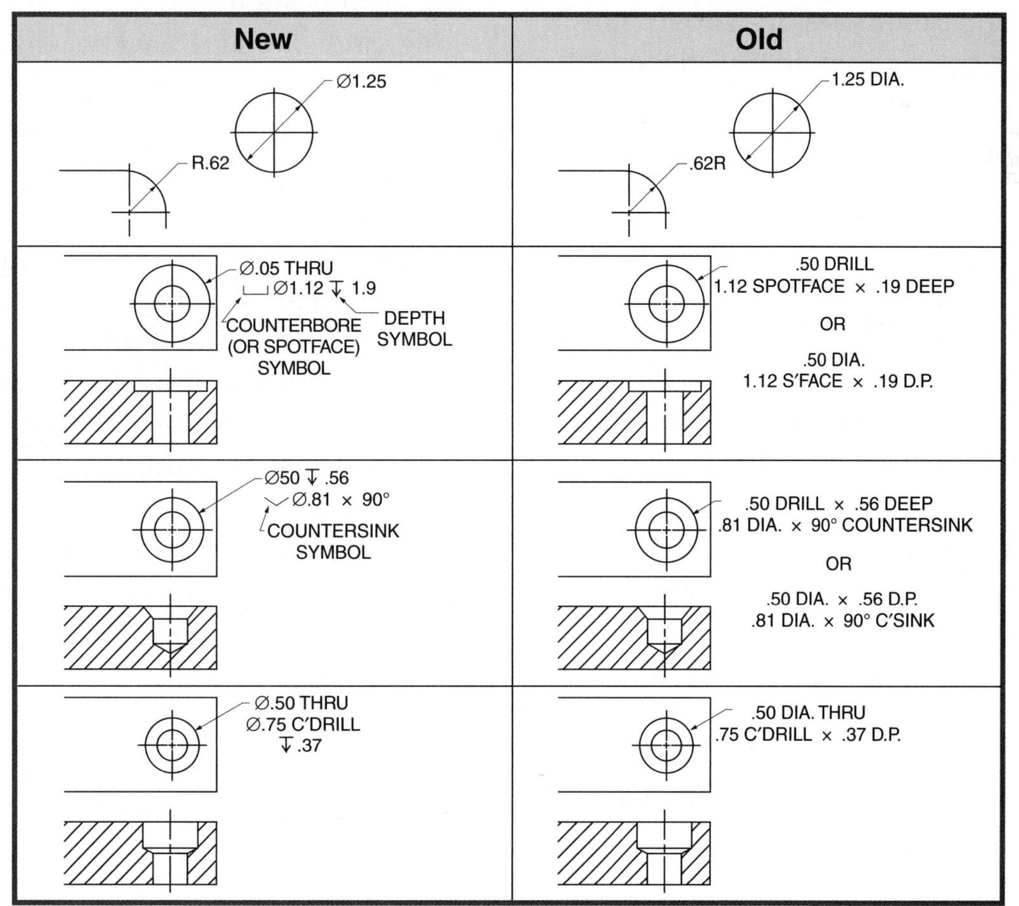

Glossary of Technical Terms

A

abrasive: A material that penetrates and cuts a material that is softer than itself. It may be natural (emery, corundum, diamond) or artificial (silicon carbide, aluminum oxide). (Ch. 7)

abrasive cutoff saw: A sawing machine that uses a rotary abrasive wheel either with or without a liquid coolant. Wet abrasive cutting uses a liquid coolant, can cut to close tolerances, and produces a fine surface finish. Dry abrasive cutting uses no coolant and is used for rapid, less critical cutting. (Ch. 11)

absolute positioning: A programming mode, indicated by the G90 code, in which the coordinate values for any point are interpreted in relation to the X0,Y0 position. (Ch. 23)

actual size: The measured size of a part after it is manufactured. (Ch. 4)

addendum: That portion of a gear tooth that projects above or outside the pitch circle. (Ch. 19)

adhesive: A material that provides one of the newer ways to join metals and keep threaded fasteners from vibrating loose. In some applications, the resulting joints are stronger than the metal itself. Adhesive-bonded joints do not require costly and time-consuming operations such as drilling, countersinking, and riveting. (Ch. 8)

all-hard blade: One of two types of power hacksaw blade (the other is a flexible-back blade). It is best for straight, accurate cutting under a variety of conditions. (Ch. 11)

alloy: A mixture of two or more metals. (Ch. 28)

Aluminum Association Designation System: A method of identifying aluminum with a four digit code plus a temper designation. The identification code and temper designation are separated by a hyphen. (Ch. 28)

American National Standards Institute (ANSI): An association that serves as a clearinghouse for nationally coordinated voluntary standards for fields ranging from information technology to building construction. Standards are established for such areas as definitions, terminology, symbols, materials, performance characteristics, procedures, and testing methods. (Ch. 4)

American National Thread System: The common thread form used in the United States, characterized by the 60° angle formed by the sides of the thread. (Ch. 7)

angularity: The condition of a surface, center plane, or axis at any specified angle from a datum plane or axis other than 90°. (Ch. 4)

annealing: A heat treatment process in which the metal is heated above its critical temperature and cooled slowly to reduce its hardness and make it easier to machine or work. (Ch. 29)

anodizing: An electrolytic process that forms a protective layer of aluminum oxide on aluminum parts. (Ch. 30)

apprentice: A student who receives on-the-job training working with a skilled machinist while also studying work-related subjects. (Ch. 2)

approved respirator: A type of mask worn over the mouth and nose that provides clean air to breathe, and prevents the inhalation of dangerous toxins. (Ch. 3)

arbor: A shaft or spindle for holding cutting tools. (Ch. 18)

arbor press: A manually operated press used on smaller work that can have a leverage of several tons. (Ch. 17)

assigned code: A code that has the same application or function on all machine controllers, regardless of application. (Ch. 23)

associate's degree: A two-year educational degree focused on preparing students for technical positions. (Ch. 2)

automatic screw machine: A variation of the lathe that was developed for high-speed production of large numbers of small parts. The machine performs a maximum number of operations, either simultaneously, or in a very rapid sequence. (Ch. 15)

automatic tool changer (ATC): A device used by some CNC machine tools that automatically changes and stores the tools. (Ch. 22)

B

bachelor's degree: A four-year educational degree focused on advanced theoretical education and skill training in specific disciplines like engineering. (Ch. 2)

backlash: Clearance designed into the mating of threaded rods and nuts to allow movement. (Ch. 22)

bacteriostat: A substance that regulates and controls the growth of bacteria. (Ch. 10)

band machining: A widely used technique that makes use of a continuous saw blade. Chip removal is rapid, and accuracy can be held to close tolerances, eliminating or minimizing many secondary machining operations. (Ch. 21)

bar puller: An attachment mounted in the turret of a turning center used to clamp and pull bar stock through the spindle for machining. (Ch. 25)

base metal: A pure, nonferrous, nonprecious metal. (Ch. 28)

bevel gear: A toothed wheel used to change the angular direction of power between shafts. The teeth are either straight or curved. (Ch. 19)

bill of materials: A listing of the numbers, names, materials, and quantities of the parts specified on a set of working drawings. (Ch. 4)

blind hole: A hole that does not go completely through the workpiece. (Ch. 7, 12)

block: In a word address format programming language, groups of words that are written on a single line and are meant to be executed together. (Ch. 23)

boring: An internal machining operation in which a single-point cutting tool is used to enlarge a hole. (Ch. 15)

boring mill: A huge machine capable of turning and boring work with diameters as large as 40′ (12 m). Work that is too large or too heavy to be turned in a horizontal position is machined on a vertical boring mill. (Ch. 15)

box annealing: A method of annealing in which the work is placed into a metal box, which is then placed into the furnace, heated to the proper temperature, and allowed to cool inside the furnace. Keeping the work inside the box during heating and cooling prevents the work from scaling or decarbonizing. (Ch. 29)

Brinell hardness test: A method of measuring a metal's resistance to deformation. In this test, a known load is applied to a metal through a steel ball of known size. The hardness reading is based on the diameter of the impression made by the steel ball. (Ch. 29)

briquetting: In the powder metallurgy process, the operation that compresses and forms metal powder into the desired shape. (Ch. 33)

broach: A long, multitoothed cutting tool with three kinds of teeth shaped to give a desired surface: rough, semifinished, and finished. (Ch. 17)

broaching: A manufacturing process for machining flat, round, and contoured surfaces, both internal and external. A broach is pushed or pulled across the work, with each tooth removing only a small portion of the material. Cutting a keyway is typically a broaching process. (Ch. 17)

broaching machine: A machine designed to push or pull a multitoothed cutter across the work, each tooth of the cutting tool (broach) removing only a small amount of material. (Ch. 1)

burnishing: The process of finishing a metal surface by compressing the surface. It is often done by tumbling the work with steel balls. (Ch. 17)

bushing: A bearing for a revolving shaft. Also, a hardened steel tube used on jigs to guide drills and reamers. (Ch. 9)

C

canned cycle: A set of commands that follows a prescribed sequence again and again until the cycle is cancelled. Also called *subroutine.* (Ch. 24)

carbon content: The amount of carbon that a material contains. In reference to steel, it is measured in percent or points. (Ch. 28)

career: An occupation requiring specialized training. (Ch. 2)

Cartesian coordinate system: A coordinate system that specifies positive (+) and negative (–) movement along X, Y, and Z axes; used in CNC programming to specify movements of the workpiece or tool. (Ch. 22)

case hardening: A heat treatment process that creates a hard shell on the surface of low-carbon steel while leaving the inner core of the metal unaffected. (Ch. 29)

center finder: A device used in a drilling machine to position the point to be drilled directly under the chuck or spindle. Also known as a "wiggler." (Ch. 12)

centerless grinding: A technique in which a workpiece is not supported between centers; rather, it is positioned on a work support blade and fed automatically between a regulating or feed wheel and a grinding wheel. (Ch. 20)

chemical blackening: A process that chemically bonds a black oxide coating to the surface of ferrous metals. The resulting coating improves corrosion and wear resistance, reduces glare from the surface, and improves the adhesion of subsequent layers of paint or other coatings. (Ch. 30)

chemical blanking: A variation of the chemical milling operation that results in the removal of metal from certain areas by chemical action. Commonly used to cut parts out of ultrathin materials. (Ch. 32)

chemical cutting fluid: Cooling and lubricating liquid that contains no oil. Because they are not actually fluids (graphite, mica, and white lead are examples), a wetting agent is often added to provide lubricating qualities. Compare with cutting fluid. (Ch. 10)

chemical machining: A category of processes that use chemicals, usually in an aqueous (with water) solution, to etch away selected portions of material. (Ch. 32)

chemical milling: A process that uses a masking technique and a chemical etchant to remove metal selectively from relatively large surface areas. (Ch. 32)

chipbreaker: A small groove cut into the face of a lathe tool near the cutting edge to break chips into small pieces. (Ch. 14)

chipless machining: A manufacturing process that forms metal wire or rod into desired shapes using a series of dies. *Also called cold forming and cold heading.* (Ch. 33)

circular interpolation: Movement in a circular pattern (in the context of CNC programming). (Ch. 23)

circularity: A form tolerance that is characterized by any given cross section taken perpendicular to the axis of a cylinder or cone, or through the common center of a sphere. (Ch. 4)

circular pitch: The distance from the center of one gear tooth to the center of the next tooth, measured along the pitch circle. (Ch. 19)

circular runout: A type of runout that provides control of single circular elements of a surface. (Ch. 4)

class of fit: Standard working tolerances for thread accuracy, indicated by the last number on a thread description. Fits for inch-based threads are as follows: Class 1, loose fit; Class 2, free fit; Class 3, medium fit; and Class 4, close fit. (Ch. 7)

climb milling: A milling technique in which the teeth of a cutting tool advance into the work in the same direction as the feed. Also known as "climb cutting." (Ch. 18)

closed-loop system: A system in which feedback is provided to the controlling mechanism; used with servo motors in CNC systems to provide feedback to the controllers regarding the position of the axis they are driving. (Ch. 22)

CNC milling center: A traditional milling machine with built-in CNC capabilities. (Ch. 22)

cold circular saw: A tool that uses a circular, toothed blade capable of producing very accurate cuts. Large cold circular saws can be used to sever round metal stock up to 27" (675 mm) in diameter. (Ch. 11)

cold finished steel: Steel that has been "pickled" or treated with a dilute acid solution to remove its oxide coating. After pickling, the steel is drawn or rolled to finished size and shape while cold. Cold finished steel is characterized by a smooth, bright finish. (Ch. 28)

cold forming: A manufacturing process that forms metal wire or rod into desired shapes using a series of dies. *Also called chipless machining and cold heading.* (Ch. 33)

cold heading: A manufacturing process that forms metal wire or rod into desired shapes using a series of dies. *Also called chipless machining and cold forming.* (Ch. 33)

column and knee milling machine: So named because of the parts that provide movement to the workpiece. It consists of a column that supports and guides the knee in vertical (up and down/Z-axis) movement, and a knee that supports the mechanism for obtaining table movements. These movements are transverse (in and out/Y-axis) and longitudinal (back and forth/X-axis). (Ch. 18)

combustible material: A solid, liquid, or gas that is capable of burning. Combustible materials are classified into four categories: Class A fires involve ordinary combustible materials (paper, wood, textiles); Class B fires involve flammable liquids and grease; Class C fires involve electrical components; and Class D fires involve flammable metals, such as magnesium and lithium. (Ch. 3)

comment code: Text in a CNC program that is enclosed in parentheses and is not read by the controller. Instead, the text appears on the controller screen as a reminder or prompt for the machine operator. (Ch. 24)

compound rest: A slide in the lathe located above a base cross-slide. The upper slide can be revolved to any required angular position. (Ch. 14)

computer numerical control (CNC): A system in which a program is used to precisely position tools or the workpiece and to carry out the sequence of operations needed to produce a part. (Ch. 1)

computer integrated manufacturing (CIM): A manufacturing system with computer-controlled machinery and adaptive tooling, which allows the system to be quickly adapted to changes in the product or the manufacturing process. *Also called a flexible manufacturing system.* (Ch. 26)

concentricity: The condition where the axes of all cross-sectional elements of a cylindrical surface are common with the axis of a datum feature. (Ch. 4)

conventional milling: A type of milling operation where the work is fed into the rotation of the cutter. Also know as "up-milling." (Ch. 18)

conversational language: An interface on CNC machines that allows the operator to select operations from menus, without having to understand G-code. (Ch. 22)

coordinate measuring machine (CMM): An instrument that makes precise measurements electronically. A CMM can be used manually or programmed to check any number of individual reference points on the object against specifications. (Ch. 27)

coordinate system: A method of locating specific points or positions in three-dimensional space. (Ch. 22)

counterboring: The process of cutting a cylindrical enlargement of a hole to a given depth and diameter to allow bolt heads to be flush with the surface of the workpiece. (Ch. 12)

countersinking: The process of chamfering a hole to receive a flathead screw. (Ch. 12)

creep grinding: A surface grinding operation that is often performed in a single pass with an unusually large depth of cut. (Ch. 20)

critical temperature: The temperature at which a metal's crystal structure changes. Most heat treatment processes require that the metal be heated to its critical temperature. (Ch. 29)

cross slide: A part of a machine tool that permits the carriage to make transverse tool movements. (Ch. 14)

cryogenic: Having or associated with extremely low temperatures. The crygogenic temperature range is generally considered to be –300°F to –460°F (–184°C to –273°C). (Ch. 33)

cutting fluid: A liquid used to cool and lubricate a cutting tool to improve the quality of the surface finish. There are four basic types of cutting fluids: mineral oils, emulsifiable (water-based) oils, chemical and semichemical cutting fluids, and gaseous fluids. (Ch. 10)

cutting speed: In reference to milling machines, the distance, measured in feet or meters, that a point (tooth) on the cutter's circumference will move in one minute. In reference to lathes, the distance the work moves past the cutting tool, expressed in feet per minute (fpm) or meters per minute (mpm). Measuring is done on the circumference of the work. In reference to drilling machines, it is the speed at which the cutting tool rotates. (Ch. 12, 14, 18)

cyanoacrylate quick setting adhesive: A bonding agent known by such names as Eastman 910™, Super Glue™, and Crazy Glue™. It is used to hold matching metal sections together while they are being machined. (Ch. 8)

cylindricity: A form tolerance identified by a radius tolerance zone establishing two perfectly concentric cylinders within which the actual surface must lie. (Ch. 4)

D

datum: A theoretically exact point, axis, line, or plane. (Ch. 4)

dedendum: The portion of a gear tooth between pitch circle and root circle; it is equal to addendum plus clearance. (Ch. 19)

depth of cut: Refers to the distance the cutter is fed into the work surface. The depth of cut varies greatly with lathe condition, material hardness, speed, feed, amount of material to be removed, and whether it is to be a roughing or finishing cut. (Ch. 14)

destructive testing: A quality control process in which the part being tested is destroyed by the testing process. (Ch. 27)

dial indicator: An instrument used for centering and aligning work on machine tools, checking for eccentricity, and inspecting. There are two types of indicators: balanced indicators take measurements on either side of a zero line, while continuous indicators read from zero in a clockwise direction. (Ch. 5)

diametral pitch: The ratio of the number of gear teeth to the number of inches of pitch diameter. (Ch. 19)

diamond-edge band: A specially designed band machining tool for cutting material that is difficult or impossible to cut with a conventional toothed blade. The diamonds are only on the front edge of the band, where the cutting is accomplished. (Ch. 21)

dielectric fluid: In electrical discharge machining, a fluid that forms a nonconductive barrier between the electrode and the work at the arc gap. (Ch. 31)

dimensions: Sizes or measurements needed to produce a part. (Ch. 4)

Direct Shell Production Casting (DSPC): A variation of the Fused Deposition Modeling process that prints a three-dimensional model using sand or ceramics instead of plastic. The finished model can be used as a mold or mold core for casting. (Ch. 26)

divider: A layout tool with pointed legs of equal length used to draw circles and arcs. (Ch. 6)

drill gage: An instrument used to measure the diameter of a drill of various series. (Ch. 12)

drilling machine: A power-driven machine that holds the material and cutting tool and brings them together to cut or enlarge a hole. (Ch. 12)

drill point gage: An instrument used to ensure a drill is correctly sharpened. (Ch. 12)

drill press: A machine that rotates a drill against stationary material with sufficient pressure to cause the drill to penetrate the material. It is primarily used for cutting round holes. (Ch. 1)

dry cycle: Cycling through all programs of a CNC machine once during setup without a part in place to check for potential problems. (Ch. 22)

dual dimensioning: A system that uses both the U.S. Conventional system of fraction or decimal dimensions and SI Metric dimensions on the same drawing. (Ch. 4)

ductility: A property of metal that permits permanent deformation by hammering, rolling, and drawing without breaking or fracturing. (Ch. 28)

dwell: To leave the cutting tool in place for a period of time after it has reached the final cutting depth. This helps to relieve pressure on the cutting tool and improves finish. (Ch. 20)

dye penetrant inspection: A nondestructive testing method in which a dye is applied to a part's surface and drawn into flaws by capillary action. The excess dye is removed and a developer is applied. The developer draws the dye back to the surface, marking the location of any flaw. (Ch. 27)

E

eddy-current inspection: A nondestructive testing method that detects flaws by measuring changes in impedance in a test coil placed near the test sample. A material without flaws will generate a different impedance than a material with flaws. (Ch. 27)

electrical discharge machining (EDM): A process in which material is eroded from workpiece by an electric arc between the electrode and workpiece. (Ch. 31)

electrical discharge wire cutting (EDWC): A machining process that removes material from the workpiece by creating an arc between a wire electrode and the workpiece. It is similar in function to band machining. (Ch. 31)

electrochemical machining (ECM): A process that removes material from the workpiece through electrolysis. In this process, an electrolyte surrounds the electrode and workpiece, which are electrically charged. (Ch. 31)

electrode: In electrical discharge machining, an negatively charged tool. When voltage builds to sufficient strength, an arc jumps from the electrode to the workpiece, eroding a small amount of metal from the workpiece. (Ch. 31)

electrohydraulic forming: A type of high-energy-rate forming in which a high-voltage discharge from a capacitor bank vaporizes an initiation wire and surrounding wire. The gases formed during the vaporization expand rapidly, applying pressure to the outer surface of the workpiece and forcing into a die. Also known as "capacitor discharge forming" and "spark forming." (Ch. 33)

electrolytic grinding: A form of electrochemical machining where electric current is passed between a metal bonded grinding wheel (the cathode) and the work (the anode) through an electrolyte to dissolve the work. The dissolved material is removed by the grinding wheel. (Ch. 20)

electromachining: A category of machining processes in which electrical current is used to remove material through erosion rather than direct physical contact. (Ch. 31)

electron beam machining (EBM): A process that uses a high-energy, highly focused beam of electrons to weld or cut materials. (Ch. 32)

electroplating: A process in which a layer of one metal is deposited on the surface of another metal by the use of an electrical current. (Ch. 30)

emulsifiable oil: Cutting fluid composed of oil droplets that are suspended in water by blending the oil with emulsifying agents and other materials. Provides increased cooling capacity over straight cutting oils with reduced misting and fogging. (Ch. 10)

encoder: A transducer in the motor that drives a CNC axis that measures the position of the moving axis and provides electronic feedback to the controller. (Ch. 22)

engineer: A trained professional who uses applied mathematics, science, and knowledge of manufacturing to design industrial products and processes. (Ch. 2)

etchant: The fluid used to erode material in a chemical machining process. Usually a strong aqueous alkaline solution. (Ch. 32)

explosive forming: A type of high-energy-rate forming in which the material is shaped by pressure resulting from the detonation of an explosive charge. The workpiece, die, and explosive charge are submerged in water, which distributes the force of the blast evenly over the outer surface of the workpiece. (Ch. 33)

external thread: A screw thread cut on an outside surface. (Ch. 16)

F

face milling: Machining large, flat surfaces parallel to the cutter face. (Ch. 18)

facing: In lathe work, cutting across the end of a workpiece to create a flat surface. (Ch. 14)

fastener: Any device used to hold two objects or parts together, such as bolts, nuts, screws, pins, keys, rivets, and chemical bonding agents or adhesives. (Ch. 8)

feature: A general term applied to a physical portion of a part. (Ch. 4)

feature control frame: A symbol used to define the geometric tolerancing characteristics of a feature. It contains the geometric characteristic symbol, allowable tolerance, and datum reference letter(s). (Ch. 4)

feed: The rate at which work moves into the cutter or the cutter moves into the work. (Ch. 12, 14, 18)

ferrous: Containing iron. (Ch. 28)

file band: A band machine accessory used to obtain a smooth, uniformly finished surface. A series of small file segments make up the file band. The segments interlock to form a continuous file. (Ch. 21)

fixed-bed milling machine: A type of milling machine that is characterized by very rigid worktable construction and support. The worktable moves only in a longitudinal (back and forth/X-axis) direction, and can vary in length. Vertical (up and down/Z-axis) and cross (in and out/Y-axis) movements are obtained by moving the cutter head. (Ch. 18)

fixture: A device for holding work rigidly while machining operations are performed. It does not guide the cutting tool. (Ch. 9)

flatness: A measure of the variation of a surface perpendicular to its plane. (Ch. 4)

flexible-back blade: A blade used on metal cutting saws when safety requirements demand a shatterproof tool. These blades are also used for cutting odd-shaped work if there is a possibility of the work coming loose in the vise. (Ch. 11)

flexible manufacturing system (FMS): A manufacturing system with computer-controlled machinery and adaptive tooling, which allows the system to be quickly adapted to changes in the product or the manufacturing process. *Also called computer integrated manufacturing.* (Ch. 26)

flutes: Two or more spiral grooves machined into a cutting tool to form the cutting edges of the drill point, facilitate easy chip removal, and to permit cutting fluid to reach the cutting point. (Ch. 12)

follower rest: Similar to a steady rest except it provides support directly in back of the cutting tool and follows along during the cut. Compare with steady rest. (Ch. 15)

foot-pound (ft·lb): The US Conventional measurement unit for torque. (Ch. 7)

form geometric tolerances: Tolerances that control the straightness, flatness, circularity, or cylindricity of a geometric shape. (Ch. 4)

form grinding: A cutting operation in which the grinding wheel is shaped to produce the required contour on the work. (Ch. 20)

friction saw: A metal-cutting tool with a blade that may or may not have teeth. The saw operates at very high speeds (20,000 surface feet or 6000 meters per minute) and actually melts its way through the metal. (Ch. 11)

Fused Deposition Modeling (FDM): A rapid prototyping technique that produces three-dimensional objects based on CAD-generated solid or surface models. A temperature-controlled head extrudes thermoplastic material layer by layer. (Ch. 26)

G

gage blocks: Precisely made steel blocks used by industry as a standard of measurement. They are made in a range of sizes and with a dimensional accuracy of ± 0.000002 (two millionths) inch, with a flatness and parallelism of ±0.000003 (three millionths) inch. Also known as "Jo-blocks." (Ch. 5)

gaging: To check parts with various gages to determine whether the pieces are made within specified tolerances. (Ch. 5)

gang milling: Using two or more milling cutters to machine several surfaces at one time. (Ch. 19)

gaseous fluid: A type of cutting fluid. Compressed air is the most commonly used. (Ch. 10)

G-codes: Preparatory codes in the ANSI/EIA 274D code format that is used to program CNC machines. (Ch. 23)

geometric dimensioning and tolerancing (GDT): The control of the size of the features of a part and the allowances (either oversize or undersize) to achieve interchangeable manufacturing. (Ch. 4)

geometric tolerance: A general term that refers to tolerances that control form, profile, orientation, or location of a feature. (Ch. 4)

glazing: A process where the grains of a grinding wheel dull before they can wear away, leaving a shiny surface and causing inefficient cutting action. (Ch. 13)

graduations: Lines that indicate measurement points on tools and machine dials. (Ch. 5)

grinding: An operation that removes material by rotating an abrasive wheel or belt against the work. (Ch. 13)

grinding machine: A machine that removes material from work by means of a rotating grinding wheel made of abrasive particles or an abrasive belt. Also known as a "grinder." (Ch. 1)

grooving: A turning operation of cutting a groove or recess into a workpiece to terminate a thread or provide adequate clearance for mating parts. (Ch. 14)

H

hardening: A heat treatment that reduces the ductility of steel. It is performed by heating metal to the proper temperature and then quenching it. (Ch. 29)

headstock: On a lathe, the structure that contains the spindle to which various work-holding attachments are fitted. (Ch. 14)

heat treatment: The controlled heating and cooling of a metal or alloy to obtain certain desirable changes in its physical characteristics. These changes include improving resistance to shock, developing toughness, and increasing wear resistance and hardness. (Ch. 29)

high-carbon steels: Steel that contains 0.6%–1.5% carbon. It is used in products that must be heat treated. (Ch. 28)

high-energy-rate forming (HERF): A manufacturing process in which a material is formed around a die by the rapid application of extreme pressures. (Ch. 33)

high-speed steels (HSS): Tool steel that has red hardness and a high resistance to abrasion. Used in some cutting tools. (Ch. 28)

honeycomb sandwich panels: A material made by bonding a honeycomb panel between two flat panels. Honeycomb sandwich panels have a very high strength-to-weight ratio and rigidity-to-weight ratio. (Ch. 28)

horizontal band saw: Frequently referred to as a cutoff machine, it is a tool with a long, continuous blade that moves in only one direction. Cutting is continuous, and the blade can run at very high speeds because it rapidly dissipates the cutting heat. (Ch. 11)

horizontal machining center (HMC): A machining center with a spindle oriented horizontally. (Ch. 22)

horizontal milling machine: A category of milling machine where the cutter is fitted onto an arbor mounted in the machine on an axis parallel with the worktable. Multiple cutters may be mounted on the arbor for some operations. (Ch. 18)

hot rolled steel: Steel that has been rolled to finished size while hot. Hot rolled steel is identifiable by its black oxide surface scale. (Ch. 28)

hydrodynamic machining (HDM): A machining process that uses a high-velocity, high-pressure stream of water to cut through materials. *Also called water jet cutting.* (Ch. 32)

I

impact machining: A machining process that removes metal by using ultrasonics and a special tool to force abrasives against the work. Also known as "slurry machining." (Ch. 32)

incremental positioning: A programming mode, indicated by the G91 code, in which the coordinate values for any point are interpreted relative to the location of the previous point. (Ch. 23)

indexable insert cutting tool: Widely used for turning and milling operations, the inserts are manufactured in a number of shapes and sizes for different turning geometries. As an edge dulls, the next edge is rotated into position until all edges are dulled; the inserts are then discarded. (Ch. 14)

internal grinding: A cutting operation done to secure a fine surface finish and accuracy on inside diameters. Work is mounted in a chuck and rotated. During the grinding operation, the revolving grinding wheel moves in and out of the hole. (Ch. 20)

internal thread: A screw thread cut on the inside surface of a piece. Internal threads are made on the lathe with a conventional boring bar and a cutting tool sharpened to the proper shape. (Ch. 16)

International System of Units (SI): The metric system of weights and measures. Abbreviated SI (Systeme International). *Also called the SI Metric system.* (Ch. 5)

J

jig: A device that guides a cutting tool and aligns it to the workpiece so that all parts produced are uniform and within specifications. (Ch. 9)

job shop: A machine shop where specialized or experimental work is machined or where the production runs are very small. (Ch. 2)

just-in-time (JIT) inventory system: An inventory management system in which parts and materials are scheduled for arrival at the time needed and not before. (Ch. 26)

K

key: A small piece of metal embedded partially in the shaft and partially in the hub to prevent rotation of the gear or pulley on the shaft. (Ch. 8)

keyway: The slot or recess in a shaft that holds the key. (Ch. 17)

knife-edge blade: A vertical band machine tool used to cut material that would tear or fray if machined by a conventional blade. Such materials include sponge rubber, cork, cloth, corrugated cardboard, and rubber. (Ch. 21)

knurling: The process of impressing diamond or straight-line patterns onto a metal surface by rolling with pressure to improve the appearance and provide better grip. The rolls are called knurls. (Ch. 15)

L

Laminated Object Manufacturing (LOM): A rapid prototyping technique that uses progressive layers of inexpensive solid sheet material, each layer bonded to the previous layer, to form the model. (Ch. 26)

laser: A device that produces a narrow and intense beam of light. Stands for Light Amplification by Stimulated Emission of Radiation. (Ch. 32)

lathe: A machine in which a workpiece in a work-holding device is rotated while a stationary cutting tool is forced against it. Some operations performed on a lathe include turning, boring, facing, thread cutting, drilling, and reaming. (Ch. 1)

lathe center: A pointed work-holding device used to accurately align the workpiece along an axis. More frequently referred to as a center. (Ch. 14)

lathe dog: A device for clamping work so that it can be machined between centers. (Ch. 14)

lay: The direction of the predominant tool marks, grain, or pattern of surface roughness. (Ch. 30)

lay out: The process of locating and scribing points for machining and forming operations. (Ch. 6)

layout dye: A coating applied to metal to make layout lines more visible. (Ch. 6)

lead: The distance a nut will advance on a screw in one revolution. (Ch. 16)

least material condition (LMC): The condition where a feature of size contains the least amount of material within the stated limits. (Ch. 4)

linear interpolation: Straight-line movement (in the context of CNC programming). (Ch. 23)

loading: A process where small pieces of metal clog the surface of the grinding wheel causing inefficient cutting action. (Ch. 13)

location geometric tolerances: Tolerances used for the purpose of locating features from datums, or for establishing coaxiality or symmetry. (Ch. 4)

low-carbon steels: Steel that contains less than 0.3%, or 30 points, or carbon. Low-carbon steel does not contain enough carbon to be hardened. However, case hardening is possible if the surface is exposed to an external source of carbon during the process. *Also called mild steel.* (Ch. 28)

M

machine control unit (MCU): Also simply called a control, the part of a CNC machine that reads the CNC program from memory and translates it into the electronic signals needed for machine operation. (Ch. 23)

machine home: For purposes of CNC programming, the location of the table and spindle on the CNC machine. (Ch. 24)

machine shield: A type of barrier on a machine that blocks flying chips and splashing cutting fluids or coolants from escaping and harming the machinist. (Ch. 3)

machine tools: That class of machines which, taken as a group, can reproduce themselves. (Ch. 1)

machining center: A CNC milling machine equipped with an automatic tool changer (ATC). (Ch. 22)

machinist: A person who is skilled in the use of machine tools and is capable of making complex machine setups. (Ch. 1)

magnetic forming: A type of high-energy-rate forming in which the magnetic force from a specially shaped induction coil provides the pressure needed to shape the material. Also known as "electromagnetic forming" and "magnetic pulse forming." (Ch. 33)

magnetic particle inspection: A nondestructive testing technique that uses iron particles and induced magnetic fields to detect flaws on or near the surface of ferromagnetic (iron-based) materials. (Ch. 27)

major diameter: The largest diameter of a thread measured perpendicular to the axis. (Ch. 16)

mandrel: A slightly tapered, hardened steel shaft that supports work machined between centers. (Ch. 15)

manual data input (MDI): A method of programming CNC programs by entering the program codes at the machine control unit. (Ch. 23)

maximum material condition (MMC): The condition where a feature contains the maximum amount of material within the stated limits. (Ch. 4)

M-codes: Miscellaneous codes in the ANSI EIA 274D code format that control machine functions in CNC programming. (Ch. 23)

medium-carbon steels: Steel that contains 0.30% to 0.60% (30 to 60 points) carbon. The carbon content is sufficient to allow partial hardening with proper heat treatment. (Ch. 28)

megahertz (MHz): A unit of measure equal to a million cycles per second. Used to measure ultrasound frequencies. (Ch. 27)

metal finish: The degree of smoothness or roughness remaining on the surface of a part after it has been machined. (Ch. 30)

metal spraying: A process in which a metal wire or powder is heated to its melting point and sprayed by air pressure to produce the desired coating on the work surface. (Ch. 30)

metrology: The science that deals with systems of measurement. (Ch. 5)

microinch: One millionth of an inch. (Ch. 30)

micrometer: One-millionth of a meter (0.000001 m). *Also called micron.* (Ch. 5, 30)

micrometer caliper: A precision tool capable of measuring to 0.001″ (0.01 mm). When fitted with a vernier scale, it will read to 0.0001″ (0.002 mm). Also known as a "mike." (Ch. 5)

micron: One-millionth of a meter (0.000001 m). *Also called micrometer.* (Ch. 5)

mild steel: Steel that contains less than 0.3%, or 30 points, or carbon. Mild steel does not contain enough carbon to be hardened. However, case hardening is possible if the surface is exposed to an external source of carbon during the process. *Also called low-carbon steel.* (Ch. 28)

milling machine: A machine that removes material from work by means of a rotary cutter. (Ch. 1)

mineral oil: Cutting fluid best suited for light-duty (low speed, light feed) operations where high levels of cooling and lubrication are not required. (Ch. 10)

minor diameter: The smallest diameter of a screw thread, measured across roots and perpendicular to axis. Also known as "root diameter." (Ch. 16)

mist coolant: Cutting fluid applied on a band machine by flooding in mist form. It is used for high-speed sawing of free machining nonferrous metals. It is also used when tough, hard-to-machine materials are cut. (Ch. 21)

modal command: A G-code command that, once activated, remains activated until the machine encounters another modal command in the CNC program. (Ch. 23)

N

newton meter (N·m): The SI Metric measurement unit for torque. (Ch. 7)

nondestructive testing: A quality control process in which the usefulness of the product is *not* impaired. Also known as "nondestructive examination" and "nondestructive inspection." (Ch. 27)

nonferrous: Containing no iron. (Ch. 28)

normalizing: A process that refines the grain structure of some metals and thereby improves their machinability. It is closely related to the annealing process. (Ch. 29)

numerical control (NC): A system (composed of a control program, a control unit, and a machine tool) that controls the actions of the machine through coded command instructions. (Ch. 1)

O

offhand grinding: A grinding operation for work that does not require great accuracy. The work is handheld and manipulated until ground to the desired shape. (Ch. 13)

offline programming: Entering a program using a computer that is not currently being used to control the operation of a CNC machine. (Ch. 23)

offset tailstock method: A method for machining external tapers on a lathe. Jobs that can be turned between centers may be taper-turned by this technique. (Ch. 16)

open-loop system: A system in which no feedback is provided to the controlling mechanism; used with stepper motors in relatively inexpensive CNC machines. (Ch. 22)

optical comparator: A gaging system in which an enlarged image of the part being inspected is projected onto a screen. The projected image can be superimposed on a grid or an accurate drawing overlay of the part to allow precise measurement of the part. (Ch. 27)

orientation geometric tolerances: Tolerances that control the relationship of features to one another. When controlling orientation tolerances, the feature is related to one or more datum features. (Ch. 4)

OSHA (Occupational Safety and Health Administration): A government agency responsible for setting and enforcing regulations regarding safety and health within the workplace. (Ch. 3)

P

parallelism: The condition of a surface or center plane equidistant from a datum plane or axis. (Ch. 4)

part catcher: A programmable device mounted to turning centers to catch finished parts as they are cut off the bar stock. (Ch. 25)

parting: The turning operation of cutting off material after it has been machined. This is one of the more difficult operations performed on a lathe. (Ch. 14)

part programmer: A skilled worker who specializes in writing programs for computer controlled machine tools. (Ch. 2)

peck drilling: A method of drilling in which the cutter is retracted at intervals to allow chips to be cleared or coolant to be flooded through the hole. (Ch. 24)

peripheral milling: A milling operation that is done when the surface being machined is parallel with the periphery of the cutter. (Ch. 18)

perpendicularity: The condition of a surface, center plane, or axis at a right angle (90°) to a datum plane or axis. (Ch. 4)

pilot hole: A smaller hole drilled prior to drilling a larger hole that greatly reduces feed pressure, improves accuracy, and allows the faster drilling. Also known as a "lead hole." (Ch. 12)

pitch: The distance from a point on one thread or gear tooth to the corresponding point on the next thread or tooth. (Ch. 16)

pitch circle: An imaginary circle located approximately half the distance from the roots and tops of gear teeth. It is tangent to the pitch circle of the mating gear. (Ch. 19)

pitch diameter: For threads, the diameter of an imaginary cylinder that would pass through the threads at such points to make the width of the thread and the width of the thread groove equal. For gears, the diameter of the pitch circle of a gear. (Ch. 16)

plain protractor: An angle-measuring tool used in layout work when angles do not need to be laid out or checked to extreme accuracy. The head is graduated from 0° to 180° in both directions for easy reading. (Ch. 6)

plain turning: Turning in which the entire length of the piece is machined to a specified diameter. (Ch. 14)

platens: Flat metal plates used to provide support behind the belt of an abrasive belt grinder. (Ch. 20)

plunge grinding: Grinding method in which work is mounted between centers and rotated while in contact with the grinding wheel. The area being ground is no wider than the wheel face. (Ch. 20)

polar coordinate system: A coordinate system in which straight-line distance and travel angle are used to specify locations or movement. (Ch. 22)

position tolerance: A tolerance used to define a zone in which the center, axis, or center plane of a feature of size is permitted to vary from true position. (Ch. 4)

post-processing: The process of translating a CAM program into a format the CNC controller can understand. (Ch. 23)

powder metallurgy (P/M): A manufacturing process used to make parts by compressing and heating metal powders. *Also called sintering.* (Ch. 33)

precision grinding: A finishing operation in which a minute amount of material is removed with each pass of the grinding wheel to generate a smooth, accurate surface. (Ch. 20)

process annealing: Removing internal stresses that have developed in parts that have been cold worked, machined, or welded. *Also called stress-relieving.* (Ch. 29)

profile geometric tolerance: A tolerance that specifies a uniform boundary along the true profile within which the elements of the surface must lie. (Ch. 4)

profile of a line tolerance: A two-dimensional or cross-sectional geometric tolerance that extends along the length of the feature. (Ch. 4)

profile of a surface tolerance: A geometric tolerance that controls the entire surface of a feature or object as a single entity. (Ch. 4)

profilometer: An electronic instrument for measuring surface roughness. (Ch. 4)

program zero position: For purposes of CNC programming, a reference point specified on the part or on the work-holding device as the zero position. (Ch. 24)

protective clothing: Clothing that is worn in a machine shop to protect the body. Safety glasses and hearing protectors are two of the most important articles because shop areas produce both noise and flying chips. Other protective clothing includes steel-toed shoes, lead aprons, caps or hairnets, and respirators. (Ch. 3)

Q

quality control (QC): A process that identifies and prevents potential product defects in the manufacturing process before they can cause injuries or damage and substandard products. (Ch. 27)

quench: To cool a metal rapidly by immersing it in a fluid or spraying a fluid on its surface. (Ch. 29)

R

radiographic inspection: A nondestructive testing method that uses X rays and gamma radiation to detect flaws (crack, pores, etc.) in an object. Also known as "X-ray inspection." (Ch. 27)

raker set: A three-tooth saw set in which one tooth is angled toward the left, another one straight, and the next one angled toward the right, alternating continuously along the length of the blade. Raker set is recommended for cutting large solids or thick plate and bar stock. (Ch. 21)

reamer: A cutting tool used to enlarge, smooth, and size a drilled hole by removing a small amount of metal. (Ch. 7)

reaming: To finish a drilled hole with a reamer. (Ch. 12)

red hardness: The ability of a metal to remain hard even when red hot. (Ch. 28)

reference line: A layout line from which all measurements are made. Also known as a "baseline." (Ch. 6)

repetitive cycle: Predefined operations specified by G-codes. Found on most turning centers; similar to the canned cycles used on CNC machining centers. (Ch. 25)

résumé: A summary of an applicant's educational and employment backgrounds. (Ch. 2)

robot: A reprogrammable, multifunctional manipulator designed to move material, parts, tools, or specialized devices through variable programmed motions for the performance of a variety of tasks. (Ch. 26)

Rockwell hardness test: The most widely used hardness testing method. A penetrator is pressed into the metal being tested under a known load. The hardness reading is based on the depth of penetration. (Ch. 29)

roller burnishing: A cold-working operation that used rollers to flatten the peaks of a metal's surface into the valleys. With the process, no honing or grinding is necessary (Ch. 30)

roughing cut: Deep cut made to remove considerable material from a workpiece. (Ch. 14)

runout geometric tolerance: A combination of geometric tolerances used to control the relationship of one or more features of a part to a datum axis. (Ch. 4)

S

safe edge: A file edge without teeth. (Ch.7)

safety equipment: Tools or equipment that help prevent or mitigate accidents in the potentially dangerous environment of a machine shop. Machine guards, fire extinguishers, eye wash stations, vacuum dust collectors, power switches (for locking off equipment), and brushes (for removing machine chips) are a few examples. (Ch. 3)

sawing machine: A machine that cuts away material from the work by means of a multitoothed saw blade. (Ch. 1)

scale drawing: A drawing made other than actual size (1:1). A drawing made one-half size would have a scale of 1:2. A scale of 2:1 would mean that the drawing is twice the size of the actual part. (Ch. 4)

scleroscope: A testing device that drops a hammer onto the test piece. The resulting rebound of the hammer determines the hardness of the material being tested. (Ch. 29)

scriber: A layout tool with a point of hardened steel used to scribe (scratch) fine, accurately straight lines into metal. (Ch. 6)

semichemical cutting fluid: Cooling and lubricating liquid that may have a small amount of mineral oil added to improve the fluid's lubricating qualities. Semichemical cutting fluids incorporate the best qualities of both chemical and emulsifiable (water-based) cutting fluids. (Ch. 10)

semiskilled worker: A worker who performs basic, routine operations that do not require a high degree of skill or training. (Ch. 2)

servomechanism: In electrical discharge machining, the drive unit used to accurately control electrode movement and to maintain the correct distance between the work and the electrode as machining progresses. (Ch. 31)

servo motor: A motor that can provide feedback to a controlling mechanism; used in higher-end CNC machines to drive axis positions. (Ch. 22)

setover: The distance a lathe tailstock is offset from the normal centerline of the machine. It is a method of taper-turning. (Ch. 16)

side milling cutter: A type of milling cutter with cutting edges located on the circumference and on one or both sides of the cutter. They are made in solid form or with inserted teeth. (Ch. 18)

SI Metric system: The metric system of weights and measures. (*SI* stands for the French words *Systeme International*). Compare with US Conventional system. (Ch. 4)

single-point cutting tool: A cutting tool with one face and one cutting edge. (Ch. 14)

sintering: In the powder metallurgy process, the operation in which the newly formed part is heated in a controlled environment furnace to give it strength. May also be used to refer to the powder metallurgy process itself. (Ch. 33)

skilled worker: A worker who has been trained, often as an apprentice, to do more complex tasks. (Ch. 2)

skills standards: Industry requirements for skilled workers and the basis for industry-recognized certification obtained through performance testing. (Ch. 1)

slitting: An operation in which a thin cutter or rotary knife is used to cut sheet metal into narrow strips. (Ch. 19)

slotting: Similar to slitting, except that the cut is made only part way through the work. The slot in a screw head is an example of slotting. (Ch. 19)

small hole EDM drilling: A process that bores small diameter holes in material by creating an arc between a hollow, spinning electrode and the workpiece. (Ch. 31)

smart tooling: Cutting tools and work-holding devices that can be easily reconfigured to produce a variety of shapes and sizes within a given part family. (Ch. 26)

spindle: On a lathe, the structure which receives tools, attachments, or the workpiece, revolves in heavy-duty bearings, and is rotated by belts, gears, or a combination of the two. (Ch. 14)

spontaneous combustion: Ignition by rapid oxidation or burning of oil without an external source of heat. (Ch. 3)

spotfacing: Machining a circular spot on the surface of a part to furnish a flat bearing surface for mounting a bolthead or nuthead. (Ch. 12)

spur gear: A wheel with teeth that run straight across the gear face, perpendicular to the sides. It is the most commonly used gear. (Ch. 19)

square: A tool used to check 90° (square) angles. It is also used for laying out lines that must be at right angles to a given edge or parallel to another edge. (Ch. 6)

statistical process control (SPC): A quality control technique in which a percentage of the products made during a production run are tested. Statistical analysis of variations detected in the products allows the manufacturer to predict and correct problems before they result in unacceptable products. (Ch. 27)

steady rest: A support for long, thin workpieces that keeps the work from springing or bending away from the cutting tool. The rest also reduces "chatter" when long shafts are machined. Compare with follower rest. (Ch. 15)

steel rule: A measuring tool, available in at least three basic types of graduations: fractional inch, decimal inch, and metric. (Ch. 5)

stepper motor: A motor that moves in small steps, or increments, and does not provide feedback to the controlling mechanism. (Ch. 22)

stereolithography: A rapid prototyping technique that uses a computer-guided low-power laser beam to harden a liquid photocurable polymer plastic into the programmed shape. (Ch. 26)

straddle milling: Using two or more milling cutters to perform several milling operations simultaneously. (Ch. 19)

straightedge: A precision tool for checking the accuracy of flat surfaces. (Ch. 6)

straightness: The measure of how closely an element of a surface or axis is to a perfectly straight line. (Ch. 4)

straight set: A two-tooth saw set in which one tooth is angled to the side and the other one straight, alternating continuously along the length of the blade. Straight set is recommended for materials like aluminum and magnesium. (Ch. 21)

stress-relieving: Removing internal stresses that have developed in parts that have been cold worked, machined, or welded. *Also called process annealing.* (Ch. 29)

subroutine: A set of commands that follows a prescribed sequence again and again until cancelled. *Also called canned cycle.* (Ch. 24)

surface gage: A scribing tool used to check whether a part is parallel to a given surface. (Ch. 6)

surface hardening: A heat treatment process used to create a medium-hard surface on high-carbon or alloy steel while leaving the inner core of the metal unaffected. (Ch. 29)

surface plate: A cast iron or granite plate, ground or lapped to a smooth flat surface, and used for precision layout and inspection. (Ch. 6)

surface roughness standards: Documents that detail how surfaces produced by machining, grinding, casting, molding, forging, or similar processes are to be measured and communicated. (Ch. 30)

swing: Indicates the largest diameter that can be turned over the ways. Along with the length of the bed, determines the size of the lathe. (Ch. 14)

Swiss-type turning center: A type of turning center or lathe that can produce elaborate detail in machined parts with a high degree of precision and accuracy. (Ch. 22)

symmetry: A relationship that indicates equal or balanced proportions on either side of a central plane or datum. (Ch. 4)

T

tailstock: A movable lathe fixture that mounts on ways to support work between centers. It can be fitted with tools for drilling, reaming, and threading. (Ch. 14)

taper: A piece that uniformly increases or decreases in diameter to assume a wedge or conical shape. (Ch. 16)

taper attachment: A guide attached to a lathe and used to accurately cut internal and external tapers. (Ch. 16)

tapping: Forming an internal screw thread in a hole or other part by means of a tap. Also, opening the pouring hole of a melting furnace to remove molten metal. (Ch. 12)

technician: A specialist in the technical details of an occupation operating between the realms of the production and engineering departments. (Ch. 2)

temper: The hardness and strength of a rolled metal. (Ch. 13)

tempering: A heat treatment process that reduces a metal's brittleness or hardness. In this process, metal is heated to a temperature below its critical temperature and then quenched. (Ch. 29)

thread cutting stop: A device used to stop the thread-cutting operation on a lathe so the tool can be removed from the work after each cut and repositioned before the next cut is started. (Ch. 16)

thread dial: A dial that meshes with the lead screw of a lathe and indicates when to engage the half-nuts, which permits the tool to follow exactly in the original cut. (Ch. 16)

threaded fastener: A bolt or similar device that uses the wedging action of the screw thread to clamp parts together. To achieve maximum strength, a threaded fastener should screw into its mating part at least a distance equal to one and one-half times the thread diameter. (Ch. 8)

three-tooth rule: In a properly selected power hacksaw blade, at least three teeth must come in contact with the work. (Ch. 11)

three-wire method: To accurately measure thread size, three wires of a specific diameter are fitted into screw threads, and a micrometer measurement is made over the wires. A mathematical formula provides the information necessary to calculate the correct measurement. (Ch. 16)

tolerance: A permissible deviation from a basic dimension. (Ch. 4)

tombstone: A common fixture-holding device used with machining centers and other CNC machine tools. They are made from heavy castings and are precisely machined. (Ch. 9)

toolmaker: A highly skilled machinist who specializes in producing the tools and tooling needed for machining operations. (Ch. 2)

tool post: Mounts the cutting tool on the carriage of the lathe. (Ch. 14)

tool rest: A part found on bench, pedestal, and belt grinding machines. Supports the workpiece during grinding operations. (Ch. 13)

tooth form: The shape of the tooth on a band machine saw. There are three basic forms: standard, skip, and hook. (Ch. 21)

tooth rest: A device that quickly and accurately positions the teeth of a gear cutter.

torque: The amount of turning or twisting force applied to a threaded fastener or part. It is measured in force units of foot-pounds (ft·lb) or the SI Metric equivalent, newton meters (N·m). Torque is the product of the force applied times the length of the lever arm. (Ch. 7)

total runout: A type of runout that provides a combined control of the circularity, straightness, angularity, and cylindricity of a surface rotated around a datum axis. (Ch. 4)

traverse grinding: A cylindrical grinding operation where a fixed amount of material is removed from the rotating workpiece as it moves past the revolving grinding wheel. Work wider than the grinding wheel; can be ground using this method. (Ch. 20)

turning: A machining process that operates on the principle of work being rotated against the edge of a cutting tool. (Ch. 1)

turning center: A CNC turning machine equipped with an automatic tool changer. (Ch. 22)

turret lathe: A lathe equipped with a rotating tower, or turret, that holds multiple tools and can be revolved to present the appropriate tool for a particular operation. Turret lathes can also be controlled by a CNC program. (Ch. 15)

twist drill: A common drill made by forging or milling rough flutes and then twisting them to a spiral shape. After twisting, the drills are milled and ground to approximate size. Finally, they are heat-treated and ground to exact size. (Ch. 12)

U

ultrasonic-assist machining: A machining process in which ultrasound waves are used to enhance the function of a conventional machine tool. (Ch. 32)

ultrasonic testing: A nondestructive testing method that uses high-frequency sound waves to detect flaws and cracks in materials. (Ch. 27)

unassigned code: A code that can be assigned to different functions on different types of machine controllers for specialty purposes. (Ch. 23)

Unified System: A thread form system adopted by the powers that make up NATO after World War II. (Ch. 7)

universal bevel protractor: An angle-measuring tool used in layout work. This tool consists of a dial, a base or stock, and a sliding blade. It is graduated into 360°. (Ch. 5)

universal tool and cutter grinder: A grinding machine designed to support cutters (primarily milling cutters) while they are sharpened to specified tolerances. Special attachments permit straight, spiral, and helical cutters to be sharpened accurately. Other attachments enable the machine to be adapted to all types of internal and external cylindrical grinding. (Ch. 20)

US Conventional system: The "English" system of weights and measures used in the United States. Compare with SI Metric system. Also known as "US Customary system." (Ch. 4)

V

V-block: A square or rectangular steel block with a 90° V-groove through the center, provided with a clamp for holding round stock for drilling, milling, and laying out operations. (Ch. 6)

ventilation: Circulation of air that brings fresh air and lowers the concentration of toxic fumes within the working area. (Ch. 3)

vernier caliper: A precision measuring instrument, used for both inside and outside measurements, that is accurate to 1/1000" (0.001") and 1/50 mm (0.02 mm). (Ch. 5)

vernier protractor: An angle-measuring tool used in layout work when angles must be extremely accurate. With this tool, angles of 1/12 of a degree (5 minutes of arc) can be precisely measured. (Ch. 6)

vertical machining center (VMC): A machining center with a vertically oriented spindle. (Ch. 22)

vertical milling machine: A type of milling machine in which the cutter is normally perpendicular (at a right angle) to the worktable. On many vertical milling machines, the spindle can be tilted to perform angular cutting operations. (Ch. 18)

vitreous enamel: A glass coating that can be fused to most metals, including steel sheet and cast iron surfaces. It forms an extremely hard coating that is smooth and easy to clean. (Ch. 30)

W

water jet cutting: A machining process that uses a high-velocity, high-pressure stream of water to cut through materials. *Also called hydrodynamic machining (HDM).* (Ch. 32)

waviness: The presence of smoothly rounded peaks and valleys caused by tool and machine vibration and chatter. Waviness is of greater magnitude than roughness. (Ch. 30)

wavy set: A saw tooth set in which several teeth are angled to the right and several to the left, alternating continuously along the blade. Wavy set is recommended for work with varying thicknesses, such as pipe, tubing, and structural materials. (Ch. 21)

ways: The flat or V-shaped bearing surface that aligns and guides the movable part of the machine. (Ch. 14)

Webster hardness tester: A portable testing device for checking the hardness of materials such as aluminum, brass, copper, and mild steel. (Ch. 29)

wheel dresser: A tool for cleaning, resharpening, and restoring the mechanical accuracy of the cutting faces of grinding wheels. (Ch. 13)

word: In a word address format programming language, a combination of letters and numerical data. (Ch. 23)

word address format: A style of programming that uses groups of letters and numbers, referred to as words, to program instructions for a CNC machine tool. (Ch. 23)

work envelope: The volume of space defined by the reach of the robot arm in three-dimensional space. (Ch. 26)

working drawing: A drawing or drawings that give a machinist the necessary information to make and assemble a mechanism. (Ch. 4)

Acknowledgments

The authors express their sincere thanks to the many organizations and manufacturers who cooperated so generously in supplying the technical information and many of the photographs used in this textbook. Any omissions from the following list are purely accidental.

3D Systems
3M Company
Alden Corp.
American Foundrymen's Society
American National Standards Institute
AMT—The Association for Manufacturing Technology
Autocon Technologies, Inc.
Baldor
Bethlehem Steel Co.
BIG Kaiser Precision Tooling Inc.
Bill Hannan
Bird-Johnson Company
Bob Walker
Bridgeport Machines, Inc.
Carboloy, Inc.
Carl Zeiss, Inc.
C. & E. FEIN GmbH
Chicago-Latrobe
Chick Machine Tool, Inc.
Cincinnati Milacron
CITCO Div., Western Atlas, Inc.
Clausing Industrial, Inc.
Cogsdill Tool Products, Inc.
CNC Software, Inc.
Coated Abrasive Manufacturers Institute
Compositek Corporation
Dapra Corporation
Delcam International
DoALL Co.
duMont Corp.
DuPont Co.
Emco-Maier Corp.
Engis Corporation
EZFeatureMILL—Engineering Geometry Systems
Fanuc Robotics North America, Inc.
Federal Products Co.
FEIN Power Tools, Inc.
Ford Motor Co.
General Dynamics Corp.
General Motors Corp.
Giddings & Lewis, Inc.
Greenfield Industries
Grizzly Industrial, Inc.
Haas Automation, Inc.
Hardinge, Inc.
Harig Div. of Bridgeport Machines Inc.
Heli-Coil Corp.
Helisys, Inc.
Hertel Cutting Technologies, Inc.
Hirata Corporation of America
Honeywell, Inc., Industrial Automation and Control
Hougen Manufacturing, Inc.
Hydromat, Inc.
IBAG North America
IBM
Idesco Corp.
Iscar Metals, Inc.
Jergens, Inc.
Jet Equipment & Tools
John Winter
Jones & Lamson Machine Co.
Jones & Shipman, Inc.
Justrite Manufacturing Company
Kennametal Inc.
Klein Tools, Inc.
K.O. Lee Co.
Komet of America, Inc.
Landis Div. of Western Atlas
LeBlond Makino Machine Tool Co.
Library of Congress, Prints & Photograph Division
Light Beam Technology, Inc.
Lockwood Products, Inc.
Lovejoy Tool Company, Inc
L. S. Starrett Co.
Lufkin Rule Co.
Lumonics Corp.
Mack Molding Co.
MAG IAS, LLC
Magnaflux Corp.
Mahr Federal Inc.
Makita USA, Inc.
Manufacturing Technology, Inc.
Marposs Corp.
Master Lock Co.
Maswerks, Inc.
Mazak Corp.
Metal Powder Industries Federation
MHI Machine Tool U.S.A., Inc.
Microphoto, Inc.
MIFCO McEnglevan Industrial Furnace Co., Inc.
Millersville University
Mitsubishi Materials USA Corporation
Mitutoyo/MTI Corp.
Morse Tool Co.
NASA
National Broach & Machine Co.
NEYTECH
Northrop-Grumman Corp.
Norton Co.
NSK America
OSG Tap & Die, Inc.
O.S. Walker Co.
Parker-Kalon
Parlec, Inc.
Peter Wolters of America, Inc.
Precision Castparts Corp.
Radyne
Rem Sales, Inc.
Renishaw, Inc.
Republic-Lagun Machine Tool Co.
Revolution Tool Company
Rofin-Sinar, Inc.
Rohm and Haas
Royal Products, Division of Curran Manufacturing Corporation
Rush Machinery, Inc.
Sandvik Coromant Co.
Sharnoa Corp.
Sharp Industries, Inc.
SpeedFam
Standard Tool Co.
Stratasys, Inc.
Sulzer
Sulzer Chemtech Ltd
Sunnen Products Company
Tibor Machine Products
Tornos-Bechler S.A.
Tri-Tool, Inc.
Union Carbide Corp.
Unison Corp.
US Air Force
US Air Force Thunderbirds
US Amada, Ltd.
US Army
US Navy
Valenite, Inc.
Vermont Machine Tool and the Bryant Grinder Division
Waldes Kohinoor Inc.
Webber Gage Div., L.S. Starrett Co.
Webster Instrument, Inc.
Weiler Corporation
Weldon Machine Tool, Inc.
Westech Products Group/ Gantrex Machine Tool Loaders
William Schotta, Millersville University
Willis Machinery and Tools Corp.
Wilson Instruments/Instron Corporation
W.J. Savage Co.
WMW Machinery Company, Inc.
Yukiwa Seiko USA, Inc.

Index

A

B

C

D

E

F

G

H

I

J

K

L

M

N

O

P

Q

R

S

T

U

V

W